# Rings with Polynomial Identities and Finite Dimensional Representations of Algebras

2010 *Mathematics Subject Classification.* Primary 16H05, 16H10, 16R20, 16R30, 16W22, 15A75, 15A72, 14L24, 16N60, 16P90.

---

For additional information and updates on this book, visit
**www.ams.org/bookpages/coll-66**

---

**Library of Congress Cataloging-in-Publication Data**

Names: Aljadeff, Eli (Eliahu), 1956– author. | Giambruno, A., author. | Procesi, Claudio, author. | Regev, Amitai, author.

Title: Rings with polynomial identities and finite dimensional representations of algebras / Eli Aljadeff, Antonio Giambruno, Claudio Procesi, Amitai Regev.

Description: [Providence, Rhode Island] : American Mathematical Society, [2020] | Series: American Mathematical Society colloquium publications, 0065-9258 ; volume 66 | Includes bibliographical references and index.

Identifiers: LCCN 2020001994 | ISBN 9781470451745 (hardcover) | ISBN 9781470456955 (ebook)

Subjects: LCSH: Polynomial rings. | PI-algebras. | Representations of algebras. | AMS: Associative rings and algebras – Algebras and orders – Separable algebra | Associative rings and algebras – Rings with polynomial identity – $T$-ideals, identities, varieties of rings and algebras. | Associative rings and algebras – Rings with polynomial identity – Semiprime p.i. rings, rings embeddable in matrices over commutative rings. | Associative rings and algebras – Rings with polynomial identity – Trace rings and invariant theory. | Associative rings and algebras – Rings and algebras with additional structure – Actions of groups and semigroups; invariant theory. | Linear and multilinear algebra; matrix theory – Basic linear algebra – Exterior algebra, Grassmann algebras. | Linear and multilinear algebra; matrix theory – Basic linear algebra – Vector and tensor algebra, theory of invariants. | Algebraic geometry – Algebraic groups – Geometric invariant theory. | Associative rings and algebras – Radicals and radical properties of rings – Prime and semiprime rings. | Associative rings and algebras – Chain conditions, growth conditions, and other forms of finiteness – Growth rate, Gelfand-Kirillov dimension.

Classification: LCC QA251.3 .A42 2020 | DDC 512/.4–dc23

LC record available at https://lccn.loc.gov/2020001994

---



∞ The paper used in this book is acid-free and falls within the guidelines
established to ensure permanence and durability.
Visit the AMS home page at `https://www.ams.org/`

10 9 8 7 6 5 4 3 2 1    25 24 23 22 21 20

American Mathematical Society
Colloquium Publications
Volume 66

# Rings with Polynomial Identities and Finite Dimensional Representations of Algebras

Eli Aljadeff
Antonio Giambruno
Claudio Procesi
Amitai Regev

AMS AMERICAN MATHEMATICAL SOCIETY

# Contents

# Preface

The idea of an identity in an algebraic structure is quite general, all the basic laws that we learn at school as the commutative, associative, and distributive laws are in fact special types of identities. Loosely speaking, an identity is a symbolic expression involving one or several operations and one or several *variables*, which is identically satisfied when the *variables* are substituted in a given algebraic structure.

In this generality, the topic is that of *universal algebra* for which we refer to the classical book of P. M. Cohn **[Coh65]**. In the present book we restrict to a special type of identities, satisfied by associative algebras, over some commutative ring $A$. We take as symbolic expressions the noncommutative polynomials, with coefficients in $A$, in some finite or infinite set of variables.

By definition, polynomials are formal linear combinations of *monomials* in a fixed *alphabet* $X$. In the noncommutative setting a monomial is just a *word* in the variables $X = \{x_1, x_2, \dots\}$. Thus a formal polynomial $p(x_1, x_2, \dots, x_n)$ is a *polynomial identity* for an algebra $R$ if $p(r_1, r_2, \dots, r_n) = 0$ whenever we substitute to the variables $x_i$ elements $r_i$ of $R$.

The simplest and, in a way, the strongest identity for an associative algebra is the commutative law $xy - yx$. The role of more complex identities for truly noncommutative algebras has been discovered by several people: Jacobson connected it with the Kurosh problem; Kaplansky proved that an algebra $R$ satisfying an identity of degree $d$ has no infinite-dimensional irreducible modules, and in fact all irreducible modules have dimension $\leq d/2$; and Amitsur and Levitzki discovered that the ring of $n \times n$ matrices over a commutative ring $A$ satisfies a kind of higher-order commutative law, the *standard identity*

$$\mathrm{St}_{2n}(x_1, x_2, \dots, x_{2n}) := \sum_{\sigma \in S_{2n}} \epsilon_\sigma x_{\sigma(1)} \cdots x_{\sigma(2n)},$$

where $S_{2n}$ denotes the symmetric group on $2n$ elements and $\epsilon_\sigma$ denotes the sign of a permutation $\sigma$.

These theorems suggest that the theory of polynomial identities (PI for short) is strongly related to finite-dimensional representations of algebras.

If an algebra $R = A\langle a_1, \dots, a_k\rangle$ is finitely generated, over a commutative domain $A$, it is easily seen that the set of representations of $R$ in an $n$-dimensional space $K^n$, where $K \supset A$ is an algebraically closed field, is an affine algebraic variety, a subvariety of the affine space of $k$-tuples of $n \times n$ matrices. The notion of isomorphism between such representations corresponds geometrically to the fact that such representations are in the same orbit under the simultaneous conjugation action of the group $\mathrm{GL}(n, K)$ of invertible matrices, on $M_n(K)^k$.

This suggests that the theory of PI should be close both to commutative algebra and to invariant theory. This we shall try to explain in this book. In parallel there is the mostly combinatorial theory of $T$-ideals; that is the ideals in the noncommutative free algebra describing identities of algebras.

The main difference between PI rings and commutative rings is that in general no analogue of the Hilbert basis theorem is available. That is, usually a finitely generated PI algebra is not Noetherian. This is the main technical difficulty and also a source of some pathological examples. The phenomenon responsible for this fact is essentially the fact that a given algebra may have irreducible representations of various dimensions. When all irreducible representations have the same dimension $d$ (and moreover the algebra satisfies all formal identities of $d \times d$ matrices) by a theorem of M. Artin, then $R$ is an Azumaya algebra, that is a possibly *nonsplit* form of the algebra of $d \times d$ matrices over a commutative ring. For such an algebra the theory is very close to the theory of commutative algebras.

Another basic technique of commutative algebra which is not really available is localization. In general, in order to construct a ring of fractions for some multiplicative system, one needs special conditions, the Ore conditions, which are often not satisfied. On the other hand representation theory of the symmetric and linear group appear in an essential way and give the main flavour to PI theory.

PI theory has developed in several steps. First the discovery of special identities, the various structure theorems, for primitive, nil, or prime algebras satisfying a PI. Then a more geometric theory strongly related to invariants of matrices. Another step is a combinatorial theory based on the representation theory of the symmetric group, investigating various problems related to the *growth* of the space of polynomial identities. The final step involves a deep structure theory analysing mostly the nil part of algebras and leading to the theorem of Razmyslov on the nilpotency of the radical of a PI algebra finitely generated over a field, and its generalizations and the theory of Kemer leading to a solution of Specht's problem, that is the finite generation of $T$-ideals (proposed by Specht in [**Spe50**]).

This last theory requires in a surprising way that we introduce superalgebras and their superidentities. In some way this is suggested by the hook theorem of Amitsur and Regev [**RA82**] and Kemer [**Kem91**]. The reason is that in the general theory the Grassmann algebra plays an important role, and a basic result of Kemer is that every PI algebra is PI equivalent to the Grassmann envelope of a finite-dimensional superalgebra; see Theorem 19.8.1.

We have the impression that Kemer's theory was not fully accessible to the PI community for more than three decades after Kemer presented his remarkable work in the mid-1980s. Fortunately, the situation has changed, and here we present a complete detailed proof (see also [**KBKR16**]).

## The plan of the book

After a brief discussion of classical finiteness problems as the Burnside and Kurosh problem, which serve as original motivations, we start our treatment by dividing the book into four main parts plus the appendix on the Golod–Shafarevich counterexample, which aims at showing typical pathologies that occur for general rings. The theme of such pathologies is somewhat skew with respect to our topics; it is a widely unexplored area where several *monsters* may appear.

**Part 1.** The first six chapters contain foundational material on representation theory and noncommutative algebra. This can be used for an introductory course in noncommutative algebra. We give detailed proofs, except for Chapters 3 and 6 where we give references to existing literature. In particular, Chapter 5 is an introduction to the theory of Azumaya algebras, which play a special role in PI theory. This theory is then completed in Chapter 10 where we apply PI theory to Azumaya algebras, giving some possibly new results. The expert reader may use this part only as reference and then start with the main topics in the remaining parts.

**Part 2.** Chapters 7 and 8 discuss mostly the combinatorial aspects of the theory, the growth theorem, the hook theorem, and Shirshov's bases. Here methods of representation theory of the symmetric group play a major role. Chapter 9 discusses in detail the PI theory of $2 \times 2$ matrices, one of the very few examples which can be fully treated.

**Part 3.** Chapter 11 begins the main body of our discussion of structure theorems for PI algebras, including the theorems of Kaplansky and Posner, the theory of central polynomials, the M. Artin theorem on Azumaya algebras, and the geometric part on the variety of semisimple representations including the foundations of the theory of Cayley–Hamilton algebras.

**Part 4.** This part is devoted first to the proof of the theorem of Razmyslov, Braun, and Kemer on the nilpotency of the nil radical for finitely generated PI algebras over Noetherian rings. We then move on to the theory of Kemer and the Specht problem, which we have split in four chapters, 17 to 20. Finally, in Chapters 21 and 22 we discuss the *PI-exponent* and codimension growth. This part uses some nontrivial analytic tools coming from probability theory.

**Appendix A.** This appendix is devoted to the counterexamples of Golod and Shafarevich to the Burnside problem.

## Differences with other books

There are several books on the topics of PI algebras; see **[Bah91]**, **[KBR05]**, **[KBKR16]**, **[Dre00]**, **[For91]**, **[GZ05]**, **[Kem91]**, **[Mue76]**, **[Pro73]**, **[Raz94]**. The main differences between the present book and the remaining literature is that, while trying to give a comprehensive treatment, we stress the invariant theory and the geometric aspects. This approach appears partly in the book **[LB08]** of Lieven Le Bruyn.

**What we do not treat.** As mentioned at the beginning, we do not treat Lie identities or identities on special nonassociative algebras, as in **[Dre74]**, **[Dre87]**, **[Dre95]**.

We do not treat various special identities (**[Dre95]** or **[Raz73]**) nor special geometric properties of the algebra of generic matrices (**[LBP87]** or **[LBVdB88]**). We treat the deep Kemer theory only in characteristic 0 and leave out the deep results of Belov and Kemer in positive characteristic; see **[Bel97]**, **[Bel00]**, **[Bel10]**, **[Kem95]**, **[Kem03]**.

We treat superalgebras in a limited way, only what is needed for Kemer's theory. In particular, the reader may look at two interesting papers connected with polynomial or trace identities, **[BR87]** by A. Berele, A. Regev, and **[Raz85]** by Razmyslov.

Finally, we do not treat the deep results of Berele for nonfinitely generated PI algebras; see **[Ber08a]**, **[Ber08c]**, **[Ber13b]**.

E. Aljadeff
A. Giambruno
C. Procesi
A. Regev

# Introduction

## 0.1. Two classical problems

*We want to motivate the theory of polynomial identities by its historical role in two classical problems of algebra—the Kurosh problem and Malćev representability question.*

**0.1.1. The Burnside and Kurosh problems.** In 1902 William Burnside asked whether a finitely generated group in which every element has finite order must necessarily be finite. The condition that all finitely generated subgroups of a given group be finite is called *local finiteness*. If a group is locally finite in particular every element $g \in G$ must have a finite period, or order, such a group is called a *torsion group*. Thus, Burnside problem can be formulated as

*Is every torsion group locally finite?*

A similar question can be formulated for algebras over a field. In this case again the condition that all finitely generated subalgebras of a given algebra be finite dimensional is called *local finiteness*. If an algebra $R$ is locally finite in particular every element $a \in R$ must be algebraic, that is it must be the root of some nontrivial polynomial; such an algebra is called *algebraic*. The analogue of Burnside problem for algebras was formulated by Kurosh **[Kur41]** as

*Is every algebraic algebra locally finite?*

It is easily seen that there is a connection between these two problems.

In particular if one can find a finitely generated algebra $R = F[a_1, \dots, a_m]$ over a field $F$ of characteristic $p > 0$ such that every element is nilpotent and $R$ is not finite dimensional (in particular a negative answer to Kurosh problem), one has a finitely generated group in which every element has order a power of $p$ and which is not finite.

In fact add 1 to $R$ constructing $F \oplus R$ and consider any element $x := 1 + a$, $a \in R$, for some power $p^h$ we have $a^{p^h} = 0$ by hypothesis. Since we are in characteristic $p$, we deduce that $x^{p^h} = 1 + a^{p^h} = 1$ so all these elements are of order a power of $p$. Then let $G$ be the subgroup generated by the elements $1 + a_i$, and we claim that $G$ is infinite; otherwise, the linear span of $G$ over $F$ is a finite-dimensional algebra which contains $1, 1 + a_i$ and hence $a_i$ for all $i$ so that it equals $R$ which is instead infinite dimensional.

This question was in fact answered in the negative in 1964 by Evgeny Golod and Igor Shafarevich, who gave a method to construct finitely generated algebras $R = F[a_1, \dots, a_m]$ over a field $F$ of characteristic $p > 0$ such that every element is nilpotent and $R$ is not finite dimensional, and hence of an infinite $p$-group that is finitely generated.

In both the group and the algebra cases, one can make a further restriction. In the group case assume that $g^n = 1$ for some fixed $n$, the local finiteness question is called the *bounded Burnside problem*. In the algebra case one can assume that each element is the root of some nontrivial polynomial of some fixed degree $n$.

Remarkably in this case, the answer is different in the two cases. For an algebra this restriction implies local finiteness, while for groups this is not usually true, but only for small $n$. In fact given $m, n$ one can construct a *universal object*, that is the free group $F_m$ on $m$ generators modulo the normal subgroup generated by all the elements $g^n$, $g \in F_m$. Call this universal group $B(m, n)$. Then the question is for which values of $m, n$ is $B(m, n)$ finite? There is a very deep and extensive literature on this question. Many cases in which the answer is positive or negative are known but there is no complete solution.

There is finally the *restricted Burnside problem* which asks if for given $m, n$ there are only finitely many finite quotients of the group $B(m, n)$. This indeed has a positive answer as shown by E. Zelmanov.

We will not discuss these problems for groups since they are not strictly connected with our theme and would deserve a separate book. The interested reader can look at the existing literature as in **[Adi79]**, **[Iva94]**, **[IO96]**, **[Kos90]**, **[Ol′91]**, **[Zel90]**, **[Zel91]**.

Now consider the algebra case. Suppose that in an algebra $R$ each element is the root of some nontrivial polynomial of some fixed degree $n$. This means that, for every $a \in R$ the elements $1, a, a^2, \dots, a^n$ are linearly dependent. Consider then the following noncommutative polynomial

$$f(x; y) := \sum_{\sigma \in S_{n+1}} \epsilon_\sigma x^{\sigma(0)} y_1 x^{\sigma(1)} y_1 x^{\sigma(2)} \dots x^{\sigma(n-1)} y_n x^{\sigma(n)}, \tag{0.1}$$

where $S_{n+1}$ denotes the group of all permutations (the symmetric group) on the set $[0, 1, \dots, n]$ and $\epsilon_\sigma = \pm 1$, the sign of the permutation.

LEMMA 0.1.1. *When we substitute to $x$ any element $a$ and to the $y_i$ elements $b_i \in R$, this expression vanishes.*

PROOF. The element $a^n$ equals a linear combination of the preceding $a^i$, by hypothesis. Making this replacement in the evaluation of $f(x; y)$, the polynomial expands in a sum of terms in which we antisymmetrize a product with two equal elements $a^i$, which therefore vanishes. □

This is an instance of a *polynomial identity* for $R$ that is *a formally nonzero noncommutative polynomial which vanishes when evaluated in $R$.*

Thus the problem arose: is every algebraic algebra which satisfies a polynomial identity locally finite?

This problem had a positive solution due to the work of Jacobson **[Jac45]**, Kaplansky **[Kap48]**, and Levitzki **[Lev43]** who started the foundations of PI theory. We will later discuss some of their structure theorems which led them to a positive solution. As for the Kurosh problem, we shall follow the combinatorial proof of Shirshov, in §8.2.1 obtained a few years later.

One should notice that there are some possible constructions of a universal algebra in which every element satisfies a polynomial of degree $n$. For this one has to work over commutative rings rather than fields, and a good example will be the algebra of generic matrices with trace, cf. §12.1.17.

**0.1.2. Malćev representability question.** One of the themes of this book is related to the following general question:

QUESTION. Find necessary and sufficient conditions that a ring $R$ can be embedded into a matrix algebra $M_n(B)$ over some commutative ring $B$.

It was already noticed by Malćev **[Mal43]** that the free algebra in two or more variables cannot be embedded in any such matrix algebra. This for us follows from the fact that such a free algebra does not satisfy any polynomial identity contrary to a matrix algebra $M_n(B)$ over some commutative ring $B$.

A complete answer to the question above is not known. We discuss several results around this question in §11.5.

Finally, we introduce the notion of *algebras with trace* in §2.3. For each $n$ one can define a formal *characteristic polynomial of degree* $n$, and we can prove in Theorem 14.2.1 a very general embedding theorem, that is an algebra with trace embeds in $n \times n$ matrices preserving the trace if and only if each element satisfies its characteristic polynomial of degree $n$.

**0.1.3. The structural approach.** As we shall see in future chapters, algebras with polynomial identities behave in quite a controlled way with respect to many standard constructions of noncommutative algebra.

A possible approach to the Kurosh theorem, as done by Kaplansky and Levitzki, is to reduce to semisimple algebras where the basic structure theorem of Kaplansky (Theorem 10.1.8) can be applied.

For this one starts by remarking that, given an algebraic algebra $R$ and an ideal $I$, we have that $R$ is locally finite if and only if both $I$ and $R/I$ are locally finite. This implies immediately that the sum of locally finite ideals is locally finite, and thus we have a maximal ideal $L(R)$ which is locally finite and $L(R/L(R)) = 0$.

Moreover, it is easily seen that every nilpotent ideal $I$ is locally finite, hence $R/L(R)$ has no nilpotent ideals. Now, if a PI algebra $S$ has no nilpotent ideals, then $S$ can be embedded in a finite direct sum of matrix algebras over rings with no nil ideals Theorem 11.2.3. In this approach then one can reduce the problem of local finiteness further to algebras over $F$ which are contained in $n \times n$ matrices $M_n(K)$ over a field $K \supset F$. In this case one has various possible options to achieve the proof; see for example Herstein **[Her94]**.

# Part 1

# Foundations

CHAPTER 1

# Noncommutative algebra

*We assume that the reader is familiar with the basic ideas of algebras, graded algebras, modules (also called* representations*), and tensor products. Also some basic commutative algebra (see* **[Eis95]** *or* **[Mat89]***) will be used. By* module *we usually mean a* lcft module.

## 1.1. Noncommutative algebras

We will usually work with associative algebras $R$, over some commutative ring $F$, with a 1 element. A ring is always thought of as an algebra over the integers $\mathbb{Z}$. Sometimes we need to analyze algebras without one, as for example algebras where every element $a$ is nilpotent, that is $a^k = 0$, for some $k$. One can pass from one class to the other in a simple way. If $R$ is any algebra, over a commutative ring $F$, one can *add* 1 to the algebra by taking the space $\bar{R} := F \oplus R$ with multiplication

$$(a, r)(b, s) = (ab, as + br + rs). \tag{1.1}$$

It is an easy exercise to verify that this is an associative multiplication, $R$ is a subalgebra of $\bar{R}$ (in fact also an ideal), and in $\bar{R}$ the element $(1, 0)$ is a unit element. This allows us to apply results for algebras with a 1 also to algebras without 1. If $R$ has already a 1, call it $e$, in order to distinguish from the added 1, then $\bar{R} = F(1, -e) \oplus R$, and now this is also a direct sum of algebras, since $(1, -e)r = r(1, -e) = 0$, $\forall r \in R$ and $(1, -e)^2 = (1, -e)$. Finally, if $R$ is an algebra, even if it has not a 1, it makes sense to use formulas as $(1 - a)b = b - ab$ in $\bar{R}$ but really in $R$.

When we work with rings with a 1 for a module, we always assume that 1 acts as the identity, otherwise we have no assumption. We usually do not fully develop this step since it is usually fairly trivial. On the other hand in the theory of $T$-ideals (§2.2.1) this distinction is quite essential and nontrivial.

Given a module $M$ over a commutative ring $A$, one has the *tensor algebra*

$$T(M) := \bigoplus_{k=0}^{\infty} M^{\otimes k}.$$

Denote by $i : M = M^{\otimes 1} \to T(M)$ the inclusion.

$T(M)$ has the universal property that any linear map $f : M \to R$, with $R$ an $A$-algebra, extends to a unique homomorphism $\bar{f} : T(M) \to R$ of algebras making the diagram commutative:

$$\begin{array}{ccc} M & \xrightarrow{\ i\ } & T(M) \\ & \searrow^{f} & \downarrow{\scriptstyle \bar{f}} \\ & & R\,. \end{array}$$

The quotient of the algebra $T(M)$ by the ideal generated by all the commutators $m_1 \otimes m_2 - m_2 \otimes m_1$ is the *symmetric algebra* $S(M) := \bigoplus_{k=0}^{\infty} S^k(M)$.

The algebra $S(M)$ has the same universal property but now for commutative algebras $R$.

If $M$ is free with basis $e_1, \dots, e_h$, then $T(M)$ is identified with the free algebra $A\langle e_1, \dots, e_h \rangle$ (Definition 2.1.13) while $S(M)$ is identified with the ring of polynomials in $k$ variables, $A[e_1, \dots, e_h]$. A linear map $f : M \to R$ is given by mapping $e_i \mapsto r_i$, and we think of the associated $\bar{f} : T(M) \to R$ as *evaluating* the polynomials (noncommutative or commutative) in $R$.

Another canonical quotient of the algebra $T(M)$ is the *exterior or Grassmann algebra* $\bigwedge M := \bigoplus_{k=0}^{\infty} \bigwedge^k(M)$, quotient of $T(M)$ by the ideal generated by the elements $m^{\otimes 2}$, $m \in M$. The multiplication is denoted by $\wedge$. If $M$ is free with basis $e_1, \dots, e_h$, then $\bigwedge M := \bigoplus_{k=0}^{h} \bigwedge^k M$ and a basis of $\bigwedge^k M$ is formed by the elements $e_{j_1} \wedge \cdots \wedge e_{j_k}$, $1 \leq j_1 < j_2 < \cdots < j_k \leq h$, with $e_i \wedge e_j = -e_j \wedge e_i$.

**1.1.1. Irreducible modules and primitive ideals.** When we speak of an *ideal* in a ring, we mean a two-sided ideal (otherwise we specify *right, left*).

DEFINITION 1.1.1. Given an inclusion of rings $A \subset B$, the *centralizer* of $A$ in $B$ is

$$A' := \{x \in B \mid xa = ax,\ \forall a \in A\}. \tag{1.2}$$

Given a module $M$ over a ring $R$, we denote by $\mathrm{End}_R(M)$ the ring of $R$-linear maps of $M$ to $M$. This is also to be thought of as the *centralizer* of the image of $R$ in $\mathrm{End}(M)$ (the ring of endomorphisms of $M$ as an abelian group). Given two modules $M, N$, denote by $\hom_R(M, N)$ the abelian group of $R$-linear maps, or homomorphisms from $M$ to $N$.

We start with a basic example. Recall the notion of *opposite* $R^{\mathrm{op}}$ of a ring $R$. This is the ring with the same additive abelian group as $R$ and opposite multiplication $a \cdot b := ba$.

LEMMA 1.1.2. *Let $R$ be a ring with a 1. If we think of $R$ as left $R$-module, then $\mathrm{End}_R(R)$ is the ring of right multiplications $\mathcal{R}_r : x \to xr$ by elements $r$ of $R$, isomorphic to the opposite ring $R^{\mathrm{op}}$.*

PROOF. Clearly, a map $\mathcal{R}_r : x \mapsto xr$ (a right multiplication) commutes with left multiplication, hence it is in $\mathrm{End}_R(R)$. Conversely, let $\phi \in \mathrm{End}_R(R)$, we claim that $\phi = \mathcal{R}_{\phi(1)}$, that is right multiplication by $\phi(1)$; in fact, $\phi(r) = \phi(r.1) = r\phi(1)$. □

If $R$ does not have a 1, this result is false since $\mathrm{End}_R(R)$ has a 1.

If $R$ is any ring and $n \in \mathbb{N}$, we denote by $M_n(R)$ the ring of $n \times n$ matrices over $R$. The following fact is easy to see. Let $M$ be a left module. If we construct a direct sum $M^{\oplus k}$ of copies of $M$, we have

$$\mathrm{End}_R(M^{\oplus k}) = M_k(\mathrm{End}_R(M)). \tag{1.3}$$

REMARK 1.1.3. In particular, $M_n(R)$ is in fact the endomorphism ring of the free module $(R^{\mathrm{op}})^n$.

In general, if we take two direct sums $\bigoplus_{i=1}^k M_i$, $\bigoplus_{j=1}^h N_j$ of $R$-modules, we have a direct sum decomposition

$$\hom_R\left(\bigoplus_{i=1}^k M_i, \bigoplus_{j=1}^h N_j\right) = \bigoplus_{i=1}^k \bigoplus_{j=1}^h \hom_R(M_i, N_j),$$

which can be interpreted in matrix language.

Given a module $M$ over a ring $R$, recall that

DEFINITION 1.1.4. $M$ is called *irreducible* if $RM \neq 0$ and $M$ has no proper submodules. $M$ is called *completely reducible* or *semisimple* if $M$ is a direct sum of irreducible submodules.

A module $M$ is *faithful* if $rM = 0$ $(r \in R)$ implies $r = 0$.

Finally, a *division ring* is a ring $R$ with a 1 in which every nonzero element is invertible.

REMARK 1.1.5. In the definition of irreducible module, we exclude the trivial case in which $RM = 0$ and $M$ is a simple abelian group.

Given an $R$-module $M$, the set $I := \{r \in R \mid rM = 0\}$ is a two-sided ideal and $M$ is a faithful $R/I$-module.

The prime example for us is when we take a vector space $V$ over a field or even a division ring $D$, and consider it as a module over $\mathrm{End}_D(V)$; it is easily seen to be irreducible. When $V$ is of finite dimension $n$, we have, choosing a basis, an isomorphism $V = D^{\oplus n}$, and from Lemma 1.1.2, Remark 1.1.3, and formula (1.3), $\mathrm{End}_D(V) = M_n(D^{\mathrm{op}})$.

When $D$ is a division ring, the ring of matrices $\mathrm{End}_D(V) = M_n(D^{\mathrm{op}})$, thought of as a module over itself, is also semisimple. In fact, $M_n(D^{\mathrm{op}})$ decomposes into the direct sum of $n$ copies of the module $V = D^n$. One way of exhibiting these copies is to consider for each $i = 1, \dots, n$, the left ideal $I_i$ of $M_n(D^{\mathrm{op}})$ formed by the matrices which are zero except on the $i$th column.

Semisimple modules are also characterized by the property that, given any submodule $N$ of $M$, one can find another submodule $P$ (a complement for $N$) with $M = N \oplus P$; see Theorem 1.2.1.

The first basic property of an irreducible module $M$ is the following.

LEMMA 1.1.6 (Schur's lemma). *If $M$ is an irreducible $R$ module, $\mathrm{End}_R(M)$ is a division ring.*

PROOF. If $a : M \to M$ is any endomorphism, its kernel and image are submodules of $M$. By irreducibility, this implies that either $a = 0$ or $a$ is an isomorphism, hence invertible. □

Let $M$ be irreducible and denote by $D := \mathrm{End}_R(M)$ the division algebra centralizer of $R$. The main theorem is

THEOREM 1.1.7 (Jacobson's density). *Let $M$ be an irreducible $R$ module, and let $D := \mathrm{End}_R(M)$.*

*Given any $k$ elements $u_1, \dots, u_k \in M$ linearly independent over $D$ and other arbitrary $k$ elements $v_1, \dots, v_k \in M$, there is an element $r \in R$ with $ru_i = v_i$, for all $i = 1, \dots, k$.*

PROOF. Consider the semisimple module $M^{\oplus k}$ and in it the element $w := (u_1, \ldots, u_k)$. The statement is equivalent to proving that $Rw = M^{\oplus k}$.

If this were not true, we could find a nonzero complement $P$ to $Rw$ in $M^{\oplus k}$ (see Theorem 1.2.1) and thus a nonzero map $\pi : M^{\oplus k} \to M^{\oplus k}$ projecting to $P$ with kernel $Rw$.

We have seen that $\pi$ is given by a matrix with entries $d_{i,j} \in D$. Hence we have $0 = \pi(w) = (\sum_j d_{1,j}u_j, \sum_j d_{2,j}u_j, \ldots, \sum_j d_{k,j}u_j)$. Since the $u_j$ are linearly independent, it follows that $d_{i,j} = 0$, for all $i, j$. Hence $\pi = 0$, a contradiction. □

If $M$ is an irreducible $R$-module and $D = \mathrm{End}_R(M)$, we have.

COROLLARY 1.1.8. *If $M$ is finite dimensional over $D$, say of dimension $k$, then the natural homomorphism $\rho : R \to \mathrm{End}_D(M) = M_k(D^{\mathrm{op}})$, defining the module structure, is surjective. In particular, if $M$ is a faithful module, then $R \simeq M_k(D^{\mathrm{op}})$.*

This corollary is usually applied to the case of $R$ a simple algebra (i.e., an algebra without nontrivial two-sided ideals) with some further restrictions. In our treatment, $R$ is usually finite dimensional over some field $F$.

In fact let us make some remarks.

PROPOSITION 1.1.9. *If $M$ is an irreducible module over a ring $R$ and $m \in M$ is a nonzero element, the map (of left $R$-modules) $\pi_m : R \to M$, $r \to rm$ is surjective, with kernel a maximal left ideal.*

PROOF. First we claim that, if $m \neq 0$, then $Rm \neq 0$. In fact, the set $X$ of elements $m \in M$ such that $Rm = 0$ is a submodule, different from $M$. Since $RM \neq 0$, we have $X \neq M$. Hence, $X$ must be equal to $(0)$.

The map $\pi_m$ is surjective since its image is a nonzero submodule. By the homomorphism theorem, $M$ is isomorphic to $R/I$ where $I = \ker(\pi_m)$ is the kernel, that is the annihilator of $m$, i.e., $I := \{r \in R \mid rm = 0\}$. Thus any left ideal $J$ with $I \subset J \subset R$ induces a submodule $J/I$ of $M$, hence the claim. □

Conversely, given a maximal left ideal $I$, the left module $R/I$ is irreducible provided that $R(R/I) \neq 0$. This condition of course is trivial for a ring with 1. For a general ring it is equivalent to the assumption that $I$ be *regular* according to

DEFINITION 1.1.10. A left ideal $I$ is *regular* if there is an $a \in R$ with $x - xa \in I$, for all $x \in R$.

We assume now that $F$ is an algebraically closed field.

PROPOSITION 1.1.11. *If $F$ is algebraically closed, then every finite-dimensional division ring $D$ over $F$ equals $F$.*

PROOF. In fact, if $a \in D$, the element $a$ generates, over $F$, a field $G \supset F$ finite dimensional over $F$. Since $F$ is algebraically closed, we have $G = F$ and $a \in F$. □

In general, for a field $F$ and an integer $n$ the question of the existence of a division ring $D$ with center $F$ and dimension $n^2$ over $F$ is a delicate issue. On the other hand one has a useful general fact (cf. **[Lam91]** or **[PY95**, page 135, Theorem (i) for infinite-dimensional algebras and Theorem (ii) for finite-dimensional division algebras]. Theorem (i) for infinite-dimensional skew fields and Theorem (ii) for finite-dimensional division algebras):

THEOREM 1.1.12. *Given any field $F$ and an integer $n$, there is a division ring $D$ with center $G \supset F$ and dimension $n^2$ over $G$.*

1.1.1.1. *Simple algebras and division rings.*

DEFINITION 1.1.13. A ring $R$ is said to be *simple* if $R^2 \neq 0$ and $R$ has no proper two-sided ideals.

In general simple rings are rather mysterious objects, for instance in **[Smo02]** it is constructed a simple nil ring. If $R$ is a simple ring with 1, then one sees easily that its center is a field. If $I$ is a maximal left ideal, which exists by Zorn's lemma, since $R$ has a 1, then $R$ acts faithfully on $R/I$ (that is, using Definition 1.1.19, $\{0\}$ is a primitive ideal).

In the theory of polynomial identities we will see that a simple PI ring is finite dimensional over its center, a theorem of Kaplansky (cf. Theorem 10.1.8). We shall need some properties of simple algebras and of division rings.

If $A \subset M_n(F)$ is a subalgebra such that $F^n$, as a module over $A$, is irreducible, then we say that $A$ is irreducible.

THEOREM 1.1.14 (Wedderburn's theorem).

(1) *If the field $F$ is algebraically closed, an $F$ subalgebra $A$ of $M_n(F)$ is irreducible if and only if it is the entire algebra $M_n(F)$ of all matrices.*

(2) *A simple algebra $R$ with 1, finite dimensional over a field $F$, is isomorphic to $M_n(D)$, where $D$ is a division ring finite dimensional over $F$. Hence, if $F$ is algebraically closed, then $D = F$.*

PROOF. (1) By Corollary 1.1.8 we have that $A = M_h(D)$, where $D \supset F$ is a division algebra centralizer of $A$ and $h$ is the dimension of $F^n$, as $D$-vector space. Hence $n = \dim_F D \cdot h$. Since $F$ is algebraically closed and $D$ is finite dimensional over $F$, we must have $F = D, h = n$.

(2) Let $I$ be a maximal left ideal of $R$ so that $R/I$ is an irreducible $R$-module. Since $R$ has no proper ideals, the statement follows from Corollary 1.1.8. □

In (2) the hypothesis that $R$ has a 1 can be removed; cf. Exercise 1.3.5.

These theorems, when applied to a finite field $F$, are to be combined with another theorem of Wedderburn which implies that if $D$ is a finite division ring, then it is a field.

REMARK 1.1.15. In part (1) of Theorem 1.1.14 the assumption that $F$ is algebraically closed is necessary; for instance, we have that the field $\mathbb{C} \subset M_2(\mathbb{R})$ is an irreducible subalgebra.

LEMMA 1.1.16. *Let $S$ be a simple algebra with a 1 and center a field $F$. If $K$ is a field extension of $F$, then $S \otimes_F K$ is a simple algebra with center $K$.*

PROOF. Take a nonzero ideal $I$ of $S \otimes_F K$; we need to show that it is equal to $S \otimes_F K$. For this it is enough to show that $I \cap S \neq 0$ since then $I \cap S$ is an ideal of $S$, hence equals $S$. Hence $1 \in I$ which implies $I = S \otimes_F K$.

So take a basis $k_i$ of $K$ over $F$ so that $S \otimes_F K = \bigoplus_i S \otimes k_i$. Take an element $u \in I$, $u \neq 0$, which, written in this basis as $u = \sum_{j=1}^m s_j \otimes k_j$, involves a minimum number $m$ of indices $j$ (which we may assume are the first $m$).

Since $S$ is simple, there are elements $a_i, b_i \in S$ such that $\sum_i a_i s_1 b_i = 1$. Consider $v := \sum_i (a_i \otimes 1) u (b_i \otimes 1) = 1 \otimes k_1 + \sum_{j=2}^m s'_j \otimes k_j$, $s'_j = \sum_i a_i s_j b_i$, and we have clearly $v \in I$, $v \neq 0$. Now, if all the elements $s'_j \in F$, we have that $v \in K$, a field. So, since $v \neq 0$, it is invertible and generates $S$ as ideal. Otherwise, there are some elements

$s'_j \notin F$, and we may assume $s'_2 \notin F$. Since $F$ is the center of $S$, there is some $a \in S$ with $[a, s'_2] = as'_2 - s'_2 a \neq 0$. Then $[a \otimes 1, u] = \sum_{j=2}^m [a, s'_j] \otimes k_j \neq 0$ is a shorter sum in $I$, giving a contradiction.

Thus we have proved that $S \otimes_F K$ is simple. As for the center, take an element $u = \sum_j s_j \otimes k_j$ in the center. If one of the elements $s_j$ say $s_1$ is not in $F$ there is an $a \in S$ with $[a, s_1] \neq 0$ and then $[a \otimes 1, u] = \sum_{j=1}^m [a, s_j] \otimes k_i \neq 0$. □

THEOREM 1.1.17. *Let $D$ be a division algebra over its center $F$, and let $K$ be a maximal subfield of $D$. If $\dim_K D = h < \infty$, as right $K$-module, we have that $D \otimes_F K = M_h(K)$ and $\dim_F K = h$.*

PROOF. If $\{u_i\}_{i \in I}$ is a basis of $D$ over $K$ and $\{v_j\}_{j \in J}$ is a basis of $K$ over $F$, we see that $\{u_i v_j\}_{i \in I,\, j \in J}$ is a basis of $D$ over $F$, hence $\dim_F D = \dim_K D \cdot \dim_F K$. If we prove that $D \otimes_F K = M_h(K)$, we have in particular that

$$\dim_F D = \dim_K M_h(K) = h^2 \implies \dim_F K = h.$$

Now consider $D$ as module over $D \otimes_F K$ by the action $(a \otimes k)b := abk$.

We clearly have that $D$ is irreducible as $D \otimes_F K$-module, and we claim that $K$ is its endomorphism ring. In fact, by Lemma 1.1.2 we have that the endomorphisms of $D$ as left $D$-module form the right multiplications $D^{\mathrm{op}}$. Any such right multiplication, which commutes also with $K$, must be in $K$, by the maximality hypothesis on $K$.

Now by the density theorem we deduce that the homomorphism of $D \otimes_F K$ to $M_h(K) = \mathrm{End}_K D$ is surjective, so it is enough to show that it is also injective. For this we apply Lemma 1.1.16, which claims that $D \otimes_F K$ is a simple algebra. □

From Theorems 1.1.14 and 1.1.17 it follows that, if $S$ is a simple algebra finite dimensional over a field $F$, then $S \simeq M_h(D)$, with $D$ of some dimension $k^2$ over its center $G \supset F$. Finally, if $\bar{G}$ is an algebraic closure of $G$, we have $S \otimes_G \bar{G} \simeq M_{hk}(\bar{G})$.

DEFINITION 1.1.18. Given $S$ a simple algebra finite dimensional over its center $G$, the number $h$ such that $\dim_G S = h^2$ is called the *degree* of $S$.

1.1.1.2. *The reduced trace.* If $S = M_n(D)$ is a matrix algebra over a division ring finite dimensional over its center $F$, we have seen the existence of *splitting fields* $K$ with

$$S \otimes_F K = M_{nh}(K), \quad h \text{ the degree of } D.$$

Then it is easy to see that for an element $a \in S$ the trace of $a \otimes 1 \in M_{nh}(K)$ lies in $F$ and it is independent of the splitting field $K$, it is called the *reduced trace* (see Theorem 2.6.15 and also §5.4.2.7 for a general treatment).

**1.1.2. The Jacobson radical.** One of the themes of noncommutative algebra is *representation* theory, that is the study of the modules of a given ring or algebra $R$. Of particular relevance are the irreducible representations. We have seen, in Proposition 1.1.9, that a maximal left ideal is the annihilator of a single nonzero element of an irreducible module. Thus the intersection of all maximal left ideals is the set of elements that annihilate all irreducible modules. That is the elements of $R$ which *do not see* any irreducible module.

DEFINITION 1.1.19. The annihilator of an irreducible module is called a *primitive ideal.*

A ring $R$ is called primitive if $\{0\}$ is a primitive ideal.

In particular, for a given irreducible module $M$ the annihilator ideal is the intersection of the annihilators of all its elements, which are maximal left ideals. Thus

DEFINITION 1.1.20. The intersection of all primitive ideals equals the intersection of all maximal left ideals $I$. This intersection is called *the Jacobson radical* of $R$ and is denoted by $J(R)$.

If $R$ is a ring with a 1, there is a simple characterization of $J(R)$.

PROPOSITION 1.1.21. *$J(R)$ is the maximal ideal $I$ of $R$ with the property that $1 - a$ is invertible for all $a \in I$.*

PROOF. First we show that $J(R)$ is also the maximal left ideal formed by elements $a$ such that $1 - a$ is left invertible. If $1 - a$ is not left invertible, $R(1 - a) \neq R$ and, by Zorn's lemma, there is a maximal left ideal $I \supset R(1 - a)$, so with $1 - a \in I, a \notin I$. Hence $a \notin I$ implies $a \notin J(R)$. Conversely, assume that $a \notin I$, for some maximal left ideal $I$, then there is a $b \in R$ such that $ba \equiv 1$ modulo $I$, that is $1 - ba \in I$. Hence $1 - ba$ is not left invertible.

Now take $a \in J(R)$ and let $b$ be such that $1 = b(1 - a)$. We then have that $1 - b = -ba \in J(R)$. Hence $b$ is also left invertible and there is an element $c$ with $cb = 1$. Hence $c = cb(1 - a) = 1 - a$, and we see that $1 - a$ is invertible. □

REMARK 1.1.22. One can also describe the same property for rings without 1, introducing the notion of *quasi-regular elements* (cf. **[Jac89]**, **[Her94]**). In fact, adding formally a 1, the statement that an element $1 - a$ has a right inverse $1 - b$ means that $a + b - ab = 0$.

So we define $a$ to be *quasi-regular* if there is a $b$ with $a + b - ab = 0$, $ab = ba$. In other words we define the *circle law*: $a \circ b := a + b - ab$ so that, if there is an element 1, $(1 - a)(1 - b) = 1 - a \circ b$, which is clearly an associative multiplication with 0 as unit element, and a quasi-regular element is an invertible element with respect to this law with 0 playing the role of 1.

PROPOSITION 1.1.23. *The Jacobson radical is an ideal whose elements are all quasi-regular and are uniquely largest with respect to this property.*

PROOF. Assume that $a \in J(R)$ does not have a right quasi-inverse. Then the left ideal $K := \{b - ba, \forall b \in R\}$ is proper and $a \notin K$. Take a maximal left ideal $I \supset K$. Again it is clear that $I$ is maximal as a proper left ideal, it is regular (Definition 1.1.10) by construction, and then the reasoning is like in the case of rings with a 1. □

Recall that

DEFINITION 1.1.24. An element $a$ of a ring $R$ is *nilpotent* if $a^n = 0$, for some positive integer $n$.

A left (resp., right) ideal $I$ is *nil* if every element of $I$ is nilpotent.

An ideal $I$ is *nilpotent* if $I^n = 0$, for some positive integer $n$.

THEOREM 1.1.25. *If $A$ is a finite-dimensional algebra over a field $F$, its Jacobson radical $J(A)$ is nilpotent.*

PROOF. The descending chain $J(A) \supseteq J(A)^2 \supseteq \cdots$ stops by finite dimensionality, say $I = J(A)^n = J(A)^{n+1}$. If $J(A)^n \neq 0$, then there is a left ideal

$L \subset I = J(A)^n$ such that the $A$-module $I/L$ is irreducible, again by finite dimensionality. But then $J(A)^{n+1} = J(A)I \subset L$, a contradiction. □

The proof can be modified if we do not assume that $A$ has a 1; see [**Jac89**, Thm. 1 on p. 38].

REMARK 1.1.26. The previous theorem is valid in the greater generality of rings with descending chain condition or Artinian rings (cf. Definition 1.3.6 and Theorem 1.3.7).

**1.1.3. Nil and nilpotent ideals.** The following proposition is straightforward, and we leave it as an exercise.

PROPOSITION 1.1.27. *If $I \subseteq J$ are ideals in $R$ such that both $I$ and $J/I$ are nil, then $J$ is nil.*

*The sum of two nil ideals is a nil ideal.*

*The sum of two left (resp., right) nilpotent ideals is nilpotent.*

*If $I$ is a left nilpotent ideal, then $IR$ is a nilpotent two-sided ideal.*

DEFINITION 1.1.28. The sum of all the nil ideals of $R$ is a nil ideal, called the *nil radical*, and denoted by $\mathcal{N}(R)$.

If $I$ is an ideal of $R$, its *nil radical* is the ideal $\sqrt{I} \supset I$ such that $\sqrt{I}/I = \mathcal{N}(R/I)$.

We call $\mathcal{N}(R)$ a *radical* since, as it is easily seen, we have $\mathcal{N}(R/\mathcal{N}(R)) = 0$. That is the ring $R$ modulo its nil radical has no nil ideals.

Clearly, any nilpotent ideal is nil but the converse is far from being true even for commutative rings. Consider now

DEFINITION 1.1.29. The sum of all the nilpotent ideals of a ring $R$ will be denoted by $N(R)$.

Contrary to the nil radical, this ideal does not behave as a radical, in the sense that it can happen that $N(R/N(R)) \neq 0$. We can thus define inductively $N_{k+1}(R)$ by

$$N_{k+1}(R) \supset N_k(R),\ N_{k+1}(R)/N_k(R) := N(R/N_k(R)). \tag{1.4}$$

This inductive construction does not end in general even after infinitely many steps, and it has to be continued by transfinite induction. Then, at some ordinal number $\lambda$, it will stop, that is $N_{\lambda+1}(R) = N_\lambda(R)$. The ideal that we obtain in this way is a nil ideal, called the *lower or Baer* nil radical, and we will denote it by $\mathcal{L}(R)$. By definition $R/\mathcal{L}(R)$ has no nilpotent ideals.

DEFINITION 1.1.30. A ring $R$ with $\mathcal{L}(R) = 0$ is called *semiprime*.

An ideal $I$ of a ring $R$ is called *semiprime* if $R/I$ is semiprime.

Notice that $R$ is semiprime if and only if $R$ has no nonzero ideal $I$ with $I^2 = 0$.

In general $\mathcal{L}(R) \subset \mathcal{N}(R)$ but the two may be different. Nil ideals for general rings may be quite strange: a famous conjecture (still unproven) of Koethe claims that if $R$ has a nonzero left nil ideal, then it also has a nonzero two-sided nil ideal. There is an extensive literature on these topics. For instance, Agata Smoktunowicz in [**Smo02**] shows the existence of a simple nil ring.

As we shall see, for rings with PI all these possible pathologies disappear. A general theorem of Amitsur will be useful (see [**Ami56**]). First we need to clarify the notion of polynomial ring $R[t]$, in the case when $R$ does not have 1.

In this case it is convenient to *add* the variable $t$. That is, one works in the usual ring of polynomials $\bar{R}[t] := \{\sum_{j=0}^{h} a_j t^j,\ a_j \in \bar{R}\}$ with coefficients in the algebra $\bar{R} := \mathbb{Z} \oplus R$ of formula (1.1) with 1 added, by taking the subring generated by $R$ and $t$, that is $R[t] := \{\sum_{j=0}^{h} a_j t^j,\ a_0 \in R,\ a_j \in \bar{R}, \forall j > 0\}$.

We also need a lemma from commutative algebra.

LEMMA 1.1.31. *Let $A$ be a commutative ring with 1. A polynomial $1 - u \in A[t]$, with $u := \sum_{j=1}^{m} b_j t^j$ is invertible if and only if the elements $b_j$ are all nilpotent.*

PROOF. If the $b_j$ are nilpotent, since $A$ is commutative, it follows that for some large $N$ all products of $N$ of the $b_j$'s are 0. Hence we see that

$$u^N = 0 \implies (1 + u + u^2 + \cdots + u^{N-1})(1 - u) = 1.$$

Conversely, if at least one element $b_j$ is not nilpotent, there is a prime ideal $P \subset A$ with $b_j \notin P$ so that in $A/P$ the image of $1 - u$ is a polynomial of positive degree which cannot be invertible. □

THEOREM 1.1.32 (Amitsur **[Ami56]**). *Let $R$ be a ring which has no nonzero nil ideals, then the Jacobson radical $J$ of the polynomial ring $R[t]$ is 0.*

PROOF. Assume by contradiction that $J \neq 0$ and pick an element

$$0 \neq r = \sum_{i=0}^{k} a_i t^{h_i} \in J, \quad h_0 < h_1 < \cdots < h_k,$$

with the least possible number $k + 1$ of nonzero coefficients $a_i$.

For all $i$ we have that $a_i r - r a_i \in J$ has at most $k$ nonzero coefficients, hence it must be 0. We deduce that the elements $a_i$ commute among each other.

We also have that $rt \in J$. Hence by Remark 1.1.22, we have that $rt$ is quasi-regular and there is an element $s \in R[t]$ with $rt \circ s = rt + s - rts = 0$ or $s = rst - rt$. Substituting $s \mapsto rst - rt$, we have that for all $n$,

$$s = r(rst - rt)t - rt = r^2 st^2 - r^2 t^2 - rt = r^3 st^3 - r^3 t^3 - r^2 t^2 - rt = \cdots$$

$$= r^{n+1} s t^{n+1} - \sum_{i=1}^{n} r^i t^i.$$

As soon as $n$ exceeds the degree $a$ of $s$, we see that $s$ coincides with the part of degree $a$ of $-\sum_{i=1}^{n} r^i t^i$. Let $R_0 \subset \bar{R}$ be the commutative subring generated by $1, a_0, \ldots, a_k$. We have that $s \in R_0[t]$, so the element $1 - rt$ has inverse $1 - s$. Then by Lemma 1.1.31 the coefficients $a_j$ are all nilpotent, in particular $a_k$ is nilpotent.

Now, for all $u_j, v_j \in R$ (or equal to 1), we also have $\sum_j u_j r v_j \in J$ has the same properties as $r$ or is 0, so $\sum_j u_j a_k v_j$ is either 0 or is the leading coefficient of $\sum_j u_j r v_j$. Hence, by the previous argument, $\sum_j u_j a_k v_j$ is nilpotent. Thus the two-sided ideal generated by $a_k$ is nil, a contradiction. □

In the same way Amitsur proves that in general $J(R[t]) = N[t]$, with $N$ some nil ideal.

### 1.1.4. Prime rings.

LEMMA 1.1.33. *The following conditions on a ring $R$ (not necessarily with a $1$) are equivalent:*

(1) *Given $a, b \in R$ both $\neq 0$, we have $aRb \neq 0$.*
(2) *Given $a, b \in R$ both $\neq 0$, we have $aR^2b \neq 0$.*
(3) *Given a right ideal $I \neq 0$ and a left ideal $J \neq 0$, we have $IJ \neq 0$.*
(4) *Given two nonzero two-sided ideals $I, J \subset R$, we have $IJ \neq 0$.*
(5) *Given two nonzero left (resp., right) ideals $I, J \subset R$, we have $IJ \neq 0$.*

PROOF. Clearly (2) implies (1). Assume (1). If $a \neq 0$, there is some $x \in R$ with $a' = ax \neq 0$. Hence $axRb \neq 0$ implies $aR^2b \neq 0$, so (1) implies (2).

(2) implies (3). In fact, take $0 \neq a \in I, 0 \neq b \in J$, we have $0 \neq aR^2b \subset IJ$.

Clearly (3) implies (4). Now assume (4), first remark that the set of elements $a \in R$ such that $aR = 0$ is a two-sided ideal, then by (4) it must be 0. Now given two left ideals $I, J$, we have that $IR$ and $JR$ are nonzero two sided ideals. If we had $IJ = 0$, we would have $IRJR \subset IJR = 0$, contradicting (4). The argument for right ideals is the same, hence (4) implies (5). Finally assume (5). Given $a \neq 0$, $b \neq 0$, we have $I = Ra$, $J = Rb$ are nonzero left ideals. Hence $IJ = RaRb \neq 0$ implies $aRb \neq 0$, so (5) implies (1). □

DEFINITION 1.1.34. A ring $R$ satisfying the previous equivalent conditions is called *prime*.

A two-sided ideal $I$ of a ring $R$ is said to be *prime* if $R/I$ is a prime ring.

Clearly the previous conditions can be interpreted as conditions on $I$.

For a set $I \subset R$, denote by $r(I) := \{x \in R \mid Ix = 0\}$ *the right annihilator of $I$*. Clearly $r(I)$ is a right ideal. Similarly, the left annihilator $\ell(I) := \{x \in R \mid xI = 0\}$ is a left ideal.

REMARK 1.1.35. If $R$ is a domain (no zero divisors), it is prime. More, if $R$ is commutative, the two notions agree. In general, if $I$ is an ideal of $R$ so that $R/I$ is a domain, we shall say that $I$ is *completely prime*.

PROPOSITION 1.1.36. *Let $R$ be a prime ring, and let $I$ be a nonzero left ideal in $R$. Then $I \cap r(I)$ is a two-sided ideal of $I$ and $I/I \cap r(I)$ is a prime ring.*

PROOF. We have that $r(I)$ is a right ideal in $R$, so $r(I) \cap I$ is also a right ideal in $I$. Now, if $a \in I$, we have $ar(I) = 0$. Hence $I \cap r(I)$ is a two-sided ideal of $I$.

Let now $a, b \in I$ and nonzero modulo $I \cap r(I)$, i.e., $a, b \notin r(I)$. This means that the left ideals $Ia, Ib$ are nonzero in $R$. Hence, since $R$ is prime, $IaIb \neq 0$. Finally, $IaIb \subset I$, but $IaIb \not\subset r(I)$, otherwise we would have that $IIaIb = 0$, contradicting item (5) of Lemma 1.1.33. □

A commutative prime ring is just an integral domain. This is not the case in the noncommutative case (e.g., a simple ring is prime), but we have several similar properties for the central part of a prime ring.

The following is quite immediate.

EXERCISE 1.1.37. Let $R$ be a prime ring, and let $\mathcal{Z}$ be its center. Assume that $\mathcal{Z} \neq 0$, then $\mathcal{Z}$ is a domain and $R$ is torsion free over $\mathcal{Z}$.

If $K$ is the field of fractions of $\mathcal{Z}$, we have that $R \subset R \otimes_{\mathcal{Z}} K$. Moreover, $R \otimes_{\mathcal{Z}} K$ is a prime ring with center $K$.

In general, if a prime ring $R$ is an algebra over a commutative ring $A$, it is faithful over $A/I$, $I = \{a \in A \mid aR = 0\}$. Then $A/I$ is a domain, $R$ is torsion free over $A/I$, and one can embed $R$ in $R \otimes_A K$, where $K$ is the field of fractions of $A$.

EXERCISE 1.1.38. Let $R \subset S$ be rings. Assume that $S$ is prime and that is $S = RC$, where $C$ is the centralizer of $S$. Prove that $R$ is prime.

In particular if $S = RZ$ with $Z$ the center of $S$, we say that $S$ is a *central extension* of $R$.

A special class of prime rings are primitive rings (Definition 1.1.19).

PROPOSITION 1.1.39. *A primitive ring $R$ is prime.*

PROOF. By definition a primitive ring $R$ has a faithful irreducible module $M$. Hence, given $a, b \neq 0$ in $R$, we have first $bM \neq 0$, by the faithfulness of the action, next $RbM = M$ since $M$ is irreducible. So finally $aRbM \neq 0$ implies $aRb \neq 0$. □

Of course when $M$ is finite dimensional over $\mathrm{End}_R(M)$, then $R$ is even simple. The next lemma is often useful.

LEMMA 1.1.40. *Let $R \subset S$ be an inclusion of rings, and let $P$ be a prime ideal of $R$. Given any ideal $I$ of $S$ with $I \cap R \subset P$, there is an ideal $Q \supset I$ maximal with the property $Q \cap R \subset P$ and $Q$ is a prime ideal.*

PROOF. The existence of $Q$ maximal is by Zorn's lemma. Assume by contradiction that $I, J$ are two ideals properly containing $Q$ with $IJ \subset Q$. Then, by the maximal property of $Q$, we have that $I \cap R, J \cap R$ are two ideals not contained in $P$ with $(I \cap R)(J \cap R) \subset IJ \cap R \subset Q \cap R = P$, contradicting the fact that $P$ is prime. □

1.1.4.1. *Semiprime rings.* By Definition 1.1.30 a ring $R$ is semiprime if there is no nonzero ideal $I \subset R$ with $I^2 = 0$. In a commutative ring this means that $R$ has no nonzero nilpotent elements, but in the noncommutative case this is clearly not the case. Any prime ring is semiprime and the converse is obviously not true even in the commutative case.

THEOREM 1.1.41. *$R$ is semiprime if and only if $\{0\}$ is an intersection of prime ideals.*

PROOF. Let $I$ be an ideal with $I^2 = 0$. If $P$ is a prime ideal, clearly $I \subset P$, so, if $\{0\}$ is an intersection of prime ideals, we have $I = 0$.

Conversely, assume that $R$ is semiprime. We need to show that, if $0 \neq a \in R$, then there is a prime ideal $P$ with $a \notin P$. We make a recursive construction setting $a := a_0$. By hypothesis, since $Ra_0R$ is a nonzero two-sided ideal, $a_0Ra_0 \neq 0$, so we may find a $b_0$ so that $a_1 = a_0b_0a_0 \neq 0$. We now repeat the construction for $a_1$ and find $b_1$ so that $a_2 = a_1b_1a_1 \neq 0$. Continuing by induction, we find a sequence $a_i, b_i$ such that $a_{i+1} = a_ib_ia_i \neq 0$, $\forall i$.

Let now $P$ be an ideal maximal with the property that $P$ does not contain any element $a_i$, this exists by Zorn's lemma, we claim that $P$ is prime. Otherwise there exist two ideals $I \supsetneq P, J \supsetneq P$ with $IJ \subset P$. By maximality we have, for some $i, j$, that $a_i \in I$, $a_j \in J$. Hence $a_h \in I$, $\forall h \geq i$, $a_k \in J$, $\forall k \geq j$. So, if we take $k \geq \max(i, j)$, we have $a_kxa_k \in IRJ \subset P$, for all $x$. In particular $a_kb_ka_k = a_{k+1} \in P$, a contradiction. □

## 1.2. Semisimple modules

The basic fact on semisimple modules is the following.

THEOREM 1.2.1. *For a module $M$ the following conditions are equivalent.*

(i) *$M$ is a sum of irreducible submodules.*
(ii) *$M$ is a direct sum of irreducible submodules.*
(iii) *Every submodule $N$ of $M$ admits a complement, i.e., a submodule $P \subset M$ such that $M = N \oplus P$.*

PROOF. This is a rather abstract theorem, and the proof is correspondingly abstract. Since we really need this only for finite-dimensional modules over some base field $F$, we give the simplified proof in this case and refer the interested reader to standard books for the general case, or we suggest the reader do it as an exercise in transfinite induction.

We remark first that, if $P, N \subset M$ are submodules and $N$ is irreducible, we either have $P \cap N = 0$ and then $P, N$ form direct sum $P \oplus N$ or $N \subset P$.

Clearly (ii) implies (i). So we prove (i) implies (ii) implies (iii) implies (ii).

(i) implies (ii). Assume (i) holds and write $M = \sum_{i=1}^{k} N_i$ for some minimal set of indices (at most the dimension of $M$ over $F$). We claim that this sum is direct and do it by induction on $k$. We have that $P := \sum_{i=1}^{k-1} N_i$ is a direct sum; by minimality we have $P \subsetneq M$, so by the previous remark $P \cap N_k = 0$, and we have the direct sum decomposition of $M$.

(ii) implies (iii). Assume (ii) and consider $M = \bigoplus_{i=1}^{k} N_i$, with the $N_i$ irreducible. Then, if $P \neq M$ there is an $N_i$ with $N_i \not\subset P$, hence $N_i \cap P = 0$ and $P \oplus N_i$ is a direct sum. We can continue in this way until we have built a direct sum of some of the $N_i$, which gives a complement to $P$.

(iii) implies (ii). Let $N_1 \subset M$ be a minimal nonzero submodule. It exists since $M$ is finite dimensional and it is irreducible since it is minimal. By (iii) we have a complement $P$ to $N_1$ so $M = P \oplus N$. If $P = 0$, we have $M$ irreducible and nothing to prove. Otherwise, let $N_2 \subset P$ be a minimal nonzero submodule, $N_2$ is irreducible and we have the direct sum $N_1 \oplus N_2 \subset M$. If $N_1 \oplus N_2 \neq M$, we take a complement and proceed in the same way. After at most $\dim_F M$ steps, we must have decomposed the entire $M$ into a direct sum of irreducible submodules. □

COROLLARY 1.2.2. *If $M$ is a semisimple module and $N \subset M$ is a submodule, then $N$ and $M/N$ are semisimple.*

PROOF. $M$ is a sum of irreducible submodules, so also $M/N$ is the sum of the images of these irreducibles which are either irreducible or zero. So $M/N$ is semisimple. Then $N$ has a complement $P$ and $N$ is isomorphic to $M/P$ so the previous argument applies. □

1.2.0.1. *Isotypic components.*

DEFINITION 1.2.3. Given a semisimple module $M$ over a ring $R$ and an irreducible module $N$, the *isotypic component of type $N$* in $M$ is the sum of all irreducible submodules of $M$ isomorphic to $N$ and denoted $I_N(M)$.

From the proof of Theorem 1.2.1 and Corollary 1.2.2, it follows that the isotypic component $I_N(M)$ is a direct sum of copies of $N$. From this it follows that, if $P \subset I_N(M)$ is an irreducible submodule, then $P$ must be isomorphic to $N$ since the projection on at least one of the summands is nonzero and hence an isomorphism.

THEOREM 1.2.4. *$M$ is the direct sum of its isotypic components. If $M, P$ are two semisimple modules and $f : M \to P$ is a morphism, then the isotypic component of type $N$ for $M$ is mapped under $f$ to the isotypic component of type $N$ for $P$.*

PROOF. We prove this by induction. Suppose that we have already constructed the isotypic components $M_1, M_2, \dots, M_k$, relative to the nonisomorphic modules $N_1, \dots, N_k$, and proved that they form a direct sum. Take another isotypic component $M_{k+1}$, then $M_{k+1} \cap M_1 \oplus M_2 \oplus \cdots \oplus M_k$ must be 0, otherwise it contains an irreducible submodule isomorphic to one of the $N_i$, $i = 1, \dots, k$, which cannot be contained in $M_{k+1}$.

The second part follows the same lines. □

Let us consider a semisimple module $M$ decomposed into its isotypic components $M_i$. Since every endomorphism of $M$ maps each $M_i$ into itself, we have $\mathrm{End}_R(M) = \bigoplus_i \mathrm{End}_R(M_i)$.

Now an isotypic component is a direct sum $N^{\oplus h}$ of $h$ copies of the same irreducible module $N$. We know that $\mathrm{End}_R(N) = D$ is a division algebra, hence

$$\mathrm{End}_R(N^{\oplus h}) = M_h(D), \quad \mathrm{End}_R(M) = \bigoplus_{i=1}^{k} M_{h_i}(D_i). \tag{1.5}$$

## 1.3. Finite-dimensional algebras

### 1.3.1. Semisimple algebras.

DEFINITION 1.3.1. A finite-dimensional algebra over a field $F$ is said to be semisimple if its Jacobson radical is zero.

PROPOSITION 1.3.2. *A finite-dimensional algebra $A$, with 1, has $J(A) = 0$ if and only if it is semisimple as a left $A$-module.*

PROOF. Assume first that $J(A) = 0$. Since by definition $J(A)$ is the intersection of the maximal left ideals and $A$ is finite dimensional, there are finitely many maximal left ideals $I_1, \dots, I_k$ such that $\bigcap_{i=1}^k I_i = 0$. Hence $A$ as $A$-module embeds into the direct sum $\bigoplus_{i=1}^k A/I_i$ which, by construction, is semisimple. Hence by Corollary 1.2.2 $A$ is semisimple as a module.

Conversely, assume that $A$ is a finite-dimensional algebra with 1 semisimple as a left module. So $A = \bigoplus N_i$, where the $N_i$ are minimal left ideals.

Then $J(A) = \bigoplus J(A)N_i = 0$. □

THEOREM 1.3.3. *A finite-dimensional algebra $A$, semisimple as a left module, is isomorphic to a direct sum $\bigoplus_{i=1}^k M_{h_i}(D_i)$ of matrix algebras over division rings.*

PROOF. By Lemma 1.1.2 have that $A^{\mathrm{op}} = \mathrm{End}_A(A)$. By Proposition 1.3.2 and by formula (1.5), $\mathrm{End}_A(A) = \bigoplus_{i=1}^k M_{h_i}(\Delta_i)$ is a direct sum of matrix algebras over division rings. Hence

$$A = \mathrm{End}_A(A)^{\mathrm{op}} = \bigoplus_{i=1}^{k} M_{h_i}(\Delta_i)^{\mathrm{op}} = \bigoplus_{i=1}^{k} M_{h_i}(\Delta_i^{\mathrm{op}}).$$

The isomorphism between $M_{h_i}(\Delta_i)^{\mathrm{op}}$ and $M_{h_i}(\Delta_i^{\mathrm{op}})$ is given by mapping a matrix to its transpose. □

COROLLARY 1.3.4. *A semisimple algebra $A$ over an algebraically closed field $F$ is isomorphic to a direct sum $\bigoplus_{i=1}^k M_{h_i}(F)$ of matrix algebras over $F$.*

PROOF. This follows from Proposition 1.1.11. □

EXERCISE 1.3.5. Prove that a finite-dimensional simple algebra $S$ over a field $F$ has a 1.

Discuss how the hypothesis that $A$ has a 1 can be removed in the previous results.

PROOF.

*Hint.* Consider the algebra with 1 $F \oplus S$ defined by formula (1.1). We remark that its Jacobson radical is 0 and then, since $S$ is an ideal, the statement follows from Theorem 1.3.3.

In general compare $J(A)$ with $J(F \oplus A)$.

□

Theorem 1.3.3 can be extended as follows.

DEFINITION 1.3.6. A ring $R$ is said to satisfy the *descending chain condition* on left ideals or to be *left Artinian*, if every descending chain of left ideals $I_1 \supset I_2 \supset \cdots \supset I_k \supset \cdots$ is stationary, that is there is a $k$ such that $I_k = I_h, \forall h \geq k$. This holds similarly for *right Artinian*.

The following is a standard general theorem for which we refer to **[Jac89]**.

THEOREM 1.3.7. *If $R$ is an Artinian ring, its Jacobson radical is nilpotent. For a ring $R$ the following conditions are equivalent.*

(1) *$R$ is semisimple Artinian.*
(2) *$R$ is isomorphic to a direct sum $\bigoplus_{i=1}^k M_{h_i}(D_i)$ of matrix algebras over division rings.*
(3) *Every module $M$ over $R$ is semisimple.*
(4) *$R$ is semisimple as a left $R$-module.*

1.3.1.1. *Change of fields.* Let $R = \bigoplus_{i=1}^t M_{h_i}(D_i)$ be a semisimple algebra finite dimensional over a field $F$, with $D_i$ division rings. Consider a finite field extension $F \subset G$. We ask if $R \otimes_F G$ is still semisimple. Since $R \otimes_F G = \bigoplus_{i=1}^t M_{h_i}(D_i \otimes_F G)$, the problem is to study $D \otimes_F G$, where $D$ is a division ring finite dimensional over $F$.

Let thus $Z$ be the center of $D$ and consider $Z \otimes_F G$. It is then a standard fact from field theory that $Z \otimes_F G$ has no nonzero nilpotent elements for all $G$, if and only if $Z \otimes_F Z$ has this property, and this is if and only if $F \subset Z$ is a separable extension. In particular, if $F \subset Z$ is a separable extension of degree $k$, we have for the algebraic closure $\bar{Z}$ of $Z$ that $Z \otimes_F \bar{Z} = \bar{Z}^k$, so that $D \otimes_F \bar{Z} = D \otimes_Z \bar{Z}^k$. By Lemma 1.1.16, we have $D \otimes_Z \bar{Z}$ is a simple algebra with center $\bar{Z}$. Hence by Corollary 1.3.4 it is of the form $M_h(\bar{Z})$, for some $h$. In particular, $\dim_F D = h^2$.

COROLLARY 1.3.8. *If $R$ is a finite-dimensional semisimple algebra over a field $F$ of characteristic 0, then for any field extension $F \subset G$ we have that $R \otimes_F G$ is a finite-dimensional semisimple algebra over $G$.*

1.3.1.2. *Idempotents.*

DEFINITION 1.3.9. An *idempotent* in a ring $R$ is an element $e$ such that $e^2 = e$. Two idempotents $e, f$ are *orthogonal* if $ef = fe = 0$.

Observe that, if $e, f$ are two orthogonal idempotents, then $e + f$ is also an idempotent.

EXERCISE 1.3.10. A finite-dimensional algebra $R$ over a field $F$ is semisimple if every left (resp., right) ideal $I$ is generated by an idempotent, that is $I = Re,\ e^2 = e$.

DEFINITION 1.3.11. An idempotent is called *primitive* if it cannot be decomposed as a sum $e = e_1 + e_2$ of two nonzero orthogonal idempotents.

EXERCISE 1.3.12. One can then verify that if $R$ is a semisimple algebra over $F$, the left ideal $Re$ is irreducible as a left $R$-module if and only if $e$ is primitive. This is equivalent to the fact that $eRe$ is a division algebra (or the field $F$ when $F$ is algebraically closed). Finally, given two primitive idempotents $e_1, e_2$ one has that $Re_1$ is not isomorphic to $Re_2$ if and only if $e_1Re_2 = 0$.

**1.3.2. Wedderburn's principal theorem.** Let $R$ be a finite-dimensional algebra over a field $F$, let $J$ be its radical, and let $\bar{R} := R/J$.

THEOREM 1.3.13. *If $\bar{R}$ is separable over $F$, there exists a subalgebra $S \subset R$ such that $R = S \oplus J$ as vector spaces.*

Of course then $S$ is isomorphic to $\bar{R}$. We have not yet defined what is a *separable* algebra $S$, and we shall develop a general theory in §5.4. For the moment assume $S$ is finite-dimensional semisimple. In characteristic 0 any finite-dimensional semisimple algebra is separable. In characteristic $p > 0$ this means that the center of $S$, which, by Theorem 1.3.3, is a direct sum of fields, is such that each of these fields is a separable extension of $F$, in turn this is equivalent to saying that $S \otimes_F S^{\mathrm{op}}$ is semisimple.

There is an immediate reduction of this statement to the special case in which $J^2 = 0$. In fact, we can prove it by induction on the nilpotence degree of $J$, that is the minimum $k$ such that $J^k = 0$.

Suppose that $J^k = 0$ and we have already proved the theorem for algebras for which $J^{k-1} = 0$, then we have the statement for the algebra $R/J^{k-1}$. So we have in this algebra a subalgebra $\bar{S}$ complement to $J/J^{k-1}$.

Let $\pi : R \to R/J^{k-1}$ be the quotient map and take $\tilde{R} := \pi^{-1}(\bar{S})$. In $R$ the subalgebra $\tilde{R}$ has the property that its radical is $\tilde{J} = J^{k-1}$ and it maps surjectively to $\bar{S}$. So $\tilde{R}$ is an algebra with radical $\tilde{J}$ with $\tilde{J}^2 = 0$ and, if we can prove the theorem for this algebra, we finally have the statement for $R$. In the end the best way to handle this theorem in general is to give to it a *cohomological flavor* but this would take us far from the theme of this book.

We shall not need this theorem in this generality; one can see for instance Jacobson **[Jac89]** or Albert **[Alb39]** who gives a direct proof. Let us prove it when $\bar{R}$ is a direct sum of matrix algebras over $F$, for instance this is the case always if $F$ is algebraically closed. Let $\bar{e}_1, \dots, \bar{e}_k$ be the unit elements of these matrix algebras. They are a list of *orthogonal idempotents*, that is $\bar{e}_i^2 = \bar{e}_i, \forall i,\ \bar{e}_i\bar{e}_j = 0,\ \forall i \neq j$. We need a simple lemma.[1]

[1] It is really a small part of Hensel's lemma.

LEMMA 1.3.14. *Let $R$ be a ring, and let $J$ be a nil ideal. Given an idempotent $\bar{e} \in R/J$, we can lift it to an idempotent $e \in R$. More generally, we can lift any $k$ orthogonal idempotents to orthogonal idempotents.*

PROOF. Let $e \in R$ be any element lifting $\bar{e}$, it is enough to do the lifting in the subring $S$ generated by $e$ which is commutative. Since $\bar{e}^2 - \bar{e} = 0$, we have that $e^2 - e = j \in J$ is nilpotent. So $(e^2 - e)^k = 0$, for some $k$, and $S$ is finite dimensional. We replace $R$ with $S$.

By induction on $k$, if $k = 1$, we are done. Otherwise, replace $e$ with the new lift $e_1 := e + (1 - 2e)j$, and set $j_1 := e_1^2 - e_1$. We have (see p. 7):

$$\begin{aligned} j_1 &= e^2 + (1-2e)^2 j^2 + 2e(1-2e)j - e - (1-2e)j \\ &= e^2 - e + (1-2e)^2 j^2 + 2ej - 4e^2 j - j + 2ej \\ &= (1-2e)^2 j^2 + 4(e - e^2)j = j^2[(1-2e)^2 - 4]. \end{aligned}$$

Hence $j_1^{k-1} = 0$, and we proceed by induction.

For the general case one can easily reduce to two orthogonal idempotents $e, f$; that is $e = e^2,\ f = f^2, 0 = ef = fe$. Having lifted one to $e$, we work in the ring $S := (1-e)R(1-e)$. We have $f = (1-e)f(1-e) \in S$ and every element of $S$ is orthogonal to $e$. Now we lift $f$ in $S$ and continue this way. □

Now let us apply this to the proof of Theorem 1.3.13 in the special case in which $\bar{R}$ is a direct sum of matrix algebras over $F$.

First of all we lift the idempotents $\bar{e}_i$ to orthogonal idempotents $e_i$ and let $e := \sum_i e_i$. Next we notice that if $R$ is an algebra with a 1, we must have $1 = e$, otherwise $1 - e$ is an idempotent in $J$ which is nil hence $1 - e = 0$. If $R$ does not have a 1, we replace $R$ with $eRe$, which has as 1 the element $e$ and $eRe/eRe \cap J = R/J$.

Next the algebras $e_i R e_i$ map to $\bar{e}_i \bar{R} \bar{e}_i = M_{n_i}(F)$, for some $n_i$.

So let us analyze the special case in which $\bar{R} = M_n(F)$ with matrix units $\bar{e}_{i,j},\ i, j = 1, \dots, n$. Next lift the orthogonal matrix idempotents $\bar{e}_{i,i},\ i = 1, \dots, n$, to orthogonal idempotents $e_{i,i}$. Finally, choose elements $e_{1,j} \in e_{1,1}Re_{j,j},\ j = 2, \dots, n$, lifting the units $\bar{e}_{1,j}$ and elements $u_{j,1} \in e_{j,j}Re_{1,1},\ j = 2, \dots, n$, lifting the units $\bar{e}_{j,1}$.

We claim that we can modify the elements $u_{j,1}$ to new elements $e_{j,1}$ such that $e_{1,j}e_{j,1} = e_{1,1}$. In fact, what we can say is that $e_{1,j}u_{j,1} = e_{1,1} + a_j,\ a_j \in e_{1,1}Je_{1,1}$. Hence $a_j^2 = 0$. Replace then $u_{j,1}$ with $e_{j,1} := u_{j,1}(e_{1,1} - a_j)$. We have now

$$e_{1,j}e_{j,1} = e_{1,j}u_{j,1}(e_{1,1} - a_j) = (e_{1,1} + a_j)(e_{1,1} - a_j) = e_{1,1}.$$

With these choices set now by definition, $e_{i,j} := e_{i,1}e_{1,j}$. Notice that if either $i$ or $j$ is the index 1, this is compatible with the definitions. We now need only show that these elements multiply as matrix units.

Take thus a product $e_{i,j}e_{h,k} = e_{i,1}e_{1,j}e_{h,1}e_{1,k}$, by construction $e_{1,j} = e_{1,j}e_{j,j}$, and $e_{h,1} = e_{h,h}e_{h,1}$. So, if $j \neq h$, we have $e_{i,j}e_{h,k} = 0$. When $j = h$ we have

$$e_{i,j}e_{j,k} = e_{i,1}e_{1,j}e_{j,1}e_{1,k} = e_{i,1}e_{1,1}e_{1,k} = e_{i,k}.$$

There are several complements to this theorem. First Malćev **[Mal42]** has shown that two different liftings of $\bar{R}$ are conjugated by an element $1 - z,\ z \in J$. Finally, one has an *equivariant* version due to Taft **[Taf57]**, see Theorem 1.3.20, that we need in the special case of superalgebras.

1.3.2.1. *The trace form.* Assume now that $R$ is a finite-dimensional algebra over a field $F$. We define a symmetric bilinear form on $R$ given by considering for each $a \in R$ the left multiplication $L_a : x \mapsto ax$ and setting

$$(a, b) := \operatorname{tr}(L_a L_b) = \operatorname{tr}(L_{ab}), \quad \text{the trace form.} \tag{1.6}$$

THEOREM 1.3.15. *If $F$ has characteristic 0, the kernel of the trace form of $R$ equals the Jacobson radical $J$.*

PROOF. First let us show that $J$ is contained in the kernel of the trace form. We need to show that, if $a \in J$, then for all $b \in R$ we have $\operatorname{tr}(L_{ab}) = 0$.

This is clear since $ab \in J$ so that $ab$ and hence also $L_{ab}$ are nilpotent and thus have trace zero.

In order to prove the converse statement, let us use Wedderburn's principal theorem $R = \bar{R} \oplus J$. It is enough to prove that the trace form, restricted to $\bar{R}$, is nondegenerate. Now, by Corollary 1.3.8 we may extend the scalars $F$ to an algebraically closed field $\bar{F}$. So assume that $F$ is algebraically closed and that $\bar{R} = \bigoplus_i M_{h_i}(F)$.

The blocks $M_{h_i}(F)$ of the decomposition of $\bar{R}$ are clearly orthogonal under the trace form since, if $a, b$ are in two different blocks , $ab = 0$. So it is enough to show that the trace form, restricted to some block $M_h(F)$, is nondegenerate. For this notice that $R$ as left module over $M_h(F)$ is isomorphic to some direct sum of some $r$ copies of the irreducible module $F^h$, so the trace of $L_a$, $a \in M_h(F)$ equals $r \operatorname{tr}(a)$, where $\operatorname{tr}(a)$ is the usual trace as matrix. Thus the trace form on $M_h(F)$ equals $\operatorname{tr}(L_{ab}) = r \operatorname{tr}(ab)$. Since we are in characteristic 0, it is then enough to verify that on $M_h(F)$ the form $\operatorname{tr}(ab)$ is nondegenerate, and this we leave to the reader. □

The trace form gives a useful criterion to determine if, for a semisimple algebra $R$ of some dimension $h$ over a field $F$ of characteristic 0, we have that $h$ elements $u_1, \ldots, u_h$ form a basis.

DEFINITION 1.3.16. The *discriminant* of $h$ elements $u_1, \ldots, u_h$ is the determinant of the $h \times h$ matrix $(u_i, u_j)$.

PROPOSITION 1.3.17. *Let $R$ be a semisimple algebra of dimension $h$ over a field $F$ of characteristic 0. Then $h$ elements $u_1, \ldots, u_h$ form a basis if and only if their discriminant is nonzero.*

1.3.2.2. *Taft's theorem.* Let us first recall another important notion, that of a *derivation*.

DEFINITION 1.3.18. Given an $R$-bimodule $M$, a linear map $d : R \to M$ is a derivation if

$$d(ab) = ad(b) + d(a)b.$$

Of particular significance are derivations of $R$ to $R$. Whenever one can justify the construction of an exponential function $e^d$, one sees that the condition of being a derivation is the *infinitesimal* condition for $e^d$ to be an automorphism.

Before stating the equivariant version due to Taft **[Taf57]** of Wedderburn's principal theorem, let us recall the notion of *equivariance*.

DEFINITION 1.3.19. Let $G$ be a group acting on two sets $X, Y$. A map $f : X \to Y$ is $G$-*equivariant* if

$$f(gx) = gf(x), \ \forall g \in G, \ \forall x \in X. \tag{1.7}$$

In particular, we shall apply this when $G$ is a linear group, $X, Y$ are two vector spaces, and $f$ is either a linear or a polynomial map.

THEOREM 1.3.20. *Let $R$ be a finite-dimensional algebra over $F$ equipped with the action of a finite group $G$ of automorphisms or anti-automorphisms whose order is not multiple of the characteristic of $F$. Then if $\bar{R} = R/J$ is separable over $F$, the subalgebra $S \subset R$ such that $R = S \oplus J$, as vector spaces, can be chosen to be $G$-stable.*

PROOF. The group $G$ clearly stabilizes $J$ and so induces a quotient action on $\bar{R}$. Let $\pi : R \to \bar{R}$ be the quotient map. We can assume, as usual, that $J^2 = 0$ since $J$ and its powers are clearly $G$-stable and the same reduction is valid. We fix a splitting $\rho : \bar{R} \to R$, that is an algebra homomorphism with $\pi\rho = 1_{\bar{R}}$ (the identity map of $\bar{R}$). We need to prove that we can choose $\rho$ to be $G$-equivariant.

Take another map $\rho + d : \bar{R} \to R$, with $\pi(\rho + d) = 1_{\bar{R}}$, that is $d : \bar{R} \to J$. This is an algebra homomorphism if and only if

$$\begin{aligned}(\rho + d)(ab) &= \rho(ab) + d(ab) = (\rho + d)(a)(\rho + d)(b)\\ &= \rho(a)\rho(b) + d(a)\rho(b) + \rho(a)d(b) \iff d(ab) = d(a)\rho(b) + \rho(a)d(b).\end{aligned}$$

In other words, $\rho + d$ is also an algebra homomorphism if and only if $d$ is a derivation in the $\bar{R}$ bimodule $J$. Observe now that, if $d_1, d_2$ are two derivations, so is $d_1 + d_2$ and also any linear combination. That is derivations form a vector space, since

$$\begin{aligned}(d_1 + d_2)(ab) &= d_1(ab) + d_2(ab) = d_1(a)\rho(b) + \rho(a)d_1(b) + d_2(a)\rho(b) + \rho(a)d_2(b)\\ &= (d_1 + d_2)(a)\rho(b) + \rho(a)(d_1 + d_2)(b).\end{aligned}$$

Let $r$ be the order of the group which is invertible by hypothesis. We claim that the *average*

$$\rho' := \frac{1}{r}\sum_{g \in G} g\rho g^{-1} : \bar{R} \to R$$

is a $G$-equivariant algebra splitting.

Clearly, by construction $\rho'$ is a $G$-equivariant map of vector spaces. Moreover, since $\pi g\rho g^{-1} = g\pi\rho g^{-1} = g1_{\bar{R}}g^{-1} = 1_{\bar{R}}$, we have $\pi\rho' = 1_{\bar{R}}$. We only need to show that $\rho'$ is an algebra homomorphism. Write

$$\rho' = \frac{1}{r}\sum_{g \in G}(\rho - (\rho - g\rho g^{-1})) = \rho - \frac{1}{r}\sum_{g \in G}(\rho - g\rho g^{-1}).$$

Since each map $g\rho g^{-1}$ is clearly an algebra splitting, each $(\rho - g\rho g^{-1})$ is a derivation, so is thus $-\frac{1}{r}\sum_{g \in G}(\rho - g\rho g^{-1})$. Hence $\rho'$ is a splitting. □

We shall apply this in particular to the case $G = \mathbb{Z}_2 := \mathbb{Z}/(2)$ in characteristic $\neq 2$ which gives to $R$ and hence to $\bar{R}$ and $S$ the structure of a *superalgebra* (see Definition 19.2.1).

**1.3.3. The double centralizer theorem.** We have seen that a semisimple algebra $R$ is isomorphic to a direct sum $\bigoplus_{i=1}^{k} M_{h_i}(D_i)$ of matrix algebras over division rings.

Each matrix algebra $M_{h_i}(D_i)$ in turn is the direct sum of $h_i$ copies of the irreducible module $N_i := D_i^{h_i}$. Any irreducible module $N$ for $R$ is isomorphic to one, and only one, of these modules since there is a surjective map from $R$ to $N$. Hence any finite-dimensional module $M$ over $R$ is a direct sum $M = \bigoplus_i N_i^{k_i}$, and it is faithful if and only if all $k_i > 0$. Otherwise it is faithful on the direct sum of those matrix blocks for which this condition is satisfied. Assume $M$ is faithful. We have seen that $\mathrm{End}_R(N_i) = D_i^{\mathrm{op}}$, and

$$S := \mathrm{End}_R M = \bigoplus_{i=1}^{k} M_{k_i}(D_i^{\mathrm{op}}).$$

It follows by symmetry

THEOREM 1.3.21 (Double centralizer). *If $M$ is a faithful semisimple $R$-module and $S := \mathrm{End}_R(M)$, then $M$ is also a semisimple $S$-module, $R = \mathrm{End}_S(M)$ and the isotypic components of the two algebras coincide. Moreover, we have an exchange between dimensions and multiplicities of the two types of irreducible modules.*

If $F$ is an algebraically closed field, then all division algebras $D_i = F$. We can describe explicitly $M, R, S$ using bases, assuming $R$ acts faithfully on $M$:

$$M = \bigoplus_{i=1}^{n} F^{h_i} \otimes F^{k_i}, \quad R = \bigoplus_{i=1}^{n} M_{h_i}(F), \quad S = \bigoplus_{i=1}^{n} M_{k_i}(F). \tag{1.8}$$

Here each summand $F^{h_i} \otimes F^{k_i}$ is an isotypic component tensor product of the irreducible module $F^{h_i}$ of the algebra $R$, through its summand $M_{h_i}(F)$ and the irreducible module $F^{k_i}$ of the algebra $S$ through its summand $M_{k_i}(F)$.

**1.3.4. Finite-dimensional group algebras.** Let $G$ be a group, and let $F$ be a field.

DEFINITION 1.3.22. An *$F$-linear* representation of $G$ is a vector space $V$ over $F$ and a homomorphism $\rho : G \to \mathrm{GL}(V)$ of $G$ into the group $\mathrm{GL}(V)$ of invertible linear transformations of $V$.

One can translate the notion of representation into one of module by introducing the *group algebra* $F[G]$ with coefficients in a field (or even a ring $F$). This is the vector space over $F$ with basis $G$ where the multiplication is defined on this basis by the group multiplication. One easily verifies that the modules over $F[G]$ are just the $F$-linear representations of the group $G$. In particular

DEFINITION 1.3.23. The action of $G$ on $F[G]$ is called the *regular representation.*

THEOREM 1.3.24 (Maschke). *If $G$ is a finite group and $|G|$ is invertible in $F$, then $F[G]$ is semisimple.*

PROOF. Let $M$ be a module, and let $N$ be a submodule. According to Theorem 1.2.1, it is enough to show that $N$ has a complement which is also a submodule. For this it is enough to construct a projection $\Pi : M \to N$, that is a linear map which is the identity on $N$, which is $G$ equivariant.

Choose any complement and so any projection $\pi : M \to N$, we then see that $\Pi := |G|^{-1} \sum_{g \in G} g\pi g^{-1}$ is the desired projection. □

By the theory of semisimple algebras if $F$ is algebraically closed, the algebra $F[G]$ decomposes into its minimal ideals which are isomorphic, as $G \times G$ representations, to $N_i \otimes N_i^*$ where $N_i$ runs over all irreducible modules.

The center of $F[G]$ is clearly formed by the elements $\sum_{g\in G} \alpha_g g$ with the property that $\alpha_g$ is constant on conjugacy classes, that is the function $g \mapsto \alpha_g$ is a *class function*. One deduces that ($F$ is algebraically closed and $|G|$ invertible):

PROPOSITION 1.3.25. *The number of irreducible representations equals the number of conjugacy classes and the dimension of the space of class functions.*

Representations of a group $G$ can also be multiplied under tensor product. If $V, W$ are two such representations, then $V \otimes W$ is a representation by acting with $G$ by $g(v \otimes w) := (gv) \otimes (gw)$.

Moreover, the dual $V^*$ of a representation is a representation by acting with $G$ by $(g\phi)(v) := \phi(g^{-1}v)$. Finally, we always have the *trivial representation*, that is the one-dimensional representation where each element of $G$ acts as the identity. [2]

Given a representation $V$ of $G$, one defines the space of invariants

$$V^G := \{v \in V \mid gv = v,\ \forall g \in G\}. \tag{1.9}$$

The space $V^G$ is thus the isotypic component of the trivial representation.

Let us denote by $V_G$ the direct sum of all the other isotypic components. The projection $\pi_G : V \to V^G$ is characterized by the fact of being a $G$-equivariant linear projection (cf. Definition 1.3.19), and we see

LEMMA 1.3.26.

$$\pi_G = \frac{1}{|G|} \sum_{g\in G} g.$$

PROOF. First of all, for all $h \in G$, we have $h\pi_G = \pi_G h = \pi_G$, from which follows that $\pi_G^2 = \pi_G$ is a projection operator and it is $G$ linear.

If $v \in V^G$, we have $\pi_G v = \frac{1}{|G|}\sum_{g\in G} gv = \frac{1}{|G|}\sum_{g\in G} v = v$. Conversely, $h\pi_G v = \frac{1}{|G|}\sum_{g\in G} hgv = \frac{1}{|G|}\sum_{g\in G} gv$ so $\frac{1}{|G|}\sum_{g\in G} gv \in V^G$, for all $v$. □

The best way to understand and work with irreducible representations of a group $G$ is via *characters*.

DEFINITION 1.3.27. Given a finite-dimensional representation $\rho : G \to \mathrm{GL}(V)$, of a group $G$, its character $\chi_\rho = \chi_V$ is the function on $G$ given by the trace $\mathrm{tr}(\rho(g))$ of the linear operators $\rho(g)$,

$$\chi_\rho(g) := \mathrm{tr}(\rho(g)). \tag{1.10}$$

A basic fact is that the character of the direct sum of two representations is the sum of the characters, and the tensor product of two representations is the product of the characters.

Let us now assume that our base field is the complex numbers.

If $G$ is finite (or more generally compact) and $V$ is a complex vector space, then the action of $G$ can be made unitary, that is we can define on $V$ a Hilbert structure (a positive Hermitian form) preserved by $G$.

This is done by choosing any positive Hermitian form $[u, v]$ and then *averaging the form*:

$$(u, v) := \frac{1}{|G|}\sum_{g\in G} [gu, gv], \qquad \text{or} \qquad \int_G [gu, gv]dg, \quad dg \quad \text{Haar measure.}$$

[2]This can be cast in the language of Hopf algebras, which we do not need.

Then, since the inverse transpose of a unitary matrix is its conjugate, it follows that $\chi_{V^*} = \bar{\chi}_V$ is the complex conjugate. Summarizing,

$$\chi_{(V\oplus W)} = \chi_V + \chi_W, \quad \chi_{(V\otimes W)} = \chi_V \chi_W, \quad \chi_{V^*} = \bar{\chi}_V. \tag{1.11}$$

For a finite group and the regular representation on the space of complex valued functions on $G$, the invariant Hilbert structure is explicitly given by[3]

$$(f_1, f_2) := \frac{1}{|G|} \sum_{g\in G} f_1(g)\bar{f}_2(g) \tag{1.12}$$

and the basic lemma (cf. with Proposition 1.3.25).

LEMMA 1.3.28. $\dim V^G = \frac{1}{|G|}\sum_{g\in G}\chi_V(g)$.

PROOF. We know that $\pi_G = \frac{1}{|G|}\sum_{g\in G} g$ is the projection to $V^G$, so $\dim V^G$ equals the trace of $\pi_G$ which is $\frac{1}{|G|}\sum_{g\in G}\chi_V(g)$. □

Now consider the representations of $G$ over $\mathbb{C}$.

THEOREM 1.3.29. *The irreducible characters of $G$ form an orthonormal basis of the class functions, for the Hilbert product* (1.12).

PROOF. Let $V_1, V_2$ be two irreducible representations of a group $G$, the space $\hom(V_1, V_2) = V_1^* \otimes V_2$ is a representation with character $\bar{\chi}_{V_1}\chi_{V_2}$. The space $\hom(V_1, V_2)^G$ equals the space of $G$-linear maps from $V_1$ to $V_2$. By Schur's lemma such a linear map is either 0 or an isomorphism, so $\hom(V_1, V_2)^G$ is 0 if the two representations are not isomorphic, otherwise it is $\mathbb{C}$.

We deduce from Lemma 1.3.28

$$\frac{1}{|G|}\sum_{g\in G}\chi_{V_1}(g^{-1})\chi_{V_2}(g) = \begin{cases} 0 & \text{if} \quad V_1 \neq V_2 \\ 1 & \text{if} \quad V_1 = V_2. \end{cases} \tag{1.13}$$ □

We can consider the group algebra $F[G]$ as a representation of $G \times G$ by left and right multiplication $(g_1, g_2).u := g_1 u g_2^{-1}$.

Then when we decompose (in characteristic 0 and over an algebraically closed field $F$) the semisimple algebra $F[G] = \bigoplus_i M_{h_i}(F)$, we have that each summand $M_{h_i}(F)$ is a minimal two-sided ideal and an isotypic component of the action. It corresponds to an irreducible representation $M_i = F^{h_i}$, and as a $G \times G$-module we have

$$F[G] = \bigoplus_i M_{h_i}(F) = \bigoplus_i M_i \otimes M_i^*. \tag{1.14}$$

Then we see that the center $Z$, of the algebra $F[G]$, has as basis over $F$ the unit elements $e_i \in M_{h_i}(F)$.

On the other hand, an element $\sum_g a_g g \in F[G]$ is in the center if and only if the function $g \mapsto a_g$ is constant on conjugacy classes, that is it is a *class function*.

We finally have the formula

$$e_i = \frac{\chi_i(1)}{|G|}\sum_{g\in G}\chi_i(g^{-1})g. \tag{1.15}$$

[3]For a Lie group one may take $L^2$ functions with respect to Haar measure.

In fact, it is clear that $e_i$ is in the center, since $\chi_i$ is a class function. Thus $e_i$ acts as a scalar on each irreducible representation, and it is thus enough to show that this scalar is 1 on $V_i$ and 0 on $V_j$, $j \neq i$. This follows immediately, computing the trace of these scalars, from formula (1.13)

$$\chi_j(e_i) = \frac{\chi_i(1)}{|G|} \sum_{g \in G} \chi_i(g^{-1})\chi_j(g) = \delta_i^j \chi_i(1). \tag{1.16}$$

1.3.4.1. *Induced representations.* When $H \subset G$ is a subgroup of a group $G$, we have the notion of *induced representation*, that is from a representation $M$ of $H$ we can construct a representation

$$\mathrm{Ind}_H^G(M) : F[G] \otimes_{F[H]} M. \tag{1.17}$$

Here $F[G]$ is considered as a right $F[H]$-module by the right action. As vector space, if $a_1H, \dots, a_kH$ are the left cosets of $H$ in $G$, we have a direct sum decomposition $\mathrm{Ind}_H^G(M) = \bigoplus_{i=1}^k a_i M$. The action of an element $g \in G$ is the following, $ga_i = a_j h$, for some $j$ and $h \in H$, and then $g(a_i m) = a_j(hm)$. Of particular importance are *permutation representations*, that is $\mathrm{Ind}_H^G(F)$ where $F$ is the trivial one-dimensional representation. In this case $\mathrm{Ind}_H^G(F)$ has as basis the cosets $aH$ and the group $G$ acts on the left by permuting the cosets.

The very special case of the *regular representation* is $F[G]$ where the basis is $G$ itself.

In general even if $M$ is an irreducible representation the induced representation is not irreducible. The way in which it decomposes is given by

THEOREM 1.3.30 (Frobenius reciprocity). *Given an irreducible representation $M$ of $H$ and an irreducible representation $N$ of $G$, its multiplicity in $\mathrm{Ind}_H^G(M)$ equals the multiplicity of $M$ in the representation $N$ (restricted to $H$).*

PROOF. This follows from the defining formula (1.17) by observing that induction transforms direct sums into direct sums, and for an irreducible $N$ we have

$$F[G] \otimes_{F[H]} M = \bigoplus_i N_i \otimes (N_i^* \otimes_{F[H]} M).$$

So we only need to understand $N_i^* \otimes_{F[H]} M$. We leave it to the reader to verify that this is isomorphic to $\hom_{F[H]}(N_i, M)$ (see also formulas (2.8) and (2.10)), so that its dimension equals the multiplicity of $M$ in the representation $N$. □

We will apply this theory to the symmetric groups in §6.1.1.

## 1.4. Noetherian rings

Noetherian rings, named in honour of Emmy Noether, constitute an important class of rings which stem from the idea of axiomatizing the Hilbert basis theorem for polynomials.

Recall that in a partially ordered set $X$ a maximal element $a$ is an element such that, if $b \in X$ and $a \leq b$, then $a = b$. In general a maximal element is not an absolute maximum. Maximal elements may or may not exist if $X$ is infinite, and one may have infinitely many maximal elements.

We leave the proof of the next proposition to the reader. It is done by induction based on Zorn's lemma.

PROPOSITION 1.4.1. *Given a module $M$ over a ring $R$ with a 1, the following conditions are equivalent.*

(1) *Every submodule $N$ is finitely generated as an $R$-module, that is $N = \sum_{i=1}^{k} Ru_i$, for some elements $u_i \in N$.*
(2) *Every ascending chain of submodules $N_1 \subset N_2 \subset \cdots N_k \subset \cdots$ stabilizes, that is there is an integer $k$ such that $N_h = N_k$ for all $h \geq k$.*
(3) *Every nonempty family of submodules $N_\alpha$ has a maximal element.*

DEFINITION 1.4.2. A module $M$ over a ring $R$ that satisfies the previous equivalent conditions is called *Noetherian*.

A ring $R$ that satisfies the previous conditions as a left module over itself is called a *left Noetherian ring*.

REMARK 1.4.3 (Noetherian induction). A very useful way of using the Noetherian property, called *Noetherian induction*, is the following. Suppose we want to prove that a certain class $\mathcal{C}$ of submodules of a Noetherian module, satisfies a given property, we then may argue by contradiction. If there is an element of the class which does not satisfy the property, there is also a maximal element which does not satisfy the property.

In several proofs, using maximality one may reach the desired contradiction.

The next proposition is also straightforward and left to the reader.

PROPOSITION 1.4.4. *Given a Noetherian module $M$, then any submodule $N$ and quotient $M/N$ is Noetherian.*

*A finite direct sum of Noetherian modules is Noetherian.*

*A finitely generated module over a Noetherian ring is Noetherian.*

There is a parallel notion of right Noetherian rings and there are simple examples of rings which are left but not right Noetherian (cf. Example 11.5.23).

In the theory of rings with polynomial identities, most of the basic objects, such as the algebras of generic matrices, are not Noetherian. Usually the Noetherian rings that appear are even finitely generated modules over commutative Noetherian rings, which is a strong restriction. Nevertheless some generalities on Noetherian rings may be useful. For general theory the reader may consult the books **[GW04]** or **[MR01]**.

DEFINITION 1.4.5. A prime ideal $P$ of a ring $R$ is *minimal*, if there is no prime ideal $Q$ with $Q \subsetneq P$.

LEMMA 1.4.6. *Given a prime ideal $P$ of a ring $R$, there is a minimal prime ideal $Q$ with $Q \subset P$.*

PROOF. The proof is by transfinite induction, that is Zorn's lemma. Take a possibly transfinite, descending chain $P_\alpha$, $\alpha \in I$ of prime ideals, that is the indexing set $I$ is well ordered, and if $\alpha < \beta \in I$, we have $P_\alpha \supset P_\beta$.

It is enough to show that we have that $\bigcap_\alpha P_\alpha$ is a prime ideal.

In fact, suppose we have two elements $a, b \in R$ with $aRb \subset \bigcap_\alpha P_\alpha$. If $a \notin \bigcap_\alpha P_\alpha$, there is some $\alpha_0$ such that $a \notin P_\alpha$, $\forall \alpha \geq \alpha_0$, then we have $b \in P_\alpha$, $\forall \alpha \geq \alpha_0$, but since the chain is descending in fact $b \in P_\alpha$, $\forall \alpha$ and so $b \in \bigcap_\alpha P_\alpha$, and this proves the claim. $\square$

PROPOSITION 1.4.7. *If $R$ is a Noetherian ring, then in $R$ there are finitely many minimal primes and there is a product of finitely many of these primes, with possible repetitions, which equals* 0.

PROOF. Let us first consider the following statement: *every two-sided ideal $I$ of $R$ contains a finite product $P_1 \cdots P_k$ of prime ideals.*

If this is not true for some ideal, by Noetherian induction, there is a maximal ideal $J$ for which this is not true. Then $J$ cannot be a prime ideal, so there are two ideals $I_1, I_2$ properly containing $J$ with $I_1 I_2 \subset J$. By maximality both $I_1, I_2$ contain a finite product $P_1 \cdots P_k$, $Q_1 \cdots Q_h$ of prime ideals and hence $J$ contains the product $P_1 \cdots P_k Q_1 \cdots Q_h$ of prime ideals, a contradiction.

Having established this, we have that $0 = P_1 P_2 \cdots P_k$ is a product of prime ideals.

If one of the $P_i$, say for instance $P_1$, is not minimal, by Lemma 1.4.6 there is a minimal prime ideal $Q_1 \subset P_1$ and clearly $0 = Q_1 P_2 \cdots P_k$, so that we may assume that all the $P_i$ are minimal primes. If $Q$ is any prime ideal, we have $Q \supset P_1 \cdots P_k$, hence $Q$ contains one of the $P_i$. Hence, if $Q$ is a minimal prime, it must be equal to one of the $P_i$, and this proves the claim. □

COROLLARY 1.4.8. *If $R$ is a Noetherian ring, then its lower nil radical $N$ (Definition 1.1.30) contains all nilpotent left or right ideals, it is nilpotent, and it is the intersection of the minimal primes.*

PROOF. We know, Theorem 1.1.41, that the lower nil radical $N$ is the intersection of all prime ideals, hence it is also the intersection of the minimal primes $P_1, \dots, P_k$. We have that a finite product of $m$ of these primes in some order and possible repetitions is 0, this product contains $N^m$ which therefore is 0.

There remains to prove that $N$ contains all left and right nilpotent ideals. For this it is enough to remark that, if $R$ is a prime ring, it does not have any nilpotent left or right ideals. □

Finally also for general Noetherian rings holds a *Hilbert basis theorem*, which is proved in exactly the same way as in the commutative case.

THEOREM 1.4.9. *If $A$ is a left Noetherian ring, so is the polynomial ring $A[x]$.*

This result is not as useful in noncommutative algebra as in the commutative case, since it does not apply to a finitely generated algebra over a field $F$, unless it is commutative.

## 1.5. Localizations

A major obstacle one encounters in noncommutative algebra, which is not present in commutative algebra, is the lack of a good theory of localization.

Abstractly, the problem is the following. We take a ring $R$, and in $R$ we consider a *multiplicative set* $S \subset R$, that is a set closed under multiplication. One then would like to construct some other ring $R_S$ with a map $j : R \to R_S$ and with the following properties:

(1) Each element $j(s)$ is invertible in $R_S$.
(2) $j$ is *universal* for the previous property, that is, given any morphism $f : R \to T$ to an algebra $T$ such that each element $f(s)$, $s \in S$, is invertible in $T$, there is a unique morphism $\bar{f} : R_S \to T$ making the diagram

commutative

$$\begin{array}{ccc} R & \xrightarrow{\ j\ } & R_S \\ & {\scriptstyle f}\searrow & \downarrow{\scriptstyle \bar{f}} \\ & & T \end{array}$$

It is quite easy to see that such a ring $R_S$ exists. One takes just the free product $R * \mathbb{Z}\langle X_S \rangle$ with the free algebra in variables $x_s$, $s \in S$, and then takes the quotient modulo the ideal generated by the elements $sx_s - 1$, $x_s s - 1$.

In fact, anticipating the material of §2.2.1.1, we can even solve this problem if we want all algebras to belong to some fixed variety of algebras $\mathcal{V}$, since we can then take the quotient of this constructed algebra $R_S$ modulo the verbal ideal of evaluations of the identities of $\mathcal{V}$.

The construction in general is so abstract that we know basically nothing about this general localized algebra. We shall never use this construction, but for a deep general theory the reader is referred to P. M. Cohn's book, **[Coh06]**.

If we work in the category of commutative rings, one easily sees that this rather abstract construction coincides with the classical construction of the *ring of fractions*. In this case all elements of $R_S$ are expressed as fractions $a/s$, $a \in R$, $s \in S$.

For a noncommutative ring $R$ one of the problems which has been analyzed is to understand under which conditions the elements of $R_S$ can be described as *fractions*.

In particular, one may consider the case in which $R$ is a domain and $S$ the set of nonzero elements. The question being under which conditions is there a division algebra of quotients.

This question is solved by Ore **[Ore31]**, and we have the notion of Ore domain.

The first obvious problem is that, when considering fractions, we have to distinguish wether we put the denominator on the left or on the right.

In general, even if fractions exist, we will have $ab^{-1} \neq b^{-1}a$.

In fact, $ab^{-1} = b^{-1}a$ implies $ba = ab$. So we are lead to discuss when is it that a fraction $ab^{-1}$ can be expressed in the form $c^{-1}d$. We see that, from $ab^{-1} = c^{-1}d$, we deduce $ca = db$. Thus

DEFINITION 1.5.1. A domain $R$ is called a *left (resp., right) Ore domain* if given any two nonzero elements $a, b \in R$ there are two nonzero elements $c, d \in R$ with $ca = db$ (resp., $ac = bd$).

This condition, for instance, is not satisfied by the free algebra in more than two variables, since two distinct variables have no common left multiple.

EXERCISE 1.5.2. A sufficient condition for a domain to be left-Ore is to be left-Noetherian.

It is not hard, although it needs a lengthy proof, to show that, if a domain satisfies the Ore conditions, then we can construct a division ring of fractions $c^{-1}d$. Basically, the idea is that one can change right fractions into left fractions and in this way define sum and multiplication of fractions (see **[Ore31]**).

1.5.0.1. *Central localization.* The localization problem has a simple answer if the multiplicative set $S$ is contained in the center $Z$ of $R$. In this case, the localized ring $Z_S$ is the usual ring of fractions of commutative algebra and we have

PROPOSITION 1.5.3. *The algebra $R \otimes_Z Z_S$ is the universal algebra $R_S$. If $R = F[a_1, \ldots, a_k]$ is finitely generated, the image $1 \otimes_Z Z_S \subset R \otimes_Z Z_S$ is the center of $R_S$.*

PROOF. We need to prove that the map $j : R \to R \otimes_Z Z_S$, $j(r) = r \otimes 1$, satisfies the universal property.

So, let $f : R \to T$ be a map such that the elements $f(s)$ become invertible. In $T$ the subring generated by $f(Z)$ and the elements $f(s)^{-1}$ is commutative and commutes with $f(R)$. We thus have by the usual properties of localization in commutative algebra a map $\bar{f} : Z_S \to T$ and then, since the two algebras $f(R)$ and $\bar{f}(Z_S)$ commute, we obtain the required map $\bar{f} : R \otimes_Z Z_S \to T$.

As for the second statement, an element $cs_1^{-1}$, $c \in R, s_1 \in S$, is in the center of $R_S$ if and only if, in $R_S$, we have $[c, a_i] = 0$, $\forall i$. This means that for some element $s \in S$ we have, in $R$, $s[c, a_i] = [sc, a_i] = 0$, $\forall i$, that is $sc \in Z$, so that $cs_1^{-1} = (cs)(ss_1)^{-1} \in Z_S$. □

PROPOSITION 1.5.4. *If $A, B$ are two algebras over a commutative ring $Z$ and $S \subset Z$ is a multiplicative set, we have*

$$(A \otimes_Z B)_S = A_S \otimes_{Z_S} B_S. \tag{1.18}$$

PROOF. This follows from the universal properties of tensor product and localization. □

## 1.6. Commutative algebra

This is a quick summary of notions that will play a role in the book for some better understanding of Azumaya algebras and for the study of the spectrum of a PI algebra. When necessary the reader should consult the standard sources for commutative algebra, such as **[Eis95]**, **[Mat89]**, **[Mat80]**, for details.

**1.6.1. Local, complete étale maps.** Here we want to collect some basic definitions.

As we already mentioned, localization in a commutative ring $A$ consists in taking a multiplicative set $S \subset A$ and forming a ring of fractions $a/s$, $s \in S$ which is denoted by $A_S$.

Localization is a useful construction for many reasons, one is that:

DEFINITION 1.6.1. If $P \subset A$ is a prime ideal, then $S := A \setminus P$ is a multiplicative set, and the corresponding localization is denoted $A_P$.

It has the important property that in $A_P$ the ideal $PA_P$ is the unique maximal ideal, and we call such a ring *local*, and $A_P/PA_P$ is the field of fractions of $A/P$.

Localization has a strong geometric meaning:

DEFINITION 1.6.2. The set of prime ideals of a commutative ring $A$ is called the *spectrum of* $A$ and is denoted by $\text{Spec}(A)$.

The set $\text{Spec}(A)$ is endowed with the *Zariski topology*.

One defines as closed sets, the sets $V(X) := \{P \in \text{Spec}(A) \mid P \supset X\}$ for any subset $X \subset A$.

A map $\phi : A \to B$ between commutative rings induces then a continuous map $\phi^* : \text{Spec}(B) \to \text{Spec}(A)$, $\phi^*(P) := \phi^{-1}(P)$.

For a localization $i_S : A \to A_S$ one has that the map $i_S^*$ is a homeomorphism of $\text{Spec}(A_S)$ with the subset of $\text{Spec}(A)$ of prime ideals $P$ with $P \cap S = \emptyset$.

If $f \in A$, $S := \{f^k \mid k \in \mathbb{N}\}$, then $\mathrm{Spec}(A_S)$ is identified with the open set complement of $V(\{f\})$.

Thus $\mathrm{Spec}(A)$ is also endowed with a *sheaf* of rings, where the sections on the open set $\mathrm{Spec}(A_S)$ is the ring $A_S$ and the *stalk* at the point $P$ is the local ring $A_P$, that is the localization at the multiplicative set $A \setminus P$. This construction is also called an *affine scheme*.

In fact, the geometric language is the following. Given $a \in A$, we say that $i_S(a)$ is the *restriction* of $a$ to the open set $\mathrm{Spec}(A_S)$. Then the sheaf property says that, given a covering by the open sets $\mathrm{Spec}(A_{S_i})$ of $\mathrm{Spec}(A)$, an element $a \in A$ is completely determined by its restrictions $i_S(a)$ to the sets $\mathrm{Spec}(A_{S_i})$.

Finally, given elements $a_i \in A_{S_i}$ they are the restrictions of an element $a \in A$ if and only if they *agree* in the intersections $\mathrm{Spec}(A_{S_i}) \cap \mathrm{Spec}(A_{S_j})$.

If $A$ is a finitely generated algebra over a field $F$ (we say then that it is *affine*), then each prime ideal is the intersection of the maximal ideals containing it so the spectrum can be replaced by the *maximal spectrum* formed only by the maximal ideals. Finally, if $F$ is algebraically closed and $\mathtt{m}$ is a maximal ideal of the ring $A = F[a_1, \dots, a_k]$, one has that $A/\mathtt{m} = F$. Thus, when we present

$$A = F[a_1, \dots, a_k] = F[x_1, \dots, x_k]/(g_1, \dots, g_h)$$

as the quotient of the polynomial ring modulo equations, one sees immediately that the maximal spectrum can be identified to the points of the affine algebraic variety formed by the subset of $F^h$ of the solutions of the equations

$$g_i(x_1, \dots, x_k) = 0, \ \forall j = 1, \dots, h.$$

In this case, of particular interest are the *affine open coverings*, that is the open coverings by the affine sets $\mathrm{Spec}(A[f_i^{-1}])$ where $f_i \in A$ are elements that generate $A$ as an ideal.

A local ring is already a simplified object but one can go further and *complete* a local ring under a topology where a set of neighbourhoods of 0 are the powers $P^k$ of the maximal ideal $P$. This ring is the inverse limit $\varprojlim A/P^k$. In some special cases, (when $A$ is regular and equicharacteristic) this ring becomes just a ring of power series, giving a theory of formal parameters around a *point* associated to $P$.

The notion of étale extension is more geometric, and we refer to Raynaud **[Ray70]** for a full treatment. It is the algebraic counterpart of the idea of *local homeomorphism*. The abstract definition comes from an idea of Grothendieck**[Gro64]**:

DEFINITION 1.6.3. A commutative $A$-algebra $B$ is said to be *formally smooth*, respectively *étale* (over $A$) if the following two conditions are satisfied:

(1) $B$ is finitely presented over $A$ (that is, it is the quotient of the polynomial ring on finitely many variables over $A$ modulo a finitely generated ideal).
(2) Given any commutative $A$-algebra $C$ and an ideal $I \subset C$ with $I^2 = 0$, we have that the map $\hom_A(B, C) \to \hom_A(B, C/I)$ is *surjective* or respectively *bijective*.

REMARK 1.6.4. If $A$ is an algebraically closed field, then to $B$ is associated an affine algebraic variety, and $B$ is formally smooth if and only if this algebraic variety is smooth and $B$ is its ring of functions (i.e., $B$ has no nilpotent elements, which one expresses by the word *reduced*).

A simple example to have in mind of an étale map is the following. Take a commutative ring $A$ and a monic polynomial $f(x) \in A[x]$. When we construct the ring $A[x]/(f(x))$, we formally *solve* the polynomial equation $f(x) = 0$ by adding to $A$ the class $\bar{x}$ of $x$ in $A[x]/(f(x))$.

The idea is that we want to exclude multiple solutions, which geometrically correspond to *ramifications*. We do this by localizing the ring at the element $f'(\bar{x})$ where $f'$ is the usual derivative.

The ring

$$A[x]/(f(x))[f'(\bar{x})^{-1}] = A[f'(\bar{x})^{-1}][x]/(f(x)) \tag{1.19}$$

is then *a simple étale extension of* $A$, the operation can be repeated giving rise to more complex étale extensions of $A$.

In fact, for a finite extension of fields the condition of being étale is the classical notion of being separable, and then, by a standard theorem, the extension is also simple.

In this case the construction of localization and completion should be complemented with a formalization of Hensel's lemma, giving rise to the notions of a Hensel ring and Henselization for which we again refer to **[Ray70]**.

DEFINITION 1.6.5. A local ring $A$ is called *Hensel ring* if, given any commutative algebra $B$ which is a finite $A$-module, then $B$ is a direct sum of local rings.

We will use repeatedly a *localization principle* which we state:

PROPOSITION 1.6.6.

(1) *If $f : M \to N$ is a morphism of two modules over a commutative ring $A$, with 1, then $f$ is injective (resp., surjective or an isomorphism), if and only if the localizations*

$$f_{\mathfrak{m}} : A_{\mathfrak{m}} \otimes_A M \to A_{\mathfrak{m}} \otimes_A N$$

*are injective (resp., surjective or an isomorphism) for all maximal ideals $\mathfrak{m}$ of $A$.*

(2) *If $a_1, \dots, a_k \in A$ are elements of $A$ generating the unit ideal $A$, then $f$ is injective (resp., surjective or an isomorphism) if and only if the localizations*

$$f_i : A[a_i^{-1}] \otimes_A M \to A[a_i^{-1}] \otimes_A N$$

*are injective (resp., surjective or an isomorphism) for all $i$.*

**1.6.2. Invariant theory.** If $G$ acts on a set $X$ and $F$ is a ring, we have that $G$ acts linearly on the space of functions from $X$ to $F$, by the formula $(gf)(x) := f(g^{-1}x)$.

Of particular importance is when $F$ is a field, $X = V$ is a vector space over $F$, then it is easily seen that $G$ acts actually also on polynomial functions on $V$, which can be identified to $S(V^*)$ the symmetric algebra on the dual space $V^*$. Moreover, polynomial functions are an algebra under multiplication and the action of $G$ is by automorphisms of this algebra.

Then the set $S(V^*)^G$ of invariants is clearly a subalgebra, and in fact a graded subalgebra.

There is an extensive theory on these invariant algebras, and we shall return to this topic in §14.1.

Of particular interest is the case in which $G$ is a *linear algebraic group*. To simplify the idea, assume $F$ algebraically closed, then we set

DEFINITION 1.6.7. A linear algebraic group $G$ is a subgroup of some matrix group $\mathrm{GL}(n, F)$ which is also a subvariety.

We use the fact that $\mathrm{GL}(n, F)$ is an affine variety being the subset of $F^{n^2+1}$ given by pairs $(X, y)$ such that $X \in M_n(F),\ y \in F$ with $\det(X)y = 1$. Then, as any affine algebraic variety, a linear algebraic group $G$ has a coordinate ring $\mathcal{O}(G)$. In particular, the coordinate ring $\mathcal{O}(\mathrm{GL}(n, F))$ is the ring $F[x_{i,j}, \det(X)^{-1}]$ of rational functions in $n^2$ variables, the coordinates of a matrix $X$, with denominator a power of the determinant $\det(X)$.

CHAPTER 2

# Universal algebra

*We shall make very moderate use of categorical language. In our case it is useful mostly to discuss some universal constructions which appear in the theory as the universal map into matrices. For a more comprehensive approach we refer the reader to the book of Mac Lane* [**ML98**].

## 2.1. Categories and functors

**2.1.1. Categories.** Recall that a Category $\mathcal{C}$ consists of a class of *objects* $\mathrm{ob}(\mathcal{C})$ and, for each pair $A, B \in \mathrm{ob}(\mathcal{C})$, a set $\hom_{\mathcal{C}}(A, B)$ of *morphisms*, together with the axioms, that morphisms compose in the usual way,

$$\hom_{\mathcal{C}}(A,B) \times \hom_{\mathcal{C}}(B,C) \to \hom_{\mathcal{C}}(A,C), \quad (f,g) \mapsto g \circ f,$$

and we have the existence of the identity $1_A \in \hom_{\mathcal{C}}(A,A)$ for each $A$ and the associative law

$$f \circ 1_A = f,\ 1_A \circ g = g, \quad h \circ (g \circ f) = (h \circ g) \circ f.$$

One usually denotes morphisms by *arrows* as

$$A \xrightarrow{f} B \xrightarrow{g} C\,.$$

One then has numerous categories which correspond to different mathematical structures, as *groups, rings, modules, topological spaces, differentiable complex or algebraic varieties, etc.*

Categories have arisen from algebraic topology and the need to connect quite different structures, a typical topic of homology theories. This is done through the notion of *functor*:

DEFINITION 2.1.1. A *covariant functor* $F : \mathcal{C} \to \mathcal{D}$ between two categories is a map between objects and morphisms.

To be precise, if $A \in \mathcal{C}$, we have $F(A) \in \mathcal{D}$, and for each pair $A, B \in \mathrm{ob}(\mathcal{C})$, a map $F : \hom_{\mathcal{C}}(A,B) \to \hom_{\mathcal{D}}(F(A), F(B))$ with

$$F(1_A) = 1_{F(A)}, \quad F(g \circ f) = F(g) \circ F(f).$$

Together with the notion of covariant functor, we also have that of *contravariant functor*, in which arrows are reversed. It is easily expressed through the notion of *opposite category* $\mathcal{C}^{\mathrm{op}}$ which is the category with the same objects of $\mathcal{C}$ where we set

$$\hom_{\mathcal{C}^{\mathrm{op}}}(A,B) := \hom_{\mathcal{C}}(B,A)$$

and reverse the arrows.

Since category theory is such a general and abstract idea, as in set theory at the beginning one has to be careful not to create logical paradoxes, as the set of all sets. With some logical provisos, due to the distinction between sets and classes, or with some restriction to small categories where the class of objects is a set, one can use functors as morphisms between categories and hence speak of the *category of all (small) categories*. In our treatment we shall mostly use, besides the category of sets, various categories of algebras and modules. For instance, the category of all rings, or the category of commutative rings will be simple examples. Of particular relevance is the category *Sets* of sets with usual maps.

2.1.1.1. *Categories of functors.* Covariant functors $F : \mathcal{C} \to \mathcal{D}$ between two given categories can also be made into a category through the notion of *natural transformation* between two functors $F, G$. This is, for all objects $A \in \mathcal{C}$ a morphism $\phi_A : F(A) \to G(A)$ satisfying the *commuting property*

$$\begin{array}{ccc} F(A) & \xrightarrow{\phi_A} & G(A) \\ {\scriptstyle F(f)}\downarrow & & {\scriptstyle G(f)}\downarrow \\ F(B) & \xrightarrow{\phi_B} & G(B) \end{array} \qquad \forall f \in \hom_{\mathcal{C}}(A, B). \tag{2.1}$$

Usually it is easy to see that the natural transformations between two functors form a set and we denote it by

$$\mathrm{Nat}(F, G).$$

DEFINITION 2.1.2. Given two categories $\mathcal{C}, \mathcal{D}$, we denote by $\mathcal{F}(\mathcal{C}, \mathcal{D})$ the category of covariant functors from $\mathcal{C}$ to $\mathcal{D}$.

EXERCISE 2.1.3. In the category of rings consider the construction of the polynomial ring $R \mapsto R[t]$. This can be considered as a functor $F$. What is $\mathrm{Nat}(F, F)$? And what if $F : R \mapsto R[t_1, t_2]$?

2.1.1.2. *The forgetful functors.* Many categories are constructed starting from a set and then adding some structure, such as algebraic operations or topologies or differentiable structures. In this case there is a very simple but useful functor from the given category $\mathcal{C}$ to the category of sets, the functor that forgets all the structure and associates to the object its underlying set. Such a functor is called *forgetful functor*. Although this seems a very trivial construction, it is quite useful when one builds universal constructions.

In the same way one may forget only part of the structure: forgetting the product in a ring gives rise to an abelian group; forgetting a differentiable structure gives rise to a topological space—the examples are many.

**2.1.2. Representable functors.** In a category $\mathcal{C}$ each object $A$ determines two special functors to sets, a covariant one $h^A = \hom_{\mathcal{C}}(A, -)$ and a contravariant $h_A = \hom_{\mathcal{C}}(-, A)$. One usually says that the functor $h^A$ is *represented* by $A$.

DEFINITION 2.1.4. A covariant functor $F : \mathcal{C} \to Sets$ from a category $\mathcal{C}$ to the category of sets is said to be *representable* if there is an object $A \in \mathrm{ob}(\mathcal{C})$ and a natural isomorphism from $F$ to the functor $h^A$ represented by $A$.

*Yoneda's lemma* (Lemma 2.1.5) is fundamental.

We start from an arbitrary covariant functor $F : \mathcal{C} \to Sets$ with values in the category of sets and a representable functor $h^A := \hom_{\mathcal{C}}(A, -)$. We want to describe the set $\mathrm{Nat}(h^A, F)$ of natural transformations between these functors and prove the following.

LEMMA 2.1.5 (Yoneda). *We have a canonical isomorphism*

$$\phi : \mathrm{Nat}(h^A, F) \xrightarrow{\cong} F(A) \tag{2.2}$$

*given by* $\varphi \mapsto \varphi_A(1_A)$.

PROOF. Consider $\varphi \in \mathrm{Nat}(h^A, F)$ a natural transformation. By definition, given an object $X \in \mathrm{ob}(\mathcal{C})$, we have a morphism

$$\varphi_X : h^A(X) = \hom_{\mathcal{C}}(A, X) \to F(X),$$

and, for each morphism $g : X \to Y$ the commutative diagram,

$$\begin{array}{ccc} \hom_{\mathcal{C}}(A, X) & \xrightarrow{\varphi_X} & F(X) \\ {\scriptstyle h^A(g)}\downarrow & & \downarrow{\scriptstyle F(g)} \\ \hom_{\mathcal{C}}(A, Y) & \xrightarrow[\varphi_Y]{} & F(Y). \end{array} \tag{2.3}$$

In particular from $\varphi_A : h^A(A) = \hom_{\mathcal{C}}(A, A) \to F(A)$, we can build the canonical element associated to $\varphi$

$$\varphi_A(1_A) \in F(A).$$

Diagram (2.3), applied to a morphism $f : A \to B$ (which by definition is an element of $h^A(B)$) gives

$$\begin{array}{ccc} 1_A \in \hom_{\mathcal{C}}(A, A) & \xrightarrow{\varphi_A} & F(A) \\ {\scriptstyle h^A(f)}\downarrow & & \downarrow{\scriptstyle F(f)} \\ f = h^A(f)(1_A) \in \hom_{\mathcal{C}}(A, B) & \xrightarrow[\varphi_B]{} & F(B). \end{array}$$

We deduce the identity

$$\varphi_B(f) = F(f)(\varphi_A(1_A)),$$

that is the element $\varphi_A(1_A)$ determines completely the value of $\varphi_B$ for all $B$.

Conversely, given an arbitrary element $u_A \in F(A)$ the formula $f \to F(f)(u_A)$ defines a natural transformation between $h^A$ and $F$.

This shows that the map $\mathrm{Nat}(h^A, F) \to F(A)$ given by $\varphi \to \varphi_A(1_A)$ is bijective. □

In particular we have an identification

$$\mathrm{Nat}(h^A, h^B) = \hom_{\mathcal{C}}(B, A), \tag{2.4}$$

which has a further interpretation.

When we associate to an object $A \in \mathrm{ob}(\mathcal{C})$, the functor $h^A$ we in fact are building a contravariant functor, from the category $\mathcal{C}$ to the category $F(\mathcal{C}, Sets)$ of set-valued functors.

In order to clarify this construction, we need some more definitions.

DEFINITION 2.1.6. A category $\mathcal{B}$ is said to be a *subcategory* of a category $\mathcal{C}$ if:

(1) The class of objects of $\mathcal{B}$ is contained in the class of objects of $\mathcal{C}$.

(2) For every pair of objects $A, B \in \text{ob}(\mathcal{B})$, we have $\hom_{\mathcal{B}}(A, B) \subset \hom_{\mathcal{C}}(A, B)$.

(3) The composition of two morphisms of $\mathcal{B}$ in $\mathcal{B}$ coincides with the composition in $\mathcal{C}$.

(4) The identity of an object in $\mathcal{B}$ coincides with the identity in $\mathcal{C}$.

(5) We say that $\mathcal{B}$ is a *full subcategory* of the category $\mathcal{C}$ if for every pair of objects $A, B \in \text{ob}(\mathcal{B})$, we have that $\hom_{\mathcal{B}}(A, B) = \hom_{\mathcal{C}}(A, B)$.

With these definitions prove:

REMARK 2.1.7. The category $\mathcal{C}$ is equivalent to the full subcategory $\mathcal{F}(\mathcal{C}, Sets)$ formed by the representable functors.

The main use of Yoneda's lemma is the following. We want to build objects in a category following the same line of ideas as for some set theoretical construction. We then follow this strategy.

We first build a set-valued functor $F$, and then we prove that it is representable. The object $A$ we want to construct is an object which represents $F$. We have that $A$ is unique up to a unique isomorphism.

COROLLARY 2.1.8. *Let $A, B$ be two objects which represent the same functor $F$. Then, applying Yoneda's lemma, we have a natural isomorphism between $h^A, h^B$ which thus is given by a unique isomorphism between $B$ and $A$.*

2.1.2.1. *Products.* Suppose we would like to define the product $A \times B$ of two objects in a category $\mathcal{C}$. The product $A \times B$ of two sets is characterized by the following property. To give a function $f : X \to A \times B$ is equivalent to give a pair of functions *the coordinates* $f_1 : X \to A,\ f_2 : X \to B$.

Thus we shall say that in a category $\mathcal{C}$ the product of two objects $A, B$ exists if the contravariant functor $\hom_{\mathcal{C}}(X, A) \times \hom_{\mathcal{C}}(X, B)$ with values in the category of sets is representable. An object which represents the functor is then the required product. All the categories that we have discussed up to now have products. In groups, rings, and modules, the product is the usual direct sum.

REMARK 2.1.9. An object $C$ represents the product functor if and only if there are two morphisms $p_1 : X \to A,\ p_2 : X \to B$ *the two projections* with the following universal property. For every object $X \in \text{ob}(\mathcal{C})$ the map $\hom_{\mathcal{C}}(X, C) \to \hom_{\mathcal{C}}(X, A) \times \hom_{\mathcal{C}}(X, B)$ given by $f \to (p_1 \circ f,\ p_2 \circ f)$ is bijective.

Not all categories have the property that two objects always have a product. Yoneda's lemma ensures that if the product exists, it is unique up to unique isomorphism.

EXAMPLE 2.1.10. A poset (partially ordered set) may be thought of as a category. Its objects are the elements of the poset, and the set of morphisms between two elements $a, b$ consists of a unique element if $a \leq b$, otherwise it is empty.

Find conditions under which there is a product in a poset.

Another useful idea is that of *initial* and *final* object in a category. For this idea consider the constant functor to sets which to each object associates the set $\{0\}$ consisting of a single element. To any morphism we associate the identity of that set. This functor can be considered either covariant or contravariant.

In the category of sets the initial object is the empty set $\emptyset$ while the final is a set with one element.

REMARK 2.1.11. Discuss in all categories you know the problem if the constant functor is representable. If it is representable as covariant functor, an object which represents this functor is said to be an *initial object;* an object which represents this as contravariant functor, is said to be a *final object* in the given category.

2.1.2.2. *Free objects.* When we have a forgetful functor $A \to f(A)$ from the category $\mathcal{C}$ to Sets, given a set $I$, a free object (in $\mathcal{C}$) on the set $I$ is an object that represents the set-valued functor $f(A)^I = \hom_{Sets}(I, f(A))$.

For instance for groups, or algebras, we have the free group, or the free algebra, $F$ in generators $x_i,\ i \in I$ corresponding to $I$. This has the property that a morphism from $F$ to $G$ is uniquely determined by giving the values of the elements $x_i$.

We will use mostly free associative algebras, but for a general theory the reader is referred to the book of Cohn **[Coh65]**.

The basic object is the *free associative monoid*[1] $W(X) = W(x_1, \dots, x_\ell)$ over a set $X := \{x_1, \dots, x_\ell\}$ of *variables* called an *alphabet.*

DEFINITION 2.1.12. $W(X)$ is the set of monomials or *free words* in these variables:

$$W(X) = \{x_{i_1} x_{i_2} \cdots x_{i_n} \mid 1 \le i_j \le \ell;\ n = 0, 1, 2, \dots\}.$$

For instance, if $X = \{a, b, c\}$, we have words such as

$$a, ba, ab, bbac, aaa, aac, acb, \quad etc.$$

We easily see that, if $X$ has $m$ elements, there are exactly $m^n$ words of length $n$ for all $n$ using the letters of $X$. The multiplication of words is just the *juxtaposition*

$$(abbc)(cgtee) = abbccgtee.$$

Often we shall need to consider infinite alphabets.

This is free in the categorical sense, that is given any map $f : X \to S$ where $S$ is an associative monoid, this extends uniquely to a homomorphism of monoids, by evaluation

$$x_{i_1} x_{i_2} \cdots x_{i_n} \mapsto f(x_{i_1}) f(x_{i_2}) \cdots f(x_{i_n}).$$

In a similar way we can construct the free associative algebra.

DEFINITION 2.1.13. The *free associative algebra* on a set $X$, called *variables*, over a commutative ring $A$, is the algebra $A\langle X\rangle$ with basis the words $W(X)$ in the set $X$ of variables where multiplication is given by juxtaposition.

Its elements are called *noncommutative polynomials* in the variables $X$.

This algebra is free in the categorical sense that it satisfies the *universal property of the free algebra*. Given any $A$-algebra $S$ and elements $a_1, \dots, a_n \in S$, that is a map of $X$ to $S$, we have a homomorphism called *evaluation* from $A\langle X\rangle$ to $S$ mapping $x_i \to a_i$ and defined by substituting in the formal polynomial to the variables the corresponding elements.

THEOREM 2.1.14. *The set of homomorphisms between the free algebra $A\langle X\rangle$ and any other $A$-algebra $S$ is in 1–1 correspondence with the set of maps from $X$ to $S$ (as sets).*

---

[1] A monoid is just a set $S$ with an associative binary multiplication $ab$.

In particular we can do this when $S = B\langle Y\rangle$ is also a free algebra in possibly a different alphabet $Y$ over the polynomial ring $B = A[t_1, \dots, t_m]$ in some commuting variables.

**Warning.** This discussion branches into two subcases. If we take the category of algebras with 1, then the free monoid has to be taken with a 1 which corresponds to the empty word, otherwise without a 1.

REMARK 2.1.15. Usually the notation $A\langle X\rangle$ refers to the free algebra with 1, it is a graded algebra with $A$ in degree 0.

The corresponding free algebra in the category of algebras without 1 is denoted $A_+\langle X\rangle$. It coincides with $A\langle X\rangle$ in all degrees $i \geq 1$ and it is 0 in degree 0.

This notation is compatible with the notation of graded algebras $S = \bigoplus_{i=0}^{\infty} S_i$ where one denoted by $S_+ := \bigoplus_{i=1}^{\infty} S_i$.

The main theme of this book will be the study of the behaviour of noncommutative polynomials with coefficients in some commutative ring $A$.

REMARK 2.1.16.

(1) Given any commutative ring extension $A \subset B$, we have

$$B \otimes_A A\langle X\rangle = B\langle X\rangle.$$

(2) The algebra $A\langle X\rangle$ has a grading determined by the degrees of its variables $x_i$.

2.1.2.3. *Coproducts, free products.* Dually one can define the *coproduct* which is the product of the objects in the opposite category. In groups and algebras the coproduct exists, it is the *free product* of the two objects.

DEFINITION 2.1.17. Given two $A$-algebras $B, C$ their free product denoted by $B *_A C$ is an $A$-algebra uniquely determined by the following universal property.

Given any $A$-algebra $U$, a homomorphism of $B *_A C$ to $U$ is uniquely determined by assigning in an arbitrary way two homomorphisms $B, C \to U$, that is $\hom_A(B * C, U) = \hom_A(B, U) \times \hom_A(C, U)$.

What we are saying thus is that $B *_A C$ *represents* the functor on $A$-algebras, $U \mapsto \hom_A(B, U) \times \hom_A(C, U)$, that is it is the coproduct of the two algebras.

In particular the identity morphism of $B *_A C$ corresponds to two morphisms $i_B : B \to B *_A C$, $i_C : C \to B *_A C$. The universal property is displayed by a diagram, for all pair of maps $f_B, f_C$ of morphisms to any algebra $R$ there is a unique map $f : B *_A C \to R$ making the following diagram commutative.

(2.5)

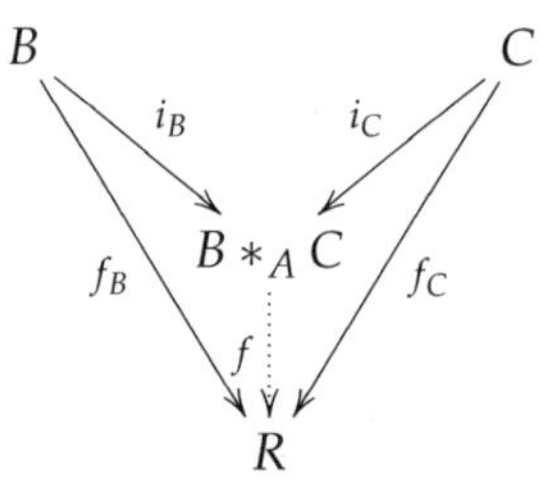

PROPOSITION 2.1.18. *Given two algebras $B, C$, the free product denoted $B *_A C$ exists, and it is unique up to a unique isomorphism.*

PROOF. For two algebras $R, S$ over $A$, take $R = A\langle X\rangle/I$ and $S = A\langle Y\rangle/J$ quotients of free algebras. Then we see that $R *_A S := A\langle X \cup Y\rangle/K$ where $K$ is the ideal generated by $I$ and $J$ is a free product, since it clearly satisfies the universal property. The fact that this construction is independent from the presentation comes again from the universal property of coproducts (the construction can be done both in the category of algebras and that of algebras with a 1). □

EXERCISE 2.1.19. In the category of algebras without 1, when the two algebras have bases $b_i$ and $c_j$, respectively, then $B *_A C$ has a sort of *concrete description*:

A basis of $B *_A C$ is formed by all elements

$$b_{i_1}c_{j_1}b_{i_2}c_{j_2}\cdots b_{i_k},\ b_{i_1}c_{j_1}b_{i_2}c_{j_2}\cdots b_{i_k}c_{j_k},\ c_{j_1}b_{i_1}c_{j_2}b_{i_2}\cdots b_{i_k},\ c_{j_1}b_{i_1}c_{j_2}b_{i_2}\cdots c_{j_k}.$$

Multiplication of two base elements $x, y$ is given by first *concatenation* $xy$ and then, if the end element of the string $x$ lies in the same algebra $B$ or $C$ of the initial element of $y$, one multiplies the two elements and expands this in the given basis.

Intrinsically for algebras (without 1) $B, C$ over a field $A = F$, the free product is also described as the direct sum of the tensor spaces

$$B, C, B \otimes C, C \otimes B, \ldots, (B \otimes C)^{\otimes k}, (B \otimes C)^{\otimes k} \otimes B, C \otimes (B \otimes C)^{\otimes k}, \ldots.$$

When we work with algebras with a 1 we complete 1 to a basis and then have as basis 1 and the previous elements in the two bases excluding 1.

If we are in the category of commutative algebras, the coproduct is the tensor product $R \otimes_A S$.

2.1.2.4. *Adjoint functors.* We complete our discussion with the important notion of *adjoint functors* which in a way summarizes all issues of representing functors. Given two categories $\mathcal{A}$, $\mathcal{B}$, one can define the product category $\mathcal{A} \times \mathcal{B}$ has having as objects the pairs of objects $(A, B)$, $A \in \mathrm{ob}(\mathcal{A})$, $B \in \mathrm{ob}(\mathcal{B})$, and as morphisms pairs of morphisms.

DEFINITION 2.1.20. Given two categories $\mathcal{A}$, $\mathcal{B}$ and two covariant functors

$$F : \mathcal{A} \to \mathcal{B},\ G : \mathcal{B} \to \mathcal{A},$$

we say that $F$, $G$ are *adjoint* ($F$ a left adjoint to $G$, resp., $G$ a right adjoint to $F$) if there is a natural isomorphism of the two covariant functors from $\mathcal{A}^{\mathrm{op}} \times \mathcal{B}$ to the category of sets

$$\hom_{\mathcal{B}}(F(A), B) \cong \hom_{\mathcal{A}}(A, G(B)).$$

We said that there is a strict connection between this notion and that of representability.

In fact $F(A)$ represents the covariant functor $B \to \hom_{\mathcal{A}}(A, G(B))$ while $G(B)$ represents the contravariant functor $A \to \hom_{\mathcal{B}}(F(A), B)$.

A basic example in algebra is in the definition of tensor product. Here we start from a ring $A$ and a right module $M_A$ and a left module ${}_AN$. Given any abelian group $G$, we define the set of $A$-bilinear maps

$$\begin{aligned}(2.6)\quad \mathcal{B}(M\times N,G)\\ &:=\{f:M\times N\to G\,|\,f(m_1+m_2,n)\\ &=f(m_1,n)+f(m_2,n),\ f(m,n_1+n_2)=f(m,n_1)+f(m,n_2),\\ &\qquad f(ma,n)=f(m,an),\ \forall m\in M,\ n\in N,\ a\in A\}.\end{aligned}$$

The abstract definition of tensor product is a consequence of the fact that this functor is representable by an abelian group $M\otimes_A N$ equipped with a universal map $(m,n)\mapsto m\otimes n$ so that we have a natural isomorphism

$$\mathcal{B}(M\times N,G)\cong \hom(M\otimes_A N,G). \tag{2.7}$$

Here hom is in the category of abelian groups. But now the additive maps $\hom(N,G)$ are in a natural way a right $A$-module by setting $(ga)(n):=g(an)$ for $g\in\hom(N,G)$, $a\in A$. To give a bilinear map $f:M\times N\to G$ is the same thing as to give an $A$-linear map $\bar f: M\to\hom(N,G)$ with $\bar f(m)(n):=f(m,n)$. Thus we have a natural isomorphism

$$\hom_A(M,\hom(N,G))\cong\mathcal{B}(M\times N,G)\cong\hom(M\otimes_A N,G). \tag{2.8}$$

This means that, given $N$, the two functors from right $A$-modules to abelian groups and in the opposite direction

$$M\mapsto M\otimes_A N,\quad G\mapsto\hom(N,G)$$

are adjoint. In fact one usually has some extra actions on $M,N$, the most general is that $M={}_BM_A$, $N={}_AN_C$ are both bimodules. In this case one can take, given a $B,C$ bimodule ${}_BG_C$, the subset

$$\begin{aligned}(2.9)\quad \mathcal{B}_{B,C}(M\times N,G):=\{f\in\mathcal{B}(M\times N,G)\,|\,f(bm,nc)=bf(m,n)c\\ \forall m\in M,\ n\in N,\ b\in B,\ c\in C\}.\end{aligned}$$

Then $M\otimes_A N$, and $\hom(N,G)$ are $B,C$ bimodules. We have the adjunction

$$\hom_{A\otimes B}(M,\hom_C(N,G))\cong\mathcal{B}_{B,C}(M\times N,G)\cong\hom_{B,C}(M\otimes_A N,G). \tag{2.10}$$

In the theory of polynomial identities our main example of adjoint functors will be discussed in §3.4, where we show that the matrix algebra $M_n(B)$, which is thought of as a functor from commutative to noncommutative rings, has a left adjoint.

EXAMPLE 2.1.21.

(1) Take the forgetful functor from abelian groups to sets. As an adjoint from sets to abelian groups, we have the construction of the free abelian group with its basis the given set.
(2) It is a similar construction for nonabelian groups.

EXERCISE 2.1.22. What is the adjoint of the functor forgetting multiplication in a ring? In an algebra? And the one that remembers only the *Lie bracket* $xy-yx$? (This part needs the notion of Lie algebra).

**2.1.3. Groups in categories.** The notion of a representable functor can be used in order to define what is a group in a category $\mathcal{C}$. Let us assume that the category has coproducts and an initial object $A$.

Our prime example is a *group* in the category of commutative algebras. In this category the coproduct of two algebras $B, C$ over $A$ is the tensor product $B \otimes_A C$ and the initial object is $A$. A group in this last category is called a *commutative Hopf algebra*.

Suppose then that we have a covariant functor $F : \mathcal{C} \to$ *Groups* to the category of groups. Forgetting the group structure, we have thus a set theoretic valued functor.

DEFINITION 2.1.23. If $F$ is representable, by an object $H$ we say that $H$ is a *group* in the category $\mathcal{C}^{\mathrm{op}}$ (or a cogroup in the category $\mathcal{C}$).

The functor $X \mapsto F(X) \times F(X)$ is thus represented by the coproduct, which we shall denote by $H \otimes H$.

We have then that group multiplication gives a natural transformation of functors $m : F(X) \times F(X) \to F(X)$, the inverse map a natural transformation of functors $i : F(X) \to F(X)$, and finally the unit element a natural transformation of functors $1 : \{1\} \to F(X)$, where $\{1\}$ denotes the constant functor.

Yoneda's lemma (Lemma 2.1.5, that is, formula (2.4)) tells us that each of these natural transformations should be interpreted as a morphism (in the opposite direction) between the objects representing the functors. Hence using the language of Hopf algebras, we get maps

$$\Delta : H \mapsto H \otimes H, \; S : H \mapsto H, \; \epsilon : H \mapsto A \tag{2.11}$$

called *comultiplication, antipode, and co-unit.*

The usual axioms of groups are

$$(ab)c = a(bc) \text{ associativity}, \quad 1a = a = a1 \text{ unit}, \quad a^{-1}a = aa^{-1} = 1 \text{ inverse},$$

and they translate in the categorical language into the corresponding commutative diagrams for the maps in formula (2.11).

EXAMPLE 2.1.24 (The general linear group). In the category of commutative rings, consider the functor $B \mapsto \mathrm{GL}(n, B)$ where by $\mathrm{GL}(n, B)$ we denote the group of invertible $n \times n$ matrices. Such an invertible matrix is given by giving the matrix entries $x_{i,j}$ of a matrix $X \in \mathrm{GL}(n, B)$ and the element $d = \det(X)^{-1}$. In other words an element of $\mathrm{GL}(n, B)$ is given by a homomorphism of the algebra

$$G_n = \mathbb{Z}[x_{i,j}, d]/(d \det(X) - 1) = \mathbb{Z}[x_{i,j}, \det(X)^{-1}] \tag{2.12}$$

of polynomials in the $n^2 + 1$ variables $x_{i,j}, d$ modulo the ideal generated by the single element $d \det(X) - 1$ (making $\det(X)$ invertible); that is, the rational functions $f(x_{i,j}) \det(X)^{-m}$ with $f(x_{i,j}) \in \mathbb{Z}[x_{i,j}]$ a polynomial with integer coefficients.

EXERCISE 2.1.25. Show that $\Delta(x_{i,j}) = \sum_{h=1}^n x_{i,h} \otimes x_{h,j}$, $S(x_{i,j}) = y_{i,j}$ are the entries of $X^{-1}$ given by Cramer's rule (the cofactors divided by the determinant) and $\epsilon(x_{i,j}) = 0$, $i \neq j$, $\epsilon(x_{i,i}) = 1$.

It is often convenient to write $x_{i,j}$ in place of the element $x_{i,j} \otimes 1$ and $y_{i,j}$ in place of $1 \otimes x_{i,j}$. Then $\Delta(x_{i,j}) = \sum_{h=1}^n x_{i,h} y_{h,j}$ is the $i, j$ entry of the product $XY$ of two generic matrices. One then writes the formulas as

$$\Delta(X) = XY, \quad S(X) = X^{-1}, \quad \epsilon(X) = 1_n.$$

This is the simplest example of a *linear algebraic group*. Later we shall study in some detail the important example of the *projective linear group* in §3.4.2.

Another important example appears when we have a group-valued functor $G$ acting on a set-valued functor $F$. Again if both functors are representable, one has by Yoneda's lemma that this action gives rise to a *co-action* between the representing objects. In the category of commutative rings, if $G$ is represented by a Hopf algebra $H$ and $F$ by a ring $A$, the co-action is a map $A \to H \otimes A$ satisfying the appropriate commutative diagrams translating the axioms of group action. The simplest examples we have in mind is for all commutative rings $B$ the action of $\mathrm{GL}(n, B)$ on $B^n$.

We have that $B^n = \hom(\mathbb{Z}[y_1, \ldots, y_n], B)$, so the coaction is the map

$$\Delta : \mathbb{Z}[y_1, \ldots, y_n] \to \mathbb{Z}[x_{i,j}, \det(X)^{-1}] \otimes \mathbb{Z}[y_1, \ldots, y_n],$$
$$\Delta(y_i) = \sum_j x_{i,j} y_j, \ \Delta(Y) = XY,$$

where $Y$ is the column vector of coordinates $y_i$.

## 2.2. Varieties of algebras

**2.2.1. $T$-ideals.** We start now to discuss the main topic of this book. We fix a commutative ring $A$ with a 1 and study associative $A$-algebras. Often we will need to restrict to the case in which $A$ is a field and for the most precise results, a field of characteristic 0.

In general we have two options, both useful: we may work with the class of algebras with a 1 or with algebras in which 1 (a 0-ary operation in the language of universal algebra) is not a datum.

Let $R$ be an associative $A$-algebra, and let $A\langle X \rangle$ be the free algebra in some set of generating variables $X$. When we are dealing with algebras with a 1, this has as its basis all words, including the empty word 1. Otherwise the role of the free algebra is that of the positive part of the free algebra with 1 which thus should be denoted by $A_+\langle X \rangle$.

By definition a homomorphism $e : A\langle X \rangle \to R$ is given by any set theoretic map from $X$ to $R$. The associated homomorphism is to be interpreted as *evaluation* of the variables $x_i$ in the elements $e(x_i) \in R$.

DEFINITION 2.2.1. An element $f(x_1, \ldots, x_n) \in A\langle X \rangle$ is a *polynomial identity* (or PI for short) for $R$ if it vanishes for all evaluations of the variables $X$ in $R$. We denote the set of these identities as $\mathrm{Id}_X(R)$ (or just $\mathrm{Id}(R)$ when $X$ is understood).

Of course a general algebra $R$ does not satisfy any identity, and $\mathrm{Id}_X(R) = 0$. Our topic is instead to study algebras which do satisfy identities.

It is clear that $\mathrm{Id}_X(R)$ is an ideal of $A\langle X \rangle$, but in fact it has a stronger property: if $\phi : A\langle X \rangle \to A\langle X \rangle$ is any homomorphism, $\phi$ is uniquely determined by evaluating the variables $X$ in $A\langle X \rangle$ itself; that is, mapping each $x \in X$ to some *polynomial* $f_x \in A\langle X \rangle$. If we compose any evaluation $e : A\langle X \rangle \to R$ with the morphism $\phi$, we have another evaluation $e \circ \phi$. Therefore if $f$ is a polynomial identity for $R$, we have $e \circ \phi(f) = 0$ for all evaluations $e$. In other words, $\phi(f) \in \mathrm{Id}_X(R)$ for all endomorphisms $\phi$ of $A\langle X \rangle$.

We denote by $T$, resp. $T_+$, the semigroup of all endomorphisms of $A\langle X\rangle$, resp. $A_+\langle X\rangle$.

DEFINITION 2.2.2. A $T$-ideal is an ideal of $A\langle X\rangle$ which is sent into itself by all endomorphisms of $A\langle X\rangle$. This holds similarly for a $T_+$-ideal of $A_+\langle X\rangle$.

REMARK 2.2.3. In particular a $T$- or $T_+$-ideal is stable under linear substitutions of the variables, and so it is a representation of the linear group of invertible substitutions of variables.

One should be aware that the notion of a $T$-ideal changes according to whether we work with algebras with a 1 or in the general case. In the first case we can substitute variables with elements which may also contain a constant term. In the other case we are only using endomorphisms of $A_+\langle X\rangle$, so a variable can only be substituted by a polynomial without constant term.

Any intersection of $T$-ideals is a $T$-ideal, hence we have the notion of a $T$-ideal generated by a set $S$ of polynomials.

Here we see a first difference when we work with algebras with a 1: the $T$-ideal generated by a polynomial of some degree $d$ may also contain polynomials of degree $< d$ obtained by substituting some variables with a 1, while for algebras without 1 this does not happen.

Let $I_1$ and $I_2$ be two ideals of $A\langle X\rangle$.

DEFINITION 2.2.4. We say that the ideal $I_1$ is $T$-equivalent to the ideal $I_2$, and we will write $I_1 \sim_T I_2$ if the greatest $T$-ideal contained in $I_1$ is equal to the greatest $T$-ideal contained in $I_2$. This is an equivalence relation.

Observe that the greatest $T$-ideal contained in the ideal $I$ is the ideal of identities of $F\langle X\rangle/I$.

Although it is easy to construct $T$-ideals, for instance by exhibiting a set of generators, an explicit description is often quite elusive. Here are two simple examples.

EXAMPLE 2.2.5. The $T$-ideal $I_2$ defining commutative algebras is defined by imposing as a polynomial identity the *commutative law* $xy - yx$. Here $A\langle X\rangle/I_2$ equals the usual ring $A[X]$ of commutative polynomials in the variables $X$.

In the category of algebras without 1, the $T$-ideal generated by a monomial $x_1x_2\cdots x_n$ equals the ideal of elements of degree $\geq n$. An algebra $R$ satisfies this identity if and only if it is nilpotent of degree $n$, that is $R^n = 0$.

We start with a simple lemma.

LEMMA 2.2.6. *Let $R$ be an algebra over a field $F$. Consider a polynomial $\sum_{i=0}^n \lambda^i a_i$, $a_i \in R$, $\lambda \in F$. If for $n+1$ distinct values of $\lambda \in F$ this takes values in a subspace $U$ of $R$, then $a_i \in U$, $\forall i$.*

PROOF. Take $n+1$ distinct elements $\lambda_j \in F$, we then have the $n+1$ linear equations

$$u_j = \sum_{i=0}^{n} \lambda_j^i a_i \in U.$$

The determinant of the matrix of coefficients $\lambda_j^i$ is the Vandermonde determinant and is nonzero, so we can solve the linear system by expressing the elements $a_i$ as linear combinations of the $u_j$. $\square$

We shall need the following.

PROPOSITION 2.2.7. *An ideal of the free algebra, over an infinite field, in infinitely many variables is a T-ideal if and only if it is stable under all automorphisms of the free algebra.*

PROOF. We only need to prove that an ideal $I$ stable under all automorphisms of the free algebra is also stable under endomorphisms. This means that if we take any polynomial $f \in I$, which we may assume depends on $x_1, \dots, x_n$, the polynomial obtained by any substitution of variables $x_i \mapsto f_i(x)$, $i = 1, \dots, n$, is still in $I$. For this we may first permute the variables $X$, an automorphism, so that the new $f$ is still in $I$, and assume that $f$ involves variables disjoint from the variables appearing in the various $f_i(x)$. Next we may reduce to analyzing a single variable substitution, say $x_1 \mapsto f_1(x)$ where $f_1$ is independent of $x_1$.

We claim then, that for every parameter $\lambda \neq 0$, the map

$$\phi : x_1 \mapsto \lambda x_1 + f_1(x),\ x_i \mapsto x_i,\ \forall i \neq 1,$$

is an automorphism of the free algebra.

In fact its inverse is $x_1 \mapsto \lambda^{-1}(x_1 - f_1(x))$, $x_i \mapsto x_i$, $\forall i \neq 1$. If we assume then that $I$ is invariant under automorphisms, we have that

$$f(\lambda x_1 + f_1(x), x_2, \dots, x_m) \in I,\ \forall \lambda \neq 0.$$

Since the field is infinite, this implies that also $f(f_1(x), x_2, \dots, x_m) \in I$ as desired, by Lemma 2.2.6. □

REMARK 2.2.8. In **[Cze71]** there is an example showing that Proposition 2.2.7 is not true for the free algebra in two variables.

If we work over a finite field, the following statement may replace Proposition 2.2.7.

DEFINITION 2.2.9. An endomorphism $\phi$ of the free algebra in infinitely many variables $X$ is *special* if there is a subset $Y \subset X$ of the variables such that $\phi(Y) = X$.

Clearly any endomorphism $\psi$ can be *finitely approximated* by special endomorphisms; that is, for any finite set $W \subset X$ of variables there is a special endomorphism $\phi$ with $\phi(w) = \psi(w)$, $\forall w \in W$. From this one easily sees that

PROPOSITION 2.2.10. *An ideal of the free algebra, over any field, in infinitely many variables is a T-ideal if and only if it is stable under all special endomorphisms of the free algebra.*

We limit now the discussion to $T$-ideals, which for $T_+$-ideals it is completely analogous.

EXERCISE 2.2.11.

(i) If $J_i$, $i \in S$, are $T$-ideals then $\bigcap_{i \in S} J_i$ and $\sum_{i \in S} J_i$ are $T$-ideals. Also the product $J_1 J_2$ of two $T$-ideals is a $T$-ideal.

Let $X \subset Y$ be two sets of variables, then

(ii) if $J \subset A\langle Y \rangle$ is a $T$-ideal, then $J \cap A\langle X \rangle$ is a $T$-ideal;

(iii) if $I \subset A\langle X \rangle$ is a $T$-ideal and $\tilde{I} \subset A\langle Y \rangle$ is the $T$-ideal generated by $I$ in $A\langle Y \rangle$, then $\tilde{I} \cap A\langle X \rangle = I$;

(iv) The nil radical $\sqrt{I}$ of a $T$-ideal is a $T$-ideal.

For (iv) we need a property that will be proved later in Corollary 11.1.8, but if the base field is infinite, it follows from Proposition 2.2.10.

THEOREM 2.2.12. *The ideal* $\mathrm{Id}_X(R)$ *of polynomial identities of* $R$ *is a* $T$*-ideal. A* $T$*-ideal* $I$ *of* $A\langle X\rangle$ *is also the ideal* $\mathrm{Id}_X(R)$ *of polynomial identities in the variables* $X$*, of* $R := A\langle X\rangle/I$.

PROOF. We have already proved the first part. As for the second part, if $f$ is a polynomial identity of $R := A\langle X\rangle/I$, then $f$ evaluated just in the quotient map $\pi$ is zero, that is $f \in I$. Conversely, if $f \in I$, consider an evaluation $e : A\langle X\rangle \to A\langle X\rangle/I$. This is given by choosing for each element $x \in X$ an element in $A\langle X\rangle/I$. Clearly we can lift this choice to a map $j$ and see that such an evaluation can be factored as

$$\begin{array}{ccc} A\langle X\rangle & \xrightarrow{\;j\;} & A\langle X\rangle \\ & \searrow^{e} & \downarrow{\scriptstyle \pi} \\ & & A\langle X\rangle/I. \end{array}$$

Since $I$ is a $T$-ideal, we have that $j(f) \in I$, therefore $e(f) = \pi(j(f)) = 0$, as desired. □

REMARK 2.2.13. In Theorem 2.2.12 we have not assumed that $X$ is infinite.

If $X$ is finite, the algebra $R := A\langle X\rangle/I$ may satisfy identities, in a larger number of variables, which are not a consequence of the elements of $I$. As an example, the reader can take identities of $n \times n$ matrices in $< 2n$ variables and prove that the standard identity in $2n$ variables cannot be deduced from those in fewer variables. This requires some care when dealing with free algebras in finitely many variables. On the other hand let $Y \subset X$ be sets of variables so $A\langle Y\rangle \subset A\langle X\rangle$. Consider any set of polynomials $S$ and evaluate in all possible ways these polynomials in $A\langle Y\rangle$ getting a $T$-ideal $I(Y)_S$ and then in $A\langle X\rangle$ getting a $T$-ideal $I(X)_S$. Then

PROPOSITION 2.2.14. $I(Y)_S = A\langle Y\rangle \cap I(X)_S$.

PROOF. Clearly $I(Y)_S \subset A\langle Y\rangle \cap I(X)_S$, conversely if $a \in A\langle Y\rangle$ is of the form $a = \sum_j b_j u_j c_j$ with $b_j, c_j \in A\langle X\rangle$ and $u_j \in A\langle X\rangle$ is an evaluation of a polynomial $s$ in $S$, we have $a = \sum_j \bar{b}_j \bar{u}_j \bar{c}_j$ where $\bar{u}$ for $u \in A\langle X\rangle$ is obtained from $u$ by setting $x = 0$, $\forall x \notin Y$. Clearly then $\bar{u}_j$ is an evaluation in $A\langle Y\rangle$ of the same polynomial $s$. □

2.2.1.1. *Varieties of algebras.* For the next discussion it is convenient to assume that $X$ is countable. In fact in general an algebra satisfying polynomial identities is such that its identities are not necessarily generated by the ones in some finitely many variables. Even if this is true (we will see a criterion when we discuss the Capelli identity (7.5)), we may need a large number of variables.

We develop the discussion for algebras with 1 and $T$-ideals in $A\langle X\rangle$. The same results hold also for $T_+$-ideals in $A_+\langle X\rangle$.

DEFINITION 2.2.15.

(1) Given a $T$-ideal $I$ of $A\langle X\rangle$, the *variety* of algebras $\mathcal{V}_I$ is the class of all algebras which satisfy all the polynomial identities of $I$.
(2) Given a $T$-ideal $I$, we say that an algebra $R$ *generates* the variety $\mathcal{V}_I$ if we have $I = \mathrm{Id}_X(R)$, the ideal of all the identities of $R$.

We should think of the variety $\mathcal{V}_I$ as a full subcategory of all algebras. Using Definition 2.2.4 we have

PROPOSITION 2.2.16. *Two ideals $I_1$ and $I_2$ of $F\langle X\rangle$ are T-equivalent if and only if $F\langle X\rangle/I_1$ and $F\langle X\rangle/I_2$ generate the same variety.*

It is essential to understand that a variety of algebras has *free algebras*.

THEOREM 2.2.17. *Given a T-ideal $I$ of $A\langle X\rangle$, the algebra $A\langle X\rangle/I$ is free on the variables $X$ in the variety $\mathcal{V}_I$.*

PROOF. The statement means that given any $R \in \mathcal{V}_I$ and a map $j : X \to R$, there is a unique morphism $A\langle X\rangle/I \to R$ extending $j$. Now, by the properties of free algebras there is a unique morphism $A\langle X\rangle \to R$ extending $j$. Since $R$ is in $\mathcal{V}_I$, this morphism vanishes on $I$ giving the required map. □

DEFINITION 2.2.18. An algebra $A\langle X\rangle/I$, quotient of the free algebra modulo a $T$-ideal, is called a *relatively free algebra*.

REMARK 2.2.19. A relatively free algebra should be thought of again as an algebra of noncommutative polynomial functions and the classes of the variables $X$ still as variables, provided we evaluate these variables not on any algebra but only on algebras of the corresponding variety.

In fact on each algebra $R$ of the variety defined by $I$, one has a homomorphism of $A\langle X\rangle$ to the algebra of polynomial maps $R^X \to R$ with kernel the polynomial identities of $R$. In particular on an algebra $R$ generating this variety, the kernel is exactly $I$, so $A\langle X\rangle/I$ is naturally a subalgebra of the algebra of polynomial maps $R^X \to R$. This is one of the interesting objects of PI theory and we shall return at several points to a study of its structure, which is usually very complex.

The proof of the following remark is left as an exercise.

REMARK 2.2.20. In fact we have free algebras in any set of variables. If $Y$ is a finite or countable set and we can embed $Y$ in $X$, one sees that $A\langle Y\rangle/(I \cap A\langle Y\rangle)$ is free on the variables $Y$ in the variety $\mathcal{V}_I$.

If $Y$ is uncountable, we embed $X \subset Y$ and $I$ generates a $T$-ideal $I_Y$ in $A\langle Y\rangle$ so that $A\langle Y\rangle/I_Y$ is free on the variables $Y$ in the variety $\mathcal{V}_I$.

The free product exists also in a variety of algebras; the definition is always the same.

DEFINITION 2.2.21. Given two $A$-algebras $B, C$ in a variety $\mathcal{V}$, the free product denoted $B *_{\mathcal{V}} C$ is an $A$-algebra in $\mathcal{V}$ uniquely determined by the following universal property. Given any $A$-algebra $U \in \mathcal{V}$, a homomorphism of $B *_{\mathcal{V}} C$ to $U$ is uniquely determined by assigning in an arbitrary way two homomorphisms $B, C \to U$, that is, $\hom_A(B *_{\mathcal{V}} C, U) = \hom_A(B, U) \times \hom_A(C, U)$.

PROPOSITION 2.2.22. *The free product $B *_{\mathcal{V}} C$ in the variety $\mathcal{V}$ exists and it is in fact the free product $B * C$ modulo the ideal $I_\Gamma \subset B * C$ generated by evaluations of the T-ideal $\Gamma$ defining $\mathcal{V}$.*

PROOF. Given an algebra $R \in \mathcal{V}$ and two maps $f_B : B \to R$, $f_C : C \to R$, one has the universal map $f : B * C \to R$ which, since $R \in \mathcal{V}$, factors through $I_\Gamma$. □

Contrary to the usual free product described by Exercise 2.1.19 for a general variety, the description is completely implicit and it is almost never possible to describe concretely. On the other hand, in the variety of commutative algebras we have seen that the free product is just the tensor product $B \otimes_A C$.

EXERCISE 2.2.23. In the category of all algebras with a 1, describe the free product $\mathbb{C} *_{\mathbb{R}} \mathbb{C}$.

Define the free product of two groups $G_1, G_2$, and then prove that the group algebra $A[G_1 * G_2]$ of this free product is the free product $A[G_1] *_A A[G_2]$ of the two group algebras.

2.2.1.2. *A characterization of varieties.* The following facts are immediate from the definition.

PROPOSITION 2.2.24. *Let $\mathcal{V}$ be a variety of algebras.*

(1) *If $R_i$, $i \in J$, is a family of algebras in $\mathcal{V}$, then $\prod_{i \in J} R_i \in \mathcal{V}$.*
(2) *If $R$ is an algebra in $\mathcal{V}$ and $S \subset R$ is a subalgebra, then $S \in \mathcal{V}$.*
(3) *If $R$ is an algebra in $\mathcal{V}$ and $S = R/I$ is a quotient algebra, then $S \in \mathcal{V}$.*

A quite remarkable fact is that the previous properties characterize varieties of algebras.

THEOREM 2.2.25. *A full subcategory $\mathcal{V}$ of the category of algebras satisfying properties* (1), (2), (3) *of Proposition* 2.2.24, *is a variety of algebras.*

PROOF. Choose a free algebra $A\langle X\rangle$ on an infinite set of variables $X$. Let $J$ be the $T$-ideal of the elements which are polynomial identities for all algebras in $\mathcal{V}$. By construction $\mathcal{V} \subset \mathcal{V}_J$; conversely, given $S \in \mathcal{V}_J$ we need to show that $S \in \mathcal{V}$. We may assume that we have chosen $X$ so that $S = A\langle X\rangle/K$ is a quotient of the free algebra $A\langle X\rangle$ modulo an ideal $K$.

Clearly we can construct a family $R_i$, $i \in I$, of algebras in $\mathcal{V}$ with the property that $J$ is the ideal of all identities of all the algebras $R_i$. Then setting $R := \prod_{i \in I} R_i$, we have that $J$ is the ideal of all identities of $R$, which, by Proposition 2.2.24(1), is in $\mathcal{V}$. Let $E$ be the set of all homomorphisms from $A\langle X\rangle$ to $R$. The set $E$ defines a morphism $j : A\langle X\rangle \to \prod_{e \in E} R$ with kernel exactly $J$. Hence again by properties (1) and (2) we have that $A\langle X\rangle/J \in \mathcal{V}$. Since $S \in \mathcal{V}_J$ and $S = A\langle X\rangle/K$ is a quotient of the free algebra, we must have $K \supset J$, hence $S$ is a quotient of $A\langle X\rangle/J \in \mathcal{V}$. We now finish the proof by applying property (3). □

COROLLARY 2.2.26. *The variety generated by an algebra $R$ is formed by all algebras $S$ which can be presented as quotients $U/I$, where $U$ is a subalgebra of a power $R^I$.*

DEFINITION 2.2.27. We say that $A$ is a generic commutative algebra if it is a commutative ring that generates the variety of all commutative $A$-algebras.

EXERCISE 2.2.28. Let $A$ be a commutative algebra, then $A$ generates the variety of all commutative $A$-algebras if and only if, given a formally nonzero polynomial $f(t) \in A[t]$, this polynomial does not vanish on $A$. If $A$ is a domain, then it is generic if and only if it is infinite.

**2.2.2. Polarization of identities.** The free algebra is *multigraded* in $X$; that is, a monomial in the variables $X$ has degree $h_i$ in the variable $x_i$ if this variable appears $h_i$ times in the monomial. A polynomial $f(x_1, x_2, \dots, x_n) \in A\langle X\rangle$ is homogeneous of degree $h_i$ in the variable $x_i$ if all the monomials appearing in $f$ have this degree.

This means also that, if we take the polynomial ring $A[t]$, substitute for $x_i$ the element $tx_i$ in $A[t]\langle X\rangle$, then $f(x_1,\dots,tx_i,\dots,x_n)=t^{h_i}f(x_1,x_2,\dots,x_n)$.

In particular

DEFINITION 2.2.29. If $f$ is of degree 1 in $x_i$, we also say that it is *linear* in $x_i$. If it is linear in all the variables that appear, then we say it is *multilinear*.

If $B$ is any commutative $A$-algebra and $R$ is a $B$-algebra, we may evaluate $f$ for $x_i\mapsto r_i\in R$. If $f$ is homogeneous of degree $h_i$ in $x_i$ and $r_i\in R$, $b_i\in B$, we have that

$$f(b_1r_1,b_2r_2,\dots,b_nr_n)=\prod_{i=1}^{n}b_i^{h_i}f(r_1,r_2,\dots,r_n).$$

If $f$ is linear in some variable $x_i$, then the evaluation in $r_1,\dots,r_n$, keeping fixed the $r_j$, $j\neq i$, is a $B$-linear map of the variable $r_i$.

We want to define the *polarizations* of a polynomial $f(x_1,x_2,\dots,x_n)\in A\langle X\rangle$.

Let us start with the concept of polarizing a single variable, say $x_1$, and for simplicity assume $f$ is homogeneous of degree $h$ in $x_1$. We then take $h$ variables $y_1,\dots,y_h$ different from the variables appearing in $f$ and substitute for $x_1$ the sum $y_1+\cdots+y_h$. The resulting polynomial decomposes then into a sum of polynomials $f_{a_1,\dots,a_h}(y_1,\dots,y_h,x_2,\dots,x_n)$ with $a_1+a_2+\cdots+a_h=h$ and is homogeneous of degree $a_i$ in $y_i$. These polynomials are the polarizations of $f$ with respect to the variable $x_1$. In particular the polynomial with all $a_i=1$ is multilinear in the $y_i$ and it is called the *full polarization* of $f$ with respect to the variable $x_1$.

Of course one can polarize all the variables obtaining a multilinear polynomial.

EXAMPLE 2.2.30. Fully polarize $x_1^2x_2x_1$ with respect to $x_1$. One must expand $(y_1+y_2+y_3)^2x_2(y_1+y_2+y_3)$. The multilinear part is then

$$y_1y_2x_2y_3+y_1y_3x_2y_2+y_2y_1x_2y_3+y_2y_3x_2y_1+y_3y_1x_2y_2+y_3y_2x_2y_1.$$

This topic will be revisited in §3.1. There we make the substitution for $x_1$ with $\sum_i\lambda_iy_i$ where $\lambda_i$ are parameters; that is, they are formal commutative indeterminates added to the base ring $A$. Then

$$f(\sum_{i=1}^{h}\lambda_iy_i,x_2,\dots,x_n)=\sum_{a_1+a_2+\cdots+a_h=h}\lambda_1^{a_1}\cdots\lambda_h^{a_h}f_{a_1,\dots,a_h}(y_1,\dots,y_h,x_2,\dots,x_n).$$

For the moment we leave it to the reader to verify

EXERCISE 2.2.31. Let $f(x_1,x_2,\dots,x_n)$ be homogeneous of degree $h$ in $x_1$. If in the full polarization of $f(x_1,x_2,\dots,x_n)$ with respect to the variable $x_1$ we set all the new variables $y_i$ equal to $x_1$, we obtain $h!f(x_1,x_2,\dots,x_n)$.

This last operation is called *restitution*.

Notice that polarization and restitution are special transformations of the free algebra induced by linear substitutions of variables.

For a general commutative ring $A$ the notion of polynomial identity and $T$-ideal with coefficients in $A$ needs some comment. First of all if $I\subset A$ is an ideal, the ideal $I\langle X\rangle\subset A\langle X\rangle$ is clearly a $T$-ideal and it defines the variety of $A$-algebras which are indeed $A/I$ algebras. We do not consider that this condition is a true polynomial identity.

DEFINITION 2.2.32. A $T$-ideal $J$ of $A\langle X\rangle$ is *proper* if it is not contained in any ideal $I\langle X\rangle$.

This means that the ideal generated by the coefficients of elements in $J$ is $A$. It then follows that there are elements $f \in J$ with the property that their coefficients generate the ideal $A$.

DEFINITION 2.2.33. A polynomial with these properties is called a *proper identity*.

In fact we usually use the following simple reduction.

PROPOSITION 2.2.34. *If an algebra $R$, over some commutative ring $A$, satisfies a homogeneous proper identity of degree $d$, then it also satisfies a multilinear identity of degree $d$ in which the coefficient of the monomial $x_1 \cdots x_d$ is equal to* 1.

PROOF. First of all by polarization we may assume that the identity is multilinear of the form $f(x_1, \dots, x_d) := \sum_{\sigma \in S_d} \alpha_\sigma x_{\sigma(1)} \cdots x_{\sigma(d)}$. By hypothesis we have some elements $\beta_\sigma \in A$ such that $\sum_{\sigma \in S_d} \alpha_\sigma \beta_\sigma = 1$. Now if we permute the variables and sum, we see that the polynomial identity

$$\sum_{\sigma \in S_d} \beta_\sigma f(x_{\sigma^{-1}(1)}, \dots, x_{\sigma^{-1}(d)}) = x_1 \cdots x_d + \cdots$$

satisfies the required condition. □

2.2.2.1. *S-ideals.* There are two notions related to $T$-ideals which sometimes are useful. This notion is also connected with the operations of polarization and restitution.

If $R$ is an algebra and $M \subset R$, a subspace, one can consider the ideal $\mathrm{Id}_X(R, M)$ of polynomials which vanish when evaluated in $M$. Then $\mathrm{Id}_X(R, M)$ is not stable under all substitutions of variables but only under linear substitutions of variables.

DEFINITION 2.2.35. An $S$-ideal is an ideal of $A\langle X\rangle$ stable under linear substitutions of variables.

If $A = F$ is an infinite field, it is easily seen that, as in Proposition 2.2.7, this is equivalent to assuming that the ideal is stable under invertible linear substitutions of variables. In other words an $S$-ideal is an ideal which is a representation of the linear group of the space $V$ spanned by the variables. Here we may order the variables as $x_1, x_2, \dots$, denote by $\mathrm{GL}(n, F)$ the subgroup of linear transformations on the first $n$ variables, and also restrict to the subgroup $\mathrm{GL}_\infty(F) = \bigcup_{i=1}^\infty \mathrm{GL}(n, F)$ of the linear group which acts only on finitely many of the variables.

### 2.2.3. Stability.

In our treatment we need to concentrate on those identities for an $A$-algebra $R$ which are satisfied by all algebras $R \otimes_A B$, for any commutative ring extension $B \supset A$. These identities we call *stable*. The main example are the multilinear identities, which thus play a special role in the theory.

PROPOSITION 2.2.36. *If $f(x_1, \dots, x_m)$ is a multilinear polynomial identity of $R$, then it is stable.*

PROOF. We compute $f$ for $x_i \mapsto \sum_j a_{i,j} \otimes b_{i,j}$, $a_{i,j} \in R$, $b_{i,j} \in B$. Since $f$ is multilinear, we have as value a linear combination

$$\sum_{j_1, \dots, j_m} f(a_{1,j_1}, a_{2,j_2}, \dots, a_{m,j_m}) \otimes b_{1,j_1} b_{2,j_2} \cdots b_{m,j_m}.$$

As $f$ is an identity of $R$, each $f(a_{1,j_1}, a_{2,j_2}, \dots, a_{m,j_m}) = 0$, and we are done. □

We can also reformulate Proposition 2.2.36 as follows.

COROLLARY 2.2.37. *Let $R \subset S$ be rings, assume that $R$ is an $A$-algebra and $B \supset A$ is a commutative ring such that $S$ is a $B$-algebra and that $S = RB$ is spanned by $R$ over $B$. Then any multilinear identity of $R$, with coefficients in $A$, is also an identity of $S$.*

As usual the theory simplifies this if we work with algebras over an infinite field $F$.

PROPOSITION 2.2.38. *Assume that $R$ is a $A$-algebra and $A$ contains an infinite field $F$. Every polynomial identity $f(x_1, \dots, x_m)$ of $R$ is stable. Moreover, all polynomials deduced from $f(x_1, \dots, x_m)$ by polarization are stable identities of $R$.*

PROOF. Take a basis $u_i,\ i \in I$, of $R$ over $F$, and one $v_j,\ j \in J$, of $A$ over $F$. Every element of $R$ can be written in the form $\sum_i \xi_i u_i,\ \xi_i \in F$ where only finitely many $\xi_j$ are nonzero. For each finite subset $K \subset I$, when we compute $f$ of total degree $d$ on the subspace spanned by the elements $u_i,\ i \in K$, we are substituting each $x_j$ with some sum $\sum_i \xi_{i,j} u_i$ with $\xi_{i,j} \in F$. But we can also consider the coordinates $\xi_{i,j}$ as variables. Hence the evaluation of $f$ that we obtain is a finite linear combination of some elements $u_h$ with coefficients some polynomials $\phi_h(\xi_{i,j})$. The elements $u_h$ are the ones which appear in the multiplications of at most $d$ of the elements $u_i,\ i \in K$, times the finitely many $v_k$ which span a subspace where lie the coefficients in $A$ of $f$.

Now $f$ is an identity in $R$ if and only if these polynomials $\phi_h(\xi_{i,j})$ vanish identically. By Lemma 2.2.6 this implies, since $F$ is infinite, that all the coefficients are 0, so the polynomial is identically 0 as a formal expression in the $\xi$.

This property is clearly satisfied also when we make an extension $S = R \otimes_A B$ since the elements $u_i$ linearly span $S$ over $B$. So, the evaluation of $f$ in elements $x_i \mapsto \sum_i b_{i,j} u_i,\ b_i \in B$, is obtained by specializing the variables $\xi_{i,j}$ to the elements $b_{i,j}$.

The second part now follows, when we polarize a polynomial in one or more variables, we are actually evaluating in $R[t_1, \dots, t_d]$. If it is a polynomial identity, it must vanish in this algebra, but this means that all of its coefficients must vanish identically. These coefficients, by definition, are the evaluations of the various polarizations. $\square$

EXERCISE 2.2.39. If a polynomial $f$ is a stable identity for an algebra $R$, then all its homogeneous components are also stable identities for $R$.

REMARK 2.2.40. If on the contrary $R$ is a finite-dimensional algebra over a finite field, then the least common multiple of the finite number of minimal polynomials satisfied by the elements of $R$, is a one-variable polynomial identity for $R$. For instance if $R = \mathbb{F}_q$ is the field with $q = p^h$ elements ($p$ a prime), then it satisfies the identity $x^q - x$.

It is clear that, if a polynomial in one-variable vanishes on a field $G$, then $G$ is finite, so a one-variable polynomial identity is never stable if $R$ is an algebra with 1.

For an algebra without 1 the only possible one-variable stable identities are multiples of $x^n$, for some $n$ (cf. Exercise 2.2.39).

We can now give, in full generality, the notion of algebra with polynomial identity, the object of this book.

DEFINITION 2.2.41. We say that an $A$-algebra $R$ is an algebra with polynomial identities, or a *PI algebra*, if $R$ satisfies a nonzero stable identity. We say that two $A$-algebras are *PI equivalent* if they satisfy the same stable identities.

Consider the set $\mathcal{M}$ of all monomials $M$ which are multilinear in the variables $x_1, \ldots, x_n$ and possibly dependent on other variables $y$; that is, each of the variables $x_i$ appears in $M$ with multiplicity 1. A polynomial $f(x_1, \ldots, x_n, y)$, multilinear in the variables $x_1, \ldots, x_n$, is by definition a linear combination of the monomials in $\mathcal{M}$.

DEFINITION 2.2.42. A polynomial $f(x_1, \ldots, x_n, y)$, multilinear in the variables $x_1, \ldots, x_n$, is *alternating* (resp., *symmetric*) in $x_1, \ldots, x_n$ if

$$f(x_1, \ldots, x_n, y) = \epsilon_\sigma f(x_{\sigma(1)}, \ldots, x_{\sigma(n)}, y) \quad f(x_1, \ldots, x_n, y) = f(x_{\sigma(1)}, \ldots, x_{\sigma(n)}, y)$$

for all permutations $\sigma$ of the variables, where $\epsilon_\sigma$ is the sign of the permutation.

REMARK 2.2.43. The symmetric group permuting the $x_i$ permutes the monomials $\mathcal{M}$ decomposing this set into orbits. In each orbit there is a monomial $a_1 x_1 a_2 \cdots x_n a_{n+1}$ where the $a_i$'s are monomials (possibly equal to 1).

A polynomial $f(x_1, \ldots, x_n, y)$ is symmetric if and only if its coefficients on a given orbit are all equal and is alternating if and only if they are equal up to the sign of the permutation passing from one element to the other of the orbit. Thus an alternating polynomial is obtained by the operator of alternation

$$\mathrm{Alt}_X(a_1 x_1 a_2 \cdots x_n a_{n+1}) := \sum_{\sigma \in S_n} \epsilon_\sigma a_1 x_{\sigma(1)} a_2 \cdots x_{\sigma(n)} a_{n+1}, \ \epsilon_\sigma \text{ the sign.}$$

A basic example of polynomial identity is when $R$ is a finitely generated module over $A$, say by $n$ elements. Then we can immediately see that $R$ satisfies any polynomial which in at least $n+1$ of its variables is multilinear and alternating, as for instance the Capelli identity, given by formula (7.5). Most of the theory then consists in analyzing how close is an algebra with polynomial identities to such a finite module.

COROLLARY 2.2.44.

(1) *Let $R$ be an algebra over an infinite field $F$, and let $f(x_1, \ldots, x_n)$ be a polynomial identity. Then all of the multihomogeneous components of $f$ are polynomial identities.*
(2) *If $f$ is homogeneous of degree $d$, its full polarization is a nontrivial multilinear identity of $R$.*
(3) *Over a field of characteristic 0, two $T$-ideals $I, J$, which contain the same multilinear elements, coincide.*
(4) *In general, over a field of characteristic 0, $I \subset J$ if and only if the multilinear elements of $I$ are contained in the multilinear elements of $J$.*

PROOF. Let us just prove the last part. Assume that $f$ is any polynomial in $I$. We fully polarize it and get a multilinear polynomial $\bar{f}$ which by hypothesis is in $J$. Since $T$-ideals are invariant both for polarization and restitution when we restitute $\bar{f}$, we see that $f \in J$. □

When we study algebras over a ring $A$ rather than a field, several difficulties may arise and we shall discuss them as they appear.

Let us remark that if $A$ is a domain which is not a finite field and $R$ is torsion free over $A$, then the field of fractions $F$ of $A$ is infinite and $R$ embeds in $R \otimes_A F$.

EXERCISE 2.2.45. Prove that the algebra $R \otimes_A F$ satisfies the same identities with coefficients in $A$ as $R$.

We can usually reduce to this case.

2.2.3.1. *Verbal ideals.*

DEFINITION 2.2.46. A verbal ideal in an algebra $R$ is an ideal generated by all the evaluations in $R$ of a set $\mathcal{S}$ of noncommutative polynomials.

REMARK 2.2.47.

- If the set $I$ of noncommutative polynomials is an ideal and $X$ is infinite, then the set of all the evaluations of $I$ in $R$ is already an ideal.
- A verbal ideal of an algebra $R$ is always the set of evaluations of a $T$-ideal $\Gamma$. We shall denote this ideal of $R$ by $\Gamma(R)$.
- If we are working with algebras over a field $F$ of characteristic 0, a verbal ideal is always the set of evaluations of the multilinear elements of a $T$-ideal.
- If we take a graded algebra $R$ and a verbal ideal generated by evaluating multilinear polynomials, then this ideal is also graded.

## 2.3. Algebras with trace

As mentioned in §2.4.3, one can develop the formalism of free algebras, identities, and $T$-ideals in much more general context.

We now develop the example of *algebras with trace*, an essential class of algebras which will play a major role in our theory.

### 2.3.1. Axioms for a trace.

DEFINITION 2.3.1. An associative algebra with trace, over a commutative ring $A$ is an associative algebra $R$ with a 1-ary operation

$$t : R \to R,$$

which is assumed to satisfy the following axioms:

(1) $t$ is $A$-linear.
(2) $t(a)b = b\,t(a), \quad \forall a, b \in R.$
(3) $t(ab) = t(ba), \quad \forall a, b \in R.$
(4) $t(t(a)b) = t(a)t(b), \quad \forall a, b \in R.$

This operation is called a *formal trace*. We denote $t(R) := \{t(a),\ a \in R\}$ the image of $t$. From the axioms it follows that $t(R)$ is a commutative algebra which we call the *trace algebra of* $R$.

REMARK 2.3.2. We have the following implications:
Axiom (1) implies that $t(R)$ is an $A$-submodule.
Axiom (2) implies that $t(R)$ is in the center of $R$.
Axiom (3) implies that $t$ is 0 on the space of commutators $[R, R]$.
Axiom (4) implies that $t(R)$ is an $A$-subalgebra and that $t$ is $t(R)$-linear.

The basic example of algebra with trace is of course the algebra of $n \times n$ matrices $M_n(B)$ over a commutative $A$-algebra $B$ with the usual trace $\mathrm{tr} : M_n(B) \to B$, thinking of $B$ as contained in $M_n(B)$ as *scalar matrices*.

REMARK 2.3.3. These axioms are for an algebra $R$ with 1. We have made no special requirements on the value of $t(1)$; of course in many important examples this is a positive integer.

If $R$ does not have a 1 or if 1 is not part of its structure, one has to add some extra axiom on $t(t(a))$ as for instance the existence of some central $x$, playing the role of $t(1)$ with

$$(5) \quad t(t(a)) = t(a)x, \quad t(xa) = xt(a), \forall a, \quad t(x) = x^2. \tag{2.13}$$

EXERCISE 2.3.4. Let $R$ be an algebra without 1 with trace satisfying the axiom (5) in (2.13). Then add 1 to $R$ and, on $A \oplus R$, define the map $t(\alpha + r) := \alpha x + t(r)$. Then $A \oplus R$ with this map $t$ is an algebra with trace with 1.

Since this last axiom is rather awkward and we also have the previous exercise, from now on when we speak of an algebra $R$ with trace, we understand that $R$ has a 1, but we make no particular restriction on $t(1)$.

Algebras with trace form a category, where objects are algebras with trace and morphisms algebra homomorphisms which commute with trace

$$t(\phi(a)) = \phi(t(a)).$$

An ideal $I$ in a trace algebra is a *trace* ideal, i.e., one which is stable under the trace, that is, if $a \in I$, then $t(a) \in I$. Then the usual homomorphism theorems are valid.

EXERCISE 2.3.5. If $R_1, t_1$ and $R_2, t_2$ are two algebras with trace, then $R_1 \oplus R_2, t_1 \oplus t_2$ is an algebra with trace and $t(R_1 \oplus R_2) = t(R_1) \oplus t(R_2)$.

REMARK 2.3.6. In §17.3.1 we shall need a slightly more general notion; that is, for some algebras without 1, we may not have $t(a) \in R$, so we need to write the axioms only for the composition $t(a)b$ with $a, b \in R$. We leave it to the reader to write down the corresponding axioms. In this case the trace algebra and the elements $t(a)$ are not in $R$ but in $\mathrm{End}(R)$.

Of course given an algebra with trace $R$, if we *forget the trace*, we just have an associative algebra. There is a simple formal construction which associates to any $A$-algebra $R$ an algebra with trace $R_T$, (a functor adjoint to the previous forgetful functor).

2.3.1.1. *The algebra $R_T$.* Consider the $A$-module $M := R/[R,R]$, quotient of $R$ by the $A$-submodule of $R$ spanned by all commutators $[a,b]$, $a, b \in R$, and let $\pi : R \to M := R/[R,R]$ be the quotient map. Then let $T := S(M) = A \oplus M \bigoplus_{i=2}^{\infty} S^i(M)$ be the symmetric algebra on the $A$ module $M$, and let $S_+(M) = \bigoplus_{i=1}^{\infty} S^i(M)$ (called *augmentation ideal*). We define on the algebra $R_T := R \otimes_A S(M)$ the trace

$$t : R_T = R \otimes_A S(M) \to S_+(M), \quad t(r \otimes a) := \pi(r)a. \tag{2.14}$$

Since $S(M)$ is a graded algebra, with $A$ in degree 0, we have that $R = R \otimes 1 \subset R \otimes_A S(M)$. The trace algebra of $R_T$ is $S_+(M)$.

Take now any algebra $U$ with trace tr, with trace algebra $\mathrm{tr}(U)$, and an algebra homomorphism $\phi : R \to U$.

PROPOSITION 2.3.7.

(1) *The homomorphism* $\phi : R \to U$ *induces a linear map* $\bar{\phi} : M = R/[R,R] \to \mathrm{tr}(U)$ *given by* $\bar{\phi}(\pi(a)) = \mathrm{tr}(\phi(a))$. *This extends to a homomorphism, still denoted by* $\bar{\phi} : S_+(M) \to \mathrm{tr}(U)$.

(2) *Then* $\phi$ *induces a unique homomorphism* $\tilde{\phi} : R_T \to U$ *of algebras with trace by the formula* $\tilde{\phi}(a \otimes b) = \phi(a)\bar{\phi}(b)$ *if* $b \in S_+(M)$ *or* $\tilde{\phi}(a \otimes 1) = \phi(a)$.

PROOF. (1) $\phi([R,R]) \subset [U,U]$, so $\mathrm{tr}(\phi([a,b]) = 0$ and induces the required linear map.

(2) We deduce a homomorphism $\bar{\phi} : S_+(M) \to \mathrm{tr}(U)$ and finally a homomorphism $\tilde{\phi}(a \otimes b) = \phi(a)\bar{\phi}(b)$ if $b \in S_+(M)$ or $\tilde{\phi}(a \otimes 1) = \phi(a)$ for which, if $b \in S_+(M)$,

$$\bar{\phi}(t(a \otimes b)) = \bar{\phi}(\pi(a)b) = \bar{\phi}(\pi(a))\bar{\phi}(b) = \mathrm{tr}(\phi(a))\bar{\phi}(b) = \mathrm{tr}(\phi(a)\bar{\phi}(b)) = \mathrm{tr}(\tilde{\phi}(a \otimes b))$$

or

$$\bar{\phi}(t(a \otimes 1)) = \bar{\phi}(\pi(a)) = \mathrm{tr}(\tilde{\phi}(a)).$$

That is, $\tilde{\phi}$ is *trace preserving*. □

In particular we may apply this to the case in which $R$ has already a trace. Take $U = R$ and take as $\phi$ the identity map $1_R : R \to R$.

EXAMPLE 2.3.8.

(1) Let $R = M_n(A)$ with the usual trace. We then have that $[R,R]$ is the submodule of trace 0 elements and $R = [R,R] \oplus Ae_{1,1}$, $M = Ax$ with $x$ the class of $e_{1,1}$ (or of any other element with trace 1). Then the symmetric algebra $S(M) = A[x]$ is the ring of polynomials in a single variable and $M = Ax$.

The associated algebra with trace $R_T := R \otimes_A S(M) = M_n(A[x])$ with trace $t$ is given by $t(a) = x \cdot \mathrm{tr}(a)$, where $\mathrm{tr}(a)$ is the usual trace of matrices. This induces the map $\bar{1}_R$ which maps $x$, the class of 1, to the trace of 1, that is $\bar{1}_R(x) = n$.

(2) $R = \mathbb{C}$ as $\mathbb{R}$ algebra, the trace $\mathrm{tr}(a + \mathrm{i}\, b) := 2a$ is the usual trace of number theory. $M = \mathbb{C}$, but it is thought of as just a two-dimensional space $\mathbb{R}x \oplus \mathbb{R}y$, with $\pi(1) = x,\ \pi(\mathrm{i}) = y$.

So $S(M) = \mathbb{R}[x,y]$, and $R_T = \mathbb{C}[x,y]$ and $\mathrm{tr}(a + \mathrm{i}\, b) = ax + by$, with $a = a(x,y), b = b(x,y) \in \mathbb{R}[x,y]$. Finally $\bar{1}_R(x) = 2,\ \bar{1}_R(y) = 0$.

(3) $R = \mathbb{C} \oplus \mathbb{C}$ as $\mathbb{R} \oplus \mathbb{R}$ algebra, the trace $\mathrm{tr}(a_1 + \mathrm{i}\, b_1, a_2 + \mathrm{i}\, b_2) := (2a_1, 2a_2)$.

**2.3.2. Algebras with trace form a category with free algebras.** We use the previous construction to exhibit the free algebras in the category of algebras with trace (with or without 1). In order to construct a free algebra, one takes first a free algebra $\mathcal{F} = A\langle X \rangle$ in indeterminates $x_i \in X$ (with or without 1). Then one has the trace algebra $\mathcal{T}_X := S(\mathcal{F}/[\mathcal{F},\mathcal{F}])$ as in the previous paragraph and finally the universal algebra with trace defined in formula (2.14).

$$\mathcal{F}_T\langle X \rangle = A\langle X \rangle \otimes \mathcal{T}_X = \mathcal{F} \otimes \mathcal{T}_X = \mathcal{T}_X\langle X \rangle, \quad \mathcal{F} = A\langle X \rangle \tag{2.15}$$

PROPOSITION 2.3.9. *The free algebra with trace in the variables X is the algebra* $\mathcal{F}_T\langle X \rangle$ *with the trace t defined in formula* (2.14).

PROOF. Given an algebra with trace $U$, a map $\phi : X \to U$ induces a homomorphism, still denoted by $\phi : \mathcal{F} \to U$, and then by Proposition 2.3.7 the unique map of trace algebras $\phi : \mathcal{F}_T\langle X\rangle \to U$. Thus $\mathcal{F}_T\langle X\rangle$ satisfies the universal property characterizing free algebras. □

We display this universal property by a commutative diagram.

$$\begin{array}{ccc} \mathcal{F} & \xrightarrow{\phi} & U \\ {\scriptstyle t}\downarrow & & \downarrow{\scriptstyle \mathrm{tr}} \\ \mathcal{F}/[\mathcal{F},\mathcal{F}] & \xrightarrow{\phi} & t(U) \end{array} \implies \begin{array}{ccc} \mathcal{T}_X\langle X\rangle = \mathcal{T}\mathcal{F} & \xrightarrow{\phi} & U \\ {\scriptstyle t}\downarrow & & \downarrow{\scriptstyle \mathrm{tr}} \\ \mathcal{T} & \xrightarrow{\phi} & t(U) \end{array}$$

We need a more *concrete* description of $\mathcal{F}_T\langle X\rangle$. Recall the *cyclic equivalence* of monomials (cf. §3.3.2). $M = AB$ is equivalent to $BA$ means that $AB = BA$ modulo $[\mathcal{F},\mathcal{F}]$, since by definition $AB - BA = [A,B]$. We leave this as

EXERCISE 2.3.10. Prove that classes of cyclic equivalence of monomials $M$, which we formally denote $t(M)$ (since in the end they are the trace of $M$), form a linear basis of $\mathcal{F}/[\mathcal{F},\mathcal{F}]$ as an $A$-module.

From this follows

THEOREM 2.3.11. *The symmetric algebra $\mathcal{T}_X := S(\mathcal{F}/[\mathcal{F},\mathcal{F}])$ over the free $A$ module $\mathcal{F}/[\mathcal{F},\mathcal{F}]$ is the commutative polynomial algebra $\mathcal{T}_X = A[t(M)]$, over $A$ in the infinitely many variables $t(M)$, $M$ running on monomials up to cyclic equivalence. The algebra $\mathcal{F}_T\langle X\rangle = A\langle X\rangle[t(M)]$.*

We shall take up this topic again in equation (3.30) where the algebra $\mathcal{T}_X$ will be replaced by an algebra defined in a characteristic free form. Then

DEFINITION 2.3.12. The commutative algebra $\mathcal{T}_X$ is called the *free ring of traces* in the variables $X$.

Then the free algebra with trace is $\mathcal{T}_X\langle X\rangle$; that is, it is the ordinary free algebra in the variables $X$ with coefficients in the free ring of traces.

The trace $t$ is the $\mathcal{T}_X$-linear map which on $\mathcal{F}$ is the projection to $\mathcal{F}/[\mathcal{F},\mathcal{F}]$.

REMARK 2.3.13. Observe in particular that every algebra endomorphism of $\mathcal{F}$ factors through $\mathcal{F}/[\mathcal{F},\mathcal{F}]$ and thus gives rise to an endomorphism of the commutative algebra $\mathcal{T}_X$ and a trace preserving endomorphism of $\mathcal{T}_X\langle X\rangle$.

A special role is played by the spaces of multilinear elements.

DEFINITION 2.3.14. We shall denote by $\mathcal{T}_X(m)$ (resp., $T_X(m)$) the space of multilinear elements in $m$ variables $x_1, \dots, x_m$, in the free ring of traces $\mathcal{T}_X$, and, respectively, in the free algebra with trace $\mathcal{T}_X\langle X\rangle$.

Let us discuss a quite remarkable fact, which will be systematically used in Chapter 12 when we discuss the role of the Cayley–Hamilton identity.

PROPOSITION 2.3.15.

(1) *The space $\mathcal{T}_X(m)$ of multilinear elements in $m$ variables in the free ring of traces $\mathcal{T}_X$ can be identified as representation of the symmetric group $S_m$, with the group algebra $A[S_m]$ and thought of as a representation of $S_m$ under the conjugation action.*

(2) *The space $T_X(m)$ of multilinear elements in $m$ variables in the free algebra with trace can be identified, as representation of the symmetric group $S_m$, with the group algebra $A[S_{m+1}]$ of the symmetric group in one more variable, where the action of $S_m \subset S_{m+1}$ is by conjugation.*

PROOF. (1) This is achieved by the map $T$ that, to a permutation

$$\sigma = (i_1 i_2 \cdots i_h) \cdots (j_1 j_2 \cdots j_k)(s_1 s_2 \cdots s_u)$$

written in its cycle decomposition, associates the multilinear trace expression

$$T_\sigma(x_1, x_2, \cdots, x_m) := t(x_{i_1} x_{i_2} \cdots x_{i_h}) \cdots t(x_{j_1} x_{j_2} \cdots x_{j_k}) t(x_{s_1} x_{s_2} \ldots x_{s_u}). \tag{2.16}$$

□

(2) The formal way of doing this, which will appear in rather concrete terms in (12.3), is the following: if $\sigma \in S_{m+1}$, we write the cycle decomposition of $\sigma$ taking in evidence the index $m+1$ as

$$\sigma = (i_1 i_2 \cdots i_h) \cdots (j_1 j_2 \cdots j_k)(s_1 s_2 \cdots s_u m+1).$$

Then associate, with a map denoted bu $\Psi$, to $\sigma$ the following multilinear trace expression

$$\psi_\sigma(x_1, x_2, \ldots, x_m) := t(x_{i_1} x_{i_2} \cdots x_{i_h}) \cdots t(x_{j_1} x_{j_2} \cdots x_{j_k}) x_{s_1} x_{s_2} \cdots x_{s_u}. \tag{2.17}$$

We have furthermore a linear isomorphism $j : T_X(m) \to \mathcal{T}_X(m+1)$ given by

$$j : a(x_1, \ldots, x_m) \mapsto t(a(x_1, \ldots, x_m) x_{m+1}),$$

making the diagram commutative

$$\begin{array}{ccc} A[S_{m+1}] & \xrightarrow{1} & A[S_{m+1}] \\ {\scriptstyle\Psi}\downarrow & & {\scriptstyle T}\downarrow \\ T_X(m) & \xrightarrow{j} & \mathcal{T}_X(m+1), \end{array}$$

that is, $T(\sigma) = t(\Psi(\sigma) x_{m+1})$. It is easily seen that this construction provides the desired formal description of multilinear elements. We have proved

PROPOSITION 2.3.16. *The maps*

$$\Psi : A[S_{m+1}] \to T_X(m), \qquad \Psi : \sigma \mapsto \psi_\sigma(x_1, x_2, \ldots, x_m), \tag{2.18}$$

$$T : A[S_m] \to \mathcal{T}_X(m), \qquad T : \sigma \mapsto T_\sigma(x_1, x_2, \ldots, x_m), \tag{2.19}$$

*are linear isomorphisms with the space of multilinear elements of degree $m$ in the free algebra with trace and, respectively, in the free trace ring.*

*For $\sigma \in S_{m+1}$ we have $T(\sigma) = t(\Psi(\sigma) x_{m+1})$.*

2.3.2.1. *Trace identities T-ideals.* As with ordinary algebras one has a notion of *trace identity* for an algebra with trace $R$ that is an element of the free algebra with trace which vanishes for all evaluations (respecting the trace) in $R$.

We also have the notion of $T$-ideal which is an ideal of $\mathcal{T}_X\langle X\rangle$ closed under all endomorphisms, compatible with the trace operation. These are induced by substitutions $X \to \mathcal{T}_X\langle X\rangle$.

DEFINITION 2.3.17. Given an algebra with trace $R$, the set of trace identities of $R$ will be denoted by $\mathrm{Tid}(R)$. It is a $T$-ideal of $\mathcal{T}_X\langle X\rangle$, the free trace algebra, and it defines a variety of algebras with trace generated by $R$.

The results of §2.2 carry over to this more general case.

In particular Theorem 2.2.17 holds for $T$-ideals of $\mathcal{T}_X\langle X\rangle$, the free trace algebra. If $R$ is a trace algebra, its relatively free algebra will be $T_X(R) := \mathcal{T}_X\langle X\rangle/\mathrm{Tid}(R)$. In particular we have the *pure trace identities* $\mathrm{Tid}(R) \cap \mathcal{T}_X$. The quotient of $\mathcal{T}_X$ by the pure trace identities will be the free trace algebra of $R$ denoted by $\mathcal{T}_X(R)$.

REMARK 2.3.18. If $R = R_1 \oplus R_2$ is a direct sum of algebras with trace, as defined by Exercise 2.3.5, we have $\mathrm{Tid}(R) = \mathrm{Tid}(R_1) \cap \mathrm{Tid}(R_2)$.

As for PI algebras the relatively free algebras satisfy $\mathcal{F}_R \subset \mathcal{F}_{R_1} \oplus \mathcal{F}_{R_1}$, so for trace algebras we also have $T_X(R) \subset T_X(R_1) \oplus T_X(R_2)$.

## 2.4. The method of generic elements

**2.4.1. The algebras of generic elements.** Consider an algebra $R$ which is a finite free module over a commutative ring $A$ with basis $u_1, \dots, u_m$. Then the (stable) polynomial identities in variables $x_\alpha$, $\alpha \in I$, of $R$ are deduced as follows. Take commutative variables $\xi_i^{(\alpha)}$, $i = 1, \dots, m$, $\alpha \in I$, and denote by $A[\xi]$ the polynomial ring in these variables.

The algebra $R \otimes_A A[\xi]$ can be thought of as the algebra of polynomial maps $F : R^I \to R$, where the multiplication is given as functions with values in the algebra $R$. On this algebra acts the group $G$ of automorphisms of the algebra $R$, by $(gF)(r_1, \dots, r_\alpha, \dots) := g(F(g^{-1}r_1, \dots, g^{-1}r_\alpha, \dots))$. We know that by evaluation the free algebra $A\langle x_\alpha\rangle$ maps to this algebra of polynomial maps and in fact it maps to the subalgebra $(R \otimes_A A[\xi])^G$ of $G$-equivariant maps (cf. Definition 1.3.19). Since $A\langle x_\alpha\rangle$ is generated by the variables $x_i$ it is natural to understand the maps associated to these elements as elements in $R \otimes_A A[\xi]$.

For this define the

$$(2.20) \qquad \textit{generic elements} \qquad \xi_\alpha := \sum_{i=1}^{m} \xi_i^{(\alpha)} u_i \in R \otimes_A A[\xi],\ \alpha \in I$$

(see Procesi **[Pro67b]**). As polynomial functions these are in fact the coordinate functions of $R^I$.

THEOREM 2.4.1. *The kernel of the map $\pi$, from the free algebra $A\langle x_\alpha\rangle$ to the algebra $R \otimes_A A[\xi]$, given by $\pi : x_\alpha \mapsto \xi_\alpha$ is the ideal of stable polynomial identities of $R$.*

PROOF. Take an element $f(x_1, \dots, x_d)$ of the free algebra $A\langle X\rangle$, when we evaluate this element in the generic elements $\xi_\alpha$, we obtain an element $\sum_{i=1}^m \phi_i(\xi_j^{(\alpha)}) u_i$ where the elements $\phi_i(\xi_j^{(\alpha)})$ are polynomials with coefficients in $A$ in the variables $\xi_j^{(\alpha)}$. Then $f$ is a stable polynomial identity if and only if these polynomials $\phi_i(\xi_j^{(\alpha)})$ vanish for all values of the variables in any commutative algebra over $A$. This means that these polynomials are identically 0, that is, they vanish on the generic elements. □

DEFINITION 2.4.2. The algebra $A\langle \xi_1, \dots, \xi_\alpha, \dots\rangle$, generated over $A$ by the generic elements, will be called the *algebra of generic elements* for the algebra $R$.

The previous theorem shows that the algebra $A\langle\xi_\alpha\rangle$ of generic elements is the relatively free algebra in the variety of algebras generated by $R$ in the variables $\xi_\alpha$ (we need to distinguish the two cases of algebras with or without 1; in the first case we take the free algebra with 1 and in the second without).

REMARK 2.4.3. Notice that this algebra is independent of the basis chosen to construct the generic elements, and one may also replace $R$ with another PI equivalent algebra.

The main example is when $R = M_n(A)$ is the algebra of $n \times n$ matrices. The generic elements can be taken with respect to the basis of matrix units $e_{i,j}$. In this case the elements $\xi_\alpha$ are matrices with entries variables $\xi^\alpha_{i,j}$, called *generic matrices.* We shall return to this basic concept in Definition 3.3.23.

We shall prove a result due to Amitsur.

THEOREM 11.3.1. *The algebra of $m \geq 2$, $n \times n$ generic matrices over a field $F$ has a total division ring of fractions, which has dimension $n^2$ over its center.*

The description of the center, due to Procesi **[Pro67b]** will be treated in §14.4.1.

2.4.1.1. *Substitutions.* Recall that $A\langle\xi_\alpha\rangle$ is a relatively free algebra in the variety generated by $R$ and also that $R \otimes A[\xi]$ is in this variety. Thus a homomorphism of $A\langle\xi_\alpha\rangle$ to $R \otimes A[\xi]$ is determined by a map $\psi : \xi_\alpha \to a_\alpha \in R \otimes A[\xi]$ or that of the free algebra $A\langle X\rangle$, $x_\alpha \to a_\alpha$. Consider such a map. We have for all $\alpha$,

$$\psi : \xi_\alpha \to \psi(\xi_\alpha) = \sum_{i=1}^{m} f_i^{(\alpha)}(\xi)u_i, \quad f_i^{(\alpha)}(\xi) \in A[\xi]. \tag{2.21}$$

We then have, denoting by $\rho : A\langle\xi_\alpha\rangle \to R \otimes_A A[\xi]$ the inclusion

THEOREM 2.4.4. *The endomorphism of $A[\xi] = A[\xi_j^{(\alpha)}]$ given by*

$$\bar{\psi}(\xi_j^{(\alpha)}) := f_i^{(\alpha)}(\xi) \tag{2.22}$$

*is the unique endomorphism of $A[\xi]$ making the following diagram commutative.*

$$\begin{array}{ccccc} A\langle X\rangle & \longrightarrow & A\langle\xi_\alpha\rangle & \xrightarrow{\rho} & R \otimes_A A[\xi] \\ & & & \searrow^{\psi} & \downarrow{\scriptstyle 1_R \otimes \bar{\psi}} \\ & & & & R \otimes_A A[\xi] \end{array} \tag{2.23}$$

PROOF. The statement is immediate from the universal property of the free algebra $A\langle\xi_\alpha\rangle$ and of the relatively free algebras $A\langle\xi_\alpha\rangle$ and $A[\xi]$. □

In particular we apply this to an endomorphism of the free algebra $A\langle X\rangle$.

Such an endomorphism is determined by a map $\psi : X \to A\langle X\rangle$ by substitution $\psi(f(x_1, \dots, x_k)) = f(\psi(x_1), \dots, \psi(x_k))$.

If $I \subset A\langle X\rangle$ is a $T$-ideal, by definition such an endomorphism $\psi$ induces an endomorphism of $A\langle X\rangle / I$. We thus have that an endomorphism of the relatively free algebra $A\langle\xi_\alpha\rangle$ is determined by a map $\psi : \xi_\alpha \to A\langle\xi_\alpha\rangle$.

We then have

COROLLARY 2.4.5. *The endomorphism of* $A[\xi]$ *given by* $\bar{\psi}(\xi_j^{(\alpha)}) := f_i^{(\alpha)}(\xi)$ *is the unique endomorphism of* $A[\xi]$ *making the following diagram commutative.*

$$\begin{array}{ccccc} A\langle X\rangle & \xrightarrow{\pi} & A\langle \xi_\alpha\rangle & \xrightarrow{i} & R\otimes_A A[\xi] \\ \psi\downarrow & & \psi\downarrow & & 1_R\otimes\bar{\psi}\downarrow \\ A\langle X\rangle & \xrightarrow{\pi} & A\langle \xi_\alpha\rangle & \xrightarrow{i} & R\otimes_A A[\xi] \end{array} \tag{2.24}$$

*If* $\psi_1, \psi_2$ *are two such endomorphisms, we have* $\overline{\psi_1\circ\psi_2} = \overline{\psi_1}\circ\overline{\psi_2}$.

**2.4.2. The algebras with trace of generic elements.** Consider an algebra $R$, a finite module over a commutative algebra $C$ that is finite dimensional over a field $F$. In most cases $C = F$ or $q$ copies of the field $F$. This last case occurs when $R = \bigoplus_{i=1}^k R_i$ is a direct sum of finite-dimensional algebras over the same field $F$. It is then convenient to consider $R$ as an algebra over $F^{\oplus k}$ and also over $F$.

We have associated to $R$ the algebra $F\langle\xi_\alpha\rangle \subset R\otimes F[\xi]$, of generic elements, formula (2.20), and Definition 2.4.2. We can map $\pi : R \to M_m(C)$ into an algebra of matrices over $C$ in several ways. One way is canonical: if $R$ has a 1 and is a free $C$ module, is the regular representation on itself relative to the left multiplication map (in case $R$ has no 1 we add it)? Of course if $R = M_n(C)$ is already the algebra of matrices (*the fundamental representation*), the regular representation is $n$ copies of the fundamental representation.

Given one of these mappings $\pi$, one also has another algebra, in which we add to $F\langle\xi_\alpha\rangle$ all the coefficients of the characteristic polynomial of all the elements $\pi(a)$, $a\in F\langle\xi_\alpha\rangle$, since $\pi$ extends to a map $\pi : R\otimes_F F[\xi] \to M_m(C[\xi])$.

DEFINITION 2.4.6. The ring $T_X(R) = T\langle\xi_\alpha\rangle$ generated by the coefficients of the characteristic polynomial of the elements of $F\langle\xi_\alpha\rangle$, is called the *commutative trace algebra of* $R$ while $F\langle\xi_\alpha\rangle T_X(R)$, is called the *noncommutative trace algebra of* $R$.

The name *trace ring* is by abuse of language, and it comes from the fact that in characteristic 0 it is generated by the traces of the algebra of generic elements.

First let $t : R \to C$ be a *trace*, in the sense of Definition 2.3.1. In this special case it is just a linear map satisfying $t(ab) = t(ba)$.

In fact we will only use four cases.

(1) If $R = M_n(F)$, then we take as $t(a) = \mathrm{tr}(a)$ the usual trace.
(2) If $R$ is finite dimensional over a field $F$ we may take $t(a) = \mathrm{tr}(L_a)$, the trace of left multiplication $L_a : R \to R$, $b \mapsto ab$ as a linear map.
(3) If $R$ is as in (2) we may take $t(a) = \mathrm{tr}(L_{\bar a})$, the trace of left multiplication $L_{\bar a} : R/J \to R/J$, $b \mapsto \bar a b$ on $R/J$, where $J$ is the Jacobson radical and $\bar a$ denotes the class of $a$ modulo $J$.
(4) If $R = \bigoplus_{i=1}^q R_i$, a direct sum of finite dimensional algebras, then $J = \bigoplus_{i=1}^q J_i$, $R/J = \bigoplus_{i=1}^q R_i/J_i$. Then we take as trace $a$ the $q$-tuple of the traces of left multiplication of $\bar a$ on each $R_i/J_i$.

We shall discuss the various cases when needed. We take one of these traces $t : R \to C$ which extends to an $C[\xi]$-linear trace $t : R\otimes F[\xi] \to C[\xi]$.

Denote again by $T_X(R) = T\langle\xi_\alpha\rangle$ the subalgebra of $C[\xi]$ generated by the elements $t(a)$ for $a \in F\langle\xi_\alpha\rangle$. Since $R$ is an $C$ module, we have that $R\otimes F[\xi]$ is an $C[\xi]$

module. We then define $F\langle\xi_\alpha\rangle T_X(R) \subset R \otimes F[\xi]$ to be the *noncommutative trace algebra*.

In particular let $\psi : F\langle\xi_\alpha\rangle \to F\langle\xi_\alpha\rangle T_X(R) \subset R \otimes F[\xi]$ be any homomorphism. By Theorem 2.4.4 this induces an endomorphism $\bar{\psi}$ of $F[\xi]$ making the diagram of formula (2.23) commutative.

PROPOSITION 2.4.7.

(1) $t : F\langle\xi_\alpha\rangle T_X(R) \to T_X(R)$.
(2) *Given a homomorphism* $\psi : F\langle\xi_\alpha\rangle \to F\langle\xi_\alpha\rangle T_X(R)$, *we have*
  (a) $t(\psi(a)) = \bar{\psi}(t(a))$.
  (b) *The map* $\bar{\psi}$ *maps* $T_X(R)$ *to itself, and* $1_R \otimes \bar{\psi}$ *maps* $F\langle\xi_\alpha\rangle T_X(R)$ *to itself extending* $\psi$.

(2.25)
$$\begin{array}{ccccc} F\langle\xi_\alpha\rangle & \xrightarrow{i} & F\langle\xi_\alpha\rangle T_X(R) & \xrightarrow{i} & R \otimes_F F[\xi] \\ & \searrow^{\psi} & \Big\downarrow{\scriptstyle 1_R\otimes\bar{\psi}} & & \Big\downarrow{\scriptstyle 1_R\otimes\bar{\psi}} \\ & & F\langle\xi_\alpha\rangle T_X(R) & \xrightarrow{i} & R \otimes_F F[\xi] \end{array}$$

PROOF. (1) If $a \in F\langle\xi_\alpha\rangle$ then $\psi(a) = \sum_k b_k\tau_k$, $b_k \in F\langle\xi_\alpha\rangle$, $\tau_k \in T_X(R)$. We have $t(\sum_k b_k\tau_k) = \sum_k t(b_k)\tau_k$ hence $t : F\langle\xi_\alpha\rangle T_X(R) \to\rangle T_X(R)$.

(2)(a) For any element $b = \sum_{k=1}^m \lambda_k u_k \in R \otimes F[\xi]$, $\lambda_k \in F[\xi]$, in the basis $u_i$ of $R$, we have $t(\sum_{k=1}^m \lambda_k u_k) = \sum_{k=1}^m \lambda_k t(u_k)$, $t(u_k) \in F$, so that

$$t((1_R \otimes \bar{\psi})(b)) = t(\sum_{k=1}^m \bar{\psi}(\lambda_k)u_k) = \sum_{k=1}^m \bar{\psi}(\lambda_k)t(u_k) = \bar{\psi}(\sum_{k=1}^m \lambda_k t(u_k)) = \bar{\psi}(t(b)).$$

In particular when $b = \psi(a)$, $a \in F\langle\xi_\alpha\rangle$,

$$t(\psi(a)) = t((1_R \otimes \bar{\psi})(a)) = \bar{\psi}(t(a))). \tag{2.26}$$

(2)(b) Observe that by linearity of the trace, one has that $t(a) \in T_X(R)$ also when $a \in F\langle\xi_\alpha\rangle T_X(R)$. We have

$$\psi(\xi_\alpha) = \sum_k b_k^{(\alpha)}\tau_k^{(\alpha)}, \ b_k^{(\alpha)} \in F\langle\xi_\alpha\rangle, \ \tau_k^{(\alpha)} \in T_X(R).$$ □

The importance of these considerations is summarized in the following

THEOREM 2.4.8.

(1) *The kernel of the map from the free algebra with trace* $\mathcal{F}_T\langle X\rangle = F\langle x_\alpha\rangle \mathcal{T}_X$ *to the algebra* $R \otimes_F F[\xi]$ *given by* $x_\alpha \mapsto \xi_\alpha$ *is the ideal of trace identities of* $R$.
(2) *The algebra* $F\langle\xi_\alpha\rangle T_X(R)$ *with the map* $t : F\langle\xi_\alpha\rangle T_X(R) \to T_X(R)$ *is an algebra with trace and it is also a relatively free algebra with trace.*

PROOF. The fact that $t : F\langle\xi_\alpha\rangle T_X(R) \to T_X(R)$ satisfies the trace axioms follows from the commutative diagram

(2.27)
$$\begin{array}{ccc} F\langle\xi_\alpha\rangle T_X(R) & \xrightarrow{t} & T_X(R) \\ {\scriptstyle i}\Big\downarrow & & {\scriptstyle j}\Big\downarrow \\ R \otimes F[\xi] & \xrightarrow{t\otimes 1} & C[\xi] \end{array}$$

in which the rows are the two trace maps and the columns are inclusions.

We have, by definition, a surjective trace map $\pi$ from the free algebra with trace to $F\langle\xi_\alpha\rangle T_X(R)$ mapping $x_\alpha$ to $\xi_\alpha$. So, by definition or by Theorem 2.2.12, the fact that $F\langle\xi_\alpha\rangle T_X(R)$ is a relatively free algebra with trace is equivalent to the fact that the kernel of $\pi$ is a $T$-ideal for algebras with trace. This is essentially a reformulation of the content of Proposition 2.4.7. □

One needs some care when working with algebras without 1 (cf. Remark 2.3.6). In this case one should not consider the ring of pure trace expressions as part of the free algebra, and so $t(M)$ need not be an element of the trace algebra of generic elements.

REMARK 2.4.9. The most important application will appear in in §17.3.1. In that section we have $R = \bigoplus_{i=1}^k R_i$ is a direct sum of some special finite-dimensional algebras over the field $F$. One takes as trace the map $t : R \to C := \bigoplus_{i=1}^k F$ given by

$$t(a_1, \dots, a_k) := (\mathrm{tr}(a_1), \dots, \mathrm{tr}(a_k)),$$

where $\mathrm{tr}(a_i)$ denotes the trace of left multiplication by $\bar{a}_i$ in $R_i/J_i$, $J_i$ the radical of $R_i$. By Exercise 2.3.5 this also satisfies the axioms of a trace.

The commutative trace algebra $T_X(R)$ is contained in $C[\xi] = F[\xi]^{\oplus k}$ as explained in Remark 2.3.18.

2.4.2.1. *Finiteness statement.* We notice an important fact, whose proof will be completed later as consequence of Shirshov's theorem (Theorem 8.2.1).

Let $R$ be an $A$-algebra with trace and a finite $A$-module. We apply the notations of Definition 2.4.6.

PROPOSITION 2.4.10. *If $X = \{x_1, \dots, x_m\}$ is finite, then $T_X(R)$ is a finitely generated $A$-algebra and $A\langle\xi_\alpha\rangle T_X(R)$ is a finite module over $T_X(R)$.*

PROOF. We claim first that it is enough to prove that $A\langle\xi_\alpha\rangle T_X(R)$ is a finite module over $T_X(R)$, say generated by some monomials $M_1, \dots, M_k$.

If this is proved, by applying formula (3.29) of Amitsur's Theorem 3.3.8 and Corollary 3.3.9, it also follows that $T_X(R)$ is a finitely generated algebra, in fact generated by the elements $\sigma_i(P)$ where $P$ are monomials in the $M_j$ of degree $\leq i$. The proof of the fact that this algebra is a finite module over $T_X(R)$ is postponed, and it will follow from Shirshov's theorem as proved in Theorem 8.2.1. □

2.4.2.2. *Field extensions.* Let $R$ be an algebra over a field $F$ of characteristic 0 and $K \supset F$ a field extension, and let $R_K := K \otimes_F R$.

We want to compare the two $T$-ideals

$$\mathrm{Id}_F(R) \subset F\langle X\rangle, \qquad \mathrm{Id}_K(R_K) \subset K\langle X\rangle = K \otimes_F F\langle X\rangle.$$

PROPOSITION 2.4.11. *We have* $\mathrm{Id}_K(R_K) = K \otimes_F \mathrm{Id}_F(R)$.

PROOF. By Corollary 2.2.44 it is enough to prove that a multilinear polynomial $f(x_1, \dots, x_m) \in K\langle X\rangle$ is in $\mathrm{Id}_K(R_K)$ if and only if it lies in $K \otimes_F \mathrm{Id}_F(R)$. Consider a basis $k_i$ of $K$ over $F$ and write

$$f(x_1, \dots, x_m) = \sum_i k_i \otimes f_i(x_1, \dots, x_m), \; f_i(x_1, \dots, x_m) \in F\langle X\rangle.$$

Clearly if $a_1, \dots, a_m \in R$, we have $f(a_1, \dots, a_m) = \sum_i k_i \otimes f_i(a_1, \dots, a_m)$.

This implies that, if $f$ is a polynomial identity of $R_K$, then $f_i$ is a PI for $R$ for all $i$. The converse is also true by multilinearity of $f$ and hence of the $f_i$. □

**2.4.3. A generalization.** In fact all the ideas and theorems of the previous paragraphs can be developed in the much more general context of *universal algebra*. In this context one may study sets with all kinds of operations. Let us quickly review the main ideas.

DEFINITION 2.4.12. Given $n \in \mathbb{N}$ and a set $X$, an $n$-ary operation $f$ on $X$ is a map $f : X^n \to X$.

Given an $n$-ary operation $f$ and $n$ operations $g_i$ which are $k_i$-ary, we can compose all these operations obtaining a $\sum_{i=1}^n k_i$-ary operation

$$X^{\sum_{i=1}^n k_i} = \prod_{i=1}^n X^{k_i} \xrightarrow{\prod_{i=1}^n g_i} X^n \xrightarrow{\ f\ } X.$$

One can then use various compositions, impose identities between these compositions (as the commutative or associative law), and build in this way very general varieties of algebras.

Lie algebras, Jordan algebras, and alternative rings are examples of this general construction. Some of the ideas we develop in this book can be applied also to these settings although each theory has its own special features.

REMARK 2.4.13. Observe that a 0-ary operation consists in just choosing a distinguished element in $X$. So for instance in a ring the elements $0, 1$ may be treated as 0-ary operations.

The most common operations are binary as the sum or the product. We will have the occasion to use also 1-ary operations, that is, special maps from $X$ to $X$.

EXAMPLE 2.4.14. A group $G$ has a 0-ary operation the neutral element 1, a 1-ary operation $g \mapsto g^{-1}$ (the inverse), and a 2-ary operation (the multiplication). The group axioms are identities between these operations.

One can show that all interesting varieties have a symbolic calculus with variables, that is, free algebras. In our setting besides commutative and associative algebras, we shall consider algebras with involution, with trace or norm and occasionally Lie algebras. In all these cases the underlying set is always a module over some commutative ring $A$. Then on top of this basic structure we build the other operations. We shall make remarks on the nature of the respective free algebras whenever we will encounter one.

## 2.5. Generalized identities

**2.5.1. Generalized identities.** The notion of generalized identity for a given $A$-algebra $R$ is not as interesting as that of polynomial identity; nevertheless let us make some remarks on this topic, since free products play an important role in Chapter 17 where these concepts find applications to ordinary polynomial identities.

For the moment this material is independent of the main theme. As usual the theory should be developed for algebras or for algebras with a 1, and we leave some details to the reader.

We form the free product $S := R *_A A\langle X\rangle$ of $R$ with the free algebra $A\langle X\rangle$ (free product in the category of $A$-algebras), this algebra is also understood as some sort of *noncommutative polynomials with coefficients in* $R$; when the role of $A$ is clear, we also write $R\langle X\rangle = R *_A A\langle X\rangle$.

By Definition 2.1.17 the free product $R\langle X\rangle$ has the property that, given any $A$-algebra $U$, a homomorphism of $R\langle X\rangle$ to $U$ is uniquely determined by assigning in an arbitrary way two homomorphisms $R \to U$, $A\langle X\rangle \to U$, that is, by the property of the free algebra, a homomorphism $j : R \to U$ and a map $X \to U$, $x_i \mapsto a_i \in U$.

In particular let us apply this to $U = R$ and $j = 1_R$ the identity. Then the resulting map $S := R\langle X\rangle \to R$ is the *evaluation* of the noncommutative polynomials in $R$ obtained by assigning to each variable $x_i$ the *value* $a_i$.

DEFINITION 2.5.1. The ideal of *generalized identities* of $R$ is the ideal of elements in $R\langle X\rangle$ vanishing under all evaluations.

If $R$ has a basis $r_i$ over $A$, one has a combinatorial description of $S := R\langle X\rangle$ in term of an explicit basis. Assume $R$ has a 1 and $r_1 = 1$, and we have as basis the elements

$$1, r_{j_1} M_1 r_{j_2} M_2 \cdots M_k r_{j_{k+1}}, \tag{2.28}$$

where the $M_i$ are monomials in the variables $X$.

The multiplication of $r_{j_1} M_1 r_{j_2} M_2 \cdots M_k r_{j_{k+1}}$ and $r_{u_1} N_1 r_{u_2} N_2 \cdots N_p r_{u_{p+1}}$ is by first juxtaposition $r_{j_1} M_1 r_{j_2} M_2 \cdots M_k r_{j_{k+1}} r_{u_1} N_1 r_{u_2} N_2 \cdots N_p r_{u_{p+1}}$ and then expanding $r_{j_{k+1}} r_{u_1} = \sum_i \alpha_i r_i$. This can be proved by verifying directly on this concrete algebra the universal properties. If $R$ is not a free module over $A$, then the elements $r_i$ can be taken as linear generators, the elements of formula (2.28) are also linear generators, over $A$.

For an interesting example the reader should look at §2.6 where we discuss generalized identities of matrices.

2.5.1.1. *$\mathcal{V}$-polynomials I.* If $R$ belongs to a variety of algebras $\mathcal{V}$, it also makes sense to construct the free product of $R$, with the relatively free algebra in the variables $X$, in the variety $\mathcal{V}$. It is easily seen, by applying the universal properties, that

REMARK 2.5.2. This free product equals $R\langle X\rangle/T(R\langle X\rangle)$ with $T(R\langle X\rangle)$ the verbal ideal (Remark 2.2.47) of $R\langle X\rangle$ obtained by evaluation of all identities in $\mathcal{V}$.

Let us denote by

$$R_{\mathcal{V}}\langle X\rangle := R\langle X\rangle/T(R\langle X\rangle). \tag{2.29}$$

We have that $R_{\mathcal{V}}\langle X\rangle$ can be thought of as an algebra of generalized polynomials on $R$; that is, given any map $X \to R$, this extends to a unique homomorphism $R_{\mathcal{V}}\langle X\rangle \to R$. In fact more is true: given any homomorphism $R \to S$ where $S \in \mathcal{V}$ and any map $X \to S$, this extends to a unique homomorphism $R_{\mathcal{V}}\langle X\rangle \to S$. In particular we may apply this when $S = R_{\mathcal{V}}\langle X\rangle$, which shows that $R_{\mathcal{V}}\langle X\rangle$ behaves as what we may call $\mathcal{V}$-polynomials, with coefficients in $R$.

DEFINITION 2.5.3. The elements of $R_{\mathcal{V}}\langle X\rangle$ will be called *$\mathcal{V}$-polynomials* with coefficients in $R$ in the variables $X$.

The expression $\mathcal{V}$-polynomials is explained since, when $\mathcal{V}$ is the variety of commutative algebras, we have indeed that $R_{\mathcal{V}}\langle X\rangle = R[X]$ is the usual commutative algebra of polynomial over $R$.

Of course if $R = A\langle Y\rangle/I$ is presented as quotient of a free algebra $A\langle Y\rangle$, then $R_{\mathcal{V}}\langle X\rangle$ is presented as quotient $\pi : A\langle X \cup Y\rangle \to R_{\mathcal{V}}\langle X\rangle$ of a free algebra $A\langle X \cup Y\rangle$

modulo the ideal $J$ generated by $I$ and by all the evaluations of the identities in $\mathcal{V}$ in $A\langle X \cup Y\rangle$. It still makes sense to consider substitutions of the variables $X$ with elements of $R$ or even of $R_{\mathcal{V}}\langle X\rangle$.

This is due to the following fact. Take a map $\phi : X \to R_{\mathcal{V}}\langle X\rangle$ and consider the mapping $\bar{\phi}$ of $A\langle X \cup Y\rangle$ to $R_{\mathcal{V}}\langle X\rangle$ which maps $X$ via $\phi$ and $Y$ via the projection $A\langle Y\rangle \to L = A\langle Y\rangle/I \to R_{\mathcal{V}}\langle X\rangle$. The map $\bar{\phi}$ factors via the ideal $J$.

$$\begin{array}{ccc} A\langle X \cup Y\rangle & \xrightarrow{\bar{\phi}} & R_{\mathcal{V}}\langle X\rangle \\ \pi\Big\downarrow & & 1\Big\downarrow \\ A\langle X \cup Y\rangle/J = R_{\mathcal{V}}\langle X\rangle & \xrightarrow{\bar{\phi}} & R_{\mathcal{V}}\langle X\rangle \end{array} \tag{2.30}$$

EXERCISE 2.5.4. Assume for simplicity that $A$ is an infinite field, and prove that the algebra $R_{\mathcal{V}}\langle X\rangle$ is graded in each of the variables in $X$.

2.5.1.2. *R-algebras.* The constructions we just made suggest that we generalize Definition 2.5.1 and develop a corresponding theory of $T$-ideals. The endomorphisms $\phi : R\langle X\rangle \to R\langle X\rangle$, which are the identity on $R$, are given by the universal property by substitutions of the variables $x_i$ in elements of $R\langle X\rangle$. Thus a $T$-ideal is an ideal of $R\langle X\rangle$ stable under all such substitutions, and we can see that $T$-ideals are the ideals of generalized identities of $R$-algebras.

DEFINITION 2.5.5. A $T$-ideal of *generalized identities* of $R$ is an ideal in $R\langle X\rangle$ stable under all substitutions of variables.

In order to understand these new $T$-ideals of generalized identities, it is useful to introduce an abstract idea—we work in the category of associative algebras with a 1.

DEFINITION 2.5.6. Given an algebra $R$, an $R$-algebra is any associative algebra $S$ with a homomorphism $i : R \to S$.

If we are working in the category of algebras over a commutative ring $A$ we assume that $R, S$ are $A$-algebras and $i$ is a homomorphism as $A$-algebras.

In this definition no hypothesis of commutativity is assumed on $R$.

Then the free product $R\langle X\rangle = R *_A A\langle X\rangle$ is the free algebra in the category of $R$-algebras. Thus the semigroup $T$ of endomorphisms of $R *_A A\langle X\rangle$ as an $R$-algebra is given exactly by the operation of substitution, of the variables $x_i \in X$ with elements in $R *_A A\langle X\rangle$, and thus the notion in Definition 2.5.5 is exactly the correct idea of $T$-ideal.

One can then define the $T$-ideal of generalized identities of an $R$-algebra as in Definitions 2.2.1 and prove the analogue of Theorem 2.2.12. One can even discuss varieties of $R$-algebras.

Later we shall need an even more general definition of $R$-algebra.

DEFINITION 2.5.7. Given an algebra $R$, an $R$-algebra is any associative algebra $S$ with a bimodule action of $R$ on $S$ satisfying

$$r(s_1s_2) = (rs_1)s_2,\ (s_1s_2)r = s_1(s_2r),\ (s_1r)s_2 = s_1(rs_2),\ \forall r \in R,\ s_1, s_2 \in S. \tag{2.31}$$

One can easily connect Definitions 2.5.6 and 2.5.7. Given an $R$-algebra $S$, according to Definition 2.5.7, we can give to $R \oplus S$ a structure of algebra by setting

$$(r_1, s_1)(r_2, s_2) = (r_1r_2, r_1s_2 + s_1r_2 + s_1s_2);$$

the axioms (2.31) are the ones necessary and sufficient to have that $R \oplus S$ is associative. Then the inclusion of $R$ in $R \oplus S$ makes this an $R$-algebra according to Definition 2.5.6.

REMARK 2.5.8. Observe that in $R\langle X\rangle$ there is a notion of degree in each variable $x_i$.

This can be defined in several ways. If $A = F$ is an infinite field, we have by definition for each variable $x_i$ a *substitution map* $x_i \mapsto \alpha x_i$, $\alpha \in F^*$, which gives an action of the multiplicative group $F^*$ on $R\langle X\rangle$. Since substituting in different variables are commuting operations, we have an action of the *torus* $(F^*)^X$ (a copy of $F^*$ for each variable). The algebra decomposes into direct sum of the multihomogeneous components associated to the characters of the torus of type $(\alpha_1, \alpha_2, \dots, \alpha_i, \dots) \mapsto \prod_i \alpha_i^{h_i}$, $h_i \in \mathbb{N}$. In fact in general giving a grading of an algebra $S$ over an infinite field is equivalent to giving a *rational* action of $F^*$ on $S$ as a group of automorphisms—here rational is in the sense of algebraic groups.

The group $F^*$ is an algebraic group with its coordinate ring the ring $F[t, t^{-1}]$ of Laurent polynomials over $F$. An action is rational if it is induced by a *comodule action* $\nabla : S \to S \otimes_F F[t, t^{-1}]$ so that, given $\alpha \in F^*$, the action of $\alpha$ is the composition of $\nabla$ and the evaluation map $F[t, t^{-1}]$ to $F$ by $t \mapsto \alpha$. In practice the comodule action $\nabla$ formalizes the idea of substitution of $x$ to $tx$ in a polynomial algebra.[2]

If the base ring is not a field, the multiplicative group of $A$ may be too small to make this a useful definition (as for instance in $\mathbb{Z}$ where $\mathbb{Z}^* = \{\pm 1\}$).

In this case one has to use the previous idea of comodule map; that is, in more elementary terms one could act with the multiplicative group of the Laurent polynomial ring $A[t_i, t_i^{-1}]$ for each variable $x_i$ and apply the substitutions $x_i \mapsto t_i x_i$. Of course this makes sense provided we extend the coefficient ring to the ring of Laurent polynomials, hence the comodule action. Since $A[t_i, t_i^{-1}]$ is a free module over $A$, for any $A$-algebra $S$ we have that $A[t_i, t_i^{-1}] \otimes S$ contains $S$; moreover, the multiplicative group of $A[t_i, t_i^{-1}]$ is now infinite and contains the free abelian group generated by the elements $t_i$ which we now use in order to define the various homogeneous degrees.

2.5.1.3. *$\mathcal{V}$-polynomials II.* If $R$ belongs to some variety $\mathcal{V}$, one can repeat verbatim the discussion of the previous paragraph taking as algebra the algebra $R_{\mathcal{V}}\langle X\rangle$ of formula (2.29). We can then define $T$-ideals in this algebra and in generalized identities with coefficients in $R$ for any algebra $S$ in the variety $\mathcal{V}$.

REMARK 2.5.9. We will apply this to the variety $\mathcal{V}_n$ of algebras satisfying a special identity, the Capelli polynomial $C_n$ (defined by formula (7.5)), and then we will denote $R_n\langle X\rangle := R_{\mathcal{V}_n}\langle X\rangle$.

## 2.6. Matrices and the standard identity

For every integer $h$ the minimal element alternating in $h$ variables is the *standard polynomial* $\mathrm{St}_h$. This is the element of the free algebra in the variables $x_1, \dots, x_h$

[2]The coordinate ring of an algebraic group is a Hopf algebra, and the language of comodules belongs to the theory of Hopf algebras.

given by the formula

(2.32) $$\mathrm{St}_h(x_1,\dots,x_h) := \sum_{\sigma\in S_h} \epsilon_\sigma x_{\sigma(1)}\cdots x_{\sigma(h)}.$$

REMARK 2.6.1. A multilinear polynomial depending only on $x_1,\dots,x_h$ and alternating in these variables is a scalar multiple of $\mathrm{St}_h(x_1,\dots,x_h)$.

We want to apply the previous theory to the special case when $R$ is a ring of matrices, in which case the theory becomes particularly explicit and simple. We shall use Theorem 10.1.4 (to be proved later); that is, the algebra of $n\times n$ matrices over any commutative ring $A$ satisfies the standard polynomial $\mathrm{St}_{2n}$.

**2.6.1. Ideals of matrix algebras.** Given a ring $B$, one denotes by $M_n(B)$ the full ring of $n\times n$ matrices over $B$. We usually write $a = (b_{i,j})$ to indicate the matrix with entries $b_{i,j}$.

If $f : B \to C$ is a ring homomorphism, we can construct

$$M_n(f) : M_n(B) \to M_n(C), \quad M_n(f)(b_{i,j}) := (f(b_{i,j})),$$

the homomorphism induced on matrices. When $B, C$ are $A$-algebras and $f$ is a morphism of $A$-algebras, we also have

(2.33) $$M_n(B) = M_n(A)\otimes_A B, \quad M_n(C) = M_n(A)\otimes_A C, \quad M_n(f) = 1_{M_n(A)}\otimes f.$$

We will use systematically the following simple lemma.

LEMMA 2.6.2. *For any ring $B$ with a 1 (not necessarily commutative), if $I$ is a two-sided ideal in $M_n(B)$, then $I = M_n(J)$ for a (unique) two-sided ideal $J$ in $B$. Furthermore $M_n(B)/I = M_n(B/J)$.*

PROOF. Let $J$ be the set of all entries of elements of $I$. We claim that $J$ is an ideal and that $I = M_n(J)$.

First we remark that, if $a = (a_{i,j}) \in I$, then the element $e_{1,i}ae_{j,1} = a_{i,j}e_{1,1} \in I$; that is, if $x$ is an entry of an element of $I$, then $xe_{1,1} \in I$.

Then, if $b \in B$, we have the constant diagonal matrix with entry $b$ given by $b1_n$ and $b1_na_{i,j}e_{1,1} = ba_{i,j}e_{1,1}$, $a_{i,j}e_{1,1}b1_n = a_{i,j}be_{1,1}$, and $J$ is an ideal. Clearly $I \subset M_n(J)$.

Conversely it is enough to show that, when $b \in J$, any element $be_{i,j} \in I$. From what we have seen $be_{1,1} \in I$ hence $be_{i,j} = e_{i,1}be_{1,1}e_{1,j} \in I$. □

If $B$ has no 1, the situation is slightly more complicated. Arguing as in the previous proof, we see only that $M_n(BIB) \subset I \subset M_n(J)$, but in general $BIB \neq I$ (give an example).

COROLLARY 2.6.3. *If $F$ is a field and $M_n(F)$ has no nontrivial two-sided ideals, it is a* simple algebra.

We always identify $R$ with the subring of $M_r(R)$ of *scalar matrices*; that is, it is diagonal and constant on the diagonal so the correspondence between ideals $I$ of $M_r(R)$ and ideals $J$ of $R$ is by $J := R\cap I$.

In this next example we assume we are in the category of algebras with a 1.

Let $F$ be a field, and let $R := M_m(F)$ be the algebra of $m\times m$ matrices. $R$ is a simple algebra (cf. Corollary 2.6.3), and thus any nonzero morphism into another algebra is an injection. So consider an injection $i : M_m(F) \to S$ into any algebra.

Let $T$ be the centralizer of $M_m(F)$ in $S$. We then have a canonical map $j : M_m(F) \otimes_F T \to S$, and we have

PROPOSITION 2.6.4. *$j : M_m(F) \otimes_F T \to S$ is an isomorphism, i.e., $S = M_m(T)$.*

PROOF. By Lemma 2.6.2 an ideal $I$ of $M_m(F) \otimes_F T$ is of the form $M_m(F) \otimes_F J$ with $J = I \cap T$ an ideal of $T$. Let us apply this to the ideal $\ker j$, since $T \subset S$, it follows that $\ker j \cap T = 0$, hence $\ker j = 0$ and $j$ is injective. So we only need to see that $j$ is surjective. Consider the matrix units $e_{i,j}$. If $s \in S$, we claim that $\sum_{i=1}^n e_{i,a} s e_{b,i} \in T$ for all $a, b$. In fact an element is in $T$ if and only if it commutes with all the matrix units and we have

$$e_{h,k} \sum_{i=1}^n e_{i,a} s e_{b,i} = e_{h,a} s e_{b,k} = \sum_{i=1}^n e_{i,a} s e_{b,i} e_{h,k}.$$

Next $1 = \sum_i e_{i,i}$ implies

$$\sum_{a,b} e_{a,b} \sum_{i=1}^n e_{i,a} s e_{b,i} = \sum_{a,b=1}^n e_{a,a} s e_{b,b} = s.$$

We have written $s$ as a matrix with entries in $T$. □

In fact the previous proof proves something more general (which will be taken up again in the theory of Azumaya algebras in Corollary 5.4.29).

EXERCISE 2.6.5. Assume that in an algebra $S$, we have a system of *matrix units* $e_{h,k}$, $h, k = 1, \dots, m$, that is, we have a morphism $i : M_m(\mathbb{Z}) \to S$ into any algebra. By Lemma 2.6.2 it follows that the kernel of $i$ is an ideal $M_n(I)$ where $I$ is an ideal of $\mathbb{Z}$, and we have an injection $\bar{i} : M_m(\mathbb{Z}/I) \to S$.

Then we have an isomorphism $j : M_m(\mathbb{Z}/I) \otimes_F T \to S$ where $T$ is the centralizer of $M_m(\mathbb{Z}/I)$ in $S$ (that is the centralizer of the matrix units).

PROPOSITION 2.6.6. *When $R = M_n(A)$, the category of $R$-algebras (over $A$) is equivalent to the category of $A$-algebras, by the functor $C \mapsto M_n(C)$. Its inverse is the functor that associates to a map $i : M_n(A) \to S$ the centralizer $C := \{a \in S \mid a i(x) = i(x) a, \forall x \in M_n(A)\}$.*

*In general if $R = M_n(B)$, where $B$ is any $A$-algebra, the category of $R$-algebras (over $A$) is equivalent to the category of $B$-algebras (over $A$).*

PROOF. This follows from Exercise 2.6.5. □

Now consider the free product algebra (cf. Definition 2.1.17) $S := M_r(A) * A\langle X \rangle$, and in this algebra define the elements

$$x_{i,j}^{(k)} := \sum_{s=1}^r e_{s,i} x_k e_{j,s} \implies \sum_{i,j} x_{i,j}^{(k)} e_{i,j} = \sum_{i,j} e_{i,i} x_k e_{j,j} = x_k. \tag{2.34}$$

The following fact is proved in **[Pro68b]** and **[Pro72a]**.

THEOREM 2.6.7. *The elements $x_{i,j}^{(k)}$ generate a free algebra $A\langle x_{i,j}^{(k)} \rangle$ and commute with $M_r(A)$. Moreover,*

$$M_r(A) * A\langle X \rangle = M_r(A\langle x_{i,j}^{(k)} \rangle). \tag{2.35}$$

PROOF. These elements commute with $M_r(A)$ since

$$e_{h,k}x_{i,j}^{(k)} = \sum_{s=1}^{r} e_{h,k}e_{s,i}x_k e_{j,s} = e_{h,i}x_k e_{j,k} = \sum_{s=1}^{r} e_{s,i}x_k e_{j,s}e_{h,k} = x_{i,j}^{(k)}e_{h,k}.$$

Next we know (by Exercise 2.6.5) that $M_r(A) * A\langle X\rangle = M_r(B)$ where $B$ is the centralizer of $M_r(A)$. We have seen that $x_{i,j}^{(k)} \in B$, and we claim they generate $B$. In fact let $C$ be the subring they generate. We have $x_i \in M_r(C)$ by formula (2.34), so since the $x_i$ and $M_r(A)$ generate $M_r(A) * A\langle X\rangle$, we must have $B = C$.

It remains to prove that the elements $x_{i,j}^{(k)}$ generate a free algebra. This comes from the universal properties.

Consider element $y_{i,j}^{(k)}$ which generates a free algebra $\mathcal{F}$, in $M_r(\mathcal{F})$, and we can define the elements $y_k := \sum_{i,j} y_{i,j}^{(k)} e_{i,j}$.

We have a map of $M_r(A)$ to $M_r(\mathcal{F})$ which is the inclusion of matrices, a map of $A\langle X\rangle$ to $M_r(\mathcal{F})$ mapping $x_i$ to $y_i$ and thus by the free product property a map of $M_r(A) * A\langle X\rangle$ to $M_r(\mathcal{F})$ which thus maps $x_{i,j}^{(k)}$ to $y_{i,j}^{(k)}$, and this proves the claim. □

As a consequence, given a noncommutative polynomial $f(y_1, \dots, y_k)$, we have that $f(x_1, \dots, x_k)$ is a matrix with entries that we call $E_{i,j}(f)(x_{i,j}^{(k)})$.

Let us call $X_{r^2}$ the set of variables $x_{i,j}^{(k)}$ so that

$$|X_{r^2}| = r^2 \cdot |X| \quad \text{and} \quad M_r(A) * A\langle X\rangle = M_r(A\langle X_{r^2}\rangle).$$

We want to understand the correspondence of Lemma 2.6.2 in the special case of $M_r(A) * A\langle X\rangle = M_r(A\langle X_{r^2}\rangle)$ and for $T$-ideals.

THEOREM 2.6.8. *The semigroup $T$ of all $M_r(A)$ algebra endomorphisms of the algebra $M_r(A) * A\langle X\rangle$ is isomorphic to the semigroup $T$ of $A$-algebra endomorphisms of $A\langle X_{r^2}\rangle$.*

*Given a $T$-ideal of generalized identities $I \subset M_r(A) * A\langle X\rangle$, $I$ is of the form $M_r(J)$ where $J$ is a $T$-ideal in $A\langle X_{r^2}\rangle$.*

*In particular if $s = ru$, the $T$-ideal of generalized identities of $M_s(A)$ equals $M_r(J)$ where $J$ is the $T$-ideal of polynomial identities of $M_u(A)$.*

PROOF. The proof is immediate by the equivalence of categories, Proposition 2.6.6, and Exercise 2.6.5. □

COROLLARY 2.6.9. *$f \in A\langle X\rangle$ is a polynomial identity of $M_r(A)$ if and only if the polynomials $E_{i,j}(f)(x_{i,j}^{(k)})$ lie all in the ideal of commutators.*

In view of Theorem 2.6.12 it seems interesting to understand possible generators of the $T$-ideal of generalized identities of $M_s(A)$ with coefficients in $M_r(A)$, $s = ru$. In fact any evaluation of such an identity will produce an $r \times r$ matrix whose entries are polynomial identities of $u \times u$ matrices.

When we say that *a ring $S$ satisfies the identities of $r \times r$ matrices*, we mean that it satisfies the identities of $M_r(\mathbb{Z})$ or, if we are working with $A$-algebras (often $A$ a field), we mean that it satisfies the identities of $M_r(A)$, provided that $A$ is generic in the category of commutative rings (cf. Definition 2.2.27). We want at this point to avoid discussing for instance the case in which $A$ is a finite field.

PROPOSITION 2.6.10. *Given $r \in \mathbb{N}$ and a ring $R$, the following statements are equivalent:*

(1) *$R$ is commutative.*
(2) *$M_r(R)$ satisfies the identities of $r \times r$ matrices.*
(3) *$M_r(R)$ satisfies the standard identity $\mathrm{St}_{2r}$.*

PROOF. Clearly (1) $\implies$ (2) $\implies$ (3), so we need only show that (3) $\implies$ (1). Given two elements $a, b \in R$ we evaluate $\mathrm{St}_{2r}$ in the $2r$ elements

$$ae_{1,1},\ be_{1,2},\ e_{2,2}, \ldots, e_{r,r},\ e_{r,1}, \tag{2.36}$$

and we claim that the evaluation is $[a,b]e_{1,1}$. In order to understand this computation, let us display these elements in the simple case $r = 3$.

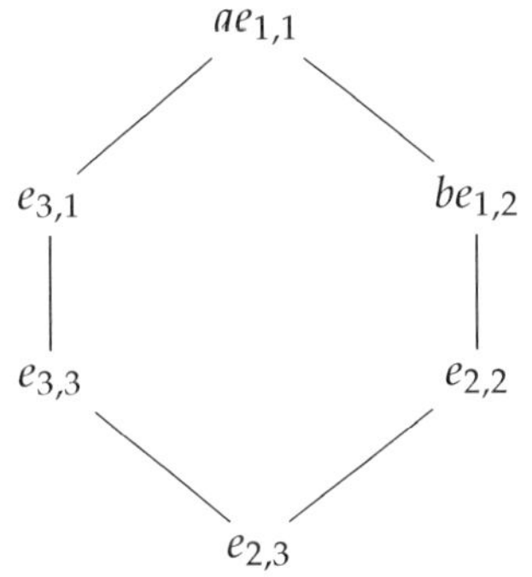


Then we see that a product of these elements is 0 unless it follows the clockwise circular order; thus we have six different nonzero products in the polynomial $\mathrm{St}_6$ which give the elements

$$ab\, e_{1,1},\ -ba\, e_{1,1},\ ab\, e_{2,2},\ -ab\, e_{2,2},\ ab\, e_{3,3},\ -ab\, e_{3,3}.$$

The value of $\mathrm{St}_6$ is thus $[a,b]e_{1,1}$. □

We now apply this to $R = A\langle X\rangle$ the free algebra in countably many variables $X = \{x_1, x_2, \ldots, x_n, \ldots\}$. Take the ideal $I$ of $M_r(R)$ generated by all evaluations of $\mathrm{St}_{2r}$ in $M_r(R)$. Then we claim

COROLLARY 2.6.11. *$I = M_r(J)$ where $J$ is the T-ideal of $R$ generated by the commutator $[x,y]$.*

PROOF. By Lemma 2.6.2 we have $I = M_r(L)$ where $L$ is an ideal of $R$ and $M_r(R)/I = M_r(R/L)$. The claim follows immediately from Proposition 2.6.10, in fact $R/J = A[x_1, \ldots, x_n, \ldots]$ is the commutative polynomial ring. So the evaluations of $\mathrm{St}_{2r}$ in $M_r(R)$ lie in $I$ and $L \subset J$. On the other hand by Proposition 2.6.10, $R/L$ is commutative so $L = J$. □

THEOREM 2.6.12. *The ideal $I \subset M_r(A) * A\langle X\rangle$ of generalized identities of $M_r(A)$ is the T-ideal of $M_r(A) * A\langle X\rangle$ generated by all evaluations of the standard identity $\mathrm{St}_{2r}$.*

PROOF. This follows from Proposition 2.6.10(3) and Theorem 2.6.8. □

It is tempting to think that this theorem may give information on ordinary polynomial identities of matrices, but this does not seem to be the case. The study of ordinary polynomial identities of matrices is a much more difficult topic with not so many explicit results.

We finally have

THEOREM 2.6.13. *Let $A, B$ be two commutative rings with 1 and $f : M_n(A) \to M_n(B)$ a (1-preserving) ring homomorphism. Then $f$ maps $A$ to $B$ and induces an isomorphism $\bar{f} : M_n(A) \otimes_A B \to M_n(B)$.*

PROOF. The kernel of $f$ is of the form $M_n(I)$ with $I$ an ideal of $A$. By Exercise 2.6.5 we have an isomorphism $M_n(B) = M_n(T)$ where $T$ is the centralizer of $M_n(\mathbb{Z}/I \cap \mathbb{Z})$ in $M_n(B)$. Since $B$ is commutative, $M_n(B)$ satisfies the standard identity $\mathrm{St}_{2n}$, so by Proposition 2.6.10 we also have that $T$ is commutative. Hence $T$ is in the center of $M_n(B)$, that is $T = B$, and the claim follows. □

COROLLARY 2.6.14.

(1) *Every $B$-linear endomorphism of $M_n(B)$ is an automorphism.*
(2) *Given $a \in M_n(A)$, let $\chi_a(t) := \det(t - a)$ be its characteristic polynomial, then*

$$\chi_{f(a)}(t) = f(\chi_a(t)). \tag{2.37}$$

This means that the trace, determinant, and all coefficients of the characteristic polynomial are *intrinsic*, they do not depend on how we present an abstract ring as matrices. In fact let $R \subset M_n(B)$ with $B$ commutative be a ring with center $A$ such that $M_n(B) = R \otimes_A B$. Assume further that $A$ is contained in $B \otimes_A B$ and $A = B \otimes 1 \cap 1 \otimes B$, then

THEOREM 2.6.15. *Under the previous hypotheses if $a \in R$ all coefficients of its characteristic polynomial are in $A$.*

PROOF. We have two homomorphisms $f_1 : b \mapsto b \otimes 1$, $f_2 : b \mapsto 1 \otimes b$ of $B$ in $B \otimes_A B$ and by hypothesis $A = \{b \in B \mid f_1(b) = f_2(b)\}$ which induce two maps $\bar{f}_1 = 1_R \otimes f_1$, $\bar{f}_2 = 1_R \otimes f_2$ of $M_n(B) = R \otimes_A B$ into $M_n(B \otimes_A B)$. If $a \in R$, we have $\bar{f}_1(a) = \bar{f}_2(a)$. So, by the previous corollary, if $a \in R$, we have $f_1(\chi_a(t)) = f_2(\chi_a(t))$, and the claim follows. □

REMARK 2.6.16. A ring $R$ with $M_n(B) = R \otimes_A B$ need not be closed under trace (as an example see generic matrices §14.4).

The condition $A = \{b \in B \mid f_1(b) = f_2(b)\}$ holds when $B$ is faithfully flat over $A$; see §5.1.0.1.

CHAPTER 3

# Symmetric functions and matrix invariants

*In this chapter we discuss, often omitting proofs, the classical theories of polarization and of symmetric functions. We devote some detail to the symbolic formulas arising from Amitsur's formula,* (3.29).

## 3.1. Polarization

**3.1.1. Commutative polynomials.** The theory of polynomial maps can be developed in various levels of generality as we shall see in Chapter 4.

In elementary algebra a polynomial function in $k$ variables over an infinite field $F$ can be uniquely written as the evaluation of a symbolic expression of the form $\sum c_{i_1,\dots,i_k} x_1^{i_1} \cdots x_k^{i_k}$, $c_{i_1,\dots,i_k} \in F$ when we substitute the *variables* $x_i$ with elements $a_i \in F$.

It is customary to denote by the symbol $F[x_1, \dots, x_k]$ the algebra of all polynomials. Here, contrary to most parts of this book, a polynomial is understood in the usual commutative algebra sense. But now clearly such a symbolic expression can also be evaluated substituting to the variables $x_i$ elements in any commutative algebra $A$ containing $F$.

An evaluation is in fact a homomorphism $\rho : F[x_1, \dots, x_k] \to A$, $\rho : x_i \mapsto a_i$, and all homomorphisms are of this form. In the language of universal algebra of Chapter 2, this means that $F[x_1, \dots, x_k]$ is the *free algebra* in $k$ variables, in the category of commutative associative $F$-algebras.

In particular we can substitute to the $x_i$ other polynomials in whichever variables we like. When we do this, we just compose homomorphisms; hence the sequence of evaluations is clearly associative and in the end all the symbolic identities that we may obtain are identities as functions on $F^k$.

DEFINITION 3.1.1.

(1) If $V$ is a vector space over a field $F$, a polynomial function on $V$ is a function $f : V \to F$ which, in a given basis, is written as polynomial in the coordinates.
(2) Given two vector spaces $V, W$, we also have the notion of *polynomial map* $f : V \to W$. This is a map such that in two given bases the coordinates of $f(v)$ are polynomials in the coordinates of $v$.

This notion is independent of the coordinates as easily seen. It is not necessary to assume that $V, W$ are finite dimensional; we only have that a polynomial by definition will depend always upon finitely many of the coordinates.

This theory is formalized in §4.1 in a categorical way, but at this point it should be just the language of functions and variables.

REMARK 3.1.2. If $A$ is any commutative ring, one has always the symbolic ring $A[x_1, \dots, x_n]$ of polynomials with coefficients in $A$ and in $n$ variables.

Each such polynomial $f(x_1, \dots, x_n)$ gives by evaluation on $A$ a *polynomial function* $\bar{f} : A^n \to A$.

For a general commutative ring it may very well be that $f \neq 0$ but $\bar{f} = 0$. The simplest example is when $A$ is a finite field with $q = p^k$ elements ($p$ a prime), where the polynomial in one variable $x^q - x$ vanishes.

On the other hand $f(x_1, \dots, x_n)$ gives by evaluation on any commutative $A$-algebra $B$ a polynomial function $\bar{f} : B^n \to B$, and it is clear that now $\bar{f}$ does not vanish for all $B$.

EXERCISE 3.1.3. Show that a nonzero polynomial $f(x_1, \dots, x_k)$ of degree $d$ with coefficients in a field $F$ with at least $d+1$ elements does not vanish as function on $F^k$.

3.1.1.1. *Polarization and multilinearization.* We want to recall the basic classical method of *polarization*. In its simplest form this is the method which associates to a quadratic form a bilinear form.

Given a polynomial function $f(v)$, on a vector space $V$ over an infinite field $F$, one can evaluate the same function also on $V \otimes_F A$ for any commutative ring $A$ (for instance by expressing it in a given basis as a polynomial in some $k$ variables). In particular we can evaluate it for $A = F[t_1, \dots, t_d]$, the ring of polynomials in $d$ variables, and the values are in $A$. In this way $f$ produces new polynomial functions in several variables by *polarization* as follows.

Fix a positive integer $d$ (in the end the degree of $f$) take $d$ copies of $V$ that is $(v_1, \dots, v_d) \in V^{\oplus d}$ as variables and auxiliary indeterminates $t_1, \dots, t_d$. Thus, we have a linear map

$$V^{\oplus d} \to V \otimes_F F[t_1, \dots, t_d], \quad (v_1, \dots, v_d) \mapsto \sum_{i=1}^{d} t_i v_i.$$

The function $f(\sum_{i=1}^d t_i v_i)$ is a polynomial both in the variables $t_i$ and in the vector variables $v_j$. We can expand this polynomial in the variables $t_i$ so that its coefficients are polynomials in the variables $v_j$ and set

$$f(\sum_{j=1}^{d} t_j v_j) := \sum_{h_1, \dots, h_d} t_1^{h_1} \cdots t_d^{h_d} f_{h_1, \dots, h_d}(v_1, \dots, v_d). \tag{3.1}$$

The first remark we should make is the *symmetry* of the operation. If $\sigma$ is a permutation of $1, 2, \dots, d$, we have $f(\sum_{i=1}^d t_i v_{\sigma(i)}) = f(\sum_{i=1}^d t_{\sigma^{-1}(i)} v_i)$,

$$\implies f_{h_1, \dots, h_d}(v_{\sigma(1)}, \dots, v_{\sigma(d)}) = f_{h_{\sigma^{-1}(1)}, \dots, h_{\sigma^{-1}(d)}}(v_1, \dots, v_d). \tag{3.2}$$

LEMMA 3.1.4. *The polynomial $f_{h_1, \dots, h_d}(v_1, \dots, v_d)$ is homogeneous of degree $h_i$ in each variable $v_i$.*

PROOF. We introduce further variables $\lambda_i$ and see that

$$\sum_{h_1,\dots,h_d} t_1^{h_1}\cdots t_d^{h_d} f_{h_1,\dots,h_d}(\lambda_1 v_1,\dots,\lambda_d v_d) = f(\sum_{j=1}^d t_j\lambda_j v_j) \tag{3.3}$$
$$= \sum_{h_1,\dots,h_d} (\lambda_1 t_1)^{h_1}\cdots(\lambda_d t_d)^{h_d} f_{h_1,\dots,h_d}(v_1,\dots,v_d)$$
$$= \sum_{h_1,\dots,h_d} t_1^{h_1}\cdots t_d^{h_d}\lambda_1^{h_1}\cdots\lambda_d^{h_d} f_{h_1,\dots,h_d}(v_1,\dots,v_d).$$

Hence $f_{h_1,\dots,h_d}(\lambda_1 v_1,\dots,\lambda_d v_d) = \lambda_1^{h_1}\cdots\lambda_d^{h_d} f_{h_1,\dots,h_d}(v_1,\dots,v_d)$, as desired. □

REMARK 3.1.5. If we just consider one variable $t$ and write $f(tv) = \sum_i t^i f_i(v)$, we have that the polynomials $f_i(v)$ are the homogeneous components of $f(v)$.

If $f$ is homogeneous of degree $d$, we see that the total degree of each of the terms $f_{h_1,\dots,h_s}(v_1,\dots,v_s)$ is also $\sum h_j = d$ equal to $d$.

We also use a compact symbolic expression for formula (3.1) which, for a homogeneous polynomial of degree $d$ is

$$f(\sum_{j=1}^d t_j v_j) = \sum_{\underline{h}\mid|\underline{h}|=d} t^{\underline{h}} f_{\underline{h}}(v_1,\dots,v_d). \tag{3.4}$$

Thus by $\underline{h}$ we mean some vector $\underline{h} = (h_1,\dots,h_d)$, $h_i\in\mathbb{N}$, by $|\underline{h}| = \sum_i h_i$ and finally $t^{\underline{h}} := t_1^{h_1}\cdots t_d^{h_d}$.

Of particular interest is the *full polarization* or *multilinearization* of $f$, that is the term $f_{1,\dots,1}(v_1,\dots,v_d)$ which is multilinear and symmetric, by formula (3.2), in $d$ vector variables.

For instance if $f(x) = x^3$ is in just one variable (i.e., $\dim(V) = 1$) the polarization is through the formula

$$\begin{aligned} f(t_1x_1+t_2x_2+t_3x_3) &= (t_1x_1+t_2x_2+t_3x_3)^3 \\ &= t_1^3x_1^3 + t_2^3x_2^3 + t_3^3x_3^3 + 3t_1^2t_2x_1^2x_2 + 3t_1^2t_3x_1^2x_3 + 3t_1t_2^2x_1x_2^2 \\ &\quad + 3t_2^2t_3x_2^2x_3 + 3t_1t_3^2x_1x_3^2 + 3t_2t_3^2x_2x_3^2 + 6t_1t_2t_3x_1x_2x_3. \end{aligned}$$

So the full polarization is $6x_1x_2x_3$. Instead in two variables $x, y$, the coordinates of a vector $v\in V$ with $\dim(V) = 2$, if $f(x,y) = x^2y$ the polarizations are obtained by substituting

$$v = (x,y) \mapsto t_1v_1 + t_2v_2 + t_3v_3 = t_1(x_1,y_1) + t_2(x_2,y_2) + t_3(x_3,y_3),$$

that is by substituting $x\mapsto t_1x_1+t_2x_2+t_3x_3$, $y\mapsto t_1y_1+t_2y_2+t_3y_3$ and computing $(t_1x_1+t_2x_2+t_3x_3)^2(t_1y_1+t_2y_2+t_3y_3)$. So the full polarization of $f(x,y) = x^2y$ is $2(x_1x_2y_3 + x_1x_3y_2 + x_2x_3y_1)$.

3.1.1.2. *Restitution.* When in $f(\sum_{j=1}^d t_jv_j)$ we set all the variables $v_i$ equal to $v$, we get the operation of *restitution*. Assuming that $f$ is homogeneous of degree $d$, we have

$$f(\sum_{j=1}^d t_jv) = (\sum_{j=1}^d t_j)^d f(v) = \sum_{h_1,\dots,h_d\mid\sum h_j=d} t_1^{h_1}\cdots t_d^{h_d} f_{h_1,\dots,h_d}(v,\dots,v).$$

One thus sees, equating coefficients on the two sides, that

$$\binom{d}{h_1,\ldots,h_d} f(v) = f_{h_1,\ldots,h_d}(v,\ldots,v). \tag{3.5}$$

In particular when we restitute the full polarization, we have

$$d!f(v) = f_{1,\ldots,1}(v,\ldots,v).$$

This formula points to the fact that this formalism works better in characteristic 0 than in positive characteristics, a fact which is the source of most difficulties in this last case.

Polarizations are clearly linear operators, in fact they are differential operators, a fact that will not play a role in our treatment. They have some complicated composition formulas with respect to the product. In fact, compute for two homogeneous polynomials $f(v), g(v)$ of degrees $a, b$

$$f(\sum_{j=1}^{d} t_j v_j) g(\sum_{j=1}^{d} t_j v_j) = \sum_{\underline{h}\,|\,|\underline{h}|=a} t^{\underline{h}} f_{\underline{h}}(v_1,\ldots,v_d) \sum_{\underline{k}\,|\,|\underline{k}|=b} t^{\underline{k}} g_{\underline{k}}(v_1,\ldots,v_d). \tag{3.6}$$

Expanding the right-hand side of (3.6), we see that the polarizations of the product of two polynomials can be expressed by universal formulas as quadratic polynomials of their respective polarizations.

If $f$ is homogeneous and $d$ equals the degree of $f$, the functions $f_{h_1,\ldots,h_d}$ are really different functions. If instead we take $d$ larger than the degree, we get repetitions of the same functions but in different variables.

This procedure can be applied also for elements of the free algebra, in fact the formalism of tensor algebra is just a modern reformulation of the classical formalism of polarization.

Polarization can be iterated: if we have a polynomial $f(v_1, v_2, \ldots, v_m)$ in several vector variables, we can polarize separately each variable.

3.1.1.3. *Symbolic polarization-multilinearization.* We have seen how polarization arises by a procedure of substitution of variables. This applies also to noncommutative polynomials, that is to the elements of the free algebra $f(x_1,\ldots,x_n) \in A\langle X\rangle$, which is analyzed in detail in §2.2.1.

We then can take an element $f(x_1,\ldots,x_n)$ and evaluate one or more of its variables in $S\langle X \cup Y\rangle$ where $S$ is any commutative $A$-algebra. In particular we can evaluate for instance the variable $x_1$ in a linear combination $\sum_{i=1}^{m} t_i y_i$, with $t_i$ commutative indeterminates. We then get an element of $A[t_1,\ldots,t_m]\langle X \cup Y\rangle$ which can be developed as a polynomial in the variables $t_i$ with coefficients in $A\langle X \cup Y\rangle$.

$$f(\sum_{j=1}^{d} t_j y_j, x_2,\ldots,x_n) = \sum_{h_1,\ldots,h_d} t_1^{h_1}\cdots t_d^{h_d} f_{h_1,\ldots,h_d}(y_1,\ldots,y_m; x_2,\ldots,x_n). \tag{3.7}$$

We can follow verbatim the discussion of the previous paragraph and see that $f_{h_1,\ldots,h_d}(y_1,\ldots,y_m; x_2,\ldots,x_n)$ is homogeneous of degree $h_i$ in each $y_i$. Furthermore, if $f$ is homogeneous of degree $d$ in $x_1$, we have again the restitution formula

$$f_{h_1,\ldots,h_d}(x_1,\ldots,x_1; x_2,\ldots,x_n) = \binom{d}{h_1,\ldots,h_d} f(x_1,\ldots,x_n). \tag{3.8}$$

Of course we may apply this procedure to all the variables.

In positive characteristic the restitution may give 0, nevertheless we claim

LEMMA 3.1.6. *Each polarized polynomial is nonzero.*

PROOF. We apply the procedure to a single monomial and to a variable $x_1$ of degree $d$ in this monomial. Since we are in a noncommutative setting, we have that the polarized noncommutative polynomial is the sum of all the distinct monomials obtained by substituting to all occurrences of $x_1$ exactly $h_i$ values of $y_i$ in all possible ways. It is clear that these noncommutative monomials are all distinct and furthermore the monomial we started from can be recovered by each of these monomials by restitution. This implies that polarizing in any way a noncommutative polynomial, we always obtain formally nonzero polynomials. □

REMARK 3.1.7. The statement of the previous lemma is not true in the commutative setting.

EXAMPLE 3.1.8. By polarizing $x^2$, we get $(x_1 + x_2)^2 = x_1^2 + x_1x_2 + x_2x_1 + x_2^2$, hence the full polarization is $x_1x_2 + x_2x_1$. If we are in characteristic 2, for the commutative polynomials we have $x_1x_2 + x_2x_1 = 0$, but in the noncommutative setting this expression is not zero.

REMARK 3.1.9. As a final remark we observe that any homogeneous polynomial can be fully polarized producing a nonzero multilinear expression.

In characteristic 0 the restitution formula (3.8) shows that the polynomial can be recovered by its full polarization.

3.1.1.4. *Polarization and representations.* The theory of polarization can be developed in an abstract and unified form, see **[DCP17]**. Here we sketch some of these ideas.

If we have a vector space $W$ of dimension $d$ over a field $F$, denote by $GL(W)$ the linear group of all invertible linear transformations on $W$.

In a basis $GL(W) = GL(d, F)$, the group of all invertible matrices.

The group $GL(W)$ acts on polynomial functions on the space $W$ by the formula $(gf)(w) := f(g^{-1}w)$. This is a linear action of automorphisms of the algebra of polynomials, the *linear changes of variables*.

In invariant theory, which we shall later encounter, we study this action when restricted to subgroups of $GL(W)$.

The transformations induced on functions by polarization and restitutions are in fact part of a more general idea related to this formalism.

When we take $d$ copies of a vector space $V^{\oplus d}$, where $d$ can also be $\infty$, we may think at this space as the tensor product $V \otimes W$ where $W$ is a $d$-dimensional vector space, in which we have chosen a basis.

The linear group $GL(W)$ acts on the tensor product $V \otimes W$, with the formula $g(v \otimes w) = v \otimes g(w)$, by acting on the second factor and hence on polynomial functions on $V \otimes W$. The polarization operators just arise as infinitesimal transformations associated to this group action. If $i \neq j$ are two indices, the operator $1 + \lambda e_{i,j}$ is in $GL(d, F)$ and transforms a polynomial $f(v_1, v_2, \dots, v_d)$ into $f(v_1, v_2, \dots, v_i - \lambda v_j, \dots, v_d)$.

Thus we have that the Lie algebra of $GL(W)$ acts as the algebra of linear polarizations. The enveloping algebra of this Lie algebra acts as differential operators commuting with $GL(V)$.

In the same way symbolic polarization is just the infinitesimal part of the action of the linear group $\mathrm{GL}(W)$ on the tensor algebra $T(W) = \bigoplus_{i=0}^{\infty} W^{\otimes i}$ which is the free algebra in the elements of a basis of $W$. The reader is referred to **[Cap02]** for a classical treatment and to **[Pro07]** for a modern one.

Usually when $W = F^{\infty}$ is infinite dimensional with basis $e_i, i = 1, \dots, \infty$, the symbol $\mathrm{GL}(W)$ will be used not for the group of all linear transformations but only for those which are given by *finite matrices*, that is that are the identity on all $e_j$ for $j$ large.

3.1.1.5. *Aronhold's method.* Let $F$ be a field of characteristic 0. We have seen two examples where we have the operations of polarization and restitution, the free algebra $F\langle x_1, \dots, x_k, \dots \rangle$ (that is the tensor algebra over $F^{\infty}$) as well as the polynomial algebra in infinitely many copies of some vector space $V$.

PROPOSITION 3.1.10. *Let $U$ be one of the two previous examples, and let $A, B \subset U$ be two subspaces closed under polarization $P$ and restitution $R$. Then $A \subset B$ (resp., $A = B$) if and only if the subspace $A^{\mathrm{mult}}$ of multilinear elements of $A$ is contained (resp., is equal) to the subspace $B^{\mathrm{mult}}$ of multilinear elements of $B$.*

PROOF. Since the subspaces are closed under polarization, they are clearly graded, that is spanned by their homogeneous elements. If $u \in A$ is homogeneous of some degree $d$, we have $u = d! R \circ P(u)$, with polarization $P$ and restitution $R$. We have by hypothesis $P(u) \in A^{\mathrm{mult}} \subset B^{\mathrm{mult}}$ so, since $B$ is closed under restitution, we have $d! u \in B$ and, since $F$ has characteristic 0, also $u \in B$. □

This proposition is the classical Aronhold's method, which was applied extensively to the space of polynomial invariants in infinitely many copies of a representation $V$ of a group $G$. In fact one can formalize this in the language of polynomial functors, as we shall see in §6.3.1.

## 3.2. Symmetric functions

**3.2.1. Elementary symmetric functions.** We give here a few definitions and facts on an important topic, symmetric functions, without proofs. For the proofs, and an extended theory, one should refer to the book of Macdonald **[Mac95]** or of Stanley **[Sta97]**, **[Sta99]**. A more limited approach directly connected to the Schur–Weyl theory is in **[Pro07]** or **[GW09]**.

The theory of symmetric functions is a classical theory developed (by Lagrange, Ruffini, Galois and others) in connection with the theory of algebraic equations in one variable and the classical question of resolution by radicals.

The main link are the formulas expressing the coefficients of a polynomial through its roots. A formal approach is the following.

Consider polynomials in commutative variables $x_1, x_2, \dots, x_n$ and an extra variable $t$ over the ring of integers.

The elementary symmetric functions $e_i := e_i(x_1, x_2, \dots, x_n)$ are implicitly defined by the formula

$$p(t) := \prod_{i=1}^{n}(1 + t x_i) := 1 + \sum_{i=1}^{n} e_i t^i. \tag{3.9}$$

More explicitly $e_i(x_1, x_2, \dots, x_n)$ is the sum of $\binom{n}{i}$ terms, the products, over all subsets of $\{1, 2, \dots, n\}$ with $i$ elements, of the variables with indices in that subset.

$$e_i = \sum_{1 \le a_1 < a_2 < \cdots < a_i \le n} x_{a_1} x_{a_2} \dots x_{a_i}. \tag{3.10}$$

If $\sigma$ is a permutation of the indices, we obviously have

$$\prod_{i=1}^{n}(1 + tx_i) = \prod_{i=1}^{n}(1 + tx_{\sigma(i)}),$$

and thus the elements $e_i$ are invariant under permutation of the variables.

Of course the polynomial $t^n p(-\frac{1}{t})$ has the elements $x_i$ as its roots.

DEFINITION 3.2.1. A polynomial in the variables $(x_1, x_2, \dots, x_n)$, invariant under permutation of these variables, is called a *symmetric function*.

The functions $e_i$ are called **elementary symmetric functions.**

We have not specified the coefficient ring of the polynomial which can be any commutative ring $A$. We use the notation of formula (1.9) which will appear systematically in this book. The ring of symmetric functions, that is the ring of invariants of the action of the symmetric group $S_n$ on the polynomial ring $A[x_1, \dots, x_n]$ is denoted by $A[x_1, \dots, x_n]^{S_n}$.

The classical theorem on symmetric functions (cf. **[Pro07]**) is

THEOREM 3.2.2. *The ring $\mathbb{Z}[x_1, \dots, x_n]^{S_n}$ is the polynomial ring $\mathbb{Z}[e_1, \dots, e_n]$.*

In fact there is a simple recursive algorithm which, given a symmetric polynomial $f(x_1, \dots, x_n)$ writes it as a polynomial in the variables $e_i$.

REMARK 3.2.3. A typical corollary of this theorem is the existence of the *discriminant* that is an explicit polynomial $\Delta(e_1, \dots, e_n) = \prod_{i \ne j}(x_i - x_j)$.

Given thus a monic polynomial with coefficients in a field $F$, one can compute the discriminant of its roots as polynomial in its coefficients. The nonvanishing of the discriminant is equivalent to the statement that the polynomial has distinct roots. Same for the *resultant* $\prod_{i,j}(x_i - y_j)$ of two polynomials.

By convention we always set $e_0 = 1$. There are several families of obviously symmetric functions. Complementary to the elementary symmetric functions are the *complete symmetric functions* $h_i$ given by the generating formula

$$\prod_{i=1}^{n} \frac{1}{1 - x_i t} = \prod_{i=1}^{n} \sum_{k=0}^{\infty} t^k x_i^k = \sum_{k=0}^{\infty} h_k(x) t^k. \tag{3.11}$$

One easily verifies that $h_k$ is the sum of all monomials of degree $k$ (from which the name complete). We have the identity (with $e_0 = h_0 = 1$)

$$(\sum_{k=0}^{\infty} h_k(x) t^k)(\sum_{k=0}^{n} (-1)^k e_k(x) t^k) = 1 \implies \sum_{k=0}^{j} (-1)^k e_k h_{j-k} = 0, \ \forall j > 0.$$

It follows that the ring $\mathbb{Z}[x_1, \dots, x_n]^{S_n}$ is also a polynomial ring $\mathbb{Z}[h_1, \dots, h_n]$ while $h_k, k > n$ is a polynomial in the preceding ones.

For us the most important will be the power sums (also known as Newton functions) and the Schur functions which will appear later:

$$\text{Newton functions} \qquad \psi_k := \sum_{i=1}^{n} x_i^k. \tag{3.12}$$

We can start with a first important fact, the explicit connection between the functions $e_i$ and the $\psi_k$. To do this, we will perform the next computations in the ring of formal power series, although the series that we will consider also have a meaning as convergent series.

Start from the identity $\prod_{i=1}^n (tx_i + 1) = \sum_{i=0}^n e_i t^i$ and take the logarithmic derivative, relative to the variable $t$, of both sides. For a function $f(t)$ it is

$$\frac{d \log(f(t))}{dt} = \frac{f'(t)}{f(t)}.$$

We use the fact that such an operator transforms products into sums to get

$$\sum_{i=1}^{n} \frac{x_i}{(tx_i + 1)} = \frac{\sum_{i=1}^n i e_i t^{i-1}}{\sum_{i=0}^n e_i t^i}.$$

The left-hand side of this formula can be developed as

$$\sum_{i=1}^{n} x_i \sum_{h=0}^{\infty} (-tx_i)^h = \sum_{h=0}^{\infty} (-t)^h \psi_{h+1}.$$

From this we get the identity[1]

$$(\sum_{h=0}^{\infty} (-t)^h \psi_{h+1})(\sum_{i=0}^{n} e_i t^i) = (\sum_{i=1}^{n} i e_i t^{i-1}) \tag{3.13}$$

which gives, equating coefficients, for all $m$

$$(-1)^m \psi_{m+1} + \sum_{i=1}^{m} (-1)^{i-1} \psi_i e_{m+1-i} = (m+1) e_{m+1}, \tag{3.14}$$

where we intend $e_i = 0$ if $i > n$.

It is clear that these formulas give recursive ways of expressing the $\psi_i$ in terms of the $e_j$ with integral coefficients. On the other hand they can also be used to express the $e_i$ in terms of the $\psi_j$, but in this case it is necessary to perform some divisions as shown by

THEOREM 3.2.4.

$$e_r = \sum_{\substack{h_1 + 2h_2 + \cdots + rh_r = r \\ h_1 \geq 0, \ldots, h_r \geq 0}} (-1)^r \prod_{j=1}^{r} \frac{(-\psi_j)^{h_j}}{h_j! j^{h_j}}. \tag{3.15}$$

PROOF. Using the Taylor expansion for $\log(1+y)$, we get

$$\sum_{i=0}^{n} (-1)^i e_i t^i = \prod_{r=1}^{n} (1 - x_r t) = \exp(-\sum_{j=1}^{\infty} \frac{\psi_j}{j} t^j). \tag{3.16}$$

Expanding and setting $e_h = 0$ for $r > n$, we deduce the formula. □

REMARK 3.2.5. Formula (3.14) for $m = n$ allows us to express $\psi_{n+1}$ as polynomial in the $\psi_i$, $i \leq n$. We shall later see how this basic identity plays a major role since it develops into the basic trace identity for matrices and from that to important information on polynomial identities of matrices, Corollary 3.3.3.

[1] Formulas found by Newton, hence the name *Newton functions* for the $\psi_k$.

EXAMPLE 3.2.6. The first expressions of $e_i$ $i = 1, 2, 3, 4$ through the $\psi_j$.

$$e_1 = \psi_1, \quad e_2 = \frac{1}{2}(\psi_1^2 - \psi_2), \quad e_3 = \frac{1}{6}(\psi_1^3 - 3\psi_1\psi_2 + 2\psi_3), \tag{3.17}$$
$$e_4 = \frac{1}{24}(\psi_1^4 - 6\psi_1^2\psi_2 + 3\psi_2^2 + 8\psi_1\psi_3 - 6\psi_4).$$

**3.2.2. Alternating functions.** Along with symmetric functions, it is important to discuss alternating (or skew-symmetric) functions. We restrict our considerations to integral polynomials.

DEFINITION 3.2.7. A function $f$ in the variables $(x_1, x_2, \dots, x_n)$, is called an alternating function if, for every permutation $\sigma$ of these variables,

$$f^\sigma = f(x_{\sigma(1)}, x_{\sigma(2)}, \dots, x_{\sigma(n)}) = \epsilon_\sigma f(x_1, x_2, \dots, x_n),$$

$\epsilon_\sigma$ being the sign of the permutation.

We have the *Vandermonde determinant* $V(x) := \prod_{i<j}(x_i - x_j)$ as a basic alternating polynomial. The main remark on alternating functions is the following.

PROPOSITION 3.2.8. *A polynomial f(x) in the variables $x = (x_1, \dots, x_n)$ is alternating if and only if it is of the form $f(x) = V(x)g(x)$, with $g(x)$ a symmetric polynomial.*

PROOF. Substitute, in an alternating polynomial $f$, to a variable $x_j$ a variable $x_i$ for $i \neq j$. We get the same polynomial if we first exchange $x_i$ and $x_j$ in $f$. Since this changes the sign, under this substitution $f$ becomes 0.

This means in turn, that $f$ is divisible by $x_i - x_j$; since $i, j$ are arbitrary, $f$ is divisible by $V(x)$. Writing $f = V(x)g$, it is clear that $g$ is symmetric. □

Let us be more formal. Let $A, S$ denote the sets of **antisymmetric** and **symmetric** polynomials. We have seen

PROPOSITION 3.2.9. *The space $A$ of antisymmetric polynomials is a free rank 1 module over the ring $S$ of symmetric polynomials generated by $V(x)$ or $A = V(x)S$.*

In particular, dividing by $V(x)$, any integral basis of $A$ gives an integral basis of $S$. In this way we will presently obtain the *Schur functions*.

In order to understand the construction, let us make a fairly general discussion. In the ring of polynomials $\mathbb{Z}[x_1, x_2, \dots, x_n]$, let us consider the basis given by the monomials (which are permuted by $S_n$).

Observe that the orbits of monomials are indexed by nonincreasing sequences of integers. To $m_1 \geq m_2 \geq m_3 \cdots \geq m_n \geq 0$ corresponds the orbit of the monomial $x_1^{m_1} x_2^{m_2} x_3^{m_3} \cdots x_n^{m_n}$.

Let $f$ be an antisymmetric polynomial, and let $(ij)$ be a transposition. Applying this transposition to $f$ changes the sign of $f$ while the transposition fixes all monomials in which $x_i, x_j$ have the same exponent. It follows that all of the monomials which have a nonzero coefficient in $f$ must have distinct exponents. Given a sequence of exponents $m_1 > m_2 > m_3 > \cdots > m_n \geq 0$, the coefficients of the monomial $x_1^{m_1} x_2^{m_2} x_3^{m_3} \cdots x_n^{m_n}$ and of $x_{\sigma(1)}^{m_1} x_{\sigma(2)}^{m_2} x_{\sigma(3)}^{m_3} \cdots x_{\sigma(n)}^{m_n}$ differ by the sign of $\sigma$. Thus,

THEOREM 3.2.10. *The functions*

$$A_{m_1>m_2>m_3>\cdots>m_n\geq 0}(x) := \sum_{\sigma \in S_n} \epsilon_\sigma x_{\sigma(1)}^{m_1} x_{\sigma(2)}^{m_2} \cdots x_{\sigma(n)}^{m_n}, \tag{3.18}$$

*form an integral basis of the space of antisymmetric polynomials.*

Of course the expression of formula (3.18) is also the value of the determinant of the $n \times n$ matrix with entries $x_i^{m_j}$.

**3.2.3. Schur functions.** We need to work often with *partitions*

DEFINITION 3.2.11.

(i) A partition $\lambda$ of a positive integer $n \in \mathbb{N}$, denoted by $\lambda \vdash n$, is a sequence $\lambda = (\lambda_1, \lambda_2, \dots)$ where the $\lambda_i \in \mathbb{N}$ are integers $> 0$, such that $\lambda_1 \geq \lambda_2 \geq \cdots > 0$ and $\sum_i \lambda_i = n$.

(ii) The number of parts of $\lambda$ is called the *length* or *height* of $\lambda$ and is denoted $\mathrm{ht}(\lambda)$.

Sometimes we also use partitions with some parts equal to 0. We denote a partition $\lambda \vdash n$, and we also write $n = |\lambda|$.

Thus $(4,2,2,1,1,1) \vdash 11$. Sometimes the order is reversed, and we would write $(1,1,1,2,2,4) \vdash 11$. We also denote $(1,1,1,2,2,4) = (1^3, 2^2, 4)$. For example, $(1^2, 3^4, 5^2) = (1,1,3,3,3,3,5,5) \vdash 24$, $\mathrm{ht}(5,2,2,1) = 4$, and $\mathrm{ht}(1^2, 3^4, 5^2) = 8$.

It is convenient to use the following conventions. Consider the sequence $\rho := (n-1, n-2, \dots, 2, 1, 0)$, $\rho \vdash \binom{n}{2}$. We clearly have

LEMMA 3.2.12. *The map*

$$\lambda = (p_1, p_2, p_3, \dots, p_n) \longrightarrow \lambda + \rho = (p_1 + n - 1, p_2 + n - 2, p_3 + n - 3, \dots, p_n)$$

*is a bijective correspondence between nonincreasing and decreasing sequences.*

Given a nonincreasing sequence $\lambda$, consider $A_{\lambda+\rho}$ the corresponding antisymmetric function. We can express it also as a determinant of the matrix $M_\lambda$ having in the $i, j$ position the element $x_j^{p_i+n-i}$.[2]

Notice that $A_\rho = V(x)$ is the Vandermonde determinant.

DEFINITION 3.2.13. The symmetric function $S_\lambda(x) := A_{\lambda+\rho}/A_\rho = A_{\lambda+\rho}/V(x)$ is called the *Schur function* associated to $\lambda$.

When there is no ambiguity, we will drop the symbol of the variables $x$ and speak of $S_\lambda$.

We can identify $\lambda$ to a *partition*, with at most $n$-parts, of the integer $\sum p_i$ and write $\lambda \vdash \sum_i p_i$. Thus we have from Theorem 3.2.10 that

THEOREM 3.2.14. *The functions $S_\lambda$, with $\lambda \vdash m$ and $\mathrm{ht}(\lambda) \leq n$, are an integral basis of the part of degree $m$ of the ring of symmetric functions in $n$ variables.*

Several interesting combinatorial facts are associated to these functions; we will see some of them in the next section. The main significance, for us, of the Schur functions is in the representation theory of the linear group, as we will see later in Theorem 6.3.21.

REMARK 3.2.15. Some special cases are important to notice. When the partition $\lambda = (1^i)$ is a unique column, the corresponding Schur function is the elementary symmetric function $e_i(x)$. When $\lambda = (i)$ is a unique row, the corresponding Schur function is the complete symmetric function $h_i(x)$.

[2] It is conventional to drop the numbers equal to 0 in a decreasing sequence.

3.2.3.1. *The Cauchy formula.* A very useful formula which has a beautiful interpretation in representation theory is the identity (see **[Pro07]** p.31)

$$\frac{1}{\prod_{i,j=1}^{n}(1-x_iy_j)}=\sum_{\lambda} S_\lambda(x_1,\dots,x_n)S_\lambda(y_1,\dots,y_n). \tag{3.19}$$

In fact this, called the *Cauchy formula*, is the first of a list of such formulas, all having interpretations in representation theory; see **[Mac95]**.

3.2.3.2. *The ring of symmetric functions.* If we set $x_n=0$ in $\mathbb{Z}[x_1,\dots,x_n]$, we see that the polynomial ring $\mathbb{Z}[e_1,\dots,e_n]$ of symmetric functions in $n$ variables maps surjectively to the polynomial ring $\mathbb{Z}[e_1,\dots,e_{n-1}]$ of symmetric functions in $n-1$ variables with kernel generated by $e_n$.

Consider the basis $S_\lambda$ of Schur functions with $\mathrm{ht}(\lambda)\le n$ of $\mathbb{Z}[e_1,\dots,e_n]$.

LEMMA 3.2.16. *Under the previous map $x_n=0$, we have that the Schur functions with $\mathrm{ht}(\lambda)=n$ map to 0, while the Schur functions with $\mathrm{ht}(\lambda)<n$ map to the corresponding basis of Schur functions for $\mathbb{Z}[e_1,\dots,e_{n-1}]$.*

In particular a Schur function with $\mathrm{ht}(\lambda)=n$ has a unique expression as polynomial with integer coefficients in the elements $e_1,\dots,e_n$.

This suggests the following

DEFINITION 3.2.17. The ring of symmetric functions is the polynomial ring $\mathbb{Z}[e_1,\dots,e_i,\dots]$, in the infinitely many variables $e_j \mid j=1,\dots,\infty$.

It has a basis over $\mathbb{Z}$ given by the Schur functions.

Over $\mathbb{Q}$ we have

$$\mathbb{Q}[e_1,\dots,e_i,\dots]=\mathbb{Q}[\psi_1,\dots,\psi_i,\dots],\ i=1,\dots,\infty.$$

It is then convenient to develop identities directly in this *stable ring*, as for instance the rule of multiplication between two Schur functions. These identities specialize to identities in each finite ring of symmetric functions.

The multiplication of two Schur functions is given by a rather complicated formula, the Littlewood–Richardson rule, which we briefly discuss in Theorem 6.4.20. Nevertheless a special case of this rule, the Pieri's rules, will be used so we need to point them out (see for instance **[Sta97]** **[Sta99]** or **[Pro07]**).

We use the language of diagrams, to be developed in §6.1.1.1.

PROPOSITION 3.2.18 (Pieri's rules).

(1) $e_i(x)S_\lambda(x)=\sum_\mu S_\mu(x)$, *where $\mu$ runs over all the partitions containing $\lambda$ and such that $\mu\setminus\lambda$ is a skew diagram with $i$ boxes and at most one box in each row.*
(2) $h_i(x)S_\lambda(x)=\sum_\mu S_\mu(x)$, *where $\mu$ runs over all the partitions containing $\lambda$ and such that $\mu\setminus\lambda$ is a skew diagram with $i$ boxes, and at most one box in each column.*

A particularly interesting case is when we multiply by $e_1(x)=h_1(x)$. Then $\mu$ runs over all diagrams in which we have added one single box to $\lambda$.

EXAMPLE 3.2.19.

$$e_2S_{3,1}=S_{3,1^3}+S_{3,2,1}+S_{4,1,1}+S_{4,2},\quad h_2S_{3,1}=S_{5,1}+S_{4,2}+S_{4,1,1}+S_{3,3}+S_{3,2,1}.$$

## 3.3. Matrix functions and invariants

**3.3.1. Matrix invariants.** Consider the ring of $n \times n$ matrices $M_n(F)$ over a commutative ring $F$ and the group $G := \mathrm{GL}(n, F)$ of invertible matrices.

The group $G$ acts on $M_n(F)$ by conjugation and hence it acts on the space of polynomial functions by the formula $f^g(A) := f(gAg^{-1})$.

Since $\det(A) = \det(gAg^{-1})$, the determinant is an invariant function. Also the characteristic polynomial $\det(\lambda - A)$ has all its coefficients invariant. We write

$$\det(\lambda - A) := \lambda^n - \sigma_1(A)\lambda^{n-1} + \sigma_2(A)\lambda^{n-2} - \cdots + (-1)^n\sigma_n(A). \tag{3.20}$$

Inside $M_n(F)$ we have the diagonal matrices which we identify to $F^n$, and inside $\mathrm{GL}(n, F)$ we have the symmetric group $S_n$, of permutation matrices, which under conjugation induces the permutation action on $F^n$.

It follows that, if we restrict a $G$-invariant function $f$ on $M_n(F)$ to $F^n$, we obtain a symmetric function. If $A = \mathrm{diag}(x_1, \dots, x_n)$ is a diagonal matrix with entries $x_i$, its characteristic polynomial is $\prod_{i=1}^n(\lambda - x_i)$; therefore the coefficients $\sigma_i(A)$ of the characteristic polynomial restrict to the elementary symmetric functions $e_i(x_1, \dots, x_n)$. When $F$ is an infinite field we have

THEOREM 3.3.1. *If $F$ is an infinite field, the restriction of the ring of $G$-invariant polynomials on $M_n(F)$ to $F^n$ is an isomorphism with the ring of symmetric polynomials.*

PROOF. The restriction map is surjective, since the $n$ elementary symmetric functions are generators for all symmetric functions, so it is enough to show that it is injective. Since the field is infinite, we may assume that it is algebraically closed. Then consider $U$ the subset of $M_n(F)$ of matrices with distinct eigenvalues, this is the affine open set where the discriminant is nonzero. By Remark 3.2.3 the discriminant is a polynomial with integer coefficients in the coefficients of the characteristic polynomial.

If $f$ is an invariant polynomial which restricted to the diagonal matrices is zero, then it is also zero when restricted to $U$ since every element of $U$ is conjugate to a diagonal matrix. But since $U$ is Zariski open nonempty, it follows that $f$ is zero. □

REMARK 3.3.2. For a linear transformation $A \in \mathrm{End}(V)$, we also have $\sigma_k(A) = \mathrm{Trace}(\bigwedge^k A)$ where $\bigwedge^k A$ is the linear transformation induced in $\bigwedge^k V$ the $k$th exterior power of the space $V$.

PROOF. By the previous argument it is enough to prove this fact when $A$ is diagonal of entries $x_1, \dots, x_n$, and then one sees that $\bigwedge^k A$ is also diagonal in the basis $e_{i_1} \wedge \cdots \wedge e_{i_k}$ with entries $x_{i_1} \cdots x_{i_k}$, $1 \le i_1 < i_2 < \cdots < i_k \le n$. □

We will use the following corollaries of these identities.

COROLLARY 3.3.3. *Let $A$ be an $n \times n$ matrix over a commutative $\mathbb{Q}$-algebra.*

(i) *We have*

$$\sigma_r(A) = \sum_{\substack{h_1+2h_2+\cdots+rh_r=r \\ h_1 \ge 0, \dots, h_r \ge 0}} (-1)^r \prod_{j=1}^r \frac{(-\mathrm{tr}(A^j))^{h_j}}{h_j!\, j^{h_j}}. \tag{3.21}$$

(ii) *If $\mathrm{tr}(A^i) = 0$, $i = 1, \dots, n$, we have $A^n = 0$.*

(iii) *If the entries of $A$ are in a ring without nilpotent elements, the converse is also true.*

PROOF. (i) formula (3.21) is a formal identity between two polynomials in the entries of $A$. As such it is universally valid if and only if it is valid when the entries of $A$ are variables. Since this is an identity between polynomials with rational coefficients, it is valid if and only if it is valid when evaluated in complex matrices. For a complex matrix $A$ we have that $\sigma_i(A)$ is the elementary symmetric function in its eigenvalues while $\operatorname{tr}(A^i)$ is the $i$th Newton function in the eigenvalues, hence the identity follows from formula (3.15).

(ii) The Cayley–Hamilton theorem states that $A$ satisfies its characteristic polynomial. Hence if $\operatorname{tr}(A^i) = 0,\ i = 1, \ldots, n$, we have that this polynomial reduces to $\lambda^n$ and so $A^n - 0$. We leave (iii) as exercise. □

The list of the first expressions of $\sigma_i(A),\ i = 1, 2, 3, 4$, in terms of the elements $\operatorname{tr}(A^i)$ is given through the formulas (3.17).

REMARK 3.3.4. Given an $n \times n$ matrix $A$ the invariant $\sigma_i(A^j)$ is a polynomial with integer coefficients in the invariants $\sigma_h(A),\ h \leq ij$. This polynomial is independent of $n$ as soon as $n \geq ij$.

PROOF. It is enough to do this when $A$ is diagonal with eigenvalues $x_1, \ldots, x_n$. Then we see that, under the identification of matrix invariants to symmetric functions, we have that $\sigma_i(A^j)$ is given by $e_i(x^j) := e_i(x_1^j, \ldots, x_n^j)$, a symmetric function of degree $ij$ with integer coefficients. This can be expressed as a polynomial with integer coefficients in the elementary symmetric functions $e_h(x_1, \ldots, x_n),\ h \leq ij$. If $n \geq ij$, we are in the stable range of symmetric functions, so the polynomial is independent of $n$. □

A possible formula is setting $\zeta_j$ a primitive $j$th root of 1.

$$
\begin{aligned}
t^{nj} &+ \sum_{h=1}^{n} (-1)^h e_h(x^j) t^{j(n-h)} \\
&= \prod_{h=1}^{n} (t^j - x_h^j) = \prod_{i=1}^{j} \prod_{h=1}^{n} (\zeta_j^i t - x_h) \\
&= \prod_{i=1}^{j} \Big( (\zeta_j^i t)^n + \sum_{h=1}^{n} (-1)^h (\zeta_j^i t)^{n-h} e_h(x) \Big).
\end{aligned}
$$

Then the coefficients are in $\mathbb{Z}$ by symmetry.

$$\text{Example:} \qquad e_2(x^2) = e_2(x)^2 - 2e_1(x)e_3(x) - 2e_4(x).$$

**3.3.2. Amitsur's formula for** $\det(x + y)$**.** We present here a remarkable identity due to Amitsur [**Ami79**], which was later rediscovered and proved in a different way by Reutenauer and Schützenberger [**RS87**].

In fact, although this formula is presented in the literature as a matrix identity, it is truly a symbolic identity in the free algebra with trace. The point is that formula (3.16) should be interpreted as a *definition* of the elements $\sigma_i(A),\ A \in \mathcal{F}_T\langle X \rangle$ the free algebra with trace over $\mathbb{Q}$ of Theorem 2.3.11.

3.3.2.1. *Symbolic formulas.*

DEFINITION 3.3.5. We define for $A \in \mathcal{F}_T\langle X\rangle$ the elements $\sigma_i(A)$ through the formula below (with $\lambda$ an auxiliary parameter) and $\sigma_0(A) = 1$:

$$\sum_{i=0}^{\infty}(-1)^i\sigma_i(A)\lambda^i := \exp(-\sum_{j=1}^{\infty}\frac{t(A^j)}{j}\lambda^j). \tag{3.22}$$

This, by taking the derivative with respect of $\lambda$, gives that

$$\sum_{i=1}^{\infty}(-1)^i i\sigma_i(A)\lambda^{i-1} = -\exp(-\sum_{j=1}^{\infty}\frac{t(A^j)}{j}\lambda^j)(\sum_{j=1}^{\infty}t(A^j)\lambda^{j-1})$$
$$= -(\sum_{i=0}^{\infty}(-1)^i\sigma_i(A)\lambda^i)(\sum_{j=1}^{\infty}t(A^j)\lambda^{j-1})$$

is equivalent to the recursive formula analogous of (3.14),

$$(-1)^m t(A^{m+1}) + \sum_{i=1}^{m}(-1)^{i-1}t(A^i)\sigma_{m+1-i}(A) = (m+1)\sigma_{m+1}(A). \tag{3.23}$$

We then reinterpret Remark 3.3.4 symbolically:

Use a primitive $h$ root of 1, $\zeta_h$ so that

$$\sum_{k=1}^{h}\zeta_h^{ik} = \begin{cases} 0 & \text{if } \; i \not\equiv 0, \bmod h \\ h. \end{cases}$$

Substitute in formula (3.22) for $A \mapsto A^h$, $\lambda \mapsto \lambda^h$.

(3.24)

$$\sum_{i=0}^{\infty}(-1)^i\sigma_i(A^h)\lambda^{hi} := \exp(-\sum_{j=1}^{\infty}\frac{t(A^{hj})}{j}\lambda^{hj})$$
$$= \exp(-\sum_{k=1}^{h}\sum_{j=1}^{\infty}\frac{t((\zeta_h^k A)^j)}{j}\lambda^j) = \prod_{k=1}^{h}(\sum_{i=0}^{\infty}(-1)^i\zeta_h^i\sigma_i(A)\lambda^i).$$

Given a sequence $\underline{a} : 0 \le a_1 \le a_2 \le \cdots \le a_h$, consider the set $C_{\underline{a}}$ of all the sequences $j_1, j_2, \ldots, j_h$, which when reordered in increasing order, equal $\underline{a}$, and let

$$m_{\underline{a}} = \sum_{j_1,j_2,\ldots,j_h \in C_{\underline{a}}} \zeta_h^{\sum_{\ell=1}^{h}\ell j_\ell}.$$

Then $m_{\underline{a}} \in \mathbb{Z}$ by Galois theory and

$$\sigma_i(A^h) = \sum_{\underline{a}:=a_1\le a_2\le\cdots\le a_h \mid \sum a_\ell = i\cdot h} m_{\underline{a}}\prod_{\ell=1}^{h}\sigma_{a_\ell}(A). \tag{3.25}$$

3.3.2.2. *Amitsur's formula.* We need some basic notions on the free algebra.

DEFINITION 3.3.6. Given a monomial $w$, denote by $\ell(w)$ its degree or *length* as a word.

We say that a monomial $w$ is *primitive* or *irreducible* if it is not of the form $w = u^k$, $k > 1$ (in particular 1 is not primitive). Every word $w$ is of the form $u^k$ with $u$ primitive in a unique way. Let $W_p$ denote the set of primitive monomials.

Consider cyclic equivalence of monomials, that is $w = ab \sim ba$. If $w$ is primitive, we have exactly $\ell(w)$ distinct elements equivalent to $w^k$ for all $k \geq 1$.

DEFINITION 3.3.7. Let $W_0$ denote a set of representatives of primitive monomials up to cyclic equivalence. We choose the first in lexicographic order which is called a *Lyndon word*, and then $W_0$ is also ordered lexicographically.

We now use a variable $\lambda$ with respect to which we write generating functions and take derivatives. We start from the identity (3.22) in the form

$$T[x,\lambda] := \sum_{j=1}^{\infty} t(x^j)\lambda^{j-1}, \qquad \Lambda[x,\lambda] := \sum_{i=0}^{\infty}(-1)^i\sigma_i(x)\lambda^i, \tag{3.26}$$

$$d\log(\Lambda[x,\lambda]) = -T[x,\lambda]d\lambda, \qquad \Lambda[x,\lambda] = \exp(-\int T[x,\lambda]d\lambda). \tag{3.27}$$

Replace $x$ with $\sum_i u_i x_i$ where the $x_i \in X$ are noncommutative variables, while the $u_i$'s are commutative variables. We work thus with the free algebra with trace in variables $x_i$ with coefficients in the polynomial ring over $\mathbb{Q}$ in the $u_i$, with basis the monomials $w$ in the $x_i$ over the trace ring.

Set $u^{\nu(w)} := \prod_{i=1}^n u_i^{h_i}$ to be the evaluation of the word $w$ under $x_i \mapsto u_i$. Thus $\nu(w) = (h_1, h_2, \dots, h_i, \dots)$ where $h_i$ is the number of occurrences of $x_i$ in the word $w$. Let $\ell(w) = |\nu(w)| = \sum_i h_i$ be the length of the monomial. We have, since formal trace is linear,

$$\begin{aligned}(\sum_i u_i x_i)^n &= \sum_{w \,|\, \ell(w)=n} u^{\nu(w)} w \implies t((\sum_i u_i x_i)^n) = \sum_{w \,|\, \ell(w)=n} u^{\nu(w)} t(w)\\
&\implies T[\sum_i u_i x_i, \lambda]d\lambda = \sum_{w\in W_p}\sum_{i=1}^{\infty} u^{i\nu(w)} t(w^i)\lambda^{i\ell(w)-1}d\lambda\\
&= \sum_{w\in W_p}\sum_{i=1}^{\infty}\lambda^{\ell(w)(i-1)} t((u^{\nu(w)}w)^i)\lambda^{\ell(w)-1}d\lambda = \sum_{w\in W_0} T[u^{\nu(w)}w, \lambda^{\ell(w)}]d(\lambda^{\ell(w)}),\end{aligned}$$

(3.28)

$$\begin{aligned}\Lambda[\sum_i u_i x_i, \lambda] &= \exp(-\sum_{w\in W_0}\int T[u^{\nu(w)}w, \lambda^{\ell(w)}]d(\lambda^{\ell(w)}))\\
&= \prod_{w\in W_0}\exp(-\int T[u^{\nu(w)}w, \lambda^{\ell(w)}]d(\lambda^{\ell(w)})) = \prod_{w\in W_0}\Lambda[u^{\nu(w)}w, \lambda^{\ell(w)}].\end{aligned}$$

Comparing degrees in $\lambda$ we have

THEOREM 3.3.8 (Amitsur [**Ami79**]). *For each integer $n > 0$ we have*

$$\sigma_n(\sum_i u_i x_i) = \sum_{\substack{(p_1<\cdots<p_k)\subset W_0,\\ j_1,\dots,j_k\in\mathbb{N},\ \sum j_i\ell(p_i)=n}} (-1)^{n-\sum j_i} u^{\sum_{i=1}^k j_i\nu(p_i)}\sigma_{j_1}(p_1)\cdots\sigma_{j_k}(p_k) \tag{3.29}$$

COROLLARY 3.3.9. *Let $f(x_1, \dots, x_k, \dots)$ be any polynomial expression (noncommutative) in the matrix variables $x_i$, with coefficients in some commutative ring $A$. We have that each $\sigma_j(f(x_1, \dots, x_k, \dots))$ can be expressed as a polynomial in the elements $\sigma_h(p)$, $p \in W_0$, with coefficients in $A$.*

PROOF. A polynomial is a linear combination $\alpha_i m_i$ of monomials. We substitute in (3.29), $u_i$ with $\alpha_i$ and $x_i$ with $m_i$, and we deduce a polynomial in the elements $\sigma_i(w)$, $w \in W$. If $w = p^j$, we apply formula (3.25) which is a polynomial with integer coefficients. □

We can thus define an *integral form* of the free algebra in variables $X$. We need to distinguish the case with or without 1. In the second case the trace algebra over $\mathbb{Q}$ is $\mathcal{F}_T\langle X\rangle = \mathbb{Q}\langle X\rangle_+ \otimes_{\mathbb{Z}} \mathcal{T}_X$ with $\mathcal{T}_X$ the polynomial ring in the variables $t(w)$, $w \in W_+$. For the integral form of trace algebra we take the ring

$$\mathcal{S}_X := \mathbb{Z}[\sigma_i(w)],\ i = 1, \dots, \infty,\ w \in W_0. \tag{3.30}$$

That these elements $\sigma_i(w)$ are algebraically independent follows easily from

PROPOSITION 3.3.10. *For a given $p \in W_0$ the elements $\sigma_i(p)$ and $t(p^i)$, with $i = 1, \dots, \infty$, generate the same algebra over $\mathbb{Q}$.*

PROOF. This follows immediately from formula (3.23) which for any $x$ gives $\sigma_i(x)$ as a polynomial with rational coefficients in the elements $\sigma_1(x^j) = t(x^j)$, and conversely $t(x^i)$ as a polynomial with integer coefficients in the elements $\sigma_j(x)$, $j = 1, \dots, i$. □

We see first that the trace algebra $\mathcal{T}_X$ is equal to $\mathbb{Q} \otimes_{\mathbb{Z}} \mathcal{S}_X$. We then have the integral form of the free algebra,

$$\mathcal{S}_X\langle X\rangle := \mathbb{Z}\langle X\rangle_+ \otimes_{\mathbb{Z}} \mathcal{S}_X, \quad \mathcal{F}_T\langle X\rangle = \mathbb{Q} \otimes_{\mathbb{Z}} \mathcal{S}_X\langle X\rangle. \tag{3.31}$$

Corollary 3.3.9 can be interpreted as

THEOREM 3.3.11. *A substitution map of the free algebra $\mathcal{F}_T\langle X\rangle$, induced by a map $\phi : X \to \mathcal{S}_X\langle X\rangle$, maps $\mathcal{S}_X\langle X\rangle$ to itself.*

DEFINITION 3.3.12. One can also work with the algebra with 1, in this case the trace algebra over $\mathbb{Q}$ is $\mathcal{F}_T\langle X\rangle = \mathbb{Q}\langle X\rangle \otimes_{\mathbb{Z}} \mathcal{T}_X[t(1)]$.

From formula (3.22), since $-\sum_{j=1}^{\infty} \frac{\lambda^j}{j} = \log(1-\lambda)$, it follows that

$$\exp(-t\sum_{j=1}^{\infty} \frac{\lambda^j}{j}) = (1-\lambda)^t \implies \sigma_i(1) = \binom{t}{i} = \frac{1}{i!}t(t-1)\cdots(t-i+1), \qquad t = t(1). \tag{3.32}$$

We then can define the integral forms with 1 as

$$\mathbb{S}\langle X\rangle := \mathbb{Z}\langle X\rangle \otimes_{\mathbb{Z}} \mathbb{S}, \quad \mathbb{S} := \mathcal{S}_X[\sigma_i(1)],\ i = 1, \dots \infty, \tag{3.33}$$

where $\mathbb{S}$ is obtained from $\mathcal{S}_X$ adding the elements $\sigma_i(1)$.

Notice that the elements $\sigma_i(1)$ form a basis over $\mathbb{Z}$ of the subalgebra $A[t]$ of $\mathbb{Q}[t]$ of integral valued polynomials on $\mathbb{Z}$ and $\mathbb{S} := \mathcal{S}_X \otimes_{\mathbb{Z}} A[t]$.

The multiplication is

$$i \le j, \ \binom{t}{i}\binom{t}{j} = \sum_{k=0}^{i} \binom{i+j-k}{k}\binom{i+j-2k}{i-k}\binom{t}{i+j-k}.$$

As a consequence the elements of the algebra $\mathcal{S}_X\langle X\rangle$ can be evaluated into matrices over any commutative ring and $\mathbb{S}\langle X\rangle$ when the ring has a 1.

In fact one could formalize this in the language of categories, one should then define the category of algebras with operations $\sigma_i$ by axiomatizing the rules used so that $\mathcal{S}_X\langle X\rangle$ is a free algebra in this category.

We may, in particular, make explicit formula (3.1) of polarization for the polynomial $\sigma_m(x)$. We group together in formula (3.29) all terms which have the same degree in each variable, that is the coefficient of $u_1^{k_1}\cdots u_m^{k_m}$.

$$\sigma_m\Big(\sum_i u_i x_i\Big) = \sum u_1^{k_1}\cdots u_m^{k_m}\sigma_{m;k_1,\dots,k_h}(x_1,\dots,x_k). \tag{3.34}$$

From Theorem 3.3.8 one deduces

THEOREM 3.3.13. *The elements $\sigma_{h_1,\dots,h_s}(x_1,\dots,x_s)$ are given by the universal polynomials with integer coefficients in the elements $\sigma_j(p)$, where $p$ is a primitive monomial in the variables $x_i$ extracted from formula* (3.29).

EXAMPLE 3.3.14. For the sum of two $2\times 2$ matrices we have

$$\begin{aligned}\det(a+b) &= \det(a)+\det(b)-\operatorname{tr}(ab)+\operatorname{tr}(a)\operatorname{tr}(b)\\ &= \sigma_2(a)+\sigma_2(b)-\sigma_1(ab)+\sigma_1(a)\sigma_1(b).\end{aligned} \tag{3.35}$$

Let us analyze formula (3.29) for the sum of three $3\times 3$ matrices, the determinant $\det(a+b+c) = \sigma_3(a+b+c)$, denoting the three variables, $a, b, c$. The possible primitive monomials appearing have length at most 3, and the corresponding Lyndon words are (ordered by degree lexicographic)

$$a,\ b,\ c,\ ab,\ ac,\ bc,\ a^2b,\ a^2c,\ ab^2,\ abc,\ acb,\ ac^2,\ b^2c,\ bc^2,$$

only elements of degree $\le 3$ may appear, hence only

$$\begin{aligned}&\sigma_i(a),\sigma_i(b),\sigma_i(c),\ i\le 3;\\ &\quad \sigma_1(ab),\sigma_1(ac),\sigma_1(bc),\sigma_1(a^2b),\sigma_1(a^2c),\sigma_1(ab^2),\sigma_1(abc),\\ &\quad \sigma_1(acb),\sigma_1(ac^2),\sigma_1(b^2c),\sigma_1(bc^2).\end{aligned}$$

We then have to write all possible products of degree 3 and sum them with signs, which gives a rather long expression:

$$\begin{aligned}\sigma_3(a+b+c) =&\sigma_1(a)\sigma_1(b)\sigma_1(c)+\sigma_1(a)(\sigma_2(b)+\sigma_2(c))\\ &+\sigma_1(b)(\sigma_2(a)+\sigma_2(c))+\sigma_1(c)(\sigma_2(a)+\sigma_2(b))\\ &-(\sigma_1(a)+\sigma_1(b)+\sigma_1(c))(\sigma_1(ab)+\sigma_1(ac)+\sigma_1(bc))\\ &+\sigma_3(a)+\sigma_3(b)+\sigma_3(c)+\sigma_1(abc)+\sigma_1(acb)+\sigma_1(a^2b)\\ &+\sigma_1(a^2c)+\sigma_1(ab^2)+\sigma_1(b^2c)+\sigma_1(ac^2)+\sigma_1(bc^2).\end{aligned} \tag{3.36}$$

We have some obvious symmetries. In the previous example we have

$$\begin{aligned}\sigma_{3;1,1,1}(a,b,c) = \sigma_1(a)\sigma_1(b)\sigma_1(c)-\sigma_1(a)\sigma_1(bc)&\\ -\sigma_1(b)\sigma_1(ac)-\sigma_1(c)\sigma_1(ab)+\sigma_1(abc)+\sigma_1(acb)&,\end{aligned} \tag{3.37}$$

$$\sigma_{3;2,1}(a,b) = -\sigma_1(a)\sigma_1(ab)+\sigma_1(b)\sigma_2(a),\quad \sigma_{3;3}(a)=\sigma_3(a).$$

Formula (3.37) in particular is the full polarization of $\sigma_3(x)$. This can be made explicit for all $m$ in the following form.

EXERCISE 3.3.15. Compare the part of formula (3.29) giving the full polarization of $\sigma_m(x)$ with the one presented in §12.1.1 as $\sum_{\sigma\in S_m}\epsilon_\sigma T_\sigma(x_1,\ldots,x_m)$ where the functions $T_\sigma$ are given by the explicit formula (12.3).

3.3.2.3. *The proof by Reutenauer and Schützenberger* **[RS87]**. We sketch this proof which is also quite instructive. It is based on a basic theorem of Lyndon and some simple formalism in the algebra of formal power series in noncommutative variables, that is the set of possibly infinite sums $\sum_{w\in W(X)} a_w w$, $a_w \in F$ ($F$ a field or coefficient ring) which has an obvious algebra structure which we omit. We restate the *generating series of words*:

$$(1-(\sum_i x_i))^{-1} = 1+\sum_{j=1}^{\infty}(\sum_{i=1}^{k} x_i)^j = \sum_{w\in W(X)} w. \tag{3.38}$$

First we choose as a canonical representative of primitive monomials up to cyclic equivalence the Lyndon words in Definition 3.3.7. A basic theorem of Lyndon (cf. **[Lot97]**) states the following.

THEOREM 3.3.16. *Every word $w$ is* uniquely factorized *as a product of powers of Lyndon words $\ell_i$:*

$$w = \ell_1^{k_1}\ell_2^{k_2}\cdots\ell_i^{k_i},\quad \ell_1>\ell_2>\cdots>\ell_i,$$

*where $\ell_1>\ell_2>\cdots>\ell_i$ in the lexicographic order.*

This is conveniently expressed by a generating series identity, in the algebra of formal power series, setting $W_0$ to be the set of Lyndon words

$$(1-(\sum_i x_i))^{-1} = 1+\sum_{j=1}^{\infty}(\sum_{i=1}^{k} x_i)^j = \sum_{w\in W(X)} w = \prod_{\ell\in W_0}(1-\ell)^{-1}, \tag{3.39}$$

where the product is over all Lyndon words in the variables $x_i$ taken in decreasing lexicographic order. Then specialize the $x_i$ in generic $n\times n$ matrices and compute the determinant of both sides. Passing to the limit $n\to\infty$, we get a formal identity

$$1-\sigma_1(\sum_i x_i)+\sigma_2(\sum_i x_i)-\cdots+(-1)^k\sigma_k(\sum_i x_i)\cdots$$

$$=\prod_\ell(1-\sigma_1(\ell)+\sigma_2(\ell)-\cdots+(-1)^k\sigma_k(\ell)+\cdots).$$

Equating the terms of degree $k$ one has Amitsur's formula for each $\sigma_k(\sum_i x_i)$.

3.3.2.4. *A symbolic calculus.* The formulas discussed in the previous section are universal formulas, which, for each $n$, can be evaluated in $n\times n$ matrices over a commutative ring. The three facts which are relevant are the following.

(i) Consider the algebra of $n\times n$ matrices $M_n(\mathbb{Z})$ (or $M_n(A)$ with $A$ a commutative ring) as naturally contained as a left upper block in $M_{n+1}(\mathbb{Z})$. To be precise, denote by $j: M_n(\mathbb{Z})\to M_{n+1}(\mathbb{Z})$ the embedding.

Then, given $a\in M_n(A)$ the characteristic polynomial of $j(a)$ is given by that of $a$ through the formula $\det(t-j(a)) = t\det(t-a)$. This means that all the coefficients $\sigma_i(a)=\sigma_i(j(a))$, $i\le n$, while $\sigma_{n+1}(j(a))=0$. Thus consider the algebra

$$M_\infty(\mathbb{Z}) = \lim_n M_n(A)$$

of infinite matrices but each of finite size. On this algebra are defined all the functions $\sigma_i(a)$ which take values in $\mathbb{Z}$ and satisfy Amitsur's formula. We then consider the evaluations of the free algebra $\mathbb{Z}_+\langle X\rangle$ or $\mathcal{S}_X\langle X\rangle_+$ in the variables $x_i$ with coefficients in $\mathbb{Z}$ without 1, in the algebra $M_\infty(\mathbb{Z})$ (notice that this algebra does not have a 1 so we need the free algebra without 1).

(ii) As we shall see (Theorem 10.1.3), every nonzero element of $\mathbb{Z}\langle X\rangle$ gives rise to a nonzero map $M_\infty(\mathbb{Z})^X \to M_\infty(\mathbb{Z})$.

The set $\mathcal{A}$ of all maps $M_\infty(\mathbb{Z})^X \to M_\infty(\mathbb{Z})$ is an algebra by pointwise sum and multiplication.

REMARK 3.3.17. In particular the coordinate maps $\xi_i : (x_1, \dots, x_i, \dots) \mapsto x_i$ can be considered as *generic matrices*, and the free algebra is isomorphic to the subalgebra of $\mathcal{A}$ generated by the generic matrices through the map $x_i \mapsto \xi_i$.

Now on $M_\infty(\mathbb{Z})$ we have the functions $\sigma_i$ which thus, by evaluation, give rise to functions of the free algebra $\mathbb{Z}_+\langle X\rangle$ to $\mathbb{Z}$ (or in case of a different ring of coefficients $A_+\langle X\rangle$ to $A$). These satisfy Amitsur's formula so we deduce an algebra generated by the functions $\sigma_i(a(\xi))$ where $a(\xi)$ is a primitive monomial in the generic matrices. By abuse of notation we identify a monomial $w \in W$ in the variables $x_i$ with its corresponding monomial in the generic matrices $\xi_i$ (this is a functional interpretation of the free algebra).

So, if $p \in W_0$, where $W_0$ denotes a set of representatives of primitive monomials up to cyclic equivalence, we also denote by $\sigma_i(p)$ the function $M_\infty(\mathbb{Z})^X \to \mathbb{Z}$ given by composing $p(\xi) = p(\xi_1, \dots, \xi_i, \dots)$ with $\sigma_i$; that is

$$\sigma_i(p) : M_\infty(\mathbb{Z})^X \xrightarrow{p(\xi)} M_\infty(\mathbb{Z}) \xrightarrow{\sigma_i} \mathbb{Z}.$$

Thus $\mathcal{S}_X^+\langle X\rangle$ of formula (3.31) gives rise to an algebra of functions from $M_\infty(\mathbb{Z})^X$ to $M_\infty(\mathbb{Z})$.

We can work first with matrices over $\mathbb{Q}$:

(iii) The final ingredient of this analysis is

THEOREM 3.3.18. *The elements $\sigma_1(w) = t(w)$, $w \in W_+$ with $w$ taken up to cyclic equivalence, or $\sigma_i(p)$, $p \in W_0 \setminus \{1\}$, are algebraically independent and generate over $\mathbb{Q}$ the same algebra of functions.*

PROOF. From Proposition 3.3.10 it is enough to prove this for the elements $\operatorname{tr}(w)$, let us sketch the proof which will be taken back in detail in Theorem 12.1.17. If we have a relation in the elements $\operatorname{tr}(w)$, first of all it is easy to see that each piece of this relation homogeneous in each variable is still a relation. We can polarize it and get a nontrivial multilinear relation of some degree $k$. We then see that no such relation exists as soon as we specialize into matrices of size $\geq k$ (see Theorem 12.1.11). Since we are working with matrices of arbitrary size the claim follows. □

COROLLARY 3.3.19. *The map of $\mathcal{S}_X\langle X\rangle$ to the algebra of functions from $M_\infty(\mathbb{Z})^X$ to $M_\infty(\mathbb{Z})$ is injective.*

So the elements $\sigma_i(p)$ and $x_i$ can be considered as functions or as purely symbolic indeterminates. The same formulas then define polynomial maps $\sigma_i$ from $\mathcal{S}_X\langle X\rangle$ to $\mathcal{S}_X$.

For any commutative ring $A$ we can extend the definition to

$$\mathbb{S}_A\langle X\rangle := A\langle X\rangle \otimes_A \mathbb{S}_A, \quad \mathbb{S}_A := A \otimes_{\mathbb{Z}} \mathbb{S}. \tag{3.40}$$

We will expand this formalism of polynomial maps in Chapter 4 where the algebra $\mathbb{S}$ appears in a different natural way; see Theorem 4.2.18.

**3.3.3. Matrix functions.** In the previous paragraph we have considered the elements $\sigma_i(p)$ as purely symbolic indeterminates, or as matrix functions but without specifying the size of the matrices.

Now we fix some number $n$ and specialize the computations to $n \times n$ matrices. First we treat the elementary case of a single matrix variable. For a given $n$ introduce $n^2$ variables $\xi_{i,j}, i, j = 1, \dots, n$, over the ring of integers, and denote by $\mathbb{Z}[\xi]$ the polynomial ring in these variables. Now take as $z$ the matrix in $M_n(\mathbb{Z}[\xi])$ with entries the commutative variables $\xi_{i,j}$. This is by definition *a generic matrix*. Now construct the functions $\sigma_i^{(n)}(z) \in \mathbb{Z}[\xi]$ from the characteristic polynomial

$$\det(t - z) = \sum_{i=0}^{n} t^{n-i}(-1)^i \sigma_i^{(n)}(z) = \sum_{i=0}^{n} t^{n-i}(-1)^i \sigma_i(z). \tag{3.41}$$

If the dimension $n$ of the matrices is understood, sometimes we will write $\sigma_i(z) := \sigma_i^{(n)}(z)$. The function $\sigma_i(z)$ is a homogeneous polynomial of degree $i$ in $\mathbb{Z}[\xi]$.n

DEFINITION 3.3.20. If $z$ is an $n \times n$ matrix and $d \in \mathbb{N}$, we set $z[d]$ to be the $dn \times dn$ matrix which is block diagonal with $d$ blocks equal to $z$.

Of particular importance is the case $d = n$. The matrix $z[n]$ is obtained from $z$ by multiplication on matrices.

We define the functions $\sigma_{i;d}(z)$ by the formula

$$\det(t - z[d]) = \det(t - z)^d = \sum_{i=0}^{dn} t^{dn-i}(-1)^i \sigma_{i;d}(z). \tag{3.42}$$

In other words we set

$$\sigma_{i;d}(z) := \sigma_i^{(dn)}(z[d]).$$

For instance

$$\sigma_{1;d}(z) = \operatorname{tr}(z[d]) = d \operatorname{tr}(z), \quad \sigma_{dn;d}(z) = \det(z[d]) = \det(z)^d.$$

From formulas (3.41) and (3.42) we have

LEMMA 3.3.21. *The functions $\sigma_{i;d}(z)$ can be expressed as polynomials with integer coefficients in the basic functions $\sigma_i(z)$. Conversely the functions $\sigma_i(z)$ can be expressed as polynomials with rational coefficients in the functions $\sigma_{i;d}(z)$.*

PROOF. What we need to prove is that if we have a polynomial $p(t) := t^n + a_1 t^{n-1} + \cdots + a_n$ and consider $p(t)^k = t^{kn} + b_1 t^{kn-1} + \cdots + b_k t^{kn-k} + \cdots + b_{kn}$, then the elements $b_i$ are polynomials in the $a_i$ with integer coefficients (this is clear) and the elements $a_i$ are polynomials in the $b_1, \dots, b_k$ with rational coefficients.

This depends upon the fact that $b_i$ is a polynomial only in $a_1, \dots, a_i$ since the maximum monomial containing some $a_i t^{n-i}$ is $t^{n(k-1)} a_i t^{n-i} = t^{nk-i} a_i$ which appears in the expansion of $b_i$ with coefficient $k$ so $b_i = k a_i + f(a_1, \dots, a_{i-1})$, and the claim follows. $\square$

EXAMPLE 3.3.22. For $z$ a $2 \times 2$ matrix and $d = 2$, the characteristic polynomial of $z[2]$ is

$$(t^2 - \sigma_1(z)t + \sigma_2(z))^2 = t^4 - 2\sigma_1(z)t^3 + (2\sigma_2(z) + \sigma_1(z)^2)t^2 - 2\sigma_1(z)\sigma_2(z)t + \sigma_2(z)^2,$$

$$\sigma_{1;2}(z) = 2\sigma_1(z), \sigma_{2;2}(z) = 2\sigma_2(z) + \sigma_1(z)^2, \sigma_{3;2}(z) = 2\sigma_1(z)\sigma_2(z),\ \sigma_{4;2} = \sigma_2(z)^2.$$

We will also need, as before, to apply this construction to the block matrices $z[h]$ and set

$$\sigma_{h_1,\dots,h_s;h}(z_1,\dots,z_s) := \sigma_{h_1,\dots,h_s}(z_1[h],\dots,z_s[h]), \tag{3.43}$$

$$\sigma_{i;h}(\sum_{j=1}^{s} t_j z_j) = \sum_{h_1,\dots,h_s \mid \sum h_j = i} t_1^{h_1}\cdots t_s^{h_s}\sigma_{h_1,\dots,h_s;h}(z_1,\dots,z_i). \tag{3.44}$$

We shall write compactly

$$\underline{h} := h_1,\dots,h_s,\ |\underline{h}| := \sum_j h_j;\quad \sigma_{\underline{h};d}(\underline{z}) := \sigma_{h_1,\dots,h_s;d}(z_1,\dots,z_s).$$

Of course in the definition only the $h_j > 0$ count and only the corresponding variables $z_j$ appear. Later we shall need to compare these functions for $n \times n$ and $n-1 \times n-1$ matrices. So if we want to stress in which matrices we are working, we write $\sigma_{\underline{h}}^{(n)}$. When $h = 1$ we drop it.

### 3.3.4. Generic matrices I.

3.3.4.1. *Matrix invariants.* We expand now for matrices the language of *generic elements* introduced in §2.4 or in Remark 3.3.17.

Instead of the algebra $\mathcal{A}$ of all maps, we want to consider *polynomial maps* from $k$-tuples of matrices to matrices. We may work over $\mathbb{Z}$. The space of polynomial functions defined over $\mathbb{Z}$ on $M_n(\mathbb{Z})^k$ will be denoted by $\mathbb{Z}[\xi] = \mathbb{Z}[\xi_{i,j}^{(u)}]$, where $u = 1,\dots,k$ and $i, j = 1,\dots,n$. For given $u$, the elements $\xi_{i,j}^{(u)}$ are the coordinate functions of the matrix $\xi_u \in M_n(\mathbb{Z})^k$.

Consider now a polynomial map $f : M_n(\mathbb{Z})^k \to M_n(\mathbb{Z})$, that is a map such that the entries of the target matrix are polynomial in the entries of the source.

Clearly two such polynomial maps $f, g : M_n(\mathbb{Z})^k \to M_n(\mathbb{Z})$ can be summed and multiplied and form an algebra; in fact this algebra is just the matrix algebra $M_n(\mathbb{Z}[\xi])$. Among the polynomial maps there are the basic *coordinate functions* (cf. Remark 3.3.17),

$$\xi_u : M_n(\mathbb{Z})^k \to M_n(\mathbb{Z}),\quad \xi_u : (a_1, a_2, \dots, a_k) \mapsto a_u,\quad u = 1,\dots,k. \tag{3.45}$$

By definition $\xi_u$ is the matrix in $M_n(\mathbb{Z}[\xi])$ with entries $\xi_{i,j}^{(u)}$.

DEFINITION 3.3.23. Let $R_{n,k} := \mathbb{Z}\langle \xi_1, \xi_2, \dots, \xi_k\rangle$ be the subring of $M_n(\mathbb{Z}[\xi])$ generated over $\mathbb{Z}$ by the generic matrices $\xi_s := (\xi_{i,j}^{(s)})$. $R_{n,k}$ is called the ring of (polynomials in) $k$, $n \times n$ *generic matrices*.

We define $\mathcal{T}_{n,k} \subset \mathbb{Z}[\xi]$ to be the ring of polynomials generated over $\mathbb{Z}$ by all the coefficients of characteristic polynomials of elements of $R_{n,k}$.

REMARK 3.3.24. The noncommutative ring $\mathcal{R}_{n,k} = \mathcal{T}_{n,k}R_{n,k} \subset M_n(\mathbb{Z}[\xi])$ of polynomial maps, has the property that each one of his elements satisfies its characteristic polynomial, a monic polynomial with coefficients in $\mathcal{T}_{n,k}$.

Given any commutative ring $A$, we have the same definitions over $A$:

$$A\langle \xi_1, \xi_2, \dots, \xi_k\rangle \subset M_n(A[\xi_{i,j}^{(s)}]). \tag{3.46}$$

REMARK 3.3.25. A homomorphism $\phi : A \to B$ between two commutative rings induces a homomorphism $\phi : A\langle \xi_1, \xi_2, \dots, \xi_k\rangle \to B\langle \xi_1, \xi_2, \dots, \xi_k\rangle$ between the two algebras of generic matrices. In general the induced map

$$\phi_B : B \otimes_A A\langle \xi_1, \xi_2, \dots, \xi_k\rangle \to B\langle \xi_1, \xi_2, \dots, \xi_k\rangle \tag{3.47}$$

is **not** an isomorphism, as examples show; see **[Sch85]**, **[ADKT00]**, and **[Kem03]**, when $A = \mathbb{Z}$ and $B = \mathbb{Z}/(p)$ for primes $p$.

EXERCISE 3.3.26. Prove that the map $\phi_B$ of formula (3.47) is an isomorphism if $B$ is flat over $A$.

REMARK 3.3.27. When calculating with these algebras, it is convenient to take infinitely many variables. So, given a set of variables $X = \{x_i\}$, $i \in I$, we denote by $\mathcal{T}_n(X), R_n(X)$ the corresponding algebras of generic matrices $\xi_i$, $i \in I$.

THEOREM 3.3.28. *The ring $\mathcal{T}_n(X)$ is generated by the coefficients $\sigma_i(p)$, $i=1,\dots,n$ of the characteristic polynomials of the primitive monomials $p \in W_0$ (cf. Definition 3.3.7) in the variables $x_i$.*

PROOF. This is an immediate consequence of the symbolic calculus developed in §3.3.2.4 and Theorem 3.3.13. □

By a deep theorem of Donkin **[Don92]** (see **[DCP17]** for an approach avoiding algebraic groups) which is beyond the scope of this book since it uses very delicate arguments of algebraic groups in prime characteristic, we have

THEOREM 3.3.29.

(1) *When the $\xi_i$ are generic matrices over a field $F$ or over $\mathbb{Z}$, the ring $\mathcal{T}_{n,k}$ coincides with the ring of* all *invariants of $k$-tuples of $n \times n$ matrices under conjugation.*
(2) *Moreover the dimension, as vector space over $F$, of $\mathcal{T}_{n,k}$ in each degree is independent of $F$ and equals the one for $\mathbb{Z}$.*

Observe that when $F$ is a finite field, by *invariant under conjugation* we mean conjugation by invertible matrices in an infinite field $G \supset F$ (not just $F$).

When working in prime characteristic, it is useful to have some more general definition, using the block diagonal matrices $x[d]$ of Definition 3.3.20. Since the map $x \mapsto x[d]$ for any $d$ is a homomorphism, we can define

DEFINITION 3.3.30. Let $R^d_{n,k} := \mathbb{Z}\langle \xi_1[d], \xi_2[d], \dots, \xi_k[d]\rangle$ be the subring of the matrix ring $M_{nd}(\mathbb{Z}[\xi])$ generated over $\mathbb{Z}$ by the matrices $\xi_s[d]$. We define then $\mathcal{T}^{(d)}_{n,k} \subset \mathbb{Z}[\xi]$ to be the ring generated by the coefficients of the characteristic polynomials of the elements of $R^d_{n,k}$. Finally, we set $\mathcal{R}^{(d)}_{n,k} := \mathcal{T}^{(d)}_{n,k} R^d_{n,k} \subset M_{nd}(\mathbb{Z}[\xi])$.

When $A$ contains $\mathbb{Q}$ (in characteristic 0) all these algebras $\mathcal{T}^{(d)}_{n,k}$ coincide and the treatment simplifies drastically; see Remark 3.3.34 and Example 3.3.35.

Let us use a fact that will be proved later in this book. By Theorem 8.2.1 it follows that $\mathcal{R}_{n,k}$ is a finitely generated module over $\mathcal{T}_{n,k}$, and in fact from Remark 8.2.3, there is a finitely generated subalgebra $B$ of $\mathcal{T}_{n,k}$ such that $\mathcal{T}_{n,k}B$ is a finitely generated module over $B$.

COROLLARY 3.3.31. *$\mathcal{T}_{n,k}$ is a finitely generated algebra over $\mathbb{Z}$.*

PROOF. By Remark 8.2.3 there is a finite list of monomials $M_1, \dots, M_N$ such that every other monomial $M$ is of the form $M = \sum_{i=1}^m \lambda_i M_i,\ \lambda_i \in B$. By Theorem 3.3.8 there is a universal formula for $\sigma_i(\sum_{i=1}^m t_i z_i)$ as an expression, homogeneous of degree $i$ in the elements $\sigma_k(N)$ where $N$ is some monomial in the $z_i$ and the degree of $\sigma_k(N)$ is $kd$ where $d$ is the degree of $N$.

It follows that $\sigma_i(M)$ lies in the ring obtained from $B$ by adding the finite number of elements $\sigma_k(N)$ where $k \le n$ and $N$ is a monomial of degree at most $n/k$ in the monomials $M_i$. □

REMARK 3.3.32. By the theorem of Donkin, Theorem 3.3.29 **[Don92]**, since $\mathcal{T}_{n,k}$ is the full ring of invariants of $k$-tuples of matrices under conjugation, finite generation follows also from general facts of invariant theory of reductive groups (cf. **[Spr77]**).

Another way of understanding $\mathcal{R}_{n,k}^{(d)}$ is the following. Embed $\mathcal{R}_{n,k} \subset M_{dn}(\mathbb{Z}[\xi])$ by mapping $\xi_u \mapsto \xi_u[d]$, and call the image $\mathcal{R}_{n,k}[d]$. Then $\mathcal{R}_{n,k}^{(d)}$ is the subalgebra of $M_{dn}(\mathbb{Z}[\xi])$ generated by $\mathcal{R}_{n,k}[d]$ and all the values of its characteristic polynomials. It is still a block diagonal algebra.

It then follows that $R_{n,k}[d]$ is a finite module over the finitely generated algebra $\mathcal{T}_{n,k}[d]$ which is a quotient of $\mathcal{T}_{dn,k}$.

The main property of these algebras is their invariance under conjugation. This should be stated in a formal way. Given any commutative ring $C$, we can form the rings

$$R_{n,k} \otimes_{\mathbb{Z}} C \subset M_n(C[\xi]), \quad \mathcal{T}_{n,k}^{(d)} \otimes_{\mathbb{Z}} C \subset C[\xi].$$

For each such $C$ consider the group $\mathrm{GL}(n, C)$ of invertible matrices with entries in $C$. This group acts by conjugation on the space of $k$-tuples of $n \times n$ matrices over $C$, and thus $\mathrm{GL}(n, C)$ also acts on the polynomial ring $\mathbb{C}[\xi]$ of the coordinates of $M_n(C)^k$.

We may also consider the $\mathrm{GL}(n, C)$-equivariant polynomial maps, that is (cf. Definition 1.3.19) polynomial maps $f : M_n(C)^k \to M_n(C)$ with

$$f(Ax_1A^{-1}, \dots, Ax_kA^{-1}) = Af(x_1, \dots, x_k)A^{-1}, \ \forall A \in \mathrm{GL}(n, C).$$

A map $f : M_n(C)^k \to M_n(C)$ is an element of $M_n(C[\xi]) = M_n(C) \otimes_C C[\xi]$.

One should verify that, the condition that $f$ is equivariant means that the element $f \in M_n(C) \otimes_C C[\xi]$ is invariant under the diagonal action of $\mathrm{GL}(n, C)$.

We usually denote by $X^G$ the invariant elements of an action of a group $G$ on a set $X$, so our equivariant maps are also denoted by $M_n(C[\xi])^{\mathrm{GL}(n,C)}$.

We then have

PROPOSITION 3.3.33. *$R_{n,k} \otimes_{\mathbb{Z}} C$ is formed by elements invariant under conjugation.*

We shall show when $C$ is a field of characteristic 0 that in fact $R_{n,k} \otimes_{\mathbb{Z}} C$ is formed by **all** the elements invariant under conjugation. This is actually also true for fields of positive characteristic due to the theorem of Donkin, Theorem 3.3.29.

REMARK 3.3.34.

(1) Clearly if $h$ divides $g$, we have that $\mathcal{T}_{n,k}^{(g)} \subset \mathcal{T}_{n,k}^{(h)}$. In characteristic 0 all these algebras are equal to $\mathcal{T}_{n,k} := \mathcal{T}_{n,k}^{(1)}$, since they can be generated by the elements $\mathrm{tr}(M)$ where $M$ runs over monomials, but not over $\mathbb{Z}$ or over a field of positive characteristic. For $k = 1$, $\mathcal{T}_{n,1}^{(d)}$ can be identified, by restriction to diagonal matrices, to the algebra generated by the coefficients of the polynomial $\prod_{i=1}^{n}(t - x_i)^d$ in the variable $t$, which are symmetric functions in the variables $x_1, \dots, x_n$.

(2) Furthermore, by the very construction, the embedding of $n \times n$ matrices as block diagonal $hn \times hn$ matrices induces two surjective maps

$$\mathcal{T}_{dn,k} \to \mathcal{T}_{n,k}^{(d)}, \quad \mathcal{R}_{dn,k} \to \mathcal{R}_{n,k}^{(d)}. \tag{3.48}$$

EXAMPLE 3.3.35. Over a field of characteristic $p$, if $d = p^k$, we have that $\mathcal{T}_{n,1}^{(p^k)}$ is generated by the elementary symmetric functions evaluated in the variables $x_i^{p^k}$, $\sigma_i(x_1^{p^k}, \dots, x_n^{p^k})$, $i = 1, \dots, n$.

3.3.4.2. *From symbolic to matrix formulas.* Given some set of variables $X$, we have introduced, in formula (3.40), the two algebras $\mathbb{S}_A\langle X\rangle := A\langle X\rangle \otimes_A \mathbb{S}_A$ and $\mathbb{S}_A := A \otimes_{\mathbb{Z}} \mathbb{S}$. We cannot interpret either of these algebras as functions on infinite matrices since this algebra does not have a 1.

For any commutative ring $A$ and $n \in \mathbb{N}$ we can instead evaluate $\mathbb{S}_A$ (resp., $\mathbb{S}_A\langle X\rangle$) as $A$-valued (resp., as $M_n(A)$-valued) functions on $M_n(A)^X$. This gives a surjective map from to the ring $\mathbb{S}_A$ to the ring $A \otimes_{\mathbb{Z}} \mathcal{T}_n(X)$ (cf. Remark 3.3.27) and one from $\mathbb{S}_A\langle X\rangle$ to $A \otimes_{\mathbb{Z}} \mathcal{R}_n(X)$ (the generic matrices with the coefficients of the characteristic polynomials added).

Thus $A \otimes_{\mathbb{Z}} \mathcal{T}_n(X)$ is a quotient of the symbolic algebra of polynomials $\mathbb{S}_A$ (3.40) and $A \otimes_{\mathbb{Z}} \mathcal{R}_n(X)$ is a quotient of $\mathbb{S}_A\langle X\rangle$.

By a deep theorem of Zubkov **[Zub94]** (see also **[DCP17]**) we have

THEOREM 3.3.36. *If $A = \mathbb{Z}$ or a field, then:*

- *The kernel of the homomorphism from the symbolic algebra $\mathbb{S} = \mathbb{Z}[\sigma_i(p)]$ to the ring of invariants $\mathcal{T}_{n,k}$ is the ideal generated by all the images of the maps $\sigma_i : A\langle X\rangle \to \mathbb{S}_A$, $\forall i > n$ and by $\sigma_i(1) = \binom{n}{i}$.*
- *The kernel of the homomorphism from $\mathbb{S}_A\langle X\rangle$ to $A \otimes_{\mathbb{Z}} \mathcal{R}_n(X)$ is the ideal generated by the previous maps and by the evaluations of the Cayley–Hamilton polynomial $\mathrm{CH}_n(x)$.*

This theorem generalizes the theorem in characteristic 0 of Procesi and Razmyslov on the Cayley–Hamilton identity, which we shall prove in Corollary 12.1.17.

## 3.4. The universal map into matrices

One of the themes of this book is related to the following general problem.

PROBLEM. Find necessary and sufficient conditions that a ring $R$ can be embedded into a matrix algebra $M_n(B)$ over some commutative ring $B$.

It was already noticed by Malćev **[Mal43]** that the free algebra in two or more variables cannot be embedded in any such matrix algebra.

In general we see that one necessary condition is that $R$ should satisfy all the polynomial identities of $n \times n$ matrices although as we shall see, in §11.5.5, this is in general not sufficient. So in this section we change the approach, and rather than asking for an embedding we just look at possible maps into matrix algebras over commutative rings.

**3.4.1. $n$-dimensional representations.** In this section for simplicity all algebras have a 1 and the homomorphisms preserve 1. Let $R$ be an $A$-algebra:

DEFINITION 3.4.1. By an *$n$-dimensional representation* over a commutative $A$-algebra $B$ one means an $A$-algebra homomorphism

$$\phi : R \to M_n(B). \tag{3.49}$$

We shall indicate by $\mathcal{R}_R^n(B)$ the set of all these representations.

It is a covariant set-valued functor on the category of commutative $A$-algebras, since, given $f : B \to C$, we can compose

$$R \xrightarrow{\ \phi\ } M_n(B) \xrightarrow{\ M_n(f)\ } M_n(C).$$

Before we prove the main statement about this functor, we develop the fundamental example of the free noncommutative algebra $A\langle x_\alpha \rangle$, $\alpha \in I$ in a set of variables $x_\alpha$, $\alpha \in I$. In this case a homomorphism $\phi : A\langle x_\alpha \rangle \to M_n(B)$ is given by choosing, for each $\alpha \in I$, a matrix $a_\alpha = (b_{i,j}^{(\alpha)}) \in M_n(B)$.

We now apply the method of generic elements developed in §2.4. For each $\alpha$ choose a set of $n^2$ variables $x_{i,j}^{(\alpha)}$, $i, j = 1, \ldots, n$.

Let $A[\xi] := A[x_{i,j}^{(\alpha)}]$ be the polynomial ring in all these variables over $A$. We have defined the *generic matrices* (in Definition 3.3.23) by considering the matrices $\xi_\alpha$ in $M_n(A[\xi])$ which, in the $i, j$ entry, has value $x_{i,j}^{(\alpha)}$.

DEFINITION 3.4.2. Let $\mathrm{j} : A\langle x_\alpha \rangle \to M_n(A[\xi])$ be defined by $\mathrm{j}(x_\alpha) = \xi_\alpha$.

The *algebra of generic matrices* is the subalgebra $A_n\langle \xi_\alpha \rangle \subset M_n(A[\xi])$, the image of the map $\mathrm{j}$ generated over $A$ by the generic matrices $\xi_\alpha$.

REMARK 3.4.3. This definition is just the generalization to any $A$ of the definition we have already introduced in Definition 3.3.23.

We have by Theorem 2.4.1 that a polynomial $f(x_1, \ldots, x_k) \in A\langle x_1, \ldots, x_k \rangle$ is in the kernel of $\mathrm{j}$ if and only if $f$ is a (stable) polynomial identity for $n \times n$ matrices.

We then have

PROPOSITION 3.4.4. *For every homomorphism $f : A\langle x_\alpha \rangle \to M_n(B)$, there exists a unique homomorphism $\bar{f} : A[\xi] \to B$ making the following diagram commutative.*

$$\begin{array}{ccccc} A\langle x_\alpha \rangle & \xrightarrow{\ \mathrm{j}\ } & A\langle \xi_\alpha \rangle & \xrightarrow{\ i\ } & M_n(A[\xi]) \\ & & {\scriptstyle f} \searrow & & \downarrow {\scriptstyle M_n(\bar{f})} \\ & & & & M_n(B) \end{array}$$

PROOF. The proof is clearly trivial. We need to set $\bar{f}(x_{i,j}^{(\alpha)}) := b_{i,j}^{(\alpha)}$ where $b_{i,j}^{(\alpha)}$ are the entries of the matrix $f(x_\alpha)$. $\square$

We can now prove (cf. Definition 2.1.4) the basic

THEOREM 3.4.5. *The functor $\mathcal{R}_R^n(-)$ is representable.*

PROOF. Take any $A$-algebra $R$ and consider a set of generators $a_\alpha$ for the algebra $R$ which is thus presented as quotient of the free noncommutative algebra $A\langle x_\alpha\rangle$:

$$\pi : A\langle x_\alpha\rangle \to R,\ \pi(x_\alpha) = a_\alpha; \quad R = A\langle x_\alpha\rangle/K,\ K := \ker \pi.$$

Let $\mathrm{j}$ be as in Definition 3.4.2, and let $I$ be the two-sided ideal in $M_n(A[\xi])$ generated by the image $\mathrm{j}(K)$.

By Lemma 2.6.2 $I = M_n(J)$ for some ideal $J$ of $A[\xi]$, and thus we have a commutative diagram

$$(3.50)\qquad \begin{array}{ccc} A\langle x_\alpha\rangle & \xrightarrow{\ \mathrm{j}\ } & M_n(A[\xi]) \\ \downarrow{\scriptstyle \pi} & & \downarrow{\scriptstyle M_n(p)} \\ R & \xrightarrow[\mathrm{j}_R]{} & M_n(A[\xi]/J) \end{array}$$

where $p : A[\xi] \to A[\xi]/J$ is the quotient map.

If $f : R \to M_n(B)$ is any map $f(a_\alpha) = (b_{i,j}^{(\alpha)})$, we have the map $f' : x_{i,j}^{(\alpha)} \to b_{i,j}^{(\alpha)}$ of $A[\xi] \to B$ constructed in Proposition 3.4.4 making the following commutative diagram.

$$\begin{array}{ccc} A\langle x_\alpha\rangle & \xrightarrow{\ \mathrm{j}\ } & M_n(A[\xi]) \\ {\scriptstyle \pi}\downarrow & & \\ R & & \downarrow{\scriptstyle M_n(f')} \\ & \searrow{\scriptstyle f} & \\ & & M_n(B) \end{array}$$

Since $f \circ \pi$ vanishes on $K$, we have that $M_n(f')$ vanishes on $\mathrm{j}(K)$ and hence $f'$ vanishes on $J$. This induces the required map $\bar{f} : A[\xi]/J \to B$ for which we have $f' = \bar{f} \circ p$. Hence $f \circ \pi = M_n(\bar{f} \circ p) \circ \mathrm{j} = M_n(\bar{f}) \circ \mathrm{j}_R \circ \pi$ and since $\pi$ is surjective, the diagram

$$\begin{array}{ccc} R & \xrightarrow{\ \mathrm{j}_R\ } & M_n(A[\xi]/J)) \\ & {\scriptstyle f}\searrow & \downarrow{\scriptstyle M_n(\bar{f})} \\ & & M_n(B) \end{array}$$

commutes. We thus have a natural isomorphism

$$(3.51)\qquad \mathcal{R}_R^n(B) \xrightarrow{\ \sim\ } \hom_A(A[\xi]/J, B)$$

of the set $\mathcal{R}_R^n(B)$ with the set of $A$-algebra homomorphisms $\hom_A(A[\xi]/J, B)$. Thus, by definition, the functor on commutative $A$-algebras $\mathcal{R}_R^n(-)$ is represented by the algebra $A[\xi]/J$. $\square$

DEFINITION 3.4.6. We shall denote by $A_n(R)$ the universal $A$-algebra $A[\xi]/J$ and shall refer to the mapping

$$R \xrightarrow{j_R} M_n(A_n(R)) \tag{3.52}$$

as the *universal n-dimensional representation*.

In Chapter 14 we shall study algebras with trace. For such algebras one may restrict to $n$-dimensional representations compatible with the trace. Also in this case one has a universal map.

Since $A_n(R)$ has been defined through a universal property, it is independent of the particular construction by generators and relations that we have used in the proof. Of course any ring is a $\mathbb{Z}$ algebra and one has the corresponding universal ring $\mathbb{Z}_n(R)$.

REMARK 3.4.7. When we change base and make an extension $B \otimes_A R$ with $B$ a commutative $A$-algebra, we have the same construction but now over the base ring $B$, and we may call the universal algebra $B_n(R)$. It is easily seen that $B_n(R) = B \otimes_A A_n(R)$.

It is then clear that

PROPOSITION 3.4.8. *An algebra $R$ can be embedded into an algebra $M_n(B)$ over a commutative algebra $B$ if and only if the universal map is an embedding.*

We shall see later (Theorem 14.2.1) a criterion for this to happen.

Notice that the construction $R \to A_n(R)$ is clearly functorial in $R$ and, for every morphism $f : R \to S$, we have the following commutative diagram.

$$\begin{array}{ccc} R & \xrightarrow{j_R} & M_n(A_n(R)) \\ \downarrow f & & \downarrow M_n(A_n(f)) \\ S & \xrightarrow{j_S} & M_n(A_n(S)) \end{array} \tag{3.53}$$

REMARK 3.4.9. The map $A_n(f)$ is uniquely determined so that if $R \xrightarrow{f} S \xrightarrow{g} U$, we have

$$A_n(g \circ f) = A_n(g) \circ A_n(f).$$

In other words

THEOREM 3.4.10. *The functor $B \to M_n(B)$ from the category $\mathcal{C}$ of commutative $A$-algebras to the category $\mathcal{N}$ of noncommutative $A$-algebras has a right adjoint*

$$\hom_{\mathcal{N}}(R, M_n(B)) = \hom_{\mathcal{C}}(A_n(R), B).$$

REMARK 3.4.11. The commutative ring $A_n(R)$ may be zero. This means that the ring $R$ does not have any $n$-dimensional representation.

Let us finish with a few properties of this construction which follow immediately from the definitions.

PROPOSITION 3.4.12.

(1) *If $R$ is a finitely generated $A$-algebra, $A_n(R)$ is also a finitely generated $A$-algebra.*

(2) *If $R_1 \to R_2$ is a surjective homomorphism, the induced map $A_n(R_1) \to A_n(R_2)$ is also surjective.*
(3) $A_n(R_1 \star_A R_2) = A_n(R_1) \otimes_A A_n(R_2)$.

REMARK 3.4.13. If $A = K$ is an algebraically closed field and $R$ is finitely generated over $K$, then $K_n(R)$ defines an affine algebraic variety whose points correspond to the representations of $R$ into $M_n(K)$.

It is possible that $K_n(R)$ is not *reduced*, that is it may have a nonzero nilpotent ideal.

For instance when $n = 1$ and $R$ is commutative, we have $R = A_1(R)$.

EXAMPLE 3.4.14. Consider the ring $U$ generated by three elements $H, X, Y$ with defining relations

$$HX - XH = Y. \tag{3.54}$$

Prove that, in this case, for each $n$, the ring $A_n(U)$ is the polynomial ring in the $2n^2$ variables $h_{ij}, x_{ij}$.

EXAMPLE 3.4.15. Consider the commutative polynomial ring $\mathbb{Z}[x, y]$ generated by two elements $x$ and $y$.

Prove that, in this case, for each $n$, the ring $A_n(\mathbb{Z}[x, y])$ is generated by the $2n^2$ variables $x_{ij}, y_{ij}$ modulo the quadratic equations $\sum_s x_{is}y_{sj} - \sum_s y_{is}x_{sj}$.

It is not known if these equations generate in general a prime ideal!

EXAMPLE 3.4.16. Consider the $\mathbb{Q}$-algebra $U$ generated by two elements $X, Y$ with defining relations

$$XY - YX = 1. \tag{3.55}$$

In this case the ring $A_n(U)$ is 0, since if $X, Y$ are two $n \times n$ matrices, we have always $\operatorname{tr}(XY - YX) = 0$ and, in characteristic 0, $\operatorname{tr}(1_n) = n \neq 0$. If instead we work over $\mathbb{Z}$, we get nontrivial rings, since the matrix equation $XY - YX = 1$ can be solved in characteristic $p > 0$, $p|n$.

### 3.4.2. The projective group.

3.4.2.1. *A localization principle.* We will use repeatedly a *localization principle* (a variation of Proposition 1.6.6) which we state for a commutative ring $A$, with 1:

PROPOSITION 3.4.17.
(1) *If $a, b \in A$ are two elements that become equal in all localizations $A_{\mathfrak{m}}$ by maximal ideals $\mathfrak{m}$, then they are equal.*
(2) *Let $f : A \to B$ be a morphism of commutative rings and assume that for all local rings $C$ we have that if*

$$\circ f : \hom(B, C) \to \hom(A, C), \quad \circ f := g \mapsto g \circ f$$

*is surjective, then $f$ is injective.*
(3) *Furthermore, if we assume that $A \subset B \subset K$ with $A$ a domain and with $K$ its field of fractions, then $A = B$.*

PROOF. (1) Let $s := a - b$, and let $I := \{x \in A \mid xs = 0\}$. If $s \neq 0$, then $I$ is a proper ideal, and let $\mathfrak{m} \supset I$ be a maximal ideal. In the localization map $j : A \to A_{\mathfrak{m}}$, the element $s$ maps to a nonzero element $j(s) = j(a) - j(b)$.

(2) Assume by contradiction that there is an $s \neq 0$ in $A$ with $f(s) = 0$.

As before let $I := \{x \in A \mid xs = 0\}$, and let $\mathfrak{m} \supset I$ be a maximal ideal. In the localization map $j : A \to A_{\mathfrak{m}}$, the element $s$ maps to a nonzero element $j(s)$.

By hypothesis there is a map $g : B \to A_{\mathfrak{m}}$ such that $j = g \circ f$ and this contradicts the hypothesis $f(s) = 0$.

(3) Since $A$ is a domain, we have $A = \bigcap_{\mathfrak{m}} A_{\mathfrak{m}}$ where $\mathfrak{m}$ runs over all maximal ideals and $A_{\mathfrak{m}}$ is the corresponding local ring. So it is enough to prove that $B \subset A_{\mathfrak{m}}$ where $\mathfrak{m}$ runs over all maximal ideals.

The inclusion $A \subset A_{\mathfrak{m}}$ by hypothesis factors though $B$ which has the same field of fractions as $A$, hence the claim. $\square$

3.4.2.2. *The universal map into matrices.* Although the definition of the universal map into matrices is quite simple, the actual nature of this map is quite deep as some basic examples will immediately show.

One of the simplest and most significant examples is to understand what is the universal map in the case in which $R = M_n(A)$ is also a ring of matrices over a commutative ring $A$. We can use Theorem 2.6.13, for a commutative ring $B$ a map $f : M_n(A) \to M_n(B)$ is given by two data: the map $f_1 : A \to B$ induced by $f$ and the map $f_2 : M_n(B) = M_n(A) \otimes_A B \to M_n(B)$ given by $f_2(r \otimes b) := f(r)b$. This last map by Corollary 2.6.14 is a $B$-linear automorphism of $M_n(B)$.

We want to understand the functor $\mathfrak{G}_n(B) := \mathrm{Aut}_B(M_n(B))$ on commutative rings $B$ and prove that it is an *affine group scheme*. This is just the geometric expression saying that it is a representable functor, the points of a commutative Hopf algebra, as we have discussed in §2.1.3. In fact it is *linear*, that is a subgroup of a linear group.

We may compare the automorphism group $\mathrm{Aut}_B(M_n(B))$ with the group of inner automorphisms of $M_n(B)$, that is $\mathrm{GL}(n, B)/B^*$, where $B^*$ denotes the group of invertible elements of $B$. Given an automorphism $\phi$ of $M_n(B)$, consider the idempotents $u_i = \phi(e_{ii})$. We have a decomposition $B^n = \bigoplus_i u_i B^n$, and it follows easily that $u_i B^n$ is a rank 1 projective module over $B$. The elements $\phi(e_{ij})$ establish isomorphisms between these rank 1 projectives. It is then clear that

LEMMA 3.4.18. *$\phi$ is inner if and only if these projectives are free.*

PROOF. Assume $\phi$ is an inner automorphism. Then there is a $g \in \mathrm{GL}(n, B)$ so that $\phi$ is conjugation by $g$. Hence $\phi(e_{i,i}) = g \circ e_{i,i} \circ g^{-1}$ so that

$$\phi(e_{i,i})B^n = g \circ e_{i,i} \circ g^{-1} B^n = g \circ e_{i,i} B^n = B g e_i.$$

Conversely assume that each $u_i B^n = B a_i$. Then the elements $a_i$ form a basis of $B^n$, and there is a $g \in \mathrm{GL}(n, B)$ so that $a_i = g e_i$.

Since the elements $u_i = \phi(e_{i,i})$ are the orthogonal idempotents, giving the decomposition $B^n = \bigoplus_i B a_i$, we must have that $u_i = g e_{i,i} g^{-1}$. Denote by $i_{g^{-1}}$ the inner automorphism induced by $g^{-1}$, and set $\psi := i_{g^{-1}} \circ \phi$. We have thus

$$\psi(e_{i,i}) = i_{g^{-1}} \circ \phi(e_{i,i}) = e_{i,i} \implies \psi(e_{i,j}) = e_{i,i}\psi(e_{i,j})e_{j,j} = \alpha_{i,j} e_{i,j},$$

and, since $\psi$ is an automorphism, we have $\alpha_{i,j} \in B^*$.

Clearly $\alpha_{i,j}\alpha_{j,k} = \alpha_{i,k}$, $\alpha_{j,i} = \alpha_{i,j}^{-1}$, hence setting $\beta_i := \alpha_{1,i}$, we have that $\alpha_{i,j} = \beta_i^{-1}\beta_j$, so that $\psi$ is conjugation by the diagonal matrix with entries $\beta_i^{-1}$. $\square$

REMARK 3.4.19. In particular if $B$ is a local ring, every projective module is free, hence every automorphism of $M_n(B)$ is inner.

Equivalence classes of rank 1 projectives form a group under tensor product called the Picard group of $B$ denoted $\mathrm{Pic}(B)$.

One can see **[Frö73]** that by mapping $\phi \in \mathrm{Aut}_B(M_n(B))$ to the class of $\phi(e_{1,1}B^n)$, we get a homomorphism $\mathrm{Aut}_B(M_n(B)) \to \mathrm{Pic}(B)$, and we have seen that its kernel is the group of inner automorphisms. This is a rather general fact.

A $B$-linear automorphism is, in particular, an invertible linear transformation so the automorphism group is a subgroup of the general linear group $\mathrm{GL}(n^2, B)$ of invertible linear transformations of $M_n(B) \simeq B^{n^2}$. We may describe it as matrices using the basis $e_{i,j}$, of $M_n(B)$, formed by the standard matrix units. We start with a simple

LEMMA 3.4.20. *$\mathrm{Aut}_B(M_n(B))$ is formed by elements of determinant* 1.

PROOF. This is clear for inner automorphisms, and one can reduce to this case by localization. If $\phi \in \mathrm{Aut}_B(M_n(B))$ for every maximal ideal $\mathfrak{m}$ of $B$, the automorphism $\phi$ induces an automorphism of $M_n(B_{\mathfrak{m}})$ which is inner and so of determinant 1. Thus the determinant of $\phi$ equals 1 in all localizations by maximal ideals, and then it must be 1. □

Given a linear map $\phi : M_n(B) \to M_n(B)$, consider its matrix relative to the basis of matrix units,

$$\phi(e_{i,j}) = \sum_{h,k} c_{i,j}^{h,k} e_{h,k}. \tag{3.56}$$

The condition that $\phi$ be an endomorphism is expressed by the quadratic relations imposed on the elements $c_{i,j}^{h,k}$, entries of an $n^2 \times n^2$ matrix,

$$\phi(e_{i,j})\phi(e_{h,k}) = \delta_h^j \phi(e_{i,k}), \quad \phi(1) = \sum_i \phi(e_{i,i}) = 1.$$

By Corollary 2.6.14 every $B$-linear endomorphism of $M_n(B)$ is an automorphism. Thus all together these algebraic restrictions on the elements $c_{i,j}^{h,k}$ define a finitely generated $\mathbb{Z}$ algebra $P_n$ which represents the functor that, to a commutative ring $B$, associates the group $\mathrm{Aut}_B(M_n(B))$ of automorphisms of $M_n(B)$. It is in fact a quotient of the coordinate ring $G_{n^2}$ 2.12 representing $\mathrm{GL}(n^2, -)$.

Since this is a group-valued functor, the ring $P_n$ has a natural structure of Hopf algebra which is made explicit since this is a subgroup of the linear group $\mathrm{GL}(n^2, -)$ (cf. formulas (2.1.25)) and makes its spectrum the form of the projective group defined over $\mathbb{Z}$.

The ring $P_n$ has a more concrete description from which we can deduce several properties, as for instance that it is a domain and a free $\mathbb{Z}$-module.

Consider the map $i_n$ from the groups $\mathrm{GL}(n, B)$ to the groups $\mathfrak{G}_n(B)$ which associates to an element $g \in \mathrm{GL}(n, B)$ the associated inner automorphism $i_g$.

Recall that $G_n := \mathbb{Z}[x_{i,j}, d^{-1}]$ is the coordinate ring of the linear group.

To the element $g = (x_{i,j}) = \sum_{i,j} x_{i,j} e_{i,j}$ with inverse $\sum_{i,j} y_{i,j} e_{i,j}$ of $\mathrm{GL}(n, G_n)$, which is the *generic element for the representable functor*, is associated the inner automorphism $i_g$ so that

$$i_g(e_{i,j}) = g e_{i,j} g^{-1} = \sum_{h,k} x_{h,i} y_{j,k} e_{h,k}.$$

This means that the comorphism of coordinate rings $i_n^* : P_n \to G_n$ maps

$$i_n^* : c_{i,j}^{h,k} \mapsto x_{h,i} y_{j,k}.$$

We have that $y_{j,k}$ is the determinant of the cofactor in place $j, k$ of the matrix $(x_{i,j})$ times the inverse of the determinant so that $x_{h,i} y_{j,k}$ is of degree 0. Denote by $G_n^0$ the span of elements of degree 0 of $G_n$. It is easy to see that $G_n^0$ is a sub-Hopf algebra of $G_n$ and it classifies a group-valued functor $\mathfrak{G}_n^0(B) := \hom(G_n^0, B)$. We want to prove

THEOREM 3.4.21. *The map $i_n^*$ is an isomorphism of $P_n$ with $G_n^0$.*

PROOF. First we need to show that $i_n^*$ is injective. Now we have for any $B$ that

$$\mathrm{GL}(n, B) = \hom(G_n, B) \mapsto \hom(P_n, B) = \mathrm{Aut}(M_n(B))$$

is surjective for $B$ local, hence the claim from (2) of Proposition 3.4.17.

We thus have an inclusion of Hopf algebras $P_n \subset G_n^0 \subset G_n$, giving natural maps

$$\mathrm{GL}(n, B) = \hom(G_n, B) \to \hom(G_n^0, B) \to \hom(P_n, B) = \mathrm{Aut}(M_n(B)).$$

The following facts follow from the definitions:

(1) The group scheme $G_m$ multiplicative group with coordinate ring $\mathbb{Z}[t^{\pm 1}]$ acts on $\mathrm{GL}_n$ by the coaction

$$\phi : G_n \mapsto G_n \otimes \mathbb{Z}[t^{\pm 1}], \ \phi(x_{i,j}) := x_{i,j} t, \ \phi(d^{-1}) = t^{-n} d^{-1}.$$

The *invariants* are exactly the algebra $G_n^0$.

(2) The map $\mathrm{GL}(n, B) = \hom(G_n, B) \to \hom(G_n^0, B)$ factors through the map $\mathrm{GL}(n, B) \to \mathrm{GL}(n, B)/B^*$ giving a factorization of group functors

$$\mathrm{GL}(n, B)/B^* \to \mathfrak{G}_n^0(B) \to \mathfrak{G}_n(B) = \mathrm{Aut}(M_n(B)).$$

(3) When $B$ is local, since $\mathrm{GL}(n, B)/B^* \to \mathrm{Aut}(M_n(B))$ is an isomorphism, we deduce that also $\mathfrak{G}_n^0(B) \to \mathfrak{G}_n(B) = \mathrm{Aut}(M_n(B))$ is an isomorphism.

Therefore, from (iii) and (3) of Proposition 3.4.17, the theorem follows once we prove that $P_n$ and $G_n^0$ have the same field of fractions.

For this it is enough to prove that $P_n \otimes \mathbb{Q} = G_n^0 \otimes \mathbb{Q}$ or $P_n \otimes \mathbb{C} = G_n^0 \otimes \mathbb{C}$.

Now these two rings are the coordinate ring of the same smooth variety: the quotient group $\mathrm{PGL}(n, \mathbb{C}) = \mathrm{GL}(n, \mathbb{C})/\mathbb{C}^*$ hence the claim. □

COROLLARY 3.4.22. *The algebra $P_n = G_n^0$ is a domain and a free abelian group.*

PROOF. In fact $G_n$ has a basis over $\mathbb{Z}$, arising from the theory of double standard tableaux (cf. **[DCP76]** or **[DCP17]**) formed by homogeneous elements. Then $P_n = G_n^0$ has as its basis those among these elements which are of degree 0. □

3.4.2.3. *Application to the universal map into matrices.* By construction the homomorphisms of $P_n$ to a commutative ring $B$ classify the automorphisms of $M_n(B)$, in particular the identity $1_P$ of $P_n$ classifies the *generic* automorphism of $M_n(P_n)$ defined by the formulas (3.56) which will be called $\mathcal{U}_n$.

The set-valued functor of maps $M_n(A) \to M_n(B)$ corresponds to pairs, a map $A \to B$, and one $P_n \to B$, so it is classified by the Hopf algebra $P_n \otimes A$. As a consequence we see

PROPOSITION 3.4.23. *The universal mapping of $M_n(A)$ to $n \times n$ matrices is, denoting by $i : A \to P_n \otimes A$, the inclusion*

$$M_n(A) \xrightarrow{M_n(i)} M_n(P_n \otimes A) \xrightarrow{\mathcal{U}_n} M_n(P_n \otimes A).$$

**Another example.** Consider $F$ an algebraically closed field, and the mappings $j : M_m(F) \to M_n(F)$. This means a representation of $M_m(F)$ is $n$-dimensional. The only representations of $M_m(F)$ are direct sums of the defining representation $F^m$, hence the set of maps $j : M_m(F) \to M_n(F)$ is empty unless $n = mk$. In this case it is a homogeneous space over the projective linear group of automorphisms of $M_n(F)$. One point is the embedding of $M_m(F)$ in $M_n(F)$ as block diagonal constant matrices which identifies $M_n(F) = M_m(F) \otimes_F M_k(F)$. The stabilizer of this point is the subgroup of automorphism of $M_k(F)$. Thus the variety associated to the universal embedding of $M_m(F)$ in $n \times n$ matrices is the homogeneous space $\mathrm{PGL}(n, F)/\mathrm{PGL}(k, F)$.

In fact one can prove that the universal ring $B := A_n(M_n(F))$ is *reduced* so it equals the coordinate ring of the homogeneous space $\mathrm{PGL}(n, F)/\mathrm{PGL}(k, F)$.

This is done by showing that the universal ring is formally smooth (see Remark 1.6.4), and we leave it to the reader as an exercise.

One uses the fact that $B$ represents the functor of maps of $M_n(F)$ to $m \times m$ matrices so that lifting of maps $B \to C/I$ to a map $B \to C$ is equivalent to lifting representations $M_n(F) \to M_m(C/I)$ to $M_n(F) \to M_m(C)$, which one does by Wedderburn's principal theorem (cf. **[Pro87]**). A general treatment as for the projective group is also possible.

**3.4.3. Symmetry of the universal map.** Given an $n$-dimensional representation of an $A$-algebra $R \xrightarrow{f} M_n(B)$ and an automorphism $\phi \in \mathfrak{G}_n(B)$ of $M_n(B)$, the composition $\phi \circ f$ is another representation, i.e., the affine group scheme $\mathfrak{G}_n(B)$ acts on the affine scheme $\mathcal{R}_R^n(-)$, or in the language of functors we act with a group-valued functor on a set-valued functor in a natural way. This again by Yoneda's lemma induces a map on the representing objects.

One can of course write this construction in terms of the coaction on rings $\delta : A_n(R) \to P_n \otimes A_n(R)$.

$$\begin{array}{ccc} R & \xrightarrow{j_R} & M_n(A_n(R)) \\ \downarrow{\scriptstyle j_R} & & \downarrow{\scriptstyle M_n(\delta)} \\ M_n(A_n(R)) & \xrightarrow{\mathcal{U}_n} & M_n(P_n \otimes A_n(R)) \end{array} \tag{3.57}$$

Let us be more concrete and assume that $A = F$ is an infinite field. Take an inner automorphism $g$ of $M_n(F)$ (this corresponds to a map $P_n \to F$). This induces an inner automorphism $g \otimes 1$ of $M_n(F_n(R)) = M_n(F) \otimes_F F_n(R)$. We then have by functoriality a map $\bar{g} : F_n(R) \to F_n(R)$ and the following commutative diagram.

$$\begin{array}{ccc} R & \xrightarrow{\mathrm{j}_R} & M_n(F) \otimes_F F_n(R) \\ \Big\downarrow \mathrm{j}_R & & \Big\downarrow 1\otimes\bar{g} \\ M_n(F_n(R)) & \xrightarrow{g\otimes 1} & M_n(F) \otimes_F F_n(R) \end{array} \tag{3.58}$$

If $g, h$ are two homomorphisms, we have

$$\begin{aligned} 1 \otimes \overline{g \circ h} \circ \mathrm{j}_R &= (g \circ h) \otimes 1 \circ \mathrm{j}_R = (g \otimes 1) \circ (h \otimes 1) \circ \mathrm{j}_R \\ &= (g \otimes 1) \circ (1 \otimes \bar{h}) \circ \mathrm{j}_R = (1 \otimes \bar{h}) \circ (1 \otimes g) \circ \mathrm{j}_R = (1 \otimes \bar{h}) \circ (1 \otimes \bar{g}) \circ \mathrm{j}_R. \end{aligned}$$

It follows that $\overline{g \circ h} = \bar{h} \circ \bar{g}$, and hence

THEOREM 3.4.24. *We have an action of the group* $\mathrm{GL}(n, F)$ *on* $M_n(F) \otimes_F F_n(R)$, *given by* $g \otimes \bar{g}^{-1}$, *and* $\mathrm{j}_R$ *maps* $R$ *into the ring of invariants* $[M_n(F) \otimes_F F_n(R)]^{\mathrm{GL}(n,F)}$.

We shall introduce later a class of algebras (with trace) for which the universal map $\mathrm{j}_R : R \to [M_n(F) \otimes_F F_n(R)]^{\mathrm{GL}(n,F)}$ is an isomorphism (see Theorem 14.2.1).

When $S = F\langle x_1, \dots, x_k\rangle$ is a free algebra in $k$ variables, the ring $F_n(S) = F[\xi]$ is just the polynomial ring of functions on $M_n(F)^k$ and the action on this polynomial ring is the one induced by conjugation, that is

$$(gf)(\xi_1, \dots, \xi_k) = f(g\xi_1 g^{-1}, \dots, g\xi_k g^{-1}).$$

If $R$ is a quotient of $F\langle x_1, \dots, x_k\rangle$, then $F_n(R)$ is the quotient of $F[\xi]$ by an invariant ideal.

In particular we have some special invariants:

PROPOSITION 3.4.25. *All the coefficients of the characteristic polynomials of all the elements* $j_R(r)$, $r \in R$, *are in* $F_n(R)^{\mathrm{GL}(n,F)}$.

We then set

DEFINITION 3.4.26. The ring $T_n(R)$, generated by the coefficients of the characteristic polynomials of all the elements $j_R(r)$, $r \in R$, will be called the *universal n trace ring of R*.

We will see after developing some invariant theory (see page 316) that

THEOREM 3.4.27. *When R is an algebra over* $\mathbb{Q}$, *the ring* $T_n(R)$ *is generated by the traces of all the elements* $j_R(r)$, $r \in R$. *By classical invariant theory, we have*

$$T_n(R) = F_n(R)^{\mathrm{GL}(n,F)}, \quad j_R(R)T_n(R) = [M_n(F) \otimes_F F_n(R)]^{\mathrm{GL}(n,F)}. \tag{3.59}$$

In some important cases, even if $R$ is not an algebra over $\mathbb{Q}$, we have that $T_n(R) = F_n(R)^{\mathrm{GL}(n,F)}$. Whether this is always the case is, for the moment, an open question.

Recall that for every morphism $f : R \to S$, we have the commutative diagram of formula (3.53).

$$\begin{array}{ccc} R & \xrightarrow{\mathrm{j}_R} & M_n(F_n(R)) \\ \Big\downarrow f & & \Big\downarrow M_n(F_n(f)) \\ S & \xrightarrow{\mathrm{j}_S} & M_n(F_n(S)) \end{array}$$

We have

PROPOSITION 3.4.28. *The morphism* $F_n(f) : F_n(R) \to F_n(S)$ *is* $\mathrm{PGL}(n, F)$ *equivariant.*

PROOF. We have from formula (3.53) that

$$1 \otimes F_n(f) \circ \mathrm{j}_R = \mathrm{j}_S \circ f$$

and, from Theorem 3.4.24 $(g \otimes \bar{g}^{-1}) \circ \mathrm{j}_R = \mathrm{j}_R$ for all $g$,

$$1 \otimes F_n(f) \circ (g \otimes \bar{g}^{-1}) \circ \mathrm{j}_R = 1 \otimes F_n(f) \circ \mathrm{j}_R = (g^{-1} \otimes \bar{g}) \circ \mathrm{j}_S \circ f$$
$$\implies (g \otimes \bar{g}^{-1}) \circ 1 \otimes F_n(f) \circ (g^{-1} \otimes \bar{g}) \circ \mathrm{j}_R = \mathrm{j}_S \circ f = 1 \otimes F_n(f) \circ \mathrm{j}_R$$
$$\implies (1 \otimes (\bar{g}^{-1} \circ F_n(f) \circ \bar{g}) \circ \mathrm{j}_R = 1 \otimes F_n(f) \circ \mathrm{j}_R \overset{3.4.9}{\implies} \bar{g}^{-1} \circ F_n(f) \circ \bar{g} = F_n(f).$$

□

We have defined a functor from the category of algebras and the category of commutative algebras equipped with an action of $\mathrm{PGL}(n, F)$ and equivariant maps.

We notice a corollary of the previous analysis:

COROLLARY 3.4.29. *The characteristic polynomial of a matrix* $a \in M_n(B)$ *is invariant under* $\mathrm{Aut}_n(B)$.

PROOF. By the localization principle (Proposition 1.6.6) and Proposition 3.4.17 it is enough to do it when $B$ is local. In this case every automorphism is inner and the proof is the usual one. □

CHAPTER 4

# Polynomial maps

*This chapter aims at giving the foundations for Corollary* 12.1.18, *which sheds light on the role of invariants of matrices in PI theory. This occurs in the formal constructions of Zubrilin, and it plays a role in* §17.5.2 *and* §18.1.3.1.

*This chapter is divided into two sections. The first part treats polynomial maps only over free modules, essentially either vector spaces over an infinite field or free abelian groups. The second section, on the Schur algebra, justifies the symbolic calculus on matrix invariants which we hinted in* §3.3.2.4. *In fact here some interesting open questions remain in the arithmetic case.*

*We follow closely the papers of Roby* **[Rob63]**, **[Rob80]** *and Ziplies* **[Zip87]**, **[Zip89]**.

## 4.1. Polynomial maps

We now present the same material of the previous sections but in a more abstract form which is useful for studying the invariants of matrices, in particular the complicated relations arising from formula (3.29) when we work with matrices of some fixed size.

We shall see that behind these computations there is an algebra of *noncommutative symmetric functions* (see §4.2, Definition 4.2.7) which is worth investigating.

Its maximal abelian quotient turns out to be the algebra $\mathbb{S} = \mathbb{Z}[\sigma_i(p)]$ defined in formula (3.33).

The connection to matrix invariants then takes a more precise form, similar to passing from the ring of formal symmetric functions to those in a fixed number of variables (see Corollary 12.1.18). This corollary essentially says that the commutative algebra of invariants of several $n \times n$ matrices is obtained as solution of a universal problem, that is the problem of classifying the *multiplicative polynomial maps* of degree $n$ from the free algebra to a commutative algebra. The map which evaluates the free algebra in generic matrices and then takes the determinant has the universal property that any multiplicative polynomial map of degree $n$ from the free algebra to a commutative algebra $A$ factors uniquely trough the determinant and a homomorphism of the algebra of invariants to $A$ (an approach of Ziplies and Vaccarino, Theorem 17.5.2, and Corollary 12.1.18).

We give a full proof of this statement only in characteristic 0. For the general characteristic-free proof, the reader is referred to the original papers or to the book of de Concini and Procesi **[DCP17]**.

**4.1.1. Definitions.** The general theory of *polynomial laws* has been developed by Roby in **[Rob63]**. We will not follow his general approach but just recall it for the reader.

DEFINITION 4.1.1. A polynomial law between two $A$-modules $N, M$ is a natural transformation $f_B : B \otimes_A N \to B \otimes_A M$ of the two functors $B \mapsto B \otimes_A N$, $B \mapsto B \otimes_A M$, from the category of commutative $A$-algebras $B$ to that of sets.

We denote by $\mathcal{P}_A(N, M)$ the polynomial laws from $N$ to $M$.

This means that *for every commutative $A$-algebra $B$*, we have a set theoretic map $f_B : B \otimes_A N \to B \otimes_A M$ of $B$-modules, and *for each morphism $h : B \to C$* of commutative algebras, we have the following commutative diagram.

$$\begin{array}{ccc} B \otimes_A N & \xrightarrow{f_B} & B \otimes_A M \\ {\scriptstyle h\otimes 1_N}\downarrow & & \downarrow{\scriptstyle h\otimes 1_M} \\ C \otimes_A N & \xrightarrow[f_C]{} & C \otimes_A M \end{array} \tag{4.1}$$

For example an $A$-linear homomorphism $f : M \to N$ clearly defines the polynomial law $f_B = id \otimes f : B \otimes_A M \to B \otimes_A N$. This is called a linear polynomial law.

As for polynomials, also for polynomial laws, there is a notion of degree.

DEFINITION 4.1.2. Let $f \in \mathcal{P}_A(M, N)$ be a polynomial law between two $A$-modules $M, N$. We say that $f$ is homogenous of degree $d \geq 0$ if for all $A$-algebras $B, b \in B$ and $z \in B \otimes_A M$, $f_B(bz) = b^d f_B(z)$.

In what follows we are going to be interested in the case in which $M, N$ are two free modules. If this is the case and $A$ is an infinite domain (for instance an infinite field), a polynomial law $f_B : B \otimes_A M \to B \otimes_A N$ is also called a polynomial map, and it is determined by a single map $f : M \to N$ which, written in coordinates, has the property that each coordinate of $f(x)$ in the basis $d_j$ of $M$ is a polynomial in the coordinates of $x$ in the basis $e_i$ of $N$, or

$$f(\sum_i x_i e_i) = \sum_\alpha x^\alpha n_\alpha;\ \alpha = (\alpha_1, \dots, \alpha_i, \dots), n_\alpha \in N, \tag{4.2}$$

where $\alpha_i \in \mathbb{N}$ are almost all 0 and $x^\alpha = \prod_i x_i^{\alpha_i}$.

There is also the restriction that if we set all except a finite number of $x_i$ equal to 0, then also all, except a finite number of $x^\alpha n_\alpha$, are equal to 0.

The polynomial law is homogeneous of degree $d$ if and only if each $\alpha$ appearing has $\sum_i \alpha_i = d$.

In our treatment we shall really only need this case. The general case needs the introduction of the algebra of *divided powers*, which the reader will find in Roby **[Rob63]**.

If $A$ is a finite field, then by definition a polynomial law $f$ is given by a map $f_G : G \otimes_A M \to G \otimes_A N$ for any field $G \supset F$, so it is still given by formula (4.2). By abuse of notation we still denote it by $f : M \to N$, although this function does not determine the law.

DEFINITION 4.1.3. Given a module $M$ and an integer $m$, we have an action of the symmetric group $S_m$ on $M^{\otimes m}$ by

$$\sigma \in S_m,\ \sigma(v_1 \otimes v_2 \otimes \cdots \otimes v_m) = v_{\sigma^{-1}(1)} \otimes v_{\sigma^{-1}(2)} \otimes \cdots \otimes v_{\sigma^{-1}(m)}. \tag{4.3}$$

The *space of symmetric tensors* $(M^{\otimes m})^{S_m}$ is

$$(M^{\otimes m})^{S_m} := \{u \in M^{\otimes m} \mid \sigma(u) = u,\ \forall \sigma \in S_m\}. \tag{4.4}$$

Given a free module $M$ with a countable or finite basis $\{e_1, e_2, \ldots, e_i, \ldots\}$ over a commutative ring $A$ and an integer $m$ the space of symmetric tensors, $(M^{\otimes m})^{S_m}$ has an explicit description since the symmetric group $S_m$ permutes the basis of $M^{\otimes m}$ formed by the tensors $e_{j_1} \otimes \cdots \otimes e_{j_m}$.

For any sequence $I = i_1 \leq i_2 \leq \cdots \leq i_m$ consider the set $\mathcal{O}_I$ of all distinct sequences obtained by permuting $I$. We set $e^I = \sum_{(j_1, j_2, \cdots, j_m) \in \mathcal{O}_I} e_{j_1} \otimes \cdots \otimes e_{j_m}$. Such a sequence is also encoded in another sequence of nonnegative integers $\alpha_1, \ldots, \alpha_k$ summing to $m$, where $\alpha_j$ counts the number of times that the number $i_s$ equals to $j$. Observe that there are $\binom{m}{\alpha_1, \ldots, \alpha_j, \ldots}$ distinct elements in the orbit $\mathcal{O}_I$.

Example $\alpha_2 = 2, \alpha_4 = 1$:

$$e_2 \otimes e_2 \otimes e_4 + e_2 \otimes e_4 \otimes e_2 + e_4 \otimes e_2 \otimes e_2, \quad \binom{3}{2,1} = 3.$$

LEMMA 4.1.4. *The $A$-module $(M^{\otimes m})^{S_m}$ is free with a basis consisting of the elements $e^I$, $|I| = m$.*

*Let $x_i$ be a sequence of elements in $A$ which are all zero except for a finite number.*

$$(4.5) \qquad (\sum_i x_i e_i)^{\otimes m} = \sum_{|I|=m} x^I e^I, \quad x^I := \prod_{h=1}^{m} x_{i_h} = x^\alpha = \prod_i x_i^{\alpha_i}.$$

PROOF. This is clear. □

PROPOSITION 4.1.5.

(1) *The map*

$$(4.6) \qquad i_M : M \to (M^{\otimes m})^{S_m},$$

*given by $i_M : a \mapsto a^{\otimes m}$, is a homogeneous polynomial map of degree $m$ and satisfies the following universal property.*

(2) *Any polynomial map $f : M \to N$, homogeneous of degree $m$ between the free modules $M$ and $N$, factors uniquely through a linear map $\bar{f}$:*

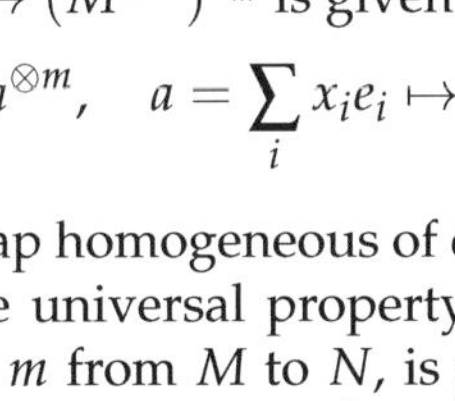


(3) *If $A$ is an infinite field, the elements $a^{\otimes m}$, $a \in M$ linearly span $(M^{\otimes m})^{S_m}$.*

PROOF. The map $i_M : M \to (M^{\otimes m})^{S_m}$ is given by formula (4.5),

$$i_M : a \mapsto a^{\otimes m}, \quad a = \sum_i x_i e_i \mapsto \sum_{|I|=m} x^I e^I,$$

so it is clearly a polynomial map homogeneous of degree $m$.

We can verify directly the universal property. By definition, a polynomial map, homogeneous of degree $m$ from $M$ to $N$, is given by an $N$-valued polynomial $\sum_{|I|=m} n_I x^I$. Thus, since the elements $e^I$ are a linear basis of $(M^{\otimes m})^{S_m}$, by assigning the values $n_I := \bar{f}(e^I) \in N$, we get a linear map $\bar{f} : (M^{\otimes m})^{S_m} \to N$ and

$$(4.7) \qquad f(\sum_i x_i e_i) = \sum_{\underline{h}} x^I n_I = \bar{f}((\sum_i x_i e_i)^{\otimes m}).$$

Finally the fact that the elements $a^{\otimes m}$ span $(M^{\otimes m})^{S_m}$ depends on the fact that otherwise there is a nonzero linear function on $(M^{\otimes m})^{S_m}$, mapping $e^I \to \lambda_I \in A$ vanishing on the elements $a^{\otimes m}$. This in turn means that there is a formally a nonzero polynomial $\sum_{\underline{h}} x^I \lambda_I$ which vanishes for all values of the $x$, which is impossible since $A$ is an infinite field. $\square$

COROLLARY 4.1.6. *Given a linear map $f : M \to N$, there is a unique linear map $\bar{f} : (M^{\otimes m})^{S_m} \to (N^{\otimes m})^{S_m}$ making the following diagram commutative.*

$$\bar{f}(a^{\otimes m}) = f(a)^{\otimes m} \qquad \begin{array}{ccc} M & \xrightarrow{i_M} & (M^{\otimes m})^{S_m} \\ {\scriptstyle f}\big\downarrow & & \big\downarrow{\scriptstyle \bar{f}} \\ N & \xrightarrow{i_N} & (N^{\otimes m})^{S_m} \end{array}$$

REMARK 4.1.7. In general, given any sequence $\underline{a} := (a_1, a_2, \dots, a_k)$ of elements $a_i \in M$, we may define, as variation of formula (4.5), the associated elements $\underline{a}^\alpha \in (M^{\otimes m})^{S_m}$ where $\alpha = (\alpha_1, \dots, \alpha_k)$, $\alpha_i \in \mathbb{N}$, and $m = \sum_{i=1}^k \alpha_i$, by introducing variables $t_i$:

$$(4.8) \qquad (\sum_i t_i a_i)^{\otimes m} = \sum_{|I|=m} t^\alpha \underline{a}^\alpha, \; t^\alpha = \prod_{i=1}^k t_i^{\alpha_i}.$$

If $M$ is an algebra with a 1, it is convenient to introduce a special notation for the element $(\underline{a}')^{\alpha'}$, for the special lists $\underline{a}' := (1, a_1, \dots, a_k)$ and the exponents $\alpha' := (m - \sum_{i=1}^k \alpha_i, \alpha_1, \dots, \alpha_k)$, under the assumption $\sum_{i=1}^k \alpha_i \le m$.

$$(4.9) \qquad (\underline{a}')^{\alpha'} := \sigma_{m;\alpha_1,\dots,\alpha_k}(a_1, \dots, a_k).$$

In particular for a single element $a$, we have

$$(4.10) \qquad (1+a)^{\otimes m} = 1 + \sum_{i=1}^m \sigma_{m;i}(a).$$

EXERCISE 4.1.8. Replacing $a$ with $\sum_i t_i a_i$, we have a *polarization formula*

$$(4.11) \qquad \sigma_{m;i}(\sum_j t_j a_j) = \sum_{\alpha_1,\dots,\alpha_k \mid \sum_j \alpha_j = m-i} \prod_j t_j^{\alpha_j} \sigma_{m;\alpha_1,\dots,\alpha_k}(a_1, \dots, a_k).$$

### 4.1.2. Multiplicative polynomial maps (according to Roby [Rob80]).

DEFINITION 4.1.9. If $M, N$ are both algebras, one says that a polynomial map $f : M \to N$ is *multiplicative* if

$$f(ab) = f(a)f(b), \; \forall a, b \in M.$$

Now notice that if $M$ is an algebra, then also $(M^{\otimes m})^{S_m}$ is an algebra, and the map $i_M : a \mapsto a^{\otimes m}$ of formula (4.6) is multiplicative.

DEFINITION 4.1.10. If $R$ is an algebra, then the algebra $(R^{\otimes m})^{S_m}$ will be denoted by $\mathcal{S}^m(R)$, and is called the *Schur algebra of $R$ of degree $m$*.

From this we immediately deduce

PROPOSITION 4.1.11. *Let $M$ and $N$ be two algebras over an infinite field $F$, and let $f : M \to N$ be a multiplicative polynomial map homogeneous of degree $m$. Then the map $\bar{f} : (M^{\otimes m})^{S_m} \to N$ is a homomorphism of algebras.*

PROOF. By definition we have

$$\bar{f}(a^{\otimes m}b^{\otimes m}) = \bar{f}((ab)^{\otimes m}) = f(ab) = f(a)f(b) = \bar{f}(a^{\otimes m})\bar{f}(b^{\otimes m}).$$

Since the elements $a^{\otimes m}$ linearly span $(M^{\otimes m})^{S_m}$ this implies that $\bar{f}$ is a homomorphism of algebras. □

In fact if $M$ and $N$ are algebras which are free modules over an infinite domain $A$, the proof goes through by passing to the quotient field $F$ of $A$.

One of our main goals in this chapter is to study the Schur algebra of a free algebra, prove Theorem 4.2.12 on its generators, and compare it with matrix invariants.

Observe that, if $R$ is graded, then also $R^{\otimes m}$ and $(R^{\otimes m})^{S_m}$ are naturally graded.

Let us give two basic examples.

EXAMPLE 4.1.12.

(1) When $R = F[t]$, the algebra $F[t]^{\otimes m} = F[t_1, \dots, t_m]$, hence $\mathcal{S}^m(F[t]) = (F[t]^{\otimes m})^{S_m}$, is by definition the algebra of symmetric polynomials in $m$ variables.

(2) When $R = \mathrm{End}(V) = M_n(F)$ (here $V$ is an $n$-dimensional vector space), we have $\mathrm{End}(V)^{\otimes m} = \mathrm{End}(V^{\otimes m})$, hence

$$(\mathrm{End}(V)^{\otimes m})^{S_m} = \mathrm{End}_{S_m}(V^{\otimes m})$$

is the centralizer of the action of the symmetric group on tensor space, a basic object discussed in §6.1.1.

Observe that, by formula (4.10), the element $\sigma_{m;i}(t) \in (F[t]^{\otimes m})^{S_m}$ coincides with the elementary symmetric function $e_i$ in the variables $t_1, \dots, t_m$.

Take an $F$-algebra $R$ and an element $a \in R$. This determines (and is equivalent to) a homomorphism $F[t] \to R$, $t \mapsto a$. Thus we have an induced homomorphism of Schur algebras and we easily see that the elementary symmetric functions $e_i(t_1, \dots, t_m)$ map to the elements $\sigma_{m;i}(a)$.

In particular if $A = F[\xi_{i,j}]$ is the algebra of polynomial functions on $m \times m$ matrices and $a = \xi = (\xi_{i,j})$ is the corresponding generic matrix, we see that the polynomial map det,

$$\det : F[t] \to A^{\mathrm{GL}(m,F)}, \ f(t) \mapsto \det(f(\xi)),$$

is multiplicative and homogeneous of degree $m$.

From Proposition 4.1.11 we then see that the map $\det : F[t] \to A^{\mathrm{GL}(m,F)}$ factors through a homomorphism $\det : (F[t]^{\otimes m})^{S_m} \to A^{\mathrm{GL}(m,F)}$.

We see that $\det(e_i)$ is the coefficient $\sigma_i(\xi)$ of the characteristic polynomial of $\xi$ which also coincides with the element $\sigma_{m;i}(\xi)$ of formula (4.10).

Therefore since $\mathcal{S}^m(F[t]) = (F[t]^{\otimes m})^{S_m}$ is the algebra of symmetric polynomials in $m$ variables, we see that the map det is an isomorphism which inverts naturally the *restriction map* given by Theorem 3.3.1.

Our goal in what follows is to apply the same method to several matrices where the restriction map to diagonal matrices is not available.

Consider the free algebra $F\langle x_1, \dots, x_n\rangle$ and the algebra $F[\xi_{i,j}^k]^{\mathrm{GL}(m,F)}$, where $i, j = 1, \dots, m$, $k = 1, \dots, n$, of conjugation invariant polynomial function on $n$, $m \times m$ matrices $\xi_k = (\xi_{i,j}^k)$. We are going to show as Corollary 12.1.18:

THEOREM 4.1.13. *Given $m \in \mathbb{N}$, for every $n = 1, \dots, \infty$,*

(1) *The following map is multiplicative:*

$$\det : F\langle x_1, \dots, x_n\rangle \to F[\xi_{i,j}^k]^{\mathrm{GL}(m,F)}, \quad f(x_1, \dots, x_n) \mapsto \det(f(\xi_1, \dots, \xi_n)).$$

(2) *The induced map*

$$\det : (F\langle x_1, \dots, x_n\rangle^{\otimes m})^{S_m} \to F[\xi_{i,j}^k]^{\mathrm{GL}(m,F)}$$

*is surjective*

(3) *Its kernel is the ideal generated by all commutators.*

This theorem, in characteristic 0, is due to Ziplies and Vaccarino and will be proved in Corollary 12.1.18.

In general the fact that the map $\det : (F\langle x_1, \dots, x_n\rangle^{\otimes m})^{S_m} \to F[\xi_{i,j}^k]^{\mathrm{GL}(m,F)}$ is surjective is proved by exhibiting generators for the Schur algebra of a free algebra (Theorem 4.2.12) and then it is a reformulation of the theorem of Donkin, Theorem 3.3.29, describing generators for the invariants, while the fact that its kernel is the ideal generated by all commutators is a reformulation of the theorem of Zubkov, see **[DCP17]** for details.

Let $R$ be an algebra, and let $\underline{a} := \{a_i,\ i \in I\}$, $\underline{b} := \{b_j,\ j \in J\}$ be two lists of elements of $R$. For two functions $\alpha : I \to \mathbb{N}$, $\beta : J \to \mathbb{N}$ with $d = \sum_i \alpha_i = \sum_j \beta_j$, we give a multiplication rule for the elements $\underline{a}^\alpha, \underline{b}^\beta \in \mathcal{S}^d(R) = (R^{\otimes d})^{S_d}$ defined in Remark 4.1.7 by formula (4.8).

Denote by $\underline{a} \circ \underline{b} := \{a_i b_j,\ (i,j) \in I \times J\}$ the *product list*. Take two sets of variables $t_i$, $i \in I$, $s_j$, $j \in J$, and consider the two elements $(\sum_{i\in I} t_i a_i)^{\otimes d}$, $(\sum_{j \in J} s_j b_j)^{\otimes d}$. We have

$$\sum_{\alpha:I\to\mathbb{N},\ |\alpha|=d,\ \beta:J\to\mathbb{N},\ |\beta|=d} t^\alpha s^\beta \underline{a}^\alpha \underline{b}^\beta = \sum_{\alpha:I\to\mathbb{N},\ |\alpha|=d} t^\alpha \underline{a}^\alpha \sum_{\beta:J\to\mathbb{N},\ |\beta|=d} s^\beta \underline{b}^\beta$$
$$\overset{(4.8)}{=} (\sum_{i\in I} t_i a_i)^{\otimes d} (\sum_{j\in J} s_j b_j)^{\otimes d} = (\sum_{(i,j)\in I\times J} t_i s_j a_i b_j)^{\otimes d}.$$

We deduce, with $ts$ denoting the list $t_i s_j$:

$$\sum_{\alpha:I\to\mathbb{N},\ |\alpha|=d,\ \beta:J\to\mathbb{N},\ |\beta|=d} t^\alpha s^\beta \underline{a}^\alpha \underline{b}^\beta = \sum_{\gamma:I\times J\to\mathbb{N},\ |\gamma|=d} (ts)^\gamma (\underline{a}\,\underline{b})^\gamma. \tag{4.12}$$

Let us define

$$M(\alpha, \beta) := \{\gamma : I \times J \to \mathbb{N} \mid \sum_{i\in I} \gamma_{i,j} = \beta_j\ \sum_{j\in J} \gamma_{i,j} = \alpha_i\}. \tag{4.13}$$

By equating coefficients in formula (4.12), we deduce the multiplication rule

THEOREM 4.1.14.

$$\boxed{\underline{a}^\alpha \underline{b}^\beta = \sum_{\gamma\in M(\alpha,\beta)} (\underline{a} \circ \underline{b})^\gamma}. \tag{4.14}$$

For the applications of the next section it is useful to cast formula (4.14) in the special case and with the notations of formula (4.9). That is we choose $\underline{a} = (1, a_1, \dots, a_h)$, $\underline{b} = (1, b_1, \dots, b_k)$ so that $\underline{a} \circ \underline{b}$ is displayed by a sequence with entries $(1, a_1, \ldots, a_h, b_1, \ldots, b_k, a_i b_j)$. To be coherent, we index the elements $\gamma_{i,j}$ with indices $i = 0, 1, \ldots, h$ and $j = 0, 1, \ldots, k$. We write then

$$\sigma_{d;\alpha_1,\ldots,\alpha_h}(a_1, \ldots, a_h)\sigma_{d;\beta_1,\ldots,\beta_k}(b_1, \ldots, b_k) = \sum_{\gamma \in M(\alpha,\beta)} \sigma_{d,\tilde{\gamma}}([\underline{a} \circ \underline{b}]_\gamma), \tag{4.15}$$

where by $\tilde{\gamma}$ we mean the subsequence of the $\gamma_{i,j} \neq 0$ and $(i,j) \neq (0,0)$, and by $[\underline{a} \circ \underline{b}]_\gamma$ we mean the subsequence of $a_1, \ldots, a_h, b_1, \ldots, b_k, a_i b_j$, where the corresponding exponent $\gamma_{i,j} \neq 0$.

It is useful to notice then that

LEMMA 4.1.15. *As soon as $d$ is sufficiently large, the right-hand side of* (4.15) *starts with the term*

$$\sigma_{d;\alpha_1,\ldots,\alpha_h,\beta_1,\ldots,\beta_k}(a_1, \ldots, a_h, b_1, \ldots, b_k).$$

*The other terms $\sigma_{d,\tilde{\gamma}}([\underline{a} \circ \underline{b}]_\gamma)$ have the property that*

$$\sum_{\gamma_{i,j} \in \tilde{\gamma}} \gamma_{i,j} = \sum_{(i,j) \neq (0,0)} \gamma_{i,j} < \sum_i \alpha_i + \sum_j \beta_j. \tag{4.16}$$

PROOF. By definition we have that

$$(i) \quad j \geq 1, \ \sum_{i=0}^{h} \gamma_{i,j} = \beta_j, \quad (ii) \quad i \geq 1, \sum_{j=0}^{k} \gamma_{i,j} = \alpha_i \tag{4.17}$$

imply

$$\sum_{(i,j) \neq (0,0)} \gamma_{i,j} = \sum_{j=1}^{k} \gamma_{0,j} + \sum_{i=1}^{h} \sum_{j=0}^{k} \gamma_{i,j} = \sum_{j=1}^{k} \gamma_{0,j} + \sum_{i=1}^{h} \alpha_i.$$

Now by formula (4.17)(i) we have that $\gamma_{0,j} \leq \beta_j$ and $\gamma_{0,j} = \beta_j$ if and only if $\gamma_{i,j} = 0$ for all $i \geq 1$. This clearly proves the claim. □

EXAMPLE 4.1.16.

$$\sigma_1(a)\sigma_1(b) = \sigma_{1,1}(a, b) + \sigma_1(ab). \tag{4.18}$$

## 4.2. The Schur algebra of the free algebra

**4.2.1. The free algebra.** Let $X = \{x_1, x_2, \ldots\}$ be a finite or countable set of variables. Consider the free algebra $A\langle X \rangle$ with $A$ a commutative ring, which has a basis consisting of the (noncommutative) monomials $W(X)$ in the variables $x_i$.

We want to study the algebras $\mathcal{S}^d_A\langle X \rangle := \mathcal{S}^d(A\langle X \rangle)$ of Definition 4.1.10 and in particular to understand the meaning of the universal property given by Proposition 4.1.5. We follow Ziplies **[Zip87]**, **[Zip89]** and Vaccarino **[Vac08]**, **[Vac09]**.

Applying Lemma 4.1.4, we deduce that the algebra $\mathcal{S}^d_A\langle X \rangle$ has a basis over $A$ indexed by functions $f : W(X) \to \mathbb{Z}^{\geq 0}$ with $\sum_{M \in W(X)} f(M) = d$. Hence we write simply $\mathcal{S}^d\langle X \rangle := \mathcal{S}^d_{\mathbb{Z}}\langle X \rangle$ and notice that $\mathcal{S}^d_A\langle X \rangle = A \otimes \mathcal{S}^d\langle X \rangle$.

It is important to distinguish the element 1 and the monomials $W(X)^+$ of positive degree. This will allow us to construct a *limit algebra* $\mathcal{S}\langle X \rangle$ (see Definition 4.2.7) as $d \mapsto \infty$.

DEFINITION 4.2.1. Denote by $\mathcal{F}$ the set of functions $f : W(X)^+ \to \mathbb{Z}^{\geq 0}$ with finite support, and set

$$|f| = \sum_{M \in W(X)^+} f(M), \ d(f) = \sum_{M \in W(X)^+} f(M) \deg(M), \tag{4.19}$$

where $\deg(M)$ is the usual degree. We denote by $\mathcal{F}_d$ the set of elements $f \in \mathcal{F}$ such that $|f| \le d$.

REMARK 4.2.2. The set $\mathcal{F}_d$ is in 1–1 correspondence with the set of functions $f : W(X) \to \mathbb{Z}^{\ge 0}$ with $\sum_{M \in W(X)} f(M) = d$ by setting $f(1) := d - |f|$.

Following Lemma 4.1.4 in order to get a basis for $\mathcal{S}^d\langle X\rangle$, we start with the symbolic element $\sum_{M\in W(X)^+} t_M M$ and expand (this is easily justified by degree considerations)

$$(4.20)\qquad (1 + \sum_{M \in W(X)^+} t_M M)^{\otimes d} = \sum_{f \in \mathcal{F}_d} t^f e_f, \quad t^f := \prod_{M \in W(X)^+} t_M^{f(M)},$$

where in particular for $f = 0$ we have $t^0 e_0 = 1$.

REMARK 4.2.3. To be precise, the element $e_f \in \mathcal{S}^d\langle X\rangle$, $f \in \mathcal{F}_d$, should be denoted by $e_f^{(d)}$.

If $f$ takes nonzero values $f_1, \dots, f_k$ on a list of monomials $A_1, \dots, A_k$, we also have that equation (4.9) gives formula (4.20) so that

$$(4.21)\qquad e_f = \sigma_{d;f_1,\dots,f_k}(A_1, \dots, A_k).$$

On the other hand the right-hand side of the previous formula has been defined also if the $A_i$'s are not all distinct.

REMARK 4.2.4. Since we will need to let $d$ go to infinity, we shall drop the symbol $d$ from $\sigma_{d;f_1,\dots,f_k}(A_1, \dots, A_k)$ and just write $\sigma_{f_1,\dots,f_k}(A_1, \dots, A_k)$.

For instance, take as alphabet $X = \{a, b\}$ and suppose $f$ is 0 on all monomials except $f(a) = 2, f(ab) = 1$. Then, if $d = 3$,

$$(4.22)\qquad e_f^{(3)} = \sigma_{2,1}(a, ab) = a \otimes a \otimes ab + a \otimes ab \otimes a + ab \otimes a \otimes a.$$

But the coefficient of $t_1t_2t_3$ in $(t_1a + t_2a + t_3ab)^{\otimes 3} = ((t_1 + t_2)a + t_3ab)^{\otimes 3}$ is

$$\sigma_{3;1,1,1}(a, a, ab) = 2e_f^{(3)}.$$

If $d = 4$, to obtain $e_f^{(4)}$ we need to insert 1 in all possible ways in $e_f^{(3)}$, as for instance $a \otimes 1 \otimes ab \otimes a$ in the previous formula obtaining four terms for each one of the three terms and thus a sum of 12 terms.

REMARK 4.2.5. If we take any algebra $R$ and a homomorphism of the free algebra $F\langle X\rangle$ to $R$ given by mapping $x_i \to a_i$, we have an induced homomorphism $\mathcal{S}^d(F\langle X\rangle) \to \mathcal{S}^d(R)$. All the formulas developed for the elements $\sigma$ of formula (4.9) constructed from monomials in $X$ specialize to formulas for their values in $R$.

4.2.1.1. *The multiplication rules.* The *multiplication rule* of the basis elements $e_f, e_g$ and in particular of an element $\sigma_q(M)$ with a basis element $e_f$ can be performed using formula (4.14). In this case we can use the property that monomials are closed under multiplication to make the formula more explicit.

As in equation (4.14), double the variables $t_M, M \in W(X)^+$ by introducing also variables $s_M$. The use of infinite variables is allowed from the fact that monomials have a degree:

$$\begin{aligned}(\sum_{f\in\mathcal{F}_d} t^f e_f)(\sum_{g\in\mathcal{F}_d} s^g e_g) &= [(1+\sum_{M\in W(X)^+} t_M M)(1+\sum_{N\in W(X)^+} s_N N)]^{\otimes d}\\ &= (1+\sum_{M\in W(X)^+}(t_M+s_M)M+\sum_{M,N\in W(X)^+} t_M s_N MN)^{\otimes d}\\ &= (1+\sum_{M\subset W(X)^+} Q_M(t,s)M)^{\otimes d},\end{aligned}$$

where

$$Q_M(t,s) = t_M + s_M + \sum_{A,\, B\in W(X)^+ \mid AB=M} t_A s_B. \tag{4.23}$$

Thus we can define, for $h, f, g \in \mathcal{F}_d$, nonnegative integers $c^h_{f,g} \in \mathbb{N}$ by

$$Q(t,s)^h := \prod_{M\in W(X)^+} Q_M(t,s)^{h(M)} = \sum_{f,g} c^h_{f,g} t^f s^g,\ c^h_{f,g} \in \mathbb{N} \tag{4.24}$$

so that

$$(1+\sum_{M\in W(X)^+} Q_M(t,s)M)^{\otimes d} = \sum_{h\in\mathcal{F}_d} Q(t,s)^h e_h$$

and we deduce

$$\sum_{f,g\in\mathcal{F}_d} t^f s^g e_f e_g = \sum_{h\in\mathcal{F}_d}\sum_{u,v} c^h_{u,v} t^u s^v e_h. \tag{4.25}$$

Thus we have shown

PROPOSITION 4.2.6. *For $h, f, g \in \mathcal{F}_d$, let the nonnegative integers $c^h_{f,g}$ be defined by formula* (4.24). *Then*

$$e_f e_g = \sum_{h\in\mathcal{F}_d} c^h_{f,g} e_h. \tag{4.26}$$

Thus formula (4.26) stabilizes as soon as $d \geq |f| + |g|$ and becomes

$$e_f e_g = \sum_{h\in\mathcal{F}} c^h_{f,g} e_h. \tag{4.27}$$

This suggests the following

DEFINITION 4.2.7. The algebra over a commutative ring $A$ with its basis the elements $e_f$, $f \in \mathcal{F}$, and its multiplication rule given by

$$e_f e_g = \sum_{h\in\mathcal{F}} c^h_{f,g} e_h$$

will be called the *free Schur algebra* in the variables $X$ with coefficients in $A$ and denoted by $\mathcal{S}_A\langle X\rangle$. The algebra $\mathcal{S}_{\mathbb{Z}}\langle X\rangle$ will be simply denoted by $\mathcal{S}\langle X\rangle$.

Consider now the augmentation $\pi : \mathbb{Z}\langle X\rangle \to \mathbb{Z}$ defined by $\pi(x_i) = 0$ for each $x_i \in X$.

This induces an algebra homomorphism $\pi_d = 1^{\otimes d} \otimes \pi : \mathbb{Z}\langle X\rangle^{\otimes d+1} \to \mathbb{Z}\langle X\rangle^{\otimes d}$ for each $d \geq 0$, which, when restricted to symmetric tensors, gives again an algebra homomorphism, still denoted by $\pi_d$,

$$\pi_d : S^{d+1}\langle X\rangle \to S^d\langle X\rangle$$

with the property that

$$\pi_d(e_f^{(d+1)}) = \begin{cases} e_f^{(d)} & \text{if } f \in \mathcal{F}_d \\ 0 & \text{if } f \in \mathcal{F}_{d+1} \setminus \mathcal{F}_d. \end{cases}$$

In particular $\pi_d$ is an isomorphism in degree $\leq d$. Thus we get an inverse system of graded algebra $(\mathcal{S}^d\langle X\rangle, \pi_d)$, and we may consider the limit algebra $\tilde{\mathcal{S}}_A\langle X\rangle$ which in each degree $d$ coincides with the degree $d$ part of all the $\mathcal{S}^m\langle X\rangle$, $m \geq d$.

THEOREM 4.2.8.

(1) *The algebras $\mathcal{S}_A\langle X\rangle$ and $\tilde{\mathcal{S}}_A\langle X\rangle$ are naturally isomorphic.*
(2) *The algebra $\mathcal{S}_A\langle X\rangle$ is an associative (but not commutative) algebra.*
(3) *For each $d$ the span of the elements $e_f$ with $|f| > d$ is an ideal $I_d$, and the algebra $S^d\langle X\rangle$ is naturally isomorphic to the quotient $\mathcal{S}_A\langle X\rangle / I_d$.*
(4) *When $X = \{x\}$ has a single element, we have that $\mathcal{S}_A\langle x\rangle$ is the ring of symmetric polynomials in an arbitrary number of variables with coefficients in $A$.*

PROOF. For each $d \geq 0$, let $\gamma_d : \tilde{\mathcal{S}}_A\langle X\rangle \to \mathcal{S}^d\langle X\rangle$ be the algebra homomorphism such that $\pi_d \circ \gamma_{d+1} = \gamma_d$. By definition it follows that $\tilde{\mathcal{S}}_A\langle X\rangle$ is a free module with basis given by the elements $\eta_f$ such that

$$\gamma_d(\eta_f) = \begin{cases} e_f^{(d)} & \text{if } f \in \mathcal{F}_d \\ 0 & \text{if } f \in \mathcal{F} \setminus \mathcal{F}_d. \end{cases}$$

So setting $\delta(e_f) = \eta_f$, we get a linear isomorphism $\delta : \mathcal{S}_A\langle X\rangle \to \tilde{\mathcal{S}}_A\langle X\rangle$.

We remark that by the definition the composition $\gamma_d \circ \delta$ is compatible with the product. It follows that $\delta$ is also compatible with the product, and point (1) follows.

After this, the proof of the remaining points is clear. □

COROLLARY 4.2.9. *The ideal $I_d$ in $\mathcal{S}_{\mathbb{Q}}\langle X\rangle$ is generated by the evaluations of $\sigma_k(x)$, $\forall k > d$ as $x \in \mathbb{Q}\langle X\rangle$.*

PROOF. This follows from part (3) of Theorem 4.2.8 and formulas (4.11) and (4.21). □

Notice that $\mathcal{S}\langle X\rangle$ is a graded algebra with $\deg e_f = d(f)$. If $X$ is finite, a simple counting argument gives

$$\sum_{n=0}^{\infty} \dim(\mathcal{S}\langle X\rangle_n)t^n = \prod_{d=1}^{\infty} \frac{1}{(1-t^d)^{|X|^d}}. \tag{4.28}$$

REMARK 4.2.10. Observe that in the algebra $\mathcal{S}\langle X\rangle$, we have *substitutional rules*, that is every map $\phi : X \to \mathbb{Z}_+\langle X\rangle$ induces a sequence of compatible homomorphisms $\mathcal{S}^d(\phi) : \mathcal{S}^d(F\langle X\rangle) \to \mathcal{S}^d(F\langle X\rangle)$ and hence a limit homomorphism $\mathcal{S}(\phi) : \mathcal{S}(F\langle X\rangle) \to \mathcal{S}(F\langle X\rangle)$.

This suggests the following notation. Take $f \in \mathcal{F}$. Let $\{M_1, \ldots, M_h\}$ be its support and, for each $i = 1, \ldots, h$, $f_i = f(M_i)$.

We denote $e_f$ by $\sigma_{f_1,\dots,f_h}(M_1,\dots,M_h)$ $(1=\sigma_0)$. In particular if for each $j$, $\phi(x_j)$ is a monomial, for each monomial $M$, also $\phi(M)$ is a monomial. We then have

$$\mathcal{S}(\phi)(\sigma_{f_1,\dots,f_h}(M_1,\dots,M_h))=\sigma_{f_1,\dots,f_h}(\phi(M_1),\dots,\phi(M_h)).$$

Notice that in view of this it makes sense to consider $\sigma_{f_1,\dots,f_h}(M_1,\dots,M_h)$ also when the monomials $(M_1,\dots,M_h)$ are not distinct. What one obtains is an integer multiple of the element $e_f$ where $f$ is defined as $f(M)=\sum_{i\mid M=M_i}f_i$. For example

$$\sigma_{1,1,1}(x,x,x)=6\sigma_3(x);\quad \sigma_{2,1}(x,x)=3\sigma_3(x).$$

4.2.1.2. *Generators for $\mathcal{S}\langle X\rangle$, $\mathcal{S}^d(\mathbb{Z}\langle X\rangle)$.* Given $f\in\mathcal{F}$, we want to exhibit an algorithm which expresses $e_f\in\mathcal{S}\langle X\rangle$ as a polynomial with integer coefficients in the elements $\sigma_q(M)$, $M\in W(X)$.

LEMMA 4.2.11. *Given $a_i\in W(X)$, each element $\sigma_{i_1,\dots,i_k}(a_1,\dots,a_k)$ is a polynomial with integer coefficients in the elements $\sigma_q(M)$ where $M$ is some monomial in the $a_i$ and with $q\le m:=\max(i_1,\dots,i_k)$.*

PROOF. We work by double induction on $k$ and of the sum $m:=\sum_{j=1}^k i_j$.

If $k=1$, the element is $\sigma_{i_1}(a_1)$ is already of the given form. If $\sum_{j=1}^k i_j=0$, the element is 1, and there is nothing to prove. Otherwise compute the product $\sigma_{i_1}(a_1)\sigma_{i_2,\dots,i_k}(a_2,\dots,a_k)$ and apply Lemma 4.1.15. Thus we get the term $\sigma_{i_1,\dots,i_k}(a_1,\dots,a_k)$ plus terms $\sigma_\gamma$ with $m'<m$, and we finish by induction. □

In fact the formula is universal in the sense that we may take the formula when $a_i$ is the variable $x_i$ and obtain the result by substitution of $x_i$ with any $a_i$.

THEOREM 4.2.12.

(1) *The algebra $\mathcal{S}\langle X\rangle$ is generated by the elements $\sigma_q(M)$ as $M$ runs over all primitive monomials (cf. Definition 3.3.6).*
(2) *In characteristic 0 the algebra $\mathcal{S}_{\mathbb{Q}}\langle X\rangle$ is generated by the elements $\sigma_1(M)$ as $M$ runs over all monomials.*

PROOF. (1) As output of the previous algorithm we have that $\mathcal{S}\langle X\rangle$ is generated by the elements $\sigma_q(M)$, $M\in W$. Now, if $M=N^k$ is not primitive, we can apply the theory of commutative symmetric functions which gives the element $\sigma_q(x^k)$ as a polynomial with integer coefficients in the elements $\sigma_j(x)$.

(2) In characteristic 0 in the ring of symmetric functions, $\sigma_q(M)$ corresponds to the elementary and $\sigma_1(M^k)$ to the Newton functions so the statement follows again from the theory of symmetric functions. □

EXAMPLE 4.2.13. Let us develop as an example the formula for the element in formula (4.22) $e_f=a\otimes a\otimes ab+a\otimes ab\otimes a+ab\otimes a\otimes a=\sigma_{2,1}(a,ab)$:

$$\sigma_1(ab)=1\otimes 1\otimes ab+1\otimes ab\otimes 1+ab\otimes 1\otimes 1,$$
$$\sigma_2(a)=a\otimes a\otimes 1+a\otimes 1\otimes a+1\otimes a\otimes a,$$
$$\begin{aligned}e_f&=\sigma_1(ab)\sigma_2(a)-\sigma_1(aba)\sigma_1(a)+\sigma_1(aba^2)\\&=\sigma_2(a)\sigma_1(ba)-\sigma_1(a^2b)\sigma_1(a)+\sigma_1(a^2ba).\end{aligned}$$

We remark that the elements $\sigma_i(p)$ are not free generators but satisfy complicated relations which have not been fully investigated.

**4.2.2. A universal commutative algebra.** Given a commutative ring $F$ and an $F$-algebra $R$ which is a free $F$-module, we take $\mathcal{S}^d(R)$ and define

DEFINITION 4.2.14. The maximal commutative quotient of $\mathcal{S}_F^d(R)$, that is the quotient modulo the ideal generated by commutators, will be denoted by $\mathbb{A}_F^d(R)$.

When $F = \mathbb{Z}$ we just write $\mathbb{A}^d(R)$. From the definition we get

PROPOSITION 4.2.15. *The mapping $a_R : R \to \mathcal{S}^d(R) \to \mathbb{A}_F^d(R)$ is a multiplicative polynomial map homogenous of degree $d$ which is* universal *with respect to multiplicative polynomial maps, homogeneous of degree $d$, $R \to U$ whose target $U$ is a commutative algebra.*

If $\phi : R \to S$ is a homomorphism of algebras, we deduce an induced homomorphism $\Phi : \mathbb{A}_F^d(R) \to \mathbb{A}_F^d(S)$ which is surjective if $\phi$ is surjective. Thus, in order to understand formal computations in $\mathbb{A}_F^d(R)$, we may as well develop them in the case in which $R$ is the free algebra $\mathbb{Z}\langle X\rangle$.

In fact we can even work with the limit algebra $\mathcal{S}\langle X\rangle$ of which $\mathcal{S}^d\langle X\rangle$ is a quotient. We will denote the maximal commutative quotient of $\mathcal{S}_F\langle X\rangle$ by $\mathbb{A}_F\langle X\rangle$ and $\mathbb{A}_{\mathbb{Z}}\langle X\rangle = \mathbb{A}\langle X\rangle$.

PROPOSITION 4.2.16. *In $\mathbb{A}\langle X\rangle$ we have $\sigma_h(ab) = \sigma_h(ba)$ for all $h$ and all pairs of elements $a, b$.*

PROOF. Let us first see that $\sigma_1(ab) = \sigma_1(ba)$. We have

$$\sigma_1(ab) = \sigma_1(a)\sigma_1(b) - \sigma_{1,1}(a,b), \quad \sigma_1(ba) = \sigma_1(b)\sigma_1(a) - \sigma_{1,1}(a,b),$$

so if we impose $\sigma_1(a)\sigma_1(b) = \sigma_1(b)\sigma_1(a)$, we deduce $\sigma_1(ab) = \sigma_1(ba)$. Observe that, by part (2) of Theorem 4.2.12, this computation implies the full proposition if $A$ contains $\mathbb{Q}$.

Let us now proceed by induction on $h$. We need a preliminary step. Let us assume that Proposition 4.2.16 holds for all $h < m$.

LEMMA 4.2.17. *Given $i_1, \ldots, i_k < m$, we have, for all $j_1, j_2, \ldots, j_k$ and $q$, that*

$$\sigma_{i_1,\ldots,i_k,q}(ab^{j_1}, \ldots, ab^{j_k}, b) = \sigma_{i_1,\ldots,i_k,q}(b^{j_1}a, \ldots, b^{j_k}a, b) \tag{4.29}$$

*in $\mathbb{A}\langle X\rangle$.*

PROOF. We start by showing that our claim holds for $q = 0$, that is,

$$\sigma_{i_1,\ldots,i_k}(ab^{j_1}, \ldots, ab^{j_k}) = \sigma_{i_1,\ldots,i_k}(b^{j_1}a, \ldots, b^{j_k}a). \tag{4.30}$$

By Theorem 4.2.12 $\sigma_{i_1,\ldots,i_k}(x_1, \ldots, x_k)$ is a polynomial with integer coefficients in elements $\sigma_q(M)$ where $q < m$ and the elements $M$ are monomials in $x_1, \ldots, x_k$. If we make in any such monomial the substitution $x_s = ab^{j_s}$ or $x_s = b^{j_s}a$, we obviously get cyclically equivalent elements. Thus our claim follows by the inductive hypothesis.

Suppose now $q > 0$ and that our lemma holds for each $q' < q$ and for all $k$. Arguing as in Lemma 4.2.11,we can write

$$\sigma_{i_1,\dots,i_k,q}(ab^{j_1},\dots,ab^{j_k},b) = \sigma_{i_1,\dots,i_k}(ab^{j_1},\dots,ab^{j_k})\sigma_q(b) - T,$$

$$\sigma_{i_1,\dots,i_k,q}(b^{j_1}a,\dots,b^{j_k}a,b) = \sigma_q(b)\sigma_{i_1,\dots,i_k}(b^{j_1}a,\dots,b^{j_k}a) - T',$$

where each addendum in $T$ (resp., $T'$) is given by a formula similar to (4.16), hence, for $T$, it is of the form $\sigma_\gamma(\underline{a})$ for $\underline{a} = ab^{j_1},\dots,ab^{j_k},ab^{j_1+1},\dots,ab^{j_k+1},b$ where $\gamma$ satisfies our inductive hypothesis. It is similar for $T'$. This implies our claim. □

Let us go back to the proof of Proposition 4.2.16. We apply Theorem 4.1.14 to $(1,a)\circ(1,b) = (1,a,b,ab)$. The solutions of $\gamma_{0,1}+\gamma_{1,1} = \gamma_{1,0}+\gamma_{1,1} = m$ are

$$\gamma_{0,1} = \gamma_{1,0} = k, \quad \gamma_{1,1} = m-k, \quad k = 0,\dots,m.$$

From formula (4.15) we get the identity.

$$\sigma_m(a)\sigma_m(b) = \sigma_m(ab) + \sum_{k=1}^{m-1}\sigma_{k,m-k,k}(a,ab,b) + \sigma_{m,m}(a,b). \tag{4.31}$$

For each $0 < k < m$, we have, by Lemma 4.2.17, that

$$\sigma_{k,m-k,k}(a,ab,b) = \sigma_{k,m-k,k}(a,ba,b) = \sigma_{k,m-k,k}(b,ba,a) \quad \text{in} \quad \mathbb{A}\langle X\rangle.$$

Exchanging $a$ and $b$ in formula (4.31) immediately implies our claim. □

Recall that $W_0$ denotes the set of Lyndon words (Definition 3.3.7).

THEOREM 4.2.18. *The algebra $\mathbb{A}\langle X\rangle$ is the polynomial ring $\mathbb{S} = \mathbb{Z}[\sigma_i(p)]$, $i = 1,\dots,\infty$, $p \in W_0$ of* §3.3.2.4.

PROOF. By Proposition 4.2.16 we have that $\mathbb{A}\langle X\rangle$ is generated over $\mathbb{Z}$ by the elements $\sigma_i(p)$, $i = 1,\dots,\infty$, $p \in W_0$. In order to prove that they are algebraically independent, it is enough to do this over $\mathbb{Q}$ and, by variable change, prove that the elements $t(M) = \sigma_1(M)$ as $M$ runs over all monomials are algebraically independent.

It is then enough to show that, for every $d \in N$ when we pass to the quotient $\mathbb{A}^d(\mathbb{Z}\langle X\rangle)$ there are no relations in degree $\leq d$.

This follows by evaluating the algebra $\mathbb{A}_{\mathbb{Q}}\langle X\rangle$ in the invariants of $d\times d$ matrices, where by Theorem 12.1.17 the relations among the elements $t(M)$ start only in degree $d+1$. □

Now we can fully justify the symbolic calculus which we hinted at in §3.3.2.4.

Given any homomorphism of the free algebra $\mathbb{Z}\langle X\rangle$ in an algebra $R$, this induces a homomorphism of $\mathbb{A}^d(\mathbb{Z}\langle X\rangle)$ in $\mathbb{A}^d(R)$.

In particular, a homomorphism of $\mathbb{Z}\langle X\rangle$ in itself is given by an arbitray map $x_i \to a_i \in \mathbb{Z}\langle X\rangle$.

Passing to the limit, we thus have a homomorphism of $\mathbb{S} = \mathbb{Z}[\sigma_i(p)]$ in itself. The value of an element $\sigma_i(p)$ under this homomorphism is computed as follows: $p$ maps to some polynomial in $\mathbb{Z}\langle X\rangle$. Then we may first apply formula (3.29) which gives the result as an expression in the elements $\sigma_i(M)$ with $M$ any polynomial. And finally if $M = p^k$, one uses the formulas for symmetric functions.

THEOREM 4.2.19. *Given any field $F$, the algebra $\mathbb{A}_F\langle X\rangle$ is the polynomial ring*

$$\mathbb{S} = F[\sigma_i(p)],\ i = 1, \dots, \infty,\ p \in W_0,$$

*of* §3.3.2.4.

PROOF. Arguing as in Theorem 4.2.18, the only point is to prove that these elements are algebraically independent. We have seen that they are algebraically independent when working over $\mathbb{Q}$ so it is enough to prove that the dimension of $\mathbb{A}_F\langle X\rangle$ over $F$ in each degree equals the dimension of $\mathbb{A}_{\mathbb{Q}}\langle X\rangle$ over $\mathbb{Q}$ in the same degree.

Now by the second part of the theorem of Donkin, Theorem 3.3.29, the algebra of invariants of $m$ copies of $n \times n$ matrices has a dimension in each degree which is independent of the characteristic. It follows then that there are no algebraic relations of degree $\leq n$ in the generators, for all fields $F$. Since the algebra of invariants of $m$ copies of $n \times n$ matrices is a quotient of $A(F\langle X\rangle)$ for all $n$, the claim follows. □

As we shall see, in Corollary 12.1.18 we have from the second fundamental theorem, Theorem 12.1.17, the rather unexpected result that $\mathbb{A}^d(\mathbb{Q}\langle X\rangle)$ is in fact isomorphic, in a canonical way, to the algebra of invariants of $d \times d$ matrices.

As for $\mathbb{A}^d(\mathbb{Z}\langle X\rangle)$ (or over a field $F$ of positive characteristic) the same statement follows from works of Donkin and Zubkov where a statement in terms of Cayley–Hamilton type of identities is developed (see **[Zub94]**, **[Zub96]**, and **[DCP17]**).

Using the approach of Ziplies and Vaccarino, we can finally explain the theorem proved by them in characteristic 0.

THEOREM 4.2.20.

(1) *The map*

$$\det : F\langle x_1, \dots, x_n\rangle \to S[X_1, \dots, X_n]^{\mathrm{GL}(m,F)}$$

*given by* $f(x_1, \dots, x_n) \mapsto \det(f(X_1, \dots, X_n))$ *is multiplicative and homogeneous of degree* $m$.

(2) *The induced map*

$$\det : (F\langle x_1, \dots, x_n\rangle^{\otimes m})^{S_m} \to S[X_1, \dots, X_n]^{\mathrm{GL}(m,F)}$$

*is surjective.*

(3) *Its kernel is the ideal generated by all commutators.*

PROOF. (1) is clear.

(2) follows from the description of $(F\langle x_1, \dots, x_n\rangle^{\otimes m})^{S_m}$ and Theorem 3.3.29.

(3) follows from the theorem of Zubkov since all the relations given in that theorem are already satisfied by $\mathbb{A}^d(F\langle X\rangle)$ which, by definition, equals the algebra $(F\langle x_1, \dots, x_n\rangle^{\otimes m})^{S_m}$ modulo the ideal generated by all commutators. □

Notice that, in characteristic 0 all the statements involved in this proof have been presented in this book. For (2) one uses the first fundamental theorem of matrix invariants (Theorem 12.1.3), and for (3) the second fundamental theorem of matrix invariants (Theorem 12.1.17). As already mentioned for a characteristic free proof, when $F = \mathbb{Z}$ or an infinite field, the reader can refer to **[DCP17]**.

4.2.2.1. *The formula* (3.29) *of Theorem* 3.3.8 *of Amitsur.* By Theorem 4.2.19 we see that formula (3.29), valid in the polynomial algebra $\mathbb{Z}[\sigma_i(M)]$, can be interpreted as a formula on the algebra $\mathbb{A}^d(F\langle X\rangle)$, where $\sigma_i : F\langle X\rangle \to \mathbb{A}^d(F\langle X\rangle)$ is indeed a polynomial map from the free algebra to this algebra, induced by the general properties of polynomial maps but which can be computed in an explicit form using the formula of Amitsur.

CHAPTER 5

# Azumaya algebras and irreducible representations

*We shall see that Azumaya algebras play a major role in the theory of polynomial identities, so here we develop their theory in some detail. We follow closely Auslander and Goldman* **[AG60a]**, **[AG60b]**, *Knus and Ojanguren* **[KO74a]**, *and Raynaud* **[Ray70]**.

In this chapter all rings are assumed to have a 1 and, on all modules, 1 acts as the identity.

## 5.1. Irreducible representations

5.1.0.1. *Flatness.* Azumaya algebras, defined in Definition 5.4.17, are a very close relative of matrix algebras over commutative rings and the simplest class of noncommutative algebras. They play a major role in the theory of polynomial identities, due to a theorem of M. Artin (Theorem 10.3.2), in particular for the study of the representations and the spectrum of a PI ring (see for instance Corollary 14.4.2). They also play an essential role in the proof of the nilpotency of the Jacobson radical in the finitely generated case; see §16.1.1. Before giving a formal definition, let us explain what we are trying to do and work out in detail an example.

Let first consider an algebra $R$ finite dimensional over its center, a field $F$. From the theory developed in Lemma 1.1.16 and §1.3.1 it follows easily that $R$ is a central simple algebra over $F$ if and only if, given an algebraic closure $\bar{F}$ of $F$, we have $R \otimes_F \bar{F} \simeq M_k(\bar{F})$. In fact there are also finite Galois extensions $L \subset F$ with $R \otimes_F L \simeq M_k(L)$. Then $R$ can be *recovered* by the action of the Galois group $G$ on $M_k(L)$ as $R = M_k(L)^G$.

Suppose now we want to take an algebra $R$ which is just a finitely generated module over its center, a commutative ring $A$. We would like to consider the property that *for some suitable commutative ring extension $\bar{A}$ of $A$*, we also have $R \otimes_A \bar{A} \simeq M_k(\bar{A})$. The problem is that it is not very clear what, for a commutative ring $A$, should be the extension $\bar{A}$? Galois theory of commutative rings is not the most suitable for our purposes.

The idea is that the ring $\bar{A}$ should be such that *no information on $R$ is lost in $R \otimes_A \bar{A}$*. The precise meaning of this statement will be made clear in §5.2 (Faithfully flat descent), a replacement of the Galois action.

For instance we would like to make sure that if $I$ is an ideal of $R$, then $I \otimes_A \bar{A} \subset R \otimes_A \bar{A}$, and if $I_1, I_2$ are two different ideals of $R$, we have that $I_1 \otimes_A \bar{A} \neq I_2 \otimes_A \bar{A}$.

It turns out that the correct hypotheses on $\bar{A}$ are that $\bar{A}$ should be *faithfully flat* over $A$. We introduce a notion coming from homological algebra:

DEFINITION 5.1.1. A right module $M$ over a ring $A$ is *flat* if, given any exact sequence of left $A$-modules

$$0 \to N \to P \to Q \to 0, \tag{5.1}$$

the sequence

$$(5.2) \qquad 0 \longrightarrow M \otimes_A N \longrightarrow M \otimes_A P \longrightarrow M \otimes_A Q \longrightarrow 0$$

is exact. Moreover, it is *faithfully flat* if the sequence (5.2) is exact if and only if the sequence (5.1) is exact.

In other words, $M$ is faithfully flat if $M$ is flat and, if $N$ is a nonzero left $A$ module, then $M \otimes_A N \neq 0$.

One of the most important examples of a flat $A$-module, for $A$ a commutative ring, is the ring $A$ localized at some multiplicative set $S$. However, as soon as $S$ contains some element $s$ which is not invertible, so that $A_S \neq A$, this is not faithfully flat. For instance $A/sA$ becomes 0 under localization as well as every module which is torsion with respect to $S$.

A simple sufficient (but not necessary) condition for a module $M$ to be faithfully flat is that $M$ is a free (or projective and faithful, see Definition 5.3.1) $A$ module.

Another important case, for $A$ commutative, is the following

EXAMPLE 5.1.2. Given elements $f_1, \ldots, f_k \in A$ such that $A = \sum_i Af_i$, the embedding $A \subset \bigoplus_{i=1}^k A[f_i^{-1}]$ is faithfully flat.

This has the following geometric meaning. $U_i := \mathrm{Spec}(A[f_i^{-1}])$ is an open set of $\mathrm{Spec}(A)$, and the condition $A = \sum_i Af_i$ means that the $U_i$ form an *open covering* of $\mathrm{Spec}(A)$. Finally the spectrum of $\bigoplus_{i=1}^k A[f_i^{-1}]$ is the disjoint union of the open sets $U_i$; this is called a *Zariski covering*. A useful notion is that of *étale covering* that is an étale map $i : A \to B$ (cf. Definition 1.6.3) such that the corresponding continuous map of their spectra $i^* : \mathrm{Spec}(B) \to \mathrm{Spec}(A)$ is surjective.

In general a flat morphism $i : A \to B$ of commutative rings is faithfully flat if and only if for every maximal ideal $m$ of $A$ we have $m = i^{-1}(B \cdot i(m))$, (cf. [**Mat80**, Theorem 3, p.28]), in particular the map $\mathrm{Spec}(B) \to \mathrm{Spec}(A)$ is surjective [**Gro60**]):

Let $\phi : A \to B$ be a homomorphism of commutative rings.

PROPOSITION 5.1.3. *The following properties are equivalent.*

(1) *$B$ is faithfully flat over $A$.*
(2) *$\phi$ is injective, and the $A$-module $B/\phi(A)$ is flat.*
(3) *The $A$-module $B$ is flat, and for every $A$-module $M$ the morphism $m \mapsto 1 \otimes m$ of $M$ in $B \otimes_A M$ is injective.*
(4) *The $A$-module $B$ is flat, and for every ideal $\mathfrak{a}$ of $A$ we have $\mathfrak{a} = \phi^{-1}(\mathfrak{a}B)$.*
(5) *The $A$-module $B$ is flat, and for every maximal ideal $\mathfrak{m}$ of $A$ there is a maximal ideal $\mathfrak{n}$ of $B$ so that $\phi^{-1}(\mathfrak{n}) = \mathfrak{m}$.*

PROOF. Let us sketch the proof of (5). If $B$ is flat, then it is faithfully flat if and only if for every nonzero $A$-module $M$ we have $B \otimes_A M \neq 0$. So if there is a module $M$ with $B \otimes_A M = 0$, by flatness we have also $B \otimes_A N = 0$ and $B \otimes_A M/N = 0$ for all submodules $N$. In particular taking $m \in M$, $m \neq 0$, the submodule $Am = A/I$ for some ideal $I$, and saying that $B \otimes_A A/I = 0$ is the same as saying that $BI = B$. Finally if $\mathfrak{m} \supset I$ is a maximal ideal, $B \otimes_A A/I = 0$ implies $B \otimes_A A/\mathfrak{m} = 0$. The claim easily follows. □

A further example is when $A$ is a Noetherian local ring and we take for $\bar{A}$ its completion (cf. [**Mat89**, Theorem 8.14 p. 62])

In this respect one simple way of defining Azumaya algebras is the following. We assume that $A$ is a commutative ring without nontrivial idempotents, which means that its spectrum is connected.

DEFINITION 5.1.4. An algebra $R$ over a commutative ring $A$ with connected spectrum is an Azumaya algebra if there exists a commutative $A$-algebra $B$ which is faithfully flat over $A$ and an integer $h$ such that

$$B \otimes_A R = M_h(B).$$

For instance if $A = F$ is a field, this immediately implies that Azumaya algebras over $F$ are the finite-dimensional central simple algebras.

If the spectrum is not connected, one may have different matrix algebras on different connected components of the spectrum.

In fact we will give a more abstract definition of Azumaya algebra, Definition 5.4.17, for which the previous statement is Corollary 5.4.38. Moreover in PI theory one is interested only in Azumaya algebras of constant rank (cf. Proposition 5.3.15) $h^2$ for some $h$.

The final characterization via polynomial identities of constant rank $h^2$ Azumaya algebras is Artin's Theorem 10.3.2. One of the most important properties that they share with matrix algebras is the fact (which will be proved in Corollary 5.4.29):

PROPOSITION 5.1.5. *If $S$ is an Azumaya algebra, the ideals of $S$ are in 1–1 correspondence with the ideals of its center $A$ by*

$$I \subset S \mapsto I \cap A, \ J \subset A \mapsto JS,$$

*and $S/J$ is an Azumaya algebra over $A/I$. In particular if $A$ is a field, $S$ is a central simple algebra over $A$.*

If $A$ is the coordinate ring of an affine algebraic variety $V$ over an algebraically closed field $K$, that is if $A$ is finitely generated over $K$ with no nilpotent elements, then $V$ corresponds to the set of maximal ideals $\mathfrak{m}$ of $A$, and we have

$$A/\mathfrak{m} = K, \quad S/\mathfrak{m}S = M_k(K).$$

In geometric terms one interprets $S$ as *the sections of a bundle of matrix algebras over $V$*; see §5.1.1.1 on page 132.

The relevance for PI theory comes from two sources.

First there is a precise connection with Definition 3.4.6 of the universal $A$-algebra $A_n(R)$ giving the *universal $n$-dimensional representation* of formula (3.52),

$$R \xrightarrow{j_R} M_n(A_n(R)).$$

It will be proved in Corollary 10.4.3 on page 283.

THEOREM 5.1.6. *$R$ is a rank $n^2$ Azumaya algebra over its center $A$ if and only if the universal $A$-algebra $A_n(R)$ is faithfully flat over $A$ and the universal $n$-dimensional representation induces an isomorphism*

$$R \otimes_A A_n(R) \simeq M_n(A_n(R)).$$

The second main source is more in the style of noncommutative algebra and it is M. Artin's Theorem 10.3.2 for algebras (generalized by Procesi for rings), which characterizes a rank $k^2$ Azumaya algebra $S$ as a ring $S$ which satisfies all polynomial identities of $k \times k$ matrices and such that, for every proper ideal $I$ of $S$, the ring $S/I$ does NOT satisfy all polynomial identities of $k-1 \times k-1$ matrices.

5.1.0.2. *Quaternion algebras.* The classical Hamilton quaternion $\mathbb{H}$ is the four-dimensional algebra over the real numbers $\mathbb{R}$ with basis $1, \mathtt{i}, \mathtt{j}, \mathtt{k}$ and multiplication table deduced from

$$\mathtt{i}^2 = \mathtt{j}^2 = \mathtt{k}^2 = -1, \ \mathtt{i} \cdot \mathtt{j} = -\mathtt{j} \cdot \mathtt{i} = \mathtt{k}.$$

Setting the *conjugate* $\overline{\mathtt{q}}$ of a quaternion $\mathtt{q} = a + b\mathtt{i} + c\mathtt{j} + d\mathtt{k}$,

$$\overline{\mathtt{q}} = a - b\mathtt{i} - c\mathtt{j} - d\,\mathtt{k}, \tag{5.3}$$

one has that the map $\mathtt{q} \mapsto \overline{\mathtt{q}}$ is an involutive antihomomorphism with the remarkable property that

$$N(\mathtt{q}) := \mathtt{q} \cdot \overline{\mathtt{q}} = a^2 + b^2 + c^2 + d^2 \in \mathbb{R},$$

which implies that, as soon as $\mathtt{q} \neq 0$, we have $N(\mathtt{q}) \neq 0$ so that $\mathtt{q}$ has an inverse, namely $\overline{\mathtt{q}} \cdot N(\mathtt{q})^{-1}$.

In general given a field $F$ and two elements $\alpha, \beta \in F$, one can construct the four-dimensional algebra $F\lceil \alpha, \beta \rceil$ over $F$ with basis $1, \mathtt{i}, \mathtt{j}, \mathtt{k}$ and multiplication table deduced from

$$\mathtt{i}^2 = \alpha, \mathtt{j}^2 = \beta, \mathtt{i} \cdot \mathtt{j} = -\mathtt{j} \cdot \mathtt{i} = \mathtt{k} \implies \mathtt{k}^2 = -\alpha\beta.$$

If the characteristic of $F$ is different from 2 this defines a noncommutative algebra. As for ordinary quaternions, one has conjugation by formula (5.3), and now the norm is

$$N(\mathtt{q}) := \mathtt{q} \cdot \overline{\mathtt{q}} = a^2 - \alpha b^2 - \beta c^2 + \alpha\beta d^2. \tag{5.4}$$

In particular, if the quadratic form of formula (5.4) has no isotropic vectors (i.e., nonzero vectors with norm 0, so in particular $\alpha\beta \neq 0$), again $F\lceil \alpha, \beta \rceil$ is a division ring still called a *quaternion algebra* with center equal to $F$. In fact it can be proved that all four-dimensional division algebras over $F$ (of characteristic $\neq 2$) have this form (a classical reference is for instance Albert [**Alb39**]).

In this case, by the standard theory one has

THEOREM 5.1.7. *The quadratic extension $F(\mathtt{i})$ of $F$ is a maximal subfield of $F\lceil \alpha, \beta \rceil$ which* splits *the algebra, that is,*

$$F\lceil \alpha, \beta \rceil \otimes_F F(\mathtt{i}) \simeq M_2(F(\mathtt{i})). \tag{5.5}$$

In particular $F\lceil \alpha, \beta \rceil$ satisfies all polynomial identities of $2 \times 2$ matrices.

Let us recall how this isomorphism is defined. One can view $F\lceil \alpha, \beta \rceil$ as a two-dimensional vector space over $F(\mathtt{i})$ with basis $1$, $\mathtt{j}$ (and $\mathtt{k} = -\mathtt{j} \cdot \mathtt{i}$) by right multiplication. Then left multiplication by $F\lceil \alpha, \beta \rceil$ identifies this as matrices, namely

$$1 \mapsto \begin{vmatrix} 1 & 0 \\ 0 & 1 \end{vmatrix}, \ \mathtt{i} \mapsto \begin{vmatrix} \mathtt{i} & 0 \\ 0 & -\mathtt{i} \end{vmatrix}, \ \mathtt{j} \mapsto \begin{vmatrix} 0 & \beta \\ 1 & 0 \end{vmatrix}, \ \mathtt{k} \mapsto \begin{vmatrix} 0 & \beta\mathtt{i} \\ -\mathtt{i} & 0 \end{vmatrix}. \tag{5.6}$$

If the formula (5.4) has an isotropic vector, then of course $F\lceil \alpha, \beta \rceil$ has zero divisors and is not a division algebra. It then may or may not be semisimple. If it is semisimple, since it is noncommutative and of dimension 4, by structure theory it must be isomorphic to the ring of $2 \times 2$ matrices over $F$.

A necessary and sufficient criterion for $F\lceil\alpha,\beta\rceil$ to be semisimple is the condition that the trace form $\mathrm{tr}(x \cdot y),\ x, y \in F\lceil\alpha,\beta\rceil$, be nondegenerate, where by tr we may take the map

$$\mathtt{tr}(\mathtt{q}) := \mathtt{q} + \overline{\mathtt{q}} = 2a.$$

By formula (5.6) we see that this is the trace as the corresponding matrices and $\frac{1}{2}$ the trace of the representation by $4 \times 4$ matrices by left multiplication. We see directly from the previous formulas that the matrix of the trace form in the basis $1, \mathtt{i}, \mathtt{j}, \mathtt{k}$ is

$$2\begin{vmatrix} 1 & 0 & 0 & 0 \\ 0 & \alpha & 0 & 0 \\ 0 & 0 & \beta & 0 \\ 0 & 0 & 0 & -\alpha\beta \end{vmatrix}$$

with determinant $-16(\alpha\beta)^2$. We deduce, since we are assuming that $F$ has characteristic not 2, that $F\lceil\alpha,\beta\rceil$ is semisimple; that is it is either a division algebra or $M_2(F)$, if and only if $\alpha\beta \neq 0$. We leave it to the reader to compute its radical in the case in which one or both $\alpha, \beta$ equals 0.

*Generic quaternion algebra.* A way of forming a *generic* quaternion algebra is thus the following. Starting from a field $K$ of characteristic $\neq 2$, let $F = K(x, y)$ be the field of rational functions and consider the algebra $F\lceil x, y\rceil$. We claim that it is a division algebra,[1] that is

LEMMA 5.1.8. *$a^2 - xb^2 - yc^2 + xyd^2 = 0$ with $a, b, c, d \in F$ implies that all these elements vanish.*

PROOF. By multiplying for some common denominator, it is clearly enough to prove this statement when $a, b, c, d \in K[x, y]$ are polynomials. In this case assume that $a^2 - xb^2 - yc^2 + xyd^2 = 0$ and set $x = 0$ getting $a(0, y)^2 - y\,c(0, y)^2 = 0$. Clearly this identity holds only if $a(0, y) = c(0, y) = 0$, that is, if both $a, c$ are divisible by $x$. In a similar way $a, b$ are both divisible by $y$. Setting $a = xy\bar{a},\ b = y\bar{b},\ c = x\bar{c}$, we deduce that

$$xy\bar{a}^2 - y\bar{b}^2 - x\bar{c}^2 + d^2 = 0.$$

By induction on the sum of the degrees of the elements $a, b, c, d$, we obtain the claim. □

We make now the main remark: If in the algebra $F\lceil x, y\rceil$ we take only the linear combinations of the basis $1, \mathtt{i}, \mathtt{j}, \mathtt{k}$ by elements $a, b, c, d \in K[x, y]$, i.e., polynomials, we still have an algebra $R$, free module over the ring $K[x, y]$. Moreover, if we invert the element $xy$ and thus consider $a, b, c, d \in K[x^{\pm 1}, y^{\pm 1}]$,

$$S := \{a + b\mathtt{i} + c\mathtt{j} + d\mathtt{k} \mid a, b, c, d \in K[x^{\pm 1}, y^{\pm 1}]\}, \tag{5.7}$$

we have, using Cramer's rule, an algebra $S$ with a special property:

THEOREM 5.1.9. *The isomorphism of formula* (5.5)

$$F\lceil x, y\rceil \otimes_F F(\mathtt{i}) \simeq M_2(F(\mathtt{i}))$$

*restricts to an isomorphism*

$$S \otimes_{K[x^{\pm 1}, y^{\pm 1}]} K[x^{\pm 1}, y^{\pm 1}][\mathtt{i}] \simeq M_2(K[x^{\pm 1}, y^{\pm 1}][\mathtt{i}]).$$

[1]This is true also in characteristic is 2, but then this is the field of rational functions in $\mathtt{i}, \mathtt{j}$.

One way of understanding this is the following.

REMARK 5.1.10. Given a commutative ring $A$, the trace on $C := M_n(A)$ induces a symmetric bilinear form $\operatorname{tr}(ab)$ whose determinant on the basis $e_{i,j}$ has determinant $\pm 1$. Thus $n^2$ elements $c_1, \dots, c_{n^2}$ of $C$ form a basis over $A$ if and only if the determinant of the matrix $\operatorname{tr}(c_i c_j)$ is invertible in $A$.

Then one easily sees that the matrix of the trace form on the basis $1, \mathrm{i}, \mathrm{j}, \mathrm{k}$ is
$$D = 4\begin{vmatrix} 1 & 0 & 0 & 0 \\ 0 & x & 0 & 0 \\ 0 & 0 & y & 0 \\ 0 & 0 & 0 & -xy \end{vmatrix}$$
with determinant $-2^8(xy)^2$. Hence for this to be a basis, we need $-2^8(xy)^2$ invertible.

One can see a consequence of this. For simplicity let us assume that $K$ is algebraically closed. Then a maximal ideal $\mathfrak{m}$ of $K[x^{\pm 1}, y^{\pm 1}]$ corresponds to a homomorphism $x \mapsto \alpha \in K^*$, $y \mapsto \beta \in K^*$ and $S/\mathfrak{m}S \simeq K\lceil \alpha, \beta \rceil = M_2(K)$. One can in fact prove (we leave it as an exercise) that all maximal ideals of $S$ are of the form $\mathfrak{m}S$.

Thus we see

COROLLARY 5.1.11. *The algebra $S$ of formula* (5.7) *is a rank* 4 *Azumaya algebra over its center $K[x^{\pm 1}, y^{\pm 1}]$.*

In fact a more careful analysis of what we have done would show that the same statement, as in Corollary 5.1.11, holds even if $K$ is only assumed to be a commutative ring where 2 is invertible.

REMARK 5.1.12. The fact that we have inverted $xy$ is essential. If we take just $R := \{a + b\mathtt{i} + c\mathtt{j} + d\mathtt{k} \mid a, b, c, d \in K[x, y]\}$, this is not an Azumaya algebra.

The reader can verify, using Remark 5.1.10, that the natural injective map
$$R \otimes_{K[x,y]} K[x, y][\mathtt{i}] \to M_2(K[x, y][\mathtt{i}])$$
is not surjective.

If we take the quotient $\bar{R}$ of $R$ by the ideal generated by $x$ (or similarly by the ideal generated by $y$), we have

$$\mathrm{i}^2 = 0,\ \mathrm{j}^2 = y,\ \mathrm{i}\mathrm{j} = \mathrm{k} = -\mathrm{j}\mathrm{i} \implies \mathrm{k}^2 = 0. \tag{5.8}$$

This algebra $\bar{R}$ is not semisimple, the elements $\mathrm{i}$, $\mathrm{k}$ span the nilpotent radical $N$ and the quotient $\bar{R}/N$ is commutative, thus as expected when the discriminant is set to 0, the algebra does not behave anymore as an Azumaya algebra, in fact modulo the ideal generated by the elements $\mathrm{i}$, $\mathrm{j}$, it is the field $F$, in particular it does not satisfy the conditions of Artin's theorem. An even more instructive example is in §14.3.2 for generic matrices.

**5.1.1. Irreducible representations.** In order to motivate our next discussion, we should make a short geometric digression. Unless otherwise specified when we speak of an *open set* in a variety or a scheme, we mean in the Zariski topology.

Let $F$ be an algebraically closed field.

Consider the set of $n$-dimensional representations of an $F$-algebra $R$, that is the set $\hom_F(R, M_n(F))$ of all $F$-linear homomorphisms $R \to M_n(F)$. For simplicity assume that $R$ is a finitely generated $F$-algebra.

By the discussion performed in §3.4 (cf. Definition 3.4.6 and Remark 3.4.7) the $n$-dimensional representations of $R$ are parametrized by the commutative $F$-algebra $F_n(R)$ which is also a finitely generated commutative $F$-algebra.

We think, using the geometric language borrowed from the theory of schemes, of an $n$-dimensional representation $\phi$ as a *point* of a *variety*, or rather as an *affine scheme* with coordinate ring $F_n(R)$. In this language, given $f \in F_n(R)$ the element $\bar{\phi}(f)$ is the *evaluation of $f$ at the point $\bar{\phi}$*.

Notice that $F_n(R)$ is not necessarily *reduced*, that is it may contain a nonzero nilpotent ideal. The simplest example, left to the reader as exercise, is when $R = F[x]/(x^2)$.

We need now to stress the fact that, over this space, the projective linear group $\mathrm{PGL}(n, F)$ acts by conjugation, and then by definition we have

REMARK 5.1.13. Two representations $\phi_1, \phi_2 : R \to M_n(F)$ are isomorphic, if and only if they are in the same orbit under the action of the group $\mathrm{PGL}(n, F)$.

By Wedderburn's theorem (Theorem 1.1.14) we have

PROPOSITION 5.1.14. *The set of irreducible representations of $R$ with values in the algebra $M_n(F)$ coincides with the set of all homomorphisms $\phi : R \to M_n(F) \to 0$ which are surjective.*

A map $\phi : R \to M_n(F)$ is surjective if and only if there exist $n^2$ elements $a_i$, $i = 1, \dots, n^2$ of $R$ which map to a basis of $M_n(F)$ over $F$. Since the symmetric bilinear form on $M_n(F)$ given by $\mathrm{tr}(ab)$ is clearly nondegenerate (cf. Theorem 1.3.15 or Remark 5.1.10), given $n^2$ elements $a_i$, $i = 1, \dots, n^2$ of $R$, they map to a basis of $M_n(F)$ if and only if the discriminant $\det(\mathrm{tr}(\phi(a_i a_j))) = \det(\mathrm{tr}(\phi(a_i)\phi(a_j)))$ is nonzero; see Definition 1.3.16.

By the universal property we have the following diagram.

$$\begin{array}{ccc} R & \xrightarrow{j_R} & M_n(F_n(R)) \\ & \underset{\phi}{\searrow} & \downarrow{\scriptstyle M_n(\bar{\phi})} \\ & & M_n(F) \end{array} \tag{5.9}$$

Then we see that

$$\det(\mathrm{tr}(\phi(a_i)\phi(a_j))) = \bar{\phi}(\det(\mathrm{tr}(j_R(a_i a_j)))).$$

Thus in this scheme the set of points in which the element $\det(\mathrm{tr}(j_R(a_i a_j)))$ is nonzero is open by definition. Hence the condition for a homomorphism to be surjective is open in this variety of all representations, since the space of surjective homomorphisms is covered by the open sets in which $n^2$ prescribed elements map to a basis (but it could be empty).

The basic example is when $R = F\langle x_1, \dots, x_m \rangle$ is the free algebra in $m$ variables $x_i$ with coefficients in $F$, then a representation is just an $m$-tuple of $n \times n$ matrices, that is a point in the $mn^2$-dimensional affine space $M_n(F)^m$.

In this case the ring $F_n(R) = F[\xi^t_{i,j}]$, $t = 1, \dots m$, $i, j = 1, \dots n$ is the ring of polynomial functions on the space $M_n(F)^m$.

Thus by Proposition 5.1.14, the irreducible representations form the open set $\mathrm{Irr}(m, n) \subset M_n(F)^m$ of $m$-tuples which generate $M_n(F)$ as a $F$-algebra.

PROPOSITION 5.1.15. *If $L$ is a field with at least $n$ elements there are two matrices $a, b \in M_n(L)$ which generate $M_n(L)$ as an $L$-algebra.*

PROOF. Take a diagonal matrix $a$ with distinct entries $d_i$ and the matrix $c$ of the cyclic permutation $e_i \mapsto e_{i+1}$, $i < n$, $e_n \mapsto e_1$ of the corresponding basis. We claim that they generate the full algebra of matrices. In fact we claim that the elements $a^i c^j$, $i, j = 0, 1, \dots, n-1$, form a basis. For this compute the discriminant, i.e., the determinant of the $n^2 \times n^2$ matrix $\operatorname{tr}(a^i c^j a^h c^k)$.

We have that $a^i c^j a^h c^k = a^i c^j a^h c^{-j} c^{j+k}$ where $a^i c^j a^h c^{-j}$ is a diagonal matrix. This has trace 0 unless $j + k$ is 0 or $n$. We see that this matrix is block diagonal with blocks the Vandermonde matrix in the elements $d_i$, hence the claim. □

COROLLARY 5.1.16. *If $n > 1$, the set $\operatorname{Irr}(m, n)$ is nonempty as soon as $m \geq 2$.*

PROOF. This follows from Proposition 5.1.15. □

As already remarked, if an $m$-tuple of matrices generates $M_n(F)$ as an $F$-algebra, there are $n^2$ monomials $M_1, \dots, M_{n^2}$ in these elements which form a basis of $M_n(F)$ as an $F$ vector space. This happens if and only if the determinant of the $n^2 \times n^2$ matrix with entries $\operatorname{Tr}(M_i M_j)$, the *discriminant* of the proposed basis, is different from 0. The set of $m$-tuples for which a particular discriminant is different from 0 is by definition an affine open set[2] of $M_n(F)^m$, and by elementary affine geometry the entire set $\operatorname{Irr}(m, n)$ is covered by a finite number of these affine open (in the Zariski topology) sets. In fact given any list of $m$ matrices, it is easy to see that if these matrices generate the algebra $M_n(F)$, then the monomials of degree $\leq n^2$ in these elements already span $M_n(F)$, hence the possible bases given by monomials have to be chosen among monomials of degree $\leq n^2$. In fact there are better estimates for this although they are probably not optimal **[Pap97]**.

When we work with representations of an algebra $R = F\langle a_1, \dots, a_k \rangle$, finitely generated by $m$ elements, the representations of dimension $n$ form a *subvariety* of $M_n(F)^m$ which may or may not intersect the open set of irreducible representations.

In fact the algebra $R = F\langle a_1, \dots, a_k \rangle$ is the quotient $R = F\langle x_1, \dots, x_k \rangle / I$ of the free algebra modulo an ideal of *relations*, the condition that $m$ matrices $\xi_1, \dots, \xi_m$ be the image of the elements $a_i$ under a homomorphism $R \to M_n(F)$ is that for all polynomial relations $f(x_1, \dots, x_n) \in I$ between the generators $a_i$, the matrix $f(\xi_1, \dots, \xi_n) = 0$. This is really what we have already seen when defining $F_n(R)$, and this condition, for a given $f$, is expressed by a system of $n^2$ polynomial equations in the coordinates $\xi_{i,j}^t$ of the matrices $\xi_i$.

When a representation is irreducible, it is clear that, under the action of the projective linear group, it has a trivial stabilizer; this is just a consequence of Schur's lemma. It is in fact a basic property, which we are going to illustrate next, that the set of $n$-dimensional irreducible representations is in fact the total space of a principal bundle under the projective group.

5.1.1.1. *A digression on bundles.* In topology one speaks of a *trivial principal bundle* when one has a continuous action of a topological group $G$ on a space $X$ which can be *trivialized*. This means that one has a homeomorphism $X = G \times Y$

[2] An *affine open set* of affine space $F^m$ is by definition the open set where a single polynomial $f(x_1, \dots, x_m)$ is different from 0.

of topological spaces, and the action is just by multiplication on the left, that is $g(h,y) = (gh,y)$.

In general we have a *topological principal bundle* when one has a continuous action of a topological group $G$ on a space $X$ which is *locally trivial*. This means that one can construct the *orbit space* $\pi : X \to X//G$, where $X//G$ is the set of all orbits of $G$ on $X$, with its quotient topology, and there is a covering $U_i$ of $X//G$ by open sets with the property that the bundle $\pi_i^{-1}(U_i) \to U_i$ is a trivial bundle.

If the group $G$ acts continuously on some space $F$, one can *change the fiber* and construct the space $(F \times X)$ of the $G$ orbits of $F \times X$ which now maps to $X//G$ with fibers homeomorphic to $F$. In particular if $G$ is the projective linear group and $F$ is the algebra of matrices, one has in this way the notion of a *matrix bundle*.

In algebraic geometry the notion needs some refinement, since the homeomorphism need not be algebraic. The simplest example of the problems involved is the map $z \mapsto w = z^2$ of $\mathbb{C}^* \to \mathbb{C}^*$. This is a principal bundle over the group $\pm 1$ but in order to make it trivial on an open set, one has to choose a square root $\sqrt{w}$ of $w$. This can be done only by cutting $\mathbb{C}^*$, and it is not a construction in the domain of algebraic geometry.

The problem has been solved by Grothendieck who generalizes the idea of topology by defining, rather than the open sets, the *coverings* which are special maps in the category of schemes. In our special case they are faithfully flat maps of commutative rings possibly with further conditions as (in the more geometric, but in a way abstract) notion of *étale topology*. We will not discuss this in any detail, and we give as reference **[Gro95]**.

The nature of this bundle $\mathrm{Irr}(m,n)$ is to be locally trivial only in the étale topology and not in the Zariski topology. The analysis of this fact leads to the theory of Azumaya algebras.

The idea is that the projective linear group is also the automorphism group of matrices (cf. §3.4.2) and so there is a dictionary which transforms facts on principal $\mathrm{PGL}(n)$ bundles into matrix bundles. In turn, over an affine scheme the global sections of a matrix bundle form an *Azumaya algebra*.

It turns out that, to some extent, Azumaya algebras are very close to commutative algebras, and they are the right objects to study deformation theory.

5.1.1.2. *Flatness.* We recall from Definition 5.1.1 the two notions of *flat* and *faithfully flat* extensions.

Of particular interest are maps of commutative rings $A \to B$, where $B$ is faithfully flat over $A$, since in this case there is the method of *faithfully flat descent* which allows us to recover modules over $A$ by modules over $B$ and *descent data*. This is a deep geometric idea of Grothendieck (a basis of his theory of *schemes*), who understood how arguments of topology could be carried out for rings by thinking of a faithfully flat map as a well-behaved covering of the spectrum.

A typical argument could be to verify that a map of $A$-modules $f : M \to N$ is injective, surjective, or an isomorphism by verifying the same statement for the map $1_B \otimes f : B \otimes_A M \to B \otimes_A N$.

A basic example comes from localization. If $A$ is a commutative ring and $S$ is a multiplicative set, the localized ring $A_S$ is flat over $A$ and the spectrum of $A_S$ is the open set of the spectrum of $A$ of prime ideals $P$ such that $P \cap S = \emptyset$. If we take several localizations $B = \bigoplus_i A_{S_i}$, then one can see that $B$ is faithfully flat over $A$ if and only if the open sets $\mathrm{Spec}(A_{S_i})$ form a covering of $\mathrm{Spec}(A)$. For instance if we take some elements $f_i$ which generate $A$ as ideal, we can take as $S_i = \{f_i^k,\ k \in \mathbb{N}\}$ and have a typical covering used in algebraic geometry. Many arguments can be done by this localization principle but some require topologies which are subtler than the Zariski topology.

In the next section we discuss some basic aspects of faithfully flat descent, which, for our purposes, is a method to deduce properties of Azumaya algebras from analogous properties of matrix algebras, which in general can be checked quite easily using matrix units calculations.

## 5.2. Faithfully flat descent

*This section is truly a complement and is not really used in the book.*

**5.2.1. Grothendieck topologies.** The ideas behind this algebraic construction come in fact from Galois theory and topology. In Galois theory if we have a finite Galois extension $F \subset H$ with $\dim_F H = m$ and a Galois group $G$, one sees that $H \otimes_F H = H \oplus \cdots \oplus H = H^{\oplus m}$ and $G$ permutes the $m$ factors in a simply transitive way.

Moreover the fact that $F = H^G$ is the field of $G$ invariants translates into the property that $F = \{h \in H \mid h \otimes 1 = 1 \otimes h \in H \otimes_F H\}$.

The analogue in topology of a finite Galois extension is a finite covering $Y \to X$ with covering group $G$. Then the previous formulation becomes the fact that if one forms the *fiber product* $Y \times_X Y$ this becomes a trivial covering over $Y$.

In algebra one can also define Galois extensions, but it is better to take a more abstract point of view, which in the end is connected with the idea of *Grothendieck topologies*, and replace covering by faithfully flat maps.

Given a map of rings $i : A \to B$, we may consider $B$ as right or left $A$-module by $ba := bi(a),\ ab := i(a)b,\ \forall a \in A,\ b \in B$. Thus we have that every left or right $A$-module $M$ gives rise to a left or right $B$-module ${}_BM = B \otimes_A M$, resp., $M_B = M \otimes_A B$. We are only interested in the case when $A, B$ are commutative and then in the case of $B$ faithfully flat as an $A$-module.

According to Grothendieck, one can axiomatize a categorical approach to topology leading to the axioms of a *site*, which is useful in the theory of schemes, or in our case just in *affine schemes*, i.e., commutative rings (actually the category opposite to commutative rings). In his approach, a topology is given by specifying for some object $X$ some special sets of maps $X_i \to X$ which play the role of *coverings* and which should satisfy some simple axioms.

In commutative rings a covering is for the spectrum $X = \mathrm{Spec}(A)$ of a commutative ring, so it has to be given by suitable maps $A \to A_i$ of commutative rings. In the Zariski topology these maps are just localizations at multiplicative sets. There are other topologies, such as the étale topology, that one can use in the same way.

The basic example to have in mind, for the étale topology, is the following. Take a commutative ring $A$, a polynomial in 1 variable $f(x) \in A[x]$, and its derivative $f'(x)$. Then an elementary étale extension of $A$ is obtained by first constructing $A[x]/(f(x))$ (which means to add a root of the equation $f(x) = 0$) and then localizing by inverting the image in $A[x]/(f(x))$ of the element $f'(x)$. This means intuitively that we want to add a *simple root* and gives an étale (open set). Usually this is not a covering, since we had to invert the derivative. When one has a covering of a space and has to define some object *locally*, the strategy is to define it on all open sets and verify that the definition is compatible on the intersections. When working with commutative rings, this can be discussed through the following complex.

Let $i : A \to B$ be a homomorphism of commutative algebras, which in particular makes $B$ into an $A$-module.

For all integers $k$ define $B^{(k)} := B \otimes_A B \otimes_A \otimes \cdots \otimes_A B$ the tensor product of $k$ copies of $B$. Define next $k+1$ morphisms $\epsilon_i : B^{(k)} \to B^{(k+1)}$, $i = 1, \cdots, k+1$, by

$$\epsilon_i(b_1 \otimes b_2 \otimes \cdots \otimes b_k) := b_1 \otimes b_2 \otimes \cdots \otimes b_{i-1} \otimes 1 \otimes b_i \otimes \cdots \otimes b_k$$

by inserting 1 between the $(i-1)$-th and the $i$th tensor factor. If $i < j$, we have $\epsilon_j \circ \epsilon_i = \epsilon_i \circ \epsilon_{j-1}$. This formalism comes of course from topology and in particular from the idea of *faces* and *simplicial objects*.

We then define a complex $\mathcal{G}(A, B)$

$$0 \longrightarrow A \xrightarrow{\ i\ } B^{(1)} \xrightarrow{\ d_1\ } B^{(2)} \ldots \longrightarrow B^{(k)} \xrightarrow{\ d_k\ } B^{(k+1)} \ldots \tag{5.10}$$

by setting

$$d_k : B^{(k)} \to B^{(k+1)}, \ d_k = -\sum_{i=1}^{k+1} (-1)^i \epsilon_i.$$

We clearly have $d_1 \circ i = 0$, so let us first verify that $d_{k+1} \circ d_k = 0$, i.e., this is a complex.

$$\begin{aligned}\sum_{i=1}^{k+2} (-1)^i \epsilon_i \circ \sum_{i=1}^{k+1} (-1)^i \epsilon_i &= \sum_{i<j\leq k+2} (-1)^{i+j} \epsilon_j \circ \epsilon_i + \sum_{j\leq i\leq k+1} (-1)^{i+j} \epsilon_j \circ \epsilon_i \\ &= \sum_{i<j\leq k+2} (-1)^{i+j} \epsilon_i \circ \epsilon_{j-1} + \sum_{j\leq i\leq k+1} (-1)^{i+j} \epsilon_j \circ \epsilon_i \\ &= -\sum_{i\leq j\leq k+1} (-1)^{i+j} \epsilon_i \circ \epsilon_j + \sum_{j\leq i\leq k+1} (-1)^{i+j} \epsilon_j \circ \epsilon_i = 0.\end{aligned}$$

In general this complex is not exact, which in topology reflects the fact that the *simplicial set* has some cohomology, but one can construct some kind of *cone* killing all cohomology. In algebraic terms we have that a standard and simple way to verify that a complex

$$\mathcal{C} : \cdots \xrightarrow{d_{i-1}} C_i \xrightarrow{d_i} C_{i+1} \xrightarrow{d_{i+1}} \cdots$$

is exact is to construct a *contracting homotopy*, that is a map in the opposite direction $s_i : C_{i+1} \to C_i$ such that for all $i$ one has $s_i d_i + d_{i-1} s_{i-1} = 1_{C_i}$.

One then has, if $a \in C_i$ is a cocycle, that is if $d_i(a) = 0$, we have

$$a = (s_i d_i + d_{i-1} s_{i-1})(a) = d_{i-1}(s_{i-1}(a)),$$

which is the condition for exactness.

LEMMA 5.2.1. *After we tensor with $B$, the complex $B \otimes_A \mathcal{G}(A, B)$ is exact, and in fact it has a contracting homotopy.*

PROOF. We define the contracting homotopy $s$ as follows:

(5.11)

$$s_k : B \otimes B^{(k+1)} \to B \otimes B^{(k)}, \ b_0 \otimes b_1 \otimes b_2 \otimes \cdots \otimes b_{k+1} \mapsto b_0 b_1 \otimes b_2 \otimes \cdots \otimes b_{k+1}.$$

Of particular importance is the beginning of this complex

$$0 \longrightarrow A \xrightarrow{\ i\ } B \xrightarrow{\ d\ } B \otimes B, \quad d(b) = b \otimes 1 - 1 \otimes b. \tag{5.12}$$

Then after tensor product with $B$ the contracting homotopy is

$$0 \longleftarrow B \xleftarrow{s_0} B\otimes B \xleftarrow{s_1} B\otimes B\otimes B, \quad s_0(b_0\otimes b_1) = b_0b_1, \tag{5.13}$$

$$\begin{aligned}(i\circ s_0 + s_1\circ d)(b_0\otimes b_1) &= b_0b_1\otimes 1 + s_1(-b_0\otimes b_1\otimes 1 + b_0\otimes 1\otimes b_1)\\ &= b_0b_1\otimes 1 - b_0b_1\otimes 1 + b_0\otimes b_1 = b_0\otimes b_1.\end{aligned}$$

We also set $s_{-1} = 0$, and we have clearly also for $k \geq 2$

$$s_k\circ 1\otimes\epsilon_i = 1\otimes\epsilon_{i-1}\circ s_{k-1}, 1 < i \leq k,\ s_k\circ 1\otimes\epsilon_1 = 1_{B\otimes B^{(k)}}$$

$$s_k(1\otimes d_k) + (1\otimes d_{k-1})s_{k-1} = -[s_k\circ\sum_{i=1}^{k+1}(-1)^i 1\otimes\epsilon_i + \sum_{i=1}^{k}(-1)^i 1\otimes\epsilon_i\circ s_{k-1}]$$

$$= s_k\circ 1\otimes\epsilon_1 - \sum_{i=2}^{k+1}(-1)^i 1\otimes\epsilon_{i-1}\circ s_{k-1} - \sum_{i=1}^{k}(-1)^i 1\otimes\epsilon_i\circ s_{k-1} = 1_{B\otimes B^{(k)}}. \qquad \square$$

From Lemma 5.2.1 and the definition of faithfully flat modules, one has immediately

THEOREM 5.2.2 (Grothendieck). *If $B$ is faithfully flat over $A$, then*

(1) *the complex* (5.10) *is exact.*
(2) *for every $A$-module $M$ the complex $\mathcal{G}(A,B)\otimes_A M$ is exact.*

PROOF. By faithful flatness of $B$ over $A$, it is enough to show that the complex $B\otimes_A(\mathcal{G}(A,B)\otimes_A M) = (B\otimes_A\mathcal{G}(A,B))\otimes_A M$ is exact. This is true since when we take the tensor product of a complex with a contracting homotopy with any module, this still has a contracting homotopy. $\square$

One of the typical uses of descent is this. Let $f : C \to B$ be a homomorphism of commutative rings, then $f$ factors through $A$, that is, there is a homomorphism $\bar{f} : C \to A$ such that $f = i\circ f$, if and only if $\epsilon_1\circ\bar{f} = \epsilon_1\circ f : C \to B\otimes_A B$.

Another typical problem of descent is to understand when, given a module $N$ over $B$, there is a module $M$ over $A$ such that $N = B\otimes_A M$.

For this assume that we have a $B$-module $N$, from $N$ we can construct two $B\otimes_A B$ modules, that is $B\otimes_A N, N\otimes_A B$.

Suppose we have a $B^{\otimes 2}$ homomorphism $\psi : N\otimes_A B \to B\otimes_A N$. Then we can construct three associated homomorphisms,

$$\psi_1 : B\otimes_A N\otimes_A B \to B\otimes_A B\otimes_A N,$$

$$\psi_2 : N\otimes_A B\otimes_A B \to B\otimes_A B\otimes_A N,$$

$$\psi_3 : N\otimes_A B\otimes_A B \to B\otimes_A N\otimes_A B,$$

inserting the identity map $1_B$ in position $1, 2, 3$.

We now define the *category of descent data*

DEFINITION 5.2.3 (Descent data). The category $\mathrm{Mod}_{A\to B}$ has as objects the pairs $(N,\psi)$ where $N$ is a $B$-module and $\psi : N\otimes_A B \to B\otimes_A N$ is an isomorphism such that

$$\psi_2 = \psi_1\circ\psi_3 : N\otimes_A B\otimes_A B \to B\otimes_A N\otimes_A B \to B\otimes_A B\otimes_A N. \tag{5.14}$$

A *morphism* between two objects $\beta : (N, \psi) \to (N', \psi')$ is a $B$-module morphism $\beta : N \to N'$ making the diagram

$$\begin{array}{ccc} N \otimes_A B & \xrightarrow{\psi} & B \otimes_A N \\ {\scriptstyle \beta\otimes 1}\downarrow & & {\scriptstyle 1\otimes\beta}\downarrow \\ N' \otimes_A B & \xrightarrow{\psi'} & B \otimes_A N' \end{array} \tag{5.15}$$

commutative.

Our goal is to show that, when $B$ is faithfully flat over $A$, the category of descent data is isomorphic to the category of $A$-modules.

For this we first can define in general a functor:

DEFINITION 5.2.4. We define a functor $F = F_A^B : \mathrm{Mod}_A \to \mathrm{Mod}_{A\to B}$ by associating to an $A$-module $M$ the $B$-module $N := B \otimes_A M$ and as an isomorphism

$$\psi : B \otimes_A M \otimes_A B \to B \otimes_A B \otimes_A M, \quad \psi(b_0 \otimes m \otimes b_1) = b_0 \otimes b_1 \otimes m. \tag{5.16}$$

To a map $\alpha : M \to M'$ of $A$-modules, one associates the map $1_B \otimes \alpha$.

It is an easy verification that $F$ is indeed a functor from $\mathrm{Mod}_A$ to $\mathrm{Mod}_{A\to B}$.

We have a *left adjoint functor* $G = G_B^A : \mathrm{Mod}_{A\to B} \to \mathrm{Mod}_A$,

$$G((N, \psi)) := \{n \in N \mid 1 \otimes n = \psi(n \otimes 1)\}, \tag{5.17}$$

since we have a natural isomorphism

$$\hom_{\mathrm{Mod_A}}(M, G((N, \psi))) \cong \hom_{\mathrm{Mod}_{A\to B}}(F(M), (N, \psi)). \tag{5.18}$$

In order to understand the next theorem, it is easier to start by verifying the behavior of these functors under *base change*. This means that, given any morphism $j : A \to C$ to a commutative ring $C$, we have by definition of base change a morphism $g \otimes i : C = C \otimes_A A \to C \otimes_A B$. This induces functors

$$I_A^C : \mathrm{Mod}_A \to \mathrm{Mod}_C, \quad M \mapsto C \otimes_A M,$$

$$I_{A\to B}^{C\to C\otimes_A B} : \mathrm{Mod}_{A\to B} \to \mathrm{Mod}_{C\to C\otimes_A B},$$

$$(N, \psi) \mapsto ((C \otimes_A B) \otimes_B N, 1 \otimes \psi) = (C \otimes_A N, 1 \otimes \psi).$$

We have natural isomorphisms

$$I_{A\to B}^{C\to C\otimes_A B} \circ F_A^B \cong F_C^{C\otimes_A B} \circ I_A^C, \qquad I_A^C \circ G_B^A \cong G_C^{C\otimes_A B} \circ I_{A\to B}^{C\to C\otimes_A B}$$

expressed by the following commutative diagrams.

$$\begin{array}{ccc} \mathrm{Mod}_A & \xrightarrow{I_A^C} & \mathrm{Mod}_C \\ {\scriptstyle F_A^B}\downarrow & & {\scriptstyle F_C^{C\otimes_A B}}\downarrow \\ \mathrm{Mod}_{A\to B} & \xrightarrow{I_{A\to B}^{C\to C\otimes_A B}} & \mathrm{Mod}_{C\to C\otimes_A B} \end{array} \qquad \begin{array}{ccc} \mathrm{Mod}_{A\to B} & \xrightarrow{I_{A\to B}^{C\to C\otimes_A B}} & \mathrm{Mod}_{C\to C\otimes_A B} \\ {\scriptstyle G_A^B}\downarrow & & {\scriptstyle G_C^{C\otimes_A B}}\downarrow \\ \mathrm{Mod}_A & \xrightarrow{I_A^C} & \mathrm{Mod}_C \end{array} \tag{5.19}$$

THEOREM 5.2.5 (Descent of modules). *When $B$ is faithfully flat over $A$, the functor $F$ is an equivalence of categories having $G$ as its inverse.*

PROOF. We have by adjoint functors two natural transformations

$$F \circ G \to 1_{\mathrm{Mod}_{A\to B}}, \quad G \circ F \to 1_{\mathrm{Mod}_A},$$

which we need to show are both isomorphisms. Take the exact sequence deduced from formula (5.12) by taking tensor product with $M$:

(5.20)

$$0 \longrightarrow M \xrightarrow{i\otimes 1_M} B \otimes_A M \xrightarrow{d\otimes 1_M} B \otimes B \otimes_A M, \quad d(b) = b \otimes 1 - 1 \otimes b.$$

By formula (5.16) the object $F(M) = (B \otimes_A M, \psi_M)$ with $\psi_M(b_0 \otimes m \otimes b_1) = b_0 \otimes b_1 \otimes m$. Thus $\psi_M(b_0 \otimes m \otimes 1) = \epsilon_2 \otimes 1_M(b_0 \otimes m)$ and

$$G(F(M)) = \{n \in B \otimes_A M \mid \epsilon_1 \otimes 1_M(n) = 1 \otimes n = \psi_M(n) = \epsilon_2 \otimes 1_M((n))\}$$
$$= \ker(d \otimes 1_M) = M.$$

We pass now to $F \circ G \to 1_{\mathrm{Mod}_{A\to B}}$, which is defined as follows. Take an element $(N, \psi)$. By construction $G((N, \psi)) \subset N$ (defined in formula (5.17)) is an $A$ submodule of $N$. Let us denote it by $M$ for simplicity, so we have the $B$-module map $\theta : B \otimes_A M \to N$ given by multiplication.

By construction we have the pair $F(G((N, \psi))) := (B \otimes_A M, \psi')$ where $\psi'$ is given by formula (5.16), $\psi'\psi(b_0 \otimes m \otimes b_1) = b_0 \otimes b_1 \otimes m, \ m \in M)$. So the first thing to do is to show that the map $\theta : (B \otimes_A M, \psi') \to (N, \psi)$ is a morphism in the category $\mathrm{Mod}_{A\to B}$, that is we need to verify the commutativity of the following diagram as in (5.15).

(5.21)
$$\begin{array}{ccc} B \otimes_A M \otimes_A B & \xrightarrow{\psi'} & N \otimes_A B \\ \downarrow{\scriptstyle \theta\otimes 1_B} & & \downarrow{\scriptstyle 1_B\otimes\theta} \\ B \otimes_A B \otimes_A M & \xrightarrow{\psi} & B \otimes_A N \end{array}$$

We have

$$\psi(\theta \otimes 1_B)(b_0 \otimes m \otimes b_1) = \psi(b_0 m \otimes b_1) = \psi((b_0 \otimes b_1)(m \otimes 1))$$
$$= (b_0 \otimes b_1)\psi(m \otimes b_1) \overset{(5.17)}{=} (b_0 \otimes b_1)(1 \otimes m) = b_0 \otimes b_1 m$$
$$= (1_B \otimes \theta)(b_0 \otimes b_1 \otimes m) \overset{(5.16)}{=} (1_B \otimes \theta)\psi'(b_0 \otimes m \otimes b_1).$$

This shows that $\theta$ gives a natural transformation $F \circ G \to 1_{\mathrm{Mod}_{A\to B}}$, which we need to show is an isomorphism if $i : A \to B$ is faithfully flat.

Consider the two homomorphisms

(5.22) $$\alpha, \beta : N \to B \otimes_A N, \ \alpha(n) := 1 \otimes n, \ \beta(n) := \psi(n \otimes 1).$$

By definition $M$ is the kernel of $\alpha - \beta$. Denote by $i_M : M \to N$ the inclusion map, by $\tau : M \otimes_A B \to B \otimes_A M$ the exchange $\tau(m \otimes b) := b \otimes m$, and we have the following diagram with exact rows.

(5.23)
$$\begin{array}{ccccccc} 0 & \longrightarrow & M \otimes_A B & \xrightarrow{i_M\otimes 1_B} & N \otimes_A B & \xrightarrow{(\alpha-\beta)\otimes 1_B} & B \otimes_A N \otimes_A B \\ & & \downarrow{\scriptstyle \theta\tau} & & \downarrow{\scriptstyle \psi} & & \downarrow{\scriptstyle 1_B\otimes\psi} \\ 0 & \longrightarrow & N & \xrightarrow{\alpha_M} & B \otimes_A N & \xrightarrow{(\epsilon_1-\epsilon_2)\otimes 1_N} & B \otimes_A B \otimes_A N \end{array}$$

Since $\psi$ is an isomorphism by assumption, if we prove that this diagram is commutative, we finally deduce that $\theta\tau$ and hence $\theta$ is an isomorphism as desired.

So let us perform this last computation. For the first square we have

$$\alpha_M\theta\tau(m\otimes b) = 1\otimes bm = (1\otimes b)(1\otimes m) = (1\otimes b)\psi(m\otimes 1)$$
$$= \psi((1\otimes b)(m\otimes 1)) = \psi(m\otimes b) = \psi(i_M\otimes 1_B)(m\otimes b).$$

For the second square we have

$$(1_B\otimes\psi)(\alpha\otimes 1_B)(n\otimes b) = 1\otimes\psi(n\otimes b)$$
$$= \epsilon_1\psi(n\otimes b)(1_B\otimes\psi)(\beta\otimes 1_B)(n\otimes b) \overset{(5.22)}{=} \psi_1(\psi(n\otimes 1)\otimes b)$$
$$= \psi_1(\psi_3(n\otimes 1\otimes b)) \overset{(5.14)}{=} \psi_2(n\otimes 1\otimes b) = \epsilon_2\psi(n\otimes b),$$

from which the claim follows. □

## 5.3. Projective modules

In order to motivate the next discussion, let us make a remark on matrices and more generally on the endomorphism ring of a projective module. Recall:

DEFINITION 5.3.1. For a module $P$ over a ring $R$ the following conditions are equivalent and, if they are satisfied, the module is called *a projective module.*

(1) $P$ is a direct summand of a free module $R^I$.
(2) Given a surjective map $q : N \to M \to 0$ of $R$-modules, the induced map $\hom_R(P, N) \to \hom_R(P, M)$, $g \mapsto q\circ g$ is surjective.

PROOF. (1) $\implies$ (2) If we have $R^I = \bigoplus_{i\in I} Re_i = P\oplus Q$, we have a projection $\pi : R^I \to P$, $e_i \mapsto p_i \in P$. Given a map $f : P \to M$, we extend it to a map still called $f$ of $R^I \to M$ by setting it equal to 0 on $Q$. Choose elements $n_i \in N$ such that $q(n_i) = f(e_i)$. We then have a mapping $\hat{f} : R^I = \bigoplus_{i\in I} Re_i \to N$ mapping $\hat{f} : e_i \mapsto n_i$ which lifts $f$. We thus define a map $P \to N$ by restriction of $\hat{f}$ to $P$.

(2) $\implies$ (1) Take a set of linear generators $p_i$ of $P$ as an $R$-module. This defines a surjective map $\pi : R^I \to P$ mapping $e_i \mapsto p_i$. The identity map $1_P$ by assumption lifts to a map $i : P \to R^I$ such that $\pi\circ i = 1_P$. This gives the required splitting. □

REMARK 5.3.2.

(1) Clearly, if $P$ is a finitely generated projective module, then it is a direct summand of a finite-dimensional free module $R^n$, where $n$ is the number of generators of $P$.
(2) If $Q$ is a direct summand of a projective module, then it is projective.
(3) If $I$ is an ideal of $R$ and $P$ is a projective $R$-module such that $IP = 0$, then $P$ is a projective $R/I$-module (this follows for instance by Definition 5.3.1(2). and the fact that for an $R/I$-module $M$ we have $\hom_R(P, M) = \hom_{R/I}(P, M)$).
(4) A projective module is flat.
(5) If $M$ is a projective module over a commutative ring $A$ and $R$ is an $A$-algebra, then $R\otimes_A M$ is projecive over $R$.

PROOF. We leave this as exercise. □

EXAMPLE 5.3.3. Let $R = M_n(A)$ be the ring of $n\times n$ matrices over a ring $A$, which acts on the set $A^n$ by multiplication. Then $A^n$ is a projecitve $R = M_n(A)$-module.

*Hint.* $M_n(A) \cong (A^n)^{\oplus n}$ is decomposed into the direct sum of its left ideals formed by a single *column*, each isomorphic to $A^n$. □

Let $P$ be a module over a ring $R$. We have that $\hom_R(P, R)$ is a right $R$-module, by setting $(fa)(p) := f(p)a$, and we can then define a mapping

$$\mu : \hom_R(P, R) \otimes_R P \to \mathrm{End}_R(P), \quad \mu(f \otimes p)(x) := f(x)p. \tag{5.24}$$

PROPOSITION 5.3.4. *The following conditions are equivalent:*

(1) *$P$ is a finitely generated projective module over $R$.*
(2) *The map $\mu$ is an isomorphism.*
(3) *The map $\mu$ is surjective.*
(4) *The image of the map $\mu$ contains $1_P$.*
(5) *There exist $n$ elements $m_i \in P$ and $n$ linear forms $\phi_i \in \hom_R(P, R)$ such that*

$$x = \sum_{i=1}^{n} \phi_i(x) m_i, \ \forall x \in P. \tag{5.25}$$

*Such a set of $m_i, \phi_j$ is called a set of* projective generators *for the module $P$.*

PROOF. (1) $\implies$ (2) If we have $R^n = \bigoplus_{i=1}^n Re_i = P \oplus Q$, we have a block decomposition

$$\hom_R(R^n, R^n) = M_n(R) = \begin{vmatrix} \hom_R(P, P) & \hom_R(Q, P) \\ \hom_R(P, Q) & \hom_R(Q, Q) \end{vmatrix}.$$

When we consider the map $\mu$ of formula (5.24) for $P = R^n$, we see by direct inspection that it is an isomorphism.

$$\hom_R(P, R) \otimes_R P = R^n \otimes_R R^n \to \mathrm{End}_R(P) = M_n(R).$$

We decompose both sides of this isomorphism in four blocks, and we have that $\mu$ restricts to the isomorphism of the four blocks,

(5.26)

$$\begin{array}{ccc} R^n \otimes \hom_R(R^n, R) & \xrightarrow{\mu} & \hom_R(R^n, R^n) = M_n(R) \\ \cong\downarrow & & \cong\downarrow \\ (P \oplus Q) \otimes_R (\hom_R(P, R) \oplus \hom_R(Q, R)) & \xrightarrow{\mu} & \begin{vmatrix} \hom_R(P, P) & \hom_R(Q, P) \\ \hom_R(P, Q) & \hom_R(Q, Q) \end{vmatrix}. \end{array}$$

(2) $\implies$ (3) $\implies$ (4) is trivial. (4) $\implies$ (5), if there is an element $u = \sum_i m_i \otimes \phi_i$ mapping, via $\mu$, to the identity map $1_P$, then for all $x \in P$ we have $x = \mu(u)x = \sum_i \phi_i(x) m_i$. This is formula (5.25).

(5) $\implies$ (1) Assume we have elements satisfying the property (5). Define a map $j : P \to R^n$, $j(p) = (\phi_1(p), \dots, \phi_n(p))$ and a map $\pi : R^n \to P$ by $\pi(a_1, \dots, a_n) := \sum_i a_i m_i$. We then have $\pi \circ \mu(p) = \sum_i \phi_i(p) m_i = p$, hence the map $\mu$ defines $P$ as direct summand of $R^n$. □

Finally one can remark that, if $P$ is a finitely generated projective $A$-module, $A$ is a commutative ring, and $m_i \in P$ and $\phi_i \in \hom_R(P, R)$ are projective generators, one has

LEMMA 5.3.5. $\mathrm{End}_A(P)$ *is generated as $A$-module by the* elementary *endomorphisms $m_i \otimes \phi_j : r \to \phi_j(r) m_i$.*

PROOF. Let $f : P \to P$ be a module endomorphism. From $r = \sum_{i=1}^{n} \phi_i(r)m_i$, we deduce $f(r) = \sum_{i=1}^{n} \phi_i(r)f(m_i)$. Since the $m_i$ are generators, one can write $f(m_i) = \sum_j \lambda_{i,j}m_j$, $\lambda \in A$ and

$$f(r) = \sum_{i=1}^{n} \sum_{j} \lambda_{i,j}\phi_i(r)m_j,$$

so $f = \sum_{i,j} \lambda_{i,j}m_j \otimes \phi_i$ as desired. □

When we localize $A$ at a prime ideal $\underline{p}$, we have

$$A_{\underline{p}} \otimes_A \mathrm{End}_A(P) = \mathrm{End}_{A_{\underline{p}}}(A_{\underline{p}} \otimes_A P).$$

On a local ring a projective module is free so $A_{\underline{p}} \otimes_A P = A_{\underline{p}}^k$ for some $k$ which in general depends on the prime ideal $\underline{p}$ (but it is a locally constant function on the spectrum). Thus, $A_{\underline{p}} \otimes_A \mathrm{End}_A(P) = M_k(A_{\underline{p}})$ is a matrix algebra, and one can express this by saying that $\mathrm{End}_A(P)$ is a *locally trivial in the Zariski topology* matrix bundle.

5.3.0.1. *Duality and trace.* Let $P$ be any $R$-module, and denote $\Omega := \mathrm{End}_R(P)$, so that $P$ is also an $\Omega$-module. We have that $\hom_R(P, R)$ is a right $\Omega$-module by setting $f\omega(p) := f(\omega p)$ and also a right $R$-module by setting $fr(p) := f(p)r$. Since $\mathrm{End}_R(P) = \Omega$ is an algebra, it is also a two-sided $\Omega$-module, and $\mu$ of formula (5.24) is a two-sided $\Omega$-map.

We define a further map, which plays the role of *trace*:

$$\tau : \hom_R(P, R) \otimes_\Omega P \to R, \quad \tau(f \otimes p) := f(p). \tag{5.27}$$

REMARK 5.3.6. When $P = A^n$ is a free module over a commutative ring $A$, we have an identification $\hom_A(A^n, A) \otimes_A A^n = \mathrm{End}(A^n) = M_n(A)$, and $\tau$ is in fact the trace.

PROPOSITION 5.3.7. *Let $P$ be a finitely generated faithful projective module over a commutative ring $A$.*

(1) *The biduality map $P \to \hom_A(\hom_A(P, A), A)$ is an isomorphism.*
(2) *The map $\mu : \hom_A(P, A) \otimes_A P \to \mathrm{End}_A(P)$ is an isomorphism.*
(3) *The trace map $\tau$ is surjective.*
(4) *The transpose map $\mathrm{End}_A(P) \to \mathrm{End}_A(\hom_A(P, A))$ is an anti-isomorphism.*
(5) $\mathrm{End}_{\mathrm{End}_A(P)}(P) = A$.

PROOF. By the principle of localization, and Propositions 1.6.6 and 3.4.17, it is enough to prove this for all localizations $A_{\underline{m}}$ with $\underline{m}$ a maximal ideal. If $A$ is local, we have that $P \simeq A^k$ for some $k$, so $\hom_A(P, A) \simeq A^k$ and $\mathrm{End}_A(P) \simeq M_k(A)$, and one then verifies directly.

In particular for (4) we have transposition of matrices. □

In fact there are examples when $R$ is noncommutative, in which case Proposition 5.3.7 does not hold.

Let us go back to the general case of an $R$-module $P$.

We also have a map $\psi : \hom_R(P, R) \to \hom_\Omega(P, \Omega)$ which associates to an element $f \in \hom_R(P, R)$ the map

$$\psi(f) \in \hom_\Omega(P, \Omega) \quad \text{given by} \quad \psi(f)(p) = \mu(f \otimes p), \tag{5.28}$$

$$\text{i.e.,} \quad \boxed{\psi(f)(p)(x) = \mu(f \otimes p)(x) = f(x)p = \tau(f \otimes x)p.}$$

We also consider the diagram

$$\begin{array}{ccc} \hom_R(P, R) \otimes_\Omega P & \xrightarrow{\psi\otimes 1} & \hom_\Omega(P, \Omega) \otimes_\Omega P \\ \tau\downarrow & & \mu\downarrow \\ R & \xrightarrow{j} & \hom_\Omega(P, P), \end{array} \tag{5.29}$$

where $j$ is the module action (and $\ker(j) = 0$ means that $P$ is a faithful $R$-module), $\tau$ the trace map, and $\mu$ is as in formula (5.24) (but for $\Omega$ instead of $R$).

Let us first verify that this is commutative. Indeed, given $f \in \hom_R(P, R)$ and $p \in P$, we have

$$(\mu \circ \psi \otimes 1)((f \otimes p))(x) = \mu((\psi(f) \otimes p))(x) \overset{(5.24)}{=} \psi(f)(x)(p)$$
$$\overset{(5.28)}{=} f(p)x = j(f(p))(x) \overset{(5.27)}{=} j(\tau(f \otimes p))(x).$$

PROPOSITION 5.3.8.

(1) *The image of $\tau$ is an ideal of $R$, denoted by $\mathfrak{T}(P)$.*
(2) *We have* $\ker(j)\mathfrak{T}(P) = 0$.

PROOF. (1) Take an element $f(p) \in \mathfrak{T}(P)$ and two elements $x, y \in R$. We have $(fy)(xp) = f(xp)y = xf(p)y$, that is, the map $\tau$ is a two-sided $R$ map.

(2) We have

$$a \in R,\ p \in P,\ f \in \hom_R(P, R) \implies af(p) = f(ap) \implies \ker(j)\mathfrak{T}(P) = 0. \qquad \square$$

LEMMA 5.3.9. *If* $\ker(j) + \mathfrak{T}(P) = R$, *then*

(1) *$\mu$ is surjective hence, by Proposition 5.3.4, $\mu$ is an isomorphism and $P$ is a finitely generated projective $\Omega$-module.*
(2) *$j$ is surjective and* $\ker(j)$ *splits as a direct summand.*

PROOF. (1) In fact this means that there is an element $w \in \hom_R(P, R) \otimes_\Omega P$ such that $\tau(w) = 1 + a,\ a \in \ker(j)$, which implies that

$$j \circ \tau(w) = 1_P = \mu \circ \phi \otimes 1(w).$$

Then we can apply Proposition 5.3.4(4) and deduce the claim.

(2) Define a map $\alpha : \hom_\Omega(P, P) \to \hom_R(P, R) \otimes_\Omega P$ by setting

$$\alpha(f) := (1 \otimes f)(w).$$

We have $\alpha(rf) := (1 \otimes rf(w)) = r(1 \otimes f(w)) = r\alpha(f)$, so $\alpha$ is a map of $R$-modules.

Let $w = \sum_h \gamma_h \otimes m_h$. We have

$$\mu \circ \psi \otimes 1 \circ \alpha(f)(p) = \mu(\sum_h \psi(\gamma_h) \otimes f(m_h))(p) = \sum_h \psi(\gamma_h)(p) f(m_h)$$
$$= f(\sum_h \psi(\gamma_h)(p) m_h) = f(p).$$

This implies that $(j \circ \tau) \circ \alpha = 1_{\hom_\Omega(P,P)}$ which implies that $j$ splits. $\square$

In order to simplify notation, set $P^* := \hom_R(P,R)$. Since now we work with right modules, there is a switch in all tensor products and, setting $\Omega^* := \mathrm{End}_R(P^*)$, the trace map is

$$\tau^* : P^* \otimes_{\Omega^*} \hom_R(P^*,R) \to R.$$

Transposition gives an antihomomorphism $\Omega = \mathrm{End}_R(P) \to \mathrm{End}_R(P^*) = \Omega^*$, so a map $P^* \otimes_\Omega \hom_R(P^*,R) \to P^* \otimes_{\Omega^*} \hom_R(P^*,R)$.

THEOREM 5.3.10. *If $\mathfrak{T}(P) = R$, then*

(1) $\ker(j) = 0$, *that is the $R$-module $P$ is faithful.*
(2) $\psi$ *is a monomorphism.*
(3) *all the maps in the diagram* (5.29) *are isomorphisms.*
(4) $\hom_R(P,R)$ *is a finitely generated projective $\Omega^*$-module.*

PROOF. (1) Since, by Proposition 5.3.8(2), $\ker(j)\mathfrak{T}(P) = 0$ if $\mathfrak{T}(P) = R$, then $\ker(j) = 0$.

(2) The kernel of $\psi$ is by definition, the set of $f$ such that $f(p)x = 0, \forall p, x \in P$. Since $\ker(j) = 0$, we have that $\psi(f)(p)(x) := f(p)x = 0, \forall x \in P$ implies $f(p) = 0$. If this is true for all $p$, we have $f = 0$.

(3) Due to Lemma 5.3.9, $P$ is a projective $\Omega$-module, hence flat, and $\psi$ is injective. Also $\psi \otimes 1$ is injective. By Lemma 5.3.9 we have that $\mu$ and $j$ are isomorphisms and, by hypothesis, $\tau$ is surjective. Now $j \circ \tau = \mu \circ \psi \otimes 1$ is injective, hence $\tau$ is an isomorphism, and then also $\psi \otimes 1$ is an isomorphism.

(4) We have the usual *biduality map* $k : P \to \hom_R(P^*,R)$. Although we do not assume that this is an isomorphism, we can essentially reproduce diagram (5.29), but for $P^*$ instead of $P$ and using $P$ in place of $\hom_R(P^*,R)$.

We deduce a commutative diagram

$$\begin{array}{ccc} P^* \otimes_\Omega P & \xrightarrow{1\otimes k} & P^* \otimes_{\Omega^*} \hom_R(P^*,R) \\ \tau\big\downarrow & & \tau^*\big\downarrow \\ R & \xrightarrow{1_R} & R \end{array} \tag{5.30}$$

from which it follows that $\tau^*$ is surjective, and we then apply Lemma 5.3.9(1). □

**5.3.1. Projective modules over commutative rings.** Consider now a finitely generated faithful projective $A$-module $P$ over a commutative ring $A$. We then have Proposition 5.3.7.

Let $\Omega := \mathrm{End}_A(P) = \hom_A(P,P)$. We have that $P$ is also a faithful projective $\Omega$ module. Consider an $\Omega$ module $M$. We have a map

$$h_M : P \otimes_A \hom_\Omega(P,M) \longrightarrow M, \quad h_M(p \otimes f) = f(p). \tag{5.31}$$

Thus, we have two functors, the first $M \mapsto \hom_\Omega(P,M)$ from $\Omega$-modules to $A$-modules, and then $N \mapsto P \otimes_A N$ from $A$-modules to $\Omega$-modules. Finally, if we combine the two functors, we have a map

$$\ell_N : N \to \hom_\Omega(P, P \otimes_A N), \quad \ell_N(n)(p) = p \otimes n. \tag{5.32}$$

PROPOSITION 5.3.11. *For all $\Omega$-modules $M$ the map $h_M$ is an isomorphism. For all $A$-modules $N$ the map $\ell_M$ is an isomorphism.*

PROOF. Since the functor $M \mapsto P \otimes_A \hom_\Omega(P, M)$ is exact and commutes with direct sums, it is enough to show the isomorphism for $M = \Omega$.

Let us first do this when $P = A^k$ is a free module so that $\Omega = M_n(A)$ is the ring of matrices. Then as the $\Omega$-module we have that $\Omega = M_n(A) = (A^k)^{\oplus k}$, so it is enough to show the isomorphism when $M = A^k$. In this case $\hom_\Omega(A^k, A^k) = A$, and the isomorphism $h_{A^k} : A^k = A^k \otimes_A A = A^k \otimes_A \hom_\Omega(A^k, A^k) \longrightarrow A^k$ is the identity.

When $P$ is just projective, we can apply the principle of localization. For every maximal ideal $\underline{m}$ of $A$ the localized module $A_{\underline{m}} \otimes_A P$ is free, so it is enough to verify that

$$\mathrm{End}_{A_{\underline{m}}}(A_{\underline{m}} \otimes_A P) = A_{\underline{m}} \otimes_A \mathrm{End}(P),$$

which is easily verified.

The same argument allows us to reduce to the case where $N = A$. When we localize, we have that $P$ becomes free. The condition that $P$ is faithful implies that for every maximal ideal $\underline{m}$ of $A$ the localized module $A_{\underline{m}} \otimes_A P$ is nonzero.

Then for a nonzero free module $P = A^k$, we have

$$(5.33) \qquad \ell_A : A \to \hom_\Omega(A^k, A^k \otimes_A A) = \hom_\Omega(A^k, A^k) = A$$

is the identity. Notice where we have used the fact that $P$ is faithful. Otherwise for some maximal ideal $\underline{m}$, the localized module is 0, so $\hom_0(0, 0) = 0$. □

We can summarize this via the statement of *Morita equivalence* (**[Mor58]**, **[Mor65]**):

THEOREM 5.3.12. *The two functors $M \mapsto \hom_\Omega(P, M)$ and $N \mapsto P \otimes_A N$ between the two categories of $\Omega$- and $A$-modules are inverse equivalences.*

*A special case.* Let now $R$ be a ring with center $Z$ and consider it first as a $Z$-module and the $Z$-linear map $\phi : \mathrm{End}_Z(R) \to R, \phi : a \mapsto a(1)$.

Notice that if $a, b \in \mathrm{End}_Z(R)$ we have $\phi(ab) = (ab)(1) = a(b(1))$. Thus $\phi$ is in fact a map of left $\mathrm{End}_Z(R)$-modules, where on $\mathrm{End}_Z(R)$ the action is by left multiplication, while $R$ is the natural action of $\mathrm{End}_Z(R)$ on $R$. Let $J_R$ be the kernel of $\phi$.

PROPOSITION 5.3.13. *The right annihilator of $J_R$ in the ring $\mathrm{End}_Z(R)$ is the right ideal $\hom_Z(R, Z)$. The ring $R$ is a projective $\mathrm{End}_Z(R)$-module if and only if $Z$ is a direct summand of $R$ as a $Z$-module.*

PROOF. To say that $f \in J_R$ (that is, $f(1) = 0$) means that $f(Z) = 0$. Therefore if $g \in \mathrm{End}_Z(R)$ is such that $f \circ g = 0$ for all $f \in J_R$, this implies that $J_R \hom_Z(R, Z) = 0$. We need to show that, conversely, if $g \in \mathrm{End}_Z(R)$ and $g \notin \hom_Z(R, Z)$, then there is some $f \in J_R$ with $f \circ g \neq 0$.

The hypothesis $g \notin \hom_Z(R, Z)$ means that there is some $r \in R$ such that $s := g(r) \notin Z$. Since $s \notin Z$ there is a $t \in R$ with $st - ts \neq 0$, so we may define $f := ad(t) : x \mapsto tx - xt$ and then $f \circ g(r) = ts - st \neq 0$.

As for the second part, if $R$ is a projective $\mathrm{End}_Z(R)$-module, the map $\phi$ splits and we have a map $h : R \to \mathrm{End}_Z(R)$ with $\phi \circ h = 1_R$. Hence $J_R$ is a direct summand of $\mathrm{End}_Z(R)$. There is thus a decomposition $1 = e + f$ where $e = h(1)$, $\phi(e) = 1$, $e$ is idempotent, and $J_R = \mathrm{End}_Z(R)(1 - e)$. The map $e$ is a projection to $Z$, but since $\phi(e) = 1$, it is the identity on $Z$.

The converse is similar. If $Z$ is a direct summand of $R$ as a $Z$-module, we have a projection $e$ from $R$ to $Z$, and we see that the map $\phi$ is an isomorphism between $\mathrm{End}_Z(R)e$ and $R$. So $R$ is a direct summand of $\mathrm{End}_Z(R)$, and hence is projective. □

5.3.1.1. *The rank of a projective module.* Let us start with a simple geometric fact.

PROPOSITION 5.3.14. *Let $A$ be a commutative ring, and suppose we decompose the spectrum* $\mathrm{Spec}(A) = U_1 \cup U_2 \cdots \cup U_k$ *where the $U_i$ are open (in the Zariski topology) and disjoint. Then there is a decomposition of $A = A_1 \oplus \cdots \oplus A_k$, so that the projection $\pi_i$ of $A$ to $A_i$ identifies $\pi_i^* : \mathrm{Spec}(A_i) \stackrel{\simeq}{\to} U_i$.*

PROOF. By induction we are reduced to the case $k = 2$. In this case we have $\mathrm{Spec}(A) = U_1 \cup U_2$ and equivalently $U_1, U_2$ are closed and disjoint. By definition of the Zariski topology, we have two ideals $I_1, I_2$ of $A$ so that

$$U_1 = \{P \in \mathrm{Spec}(A) \mid P \supset I_1\}, \quad U_2 = \{P \in \mathrm{Spec}(A) \mid P \supset I_2\}.$$

The fact $U_1 \cap U_2 = \emptyset$ implies $I_1 + I_2 = A$, and hence we have a direct sum $A/I_1I_2 = A/(I_1 + I_1I_2) \oplus A/(I_2 + I_1I_2)$, or we have an idempotent $e$ modulo $I_1I_2$ such that $Ae + I_1I_2 = I_1 + I_1I_2$, $A(1-e) + I_1I_2 = I_2 + I_1I_2$.

The fact $U_1 \cup U_2 = \mathrm{Spec}(A)$ implies $I_1I_2$ is a nil ideal. Then the idempotent $e$ modulo $I_1I_2$ can be lifted to an idempotent $e_1$ of $A$ (by Lemma 1.3.14) such that $I_1 = A(1-e_1)$, $I_2 = Ae_1$, and the claim follows. □

This proposition is not true when $A$ in noncommutative as the simplest examples of finite-dimensional algebras are indecomposable, as for instance triangular matrices (even $2 \times 2$) show. The problem is that when we lift a central idempotent to an idempotent, we usually loose the condition of being central.

PROPOSITION 5.3.15. *Let $P$ be a finitely generated projective module over a commutative ring $A$:*

(1) *For every prime ideal $\underline{p}$ of $A$ the localization $P_{\underline{p}} := P \otimes_A A_{\underline{p}}$ is free of some rank $n(\underline{p})$ (or 0, i.e., $n(\underline{p}) = 0$).*
(2) *The rank is a locally constant function on the spectrum of $A$.*
(3) *There is a decomposition $A = \bigoplus_i A_i$, $A_i = Ae_i$, $e_ie_j = \delta^i_j e_j$ orthogonal idempotents, so that $P = \bigoplus_i P_i$, $P_i = P \otimes_A A_i = e_iP$ and $P_i$ is projective over $A_i$ of constant rank.*

PROOF. Let $m_1, \dots, m_k$, $\phi_1, \dots, \phi_k$ be projective generators for $P$. The first remark is that the elements $m_1, \dots, m_k$ form a basis of $P$ over $A$ if and only if the determinant of the matrix with entries $\phi_i(m_j)$ is different from 0.

In fact saying that the elements $m_1, \dots, m_k$ form a basis of $P$ over $A$ means that the map $\pi : A^k \to P$, $\pi(a_1, \dots, a_k) \sum_i a_im_i$, is an isomorphism. We also have the map in the opposite direction $j : P \to A^k$ mapping $m_i \mapsto (\phi_1(m_i), \dots, \phi_k(m_i))$, and $\pi \circ j = 1_P$. Since $\pi$ is an isomorphism, this means that $A^k = j(P)$, that is the elements $(\phi_1(m_i), \dots, \phi_k(m_i))$ form a basis of $A^k$, and this is equivalent to the fact that the determinant of the matrix with entries $\phi_i(m_j)$ is different from 0.

Since a free module finitely generated by $k$ elements has rank $\leq k$, the rank function for $P$ with $k$ projective generators can have at most $k+1$ values (it is possible that it is 0 at some point). What we need to show is that the set of points in the spectrum of $A$ where the rank of $P$ is a fixed number $n$ is open. In fact suppose that $\underline{p}$ is a prime ideal of $A$, such that when we localize $A$ at $\underline{p}$, getting the local ring $A_{\underline{p}}$, the module $A_{\underline{p}} \otimes_A P$ is free over $A_{\underline{p}}$ of rank $n$. Then there is a set of projective generators $p_1, \dots, p_n \in P, \phi_1, \dots . \phi_n$ such that $p_1, \dots, p_n \in P$ is a basis of $A_{\underline{p}} \otimes_A P$ over $A_{\underline{p}}$. When we write these elements, the matrix with entries $\phi_i(m_j)$ whose determinant is not zero, we see by basic localization arguments that there is an element $f \in A$ so that all generators are for $A[f^{-1}]$ and the determinant is still nonzero. This means that $p_1, \dots, p_n \in P$ is a basis of $A[f^{-1}] \otimes_A P$ over $A[f^{-1}]$, hence it is a basis of $A_{\underline{q}} \otimes_A P$ over $A_{\underline{q}}$ for all prime ideals $\underline{q}$ in the open neighborhood of $P$ where $f \neq 0$. □

COROLLARY 5.3.16. *If the spectrum of a commutative ring $A$ is connected, every projective module $P \neq \{0\}$ over $A$ has a constant rank $k$, and it is faithful.*

PROOF. Let us show that $P$ is faithful. Let $a$ be an element of $A$ with $aP = 0$. For all prime ideals $\underline{p}$ we have $A_{\underline{p}} \otimes_A P \simeq A_{\underline{p}}^k$, so $a$ maps to 0 in $A_{\underline{p}}$ for all $\underline{p}$, hence $a$ is 0. □

## 5.4. Separable and Azumaya algebras

**5.4.1. Separable algebras.** Although we do not really need all the general properties of Azumaya algebras, we feel it is useful to set them in their proper framework, which is that of *separable algebras*. This is a quite general notion that extends the theory of finite separable field extensions. We follow very closely the basic papers of Auslander and Goldman **[AG60a]**, **[AG60b]**.

If $R$ is an $A$-algebra, we denote by $R^{\text{op}}$ its opposite, which acts on $R$ by right multiplications. We also set

$$R^e := R \otimes_A R^{\text{op}} \quad \text{the envelope of} \quad R.$$

$R$ is naturally an $R^e$-module by $(a \otimes b)r := arb$, in other words

DEFINITION 5.4.1. There is a natural homomorphism $\eta_R : R^e \to \text{End}_A(R)$ of algebras which associates to the element $a \otimes b$ the endomorphism of $R$ as $A$-module, given by $r \mapsto arb$.

If $R$ is an algebra with a 1, we clearly have that $R$, as the $R^e$-module is generated by 1, so we have the surjective $R^e$-module map

$$\phi : R^e = R \otimes_A R^{\text{op}} \to R, \ \phi : x \otimes y \mapsto xy = \eta_R(x \otimes y)(1). \tag{5.34}$$

LEMMA 5.4.2. *The kernel of $\phi$ is the left ideal $J_R$ of $R \otimes_A R^{\text{op}}$ generated by the elements $x \otimes 1 - 1 \otimes x$.*

PROOF. Clearly, $\phi(x \otimes 1 - 1 \otimes x) = x - x = 0$; conversely, assume that an element $\sum_i x_i \otimes y_i$ is in $\ker(\phi)$, hence $0 = \sum_i x_i y_i$. This implies

$$-\sum_i x_i(y_i \otimes 1 - 1 \otimes y_i) = \sum_i x_i \otimes y_i.$$ □

Let $M$ be any $R^e$-module, which we may also consider as an $R, R^{\text{op}}$ bimodule where $R^{\text{op}}$ acts on the right over $M$. That is, we also write

$$(a \otimes b)m = amb.$$

Consider the subset

$$M^R := \{m \in M \,|\, xm = mx, \ \forall x \in R, \iff (x \otimes 1 - 1 \otimes x)m = 0\}.$$

In other words, using the left $R^e$-module notation, we have that $M^R$ is the set of elements of $m \in M$ with $J_R m = 0$.

One can identify, as for all modules $M \cong \hom_{R^e}(R^e, M)$. Then the set $M^R$ of elements of $m \in M$ with $J_R m = 0$ is

$$M^R \cong \hom_{R^e}(R^e/J_R, M) = \hom_{R^e}(R, M). \tag{5.35}$$

We have two particularly useful instances of the previous construction.

The first instance is when $M = R$. In this case we have that $Z := R^R = \hom_{R^e}(R, R)$ is the center of $R$, which we shall denote by $Z$.

The second instance in which $M = R^e$ and then $(R^e)^R = \hom_{R^e}(R, R^e)$ is the right annihilator $K$ of the left ideal $J_R$, hence it is a right ideal.

LEMMA 5.4.3. *Under the map $\eta_R : R^e \to \mathrm{End}_A(R)$ the right annihilator $K$ of the left ideal $J_R$ maps to* $\hom_Z(R, Z)$.

PROOF. Let $\sum_i x_i \otimes y_i \in K$, hence $(x \otimes 1 - 1 \otimes x)\sum_i x_i \otimes y_i = 0,\ \forall x \in R$. This implies, applying the homomorphism $\eta_R$, that

$$x \sum_i x_i r y_i - \sum_i x_i r y_i x = 0,\ \forall x, r \in R.$$

Hence $\sum_i x_i r y_i \in Z$, that is $\eta_R(K) \subset \hom_A(R, Z)$. On the other hand if $r \in R$ and $a \in Z$ we have that $\sum_i x_i a r y_i = a \sum_i x_i r y_i$, and hence each element of $\eta_R(K)$ is a $Z$ homomorphism. □

Then according to Auslander and Goldman **[AG60a]**, **[AG60b]**, we make the following

DEFINITION 5.4.4. An algebra $R$ over a commutative ring $A$ is called a *separable algebra* if $R$ is projective as $R^e = R \otimes_A R^{\mathrm{op}}$-module.

EXAMPLE 5.4.5 (Exercise). The algebra $M_n(A)$ of $n \times n$ matrices over a commutative ring $A$ is separable, $M_n(A)^e \cong M_{n^2}(A) \cong \mathrm{End}_A(M_n(A))$.

*Hint.* The isomorphism $M_n(A)^e \cong M_{n^2}(A) \cong \mathrm{End}_A(M_n(A))$ is verified using matrix units showing that the elements $e_{i,j} \otimes e_{h,k}$ correspond to matrix units of $\mathrm{End}_A(M_n(A)) = M_{n^2}(A)$.

The property of being projective is then a special case of the fact that $A^k$ is a projective $M_k(A)$-module for all $k$ (cf. Example 5.3.3). □

Let us first deduce some facts about separable algebras. So in the remainder of this section we assume that $R$ is separable.

By Lemma 5.4.2 we have an exact sequence

$$0 \to J_R \longrightarrow R \otimes_A R^{\mathrm{op}} \xrightarrow{\phi} R \to 0 \tag{5.36}$$

of $R^e$-modules, which splits if and only if $R$ is projective as an $R^e$-module. Hence $R$ is a separable algebra if and only if we have a module map $\rho : R \to R^e$ which is composed with $\phi$ as the identity of $R$.

This is equivalent to the fact that for every $R^e$-module $M$ the sequence, associated to the split exact sequence (5.36)

$$0 \longrightarrow \hom_{R^e}(M, J_R) \longrightarrow \hom_{R^e}(M, R^e) \longrightarrow \hom_{R^e}(M, R) \longrightarrow 0 \tag{5.37}$$

is split exact.

PROPOSITION 5.4.6. *If $R$ is separable and $\phi \circ \rho = 1_R$, we have a decomposition of $R^e = \rho(R) \oplus J_R = \rho(R) \oplus \ker(\phi)$ as direct sum of two left ideals.*

REMARK 5.4.7. In concrete terms the module map $\rho : R \to R^e$ maps 1 to some element $u = \sum_i x_i \otimes y_i$ with $\sum_i x_i y_i = 1$ and $J_R \sum_i x_i \otimes y_i = 0$. These conditions are in fact equivalent to the existence of the map $\rho$ and so to the separability of $R$.

LEMMA 5.4.8. *Let $S = I \oplus J$ be a decomposition of a ring $S$ with a 1 into a direct sum of two left ideals, and let $1 = e + f,\ e \in I,\ f \in J$. Then $e, f$ are orthogonal idempotents, i.e., $e = e^2,\ f = f^2,\ ef = fe = 0$.*

PROOF. We have $e = e^2 + ef \implies 0 = e - e^2 + ef$. Since $e - e^2 \in I,\ ef \in J$, we must have $e - e^2 = ef = 0$. The argument is similar with $f$. □

If $M$ is an $S$-module, we have an isomorphism $M \to \hom_S(S, M)$ by mapping $m$ to the homomorphism which maps $s \mapsto sm$. The inverse $\hom_S(S, M) \to M$ is defined by associating, to a homomorphism $f \in \hom_S(S, M)$, the element $m := f(1)$.

When we have a decomposition $S = I \oplus J = Se \oplus Sf$ with $1 = e + f$, we have the decomposition $M = eM \oplus fM$ and a corresponding decomposition, $\hom_S(S, M) = \hom_S(Se, M) \oplus \hom_S(Sf, M)$. Then in the given isomorphism,

$$\begin{aligned} \hom_S(Se, M) &= eM = \{m \in M \mid fm = 0\}, \\ \hom_S(Sf, M) &= fM = \{m \in M \mid em = 0\}. \end{aligned} \tag{5.38}$$

COROLLARY 5.4.9.

(a) *The elements $v := \sum_i x_i(y_i \otimes 1 - 1 \otimes y_i)$ and $u = \rho(1) := \sum_i x_i \otimes y_i$ of Remark 5.4.7 are orthogonal idempotents giving the decomposition*

$$R^e = \rho(R) \oplus J_R,\ \rho(R) = R^e u,\ J_R = R^e(1 - u) \tag{5.39}$$

*as the direct sum of two left ideals.*

(b) *The element*

$$\eta_R(\rho(1)) = \eta_R(u) \in \operatorname{End}_A(R),\ \eta_R(u)(r) := \sum_i x_i r y_i \tag{5.40}$$

*is a $Z$-linear projection to the center $Z$ of $R$ and $Z = \phi(\rho(1)R^e)$.*

(c) *The center $Z$ of $R$ is a direct summand as $Z$-module.*

PROOF. (a) is a consequence of Proposition 5.4.6 and Lemma 5.4.8 since

$$1 = \sum_i x_i y_i \otimes 1 = \sum_i x_i(y_i \otimes 1 - 1 \otimes y_i) + \sum_i x_i \otimes y_i,$$

and $\sum_i x_i(y_i \otimes 1 - 1 \otimes y_i) \in J_R$ while $u = \sum_i x_i \otimes y_i \in \rho(R)$.

(b) Since $u$ is idempotent and $\eta_R$ is a homomorphism to $\operatorname{End}_Z(R)$, we have that $\eta_R(u) : R \to R$ is a projection. Since $J_R u = 0$ we have, by Lemma 5.4.3, that the image of $\eta_R(u)$ is in $Z$. Now $\eta_R(u)(r) = \sum_i x_i r y_i$ implies $\eta_R(u)(1) = 1$, and if $r \in Z$ also $\eta_R(u)(r) = \sum_i x_i r y_i = r \sum_i x_i y_i = r$, so $\eta_R(u)$ is a $Z$-linear projection of $R$ to $Z$. The identity $Z = \phi(\rho(1)R^e)$ follows from formula (5.34), since

$$\phi(u \cdot x \otimes y) = \eta_R(u \cdot x \otimes y)(1) = \eta_R(u) \circ \eta_R(x \otimes y)(1) \in Z$$

and $\phi(u) = 1$.

(c) This is a consequence of (b). □

We identify $\rho(R) = R^e u$ with $R$ and write $R^e = R^e u \oplus J_R = R \oplus J_R$. Thus given any $R^e$-module $M$ (i.e., an $R, R$ bimodule), we have

$$M = \hom_{R^e}(R^e, M) = \hom_{R^e}(R, M) \oplus \hom_{R^e}(J_R, M). \tag{5.41}$$

LEMMA 5.4.10. *We have that*

$$\begin{aligned} \hom_{R^e}(R, M) &= \{m \in M \mid um = m\} = uM \\ &= \{m \in M \mid J_R m = 0\} = \{m \in M \mid xm = mx, \ \forall x \in R\}. \end{aligned} \tag{5.42}$$

PROOF. From formula (5.38), in the isomorphism $M = \hom_{R^e}(R^e, M)$ an element $m \in \hom_{R^e}(R, M)$ if and only if $J_R m = 0$, and since $J_R$ is generated by the elements $x \otimes 1 - 1 \otimes x$, this is equivalent to the given condition. □

If $M, N$ are two $R$-modules, we have that the $A$-module $\hom_A(N, M)$ is naturally an $R^e$-module by setting

$$(a \otimes b)(f)(n) := a(f(bn)). \tag{5.43}$$

Notice that if $f \in \hom_A(N, M)$, then $(x \otimes 1 - 1 \otimes x)f = 0$, and this says that for all $n \in N$ we have $0 = (x \otimes 1 - 1 \otimes x)(f)(n) = xf(n) - f(xn)$. Thus we have the following

COROLLARY 5.4.11.

$$\hom_R(N, M) = \hom_{R^e}(R, \hom_A(N, M)). \tag{5.44}$$

In particular suppose that $N$ is an $R$ submodule of $M$. If $N$ is a direct summand of $M$ as an $A$-module, there is an $A$-linear map $\pi : M \to N$ which is the identity on $N$. The $A$-module $\hom_A(M, N)$ is also an $R^e$-module by setting $(x \otimes y)\pi(m) := x\pi(ym)$. In other words define $\pi' := u\pi$,

$$\pi'(m) = \sum_i x_i \pi(y_i m) \implies \pi'(n) = \sum_i x_i \pi(y_i n) = \sum_i x_i y_i n = n, \forall n \in N. \tag{5.45}$$

COROLLARY 5.4.12. *If $N$ is an $R$ submodule of $M$ and a direct summand of $M$ as an $A$-module, then $N$ is also a direct summand of $M$ as $R$-module. In particular if $A$ is a field, every $R$-module is semisimple, hence $R$ is a direct sum of matrix algebras over division rings.*

PROOF. Consider the map $\pi'$ of formula (5.45). We have seen that it is the identity on $N$, so we need to see that it is an $R$-linear map, but this means that $J_R \pi' = 0$. This is true since $J_R u = 0$.

If $A$ is a field, then every $A$ submodule is a direct summand, so every module over $R$ is semisimple and this is equivalent for $R$ to be semisimple Artinian (cf. Theorem 1.3.7). A more precise result is Theorem 5.4.15. □

We have previously explained that $\hom_{R^e}(R, R)$ can be identified to the center $Z$ of $R$.

PROPOSITION 5.4.13.

(1) *If $\pi : R \to S$ is a surjective homomorphism of $A$-algebras and $R$ is separable, so also is $S$. The center of $S$ is the image of the center of $R$.*
(2) *If $R$ is a separable $A$-algebra and $f : A \to B$ is a morphism of commutative rings, then $B \otimes_A R$ is a separable $B$-algebra. The center of $B \otimes_A R$ is $B \otimes_A Z$, where $Z$ is the center of $R$.*

(3) *If $R_1, R_2$ are both separable algebras over two commutative $A$-algebras $A_1, A_2$ with centers $Z_1, Z_2$, we have that $\Gamma := R_1 \otimes_A R_2$ is either 0 or separable over $A_1 \otimes_A A_2$ with center $Z_1 \otimes_A Z_2$.*

PROOF. (1) Since $R$ is separable, we have an element $\sum_i x_i \otimes y_i \in R^e$ with $\sum_i x_i y_i = 1$ and $J_R \sum_i x_i \otimes y_i = 0$ with $J_R$ as in Lemma 5.4.2. We take the image in $S$ of this element $\sum_i \pi(x_i) \otimes \pi(y_i) \in S^e$ with $\sum_i \pi(x_i)\pi(y_i) = 1$. Since $\pi$ is surjective and $J_R$ is the left ideal of $R^e$ generated by the elements $x \otimes 1 - 1 \otimes x$, we have that $\pi \otimes \pi$ maps $J_R$ surjectively to the corresponding left ideal $J_S$ of $S^e$ and hence $J_S \sum_i \pi(x_i) \otimes \pi(y_i) = \pi(J_R \sum_i x_i \otimes y_i) = 0$. Thus, we have the condition that $S$ is separable as from Remark 5.4.7. As for the center, this follows from Corollary 5.4.9(b).

(2) This has a similar proof.

(3) Set $\Gamma := R_1 \otimes_A R_2$. We have

$$\Gamma^e = \Gamma \otimes_{A_1 \otimes_A A_2} \Gamma^{\mathrm{op}} = (R_1 \otimes_{A_1} R_1^{\mathrm{op}}) \otimes_A (R_2 \otimes_{A_2} R_2^{\mathrm{op}}),$$

from which it easily follows that if $R_i$ is projective as an $R_i^e$-module, $i = 1, 2$, then $R_1 \otimes_A R_2$ is projective as an $(R_1 \otimes_A R_2)^e$-module.

In fact as element $u \in (R_1 \otimes_A R_2)^e = (R_1)^e \otimes_A (R_2)^e$, we can clearly take $u_1 \otimes u_2$ where $u_i \in R_i^e$ satisfies the properties of Remark 5.4.7.

If $\rho_i$ is a splitting of $\phi_i$, $i = 1, 2$, we have that $\rho := \rho_1 \otimes \rho_2$ is a splitting of $\phi := \phi_1 \otimes \phi_2$. Form formula (5.40) we have

$$Z_i = \phi_i(\rho_i(1)R_i^e) \implies Z_1 \otimes_A Z_2 = \phi(\rho(1)\Gamma^e). \qquad \square$$

We want to characterize separable algebras over a field, and we start with

LEMMA 5.4.14. *Let $K \subset C$ be an extension of fields. Then this is a finite separable extension, in the sense of Galois theory, if and only if for any field extension $K \subset G$, we have that $G \otimes_K C$ is a finite direct sum of fields.*

PROOF. In one direction it is immediate: if $K \subset C$ is a finite separable extension, we have $C = K[x]/(f(x))$, where $f(x)$ is a separable polynomial. This implies that, for any field extension $K \subset G$, the polynomial $f(x)$ factors, over $G$, as $f(x) = \prod_{i=1}^{j} f_i(x)$ into distinct separable irreducible polynomials. Hence $G \otimes_K C = G[x]/(f(x)) = \bigoplus_{i=1}^{h} G[x]/(f_i(x))$, and each $G[x]/(f_i(x))$ is a field (finite and separable over $G$).

In the other direction, if $C$ is not algebraic over $K$, there is a trascendental element $a \in C$, and taking $G = K(x)$ the field of rational functions, we have that $G \otimes_K C$ is an integral domain but the element $x + a$ is not invertible.

So $C$ is algebraic over $K$. If it has an element $a$ which is not separable, we are in characteristic $p > 0$ and one sees that taking $G = K(a)$, the ring $K(a) \otimes_K C$ contains a nonzero nilpotent element. Finally, if $C$ is algebraic separable but infinite-dimensional, taking as $G$ the algebraic closure $\bar{K}$ of $K$, we have that $\bar{K} \otimes_K C$ has infinitely many maximal ideals corresponding to the infinitely many embeddings of $C$ in $\bar{K}$. $\square$

THEOREM 5.4.15. *Let $R$ be a $K$-algebra, and let $K$ be a field. Then $R$ is a separable $K$-algebra if and only if $R$ is finite dimensional over $K$ and semisimple with its center a separable extension of $K$.*

PROOF. If $R$ is separable, then for any field extension $K \subset G$, we also have that $G \otimes_K R$ is a separable $G$ algebra. Then by Corollary 5.4.12, $G \otimes_K R$ is semisimple Artinian. In particular $R = \bigoplus_i M_{h_i}(D_i)$ is a direct sum of matrix algebras over division rings. Let $C_i$ be the center of $D_i$, and let $K_i \supset C_i$ be a maximal subfield of $D_i$. We take $G = K_i$, and we have that $K_i \otimes_K M_{h_i}(D_i)$ is semisimple. We then consider the space $D_i^{h_i}$ as a module over $S := K_i \otimes_K M_{h_i}(D_i)$ by left multiplication by $M_{h_i}(D_i)$ and by right multiplication by $K_i$. We clearly have that $D_i^{h_i}$ is an irreducible module and $\mathrm{End}_S(D_i^{h_i}) = K_i$. Since $K_i \otimes_K M_{h_i}(D_i)$ is semisimple, we must have that $D_i^{h_i}$, and hence $D_i$ is finite dimensional over $K_i$, and hence $K_i$ is finite-dimensional over the center $C_i$ of $D_i$. Finally, we must have that $G \otimes_K \bigoplus_i C_i = \bigoplus_i G \otimes_K C_i$ must be a finite direct sum of fields for all $G$. Hence each $G \otimes_K C_i$ must be a finite direct sum of fields for all $G$, and this is indeed equivalent to the fact that all the $C_i$ are finite separable field extensions of $K$ by Lemma 5.4.14. □

COROLLARY 5.4.16. *Let $R$ be a separable algebra over its center $Z$, and let $M$ be a maximal ideal of $R$. Then $M = \mathfrak{m}R$, where $\mathfrak{m}$ is a maximal ideal of $Z$.*

PROOF. Let $\mathfrak{m} := M \cap Z$ so that $M \supset \mathfrak{m}R$, and consider the two algebras $R/M$, $R/\mathfrak{m}R$, both separable algebras over their center which is $Z/\mathfrak{m}$, by Proposition 5.4.13(1). Since $R/M$ is a simple algebra, its center is a field, so $\mathfrak{m}$ is a maximal ideal of $Z$ and then $R/\mathfrak{m}R$ is a separable algebra over its center a field. By Theorem 5.4.15 we have then that $R/\mathfrak{m}R$ is also a simple algebra, hence $R/M = R/\mathfrak{m}R$, that is $M = \mathfrak{m}R$. □

### 5.4.2. Azumaya algebras.

DEFINITION 5.4.17. An algebra $R$ over a commutative ring $A$ is called an *Azumaya algebra* or a *maximally central algebra*, cf. **[Azu51]** over $A$ if:

(1) $R$ is a finitely generated projective $A$-module.

(2) The natural map $\eta_R : R \otimes_A R^{\mathrm{op}} \to \mathrm{End}_A(R)$ given by $\eta_R(a \otimes b)(r) := arb$ is an isomorphism.

REMARK 5.4.18. Once we know that $R$ is finitely generated projective to verify (2), it is enough to see that $\eta_R : R^e \to \mathrm{End}_A(R)$ is surjective. In fact, since $R^e$, $\mathrm{End}_A(R)$ are both finitely generated projective of the same rank (the square of the rank of $R$), one can reduce to the local case remarking that a surjective map between two free modules (over a commutative ring) of the same rank is necessarily an isomorphism.

REMARK 5.4.19. By Theorem 5.4.15 an Azumaya algebra over a field $F$ is a finite-dimensional central simple algebra with center $F$.

Conversely,

EXERCISE 5.4.20. A finite-dimensional simple algebra is Azumaya over its center.

We leave it to the reader to verify that when $A$ is a commutative ring, the algebra $M_n(A)$ is an Azumaya algebra.

After this, the simplest example of Azumaya algebra is the ring $\mathrm{End}_A(P)$ where $P$ is a finitely generated projective module. When $P = A^n$ is free, then of course $\mathrm{End}_A(P) = M_n(A)$.

This can be proved by the principle of localization. Given any maximal ideal $\underline{m}$ of $A$, we have that the localized module $P_{\underline{m}} = P \otimes_A A_{\underline{m}}$ is a projective module over a local ring. Hence by Nakayama's Lemma it is a free module, so that the map $\eta_R$ is an isomorphism after localization, but then since this is true for all maximal ideals, $\eta_R$ is an isomorphism.

5.4.2.1. *A typical example.* Let us recall a

DEFINITION 5.4.21. A ring $R$ is a left (resp., right) order in a ring $S$, if $R \subset S$ and every element of $S$ is of the form $ab^{-1}$ (resp., $b^{-1}a$) with $a, b \in R$ and $b$ invertible in $S$. Furthermore, $R$ is a *central order* if in the fractions $ab^{-1}$ one can take $b$ to be in the center of $R$.

We have already mentioned (see the discussion on Ore domains in Definition 1.5.1) that this setting is rather special in noncommutative algebra. A very important theorem in this direction is the theorem of Posner, Theorem 11.2.6 (see **[Pos60]**), stating that a prime PI ring is always a central order in a finite-dimensional central simple algebra.

A typical way of constructing prime Azumaya algebras is the following.

Let $R$ be a ring which is an order in a simple algebra $S$ finite-dimensional over its center $F$, say $\dim_F S = n^2$. Assume that, for every $r \in R$ the reduced trace $\operatorname{tr}(r) \in R$ (cf. §1.1.1.2).

Let $r_1, \dots, r_{n^2}$ be a basis of $S$ over $F$ formed by elements in $R$, and finally let $d := \det(\operatorname{tr}(r_i r_j))$ be the discriminant of this basis, then

PROPOSITION 5.4.22. *$U := R[d^{-1}]$ is Azumaya over its center $Z$ which has $F$ as a field of fractions.*

PROOF. Given any element $s \in S$, we have $s = \sum_i \lambda_i r_i$, for some $\lambda_i \in F$.

One can compute the coefficients $\lambda_i$ by solving the system of linear equations $\operatorname{tr}(r_j s) = \sum_i \lambda_i \operatorname{tr}(r_j r_i)$, where by Cramer's rule the denominator of the formulas is $d$. Thus if $R$ is stable under reduced trace, we deduce that after inverting $d$, the elements $r_i$ are a basis of $U$ over its center $Z$ (so it is a free $Z$-module), which has as ring of fractions $F$ since we are assuming that $R$ is an order in $S$. The fact that the map $j : U^e \to \operatorname{End}_A(U)$ is an isomorphism can be proved by again computing the trace form. In fact we have that the trace of the endomorphism $j(r_i \otimes r_j)$ is $\operatorname{tr}(r_i)\operatorname{tr}(r_j)$. It follows that the matrix of the trace form $M'$ of the $n^4$ endomorphisms $r_i \otimes r_j$ is the tensor product $M \otimes M$ where $M$ is the matrix of the reduced trace form of $S$ computed on the basis $s_i$. Therefore the determinant of $M'$ is $d^{2n^2}$.

By construction $d^{2n^2}$ is invertible in the ring $A$, hence these $n^4$ elements form a basis of $\operatorname{End}_A(U)$, and so $j$ is an isomorphism. □

EXERCISE 5.4.23. Let $R$ be an order in a central simple algebra $S$ closed under reduced trace. Then $R$ is Azumaya over its center $Z$ if and only if the ideal of $Z$ generated by the discriminants of the bases of $S$ contained in $R$ is the entire $Z$.

5.4.2.2. *Structure theorems.* Let $R$ be Azumaya over $Z$ and consider an $R^e$-module $M$ (i.e., an $R - R$ bimodule).

We have seen (formula (5.42)) that inside $M$ the subset

$$uM = M^R := \{m \in M \mid xm = mx, \forall x \in R\}$$

of the elements annihilated by all $x \otimes 1 - 1 \otimes x$ can be identified to $\hom_{R^e}(R, M)$.

The multiplication map $R^e \otimes_Z M^R \to M$ is 0 on $J_R \otimes_Z M^R$ where $J_R$ is the left ideal of $R^e$ generated by the elements $x \otimes 1 - 1 \otimes x$ and $R^e / J_R = R$, hence we have a map $R \otimes_Z M^R \to M$. One of the main theorems is

THEOREM 5.4.24. *For every $R^e$-module $M$ the map $R \otimes_Z M^R \to M$ is an isomorphism.*

We will prove this theorem as a part of Theorem 5.4.27. In order to do this, we need to present several characterizations of Azumaya algebras.

Let $R$ be an algebra with center $Z$, and let $R^e = R \otimes_Z R^{\mathrm{op}}$. We use the commutative diagram (5.29), where $R$ is replaced by $R^e$, $P$ is replaced by $R$, $j = \phi$, and finally $\Omega := \operatorname{End}_{R^e}(R) = Z$.

$$\begin{array}{ccc} \hom_{R^e}(R, R^e) \otimes_Z R & \xrightarrow{\psi \otimes 1} & \hom_Z(R, Z) \otimes_Z R \\ \tau \downarrow & & \downarrow \mu \\ R^e & \xrightarrow{\phi} & \hom_Z(R, R) \end{array} \tag{5.46}$$

THEOREM 5.4.25. *Let $R$ be an algebra with center $Z$, and let $R^e = R \otimes_Z R^{\mathrm{op}}$. Then the following properties are equivalent:*

(a) *$R$ is separable over $Z$.*
(b) *$R^e (R^e)^R = R^e$.*
(c) *In the diagram* (5.46) *all maps are isomorphisms.*
(d) *$R$ is an Azumaya algebra over $Z$.*
(e) *The map $R^e \to \operatorname{End}_Z(R)$ is an isomorphism, and $Z$ is a direct summand of $R$ as $Z$-module.*

PROOF. (a) $\implies$ (b) If $R$ is separable over $Z$, we have by Proposition 5.4.13(3) that also $R^e$ is separable over $Z$. Furthermore $(R^e)^R = uR^e$, where the element $u := \sum_i x_i \otimes y_i$ of Remark 5.4.7 is an idempotent by Corollary 5.4.9.

So $R^e (R^e)^R = R^e u R^e$ is a two-sided ideal of $R^e$. If $R^e u R^e \neq R^e$, there is a maximal ideal $M \supset R^e u R^e$, and by Corollary 5.4.16 we have that $M = \mathfrak{m} R^e$ with $\mathfrak{m}$ a maximal ideal of $Z$.

From Corollary 5.4.9(b) we have $Z = \phi(uR^e)$ and $\phi(uR^e) \subset \phi(\mathfrak{m} R^e) = \mathfrak{m} R$. This contradicts the fact that $\phi(\mathfrak{m} R^e) = \mathfrak{m} R$ has the property that $Z \cap \mathfrak{m} R = \mathfrak{m}$.

(b) $\implies$ (c) Recall that $(R^e)^R \overset{(5.35)}{=} \{a \in R^e \,|\, (x \otimes 1)a = (1 \otimes x)a, \ \forall x\}$ is the (isomorphic) image of the space $\hom_{R^e}(R^e / J_R, R^e) = \hom_{R^e}(R, R^e)$ under the map $f \mapsto f(1)$.

The trace map $\tau(f \otimes r) = f(r) = (r \otimes 1)(f(1)) = (1 \otimes r)(f(1))$ has as image a two-sided ideal, Proposition 5.3.8(i), $R \otimes 1 \cdot (R^e)^R = 1 \otimes R \cdot (R^e)^R = R^e \cdot (R^e)^R$. Hence hypothesis (b) is that $\tau$ is surjective, and then we apply Theorem 5.3.10(3) to deduce that all maps of diagram (5.46) are isomorphisms.

(c) $\implies$ (d) From this the first condition of being Azumaya of Definition 5.4.17 follows from Proposition 5.3.4 applied to $\tau$, while the second condition is the fact that $\phi$ is an isomorphism.

(d) $\implies$ (e) Since $R$ is Azumaya, the map $R^e \to \operatorname{End}_Z(R)$ is an isomorphism by hypothesis. The fact that $Z$ is a direct summand of $R$ as $Z$-module follows from Proposition 5.3.13.

(e) $\implies$ (a) This follows from Proposition 5.3.13. □

THEOREM 5.4.26. *Let $R$ be an $A$-algebra with center $Z$. Then $R$ is separable over $A$ if and only if $R$ is separable over $Z$ and $Z$ is separable over $A$.*

PROOF. If $R$ is separable over $A$ since the $R \otimes_A R^{\mathrm{op}}$-module structure of $R$ factors through $R \otimes_Z R^{\mathrm{op}}$, we have that $R$ is projective as a $R \otimes_Z R^{\mathrm{op}}$-module (Remark 5.3.2(3)), so $R$ is separable over $Z$. Moreover, by part (d) of Theorem 5.4.25, $R$ is projective over $Z$, and hence $R \otimes_A R^{\mathrm{op}}$ is projective over $Z \otimes_A Z$. Finally since $Z$ is a direct summand of $R$ as $Z$-module and $R$ is a direct summand of $R \otimes_A R^{\mathrm{op}}$ as an $R \otimes_A R^{\mathrm{op}}$-module, so $R$ and finally also $Z$ are direct summands of $R \otimes_A R^{\mathrm{op}}$ as $Z \otimes_A Z$-module, and hence they are projective.

Assume now that $R$ is separable over $Z$ and $Z$ is separable over $A$. We have a split exact sequence $0 \to J_0 \to Z \otimes_A Z \to Z \to 0$ of $Z \otimes_A Z$-modules from which, tensoring with $(R \otimes_A R^{\mathrm{op}}) \otimes_{Z \otimes_A Z}$, we deduce a split exact sequence,

$$(5.47)\quad 0 \longrightarrow J_1 \longrightarrow R \otimes_A R^{\mathrm{op}} \longrightarrow (R \otimes_A R^{\mathrm{op}}) \otimes_{Z \otimes_A Z} Z \longrightarrow 0.$$

Now we claim that $(R \otimes_A R^{\mathrm{op}}) \otimes_{Z \otimes_A Z} Z = R \otimes_Z R^{\mathrm{op}}$ as an $R \otimes_A R^{\mathrm{op}}$-module; hence $R \otimes_Z R^{\mathrm{op}}$ is projective as $R \otimes_A R^{\mathrm{op}}$-module.

Since $R$ is a direct summand of $R \otimes_Z R^{\mathrm{op}}$ as an $R \otimes_A R^{\mathrm{op}}$-module, it is projective, and hence $R$ is a separable $A$-algebra. □

5.4.2.3. *Two-sided modules over Azumaya algebras.* Let $R$ be an algebra with center $Z$ and $R^e = R \otimes_Z R^{\mathrm{op}}$. We then construct for every $R^e$-module $M$ the $Z$-module $M^R = \hom_{R^e}(R, M)$ as in formula (5.35). We then have a map $g : R \otimes_Z M^R \to M$ defined by $g(r \otimes m) = (r \otimes 1)m$ (or $(r \otimes 1)m = rm$ if we consider the $R^e$-module $M$ as an $R - R$ bimodule). We have thus two functors $M \mapsto M^R$ from $R^e$-modules to $Z$-modules, and $N \mapsto R \otimes_Z N$ from $Z$-modules to $R^e$-modules. Finally, if we combine the two functors, we have a map, which is analogous to the one of formula (5.32),

$$(5.48)\qquad \ell_N : N \to \hom_{R^e}(R, R \otimes_Z N), \quad \ell_N(n)(r) = r \otimes n.$$

THEOREM 5.4.27. *For an algebra $R$ with center $Z$, the following conditions are equivalent.*

(a) *$R$ is an Azumaya algebra.*
(b) *The map $g : R^e \otimes_Z (R^e)^R \to R^e$ is an epimorphism.*
(c) *For every $R^e$-module $M$, the map $g : R \otimes_Z M^R \to M$ is an isomorphism.*

PROOF. Conditions (a) and (b) are equivalent by Theorem 5.4.25. Clearly, (c) implies (b); hence we only need to show that (a) implies (c).

If $R$ is an Azumaya algebra, we have that $R$ is a finitely generated, faithful, projective $Z$-module. Moreover $R^e = \Omega = \mathrm{End}_Z(R)$, so we can use Proposition 5.3.11 applied to $P = R$, which gives that $g = h_M$ (as in formula (5.31)) is an isomorphism. □

This theorem together with Proposition 5.3.11 and Theorem 5.3.12 has a number of important consequences.

PROPOSITION 5.4.28. *If $R$ is an Azumaya algebra over a commutative ring $A$ and $B$ is a commutative $A$-algebra, then $R_B := R \otimes_A B$ is an Azumaya algebra over $B$. Conversely, if $B$ is faithfully flat over $A$ and $R_B$ is Azumaya over $B$, then so is $R$ an Azumaya algebra over $A$.*

PROOF. If $R$ is $A$ projective, then $R_B$ is $B$ projective, and clearly we have the maps

$$R_B^e \xrightarrow{\cong} (R^e)_B \to (\mathrm{End}_A(R))_B \xrightarrow{\cong} \mathrm{End}_B(R_B),$$

which give the required isomorphism for $R_B$ given the one for $R_A$. We then need to show that $R_B$ is faithful over $B$ so that from the previous isomorphism, we have that $B$ is the center of $R_B$. This follows from the fact that $A$ is a direct summand of $R$. Conversely, if $B$ is faithfully flat, an isomorphism over $B$ comes from one over $A$ and similarly for the finitely generated projective condition. □

Using Proposition 5.3.15, we have a decomposition $A = \bigoplus_i A_i$, $A_i = Ae_i$ so that $R = \bigoplus R_i$, $R_i = Re_i$, and $R_i$ has constant rank, which must be a square $n_i^2$.

COROLLARY 5.4.29. *Let $R$ be an Azumaya algebra over its center $Z$. Then:*

(1) *The ideals of $R$ are in 1–1 correspondence with the ideals of $Z$ by $I \mapsto I \cap Z$, so that $I = R(I \cap Z)$ and for all ideals $\underline{a}$ of $Z$, we have $R\underline{a} \cap Z = \underline{a}$.*
(2) *Given a homomorphism $j : R \to S$ where $S$ is any $Z$-algebra, we have an algebra isomorphism $S \simeq R \otimes_Z S^R$ where*
$$S^R := \{s \in S \mid sj(r) = j(r)s, \ \forall r \in R\}$$
*is the centralizer of $j(R)$ in $S$.*
(3) *$S$ is Azumaya over its center $Z'$ if and only if $S^R$ is Azumaya over $Z'$. In this case if $R$ and $S$ have the same rank $n^2$ over $Z$ and over $Z'$, respectively, then $S^R = Z'$.*

PROOF. (1) An ideal $I$ of $R$ is in particular a two-sided module and $I^R = I \cap Z$ is an ideal of $Z$; hence Theorem 5.4.27(c) implies that $I = RI^R$. Conversely, if $\mathfrak{a}$ is an ideal of $Z$, we have by Proposition 5.4.28 that $R \otimes_Z Z/\mathfrak{a} = R/\mathfrak{a}R$ is Azumaya over $Z/\mathfrak{a}$, which implies that $\mathfrak{a}R \cap Z = \mathfrak{a}$.

(2) Since $S$ is a $Z$-algebra, we may consider $S$ as an $R \otimes_Z R^{\mathrm{op}}$-module by left and right multiplication and apply Theorem 5.4.27(c) where $S^R$ is now the centralizer of $R$ in $S$. We have thus $S = R \otimes_Z S^R$.

(3) Denote by $Z'$ the center of $S$. We have $S = R \otimes_Z S^R = (R \otimes_Z Z') \otimes_{Z'} S^R$, so since $R \otimes_Z Z'$ is Azumaya over $Z'$, we are reduced to the case $Z = Z'$. Then in one direction this is Proposition 5.4.13(3).

Conversely, assume that $S = R \otimes_Z B$ is Azumaya with center $Z$ for some $Z$-algebra $B \subset S$. So $R \otimes_Z B$ is a finitely generated projective module over $Z$, and it is also projective over $S^e = S \otimes_Z S^{\mathrm{op}} = R^e \otimes_Z B^e$.

Now $R^e$ is a projective over $Z$ and $Z$ is a direct summand, so $B^e$ is also a direct summand. $S^e$ is projective over $B^e$ hence also $R \otimes_Z B$ is projective over $B^e$. Since $B$ is a direct summand, also $B$ is projective over $B^e$, and the two conditions for $B$ to be Azumaya over $Z$ are verified.

The last statement follows from the fact that the rank of $R \otimes_Z S^R = (R \otimes_Z Z') \otimes_{Z'} S^R$ is the product of the two ranks of $R$ over $Z$ which is equal to that of $R \otimes_Z Z'$ over $Z'$ and the rank of $S^R$ over $Z'$. So finally the rank of $S^R$ as projective module over $Z'$ is 1, and the claim follows since $Z'$ is a direct summand of $S^R$. □

COROLLARY 5.4.30. *Let $R_1, R_2$ be two Azumaya algebras over the same center $Z$ and of the same rank. Given a $Z$ homomorphism $\phi : R_1 \to R_2$, then $\phi$ is an isomorphism.*

COROLLARY 5.4.31. *If $R$ is an Azumaya algebra over its center $Z$, it is a central simple algebra if and only if $Z$ is a field.*

As a corollary, given elements $a_i \in A$ which generate the unit ideal, consider the localizations $A[1/a_i]$. Since $\bigoplus_i A[1/a_i]$ is faithfully flat over $A$ (Example 5.1.2), then $R$ is an Azumaya algebra over $A$ if and only if the algebras $R \otimes A[1/a_i]$ are Azumaya. In other words the Azumaya property is a local property (in the Zariski and étale topology of the base). One could now define a sheaf of Azumaya algebras over a scheme, etc., but we will not need all these facts.

5.4.2.4. *Central simple algebras.*

THEOREM 5.4.32. *If $K$ is a field and $R$ is an Azumaya algebra over $K$, then $R$ is a finite-dimensional central simple algebra over $K$ (cf. Corollary 5.4.31). Thus if $K$ is an algebraically closed field, $R = M_n(K)$ for some $n$.*

PROOF. Since $R$ is a finite-dimensional vector space over $K$ of some dimension $m$, we have that $R \otimes_K R^{\mathrm{op}}$ is isomorphic to the algebra $M_m(K)$. A two-sided ideal of $R$ is just an $R \otimes_K R^{\mathrm{op}}$ submodule. Since by assumption this is the entire ring of endomorphisms, it follows that $R$ has no nonzero two-sided ideals, hence it is simple.

We need to show that $K$ is the center of $R$. The center of $R$ can be identified with the endomorphisms of $R$ which commute with $R \otimes_K R^{\mathrm{op}}$ which is the algebra $M_m(K)$, clearly these are just $K$.

If $K$ is algebraically closed, a finite-dimensional central simple algebra is a matrix algebra over its center. □

COROLLARY 5.4.33. *Let $R$ be an algebra with its center a local ring $A$ with maximal ideal $\underline{m}$. Then $R$ is an Azumaya algebra over $A$ if and only if:*

(1) *$R$ is a projective, hence free, module over $A$.*
(2) *$\underline{m}R$ is a maximal ideal of $R$ and $A/\underline{m}$ is the center of $R/\underline{m}R$.*

PROOF. If $R$ is Azumaya, it is separable with center $A$; hence $R/\underline{m}R$ is separable with center $A/\underline{m}$ a field, by Proposition 5.4.13. Then by Theorem 5.4.32 $R/\underline{m}R$ is a central simple algebra over $A/\underline{m}$, so $\underline{m}R$ is a maximal ideal of $R$.

Conversely, if $\underline{m}R$ is a maximal ideal of $R$, then $R/\underline{m}R$ is a simple algebra and is central over $A/\underline{m}$, so it is an Azumaya algebra. Then consider the diagram

$$\begin{array}{ccc} R^e & \xrightarrow{j} & \operatorname{End}_A(R) \\ \downarrow & & \downarrow \\ (R/\underline{m}R)^e & \xrightarrow{j} & \operatorname{End}_A(R/\underline{m}R). \end{array} \tag{5.49}$$

Since we are assuming that $R$ is a free module over $A$ of some rank $n$, we have that $j : R^e \to \operatorname{End}_A(R)$ is a map between two free modules over $A$ of rank $n^2$. So by Nakayama's lemma this map is an isomorphism as soon as it is an isomorphism when reducing modulo $\underline{m}$.

Now the equality $\operatorname{End}_A(R/\underline{m}R) = \operatorname{End}_A(R)/\underline{m}\operatorname{End}_A(R)$ depends on the fact that choosing a basis of $R$ over $A$, $\operatorname{End}_A(R) = M_n(A)$, $\operatorname{End}_A(R/\underline{m}R) = M_n(A/\underline{m})$. □

Theorem 5.4.32 already shows the existence of nontrivial Azumaya algebras, that is not isomorphic to $\mathrm{End}_A(P)$, the finite-dimensional division algebras. To understand the general case we develop some basic facts.

5.4.2.5. *Complete local rings.* The idea of faithful flatness can be used in various ways, the simplest is to analyze locally the structure of Azumaya algebras. One can take either the formal or the local algebraic approach. In the first approach, we let $A$ be a complete local ring with an algebraically closed residue field $K$.

THEOREM 5.4.34. *Let $A$ be a complete local ring with maximal ideal $\underline{m}$ and an algebraically closed residue field $K = A/\underline{m}$. An $A$-algebra $R$ is an Azumaya algebra if and only if:*

(1) *it is a free $A$-module of some rank $n^2$ and $R/\underline{m}R = M_n(K)$.*

(2) *moreover, fixing $n$, this is equivalent to $R$ being isomorphic to $M_n(A)$.*

(3) *more generally, if $A$ is a complete local ring with residue field $F$, $R$ is an Azumaya algebra over $A$, then $R/\underline{m}R = M_h(D)$ where $D$ is a $k^2$-dimensional division algebra over its center $F$, $R$ has rank $n^2$ with $n = hk$ and moreover $R = M_h(S)$ where $S$ is a rank $k^2$ Azumaya algebra over $A$ and $S/\underline{m}S = D$.*

PROOF. Condition (1) is clearly necessary and sufficient by Corollary 5.4.33 and the fact that a central simple algebra over an algebraically closed field $K$ is of the form $M_n(K)$. We prove then (2).

The idea is to lift a set of matrix units $e_{i,j}$ modulo the various powers of the maximal ideal and, by Nakayama's lemma, obtain always a basis of matrix units over $A/\underline{m}^k$ for all $k$, then pass to the limit by completeness.

In fact we claim that it suffices to lift the orthogonal idempotents $e_{ii}$ to orthogonal idempotents $u_{ii}$ which add to 1, and this we can do by Lemma 1.3.14, induction and completeness.

Suppose then we have lifted the $e_{ii}$ to orthogonal idempotents $u_{ii}$ which add to 1, so that $R = \bigoplus_{ij} u_{ii}Ru_{jj} = \bigoplus_i Ru_{ii}$. Next since $R$ is a free $A$-module, we have that $Ru_{ii}$ is a free module over $A$ for every $i$. By reducing mod $\underline{m}$, we see that its rank is $n$. Then it is enough to show that $R = R \otimes 1 \subset R^e = \mathrm{End}_A(R)$ maps isomorphically to $\mathrm{End}_A(Ru_{11}) = M_n(A)$. In fact when we have a module $M$ over some ring $R$, we decompose it as $M = N \oplus P$ and let $e$ be the idempotent projecting $M$ to $N$ with kernel $P$. We have that $\mathrm{End}_R(N) = e\,\mathrm{End}_R(M)e$.

In our case the idempotent is $1 \otimes u_{11} \in R^e$, and we see that $u_{ii}Ru_{ii}$ is a free rank 1 module over $A$ for every $i$ (again by reducing mod $\underline{m}$). Then

$$1 \otimes u_{11}R^e 1 \otimes u_{11} = R \otimes u_{11}R^{\mathrm{op}}u_{11}.$$

Next $u_{11}R^{\mathrm{op}}u_{11} = A$, and the claim follows.

Condition (3) is similar to (2), and we leave it to the reader. □

In particular this result can also be used when $A/\underline{m}$ is a finite field $\mathbb{F}$, in which case $D = \mathbb{F}$ by Wedderburn's theorem, so for instance an Azumaya algebra over a ring of $p$-adic numbers is a matrix algebra. See also Knus, Ojanguren [**KO74a**, pp. 100–105].

5.4.2.6. *The faithfully flat and étale splitting.* When working with Azumaya algebras over a connected spectrum, we have a fixed rank which will be some number $n^2$. In general we can decompose $A = \bigoplus A_i$ and $R = \bigoplus R_i$ so that $R_i$ is Azumaya over $A_i$ of constant rank $n_i^2$ (see Proposition 5.3.15).

We use a method of *reduction to finitely generated commutative rings*. If a commutative ring $B$ is finitely generated as a ring, then it is Noetherian and, given $\mathfrak{m}$ a maximal ideal of $B$, the quotient $B/\mathfrak{m}$ is a finite field.

The first example of the method is this.

THEOREM 5.4.35. *Let $P$ be a finitely generated projective module over a commutative ring $A$. Then there is a finitely generated commutative ring $B \subset A$ and a finitely generated projective module $\bar{P}$ over $B$ with $P = \bar{P} \otimes_B A$.*

PROOF. $P$ is finitely projective over $A$, say generated by $k$ elements. So there is a map $\pi : A^k \to P$, surjective, and the map splits, so there is an inclusion of $\phi : P \to A^k$ with $\pi \circ \phi = 1_P$. In fact in the language of projective generators if $A^k = \bigoplus_{i=1}^k Ae_i$, we have $\pi(e_i) = m_i$, $\phi(x) = \sum_{i=1}^k \phi_i(x)e_i$,

$$x = \pi \circ \phi(x) = \pi(\sum_{i=1}^{k} \phi_i(x)e_i) = \sum_{i=1}^{n} \phi_i(x)m_i, \ \forall x \in P. \tag{5.50}$$

So the entries of the matrix $\pi$ are $\phi_i(m_j)$ and the condition (5.50) is that $\pi^2 = \pi$.

$$m_j = \sum_{i=1}^{n} \phi_i(m_j)m_i \implies \phi_\ell(m_j) = \sum_{i=1}^{n} \phi_i(m_j)\phi_\ell(m_i).$$

Define $B$ as any ring containing the entries $\psi_i(m_j)$ of $\pi$. Then the same matrix gives rise to a projection $\pi + B : B^k \to B^k$ and, letting $\bar{P} = Im(\pi_B)$, we have that $\bar{P}$ is a finitely generated projective module over $B$ and $\pi = 1_A \otimes_B \pi_B$, so $P = A \otimes_B \bar{P}$. □

In the second approach we have

PROPOSITION 5.4.36. *Let $R$ be an Azumaya algebra free of rank $n^2$ over its center $A$. Then there is a finitely generated ring $B \subset A$ and a $B$ subalgebra $U \subset R$ so that $U$ is an Azumaya algebra free of rank $n^2$ over its center $B$ and $R = U \otimes_B A$.*

PROOF. Let $m_1, \dots, m_{n^2}$ be a basis of $R$ over $A$, so that $m_i \otimes m_j$ is a basis of $R^e = R \otimes_A R^{\mathrm{op}}$ over $A$.

The space $\mathrm{End}_A(R)$ has as its basis the elementary matrices $e_{i,j} : m_h \mapsto \delta_h^j m_i$ and also, by hypothesis, the isomorphism $\psi : R^e \to \mathrm{End}_A(R)$ gives a basis of elements $\psi(m_i \otimes m_j) : x \mapsto m_i x m_j$.

By hypothesis there are elements $\lambda_{i,j}^h$ in $A$ giving in $R$ a multiplication table

$$m_i m_j = \sum_h \lambda_{i,j}^h m_h$$

and elements $\mu_{i,j}^{h,k} \in A$ with

$$e_{i,j} = \sum_{h,k} \mu_{i,j}^{h,k} \psi(m_h \otimes m_k).$$

If $B$ is the subring of $A$ generated by the elements $\lambda_{i,j}^h, \mu_{i,j}^{h,k} \in A$, we clearly have that the subspace $U := \sum_h Bm_h$ is an algebra and that the corresponding map $\psi : U^e \to \mathrm{End}_A(U)$ is an isomorphism. Finally clearly $R = U \otimes_B A$. □

Let $R$ be a rank $n^2$ Azumaya algebra over its center $A$ which is finitely generated over $\mathbb{Z}$. Let $\mathfrak{m}$ be a maximal ideal of $A$ so that $A/\mathfrak{m}$ is a finite field $F$ and $R/\mathfrak{m}R$ is isomorphic to $M_n(F)$ since by Wedderburn's theorem a finite division algebra is commutative. Let $A_\mathfrak{m}$ be the localization at $\mathfrak{m}$, and let $\hat{A}_\mathfrak{m}$ be its completion. Then the map $i : A_\mathfrak{m} \to \hat{A}_\mathfrak{m}$ is a faithfully flat inclusion (see **[Mat89**, Theorem 8.14]) and $R_{A_\mathfrak{m}} \otimes_{A_\mathfrak{m}} \hat{A}_\mathfrak{m}$ is isomorphic to $M_n(\hat{A}_\mathfrak{m})$ (Theorem 5.4.34(3)).

Let $G \supset F$ be the finite field of degree $n = [G : F]$. Then there is a generator $\bar{a} \in G$ with $G = F[\bar{a}]$, and $\bar{a}$ satisfies a monic irreducible polynomial $g(x) \in F[x]$ of degree $n$. Moreover, if $g'(x)$ denotes the derivative of $g(x)$, we have $g(\bar{a}) \neq 0$. Moreover, $G$ embeds in $M_n(F)$, so we consider $M_n(F)$ as a module over $G$ by right multiplication, and we let $\bar{b}_1, \dots, \bar{b}_n$ be a basis. We may take $a, b_1, \dots, b_n \in R$ reducing, modulo $\mathfrak{m}$ to $\bar{a}, \bar{b}_1, \dots, \bar{b}_n$. We have the isomorphism

$$\bar{j} : M_n(F) \otimes_F G \to \mathrm{End}_G(M_n(F)), \quad \bar{j}(u \otimes g)(x) := uxg.$$

Let $a \in R$ map to $\bar{a}$, then the same $a$ thought of as element of $R_{A_\mathfrak{m}} \subset M_n(\hat{A}_\mathfrak{m})$ satisfies its characteristic polynomial $\chi_a(x)$ monic of degree $n$, with coefficients in $A_\mathfrak{m}$ by Theorem 2.6.15, which reduces modulo $\mathfrak{m}$ to $g(x)$. So let

$$S_\mathfrak{m} := A_\mathfrak{m}[x]/\chi_a(x) = A_\mathfrak{m}[a] \subset R_{A_\mathfrak{m}}$$

be a commutative ring free over $A_\mathfrak{m}$ with basis $1, a, \dots, a^{n-1}$. The ideal $S_\mathfrak{m}\mathfrak{m}$ is maximal and $S_\mathfrak{m}/S_\mathfrak{m}\mathfrak{m} = G$. By the going-up lemma $S_\mathfrak{m}$ is also local. In particular the derivative $\chi'_a(x)$ computed in $a$ is invertible modulo $\mathfrak{m}S_\mathfrak{m}$, and so it is invertible. In other words the extension $A_\mathfrak{m} \subset S_\mathfrak{m}$ is a simple étale extension (cf. formula (1.19)).

Since $S_\mathfrak{m} \subset R_{A_\mathfrak{m}}$, we consider $R_{A_\mathfrak{m}}$ as $S_\mathfrak{m}$ module by multiplication on the right. We have a map

$$j : R_{A_\mathfrak{m}} \otimes_{A_\mathfrak{m}} S_\mathfrak{m} \to \mathrm{End}_{S_\mathfrak{m}}(R_{A_\mathfrak{m}}), r \otimes b(x) := rxb,$$

and modulo $\mathfrak{m}$ the map $j$ is the isomorphism $\bar{j}$. So using Nakayama's lemma, we deduce that $R_{A_\mathfrak{m}} \otimes_{A_\mathfrak{m}} S_\mathfrak{m}$ is isomorphic to $M_n(S_\mathfrak{m}) = \mathrm{End}_{S_\mathfrak{m}}(R_{A_\mathfrak{m}})$.

Furthermore the elements $b_j a^i$, $i = 0, \dots, n-1$ and $j = 1, \dots, n$ form a basis of $R_{A_\mathfrak{m}}$ over $A_\mathfrak{m}$ and $b_j$, $j = 1, \dots, n$ form a basis of $R_{A_\mathfrak{m}}$ over $S_\mathfrak{m}$.

PROPOSITION 5.4.37. *There is an $f \notin \mathfrak{m}$ so that*

(1) *$a$ satisfies a monic polynomial $u(x)$ with coefficients in $A[f^{-1}]$ which equals $\chi_a(x)$ in $A_\mathfrak{m}$.*
(2) *Setting $S_f := A[f^{-1}][a]$, we have that $S_f$ is an algebra and a free module with basis $1, a, a^2, \dots, a^{n-1}$ over $A[f^{-1}]$.*
(3) *$R[f^{-1}] = R \otimes_A A[f^{-1}]$ is a free module over $S_f$, acting on the right with basis $b_1, \dots, b_n$.*
(4) *The analogue of the map $j$,*

$$\tilde{j} : R[f^{-1}] \otimes_{A[f^{-1}]} S_f \to \mathrm{End}_{S_f}(R[f^{-1}]), \quad j(r \otimes b)(x) := rxb, \forall x \in R[f^{-1}], \tag{5.51}$$

*is an isomorphism.*
(5) *Furthermore, we may also assume that $u(a)$ is invertible in $S_f$, so $S_f$ is a simple étale extension.*

PROOF. First there is an element $k \in A \setminus \mathfrak{m}$ so that the $n^2$ elements $a^i b_j$, $i = 0, \dots, n-1$ and $j = 1, \dots, n$ form a basis of $R \otimes_A A[k^{-1}]$ over $A[k^{-1}]$.

Next there is $\ell \in A \setminus \mathfrak{m}$ so that the polynomial $\chi_a(x) = j_\ell(u(x))$ with $u(x) \in A[\ell^{-1}][x]$ and $j_\ell : A[\ell^{-1}] \to A_\mathfrak{m}$ is the further localization.

The fact that $\chi_a(a) = 0$ in $R_{A_\mathfrak{m}}$ means that there is a further $h \in A \setminus \mathfrak{m}$ so that $u(a)h = 0$ in $R[\ell^{-1}]$, that is setting $f := k \cdot h \cdot \ell$, we have $u(a) = 0$ in $R[f^{-1}]$.

The map $\tilde{j}$ of formula (5.51) is a morphism of two Azumaya algebras of rank $n^2$ over $S_f$, so it is an isomorphism by Corollary 5.4.30. □

Observe that

$$R[f^{-1}] \otimes_{A[f^{-1}]} S_f = (R \otimes_A A[f^{-1}]) \otimes_{A[f^{-1}]} S_f = R \otimes_A S_f.$$

Now for each maximal ideal $\mathfrak{m}_i$, we may choose one such $f_i \notin \mathfrak{m}_i$, and then the ideal generated by all these $f_i$ is $A$. So there is a finite list of $f_i$, $i = 1, \dots, p$, so that there are elements $a_i \in A$, $i = 1, \dots, p$, with $\sum_{i=1}^p a_i f_i = 1$.

It follows that

COROLLARY 5.4.38. *If $R$ is an Azumaya algebra of rank $n^2$ over a finitely generated ring, then we have an étale covering $C := \bigoplus_{i=1}^p S_{f_i}$ of $A$ with $C \otimes_A R \simeq M_n(C)$.*

PROOF. It follows from Proposition 5.4.37 that

$$C := \bigoplus_{i=1}^{p} S_{f_i}, \quad R \otimes_A C = M_n(C).$$

Furthermore, the extension $A \subset C$ is faithfully flat and étale. □

THEOREM 5.4.39. *Let $R$ be a rank $n^2$ Azumaya algebra over its center $A$. Then there is an étale covering $\eta : A \to S$ (i.e., étale and faithfully flat) so that $S \otimes_A R \simeq M_n(S)$.*

PROOF. First there is an affine covering $A \subset \bigoplus_i A[f_i^{-1}]$ so that each algebra $R_i := A[f_i^{-1}] \otimes_A R$ is a free module over $A[f_i^{-1}]$ of rank $n^2$, so we are reduced to the case $R$ a free module of rank $n^2$ over its center $A$.

By Proposition 5.4.36 we have $R = A \otimes_B U$ and $U$ satisfies the hypotheses of Corollary 5.4.38, so we have the extension $B \subset C$ is faithfully flat and étale with $C \otimes_B U \simeq M_n(C)$.

Then setting $S := C \otimes_B A$, we have that $S$ is faithfully flat and étale over $A$ and

(5.52)
$$S \otimes_A R \simeq (C \otimes_B A) \otimes_A A \otimes_B U \simeq A \otimes_B (C \otimes_B U) \simeq A \otimes_B M_n(C) = M_n(S).$$

□

5.4.2.7. *Reduced trace and characteristic polynomial.* Let $R$ be a rank $n^2$ Azumaya algebra over its center $Z$. Take a faithfully flat splitting $R \otimes_Z B \cong M_n(B)$.

THEOREM 5.4.40. *Given an element $a \in R$ the coefficients of the characteristic polynomial of the matrix $a \otimes 1 \in M_n(B)$ lie in $Z$ and are independent of the faithfully flat splitting.*

PROOF. For the first statement we use Theorem 2.6.15. For the second if we have two faithfully flat embeddings, we combine them in a unique embedding $Z \to B_1 \otimes_Z B_2$. □

We can thus define the *reduced trace*, the *norm*, and the *reduced characteristic polynomial* $\chi_a(t) \in Z[t]$ of an element $a \in R$ as the trace, the determinant, and the characteristic polynomial of the matrix $a \in M_n(B)$, and we have $\chi_a(a) = 0$.

We now can see that the reduced characteristic polynomial behaves well under morphisms of rank $n^2$ Azumaya algebras.

PROPOSITION 5.4.41. *Given a homomorphism $\phi : R \to S$ of rank $n^2$ Azumaya algebras and an element $a \in R$, we have that $\phi(\chi_a(t)) = \chi_{\phi(a)}(t)$.*

PROOF. Let $Z$ be the center of $R$, and let $Z'$ be the center of $S$. By Corollary 5.4.29 we have $S = Z' \otimes_Z R$.

Take a faithfully flat map $Z \to B$ splitting $R$, that is $B \otimes_Z R = M_n(B)$. Then the map $Z' \to B \otimes_Z Z'$ is faithfully flat and

$$(B \otimes_Z Z') \otimes_{Z'} S = B \otimes_Z (Z' \otimes_Z R) = (B \otimes_Z Z') \otimes_Z R = M_n(B \otimes_Z Z').$$

The statement for matrices is the usual one given in Corollary 2.6.14. □

EXERCISE 5.4.42. Let $R$ be a rank $n^2$ Azumaya algebra over its center $Z$. Prove that the trace map to $Z$ is surjective.

*Hint*. The image of the trace is an ideal $I$ of $Z$. Deduce a contradiction if $I \neq Z$.

**5.4.3. Using faithfully flat descent.** One can use ideas of faithfully flat descent to derive properties of $R$. By way of example we reprove various properties of Azumaya algebras using descent.

PROPOSITION 5.4.43. *Assume $R$ is Azumaya over $A$ and $A \mapsto B$ is a faithfully flat extension of $A$ such that $B \otimes_A R = M_n(B)$.*

(1) *Let $I$ be a two-sided ideal of $R$, and let $J = R \cap A$. Then $I = JR$ and $R/I = R \otimes_A A/J$ is Azumaya.*
(2) *$A$ is the center of $R$.*
(3) *There is an $A$-linear map* $\mathrm{tr} : R \to A$, *the* reduced trace, *which coincides with the trace in any splitting. Conceived as an endomorphism $\mathrm{tr} : R \to A \subset R$, it is associated to a canonical element $\xi = \sum_j r_j \otimes s_j \in R^e$, $\mathrm{tr}(r) = \sum_j r_j r s_j$.*
(4) *$R$ is a projective $R^e$-module.*
(5) *The functor $N \to R \otimes_A N$ is an equivalence between the category of $A$ modules and that of $R^e$-modules. Given an $R^e$-module $M$, the $A$-module*

$$M^R := \{x \in M | (1 \otimes r - r \otimes 1)x = 0, \ \forall r \in R\}$$

*equals* $\hom_{R^e}(R, M)$ *and gives the isomorphism $M = R \otimes_A M^R$.*
(6) *If $R \subset U$ with $U$ any $A$-algebra, then $U = R \otimes_A R'$ where $R'$ is the centralizer of $R$ in $U$.*

PROOF. (1) To prove $I = JR$, one extends scalars to $B$ and it suffices to prove that $I_B = J_B R_B$; this is Lemma 2.6.2.

(2) Let $B$ be faithfully flat over $A$, and let $R_B = M_n(B)$. If $Z$ is the center of $R$, then $A \subset Z$ and $Z_B$ is contained in the center of $R_B$, which is $B$.

(3) One shows that $\mathrm{tr} : M_n(B) = R \otimes_A B \to B$ maps $R$ to $A$ by verifying the faithfully flat descent criterion using that the two isomorphisms of $R \otimes_A (B \otimes_A B)$ with $M_n(B \otimes_A B)$ are conjugate by an automorphism that leaves the trace invariant.

(4) Since $R^e$ is identified to $\operatorname{End}(R)$, it is enough to show that a finitely generated projective module $P$ over a commutative ring $A$ is also projective over $\operatorname{End}(P)$.

The property is easily seen to be local on the spectrum of $A$, and thus one can reduce to the case $A$ local and $P$ free of some rank $k$. Then $P = A^k$, hence $\operatorname{End}(P) = M_k(A)$, and $P$ is clearly projective over the ring of matrices which splits as direct sum of the spaces of columns, each isomorphic to $P$. Hence (4).

(5) Let $J$ be the left ideal in $R^e$ which annihilates the element $1 \in R$. We claim that$J$ is generated by the elements $\{1 \otimes r - r \otimes 1 | r \in R\}$. In fact if $\sum_i a_i b_i = (\sum_i a_i \otimes b_i)1 = 0$, we have $\sum_i a_i \otimes b_i = \sum_i a_i \otimes 1(1 \otimes b_i - b_i \otimes 1)$.

From (3) since the trace is surjective, we can split $R = A \oplus \bar{R}$ as direct sum decomposition of $A$-modules.

Since $R = R^e/J_R$, the identification $\hom_{R^e}(R, M) = M^R$ is by $\phi \to \phi(1)$.

One has to prove that the natural map $R \otimes_A M^R = R \otimes_A \hom_{R^e}(R, M) \to M$ is an isomorphism. As usual it suffices to show this, using a faithfully flat extension, for $R = M_n(B)$, and we leave it to the reader.

(6) is a special case of (5). □

Finally, there is a geometric point of view—we view $R$ as a *torsor*. We will take this up in §10.4.1; see the discussion on page 283.

5.4.3.1. *Braun's theorem.* Recall that a simple algebra $S$ is an algebra with no two-sided ideals. If $S$ is an algebra with a 1, then its center is clearly a field.

LEMMA 5.4.44. *Let $R, S$ be two $F$-algebras with 1, where $F$ is a field, and assume $S$ is simple with center $F$. Then every nonzero ideal $I$ of $R \otimes_F S$ is of the form $I = J \otimes_F S$ for some ideal $J$ in $R$.*

PROOF. Let $J := \{x \in R \,|\, x \otimes 1 \in I\}$. We have that $J$ is an ideal of $R$ and $I \supseteq J \otimes S$, and we claim that $I = J \otimes S$.

Replacing $R$ with $R/J$, we see that it is enough to prove, by contradiction, that if $I \neq 0$, then there is an element $a \in R$, $a \neq 0$ with $a \otimes 1 \in I$.

Take a shortest (with $k$ minimal) expression $x = \sum_{i=1}^k a_i \otimes b_i \in I$; clearly the elements $a_i$ are linearly independent over $F$. Since $S$ is simple, we have some elements $u_j, v_j \in S$ so that $\sum_j u_j b_1 v_j = 1$, so (identifying $S$ and $1 \otimes S$)

$$\sum_j u_j x v_j = a_1 \otimes 1 + \sum_{i=2}^k a_i \otimes c_i, \; c_i = \sum_j u_j b_i v_j,$$

and we may assume that $b_1 = 1$. We claim that $k = 1$; otherwise, consider $c_2$. If $c_2 \in F$, then we can write $a_1 \otimes 1 + a_2 \otimes c_2 = (a_1 + c_2 a_2) \otimes 1$ and $x$ is not a shortest expression. So $c_2 \notin F$, and since $F$ is the center, there is a $y \in S$ with $[c_2, y] \neq 0$ then $[x, 1 \otimes y] \in I$ and $[x, 1 \otimes y] = \sum_{i=2}^k a_i \otimes [c_i, y] \neq 0$ is a shorter expression, giving a contradiction. □

COROLLARY 5.4.45. *If $R, S$ are two $F$-algebras, over a field $F$, both simple and $S$ has center $F$, then $R \otimes_F S$ is simple.*

We can now give the proof by Dicks **[Dic88]** of Braun's characterization of Azumaya algebras. The notations $\eta = \eta_R : R^e \to \operatorname{End}_A(R), \phi(x) = \eta_R(x)(1)$, $J_R = \ker \phi$ are as in §5.4.1.

THEOREM 5.4.46 (Braun [**Bra84**]). *A ring $R$ is an Azumaya algebra over its center $A$ if and only if:*
(1) *there exist elements $a_i, b_i \in R$ with $\sum_i a_i b_i = 1$, and are such that*
(2) $\sum_i a_i r b_i \in A, \ \forall r \in R$.

PROOF. If $R$ is an Azumaya algebra over $A$, then it is separable and the statement follows from Corollary 5.4.9 and formula (5.40).

Conversely, let $e := \sum_i a_i \otimes b_i$ satisfy the conditions of the theorem,

$$\eta(e) \in \hom(R, A), \ \eta(e)(1) = 1.$$

By Remark 5.4.7 it is sufficient to prove that $J_R \cdot e = 0$.

What we do have is that $\eta(J_R \cdot e) = 0$. In fact $\eta(j \cdot e)(r) = \eta(j)\eta(e)(r)$ and $\eta(e)(r) \in A$ while $\eta(j)$ is $A$ linear. So that, if $j \in J_R$,

$$\eta(j \cdot e)(r) = \eta(j)\eta(e)(r) = \eta(e)(r)\eta(j)(1) = \eta(e)(r)\phi(j) = 0$$

since $J_R = \ker \phi$.

We claim first that the two-sided ideal $I = R^e e R^e$, of the algebra $R^e$, generated by $e$ is $R^e$. Let $N := \{r \in R \,|\, r \otimes 1 \in I\}$. Clearly $N$ is a two-sided ideal of $R$. If $I \neq R^e$, then $N \neq R$, so there is a maximal ideal $M \supset N$ in $R$, so $R/M$ and $(R/M)^{\mathrm{op}}$ are both simple algebras. If $F$ is the common center, then we have a map

$$R^e = R \otimes_A R^{\mathrm{op}} \to R/M \otimes_F R/M^{\mathrm{op}}.$$

Under this map the element $e$ maps to an element $\bar{e}$ with the same properties.

In particular $\bar{e} \neq 0$, so the ideal $I$ maps to a nonzero ideal which, since the ring $R/M \otimes_F R/M^{\mathrm{op}}$ is simple (Corollary 5.4.45), must coincide with $R/M \otimes_F R/M^{\mathrm{op}}$. In particular there is some element $\sum_j c_j \otimes d_j \in I$ mapping to 1. Then, since, by the hypothesis on $e$, $\sum_i a_i d_j b_i \in A, \ \forall j$, we have

$$\sum_j c_j \otimes \sum_i a_i d_j b_i = \sum_i \sum_j c_j \otimes a_i d_j b_i = \sum_i 1 \otimes a_i (\sum_j c_j \otimes d_j) 1 \otimes b_i \in I,$$

$$\sum_j c_j \otimes \sum_i a_i d_j b_i = \sum_j c_j \sum_i a_i d_j b_i \otimes 1 \implies \sum_j c_j \sum_i a_i d_j b_i \in N \subset M.$$

This is a contradiction since modulo $M$ this element is 0, but $\sum_j c_j \otimes d_j$ maps to 1, so $\sum_j c_j \otimes \sum_i a_i d_j b_i$ maps to $1 \otimes \sum_i \bar{a}_i \bar{b}_i = 1$. Thus $R^e e R^e = R^e$. Now take $u, v \in R^e$ and compute in $\mathrm{End}(R \otimes R)$:

$$1 \otimes \eta(uev) = 1 \otimes \eta(u)\eta(e)\eta(v) \in \mathrm{End}(R \otimes R) = \mathrm{End}(R \otimes R^{\mathrm{op}}).$$

Apply $1 \otimes \eta(uev)$ to an element $x = \sum_k g_k \otimes h_k \in R \otimes R$. Since $\eta(e)\eta(v)(h_k) \in A$, we have $\eta(u)\eta(e)\eta(v)(h_k) = \eta(e)\eta(v)(h_k)\eta(u)(1)$, so

$$(1 \otimes \eta(uev))(x) = \sum_k g_k \otimes \eta(u)\eta(e)\eta(v)(h_k) = \sum_k g_k \eta(e)\eta(v)(h_k) \otimes \eta(u)(1).$$

We claim that $\sum_k g_k \eta(e)\eta(v)(h_k) \in \eta(x)(R)R$ for all $v$. In fact we may assume $v = t \otimes z$ so $\eta(v)(h_k) = t h_k z$, and identifying $x \in R \otimes R^{\mathrm{op}}$,

$$\sum_k g_k \eta(e)\eta(v)(h_k) = \sum_k g_k \eta(e) t h_k z = \sum_i \sum_k g_k a_i t h_k z b_i = \sum_i \eta(x)(a_i t) z b_i.$$

Take now $x \in J_R e$, so $\eta(x) = 0$, and from the previous argument we have that

$$(1 \otimes \eta(uev))(x) \in \eta(x)(R)R \otimes R = 0.$$

Since $R^e e R^e = R^e$, we have $1 = \sum_i u_i e v_i$, and we deduce $(1 \otimes \eta(1))(x) = x = 0$. □

Further properties of Azumaya algebras will be discussed in §10.3.

5.4.3.2. *The Brauer group.* Although we will not need this, we should recall that the main motivation to introduce Azumaya algebras is in the theory of the Brauer group over commutative rings or even schemes generalizing the classical theory of Brauer over fields.

We have seen that if $R, S$ are two Azumaya algebras over the same commutative ring $A$, so is $R \otimes_A S$.

Next one defines an equivalence on Azumaya algebras given by $R \equiv S$ if there exist two finitely generated projective modules $P_i$ such that $R \otimes_A \mathrm{End}_A(P_1) = S \otimes_A \mathrm{End}_A(P_2)$. In this equivalence in particular all algebras $\mathrm{End}_A(P)$ for $P$ projective are equivalent. Next one sees that the equivalence classes of Azumaya algebras form a commutative group under the tensor product, in which the class of $\mathrm{End}(P)$ is the identity element and $R^{\mathrm{op}}$ gives an inverse to $R$, since $R \otimes_A R^{\mathrm{op}} \cong \mathrm{End}_A(R)$.

This group, denoted $\mathrm{Br}(A)$ is functorial in $A$, in the classical case when $A$ is a field. It can be described by the theory of crossed products by Galois cohomology, which in the general case it can be described as an étale cohomology group.

CHAPTER 6

# Tensor symmetry

*In Remark* 2.2.3 *we saw that a T-ideal is in particular an ideal of the tensor algebra* $T(V)$ *which is stable under the action of the linear group* $\mathrm{GL}(V)$. *Thus it is essential to recall the structure of* $T(V)$ *as representation of* $\mathrm{GL}(V)$, *which is the object of this chapter. We recall the representation theory of* $\mathrm{GL}(V)$ *and of the symmetric group, the two go together. The chapter is mostly without proofs. For the proofs we refer to* **[Pro07]** *or* **[GW09]**.

## 6.1. Schur–Weyl duality

**6.1.1. Symmetry on tensor algebras.** We want to recollect some basic facts without proofs, see for instance **[Pro07]** or **[GW09]**. If $V$ is a finite-dimensional vector space over a field $F$, the tensor power $V^{\otimes n}$ is a representation both of the group $\mathrm{GL}(V)$ of invertible linear transformations of $V$, and of the symmetric group $S_n$.

The group $\mathrm{GL}(V)$ acts by $g(v_1 \otimes v_2 \otimes \cdots \otimes v_i) = g(v_1) \otimes g(v_2) \otimes \cdots \otimes g(v_n)$ while $\sigma \in S_n$ acts by $\sigma(v_1 \otimes v_2 \otimes \cdots \otimes v_n) = v_{\sigma^{-1}(1)} \otimes v_{\sigma^{-1}(2)} \otimes \cdots \otimes v_{\sigma^{-1}(n)}$ (cf. Definition 4.1.3). The two actions commute. If $F$ is infinite, we have

THEOREM 6.1.1. *The algebras* $A, B$ *spanned respectively by the elements* $g^{\otimes i}$, *with* $g \in \mathrm{GL}(V)$ *and* $\sigma \in S_n$, *are the centralizers of each other.*

PROOF. Let us sketch the proof in characteristic 0.

First one notices that one identifies $\mathrm{End}(V^{\otimes n}) = \mathrm{End}(V)^{\otimes n}$, and then one easily sees that $\mathrm{End}_{S_n}(V^{\otimes n}) = (\mathrm{End}(V)^{\otimes n})^{S_n}$ coincides with the symmetric tensors of $\mathrm{End}(V)^{\otimes n}$. This is the Schur algebra defined and discussed in §4.1.2, and if $F$ is infinite, it is easily seen to be generated by the tensors $g^{\otimes n}$, $g \in \mathrm{GL}(V)$. Then the full statement follows from the double centralizer Theorem 1.3.21. □

In fact this theorem can be viewed also as the *first fundamental theorem* (*FFT*) of invariant theory of the group $\mathrm{GL}(V)$.

THEOREM 6.1.2 (FFT of the group $\mathrm{GL}(V)$). *The ring of polynomial invariants on the space* $V^{\oplus h} \oplus (V^*)^{\oplus k}$ *of h vectors* $v_1, \dots, v_h \in V$ *and of k forms* $\phi_1, \dots, \phi_k \in V^*$ *is generated by the elements* $\langle \phi_i \mid v_j \rangle = \phi_i(v_j)$.

This theorem is classical in characteristic 0 and was proved in **[DCP76]** for all characteristics using combinatorial methods in invariant theory.

In characteristic 0 both algebras $A, B$ are semisimple and satisfy thus all the statements of the double centralizer theorem, so we have some formula like (1.8). The precise description of this formula—the isotypic components and irreducible representations—will be given in formula (6.8). In this case the division rings reduce all to the field $F$, and we can in fact work in an *arithmetic way* with $F = \mathbb{Q}$.

In invariant theory usually, besides a first, there is also a *second fundamental theorem* (*SFT*) describing the ideal of relations among the generators of the ring of invariants. In this case we have two possible equivalent formulations of this theorem:

THEOREM 6.1.3 (SFT of the group $GL(V)$). *Let $n = \dim V$.*

(1) *The ideal of relations between the invariants $x_{i,j} := \langle \phi_i \,|\, v_j \rangle = \phi_i(v_j)$ is 0 if $n \geq \min(h,k)$; otherwise, it is generated by the determinants of the $n+1 \times n+1$ minors of the $h \times k$ matrix in the elements $x_{i,j}$.*
(2) *The kernel of the map $j : F[S_m] \to \mathrm{End}(V)^{\otimes m}$ is 0 if $n \geq m$; otherwise, it is the ideal $I_{n+1}$ generated by the antisymmetrizer $\sum_{\sigma \in S_{n+1}} \epsilon_\sigma \sigma$ (where $S_{n+1} \subset S_m$).*

One can approach this theory from two complementary angles: either develop first the representation theory of the symmetric group and deduce from it one of the linear group, or vice versa.

6.1.1.1. *Partitions, Young diagrams, and tableaux.* We recall the representation theory of the symmetric group in characteristic 0. For a systematic treatment the reader can refer to **[Sag01]** and **[JK81]**.

We have defined partitions in Definition 3.2.11. In this chapter partitions are represented by their *Young diagrams*.

Formally, such a diagram can be defined as a suitable set of integer points in the plane, but we shall use a less formal and more pictorial approach.

For example, the Young diagram of the partition $\lambda = (5, 2, 2, 1)$ of 10 is the diagram

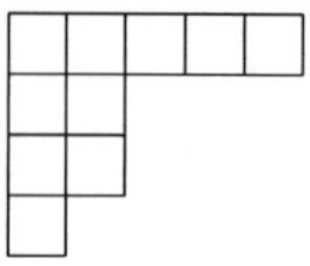

5 boxes
2 boxes
2 boxes
1 box,

consisting of ten boxes, arranged corresponding to the parts of $\lambda$.

Let $\lambda = (\lambda_i)$ be a partition. The diagram $\check{\lambda}$, corresponding to the *conjugate* or *dual* partition $\check{\lambda} = (\check{\lambda}_j)$, is obtained by flipping the diagram of $\lambda$ along its main diagonal, that is, by interchanging rows and columns.

For example, for $\lambda = (5, 2^3, 1)$, the diagrams of $\lambda$ and of $\check{\lambda}$ are given by

$\lambda =$ 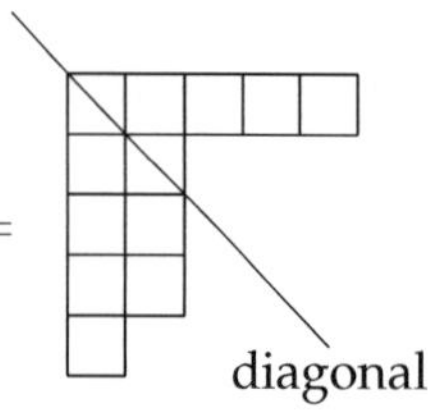

and $\check{\lambda} =$ 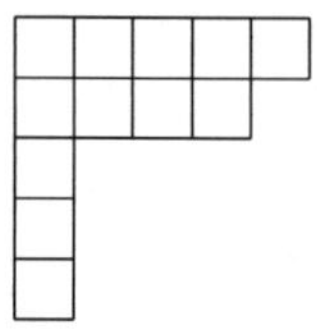 ,

and so the conjugate partition is $\check{\lambda} = (5, 4, 1^3)$. Thus $\lambda_i$ is the length of the $i$th row in the diagram of $\lambda$, while $\check{\lambda}_j$ is the length of the $j$th column in the diagram of $\lambda$. In particular the length or height $\mathrm{ht}(\lambda) = \check{\lambda}_1$.

There are at least three useful partial orders on partitions, one is the simple geometric inclusion of diagrams, which we denote by $\lambda \supseteq \mu$. Formally, if $\lambda, \mu$ have rows $\lambda_i, \mu_i$, then $\lambda \supseteq \mu$ if and only if $\mu_i \leq \lambda_i$ for all $i$.

In this case we can also define the *skew diagram* $\lambda \setminus \mu$ formed by all the boxes of $\lambda$ not contained in $\mu$.

The second is the lexicographic order.

The third subtler order is the *dominance order*, and it is related to various geometric and combinatorial theories.

DEFINITION 6.1.4. We say that $\lambda \geq \mu$ in the dominance order, if for all $i$ we have $\sum_{j=1}^{i} \lambda_j \geq \sum_{j=1}^{i} \mu_j$. Intuitively, if $\lambda, \mu$ are partitions of the same $n$, then $\lambda \geq \mu$ if and only if $\lambda$ is obtained from $\mu$ by *moving* boxes of $\mu$ from lower rows to upper rows.[1]

**6.1.2. Tableaux.** Let $\lambda$ be a partition of $n$.

DEFINITION 6.1.5. A bijective map from the set of boxes of the Young diagram of $\lambda$ to the interval $(1, 2, 3, \ldots, n-1, n)$ is called a Young *tableau* of shape $\lambda$.

It can be thought of as a *filling* of the diagram with numbers. The given partition $\lambda$ is called the *shape* of the tableau.

In this definition each number $1, \ldots, n$, is precisely being used once. In representation theory also another type of tableaux is useful, as we shall see later. For example, a Young tableau of the partition $\lambda = (5, 3, 2)$ of 10 is given by

$$T = \begin{array}{|c|c|c|c|c|} \hline 7 & 2 & 4 & 8 & 3 \\ \hline 5 & 10 & 6 & \multicolumn{2}{c}{} \\ \cline{1-3} 1 & 9 & \multicolumn{3}{c}{} \\ \cline{1-2} \end{array} \; .$$

DEFINITION 6.1.6. A Young tableau $T$ is *standard* if the numbers in $T$ increase along the rows and down the columns; see Example 6.1.7.

We denote the number of standard tableaux of shape $\lambda$ by $f_\lambda$. We will give two formulas for calculating $f_\lambda$, (6.23) and (6.25).

EXAMPLE 6.1.7. Given a partition $\lambda$ of $n$, there are $n!$ tableaux of shape $\lambda$. So for $\lambda = (2, 1)$, there are six tableaux of shape $\lambda$:

$$\begin{array}{|c|c|} \hline 1 & 2 \\ \hline 3 & \multicolumn{1}{c}{} \\ \cline{1-1} \end{array}\, , \quad \begin{array}{|c|c|} \hline 1 & 3 \\ \hline 2 & \multicolumn{1}{c}{} \\ \cline{1-1} \end{array}\, , \quad \begin{array}{|c|c|} \hline 2 & 1 \\ \hline 3 & \multicolumn{1}{c}{} \\ \cline{1-1} \end{array}\, , \quad \begin{array}{|c|c|} \hline 2 & 3 \\ \hline 1 & \multicolumn{1}{c}{} \\ \cline{1-1} \end{array}\, , \quad \begin{array}{|c|c|} \hline 3 & 1 \\ \hline 2 & \multicolumn{1}{c}{} \\ \cline{1-1} \end{array}\, , \quad \begin{array}{|c|c|} \hline 3 & 2 \\ \hline 1 & \multicolumn{1}{c}{} \\ \cline{1-1} \end{array}\; .$$

Only the first two of these six tableaux are standard, hence $f_{(2,1)} = 2$.

This notation is also called the *English notation* we also have the *French notation*. In this notation a Young diagram may be identified to a finite set $D$ of points $(a, b) \in \mathbb{N}^2$ such that, if $(a, b) \in D$ and $a' \leq a, b' \leq b,\ a', b' \in \mathbb{N}$, then $(a', b') \in D$.

[1]One can give it a funny economic interpretation—the rows are distributions of wealth and this is like the opposite of Robin Hood, take from the poor to give to the rich.

EXAMPLE 6.1.8. Let $\lambda = (4,3,1,1)$. We omit drawing the boxes:

$$\text{French notation} \quad \begin{array}{cccc} 3 & & & \\ 1 & & & \\ 5 & 2 & 7 & \\ 4 & 9 & 6 & 8 \end{array} \quad , \quad \begin{array}{cccc} 7 & & & \\ 4 & & & \\ 3 & 6 & 8 & \\ 1 & 2 & 5 & 9. \end{array}$$

The first is not standard while the second is standard.

In representation theory it is important to also have a more general notion of tableau as a filling $T$ of a Young diagram, with $n \leq m$ boxes, with some or all of the numbers $1, 2, \ldots, m$.

DEFINITION 6.1.9. We say that a tableau $T$ is a *row* (resp., *column*) *semistandard* tableau on indices $1, 2, \ldots, m$ if the columns (resp., rows) are strictly increasing while the rows (resp., columns) are weakly increasing.

For example, $n = m = 3$ for $\lambda = (2,1)$. Besides the previous standard, we have the column semistandard tableaux:

$$\begin{array}{|c|c|}\hline 1 & 2 \\ \hline 1 \\ \cline{1-1}\end{array}\,,\quad \begin{array}{|c|c|}\hline 1 & 3 \\ \hline 1 \\ \cline{1-1}\end{array}\,,\quad \begin{array}{|c|c|}\hline 2 & 3 \\ \hline 2 \\ \cline{1-1}\end{array}\,.\quad \begin{array}{|c|c|}\hline 2 & 3 \\ \hline 3 \\ \cline{1-1}\end{array}\,,\quad \begin{array}{|c|c|}\hline 1 & 3 \\ \hline 3 \\ \cline{1-1}\end{array}\,,\quad \begin{array}{|c|c|}\hline 1 & 2 \\ \hline 2 \\ \cline{1-1}\end{array}\,.$$

Occasionally we may have to use some *skew tableaux*, that is fillings of a skew diagram.

For a general tableau we use the notion of

DEFINITION 6.1.10. The *content* of a tableau $T$ is the sequence $h_j$ where $h_j$ counts the number of occurrences of the index $j$ on $T$.

6.1.2.1. *Conjugacy classes.* There are several approaches to the study of $V^{\otimes n}$ as representation both of $GL(V)$ and of the symmetric group. Let us recall the approach through the symmetric group.

One thinks of a permutation $\sigma$ on $1, 2, \ldots, n$ as a $\mathbb{Z}$ action, where $m \in \mathbb{Z}$ acts by $\sigma^m$. As any group action this decomposes the set into orbits, which are just the *cycles* of the permutation $\sigma$. Saying that two permutations are conjugate is the same as saying that the two actions are isomorphic.

Thus the theory of cycles implies that the conjugacy classes of $S_n$ are in 1–1 correspondence with the isomorphism classes of $\mathbb{Z}$ actions on $1, 2, \ldots, n$. These are given in turn by the cycle decomposition of a permutation, the lengths $\mu := k_1, k_2, \ldots, k_n$ of the cycles is described by a partition $\mu \vdash n$ of $n$.

We shall denote by $C(\mu)$ the conjugacy class in $S_n$ formed by the permutations decomposed in cycles of length $k_1, k_2, \ldots, k_n$, hence $S_n = \bigsqcup_{\mu \vdash n} C(\mu)$.

Consider the group algebra $R := \mathbb{Q}[S_n]$ of the symmetric group. We wish to work over $\mathbb{Q}$ since the theory has this really more arithmetic flavor.

We will (implicitly) exhibit a decomposition of $R$ as a direct sum of matrix algebras over $\mathbb{Q}$:

$$R = \mathbb{Q}[S_n] := \bigoplus_{\mu \vdash n} M_{d(\mu)}(\mathbb{Q}). \tag{6.1}$$

The numbers $d(\mu)$ can be computed in several ways from the partition $\mu$.

From the theory of group characters (cf. §1.3.4),

$$R_{\mathbb{C}} := \mathbb{C}[S_n] := \sum_i M_{n_i}(\mathbb{C}),$$

where the number of summands is equal to the number of conjugacy classes, hence the number of partitions of $n$. For every partition $\lambda \vdash n$ one can construct a primitive idempotent $e_\lambda$ in $R$, called a *Young symmetrizer* so that $R = \bigoplus_{\lambda \vdash n} Re_\lambda R$ and $\dim_{\mathbb{Q}} e_\lambda R e_\lambda = 1$. In this way the left ideals $Re_\lambda$ exhaust all irreducible representations.

## 6.2. The symmetric group

**6.2.1. Young symmetrizers.** The construction of Young symmetrizers is the following. For a partition $\lambda \vdash n$, construct its *Young diagram* and the corresponding tableaux.

Denote by $B$ the set of boxes of the Young diagram corresponding to $\lambda$, so that a tableau is a bijective map $T : B \to (1, 2, \dots, n-1, n)$.

The symmetric group $S_n$ acts on tableaux by composition:

$$\sigma T : B \xrightarrow{\ T\ } (1, 2, \dots, n-1, n) \xrightarrow{\ \sigma\ } (1, 2, \dots, n-1, n).$$

A tableau induces two partitions on $(1, 2, \dots, n)$, the *row partition* defined by: $i, j$ are in the same part if they appear in the same row of $T$ and similarly for the *column partition*.

To a partition $\pi$ of $(1, 2, \dots, n)$ one associates the subgroup $S_\pi$ of the symmetric group of permutations which preserve the partition. It is isomorphic to the product of the symmetric groups of all the parts of the partition. To a tableau $T$ one associates two subgroups $\mathrm{R}_T, \mathrm{C}_T$ of $S_n$.

(1) $\mathrm{R}_T$ is the group preserving the row partition.
(2) $\mathrm{C}_T$ the subgroup preserving the column partition.

It is clear that $\mathrm{R}_T \cap \mathrm{C}_T = 1$ since each box is an intersection of a row and a column.

Notice that, if $\sigma \in S_n$, the row and column partitions associated to $\sigma T$ are obtained by applying $\sigma$ to the corresponding partitions of $T$, thus,

$$\mathrm{R}_{\sigma T} = \sigma \mathrm{R}_T \sigma^{-1}, \quad \mathrm{C}_{\sigma T} = \sigma \mathrm{C}_T \sigma^{-1}.$$

For example let

$$T = \begin{array}{|c|c|c|c|c|} \hline 1 & 2 & 4 & 8 & 11 \\ \hline 3 & 5 & 6 \\ \cline{1-3} 7 & 9 \\ \cline{1-2} 10 & 12 \\ \cline{1-2} 13 \\ \cline{1-1} \end{array}$$

then

$$\begin{aligned} R_T &= S_{\{1,2,4,8,11\}} \times S_{\{3,5,6\}} \times S_{\{7,9\}} \times S_{\{10,12\}}, \\ C_T &= S_{\{1,3,7,10,13\}} \times S_{\{2,5,9,12\}} \times S_{\{4,6\}}. \end{aligned}$$

We define two elements in $R = \mathbb{Q}[S_n]$:

$$(6.2)\qquad s_T = \sum_{\sigma \in R_T} \sigma \quad \text{the } \textit{symmetrizer on the rows},$$

$$(6.3)\qquad a_T = \sum_{\sigma \in C_T} \epsilon_\sigma \sigma \quad \text{the } \textit{antisymmetrizer on the columns},$$

where $\epsilon_\sigma$ denotes the sign of the permutation. We have the two identities:

$$s_T^2 = \prod_i h_i! s_T, \qquad a_T^2 = \prod_i k_i! a_T,$$

where the $h_i$ are the lengths of the rows and $k_i$ are the length of the columns.

$$p s_T = s_T = s_T\, p,\ \forall p \in R_T; \qquad q a_T = a_T\, q = \epsilon_q a_T,\ \forall q \in C_T.$$

It is then an easy exercise to check

PROPOSITION 6.2.1. *The left ideal $\mathbb{Q}[S_n]s_T$ has as its basis the elements $gs_T$ as $g$ runs over a set of representatives of the cosets $gR_T$ and it equals, as representation, the permutation representation on such cosets.*

*The left ideal $\mathbb{Q}[S_n]a_T$ has as its basis the elements $ga_T$ as $g$ runs over a set representatives of the cosets $gC_T$ and it equals, as representation, the representation induced to $S_n$ by the sign representation of $C_T$.*

Now the remarkable fact comes: Consider the product,

$$c_T := s_T a_T = \sum_{p \in R_T,\, q \in C_T} \epsilon_q pq.$$

THEOREM 6.2.2. *For each $T$ there exists a positive integer $p(T)$ such that the element $e_T := \frac{c_T}{p(T)}$ is a primitive idempotent.*

$$(6.4)\qquad e_T^2 = e_T, \quad e_T\mathbb{Q}[S_n]e_T = \mathbb{Q}e_T.$$

*If $T_1$, $T_2$ are tableaux associated to two different partitions, then*

$$(6.5)\qquad e_{T_1}\mathbb{Q}[S_n]e_{T_2} = 0.$$

DEFINITION 6.2.3. The idempotent $e_T := \frac{c_T}{p(T)}$ is called the *Young symmetrizer* relative to the given tableau.

REMARK 6.2.4. Often it is easier to compute with $c_T$ instead of $e_T$. Then we have $c_T^2 = p(T)c_T$, and we express this property by saying that $c_T$ is an *essential idempotent*.

We have thus, for a given $\lambda \vdash n$, several conjugate primitive idempotents, associated to tableaux of row shape $\lambda$. Each of them generates a left ideal which is an irreducible $S_n$-module associated to $\lambda$ and which will be denoted by $M_\lambda$. The integer $p(T)$ depends only on the shape $\lambda$ of $T$, and thus we will denote it by $p(T) = p(\lambda)$.

By formula (6.4) $\mathbb{Q}[S_n]e_T$ is irreducible, and by formula (6.5) and Example 1.3.12, these irreducibles are nonisomorphic for different partitions; therefore we have

THEOREM 6.2.5. *The modules $M_\lambda$ as $\lambda \vdash n$ runs over the partitions of $n$ form the full list of irreducible $S_n$-modules.*

The basic properties of these elements are given in the following theorem.

THEOREM 6.2.6. *Let $\lambda \vdash n$ be a partition with the corresponding minimal two-sided ideal $I_\lambda \subseteq \mathbb{Q}[S_n]$. Let $T = T_\lambda$ be a tableau of shape $\lambda$, and let $\mathbb{Q}[S_n]e_T$ be the left ideal of $\mathbb{Q}[S_n]$ generated by $e_T$.*

(1) *Then $e_T\mathbb{Q}[S_n]e_T = \mathbb{Q}e_T$ and $\mathbb{Q}[S_n]e_T$ is a minimal left ideal in $\mathbb{Q}[S_n]$.*
(2) *$I_\lambda$ is the direct sum of the minimal left ideals $\mathbb{Q}[S_n]e_{T_\lambda}$ where the tableaux $T_\lambda$ are standard.*
(3) *Similar statements for $\bar{c}_T = a_T s_T$, $\bar{e}_T = a_T s_T p(T)^{-1}$ and for right ideals.*

Define the involution $u : S_n \to S_n$ via $u(\sigma) = \sigma^{-1}$ and extend by linearity to $u : \mathbb{Q}[S_n] \to \mathbb{Q}[S_n]$. Then the anti-isomorphism $u$ satisfies the following

COROLLARY 6.2.7.

$$u(a_T) = a_T,\ u(s_T) = s_T,\ u(e_T) = \bar{e}_T. \tag{6.6}$$

*Let $\lambda \vdash n$ with the corresponding $I_\lambda \subseteq \mathbb{Q}[S_n]$ as above, then $u(I_\lambda) = I_\lambda$.*

PROOF. Clearly $u(R_{T_\lambda}) = R_{T_\lambda}$ and $u(C_{T_\lambda}) = C_{T_\lambda}$. Since $\operatorname{sgn}(\sigma) = \operatorname{sgn}(\sigma^{-1})$, it follows that $u(s_{T_\lambda}) = s_{T_\lambda}$ and $u(a_{T_\lambda}) = a_{T_\lambda}$. Thus $u(\mathbb{Q}[S_n]e_{T_\lambda}) = u(e_{T_\lambda})u(\mathbb{Q}[S_n]) = \bar{e}_{T_\lambda}\mathbb{Q}[S_n]$. By Theorem 6.2.6 this implies that $u : I_\lambda \to I_\lambda$.

The proof now follows since $u$ is an anti-isomorphism. □

REMARK 6.2.8.

(1) Given a partition $\lambda$, we have thus a corresponding irreducible representation, which we will often denote by $M_\lambda$ and which can be constructed as $\mathbb{Q}[S_n]e_{T_\lambda}$.
(2) We then have by formula (6.4) and by Definition 6.2.3 that $e_{T_\lambda}M_\lambda = c_{T_\lambda}M_\lambda$ is nonzero and in fact it is a one-dimensional subspace.
    If $g \in R_{T_\lambda}$, we have $ge_{T_\lambda} = e_{T_\lambda}$. Thus, for an element $f \in e_{T_\lambda}M_\lambda$, we have $gf = f$, $\forall g \in R_{T_\lambda}$, that is $f$ is symmetric under the row subgroup $R_{T_\lambda}$.
(3) Similarly we can find another element $f' \in u(e_{T_\lambda})M_\lambda$ image of some element under the antisymmetrizer $a_{T_\lambda}$ so $gf = \epsilon_g f$, $\forall g \in C_{T_\lambda}$, that is $f'$ is antisymmetric under the column subgroup $a_{T_\lambda}$.

Define instead the automorphism $v$ of $\mathbb{Q}[S_n]$ via $v(\sigma) = \epsilon_\sigma \sigma$.

COROLLARY 6.2.9. *Let $\lambda \vdash n$ with the corresponding $I_\lambda \subseteq \mathbb{Q}[S_n]$ as above. Then $v(I_\lambda) = I_{\check{\lambda}}$, where $\check{\lambda}$ is the conjugate partition.*

PROOF. Clearly $v(s_{T_\lambda}) = \sum_{\sigma \in R_{T_\lambda}} \epsilon_\sigma \sigma = \sum_{\sigma \in C_{T_{\check{\lambda}}}} \epsilon_\sigma \sigma = a_{T_{\check{\lambda}}}$ and $v(a_{T_\lambda}) = s_{T_{\check{\lambda}}}$. We then apply Corollary 6.2.7 using formula (6.6). □

6.2.1.1. *Young's rule.* There is a precise description of how the two left ideals $\mathbb{Q}[S_n]s_T$, $\mathbb{Q}[S_n]a_T$ decompose into irreducible representations of $S_n$. First we remark that the automorphism of $\mathbb{Q}[S_n]$, $\tau : \sigma \mapsto \epsilon_\sigma \sigma$ exchanges symmetrizers with antisymmetrizers and the module $M_\lambda$ with $M_{\check{\lambda}}$ (conjugate partition). Hence it is enough to describe $\mathbb{Q}[S_n]s_T$ for $T$ a tableau of some row shape $\lambda = (\lambda_1, \lambda_2, \dots, \lambda_u)$. Here one has that:

THEOREM 6.2.10 (Young's rule (see **[JK81**, 2.8.5])). *The multiplicity of an irreducible $M_\mu$ into $\mathbb{Q}[S_n]s_T$ is given by the* Kostka number *$K_{\mu,\lambda}$ defined to be the number of row semistandard (Definition 6.1.9) tableaux of shape $\mu$ and content $\lambda$.*

Here content is defined in Definition 6.1.10. Of course there is a bijection between the set of row semistandard tableaux of shape $\mu$ filled with $\lambda_j$ entries of $j$ and the set of column semistandard tableaux of (conjugate) shape $\check{\mu}$ filled with $\lambda_j$ entries of $j$.

REMARK 6.2.11. The number $K_{\mu,\lambda}$ is different from 0 if and only if $\mu \geq \lambda$ in the dominance order.

PROOF. Let $\mu_1 \geq \cdots \geq \mu_r$ be the rows of $\mu$. By definition of row-semistandard tableau $T$ it follows that an index $i$ in $T$ must be in one of the first $i$ rows, thus we must have for all $i$ that $\lambda_1 + \lambda_2 + \cdots + \lambda_i \leq \mu_1 + \mu_2 + \cdots + \mu_i$.

Now if this condition is satisfied, we see that filling the diagram of shape $\mu$ from left to right and from top to bottom with the numbers $1, 2, \ldots$ with multiplicity $\lambda_1, \lambda_2, \ldots, \lambda_u$ produces a (row)-semistandard tableau of shape $\mu$ and content $\lambda$. □

For example, let

$$\lambda_1 = (4,4,1,1,1,1,1),\ \lambda_2 = (3,3,3,2,1,1), \quad \text{and} \quad \mu = (5,3,2,2,1).$$

The previous procedure gives the corresponding row-semistandard tableaux.

$T_1 =$

| | | | | |
|---|---|---|---|---|
| 1 | 1 | 1 | 1 | 2 |
| 2 | 2 | 2 | | |
| 3 | 4 | | | |
| 5 | 6 | | | |
| 7 | | | | |

$T_2 =$

| | | | | |
|---|---|---|---|---|
| 1 | 1 | 1 | 2 | 2 |
| 2 | 3 | 3 | | |
| 3 | 4 | | | |
| 4 | 5 | | | |
| 6 | | | | |

.

## 6.3. The linear group

**6.3.1. Schur functors.** When we act with the symmetric group on a tensor power $V^{\otimes n}$ of a vector space $V$ of some dimension $m = \dim_F V$, the map $F[S_n] \to \mathrm{End}(V^{\otimes n})$ from the group algebra of $S_n$ to the endomorphisms of the tensor power is injective if and only if $n \leq m$. As soon as $n > m$, we have (see **[Pro07**, §6.1])

THEOREM 6.3.1. *The kernel of* $F[S_n] \to \mathrm{End}(V^{\otimes n})$ *is the two-sided ideal generated by an antisymmetrizer* $\sum_{\sigma \in S_{m+1}} \epsilon_\sigma \sigma$ *on (any)* $m+1$ *elements. In characteristic* 0 *this ideal is the sum of all the isotypic components relative to the Young diagrams of height* $> m$ *(by Young's rule). If* $e_T$ *is a Young symmetrizer relative to a partition* $\lambda$ *of* $n$ *of height* $k$*, then* $e_T V^{\otimes n}$ *is an irreducible* $\mathrm{GL}(V)$*-module which is nonzero if and only if* $k \leq m$*; its isomorphism type depends only upon* $\lambda$ *and not* $T$*.*

If $W$ is another vector space and $f : V \to W$ a linear map, we have an induced map $f^{\otimes n} : V^{\otimes n} \to W^{\otimes n}$ which induces a map

$$f^{\otimes n} : e_T V^{\otimes n} \to e_T W^{\otimes n}. \tag{6.7}$$

It is clear that formula (6.7) gives a functor, and that this construction, up to natural isomorphism, depends only upon $\lambda$. So we set

DEFINITION 6.3.2. We denote $S_\lambda(V) := e_T V^{\otimes n}$ and call it a *Schur functor*.

One verifies that to the partition $(1^k)$, $k \leq \dim_F V$ (a single column of length $k$) is associated the exterior power $S_{(1^k)}(V) = \bigwedge^k V$ while to the partition $(k)$, (a single row of length $k$) is associated the symmetric power $S_{(k)}(V) = S^k(V)$.

If $V \subset W$ is an inclusion of vector spaces, we have that $S_\lambda(V) \subset S_\lambda(W)$.

Then the rational irreducible representations of $\mathrm{GL}(V)$ (cf. Definition 14.1.2) are obtained as follows. We have the one-dimensional representations given by the integer powers of the determinant $\det^i$. We take all possible partitions $\lambda$ of height $\leq \dim_F V - 1$ and construct the corresponding irreducible representations $S_\lambda(V) \otimes \det^i$, $i \in \mathbb{Z}$, then

THEOREM 6.3.3. *The irreducible representations* $S_\lambda(V) \otimes \det^i$, $i \in \mathbb{Z}$, *with* $\mathrm{ht}(\lambda) \leq \dim_F V - 1$, *form the complete list of the rational irreducible representations of* $\mathrm{GL}(V)$.

A partition $\lambda$ of height $\dim_F V$ can be decomposed into a partition $\mu$ of height $< \dim_F V$ and $i$ rows of length equal to $\dim_F V$. For the corresponding irreducible representation, we have that $S_\lambda(V) = S_\mu(V) \otimes \det^i$.

Of course this theorem holds only in characteristic 0; in positive characteristic, many problems are still open.

In our computations we only use *polynomial representations*, that is representations of $\mathrm{GL}(V)$ in which the coordinates are polynomials in the matrix variables. These are exactly the direct sums of the irreducible representations $S_\lambda(V)$, so the inverse of the determinant does not appear.

Putting together these facts we see that (cf. Theorem 6.2.5)

THEOREM 6.3.4. *The decomposition of* $V^{\otimes n}$ *with respect to* $\mathrm{GL}(V) \times S_n$ *is*

$$V^{\otimes n} = \bigoplus_{\lambda \vdash n,\, ht(\lambda) \leq \dim_F V} M_\lambda \otimes S_\lambda(V). \tag{6.8}$$

REMARK 6.3.5. Notice that under the action of the symmetric group $S_n$ the space $V^{\otimes n}$ contains always the symmetric tensors which correspond to the partition $\lambda$ with just one row equal to $n$. In this case since, $M_\lambda$ is the trivial representation of dimension 1, we see that the symmetric tensors form an irreducible representation of $\mathrm{GL}(V)$.

If $\dim V \geq n$ under the action of the symmetric group $S_n$, the space $V^{\otimes n}$ contains also the nonzero space of antisymmetric tensors, which can be identified to the exterior power $\bigwedge^n V$ and which correspond to the partition $\lambda$ with just one column equal to $n$. In this case since $M_\lambda$ is the sign representation of dimension 1, we see that also $\bigwedge^n V$ forms an irreducible representation of $\mathrm{GL}(V)$. All other isotypic components correspond to irreducible representations $M_\lambda$ of the symmetric group of dimension $> 1$, so they are not irreducible but decompose as direct sum in as many irreducibles as the dimension of $M_\lambda$.

In a precise way the theory of Young symmetrizers tells us that we can *glue together* the symmetric and exterior powers of $V$ in order to construct all the Schur functors $S_\lambda(V)$.

6.3.1.1. *Highest weight vectors.* A further important fact comes from the theory of highest weight vectors. Let us fix a basis $x_1, \ldots, x_m$ of $V$, and let $U_m$ be the group of upper triangular matrices with a 1 in the diagonal. In the general theory of algebraic groups this is the unipotent radical of the group $B_m$ of upper triangular matrices which is a maximal connected solvable subgroup i.e., a *Borel subgroup*.

If $D$ is the group of diagonal matrices $d$ with entries $d = \mathrm{diag}(d_1, \dots, d_m)$, we have a semidirect product decomposition $B_m = DU_m$. A vector $v$ in a representation of $\mathrm{GL}(V) = \mathrm{GL}(m, F)$ is a *weight vector* for $D$ if it is an eigenvector under the action of $D$. In this case there exist integers $h_1, \dots, h_m$ such that $d \cdot v = \prod_{i=1}^m d_i^{h_i} v$. The $m$-tuple $h_1, \dots, h_m$ is the *weight* of the vector $v$.

DEFINITION 6.3.6. An element $u$ of a rational representation of $\mathrm{GL}(V)$ which is a weight vector for $B_m$ of weight $\lambda$ is called a *highest weight vector*.

In the representation $V^{\otimes k}$ a basis of weight vectors is given by the decomposable tensors $x_{i_1} \otimes \cdots \otimes x_{i_k}$ which we also call *monomials* and write simply as words $x_{i_1} \cdots x_{i_k}$. The weight counts the number of times that each index $i$ appears in the factors $x_j$.

In particular when $k = m$ the space of multilinear noncommutative polynomials in the variables $x_1, \dots, x_m$ is the span of the *multilinear tensors*, that is the tensors of weight $\prod_{i=1}^m d_i$.

THEOREM 6.3.7.

(1) *For every partition $\lambda$ of height $\leq m = \dim_F V$, the subspace $S_\lambda(V)^{U_m}$ is one-dimensional.*
(2) *If $u_\lambda$ is a nonzero vector in $S_\lambda(V)^{U_m}$, $u_\lambda$ is a weight vector of weight $h_1, \dots, h_n$, where the $h_i$ are the rows of the partition $\lambda$.*
(3) *Moreover if the height of $\lambda$ is $h < m$, we have that $S_\lambda(V)^{U_m} = S_\lambda(V_h)^{U_m}$, where $V_h$ is the subspace spanned by $x_1, \dots, x_h$.*

$$(V^{\otimes n})^{U_m} = \bigoplus_{\lambda \vdash n,\, ht(\lambda) \leq m = \dim_F V} M_\lambda \otimes S_\lambda(V)^{U_m} \tag{6.9}$$

*is the direct sum of all irreducible representations $M_\lambda$ of the symmetric group $S_n$, with $\lambda \vdash n$, $\mathrm{ht}(\lambda) \leq m = \dim_F V$ each appearing with multiplicity* 1.

(4) $(S_\lambda(V) \otimes \bigwedge^m V)^{U_m} = (S_\lambda(V))^{U_m} \otimes \bigwedge^m V$.
(5) *A vector $u$ of a rational representation of $\mathrm{GL}(V)$ of weight $\lambda$ is a highest weight if and only if it is killed by all elements $e_{i,j}, i < j$. Then it generates, over $\mathrm{GL}(V)$, an irreducible representation isomorphic to $S_\lambda(V)$, which is also generated by $u$ under the action of the enveloping algebra of the Lie algebra of strictly lower triangular matrices, that is the algebra of matrices with basis $e_{i,j}, i > j$.*

EXAMPLE 6.3.8. In the language of free algebra, the vector space $V$ has as its basis the variables $x_1, x_2, \dots, x_m$ and $V^{\otimes n}$ is the span of noncommutative polynomials in these variables of degree $n$. The weight of a monomial is just the vector of degrees in the variables.

We remark that the standard polynomial (2.32) $\mathrm{St}_k(x_1, x_2, \dots, x_k)$ is a highest weight vector for $\bigwedge V^k$. Thus the space $S_\lambda(V)^{U_m}$, where $\lambda$ has columns $k_1, \dots, k_n$, $\sum k_i = n$, contains then the products in all orders

$$\prod_{i=1}^n \mathrm{St}_{k_i}(x_1, x_2, \dots, x_{k_i}).$$

In fact it is the span of the transforms under the symmetric group $S_n$ which acts on polynomials by *exchanging places*. Clearly the degree of these elements in the variable $x_i$ is the length $h_i$ of the $i$th row.

REMARK 6.3.9. A dual result is obtained when $m = n$. In the space $V^{\otimes n}$ we consider the subspace $V^{\otimes n}_{\text{mult}}$ of *multilinear elements* spanned by the $n!$ tensors $x_{\sigma(1)} \otimes \cdots \otimes x_{\sigma(n)} = x_{\sigma(1)} \cdots x_{\sigma(n)}$ as $\sigma$ runs over the symmetric group $S_n$. This is the space of vectors of weight $\prod_{i=1}^n d_i$. Then $V^{\otimes n}_{\text{mult}}$ is stable under two different and commuting copies of the symmetric group $S_n$. One is the symmetric group $S_n$ which acts on $V^{\otimes n}$ in the usual way by permuting the tensor. The other symmetric group $S_n$ is the subgroup of $\mathrm{GL}(V)$ which permutes the basis elements $x_1, \dots, x_m$.

REMARK 6.3.10. The map $\sigma \mapsto x_{\sigma(1)} \otimes \cdots \otimes x_{\sigma(n)} = x_{\sigma(1)} \cdots x_{\sigma(n)}$ identifies the space $V^{\otimes n}_{\text{mult}}$ to the group algebra of the symmetric group, and the two actions correspond to the right (the permutation of tensors) and left (the action of the subgroup $S_n$ of $\mathrm{GL}(V)$) action.

From the decomposition (6.8), since $M_\lambda$ is self-dual, we deduce

COROLLARY 6.3.11.

$$\bigoplus_{\lambda \vdash n} M_\lambda \otimes M_\lambda^* = F[S_n] = V^{\otimes n}_{\text{mult}} = \bigoplus_{\lambda \vdash n} M_\lambda \otimes S_\lambda(V)_{\text{mult}}. \tag{6.10}$$

*And $S_\lambda(V)_{\text{mult}}$ is also isomorphic to the representation $M_\lambda^* = M_\lambda$.*

This corresponds to the usual decomposition of the group algebra of the symmetric group.

6.3.1.2. *Interpretation by identities.*

REMARK 6.3.12. The above argument also gives another interpretation for $\pi\sigma$, where $\pi, \sigma \in S_n$. Write $\pi \equiv M_\pi(x_1, \dots, x_n) = y_1 \cdots y_n$. Namely, we have again $y_i = x_{\pi(i)}$ while $y_{\sigma(j)} = x_{\pi(\sigma(j))}$. Then we just saw that

$$\pi\sigma \equiv (y_1 \cdots y_n)\sigma = y_{\sigma(1)} \cdots y_{\sigma(n)}. \tag{6.11}$$

Thus,

(1) When a permutation $\pi \in S_n$ acts on a monomial from the left, it substitutes the variables according to $\pi$.
(2) When a permutation $\sigma \in S_n$ acts on a monomial from the right, it permutes the *places* of the variables according to $\sigma$.

DEFINITION 6.3.13. We denote by $V_n \simeq F[S_n]$ the span of the multilinear monomials in $n$-variables $x_1, \dots, x_n$.

EXAMPLE 6.3.14. Let $f(x) = f(x_1, \dots, x_n) = \sum_{\sigma \in S_n} \alpha_\sigma x_{\sigma(1)} \cdots x_{\sigma(n)} \in V_n$, then denote

$$\begin{aligned} &f^*(x_1, \dots, x_n; x_{n+1}, \dots, x_{2n-1}) \\ &\quad = \sum_{\sigma \in S_n} \alpha_\sigma x_{\sigma(1)} x_{n+1} x_{\sigma(2)} x_{n+2} \cdots x_{\sigma(n-1)} x_{2n-1} x_{\sigma(n)}, \end{aligned} \tag{6.12}$$

namely, the variables $x_{\sigma(1)}, x_{\sigma(2)}, \dots, x_{\sigma(n)}$ are separated by the other variables $x_{n+1}, x_{n+2}, \dots, x_{2n-1}$. Then by Remark 6.3.12 with $\eta$ replacing $\sigma$,

$$f^*(x) = (f(x_1, \dots, x_n) x_{n+1} \cdots x_{2n-1})\eta,$$

where $\eta \in S_{2n-1}$ is the permutation

$$\eta = \begin{pmatrix} 1 & 2 & 3 & 4 & \cdots & 2n-1 \\ 1 & n+1 & 2 & n+2 & \cdots & n \end{pmatrix}. \tag{6.13}$$

As we shall see in §7.3.1, if we consider $d \times d$ matrices, they satisfy the standard identity $\mathrm{St}_{2d}$. Then if we take $2d \leq n \leq d^2$ and perform the previous construction to $f = \mathrm{St}_n$, we have that $f$ is an identity while $f^*$ is not an identity of $d \times d$ matrices.

COROLLARY 6.3.15. *Let $f$ be a multilinear polynomial of degree $n$. Then $f$ is equivalent to a system of polynomials of the form $e_{T_\lambda} g_i$, with $g_i = g_i(x_1, \ldots, x_n)$ multilinear polynomials in the variables $x_1, \ldots, x_n$, $\lambda$ is a partition of $n$, and $T_\lambda$ is a Young tableau associated to $\lambda$.*

PROOF. Here equivalent means that the $S_n$ submodule $N = F[S_n] \cdot f$, generated by $f$ by permuting the variables, which is thus made of polynomials deduced from $f$, has a basis of elements of the type $e_{T_\lambda} g$. Now $N$ is a direct sum of submodules of type $M_\lambda$. Given a tableau $T$ of shape $\lambda$, we have by Theorem 6.2.6 that $\dim e_T M_\lambda = 1$, since $e_T$ is idempotent there is an $a \in M_\lambda$, $a \neq 0$ such that $e_T a = a$. Then $M_\lambda$ has a basis of elements $\sigma a$ for $\sigma$ some set of permutations. Finally each $\sigma e_T a = \sigma e_T \sigma^{-1} \sigma a = e_{\sigma T} \sigma a$ is of the desired type. $\square$

**6.3.2. A combinatorial approach.** *This small section with no proofs is a* challenge *to the reader.*

The tensor power $V^{\otimes n}$ has a basis given by all *words* of length $n$ in the elements of a basis $x_1, \ldots, x_m$ of $V$. By the Robinson–Schensted correspondence (see §7.1.1.2) such a set of words is in 1–1 correspondence with pairs of tableaux $(P, Q)$ of the same shape $\lambda$, where $Q$ is standard while $P$ is (row)-semistandard and filled with the numbers $1, 2, \ldots, m$. This combinatorial correspondence has a counterpart in representation theory.

THEOREM 6.3.16. *$M_\lambda$ has a basis indexed by standard tableaux of shape $\lambda$. $S_\lambda(V)$ has a basis indexed by semistandard tableaux of shape $\lambda$ filled with the numbers* $1, 2, \ldots, m$.

In particular the dimension of $M_\lambda, S_\lambda(V)$ equals the number of corresponding tableaux. For these numbers there exist several formulas, some of which we review in the next section.

But first we want to discuss a combinatorial basis of $V^{\otimes n}$ indexed by all pairs $(P_\lambda, Q_\lambda)$, where $\lambda \vdash n$ is a partition of height $\leq \dim V$, $P_\lambda$ is a standard tableau, and $Q_\lambda$ is a semistandard tableau with entries the numbers $1, 2, \ldots, \dim_F V$.

The basis is best described for $(V^*)^{\otimes n}$, the multilinear polynomial functions, and it is valid in all characteristics and even over $\mathbb{Z}$. It is based on the theory of *double standard tableaux*.

In our setting this theory can be formulated as follows. Given a tableau $T$ with possible repetitions on columns but not on rows, we associate to $T$ the noncommutative polynomial $\mathrm{St}_T$ product of the standard polynomials in the variables of a row. For instance,

$$T = \begin{array}{|c|c|c|c|c|} \hline 7 & 2 & 4 & 8 & 3 \\ \hline 5 & 10 & 6 \\ \cline{1-3} 5 & 3 \\ \cline{1-2} \end{array}$$

corresponds to $\mathrm{St}_T = \mathrm{St}_5(x_7, x_2, x_4, x_8, x_3)\,\mathrm{St}_3(x_5, x_{10}, x_6)\,\mathrm{St}_2(x_5, x_3)$.

On the other hand, to a tableau $T$ with no repetitions, we associate the permutation $\eta_T$ reading the rows from top to bottom. Finally to a pair of tableaux $T_1, T_2$, the first with no repetitions, we associate the element $\mathrm{St}_{T_2}\,\eta_{T_1}$. We call this a *double tableau*.

EXAMPLE 6.3.17.

$$\begin{array}{|c|c|} \hline 1 & 3 \\ \hline 2 \\ \cline{1-1} \end{array}\ , \quad \begin{array}{|c|c|} \hline 1 & 3 \\ \hline 1 \\ \cline{1-1} \end{array}\ ,$$

the permutation is $\eta = \eta_{T_1} = (1, 3, 2)$, the polynomial $\mathrm{St}_{T_2}$ is $(x_1x_3 - x_3x_1)x_1$, so the associated polynomial ($\eta_{T_1}$ acts on the right cf. Remark 6.3.12 ) is

$$\mathrm{St}_{T_2}\,\eta_{T_1} = x_1x_1x_3 - x_3x_1x_1.$$

We can interpret the theory of double standard tableaux (cf. **[DCP17]**) by

THEOREM 6.3.18. *The polynomials $\mathrm{St}_{T_2}\,\eta_{T_1}$ as the pair $T_1, T_2$ runs over standard (resp., semistandard) tableaux with entries $1, 2, \ldots, m$ of all shapes $\lambda \vdash n$ are called* double standard tableaux *and form a basis over $\mathbb{Z}$ of the noncommutative polynomials of degree $n$ in $m$ variables.*

Of course the multilinear polynomials correspond to a pair of tableaux which are standard. One of the interesting points of this description is the connection with representation theory. The span $V_{\geq\lambda}$ of all double tableaux of shape $\geq \lambda$, where $\geq$ is the dominance order of Definition 6.1.4, is stable under both actions of $S_n, \mathrm{GL}(V)$ and has as its basis all double standard tableaux of shape $\geq \lambda$.

In characteristic 0 we have a decomposition,

$$V_{\geq\lambda} = M_\lambda \otimes S_\lambda(V) \oplus V_{>\lambda} = \bigoplus_{\mu\geq\lambda} M_\mu \otimes S_\mu(V).$$

6.3.2.1. *Characters.* From the basic formula (6.8) one can draw several important consequences. One may take an element $(\sigma, g) \in S_n \times \mathrm{GL}(V)$ and compute its trace on the representation $V^{\otimes n}$, with $n = \dim V$.

One obtains that

$$\mathrm{tr}(\sigma \circ g^{\otimes n}) = \sum_{\lambda\vdash n,\ \mathrm{ht}(\lambda)\leq \dim_F V} \chi_\lambda(\sigma) S_\lambda(g), \tag{6.14}$$

where $\chi_\lambda(\sigma)$ is the trace, that is the character, of $\sigma$ acting on the module $M_\lambda$ while $S_\lambda(g)$ is the trace, that is the character, of $g$ acting on the module $S_\lambda(V)$.

In order to compute $\mathrm{tr}(\sigma \circ g^{\otimes n})$, we need a simple but fundamental character computation.

Consider $n$ matrices $X_1, X_2, \dots, X_n \in \operatorname{End}(V)$. We can form the linear operator $X_1 \otimes X_2 \otimes \cdots \otimes X_n : V^{\otimes n} \to V^{\otimes n}$, take $\sigma \in S_n$, which is also a linear operator on $V^{\otimes n}$, and we have

LEMMA 6.3.19.

$$\begin{aligned} \operatorname{tr}(\sigma^{-1} \circ X_1 \otimes X_2 \otimes \cdots \otimes X_n) & \\ = \operatorname{tr}(X_{i_1} X_{i_2} \cdots X_{i_h}) \operatorname{tr}(X_{j_1} X_{j_2} \cdots X_{j_k}) \cdots \operatorname{tr}(X_{s_1} X_{s_2} \cdots X_{s_m}), & \end{aligned} \tag{6.15}$$

*where* $\sigma = (i_1 i_2 \cdots i_h)(j_1 j_2 \cdots j_k) \cdots (s_1 s_2 \cdots s_m)$ *is the cycle decomposition of* $\sigma$.

PROOF. Since the identity is multilinear, it is enough to prove it on the decomposable tensors of $\operatorname{End}(V) = V \otimes V^*$, so we may assume $X_i := u_i \otimes \phi_i$.

We need to use the rules

$$u \otimes \phi \circ v \otimes \psi = u \otimes \langle \phi \,|\, v \rangle \psi, \quad \operatorname{tr}(u \otimes \phi) = \langle \phi \,|\, u \rangle.$$

From these rules Lemma 6.3.19 easily follows. □

In particular we have $\operatorname{tr}(\sigma \circ X^{\otimes n}) = \prod \operatorname{tr}(X^{k_i})$, where the $k_i$ are the lengths of the cycles of $\sigma$. From the decomposition (6.8) we deduce

$$\operatorname{tr}(\sigma \circ X^{\otimes n}) = \prod \operatorname{tr}(X^{k_i}) = \sum_{\lambda \vdash n,\ \operatorname{ht}(\lambda) \le n = \dim V} \chi_\lambda(\sigma) \operatorname{tr}(S_\lambda(X)). \tag{6.16}$$

Call $\mu_\sigma$ the partition given by the lengths $k_i$ of the cycles of $\sigma$.

$$\prod \operatorname{tr}(X^{k_i}) = \prod_i \psi_{k_i}(x_1, \dots, x_m) := \Psi_{\mu_\sigma}(x_1, \dots, x_m),$$

where $\psi_k(x_1, \dots, x_m) = \sum_{i=1}^m x_i^k$. Thus formula (6.16) gives

THEOREM 6.3.20.

$$\Psi_{\mu_\sigma}(x_1, \dots, x_m) = \sum_{\lambda \vdash n,\ \operatorname{ht}(\lambda) \le n = \dim V} \chi_\lambda(\sigma) S_\lambda(x_1, \dots, x_m). \tag{6.17}$$

This formula can be effectively used to compute the characters of the symmetric group; see §6.4.5.

From this and Cauchy's formula, (3.19), one then can deduce (see for instance **[Pro07]**) the characters of the linear group:

THEOREM 6.3.21. *If* $X$ *has eigenvalues* $x_1, \dots, x_m$, *then* $\operatorname{tr}(S_\lambda(X))$ *equals the Schur function* $S_\lambda(x_1, \dots, x_m)$ *defined in* §3.2.3.

Special cases of this formula are

COROLLARY 6.3.22. *The character of* $\bigwedge^i(V)$ *is the elementary symmetric function* $e_i(x)$ *and the one of* $S^i(V)$ *is the complete symmetric function* $h_i(x)$.

Now that we established the connection between Schur functions and polynomial irreducible representations of $\mathrm{GL}(V)$, we see that the character of the tensor product $S_\lambda(V) \otimes S_\mu(V)$ is the Schur function $S_\lambda(x) S_\mu(x)$. So, if we want to understand how the representation $S_\lambda(V) \otimes S_\mu(V)$ decomposes into irreducible representations, we need to compute $S_\lambda(x) S_\mu(x)$ as a linear combination of Schur functions.

In particular Pieri's rule (Proposition 3.2.18) gives information on tensor products by exterior or symmetric powers.

The Cauchy formula also has an interpretation as

$$S(V \otimes W) = \bigoplus_{\lambda} S_\lambda(V) \otimes S_\lambda(W). \tag{6.18}$$

REMARK 6.3.23. If $\dim V = \dim W = n$, we have the one-dimensional space $S_{1^n}(V) \otimes S_{1^n}(W) = \bigwedge^n(V) \otimes \bigwedge^n(W)$. Moreover if a partition $\mu$ is obtained from a partition $\lambda$ by adding a column of length $n$, we have

$$\begin{aligned} S_\mu(V) \otimes S_\mu(W) &= S_\lambda(V) \otimes S_\lambda(W) \cdot \bigwedge^n(V) \otimes \bigwedge^n(W) \\ &= S_\lambda(V) \otimes S_\lambda(W) \otimes \bigwedge^n(V) \otimes \bigwedge^n(W). \end{aligned} \tag{6.19}$$

Finally consider $V$ of dimension $n$ and $S(V^* \otimes F^n)$ the algebra of polynomial functions on $n$ copies of $V$. The one-dimensional space $\bigwedge^n(V^*) \otimes \bigwedge^n(F^n)$ is generated by the determinant function on $V^n$, $(v_1, \dots, v_n) \mapsto v_1 \wedge \cdots \wedge v_n \in \bigwedge^n V = F$, where we have identified (a trivialization) $\bigwedge^n V = F$.

A typical application of this formula is in classical invariant theory.

Consider a group $G$ acting on a vector space $V$. Then for any other vector space $U$ the group $G$ acts on $V \otimes U$, and we have

$$S[V \otimes U]^G = \bigoplus_{\lambda \in \Pi_k} S_\lambda[V]^G \otimes S_\lambda[U]. \tag{6.20}$$

Taking a basis of $U$, for instance $U = F^\infty$, we have that $S[V \otimes F^\infty]^G = S[V^{\oplus\infty}]^G$ is the ring of invariants of countably many copies of the representation $V^*$, a graded representation of $\mathrm{GL}_\infty(F)$ of height $\leq k$.

COROLLARY 6.3.24. *For $S[V \otimes F^\infty]^G$ we have $p_\lambda = \dim S_\lambda[V]^G$. In particular, $p_\lambda = 0$ if the height of $\lambda$ is $> \dim V$.*

## 6.4. Characters

**6.4.1. Degree formulas for $\mathrm{GL}(V)$ and $S_n$.** The dimension of the irreducible representation $M_\lambda$ associated to a partition $\lambda$ is clearly $f_\lambda := \chi_\lambda(1)$.

A remarkable combinatorial fact is given in Theorem 6.3.16, i.e., the existence of a basis of $M_\lambda$ indexed by standard tableaux of shape $\lambda$. In fact there are several different ways of constructing these types of bases. Thus we have the following remarkable theorem.

THEOREM 6.4.1. *The degree $f_\lambda = \chi_\lambda(1)$ equals the number of standard tableaux of shape $\lambda$:*

$$\deg(\chi_\lambda) = \chi_\lambda(1) := f_\lambda. \tag{6.21}$$

Similarly the module $S_\lambda(F^m)$ has a basis indexed by semistandard tableaux which are also weight vectors for the diagonal matrices. The weight $w(T)$ of a tableau $T$ is given by the integers $h_1, \dots, h_m$, with $h_i$ the number of occurrences of $i$ in $T$. One deduces

$$S_\lambda(x_1, \dots, x_m) = \sum_{T \text{ semistandard}} x^{w(T)}.$$

**Weyl's dimension formula.** We start with a dimension formula for the space $S_\lambda(V)$, $\dim V = n$, $\lambda \vdash n$. Of course its value is $S_\lambda(1,1,\dots,1)$ which we want to compute from the determinantal formulas giving $S_\lambda(x) = A_{\lambda+\rho}(x)/V(x)$.

As usual let $\lambda := (\lambda_1, \lambda_2, \dots, \lambda_n)$ and $l_i := \lambda_i + n - i$. Of course we cannot substitute the number 1 directly for the $x_i$ else we get $0/0$. Thus we first substitute to $x_i \to x^{i-1}$ and then take the limit as $x \to 1$. Under the previous substitution we see that $A_{\lambda+\rho}$ becomes the Vandermonde determinant of the elements $x^{l_i}$, hence

$$S_\lambda(1, x, x^2, \dots, x^{n-1}) = \prod_{1 \le i < j \le n} \frac{x^{l_i} - x^{l_j}}{x^{n-i} - x^{n-j}} = \prod_{1 \le i < j \le n} \frac{\frac{x^{l_i} - x^{l_j}}{x-1}}{\frac{x^{n-i} - x^{n-j}}{x-1}}.$$

If $a > b$, we have

$$\frac{x^a - x^b}{x-1} = x^b(x^{a-b-1} + x^{a-b-2} + \dots + 1) \implies \lim_{x \to 1} \frac{x^a - x^b}{x-1} = a - b.$$

Hence, we deduce *Weyl's dimension formula*,

$$\text{(6.22)} \quad \dim S_\lambda(V) = S_\lambda(1,1,1,\dots,1) = \prod_{1 \le i < j \le n} \frac{l_i - l_j}{j - i} = \prod_{1 \le i < j \le n} \frac{\lambda_i - \lambda_j + j - i}{j - i}.$$

**Young–Frobenius and hook formula.** We bring here two useful formulas for computing $f_\lambda$; see **[Bo70]** and **[Mac95, Sag01]**.

THEOREM 6.4.2 (Young–Frobenius formula). *Let $\lambda = (\lambda_1, \dots, \lambda_k)$ be a partition of $n$, and define $\ell_j := \ell_j(\lambda) = \lambda_j + k - j$ for $1 \le j \le k$. Then the number of standard tableaux of shape $\lambda$ is given by*

$$\text{(6.23)} \qquad f_\lambda = \frac{n! \cdot \Delta_k(\ell_1, \dots, \ell_k)}{\ell_1! \cdots \ell_k!},$$

*where $\Delta_k(\ell_1, \dots, \ell_k) = \prod_{1 \le i < j \le k} (\ell_i - \ell_j)$.*

For example, let $\lambda = (2,1)$, then $(\ell_1, \ell_2) = (3,1)$, $\Delta_2(3,1) = 3 - 1 = 2$, so

$$f_{(2,1)} = \frac{3!}{3! \cdot 1!}(3-1) = 2.$$

The same partition $\lambda$ can be written as $\lambda = (2,1,0)$. Then $(\ell_1, \ell_2, \ell_3) = (4,2,0)$, $\Delta_3(4,2,0) = (4-2)(4-0)(2-0) = 16$, so

$$f_{(2,1,0)} = \frac{3!}{4! \cdot 2! \cdot 0!} 16 = 2.$$

In general, adding a tail of zeroes does not affect the value $f_\lambda$ as calculated by the Young–Frobenius formula.

Let $\lambda$ be a partition of $n$. The hook formula for calculating $f_\lambda$ is due to Frame, Robinson, and Thrall **[Sag01]**. We begin with

DEFINITION 6.4.3. Let $\lambda$ be a partition of $n$, and let $x = (i,j) \in \lambda$ be a box in the corresponding Young diagram. The *hook number* $h(x) = h(i,j)$ is defined as

$$\text{(6.24)} \qquad h(i,j) = \lambda_i + \check{\lambda}_j - i - j + 1.$$

EXAMPLE 6.4.4. Note that the box $x$ defines a hook in the diagram $\lambda$, and $h(x)$ equals the length (number of boxes) of this hook:

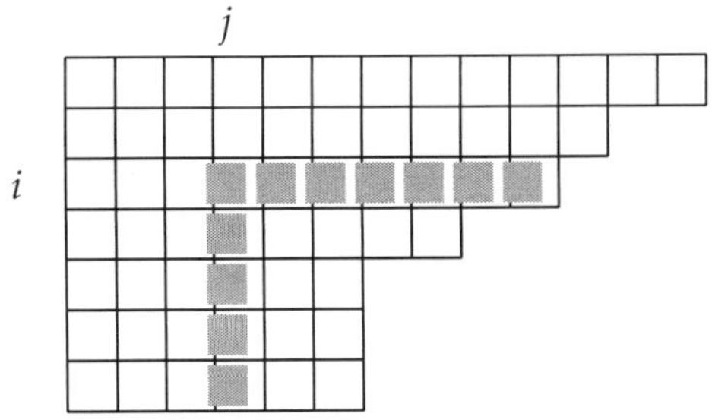


In this figure, we have $\lambda=(13,11,10,8,6^3)$ with $x=(3,4)$. Then $\check{\lambda}=(7^6,4^2,3^2,2,1^2)$ and $h(x)=\lambda_3+\check{\lambda}_4-3-4+1=10+7-6=11$.

Here is another example. In the following diagram of shape $\lambda=(8,3,2,1)$, each hook number $h(x)$ is written inside its box in the diagram $\lambda$:

| | | | | | | | |
|---|---|---|---|---|---|---|---|
| 11 | 9 | 7 | 5 | 4 | 3 | 2 | 1 |
| 5 | 3 | 1 | | | | | |
| 3 | 1 | | | | | | |
| 1 | | | | | | | |

EXAMPLE 6.4.5. Let $\lambda$ be the $u\times v$ rectangle, namely $\lambda=(u^v)$. Then the sum of all hook numbers in the diagram $\lambda$ is

$$\begin{aligned}\sum_{x\in\lambda} h(x) &= \sum_{i=1}^{v} \text{the sum of the hook numbers in the } (v-i)\text{-th row of } \lambda\\ &= \sum_{i=1}^{v}((1+\cdots+u)+(v-i)u)=uv(u+v)/2.\end{aligned}$$

The calculation done in this example will be applied in Theorem 7.5.15.

Let $\lambda$ be a partition. The hook formula expresses $f_\lambda$, the number of standard tableaux of shape $\lambda$, by the hook numbers of $\lambda$.

THEOREM 6.4.6 (The hook formula). *Let $\lambda\vdash n$ be a partition of $n$, and let $f_\lambda$ be the number of standard tableaux of shape $\lambda$. Then*

$$f_\lambda=\frac{n!}{\prod_{x\in\lambda}h(x)}. \tag{6.25}$$

EXAMPLE 6.4.7. Let $\lambda=(2,1)$, then the hook formula says that

$$f_{(2,1)}=\frac{3!}{3\cdot1\cdot1}=2;$$

compare with Example 6.1.7.

For a second example, let $\lambda$ be a so-called $(1,1)$-*hook partition* of $n$, that is, a partition of $n$ of the form $\lambda=(n-k+1,1,\dots,1)=(n-k+1,1^{k-1})$ for some $1\le k\le n$. Applying the hook formula is particularly convenient in this case, and it follows that

$$f_{(n-k+1,1^{k-1})}=\frac{n!}{n\cdot(k-1)!\cdot(n-k)!}=\binom{n-1}{k-1}.$$

EXERCISE 6.4.8. Describe the standard tableaux for a $(1,1)$-hook partition.

DEFINITION 6.4.9. Let $k, \ell \geq 0$ be integers, then denote

$$H(k,\ell;n) = \{\lambda = (\lambda_1, \lambda_2, \dots) \vdash n \mid \lambda_{k+1} \leq \ell\}.$$

In particular $H(k,0;n) = \{\lambda \vdash n \mid \mathrm{ht}(\lambda) \leq k\}$. We also define

$$H(k,\ell) = \bigcup_{n\geq 0} H(k,\ell;n). \tag{6.26}$$

This is the set of all partitions which *do not contain* the rectangle $k+1, \ell+1$ or equivalently the box of coordinates $k+1, \ell+1$. The union of all these diagrams is made of a horizontal infinite strip of height $k$ and a vertical infinite strip of width $\ell$. This is the *$k, \ell$ hook*.

These subsets of partitions play important roles in PI theory. For example, the sum

$$\sum_{\lambda\in H(k,0;n)} (f_\lambda)^2$$

plays an important role in the proof of Lemma 7.1.11, which is a key step in proving the exponential bound for the codimensions; see §7.1. These subsets $H(k,\ell;n)$ also play important roles in the representation theory of Lie algebras and of Lie superalgebras; see for example **[BR87]**.

**6.4.2. $S_n$ characters—some basic facts.** Some main facts from the character theory of $S_n$ are reviewed here, with emphasis on results that will be applied in later sections. The reader is advised to look at some textbooks on the subject, see for example **[Bo70, JK81, Mac95, Sag01]**.

From Theorem 6.2.6 the regular representation $\mathbb{Q}[S_n]$ of $S_n$ decomposes as a direct sum $\mathbb{Q}[S_n] = \bigoplus_{\lambda\vdash n} I_\lambda$ with $I_\lambda$ minimal *two-sided* ideals.

We have $\dim I_\lambda = (\deg(\chi_\lambda))^2 = (f_\lambda)^2$. By computing dimensions this implies

$$n! = \sum_{\lambda\vdash n} (f_\lambda)^2. \tag{6.27}$$

Furthermore, if $J_\lambda \subseteq I_\lambda$ is a minimal *left* ideal of $\mathbb{Q}[S_n]$, then $J_\lambda$ is a subrepresentation of the regular representation of $S_n$, with $S_n$-character $\chi_\lambda$,

$$\chi_{regular}(J_\lambda) = \chi_\lambda, \quad \text{hence } \dim J_\lambda = f_\lambda.$$

Also, $I_\lambda$ equals the sum of its minimal left ideals, and is a subrepresentation of the regular representation of $S_n$, with corresponding $S_n$-character,

$$\chi_{regular}(I_\lambda) = f_\lambda \chi_\lambda, \qquad \text{hence} \qquad \chi_{regular}(\mathbb{Q}[S_n]) = \sum_{\lambda\vdash n} f_\lambda \chi_\lambda, \tag{6.28}$$

which is the decomposition of the character of the regular representation of $S_n$ into irreducibles.

**6.4.3. Branching and induction of $S_n$ characters.** As usual, $S_{n-1}$ is viewed as a subgroup of $S_n$ fixing the integer $n$, i.e., $\sigma \in S_{n-1}$ if and only if $\sigma \in S_n$ and $\sigma(n) = n$. Since $S_{n-1} \subset S_n$, any $S_n$ character is also an $S_{n-1}$ character. If $\lambda \vdash n$, then the irreducible $S_n$ character $\chi_\lambda$ restricts to an $S_{n-1}$ character, and we denote its restriction by $\chi_\lambda \downarrow_{S_{n-1}}$.

At the same time we can induce $\chi_\lambda$ to an $S_{n+1}$ character, and we denote this induction by $\chi_\lambda \uparrow^{S_{n+1}}$

In order to describe restriction and induction, we need some combinatorics of diagrams.

A *corner* in a partition $\lambda$ is a box in the diagram $\lambda$ which is rightmost in its row and lowest in its column. For example the corners of $\lambda = (5,2,2,1)$ are the boxes $(1,5)$, $(3,2)$, and $(4,1)$.

By adding a box to $\lambda$ such that the result is a partition, we obtain a new corner. The new corners for $\lambda = (5,2,2,1)$ are $(1,6)$, $(2,3)$, $(4,2)$, and $(5,1)$.

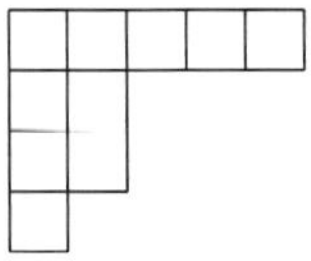

DEFINITION 6.4.10. Let $\lambda$ be a partition of $n$, and denote by $\lambda^-$ the set of partitions of $n-1$ which are obtained from $\lambda$ by the removal of one corner. Similarly $\lambda^+$ denotes the set of partitions of $n+1$ which are obtained from $\lambda$ by the addition of one box. For example,

$$\begin{aligned} (5,5,3,2)^- &= \{(5,4,3,2),\ (5,5,2,2),\ 5,5,3,1)\}, \\ (5,5,3,2)^+ &= \{(6,5,3,2),(5,5,4,2),(5,5,3,3),(5,5,3,2,1)\}. \end{aligned}$$

The following branching theorem can be proved for the irreducible characters $\chi_\lambda$.

THEOREM 6.4.11. *Let $\lambda$ be a partition of $n$, then*

$$\chi_\lambda \downarrow_{S_{n-1}} = \sum_{\nu \in \lambda^-} \chi_\nu, \qquad \textit{hence} \qquad f_\lambda = \sum_{\nu \in \lambda^-} f_\nu. \tag{6.29}$$

$$\chi_\lambda \uparrow^{S_{n+1}} = \sum_{\nu \in \lambda^+} \chi_\nu, \qquad \textit{hence} \qquad (n+1) f_\lambda = \sum_{\nu \in \lambda^+} f_\nu. \tag{6.30}$$

Let $\mu \vdash m$, $\lambda \vdash n$, $m < n$ such that $\mu \subseteq \lambda$ (namely $\mu_i \leq \lambda_i$ for all $i$). We see then that when we restrict the representation $M_\lambda$ from $S_n$ to $S_m$, we can do it by successive restrictions of $S_m$ to $S_{m-1}$ to $S_{m-2}$ and so on. We can keep track at each step of how a given irreducible representation decomposes by marking the case we remove.

If we do it by decreasing markings, we see that the multiplicity of $M_\mu$ in the restriction of $M_\lambda$ equals the number of *skew standard tableaux* $\lambda/\mu$.

For example, there are four standard skew Young tableaux of the pair of partitions $\lambda = (5,3,2)$, $\mu = (3,3)$. We show two of them given by

$T_1 =$ 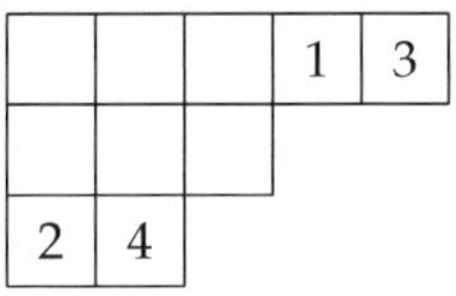

, $T_2 =$ 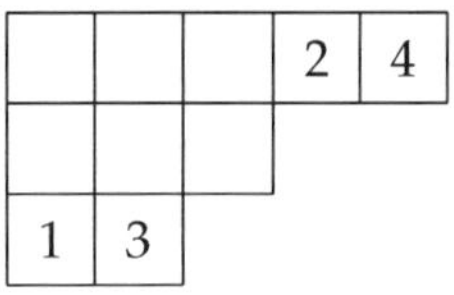

.

COROLLARY 6.4.12. *If $\mu \subset \lambda$, we have $f_\mu \leq f_\lambda$.*

PROOF. By (6.29)

$$f_\lambda = \sum_{\nu \in \lambda^-} f_\nu = f_\mu + \text{ some other summands,}$$

and the proof follows.

Here is another elementary proof. Given a SYT (standard Young tableau) $T_\mu$ of shape $\mu$, fill the skew shape $\lambda \setminus \mu$ with $m+1, m+2, \ldots, n$ by first filling, from left to right, the first row of $\lambda \setminus \mu$, then the second row, etc. The result is an SYT $T_\lambda$ of shape $\lambda$ (with $T_\mu \subseteq T_\lambda$). This gives a bijection of the set of $f_\mu$ SYTs of shape $\mu$, with a subset of $f_\lambda$ SYTs of shape $\lambda$. □

Formula (6.30) implies the following theorem.

THEOREM 6.4.13. *Let $\mu \vdash n$ with $I_\mu \subseteq \mathbb{Q}[S_n]$ the corresponding two-sided ideal, and let $n \leq m$. Then with the notations of page 166:*

$$\mathbb{Q}[S_m] I_\mu \mathbb{Q}[S_m] = \bigoplus_{\lambda \vdash m,\ \mu \subseteq \lambda} I_\lambda.$$

Another consequence of Theorem 6.4.11 is the following. From formula (6.27) it follows that

$$\sum_{\lambda \vdash n} f_\lambda \geq \sqrt{\sum_{\lambda \vdash n} (f_\lambda)^2} = \sqrt{n!}.$$

Hence $\sum_{\lambda \vdash n} f_\lambda$ grows to infinity faster than any exponent $q^n$. However, restricting $\lambda$ to the $k$-strip $H(k, 0; n)$, Theorem 6.4.11 implies the following exponential bounds.

COROLLARY 6.4.14.

$$\sum_{\lambda \in H(k,0;n)} f_\lambda \leq k^n, \quad \Longrightarrow \quad \sum_{\lambda \in H(k,0;n)} (f_\lambda)^2 \leq k^{2n}.$$

PROOF. $F(n,k) := \sum_{\lambda \in H(k,0;n)} f_\lambda$ counts the number of standard tableaux, with $n$ boxes and at most $k$ rows. The proof is by induction on $n$, and we assume $F(n-1,k) \leq k^{n-1}$. Such tableaux are obtained from standard tableaux, with $n-1$ boxes and at most $k$ rows by adding $n$ in a rim box in one of the $k$ rows. This is done in at most $k$ ways, hence $F(n,k) \leq kF(n-1,k)$, and the claim follows. □

Notice that if we only use *staircase tableaux*, that is with $\lambda_i = \lambda_1 - i + 1,\ i = 2, \ldots, k,\ \lambda_j = 0, \forall j > k$, we can also deduce a lower bound

$$F(n,k) \geq (k!)^{n/k}.$$

REMARK 6.4.15. The estimate $\sum_{\lambda \in H(k,0;n)} (f_\lambda)^2 < k^{2n}$ follows also immediately from representation theory.

We claim that $\sum_{\lambda \in H(k,0;n)} (f_\lambda)^2$ is the dimension of the algebra $\Sigma_n(k)$ spanned by the symmetric group $S_n$ acting on $V^{\otimes n}$ where $\dim V = k$.

In fact the second fundamental theorem, Theorem 6.1.3, says that this algebra is the image of the group algebra of the symmetric group modulo the ideal generated by any antisymmetrizer on $k+1$ elements. So this ideal is the sum of the simple ideals $I_\lambda, \mathrm{ht}(\lambda) > k$ and $\Sigma_n(k)$ is the direct sum of the ideals $I_\lambda, \mathrm{ht}(\lambda) \leq k$.

Clearly $\dim \Sigma_n(k) < \dim \mathrm{End}(V^{\otimes n}) = k^{2n}$.

Notice also that $\sum_{n=0}^{\infty} \dim \Sigma_n(k) t^n$ is the constant term with respect to the variables $x_i$ which are bound to $\prod_i x_i = 1$ of the series

(6.31)

$$E_k(x,t) := \frac{1}{k!} \frac{\prod_{1 \le i < j \le k}(2 - (x_i x_j^{-1} + x_j x_i^{-1}))}{1 - (\sum_{i,j=1}^{k} x_i x_j^{-1})t}$$

$$= \frac{1}{k!} \sum_{n=0}^{\infty} \prod_{1 \le i < j \le k} (2 - (x_i x_j^{-1} + x_j x_i^{-1}))(k + \sum_{1 \le i < j \le k} (x_i x_j^{-1} + x_j x_i^{-1}))^n t^n.$$

This follows from the fact that $(\sum_{i,j=1}^{k} x_i x_j^{-1}) = k + \sum_{1 \le i < j \le k}(x_i x_j^{-1} + x_j x_i^{-1})$ is the character $\chi(g)$ of the action of $\mathrm{SU}(k)$ (the special group of $k \times k$ unitary matrices with $\det = 1$) on $V \otimes V^* = \mathrm{End}(V)$ restricted to the diagonal matrices, so

$$\chi(g)^n = (\sum_{i,j=1}^{k} x_i x_j^{-1})^n = (k + \sum_{1 \le i < j \le k} (x_i x_j^{-1} + x_j x_i^{-1}))^n$$

is the character of the action of $\mathrm{SU}(k)$ on $\mathrm{End}(V)^{\otimes n}$, and the invariants of this action give the algebra $\Sigma_n(k)$.

The dimension of the invariants is $\int_{\mathrm{SU}(k)} \chi(g)^n dg$, where $dg$ is the normalized Haar measure, and by Weyl's integration formula

$$\int_{\mathrm{SU}(k)} \chi(g)^n dg = \frac{1}{k!} \int_T \chi(t)^n V(x) \bar{V}(x) dt,$$

where $T$ is the compact torus of diagonal matrices with entries complex numbers $x_i$ of absolute value 1 and $\prod_i x_i = 1$. $V(x_1, \dots, x_k)$ is the Vandermonde determinant. So

$$V(x_1, \dots, x_k) V(x_1^{-1}, \dots, x_k^{-1}) = \prod_{i<j} (2 - x_i x_j^{-1} - x_j x_j^{-1}).$$

On such a torus the integral of a Fourier series is the constant term.

**6.4.4. Kronecker products of $S_n$ characters.** Given any group $G$, we have seen in §1.3.4 that given two representations $M, N$ of $G$, the tensor product $M \otimes N$ is also a representation of $G$ by the formula $g(m \otimes n) := gm \otimes gn$. The character of the tensor product is also called Kronecker product of the characters $\chi_1, \chi_2$ and is the function $\chi_1 \otimes \chi_2 : G \longrightarrow F$ given by $(\chi_1 \otimes \chi_2)(g) = \chi_1(g) \cdot \chi_2(g)$. Clearly $\deg(\chi_1 \otimes \chi_2) = \deg(\chi_1) \cdot \deg(\chi_2)$. We now apply this to the symmetric group.

Since the characteristic of the base field is zero, every character is a sum of irreducible characters. In particular this applies to the character $\chi_1 \otimes \chi_2$. For $\lambda, \mu \vdash n$, the problem of decomposing $\chi_\lambda \otimes \chi_\mu$ is extremely involved, and so far there are no known effective algorithms for computing that decomposition; see for example **[Dvi93]**, **[GR85]**.

Consider now the conjugation representation of $S_n$, that is where $S_n$ acts on $\mathbb{Q}[S_n]$ by conjugation. Let $\lambda \vdash n$, and let $I_\lambda \subseteq \mathbb{Q}[S_n]$ be the corresponding minimal two-sided ideal. Then $I_\lambda$ is an $S_n$ submodule for the conjugation action $*$: if $x \in \mathbb{Q}[S_n]$ and $\sigma \in S_n$, then $\sigma * x = \sigma x \sigma^{-1}$. We denote the corresponding $S_n$ character by $\chi_{\mathrm{conj}}(I_\lambda)$. Also let $\chi_\lambda$ be the irreducible $S_n$ character corresponding to $\lambda$. Then since $I_\lambda = M_\lambda \otimes M_\lambda$, we have

$$\chi_{\mathrm{conj}}(I_\lambda) = \chi_\lambda \otimes \chi_\lambda. \tag{6.32}$$

**6.4.5. The Murnaghan–Nakayama rule.** Let $\lambda, \mu \vdash k$, and let $C(\mu) \subseteq S_k$ be the conjugacy class of $S_k$ indexed by $\mu$. We denote

$$\chi_\lambda(\mu) := \chi_\lambda(\pi),$$

where $\pi$ is any permutation in $C(\mu)$. The Murnaghan–Nakayama rule is an algorithm for computing $\chi_\lambda(\mu)$. The main point in this rule is Theorem 6.3.20 and formula (6.17).

Since for a partition $\mu = (k_1, \dots, k_u)$ we have $\Psi_\mu = \prod_j \psi_{k_j}$, formula (6.17) can be computed recursively if one knows the multiplication rule between a Newton function $\psi_k$ and the Schur functions. This is the Murnaghan–Nakayama rule. A proof of this rule can be found, for example, in **[Mac95, Sag01]** or **[Pro07]**. We now describe that algorithm.

THEOREM 6.4.16. *$\psi_k(x)S_\lambda(x) = \sum \pm S_\mu$, where $\mu$ runs over all diagrams, such that by removing a connected set of $k$ boxes* (a rim-hook) *of the rim of $\mu$ we have $\lambda$.*

*The sign to attribute to $\mu$ is $+1$ if the number of rows modified from $\lambda$ is odd; otherwise it is $-1$.*

This property says that the new diagrams $\mu$ are the diagrams containing the diagram $\lambda$ and such that their difference is a (*connected*) *rim-hook*, made of $k$ boxes of the rim of $\mu$. Intuitively, it is like a *Slinky*.[2] So, if we write the diagram in French notation, one may think of a slinky made of $k$ boxes $\bullet$, *sliding down the diagram* in all possible ways.

The sign to attribute to such a configuration is $+1$ if the number of rows occupied is odd; otherwise, it is $-1$.

For instance $\psi_3 S_{(3,2,1)} = S_{(3,2,1^4)} - S_{(3,2^3)} - S_{(3^3)} - S_{(4^2,1)} + S_{(6,2,1)}$ can be visualized in French notation as

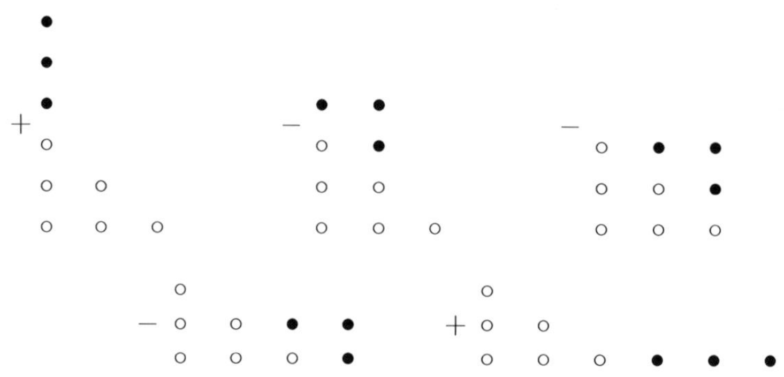

Formally, one can define a $k$-slinky as a walk in the plane $\mathbb{N}^2$ made of $k$-steps, and each step is either one step down or one step to the right. The sign of the slinky is $-1$ if it occupies an even number of rows, and $+1$ otherwise.

When $\mu = (1^n)$, the slinky is one box, and one recovers the standard tableaux.

EXAMPLE 6.4.17. Let $\sigma \in S_k$ be a $k$-cycle, then $\sigma \in C(k)$, so $\chi_\lambda(\sigma) = \chi_\lambda((k))$ and $\Psi_{(k)} = \psi_k$, and formula (6.17) becomes

$$\psi_k = \sum \chi_\lambda((k))S_\lambda.$$

[2] This was explained to me by A. Garsia and refers to a spring toy sold in novelty shops.

Then $\psi_k = \psi_k \cdot 1 = \psi_k \cdot S_{(0)}$,

$$\chi_\lambda(\sigma) = \begin{cases} (-1)^i & \text{if } \lambda = (k-i, 1^i),\ i = 0, \dots, k-1, \\ = 0 & \textit{otherwise}. \end{cases} \tag{6.33}$$

PROOF. If $\chi_\lambda((k)) \neq 0$, then necessarily $\lambda$ is itself a rim-hook of length $k$, namely $\lambda = (k-i, 1^i)$ for some $i = 0, \dots, k-1$. The height of $(k-i, 1^i)$ is $i+1$, hence

$$\psi_k = \sum_i (-1)^i S_{(k-i,1^i)} \implies \chi_\lambda(\sigma) = \chi_{(k-i,1^i)}((k)) = (-1)^{(i+1)+1} = (-1)^i. \quad \square$$

**6.4.6. Frobenius character.** From the theory of Young symmetrizers it follows that the characters $\beta_\lambda$ of the permutation representations which are associated to the Young subgroups, $S_\lambda = S_{\lambda_1} \times S_{\lambda_2} \times \cdots \times S_{\lambda_k}$, that is the subgroup preserving the rows of a tableau of shape $\lambda = (\lambda_1, \lambda_2, \dots, \lambda_k)$, form a basis of the characters over $\mathbb{Z}$. The change of basis from this to the irreducible characters is triangular and in fact the irreducible representations $M_\mu$ appearing in the permutation representation $F[S_n/S_\lambda]$ are the ones for which $\mu \geq \lambda$ in the *dominance order*, (cf. Definition 6.1.4). The formulas are best cast by introducing the *Frobenius character* $\mathbb{F}(M)$ of a representation $M$ of $S_n$. By definition, $\mathbb{F}(M)$ is a symmetric function. If $M = \bigoplus_\lambda m_\lambda M_\lambda$, we set

$$\mathbb{F}(M) = \sum_\lambda m_\lambda S_\lambda(X).$$

We then have that the Frobenius character of $F[S_n/S_\lambda]$ is the symmetric function $h^\lambda$ equal to the sum of all monomials with exponents the numbers $\lambda_i$. Thus the multiplicities of the irreducible representations $M_\mu$ appearing in the permutation representation $F[S_n/S_\lambda]$ are the coefficients (called Kostka numbers) of the expansion of $h^\lambda$ in the basis of Schur functions.

Another useful formula involving the Frobenius character is the one for the following induced representation.

Let $n = h + k$, and let $\lambda \vdash h$, $\mu \vdash k$ be two partitions of $h, k$. To these two partitions we have associated the irreducible representations $M_\lambda$, $M_\mu$ of $S_h, S_k$ and hence the irreducible representation $M_\lambda \otimes M_\mu$ of $S_h \times S_k$. We then have the formula for the induced representation

$$\mathbb{F}(\mathrm{Ind}_{S_h \times S_k}^{S_{h+k}} M_\lambda \otimes M_\mu) = \mathbb{F}(M_\lambda)\mathbb{F}(M_\mu) = S_\lambda(X) S_\mu(X). \tag{6.34}$$

Its Frobenius character is given by the product of the corresponding Schur functions in the ring of symmetric functions.

DEFINITION 6.4.18. The representation $\mathrm{Ind}_{S_h \times S_k}^{S_{h+k}} M_\lambda \otimes M_\mu$ is also called the *outer tensor product* of $M_\lambda$ and $M_\mu$ and is denoted by $M_\lambda \hat{\otimes} M_\mu$. We use the same notation for characters $\chi_\lambda \hat{\otimes} \chi_\mu$

The rule of multiplication of two Schur functions $S_\alpha S_\beta = \sum_\gamma c_{\alpha,\beta}^\gamma S_\gamma$ is a rather difficult combinatorial rule, based on the notion of a *lattice permutation*.

DEFINITION 6.4.19. A word $w$ in the numbers $1, \dots, r$ is called a lattice permutation if, for each initial subword (a prefix) $a$, i.e., such that $w = ab$, we have that, setting $k_i$ to be the number of occurrences of $i$ in $a$, we have $k_1 \geq k_2 \geq \cdots \geq k_r$.

A reverse lattice permutation is a word $w$ such that the opposite word $w^o$,[3] is a lattice permutation.

If we read a skew standard tableau from left to right and from up to bottom, we display its entries as a *row word*.

THEOREM 6.4.20 (The Littlewood–Richardson rule).

(i) *A partition $\gamma$ which appears with $c^{\gamma}_{\alpha,\beta} > 0$ must contain $\alpha$ and $\beta$.*

(ii) *The multiplicity $c^{\gamma}_{\alpha,\beta}$ equals the number of skew semistandard tableaux of shape $\gamma \setminus \alpha$ and content $\beta$ whose associated row word is a reverse lattice permutation.*

Part (i) is quite easy, and it follows from the branching rule (6.30). The harder part (ii) will not be used in this book.

[3]The opposite word is just the word read from right to left.

# Part 2

# Combinatorial aspects of polynomial identities

CHAPTER 7

# Growth

*In this chapter we start the combinatorial approach to polynomial identities. The main results treated are the theorems of Regev on exponential growth of codimensions (Theorem 7.1.6) with its application to the tensor product of two PI algebras. Then we present the hook theorem (Theorem 7.5.1) of Amitsur and Regev and, independently, Kemer.*

## 7.1. Exponential bounds

**7.1.1. Codimensions.** Assume now that $A$ is an algebra over a field $F$ of characteristic 0, by identifying the free algebra $F\langle X\rangle$ to the tensor algebra $T(V)$, where $V$ is the vector space having $X$ as its basis. We have seen in Remark 2.2.3 that any $T$-ideal $I$ of $T(V)$ is in particular a representation of $\mathrm{GL}(V)$.

The homogeneous component of $I$ (resp., of $T(V)/I$) of degree $m$, decomposes into isotypic components, and for each partition $\lambda \vdash m$ the corresponding component is a direct sum of representations isomorphic to $S_\lambda(V)$.

With the notations of §6.3.1.2 we identify elements of the group algebra $\mathbb{Q}[S_n]$ with multilinear polynomials $\mathbb{Q}[S_n] \simeq V_n$, namely

$$\sum_{\sigma\in S_n} a_\sigma \sigma \equiv \sum_{\sigma\in S_n} a_\sigma x_{\sigma(1)} \cdots x_{\sigma(n)}.$$

In Remark 6.3.10 we have clarified the role of left and right action.

For a $T$-ideal $I = \mathrm{Id}(A)$, the space $I \cap V_n$ is stable under left action so it is a representation of the symmetric group $S_n$. In characteristic 0 this space decomposes into the direct sum of isotypic components relative to the partitions of $n$. This allows the application of the theory of $S_n$-characters to the study of PI algebras with the introduction of the cocharacter sequence $\chi_n(A)$ of a PI algebra $A$.

Assume $X$ is infinite or, equivalently, $V$ is infinite dimensional.

DEFINITION 7.1.1.

(1) The character of the representation of $S_n$ in the space $V_n/I \cap V_n = V_n/(\mathrm{Id}(A) \cap V_n)$ is called the *cocharacter* and is denoted by $\chi_n(A)$ or $\chi_n(I)$.
(2) The cocharacter $\chi_n(A)$ decomposes as $\sum_{\lambda\vdash n} m_\lambda(A)\chi_\lambda$, where $m_\lambda(A)$ is the multiplicity of the irreducible representation $M_\lambda$ of the symmetric group $S_n$ in $V_n/(\mathrm{Id}(A) \cap V_n)$.
(3) The sum $\sum_{\lambda\vdash n} m_\lambda(A) \in \mathbb{N}$ is the number of irreducible representations decomposing $V_n/(\mathrm{Id}(A) \cap V_n)$ and is called the *colength*.

According to Theorem 6.3.7 and the discussion following this theorem, in particular Corollary 6.3.11, if $|\lambda| = n$, that is $S_\lambda(V) \subset V^{\otimes n}$, this multiplicity can be computed in two different and both useful ways.

PROPOSITION 7.1.2. *The following numbers are equal:*

(1) *The multiplicity $m_\lambda$ of $S_\lambda(V)$ in $T(V)/I$,*
(2) *the multiplicity $m_\lambda(I)$ in the cocharacter of $I$,*
(3) *the dimension of the space of $U_V$ invariants in $T(V)/I$ of weight $\lambda$.*

Here $U_V$ is the group of upper triangular matrices with 1 on the diagonal, well defined as soon as we have chosen a basis $v_1, v_2, \dots$ of $V$.

In view of this proposition we see

COROLLARY 7.1.3. *The Frobenius character $\sum_\lambda m_\lambda S_\lambda$ in the ring of symmetric functions equals the graded Hilbert–Poincaré series of $T(V)/I$ (see (8.12)).*

A coarser invariant that one can define also in positive characteristic is

DEFINITION 7.1.4. The dimension of the space $V_n/(\mathrm{Id}(A) \cap V_n)$, of multilinear elements in $n$-variables of $T(V)/I$, is denoted by $c_n(A)$, and is called the *codimension* of $A$ in degree $n$.

Since $\dim V_n = n!$, saying that $A$ satisfies a polynomial identity of degree $n$ is then clearly equivalent to saying that $c_n(A) < n!$.

REMARK 7.1.5. There exist $c_n(A)$ monomials $K_i(x_1, \dots, x_n) \in V_n$, $1 \le i \le c_n(A)$ and coefficients $\{\alpha_{i,\sigma} \in F \mid \sigma \in S_n\}$ such that for any $\sigma \in S_n$, and for any $a_1, \dots, a_n \in A$,

$$M_\sigma(a_1, \dots, a_n) = \sum_{i=1}^{d_n} \alpha_{i,\sigma} K_i(a_1, \dots, a_n).$$

The following theorem shows that the codimensions are always exponentially bounded. This implies that, in a sense, $T$-ideals in $F\langle X\rangle$ are always large. The exponential bound theorem has found several applications in PI theory. Originally it was proved, in **[Reg72]**, in order to prove Theorem 7.2.2 below.

THEOREM 7.1.6 (**[Reg72]**). *Let $A$ be a PI algebra over a field $F$ of an arbitrary characteristic. Then there exists $0 < \alpha \in \mathbb{R}$ such that*

$$c_n(A) \le \alpha^n.$$

*In fact, if $A$ satisfies an identity of degree $d$, then for all $n$,*

$$c_n(A) \le (d-1)^{2n}. \tag{7.1}$$

The proof is given below. We remark that the theorem holds for a PI algebra defined over any commutative ring—and in any characteristic, provided the identity is multilinear and with coefficients $\pm 1$.

For a proof based on the Dilworth theorem **[Dil50]**, see **[Dre00]**, **[Lat72]**, **[Row80]**. The proof presented here is based on classical, well known—and rather deep—results in combinatorics and in the representation theory of $S_n$.

This theorem can be refined as in **[GZ03a]** by Giambruno and Zaicev and will be presented in Chapter 21.

THEOREM 7.1.7. *We have*

$$\lim_{n\to\infty} c_n(A)^{1/n} = k \in \mathbb{N}.$$

*The integer $k$ is called the* exponent *of $A$.*

Introduce the left lexicographic order on permutations, hence on multilinear monomials: Given $\sigma, \pi \in S_n$, we set $\sigma < \pi$ if for some $1 \le i \le n$, $\sigma(1) = \pi(1), \dots, \sigma(i-1) = \pi(i-1)$ and $\sigma(i) < \pi(i)$. For example when $n = 3$,

$$x_1x_2x_3 < x_1x_3x_2 < x_2x_1x_3 < x_2x_3x_1 < x_3x_1x_2 < x_3x_2x_1;$$

see **[Lat72]**.

DEFINITION 7.1.8. Let $0 < d$ be an integer, and let $\sigma \in S_n$. Then $\sigma$ is called *d-bad* if $\sigma$ has a descending subsequence of length $d$, namely, if there exist $1 \le i_1 < i_2 < \cdots < i_d \le n$ such that $\sigma(i_1) > \sigma(i_2) > \cdots > \sigma(i_d)$. Otherwise $\sigma$ is called *d-good*.

Let $\sigma \in S_n$, then the monomial $x_{\sigma(1)} \cdots x_{\sigma(n)}$ is said to be $d$-good if and only if $\sigma$ is $d$-good.

REMARK 7.1.9. $\sigma$ is $d$-good if any descending subsequence of $\sigma$ is of length $\le d-1$. If $\sigma$ is $d$-good, then $\sigma$ is $d'$-good for any $d' \ge d$.

Every permutation is 1-bad.

Denote by $g_d(n)$ the number of $d$-good permutations in $S_n$. For example $g_2(3) = 1$, $g_3(3) = 5$, and $g_4(3) = 6$: $(123)$ is 1-bad but not 2-bad, hence it is $d$-good for any $d \ge 2$. Similarly, each of $(132)$, $(213)$, $(231)$, and $(312)$ is 2-bad but not 3-bad, hence is $d$-good for any $d \ge 3$; $(321)$ is 3-bad, and finally every $\sigma \in S_3$ is $d$-good for $d \ge 4$.

7.1.1.1. *The Latyshev–Shirshov lemma.* The Latyshev–Shirshov lemma relates codimensions with the numbers $g_d(n)$ of $d$-good permutations in $S_n$; see also Remark 8.1.9 below.

LEMMA 7.1.10 (**[Lat72]**). *Let $A$ be a PI algebra satisfying a proper identity* (*Definition* 2.2.33) *homogeneous of degree $d$. Then modulo* $\mathrm{Id}(A) \cap V_n$*, the module $V_n$ is spanned by the $d$-good monomials (i.e., $d$-good permutations). Thus $c_n(A) \le g_d(n)$.*

PROOF. The lemma follows once we prove the following: Let $\sigma \in S_n$ be $d$-bad. Then modulo $\mathrm{Id}(A) \cap V_n$, $\sigma$ is a linear combination of permutations (i.e., monomials) that are strictly smaller than $\sigma$ in the above left lexicographic order.

By Proposition 2.2.34 this identity can be taken to be multilinear and of the form

$$x_1 \cdots x_d - \sum_{1 \ne \pi \in S_d} \alpha_\pi x_{\pi(1)} \cdots x_{\pi(d)} \in \mathrm{Id}(A).$$

Therefore, for any polynomials $a_0, h_1, \dots, h_d$, by substituting $x_i \to h_i$ and by left multiplying by $a_0$, also

$$a_0 h_1 \cdots h_d - \sum_{1 \ne \pi \in S_d} \alpha_\pi \cdot a_0 h_{\pi(1)} \cdots h_{\pi(d)} \in \mathrm{Id}(A). \tag{7.2}$$

Since $\sigma$ is assumed to be $d$ bad, there are $d$ indices $1 \le i_1 < i_2 < \cdots < i_d \le n$ such that $\sigma(i_1) > \sigma(i_2) > \cdots > \sigma(i_d)$. Corresponding to the indices $i_1, \dots, i_d$, we parenthesize the monomial $x_{\sigma(1)} \cdots x_{\sigma(n)}$ as follows:

$$x_{\sigma(1)} \cdots x_{\sigma(n)} = a_0 (x_{\sigma(i_1)} \cdots)(x_{\sigma(i_2)} \cdots) \cdots (x_{\sigma(i_d)} \cdots x_{\sigma(n)}),$$

where $a_0 = x_{\sigma(1)} \cdots x_{\sigma(i_1 - 1)}$. Denote $h_j = (x_{\sigma(i_j)} \cdots)$, $j = 1, \dots d$, then

$$x_{\sigma(1)} \cdots x_{\sigma(n)} = a_0 h_1 \cdots h_d.$$

With these $a_0$ and $h_j$'s, the left-hand side of (7.2) is also in $V_n$—hence, it is in $\mathrm{Id}(A) \cap V_n$. Thus, modulo $\mathrm{Id}(A) \cap V_n$, $x_{\sigma(1)} \cdots x_{\sigma(n)}$ is a linear combination of the monomials $a_0 h_{\pi(1)} \cdots h_{\pi(d)}$, $\pi \in S_d$ and $\pi \neq 1$. Finally, note that in the above left lexicographic order,

$$x_{\sigma(1)} \cdots x_{\sigma(n)} = a_0 h_1 \cdots h_d > a_0 h_{\pi(1)} \cdots h_{\pi(d)}$$

whenever $\pi \neq 1$. This shows that modulo $\mathrm{Id}(A)$ a $d$-bad monomial can be replaced by lexicographically lower monomials. We repeat this algorithm until all monomials are $d$-good, and this completes the proof of the lemma. □

7.1.1.2. *The RSK and d-good permutations.* The RSK correspondence[1] is a combinatorially defined bijection $\sigma \longleftrightarrow (P_\lambda, Q_\lambda)$ (see **[Knu98]**, **[Sta99]**) between $\sigma \in S_n$ and pairs $P_\lambda, Q_\lambda$ of standard tableaux of same shape $\lambda$, where $\lambda \vdash n$.

In fact more generally it associates to a word, in the free monoid, and a pair of tableaux, one standard and the other semistandard filled with the letters of the word.

The correspondence is based on a simple *game* of inserting a letter. We have some letters piled up so that lower letters appear below higher letters, and we want to insert a new letter $x$. If $x$ fits on top of the pile, we place it there; otherwise, we go down the pile until we find a first place where we can replace the existing letter with $x$. We do this and expel that letter. If we have a second pile of letters, then we try to place that letter there, and so on.

So let us pile inductively the word *strange*.

$$e \mapsto e,\ g \mapsto \begin{array}{c} g \\ e \end{array},\ n \mapsto \begin{array}{c} n \\ g \\ e \end{array},\ a \mapsto \begin{array}{cc} n & \\ g & \\ a & e \end{array},\ r \mapsto \begin{array}{cc} r & \\ n & \\ g & \\ a & e \end{array},\ t \mapsto \begin{array}{cc} t & \\ r & \\ n & \\ g & \\ a & e \end{array},\ s \mapsto \begin{array}{cc} s & \\ r & \\ n & \\ g & t \\ a & e \end{array}.$$

Notice that, as we proceed, we can keep track of where we have placed the new letter. We do this by filling a corresponding tableau:

$$\begin{array}{cc} 6 & \\ 5 & \\ 3 & \\ 2 & 7 \\ 1 & 4 \end{array},\quad \begin{array}{cc} s & \\ r & \\ n & \\ g & t \\ a & e \end{array}.$$

It is not hard to see that from the two tableaux one can *decrypt* the word we started with by giving the bijective correspondence.

Assume now that $\sigma \longleftrightarrow (P_\lambda, Q_\lambda)$, where $P_\lambda, Q_\lambda$ are standard tableaux given by the RSK correspondence. By a classical theorem of Schensted **[Sch61]**, $\mathrm{ht}(\lambda)$ equals the length of a longest decreasing subsequence in the permutation $\sigma$. Hence $\sigma$ is $d$-good if and only if $\mathrm{ht}(\lambda) \leq d-1$.

Recall that $H(k, 0; n) = \{\lambda \vdash n \mid \mathrm{ht}(\lambda) \leq k\}$.

LEMMA 7.1.11. *We have*

$$g_d(n) = \sum_{\lambda \in H(d-1,0;n)} (f_\lambda)^2.$$

[1] Robinson, Schensted, Knuth

PROOF. Apply the RSK correspondence and the theorem of Schensted.

Now there are $f_\lambda$ standard tableaux $P_\lambda$ and $f_\lambda$ standard tableaux $Q_\lambda$, hence there are $(f_\lambda)^2$ pairs $(P_\lambda, Q_\lambda)$ and therefore $(f_\lambda)^2$ permutations in $S_n$ corresponding to $\lambda$.

The proof now follows by summing over all $\lambda \in H(d-1, 0; n)$. $\square$

We can now complete the proof of Theorem 7.1.6.

PROOF. By Lemma 7.1.10 $c_n(A) \leq g_d(n)$, and by Lemma 7.1.11 we have that $g_d(n) = \sum_{\lambda \in H(d-1,0;n)} (f_\lambda)^2$. Finally, by Corollary 6.4.14 we have the estimate $\sum_{\lambda \in H(d-1,0;n)} (f_\lambda)^2 \leq (d-1)^{2n}$, and the proof follows. $\square$

REMARK 7.1.12. The exponential bound on the codimensions was proved—by a rather long argument—in **[Reg72]**, in order to prove Theorem 7.2.2 below. Latyshev simplified that proof considerably by introducing "$d$-good permutations" and by proving Lemma 7.1.10. The final step in Latyshev's proof of Theorem 7.1.6, that of bounding $g_d(n)$ by $(d-1)^{2n}$, is done in **[Lat72]** by applying a combinatorial theorem called the Dilworth theorem **[Dil50]**; see also **[Hal86]**. For an exposition of Latyshev's proof, see **[Row80**, Section 6.1]. Our proof here of bounding $g_d(n)$ by $(d-1)^{2n}$ applies the Schur–Weyl duality of §6.1, as well as the RSK correspondence.

## 7.2. The $A \otimes B$ theorem

Let us first define PI equivalence.

DEFINITION 7.2.1. We say that two algebras $A, A'$ are PI equivalent if they satisfy the same polynomial identities.

As a first application of the exponential bound on the codimensions, we have

THEOREM 7.2.2 (**[Reg72]**). *Let $A$ and $B$ be two PI algebras. Then $A \otimes B$ is also PI.*

Before we prove this theorem, let us give a theorem by Leron and Vapne (former name of Regev) **[LV70]** showing that the identities of the tensor product depend only on the identities of the two factors.

THEOREM 7.2.3 (**[LV70]**). *Let $A, A', B, B'$ be algebras over a field. Then if $A$ is PI equivalent to $A'$ and $B$ is PI equivalent to $B'$, we have that $A \otimes B$ is PI equivalent to $A' \otimes B'$.*

PROOF. Let $U_1 = F\langle X \rangle / I_1$, $U_2 = F\langle X \rangle / I_2$, respectively, be the relatively free algebras in countably many variables for $A, A'$ and $B, B'$ so that these algebras are quotients of their relatively free algebras. $I_1, I_2$ are the two $T$-ideals of identities of the two pairs of algebras.

It is then enough to show that $A \otimes B$ (and hence also $A' \otimes B'$) is PI equivalent to $U_1 \otimes U_2$.

In fact it is enough to do it in two steps and prove for instance that $A \otimes B$ is PI equivalent to $U_1 \otimes B$ and then, in the same way, that $U_1 \otimes B$ is PI equivalent to $U_1 \otimes U_2$.

Now let $I$ be the set of homomorphisms of $U_1$ to $A$. We deduce a mapping $j : U_1 \to A^I$ given by evaluations of all these homomorphisms. Since $I_1$ is the ideal of identities of $A$, we have that $j : U_1 \to A^I$ is injective, so $U_1 \otimes B$ injects into $(A^I) \otimes B$, which in turn injects into $(A \otimes B)^I$.

Therefore $U_1 \otimes B$ satisfies all PIs of $A \otimes B$. But now $A$ is a quotient of $U_1$, so we have also the converse, $A \otimes B$ satisfies all PI's of $U_1 \otimes B$, and the claim follows. $\square$

We can slightly strengthen this result in

COROLLARY 7.2.4. *Let $A, A', B$ be algebras over a field. Then if $A$ satisfies all the PI of $A'$, we have that $A \otimes B$ satisfies all the PI of $A' \otimes B$.*

PROOF. Let $U_1 = F\langle X\rangle/I_1$, $U_2 = F\langle X\rangle/I_2$, respectively, be the relatively free algebras in some large number of variables for $A$ and $A'$. Here $I_1, I_2$ are the two $T$-ideals of identities of these two algebras. The hypothesis is that $I_1 \supset I_2$ or that $U_1$ is a quotient of $U_2$.

We know by the previous result that $A \otimes B$ is PI equivalent to $U_1 \otimes B$ and $A' \otimes B$ is PI equivalent to $U_2 \otimes B$.

Since $U_1$ is a quotient of $U_2$, we have that $U_1 \otimes B$ is a quotient of $U_2 \otimes B$, and the claim follows. $\square$

COROLLARY 7.2.5. *If $A, A'$ are PI equivalent and $k \in \mathbb{N}$, then $M_k(A)$, $M_k(A')$ are PI equivalent. If $A$ satisfies all the PI of $A'$, then $M_k(A)$ satisfies all the PI of $M_k(A')$.*

We now concentrate on Theorem 7.2.2. Recall that the codimensions $c_n(A)$ of an $F$-algebra $A$ are defined whether or not $A$ is PI, and $A$ is PI if and only if there exists $n$ such that $c_n(A) < n!$. This is similar for cocharacters, where we can also define a partial order. If $\chi_1, \chi_2$ are two $S_n$ characters, we say that $\chi_1 \leq \chi_2$ if and only if $\chi_2 - \chi_1 \geq 0$, that is it is the character of a representation. In other words if $\chi_1 \leq \chi_2$ are characters of two representations $V_1, V_2$; saying that $\chi_1 \leq \chi_2$ means that $V_1$ is isomorphic to a subrepresentation of $V_2$. We first prove

PROPOSITION 7.2.6. *For any algebras $A$ and $B$, their codimension sequences satisfy*

$$c_n(A \otimes B) \leq c_n(A) \cdot c_n(B),$$

*which has a cocharacters refinement, namely*

$$\chi_n(A \otimes B) \leq \chi_n(A) \otimes \chi_n(B). \tag{7.3}$$

PROOF. Let $A$ be an algebra. For every $n \in \mathbb{N}$, we have the multiplication map

$$m : A^{\otimes n} \to A \quad \text{given by} \quad a_1 \otimes a_2 \otimes \cdots \otimes a_n \mapsto a_1 a_2 \cdots a_n.$$

On $A^{\otimes n}$ we have the usual action of the symmetric group (Section 6.1.1). This induces an action of the symmetric group on the space of linear maps $\hom_F(A^{\otimes n}, A)$, that is the multilinear maps from $A \times \cdots \times A$ to $A$:

$$\sigma f(a_1 \otimes a_2 \otimes \cdots \otimes a_n) = f(\sigma^{-1}(a_1 \otimes a_2 \otimes \cdots \otimes a_n)) = f(a_{\sigma(1)} \otimes a_{\sigma(2)} \otimes \cdots \otimes a_{\sigma(n)}).$$

We have also a map $e_A : F[S_n] \mapsto \hom(A^{\otimes n}, A)$,

$$e_A(\sigma)(a_1 \otimes a_2 \otimes \cdots \otimes a_n) := m(\sigma^{-1}(a_1 \otimes a_2 \otimes \cdots \otimes a_n)) = a_{\sigma(1)} a_{\sigma(2)} \cdots a_{\sigma(n)}.$$

If we identify $F[S_n]$ with the multilinear polynomials in $n$ variables, we see that the kernel of $e_A$ is the space of multilinear identities of degree $n$ for $A$.

Recall that, if $j_1 : V_1 \to V_2$, $j_2 : W_1 \to W_2$ are injective maps of vector spaces, then $j_1 \otimes j_2 : V_1 \otimes V_2 \to W_1 \otimes W_2$ is injective. Thus, given two algebras $A, B$, we have an injective map

$$\eta : \hom(A^{\otimes n}, A) \otimes \hom(B^{\otimes n}, B) \to \hom(A^{\otimes n} \otimes B^{\otimes n}, A \otimes B)$$
$$= \hom((A \otimes B)^{\otimes n}, A \otimes B),$$

$$\eta(f \otimes g)(a_1 \otimes b_1) \otimes \cdots \otimes (a_n \otimes b_n) = f(a_1 \otimes a_2 \otimes \cdots \otimes a_n) \otimes g(b_1 \otimes b_2 \otimes \cdots \otimes b_n).$$

We then have a map,

$$F[S_n] \otimes F[S_n] \to \hom((A \otimes B)^{\otimes n}, (A \otimes B)), \sigma \otimes \tau \mapsto \eta(e_A(\sigma) \otimes e_B(\tau)).$$

This map induces an injective map

$$F[S_n]/\ker e_A \otimes F[S_n]/\ker e_B \to \hom((A \otimes B)^{\otimes n}, (A \otimes B)).$$

When we combine it with the diagonal map $\delta : F[S_n] \to F[S_n] \otimes F[S_n], \sigma \mapsto \sigma \otimes \sigma$, we see that $e_A \otimes e_B \circ \delta = e_{A \otimes B}$. Then we have an embedding

$$F[S_n]/\ker e_{A\otimes B} \to F[S_n]/\ker e_A \otimes F[S_n]/\ker e_B. \tag{7.4}$$

This is $S_n$ equivariant, so it gives the inclusion of cocharacters.

Notice that formula (7.4) gives an explicit algorithm of how to describe the multilinear polynomial identities of $A \otimes B$ from the ones of $A, B$, and is a more concrete proof of Theorem 7.2.3. □

REMARK 7.2.7. The previous construction gives a possible algorithm to deduce the multilinear identities of $A \otimes B$ from those of $A, B$ in a degree $n$ as follows. Take two sets of free variables $x_1, \dots, x_n$ and $y_1, \dots, y_n$ so that the $x$ commute with the $y$. Given an element $f := \sum_{\sigma \in S_n} a_\sigma \sigma$, construct the element

$$F(x, y) := \sum_{\sigma \in S_n} a_\sigma x_{\sigma(1)} y_{\sigma(1)} \cdots x_{\sigma(n)} y_{\sigma(n)} = \sum_{\sigma \in S_n} a_\sigma x_{\sigma(1)} \cdots x_{\sigma(n)} y_{\sigma(1)} \cdots y_{\sigma(n)}.$$

The element $f$ corresponds to a PI of $A \otimes B$ if and only if there exist PI $f_i(x)$ of $A$ and $g_i(y)$ of $B$ and multilinear polynomials $h_i(x), k_i(y)$ such that

$$F(x, y) = \sum_i f_i(x) k_i(y) + h_i(x) g_i(y).$$

We are now ready to prove Theorem 7.2.2.

PROOF OF THEOREM 7.2.2. We have remarked that in order to prove that $A \otimes B$ is PI, we must show that $c_n(A \otimes B) < n!$ for some $n$.

By Theorem 7.1.6 there exist $0 < \alpha \in \mathbb{R}$ such that for all $n$, $c_n(A) \leq \alpha^n$. Similarly $c_n(B) \leq \beta^n$ for some $0 < \beta \in \mathbb{R}$. Let $n$ be large enough so that $(\alpha\beta)^n < n!$. Then $c_n(A \otimes B) \leq c_n(A) \cdot c_n(B) \leq (\alpha\beta)^n < n!$, which completes the proof. □

REMARK 7.2.8.

(1) The exponential bound of the codimensions, together with techniques of Young tableaux and Young symmetrizers, allow one to construct explicit identities for a PI algebra, provided we know a degree of an identity of such algebra.
(2) If $A$ satisfies an identity of degree $d$ and $B$ satisfies an identity of degree $h$, then $A \otimes B$ satisfies an identity of degree $\leq e(d-1)^2(h-1)^2 + 1$, where $e = 2.718 \cdots$ is the classical Neper's constant.

Indeed, the classical inequality $(n/e)^n < n!$ implies that $\alpha^n < n!$ whenever $e\alpha < n$. Thus, with $\alpha = (d-1)^2(h-1)^2$, if $e(d-1)^2(h-1)^2 < n$, then by formula (7.1),

$$c_n(A \otimes B) \leq c_n(A)c_n(B) \leq ((d-1)^2(h-1)^2)^n < n!,$$

which implies the remark.

## 7.3. Cocharacters of a PI algebra

Note that in general $\mathrm{Id}(A) \cap V_n$ is not a right ideal in $V_n$, for instance the $T$-ideal of identities of the Grassmann algebra gives an example (see 19.1).

**7.3.1. Alternating, standard, and Capelli polynomials.** Following Razmyslov, we now introduce the *Capelli polynomials*,

(7.5)
$$C_m(x_1, x_2, \dots, x_m; y_1, y_2, \dots, y_{m+1}) := \sum_{\sigma \in S_m} \epsilon_\sigma y_1 x_{\sigma(1)} y_2 x_{\sigma(2)} \cdots y_m x_{\sigma(m)} y_{m+1}.$$

In fact this is only an analogy of the classical Capelli identity, which is instead an identity of differential operators (see **[Pro07]**).

Sometimes it is better to define the Capelli polynomial as

$$C'_m(x_1, x_2, \dots, x_m; y_1, y_2, \dots, y_{m-1}) := \sum_{\sigma \in S_m} \epsilon_\sigma x_{\sigma(1)} y_1 x_{\sigma(2)} \cdots y_{m-1} x_{\sigma(m)}. \tag{7.6}$$

$$C_m(x_1, x_2, \dots, x_m; y_1, y_2, \dots, y_{m+1}) = y_1 C'_m(x_1, x_2, \dots, x_m; y_2, \dots, y_m) y_{m+1}.$$

The element of formula (7.5) can be deduced from (7.6), and when we work with algebras with a 1, also the converse is true.

By construction $C_m$ is alternating in the variables $x_i$. Moreover we have

LEMMA 7.3.1.

(1) *Denote* $x_{n+j-1} = y_j$, $2 \leq j \leq 2n$. *Then the Capelli polynomial* $C_n(x; y)$ *satisfies*

$$C_n(x_1, \dots, x_n; y_1, \dots, y_{n+1}) = y_1(St_n(x_1, \dots, x_n) y_2 \cdots y_{n+1})\eta,$$

*where* $\eta$ *is given by* (6.13).

(2) *Both the standard polynomial* $\mathrm{St}_n(x_1, \dots, x_n)$ *and the Capelli polynomial*

$$C_n(x_1, \dots, x_n : y_1, \dots, y_{n-1})$$

*are alternating in* $x_1, \dots, x_n$.

(3) *Given any polynomial* $f(x_1, \dots, x_n, Y)$ *multilinear and alternating in the variables* $x_1, \dots, x_n$ *and possibly depending on other variables* $Y$, *we have*

$$f(x_1, \dots, x_n, Y) = \sum_{i,\, a_1, \dots, a_{n+1}} \lambda_i C_n(x_1, \dots, x_n : a_1, \dots, a_{n+1}), \tag{7.7}$$

*where* $\lambda_i \in F$ *and* $a_j$ *are monomials in* $Y$.

PROOF. Part (1) is a special case of Example 6.3.14. The remaining statements follow immediately from Remark 2.2.43. In part (3). we have to allow the possibility that some of the monomials $a_j = 1$. □

As a corollary we can now prove

PROPOSITION 7.3.2. *Let $A$ be an $F$-algebra (not necessarily with a 1) with $\dim A \leq k$. Then $A$ satisfies the polynomial identity $C_{k+1}(x;y) = 0$ and also all the identities deduced from $C_{k+1}(x;y)$ setting some variables $y_i$ equal to 1.*

PROOF. When we evaluate $C_{k+1}(x;y)$ in elements $a_1, \dots, a_{k+1}, b_1, \dots, b_k \in A$, we see that as a function of the $a_i$, we have a multilinear and antisymmetric function in a number of variables higher than the dimension, and any such function is identically 0. □

DEFINITION 7.3.3. The identities deduced from $C_k(x;y)$ setting some variables $y_i$ equal to 1 will be called the *Capelli list* of rank $k$ and is denoted by $\mathcal{C}_k(x;y)$.

The $T$-ideal generated by the list $\mathcal{C}_k(x;y)$ will be denoted by $\mathbb{I}_k$ and called the $k$th-Capelli ideal.

The main property of the Capelli polynomial is for $T$-ideals. For this we need to distinguish $T$-ideals in the category of algebras with a 1, or *without* 1. In this case we are not allowed to substitute 1 to a variable. Also the finite list of polynomials obtained from $C_k(x;y)$ by specializing some of the variables $y_i$ to 1 is not contained in the $T$-ideal generated by $C_k(x;y)$. On the other hand we have

EXERCISE 7.3.4. The $T$-ideal in the category of algebras without a 1 generated by the Capelli polynomial $C_k(x;y)$ contains the Capelli list $\mathcal{C}_{2k+1}(x;y)$ so it contains $\mathbb{I}_{2k+1}$.

We identify the free algebra $F\langle X\rangle$ in countably many variables $x_i$ with the tensor algebra $T(V)$, where $V$ is the vector space with basis the $x_i$.

THEOREM 7.3.5. *In characteristic 0 the kth-Capelli ideal $\mathbb{I}_k$ contains all the isotypic components of type $S_\lambda(V)$ for all partitions $\lambda$ of height $\geq k$.*

PROOF. It is enough to show that this $T$-ideal contains the multilinear parts of such representations. Such a multilinear part by Remark 6.3.9 is under the left and right actions of the two symmetric groups isomorphic to $M_\lambda \otimes M_\lambda$. Using Remark 2.2.43, it is thus enough to prove that $M_\lambda$ is spanned by elements which are antisymmetric with respect to a symmetric group in at least $k$ variables.

Now $M_\lambda$ is spanned by elements of the form $\bar{e}_T g$ for $T$ some tableau of shape $\lambda$ where the Young symmetrizer $\bar{e}_T = a_T s_T$ is discussed in Theorem 6.2.6. In particular it is spanned by elements of the form $a_T h,\ h = s_T g$ for some $h$ where $a_T$ is the antisymmetrizer. Since the height is $\geq k$, the subgroup preserving the columns contains a symmetric group $S_a,\ a \geq k$, and the antisymmetrizer $a_T$ is factored as product of antisymmetrizers on the columns, one of which gives the desired property. □

COROLLARY 7.3.6. *If an algebra $R$ satisfies the Capelli polynomial $C_k(x;y)$ (or the Capelli list), then in its cocharacter $\chi_m(R)$ the multiplicities $m_\lambda(R)$ vanish for all partitions $\lambda$ of height $\geq k$.*

In all characteristics and over $\mathbb{Z}$ one may develop a combinatorial theory of tableaux and show that it contains the span of double standard tableaux of height $\geq k$, but we will not discuss this in this book. This result implies an important

COROLLARY 7.3.7. *Consider a free algebra $R = T(V)/\mathbb{I}_k$ in infinitely many variables over a field of characteristic 0 modulo the kth-Capelli ideal $\mathbb{I}_k$. Then any T-ideal J of R is generated, as T-ideal by its intersection with the free algebra in $k-1$ variables.*

PROOF. From Theorem 7.3.5 the algebra $R$ is a direct sum of copies of irreducible representations $S_\lambda(V)$ for partitions $\lambda$ of height $t \leq k-1$. So also $J$ is a direct sum of copies of irreducible representations $S_\lambda(V)$ for partitions $\lambda$ of height $t \leq k-1$.

Under the linear group $\mathrm{GL}(V)$, that is under linear substitutions of the variables $x_i$, each representation $S_\lambda(V)$ is generated by its highest weight vector, which is in the algebra generated by the first $t$ variables $x_i$, where $t$ is the height of $\lambda$, thus the claim follows. □

REMARK 7.3.8. If we work with algebras without 1, we have the same statement, for $T(V) = \bigoplus_{k=1}^{\infty} V^{\otimes k}$, but using as identities all polynomials of the list $C_k$.

Then we reformulate the corollary as

THEOREM 7.3.9. *Let I be a T-ideal in the free algebra $F\langle X\rangle$ containing the list $C_k$. Consider $I_h := I \cap F\langle x_1, \dots, x_h\rangle$ with $h \geq k-1$. Then I is the ideal of polynomial identities of the finitely generated algebra $F\langle x_1, \dots, x_h\rangle/I_h$.*

PROOF. The ideal $I_h = I \cap F\langle x_1, \dots, x_h\rangle$ is still a $T$-ideal in the free algebra $F\langle x_1, \dots, x_h\rangle$, hence it is the ideal of polynomial identities, in the variables $x_1, \dots, x_h$, of the algebra $F\langle x_1, \dots, x_h\rangle/I_h$. Thus $I$ is contained in the ideal $J$ of polynomial identities of this algebra.

We claim that $I$ and $J$ coincide. Since both contain the Capelli list, it is enough to prove that $I \cap F\langle x_1, \dots, x_{k-1}\rangle = J \cap F\langle x_1, \dots, x_{k-1}\rangle$. This is the case since they are both the ideal of polynomial identities of the same algebra $F\langle x_1, \dots, x_{k-1}\rangle/I_{k-1}$ in the variables $x_1, \dots, x_{k-1}$. □

We can make Proposition 7.1.2 more precise. Let $V_{k-1}$ be the subspace of $V$ with basis $x_1, \dots, x_{k-1}$, then

THEOREM 7.3.10. *If the T-ideal I contains a Capelli list $C_k$, the multiplicity $m_\lambda(I)$ of $S_\lambda(V)$ in $T(V)/I$ equals the multiplicity of $S_\lambda(V_{k-1})$ in $T(V_{k-1})/I \cap T(V_{k-1})$.*

PROOF. In fact by part (3) of Proposition 7.1.2, the multiplicity $m_\lambda(I)$ of $S_\lambda(V)$ in $T(V)/I$ equals the dimension of the space of $U_V$ invariants in $T(V)/I$ of weight $\lambda$. But under our hypotheses the height of $\lambda$ is $\leq k$, so the $U_V$ invariants are contained in $S_\lambda(V_{k-1})$. □

We shall need the following properties of the standard polynomials $\mathrm{St}_n$.

LEMMA 7.3.11.

(1)

$$\mathrm{St}_n(x_1, \dots, x_n) = \sum_{i=1}^{n} (-1)^{i-1} x_i \cdot \mathrm{St}_{n-1}(x_1, \dots, x_{i-1}, x_{i+1}, \dots, x_n). \tag{7.8}$$

(2) *Let the element $a$ commute with $x_2, x_3, \ldots, x_n$, for example let $a = 1$.*
(a) *If $n = 2k+1$, then*

$$\mathrm{St}_{2k+1}(a, x_2, \ldots, x_{2k+1}) = a \cdot \mathrm{St}_{2k}(x_2, \ldots, x_{2k+1}).$$

(b) *If $n = 2k$, then*

$$\mathrm{St}_{2k}(a, x_2, \ldots, x_{2k}) = 0.$$

PROOF. (1) is immediate from the definition.

(2) To prove (2), substitute $x_1$ with $a$ in formula (7.8).

If $n$ is odd, the terms $i \neq 1,\ x_i \cdot \mathrm{St}_{n-1}(a, \ldots, x_{i-1}, x_{i+1}, \ldots, x_n) = 0$ by induction.

If $n$ is even,

$$i \neq 1,\ x_i \cdot \mathrm{St}_{n-1}(a, \ldots, x_{i-1}, x_{i+1}, \ldots, x_n) = a x_i \cdot \mathrm{St}_{n-2}(\ldots, x_{i-1}, x_{i+1}, \ldots, x_n),$$

by induction. So formula (7.8) becomes

$$a \cdot \mathrm{St}_{2k}(x_2, \ldots, x_{2k+1}) - a \sum_{i=2}^{n} (-1)^{i-2} x_i \cdot \mathrm{St}_{n-2}(\ldots, x_{i-1}, x_{i+1} = 0. \qquad \square$$

COROLLARY 7.3.12. *If an algebra with 1 satisfies a standard identity, then the minimal degree of such an identity is even.*

**7.3.2. The multilinear cocharacters.** By Definition 7.1.1 of the cocharacters and the codimensions of an algebra $A$, we have

$$\chi_n(A) = \sum_{\lambda \vdash n} m_\lambda(A) \chi_\lambda \quad \text{hence} \quad c_n(A) = \sum_{\lambda \vdash n} m_\lambda(A) f_\lambda, \tag{7.9}$$

where $f_\lambda = \chi_\lambda(1)$ is the dimension of the corresponding irreducible representation of $S_n$.

For most PI algebras $A$, the computation of the exact values of the multiplicities $m_\lambda(A)$ seems to be too complicated. Consider for example the cocharacters $\chi_n(M_k(F))$ of the $k \times k$ matrices. In the case $k = 2$, $\chi_n(M_2(F))$ has been fully determined; see Theorem 7.3.13 below. However, recent results indicate that the problem of determining $\chi_n(M_k(F))$, $k \geq 3$, seems to be far too complicated; see [**Ber97, DG03**].

For $A$ a finite-dimensional algebra (or an algebra satisfying a Capelli identity) we shall see, in Theorem 18.3.1, that in fact $m_\lambda(A)$ has a combinatorial structure, somewhat similar but much more complicated than that of Theorem 7.3.13 for $2 \times 2$ matrices.

One can in general try to estimate these $m_\lambda(A)$ asymptotically, as $n$ goes to infinity. For example, in §8.2 we prove that *always*, the colength sequence $\sum_\lambda m_\lambda(A)$ is polynomially bounded.

We now quote the formulas for the cocharacters of the $2 \times 2$ matrices $M_2(F)$, see §9.3.

THEOREM 7.3.13 ([**Dre81, For84**]). *Let $\chi_n(M_2(F))$ be the nth cocharacter of the algebra $M_2(F)$. If $\mathrm{ht}(\lambda) \geq 5$, then $m_\lambda(M_2(F)) = 0$. Hence*

$$\chi_n(M_2(F)) = \sum_{\lambda = (\lambda_1, \lambda_2, \lambda_3, \lambda_4) \vdash n} m_\lambda(M_2(F)) \cdot \chi_\lambda,$$

*and we have that*

(1) *For most partitions* $\lambda = (\lambda_1, \lambda_2, \lambda_3, \lambda_4) \vdash n$,
$m_\lambda(M_2(F)) = (\lambda_1 - \lambda_2 + 1)(\lambda_2 - \lambda_3 + 1)(\lambda_3 - \lambda_4 + 1)$.

*The only exceptions are the following:*

(2) $m_{(n)}(M_2(F)) = 1$.

(3) $m_{(\lambda_1,\lambda_2)}(M_2(F)) = (\lambda_1 - \lambda_2 + 1)\lambda_2$, *provided* $\lambda_2 > 0$.

(4) $m_{(\lambda_1,1,1,\lambda_4)}(M_2(F)) = \lambda_1(2 - \lambda_4) - 1$ *(here* $\lambda_4 \leq 1$*).*

## 7.4. Proper polynomials

Let $A$ be an associative algebra over a field $F$, and denote by $A^{(\mathrm{Lie})}$ the algebra $A$ regarded as a Lie algebra under the Lie bracket $[a, b] = ab - ba$.

Recall that if $L$ is a Lie algebra over a field $F$, a unitary associative algebra $\mathcal{U} = \mathcal{U}(L)$ over $F$ is a universal enveloping algebra of $L$ if the following holds. There exists a canonical homomorphism $\epsilon : L \to \mathcal{U}^{(\mathrm{Lie})}$ such that for any associative algebra $A$ and for any homomorphism of Lie algebras $\varphi : L \to A^{(\mathrm{Lie})}$, there exists a unique homomorphism of associative algebras $\psi : \mathcal{U} \to A$ such that $\psi(\epsilon(x)) = \varphi(x)$, for all $x \in L$.

The existence of an enveloping algebra of a Lie algebra is just given by generators and relations as a quotient of the tensor algebra $T(L)$ modulo the ideal generated by the elements $x \otimes y - y \otimes x - [x, y]$ for all $x, y \in L$. The deeper fact is that $\epsilon$ is in fact injective and this is assured by the more general Poincaré–Birkhoff–Witt theorem. Clearly $\mathcal{U}$ is generated as an associative algebra by $\epsilon(L)$.

THEOREM 7.4.1 (PBW theorem). *For any Lie algebra $L$ there exists a universal enveloping algebra $\mathcal{U}(L)$, and the canonical homomorphism $\epsilon : L \to \mathcal{U}(L)$ is injective. If we choose a linearly ordered basis $B$ of $L$ and we identify $\epsilon(x) \in \mathcal{U}(L)$ with $x$, for all $x \in L$, then*

$$\{1, b_1 \cdots b_n \mid b_i \in B, \quad b_1 \leq \cdots \leq b_n,\ n = 1, 2, \ldots\}$$

*is a basis of $\mathcal{U}(L)$.*

As a consequence, by applying the universal properties (cf. **[Reu93]**) we easily get

COROLLARY 7.4.2. *If $F\langle X\rangle$ is the free associative algebra over $F$ on a set $X$, then the Lie subalgebra of $F\langle X\rangle^{(\mathrm{Lie})}$ generated by $X$ is the free Lie algebra $\mathcal{L}(X)$ on $X$.*

*$F\langle X\rangle$ is the enveloping algebra of $\mathcal{L}(X)$.*

An element of $\mathcal{L}(X) \subseteq F\langle X\rangle$ will be called a *Lie polynomial.*

DEFINITION 7.4.3. For any $n \geq 2$, define inductively

$$[x_1, \ldots, x_n] := [[x_1, \ldots, x_{n-1}], x_n]. \tag{7.10}$$

This is called a *left normed Lie monomial* of length $n$.

For instance $[x_1, x_2, x_3] = [[x_1, x_2], x_3]$.

The importance of left-normed monomials is due to the fact that a homogeneous Lie polynomial of degree $k$, $f = f(x_1, \ldots, x_n)$, can be written as a linear combination of left-normed monomials $[x_{i_1}, \ldots, x_{i_k}]$.

In fact there is a deep combinatorial theory dealing with this structure. We do not need this but refer the interested reader to **[Reu93]**.

As an application of Corollary 7.4.2, we can write a new basis of the free associative algebra as follows.

Fix an ordering of a basis of $\mathcal{L}\langle X\rangle$ consisting of the variables $x_1, x_2, \dots$ and left-normed commutators such that the variables precede the commutators and shorter commutators precede longer ones

$$x_1, x_2, \dots, [x_{i_1}, x_{i_2}], [x_{j_1}, x_{j_2}], \dots, [x_{k_1}, x_{k_2}, x_{k_3}], \dots.$$

PROPOSITION 7.4.4. *A basis of the vector space $F\langle X\rangle$ is given by the polynomials*

$$x_1^{a_1} \cdots x_n^{a_n} [x_{i_1}, x_{i_2}]^b \cdots [x_{j_1}, \dots, x_{j_m}]^c, \tag{7.11}$$

*where $a_1, \dots, a_n, b, c$ are nonnegative integers and $[x_{i_1}, x_{i_2}] < \cdots < [x_{j_1}, \dots, x_{j_m}]$ in the ordering of the basis of $\mathcal{L}\langle X\rangle$.*

DEFINITION 7.4.5. A polynomial $f \in F\langle X\rangle$ is called a proper polynomial if it is a linear combination of products of left-normed commutators.

Hence a proper polynomial $f$ is of the form

$$f(x_1, \dots, x_n) = \sum \alpha_{i_1,\dots,j_l} [x_{i_1}, \dots, x_{i_k}] \cdots [x_{j_1}, \dots, x_{j_l}]$$

with $\alpha_{i_1,\dots,j_l} \in F$.

Observe that if $f(x_1, \dots, x_n)$ is a proper polynomial, then also all of its multihomogeneous components are proper. Setting any of the variables equal to 1, we obtain 0. If we substitute for $x_1 \mapsto x_1 + y_1$, the resulting polynomial is also proper, so finally we deduce that, for a proper polynomial $f(1 + x_1, \dots, x_n) = f(x_1, \dots, x_n)$.

The relevance of proper polynomials is given in

THEOREM 7.4.6. *If $A$ is a unitary algebra over an infinite field, then $\operatorname{Id}(A)$ is generated by the proper polynomials. If $\operatorname{char} F = 0$, then $\operatorname{Id}(A)$ is generated by the multilinear proper polynomials.*

PROOF. Let $f = f(x_1, \dots, x_n) \in \operatorname{Id}(A)$ and assume, as we may, that $f$ is multihomogeneous. By the above discussion we write $f$ as

$$f = \sum \alpha_{\underline{a}} x_1^{a_1} \cdots x_n^{a_n} w_{\underline{a}}(x_1, \dots, x_n), \quad \underline{a} := \{a_1, \dots, a_n\}, \tag{7.12}$$

where $w_{\underline{a}}(x_1, \dots, x_n)$ is a linear combination of products of commutators. We want to show that all the polynomials $w_{\underline{a}}(x_1, \dots, x_n)$ are identities.

Since $1 \in A$, then $f(1 + x_1, x_2, \dots, x_n) \in \operatorname{Id}(A)$. By the previous remarks $w_{\underline{a}}(1 + x_1, \dots, x_n) = w_{\underline{a}}(x_1, \dots, x_n)$, hence

$$f(1 + x_1, x_2, \dots, x_n) = \sum \alpha_{\underline{a}} \sum_{i=0}^{a_1} \binom{a_1}{i} x_1^{a_1 - i} x_2^{a_2} \cdots x_n^{a_n} w_{\underline{a}}(x_1, \dots, x_n).$$

Let $c$ be the maximum of the exponents $a_1$ appearing in $f$. If $c = 0$, we pass to the maximum of the exponents $a_2$ and so on, thus we make a double induction. Since $\operatorname{Id}(A)$ is multihomogeneous, we obtain that

$$\sum \alpha_{\underline{b}} x_2^{a_2} \cdots x_n^{a_n} w_{\underline{b}}(x_1, \dots, x_n) \in \operatorname{Id}(A), \tag{7.13}$$

where $\underline{b}$ runs over all the indices $\underline{a}$ with $a_1 = c$.

By induction all these $w_{\underline{b}}(x_1, \dots, x_n) \in \operatorname{Id}(A)$, then subtracting from $f$ the element $x_1^c \sum \alpha_{\underline{b}} x_2^{a_2} \cdots x_n^{a_n} w_{\underline{b}}(x_1, \dots, x_n)$, we have an identity in which the maximum exponent of $x_1$ is decreased, and we still finish by induction. □

Recall that $V_n$ denotes the space of multilinear polynomials in the $n$ variables $x_1, \dots, x_n$. We denote by $P_n$ the subspace of proper multilinear polynomials. For an algebra $A$ with 1, let us consider the two spaces

$$P_n/P_n \cap \mathrm{Id}(A) \subset V_n/V_n \cap \mathrm{Id}(A)$$

and choose for each $k$ a list of polynomials $f_{i,k}(x_1, \dots, x_k) \in P_k$ which form a basis modulo $P_k \cap \mathrm{Id}(A)$.

Next given $k \leq n$ and a subset $S = \{i_1 < i_2 < \cdots < i_{n-k}\} \subset \{1, \dots, n\}$ denote by

$$X_S := x_{i_1} x_{i_2} \cdots x_{i_{n-k}}, \quad f_{i,k}(S') := f_{i,k}(x_{j_1}, x_{j_2}, \dots, x_{j_k}),$$

where $j_1 < j_2 < \cdots < j_k$ is the complement $S'$ to $S = i_1 < i_2 < \cdots < i_{n-k}$.

LEMMA 7.4.7. *The polynomials $X_S f_{i,k}(S')$ as $S$ run over all subsets of $\{1, \dots, n\}$ and $f_{i,k}$ runs over the chosen basis elements for $P_k/P_k \cap \mathrm{Id}(A)$, where $k + |S| = n$ form a basis of $V_n$ modulo $V_n \cap \mathrm{Id}(A)$.*

PROOF. Consider any multilinear polynomial $f$. By the previous discussion we have that $f$ can be uniquely written in the form $f = \sum_S X_S g_S$ where $g_S$ is a multilinear proper polynomial in the variables $x_i$, $i \in S'$ the complement of $S$.

We first claim that $f \in \mathrm{Id}(A)$ if and only if $g_S \in \mathrm{Id}(A)$ for all $S$. In one direction this is clear, so we only need to show that if $f \in \mathrm{Id}(A)$, then $g_S \in \mathrm{Id}(A)$ for all $S$.

For this by contradiction we may reduce to the case that, for all the $S$ appearing, $g_S \notin \mathrm{Id}(A)$. So choose one such $S_0 = i_1 < i_2 < \cdots < i_{n-k}$ with $n-k$ maximal and set all $x_i = 1$ for $i \in S_0$. Then all the $S \neq S_0$ appearing do not contain all elements in $S_0$, hence the polynomial $g_S$ contains one of the variables $x_i$, $i \in S_0$, and so vanishes. Thus $f$ specializes to $g_{S_0}$ which is a PI of $A$, a contradiction.

Now the claim follows. □

At this point we can see that for all algebras with 1, one has the same phenomenon observed for $2 \times 2$ matrices in formula (9.61). The result is due to Drensky **[Dre81]**.

THEOREM 7.4.8. *As a representation of $S_n$ we have the isomorphism*

$$V_n/V_n \cap \mathrm{Id}(A) \simeq \bigoplus_{k=2}^{n} \mathrm{Ind}_{S_{n-k} \times S_k}^{S_n} P_k/P_k \cap \mathrm{Id}(A). \tag{7.14}$$

PROOF. This is just an interpretation of the previous statement.

The $S_n$ equivariant embedding of $\mathrm{Ind}_{S_{n-k} \times S_k}^{S_n} P_k/P_k \cap \mathrm{Id}(A)$ in $V_n/V_n \cap \mathrm{Id}(A)$ is induced by mapping $f \in P_k/P_k \cap \mathrm{Id}(A)$ to $V_n/V_n \cap \mathrm{Id}(A)$ by

$$\sigma(x_1 x_2 \cdots x_{n-k}) f(x_{n-k+1}, \dots, x_n) \in V_n/V_n \cap \mathrm{Id}(A),$$

where by $\sigma(x_1 x_2 \cdots x_{n-k})$ we denote the symmetrization of this monomial and $S_{n-k}$ acts trivially on $P_k/P_k \cap \mathrm{Id}(A)$.

The fact that this is an embedding and that these embeddings form a direct sum exhausting $V_n/V_n \cap \mathrm{Id}(A)$ comes from Lemma 7.4.7 and the algorithm of the PBW basis. □

DEFINITION 7.4.9. The character of $S_n$ on $P_n/P_n \cap \mathrm{Id}(A)$ will be called the *proper cocharacter* and is denoted by $\bar{\chi}_n(A)$.

The dimension of $P_n/P_n \cap \mathrm{Id}(A)$ will be called the *proper codimension* and denoted by $\bar{c}_n(A)$ and by $\bar{\chi}_n(A) = \sum_{\lambda\vdash n} \overline{m}_\lambda(A)\chi_\lambda$ the *proper cocharacter* of $A$.

Thus for an algebra $A$ with 1 the relation between usual and proper cocharacter and codimension is given by the formula

$$(7.15)\qquad \chi_n(A) = \sum_{k=2}^{n} \mathrm{Ind}_{S_{n-k}\times S_k}^{S_n} \bar{\chi}_n(A), \quad c_n(A) = \sum_{k=2}^{n} \binom{n}{k} \bar{c}_k(A).$$

One may express the property of the previous formula by saying that the cocharacters are *Young derived*.

In terms of Frobenius character using Corollary 7.1.3, one has that the generating function of the corresponding Frobenius characters is

$$(7.16)\qquad \sum_\lambda m_\lambda S_\lambda = (\sum_\lambda \overline{m}_\lambda S_\lambda)(\sum_{i=2}^{\infty} S_{(i)}).$$

## 7.5. Cocharacters are supported on a $(k,\ell)$ hook

**7.5.1. The $(k,\ell)$ hook theorem.** The assumption that $\mathrm{char}(F) = 0$ is crucial and assumed in this section.

Recall from Definition 6.4.9 that

$$H(k,\ell;n) = \{\lambda = (\lambda_1,\lambda_2,\dots) \vdash n \mid \lambda_{k+1} \le \ell\} \text{ and } H(k,\ell) = \bigcup_{n\ge 0} H(k,\ell;n).$$

Given a PI algebra $A$, let $\chi_n(A)$ be its $n$th cocharacter. We say that the cocharacters of $A$ are supported on the $(k,\ell)$ hook if for all $n$,

$$(7.17)\qquad \chi_n(A) = \sum_{\lambda\in H(k,\ell;n)} m_\lambda(A)\chi_\lambda.$$

We denote this by $\chi(A) \subseteq H(k,\ell)$. Geometrically, $H(k,\ell)$ contains those partitions whose Young diagram is contained in the (infinite) $(k,\ell)$ hook.

In this section we prove that for any PI algebra $A$ there exist numbers $k,\ell$ such that the cocharacters of $A$ are supported on the $(k,\ell)$ hook $H(k,\ell)$. We calculate these $k$ and $\ell$ explicitly. This is Theorem 7.5.1, or equivalently Theorem 7.5.7 below. This theorem is of fundamental importance in PI theory, and it was proved, independently, in **[RA82]** and in **[Kem91]**.

THEOREM 7.5.1 (**[RA82]**, **[Kem91]**). *Let $A$ be any PI algebra. Then there exist $k,\ell$ such that its cocharacters $\chi_n(A)$ are supported on the $(k,\ell)$ hook.*

*Explicitly, let $A$ satisfy an identity of degree $d$, and let $k,\ \ell \ge e\cdot(d-1)^4 - 1$, where $e = 2.718281828\cdots$. Then $\chi(A) \subseteq H(k,\ell)$.*

A special case occurs when $\ell = 0$. In this case the hook is in fact a *strip* formed by the diagrams with at most $k$ rows. This is the case, by Theorem 7.3.5, when the PI algebra $A$ satisfies the Capelli list $\mathcal{C}_{k+1}$. On the other hand we shall see that for instance the Grassmann algebra does not satisfy any Capelli list, and in fact its cocharacters are explicitly described as belonging to $H(1,1)$ (Theorem 19.1.5).

The hook theorem is based on

DEFINITION 7.5.2. Let $A$ be a PI algebra. The multilinear polynomial $g \in V_n$ is a *strong* identity of $A$ if for every $n \le m$ we have $F[S_m]\cdot g\cdot F[S_m] \subseteq \mathrm{Id}(A)$.

REMARK 7.5.3. For a multilinear polynomial $g \in V_n$ the condition that $F[S_n] \cdot g \cdot F[S_n] \subseteq \mathrm{Id}(A)$ is not sufficient to make it a strong identity of $A$. For instance the standard identity $S_{2n}$ spans the two-sided ideal of the sign representation but it is not a strong identity for $n \times n$ matrices (prove it using formula (6.12)). We will see, in Proposition 7.5.14, that a sufficient condition for $g \in V_n$ to be a strong identity of $A$ is that $F[S_m] \cdot g \cdot F[S_m] \subseteq \mathrm{Id}(A)$, $\forall m,\ n \leq m \leq 2n-1$.

Assume there is a strong identity $g \in V_n$ for $A$, let $I(g) = \bigoplus_i I_{\lambda_i}$ be the ideal generated by $g$ in $F[S_n]$, then

PROPOSITION 7.5.4. *For all i and for all $m \geq n$ and partitions $\mu \vdash m$ with $\lambda_i \subset \mu$, we have that the ideal $I_\mu$ of $F[S_m]$ is formed by strong polynomial identities of $A$.*

PROOF. In fact, by Theorem 6.4.13, $F[S_m] I_{\lambda_i} F[S_m] = \bigoplus_{\lambda_i \subset \mu \vdash m} I_\mu$. □

We then have

PROPOSITION 7.5.5. *The following two statements are equivalent.*

(1) *The hook theorem holds for the $(k, \ell)$ hook.*
(2) *The multilinear polynomials in the $(k+1)^{\ell+1}$ rectangle are strong identities.*

PROOF. This is basically tautological, since for $m \geq (k+1) \cdot (\ell+1)$ the ideal $I_{(k+1)^{\ell+1}}$ generated by the rectangle $(k+1) \times (\ell+1)$ is formed by all diagrams containing this rectangle. These are exactly the diagrams not contained in the $(k, \ell)$ hook. □

COROLLARY 7.5.6. *If there is a strong identity $g$, then the hook theorem follows.*

PROOF. If there is a strong identity $g$, by Remark 7.5.3 any rectangle containing some $\lambda$ with $I_\lambda \subset I(g)$ is formed of strong identities, hence the hook theorem follows. □

In fact, we prove the following version of the $(k, \ell)$ hook theorem, which is essentially equivalent to Theorem 7.5.1.

THEOREM 7.5.7. *Let $\mathrm{char}(F) = 0$, let $A$ satisfy an identity of degree $d$, and let $0 < u, v \in \mathbb{Z}$ satisfy*

$$\frac{uv}{u+v} \cdot \frac{2}{e} \geq (d-1)^4. \tag{7.18}$$

*For example, choose $u = v \geq e \cdot (d-1)^4$. Let $k = u-1$ and $\ell = v-1$, then $\chi(A) \subseteq H(k, \ell)$.*

*Equivalently, let $u, v$ satisfy (7.18). Let $(u^v)$ be the $u \times v$ rectangle, and let $\lambda \supseteq (u^v)$. Then $I_\lambda \subseteq \mathrm{Id}(A)$, where $I_\lambda \subseteq F[S_{|\lambda|}]$ is the minimal two-sided ideal in $F[S_{|\lambda|}]$ corresponding to $\lambda$.*

The starting point toward proving the $(k, \ell)$ hook theorem, Theorem 7.5.7, is

LEMMA 7.5.8. *Let $A$ be an $F$-algebra, and let $\lambda$ be a partition of $n$. If $f_\lambda > c_n(A)$, then $I_\lambda \subseteq \mathrm{Id}(A) \cap V_n$.*

PROOF. Let $J_\lambda \subseteq I_\lambda$ be a minimal left ideal in $I_\lambda$, so $\dim J_\lambda = f_\lambda > c_n(A)$. Since $J_\lambda$ is minimal, either $J_\lambda \subseteq \mathrm{Id}(A) \cap V_n$ or $J_\lambda \cap \big(\mathrm{Id}(A) \cap V_n\big) = 0$.

If $J_\lambda \cap \big(\mathrm{Id}(A) \cap V_n\big) = 0$, then $c_n(A) = \dim V_n / \big(\mathrm{Id}(A) \cap V_n\big) \geq \dim J_\lambda = f_\lambda$, a contradiction. Since $I_\lambda$ is the sum of these $J_\lambda$, the claim follows. □

Let $c_n(A) \le \alpha^n$ for all $n$. The next lemma determines rectangles $\mu = (u^v) \vdash n$ such that $\alpha^n < f_\mu$.

LEMMA 7.5.9. *Let $0 < u, v$ be integers, and let $\mu = (u^v) \vdash u \cdot v$ be the $u \times v$ rectangle. Also denote $u \cdot v = n$. Then*

$$\left(\frac{uv}{u+v}\right)^n \cdot \left(\frac{2}{e}\right)^n < f_\mu \qquad (\textit{where} \quad e = 2.718281828\cdots).$$

*In particular, if $\alpha \le \frac{uv}{u+v} \cdot \frac{2}{e}$, then $\alpha^n < f_\mu$.*

PROOF. Recall from Example 6.4.5 that the sum of hook numbers for the $(u^v)$ rectangle is

$$\sum_{x\in\mu} h_x = uv(u+v)/2 = n(u+v)/2.$$

Since the geometric mean is bounded by the arithmetic mean,

$$\left(\prod_{x\in\mu} h_x\right)^{1/n} \le \frac{1}{n}\sum_{x\in\mu} h_x = \frac{u+v}{2}, \qquad \text{hence} \qquad \left(\frac{2}{u+v}\right)^n \le \left(\frac{1}{\prod_{x\in\mu} h_x}\right).$$

Together with $n = uv$ and with the classical inequality $(n/e)^n < n!$, this implies that

$$\left(\frac{uv}{u+v}\right)^n \cdot \left(\frac{2}{e}\right)^n = \left(\frac{n}{e}\right)^n \cdot \left(\frac{2}{u+v}\right)^n < \frac{n!}{\prod_{x\in\mu} h_x} = f_\mu. \qquad \square$$

From the previous two lemmas it follows that, under the hypothesis of Lemma 7.5.9, the ideal $I_{(u^v)}$ is all formed of polynomial identities. This is not enough since we need to find a rectangle formed by *strong* polynomial identities. This requires two further steps.

PROPOSITION 7.5.10. *Let $A$ be a PI algebra satisfying an identity of degree $d$. Choose natural numbers $u$ and $v$ such that*

$$\frac{uv}{u+v} \cdot \frac{2}{e} \ge (d-1)^4. \tag{7.19}$$

*For example, choose $u = v \ge e \cdot (d-1)^4$. Let $n = uv$, and let $\mu = (u^v)$ be the $u \times v$ rectangle.*

*Let $n \le m \le 2n$, and let $\lambda \vdash m$ be any partition of $m$ which contains $\mu$: $(u^v) \subseteq \lambda$. Then the elements of the corresponding two-sided ideal $I_\lambda \subseteq F[S_m]$ are identities of $A$: $I_\lambda \subseteq \mathrm{Id}(A) \cap V_m$. Thus for $n \le m \le 2n$,*

$$\bigoplus_{\substack{\lambda\vdash m\\ \mu\subseteq\lambda}} I_\lambda \subseteq \mathrm{Id}(A).$$

PROOF. By Theorem 7.1.6 $c_m(A) \le (d-1)^{2m} \le (d-1)^{4n}$ since $m \le 2n$. By assumption $(d-1)^4 \le \frac{uv}{u+v} \cdot \frac{2}{e}$. By Lemma 7.5.9 $\left(\frac{uv}{u+v} \cdot \frac{2}{e}\right)^n < f_\mu$ and since $\mu \subseteq \lambda$, by Corollary 6.4.12 $f_\mu \le f_\lambda$. Thus

$$c_m(A) \le (d-1)^{2m} \le (d-1)^{4n} \le \left(\frac{uv}{u+v} \cdot \frac{2}{e}\right)^n < f_\mu \le f_\lambda,$$

and the proof follows by Lemma 7.5.8. $\square$

In order to complete the proof, we shall need the following construction, which is due to Amitsur.

**7.5.2. Existence of strong identities.** The natural embedding of the two groups $S_n \subset S_{n+1}$ (via $\sigma(n+1) = n+1$ for $\sigma \in S_n$) induces the embedding of spaces $V_n \subset V_{n+1}$: $f(x_1, \dots, x_n) \mapsto f(x_1, \dots, x_n) \cdot x_{n+1}$.

More generally, for any $n < m$, we have the inclusion $V_n \subset V_m$ via the map $f(x_1, \dots, x_n) \mapsto f(x_1, \dots, x_n) \cdot x_{n+1} \cdots x_m$.

With the notations of Example 6.3.14, let

$$f(x) = f(x_1, \dots, x_n) = \sum_{\sigma \in S_n} \alpha_\sigma x_{\sigma(1)} \cdots x_{\sigma(n)} \in V_n,$$

and build as in formula (6.12)

$$\begin{aligned} f^*(x_1, \dots, x_n; x_{n+1}, \dots, x_{2n-1}) & \\ &= \sum_{\sigma \in S_n} \alpha_\sigma x_{\sigma(1)} x_{n+1} x_{\sigma(2)} x_{n+2} \cdots x_{\sigma(n-1)} x_{2n-1} x_{\sigma(n)} \\ &= (f(x_1, \dots, x_n) x_{n+1} \cdots x_{2n-1}) \eta, \end{aligned} \tag{7.20}$$

where $\eta \in S_{2n-1}$ is the permutation as in formula (6.13),

$$\eta = \begin{pmatrix} 1 & 2 & 3 & 4 & \cdots & 2n-1 \\ 1 & n+1 & 2 & n+2 & \cdots & n \end{pmatrix}. \tag{7.21}$$

REMARK 7.5.11. Let $L \subseteq \{x_{n+1}, \dots, x_{2n-1}\}$, and denote by $f_L^*$ the polynomial obtained from $f^*$ by substituting $x_j \to 1$ for all $x_j \in L$. Rename the variables in $\{x_{n+1}, \dots, x_{2n-1}\} \setminus L$ to be $\{x_{n+1}, \dots, x_{n+q}\}$ (where $q = n - 1 - |L|$) and denote the resulting polynomial by $f_L^*$. Then similar to (7.20), there exists a permutation $\rho \in S_{n+q}$ such that $f_L^* = (f(x_1, \dots, x_n) x_{n+1} \cdots x_{n+q}) \rho$.

Note that if $1 \in A$ and $f^* \in \mathrm{Id}(A)$, then also $f_L^* \in \mathrm{Id}(A)$ for any such $L$, and in particular $f \in \mathrm{Id}(A)$. The converse is not true: as shown in the next example, it is possible that $f \in \mathrm{Id}(A)$, but $f^* \notin \mathrm{Id}(A)$.

EXAMPLE 7.5.12. Let $\mathrm{St}_n$ be the standard polynomial. Then $\mathrm{St}_n^*$ is the Capelli polynomial $\mathrm{St}_n^* = C_n(x; y)$. Consider for example $A = M_3(F)$. Since $\dim A = 9$, by Proposition 7.3.2 $A$ satisfies the Capelli identity $C_{10}(x; y)$. By formula (10.10) $A$ does *not* satisfy $C_m(x; y)$ for any $m \leq 9$. Thus, by the Amitsur–Levitzki theorem, Theorem 10.1.4, $A$ satisfies the standard polynomial $\mathrm{St}_6$, but it does not satisfy $\mathrm{St}_6^* = C_6(x; y)$.

PROPOSITION 7.5.13. *Let $A$ be a PI algebra satisfying an identity of degree $d$. Choose natural numbers $u$ and $v$ as in formula (7.19). Let $n = uv$, and let $\mu = (u^v)$ be the $u \times v$ rectangle. If $f \in I_\mu$, then for any subset $L \subseteq \{n+1, \dots, 2n-1\}$, $f_L^* \in \mathrm{Id}(A)$. In particular $f, f^* \in \mathrm{Id}(A)$.*

PROOF. Notice that by Theorem 6.4.13, the two-sided ideal generated in $V_m$ by $I_\mu$ is

$$V_m I_\mu V_m = F[S_m] I_\mu F[S_m] = \bigoplus_{\substack{\lambda \vdash m \\ \mu \subseteq \lambda}} I_\lambda.$$

Proposition 7.5.10 implies that $F[S_m] I_\mu F[S_m] \subseteq \mathrm{Id}(A)$ for any $n \leq m \leq 2n-1$. In particular, for such $m$, if $f \in I_\mu$ and $\rho \in S_m$, then $f\rho \in \mathrm{Id}(A)$. By the definition of $f^*$ in formula (7.20) and of $f_L^*$ in Remark 7.5.11, the claim follows. □

PROPOSITION 7.5.14. *Let $A$ be a PI algebra, let $I \subseteq V_n$ be a two-sided ideal in $V_n$, and assume that for any $L \subseteq \{n+1,\dots,2n-1\}$, $f_L^* \in \mathrm{Id}(A)$ (hence in particular $f \in \mathrm{Id}(A)$). Then for any $m > n$, $F[S_m]IF[S_m] \subseteq \mathrm{Id}(A)$.*

PROOF. Since $F[S_m]I \subseteq \mathrm{Id}(A)$, it suffices to prove the following

CLAIM. If $f \in I$ and $\pi \in S_m$, then $f\pi := (f(x_1,\dots,x_n)x_{n+1}\cdots x_m)\pi \in \mathrm{Id}(A)$.

Let $f = \sum_{\sigma\in S_n} a_\sigma \sigma(x_1\cdots x_n)$, then $f\pi = \sum_{\sigma\in S_n} a_\sigma\sigma((x_1\cdots x_n\cdots x_m)\pi)$ since the left and right actions commute.

Consider the positions of $x_1,\dots,x_n$ in the monomial $(x_1\cdots x_n\cdots x_m)\pi$. There exists $\tau \in S_n$ such that

$$\begin{aligned}(x_1\cdots x_n\cdots x_m)\pi &= g_0x_{\tau(1)}g_1x_{\tau(2)}g_2\cdots g_{n-1}x_{\tau(n)}g_n\\ &= \tau(g_0x_1g_1x_2g_2\cdots g_{n-1}x_ng_n),\end{aligned}$$

where each $g_j$ is $=1$ or is a monomial in some of the variables $x_{n+1},\dots,x_m$. It follows that

$$\begin{aligned}f\pi &= \sum_{\sigma\in S_n} a_\sigma\sigma\tau(g_0x_1g_1x_2g_2\cdots g_{n-1}x_ng_n)\\ &= \sum_{\sigma\in S_n} a_\sigma(g_0x_{\sigma\tau(1)}g_1x_{\sigma\tau(2)}g_2\cdots g_{n-1}x_{\sigma\tau(n)}g_n).\end{aligned}$$

Now $\tau(x_1x_2\cdots x_n) = (x_1x_2\cdots x_n)\tau$. Hence,

$$\sum_{\sigma\in S_n} a_\sigma\sigma\tau(x_1x_2\cdots x_n) = \sum_{\sigma\in S_n} a_\sigma\sigma(x_1x_2\cdots x_n)\tau = f\tau.$$

Since $f \in V_n$ and $\tau \in S_n$, $f\tau \in I$, hence by assumption $(f\tau)^* \in \mathrm{Id}(A)$.

$$\Longrightarrow f\pi = g_0((f\tau)^*(x_1,\dots,x_n;g_1,\dots,g_{n-1}))g_n \in \mathrm{Id}(A).$$

Since, if $L$ is the subset of the indices $1,\dots,n-1$ where $g_i \neq 1$, we have that $(f\tau)^*(x_1,\dots,x_n;g_1,\dots,g_{n-1})$ is obtained from $(f\tau)_L^* \in \mathrm{Id}(A)$ by substituting some of the variables with the elements $g_i$. □

Recall that the ceiling of a real number $x$, denoted by $\lceil x\rceil$, is defined to be least integer greater than or equal to $x$.

By Proposition 7.5.10 and Lemma 6.11, we have proved

THEOREM 7.5.15. *Let $\mathrm{char}(F) = 0$. Then every PI algebra $A$ satisfies nontrivial strong identities. Explicitly, let $A$ satisfy an identity of degree $d$. Let $u, v \in \mathbb{N}$ be such that*

$$\frac{uv}{u+v}\cdot\frac{2}{e} \geq (d-1)^4,$$

*and let $\mu = (u^v)$ be the $u \times v$ rectangle. Then every $g \in I_\mu$ is a strong identity of $A$. The degree of such strong identity $g$ is $uv$. We can choose for example $u = v = \lceil e\cdot(d-1)^4\rceil$, so $h = \deg(g) = \lceil e\cdot(d-1)^4\rceil^2 \approx e^2(d-1)^8$.*

Finally:

PROOF OF THEOREM 7.5.7. This follows now from Corollary 7.5.6. □

Another application of strong identities is given in §7.6.

**7.5.3. Amitsur–Capelli polynomials.** Let $\lambda \vdash n$ with corresponding irreducible $S_n$ character $\chi_\lambda$. To $\lambda$ also corresponds the minimal two-sided ideal $I_\lambda \subseteq \mathbb{Q}[S_n]$ with its central idempotent $E_{(\lambda)}$. It is well known (see for example **[CR62**, page 236]) that

$$E_{(\lambda)} = \frac{1}{n!} \sum_{\sigma \in S_n} \chi_\lambda(\sigma)\sigma.$$

Let $(u^v)$ denote the $u \times v$ rectangle, with $S_{uv}$ the corresponding symmetric group, so

$$E_{(u^v)} = \frac{1}{(uv)!} \sum_{\sigma \in S_{uv}} \chi_{(u^v)}(\sigma)\sigma.$$

The polynomials $E^*_{(u^v)}$, defined by formula (6.12), were introduced by S. Amitsur **[RA82]**, and we call them the Amitsur–Capelli polynomials. They characterize the condition $\chi(A) \subseteq H(k, \ell)$ (see (7.17)). This is

THEOREM 7.5.16. *Let* $\mathrm{char}(F) = 0$, *and let* $A$ *be a PI algebra. Let* $k, \ell \geq 0$ *integers, denote* $u = k+1, v = \ell+1$, *and let* $\mu = (u^v) = ((k+1)^{\ell+1}) \vdash n$ *be the* $u \times v$ *rectangle partition, where* $n = uv$. *The following are equivalent.*

(1) $\chi(A) \subseteq H(k, \ell)$.
(2) $E^*_\mu = E^*_{((k+1)^{\ell+1})} \in \mathrm{Id}(A)$.

PROOF. (1) Assume first that $\chi(A) \subseteq H(k, \ell)$, then for all $m > n$

$$F[S_m] \cdot I_\mu \cdot F[S_m] \subseteq \mathrm{Id}(A),$$

and in particular $E^*_\mu \in \mathrm{Id}(A)$ since $E^*_\mu = E_\mu \eta$, where $\eta \vdash 2n - 1$, see (7.21).

(2) Conversely, assume $E^*_\mu \in \mathrm{Id}(A)$. Let $f \in I_\mu$, then $f = aE_\mu$ for some $a \in \mathbb{Q}[S_n]$. Since $f^* = f\eta$, we get

$$f^* = (aE_\mu)^* = (aE_\mu)\eta = a(E_\mu \eta) = aE^*_\mu.$$

By assumption $E^*_\mu \in \mathrm{Id}(A)$, which implies that also $f^* = aE^*_\mu \in \mathrm{Id}(A)$.

Thus $I = I_\mu$ satisfies the condition in Proposition 7.5.14, hence for any $n < m$,

$$F[S_m] \cdot I_\mu \cdot F[S_m] \subseteq \mathrm{Id}(A),$$

which implies that $\chi(A) \subseteq H(k, \ell)$. □

Clearly, Theorems 7.5.1 and 7.5.16 together imply

COROLLARY 7.5.17. *Let the algebra* $A$ *satisfy an identity of degree* $d$. *Assume* $k, \ell > e(d-1)^4 - 1$, *then* $A$ *satisfies the Amitsur–Capelli identity,* $E^*_{((k+1)^{\ell+1})}$, *namely* $E^*_{((k+1)^{\ell+1})} \in \mathrm{Id}(A)$.

## 7.6. Application: A theorem of Kemer

As an application of strong identities (Definition 7.5.2) we now prove the following important theorem of Kemer which has been used in the first proofs of the Razmyslov–Kemer–Braun theorem of the nilpotency of the Jacobson radical of a finitely generated PI algebra.

As in the previous section, the assumption that the characteristic is zero is crucial here.

THEOREM 7.6.1 (**[Kem80]**). *Let $A = F\langle a_1, a_2, \dots, a_r\rangle$ be a finitely generated PI algebra, satisfying a strong identity $p$ of degree $h$, and let $n \geq r^h + h$. Then $A$ satisfies the Capelli identity $C_n(x; y)$.*

The proof is given below. We begin with a few remarks.

REMARK 7.6.2. (1) Let $A$ satisfy an identity of degree $d$. Then, by Theorem 7.5.15, $A$ satisfies a strong multilinear identity

$$p = \sum_{\sigma \in S_h} \beta_\sigma \sigma = \sum_{\sigma \in S_h} \beta_\sigma x_{\sigma(1)} \cdots x_{\sigma(h)}, \ \beta_1 = 1, \tag{7.22}$$

of degree $h$ (and we can take $h = \lceil e^2(d-1)^8 \rceil$ the integral part of $e^2(d-1)^8$).

Assume $A$ satisfies the strong multilinear identity (7.22).

LEMMA 7.6.3. *Let $f(x_1, \dots, x_h; Y)$ be any multilinear polynomial in the variables $x_1, \dots, x_h$ and $Y$ (where $Y$ is any set of additional variables). Then*

$$\sum_{\sigma \in S_h} \beta_\sigma f(x_{\sigma(1)}, \dots, x_{\sigma(h)}; Y) \in \mathrm{Id}(A). \tag{7.23}$$

PROOF. Clearly it is enough to prove this when $f = M$ is a multilinear monomial identified with a permutation $\eta \in S_m$ for some $m \geq h$. Then the monomial $M(x_{\sigma(1)}, \dots, x_{\sigma(h)}; Y)$
$= \sigma\eta$, so (see (7.22))

$$\sum_{\sigma \in S_h} \beta_\sigma M(x_{\sigma(1)}, \dots, x_{\sigma(h)}; Y) = \sum_{\sigma \in S_h} \beta_\sigma \sigma\eta = p\eta \in F[S_m]pF[S_m] \subseteq \mathrm{Id}(A),$$

since $p$ is a strong identity of $A$. □

Let $A = F\langle a_1, a_2, \dots, a_r\rangle$ be a PI algebra, satisfying a strong multilinear identity $p$ of degree $h$ as in formula (7.22).

For $h \leq n$, let $f(x_1, \dots, x_n; Y)$ be a multilinear polynomial in $x_1, \dots, x_n$ and possibly some extra variables $Y$. We denote by $\bar{Y}$ any specialization of $Y$ in $A$, and consider $\Delta = f(v_1, \dots, v_n; \bar{Y}) \in A$ with $v_i$, $1 \leq i \leq n$ words in the generators $a_1, a_2, \dots$. Then

LEMMA 7.6.4. *$\Delta$ is a linear combination of terms $\Delta' = f(v'_1, \dots, v'_n; \bar{Y})$ (same $\bar{Y}$) where at most $h-1$ among the $v_i$'s have length $\geq h$ as words in $a_1, a_2, \dots$.*

PROOF. If there are at least $h$ among the $v_i$'s with $|v_i| \geq h$, we may assume without loss of generality that $\Delta = f(v_1, \dots, v_n; \bar{Y})$ where $|v_1|, \dots, |v_h| \geq h$ and also up to changing name to the variables that $|v_1| \geq |v_2| \geq \cdots \geq |v_n|$. Write $v_i = w_i u_i$ where $|u_i| = i$, $1 \leq i \leq h$.

Since we assume the strong identity $p$, we may apply formula (7.23) to $\Delta$. Thus in $A$ we have that the polynomial $\Delta$ is a linear combination of terms $\Delta_\sigma$ of the form $f(w_1 u_{\sigma(1)}, \dots, w_h u_{\sigma(h)}; \bar{Y})$ where $1 \neq \sigma \in S_h$ ($\Delta$ itself corresponds to $\sigma = 1$).

Denote $w_i u_{\sigma(i)} = v'_i$, $1 \leq i \leq h$ and $v'_i = v_i$ for $i > h$. The new vector of lengths, after reordering the $v'_i$ so that they are again nonincreasing, is clearly strictly larger lexicographically than the vector $|v_i|$ but $\sum_i |v_i| = \sum_j |v'_j|$, and the proof of the lemma follows by induction. □

We are now ready to prove Theorem 7.6.1.

PROOF. We can assume that $r \geq 2$. Consider $C_n(v_1, \ldots, v_n; \bar{Y})$ where $v_i \in A$ is a monomial in the generators and the elements in $\bar{Y}$ are in $A$.

By Lemma 7.6.4 we can assume that at most $h-1$ among the $v_i$'s have length $\geq h$, hence at least $n-(h-1)$ among the $v_i$'s have length $\leq h-1$ as words in the generators $a_1, \ldots, a_r$. The number of words of length $q$ is $\leq r^q$. Hence the number of words of length $\leq h-1$ is

$$\leq 1+r+r^2+\cdots+r^{h-1} = \frac{r^h-1}{r-1} \leq r^h - 1 < r^h.$$

But we have at least $n-(h-1)$ such words (namely of length $\leq h-1$) among $v_1, \ldots, v_n$, and $n-(h-1) > r^h$ (since by assumption $n \geq r^h + h$). It follows that there must be repetitions among $v_1, \ldots, v_n$, so $C_n[v_1, \ldots, v_n; \bar{Y}] = 0$. $\square$

CHAPTER 8

# Shirshov's Height Theorem

*Theorem 8.1.7 is Shirshov's height theorem* **[Šir57a, Šir57b, Šir58]**. *This theorem is a powerful tool in the study of PI algebras. It was published—in Russian—around the year 1957, but was ignored in the West for several years; see* **[Pro73**, page 151**]**. *For additional expositions of Shirshov's height theorem—see for example* **[KBR05]**, **[Row88]**, **[Reu86]**, *relating to Lyndon words or* **[Pir96]**, *using infinite words, or* **[dLV99]**.

*In Section 8.2 we bring some applications of this theorem, including Berele's theorem* **[BR83]** *about the multiplicities in cocharacters being polynomially bounded. Shirshov's height theorem is one of the main ingredients in the proof of that theorem of Berele; see Theorem 8.3.14.*

## 8.1. Shirshov's height theorem

### 8.1.1. Shirshov's lemma.

8.1.1.1. *Words and the lexicographic order.* Let us recall some basic notions as in §4.2.1. Let $x_1, \ldots, x_\ell$ be associative and noncommutative variables, and let $W(X) = W(x_1, \ldots, x_\ell)$ be the set of monomials or *words* in these variables,

$$W(X) = \{x_{i_1} x_{i_2} \cdots x_{i_n} \mid 1 \leq i_j \leq \ell;\ n = 0, 1, 2, \ldots\}.$$

The set $X = \{x_1, \ldots, x_\ell\}$ is also called *alphabet*, the variables $x_1, \ldots, x_\ell$ are also called *letters*, $W(X)$ is called the *free, associative,* monoid generated by these letters, and the monomials in $W(X)$ are also called *words*.

The word $w = x_{i_1} \cdots x_{i_n}$ will be said to have *length n*, written $|w| = n$. If $w = w_1 w_2 w_3$ in $W(X)$, we call the $w_i$ *subwords* of $w$; $w_1$ is an *initial* and $w_3$ is a *terminal* subword of $w$. A subword $v$ of $w$ has *multiplicity* $k$ in $w$ if there are $k$ nonoverlapping occurrences of $v$ in $w$. For example, the subword $x_1x_2x_1$ has multiplicity 2 in $x_1x_2x_1x_1x_2x_1$ but multiplicity 1 in $x_1x_2x_1x_2x_1$. If the occurrences are consecutive, we write $v^k$ to denote $v \cdots v$, i.e., $v$ repeating $k$ times. The *empty word*, denoted $\emptyset$, is the unit element 1 in $W(X)$.

One of the main tools here is the *lexicographic order* $\succ$ on monomials (we write $w_1 \prec w_2$ if and only if $w_2 \succ w_1$).

DEFINITION 8.1.1. First, the variables are ordered by the following total order:

$$x_\ell \succ x_{\ell-1} \succ \cdots \succ x_2 \succ x_1.$$

For two monomials $w_1 = x_{i_1} v_1$ and $w_2 = x_{i_2} v_2$, we set $w_1 \prec w_2$ if $i_1 \prec i_2$, or if $i_1 = i_2$, we have inductively, $v_1 \prec v_2$.

REMARK 8.1.2. This is a partial order, as we do not compare a monomial with the empty monomial 1. For example, $x_1^2 x_2^3 x_3^2 \prec x_1^2 x_2 x_3$, but $x_1x_2$ and $x_1x_2^2$ are not comparable. Two monomials $v \neq w$ with $|v| \leq |w|$ are *comparable* if and only if $v$

is *not* an initial subword of $w$. Thus, any two monomials of the same length are comparable.

REMARK 8.1.3. Let $w_1 \succ w_2$, then $uw_1 \succ uw_2$ and $w_1u \succ w_2v$ for any words $u$, $v$. For the second statement we use the fact that $w_2$ is not an initial of $w_1$.

LEMMA 8.1.4. *Let* $w_1 \succ w_2 \succ \cdots \succ w_d$, *and let* $1 \neq \pi \in S_d$, *a permutation. Then* $w_1w_2\cdots w_d \succ w_{\pi(1)}w_{\pi(2)}\cdots w_{\pi(d)}$.

PROOF. Since $\pi \neq 1$, there exists $r$, $1 \leq r \leq d$ such that $\pi(i) = i$ for all $i$ with $1 \leq i \leq r-1$ and $\pi(r) \neq r$. Then necessarily $r < \pi(r)$, so $w_r \succ w_{\pi(r)}$. Since in the lexicographic order monomials are compared from left to right, the proof now follows from Remark 8.1.3. □

DEFINITION 8.1.5. A *d-decomposition* of a word $w$ is a factorization

$$w = w_1w_2\cdots w_d$$

such that, for every permutation $1 \neq \pi \in S_d$,

$$w_1w_2\cdots w_d \succ w_{\pi(1)}w_{\pi(2)}\cdots w_{\pi(d)}.$$

A word $w$ is called *d-decomposable* (with respect to $\succ$) if it contains a subword which has a $d$-decomposition. A word which is not $d$-decomposable is called *d-indecomposable*.

If $w$ contains a subword $w_1a_1w_2a_2\cdots a_{d-1}w_d$ such that $w_1 \succ w_2 \succ \cdots \succ w_d$, then, by Remark 8.1.3 and Lemma 8.1.4, $w$ is $d$-decomposable. The converse need not be true. As shown in Example 8.1.6, it is possible that for every permutation $1 \neq \pi \in S_d$,

$$w_1w_2\cdots w_d \succ w_{\pi(1)}w_{\pi(2)}\cdots w_{\pi(d)},$$

*without* having $w_1 \succ w_2 \succ \cdots \succ w_d$. Note also that every word is 1-decomposable, hence there are no 1-indecomposable words.

EXAMPLE 8.1.6.

(1) On the alphabet $\{x_1, x_2\}$ (with $x_2 \succ x_1$) consider $w_1 = x_1x_2$ and $w_2 = x_1$, then $w_1w_2 \succ w_2w_1$ but $w_1 \not\succ w_2$ (in fact, $w_1$ and $w_2$ are not comparable).
 Similarly let $w_1 = x_1x_2x_2$, $w_2 = x_1x_2$ and $w_3 = x_1$. Then we have $w_1w_2w_3 \succ w_{\pi(1)}w_{\pi(2)}w_{\pi(3)}$ for any $1 \neq \pi \in S_3$, but $w_1, w_2$ and $w_3$ are not comparable.
(2) If $\ell = 1$, the only words are $x_1^k$, and two different words are not comparable; hence for any $d > 1$, any word is $d$-indecomposable.

It is instructive to see which are the 2-indecomposable words. They are words in which if $i < j$ any occurrence of $x_i$ is to the left of all occurrences of $x_j$, that is a 2-indecomposable word is an *ordered monomial* of type

$$x_1^{h_1}x_2^{h_2}\cdots x_\ell^{h_\ell}. \tag{8.1}$$

Then the main task is to understand indecomposable words and prove

THEOREM 8.1.7 (Shirshov's height theorem). *Given $d$ and $\ell$, there exists an $h \in \mathbb{N}$, dependent on $d, \ell$ such that every $d$-indecomposable word $w$ in $\ell$ letters is of the form*

$$w = u_1^{k_1}u_2^{k_2}\cdots u_j^{k_j}, \quad j \leq h, \quad |u_i| \leq d-1. \tag{8.2}$$

For $d = 2$ this follows from formula (8.1) with $h = \ell$. The next few sections are devoted to the proof of this theorem .

8.1.1.2. *Representations of words in an algebra.* Our main use of the definitions of decompositions and indecomposable words will be to PI algebras.

The commutative law implies that every monomial in some variables $x_i$ can be reordered to be of the form of formula (8.1), and a similar statement holds for higher polynomial identities.

We consider an algebra $A$ finitely generated over a commutative ring $F$ by generators $\underline{a} := \{a_1, \ldots, a_\ell\}$.

Let $F\langle X\rangle = F\langle x_1, \ldots, x_\ell\rangle$ be the free associative $F$-algebra in $\ell$ variables.

The choice of the generators $\underline{a} := \{a_1, \ldots, a_\ell\}$ of $A$ yields the homomorphism $\Phi : F\langle X\rangle \to A$, given by $x_i \mapsto a_i \in A$, which is surjective. In order to stress both the word and the generators, we write $w(\underline{a}) := \Phi(w)$ for the image of a word $w$ under $\Phi$. For a set of words $U \subseteq W(X)$, we write $U(\underline{a})$ for $\{w(\underline{a}) : w \in U\}$.

DEFINITION 8.1.8. The *content* or *weight* of a word $w$ is the commutative monomial associated to $w$; equivalently it is the integer valued vector which gives the multiplicity of each variable $x_i$ in the monomial.

Definition 8.1.5 motivates the following important remark, which is one of the main motivations for proving Shirshov's height theorem. Recall that, by Proposition 2.2.34 a PI algebra always satisfies a multilinear polynomial identity of the form given by the next formula (8.3).

REMARK 8.1.9. Let

$$f = y_1 \ldots y_d - \sum_{1 \neq \pi \in S_d} \alpha_\pi y_{\pi(1)} \ldots y_{\pi(d)} \in \mathrm{Id}(A) \tag{8.3}$$

be a polynomial identity of the algebra $A$, and let $w = w_1 w_2 \cdots w_d$ be a $d$-decomposition of a word $w$. By applying the substitutions $y_i \to w_i$, we deduce that

$$w_1 \cdots w_d - \sum_{1 \neq \pi \in S_d} \alpha_\pi w_{\pi(1)} \cdots w_{\pi(d)} \in \mathrm{Id}(A).$$

Evaluating this identity of $A$, in the elements $\underline{a}$ we have

$$(w_1 \cdots w_d)(\underline{a}) = \sum_{1 \neq \pi \in S_d} \alpha_\pi (w_{\pi(1)} \cdots w_{\pi(d)})(\underline{a}).$$

By assumption $w_1 \cdots w_d \succ w_{\pi(1)} \cdots w_{\pi(d)}$ for any $1 \neq \pi \in S_d$. In other words, modulo the identities of $A$, we can replace $w_1 \cdots w_d(\underline{a})$ by the linear combination $\sum_{1 \neq \pi \in S_d} \alpha_\pi w_{\pi(1)} \cdots w_{\pi(d)}(\underline{a})$, a combination of words of the same length in the same elements $(\underline{a})$ and with the same content which are smaller in the lexicographic order.

Take a Z algebra $A$ which satisfies a multilinear PI of degree $d$, with coefficients in Z and one coefficient equal to 1.

COROLLARY 8.1.10. *If $A$ is generated, over Z, by some elements $\underline{a}$, then $A$ is spanned linearly, over Z, by the monomials $w(\underline{a})$, where $w$ runs over the words which are $d$-indecomposable. More precisely any word $w(\underline{a})$ equals a linear combination, with coefficients in Z, of $d$-indecomposable words of the same length and with the same content.*

PROOF. $A$ is spanned by the monomials $w(\underline{a})$ where $w$ runs over all the words. If $w$ is $d$-decomposable, then by the previous remark $w(\underline{a})$ is a linear combination of words of the same length and which are smaller in the lexicographic order. Since the words of a given length form a finite set, this replacement algorithm stops after a finite number of steps and it terminates only when all words appearing are $d$-indecomposable. □

Therefore by the characterization of $d$-indecomposable words given by Theorem 8.1.7, we finally have the main

COROLLARY 8.1.11. *If $A$ satisfies the conditions of Corollary 8.1.10, there exists an integer $h$ such that $A$ is spanned over $F$ by the set $u_1(\underline{a})^{k_1}u_2(\underline{a})^{k_2}\cdots u_j(\underline{a})^{k_j}$, with $j \leq h$, $|u_i| \leq d-1$. More precisely, a monomial $u(a)$ is a linear combination of elements of the form $u_1(\underline{a})^{k_1}u_2(\underline{a})^{k_2}\cdots u_j(\underline{a})^{k_j}$, with $j \leq h$, $|u_i| \leq d-1$ of the* same degree *that is $|u| = \sum_i k_i|u_i|$ and of the same content.*

DEFINITION 8.1.12. The elements $u(\underline{a})$ with $|u| \leq d-1$ are called a *Shirshov basis* for $A$, or a Shirshov basis associated to the generators $\underline{a}$.

8.1.1.3. *Primitive words and periods.* Here by *word* we always mean a nonempty one.

DEFINITION 8.1.13. A nonempty word $w$ is called *primitive* if it is not of the form $w = u^k$ for suitable $k > 1$.

REMARK 8.1.14. Every nonempty word $w$ is uniquely of the form $w = u^k$ with $u$ primitive. The word $u$ is called the *period* of $W$ and the number $|u|$ is called the *periodicity* of $w$. We have $|w| = |u|k$.

DEFINITION 8.1.15. The cyclic shift $\mathcal{CS}$ on nonempty words $w$ is defined by $\mathcal{CS}(ux) = xu$, for any word $w = ux$ with $x$ a letter.

Note that for words $u, v$, $\mathcal{CS}^{|v|}(uv) = vu$ in particular $\mathcal{CS}^{|v|}(v^k) = v^k$, $\forall k \in \mathbb{N}$.

LEMMA 8.1.16. *Let $1 \leq k < |w|$ and assume $w = uv = vu$ where $|v| = k$ (equivalently, $\mathcal{CS}^k(w) = w$), then $u, v, w$ have the same period $\hat{u}$. The periodicity $|\hat{u}|$ divides $k$.*

PROOF. By assumption $|u| \geq 1$, and we may assume $|u| \leq |v|$. Since $uv = vu$, $u$ is an initial subword of $v$, so we write $v = uv'$. If $v' = 1$, we are clearly done. Otherwise $uuv' = w = vu = uv'u$, so $uv' = v'u$ and is of length $< |w|$. Thus, by induction on $|w|$, we have $u, v', uv'$ have the same period $\hat{u}$, and $|\hat{u}|$ dividing both $|u|$ and $|v'|$. Writing $u = \hat{u}^i$ and $v' = \hat{u}^m$, we have $v = \hat{u}^{i+m}$ (so $j = i + m$) and $w = \hat{u}^{i+j}$. □

COROLLARY 8.1.17. *The periodicity of a word $v$ is the minimum positive exponent $h$ such that $v = \mathcal{CS}^h(v)$. Moreover, $\mathcal{CS}^i(v) = \mathcal{CS}^j(v)$, if and only if $i \cong j$ modulo $h$.*

PROOF. By elementary group theory it is enough to show the first statement. If $v = u^k$ with $u$ primitive and $|u| = h$, we clearly have $v = \mathcal{CS}^h(v)$. If for $i < h$, we had $v = \mathcal{CS}^i(v)$, it would mean that $v = ab = ba$ with $|b| = i$. Then by the previous lemma the periodicity of $v$ equals that of $b$ and divides $i$ an absurdity. □

8.1.1.4. *d-indecomposable words.* We start with

LEMMA 8.1.18. *Suppose $w = w_1 w_2 \cdots w_d$ is a d-decomposition, and take a word $w_0 \succ w_1, w_2, \dots, w_d$ which is higher of all these factors. Then $w_0 w = w_0 w_1 w_2 \dots w_d$ is a $d+1$-decomposition of $w_0 w$.*

PROOF. This is immediate from Remark 8.1.3. □

Let $u$ be a word containing two subwords $c_1, c_2$ (not necessarily disjoint) namely $u = a_1 c_1 r_1 = a_2 c_2 r_2$, and such that $c_1 \succ c_2$. By Remark 8.1.3, $c_1 r_1 \succ c_2 r_2$. Hence, with $b_i = c_i r_i$, $i = 1, 2$, we have $u = a_1 b_1 = a_2 b_2$, and $b_1 \succ b_2$. Thus, if $u$ contains $d$ different and lexicographically comparable subwords $c_i$, then we can write $u = a_i b_i$, $i = i, \dots, d$ such that $b_1 \succ \cdots \succ b_d$.

For example, recall that $x_2 \succ x_1$, and let $u = x_1^3 x_2^3$. Then

$$u = a_1 \cdot b_1 = x_1^3 \cdot x_2^3, \quad u = a_2 \cdot b_2 = x_1^2 \cdot x_1 x_2^3, \quad u = a_3 \cdot b_3 = x_1 \cdot x_1^2 x_2^3,$$

and $b_1 \succ b_2 \succ b_3$.

Recall that $u$ has multiplicity $k$ in a word $w$ if it has $k$ nonoverlapping occurrences in $w$.

LEMMA 8.1.19. *If a subword u has multiplicity $\geq d$ in w, and u contains d different (though not necessarily disjoint) and lexicographically comparable subwords, then w is d-decomposable.*

PROOF. By assumption we can write $w = s_0 u s_1 u \cdots s_{d-1} u s_d$, and, by the previous remark, $u = a_i b_i$, $i = i, \dots, d$ such that $b_1 \succ \cdots \succ b_d$. Then we can write $w = s_0 a_1 (b_1 s_1 a_2)(b_2 s_2 a_3) \cdots (b_{d-1} s_{d-1} a_d)(b_d s_d)$, and $b_1 s_1 a_2 \succ b_2 s_2 a_3 \succ \cdots \succ b_{d-1} s_{d-1} a_d \succ b_d s_d$. Hence $w$ is $d$-decomposable by the remark following Definition 8.1.5. □

LEMMA 8.1.20. *If v is a primitive word and $d \leq |v|$, then $v^{2d}$ is d-decomposable.*

PROOF. Since $v$ is primitive, it has period $|v|$ which by hypothesis is $\geq d$, hence the $d$ words $\mathcal{CS}^j(v)$, $j = 0, 1, \dots d-1$, are distinct and of the same length, thus comparable. Each $\mathcal{CS}^j(v) = b_j a_j$, $v = a_j b_j$, is a subword of $v^2$. Hence, by Lemma 8.1.19 with $w = v^{2d}$ and $u = v^2$, it follows that $v^{2d}$ is $d$-decomposable. □

8.1.1.5. *d-indecomposable words and a lemma of Shirshov.* Given the integers $\ell$, $k$ and $d$, consider $d$-indecomposable words on an alphabet of $\ell$ letters, $\{x_1, \dots, x_\ell\}$. For a given $k \in \mathbb{N}$, we look for subwords of the form $v^k$. The next lemma of Shirshov asserts that if such a $d$-indecomposable word is long enough, then it contains such a subword $v^k$. Moreover if $k \geq 2d$, we have that $v$ has length $|v| \leq d-1$.

We can now prove Shirshov's lemma.

PROPOSITION 8.1.21 (Shirshov's lemma).

(i) *For every choice of $\ell, k, d \in \mathbb{N}$ there is $\mathcal{N} = \mathcal{N}(\ell, k, d) \in \mathbb{N}$ such that every d-indecomposable word w of length $\geq \mathcal{N}$ in $\ell$ letters must contain a nonempty subword of the form $v^k$. Equivalently, if $|w| \geq \mathcal{N}(\ell, k, d)$ and w does not contain any nonempty subword of the form $v^k$, then w is d-decomposable.*

(ii) *If w is indecomposable and $k \geq 2d$, then a subword of the form $v^k$ with v primitive has length $1 \leq |v| \leq d-1$.*

PROOF. Let us show that (i) implies (ii). In fact by (i) $w$ contains a nonempty subword $v^k$, and we may assume $v$ primitive, hence if $|v| \geq d$, by Lemma 8.1.20, $v^{2d}$ is $d$-decomposable. But $w$ contains $v^k$, therefore it contains $v^{2d}$ since $k \geq 2d$, hence also $w$ is $d$-decomposable, a contradiction.

So our main task is to prove part (i). We keep $k$ fixed and work by double induction, first on $d$ and then on $\ell$.

If $d = 1$, then, for any $\ell$ and $k$, the assertion holds for any $\mathcal{N}$, $\mathcal{N} = \mathcal{N}(\ell, k, 1)$, since there are no 1-indecomposable words.

If $d = 2$ the indecomposable words are the ordered monomials, (8.1), so if the degree of such a monomial is $\geq (k-1)\ell + 1$, then at least one of the exponents $k_i \geq k$, thus clearly $\mathcal{N}(\ell, k, 2) = (k-1)\ell + 1$.

If $\ell = 1$, then just take $\mathcal{N} = k$, and $v = x_1$. Then every word $w$ of length $|w| \geq \mathcal{N} = k$ contains $x_1^k$, and in this case all words $w$ are $d$-indecomposable.

We assume now that we are given $\ell, k, d$ with $2 \leq \ell, 3 \leq d$. By induction on $d$ we assume that for every number $q$, we have a corresponding $\mathcal{N}(q, k, d-1)$. And by induction on $\ell$ we also have $\mathcal{N}(\ell - 1, k, d)$, both satisfying the lemma.

Assume $w$ is any word which does not contain any subword of the form $v^k$. We shall calculate $\mathcal{N} = \mathcal{N}(\ell, k, d)$ large enough such that, $w$ must be $d$-decomposable whenever $|w| \geq \mathcal{N}$.

Recall that $x_\ell \succ \cdots \succ x_1$. Given $w$, we list each occurrence of $x_\ell$ in $w$ separately, i.e.,

$$w = v_0 x_\ell^{t_1} v_1 x_\ell^{t_2} v_2 \cdots x_\ell^{t_m} v_m, \tag{8.4}$$

where $v_i$ are words in $x_1, \ldots, x_{\ell-1}$. The words $v_0$ and $v_m$ could be 1, but the other $v_i$ are not equal to 1. By the induction hypothesis on $\ell - 1$ we are done, unless each $|v_i| < \mathcal{N}(\ell - 1, k, d)$. Also, by hypothesis on $w$, $t_i < k$ for each $i$.

Form a new alphabet $X'$ as follows:

$$X' = \{v x_\ell^t \mid 1 \leq t \leq k-1,\ v \in W(x_1, \ldots, x_{\ell-1}) \text{ and } 1 \leq |v| < \mathcal{N}(\ell - 1, k, d)\}.$$

Thus, every new letter $x' \in X'$ is a word in $X$ which starts with some $x_j$ where $1 \leq j \leq \ell - 1$. Notice that the natural map from $i : W(X') \to W(X)$ is injective by the special form of the elements in $X'$. Moreover one can compare the length of $w \in W(X')$ with $i(w) \in W(X)$.

Since for every $x' \in X'$ we have $|i(x')| \leq k - 2 + \mathcal{N}(\ell - 1, k, d)$, we have

$$\forall w \in W(X'), \quad |i(w)| \leq (k - 2 + \mathcal{N}(\ell - 1, k, d))|w|. \tag{8.5}$$

We totally order this new alphabet $X'$ as follows: $v_1 x_\ell^{t_1} \succ' v_2 x_\ell^{t_2}$, if and only if either $v_1 x_\ell^{t_1} \succ v_2 x_\ell^{t_2}$ in the old lexicographic order $\succ$, or $v_2 x_\ell^{t_2}$ is an initial subword of $v_1 x_\ell^{t_1}$ (in which case $v_1 = v_2$ and $t_1 > t_2$). In particular, if $a, b \in X'$ are such that $i(a) \succ i(b)$, then $a \succ' b$. This is a total order on $X'$, which induces the corresponding lexicographic order on the monoid $W(X')$ generated by $X'$, denoted also by $\succ'$. We may apply all the notions and remarks associated to such an order.

The main property of this embedding, from which Shishov's lemma follows is

LEMMA 8.1.22. *A $q$-decomposition $w = w_1 \cdots w_q$, where $w_1, \ldots, w_q \in W(X')$, in the $\succ'$ order gives rise to a $q$-decomposition $i(w) = i(w_1) \cdots i(w_q)$ (for the original order $\succ$).*

Before we prove this fact, let us see how from this follows Shishov's lemma, Lemma 8.1.21. Denote $\ell' := |X'|$ (in fact $\ell' = [\sum_{i=1}^{\mathcal{N}(\ell-1,k,d)-1}(\ell-1)^i](k-1)$). We claim that we can estimate

$$\mathcal{N}(\ell,k,d) \leq \mathcal{N}' := (k-2+\mathcal{N}(\ell-1,k,d))[\mathcal{N}(\ell',k,d-1)+2].$$

In fact take our word $w = v_0 x_\ell^{t_1} v_1 x_\ell^{t_2} v_2 \cdots x_\ell^{t_m} v_m$ of formula (8.4). We can write it as $w = v_0 x_\ell^{t_1} i(u) v_m$ where $u = x'_{i_1} \cdots x'_{i_{m-1}}$ with $x'_{i_j} = v_j x_\ell^{t_{j+1}}$.

If $|w| > \mathcal{N}'$, we have by formula (8.5) that $|u| > \mathcal{N}(\ell',k,d-1)$. Since we are assuming that $w$ does not contain any nonempty subword of type $v^k$, clearly the same holds for $u$ which is therefore $(d-1)$-decomposable.

So if $w_1 \cdots w_{d-1}$ is a $(d-1)$-decomposition of a subword of $u$, we have that

$$w = v_0 x_\ell^{t_1} w_0 i(w_1) \cdots i(w_{d-1}) z,$$

and by Lemma 8.1.22 $i(w_1) \cdots i(w_{d-1})$ is a $(d-1)$-decomposition of a subword of $w$.

We then have by the definitions that $x_\ell^{t_1} w_0 \succ i(w_1), \ldots, i(w_{d-1})$ hence by Lemma 8.1.22 $x_\ell^{t_1} w_0 i(w_1) \cdots i(w_{d-1})$ is a $d$-decomposition of a subword of $w$ contradicting the hypothesis that $w$ is $d$-indecomposable. □

Let us now prove Lemma 8.1.22.

The relation between $\succ$ and $\succ'$ is now analyzed in the next two claims.

CLAIM 1. If $f, g \in W(X')$ and $i(f) \succ i(g)$, then $f \succ' g$.

Indeed, write $f = a_1 \cdots a_p$, $g = b_1 \cdots b_q$, where $a_i, b_i \in X'$. Since $i(f) \succ i(g)$, $g$ is not an initial of $f$, hence for some $n$, $a_i = b_i$ for $0 \leq i \leq n-1$, and $i(a_n) \succ i(b_n)$, so $a_n \succ' b_n$, and therefore $f \succ' g$ by Remark 8.1.3.

CLAIM 2. Let $f, g \in W(X')$ and let $f \succ' g$. Then either $i(f) \succ i(g)$ or $i(g)$ is an initial of $i(f)$ ("$X$-initial").

Indeed, assume $i(g)$ is not an $X$-initial of $i(f)$. Since $f \succ' g$, $g$ is not an initial of $f$ in the alphabet $X'$, so $f$ and $g$ can be decomposed as follows: $f = uar$, $g = ubs$, with $u, r, s \in W(X')$, $a, b \in X'$, and $a \succ' b$. Let $a = v_1 x_\ell^{t_1}$ and $b = v_2 x_\ell^{t_2}$.

CASE 1. $v_1 \succ v_2$. Then by Remark 8.1.3 $f \succ g$.

CASE 2. $v_1 = v_2 = v$. Then $t_1 > t_2$. Rewrite $f = uvx_\ell^{t_1} r = uvx_\ell^{t_2} x_\ell^{t_1-t_2} r$ and $g = uvx_\ell^{t_2} s$. Since $g$ is not an $X$-initial of $f$, $s \neq 1$. Now $s$ starts with some $x_i$ where $1 \leq i \leq \ell - 1$, which implies that $x_\ell^{t_1-t_2} r \succ s$, and finally $f \succ g$.

PROOF OF LEMMA 8.1.22. We have $w_1 \cdots w_q \succ' w_{\pi(1)} \cdots w_{\pi(q)}$ for every $1 \neq \pi \in S_q$, by assumption. (Hence also $w_1 \cdots w_q \neq w_{\pi(1)} \cdots w_{\pi(q)}$). Since both words have the same length and are different, one cannot be the initial of the other. By the above Claim 2 it follows that $i(w_1) \cdots i(w_q) \succ i(w_{\pi(1)}) \cdots i(w_{\pi(q)})$, hence $i(w)$ is $q$-decomposable in the original order $\succ$. □

Let $\mathcal{N} = \mathcal{N}(\ell, 2d, d) \gg 2d$ be the number determined in Proposition 8.1.21.

COROLLARY 8.1.23. *We can write every d-indecomposable word w in the form*

$$w = s_0 u_1^{k_1} v_1 s_1 u_2^{k_2} v_2 s_2 u_3^{k_3} v_3 \cdots s_{t-1} u_t^{k_t} v_t w', \tag{8.6}$$

*where* $|s_i| \leq \mathcal{N} - 2d, k_i \geq 2d,\ |u_i| = |v_i| < d, u_i \neq v_i,\ |w'| < \mathcal{N}$. *Or*

$$w = s_0 u_1^{k_1} v_1 s_1 u_2^{k_2} v_2 s_2 u_3^{k_3} v_3 \cdots s_{t-1} u_t^{k_t} w', \quad |w'| < |u_t|. \tag{8.7}$$

PROOF. In fact if $|w| < \mathcal{N}$, we take $w = w'$, otherwise write $w = w_1 w_2$ with $|w_1| = \mathcal{N}$, by Shirshov's lemma, Lemma 8.1.21, write $w_1 = s_0 u_1^{h_1} x$ with $h_1 \geq 2d$ and $1 \leq |u_1| < d$. Since $|u_1^{h_1}| \geq 2d$ we have necessarily $|s_0| \leq \mathcal{N} - 2d$. We then can write $w = s_0 u_1^{k_1} \tilde{w}$ with $k_1 \geq h_1$ and $u_1$ is not an initial word in $\tilde{w}$. If $|\tilde{w}| < |u_1|$, we take $t = 1$ and $w' = \tilde{w}$, otherwise we can write $\tilde{w} = v_1 \bar{w}$ where $|u_1| = |v_1|$, $u_1 \neq v_1$ and then repeat the procedure on $\bar{w}$. □

### 8.1.2. Proof of Shirshov's height theorem, Theorem 8.1.7.

LEMMA 8.1.24. *Let a, b be two different and comparable words (i.e., none is the initial of the other). Then* $a^{d-1}b$ *contains d different and comparable words.*

PROOF. (1) If $a \succ b$, then $ab \succ b$, hence $a^{d-1}b \succ \cdots \succ a^2 b \succ ab \succ b$.
(2) If $b \succ a$, then $b \succ ab$, hence $b \succ ab \succ a^2 b \succ \cdots \succ a^{d-1}b$.
Thus by Lemma 8.1.19, any word $w$ in which a word $u = a^{d-1}b$ has multiplicity $\geq d$ is $d$-decomposable. □

For a word $w$ let us define the $d$-height $h_d(w)$ to be the minimum $j$ such that $w$ can be written in the form $w = u_1^{k_1} u_2^{k_2} \cdots u_j^{k_j}$, $|u_i| \leq d - 1$.

Write a word $w$ as $a_1 a_2 \cdots a_k b$, where $|a_i| = d - 1$ and $0 \leq r = |b| < d - 1$. We have $|w| = k(d-1) + r$, $0 \leq r < d - 1$, is the division with remainder, and the integer part of $|w|/(d-1)$ equals to $r$. We have a trivial estimate for every word

$$\text{(i) } h_d(w) \leq |w|/(d-1) + 1 \quad \text{and} \quad \text{(ii) } h_d(w_1 w_2) \leq h_d(w_1) + h_d(w_2). \tag{8.8}$$

Let now $\mathcal{N} = \mathcal{N}(\ell, 2d, d)$ be the number determined in Proposition 8.1.21. Then Theorem 8.1.7 follows from the more precise estimate:

THEOREM 8.1.25. *If w is a d-indecomposable word, then*

$$h_d(w) \leq 3\ell^{2d} \mathcal{N}. \tag{8.9}$$

PROOF. We write $w$ in the form given by (8.6) or (8.7) as in Corollary 8.1.23.

Since the lengths of $u_i$ and $v_i$ are $\leq d - 1$, each of the terms $u_i^{k_i}$, $v_i$ contributes 1 to the height $h_d(w)$.

By (8.8)(i) we have $h_d(s_i) \leq (\mathcal{N} - 2d)/(d-1) + 1 \leq \mathcal{N}/(d-1) - 1$ and $h_d(w') \leq (\mathcal{N} - 1)/(d-1) + 1$, thus by (8.8)(ii) we arrive at the estimate

$$h_d(w) \leq 2t + 2t(\mathcal{N}/(d-1) - 1) + \mathcal{N}/(d-1) + 1 = (2t+1)\mathcal{N}/(d-1) + 1. \tag{8.10}$$

So the main point is to estimate $t$. We show next that in either cases (8.6) or (8.7),

$$t < d\,\ell^{2d}. \tag{8.11}$$

Note that the number of words of length $r$ in the alphabet $\{x_1, \ldots, x_\ell\}$ is $\ell^r$. So the number of words of length $< d$ is $\sum_{r=0}^{d-1} \ell^r = (\ell^d - 1)/(\ell - 1) < \ell^d$ (since by assumption $\ell \geq 2$). Here, for each $i$, $|u_i| = |v_i| = r_i$ for some $1 \leq r_i \leq d-1$, thus the number of possible different words $u_i^{d-1}v_i$ is $< \ell^{2d}$.

Note also that each $u_i^{k_i}v_i$ contains the subword $u_i^{d-1}v_i$.

Thus, if $t \geq d\,\ell^{2d}$, some $u_i^{d-1}v_i$ repeats at least $d$ times in $w$, therefore by Lemma 8.1.24 $w$ is $d$-decomposable, a contradiction. Combining estimates (8.10) and (8.11), we conclude (assume $\mathcal{N} > d$)

$$h_d(w) \leq (2t+2)(\mathcal{N}/(d-1)) \leq 2d\,\ell^{2d}(\mathcal{N}/(d-1)) \leq 3\ell^{2d}\mathcal{N}.$$

This completes the proof. □

## 8.2. Some applications of Shirshov's height theorem

**8.2.1. Algebraic and PI imply local finite-dimensionality.** We treat the Kurosh problem in a general setting of an algebra $R$, not necessarily with a 1 over a commutative ring $A$ (with a 1).

Recall that given an $A$-algebra $R$, an element $a \in R$ is *integral* over $A$ if there exists a nonzero monic polynomial $g(x) = g_a(x)$ over $A$ (in one variable) such that $g_a(a) = 0$. The algebra $R$ is *integral* over $A$ if every element $a \in R$ is integral.

When $A$ is a field, the condition of being monic can be dropped, and one uses the term *algebraic* rather than integral. For example, any algebra $R$ which is finite dimensional over a field $A$ is algebraic. In fact let $\dim R = d$, and let $a \in R$. Then the sequence of powers $1, a, a^2, \dots, a^d$ (or $a, a^2, \dots, a^d, a^{d+1}$ if $R$ does not have a 1) is linearly dependent, so $\sum_{j=0}^{d} \alpha_j a^j = 0$. Thus $g(a) = 0$ where $g(x) = \sum_{j=0}^{d} \alpha_j x^j$.

As a first application of Shirshov's height theorem, we now show that if $R$ is PI and also integral over $A$, one obtains a combinatorial proof of the Kurosh problem.

THEOREM 8.2.1. *A PI algebra which is finitely generated and integral over a commutative ring $A$ is a finitely generated module over $A$.*

In fact, in this approach we can prove a stronger statement.

THEOREM 8.2.2.

(1) *Let $R = A\langle a_1, \dots, a_\ell\rangle = A\langle \underline{a}\rangle$ be a finitely generated PI algebra satisfying an identity of degree $d$. If the monomials in the $a_i$ of degree $< d$ are all integral, then $R$ is a finitely generated module over $A$.*
(2) *If in $R$ every monomial in the $a_i$ of degree $< d$ is nilpotent, then $R$ is nilpotent.*

PROOF. (1) The case $\ell = 1$: In this case $R = A\langle a\rangle$ and $a$ is integral. We just saw that in this case, $R$ is a finitely generated $A$-module.

By Corollary 8.1.10 and Shirshov's height theorem (Theorem 8.1.7) there exists an integer $h$ such that $R$ is spanned by the set $\{w_1(\underline{a})^{k_1} \cdots w_j(\underline{a})^{k_h} \mid |w_i| < d,\ j \leq h\}$ over $A$. In fact more is true, that is a given monomial $w(a)$ is a linear combination of monomials of the previous type of *the same degree* $\sum_i k_i|w_i| = |w|$ as $w(a)$.

Now each $w_i(\underline{a})$, with $|w_i| < d$, is by assumption integral of some degree $q_i$, hence each $w_i(\underline{a})^{k_i}$ is a linear combination of elements $w_i(\underline{a})^c$ with $c \leq q_i - 1$. It follows that $R$ is spanned over $A$ by the finite set

$$\{w_1(\underline{a})^{c_1} \cdots w_h(\underline{a})^{c_h} \mid |w_i| \leq d-1, \quad c_i \leq q_i - 1\}.$$

(2) Let $k$ be such that all monomials $w_i(a)$ in the generators $a_1, \dots, a_\ell$ with the length $w_i$ less than or equal to $d-1$ have $w_i(a)^k = 0$.

Then, with $h$ as before, take a monomial $w(a)$ of degree $> h(k-1)(d-1)$. When we write it as a linear combination of words of type (8.2), we have that in each one of these words $h(d-1)\max(k_i) \geq \sum_j (d-1)k_j > h(d-1)(k-1)$, so we have that one of the exponents $k_i \geq k$, which implies that each of these words equals to 0. $\square$

REMARK 8.2.3. We have a subalgebra $B$ of $A$, finitely generated by the coefficients of the monic polynomials satisfied by the elements $w_i(\underline{a})$, so that in fact the subalgebra $B\langle a_1, \ldots, a_\ell\rangle$ is a finitely generated module over $B$.

A useful fact is also

PROPOSITION 8.2.4. *Let $R = A\langle a_1, \ldots, a_\ell\rangle$ be a finitely generated algebra over a commutative Noetherian ring $A$ and also a finite module over its center $Z$. Then $Z$ is also a finitely generated algebra over $A$.*

PROOF. Let $u_1, \ldots, u_m$ be linear generators of $R$ over $Z$.

We have $u_i u_j = \sum_k \alpha_{i,j}^k u_k$, and $a_i = \sum_k \beta_i^k u_k$ with $\alpha_{i,j}^k, \beta_i^k \in Z$. Then the algebra $R$ is also a finite module over the finitely generated subalgebra $B = A[\alpha_{i,j}^k, \beta_i^k] \subset Z$. Thus by the usual Noetherian property $Z$ also is a finite module over $B$. $\square$

One should compare this result with the Dubnov–Ivanov–Nagata–Higman theorem, discussed in §12.2.5 which states that an algebra $R$ in which every element is nilpotent of fixed degree $n$ is nilpotent, provided the characteristic is $> n$ (see **[For90]** for a survey).

**8.2.2. The trace ring.** There is an important consequence of this theory. Let $S := M_m(B)$ be the ring of $m \times m$ matrices over some commutative ring $B$. Consider a ring $A \subset B$, (for instance $\mathbb{Z}1$) and the subalgebra $R = A\langle a_1, \ldots, a_k\rangle$ of $S$ generated by some elements $a_i \in S$. Consider the ring $T$, the *trace ring of $R$* generated over $A$ by all the coefficients of the characteristic polynomials of all monomials in the elements $a_i$.

THEOREM 8.2.5. *$T$ is a finitely generated algebra over $A$ and $TR = T\langle a_1, \ldots, a_k\rangle$ is finitely generated as module over $T$.*

PROOF. Consider first the ring $T'$ generated over $A$ by all the coefficients of the characteristic polynomials of all monomials in the elements $a_i$ of degree $\leq 2m-1$. This ring is finitely generated over $A$, and all monomials of degree $\leq 2m-1$ in the elements $a_i$ satisfy a monic polynomial with coefficients in $T'$, that is their characteristic polynomial, by construction. The ring $RT$ satisfies the standard identity of degree $d = 2m$, thus, by Theorem 8.2.2, the ring $RT'$ is a finitely generated module over $T'$ spanned by the monomials of some bounded degree $\leq N$ in the elements $a_i$. Now if we take any element $u \in R$, write $u = \sum_j t_j M_j$ where $t_j \in T'$ and $M_j$ is a monomial of degree $\leq N$. By Amitsur's theorem (Theorem 3.3.8) the characteristic polynomial of $u$ is a specialization of the characteristic polynomial of $\sum_i \lambda_i \xi_i$ where the $\lambda_i$ are indeterminates and $\xi_i$ are generic matrices. Thus it is a polynomial in the elements $t_j$ and the coefficients of the characteristic polynomials of monomials of degree $\leq m$ in the $M_j$.

We thus deduce that $T$ is generated by all the coefficients of the characteristic polynomials of all monomials in the elements $a_i$ of degree $\leq mN$, this ring is finitely generated over $A$. $\square$

REMARK 8.2.6. This theorem applies in particular to $R = \mathbb{Z}\langle\xi_1,\dots,\xi_k\rangle$ the ring of generic $m \times m$ matrices over $\mathbb{Z}$.

In fact when we work over $\mathbb{Q}$, that is in characteristic 0, we have a more precise result which will be proved in the Razmyslov theorem (Theorem 12.2.13). That is $T$ is generated by the elements $\mathrm{tr}(M)$ where $M$ runs over monomials of degree $\leq m^2$ and $RT$ is also spanned over $T$ by monomials of degree $< m^2$. In fact this is conjecturally not the best estimate.

REMARK 8.2.7. By the Donkin theorem (Theorem 3.3.29), the ring generated by the coefficients of the characteristic polynomials of all monomials in the generic matrices $\xi_i$, when the coefficient ring is $\mathbb{Z}$ or a field $F$, is in fact the full ring of invariants under conjugation.

When $F$ is a field of characteristic 0, this follows fairly easily by classical invariant theory, but in positive characteristic or over $\mathbb{Z}$ one needs the representation theory of the linear and symmetric group which are no more linearly reductive (see 14.1.6), and thus the representation theory is much more complex.

## 8.3. Gel'fand–Kirillov dimension

**8.3.1. Dimension.** The idea of dimension in mathematics is quite general—in an algebraic approach this usually reflects in some form some maximum number of *independent parameters*.

For an associative algebra $A$ with a 1 which is finitely generated, let $V \subset A$ be a finite-dimensional subspace such that $A$ is generated by $V$. Let $V_n$ denote the span of all products of $n$ elements of $V$ (for $n = 0$ this is the multiples of 1), and set $d_V(n) := \dim_F V_n$.

It is useful to pass to the associated graded algebra

$$\mathrm{Gr}(A) := \bigoplus_{i=0}^{\infty} A_i,\ A_i := V_i/V_{i-1},$$

where by definition if $A$ has a 1, we set $A_0 = F$.

Then one can compute $d_V(n) = \sum_{i=1}^{n} \dim_F V_i/V_{i-1}$.

Consider a multigraded vector space $M = \bigoplus_{\underline{i}\in\mathbb{Z}^k} M_{\underline{i}}$, over a field $F$, such that $\dim_F M_{\underline{i}} < \infty, \forall \underline{i}$. It is convenient to express the dimension of $M$ in each multidegree as a formal series,

$$\mathrm{h}_M(\underline{t}) := \sum_{\underline{i}\in\mathbb{Z}^k} \dim_F(M_{\underline{i}})\underline{t}^{\underline{i}},\quad \underline{i} = (i_1,\dots,i_k),\ \underline{t}^{\underline{i}} := \prod_{j=1}^{k} t_i^{i_j}, \tag{8.12}$$

called a *graded dimension* or a Hilbert–Poincaré series.

Usually this is applied to the case where in fact $M_{\underline{i}} = 0$ if $\underline{i} \notin \mathbb{N}^k$ or at least $M_{\underline{i}} = 0$ for all but a finite number of $\underline{i} \notin \mathbb{N}^k$.

Under this hypothesis this has the advantage that it is compatible with the direct sum and tensor product of graded spaces, that is

$$\mathrm{h}_{M_1\oplus M_2}(\underline{t}) = \mathrm{h}_{M_1}(\underline{t}) + \mathrm{h}_{M_2}(\underline{t}),\quad \mathrm{h}_{M_1\otimes M_2}(\underline{t}) = \mathrm{h}_{M_1}(\underline{t})\mathrm{h}_{M_2}(\underline{t}). \tag{8.13}$$

One main example is when $M$ is a finitely generated graded module over a commutative finitely generated graded algebra $A = F[a_1,\dots,a_m]$, $\deg a_i = h_i$.

Of course $M$ is also a module over the polynomial ring $F[y_1,\dots,y_m]$, with $\deg y_i = h_i$. In order to understand this case, let us remark that $F[y_1,\dots,y_m] = F[y_1]\otimes\cdots\otimes F[y_k]$. Moreover if $y$ is a variable of degree $h$, we have

$$\mathrm{h}_{F[y]}(t) = \sum_i t^{hi} = \frac{1}{1-t^h} \overset{(8.13)}{\Longrightarrow} \mathrm{h}_{F[y_1,\dots,y_m]}(t) = \prod_{i=1}^m \frac{1}{1-t^{h_i}}. \tag{8.14}$$

EXAMPLE 8.3.1.

$$\frac{1}{(1-t)^h} = \sum_{n=0}^{\infty} \binom{h+n-1}{n} t^n. \tag{8.15}$$

In fact the function of formula (8.15) is the generating function for the dimension of the polynomial ring in $h$ variables, so let us observe that the number of monomials in the variables $y_i$ of degree $n$ is

$$\binom{h+n-1}{n} = \frac{(h+n-1)(h+n-2)\cdots(n+1)}{(h-1)!},$$

a polynomial in $n$ of degree $h-1$.

This can also be verified in a simple combinatorial way. We have to count the number of all possible choices of $h$ numbers $k_i$, $i=1,\dots,h$ with $0\le k_i$ and $\sum_i k_i = n$ or equally the number of all possible choices of $h-1$ numbers $k_i$, $i = 1,\dots,h-1$ with $0\le k_i$ and $\sum_i k_i \le n$. If we choose $h-1$ distinct elements in the set $1,2,\dots,n+h-1$, we list them in increasing order and see they can be uniquely written as $1+k_1, 2+k_1+k_2,\dots,h+k_1+k_2+\cdots+k_{h-1}$ with $0\le k_i$ and $h-1+k_1+k_2+\cdots+k_{h-1}\le h+n-1 \iff k_1+k_2+\cdots+k_{h-1}\le n$.

We then have the following

THEOREM 8.3.2. *If $M$ is a graded finitely generated module over the polynomial ring $F[y_1,\dots,y_m]$, $\deg y_i = h_i$, we have*

$$\mathrm{h}_M(t) = \frac{q(t,t^{-1})}{\prod_{i=1}^m (1-t^{h_i})}, \quad q(t,t^{-1})\in\mathbb{Z}[t,t^{-1}]. \tag{8.16}$$

PROOF. The proof is by induction on $m$. If $m=0$, then $M$ is finite dimensional, it has a basis $v_1,\dots,v_u$ of vectors with some degrees $c_i\in\mathbb{Z}$, and its Hilbert series is $\sum_{i=1}^u t^{c_i}\in\mathbb{Z}[t,t^{-1}]$. Otherwise consider the exact sequence

$$0 \longrightarrow M_0 \longrightarrow M \xrightarrow{y_m} M \longrightarrow M_1 \longrightarrow 0,$$

the middle map is multiplication by $y_m$. We have that $y_m$ acts as 0 on the two spaces $M_0 = \ker y_m$, and $M_1 = M/y_mM$, hence $M_0, M_1$ are both finitely generated modules over $F[y_1,\dots,y_{m-1}]$ for which we apply induction.

One easily sees that

$$0 = \mathrm{h}_{M_0}(t) - \mathrm{h}_M + t^{h_m}\mathrm{h}_M(t) - \mathrm{h}_{M_1}(t) \implies \mathrm{h}_M(t) = \frac{1}{1-t^{h_m}}(\mathrm{h}_{M_0} - \mathrm{h}_{M_1}),$$

which by induction implies the claim. □

Notice that if $M$ is generated by elements in degree $\ge 0$, then the Laurent polynomial $q(t,t^{-1}) = q(t)\in\mathbb{Z}[t]$ is a polynomial.

DEFINITION 8.3.3. The function $c_k$ with generating series

$$\prod_{j=1}^{N}(1-t^{h_j})^{-1}=\prod_{j=1}^{N}(\sum_{i=0}^{\infty}t^{ih_j})=\sum_{k=0}^{\infty}c_kt^k$$

is classically known as a *partition function*.

We see that $c_k$ is a positive integer which counts in how many ways the integer $k$ can be written in the form $k=\sum_{j=1}^{N}i_jh_j,\ i_j\in\mathbb{N}$.

Such a function $c_k$ in fact coincides, on the set of positive integers, with a *quasi-polynomial* of degree $N-1$.

DEFINITION 8.3.4. A function $f:\mathbb{Z}\to\mathbb{C}$ is a *quasi-polynomial* if there is a modulus $m\in\mathbb{N}$ such that $f$ is a polynomial on each coset modulo $m$.

We now give the following definition of dimension.

DEFINITION 8.3.5. We define as *dimension* for a rational function as in formula (8.16) to be the order of the pole of this function at $t=1$. For a graded module $M$, we take as dimension written $\dim M$ to be the dimension of $\mathtt{h}_M(t)=\sum_{k=0}^{\infty}c_kt^k,\ c_k:=\dim_F M_k$.

PROPOSITION 8.3.6.

(1) *Let* $H(t)=\sum_{k=0}^{\infty}c_kt^k$ *be equal to a rational function* $\frac{p(t)}{\prod_{i=1}^{m}(1-t^{h_i})}$ *with* $p(t)\in\mathbb{Z}[t]$ *some polynomial. For* $k$ *sufficiently large the function* $c_k$ *is a quasi-polynomial, in fact a polynomial on each coset modulo the least common multiple* $m$ *of the* $h_i$.

(2) *If* $c_k\geq 0$ *for* $k\gg 0$, *then the dimension of* $H(t)$ *equals the (maximum) degree of these polynomials plus* 1.

(3) *If we set* $d(k):=\sum_{i=0}^{k}c_i$, *we have* $\sum_{k=0}^{\infty}d(k)t^k=\frac{H(t)}{1-t}$ *which has at* 1 *a pole of order* $\lim_{k\to\infty}\frac{\log(d(k))}{\log(k)}=d$.

PROOF. (1) Let $m$ be the least common multiple of the integers $h_i$ so that

$$m=h_ik_i,\implies(1-t^m)=(1-t^{h_i})(\sum_{j=0}^{k_i-1}t^{h_ij}).$$

It then follows that $H(t)$ can be written in the form $\frac{P(t)}{(1-t^m)^d}$ with $P(t)\in\mathbb{Z}[t]$ some polynomial. Since for $i>0$ we have

$$\frac{t^m}{(1-t^m)^i}=\frac{1-(1-t^m)}{(1-t^m)^i}=\frac{1}{(1-t^m)^i}-\frac{1}{(1-t^m)^{i-1}},$$

it then follows that $H(t)$ has an expansion in partial fractions

$$H(t)=\sum_{i=1}^{d}\frac{p_i(t)}{(1-t^m)^i}+q(t)$$

with all the $p_i(t)$ polynomials of degree $<m$ in $t$ and $q(t)$ a polynomial.

Then we remark that the generating function of

$$\frac{t^j}{(1-t^m)^i}\overset{(8.15)}{=}\sum_{k=0}^{\infty}\binom{i+k-1}{i-1}t^{mk+j},\ 0\leq j<m,$$

gives a polynomial on the coset $\mathbb{Z}m + j$, with positive values, of degree $i-1$.

(2) It follows that after a finite number of steps (given by the polynomial $q(t)$) the fucntion $c_k$ is a polynomial on each coset, and of degree $d-1$ on some cosets where the coefficients of $p_d(t)$ are different from 0. If $c_k$ is definitely positive, it follows that all the coefficients of $p_d(t)$ are nonnegative, in particular $p_d(1) \neq 0$ and the order of the pole at 1 of $H(t)$ is clearly $d$.

(3) On the other hand,

$$d(k) := \sum_{i=0}^{k} c_i \implies \sum_{k=0}^{\infty} d(k)t^k = (\sum_{k=0}^{\infty} c_k t^k)(\sum_{k=0}^{\infty} t^k) = \frac{H(t)}{1-t}$$

is clearly a quasi-polynomial of degree $d$ on each coset. Then given a polynomial $d(k) = a\prod_{i=1}^{d}(k-b_i)$, we have

$$\log(d(k)) = \log a + \sum_{i=1}^{d} \log(k-b_i) \implies \lim_{k\to\infty} \frac{\log(d(k))}{\log(k)} = d. \tag{8.17}$$

Hence also if $d(k) := \sum_{i=0}^{k} c_i$ is a quasi-polynomial of degree $d$, we have

$$\lim_{k\to\infty} \frac{\log(d(k))}{\log(k)} = d. \qquad \square$$

In fact for the partition function one has more precise results and there is an extensive literature on such functions (cf. **[DCP11]**).

For a finitely generated commutative algebra $R$, the dimension can be defined as the dimension of the associated graded algebra, as computed by the Hilbert series. It is easy to see that this dimension is independent from the associated graded algebra, which instead depends on the choice of the generators. The dimension has a geometric meaning and equals the dimension of its associated affine variety $V(R)$ (which can be defined in several different ways), in particular as *Krull dimension*; see Definition 15.1.7. In particular if a finitely generated commutative algebra $R$ over a field $F$ is also an integral domain, it easily follows that its dimension equals the transcendence degree over $F$ of its field of fractions.

For a module $M$ we have the notion of *support of* $M$, that is the set of point $p \in V(R)$ where $M$ is not zero, that is $M \otimes_R R_p \neq 0$ where $R_p$ is the local ring at $p$. Then the dimension of $M$ equals the dimension of its support. The support is computed as follows

REMARK 8.3.7. It is well known that a finitely generated module $M$ over a commutative Noetherian ring $R$ has a finite filtration $0 \subset M_1 \subset M_2 \subset \cdots \subset M_k = M$ such that $M_{i+1}/M_i = R/P_i$ where $P_i$ is a prime ideal.

If $R, M$ are graded and $R$ is finitely generated, the $M_i$ can be taken graded and the Hilbert series is the sum of the Hibert series of the $R/P_i$, the dimension is the maximum of the dimensions of the $R/P_i$.

In the language of algebraic geometry, $R$ defines the algebraic variety $V(R)$ and $P_i$ some subvarieties with coordinate rings $R/P_i$. One sees by induction that $M$ is *supported* on the union of these subvarieties, and thus its dimension is the dimension of its support.

If $M$ is torsion free over a domain $R$, then it contains a free module $R^k \subset M$ so that $M/R^k$ is supported in a proper subvariety so it has lower dimension and the dimension of $M$ equals that of $R$.

The concept of dimension as presented for commutative algebras can be generalized; a useful approach is due to Gel'fand and Kirillov [**GK66**]. Even starting from a graded finitely generated associative algebra $A$ its Hilbert series need not be a rational function, or one of type (8.16). For such algebras one can try to use as definition the formula in Proposition 8.3.6(3). For instance for the free algebra in $m$ generators, its dimension in degree $n$ is $m^n$ and one obtains as dimension $\infty$. In general, mimicking Proposition 8.3.6(3), one proceeds as follows.

For an associative algebra $A$ with a 1, over a field $F$, let $V \subset A$ be a finite-dimensional subspace with $1 \in V$. Let $V_n$ denote the span of all products of $n$ elements of $V$ and set $d_V(n) := \dim_F V_n$.

DEFINITION 8.3.8 (Gel'fand–Kirillov dimension (GK dimension)).

$$\text{Dim } A := \sup_V \limsup_{n\to\infty} \log d_V(n)/\log n. \tag{8.18}$$

PROPOSITION 8.3.9. *If $A$ is generated by $V$, then*

$$\text{Dim } A = \limsup_{n\to\infty} \log d_V(n)/\log n.$$

PROOF. Let $W$ be any finite-dimensional subspace of $A$. Then for some $k$ we have $W \subset V_k$, hence $W_n \subset V_{kn}$. We have

$$\begin{aligned}\limsup_{n\to\infty} \log d_W(n)/\log n &\leq \limsup_{n\to\infty} \log d_V(kn)/\log n \\ &= \limsup_{n\to\infty} \log d_V(kn)/\log(kn) \leq \limsup_{n\to\infty} \log d_V(n)/\log n.\end{aligned}$$

Thus, finally, the sup (over all $W$) of the numbers $\limsup_{n\to\infty} \log d_W(n)/\log n$ equals, as claimed, $\limsup_{n\to\infty} \log d_V(kn)/\log n$. □

If $A$ is generated by $V$, then it is useful to pass to the associated graded algebra

$$\text{Gr}(A) := \bigoplus_{i=0}^{\infty} A_i, \; A_i := V_i/V_{i-1},$$

where by definition if $A$ has a 1, we set $A_0 = F$. Then one can compute $d_V(n) = \sum_{i=1}^n \dim_F V_i/V_{i-1}$.

PROPOSITION 8.3.10. *If the Hilbert series of* $\text{Gr}(A)$ *is a rational function of the type given by formula* (8.16), *then the GK dimension of $A$ equals the order of the pole at* 1 *of this rational function.*

PROOF. This follows from Proposition 8.3.6(3) and the definition of GK dimension. □

We leave to the reader to prove the following fact which will be used later:

PROPOSITION 8.3.11. *Let $A$ be a commutative domain over a field $F$, and let $R$ be an algebra which is a finite torsion free module over $A$. Then the GK dimension of $A$ and $R$, given by formula* (8.18), *coincides with the transcendence degree over $F$ of the field of fractions of $A$.*

In noncommutative algebra usually things are much more complicated than in the commutative case. The function $\dim_F V_i/V_{i-1}$ may have exponential growth, as for the free algebra. This as we shall see does not happen for PI algebras, where on the other hand the Hilbert series need not be a rational function (cf. examples in [**BBL97**]). The dimension need not be an integer, in fact in [**BK76**] they exhibit algebras of any real dimension $\gamma \geq 2$. For a finitely generated PI algebra, the Krull dimension and the GK dimension are both finite, but the GK dimension is almost never equal to the Krull dimension which instead equals the GK dimension of $R/J$ where $J$ is the nilpotent radical.

EXERCISE 8.3.12. Consider in the free algebra $F\langle x,y\rangle$ the ideal $Y$ generated by $y$, so that $F\langle x,y\rangle/Y = F[x]$. Prove that the algebra $F\langle x,y\rangle/Y^n$ is PI, with radical $Y/Y^n$ and GK dimension $n$.

*Hint.* Compute the dimension in each degree, $k$ as $\sum_{i=0}^{\min(n-1,k)} \binom{k}{i}$.

We shall see in Corollary 18.1.4 that the Hilbert series is a rational function for the relatively free algebras in finitely many variables for any variety of PI algebras, and we shall study its dimension.

Let $A = F\langle a_1, \dots, a_\ell\rangle$ be a finitely generated algebra satisfying a polynomial identity of degree $d$. Let $U(d,\ell)$ denote the set of monomials of degree $\leq d-1$ in $a_1, \dots, a_\ell$. By Shirshov's height theorem (Theorem 8.1.7) there exist $h$ (depending only on $\ell$ and the degree $d$ of a minimal identity) such that $A$ is spanned by the elements

$$\{u_1^{k_1} \cdots u_j^{k_j} \mid u_i \in U(d,k),\ j \leq h\}. \tag{8.19}$$

In fact a monomial in the generators $a_i$ of some degree $n$ can be expressed as a linear combination of elements of type (8.19) of the same degree $n$.

PROPOSITION 8.3.13. *The number of monomials of type* (8.19) *of degree n is bounded by* $Cn^h$, *where C is a fixed constant. The GK dimension of A is* $\leq h$.

PROOF. The second part is clearly a consequence of the first, since by (8.19) if $V = \operatorname{span}\{a_1, \dots, a_\ell\}$, we see that $\dim_F V_n$ is less or equal that the number of monomials of formula (8.19).

This number is less than or equal to the sum over all possible sequences $u_1, \dots, u_j$ of the dimension of the polynomials of degree $\leq n$ of the polynomial ring in $j \leq h$ variables so it is less than or equal to a constant times the dimension of the space of polynomials of degree $\leq n$ of the polynomial ring in $h$ variables. The claim follows from Example 8.3.1 and Proposition 8.3.6. □

In general an algebra of finite GK dimension may be far from PI, as for instance the Weyl algebras or most universal enveloping algebra of Lie algebras. An exception is when the GK dimension is 1. Then by a result of L. W. Small, J. T. Stafford, and R. B. Warfield [**SSW85**], one has that a finitely generated algebra over a field of GK dimension 1 satisfies a polynomial identity.

**8.3.2. Cocharacter-multiplicities are polynomially bounded.** Here we apply Shirshov's height theorem (Theorem 8.1.7) to prove the following theorem of A. Berele [**BR83**]. Recall that $H(k,0;n)$ is the set of partitions $\lambda$ of $n$ satisfying $\mathrm{ht}(\lambda) \leq k$.

Let $A$ be a finitely generated PI algebra with $\chi_n(A) = \sum_{\lambda \vdash n} m_\lambda(A)\chi_\lambda$ its cocharacter. From Theorem 7.6.1 one has that it satisfies a Capelli identity, so all its cocharacters are contained in a strip. From the next theorem it follows that all the multiplicities $m_\lambda(A)$ in its cocharacters are polynomially bounded. In fact we shall prove a much stronger theorem stating that the generating series $\sum_{\lambda,\ ht(\lambda)\le k} m_\lambda t^\lambda$ is a nice rational function; see Theorem 18.3.1.

In §8.3.3 we extend Theorem 8.3.14 to show that for any PI algebra $A$, all its cocharacter-multiplicities are polynomially bounded; see Theorem 8.3.15.

THEOREM 8.3.14. *For any fixed $k$ the multiplicities $\{m_\lambda(A) \mid \lambda \in H(k,0;n)\}$ are polynomially bounded: there exist a constant $C = C(k)$ and a power $p = p(k)$ such that for all $n$ and for all $\lambda \in H(k,0;n)$, $m_\lambda(A) \le Cn^p$.*

PROOF. Given $k$, choose a vector space $V$ with $\dim_F V = k$. By Corollary 6.3.11 and Proposition 7.1.2 it follows that as representation of $\mathrm{GL}(V)$, we have

$$\frac{V^{\otimes n}}{V^{\otimes n} \cap \mathrm{Id}(A)} \simeq \bigoplus_{\lambda \in H(k,0;n)} m_\lambda(A) \cdot S_\lambda(V)$$

with the same multiplicities $m_\lambda(A)$ as in the cocharacter of $A$—whenever $\lambda \in H(k,0;n)$. Thus, in order to prove Theorem 8.3.14, it suffices to show that the sequence of dimensions $\dim_F \left(V^{\otimes n}/(V^{\otimes n} \cap \mathrm{Id}(A))\right)$ is polynomially bounded.

Recall that $T(V) = \bigoplus_n V^{\otimes n} \equiv F\langle x_1, \dots, x_k\rangle$. Let $I = T(V) \cap \mathrm{Id}(A)$, which is a two-sided ideal in $T(V)$.

Since $\mathrm{Id}(A)$ is homogeneous, $I = \bigoplus_n I_n$ where $I_n = V^{\otimes n} \cap \mathrm{Id}(A)$. Then

$$\frac{T(V)}{I} \cong \bigoplus_n \frac{V^{\otimes n}}{I_n}, \quad \text{namely} \quad \frac{T(V)}{T(V) \cap \mathrm{Id}(A)} \cong \bigoplus_n \frac{V^{\otimes n}}{V^{\otimes n} \cap \mathrm{Id}(A)}.$$

The graded algebra $\frac{T(V)}{I}$ satisfies a PI and it is finitely generated by a basis $x_i$, $i = 1, \dots, k$ of $V$. Therefore the claim follows from Proposition 8.3.13. □

**8.3.3. Polynomially bounded multiplicities—the general case.** The following theorem extends Theorem 8.3.14 and is proved in [**BR83**].

THEOREM 8.3.15 ([**BR83**]). *For any PI algebra $A$, all its cocharacter-multiplicities $m_\lambda(A)$ are polynomially bounded: there exist a constant $C$ and a power $p$ such that for all $n$ and $\lambda \vdash n$, $m_\lambda(A) \le Cn^p$.*

This follows from Lemma 19.4.20 which gives the more precise Theorem 19.4.21 that the colength sequences (Definition 7.1.1(3)) are polynomially bounded.

This will be a part of a general theory, to be developed in Chapters 19 and 20, leading to the fact that any PI algebra is PI equivalent to the Grassmann envelope of a finite-dimensional superalgebra (Theorem 19.8.1).

CHAPTER 9

# $2 \times 2$ matrices

*We want to discuss in detail the example of $2 \times 2$ matrices. This is the only case for matrix algebras, apart from the trivial case of scalars, where the description of the free algebra (Definition 2.2.18), of the trace algebra (Definition 2.4.6), and their centers in any number of variables can be completely performed.*

*Most of our computations hold in all characteristics not 2 or 3, although the theory of invariant ideals is developed only in characteristic 0.*

## 9.1. $2 \times 2$ matrices

**9.1.1. A special case.** We start from the very special case of $m = 2$, $2 \times 2$ matrices (cf. **[FHL81]**), which we leave as exercise. The reader will be able to perform it after reading some of this chapter.

EXERCISE 9.1.1. The ring of invariants of two $2 \times 2$ matrices $X, Y$ is the polynomial ring in five generators

$$\operatorname{tr}(X),\ \det(X),\ \operatorname{tr}(Y),\ \det(Y),\ \operatorname{tr}(XY). \tag{9.1}$$

The ring of equivariant maps of two $2 \times 2$ matrices $X, Y$ to $2 \times 2$ matrices is a free module over the ring of invariants with basis

$$1, X, Y, XY. \tag{9.2}$$

The multiplication table is given by Cayley and Hamilton (cf. formula (12.17)):

$$X^2 = \operatorname{tr}(X)X - \det(X), \quad Y^2 = \operatorname{tr}(Y)Y - \det(Y),$$

$$YX = -XY + \operatorname{tr}(X)Y + \operatorname{tr}(Y)X + \operatorname{tr}(XY) - \operatorname{tr}(X)\operatorname{tr}(Y). \tag{9.3}$$

As soon as we want to treat $m > 2$, $2 \times 2$ matrices $X_1, \dots, X_m$ the picture is much more complicated, but it can be treated by combinatorial methods as we plan to show in this chapter.

**9.1.2. The plan of the chapter.** A basic question of symbolic calculus is that of finding a *normal form* for the symbolic expressions involved. In PI theory this means to describe some explicit linear basis of a given relatively free algebra and also the rule to express any given expression in terms of this basis.

Such a rule gives in particular a rule to multiply the given basis elements.

The simplest example is for the relatively free algebras of commutative algebras. These are the polynomial rings which have the usual basis of ordered monomials. In this case the commutative law tells us immediately how to reorder any monomial.

It turns out that very few other examples can be fully developed; an example is the Grassmann algebra, Theorem 19.1.5. A major problem is to develop a similar calculus for matrices, that is to find an explicit basis for the ring of generic matrices.

This appears to be a basically impossible task for $n \times n$ matrices except for the case $n = 2$; see Corollaries 9.1.21 and 9.1.24. In this case the bases are deduced from methods of representation theory; cf. the final formulas (9.58) and (9.59).

In particular behind the success of the method are two facts. One is the general fact that irreducible polynomial representations of the linear group $\mathrm{GL}(m, F)$ have bases indexed by semistandard tableaux. This we apply to the decomposition of the free algebra in $m$ variables as representation of the group $\mathrm{GL}(m, F)$ which acts linearly on the variables. This of course is quite general and holds also for other relatively free algebras. What makes the case of $2 \times 2$ matrices special is the fact that the algebra generated by $m$ generic $2 \times 2$ matrices, with trace 0 which we denote by $\tilde{V}_m$, has the property that the irreducible representations of $\mathrm{GL}(m, F)$ which in fact appear with multiplicity 1. This makes an enormous difference since it makes all the structure combinatorial.

The phenomenon of having algebras with the action of some algebraic group under which they decompose into irreducible representations with multiplicity 1 belongs to a very large and complex area of algebraic groups, called the theory of *spherical varieties*; the computations here are clearly inspired by computations of this type. These computations have a deep geometric meaning; see **[Bri12]**, **[DCEP82]**, **[DCEP80]**. Our case is one of the simplest of this theory, so we hide the geometry behind it. Except for a twist which is unusual in algebraic groups, our algebras are noncommutative, so the rules of multiplication are more complicated and not entirely geometric.

The final point is that we can also describe explicitly the *conductor* of the algebra of generic matrices inside its trace algebra, that is the maximal common two-sided ideal; see Theorem 9.3.1.

The multiplicity 1 phenomenon depends on two special properties of $2 \times 2$ matrices. The first, if $A, B$ are two $2 \times 2$ matrices of trace 0, we have $\mathrm{tr}(AB) = AB + BA$, a phenomenon unique to this case. From this in Corollary 9.1.16 it follows that the algebra $\tilde{V}_m$ is already closed under trace. The second is that the three-dimensional space of trace 0 matrices, as representation of $\mathrm{GL}(3, F)$, under conjugation is really the fundamental representation of $\mathrm{SO}(3, F)$ for which the invariant theory is classical.

In order to apply this to the algebra of generic $2 \times 2$ matrices, which is connected in a simple but nontrivial way to $\tilde{V}_m$, we need some control on ideals of $\tilde{V}_m$ which are invariant under linear substitution of variables. In fact we could use only a piece of this theory, but we have decided to discuss it in full (although this occupies several pages of lengthy computations in §9.2) since it can then be applied to several possible developments, as for instance the study of $T$-ideals containing the $T$-ideal of identities of $2 \times 2$ matrices.

As we shall see in Chapter 17 and in particular in Theorem 18.1.2, for a general relatively free algebra associated to a finite-dimensional algebra there is a natural sequence of $T$-ideals which plays the role of a sequence of conductors. If one could have a grasp on these ideals similar to what we do here, one would have a complete understanding of these free algebras. It is in fact rather clear that the complexity of the problem increases in an untreatable way, so even for $3 \times 3$ matrices we may never be able to attain such a control.

For $2 \times 2$ matrices instead this sequence is quite simple. It has only three steps: the entire ring $R_m$ of generic matrices, the ideal $[R_m, R_m]$ generated by commutators, and finally the conductor in its trace algebra which is the ideal $[[R_m, R_m], R_m]$ generated by triple commutators. The graded pieces of this simple filtration have a very precise combinatorial description, given in formula (9.59), which allows us to compute the basic invariants of this setting, the Hilbert series, the codimensions and the cocharacters.

### 9.1.3. Some classical invariant theory.

9.1.3.1. *Notations.* For simplicity let us use the following notations: $F$ is a field of characteristic $\neq 2, 3$; and denote by $M := M_2(F)$ the space of $2 \times 2$ matrices over $F$ and by $M_0$ the space of trace 0 matrices. As usual the linear group $G := \mathrm{GL}(2, F)$ acts by conjugation on these two spaces. Therefore it acts also on the spaces of $m$-tuples, $M^m, M_0^m$ of matrices (resp., trace 0 matrices), and thus on polynomial functions on these spaces. We shall use systematically the algebras of invariant polynomial functions, called for short *invariants of matrices.*

Our goal is to describe the ring $F\langle X_1, X_2, \dots, X_m\rangle$ of $m$ generic $2 \times 2$ matrices, which for simplicity we shall denote by $R_m$.

In order to do this, we first study the algebras $T_m = S[(M^m)^*]^G$ of invariants of $m$-tuples of matrices (where as usual the symmetric algebra $S[(M^m)^*]$ is the algebra of polynomial functions on $M^m$), and the *trace algebra* $V_m$, that is the algebra of $G$ equivariant maps $M^m \to M$ or $V_m = (S[(M^m)^*] \otimes M)^G$ (see §3.3.4.1). Since we assume that the characteristic of $F$ is not 2, we have a decomposition $M = F \oplus M_0$ where as usual we identify $F$ with the scalar matrices. Thus a matrix $X$ is decomposed as $X = \mathrm{tr}(X)/2 + Y$ with $\mathrm{tr}(Y) = 0$, we use the notation $X_i = \mathrm{tr}(X_i)/2 + Y_i = t_i + Y_i$. The main computation is in describing the ring generated by $m$ generic trace 0 matrices $Y_i$ which turns out to have a very special structure, cf. Corollary 9.1.16. Summarising, we have

DEFINITION 9.1.2.

(1) $R_m$ denotes the algebra of $m$ generic $2 \times 2$ matrices.
(2) $T_m := S[(M^m)^*]^G$ denotes the commutative algebra of invariants of $m$ generic $2 \times 2$ matrices.
(3) $V_m := T_m R_m = (S[(M^m)^*] \otimes M)^G$ denotes the noncommutative algebra of $m$ generic $2 \times 2$ matrices and their traces.
(4) $T_m^0 := S[(M_0^m)^*]^G$ denotes the commutative algebra of invariants of $m$ generic $2 \times 2$ matrices of trace 0.
(5) $\tilde{V}_m := [S[(M_0^m)^*] \otimes M]^G$ is the main object of this paragraph and is described in Corollary 9.1.16. Since $M = M_0 \oplus F$, we decompose

$$\tilde{V}_m := V_m^0 \oplus T_m^0, \quad V_m^0 := [S[(M_0^m)^*] \otimes M_0]^G. \tag{9.4}$$

Since $M^m = M_0^m \oplus F^m$ where $F^m$ is the trivial representation, we deduce that

COROLLARY 9.1.3. *The ring $T_m$ of invariants of $m$ matrices $X_i \in M$ is the polynomial ring $T_m^0[t_1, \dots, t_m]$ in the $m$ variables $t_i := \mathrm{tr}(X_i)/2$ over the ring $T_m^0$ of invariants of $m$-matrices $Y_i$ of trace 0.*

We have stated the equality $V_m = T_m R_m = (S[(M^m)^*] \otimes M)^G$ for the algebra $V_m$. This is a general fact proven in Theorem 12.1.5 for all matrices in characteristic 0, that is:

*The algebra of equivariant maps from matrices to matrices equals the algebra generated by the generic matrices and their traces.*

In characteristic not 2 and for $2 \times 2$ matrices, the same statement holds by the Donkin theorem, Theorem 3.3.29, and the formula for the determinant of a matrix.

In our case we can see it explicitly. Take $S[(M^m)^*] \otimes M = M_2(S[(M^m)^*])$, that is the ring of $2 \times 2$ matrices with entries the ring of polynomial functions on $m$-tuples of $2 \times 2$ matrices. This matrix algebra acts in a diagonal way on the group $G := \mathrm{PGL}(2, F)$ of inner automorphisms of $2 \times 2$ matrices, and thus our claim is that $V_m = (S[(M^m)^*] \otimes M)^{\mathrm{PGL}(2,F)}$.

We also have the decompositions

$$S[(M^m)^*] \otimes M = S[(M^m)^*] \otimes M_0 \oplus S[(M^m)^*],$$

$$S[(M^m)^*] = S[(M_0^m)^*] \otimes F[t_1, \dots, t_m].$$

These induce a reduction in the computation of invariants, that is one has to study the space $V_m^0 := [S[(M_0^m)^*] \otimes M_0]^G$ and

$$V_m = [V_m^0 \oplus T_m^0] \otimes_F F[t_1, \dots, t_m]. \tag{9.5}$$

Notice though, that $V_m^0$ contrary to $V_m$ is not closed under product but only with respect to Lie bracket $[\,,\,]$. On the other hand,

$$\tilde{V}_m := [S[(M_0^m)^*] \otimes M]^G = V_m^0 \oplus T_m^0$$

is again an associative algebra and $V_m = \tilde{V}_m[t_1, \dots, t_m]$ is the polynomial ring over $\tilde{V}_m$.

The crucial object to analyze is thus $\tilde{V}_m$, which turns out to have quite a remarkable structure as explained in the next paragraph. In the last section, finally, we use all the combinatorics developed in order to understand how the ring of generic matrices $R_m$ sits inside $V_m$.

We will prove that the algebra $\tilde{V}_m = F\langle Y_1, \dots, Y_m\rangle$ is generated by $m$ generic $2 \times 2$ matrices $Y_i$ of trace 0, that is that this algebra is closed under trace; see Corollary 9.1.16. As a representation of $\mathrm{GL}(m, F)$, if the characteristic of $F$ is 0, we have the decomposition of $\tilde{V}_m$ in irreducible representations $L_\lambda$ of $\mathrm{GL}(m, F)$ with $\lambda$ in the set $\mathcal{P}^3$ formed by all partitions with at most three columns (Corollary 9.1.21):

$$\tilde{V}_m = \bigoplus_{\lambda \in \mathcal{P}^3} L_\lambda. \tag{9.6}$$

DEFINITION 9.1.4. By $L_\lambda = L_\lambda(m) := S_{\check{\lambda}}(W)$ we denote for simplicity the irreducible representation $S_{\check{\lambda}}(W)$, $\dim W = m$ of Definition 6.3.2.

The reason to pass to a notation with conjugate partitions depends on the display of a basis of $\tilde{V}_m$ which is indexed by column semistandard tableaux with at most three columns (Corollary 9.1.21).

As a consequence the trace algebra $V_m = R_m T_m$ associated to the algebra $R_m = F\langle X_1, \dots, X_m\rangle$ of generic $2 \times 2$ matrices is just the polynomial ring $V_m = \tilde{V}_m[t_1, \dots, t_m]$, $t_i := \mathrm{tr}(X_i)/2$, $X_i = Y_i + t_i$ (cf. Definition 9.1.2). Furthermore we have a precise description of the conductor of the inclusion of the algebra $R_m$ of generic $2 \times 2$ matrices into the algebra $V_m$ (Theorem 9.3.1). From this we deduce a precise description of the algebra of generic $2 \times 2$ matrices $R_m$ (Corollary 9.3.4).

The fact that, in this case, we will produce a complete description is due to the following simple fact (left to the reader):

*The space $M_0$ of trace 0 matrices is equipped with the nondegenerate symmetric bilinear form $(X, Y) := \operatorname{tr}(XY)$.*

*The linear group $\mathrm{GL}(2, F)$, acting by conjugation on the space $M_0$, induces the full special orthogonal group (of this form) on this three-dimensional space.*

It follows that the ring of invariants of $m$-matrices of trace 0 coincides with the ring of invariants of $\mathrm{SO}(3, F)$ on $m$ copies of its defining representation. In characteristic not 2 this has a simple classical description, and in characteristic 0 this is even more precise.

Let us recall the general theorem in characteristic 0 and its generalization in characteristic different from 2.

If $W$ is a vector space of dimension $n$ over a field $F$ equipped with a nondegenerate symmetric form $(u, v)$ (a scalar product), we have the orthogonal group $O(W)$ of linear transformations of $W$ preserving this form and the special orthogonal group $\mathrm{SO}(W)$, the subgroup of $O(W)$ of elements of determinant 1. We act with these groups on the space $W^m = W \otimes_F F^m = \{(v_1, \ldots, v_m)\}$, $v_i \in W$ of $m$-tuples of elements of $W$, and we consider the respective rings of invariants, that is the ring of polynomial functions on these spaces invariant under the action of the corresponding group. The classical theorem (see for instance **[Wey97]** or **[Pro07**, page 390], or **[DCP76]** for a characteristic free, not 2, treatment) is

THEOREM 9.1.5. *The ring of orthogonal invariants of $W^m$ is generated by the scalar products $(v_i, v_j)$, $i, j \leq m$. The ring of special orthogonal invariants of $W^m$ is generated by the scalar products $(v_i, v_j)$, $i, j \leq m$, and by the determinants*

$$\det(v_{i_1}, v_{i_2}, \ldots, v_{i_n}) = v_{i_1} \wedge v_{i_2} \wedge \cdots \wedge v_{i_n}, \ 1 \leq i_1 < i_2 < \cdots < i_n \leq m.$$

For characteristic 2, see **[DF04]**.

9.1.3.2. *Tableaux.* In our case we have the following interpretation. Let us denote by the symbol $(Y_1, \ldots, Y_m) \in (M_0)^m$ an $m$-tuple of trace 0, $2 \times 2$ matrices and by $X_i = Y_i + t_i$ matrices with trace $t_i/2$.

Notice first that, from formula (9.3) if $Y$ has trace 0 and $X$ is any $2 \times 2$ matrix, we have by (9.3) that

$$YX + XY - \operatorname{tr}(X)Y = \operatorname{tr}(XY). \tag{9.7}$$

This gives the recursive formula

(9.8)

$$\operatorname{tr}(Y_1Y_2 \cdots Y_{n-1}Y_n) = Y_1Y_2 \cdots Y_{n-1}Y_n + Y_nY_1Y_2 \cdots Y_{n-1} - \operatorname{tr}(Y_1Y_2 \cdots Y_{n-1})Y_n.$$

EXERCISE 9.1.6. $(Y_i, Y_j) := \operatorname{tr}(Y_iY_j) = Y_iY_j + Y_jY_i$ and

$$\begin{aligned} Y_1 \wedge Y_2 \wedge Y_3 &= tr(Y_1Y_2Y_3) \\ &= Y_3Y_1Y_2 - Y_2Y_1Y_3 = Y_1Y_2Y_3 - Y_3Y_2Y_1 = Y_2Y_3Y_1 - Y_1Y_3Y_2. \end{aligned} \tag{9.9}$$

In particular $\operatorname{tr}(Y_1Y_2Y_3)$ is an antisymmetric function.

$$Y_i = \begin{vmatrix} a_i & b_i \\ c_i & -a_i \end{vmatrix}, \ i = 1, 2, 3, \quad Y_1 \wedge Y_2 \wedge Y_3 = \det \begin{vmatrix} a_1 & b_1 & c_1 \\ a_2 & b_2 & c_2 \\ a_3 & b_3 & c_3 \end{vmatrix}.$$

From Theorem 9.1.5 and Exercise 9.1.6 then follows

THEOREM 9.1.7. *The ring $T_m^0$ is generated by the elements* $\operatorname{tr}(Y_iY_j), \operatorname{tr}(Y_iY_jY_k)$. *The ring $T_m$ is generated by the elements* $\operatorname{tr}(X_i)$, $\operatorname{tr}(X_iX_j)$, $\operatorname{tr}(X_iX_jX_k)$.

THEOREM 9.1.8. *The space $\tilde{V}_m$ is generated as a module over $T_m^0$ by the elements* $1$, $Y_i$, $Y_iY_j$, *or* $1, Y_i, [Y_i, Y_j]$. *The space $V_m$ is generated as a module over $T_m$ by the elements* $1, X_i, X_iX_j$.

PROOF. Since $X_i = Y_i + 1/2\operatorname{tr}(X_i)$, the two statements are equivalent. As for the first, by formula (9.9) one deduces

$$2Y_1Y_2Y_3 = Y_1\operatorname{tr}(Y_2Y_3) - \operatorname{tr}(Y_1Y_3)Y_2 + Y_3\operatorname{tr}(Y_1Y_2) + \operatorname{tr}(Y_1Y_2Y_3). \qquad \square$$

These invariants satisfy some relations, but the most interesting fact is that these rings admit special bases indexed by standard tableaux. This has to do with the representation theory of $\mathrm{GL}(m, F)$.

The group $\mathrm{GL}(m, F)$ acts naturally on both rings, which acts on the vector space $W^m = W \otimes_F F^m$ by $A(v \otimes u) = v \otimes A(u)$, $A \in \mathrm{GL}(m, F)$ commuting with the action of $\mathrm{GL}(W)$. In characteristic 0 the ring of invariants decomposes into a direct sum of irreducible representations of $\mathrm{GL}(m, F)$, and we have a remarkable fact (cf. **[Pro07**, page 409**]**).

THEOREM 9.1.9. *The ring of orthogonal invariants of $W^m$, as a representation of the group $\mathrm{GL}(m, F)$, is the direct sum of the Schur modules $S_\lambda(F^m)$, each appearing only once, where $\lambda$ is a partition of height $\leq n$ and its columns appear each an even number of times.*

*The ring of special orthogonal invariants of $W^m$, as a representation of $\mathrm{GL}(m, F)$, is the direct sum of the Schur modules $S_\lambda(F^m)$, each appearing only once, where $\lambda$ is a partition of height $\leq n$ and its columns of length $< n$ appear each an even number of times.*

It is well known that representations of the linear group have bases indexed by (semi-)standard tableaux. In fact the theory has an extension over fields of positive characteristic, and the two algebras of invariants of Theorem 9.1.9 have a basis indexed by suitable standard tableaux.

Let us explain this theory for invariants under the special orthogonal group of copies of a three-dimensional space $V$ equipped with a scalar product.

In our case, $V$ is the space of traceless $2 \times 2$ matrices, for which the scalar product is by definition $(A, B) = \operatorname{tr}(AB)$.

In all characteristics different from 2 this ring, denoted by $T_m^0$, has a combinatorial basis, over $F$, indexed by *standard tableaux* of row shape $3^k2^{2p}1^{2n}$ or of column shape $k + 2p + 2n, k + 2p, k$; see **[Pro07**, page 542 and following**]** or **[DCP76]**.

DEFINITION 9.1.10. We denote by $\mathcal{P}^3$ the partitions of type $3^k2^i1^j$. These are given by three columns $h_1 = k + i + j$, $h_2 = k + i$, $h_3 = k$. If they are such that $j = h_1 - h_3$, $i = h_2 - h_3$ are even, they will also be called *even partitions* and denoted $\mathcal{P}^{3,\mathrm{even}}$.

The other partitions with three columns will be called *odd* and denoted $\mathcal{P}^{3,\mathrm{odd}}$.

If we want to stress that a given partition is a partition of some integer $n$, we write $\mathcal{P}^3_n, \mathcal{P}^{3,\mathrm{even}}_n, \mathcal{P}^{3,\mathrm{odd}}_n$.

We start by defining three types of special tableaux, representing elements of $T^0_m$, of shapes $1^2, 2^2, 3$, respectively.

(A) $\begin{vmatrix} i \\ j \end{vmatrix} := \operatorname{tr}(Y_iY_j) = Y_iY_j + Y_jY_i, \quad \begin{vmatrix} i \\ j \end{vmatrix} = \begin{vmatrix} j \\ i \end{vmatrix}.$

(B)

$$\begin{vmatrix} i & h \\ j & k \end{vmatrix} := \det \begin{vmatrix} \begin{vmatrix} i \\ j \end{vmatrix} & \begin{vmatrix} i \\ k \end{vmatrix} \\ \begin{vmatrix} h \\ j \end{vmatrix} & \begin{vmatrix} h \\ k \end{vmatrix} \end{vmatrix} = \begin{vmatrix} i \\ j \\ h \\ k \end{vmatrix} - \begin{vmatrix} i \\ k \\ h \\ j \end{vmatrix}$$

$$= \operatorname{tr}(Y_iY_j)\operatorname{tr}(Y_hY_k) - \operatorname{tr}(Y_iY_k)\operatorname{tr}(Y_hY_j).$$

(C) $\begin{vmatrix} i & j & k \end{vmatrix} := \det(Y_i, Y_j, Y_k) = Y_i \wedge Y_j \wedge Y_k = \operatorname{tr}(Y_iY_jY_k).$

In (C), the determinant $Y_1 \wedge Y_2 \wedge Y_3$ is of the three vectors, $Y_1, Y_2, Y_3$ and is from Exercise 9.1.6.

The elements of type (B) are not necessary for generating the algebra $T^0_m$ but are introduced in order to give an explicit basis of $T^0_m$ (which is connected with the SFT, which we shall not discuss).

When we take a product of $k$ elements of type (C), $p$ elements of type (B), and $n$ of type (A), we display it as a tableau of shape $3^k2^{2p}1^{2n}$ by placing these symbols one on top of the other; compare this with the discussion in §6.3.2. As an example (reading from bottom to top):

$$\begin{array}{lll} 1 & 2 & 3 \\ 1 & 3 & 4 \\ 1 & 3 & \\ 2 & 4 & \\ 1 & & \\ 2 & & \end{array}\quad , \qquad \begin{array}{lll} 1 & 3 & 4 \\ 1 & 3 & 4 \\ 1 & 3 & \\ 2 & 3 & \\ 2 & 3 & \\ 2 & 4 & \end{array} \tag{9.10}$$

represent, respectively, the products

$$\operatorname{tr}(Y_1Y_2)[\operatorname{tr}(Y_1Y_2)\operatorname{tr}(Y_3Y_4) - \operatorname{tr}(Y_1Y_4)\operatorname{tr}(Y_2Y_3)]\operatorname{tr}(Y_1Y_3Y_4)\operatorname{tr}(Y_1Y_2Y_3)$$

$$[\operatorname{tr}(Y_2^2)\operatorname{tr}(Y_3Y_4) - \operatorname{tr}(Y_2Y_4)\operatorname{tr}(Y_2Y_3)][\operatorname{tr}(Y_1Y_2)\operatorname{tr}(Y_3^2) - \operatorname{tr}(Y_1Y_3)\operatorname{tr}(Y_2Y_3)]\operatorname{tr}(Y_1Y_3Y_4)^2.$$

DEFINITION 9.1.11. A general tableau is by definition a way of displaying a product of the tableaux of the previous types (A), (B), (C).

Of course this depends on the order of the elements in the product, but given a tableau of the shape of a partition in $\mathcal{P}^{3,\mathrm{even}}$, that is in which the number of rows of length 2, resp., 1 are even, we have a corresponding product of invariants.

Here we call such a tableau *standard* (rather than semistandard as in §6.3.2) if the indices appearing are strictly increasing on the rows and nondecreasing on the columns. In our example (9.10) the second is standard, and the first is not. Then one of the main facts, proved in **[DCP76]** in general, is

THEOREM 9.1.12. *The standard tableaux, filled with indices in* $1, \dots, m$ *of all shapes* $3^k2^{2p}1^{2n}$ *give a linear basis of* $T^0_m$ *over* $F$.

**9.1.4. The straightening algorithm.** Let us sketch the algorithm which allows us to express any tableau as a linear combination of standard ones.

First we order the tableaux by shape and for the same shape lexicographically, reading each row from left to right and from up to down.

The main step is to describe some *local laws* which replace a tableau which is a product of two tableaux of types (A), (B), (C), which is not standard at some point with a linear combination of tableaux of higher shape (that is with some boxes raised) or lexicographically inferior. This gives a recursive algorithm on all tableaux.

The introduction of (B) is used to *straighten* a tableau with a single column with entries $i, j, h, k$ which are not weakly increasing, expressing a product of the type $\operatorname{tr}(Y_iY_j)\operatorname{tr}(Y_hY_k)$. As example $\operatorname{tr}(Y_1Y_4)\operatorname{tr}(Y_2Y_3)$,

$$\operatorname{tr}(Y_1Y_4)\operatorname{tr}(Y_2Y_3) = \begin{vmatrix}1\\4\\2\\3\end{vmatrix} = \begin{vmatrix}1\\2\\3\\4\end{vmatrix} - \begin{vmatrix}1&3\\2&4\end{vmatrix}.$$

Notice that this symbol (B) is symmetric by exchange of rows and antisymmetric by exchanging indices in every row. Verify that

$$\begin{vmatrix}i&j\\h&k\end{vmatrix} = \begin{vmatrix}h&k\\i&j\end{vmatrix} = -\begin{vmatrix}j&i\\h&k\end{vmatrix} = -\begin{vmatrix}i&j\\k&h\end{vmatrix} = \begin{vmatrix}j&i\\k&h\end{vmatrix}, \tag{9.11}$$

and we have further the first *straightening law*

$$\begin{vmatrix}i&j\\h&k\end{vmatrix} = \begin{vmatrix}i&h\\j&k\end{vmatrix} + \begin{vmatrix}i&k\\h&j\end{vmatrix}. \tag{9.12}$$

So up to sign, using (9.11), we can rearrange the left-hand side of formula (9.12) to have $i < j,\ h < k$. Then using the row exchange, we may assume that $i \le h$, if $j \le k$ the tableau is standard; otherwise we apply the straightening law (9.12) and write it as a linear combination of standard tableaux.

$$\begin{vmatrix}1&4\\2&3\end{vmatrix} = \begin{vmatrix}1&2\\4&3\end{vmatrix} + \begin{vmatrix}1&3\\2&4\end{vmatrix} = -\begin{vmatrix}1&2\\3&4\end{vmatrix} + \begin{vmatrix}1&3\\2&4\end{vmatrix}. \tag{9.13}$$

REMARK 9.1.13. Essentially all strainghtening laws follow from a basic rule of alternation for Plücker coordinates.

(a) Suppose we consider $m$ vectors $v_1, \dots, v_m$ of dimension $q$, which in a basis we may consider as the columns of a $q \times m$ matrix $X$. Given any $q$ of these vectors $v_{i_1}, \dots, v_{i_q}$, we denote by a row tableau

$$\text{Plücker coordinates} \quad |i_1, \dots, i_q| := v_{i_1} \wedge \cdots \wedge v_{i_q}$$

the determinant of the maximal minor of $X$ extracted from the corresponding columns. Then we consider the algebra generated by these elements and a product we display as a tableau. Given some two consecutive rows, if one of the top indices is bigger than the bottom index, *a violation of being standard*, then we use the fact that if we antisymmetrize in a tableau the $q + 1$ underbraced indices of two consecutive rows, we get zero since we are alternating $q+1$ vectors of dimension $q$.

$$\left|\begin{matrix} i_1, i_2 \dots, \underbrace{i_k,\ i_{k+1}, \dots, i_p, \dots, i_q} \\ \underbrace{j_1, j_2, \dots, j_k},\ j_{k+1}, \dots, j_p, \dots, j_q \end{matrix}\right|, \quad i_k > j_k.$$

Since by construction each row is already antisymmetric, the required process of antisymmetrization gives a sum of $\binom{q+1}{k}$ terms in which some of the indices below rise above and some of the indices above fall below, giving rise to an expression of the given tableau by lexicographically lower terms, which then allows us to proceed by induction.

(b) General rules for the product of two determinants of minors of an $a \times b$ matrix $X$ is reduced to the previous case. One adds to $X$ the identity matrix $Y = X\ 1_a$ and remarks that the determinant of a $k \times k$ minor of $X$ equals up to sign the determinant of a maximal minor of $Y$ in which the $a-k$ columns are from $1_a$ (see **[DCP76]**).

This method applies to products of two of the basic tableaux when the first is of type (C), since in each case we alternate on four elements.

$$\begin{matrix} \underbrace{i_1\ \ i_2\ \ i_3} \\ \underbrace{i_4}\ \ i_5 \\ i_6\ \ i_7 \end{matrix} = 0, \qquad \begin{matrix} i_1\ \underbrace{i_2\ i_3} \\ \underbrace{i_4\ i_5} \\ i_6\ i_7 \end{matrix} = 0, \qquad \begin{matrix} \underbrace{i_1\ \ i_2\ \ i_3} \\ \underbrace{i_4} \\ i_5 \end{matrix} = 0.$$

For a product of two elements of type (B) one has, using Remark 9.1.13(b),

$$\begin{vmatrix} a & b \\ c & d \\ h & k \\ i & j \end{vmatrix} + \begin{vmatrix} a & b \\ d & h \\ c & k \\ i & j \end{vmatrix} - \begin{vmatrix} a & b \\ c & h \\ d & k \\ i & j \end{vmatrix} - 2\begin{vmatrix} c & d & h \\ a & b & j \\ i & & \\ k & & \end{vmatrix} + 2\begin{vmatrix} c & d & h \\ a & b & i \\ j & & \\ k & & \end{vmatrix} = 0.$$

So if $h < c$, one has an expression of the first tableu in terms of either tableaux of the same shape but lexicographically inferior, or tableaux of a *higher shape*. Repeating this kind of argument, one has the required expression as a sum of standard tableaux. A similar expression (with nine terms) treats the case $k < d$.

$$0 = \begin{vmatrix} a & b \\ c & d \\ h & k \\ i & j \end{vmatrix} - \begin{vmatrix} a & b \\ c & h \\ d & k \\ i & j \end{vmatrix} + \begin{vmatrix} a & b \\ c & k \\ d & h \\ i & j \end{vmatrix} - 2\begin{vmatrix} a & b & j \\ c & d & h \\ i & & \\ k & & \end{vmatrix} + 2\begin{vmatrix} a & b & i \\ c & d & h \\ j & & \\ k & & \end{vmatrix} + 2\begin{vmatrix} a & b & j \\ c & d & k \\ i & & \\ h & & \end{vmatrix}$$

$$-2\begin{vmatrix} a & b & i \\ c & d & k \\ j & & \\ h & & \end{vmatrix} - 2\begin{vmatrix} a & b & j \\ c & h & k \\ i & & \\ d & & \end{vmatrix} + 2\begin{vmatrix} a & b & i \\ c & h & k \\ j & & \\ d & & \end{vmatrix}.$$

In this case one should notice the appearence of the factor 2, which makes our notation different from **[DCP76]**, and which is due to the fact that since the scalar product is not written in an orthonormal basis we have

$$\det(\operatorname{tr}(x_i y_j)) = -2\operatorname{tr}(x_1x_2x_3)\operatorname{tr}(y_1y_2y_3).$$

Finally, for type (B) times type (A),

$$-\begin{vmatrix} a & b\\ c & d\\ h & \\ j & \end{vmatrix} - \begin{vmatrix} a & b\\ d & h\\ c & \\ j & \end{vmatrix} + \begin{vmatrix} a & b\\ c & h\\ d & \\ j & \end{vmatrix} - 2\begin{vmatrix} c & d & h\\ a & b & j\end{vmatrix} = 0.$$

EXERCISE 9.1.14. Prove, for example (9.10), using most of the formulas,

$$\begin{array}{ccc} 1 & 2 & 3\\ 1 & 3 & 4\\ 1 & 3 & \\ 2 & 4 & \\ 1 & & \\ 2 & & \end{array} = \begin{array}{ccc} 1 & 2 & 3\\ 1 & 3 & 4\\ 1 & 3 & \\ 1 & 4 & \\ 2 & & \\ 2 & & \end{array} - \begin{array}{ccc} 1 & 2 & 3\\ 1 & 2 & 4\\ 1 & 3 & \\ 1 & 3 & \\ 2 & & \\ 4 & & \end{array} + \begin{array}{ccc} 1 & 2 & 3\\ 1 & 2 & 3\\ 1 & 3 & \\ 1 & 4 & \\ 2 & & \\ 4 & & \end{array} + 2\begin{array}{ccc} 1 & 2 & 3\\ 1 & 2 & 3\\ 1 & 2 & 4\\ 1 & 3 & 4 \end{array}.$$

9.1.4.1. *The algebra* $\tilde{V}_m$. In fact the previous theory is strictly connected with the structure of $T_m^0$ as representation of the group $\mathrm{GL}(m, F)$ induced by the standard action of this group $g(x \otimes v) = x \otimes gv$ on the space $(M_0)^m = M_0 \otimes_F F^m$ of $m$-tuples of traceless matrices.

We use then the dominance order of partitions, Definition 6.1.4. We filter in each degree $k$ the homogeneous elements of $T_m^0$ of degree $k$. The filtration is indexed by partitions $3^h 2^{2p} 1^{2n} := \lambda \vdash k$ (i.e., $3h + 4p + 2n = k$). The corresponding subspace $T_{m,\lambda}^0$ is spanned by all tableaux of shape $\mu \geq \lambda$, $\mu \in \mathcal{P}_k^{3,\text{even}}$, and one can prove (cf. **[DCP76]**) that it is $\mathrm{GL}(m, F)$ stable and has as its basis the standard tableaux of these shapes.

It turns out that this is a *good filtration* in the sense of Donkin. Denote by $T_{m,>\lambda}^0$ the subspace spanned by all tableaux of shape $\mu > \lambda$, $\mu \in \mathcal{P}_k^{3,\text{even}}$. This space is also a representation. In characteristic 0 we have complete reducibility, and we can write everything as direct sum of the irreducible representation $L_\lambda := S_\lambda(F^m)$, so that

$$T_{m,\lambda}^0 = S_\lambda(F^m) \oplus T_{m,>\lambda}^0 = L_\lambda \oplus T_{m,>\lambda}^0.$$

From the general theory of invariants of the special orthogonal group, as in **[DCP76]**, we have in our case

THEOREM 9.1.15. *If $F$ is a field of characteristic 0, we have the decomposition in irreducible representations of* $\mathrm{GL}(m, F)$:

$$T_m^0 = \bigoplus_{k=0}^{\infty} \bigoplus_{\lambda \in \mathcal{P}_k^{3,\text{even}}} L_\lambda \tag{9.14}$$

Where $\bigoplus_{\lambda \in \mathcal{P}_k^{3,\text{even}}} L_\lambda$ is the part in degree $k$ of $T_m^0$.

Our next task is to apply this theory to generic matrices. For this it is important to reinterpret invariants in a noncommutative language, as given by Exercise 9.1.6 and (9.8).

COROLLARY 9.1.16. *The algebra* $\tilde{V}_m = V_m^0 \oplus T_m^0$ *is generated by the coordinates* $Y_i$. *That is* $\tilde{V}_m = F\langle Y_1, \dots, Y_m\rangle$ *is the algebra of generic trace* 0 *matrices.*

PROOF. In fact the previous formulas tell us that the generators of the invariants are polynomials in the matrix variables $Y_i$. □

COROLLARY 9.1.17. *The identity map on the generators $Y_i$ extends to an involutory antiautomorphism $a \mapsto a^*$ of $\tilde{V}_m = F\langle Y_1, \dots, Y_m \rangle$ which is the identity on its center.*

PROOF. Since $Y_1, \dots, Y_m$ are generic matrices of trace 0, also their transposes $Y_1^t, \dots, Y_m^t$ are generic matrices of trace 0. Therefore, we have an isomorphism $j : F\langle Y_1^t, \dots, Y_m^t \rangle \simeq F\langle Y_1, \dots, Y_m \rangle$. We also have an antihomomorphism, given by transposition $\tau : F\langle Y_1, \dots, Y_m \rangle \to F\langle Y_1^t, \dots, Y_m^t \rangle$, so $j \circ \tau$ is the required involutory antiautomorphism. The fact that this is the identity on its center follows from the fact the center is generated by the elements $\mathrm{tr}(X)$, $X \in \tilde{V}_m$ and that $\mathrm{tr}(X) = \mathrm{tr}(X^t)$. □

9.1.4.2. *Odd tableaux.* Now we need to interpret also the elements of $V_m^0$ as tableaux. We do this by a standard trick, that will be used later in several places (cf. for instance formula (12.7)). Let us start from the simple

LEMMA 9.1.18. *Given an odd partition $\lambda$ of columns $h + a, h + b, h$ there is a unique way to add to its diagram a single box and make it even with at most three columns.*

PROOF. The fact that $\lambda$ is odd means that either $a$ or $b$, or both, is odd. If $a$ is odd and $b$ is even, we change $a$ to $a + 1$, that is we put the box on the first column. If $a$ is even and $b$ is odd, we change $b$ to $b + 1$, that is we put the box on the second column. If both $a, b$ are odd, we change $h$ to $h + 1$, that is we put the box on the third column. □

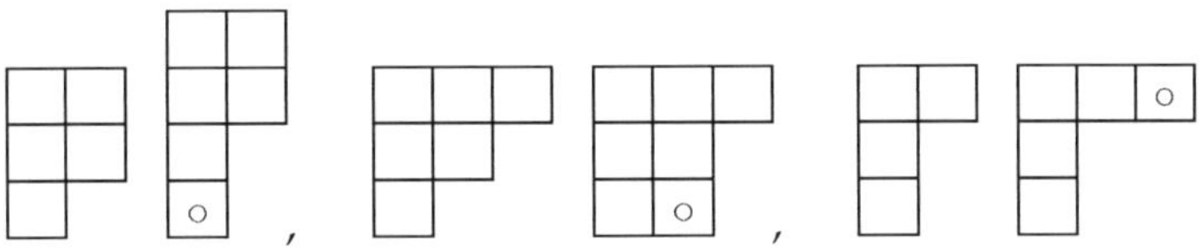

This means that if we have a tableau $A$ (in the indices $1, \dots, m$) with odd shape, there is a unique tableau $B$ of even shape and with $m + 1$ in the added box. Notice that $A$ is standard if and only if $B$ is standard.

Let us now analyze the space $V_m^0$ of equivariant maps $f : M_0^m \to M_0$. Given such a map from $m$-tuples $Y_i$ of trace 0 matrices to trace 0 matrices, we construct an invariant $\mathtt{f}$ of $m + 1$-tuples of matrices by the formula

$$\mathtt{f}(Y_1, \dots, Y_m, Y_{m+1}) := \mathrm{tr}(f(Y_1, \dots, Y_m) Y_{m+1}).$$

Clearly, the invariant $\mathtt{f}$ is linear in $Y_{m+1}$. Conversely, since the bilinear form on $M_0$, $\mathrm{tr}(Y_1 Y_2)$, is nondegenerate, given any invariant $\mathtt{f}(Y_1, \dots, Y_m, Y_{m+1})$ linear in $Y_{m+1}$, we have a unique map $f : M_0^m \to M_0$ satisfying the relation

$$\mathtt{f}(Y_1, \dots, Y_m, Y_{m+1}) = \mathrm{tr}(f(Y_1, \dots, Y_m) Y_{m+1}).$$

By nondegeneracy, it is easily seen that $f$ is an equivariant polynomial map.

We may thus identify the space $V_m^0$ with the space of invariants of $m + 1$ traceless matrices, linear in the extra variable $Y_{m+1}$ by associating to an element $u \in V_m^0$, the invariant $\mathrm{tr}(u Y_{m+1})$. We then have a theory of standard tableaux for $V_m^0$, by restricting the standard tableau of $T_{m+1}^0$ in which $m + 1$ appears exactly once and which displays such a tableau as odd tableau by removing the box in which $m + 1$ appears. More formally if $A$ is an odd tableau and $B$ is the associated even tableau with $m + 1$ added, we have

DEFINITION 9.1.19. We define the equivariant map encoded by the tableau $A$ by setting $B := \mathrm{tr}(AY_{m+1})$.

If we do this, we see that we introduce two basic odd tableaux, $|i|$, $|i\ j|$, which by convention are defined by

$$\mathrm{tr}(|i|Y_{m+1}) = \begin{vmatrix} i \\ m+1 \end{vmatrix}, \quad \mathrm{tr}(|i\ j|Y_{m+1}) = |i, j, m+1|.$$

LEMMA 9.1.20.

$$|i| := Y_i, \quad |i \quad j| := Y_iY_j - \mathrm{tr}(Y_iY_j)/2 = [Y_i, Y_j]/2. \tag{9.15}$$

PROOF. We have that $\begin{vmatrix} i \\ m+1 \end{vmatrix} := \mathrm{tr}(Y_iY_{m+1})$ is associated with the map $Y_i$.

$$|i \quad j \quad m+1| = \mathrm{tr}(Y_iY_jY_{m+1}) = \mathrm{tr}((Y_iY_j - \mathrm{tr}(Y_iY_j)/2)Y_{m+1}),$$
$$\text{with} \quad (Y_iY_j - \mathrm{tr}(Y_iY_j)/2) \in V_m^0$$

is associated thus

$$|i\ j| := Y_iY_j - \mathrm{tr}(Y_iY_j)/2 = [Y_i, Y_j]/2.$$ □

We have a first multiplication rule,

$$|i||j| = Y_iY_j = \frac{1}{2}(Y_iY_j + Y_jY_i - Y_jY_i + Y_iY_j) = \frac{1}{2}\begin{vmatrix} i \\ j \end{vmatrix} + |i \quad j|. \tag{9.16}$$

COROLLARY 9.1.21.

(1) *If $F$ is a field of characteristic 0, we have the decomposition in irreducible representations of $\mathrm{GL}(m, F)$:*

$$V_m^0 = \bigoplus_{k=0}^{\infty} \bigoplus_{\lambda \in \mathcal{P}_k^{3,\mathrm{odd}}} L_\lambda \tag{9.17}$$

(2) *The algebra $\tilde{V}_m = V_m^0 \oplus T_m^0$ has a basis indexed by all possible standard tableaux, with entries $1, 2, \ldots, m$, of diagrams with at most three columns.*

PROOF. This follows from the constructions made. □

REMARK 9.1.22. The space $V_m^0 \subset \tilde{V}_m$ is the subspace of $\tilde{V}_m$ formed by the trace 0 elements. Thus it is in fact a Lie subalgebra. One can prove, by induction, that $V_m^0$ is the *Lie subalgebra generated by the elements* $Y_i$. We leave this to the reader.

Observe that an odd tableau is naturally the product of an even tableau times one of the three basic ones,

$$|i|, \quad |i\ j|, \quad \begin{vmatrix} i & j \\ h & \end{vmatrix} \quad \mathrm{tr}\left(\begin{vmatrix} i & j \\ h & \end{vmatrix} Y_{m+1}\right) = \begin{vmatrix} i & j \\ h & m+1 \end{vmatrix}.$$

In order to decode the map associated to an odd tableau according to Definition 9.1.19, we only need to decode the third one, which is

$$\begin{vmatrix} i & j \\ h & m+1 \end{vmatrix} = \mathrm{tr}(Y_iY_h)\,\mathrm{tr}(Y_jY_{m+1}) - \mathrm{tr}(Y_jY_h)\,\mathrm{tr}(Y_iY_{m+1}).$$

$$\begin{vmatrix} i & j \\ h & \end{vmatrix} = \begin{vmatrix} i \\ h \end{vmatrix} |j| - \begin{vmatrix} j \\ h \end{vmatrix} |i| = \begin{vmatrix} i \\ h \\ j \end{vmatrix} - \begin{vmatrix} j \\ h \\ i \end{vmatrix} = \mathrm{tr}(Y_iY_h)Y_j - \mathrm{tr}(Y_jY_h)Y_i. \tag{9.18}$$

In order to have a complete calculus, one needs to have the multiplication rules between basic tableaux and the straightening laws. Using these, one has that the algebra $\tilde{V}_m^0$ has a basis of standard tableaux which we know how to multiply. For instance we have

$$|i|\,|j\ \ h| = \frac{1}{2}Y_i[Y_j, Y_h] = |i\ \ j\ \ h| + \begin{vmatrix} i & j \\ h & \end{vmatrix}. \tag{9.19}$$

Notice that

$$([Y_h, Y_k][Y_i, Y_j] + [Y_i, Y_j][Y_h, Y_k]) = -2\begin{vmatrix} i & j \\ h & k \end{vmatrix}$$

$$\operatorname{tr}([Y_i, Y_j][Y_h, Y_k]) = \operatorname{tr}([Y_h, Y_k][Y_i, Y_j]) = -2\begin{vmatrix} i & j \\ h & k \end{vmatrix} \tag{9.20}$$

$$Y_i^2 = \frac{1}{2}\begin{vmatrix} i \\ i \end{vmatrix} \qquad [Y_i, Y_j]^2 = -\begin{vmatrix} i & j \\ i & j \end{vmatrix}$$

$$[Y_i, Y_j][Y_h, Y_k] = 2\operatorname{tr}(Y_iY_jY_h)Y_k - 2\operatorname{tr}(Y_iY_jY_k)Y_h - \begin{vmatrix} i & j \\ h & k \end{vmatrix}$$

$$[i, j][h, k] = 2\begin{vmatrix} i & j & h \\ k & & \end{vmatrix} - 2\begin{vmatrix} i & j & k \\ h & & \end{vmatrix} - \begin{vmatrix} i & j \\ h & k \end{vmatrix} \tag{9.21}$$

$$[[Y_i, Y_j], [Y_h, Y_k]] = 2(\operatorname{tr}(Y_iY_jY_h)Y_k + \operatorname{tr}(Y_iY_jY_k)Y_h - \operatorname{tr}(Y_hY_kY_i)Y_j - \operatorname{tr}(Y_hY_kY_j)Y_i)$$

REMARK 9.1.23. If $A$ is an odd tableau, $B$ is the associated even tableau, and $m+1$ is on the first column, then $A = EY_i$ where $i$ is the last index of $A$ on the first column and $E$ is the even tableau obtained from $A$ by removing the last box containing $i$. Similar analysis holds in the other two cases. If $m+1$ is on the second column, then we have $A = EG$, where $E$ is the even tableau in which we remove the 2, 1 hook $\begin{vmatrix} i & j \\ h & m+1 \end{vmatrix}$ in which is placed $m+1$ and $G$ is associated to the hook $\begin{vmatrix} i & j \\ h & \end{vmatrix}$. If $m+1$ is on the third column, then we have $A = E|i\ j|$, where $E$ is the even tableau in which we remove the first row of length 2 and $|i\ j|$ is the tableau in that row.

9.1.4.3. *Summarizing.* The space $V_m^0$ has a basis in degree $k$ indexed by standard tableaux of shapes $\lambda$ in $\mathcal{P}_{k+1}^{3,\text{even}}$ in the indices $1, 2, \ldots, m, m+1$ in which the index $m+1$ appears exactly once. Since the tableau is standard, this index appears in an external corner of $\lambda$, which thus can be either the last box of the last row with 3, or the last box of the last row with 2, or the last box of the last row with 1 element. The partition obtained by removing that corner from $\lambda$ is one of the corresponding three types (cf. example on page 241)

$$3^q2^{2p+1}1^{2n},\ 3^q2^{2p+1}1^{2n+1},\ 3^q2^{2p}1^{2n+1}.$$

These form all possible partitions with at most three columns in which at least one of the exponents of 2 or 1 is odd.

COROLLARY 9.1.24. *$T_m^0$ has a basis indexed by standard tableaux of shapes $\lambda$ in $\mathcal{P}^{3,\text{even}}$ filled with the indices $1, 2, \ldots, m$. $V_m^0$ has a basis indexed by standard tableaux of shapes $\lambda$ in $\mathcal{P}^{3,\text{odd}}$ filled with the indices $1, 2, \ldots, m$. $\tilde{V}_m$ has a basis indexed by standard tableaux of shapes $\lambda$ in $\mathcal{P}^3$ filled with the indices $1, 2, \ldots, m$.*

In characteristic 0 we deduce a decomposition into irreducible representations

$$T_m^0 = \bigoplus_{\lambda\in\mathcal{P}^{3,\mathrm{even}}} L_\lambda,\ V_m^0 = \bigoplus_{\lambda\in\mathcal{P}^{3,\mathrm{odd}}} L_\lambda,\ \tilde{V}_m = \bigoplus_{\lambda\in\mathcal{P}^{3}} L_\lambda, \tag{9.22}$$

then all these formulas can be fed into formula (9.5) in order to give a precise description of $V_m$.

9.1.4.4. *The defining ideal.* We want to give a presentation of the algebra of $m$ generic trace 0 matrices, following Drensky and Koshlukov **[DK87]** who prove the following general fact.

One starts from the fact that given two trace 0 matrices $x, y$, one has the identity $[x^2, y] = 0$. Then consider the free algebra $F\langle x_1, \dots, x_m\rangle$ identified to the tensor algebra $T(V)$, $\dim V = m$. We see that he algebra of $m$ generic trace 0 matrices is a quotient of the algebra,

$$C(V) := T(V)/I, \quad I \text{ the ideal generated by } \ [v^2, w],\ v, w \in V.$$

Clearly, the ideal $I$ is stable under the action of $\mathrm{GL}(V)$ so that this action passes to $C(V)$.

THEOREM 9.1.25. *As representation of* $\mathrm{GL}(V)$, *we have*

$$C(V) = \bigoplus_\lambda S_\lambda(V). \tag{9.23}$$

*That is it is the direct sum of all the irreducible polynomial representations of* $\mathrm{GL}(V)$ *each with multiplicity* 1.

PROOF. The elements $v^2 \in V^{\otimes 2}$ span the irreducible representation $S^2(V)$ so $C(V)$ is the algebra over $A = S(S^2(V))$, with generators the space $V$ and with the relations $v^2 = v^{\otimes 2} \in A$. This algebra is in fact the Clifford algebra for the free $A$-module $V \otimes A$ with the *universal quadratic form* $v^2 = v^{\otimes 2} \in A$.

As any Clifford algebra this can be filtered by assigning degree 0 to $A$ and 1 to $V$ so that the associated graded algebra is $\bigwedge V \otimes S(S^2(V))$. This last algebra is clearly isomorphic to $C(V)$ as representation of $\mathrm{GL}(V)$.

Now the theorem follows from two standard facts. The first is the *Plethysm formula* that $S(S^2(V)) = \bigoplus_\lambda S_{\lambda\ \mathrm{even}}(V)$, which is just one of Cauchy's identities for symmetric functions (even means even rows)

$$\frac{1}{\prod_{i,j=1}^m (1 - x_i x_j)} = \sum_{\lambda\ \mathrm{even}} S_\lambda(x_1, \dots, x_m).$$

The second is Pieri's rule for $\bigwedge^i V \otimes S_\lambda = \bigoplus_\mu S_\mu$, where $\mu$ is obtained from $\lambda$ by adding $i$ boxes but at most 1 on each row. □

COROLLARY 9.1.26. *The algebra* $\tilde{V}_m$ *of* $m$ *generic trace* 0 *matrices is the quotient of the algebra* $C(V)$ *modulo the ideal generated by the standard identity in four variables* $\mathrm{St}(x_1, x_2, x_3, x_4)$.

PROOF. We have already remarked that $\tilde{V}_m$ is a quotient of $C(V)$ and also of $\mathrm{St}(x_1, x_2, x_3, x_4)$. So it is enough to remark that the ideal in $C(V)$ generated by $\mathrm{St}(x_1, x_2, x_3, x_4)$ is the direct sum of all $S_\lambda(V)$ with height $\geq 4$ while $\tilde{V}_m$ is the direct sum of all $S_\lambda(V)$ with height $\leq 3$. □

9.1.4.5. *The highest weight vectors.* The theory has been explained in Theorem 6.3.7.

REMARK 9.1.27. On the algebra $\tilde{V}_m$, the group $G$ acts as a group of isomorphisms; its Lie algebra $M_m(F)$ acts by derivations.

If $a, b \in \tilde{V}_m$ are elements of some weights $w_1, w_2$, then $ab$ has weight $w_1 w_2$. It will be more convenient to use the weights for the Lie algebra, which is still the space of diagonal matrices, then the weight of $ab$ is weight $w_1 + w_2$.

Denote the diagonal matrices by $(d_1, \dots, d_m)$. Then the weight of $Y_i$ is $d_i$.

An elementary matrix $e_{h,k}$ induces the derivation on $\tilde{V}_m$ which on the generators acts by

$$e_{h,k} Y_i = \delta_i^k Y_h.$$

If $A$ has some weight $w$, then $e_{h,k}A$, if nonzero, has weight $w + d_h - d_k$.

If $u$ is a weight vector, then $u$ is a highest weight vector if and only if $\forall i < m$ we have $e_{i,i+1}u = 0$. In this case $u$ generates the irreducible representation $L_\lambda$ and the weight is $\sum_i h_i d_i$ where $(h_1, \dots, h_m)$ are the columns of $\lambda$.

In our case, denote by $K_\lambda \in L_\lambda$ a highest weight vector, one easily sees that it is a multiple of the tableau with 1, 2, 3, respectively, on the corresponding columns.

By definition an element $K_\lambda \in L_\lambda$ is characterized, up to a scalar normalization, by the fact that under the action of the subgroup $B$ of upper triangular matrices, it transforms under the character $\chi_\lambda$ taking a diagonal matrix $\mathrm{diag}(d_1, \dots, d_m)$ to $d_1^{h_1} d_2^{h_2} d_3^{h_3}$ where $h_1, h_2, h_3$ are the lengths of the columns. Remark that $K_\lambda$ lies already in the representation for $m = 3$ and each of the representations $L_\lambda$ appear already for $m = 3$ (but not for $m = 2$). Since the group action is by automorphisms and the algebra $V_m$ is a domain, it follows that the product of two highest weight vectors is a highest vector with highest weight the sum of the highest weights. The invariant algebra $\tilde{V}_m^{U_m}$ has a basis of the highest weight vectors $K_\lambda$, $\lambda \in \mathcal{P}^3$.

We choose as highest weight vectors of $L_1, L_2, L_3$, respectively,

$$K_1 := Y_1, \quad K_2 := [Y_1, Y_2], \quad K_3 := \frac{\mathrm{St}_3(Y_1, Y_2, Y_3)}{3} = \mathrm{tr}(Y_1 Y_2 Y_3).$$

$K_3$ is in the center as well as

$$K_{1^2} := -K_1^2 = -Y_1^2 = \det(Y_1), \; K_{2^2} := -K_2^2 = -[Y_1, Y_2]^2 = \det([Y_1, Y_2]).$$

Thus the product

$$K_\lambda := \mathrm{tr}(Y_1 Y_2 Y_3)^h [Y_1, Y_2]^{2p} Y_1^{2n}, \quad \lambda = 3^h 2^{2p} 1^{2n},$$

is a highest weight vector of $L_{3^h 2^{2p} 1^{2n}}$.

Finally the polynomial

(9.24)

$$Y_1[Y_1, Y_2] = Y_1^2 Y_2 - Y_1 Y_2 Y_1 = -[Y_1, Y_2] Y_1, \; [Y_1, [Y_1, Y_2]] = 2(Y_1^2 Y_2 - Y_1 Y_2 Y_1)$$

is a highest weight vectors for $L_{2,1}$.

Next any partition can be decomposed by taking out the largest even partition so that the highest weight vectors can be chosen for all types to be

(9.25) $$K_\lambda, \quad K_\lambda Y_1, \quad K_\lambda [Y_1, Y_2], \quad K_\lambda Y_1 [Y_1, Y_2], \; \lambda = 3^h 2^{2p} 1^{2n}.$$

In other words

COROLLARY 9.1.28. *The algebra of $U_m$ invariants with basis the highest weight vectors is a quaternion algebra over the polynomial ring in the three variables $K_1^2, K_2^2, K_3$ with basis $1, K_1, K_2, K_1K_2 = -K_2K_1$.*

*For the antiautomorphism of Corollary 9.1.17 we have*

$$K_1^* = K_1,\ K_2^* = -K_2,\ K_3^* = -K_3. \tag{9.26}$$

*It equals multiplication by $-1$ on $L_\lambda$, $\lambda = 3^{2h}2^{2p+1}1^n, 3^{2h+1}2^{2p}1^n$ and $1$ on the rest $\lambda = 3^{2h+1}2^{2p+1}1^n, 3^{2h}2^{2p}1^n$.*

## 9.2. Invariant ideals

*This long section is not strictly needed for the study of the algebra of generic matrices. A special case of Proposition 9.2.3 would be sufficient, nevertheless we feel that this is an interesting topic which is connected with the study of T-ideals which contain the identities of $2 \times 2$ matrices.*

**9.2.1. The basic theorem.** In order to state the main theorem of this section we need a combinatorial definition which originates from the study of determinantal ideals **[DCEP80]**.

DEFINITION 9.2.1. We say that a set $\mathcal{I} \subset \mathcal{P}^3$ is an *ideal* of partitions, if for every $\lambda \in \mathcal{I}$ and for every $\mu \supset \lambda$, $\mu \in \mathcal{P}^3$, we have that $\mu \in \mathcal{I}$.

Then the main theorem is

COROLLARY 9.2.2. *Any left or right ideal $I$ of $\tilde{V}_m$ invariant under all automorphisms induced by $\mathrm{GL}(m, F)$ is a two-sided ideal, and it is of the form*

$$I = \bigoplus_{\lambda \in \mathcal{I}} L_\lambda, \quad \textit{with } \mathcal{I} \textit{ an ideal of partitions.} \tag{9.27}$$

PROOF. The proof is based upon formula (9.28), the content of Proposition 9.2.3, which unfortunately has a very long and computational proof.

Since the irreducible representations $L_\lambda$ are all distinct, any subspace $I$ of $\tilde{V}_m$ invariant under all automorphisms induced by $\mathrm{GL}(m, F)$ must be a direct sum $\bigoplus_{\lambda \in \mathcal{I}} L_\lambda$ where $\mathcal{I}$ is a set of partitions. Moreover if $I$ is an ideal, we have that $\mathcal{I}$ is an ideal of partitions by formula (9.28).

Observe that $I$ is generated as ideal by the representations $L_\lambda$ where $\lambda$ is minimal for the order of inclusion in the set $\mathcal{I}$. □

**9.2.2. The basic multiplication.** The next computation may hold in a more general setting, but we perform only in characteristic 0 for $\tilde{V}_m = \bigoplus_{\lambda \in \mathcal{P}^3} L_\lambda$; see **[Kos88]** or **[Pro84]**.

PROPOSITION 9.2.3. *Let $\lambda \in \mathcal{P}_k^3$. Then*

$$L_1 L_\lambda = L_\lambda L_1 = \bigoplus_{\mu \supset \lambda,\ \mu \in \mathcal{P}_{k+1}^3} L_\mu. \tag{9.28}$$

PROOF. By Corollary 9.1.17 it is enough to prove this for $L_1 L_\lambda$. This is the image under multiplication that we denote by

$$j_1 : L_1 \otimes L_\lambda \to L_1 L_\lambda,\ j_1(a \otimes b) := ab,$$

which is a $G$ equivariant map. We could also define the two other maps,

$$j_2 : a \otimes b \mapsto ba,\ j_3 : a \otimes b \mapsto [a,b],\ j_3 = j_1 - j_2,$$

which one can deduce from $j_1$ and the $\star$ operator.

The idea of the proof is simple. By Pieri's rule (Proposition 3.2.18 and Corollary 6.3.22) the representation $L_1 \otimes L_\lambda,\ \lambda \vdash k$ is the direct sum of all the representations $\bigoplus_{\mu \supset \lambda,\ \mu \vdash k+1} L_\mu$.

Of course in the image of the map, $j_1$ can appear only the representations with $\mu \in \mathcal{P}^3_{k+1}$, and we need to prove that *all of them appear*. In order to do this we first compute the highest weight vectors $K_\mu,\ \mu \in \mathcal{P}^3$, of the representations appearing in $L_1 \otimes L_\lambda,\ \lambda \vdash k$, and then we must prove that $j_1(K_\mu) \neq 0$. Then $j_1(K_\mu)$ is a highest weight vector of a representation appearing in $L_1 L_\lambda$.

The proof of these two facts needs a lot of explicit computations; we will do some cases and leave the others to the reader.

STEP 1 (applying Pieri's rule). The highest weight vectors $K_\mu,\ \mu \in \mathcal{P}^3$, appearing in $L_1 \otimes L_\lambda$ already appear for $m = 3$, so we may assume that $m = 3$.

Let $\lambda$ have columns $(h_1, h_2, h_3)$. When $m = 3$ the $h_3$ rows of length 3 just multiply the representation $L_{h_1-h_3,h_2-h_3}$ by the one-dimensional representation $\bigwedge^3 V$ of the three-dimensional space $V$ spanned by $Y_1, Y_2, Y_3$. So $L_\lambda = L_\mu \otimes (\bigwedge^3 V)^k$ where $\mu = (k_1, k_2),\ k_1 = h_1 - h_3, k_2 = h_2 - h_3$ and, by Exercise 9.1.6,

$$L_1 \otimes L_\lambda = (L_1 \otimes L_\mu) \otimes (\bigwedge^3 V)^k \implies L_1 L_\lambda = L_1 L_\mu \operatorname{tr}(Y_1 Y_2 Y_3)^k.$$

We may thus replace $L_\lambda$ with $L_\mu$ or, by change of notation, it is enough to study $j_1(L_1 \otimes L_\lambda)$ when $\lambda$ has only two columns $h_1, h_2$.

When we add one single box to a partition $\lambda = 2^p 1^n$ of the previous type, with columns $h_1 = p + n, h_2 = p$, we can get only the new partitions

$$\lambda_1 := h_1 + 1, h_2,\ \lambda_2 := h_1, h_2 + 1 \text{ if } h_1 > h_2,\ \lambda_3 := h_1, h_2, 1 \text{ if } h_2 > 0.$$

In other words we get $\lambda_2$ if $n > 0$ and $\lambda_3$ if $p > 0$.

$\lambda =$ 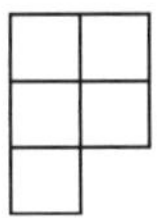, $\lambda_1 =$ 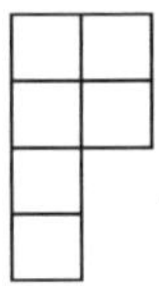, $\lambda_2 =$ 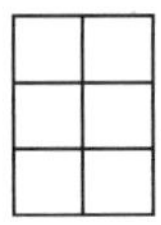, $\lambda_3 =$ 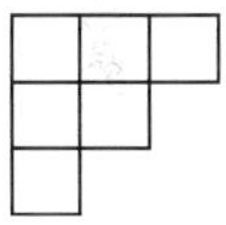

By the previous analysis we then have that

$$\begin{cases} L_1 \otimes L_\lambda = L_{\lambda_1} \oplus L_{\lambda_2} \oplus L_{\lambda_3} & \text{if } h_1 > h_2 > 0 \\ L_1 \otimes L_\lambda = L_{\lambda_1} \oplus L_{\lambda_3} & \text{if } h_1 = h_2 > 0 \\ L_1 \otimes L_\lambda = L_{\lambda_1} \oplus L_{\lambda_2} & \text{if } h_1 > 0, h_2 = 0 \\ L_1 & \text{if } h_1 = h_2 = h_3 = 0. \end{cases}$$

STEP 2 (the three highest weight vectors of the $L_{\lambda_i}$). Call $\mathrm{gl}(V)$ the Lie algebra of linear transformations of the three-dimensional vector space $V$, and the diagonal matrices form an abelian subalgebra (a Cartan algebra), $\mathfrak{t}$, with basis $e_{i,i},\ i = 1, \ldots, 3$. For a representation of this Lie algebra, by definition, a weight vector is an eigenvector of $\mathfrak{t}$ and it is a highest weight vector if and only if it is killed by $e_{1,2}, e_{2,3}$, elements which generate the Lie algebra of strictly upper triangular $3 \times 3$ matrices (with $e_{1,3} = [e_{1,2}, e_{2,3}]$).

By definition a weight vector is an eigenvector of $\mathfrak{t}$. The corresponding eigenvalue is a linear map on $\mathfrak{t}$. We set $d_i$ to be the dual basis, so we write a weight as $\sum_{i=1}^3 \alpha_i d_i$, and saying that $u$ is a vector of weight $\sum_{i=1}^3 \alpha_i d_i$ means that $e_{i,i}u = \alpha_i u$.

We can identify the highest weight of $L_\lambda$, to a partition $\lambda = \sum_i h_i d_i$, the coefficient of $d_i$ is the length of the $i$th column. If $\lambda = h_1 d_1 + h_2 d_2$, we have

$$\lambda_1 = \lambda + d_1,\ \lambda_2 = \lambda + d_2,\ \lambda_3 = \lambda + d_3.$$

REMARK 9.2.4. The computational rules we use are the following.

(1) $e_{i,i}K_\lambda = h_i K_\lambda, \quad e_{i,j}K_\lambda = 0$, if $i < j$.

(2) The elements $e_{i,j}$ as operators satisfy the usual commutation relations as in matrices (but not the matrix multiplication rules).

(3) The space $L_\lambda$ is generated by $K_\lambda$, as a module over the algebra generated by the $e_{i,j},\ i > j$.

(4) An element $Y_1 \otimes A + Y_2 \otimes B + Y_3 \otimes C \in L_1 \otimes L_\lambda$ has weight $\mu$ if and only if $A, B, C$ have weights $\mu - d_1,\ \mu - d_2,\ \mu - d_3$.

LEMMA 9.2.5. *The three possible highest weight vectors in $L_1 \otimes L_\lambda$ are the following.*

(1) $V_1 := Y_1 \otimes K_\lambda$ *for* $L_{\lambda_1}$.

(2) *If* $(h_1 - h_2) \neq 0$ *for* $L_{\lambda_2}$,

$$V_2 := -(h_1 - h_2)^{-1} Y_1 \otimes e_{2,1}K_\lambda + Y_2 \otimes K_\lambda. \tag{9.29}$$

(3) *If* $h_2 > 0$ *for* $L_{\lambda_3}$, *we have* $V_3 :=$

$$Y_1 \otimes (-(1 + h_2)e_{3,1} + e_{3,2}e_{2,1})K_\lambda - (h_1 + 1)Y_2 \otimes e_{3,2}K_\lambda + h_2(h_1 + 1)Y_3 \otimes K_\lambda. \tag{9.30}$$

PROOF. (1) We see that $Y_1 \otimes K_\lambda$ has weight $\lambda_1$, and we immediately check that it is killed by $e_{1,2}, e_{2,3}$.

(2) As for $\lambda_2 = \lambda + d_2$, we should look for a vector of type $Y_1 \otimes A + Y_2 \otimes B$ where $B$ has weight $\lambda$ and $A$ has weight $\lambda + (d_2 - d_1)$. This implies (by Remark 9.2.43) that we may take $B = K_\lambda$ and for $A$ some multiple of $e_{2,1}K_\lambda$. Take such an element

$$K_x := xY_1 \otimes e_{2,1}K_\lambda + Y_2 \otimes K_\lambda.$$

We need to impose the condition that this element is killed by $e_{1,2}, e_{2,3}$. Clearly $e_{2,3}$ kills both summands while $e_{1,2}Y_1 = 0,\ e_{1,2}Y_2 = Y_1$ so acts as

$$\begin{aligned} e_{1.2}K_x &= xY_1 \otimes e_{1,2}e_{2,1}K_\lambda + Y_1 \otimes K_\lambda = xY_1 \otimes [e_{1,2}, e_{2,1}]K_\lambda + Y_1 \otimes K_\lambda \\ &= xY_1 \otimes (h_1 - h_2)K_\lambda + Y_1 \otimes K_\lambda = 0 \iff x = -(h_1 - h_2)^{-1}. \end{aligned}$$

We remark that at the end we have used the fact that $(h_1 - h_2) \neq 0$. We deduce that a highest weight vector for $\lambda_2$ is the one given by formula (9.29).

(3) As for $\lambda_3$ we look for a vector of type $Y_1 \otimes A + Y_2 \otimes B + Y_3 \otimes C$ where $B$ has weight $\lambda + d_3 - d_2$, $A$ has weight $\lambda + (d_3 - d_1)$, and $C$ has weight $\lambda$. This implies by simple considerations that we may take $C = K_\lambda$ and for $B$ some multiple $xe_{3,2}K_\lambda$ and finally for $A = (ye_{3,1} + ze_{3,2}e_{2,1})K_\lambda$. Set

$$K_{x,y,z} := Y_1 \otimes (ye_{3,1} + ze_{3,2}e_{2,1})K_\lambda + Y_2 \otimes xe_{3,2}K_\lambda + Y_3 \otimes K_\lambda.$$

We need to impose the condition that $K_{x,y,z}$ is killed by $e_{1,2}, e_{2,3}$.

$$\begin{aligned} e_{1,2}K_{x,y,z} &= Y_1 \otimes e_{1,2}(ye_{3,1} + ze_{3,2}e_{2,1})K_\lambda + Y_1 \otimes xe_{3,2}K_\lambda \\ &= Y_1 \otimes \theta K_\lambda,\ \theta = [ye_{1,2}e_{3,1} + ze_{1,2}e_{3,2}e_{2,1} + xe_{3,2}]. \end{aligned}$$

Now $e_{1,2}e_{3,1} = -e_{3,2} + e_{3,1}e_{1,2}$ and

$$e_{1,2}e_{3,2}e_{2,1} = e_{3,2}e_{1,2}e_{2,1} = e_{3,2}[e_{1,1} - e_{2,2}] + e_{3,2}e_{2,1}e_{1,2}.$$

So $\theta$ applied to $K_\lambda$ gives

$$((x-y)e_{3,2} + ze_{3,2}[e_{1,1} - e_{2,2}])K_\lambda = (x - y + z(h_1 - h_2))e_{3,2}K_\lambda,$$

which is 0 if and only if $x - y + z(h_1 - h_2) = 0$.

$$e_{2,3}K_{x,y,z} = Y_1 \otimes e_{2,3}(ye_{3,1} + ze_{3,2}e_{2,1})K_\lambda + Y_2 \otimes e_{2,3}xe_{3,2}K_\lambda + Y_2 \otimes K_\lambda.$$

As before we see that $xe_{2,3}e_{3,2}K_\lambda = x[e_{2,3}, e_{3,2}]K_\lambda = h_2xK_\lambda$, so we need to take $x = -h_2^{-1}$ and we see the need that $h_2 \neq 0$.

Next

$$\begin{aligned} e_{2,3}(ye_{3,1} + ze_{3,2}e_{2,1}) &= y[e_{2,3}, e_{3,1}] + ye_{3,1}e_{2,3} + z[e_{2,3}, e_{3,2}]e_{2,1} + ze_{3,2}e_{2,1}e_{2,3} \\ &= ye_{2,1} + ye_{3,1}e_{2,3} + z[e_{2,2} - e_{3,3}]e_{2,1} + ze_{3,2}e_{2,1}e_{2,3} \\ &= ye_{2,1} + ze_{2,1} + ze_{2,1}e_{2,2} - ze_{2,1}e_{3,3} + (ye_{3,1} + ze_{3,2}e_{2,1})e_{2,3} \end{aligned}$$

applied to $K_\lambda$ gives

$$(y + z(1 + h_2))e_{2,1}K_\lambda.$$

So finally we have the equations

$$y + z(1 + h_2) = -h_2^{-1} - y + z(h_1 - h_2) = 0,$$

$$-h_2^{-1} + z(h_1 + 1) = 0 \implies z = \frac{1}{h_2(h_1+1)}, \quad y = -\frac{1 + h_2}{h_2(h_1 + 1)}.$$

If we multiply by $h_2(h_1 + 1)$, we have thus the highest weight vector given by formula (9.30). $\square$

STEP 3 (apply the map $j_1$). We now need to apply these formulas in order to compute the image of these three highest weight vectors given by Lemma 9.2.5 under one of the three maps $L_1 \otimes L_\lambda \to \tilde{V}_m$, and verify if it is nonzero.

$$j_1 : a \otimes b \mapsto ab, \; j_2 : a \otimes b \mapsto ba, \; j_3 : a \otimes b \mapsto [a, b].$$

We shall work in detail $j_1$, then $j_2, j_3$ can be left as exercise using Corollary 9.1.28. Let $\lambda = 1^a2^b$ so that $h_1 = a + b$, $h_2 = b$. We have thus $K_\lambda = Y_1^a[Y_1, Y_2]^b$ and, from the formulas in Lemma 9.2.5,

$$j_1(Y_1 \otimes K_\lambda) = Y_1^{a+1}[Y_1, Y_2]^b, \tag{9.31}$$

$$j_1(-a^{-1}Y_1 \otimes e_{2,1}K_\lambda + Y_2 \otimes K_\lambda) = -a^{-1}Y_1e_{2,1}(Y_1^a[Y_1, Y_2]^b) + Y_2Y_1^a[Y_1, Y_2]^b,$$

$$\begin{aligned} j_1(Y_1 \otimes (-(1+b)e_{3,1} + e_{3,2}e_{2,1})K_\lambda - (a+b+1)Y_2 \otimes e_{3,2}K_\lambda + b(a+b+1)Y_3 \otimes K_\lambda)) \\ = Y_1(-(1+b)e_{3,1} + e_{3,2}e_{2,1})Y_1^{a+1}[Y_1, Y_2]^b \\ - (a+b+1)Y_2e_{3,2}Y_1^{a+1}[Y_1, Y_2]^b \\ + b(a+b+1)Y_3Y_1^{a+1}[Y_1, Y_2]^b. \end{aligned}$$

So our task is to compute explicit formulas for the right-hand side of these three elements—in fact only for the last two—and show that they are not 0.

We need to start with

LEMMA 9.2.6. *Let us give two elements $u, v$ in a ring so that $u^2$ is central and give a commuting variable $x$. Then the coefficient of $x$ in the binomial $(u+xv)^a$ is the element $Z_a$ given by the formula*

$$Z_a = \begin{cases} \text{if } a=1 & v \\ \text{if } a=2b & bu^{2b-2}(uv+vu) \\ \text{if } a=2b+1,\ b>0 & u^{2b-2}(buvu+(b+1)vu^2). \end{cases} \tag{9.32}$$

PROOF. By induction we see that by definition

$$(u+xv)^a = u^a + xZ_a + O(x^2)$$
$$\implies (u+xv)^{a+1} = u^{a+1} + xZ_{a+1} + O(x^2) = u^{a+1} + x(Z_a u + u^a v) + O(x^2),$$

thus we have the inductive formula $Z_{a+1} = Z_a u + u^a v$. We then see that $Z_1 = v$ by definition $Z_2 = vu + uv$ and by induction, if $a = 2b+1$ is odd and bigger than 2, we have

$$\begin{aligned} Z_{2b+1} &= Z_{2b}u + u^{2b}v = bu^{2b-2}(uv+vu)u + u^{2b+1}v = bu^{2b-2}uvu + bu^{2b}v + u^{2b}v \\ &= u^{2b-2}(buvu + (b+1)vu^2), \\ Z_{2b+2} &= Z_{2b+1}u + u^{2b+1}v = u^{2b-2}(buvu+(b+1)vu^2)u + u^{2b+1}v \\ &= bu^{2b-2}uvu^2 + (b+1)u^{2b}vu + u^{2b+1}v = (b+1)u^{2b}(uv+vu). \end{aligned}$$

□

We are now ready to compute

$$e_{2,1}K_\lambda,\ e_{3,1}K_\lambda,\ e_{3,2}K_\lambda,\ e_{3,2}e_{2,1}K_\lambda,\ K_\lambda = Y_1^a[Y_1, Y_2]^b.$$

In general to compute $e_{i,j}A$, where $A$ is some expression in the $Y_i$, since $e_{i,j}$ is a derivation (cf. Remark 9.1.27), we must consider the isomorphism induced by the invertible linear transformation $\tau : Y_j \mapsto Y_j + xY_i,\ Y_h \mapsto Y_h,\ \forall h \neq j$, with $x$ a parameter, and take the coefficient of $x$.

When we do this to $Y_1^a[Y_1, Y_2]^b$, we get 0 unless $j = 1, 2$. Since $e_{i,j}$ is a derivation, we have

$$e_{i,j}(Y_1^a[Y_1, Y_2]^b) = e_{i,j}(Y_1^a)[Y_1, Y_2]^b + Y_1^a e_{i,j}([Y_1, Y_2]^b).$$

We have three cases, we have to expand

$$\text{if } j=1,\ (Y_1 + xY_i)^a, \quad [Y_1 + xY_i, Y_2]^b, \text{ if } j=2,\ [Y_1, Y_2 + xY_i]^b.$$

Since $[Y_1, Y_2 + xY_i]^b = ([Y_1, Y_2] + x[Y_1, Y_i])^b$ and $Y_1^2,\ [Y_1, Y_2]^2$ are central, we can apply formula (9.32) and we deduce

$$e_{2,1}Y_1^a = \begin{cases} \text{if } a=1 & Y_2 \\ \text{if } a=2b & bY_1^{2b-2}(Y_1Y_2 + Y_2Y_1) \\ \text{if } a=2b+1,\ b>0 & Y_1^{2b-2}(-bY_1[Y_1, Y_2] + (2b+1)Y_2Y_1^2), \end{cases} \tag{9.33}$$

$$e_{3,1}Y_1^a = \begin{cases} \text{if } a=1 & Y_3 \\ \text{if } a=2b & bY_1^{2b-2}(Y_1Y_3 + Y_3Y_1) \\ \text{if } a=2b+1,\ b>0 & Y_1^{2b-2}(-bY_1[Y_1, Y_3] + (2b+1)Y_3Y_1^2), \end{cases} \tag{9.34}$$

(9.35)
$$e_{3,1}[Y_1,Y_2]^b=\begin{cases}\text{if } b=1 & [Y_3,Y_2]\\ \text{if } b=2c & c[Y_1,Y_2]^{2c-2}([Y_1,Y_2][Y_3,Y_2]+[Y_3,Y_2][Y_1,Y_2])\\ \text{if } b=2c+1, & c>0\\ \multicolumn{2}{l}{[Y_1,Y_2]^{2c-2}(c[Y_1,Y_2][Y_3,Y_2][Y_1,Y_2]+(c+1)[Y_3,Y_2][Y_1,Y_2]^2),}\end{cases}$$

(9.36)
$$e_{3,2}[Y_2,Y_1]^b=\begin{cases}\text{if } b=1 & [Y_3,Y_1]\\ \text{if } b=2c & c[Y_2,Y_1]^{2c-2}([Y_2,Y_1][Y_3,Y_1]+[Y_3,Y_1][Y_2,Y_1])\\ \text{if } b=2c+1, & c>0\\ \multicolumn{2}{l}{[Y_2,Y_1]^{2c-2}(c[Y_2,Y_1][Y_3,Y_1][Y_2,Y_1]+(c+1)[Y_3,Y_1][Y_2,Y_1]^2).}\end{cases}$$

From formula (9.33) we further deduce by the same method

$$(9.37)\quad e_{3,2}e_{2,1}Y_1^a=\begin{cases}\text{if } a=1 & Y_3\\ \text{if } a=2b & bY_1^{2b-2}(Y_1Y_3+Y_3Y_1)\\ \text{if } a=2b+1,\ b>0 & Y_1^{2b-2}(-bY_1[Y_1,Y_3]+(2b+1)Y_3Y_1^2).\end{cases}$$

Let $V_2$ be the highest weight vector of weight $\lambda_2$ given by formula (9.29) and $j_1(V_2)=-a^{-1}Y_1e_{2,1}(Y_1^a[Y_1,Y_2]^b)+Y_2Y_1^a[Y_1,Y_2]^b$ given by formula (9.31).

Since $[Y_1+xY_2,Y_2]=[Y_1,Y_2]$, we see immediately that $[Y_1,Y_2]^b$ is a factor killed by $e_{2,1}$, so it can be pulled out of the sum and it is enough to compute when $b=0$. We need to apply formula (9.33) and we have

(9.38)
$$=\begin{cases}\text{if } a=1 & -[Y_1,Y_2]\\ \text{if } a=2c & -\frac{1}{2}Y_1^{2c-2}(Y_1Y_2+Y_2Y_1)+Y_2Y_1^{2c}=-\frac{1}{2}Y_1^{2c-2}[Y_1,Y_2]\\ \text{if } a=2c+1,\ c>0 & -a^{-1}Y_1^{2c-1}(-cY_1[Y_1,Y_2]+(2c+1)Y_2Y_1^2)+Y_2Y_1^a\\ & =\frac{c}{2c+1}Y_1^{2c}[Y_1,Y_2].\end{cases}$$

For all $\lambda$ we thus have that $j_1(V_2),j_2(V_2)$ are both nonzero.

Let $V_3$ be the highest weight vector of weight $\lambda_3$ given by formula (9.30),

$$Y_1\otimes(-(1+h_2)e_{3,1}+e_{3,2}e_{2,1})K_\lambda-(h_1+1)Y_2\otimes e_{3,2}K_\lambda+h_2(h_1+1)Y_3\otimes K_\lambda.$$

We need thus to compute the element $j_1(V_3)$ and prove that it is nonzero,

$$(9.39)\qquad Y_1(-(1+h_2)e_{3,1}+e_{3,2}e_{2,1})K_\lambda-(h_1+1)Y_2e_{3,2}K_\lambda+h_2(h_1+1)Y_3K_\lambda.$$

This turns out to be a very lengthy computation. Let us do in detail only the case $h_1=2b+2d,\ h_2=2d$; the other cases are similar. We split the computation of (9.39) into its three summands, which we call $A_1,A_2,A_3$. For the first summand we need to compute the two terms arising from $e_{3,1}K_\lambda$ and $e_{3,2}e_{2,1}K_\lambda$.

We start by computing $e_{3,2}e_{2,1}K_\lambda$, recall that for two trace 0, $2\times 2$ matrices $A,B$, we have $AB+BA=\operatorname{tr}(AB)$ so,

$$\begin{aligned}e_{3,2}e_{2,1}(Y_1^{2b}[Y_1,Y_2]^{2d})&=e_{3,2}bY_1^{2b-2}(Y_1Y_2+Y_2Y_1)[Y_1,Y_2]^{2d}\\&=bY_1^{2b-2}e_{3,2}(Y_1Y_2+Y_2Y_1)[Y_1,Y_2]^{2d}=bY_1^{2b-2}e_{3,2}\operatorname{tr}(Y_1Y_2)[Y_1,Y_2]^{2d},\end{aligned}$$

$$(9.40)\qquad \begin{aligned} e_{3,2}\operatorname{tr}(Y_1Y_2)[Y_1,Y_2]^{2d} &= \operatorname{tr}(Y_1Y_3)[Y_1,Y_2]^{2d} + \operatorname{tr}(Y_1Y_2)e_{3,2}[Y_1,Y_2]^{2d}\\ &= \operatorname{tr}(Y_1Y_3)[Y_1,Y_2]^{2d} + \operatorname{tr}(Y_1Y_2)d[Y_2,Y_1]^{2d-2}\operatorname{tr}([Y_2,Y_1][Y_3,Y_1]), \end{aligned}$$

$$\begin{aligned} &e_{3,2}e_{2,1}(Y_1^{2b}[Y_1,Y_2]^{2d})\\ &\quad = Y_1^{2b-2}[Y_1,Y_2]^{2d-2}(b\ \operatorname{tr}(Y_1Y_3)[Y_1,Y_2]^2 + bd\ \operatorname{tr}(Y_1Y_2)\operatorname{tr}([Y_1,Y_2][Y_1,Y_3])). \end{aligned}$$

Next we compute $e_{3,1}K_\lambda$,

$$\begin{aligned} e_{3,1}(Y_1^{2b}[Y_1,Y_2]^{2d}) &= e_{3,1}(Y_1^{2b})[Y_1,Y_2]^{2d} + Y_1^{2b}e_{3,1}([Y_1,Y_2]^{2d})\\ &= bY_1^{2b-2}\operatorname{tr}(Y_1Y_3)[Y_1,Y_2]^{2d} + Y_1^{2b}d[Y_1,Y_2]^{2d-2}\operatorname{tr}([Y_1,Y_2][Y_3,Y_2])\\ &= Y_1^{2b-2}[Y_1,Y_2]^{2d-2}(b\ \operatorname{tr}(Y_1Y_3)[Y_1,Y_2]^2 + Y_1^2 d\ \operatorname{tr}([Y_1,Y_2][Y_3,Y_2])). \end{aligned}$$

Summing these two computations, we have

$$\begin{aligned} &-(1+2d)(b\ \operatorname{tr}(Y_1Y_3)[Y_1,Y_2]^2 + Y_1^2 d\ \operatorname{tr}([Y_1,Y_2][Y_3,Y_2]))\\ &\qquad + b\ \operatorname{tr}(Y_1Y_3)[Y_1,Y_2]^2 + bd\ \operatorname{tr}(Y_1Y_2)\operatorname{tr}([Y_1,Y_2][Y_1,Y_3])\\ &\quad = d(-2b\ \operatorname{tr}(Y_1Y_3)[Y_1,Y_2]^2 - (1+2d)Y_1^2\ \operatorname{tr}([Y_1,Y_2][Y_3,Y_2])\\ &\qquad\qquad + b\ \operatorname{tr}(Y_1Y_2)\operatorname{tr}([Y_1,Y_2][Y_1,Y_3])), \end{aligned}$$

and we deduce that

$$\begin{aligned} -(1+h_2)e_{3,1} + e_{3,2}e_{2,1})K_\lambda &= Y_1^{2b-2}[Y_1,Y_2]^{2d-2}d\,X_1,\\ X_1 &= -(1+2d)Y_1^2\ \operatorname{tr}([Y_1,Y_2][Y_3,Y_2])\\ &\quad + b(\ \operatorname{tr}(Y_1Y_2)\operatorname{tr}([Y_1,Y_2][Y_1,Y_3]) - 2\ \operatorname{tr}(Y_1Y_3)[Y_1,Y_2]^2). \end{aligned}$$

The first term $A_1$ of (9.39) contributes thus $Y_1^{2b-2}[Y_1,Y_2]^{2d-2}d\,Y_1X_1$.

As for the second term $A_2$ of (9.39) we remark first that

$$e_{3,2}Y_1^{2b}[Y_1,Y_2]^{2d} = Y_1^{2b}e_{3,2}[Y_1,Y_2]^{2d}$$

and then apply formula (9.36)

$$\begin{aligned} A_2 &= -(2b+2d+1)Y_2Y_1^{2b}(d[Y_2,Y_1]^{2d-2}\operatorname{tr}([Y_2,Y_1][Y_3,Y_1]))\\ &= -d(2b+2d+1)Y_1^{2b-2}[Y_1,Y_2]^{2d-2}Y_2X_2,\ X_2 = Y_1^2\operatorname{tr}([Y_2,Y_1][Y_3,Y_1]). \end{aligned}$$

The third term gives $2d(2b+2d+1)Y_1^{2b-2}[Y_1,Y_2]^{2d-2}Y_3Y_1^2[Y_1,Y_2]^2$. So we deduce that $j_1(V_3) = Y_1^{2b-2}[Y_1,Y_2]^{2d-2}d\,R$,

$$R = Y_1X_1 - (2b+2d+1)Y_2X_2 + 2d(2b+2d+1)Y_3Y_1^2[Y_1,Y_2]^2.$$

The final formula to compute is the value of $R$:

$$\begin{aligned} R &= \boxed{-(1+2d)Y_1^3\operatorname{tr}([Y_1,Y_2][Y_3,Y_2]))}\\ &\quad + b\boxed{Y_1\operatorname{tr}(Y_1Y_2)\operatorname{tr}([Y_1,Y_2][Y_1,Y_3]) - 2Y_1\operatorname{tr}(Y_1Y_3)[Y_1,Y_2]^2}\\ &\quad + (2b+2d+1)\boxed{Y_2Y_1^2\operatorname{tr}([Y_1,Y_2][Y_3,Y_1]) + 2Y_3Y_1^2[Y_1,Y_2]^2}. \end{aligned}$$

We need some formulas:

$$(9.41)\qquad Y_1^2 = \frac{1}{2}\begin{vmatrix}1\\1\end{vmatrix},\qquad \operatorname{tr}([Y_3,Y_2][Y_1,Y_2]) = 2\begin{vmatrix}1 & 2\\2 & 3\end{vmatrix},$$

$$\left|\begin{matrix}1&2\\2&3\\1&\end{matrix}\right| = \left|\begin{matrix}1&2\\1&3\\2&\end{matrix}\right| - \left|\begin{matrix}1&2\\1&2\\3&\end{matrix}\right| - 2\left|\begin{matrix}1&2&3\\1&2&\end{matrix}\right|. \tag{9.42}$$

We now associate to the various monomials their expression through tableaux using formulas (9.41) and (9.42). For the first line we obtain:

$$Y_1^3\operatorname{tr}([Y_1,Y_2][Y_3,Y_2]) = \left|\begin{matrix}1&2\\2&3\\1&\\1&\\1&\end{matrix}\right| = -2\left|\begin{matrix}1&2&3\\1&2&\\1&&\\1&&\end{matrix}\right| + \left|\begin{matrix}1&2\\1&3\\1&\\1&\\2&\end{matrix}\right| - \left|\begin{matrix}1&2\\1&2\\1&\\1&\\3&\end{matrix}\right|.$$

The second line consists of two terms. The first term gives

$$Y_1\operatorname{tr}(Y_1Y_2) = 2Y_2Y_1^2 - Y_1[Y_1,Y_2] = \left|\begin{matrix}1\\1\\2\end{matrix}\right| - 2\left|\begin{matrix}1&2\\1&\end{matrix}\right|, \tag{9.43}$$

$$Y_1\operatorname{tr}(Y_1Y_2)\operatorname{tr}([Y_1,Y_2][Y_1,Y_3]) = -2\left(\left|\begin{matrix}1\\1\\2\end{matrix}\right| - 2\left|\begin{matrix}1&2\\1&\end{matrix}\right|\right)\left|\begin{matrix}1&2\\1&3\end{matrix}\right| \tag{9.44}$$

$$= -2\left(\left|\begin{matrix}1&2\\1&3\\1&\\1&\\2&\end{matrix}\right| - 2\left|\begin{matrix}1&2\\1&3\\1&2\\1&\end{matrix}\right|\right), \quad -\left|\begin{matrix}1&2\\1&3\\1&2\\1&\end{matrix}\right| + \left|\begin{matrix}1&2\\1&2\\1&3\\1&\end{matrix}\right| = 2\left|\begin{matrix}1&2&3\\1&2&\\1&&\\1&&\end{matrix}\right|. \tag{9.45}$$

The second term gives

$$Y_1\operatorname{tr}(Y_1Y_3)[Y_1,Y_2]^2 = -\left(\left|\begin{matrix}1\\1\\3\end{matrix}\right| - 2\left|\begin{matrix}1&3\\1&\end{matrix}\right|\right)\left|\begin{matrix}1&2\\1&2\end{matrix}\right| = -\left|\begin{matrix}1&2\\1&2\\1&\\1&\\3&\end{matrix}\right| + 2\left|\begin{matrix}1&2\\1&2\\1&3\\1&\end{matrix}\right|. \tag{9.46}$$

The last line has also two terms which by the same type of computations are

$$Y_2Y_1^2\operatorname{tr}([Y_1,Y_2][Y_3,Y_1]) = \left|\begin{matrix}1&2\\1&3\\1&\\1&\\2&\end{matrix}\right|, \quad 2Y_3Y_1^2[Y_1,Y_2]^2 = -\left|\begin{matrix}1&2\\1&2\\1&\\1&\\3&\end{matrix}\right|,$$

substituting in the formula we finally have

$$2(1+2d)\begin{vmatrix}1&2&3\\1&2&\\1&&\\1&&\end{vmatrix}-(1+2d)\begin{vmatrix}1&2\\1&3\\1&\\1&\\2&\end{vmatrix}+(1+2d)\begin{vmatrix}1&2\\1&2\\1&\\1&\\3&\end{vmatrix}$$

$$-2b\left(\begin{vmatrix}1&2\\1&3\\1&\\1&\\2&\end{vmatrix}-2\begin{vmatrix}1&2\\1&2\\1&3\\1&\end{vmatrix}-4\begin{vmatrix}1&2&3\\1&2&\\1&&\\1&&\end{vmatrix}-\begin{vmatrix}1&2\\1&2\\1&\\1&\\3&\end{vmatrix}+2\begin{vmatrix}1&2\\1&2\\1&3\\1&\end{vmatrix}\right)$$

$$+(2b+2d+1)\begin{vmatrix}1&2\\1&3\\1&\\1&\\2&\end{vmatrix}-(2b+2d+1)\begin{vmatrix}1&2\\1&2\\1&\\1&\\3&\end{vmatrix},$$

which finally gives the nonzero highest weight vector

$$2(1+4b+2d)\begin{vmatrix}1&2&3\\1&2&\\1&&\\1&&\end{vmatrix}=j_1(V_3)\neq 0,$$

which is nonzero, completing the proof of Proposition 9.2.3. □

## 9.3. The structure of generic $2\times 2$ matrices

**9.3.1. The commutators' ideals.** We need a simple remark—the algebra $\tilde{V}_m$ is generated by the elements $Y_i$, so when we pass modulo the ideal generated by the commutators, that is we take the maximal abelian quotient, we still have an algebra generated by the classes of the $Y_i$ which we may denote by $y_i$. We claim that these elements are algebraically independent. For this it is enough to show that there is a homomorphism of $\tilde{V}_m$ into a polynomial ring in some variables $u_i$ such that $Y_i \mapsto u_i$.

This can be defined by specializing the generic trace 0 matrices,

$$Y_i \mapsto u_i := \begin{vmatrix} y_i & 0\\ 0 & -y_i\end{vmatrix}.$$

It follows then that also the maximal abelian quotient of $V_m$ is a polynomial ring in the $2m$ variables $y_i, t_i$ and the image of the generic matrix $X_i$ is $y_i+t_i$, and gives the maximal abelian quotient of $R_m$.

9.3.1.1. *The algebra $R_m$ of generic matrices.* In this section let us simplify the discussion by assuming that we are in characteristic 0. This is not essential for the results, and for the details one can look at **[Pro84]**. We start from the *Wagner identity*

(9.47) $$[x,y]^2=-\det([x,y]) \implies [[x,y]^2,z]=0,\quad \text{Wagner identity},$$

or better a multilinear form (cf. Exercise 9.1.6 and formula (9.20) which gives a tableau interpretation),

$$[x_1,x_2][x_3,x_4]+[x_3,x_4][x_1,x_2]=\operatorname{tr}([x_1,x_2][x_3,x_4])=-2\begin{vmatrix}1&2\\3&4\end{vmatrix}, \tag{9.48}$$

so that $[x_1,x_2][x_3,x_4]+[x_3,x_4][x_1,x_2]$ is a central polynomial with value

$$\begin{aligned}\operatorname{tr}([x_1,x_2][x_3,x_4])&=\operatorname{tr}([x_1,x_2]x_3x_4)-\operatorname{tr}([x_1,x_2]x_4x_3)\\&=\operatorname{tr}([x_1,x_2]x_3x_4)-\operatorname{tr}(x_3[x_1,x_2]x_4)=\operatorname{tr}([[x_1,x_2],x_3]x_4).\end{aligned}$$

Introducing a new variable $z$, we have (cf. Razmyslov's transform, Theorem 10.2.7)

$$\begin{aligned}\operatorname{tr}\Big(z([x_1,x_2][x_3,x_4]+[x_3,x_4][x_1,x_2])\Big)&=\operatorname{tr}(z)\operatorname{tr}([[x_1,x_2],x_3]x_4)\\&=\operatorname{tr}(\operatorname{tr}(z)[[x_1,x_2],x_3]x_4)=\operatorname{tr}(g(z,x_1,x_2,x_3)x_4),\end{aligned}$$

$$g(z,x_1,x_2,x_3)=[z[x_1,x_2]+[x_1,x_2]z,x_3]. \tag{9.49}$$

Hence the main formula,

$$[z[x_1,x_2]+[x_1,x_2]z,x_3]=\operatorname{tr}(z)[[x_1,x_2],x_3]. \tag{9.50}$$

We now recall that by formula (9.22) in characteristic 0 the algebra $\tilde V_m$ of generic trace 0 matrices has the canonical decomposition into $\mathrm{GL}(m,F)$ irreducible representations $\tilde V_m=\bigoplus_{\lambda\in\mathcal P^3}L_\lambda$. In particular let us look at the first three degrees.

In degree 1 we have only $L_1$ which is the span of the generic matrices $Y_i$. In degree 2 we have $L_2\oplus L_{1,1}$ and we have $L_2\subset V_m^0$, $L_{1,1}\subset T_m^0$.

We claim that

$$[L_1,[L_1,L_1]]=L_{2,1}, \tag{9.51}$$

in fact since $L_{2,1}$ is irreducible and clearly $[L_1,[L_1,L_1]]\neq 0$, it is enough to see that $[L_1,[L_1,L_1]]\subset L_{2,1}$.

Next observe that $[L_1,[L_1,L_1]]\subset V_m^0$ since it is formed by trace 0 elements. But $V_m^0$ in degree 3 coincides with $L_{2,1}$.

We have now a first main theorem,

THEOREM 9.3.1.

(1)

$$L_{2,1}V_m=V_mL_{2,1}\subset R_m \tag{9.52}$$

(2) $L_{2,1}V_m=I[t_1,\dots,t_m]$, *where* $I:=\bigoplus_{\lambda\in\mathcal P^3,\lambda\supset 2,1}L_\lambda,\quad t_i=\frac{\operatorname{tr}(X_i)}{2}$.

PROOF. (1) We have that $[X_i,[X_j,X_k]]=[Y_i,[Y_j,Y_k]]\in L_{2,1}\cap R_m$. Since $L_{2,1}$ is an irreducible representation of $\mathrm{GL}(m,F)$, we have $L_{2,1}\subset R_m$. By formula (9.50) follows that for every $z\in R_m$, we have that $\operatorname{tr}(z)[X_i,[X_j,X_k]]$ is the evaluation of the polynomial $g$ of formula (9.49), hence by applying $\mathrm{GL}(m,F)$, we have $\operatorname{tr}(z)L_{2,1}\subset R_m$. But now formula (9.3) shows by induction that all products $\operatorname{tr}(z_1)\operatorname{tr}(z_2)\cdots\operatorname{tr}(z_k)$ lie in the subspace of $S$ of linear combinations of elements $\operatorname{tr}(z)$ with elements in $R_m$, hence it follows that also $r(z_1)\operatorname{tr}(z_2)\cdots\operatorname{tr}(z_k)L_{2,1}\subset R_m$.

(2) Since $V_m=\tilde V_m[t_1,\dots,t_m]$, it is enough to remark that $I:=L_{2,1}\tilde V_m$ is the direct sum $\bigoplus_{\lambda\in\mathcal P^3,\lambda\supset 2,1}L_\lambda$. Since $\tilde V_m$ is generated, as algebra, by $L_1$, this statement follows from Proposition 9.2.3. □

REMARK 9.3.2. The space

$$L_{2,1}V_m = I[t_1, \dots, t_m] = \bigoplus_{\lambda \in \mathcal{P}^3, \lambda \supset 2,1} L_\lambda[t_1, \dots, t_m] \tag{9.53}$$

is a common ideal of $V_m$ and $R_m$.

EXERCISE 9.3.3. Prove that $L_{2,1}V_m$, as an ideal of $R_m$, is generated by $L_{2,1}$ and by the elements $[X_i, X_j][X_h, X_k]$ (cf. formula (9.21)).

Since $L_{2,1}V_m$ is described explicitly by formula (9.53), we have the description

$$V_m/L_{2,1}V_m = K[t_1, \dots, t_m],\ K := \bigoplus_{\lambda \in \mathcal{P}^3, \lambda \not\supset 2,1} L_\lambda. \tag{9.54}$$

The next step is to understand $R_m/L_{2,1}V_m \subset V_m/L_{2,1}V_m$.

Now a partition $\lambda \not\supset 2,1$ can be of three types, the partition 3 or 2 or $1^k$.

Let $[R_m, R_m]$ (and $[V_m, V_m]$) denote the two-sided ideal generated by the commutators in $R_m$, resp., in $V_m$. We have that $R_m/[R_m, R_m]$ is a polynomial ring in $m$ variables, the classes of the generic matrices $X_i$, so we need to understand $[R_m, R_m]/L_{2,1}V_m \subset [V_m, V_m]/L_{2,1}V_m$. We have that $[V_m, V_m]$ is generated by $L_2$ so that

(9.55)

$$[V_m, V_m] = \bigoplus_{\lambda \in \mathcal{P}^3, \lambda \supset 2} L_\lambda[t_1, \dots, t_m],\ [V_m, V_m]/L_{2,1}V_m = (L_2 \oplus L_3)[t_1, \dots, t_m].$$

Consider first $L_3$. By Proposition 9.2.3 we have $L_1L_3 = L_3L_1 = L_{3,1} \subset L_{2,1}$, so in the algebra $V_m/L_{2,1}V_m$ we have

$$Y_ia = aY_i = 0, \implies X_ia = aX_i = t_ia,\ \forall a \in L_3.$$

We have that $L_2$ is the span of the commutators $[X_i, X_j]$.

From formula (9.19) and Corollary 9.1.17 we also have, modulo $L_{2,1}V_m$,

$$X_i[X_h, X_k] = t_i[X_h, X_k] + |i, h, k| = [X_h, X_k]X_i, \tag{9.56}$$

$$X_jX_i[X_h, X_k] = t_jt_i[X_h, X_k] + t_j|i, h, k| + t_i|j, h, k|. \tag{9.57}$$

We deduce

COROLLARY 9.3.4. *The ideal $[R_m, R_m]$ of $R_m$ generated by the commutators contains $L_{2,1}V_m$. Furthermore*

$$[R_m, R_m]/L_{2,1}V_m = R_mL_2 \overset{j}{\simeq} F[t_1, \dots, t_m] \otimes L_2. \tag{9.58}$$

PROOF. From formula (9.56) it follows that, modulo $L_{2,1}V_m$, the left and right action of $R_m$ on the commutators coincide, and from formula (9.57) it follows that this action factors through the commutators. The isomorphism $j$ follows by passing modulo $L_3[t_1, \dots, t_m]$ and from the fact that $V_m/L_{2,1}V_m$ is a free $F[t_1, \dots, t_m]$-module given by formula (9.54). □

9.3.1.2. *Hilbert series and cocharacters.* Thus we have this canonical filtration $R_m \supset [R_m, R_m] \supset L_{2,1}V_m$ and an explicit description of the factors so we can perform several computations:

$$R_m/[R_m, R_m] = F[t_1, \dots, t_m], \quad [R_m, R_m]/L_{2,1}V_m = F[t_1, \dots, t_m] \otimes L_2,$$

$$L_{2,1}V_m = F[t_1, \dots, t_m] \otimes \Big( \bigoplus_{\lambda \in \mathcal{P}^3, \lambda \supset 2,1} L_\lambda \Big). \tag{9.59}$$

First of all let us compute the Hilbert series $H_{R_m}(t) = \sum_{i=0}^{\infty} \dim_F((R_m)_i)$. Recall that the Hilbert series of the polynomial ring $F[t_1, \dots, t_m]$ is $= \frac{1}{(1-t)^m}$, the Hilbert series of a direct sum is the sum and of a tensor product is the product, so we have clearly

$$\begin{aligned} H_{R_m}(t) &= H_{R_m/[R_m,R_m]}(t) + H_{[R_m,R_m]/L_{2,1}V_m}(t) + H_{L_{2,1}V_m}(t) \\ &= \frac{1}{(1-t)^m}\Big(1 + \binom{m}{2} t^2 + H_{L_{2,1}V_m]}(t)\Big). \end{aligned}$$

For $H_{L_{2,1}V_m}(t)$ recall Weyl's dimension formula. This is a general formula for the dimension of an irreducible representation of a semisimple group. In our case we have for $\lambda = 3^a 2^b 1^c$ a partition with three columns $h_1 = a+b+c, h_2 = a+b, h_3 = a$, the condition that $\lambda \supset 2,1$ is that $a+b+c \geq 2,\ a+b \geq 1$, and we have

$$\begin{aligned} d_\lambda &= \prod_{m \geq i > j \geq 1} \frac{h_j - h_i + i - j}{i - j} \\ &= (h_1 - h_2 + 1)(h_2 - h_3 + 1)\frac{h_1 - h_3 + 2}{2} \prod_{i=4}^{m} \Big(\frac{h_3 + i - 3}{i - 3}\frac{h_2 + i - 2}{i - 2}\frac{h_1 + i - 1}{i - 1}\Big) \\ &= (c+1)(b+1)\frac{b + c + 2}{2} \prod_{i=4}^{m} \Big(\frac{a + i - 3}{i - 3}\frac{a + b + i - 2}{i - 2}\frac{a + b + c + i - 1}{i - 1}\Big), \end{aligned}$$

the Hilbert series has the form

$$\sum_{\lambda \supset 2,1} d_\lambda t^{|\lambda|} = \sum_{a,b,c,\ a+b+c \geq 2,\ a+b \geq 1} d_{3^a 2^b 1^c} t^{3a+2b+c},$$

and it is easily seen that this series can be written as a rational function

$$\frac{p_m(t)}{(1-t)^{d_1+1}(1-t^2)^{d_2+1}(1-t^3)^{d_3+1}},$$

where the $d_i$ are the degrees in $c, b, a$, respectively, of the polynomial $d_\lambda$,

$$d_1 = m - 1, \quad d_2 = 2(m-2), \quad d_3 = 3(m-3).$$

As for the cocharacters one may discuss it for the various algebras constructed. The most difficult case is for polynomial identities of matrices, that is the description as representation of $S_m$ of the multilinear part of $R_m$ as function of $m$. This is a sum in which each of the three factors which are described in formulas (9.59) contribute with their respective multilinear parts. The first is $R_m/[R_m, R_m] = F[t_1, \dots, t_m]$ equals the polynomial ring in $m$ variables for each $m$ contributes the trivial representation given by the commutative monomial $t_1 t_2 \cdots t_m$.

For the other two factors we need to understand the multilinear part of some representation $F[t_1, \dots, t_m] \otimes L_\lambda(m)$.

By Definition 9.1.4 the symbol $L_\lambda = L_\lambda(m) = S_{\check{\lambda}}(V)$, $\dim V = m$.

In degree $k = |\lambda|$, the multilinear part of $L_\lambda(k)$ is the irreducible representation $M_{\tilde{\lambda}}$ of the symmetric group associated to the conjugate partition $\tilde{\lambda}$. In general thus $L_\lambda(m)$ will contribute to the cocharacter $\chi_m$, that is to multilinear part of $F[t_1, \dots, t_m] \otimes L_\lambda(m)$, only in degrees $m \geq k$.

PROPOSITION 9.3.5. *The multilinear part of $L_\lambda[t_1, \dots, t_m]$ as representation of $S_m$ is nonzero only in degrees $m \geq k = |\lambda|$. In this case it is the one induced*

$$\mathrm{Ind}_{S_k \times S_{m-k}}^{S_m} M_{\tilde{\lambda}} \otimes I_{m-k},$$

*where $I_{m-k}$ is the trivial representation of $S_{m-k}$.*

PROOF. The multilinear part of $L_\lambda[t_1, \dots, t_m]$ is the direct sum of the $\binom{m}{k}$ spaces given by a monomial in $m-k$ of the variables $t_i$ and the part of $L_\lambda$ which is multilinear in the complementary indices. But these spaces are permuted by the symmetric group $S_m$ starting from $M_{\tilde{\lambda}} t_{k+1} t_{k+2} \cdots t_m$ which equals $M_{\tilde{\lambda}} \otimes I_{m-k}$ as representation of $S_k \times S_{m-k}$. □

This implies an explicit description of the cocharacter in degree $m$. We have the trivial representation $I_m$ coming from the first factor, the one induced from $M_{1,1} \otimes I_{m-2}$ coming from the second factor, and finally from the third factor the ones induced by $M_{\tilde{\lambda}} \otimes I_{m-k}$ where $\tilde{\lambda}$ is a partition of $k \leq m$ which has at most three rows and must contain the hook 2, 1. Let us denote by $C_k$ this set of partitions.

THEOREM 9.3.6. *The final formula for the cocharacter of 2 × 2 matrices is then*

$$I_m \oplus \mathrm{Ind}_{S_2 \times S_{m-2}}^{S_m} M_{1,1} \otimes I_{m-2} \oplus \sum_{k=3}^{m} \sum_{\lambda \in C_k} \mathrm{Ind}_{S_k \times S_{m-k}}^{S_m} M_{\tilde{\lambda}} \otimes I_{m-k}. \tag{9.60}$$

This formula is also connected to the following explicit basis of the space of multilinear elements modulo the identities of 2 × 2 matrices. For a subset $S \subset \{1, \dots, m\}$ with $t$ elements $i_1 < i_2 \cdots < i_t$, let $\mathfrak{m}_S := x_{i_1} \cdots x_{i+t}$ be the corresponding ordered monomial. Let $\mathcal{T}$ denote the set of standard tableaux $T$ of size $k$ containing the little hook 2, 1 with at most three columns and with indices in a subset $S(T) \subset \{1, \dots, m\}$.

To each such tableau we can associate, using the theory developed, a multilinear noncommutative polynomial in the variables $x_i$, $i \in S$ which we denote by $\mathfrak{p}_T$. If $S \subset \{1, \dots, m\}$, denote by $S^c$ its complement in $\{1, \dots, m\}$. Then a basis is given by

$$x_1 \cdots x_m, \; [x_i, x_j] \mathfrak{m}_{\{i,j\}^c}, \; i < j, \quad \mathfrak{p}_T \mathfrak{m}_{S(T)^c}, \; T \in \mathcal{T}.$$

There is moreover a standard filtration of subspaces spanned by the obvious subsets of this basis which is stable under the symmetric group.

We can recover the formulas given by Drensky **[Dre90]** as follows. In formula (9.60) we have the three summands:

(1)
$$Q_1 = I_m,$$

(2)
$$Q_2 = \mathrm{Ind}_{S_2 \times S_{m-2}}^{S_m} M_{1,1} \otimes I_{m-2},$$

(3)
$$Q_3 = \sum_{k=3}^{m} \sum_{\lambda \in C_k} \mathrm{Ind}_{S_k \times S_{m-k}}^{S_m} M_{\tilde{\lambda}} \otimes I_{m-k}.$$

Here $C_k$ are the conjugates of partitions of $k$ with $\leq 3$ rows containing $\mu = (2,1)$.

REMARK 9.3.7.

(1) The computation of $\text{Ind}_{S_k \times S_{m-k}}^{S_m} M_{\tilde{\lambda}} \otimes I_{m-k}$ is done using its Frobenius character and formula (6.34). In this case we need to understand the multiplication of the two Schur functions $S_{\tilde{\lambda}} S_{m-k} = S_{\tilde{\lambda}} h_{m-k}$ which is given by one of Pieri's rules, Proposition 3.2.18.

(2) When $k \geq 4$ all partitions of $k$ with $\leq 3$ rows contain $\mu = (2,1)$, except the unique row partition. While for $k = 3$ we miss the row and the column partition.

We denote by $m_\lambda = m_\lambda(M_2(F))$ the corresponding multiplicity in the cocharacter of $M_2(F)$. Trivially, $m_{(m)} = 1$ is the contribution of $Q_1$. Next, we have the two parts partitions.

LEMMA 9.3.8. *Let* $\lambda = (\lambda_1, \lambda_2)$ *with* $\lambda_2 \geq 1$. *Then* $m_\lambda = (\lambda_1 - \lambda_2 + 1)\lambda_2$.

PROOF. Let $m = \lambda_1 + \lambda_2$. If $\lambda_2 \geq 2$, there is no contribution to $m_\lambda$ from the summands $Q_1, Q_2$ when we apply Pieri's formula, only from $Q_3$.

A partition $\mu \in \bigcup_{k \geq 3} C_k$ contributes to $m_\lambda$ if and only if $S_\lambda$ appears in the product $S_\mu h_{m-|\mu|}$, and this happens if and only if $\mu = (r+2, u)$ with the restrictions $\lambda_2 \leq r+2 \leq \lambda_1$ and $1 \leq u \leq \lambda_2$. The number of choices of such pairs is then $m_\lambda = (\lambda_1 - \lambda_2 + 1)\lambda_2$.

If $\lambda_2 = 1$, we have a contribution from $Q_2$ since in this case Pieri's rule gives $h_{m-k} S_{1,1} = S_{1+m-k,1} + S_{m-k,1,1}$. From the term $Q_3$ we have a contribution from each $\mu = (r, 1)$ with $2 \leq r \leq \lambda_1$. A total of $m_\lambda = \lambda_1$. □

LEMMA 9.3.9. $m_{(\lambda_1,1,1)} = 2\lambda_1 - 1$ *and* $m_{(\lambda_1,1,1,1)} = \lambda_1$.

PROOF. First $m_{(\lambda_1,1,1)}$. Clearly $Q_2$ contributes one summand from $h_{\lambda_1} S_{1,1} = S_{1+\lambda_1,1} + S_{\lambda_1,1,1}$. From $Q_3$ we get contributions from $S_{r,1} S_{m-r-1}$, $2 \leq r \leq m-2$, so $m-3$ cases; and from $S_{r,1,1} S_{m-r-2}$, again $2 \leq r \leq m-2$, so $m-3$ cases.

Together we get $1 + 2(m-3) = 2m - 5 = 2(m-2) - 1 = 2\lambda_1 - 1$ cases, so $m_{(\lambda_1,1,1)} = 2\lambda_1 - 1$.

Now $m_{(\lambda_1,1,1,1)}$ is obtained only from $S_{r,1,1} S_{m-r-2}$, $2 \leq r \leq m-3$, so $m-4$ cases. Hence $m_{(\lambda_1,1,1,1)} = (m-3,1,1,1) = m - 4 = \lambda_1 - 1$. □

LEMMA 9.3.10. *For all other* $\lambda$ *with* $\text{ht}(\lambda) \leq 4$,

$$m_\lambda = (\lambda_1 - \lambda_2 + 1)(\lambda_2 - \lambda_3 + 1)(\lambda_3 - \lambda_4 + 1).$$

PROOF. We can assume that $\lambda = (\lambda_1, \ldots, \lambda_4)$ with $\lambda_1 \geq \lambda_2 \geq 2$ and $\lambda_3 \geq 1$. So only $Q_3$ contributes to $m_\lambda$. Thus $\mu \in C_k$ contributes to $m_\lambda$ if and only if $\lambda_{r+1} \leq \mu_r \leq \lambda_r$ (so $\lambda_r - \lambda_{r+1} + 1$ such $\mu_r$'s, $1 \leq r \leq 3$). The proof clearly follows. □

REMARK 9.3.11. For each $k \geq 2$ there is a subset $\Omega(k) \subseteq \text{Par}(k)$ of the partitions of $k$, such that the $n$th cocharacter $\chi_n(M_2(F))$ satisfies

$$\chi_n(M_2(F)) = 1 + \sum_{k \geq 2}^{n} \sum_{\lambda \in \Omega(k)} \text{Ind}_{S_k \times S_{n-k}}^{S_n} M_\lambda \otimes I_{n-k}. \tag{9.61}$$

In fact the reader can verify that the sum computes the cocharacter when restricted to *proper polynomials*, cf. Definition 7.4.5.

Call $\{\sum \chi_\lambda \mid \lambda \in \Omega(k)\}$ (cf. Remark 9.3.7) **the $k$th reduced cocharacter of** $M_2(F)$, and $\sum_{\lambda\in\Omega(k)} f_\lambda$ **the $k$th reduced codimension of** $M_2(F)$. Note that formula (9.61) gives for the codimension the formula,

$$c_n(M_2(F)) = 1 + \sum_{k\geq 2}^{n} \binom{n}{k} \sum_{\lambda\in\Omega(k)} f_\lambda. \tag{9.62}$$

We want to make this formula more explicit, so we need a few manipulations which will lead to the final formula (9.66) (compare with Drensky[**Dre00**, Theorem 4.3.12(ii)]).

9.3.1.3. *The generating function $\sum_{n=0}^{\infty} c_n(M_2(F))t^n$. Catalan and Motzkin numbers.* Let $C_n := \frac{1}{n+1}\cdot\binom{2n}{n}$ denote the $n$th Catalan number, $C_0 = 1,\ C_1 = 1,\ C_2 = 2,\ C_3 = 5$, etc.: $1, 1, 2, 5, 14, 42, 132, \ldots$. The generating function is well known:

$$\sum_{n=0}^{\infty} C_n t^n = \frac{1-\sqrt{1-4t}}{2t} = \frac{2}{1+\sqrt{1-4t}}. \tag{9.63}$$

Let $\mathtt{M}_n := \sum_{j\geq 0} \frac{n!}{j!(j+1)!(n-2j)!}$ (see [**Reg84**]) be the $n$th Motzkin number:

$$\mathtt{M}_0 = 1, \mathtt{M}_1 = 1, \mathtt{M}_2 = 2, \mathtt{M}_3 = 4, etc.:\ 1, 1, 2, 4, 9, 21, 51, \ldots.$$

The following properties are fairly well known (see [**Reg81**]):

$$\mathtt{M}_n = \sum_{\lambda\vdash n,\ \mathrm{ht}(\lambda)\leq 3} f_\lambda, \quad \sum_{j=0}^{n}\binom{n}{j}\cdot \mathtt{M}_j = C_{n+1}. \tag{9.64}$$

By formula (9.64) and Remark 9.3.7 we deduce

$$\sum_{\lambda\in\Omega(k)} f_\lambda = \begin{cases} \mathtt{M}_k - 1 & \text{if} \quad k\neq 3 \\ \mathtt{M}_k - 2 & \text{if} \quad k = 3. \end{cases}$$

If $k=0$ or $k=1$, then $\mathtt{M}_k = 1$, $\mathtt{M}_k - 1 = 0$, hence in (10.49) we can write

$$\begin{aligned} c_n(M_2(F)) &= 1 + \sum_{k=0}^{n}\binom{n}{k}\cdot(\mathtt{M}_k - 1) - \binom{n}{3} \\ &= 1 + \sum_{k=0}^{n}\binom{n}{k}\cdot \mathtt{M}_k - \sum_{k=0}^{n}\binom{n}{k} - \binom{n}{3} = 1 + C_{n+1} - 2^n - \binom{n}{3} \\ &= \frac{1}{n+2}\cdot\binom{2n+2}{n+1} + 1 - 2^n - \binom{n}{3}. \end{aligned} \tag{9.65}$$

It follows from formulas (9.63) and (9.65) the final formula,

$$\sum_{n=0}^{\infty} c_n(M_2(F))t^n = \frac{1-\sqrt{1-4t}}{2t^2} - \frac{1}{t} + \frac{1}{1-t} - \frac{1}{1-2t} - \frac{t^3}{(1-t)^4}. \tag{9.66}$$

REMARK 9.3.12. formula (9.66) shows in particular that $\sum_{n=0}^{\infty} c_n(M_2(F))t^n$ is an algebraic function. One may thus ask for which PI algebras the generating function of codimension is algebraic.

By formula (9.66) together with the results in [**BR87**], the generating function $\sum_n c_n(M_k(F))t^n$ is NOT algebraic for $k\geq 3$ odd.

*Conjecture.* The above is not algebraic for all $k\geq 3$.

9.3.1.4. *Central polynomials.* Central polynomials are just the center of $R_m$ as soon as $m > 1$. In order to understand this center, let us first analyze the center of $V_m = \tilde{V}_m[t_1, \dots, t_m]$.

This is $T_m = \mathcal{T}_m^0[t_1, \dots, t_m]$, since $T_m^0$ is the center of $\tilde{V}_m = V_m^0 \oplus T_m^0$.

We deduce

THEOREM 9.3.13. *The center of $R_m$ equals $\mathcal{Z}'$, where* (cf. formula (9.14))

$$\mathcal{Z}' = \bigoplus_{\lambda \in \mathcal{P}^{3,\text{even}},\ \lambda \supset 2,1} L_\lambda[t_1, \dots, t_m]. \tag{9.67}$$

PROOF. We have that $\mathcal{Z}' = \mathcal{T}_m \cap L_{2,1} V_m \subset R_m$ is an ideal of $\mathcal{T}_m$, the part of the center of $V_m$ contained in the ideal $L_{2,1} V_m \subset R_m$, so we only need to show that there is no other term in the center of $R_m$. We know that central polynomials of $n \times n$ matrices are identities for $n - 1 \times n - 1$ matrices (Proposition 10.2.2) so they are contained in $[R_m, R_m]$.

Consider $[R_m, R_m]/L_{2,1} V_m \subset [V_m, V_m]/L_{2,1} V_m$. By formula (9.55) the image of the center of $V_m$ in $[V_m, V_m]/L_{2,1} V_m$ is $L_3[t_1, \dots, t_m]$ which we have seen is complementary to the image of $R_m$ which is instead $L_2[t_1, \dots, t_m]$, hence the center of $R_m$ is contained in $L_{2,1} V_m$ so equals $\mathcal{Z}'$. □

COROLLARY 9.3.14. *Central polynomials for $2 \times 2$ matrices are in the conductor of $R_m$ in $V_m$.*

REMARK 9.3.15. One can compute the dimensions and cocharacters for central polynomials in a way similar to that of the previous section. Observe that $\mathcal{Z}'$ is an ideal of the finitely generated commutative ring $T_m$, its associated variety is formed by the complement of the irreducible representations.

*Central cocharacters and codimensions.* So the analogue of $C_k$ would be $\bar{C}_k = \mathcal{P}_k^{3,\text{even}}$, and the central cocharacter $\chi_n^Z(M_2(F))$ is

$$\chi_n^Z(M_2(F)) = \sum_{k \ge 3}^{n} \sum_{\lambda \in \mathcal{P}_k^{3,\text{even}}} \operatorname{Ind}_{S_k \times S_{n-k}}^{S_n} M_\lambda \otimes I_{n-k}.$$

9.3.1.5. *A noncommutative presentation.* Both algebras $R_m$ and $\tilde{V}_m$ are quotients of the free algebra $F\langle x_1, \dots, x_m \rangle$, so it is a natural question to describe the ideal modulo which they are quotients. For $R_m$ this is a $T$-ideal, and a theorem of Razmyslov and Drensky implies that it is generated, as a $T$-ideal, by the standard polynomial $\mathrm{St}_4$ and the Wagner identity (9.47).

For the details see **[Raz73]**, **[Raz94]**, and Drensky **[Dre81]**.

# Part 3

# The structure theorems

CHAPTER 10

# Matrix identities

*In this chapter we start from the classical theorem of Amitsur and Levitzki, Theorem 10.1.4. We then construct central polynomials for matrices using the two approaches of Formanek and Razmyslov in §10.2. We close the chapter by using central polynomials in order to prove the basic characterization of Azumaya algebras, Theorem 10.3.2, in terms of polynomial identities, due to M. Artin.*

## 10.1. Basic identities

**10.1.1. The Amitsur–Levitzki theorem.** The simplest algebras that satisfy polynomial identities are finite modules over commutative rings. We start with

PROPOSITION 10.1.1. *If $M$ is a module generated by $n$ elements over a ring $B$, then $\mathrm{End}_B(M)$ is a quotient of a subring of $M_n(B^{\mathrm{op}})$.*

PROOF. Let $u_1, \dots, u_n$ be linear generators of $M$ over $B$, so that $M = \sum_{i=1}^n Bu_i$. Consider the free $B$-module $\tilde{M}$ on generators $\tilde{u}_i$ which maps surjectively to $M$ by $\pi : \tilde{u}_i \mapsto u_i$. Let $K$ be the kernel of the map $\pi$.

An endomorphism $\phi$ of $\tilde{M}$ induces an endomorphism $\bar{\phi}$ of the module $M$, with $\pi \circ \phi = \bar{\phi} \circ \pi$, if and only if $\phi(K) \subset K$. The set of endomorphisms $\phi$ of $\tilde{M}$ with $\phi(K) \subset K$ is clearly a subalgebra $S$ of $\mathrm{End}_B(\tilde{M})$. Moreover the map $\phi \mapsto \bar{\phi}$ of $S$ to $\mathrm{End}_B(M)$ is a homomorphism.

Now take an endomorphism $\psi$ of $M$ with $\psi(u_i) = \sum_j \alpha_{i,j} u_j$ and define $\phi(\tilde{u}_i) = \sum_j \alpha_{i,j} \tilde{u}_j$ so that $\pi \circ \psi = \phi \circ \pi$. We then see that if $k \in K$, $\pi(k) = 0$, we have $\pi \circ \psi(k) = \phi \circ \pi(k) = 0$. Thus $\psi \in S$ induces on $M$ the endomorphism $\phi$, and this means that $\mathrm{End}_B(M)$ is a quotient of the subalgebra $S$ of $\mathrm{End}_B(\tilde{M}) = M_n(B^{\mathrm{op}})$, hence the claim. □

COROLLARY 10.1.2. *If $R$ is an algebra with a 1 which is a module generated by $n$ elements over a commutative algebra $B$, then $R$ satisfies the identities of $n \times n$ matrices, over $B$.*

PROOF. By using left multiplication, we identify $R$ with a subalgebra of the algebra $\mathrm{End}_B(R)$, so the claim follows from Proposition 10.1.1. □

If $R$ does not have a 1, then by adding a 1, one has a module generated by $n + 1$ elements.

The identities of $n \times n$ matrices are a rather difficult and deep subject. We start with a commutative ring $B$ and the algebra of $n \times n$ matrices over $B$.

THEOREM 10.1.3. *$M_n(B)$ does not satisfy any multilinear identity of degree $< 2n$.*

PROOF. The fact that $M_n(B)$ does not satisfy any multilinear identity of degree $k < 2n$ can be seen as follows. One takes such a proposed identity and up to reordering the variables and multiplying it if necessary by further variables, may assume that it is of degree $2n-1$ and of the form $f(x) := a\, x_1 x_2 \cdots x_{2n-1} + M$ where $a \neq 0$ and $M$ collects the other terms.

Then one observes that the sequence of $2n-1$ matrix units

$$e_{1,1}, e_{1,2}, e_{2,2}, e_{2,3}, e_{3,3}, \dots, e_{n-1,n}, e_{n,n}$$

gives product 0 unless it is multiplied in the order in which it has been presented, and so, when one evaluates $f(x)$ in this sequence, all monomials vanish except the first one which gives $a\, e_{1,n} \neq 0$. □

As for degree $2n$, the answer is given by the theorem of Amitsur and Levitzki, a cornerstone of the theory of polynomial identities. It states:

THEOREM 10.1.4. *The algebra of $n \times n$ matrices over any commutative ring $A$ satisfies the standard polynomial* $\mathrm{St}_{2n}$*; see formula* (2.32).

This theorem has received several proofs, after the original proof **[AL50]**, which is a direct verification. These include a proof by Swan using graphs **[Swa63]**; one by Kostant relating it to Lie algebra cohomology **[Kos58]**; a proof by Razmyslov as a consequence of the theory of trace identities and of the Cayley–Hamilton theorem (and independently by Formanek unpublished) **[Raz74b]**; one by Szigeti , Tuza and Révész **[STR93]**, who again apply graph theory in a different way constructing several matrix identities. Finally, there is a very simple proof by Rosset using Grassmann variables **[Ros76]**, which inspired the proof given here.

The proof presented here shows that the Amitsur–Levitzki theorem is the Cayley–Hamilton identity for the *generic Grassmann matrix.*

In this formulation the theorem is the first step for the general theory of alternating equivariant maps (with respect to conjugation) from matrices to matrices (see **[BPŠ15]** and **[DCPP15]**).

LEMMA 10.1.5. *It is enough to prove the theorem for matrices over $\mathbb{Q}$.*

PROOF. The identity has coefficients $\pm 1$ and it is equivalent to the identical vanishing of $n^2$ polynomials with integer coefficients in $2n^3$ commutative variables, the entries of $2n$ $n \times n$ matrices.

A polynomial with integer coefficients vanishes identically if and only if it vanishes when computed in $\mathbb{Q}$, hence the claim. □

We will assume we are in this setting from now on.

**10.1.2. Antisymmetry.** By the *antisymmetrizer* we mean the operator that maps a multilinear expression $f(x_1, \dots, x_h)$ into the antisymmetric expression

$$\sum_{\sigma \in S_h} \epsilon_\sigma f(x_{\sigma(1)}, \dots, x_{\sigma(h)}).$$

Such an operation can be applied to different settings. For instance, if $\phi_1, \dots, \phi_h$ are linear forms on $V$, by antisymmetrizing the function $\phi_1(v_1) \cdots \phi_h(v_h)$ one has the function $\phi_1 \wedge \cdots \wedge \phi_h$ which on $h$ vectors $(v_1, \dots, v_h)$ takes as value $\det(\phi_i(v_j))$. Instead, applying the antisymmetrizer to the noncommutative monomial $x_1 \cdots x_h$, we get the standard polynomial of degree $h$,

$$\mathrm{St}_h(x_1, \dots, x_h) = \sum_{\sigma \in S_h} \epsilon_\sigma x_{\sigma(1)} \cdots x_{\sigma(h)}.$$

Up to a scalar multiple this is the only multilinear antisymmetric noncommutative polynomial of degree $h$.

Let $A$ be any $F$-algebra (not necessarily associative) with basis $e_i$, and let $V$ be a finite-dimensional vector space over a field $F$.

The space $\mathrm{Alt}^k(V, A)$ of multilinear antisymmetric functions from $V^k$ to $A$ is given by functions $G(v_1, \dots, v_k) = \sum_i G_i(v_1, \dots, v_k)e_i$ with $G_i(v_1, \dots, v_k)$ multilinear antisymmetric functions from $V^k$ to $F$. Moreover, if $A$ is infinite dimensional, only finitely many $G_i(v_1, \dots, v_k)$ appear for any given $G$.

In other words $G_i(v_1, \dots, v_k) \in \bigwedge^k V^*$, and this space $\mathrm{Alt}^k(V, A)$ can be identified with $\bigwedge^k V^* \otimes A$. Using the algebra structure of $A$, we have a wedge product of these functions:

For $G \in \bigwedge^h V^* \otimes A$, $H \in \bigwedge^k V^* \otimes A$, we define

$$\begin{aligned}(G \wedge H)(v_1, \dots, v_{h+k}) &:= \frac{1}{h!k!} \sum_{\sigma \in S_{h+k}} \epsilon_\sigma G(v_{\sigma(1)}, \dots, v_{\sigma(h)}) H(v_{\sigma(h+1)}, \dots, v_{\sigma(h+k)}) \\ &= \sum_{\sigma \in S_{h+k}/S_h \times S_k} \epsilon_\sigma G(v_{\sigma(1)}, \dots, v_{\sigma(h)}) H(v_{\sigma(h+1)}, \dots, v_{\sigma(h+k)}).\end{aligned}$$

As an example we have for the standard polynomials $\mathrm{St}_a \wedge \mathrm{St}_b = \mathrm{St}_{a+b}$.

With this multiplication the algebra of multilinear antisymmetric functions from $V$ to $A$ is isomorphic to the tensor product algebra $\bigwedge V^* \otimes A$. We shall denote by $\bigwedge$ the product in this algebra.

Assume now that $A$ is an associative algebra and $V \subset A$. The inclusion map $X : V \to A$ is of course antisymmetric, since the symmetric group on one variable is trivial, hence $X \in \bigwedge V^* \otimes A$. By iterating the definition of wedge product, we have the important fact:

PROPOSITION 10.1.6. *As a multilinear function $X^k : V^k \to A$, each power $X^k := X^{\wedge k}$ equals the standard polynomial* $\mathrm{St}_k$ *computed in $V$.*

We apply this to $V = A = M_n(F) := M_n$; the group $G = \mathrm{PGL}(n, F)$ acts on this space and hence on functions by conjugation, and it is interesting to study the *invariant algebra*, i.e., the algebra of $G$-equivariant maps

$$A_n := (\textstyle\bigwedge M_n^* \otimes M_n)^G. \tag{10.1}$$

This can be fully described; see **[BPŠ15]**.

In the natural basis $e_{ij}$ of matrices and the coordinates $x_{ij}$, the element $X \in A_n$ (cf. formula (10.1)) is the identity element of $M_n^* \otimes M_n$ and can be thought of as the *generic Grassmann matrix* $X = \sum_{h,k} x_{hk} e_{hk}$.

Hence in this language the Amitsur–Levitzki theorem is the single identity $X^{2n} = 0$.

PROOF OF THE AMITSUR–LEVITZKI THEOREM $X^{2n} = 0$. Since $X$ is an element of degree 1, we have that $X^2$ is in $\bigwedge^2 M_n^* \otimes M_n \subset M_n(\bigwedge^{\mathrm{even}} M_n^*)$.

We have that $X^2$ is an $n \times n$ matrix over a commutative ring, the even part of the Grassmann algebra, hence in order to prove that $X^{2n} = (X^2)^n = 0$ it is enough to show that $\mathrm{tr}((X^2)^i) = \mathrm{tr}(X^{2i}) = 0$ for $i \le n$ (cf. Corollary 3.3.3).

Now it is immediate that given two homogeneous elements $a, b \in M_n(\bigwedge M_n^*)$, we have $\mathrm{tr}(ab) = (-1)^{\deg(a)\deg(b)} \mathrm{tr}(ba)$.

So, for an odd element $a$, we have

$$\operatorname{tr}(a^{2k}) = \operatorname{tr}(aa^{2k-1}) = -\operatorname{tr}(a^{2k-1}a) = -\operatorname{tr}(a^{2k}).$$

Since $X$ is an element of degree 1 we have $\operatorname{tr}((X^2)^i) = \operatorname{tr}(X^{2i}) = 0$. $\square$

**10.1.3. Uniqueness.** The Amitsur–Levitzki theorem has the following additional part.

THEOREM 10.1.7.

(1) *Given a commutative ring $A$, the algebra $M_k(A)$ does not satisfy any identity of degree $\leq 2k-1$.*
(2) *Let $f(x_1, \dots, x_{2k})$ be a multilinear identity of $M_k(A)$ of degree $2k$, then, for some $\alpha \in A$, $f(x_1, \dots, x_{2k}) = \alpha \cdot \mathrm{St}_{2k}(x_1, \dots, x_{2k})$.*

PROOF. (1) This part has already been proved.

(2) Let

$$f(x_1, \dots, x_{2k}) = \sum_{\sigma \in S_{2k}} \alpha_\sigma x_{\sigma(1)} \cdots x_{\sigma(2k)}$$

be a multilinear identity of $M_k(A)$. Let $\sigma \in S_{2k}$, denote $x_{\sigma(1)} \cdots x_{\sigma(2k)} = y_1 \cdots y_{2k} = M(y)$, and $\alpha_\sigma = \alpha$, so

$$\alpha M(y) = \alpha y_1 \cdots y_{r-1} \cdot y_r \cdot y_{r+1} \cdots y_{2k}$$

is a summand of $f$. Then, with the appropriate coefficient $\beta$, $f$ also has the summand

$$\beta N(y) = \beta y_1 \cdots y_{r-1} \cdot y_{r+1} \cdot y_r \cdots y_{2k},$$

and we compare $\alpha$ with $\beta$.

CLAIM. With these notations $\beta = -\alpha$ (which implies Theorem 10.1.7(2)). To simplify the exposition, we let $2k = 6$, and we work with examples.

*The case $r$ is odd.* Let $r = 3$ and compare $\alpha M(y) = \alpha y_1 y_2 y_3 y_4 y_5 y_6$ with $\beta N(y) = \beta y_1 y_2 y_4 y_3 y_5 y_6$. Substitute

$$(y_1, \dots, y_6) \to (\bar{y}_1, \dots, \bar{y}_6) = (e_{1,1}, e_{1,2}, e_{2,2}, e_{2,2}, e_{2,3}, e_{3,3}),$$

then $M(\bar{y}) = N(\bar{y}) = e_{1,1} \cdot e_{1,2} \cdot e_{2,2} \cdot e_{2,2} \cdot e_{2,3} \cdot e_{3,3} = e_{1,3}$, and these are the only products of $\bar{y}_1, \dots, \bar{y}_6$ which equal $e_{1,3}$, while any other product equals $e_{i,j}$ with $(i,j) \neq (1,3)$. Thus

$$0 = f(\bar{y}) = (\alpha + \beta)e_{1,3} + \sum_{(i,j) \neq (1,3)} \gamma_{i,j} e_{i,j},$$

hence $\beta = -\alpha$.

*The case $r$ is even.* Let $r = 4$ and now compare $\alpha M(y) = \alpha y_1 y_2 y_3 y_4 y_5 y_6$ with $\beta N(y) = \beta y_1 y_2 y_3 y_5 y_4 y_6$. Substitute

$$(y_1, \dots, y_6) \to (\bar{y}_1, \dots, \bar{y}_6) = (e_{1,2}, e_{2,2}, e_{2,3}, e_{3,3}, e_{3,3}, e_{3,1}),$$

then $M(\bar{y}) = N(\bar{y}) = e_{1,1}$, and these are the only products of $\bar{y}_1, \dots, \bar{y}_6$ which equal $e_{1,1}$, while any other product equals $e_{i,j}$ with $(i,j) \neq (1,1)$. Thus

$$0 = f(\bar{y}) = (\alpha + \beta)e_{1,1} + \sum_{(i,j) \neq (1,1)} \gamma_{i,j} e_{i,j},$$

hence $\beta = -\alpha$.

The cases of general $r$ and $k$ are done exactly the same way. It proves the claim that $\beta = -\alpha$, and this implies that $f(x_1, \dots, x_{2k}) = \alpha \operatorname{St}_{2k}(x_1, \dots, x_{2k})$ for some $\alpha \in A$. □

**10.1.4. Kaplansky's theorem.** The starting point of structure theory of rings with polynomial identities is due to Kaplansky. Suppose that an algebra $R$ over a commutative ring $A$ satisfies a proper multilinear identity $f$ of degree $d$, with coefficients in $A$ (generating $A$ as ideal).

Let $M$ be an irreducible $R$-module, and let $D := \hom_R(M, M)$ be the centralizer of $R$ which, by Schur's lemma, is a division ring with center a field $F$, then:

THEOREM 10.1.8 (Kaplansky theorem).

$$\dim_F D = h^2 < \infty, \quad \dim_{DM} = k, \quad 2hk \leq d. \tag{10.2}$$

PROOF. It is in two steps.

(1) *Assume that $D = K$ is commutative.* We apply the Jacobson density theorem to prove that $\dim_K M \leq \frac{d}{2}$. If we had $\dim_K M > \frac{d}{2}$ we could find $k > \frac{d}{2}$ linearly independent elements over $K$ and a subring $S$ of $R$ mapping surjectively to the ring of all endomorphisms, equal to $M_k(K)$, of the subspace generated by these elements. This implies that the matrix ring $M_k(K)$ satisfies the identity $f$, because if an identity is satisfied by a ring it is satisfied by its subrings and quotients of them. Now we have seen that the matrix ring $M_k(K)$ does not satisfy any identity of degree $< 2k$, so we must have $d \geq 2k$ a contradiction

(2) *Reduction to $D$ is commutative.* Let $K$ be a maximal commutative subfield of $D$. Then $K$ equals the centralizer of $K$ in $D$, and if $RK$ denotes the $K$-algebra of operators of $M$ spanned by $K$, $R$, we clearly have that $M$ is irreducible under $RK$ and $\operatorname{End}_{RK} M = K$. Since the identity we start from is multilinear, $RK$ satisfies the same identity, cf. Lemma 2.2.37. By the commutative case we can deduce that $2 \dim_K M \leq d$.

Clearly $\dim_K M = \dim_K D \dim_{DM}$, hence $\dim_K D < \infty$, and by Theorem 1.1.17 in this case $\dim_F D = [\dim_K D]^2$, and the claim follows. □

COROLLARY 10.1.9.

(1) *A primitive ring satisfying a PI is a simple algebra finite dimensional over its center.*
(2) *A simple ring satisfying a PI is finite dimensional over its center.*
(3) *If a ring $R$ satisfies a PI, its Jacobson radical $J(R)$ is the intersection of all maximal two-sided ideals of $R$.*

PROOF. (1) A primitive ring $R$ by definition has a faithful irreducible module which, by the previous theorem, is a vector space $V$ of finite dimension $h$ over a division ring $D$ finite dimensional over its center. Then $R = M_h(D)$ by the Jacobson density theorem and $R = M_h(D)$ is simple.

(2) If the ring has a 1, then it is primitive, and we can use (1). Otherwise observe that $S$ is an irreducible module over $S \otimes S^{\mathrm{op}}$ which, by Theorem 7.2.2 is a PI algebra. Hence by the Kaplansky theorem, Theorem 10.1.8, we have that $S$ is finite dimensional over a field $F$ commuting with the two left and right actions of $S$. Thus $S$ is a finite-dimensional algebra, and we may apply Theorem 1.1.14.

(3) The Jacobson radical is the intersection of all primitive ideals, so the second claim follows once we observe that conversely every maximal ideal of a PI algebra is primitive. In other words a simple PI algebra $R$ is a primitive ring. □

## 10.2. Central polynomials

10.2.0.1. *Central polynomials for $n \times n$ matrices.* A very interesting fact is the existence of *central polynomials* for $n \times n$ matrices. By this we mean

DEFINITION 10.2.1. A central polynomial for $n \times n$ matrices is a noncommutative polynomial without constant term which does not vanish on $n \times n$ matrices but takes only scalar values.

For $2 \times 2$ matrices the polynomial $[x, y]^2$ is clearly central but for a long time no such polynomial was known for $n > 2$. The existence of central polynomials for all $n$ was conjectured by Kaplansky **[Kap70]** and **[Kap57]**, and proved independently by Formanek **[For72]** and Razmyslov **[Raz74a]**, **[Raz73]**.

The existence of central polynomials is in fact equivalent to the existence of a large center of the ring of generic matrices $\mathbb{Z}\langle \xi_\alpha \rangle$. We will see this in Proposition 11.2.5.

There is a simple but useful remark on such elements. Let $F$ be a field.

PROPOSITION 10.2.2. *If $f \in F\langle x_\alpha \rangle$ is a central polynomial of $n \times n$ matrices, then it is a polynomial identity for any proper subalgebra of $n \times n$ matrices.*

PROOF. First we show that it is an identity for $k \times k$ matrices $k < n$, we may do it for $k = n - 1$. We can embed the ring of $n - 1 \times n - 1$ matrices into the ring of $n \times n$ matrices as an upper block with the entries $x_{n,j} = x_{j,n} = 0,\ \forall j$. Clearly no scalars are contained in this subring, except 0. By hypothesis, $f$ evaluated in $n \times n$ matrices takes scalar values, hence it must vanish on this subalgebra.

Now, by Wedderburn's theorem, Theorem 1.1.14, a proper subalgebra $R$ of $n \times n$ matrices over an algebraically closed field $F$ is not irreducible. That is it has a proper stable invariant subspace $U \subset F^n$, $0 < p := \dim_F U < n$. Then fixing a basis of $F^n$ so that its first $p$ elements form a basis of $U$, we see that in this basis $R$ is contained in the algebra of matrices which are upper triangular with a two-blocks decomposition. When we evaluate the central polynomial, we see by the previous analysis that we have an element which is 0 on the two diagonal blocks. Since the value is a scalar, it must be identically 0. □

**10.2.1. Formanek's polynomials.** The construction of Formanek starts from a remark. Assume that we take a polynomial $\phi(X, Y_1, \dots, Y_n)$ linear in the variables $Y_i$ and homogeneous in $X$, which is not a polynomial identity for $n \times n$ matrices. First of all we remark that a sufficient condition for an $n \times n$ matrix $A$ to be diagonalizable (in an algebraic closure of $F$) is to have $n$ distinct eigenvalues. This condition is that its discriminant (a polynomial in the coefficients of its characteristic polynomial) be different from 0. Since of course this is a nonzero polynomial in the entries of $A$, clearly a generic matrix satisfies this condition.

LEMMA 10.2.3. *$\phi(X, Y_1, \dots, Y_n)$ is a central polynomial if and only if it takes a scalar value when $X$ is computed in a diagonal matrix* $\operatorname{diag}(x_1, \dots, x_n)$ *and the $Y_i$ are computed into elementary matrices $e_{h_i,k_i}$.*

PROOF. The condition that $\phi(X, Y_1, \dots, Y_n)$ takes values in the center is the equation $[\phi(X, Y_1, \dots, Y_n), Z] = 0$. This is satisfied for all matrices if and only if it is satisfied for all $Y_i$ and $X$ evaluated in generic matrices. A generic matrix can be diagonalized (over some algebraic closure of the field of rational functions in its entries). Since the condition is invariant under conjugation, it is clear that

$\phi(X, Y_1, \ldots, Y_n)$ is a central polynomial if and only if it takes a scalar value when $X$ is computed in a diagonal matrix $\mathrm{diag}(x_1, \ldots, x_n)$ and the $Y_i$ in arbitrary matrices. But since the polynomial is linear in the $Y_i$, this is also equivalent to assuming the statement when the $Y_i$ are computed into elementary matrices $e_{h_i,k_i}$. □

Now take a monomial $X^{m_1}Y_1X^{m_2}Y_2 \cdots X^{m_n}Y_nX^{m_{n+1}}$ and compute it on a diagonal matrix $X = \mathrm{diag}(x_1, \ldots, x_n)$ and $Y_i = e_{h_i,k_i}$. One has $X^{m_i}Y_i \mapsto x_{h_i}^{m_i}e_{h_i,k_i}$ which implies

$$X^{m_1}Y_1X^{m_2}Y_2 \cdots X^{m_n}Y_nX^{m_{n+1}} \mapsto \begin{cases} x_{k_n}^{m_{n+1}} & \prod_{i=1}^{n} x_{h_i}^{m_i} e_{h_1,k_n} \\ \text{if} & k_i = h_{i+1},\ \forall i < n \\ 0 & \text{otherwise.} \end{cases} \tag{10.3}$$

Then associate to a polynomial

$$g(x_1, \ldots, x_{n+1}) = \sum_{\underline{m}:=m_1,\ldots,m_{n+1}} c_{\underline{m}} x_1^{m_1} \cdots x_{n+1}^{m_{n+1}}$$

a corresponding noncommutative polynomial

$$\bar{g}(X, Y_1, \ldots, Y_n) := \sum_{m_1,\ldots,m_{n+1}} c_{\underline{m}} X^{m_1}Y_1X^{m_2}Y_2 \cdots X^{m_n}Y_nX^{m_{n+1}}. \tag{10.4}$$

Now take

$$g(x_1, \ldots, x_{n+1}) = \prod_{2 \le i \le n} (x_1 - x_i)(x_{n+1} - x_i) \prod_{2 \le i < j \le n} (x_i - x_j)^2. \tag{10.5}$$

This polynomial has the property that, when we set $x_{n+1} = x_1$, we obtain that $g(x_1, \ldots, x_n, x_1) = \Delta^2$, $\Delta = \prod_{1 \le i < j \le n}(x_i - x_j)^2$. Call $\phi(X, Y_1, \ldots, Y_n)$ the corresponding noncommutative polynomial from formula (10.4).

THEOREM 10.2.4. *The polynomial*

$$\begin{aligned} G &:= G(X, Y_1, \ldots, Y_n) \\ &= \phi(X, Y_1, \ldots, Y_n) + \phi(X, Y_2, \ldots, Y_n, Y_1) + \cdots + \phi(X, Y_n, Y_1, \ldots, Y_{n-1}) \end{aligned}$$

*is a nonzero central polynomial for $n \times n$ matrices.*

PROOF. We apply Lemma 10.2.3 and thus compute $G$ for $X$ diagonal and the $Y_i \mapsto e_{h_i,k_i}$.

We see that $\phi(X, Y_1, \ldots, Y_n) = 0$ unless the evaluations $Y_i \mapsto e_{h_i,k_i}$ satisfy $k_i = h_{i+1}$, $\forall i < n$, in which case we have as value $g(x_{h_1}, \ldots, x_{h_n}, x_{k_n})e_{h_1,k_n}$.

Now notice that $g(x_{h_1}, \ldots, x_{h_n}, x_{k_n}) = 0$ unless the elements $h_1, \ldots, h_n$ are all distinct and, furthermore, $k_n = h_1$.

In this case $g(x_{h_1}, \ldots, x_{h_n}, x_{k_n}) = \Delta^2$ and $\phi(X, Y_1, \ldots, Y_n) \mapsto \Delta^2 e_{h_1,h_1}$.

But the same argument shows that $\phi(X, Y_2, \ldots, Y_n, Y_1) = \Delta^2 e_{h_2,h_2}$, and so on, so that $G(X, Y_1, \ldots, Y_n) \mapsto \Delta^2 \sum_h e_{h,h} = \Delta^2 1_n$. □

**10.2.2. Capelli and matrices.** We now follow Razmyslov in the construction of central polynomials.

10.2.2.1. *The Razmyslov transform.* A multilinear polynomial $f = f(x_1, \ldots, x_m)$ takes values in the center, of some algebra $A$, if and only if $[f(x_1, \ldots, x_m), u]$ is a polynomial identity of $A$. If $A$ is the algebra of $n \times n$ matrices over a field $F$, then this is equivalent to the fact that $\mathrm{tr}([f(x_1, \ldots, x_m), u]v) = 0$ for all $v$.

Fixing a variable $x_i$, write $f(x_1, \ldots, x_m) = \sum_j a_j x_i b_j$ so

$$\mathrm{tr}([f(x_1, \ldots, x_m), u]\, v) = \mathrm{tr}(\sum_j a_j x_i b_j uv - \sum_j u a_j x_i b_j v)$$

$$= \mathrm{tr}(x_i \left( \sum_j b_j uva_j - \sum_j b_j vua_j \right)) = \mathrm{tr}(x_i \sum_j b_j [u, v] a_j),$$

getting

LEMMA 10.2.5. *A multilinear polynomial $f(x_1, \ldots, x_m) = \sum_j a_j x_i b_j$ takes values in the center of the algebra of $n \times n$ matrices over a field $F$ if and only if the polynomial $\sum_j b_j [u, v] a_j$ is a PI algebra of $n \times n$ matrices.*

So one way of constructing central polynomials is by this method. In particular one may formalize this $n \times n$

DEFINITION 10.2.6. A multilinear polynomial $f(x_1, \ldots, x_m) = \sum_j a_j x_i b_j$ is a *weak identity* of the algebra of $n \times n$ matrices over a field $F$ for the variable $x_i$ if it vanishes when the variables are evaluated in matrices and the variable $x_i$ in a trace 0 matrix, $\mathrm{tr}(x_i) = 0$.

Take a multilinear polynomial $f(x_1, \ldots, x_m)$ and write it as $\sum_i a_i x_1 b_i$. Take another variable $z$ and compute on $n \times n$ matrices. We have the identity

$$\mathrm{tr}(z f(x_1, \ldots, x_m)) = \mathrm{tr}(\sum_i b_i z a_i x_1).$$

We define thus the Razmyslov transform of $f(x_1, \ldots, x_m)$ to be

$$Rf(x_1, \ldots, x_m) := \sum_i b_i x_1 a_i.$$

One has $R^2 f = f$ and

$$\mathrm{tr}(z f(x_1, \ldots, x_m)) = \mathrm{tr}((\sum_i b_i z a_i) x_1) = \mathrm{tr}(x_1 Rf(z, x_2, \ldots, x_n)).$$

THEOREM 10.2.7. *$f$ is a polynomial identity of $n \times n$ matrices if and only if $Rf$ is a polynomial identity. $f$ is a central polynomial of $n \times n$ matrices if and only if $Rf$ is a weak identity for the variable $x_1$ and not a polynomial identity.*

Notice that if $n$ is invertible in $F$, any matrix $x$ can be written as $a + n^{-1}\,\mathrm{tr}(a)$ with $\mathrm{tr}(a) = 0$ so that if $f(x_1, \ldots, x_m) = \sum_j a_j x_i b_j$ is a weak identity, one has that $f(x_1, \ldots, x_m) = n^{-1}\,\mathrm{tr}(x_i) \sum_j a_j b_j$.

There are several ways of finding weak identities, for instance by Halpin **[Hal83b]**. We do not know in general the minimum degree $d_n$ of such weak identities and hence of central polynomials. Formanek conjectured that $d_n = (n^2 + 3n - 2)/2$. Special results are due to V. Drensky and A. Kasparian **[DK85]**, who exhibit a central polynomial of degree 8 for $3 \times 3$ matrices and to V. Drensky and P. Cattaneo, and G. Maria **[DPC94]**, who exhibit a central polynomial of degree 13 for $4 \times 4$ matrices, thus agreeing with Formanek's conjecture. Finally,

Drensky constructs for each $n$ a central polynomial of degree $(n-1)^2+4$ **[Dre95]**. See also **[DR93]**.

10.2.2.2. *The use of the Capelli polynomial.* We need a standard fact from commutative algebra. Let $v_1 = (v_{1,1}, \dots, v_{1,m}), \dots, v_m = (v_{m,1}, \dots, v_{m,m})$ be $m$ vector variables, i.e., with the $v_{i,j}$ indeterminates.

PROPOSITION 10.2.8. *The determinant* $\det(v_1, \dots, v_m)$ *is a polynomial in the* $m^2$ *variables* $v_{i,j}$ *with coefficients* $\pm 1$ *irreducible over* $\mathbb{Z}$ *and also over any field* $F$.

PROOF. The formula for the determinant shows that it is a polynomial in the $m^2$ variables $v_{i,j}$ with coefficients $\pm 1$.

We show that it is irreducible by induction, for $m = 1$ being trivial. In fact by expanding on the first row, we have

$$\det(v_1, \dots, v_m) = \sum_{i=1}^{m} (-1)^{i+1} v_{1,i} D_{1,i} := v_{1,1} A + B$$

with $D_{1,i}$ the determinant of the corresponding cofactor. Now both $A$ and $B$ are polynomials in the variables $v_{i,j}$ excluding $v_{1,1}$.

Now, if $R$ is a unique factorization domain, it is easy to see that a polynomial $ax+b$, $a, b \in R$ and with $a$ irreducible is irreducible unless $a$ divides $b$. In our case by induction $A = D_{1,1}$ is irreducible so it is enough to show that it does not divide $B$, that is we do not have $B = AC$, with $C$ necessarily of degree 1. But expanding $D_{1,2}$ with respect to the first column, we see that $B$ contains terms homogeneous of degree 1 in both variables $v_{1,2}, v_{2,1}$. Since $D_{1,1}$ does not contain any of these variables and $C$ is homogeneous of degree 1, we have a contradiction. □

Consider a polynomial function $f(v_1, \dots, v_m, w) \in F[v_{i,j}, w]$ in $m+k$ vector variables $v_i := (v_{i,1}, \dots, v_{i,m})$ and other variables $w_j$.

LEMMA 10.2.9. *If* $f$ *is multilinear and antisymmetric in the variables* $v_i$, *then it is a multiple* $\det(v_1, \dots, v_m) g(w_1, \dots, w_k)$ *with* $g(w_1, \dots, w_k)$ *still a polynomial. If* $f$ *has integer coefficients, then* $g$ *is with integer coefficients.*

PROOF. We may first assume that $F$ is algebraically closed. We have that $\det(v_1, \dots, v_m)$ is an irreducible polynomial and the variety in which it vanishes is the set of $m$-tuples $v_1, \dots, v_m$ which are linearly dependent.

Clearly, by the hypotheses made, $f$ vanishes on this hypersurface, and so it is a multiple of $\det(v_1, \dots, v_m)$. Since $\det(v_1, \dots, v_m)$ is an irreducible polynomial with coefficients $\pm 1$, if $f$ has integer coefficients, then $\det(v_1, \dots, v_m)$ appears as an irreducible factor in the factorization over $\mathbb{Z}$ of $f$, and so $g$ is the product of the remaining factors, hence with integer coefficients. □

We now draw a consequence of the previous lemma which will be useful when discussing the Capelli polynomial.

Let us consider an $n$-dimensional vector space $V$ over some field $F$. We trivialize $\bigwedge^n V = F$ so that, if $v_1, \dots, v_n$ are $n$ vectors on $V$, we can speak of the determinant $\det(v_1, \dots, v_n) := v_1 \wedge \cdots \wedge v_n \in \bigwedge^n V = F$.

Consider a multilinear map $f(v_1, \dots, v_n)$ from $n$ copies of $V$ to some other vector space $W$.

PROPOSITION 10.2.10. *Assume that $f$ is antisymmetric, and let $A : V \to V$ be any linear map. Then we have*

$$f(Av_1,\ldots,Av_n) = \det(A)f(v_1,\ldots,v_n). \tag{10.6}$$

PROOF. If $W$ has a basis $u_1,\ldots,u_m$, we have in coordinates that $f(v_1,\ldots,v_n) = \sum_{i=1}^m f_i(v_1,\ldots,v_n)u_i$ where each coordinate $f_i(v_1,\ldots,v_n)$ is a polynomial satisfying the conditions of Lemma 10.2.9, hence $f_i(v_1,\ldots,v_n) = \alpha_i \det(v_1,\ldots,v_n)$ for some $\alpha_i \in F$.

Equation (10.6) is the usual rule $\det(Av_1,\ldots,Av_n) = \det(A)\det(v_1,\ldots,v_n)$ of determinants. □

COROLLARY 10.2.11.

$$f((t-A)v_1,\ldots,(t-A)v_n) = \chi_A(t)f(v_1,\ldots,v_n),$$

*where $\chi_A(t) = t^n + \sum_{i=1}^n (-1)^i \sigma_i(A)t^{n-i}$ is the characteristic polynomial. In particular equating coefficients in $t$, we have*

$$\sigma_i(A)f(v_1,\ldots,v_n) = \sum_{1\le j_1<j_2<\cdots<j_i\le n} f(v_1,\ldots,Av_{j_1},\ldots,Av_{j_i},\ldots,v_n). \tag{10.7}$$

$$\sum_{i=1}^n f(v_1,\ldots,Av_i,\ldots,v_n) = \operatorname{tr}(A)f(v_1,\ldots,v_n).$$

We apply this to elements of the free algebra $f(x_1,\ldots,x_k,y)$ which are multilinear and alternating in the variables $x_i$ and may depend on other variables.

Out of such a polynomial, for each $i \le k$, we can construct a new polynomial with a further variable $z$

(10.8)

$$(S_{i,z}f)(z,x_1,\ldots,x_k,y) := \sum_{1\le j_1<j_2<\cdots<j_i\le k} f(x_1,\ldots,zx_{j_1},\ldots,zx_{j_i},\ldots,x_k,y).$$

The polynomial $(S_{i,z}f)(z,x_1,\ldots,x_k,y)$ is still multilinear and alternating in the variables $x_i$, and it depends on $z$ as a homogeneous variable of degree $i$. This implies in particular that we can iterate this construction and build a series of polynomials

$$(S_{i_1,z_1}\cdots S_{i_r,z_r}f)(z_1,\ldots,z_r,x_1,\ldots,x_k,y).$$

We now choose an algebra $R$ and a subalgebra $B$ of dimension $k$. Fix an evaluation of the polynomial of formula (10.8) in the variables $y$, and then evaluate the variables $x_i, z$ in $B$.

We have that, since $B$ is a subalgebra, multiplication by $z$ is a linear map, which may play the role of the operator $A$, thus Corollary 10.2.11 gives

COROLLARY 10.2.12.

(1) *When we evaluate $(S_{i,z}f)(z,x_1,\ldots,x_k,y)$ in the algebra $R$, if we evaluate the variables $x_i$ and $z$ in $B$, we see that*

$$(S_{i,z}f)(z,x_1,\ldots,x_k,y) = \sigma_i(z)f(x_1,\ldots,x_k,y), \tag{10.9}$$

$$(S_{i_1,z_1}\cdots S_{i_r,z_r}f)(\underline{z},\underline{x},y) = \prod_{j=1}^r \sigma_{i_j}(z_j)f(\underline{x},y),$$

*where $(-1)^i\sigma_i(z)$ is the ith coefficient of the characteristic polynomial of the multiplication of $z$ on $B$.*

(2) *If $I$ is an ideal of $R$, with $\dim_F R/I = k$ and $f$ vanishes whenever a variable $x_i$ is computed in $I$, then again formula* (10.9) *holds, where now $(-1)^i\sigma_i(z)$ is the ith coefficient of the characteristic polynomial of the multiplication of $z$ on $R/I$.*

PROOF. The first part is just a special case of Corollary 10.2.11. As for the second part the hypothesis implies that $f$ induces, for each evaluation of the variables $y$, a multilinear and antisymmetric map from the vector space $A/I$ to $A$ so again a special case of Corollary 10.2.11. □

We apply now this corollary to matrices and the Capelli polynomial, (7.6), $C_{n^2}$ computed on $n \times n$ matrices over $\mathbb{Z}$ or any commutative ring $F$.

PROPOSITION 10.2.13. *The Capelli polynomial $C_{n^2}$ is not a polynomial identity of $M_n(F)$.*

PROOF. Indeed, given any pair $i, j$, order the $n^2$ matrix units so that the first is some $e_{i,h}$ and the last some $e_{k,j}$.

Observe that for every matrix unit $a$, we have $e_{i_1 j_1} a e_{i_2 j_2} = 0$ unless $a = e_{j_1 i_2}$, in which case $e_{i_1 j_1} a e_{i_2 j_2} = e_{i_1 j_2}$. Hence there exists a unique evaluation of the $y_i$'s in matrix units $\bar{y}_i$'s such that $\bar{x}_1\bar{y}_1\bar{x}_2\bar{y}_2\cdots\bar{x}_{n^2} = e_{i,j}$. Now the same argument shows that $\bar{y}_{i-1}\bar{x}_{\sigma(i)}\bar{y}_i = 0$ unless $\sigma(i) = i$; therefore

$$\bar{x}_{\sigma(1)}\bar{y}_1\bar{x}_{\sigma(2)}\bar{y}_2\cdots\bar{x}_{\sigma(n^2)} = 0$$

if $\sigma \neq 1$. It follows that

$$C_{n^2}(\bar{x}_1,\dots,\bar{x}_{n^2},\bar{y}_1,\dots,\bar{y}_{n^2-1}) = e_{i,j}. \tag{10.10}$$

□

We then have a matrix valued polynomial map such that each entry is multilinear and alternating in the $n^2$ matrix variables $x_i$. We can thus apply Lemma 10.2.9 and Proposition 10.2.13 to deduce

PROPOSITION 10.2.14. *When computed on $n \times n$ matrices over $\mathbb{Z}$, we have*

$$C_{n^2}(x_1,\dots,x_{n^2},\underline{y}) = \det(x_1,\dots,x_{n^2})B(y_1,\dots,y_{n^2-1}), \tag{10.11}$$

*where $B(y_1,\dots,y_{n^2-1})$ is some equivariant polynomial map with integer coefficients. This is nonzero in any matrix algebra $M_n(A)$ for all commutative rings $A$.*

Observe that if $z \in M_n(F)$, the linear map $z(n) : x \mapsto xz$ has determinant $\det(z)^n$. Hence by Proposition 10.2.10

$$\det(x_1 z,\dots,x_{n^2}z) = \det(zx_1,\dots,zx_{n^2}) = \det(z)^n\det(x_1,\dots,x_{n^2}).$$

We deduce

PROPOSITION 10.2.15. *When computed on $n \times n$ matrices over $\mathbb{Z}$, we have*

$$C_{n^2}(zx_1,\dots,zx_{n^2},\underline{y}) = C_{n^2}(x_1 z,\dots,x_{n^2}z,\underline{y}) = \det(z)^n C_{n^2}(x_1,\dots,x_{n^2},\underline{y}). \tag{10.12}$$

Of course this identity over $\mathbb{Z}$ induces an identity on matrices over any commutative ring $A$.

We then can introduce in a noncommutative way the functions of §3.3.3. We claim, following the lines of Remark 10.2.12,

PROPOSITION 10.2.16. *For each $\underline{h} := h_1, \dots, h_i$ there exists a noncommutative polynomial $C_{\underline{h}}(x, y, z)$ with integer coefficients, deduced from the Capelli polynomial $C_{n^2}$, such that as functions on matrices,*

$$C_{\underline{h}}(x, y, z_1, \dots, z_i) = \sigma_{h_1,\dots,h_i;n}(z_1, \dots, z_i) C_{n^2}(x, y). \tag{10.13}$$

*In the previous formula we can replace the $\sigma_{i;n}$ with their expressions as universal polynomials with integer coefficients in the $\sigma_i$; cf. Lemma* 3.3.21.

PROOF. We have, when we evaluate on matrices $x, y, z$,

$$\begin{aligned} C_{n^2}(x_1 z, \dots, x_{n^2} z, y) &= \det(x_1 z, \dots, x_{n^2} z) B(y_1, \dots, y_{n^2-1}) \\ &= \det(z)^n \det(x_1, \dots, x_{n^2}) B(y_1, \dots, y_{n^2-1}) = \det(z)^n C_{n^2}(x_1, \dots, x_{n^2}, y). \end{aligned}$$

Now substitute to $z$ the element $t - \sum_{i=1}^{n^2} t_i z_i$, and develop both sides as polynomials in the commuting variables $t, t_i$. By homogeneity we get a sum where the monomials are of the form $t^i \prod_{j=1}^{n^2} t_j^{h_j}$, $i + \sum_j h_j = n^2$. In the left-hand side we are developing

$$C_{n^2}(x_1(t - \sum_{i=1}^{n^2} t_i z_i), \dots, x_{n^2}(t - \sum_{i=1}^{n^2} t_i z_i), y)$$

in the free algebra in the variables $x, z$ with coefficients in the polynomial ring in the variables $t, t_i$. Hence we have that the coefficients of these monomials are noncommutative polynomials, which we call $C_{\underline{h}}(x, y, z_1, \dots, z_i)$. On the right we are developing the formula inside the algebra of matrix functions and have the corresponding expressions in the $\sigma$. □

In particular notice that, the trace of the multiplication map of a matrix $z$ on $n \times n$ matrices equals $n\ \mathrm{tr}(z)$, hence from formula (10.7), as functions on $n \times n$ matrices, we have

$$\bar{C}(n) := \sum_{i=1}^{n^2} C_{n^2}(x_1, \dots, x_i z, \dots x_{n^2}, y) = n\ \mathrm{tr}(z) C_{n^2}(x_1, \dots, x_{n^2}, y). \tag{10.14}$$

$$\sum_{1 \le i_i < \dots < i_q \le n^2} C_{n^2}(x_1, \dots, z x_{i_1}, \dots, z x_{i_q}, \dots x_{n^2}, y) = \sigma_{q;n}(z) C_{n^2}(x_1, \dots, x_{n^2}, y). \tag{10.15}$$

Thus the left-hand side of (10.14) is a weak identity for $n \times n$ matrices and, if the characteristic does not divide $n$, it is not a polynomial identity. Applying Theorem 10.2.7, relative to the Razmyslov transform of (10.14), we have

THEOREM 10.2.17. *The polynomial $R\bar{C}(n)(x_1, \dots, x_{n^2}, y, z)$ takes values in the center, and if the characteristic does not divide n, it is not a polynomial identity.*

A very interesting class of central polynomials will be discussed in §12.3.

10.2.2.3. *A second construction.* There is another construction which produces a central polynomial in all characteristics. Razmyslov argues by constructing a kind of *noncommutative discriminant* as follows.

We start by remarking that in $n \times n$ matrices over a field $F$, the form $\mathrm{tr}(xy)$ is nondegenerate and, if $e_i$, $i = 1, \dots, n^2, e_j^*$, $j = 1, \dots, n^2$ are dual bases for this form, that is:

LEMMA 10.2.18. *For any scalar factor $\alpha \in F$ if $\mathrm{tr}(e_i e_j^*) = \alpha \delta^i_j$, we have the identity*

$$\sum_{i=1}^{n^2} e_i a e_i^* = \alpha \,\mathrm{tr}(a) \tag{10.16}$$

*for all matrices a (we identify the scalars with the multiples of the identity).*

PROOF. It is enough to prove this for $\alpha = 1$. A possible choice of dual bases is $e_{i,j}$ and $e_{j,i}$, and for any matrix $a = (a_{i,j})$, we have that

$$\sum_i \sum_j e_{i,j} a e_{j,i} = \sum_i \sum_j a_{j,j} e_{i,i} = \sum_i \mathrm{tr}(a) e_{i,i} = \mathrm{tr}(a).$$

We claim that the expression $\sum_{i=1}^{n^2} e_i a e_i^*$ is independent of the dual bases chosen. In fact, if $f_i, f_i^*$ is another pair of dual bases, we have

$$f_i = \sum_j c_{i,j} e_j, \quad f_i^* = \sum_j d_{i,j} e_j^*,$$

$$\delta^i_j = \mathrm{tr}(f_i f_j^*) = \mathrm{tr}((\sum_k c_{i,k} e_k)(\sum_h d_{j,h} e_h^*)) = \sum_k c_{i,k} d_{j,k},$$

hence if $C$ is the matrix with entries $c_{i,j}$ and $D$ is the one with entries $d_{i,j}$, we have $CD^t = 1$. Hence also $D^t C = 1$ and $\sum_i c_{i,j} d_{i,h} = \delta^j_h$, then

$$\sum_{i=1}^{n^2} f_i a f_i^* = \sum_{i=1}^{n^2} (\sum_j c_{i,j} e_j) a (\sum_h d_{i,h} e_h^*) = \sum_{j,h} \sum_i c_{i,j} d_{i,h} e_j a e_h^* = \sum_i e_i a e_i^*. \qquad \square$$

Given a noncommutative polynomial $f$ in several variables which is linear in a given variable $x_i$, write it in the form $f = \sum_k a_k x_i b_k$. Consider it as a function on matrices, and set $f_i := \sum_k b_k a_k$. We have

$$\mathrm{tr}(f) = \mathrm{tr}(x_i f_i).$$

Suppose that $f = \sum_k a_k x_i b_k$ is a polynomial which depends also upon another variable $x_j$. Assume that $f$ changes sign when one exchanges $x$ with $x_j$. Then when we substitute in $f$, $x_i$ with $x_j$ we get $\sum_k a_k x_j b_k = 0$, and we also deduce

$$\mathrm{tr}(x_j f_i) = \mathrm{tr}(\sum_k a_k x_j b_k) = 0.$$

We apply this to the Capelli polynomial,

$$\mathtt{C}_{n^2}(x_1, \dots, x_{n^2}, y_1, \dots, y_{n^2}) := C_{n^2}(x, y) y_{n^2} \tag{10.17}$$

Then for each variable $x_i$, $i = 1, \dots, n^2$, we construct, by the previous procedure, a polynomial $\mathtt{C}_{n^2,i}(x_1, \dots, x_{i-1}, x_{i+1}, \dots, x_{n^2}, y_1, \dots, y_{n^2})$ (omitting the variable $x_i$), and by the previous remarks we have

$$\delta^i_j \,\mathrm{tr}(\mathtt{C}_{n^2}) = \mathrm{tr}(x_j \mathtt{C}_{n^2,i}), \ \forall i, j = 1, \dots, n^2. \tag{10.18}$$

We want to combine the previous formula with Lemma 10.2.18. For this we define and compute
(10.19)

$$h(x_1, \dots, x_{n^2}, y_0, y_1, \dots, y_{n^2}) := \sum_{i=1}^{n^2} x_i y_0 \mathtt{C}_{n^2,i}(x_1, \dots, x_{i-1}, x_{i+1}, \dots, y_1, \dots, y_{n^2}).$$

It follows that:

THEOREM 10.2.19. *The polynomial* $h(x,y)$ *of formula* (10.19), *evaluated in* $n \times n$ *matrices takes value* $h = \mathrm{tr}(y_0)\,\mathrm{tr}(\mathtt{C}_{n^2})$ *in the center and it does not vanish identically.*

PROOF. We can apply Lemma 10.2.18 to $e_i = x_i$ and $e_i^* = \mathtt{C}_{n^2,i}$ which, by formula (10.18) are dual bases up to the scalar factor $\alpha = \mathrm{tr}(\mathtt{C}_{n^2})$.

From formula (10.10) when $i = j$, we deduce that this scalar $\alpha = \mathrm{tr}(\mathtt{C}_{n^2})$ is not identically 0. Formula (10.16), applied to these elements gives formula (10.19). □

When $y_0 = 1$, we deduce again a noncommutative formula for $n\ \mathrm{tr}(\mathtt{C}_{n^2})$.

We finish with the following fact which will be used in §10.3.1.1 in order to prove the basic theorem of Artin on Azumaya algebras.

PROPOSITION 10.2.20. *The polynomial* $h(x,y)$ *of formula* (10.19) *equals*

$$\sum_{j=1}^{n^2}(-1)^{(n+1)(j+1)}\mathtt{C}_{n^2}(x_1,\dots,x_{n^2},y_0y_j,y_{j+1},\dots,y_{n^2},y_1,\dots,y_{j-1}), \tag{10.20}$$

*is alternating in* $x_1,\dots,x_{n^2}$ *and lies in the T-ideal generated by* $C_{n^2}$.

PROOF. We have, from the defining formulas (7.5) or (10.17)

$$\mathtt{C}_{n^2,i}(x,y) = \sum_{j=1}^{n^2}\sum_{\sigma\in S_{n^2}\,|\,\sigma(j)=i}\epsilon_\sigma y_j x_{\sigma(j+1)}\cdots x_{\sigma(n^2)}y_{n^2}x_{\sigma(1)}y_1\cdots x_{\sigma(j-1)}y_{j-1}$$

$$\implies h(x,y) = \sum_j\sum_\sigma \epsilon_\sigma x_{\sigma(j)}y_0y_jx_{\sigma(j+1)}\cdots x_{\sigma(n^2)}y_{n^2}x_{\sigma(1)}y_1\cdots x_{\sigma(j-1)}y_{j-1},$$

so it is enough to analyze, for a given $j$, $1 \le j \le n^2$,

$$h_j(x,y) := \sum_{\sigma\in S_{n^2}}\epsilon_\sigma x_{\sigma(j)}y_0y_jx_{\sigma(j+1)}\cdots x_{\sigma(n^2)}y_{n^2}x_{\sigma(1)}y_1\cdots x_{\sigma(j-1)}y_{j-1}. \tag{10.21}$$

If $\tau$ is the permutation $(j,\dots,n^2,1,2,\dots,j-1)$, we have

$$\begin{aligned} h_j(x,y) &:= \sum_\sigma \epsilon_\sigma x_{\sigma\tau(1)}y_0y_jx_{\sigma\tau(2)}\cdots x_{\sigma\tau(n^2-j+1)}y_{n^2}x_{\sigma\tau(n^2-j+2)}y_1\cdots x_{\sigma\tau(n^2)}y_{j-1}\\ &= \epsilon_\tau\sum_\sigma \epsilon_\sigma x_{\sigma(1)}y_0y_jx_{\sigma(2)}\cdots x_{\sigma(n^2-j+1)}y_{n^2}x_{\sigma(n^2-j+2)}y_1\cdots x_{\sigma(n^2)}y_{j-1} \qquad (10.22)\\ &= (-1)^{(n+1)(j+1)}\mathtt{C}_{n^2}(x_1,\dots,x_{n^2},y_0y_j,y_{j+1},\dots,y_{n^2},y_1,\dots,y_{j-1}). \end{aligned}$$

It follows that $h_j(x,y)$, and hence $h(x,y) = \sum_{j=1}^{n^2} h_j(x,y)$ is antisymmetric and is deduced from $C_{n^2}$. □

## 10.3. The theorem of M. Artin on Azumaya algebras

**10.3.1. The theorem of M. Artin.** There is a remarkable theorem, due to M. Artin [**Art69**] for algebras over a field, which was extended by Procesi in general [**Pro72c**], characterizing Azumaya algebras of rank $n^2$ over their center, giving a strong link between PI algebras and commutative algebras. Here, the letter "c" *is a central polynomial for* $n \times n$ *matrices*, we mean that $c$ is a polynomial with integer coefficients, and no constant term, which takes values in the center and does not vanish on $M_n(A)$ for all commutative rings $A$. In the previous paragraph we have seen examples of these polynomials. We start with a simple

LEMMA 10.3.1. *Consider a ring* $R$ *with a* 1 *which satisfies all the polynomial identities of* $n \times n$ *matrices. For* $R$ *the following conditions are equivalent:*

(1) *The ring $R$ does not have any quotient which satisfies all the polynomial identities of $n-1 \times n-1$ matrices.*
(2) *$R$ does not have any quotient which is a simple algebra of degree $< n$.*
(3) *If $c$ is a central polynomial for $n \times n$ matrices, the evaluations of $c$ in $R$ generate $R$ as ideal.*

PROOF. Clearly (1) implies (2). Next assume (2) and let $I$ be the ideal generated by the evaluations of $c$ in $R$. If $I \neq R$, then there is a maximal ideal $M \supset I$, and by (2) we have that $R/M$ is a simple algebra of degree $n$. Then $c$ does not vanish in $R/M$, a contradiction, and (3) holds. Finally, assuming (3), if $R/I$ satisfies all the polynomial identities of $n-1 \times n-1$ matrices, then $c$ vanishes on $R/I$, that is the ideal of the evaluations of $c$ is contained in $I$, a contradiction. □

THEOREM 10.3.2 (Artin **[Art69]** and Procesi **[Pro72c]**). *A ring $R$, with a 1, is an Azumaya algebra of rank $n^2$ over its center if and only if the following holds:*

(P) *$R$ satisfies all the polynomial identities of $n \times n$ matrices and moreover it satisfies the equivalent conditions of Lemma* 10.3.1.

The conditions (P) are necessary. By Theorem 5.4.39 an Azumaya algebra $R$ of rank $n^2$ over its center $A$ embeds into $n \times n$ matrices over some commutative ring $B$, so it satisfies the identities of $n \times n$ matrices. Moreover by Corollary 5.4.29 every ideal is of the form $\underline{m}R$ with $\underline{m}$ an ideal of $A$, and $R/\underline{m}R$ is Azumaya with center $A/\underline{m}$. Hence the simple quotients of $R$ are $R/\underline{m}R$ with $\underline{m}$ a maximal ideal of $A$, and $R/\underline{m}R$ has rank $n^2$ over its center $A/\underline{m}$.

The converse is of course very subtle since assuming (P) one has to exhibit enough elements in the center $A$ of $R$ to prove the Azumaya property. We shall follow Schelter's proof **[Sch77b]**.

10.3.1.1. *Schelter's proof.* Take the multilinear polynomial, defined in formula (10.19), of Razmyslov $h(x_1, \dots, x_{n^2}, y_0, y_1, \dots, y_{n^2})$. This is a central polynomial for $n \times n$ matrices, taking the value $\operatorname{tr}(y_0)\operatorname{tr}(C_{n^2}(x_1, \dots, x_{n^2}, y_1, \dots, y_{n^2}))$.

By Proposition 10.2.20 the polynomial $h$ is multilinear and alternating in the variables $x_1, \dots, x_{n^2}$, so we can deduce, adding another variable and alternating $h(x_1, \dots, x_{n^2}, \underline{y})x$ in the $n^2+1$ variables $x_1, \dots, x_{n^2}, x$, the identity

$$h(x_1, \dots, x_{n^2}, \underline{y})x - \sum_{j=1}^{n^2} h(x_1, \dots, x_{j-1}, x, x_{j+1}, \dots, x_{n^2}, \underline{y})x_j = 0 \tag{10.23}$$

on $n \times n$ matrices.

Using this as basic tool, we strengthen the statement of Theorem 10.3.2 as

THEOREM 10.3.3. *A ring $R$, with a 1, is an Azumaya algebra of rank $n^2$ over its center if and only if the following holds:*

(P) *$h(x_1, \dots, x_{n^2}, \underline{y})$ takes values in the center, formula* (10.23) *is satisfied in $R$, (that it the left side is a polynomial identity of $R$), and moreover $R$ does not have any simple quotient of degree $< n$.*

PROOF. The conditions (P) are necessary since an Azumaya algebra of rank $n^2$ over its center can be split to an algebra $M_n(A)$ with $A$ commutative, where $h$ is central and (10.23) is an identity.

Conversely, assume conditions (P). We are going to show that $R$ is a projective module over its center $A$ and that the map $\eta_R : R^e \to \operatorname{End}_A(R)$ is surjective; this, by Remark 5.4.18, is one of the characterizations of Azumaya algebras. Consider all possible evaluations of the central polynomial $h$ in $R$.

The ideal generated by all these evaluations must be the entire $R$, otherwise it is contained in a maximal ideal $\underline{m}$ and $R/\underline{m}$ is a simple algebra of degree $\geq n$ where $h$ vanishes. This is a contradiction, since $h$ is not a polynomial identity of matrices $M_k(F)$ for $k \geq n$ and hence also not a polynomial identity of any simple algebra of degree $\geq n$.

In order to show that $R$ is a projective module over its center $A$, we exhibit a set of projective generators, cf. Proposition 5.3.4(5). Since, by assumption, $h$ takes values in the center $A$ of $R$, we have $1 = \sum_{i=1}^{k} h_i t_i$ where the elements $h_i$ are evaluations of $h$ in elements $x_j \to a_j^{(i)}$, $y_j \to b_j^{(i)}$, $j = 0, 1, \ldots, n^2$, and $t_i \in R$.

$$1 = \sum_i h(a_j^{(i)}, b_j^{(i)}) t_i \implies r = \sum_i h(a_j^{(i)}, b_j^{(i)}) r t_i, \quad \forall r \in R. \tag{10.24}$$

We claim that the $kn^2$ elements $a_j^{(i)} t_i \in R$ can be completed to a set of projective generators for $R$ as an $A$-module, cf. Proposition 5.3.4(5), and formula (5.50).

By definition this means that we need to find $kn^2$ linear forms: $f_j^{(i)} \in R^* = \hom_A(R, A)$ with $r = \sum_{j,i} f_j^{(i)}(r) a_j^{(i)} t_i$ for all $r \in R$.

Take as linear forms the elements $f_j^{(i)}$ defined as

$$f_j^{(i)}(x) := h(a_1^{(i)}, \ldots, a_{j-1}^{(i)}, x, a_{j+1}^{(i)}, \ldots, a_{n^2}^{(i)}, b_0^{(i)}, \ldots, b_{n^2}^{(i)}). \tag{10.25}$$

Thus $x$ is not evaluated at all in $h$ and when we evaluate it, we obtain a linear form $r \mapsto f_j^{(i)}(r)$ for $x = r \in R$.

By the identity of formula (10.23) (which by hypothesis holds), we get for $x \mapsto r \in R$

$$h(a_1^{(i)}, \ldots, a_{n^2}^{(i)}, b_0^{(i)}, \ldots, b_{n^2}^{(i)}) r = \sum_{j=1}^{n^2} f_j^{(i)}(r) a_j^{(i)}$$

$$\implies r h(a_1^{(i)}, \ldots, a_{n^2}^{(i)}, b_0^{(i)}, \ldots, b_{n^2}^{(i)}) t_i = \sum_{j=1}^{n^2} f_j^{(i)}(r) a_j^{(i)} t_i.$$

Summing as in formula (10.24), we have $r = \sum_{j,i} f_j^{(i)}(r) a_j^{(i)} t_i$ which is the condition that the $kn^2$ elements $a_j^{(i)} t_i \in R$ and the $kn^2$ elements $f_j^{(i)}(x) \in \hom_A(R, A)$ are projective generators, by Proposition 5.3.4.

Now we only need to prove that the mapping $\eta_R : R^e \to \mathrm{End}_A(R)$ is surjective (by Remark 5.4.18). For this, by Lemma 5.3.5, it is enough to show that the elementary endomorphisms

$$a_j^{(i)} t_i \otimes f_h^{(k)} : r \mapsto f_h^{(k)}(r) a_j^{(i)} t_i \tag{10.26}$$

associated to the projective generators which we have found, are in the image of $\eta_R$.

For this we remark that $f_j^{(i)}(r)$ is the evaluation of a noncommutative polynomial, linear in $x$. Hence there exist some (explicit) elements $d_{j,s}^{(i)}, e_{j,s}^{(i)} \in R$, deduced from formula (10.25), for which

$$f_j^{(i)}(x) = \sum_s d_{j,s}^{(i)} x e_{j,s}^{(i)} \implies f_j^{(i)}(r) = \sum_s d_{j,s}^{(i)} r e_{j,s}^{(i)}, \ \forall r \in R.$$

Thus for the elementary endomorphisms of formula (10.26) we have

$$f_h^{(k)}(r) a_j^{(i)} t_i = \sum_s d_{h,s}^{(k)} r e_{h,s}^{(k)} a_j^{(i)} t_i = \eta_R(\sum_s d_{h,s}^{(k)} \otimes e_{h,s}^{(k)} a_j^{(i)} t_i)(r),$$

which proves the claim. □

REMARK 10.3.4. Looking at the proof of the theorem, one sees that the condition that $R$ has no simple quotients of degree $< n$ can be replaced by the condition that the ideal generated by all evaluations of $h$ is $R$.

We have useful corollaries for this theorem.

COROLLARY 10.3.5. *Let $R$ be a ring, with center $\mathcal{Z}$, which satisfies all the identities of $n \times n$ matrices. Let $\lambda \in \mathcal{Z} \subset R$ be a nonzero evaluation of a central polynomial for $n \times n$ matrices, with no constant term, $c(x_1, \dots, x_m)$. Then the localized algebra $R[\lambda^{-1}]$ is a rank $n^2$ Azumaya algebra, with center $\mathcal{Z}[\lambda^{-1}]$.*

PROOF. Clearly $R[\lambda^{-1}]$ satisfies the identities of $n \times n$ matrices. We have seen, Proposition 10.2.2, that $c(x_1, \dots, x_m)$ is a PI for $(n-1) \times (n-1)$ matrices, so once we invert an evaluation of $c$, this polynomial is nonzero in every nonzero quotient. So no quotient of $R[\lambda^{-1}]$ satisfies the identities of $(n-1) \times (n-1)$ matrices, and thus $R[\lambda^{-1}]$ satisfies the conditions of Theorem 10.3.2. □

COROLLARY 10.3.6. *If $R = F\langle a_1, \dots, a_k \rangle$ is a rank $n^2$ Azumaya algebra, its center is finitely generated as algebra over $F$.*

PROOF. The proof follows along the same lines as that of Proposition 11.4.2, and we leave it as exercise. □

Suppose only that the polynomial $h$ takes values in the center $\mathcal{Z}$ of $R$ and that $R$ satisfies the PI of formula (10.23).

Let $I$ be the ideal of $R$ generated by the evaluations of $h$ in $R$ and $J := I \cap \mathcal{Z}$. If we take any element $\lambda \in J$, the localized algebra $R[\lambda^{-1}]$ has now the property that the ideal of $R[\lambda^{-1}]$ generated by the evaluations of $h$ in $R[\lambda^{-1}]$ equals $R[\lambda^{-1}]$. Thus by Remak 10.3.4, $R[\lambda^{-1}]$ satisfies the properties (P) of Theorem 10.3.3, hence it is Azumaya.

Assume that $R$ satisfies the identities of $n \times n$ matrices, and let $\mathcal{Z}$ be the center of $R$. Let $P$ be a prime ideal of $R$ such that $R/P$ has ring of fractions a simple algebra of dimension exactly $n^2$ over its center (in general it may be a lower dimension), that is $R/P$ has PI degree $n$. Let $S := \mathcal{Z} \setminus P$ be the elements in the center $\mathcal{Z}$ which are not in $P$. Then

COROLLARY 10.3.7. *$R \otimes_{\mathcal{Z}} \mathcal{Z}_S$ is a rank $n^2$ Azumaya algebra.*

*$R \otimes_{\mathcal{Z}} \mathcal{Z}_S$ is a local ring with maximal ideal $M$ and, if $j : R \to R \otimes_{\mathcal{Z}} \mathcal{Z}_S$, we have $j^{-1}(M) = P$, and $R \otimes_{\mathcal{Z}} \mathcal{Z}_S / M = Q(R/P)$ is the full ring of fractions of $R/P$.*

PROOF. By hypothesis, $R/P$ has PI degree $n$, so there is an evaluation $\lambda$ of a central polynomial which is in $S$. Thus $R \otimes_{\mathcal{Z}} \mathcal{Z}_S = R[\lambda^{-1}] \otimes_{\mathcal{Z}[\lambda^{-1}]} \mathcal{Z}_S$ is Azumaya with center $\mathcal{Z}_S$, by Corollary 10.3.5 and Proposition 5.4.28.

The fact that $j^{-1}(M) = P$ and $R \otimes_Z \mathcal{Z}_S/M = Q(R/P)$ follows as for the commutative case, and we leave to it the reader. $\square$

What is remarkable about this corollary is that it gives a localization at a prime ideal $P$, something which rarely exists in a noncommutative ring.

If on the other hand $P$ is a prime ideal of $R$ such that $R/P$ has PI degree $< n$, that is its ring of fractions is a simple algebra of dimension $< n^2$, then the central polynomials for $n \times n$ matrices vanish in $R/P$.

Thus the ideal $\mathcal{I}$ of $R$, generated by the evaluations of central polynomials for $n \times n$, behaves as some sort of *discriminant ideal*. The prime ideals $P \supset \mathcal{I}$ are exactly those for which the PI degree $< n$, the prime ideals $P \not\supset \mathcal{I}$ are exactly those for which the PI degree equals $n$, and they admit a central localization which is a rank $n^2$ Azumaya algebra.

REMARK 10.3.8. We have seen in Schelter's proof that we do not need to assume that $R$ satisfies all identities of $n \times n$ matrices, but only those coming from the central polynomial $h$. This is useful for further results on the nil radical.

## 10.4. Universal splitting

**10.4.1. Some complements.** What is remarkable is that for a ring $R$ satisfying the hypotheses of Artin's theorem, the center appears as a construction and not as given. This means that one can treat Azumaya algebras without making explicit reference to their centers.

PROPOSITION 10.4.1. *Let $R$ be a rank $n^2$ Azumaya algebra with center $A$. Then every element $a \in A$ is the sum of values of any given multilinear central polynomial $f(x_1, \ldots, x_k)$ for $n \times n$ matrices.*

PROOF. We first claim that the set $I$ of elements $a \in A$, which are sums of values of $f$, is an ideal of $A$. In fact if $a = f(u_1, \ldots, u_k)$ and $c \in A$ we have $ca = f(cu_1, \ldots, u_k)$.

If $I \neq A$, the algebra $R/IR$ is still a nonzero rank $n^2$ Azumaya algebra with center $A/I$, and central polynomials take nonzero values in $R/IR$, a contradiction. $\square$

PROPOSITION 10.4.2. *Let $\phi : R \to S$ be a map of rings with 1. Assume that $R$ is a rank $n^2$ Azumaya algebra with center $A$ and $S$, with center $B$, satisfies the identities of $n \times n$ matrices. Then $\phi(A) \subset B$, $S$ is a rank $n^2$ Azumaya algebra over $B$ and $S = R \otimes_A B$.*

PROOF. First $S$ is a rank $n^2$ Azumaya algebra, in fact it satisfies the hypotheses of Artin's theorem since if we had a nonzero quotient $S/J$ satisfying the identities of $n-1 \times n-1$ matrices, the quotient $R/J \cap R$ also would satisfy these identities, a contradiction.

Next by the previous proposition every element of $A$ is a sum of values $a = f(u_1, \ldots, u_k) \in A$, each the value of some multilinear polynomial $f$ for $n \times n$ matrices.

Of course such a polynomial $f(x_1, \ldots, x_k)$ takes values also in the center of $S$, so $\phi(a) = f(\phi(u_1), \ldots, \phi(u_k)) \in B$ and $\phi(A) \subset B$. The final identity $S = R \otimes_A B$ is from Corollary 5.4.29. $\square$

Given a ring $R$, in §3.4.1 we have considered the set-valued functor $B \mapsto \mathcal{R}_R^n(B)$, on the category $\mathcal{C}$ of commutative rings, formed by all homomorphisms $\phi : R \to M_n(B)$. We have shown that it is representable giving rise to a universal map into $n \times n$ matrices over a commutative ring $j : R \to M_n(A_n(R))$, i.e., $\hom_{\mathcal{R}}(R, M_n(B)) \simeq \hom_{\mathcal{C}}(A_n(R), B)$.

We have introduced in §3.4.2 the *projective group*, which as a group-valued functor of the category of commutative rings, is $B \mapsto \mathrm{Aut}_B(M_n(B))$. This functor is represented by the Hopf algebra $P_n$. In §3.4.3 we have studied the action

$$\mathrm{Aut}_B(M_n(B)) \times \mathcal{R}_R^n(B) \to \mathcal{R}_R^n(B). \tag{10.27}$$

If $R = M_n(\mathbb{Z})$, then the set of homomorphisms $\phi : M_n(\mathbb{Z}) \to M_n(B)$ as a functor on the category of commutative rings can be identified with $\mathrm{Aut}_B(M_n(B))$ by applying the action (10.27) to the trivial map $M_n(\mathbb{Z}) \to M_n(B)$, $e_{i,j} \mapsto e_{i,j}$, Proposition 3.4.23.

Assume now that $R$ is a rank $n^2$ Azumaya algebra over its center $Z$.

Given $\phi : R \to M_n(B)$ a homomorphism, we have seen, in Proposition 10.4.2, that $\phi : Z \to B$, and this gives rise to a isomorphism $\phi : B \otimes_Z R \to M_n(B)$.

In particular we have

COROLLARY 10.4.3. *Let $R$ be a rank $n^2$ Azumaya algebra with center $Z$. The universal map into matrices $j : R \to M_n(A_n(R))$ is injective. We further have $j(Z) \subset A_n(R)$ so $M_n(A_n(R)) = A_n(R) \otimes_Z R$. $A_n(R)$ is faithfully flat over $Z$.*

PROOF. The main point is to prove that $j$ is injective. This follows from the fact that, by Theorem 5.4.39, there is some embedding of $R$ into $n \times n$ matrices over some commutative ring.

Finally let $B \supset Z$ be a faithfully flat splitting, i.e., $M_n(B) = B \otimes_Z R$. Then the map $1_B \otimes j$,

$$M_n(B) = B \otimes_Z R \xrightarrow{1_B \otimes j} B \otimes_Z M_n(A_n(R)) = M_n(B \otimes_Z A_n(R)),$$

is the universal map of $M_n(B)$ to $n \times n$ matrices. So we have an isomorphism

$$B \otimes_Z A_n(R) \simeq B \otimes_{\mathbb{Z}} P_n = B \otimes_Z (Z \otimes_{\mathbb{Z}} P_n).$$

Since by Corollary 3.4.22 the algebra $Z \otimes_{\mathbb{Z}} P_n$ is a free $Z$-module, it follows that $B \otimes_{\mathbb{Z}} P_n$ is faithfully flat over $Z$. So both $B$ and $B \otimes_Z A_n(R)$ are faithfully flat over $Z$. From this it easily follows that also $A_n(R)$, the universal splitting, is faithfully flat over $Z$. $\square$

If $\mathcal{R}_R^n(B) = \hom(R, M_n(B)) \simeq \hom_Z(A_n(R), B) \neq \emptyset$, each element of $\mathcal{R}_R^n(B)$ gives an isomorphism $B \otimes_Z R \simeq M_n(B)$. We say that $B$ *splits* $R$.

The action of $\mathrm{Aut}_B(M_n(B))$ on $\mathcal{R}_R^n(B)$ is simply transitive. That is, by definition, we have that $A_n(R)$ is a *torsor* for $Z \otimes_{\mathbb{Z}} P_n$ (in the opposite category of that of $Z$ commutative algebras).

DEFINITION 10.4.4. Let $\mathcal{C}$ be a category, and let $G(-), F(-)$ be a covariant group valued and a set valued functor, respectively, together with a functorial group action $\mu : G(-) \times F(-) \to F(-)$.

We say that $F$ is a *torsor* over $G$ if for all $X \in \mathcal{C}$ given $x, y \in F(X)$ there is a unique element $g \in G(X)$ with $gy = x$.

If $G(X) = \hom_{\mathcal{C}}(\mathtt{G}, X)$, $F(X) = \hom_{\mathcal{C}}(\mathtt{F}, X)$ are represented by two objects $\mathtt{G}$, $\mathtt{F}$, and $\mathcal{C}^{op}$ has products and a final object, then, by Yoneda's lemma we have that $\mathtt{G}$ is a group-like object (Definition 2.1.23) of $\mathcal{C}^{\mathrm{op}}$. The axioms on the group action are reflected in axioms on the map $\mathtt{G} \times \mathtt{F} \to \mathtt{F}$, in $\mathcal{C}^{\mathrm{op}}$. We say that $\mathtt{F}$ is a torsor over the group-like object $\mathtt{G}$.

Notice again that when $F(X) = \emptyset$, the condition is void.

The previous condition can be conveniently reformulated as *the natural transformation of functors, or the map of their representing objects:*

$$(10.28)\qquad \begin{array}{ccccc} G(X) \times F(X) & \xrightarrow{1\times\delta} & G(X) \times F(X) \times F(X) & \xrightarrow{\mu\times 1} & F(X) \times F(X) \\ \mathtt{G} \times \mathtt{F} & \xrightarrow{1\times\delta} & \mathtt{G} \times \mathtt{F} \times \mathtt{F} & \xrightarrow{\mu\times 1} & \mathtt{F} \times \mathtt{F} \end{array}$$

*is an isomorphism* (*with $\delta$ the diagonal*).

In terms of algebras we have that the action gives a map (*the coaction*)

$$\eta : A_n(R) \to (P_n \otimes Z) \otimes_Z A_n(R) = P_n \otimes A_n(R),$$

making $A_n(R)$ a comodule over the Hopf algebra $P_n$. From formula (10.28) the property of being a torsor can be stated in terms of the coaction of $Z \otimes_{\mathbb{Z}} P_n$ over $A_n(R)$, by the fact that the map $1 \otimes m \circ \eta \otimes 1$ ($m$ the multiplication $m(a \otimes b) := ab$) is an isomorphism,

$$1 \otimes m \circ \eta \otimes 1 : A_n(R) \otimes_Z A_n(R) \xrightarrow{\eta\otimes 1} P_n \otimes A_n(R) \otimes_Z A_n(R) \xrightarrow{1\otimes m} P_n \otimes A_n(R).$$

This formula at the level of points means exactly that given two points $(x, y)$ in $\hom(A_n(R), A)$ there is a unique $g \in \mathfrak{G}(A)$ with $x = gy$, $(x, y) = (gy, y)$.

We leave it to the reader to show that

$$Z = \{a \in A_n(R) \mid \eta(a) = 1 \otimes a\},$$

which can be interpreted as $Z$, is the ring of invariants under the action of the projective group. Moreover recall the general commutative diagram (3.57)

$$(10.29)\qquad \begin{array}{ccc} R & \xrightarrow{j_R} & M_n(A_n(R)) \\ \downarrow{\scriptstyle j_R} & & \downarrow{\scriptstyle M_n(\eta)} \\ M_n(A_n(R)) & \xrightarrow{U_n} & P_n \otimes_{\mathbb{Z}} M_n(A_n(R)). \end{array}$$

EXERCISE 10.4.5. If $R$ is a rank $n^2$ Azumaya algebra, we have that

$$(10.30)\qquad R = \{a \in M_n(A_n(R)) \mid U_n(a) = 1 \otimes a\}.$$

Now given a ring $R$, let $j : R \to M_n(A_n(R))$ be the universal map of $R$ to $n \times n$ matrices over a commutative ring. We have defined in Definition 3.4.26 the ring $T_n(R)$ as the subring of $A_n(R)$ generated by all the coefficients of the characteristic polynomial of elements of $R$, called the universal $n$-trace ring of $R$.

By Amitsur's theorem, Theorem 3.3.8, we have that the coefficients of the characteristic polynomial of elements of the ring $j(R)T_n(R)$ still lie in $T_n(R)$.

Consider then an evaluation $a$ of a central polynomial for $n \times n$ matrices in $R$. We have that $a^n = \det(a) \in T_n(R)$. If $R$ is an algebra over a field of characteristic 0, then $a = n^{-1}\operatorname{tr}(a) \in T_n(R)$.

In general let $\tilde{T}_n(R)$ be the ring generated by $T_n(R)$ and all evaluations in $R$ of central polynomials. If we localize $j(R)\tilde{T}_n(R)$ in $a$, we get the ring $j(R)\tilde{T}_n(R)[a^{-1}]$.

PROPOSITION 10.4.6. *The ring $j(R)\tilde{T}_n(R)[a^{-1}]$ is a rank $n^2$ Azumaya algebra, with center $\tilde{T}_n(R)[a^{-1}]$.*

PROOF. By Corollary 10.3.5 $j(R)\tilde{T}_n(R)[a^{-1}]$ is a rank $n^2$ Azumaya algebra. We only need to show that its center is $\tilde{T}_n(R)[a^{-1}]$. By Proposition 10.4.1 the center of $j(R)\tilde{T}_n(R)[a^{-1}]$ is given by all evaluations of multilinear central polynomials for $n \times n$ matrices. This is the ring generated by $\tilde{T}_n(R)[a^{-1}]$, and all evaluations of central polynomials in $R$ which by definition lie in $\tilde{T}_n(R)$, and the claim follows. □

**10.4.2. The conductor.** We want to understand now another important property of the Capelli polynomial $C_{n^2}$ and also of central polynomials—the *conductor property*.

DEFINITION 10.4.7. Given an inclusion $A \subset B$ of rings, the *conductor* of this inclusion is the maximal ideal $J$ of $A$ which is also an ideal of $B$.

In general the conductor is 0, so when it is nonzero this gives some nontrivial relationship between the two rings.

For each $n$ fixed, we want to apply this idea to the inclusion of the ring of $n \times n$ generic matrices $A = F\langle \xi_i \rangle$ inside the trace ring $B = \mathcal{T}_n\langle \xi_i \rangle F\langle \xi_i \rangle$.

Here $\mathcal{T}_n\langle \xi_i \rangle = T_A$, defined in Definition 2.4.6, is by Theorem 3.3.28 the central ring $T_A := F\langle \sigma_i(p) \rangle$ generated by the elements $\sigma_i(p)$, with $p \in W_0$ Lyndon words in the variables $\xi_j$. We assume $F$ to be a field or the integers.

The conductor of $F\langle \xi_i \rangle$ in $\mathcal{T}_n\langle \xi_i \rangle F\langle \xi_i \rangle$ is clearly

$$\{a \in F\langle \xi_i \rangle \mid a\mathcal{T}_n\langle \xi_i \rangle \subset F\langle \xi_i \rangle\}.$$

THEOREM 10.4.8.

(1) *Let $A := F\langle \xi_i \rangle_{i \in I}$ in possibly infinitely many variables, and let $F$ be a field or the integers $\mathbb{Z}$. The set $J := \{x \in A \mid x \det(r) \in A, \ \forall r \in A\}$, is the conductor of $A$ in $B = AT_A$.*

(2) *If $x \in A$ depends on $\xi_1, \dots, \xi_k$, then $x \in J$ if and only if $x \det(\xi_{k+1}) \in A$.*

PROOF. Certainly if $x$ satisfies (2) substituting, for any $r \in A$, $\xi_{k+1} \mapsto r$, we see that $x$ is in $J$. So we need to prove (1).

(1) We have that $J$ is an ideal of $A$ and clearly contains the conductor. Take $u \in J$ depending on some variables $\xi_1, \dots, \xi_k$. Let us add a central variable $y$ to $A$. In $A[y]$ we have $u \det(y - \xi_{k+1}) \in A[y]$, and this implies that the various coefficients of the powers of $y$, that is $u\sigma_i(\xi_{k+1})$ are also in $A$. Now we claim that in fact $u \det(r) \in J$. For this we need to show that $u \det(r) \det(r') \in A$ for all $r' \in A$. This is clear since $u \det(r) \det(r') = u \det(r \cdot r') \in A$.

Finally, since $AT_A$ is obtained from $A$ by adding the elements $\sigma_i(r), r \in R$, it follows that $J$ is also an ideal of $AT_A$, that is $J$ is in the conductor of $A$ in $AT_A$. Since $J$ contains the conductor, it follows that $J$ coincides with the conductor. □

We have several classes of examples of polynomials in the conductor of the ring $A$ of generic matrices in $AT_A$.

First of all when $F$ is a field of characteristic 0, it is useful to weaken the conditions of Theorem 10.4.8.

PROPOSITION 10.4.9. *When $F$ is a field of characteristic 0, then $x \in A$ is in $J$ if and only if $x \det(\xi_{k+1})^m \in A$ for some integer $m > 0$.*

PROOF. The argument of Theorem 10.4.8 applied verbatim implies that for all the elements $\sigma_{i;m}(r)$, $r \in A$, obtained by polarizing $\det(z)^n$ (cf. formula (3.42)) we have $x\sigma_{i;m}(r) \in A$. But, since we are in characteristic 0, by Lemma 3.3.21 the algebra obtained by polarizing $\det(z)$ or $\det(z)^n$ is the same, so from that it follows also that $x \det(\xi_{k+1}) \in A$ (or use formula (10.9)). □

PROPOSITION 10.4.10. *If F is a field of characteristic* 0, *then the Capelli polynomial* $C_{n^2}(x, y)$ *is in the conductor.*

PROOF. This follows from Propositions 10.4.9 and 10.2.14. □

In all characteristics central elements in the conductor can be constructed as follows.

THEOREM 10.4.11. *If* $c(\xi_1, \dots, \xi_k) \in A$ *is a central polynomial with no constant term, there is an exponent m such that* $c(\xi_1, \dots, \xi_k)^m$ *is in the conductor of A in* $AT_A$.

PROOF. By Corollary 10.3.5, we have that $A[c^{-1}]$ is an Azumaya algebra of rank $n^2$. Then we have that the determinant $\det(\xi_{k+1})$ of $\xi_{k+1}$ is in $A[c^{-1}]$ (Theorem 5.4.40), hence for some power $m$ we have that $p_m(\xi_1, \dots, \xi_{k+1}) := c^m \det(\xi_{k+1})$ lies in $A$. By Theorem 10.4.8(2), the claim follows. □

In characteristic 0 we shall see some very explicit central polynomials in the conductor which have a very special meaning; Theorem 12.3.18.

CHAPTER 11

# Structure theorems

*This chapter is devoted to the classical structure theory of PI algebras satisfying additional conditions, such as being nil, Levitzki's and Amitsur's Theorem 11.1.4, or on the opposite primitive, Kaplansky's Theorem 10.1.8, or prime, Posner's Theorem 11.2.6. In this chapter unless specified, by polynomial identity we shall mean always a stable identity.*

## 11.1. Nil ideals

**11.1.1. The theorems of Levitzki and Amitsur.** One of the first inputs of PI theory was the goal to give a positive answer to the Kurosch problem under special conditions. Recall that this problem asks whether an algebra in which every element is algebraic is *locally finite*, that is finitely generated subalgebras are finite dimensional.

We now know, thanks to the work of Golod and Shafarevitch **[GŠ64]** that in general the answer is negative, even in the special case in which the algebraic equation is the condition that each element is nilpotent; see Appendix A.

If one assumes that every element satisfies an algebraic equation of fixed degree $n$, we have seen in Lemma 0.1.1 that the algebra satisfies the polynomial identity (0.1). Therefore a natural question is whether the existence of a PI implies a positive answer to the Kurosch problem. We have already deduced this from Shirshov's theory (cf. Theorem 8.2.2), but in fact this was rapidly achieved, with a positive answer, by the work of Levitzki and Kaplansky, and we start to review the first ideas in this context.

**11.1.2. Locally nilpotent radical.** We start by reproducing the results of Levitzki in **[Lev46]**. In that paper he uses the expression "semi-nilpotent" instead of "locally nilpotent", which is used here.

Recall that we have introduced for a general ring $R$ two nil radicals, the full nil radical $\mathcal{N}(R)$ is the maximal ideal formed of nilpotent elements and the lower nil radical $\mathcal{L}(R)$ is obtained by a transfinite induction, cf. §1.1.3.

For PI rings we shall see that these two radicals coincide (Corollary 11.1.5), and in fact this follows from a positive solution of Kurosch's problem in the case of nil PI algebras. It is useful for this to introduce a third nil radical, $\mathcal{L}_n(R)$, intermediate between the two previously defined nil radicals. For PI algebras then all these nil radicals coincide.

Recall that a ring is locally nilpotent if every finitely generated subring is nilpotent.

THEOREM 11.1.1. *The sum $I = I_1 + I_2$ of two locally nilpotent left ideals is also locally nilpotent.*

PROOF. Take a subring generated by finitely many elements $s_i = a_i + b_i$, with $a_i \in I_1$, $b_i \in I_2$. This is contained in the subring generated by all the elements $a_i, b_i$, so it is enough to prove that every subring generated by finitely many elements $r_i$ which are either in $I_1$ or in $I_2$ is nilpotent. If by contradiction this were not true, we could find a minimum number $m$ of elements $r_1, \dots, r_m$ which generate a subring which is not nilpotent; we may assume $r_1 \in I_1$. So by hypothesis $r_1^k = 0$ for some $k > 1$ and the subring generated by $r_2, \dots, r_m$ is nilpotent of some exponent $\ell$, so it is spanned by finitely many monomials $w_1, \dots, w_s$ (in $r_2, \dots, r_m$) of some degree $< \ell$.

Now consider the elements $w_i r_1^j$, $1 \leq j < k$. They are finite in number and belong to the left ideal $I_1$, so they generate a nilpotent subring. This means that there is some $h$ such that any product of $h$ out of these monomials equals 0. On the other hand every monomial in $r_1, \dots, r_m$ is a linear combination of monomials of type either $w_i$ or of products of $w_i x_1^j$, so we have that every monomial of degree at least $(k-1)\ell h$ in $r_1, \dots, r_m$ equals 0. □

Then we deduce the existence of a *locally nilpotent radical.*

THEOREM 11.1.2. *The sum of all locally nilpotent left ideals of a ring R is a locally nilpotent two-sided ideal $\mathcal{L}_n(R)$ equal to the sum of all locally nilpotent right ideals. Moreover $\mathcal{L}_n(R)$ is a* radical *in the sense that*

$$\mathcal{L}_n(R/\mathcal{L}_n(R)) = 0.$$

PROOF. By the previous theorem $\mathcal{L}_n(R)$ is a locally nilpotent left ideal. Now given any locally nilpotent left ideal $I$, we claim that for all $a \in R$ we have that $Ia$ is also a locally nilpotent left ideal.

In fact given finitely many elements $r_i a$, consider the elements $a r_i \in I$ which generate a nilpotent subring. Thus there is an $h > 0$ so that the product of any $h$ of these elements equals 0, but then a product of $h+1$ elements $r_i a$ also equals 0. This implies that $\mathcal{L}_n(R)$ is a two-sided ideal. If $I$ is a left ideal containing $\mathcal{L}_n(R)$ and is locally nilpotent modulo $\mathcal{L}_n(R)$, we claim that it is locally nilpotent, hence contained in $\mathcal{L}_n(R)$. This will prove that $\mathcal{L}_n(R/\mathcal{L}_n(R)) = 0$.

As for the previous statement, consider elements $r_1, \dots, r_m \in I$, we have that there is an $h$ such that all the monomials $M_\alpha$ of degree $h$ in these elements lie in $\mathcal{L}_n(R)$. This is a finite set of elements so there is a $k$ such that all monomials of degree $k$ in the $M_\alpha$'s equal 0, hence all monomials of degree $kh$ in the $r_1, \dots, r_m$ are 0 as required. □

11.1.2.1. *Locally nilpotent property revisited.* We want to prove now, without the use of Shirshov's bases,

THEOREM 11.1.3. *If R is a nil ring satisfying a polynomial identity $f(x_1, \dots, x_d)$, with coefficients $\pm 1$, then it is locally nilpotent.*

PROOF. By Theorem 11.1.2 it is equivalent to prove that $R = \mathcal{L}_n(R)$. In other words assuming that $\mathcal{L}_n(R) = 0$, we need to show that $R = 0$.

We go by induction on the degree of the identity which we may assume to be multilinear of the form $f(x_1, \dots, x_d) = x_1 g(x_2, \dots, x_d) + h(x_1, \dots, x_d)$, where $h$ is a linear combination of monomials which does not start with $x_1$. Take $a \in R$ with $a \neq 0$, $a^2 = 0$.

Now consider the left ideal $Ra$, and let $J := \{x \in Ra \mid Rax = 0\}$ be its right annihilator as subring. We claim that $Ra$ is locally nilpotent. In fact substitute in

$f$ for $x_1 \mapsto a$, $x_i \mapsto r_i a$, $i \geq 2$. We then have that every monomial not starting with $x_1$ contains the element $a^2 = 0$, hence $h(a, r_2 a, \ldots, r_d a) = 0$ so that also $ag(r_2 a, \ldots, r_d a) = 0$, that is $g(r_2 a, \ldots, r_d a) \in J$. This implies that $Ra/J$ satisfies the polynomial identity $g$ of lower degree $d-1$ so, by induction is locally nilpotent.

So take any finitely generated subring $S$ of $Ra$, by induction $S$ is nilpotent modulo $J$, but then $S^k \subset J$ implies $S^{k+1} = 0$. This implies that the left ideal $Ra$ is locally nilpotent, hence by Theorem 11.1.1 it is contained in $\mathcal{L}_n(R)$. Since by hypothesis this equals 0, we have that $Ra = 0$. But the set of $a \in R$ such that $Ra = 0$ is an ideal $K$ with $K^2 = 0$, hence by hypothesis $K \subset \mathcal{L}_n(R) = 0$, and we reach a contradiction having assumed that $a \neq 0$. □

In fact a more careful analysis of the proof shows that we do not really need to have an identity with coefficients $\pm 1$ but just a proper identity.

This in fact is mostly a nuisance and not a serious problem since one has that every PI algebra satisfies a power of some standard identity which when made multilinear has in fact coefficients $\pm 1$; see Corollary 11.3.5.

Recall that $N(R)$ denotes the sum of all nilpotent ideals of $R$; cf. Definition 1.1.29. The first substantial theorem of Levitzki **[Lev43]** as generalized by Amitsur **[Ami51]** is

THEOREM 11.1.4. *Let $R$ satisfy a proper multilinear identity of degree $d$. Then if $T \subset R$ is a nil subring, we have $T^{[d/2]} \subset N(R)$.*

PROOF. Let us take first $T$ nilpotent. If $T \subset N(R)$, there is nothing to prove. Otherwise set $n$ to be such that $T^n \not\subset N(R)$, $T^{n+1} \subset N(R)$. We need to show that $n < [d/2]$. Consider the $2n+1$ subrings

$$A_{2j-1} := T^{n-j+1} R T^{j-1}, \quad j = 1, 2, \ldots, n+1, \qquad (11.1)$$
$$A_{2j} := T^{n-j+1} R T^{j}, \quad j = 1, 2, \ldots, n,$$

and form the products

$$B_i : A_1 \cdot A_2 \cdots A_i, \quad i = 1, 2, \ldots, 2n+1.$$

It follows that

$$B_{2j-1} := (T^n R)^{2j-1} T^{j-1}, \quad j = 1, 2, \ldots, n+1, \qquad (11.2)$$
$$B_{2j} := (T^n R)^{2j} T^{j}, \quad j = 1, 2, \ldots, n.$$

From (11.1) we deduce $A_s A_t \subset R T^{n+1} R$ for $s > t$. Hence, for every $r \leq 2n+1$ and any permutation $i_1, i_2, \ldots, i_r$ different from the identity of the numbers $1, 2, \ldots, r$, we have $A_{i_1} \cdot A_{i_2} \cdots A_{i_r} \subset R T^{n+1} R$. Assume that $n \geq [d/2]$, that is $d \leq 2n+1$, and take the multilinear identity which we write in the form

$$\beta x_1 x_2 \cdots x_d = \sum_{\sigma \neq 1} \alpha_\sigma x_{\sigma(1)} x_{\sigma(2)} \cdots x_{\sigma(d)} \quad \text{on } R.$$

We deduce, taking $a_i \in A_i$, that

$$\beta a_1 a_2 \cdots a_d \in R T^{n+1} R \implies \beta A_1 A_2 \cdots A_d \subset R T^{n+1} R.$$

Repeating the argument for all the monomials of the identity since the identity is proper, we deduce

$$a_1 a_2 \cdots a_d \in R T^{n+1} R \implies B_d = A_1 A_2 \cdots A_d \subset R T^{n+1} R.$$

If $d = 2q - 1$, we have $B_d = (T^nR)^{2q-1}T^{q-1}$; if $d = 2q$, we have $B_d = (T^nR)^{2q}T^q$. In the first case multiply (11.2) by $T^{n-q+1}R$ and get $(T^nR)^{2q} \subset RT^{n+1}R$. In the second case multiply by $T^{n-q}R$ and get $(T^nR)^{2q+1} \subset RT^{n+1}R$. In either case we observe that, since $RT^{n+1}R$ is a nilpotent ideal by assumption, we also have that $RT^nR$ is a nilpotent ideal.

This contradicts the minimality of the choice of $n$.

Now we pass to the general case and assume just $T$ to be nil. By Theorem 11.1.3 every finitely generated subring $U$ of $T$ is nilpotent, thus we can apply the previous analysis and deduce that $U^{[d/2]} \subset N(R)$. Since this is true for any finitely generated $U$, it follows that $T^{[d/2]} \subset N(R)$, as desired. □

This theorem has an important corollary:

COROLLARY 11.1.5. *The nil radical of a PI ring $R$ coincides with the lower nil radical, which in turn coincides with $N_2(R)$ (defined by formula* (1.4)*).*

In particular a semiprime PI ring has no nonzero nil left ideals. Furthermore there are examples of PI rings in which $N_2(R) \neq N(R)$.

EXAMPLE 11.1.6. The following example is also due to Amitsur.

Take a commutative ring $A$ with a 1, so that its nil radical $N = N(A) = \sum_\alpha I_\alpha$ with $I_\alpha$ nilpotent is not nilpotent, and consider $M_n(A)$, the matrix algebra which satisfies the standard identity $\mathrm{St}_{2n}$. The nil radical of $M_n(A)$ is clearly $M_n(N)$ which is a sum of nilpotent ideals $M_n(I_\alpha)$.

We take as $R$ the subring of all matrices with entries $\alpha_{i,j} \in N$, $\forall i \geq j$ (no restriction on $\alpha_{i,j}$, $i < j$). Modulo $M_n(N)$ the ring $R$ becomes the ring of strictly upper triangular matrices with entries in $A/N$, a ring nilpotent of index $n$.

So $R^n \subset M_n(N)$. Now take the element $e_{1,2} \in R$, and we claim that the ideal $I$ that it generates is not nilpotent. We have $e_{1,2}(\rho e_{2,1}) = \rho e_{1,1} \in I$, $\forall \rho \in N$, and these elements form a ring isomorphic to $N$, nil but not nilpotent. This implies that $e_{1,2} \notin N(R)$. We have thus a nil ring satisfying an identity of degree $2n$ so that every element $x$ raised to $n$ is in $N(R)$, but we have elements $x$ which do not generate a nilpotent ideal. This gives the required example, in fact we have $N(R) = M_n(N(A))$ (verify it).

DEFINITION 11.1.7. If $I$ is an ideal of a PI ring $R$, we denote by $\sqrt{I}$, and call it the nil radical of $I$, the ideal such that

$$\sqrt{I} \supset I, \ \sqrt{I}/I = \mathcal{L}(R/I).$$

Now consider the free algebra $A\langle X\rangle$ in infinitely many variables over a commutative ring $A$. A $T$-ideal is an ideal $I$ of the free algebra $A\langle X\rangle$ closed under substitutions, and the Definition 2.2.32 implies that $I$ is proper if it contains a proper polynomial identity. We generalize the result, of Exercise 2.2.11(iv), where $A$ is an infinite field.

COROLLARY 11.1.8. *The nil radical $J := \sqrt{I}$ of a proper $T$-ideal is a $T$-ideal.*

PROOF. Let $I$ be a proper $T$-ideal. The nil radical $J$ contains all left ideals or right ideals which are formed by nil elements modulo $I$ (because the quotient $A\langle X\rangle/I$ is PI). Take an element $a \in J$. We have for a variable $z$ which does not appear in $a$ that $za \in J$, hence $(za)^k \in I$ for some $k \in \mathbb{N}$. Now take some substitution of variables which maps $a$ to some element $b$. We need to prove that $b$ is in the nil radical of $I$, so we need to prove that it generates a nil left ideal modulo $I$ or that $ub$ is nil for all $u$. We then complete the substitution so that $z$ maps to $u$ and then $za$ maps to $ub$. Since $(za)^k \in I$ and $I$ is a $T$-ideal, we have that it is closed under substitutions so $(ub)^k \in I$, and the left ideal generated by $b$ is nil modulo $I$. $\square$

We need next to understand semiprime proper $T$-ideals. This will be done when $A$ is a field in Corollary 11.3.4.

## 11.2. Semisimple and prime PI algebras

**11.2.1. Kaplansky's theorem.** The starting point of structure theory of rings with polynomial identities is due to Kaplansky. Suppose that an algebra $R$ over a commutative ring $A$ satisfies a proper multilinear identity $f$ of degree $d$, with coefficients in $A$ (generating $A$ as ideal).

Let $M$ be an irreducible $R$-module, $D := \hom_R(M, M)$ the centralizer of $R$ which, by Schur's lemma, is a division ring with center a field $F$, then

THEOREM 11.2.1 (Kaplansky's theorem). $\dim_F D = h^2 < \infty$, $\dim_{DM} = k$, *and* $2hk \leq d$.

PROOF. This is in two steps.

(1) Reduction to $D$ commutative. Let $K$ be a maximal commutative subfield of $D$. Then $K$ equals the centralizer of $K$ in $D$, and if $RK$ denotes the $K$-algebra of operators of $M$ spanned by $K, R$, we clearly have that $M$ is irreducible under $RK$ and $\mathrm{End}_{RK} M = K$. Since the identity we start from is multilinear, $RK$ satisfies the same identity, cf. Lemma 2.2.37. By the commutative case we can deduce that $2\dim_K M \leq d$.

Clearly, $\dim_K M = \dim_K D \dim_{DM}$, hence $\dim_K D < \infty$, and by Theorem 1.1.17 in this case $\dim_F D = [\dim_K D]^2$, and (1) is proved.

(2) To prove the commutative case $D = K$, one has to apply the Jacobson density theorem. If we had $\dim_K M > \frac{d}{2}$, we could find $k > \frac{d}{2}$ linearly independent elements over $K$ and a subring $S$ of $R$ mapping surjectively to the ring of all endomorphisms, equal to $M_k(K)$, of the subspace generated by these elements. This implies that the matrix ring $M_k(K)$ satisfies the identity $f$, because if an identity is satisfied by a ring, it is satisfied by its subrings and quotients of them. Now we have seen that the matrix ring $M_k(K)$ does not satisfy any identity of degree $< 2k$, so we must have $d \geq 2k$, a contradiction. $\square$

COROLLARY 11.2.2. *A primitive ring satisfying a PI is a simple algebra finite dimensional over its center. If a ring $R$ satisfies a PI, its Jacobson radical $J(R)$ is the intersection of all maximal two-sided ideals of $R$.*

PROOF. A primitive ring $R$ by definition has a faithful irreducible module which, by the previous theorem, is a vector space $V$ of finite dimension $h$ over a division ring $D$ finite dimensional over its center. Then $R = M_h(D)$ by the Jacobson density theorem and $R = M_h(D)$ is simple.

The Jacobson radical is the intersection of all primitive ideals, so the second claim follows once we observe that conversely every maximal ideal of a PI algebra is primitive. In other words a simple PI algebra $R$ is a primitive ring. This is clear if the algebra has a 1 since then a simple algebra with a 1 has Jacobson radical equal to $(0)$, so it is a primitive ring. Otherwise we need to apply the next theorem of Posner, $R$ is prime and its center is nonzero so it is a field, and then $R$ is a finite-dimensional central simple algebra. □

11.2.1.1. *Amitsur's embedding theorem.*

THEOREM 11.2.3. *Let $R$ be a semiprime ring satisfying a PI of degree $d$. Let $n := [d/2]$ be the integral part of $d/2$.*

(1) *Then there exist $m \leq n$ semiprime commutative rings $A_1, \ldots, A_m$ and an embedding*

$$i : R \to \bigoplus_{i=1}^{m} M_{h_i}(A_i), \ 1 \leq h_1 < h_2 < \cdots < h_m \leq d/2.$$

(2) *Moreover, $R$ satisfies the same polynomial identities as $\bigoplus_{i=1}^{m} M_{h_i}(A_i)$.*
(3) *If $R$ is prime, then one can embed $R$ in just one matrix algebra.*
(4) *If $R$ is a finitely generated algebra over some commutative ring $A$, we can take all the $A_i$ as finitely generated over $A$.*

PROOF. By passing to $R[x]$ and using Theorem 1.1.32, we may assume that $R$ is semisimple, and by Corollary 11.2.2 we have $0 = \bigcap_\alpha M_\alpha$ where $M_\alpha$ runs over all primitive ideals of $R$ (which since $R$ is PI are in fact maximal; Theorem 11.2.1). Now each algebra $R/M_\alpha$ is a simple algebra over some center $F_\alpha$ of dimension $h_\alpha^2$, $h_\alpha \leq d/2$, and if $\bar{F}_\alpha$ is an algebraic closure of $F_\alpha$, we have $R/M_\alpha \otimes_{F_\alpha} \bar{F}_\alpha = M_{h_\alpha}(\bar{F}_\alpha)$. Thus we define $A_i := \prod_\alpha \bar{F}_\alpha$ with $h_\alpha = i$, and we clearly have

$$R \subset \prod_\alpha R/M_\alpha \subset \prod_\alpha M_{h_\alpha}(\bar{F}_\alpha) = \bigoplus_{i=1}^{m} M_i(A_i).$$

The fact that $R$ satisfies the same polynomial identities as $\bigoplus_{i=1}^{m} M_i(A_i)$ comes from the construction.

Let $I_i$ be the kernel of the projection to $M_i(A_i)$. We have $\prod_i I_i \subset \bigcap_i I_i = 0$, hence if $R$ is prime, we have that one of the ideals $I_i$ is 0 and $R$ embeds into $M_i(A_i)$. Finally if $R$ is generated by some elements $a_j$, it is enough to take as $A_i$ the ring generated by the entries of all the matrices associated to the elements $a_j$. □

If $R$ is a prime ring, then it has a characteristic, either 0 or $p$ a prime number.

By Theorem 11.2.3(3) we have $R \subset M_i(A)$ with $A$ a commutative algebra, and $R$ satisfies the same (stable) polynomial identities as $M_i(A)$. In particular $i$ is the minimum integer such that $R$ satisfies the standard identity $S_{2i}$.

DEFINITION 11.2.4.

(1) The minimum $i$ is called the *PI degree* of the prime PI ring $R$, denoted PI-deg$(R)$.
(2) If $R$ is any ring and $P$ a prime ideal of $R$, we define the *PI degree* of $P$ to be the PI degree of $R/P$ if $R/P$ satisfies a polynomial identity, otherwise we say that the PI degree of $P$ is $\infty$.

### 11.2.2. Posner's theorem.

11.2.2.1. *Posner's theorem.* We arrive now at the main result. The statement of Posner's theorem is in fact stronger than the original statement in **[Pos60]**, which only stated that a prime PI ring has a ring of quotients which is a central simple algebra. In our setting, we actually prove that the denominators can be taken from the center (cf. Definition 5.4.21). Let $R$ be a prime ring satisfying a polynomial identity of degree $d$, and let $\mathcal{Z}$ be the center of $R$.

PROPOSITION 11.2.5. *If $I \subset R$ is a nonzero two-sided ideal, then $\mathcal{Z} \cap I \neq 0$. In particular $\mathcal{Z} \neq 0$. In fact if the PI degree of $R$ is $i$ (cf. Definition 11.2.4) we may replace $\mathcal{Z}$ with its subring obtained by evaluations of multilinear central polynomials for $i \times i$ matrices.*

PROOF. By the embedding theorem, Theorem 11.2.3, and the fact that $R$ is prime, it follows that $R$ embeds into a product $M_i(A_i) = \prod_\alpha M_i(\bar{F}_\alpha)$ for some $i \leq d/2$.

We have $M_i(\bar{F}_\alpha) = R[x]/M_\alpha \otimes_{F_\alpha} \bar{F}_\alpha$ and $M_\alpha$ some primitive ideal of $R[x]$. In particular for some $\alpha_0$ we have that the image of the ideal $I[x]$ of $R[x]$ in the factor $M_i(\bar{F}_{\alpha_0})$ is not zero. This means that $I[x] \not\subset M_{\alpha_0}$, hence, since $I[x]$ is an ideal, the image of $I[x]$ is $R/M_{\alpha_0}$. Multiplication by $x$ commutes with multiplication by any element of $R/M_{\alpha_0}$, so it is multiplication by some element of $\bar{F}_\alpha$, and the image of $I$ spans $M_i(\bar{F}_\alpha)$ over $\bar{F}_\alpha$. We then can evaluate one of the nontrivial multilinear central polynomials for $i \times i$ matrices on $I$ so that its value is nonzero. This is in the center of $R$ since such a polynomial evaluated in $M_i(A_i)$ takes central values. □

Since $\mathcal{Z} \neq 0$, we have that it is a domain and $R$ is torsion free over $\mathcal{Z}$. Thus let $K$ be the field of fractions of $\mathcal{Z}$. Since $R$ is torsion free over $\mathcal{Z}$, the ring $R$ embeds in the algebra $Q = R \otimes_{\mathcal{Z}} K$. It is clear that $K$ is the center of $Q$, and $Q$ satisfies the identities of $R$ (Exercise 2.2.45).

THEOREM 11.2.6 (Posner). *$Q$ is a finite-dimensional central simple algebra over its center $K$ satisfying the same identities as $R$, and $Q$ is a full ring of quotients of $R$.*

PROOF. Since $Q$ is a PI ring, it is enough, by (2) of Corollary 11.2.2, to prove that $Q$ is a simple algebra. If $J \neq 0$ is an ideal of $Q$, we have that $J \cap R$ is a nonzero ideal of $R$. Thus by Proposition 11.2.5 we have that $J \cap Z \neq 0$, and this implies $J = Q$ since the nonzero elements of $Z$ are invertible in $Q$. □

The following simple consequence will be used later.

COROLLARY 11.2.7. *Let $R$ be a prime PI ring, and let $S$, with $\mathcal{Q}(R) \supset S \supset R$, be any ring contained in the ring of fractions $\mathcal{Q}(R)$ of $R$. If $I$ is a nonzero ideal of $S$, then $I$ intersects the center of $R$, in fact it contains nonzero evaluations of central polynomials.*

PROOF. It is enough to see that $I \cap R \neq 0$. Now take any element $a \in I$, $a \neq 0$, we have $a = \alpha^{-1}b$, $b \in R$, and $\alpha$ in the center of $R$. Clearly $b \in I \cap R$. □

COROLLARY 11.2.8. *The PI degree of a prime PI ring $R$ over a field $F$ equals the number $i$ such that* $\dim_K Q = i^2$, *where $Q = R \otimes_{\mathcal{Z}} K$ with $\mathcal{Z}$ the center of $R$.*

PROOF. If $\bar{K}$ is an algebraic closure of $K$, we have

$$R \otimes_{\mathcal{Z}} \bar{K} \simeq Q \otimes_K \bar{K} = M_i(\bar{K}).$$ □

COROLLARY 11.2.9. *The Gelfand–Kirillov dimension of a prime PI algebra $R$ over a field $F$ equals the Gelfand–Kirillov dimension of its center $\mathcal{Z}$ which equals the transcendence degree of the center $G$ of its full ring of quotients over $F$.*

PROOF. This follows from Proposition 8.3.11 and the previous two results, since clearly

$$\text{Dim } \mathcal{Z} \leq \text{Dim } R \leq \text{Dim } Q = \text{Dim } G = \text{Dim } \mathcal{Z}. \qquad \square$$

Let $R$ be a prime PI algebra, and let $Q(R)$ be its simple algebra of fractions, of some degree $d$. Let $T_R$ be the subring of the center of $Q(R)$ generated by all coefficients of the characteristic polynomial of all the elements $r \in R$. Assume that $R$ has PI degree $d$.

PROPOSITION 11.2.10. *If $a := c(r_1, \dots, r_k) \in R$ is the nonzero evaluation of a central polynomial $c(x_1, \dots, x_k)$ with no constant term for $d \times d$ matrices, then*

(i) *$R[a^{-1}]$ is an Azumaya algebra.*
(ii) *there is an exponent $m$ such that $a^m RT_R \subset R$.*
(iii) *$RT_R[a^{-1}] = R[a^{-1}]$.*

PROOF. (i) This is a special case of Corollary 10.3.5.

(ii) This is now a corollary of Theorem 10.4.8. In fact for $I$ large enough the algebra $R$ is a quotient of $S = \mathbb{Z}\langle \xi_i \rangle_{i \in I}$ in such a way that $T_R$ is a quotient of $T_S$ and similarly $RT_R$ is a quotient of $ST_S$. This is easily seen by splitting $Q(R)$ and thus embedding $R$ in a ring of $d \times d$ matrices over a commutative ring $A$ and then choosing a set $r_i$, $i \in I$, of generators of $R$ as ring.

Once this is done, the same power $m$ for which $p_m(\xi_1, \dots, \xi_{k+1}) = c^m \det(\xi_{k+1})$ specializes to $p_m(r_1, \dots, r_{k+1}) = c(r_1, \dots, r_k)^m \det(r_{k+1})$ which implies the conductor property as in Theorem 10.4.8.

(iii) This follows from (ii). $\square$

### 11.2.3. The *spectrum* of a PI ring.

11.2.3.1. *The spectrum of a ring.* In affine algebraic geometry the prime ideals of the ring $\mathbb{C}[x_1, \dots, x_n]$ of polynomials, correspond to the irreducible affine subvarieties of $\mathbb{C}^n$. In general commutative algebra prime ideals play a major role and are usually equipped with the *Zariski topology*, which we have discussed in §1.6.1. In view of the previous results, some parts of this theory can be carried out to general noncommutative rings and, by Theorem 11.2.6, to stronger results in PI theory. Let us establish some basic definitions. As in commutative algebra one has

DEFINITION 11.2.11.

(1) The spectrum $\text{Spec}(R)$ of a ring $R$ is the set of all prime ideals $P$ and the maximal spectrum is the set of all maximal ideals.
(2) As in the commutative case we can endow the spectrum of the *Zariski topology* where, given any set $A \subset R$, we set

$$V(A) := \{P \in \text{Spec}(R) \mid P \supset A\}.$$

(3) The sets $V(A)$ are the closed sets of the topology by definition. Note that $V(A) \cup V(B) = V(ARB)$.

As in commutative algebra, by Theorem 1.1.41 the map $A \mapsto V(A)$ is a 1–1 correspondence between semiprime ideals (Definition 1.1.30) and closed subsets of $\mathrm{Spec}(R)$. Furthermore we can, as in commutative algebra, define *irreducible* to be a closed set $V$ which is not equal to $V_1 \cup V_2$ for two proper closed subsets $V_1, V_2 \subsetneq V$. One then sees that under the previous correspondence, irreducible closed subsets correspond to prime ideals.

In general a map $\pi : R \to S$ does not induce a map of spectra as very simple examples show (consider for instance the embedding of triangular matrices into matrices), but we have good behavior under special maps, called *extensions*:

DEFINITION 11.2.12.

(1) We say that a map of rings $\pi : R \to S$ is an extension, if $S = \pi(R)S^R$ where $S^R = \{s \in S \mid s\pi(r) = \pi(r)s, \ \forall r \in R\}$ is the centralizer of $\pi(R)$ in $S$.
(2) We say that an extension $R \subset S$ is *finite* if $S$ is a finitely generated right (or left) $R$-module.
(3) An extension $\pi : R \to S$ is *central* if $S$ is generated over $\pi(R)$ by the center of $S$.

EXAMPLE 11.2.13. The prime example we have in mind is when $R \subset S = M_n(R)$, in which case $S^R = M_n(\mathcal{Z})$ where $\mathcal{Z}$ is the center of $R$.
This is a finite extension.

PROPOSITION 11.2.14. *If $\pi : R \to S$ is an extension, the map $\pi^* : P \mapsto \pi^{-1}(P)$ is a continuous map $\pi^* : \mathrm{Spec}(S) \to \mathrm{Spec}(R)$.*

PROOF. By reducing modulo $P$, we may assume that $S$ is a prime ring and $R \subset S$. We need to verify that $R$ is prime. Let us take the criterion $aRb = 0, a, b \in R$. Then we see that $aSb = aRS^Rb = aRbS^R = 0$, hence since $S$ is prime either $a = 0$ or $b = 0$, as wished. □

When $S = R/I$, the map $\pi^* : \mathrm{Spec}(R/I) \to \mathrm{Spec}(R)$ is a homeomorphism onto the closed subset $V(I)$, and it is a *closed embedding*.

In commutative algebra there is also the very useful notion of *open embedding*, that is when $S = R_T$ is obtained from $R$ by localization, that is inverting the elements of a multiplicative set $T$.

Then the map $\pi^* : \mathrm{Spec}(R_T) \to \mathrm{Spec}(R)$ is a homeomorphism onto the open subset $\mathrm{Spec}(R) \setminus V(T)$, and it is an *open embedding*.

Localization is one of the tools not easily available in noncommutative rings. The previous statement holds and the proof is the same, in general, if $T$ is formed by central elements.

REMARK 11.2.15. Assume that $R \subset S$ is an extension (resp., central). Then,

(1) $S^R \subset S$ is also an extension, we have $S = RS^R$, and if $S$ is prime, so is $R$ and $S^R$.
(2) The center of $R$ is contained in the center of $S$ which equals the center of $S^R$.
(3) If $I$ is an ideal of $S$, then $R/R \cap I \subset S/I$ is an extension (resp., central).
(4) If $R$ is an Azumaya algebra with center $A$, then $S = R \otimes_A S^R$.
(5) If $R$ is a finite-dimensional central simple algebra with center $F$, then $S = R \otimes_F S^R$.

(6) If $R$ and $S$ are both finite-dimensional central simple algebras, then $S^R$ is a central simple algebra.

(7) If $R_1 \subset R_2$ and $R_2 \subset R_3$ are extensions, then $R_1 \subset R_3$ is also an extension.

PROOF. All statements are immediate except perhaps (4), (5) and (6) which follow from the general statement on Azumaya algebras in Corollary 5.4.29. As for (7) we have that $R_3 = R_2 R_3^{R_2} = R_1 R_2^{R_1} R_3^{R_2}$ and $R_2^{R_1} R_3^{R_2} \subset R_3^{R_1}$, so $R_3 = R_1 R_3^{R_1}$. □

11.2.3.2. *The case of PI rings.* For a ring $R$ which satisfies a PI of some degree $d$, every prime ideal $P$ has a degree $i \leq [\frac{d}{2}]$, that is the PI-degree of $R/P$ so that the spectrum decomposes into

$$(11.3)\quad \operatorname{Spec}(R) = \bigcup_{i=1}^{\frac{d}{2}} \operatorname{Spec}_i(R),\ \operatorname{Spec}_i(R) := \{P \in \operatorname{Spec}(R) \mid \text{PI-deg}(R/P) = i\}.$$

Moreover, for each $j$ the subset $\bigcup_{i \leq j} \operatorname{Spec}_i(R)$ is closed. In fact it is $V(I_j(R))$, where $I_j(R)$ is the verbal ideal of $R$ generated by evaluating all the PIs of $j \times j$ matrices. Finally the quotient map $\pi_j : R \to R/I_j(R)$ induces a homeomorphism between $\bigcup_{i \leq j} \operatorname{Spec}_j(R/I_j(R))$ and $\bigcup_{i \leq j} \operatorname{Spec}_j(R)$.

Consider then $I^{(j)}(R) := \bigcap_{P \in \operatorname{Spec}_j(R)} P$.

LEMMA 11.2.16.

(1) *The ring* $R/I^{(j)}(R) = R/\bigcap_{P \in \operatorname{Spec}_j(R)} P$ *is semiprime, and it embeds into a ring* $M_j(A)$ *of* $j \times j$ *matrices over a commutative ring* $A$ *with no nilpotent elements.*

(2) *If* $R$ *is prime of degree* $j$*, then* $I^{(j)}(R) = \{0\}$.

PROOF. (1) For each $P \in \operatorname{Spec}_j(R)$ by Posner's theorem, we have $R/Q$ has a quotient $Q(R/P)$ which is a central simple algebra of degree $j$ so it embeds into a ring $M_j(F_P)$ of $j \times j$ matrices over a field $F_P$, and $M_j(F) = (R/P)F_P$. Thus $R/\bigcap_{P \in \operatorname{Spec}_j(R)} P$ embeds into $M_j(A)$ with $A := \prod_{P \in \operatorname{Spec}_j(R)} F_P$.

(2) In this case $\{0\} \in \operatorname{Spec}_j(R)$, and the claim follows. □

**A remarkable property.** Each of the strata $\operatorname{Spec}_j(R) = \operatorname{Spec}_j(R/I_j(R))$ is homeomorphic to an open set of the spectrum of a *commutative ring!*, the center of $R/I_j(R)$.

In order to see this, we may assume that $I_j(R) = \{0\}$, that is $R$ satisfies the PIs of $j \times j$ matrices.

Let us denote by $\mathcal{Z}$ the center of $R$ and by $\mathcal{C}$ the ideal of $\mathcal{Z}$ generated by the evaluations of all central polynomials $c$ for $j \times j$ matrices. Under these hypotheses consider the map from $\operatorname{Spec}_j(R)$ to $\operatorname{Spec}(\mathcal{Z})$ induced by the inclusion. We have

THEOREM 11.2.17.

(1) *For each* $P \in \operatorname{Spec}_j(R)$*, let* $\pi(P) := P \cap \mathcal{Z}$*. The algebra* $R \otimes_{\mathcal{Z}} \mathcal{Z}_{\pi(P)}$ *is an Azumaya algebra of degree* $j$ *over* $\mathcal{Z}_{\pi(P)}$.

(2) $\operatorname{Spec}_j(R) = \operatorname{Spec}(R) \setminus V(\mathcal{C})$.

(3) *The map* $P \mapsto P \cap \mathcal{Z}$ *from* $\operatorname{Spec}_j(R)$ *to* $\operatorname{Spec}(\mathcal{Z})$ *is a homeomorphism with* $\operatorname{Spec}(\mathcal{Z}) \setminus V(\mathcal{C})$.

PROOF. This is truly a consequence of Artin's theorem. The point is the following. If $P \in \mathrm{Spec}_j(R)$, there is an evaluation of a central element nonzero in $R/P$. In other words there is an element $c \in \mathcal{Z} \setminus \mathcal{C}$ with $c \notin P$.

This means that

$$\mathrm{Spec}_j(R) = \bigcup_{c \in \mathcal{Z} \setminus \mathcal{C}} (\mathrm{Spec}(R) \setminus V(c)). \tag{11.4}$$

But $\mathrm{Spec}(R) \setminus V(c)$ is the open set homomorphic to $\mathrm{Spec}(R[c^{-1})$ which is an Azumaya algebra of degree $j$ with center $\mathcal{Z}[c^{-1}])$. Hence $\mathrm{Spec}(R[c^{-1}])$ is homomorphic to $\mathrm{Spec}(\mathcal{Z}[c^{-1}])$ by Corollary 5.4.29. □

PROPOSITION 11.2.18. *Let $R \subset S$ be an inclusion of rings. Assume that $R$ is a PI ring and $S$ is a finite right $R$-module, that is $S = \sum_{i=1}^k s_i R$. Then $S$ is also a PI ring.*

PROOF. Consider $S$ has a module over $R^{\mathrm{op}}$ by right multiplication. By using left multiplication, we identify $S$ with a subalgebra of $\mathrm{End}_{R^{\mathrm{op}}}(S)$. By Proposition 10.1.1 $S$ is isomorphic to a homomorphic image of a subring $\hat{S} \subset M_n(R)$. By the tensor product theorem a ring of $n \times n$ matrices over a PI ring is also PI. □

PROPOSITION 11.2.19. *Let $R \subset S = RS^R$ be an extension, and let $S$ be a prime PI ring. Then:*

(1) *We have an embedding of rings of fractions $\mathcal{Q}(R) \subset \mathcal{Q}(S)$ which is an extension. Similarly for $\mathcal{Q}(S^R) \subset \mathcal{Q}(S)$.*
(2) *$\mathcal{Q}(S) = \mathcal{Q}(R) \otimes_K \mathcal{Q}(S^R)$.*
(3) *The PI degree of $S$ is the product of the PI degrees of $R$ and $S^R$. In particular the PI degree of $R$ equals that of $S$ if and only if the extension is central.*

PROOF. Assume $S$ is prime PI, and then of course also $R$ and $S^R$ are prime PI.

Let $A \subset B$ be the inclusion of the center of $R$ in that of $S$ which, by Remark 11.2.15(2), equals the center of $S^R$. Let $F \subset G$ be the inclusion of their respective fields of fractions. We have $\mathcal{Q}(R) = F \otimes_A R$, $\mathcal{Q}(S) = G \otimes_B S = G \otimes_{F \otimes_A B} (F \otimes_A S)$. Now $F \otimes_A S$ contains $\mathcal{Q}(R)$ and is an extension; moreover the centralizer of $\mathcal{Q}(R)$ in $F \otimes_A S$ is clearly $F \otimes_A S^R$ so by (4) we have

$$F \otimes_A S = \mathcal{Q}(R) \otimes_F (F \otimes_A S^R) \implies \mathcal{Q}(S)^{\mathcal{Q}(R)} = G \otimes_B S^R.$$

By the theory of central simple algebras also $G \otimes_B S^R$ is a simple algebra with center $G$. The PI degree of the tensor product $S_1 \otimes_F S_2$ of two central simple algebras $S_1, S_2$ with centers $F, G$ with $F \subset G$ is the product of the two PI degrees since if $\bar{G}$ is an algebraic closure of $G$, we have

$$(S_1 \otimes_F S_2) \otimes_G \bar{G} = (S_1 \otimes_F \bar{G}) \otimes_{\bar{G}} (S_2 \otimes_G \bar{G}) = M_{n_1}(\bar{G}) \otimes_{\bar{G}} M_{n_2}(\bar{G}) = M_{n_1 n_2}(\bar{G}).$$

□

## 11.3. Generic matrices

**11.3.1. Properties of generic matrices.** Let $j : A\langle x_\alpha \rangle_{\alpha \in I} \to M_n(A[\xi])$ be the universal map into generic matrices defined in Definition 3.4.2. By Theorem 2.4.1 the kernel of $j$ is the $T$-ideal of stable polynomial identities of $n \times n$ matrices.

By Definition 3.3.23 its image is the algebra $A_n\langle \xi_\alpha \rangle$ of generic matrices. As soon as $I$ has at least two elements, this is a rather nontrivial construction and its structure is not completely understood for $n > 2$.

We shall use Theorem 1.1.12. Given any field $F$ and a number $n \in \mathbb{N}$, there is a division algebra $\Delta$ containing $F$ (but with center $G$ possibly larger) and such that $\Delta \otimes_G \bar{G} = M_n(\bar{G})$. Notice that $\Delta$ is PI equivalent (as $F$-algebra) to $M_n(\bar{G})$.

THEOREM 11.3.1. *Assume that $A$ is a domain with field of fractions $F$ and that $I$ is a set with at least two elements:*

(1) *The algebra of generic matrices $A\langle \xi_\alpha \rangle_{\alpha \in I} \subset F\langle \xi_\alpha \rangle = F \otimes_A A\langle \xi_\alpha \rangle$ is also a domain.*
(2) *Let $C$ be the center of $A\langle \xi_\alpha \rangle$, and let $K$ be its field of fractions: $D := A\langle \xi_\alpha \rangle \otimes_C K$ is a division algebra of dimension $n^2$ over its center $K$.*
(3) $D \otimes_K F(\xi_{i,j}^\alpha) = M_n(F(\xi_{i,j}^\alpha))$.

PROOF. (1) By definition $A\langle \xi_\alpha \rangle_{\alpha \in I} \subset M_n(A[\xi])$ is torsion free with respect to $A$, so we may assume that $A = F$ is a field. In order to prove that the ring $F\langle \xi_\alpha \rangle$ is a domain by the method of generic elements, take a division algebra $\Delta$ containing $F$ such that $\Delta \otimes_G \bar{G} = M_n(\bar{G})$. Then if we take a basis $u_1, \dots, u_{n^2}$ of $\Delta$, we can replace the algebra $F\langle \xi_\alpha \rangle$ by $F\langle \eta_\alpha \rangle$ where the $\eta_\alpha$ are generic elements in this basis, so they are contained in a polynomial ring over $\Delta$ which is also a domain. Hence, since $\Delta$ and $M_n(F)$ are PI equivalent, the two relatively free algebras $F\langle \xi_\alpha \rangle$ and $F\langle \eta_\alpha \rangle$ are isomorphic and the last is a domain being contained in a domain.

(2) Now a domain is also a prime ring, so that if $C$ is its center and $K$ is its field of fractions, then $D := F\langle \xi_\alpha \rangle \otimes_C K$ is a division algebra of some dimension $m^2$ over its center $K$, by Theorem 11.2.6. We only need to show that $m = n$. This follows from part (3).

(3) We have the following commutative diagram of inclusions.

$$\begin{array}{ccc} F\langle \xi_\alpha \rangle & \longrightarrow & M_n(F[\xi_{i,j}^\alpha]) \\ \downarrow & & \downarrow \\ D = F\langle \xi_\alpha \rangle \otimes_C K & \longrightarrow & M_n(F(\xi_{i,j}^\alpha)) \\ \downarrow & & 1\downarrow \\ D \otimes_K F(\xi_{i,j}^\alpha) & \xrightarrow[\simeq]{i} & M_n(F(\xi_{i,j}^\alpha)) \end{array} \quad (11.5)$$

We have that $D \otimes_K F(\xi_{i,j}^\alpha)$ is a simple algebra (Lemma 1.1.16). By Proposition 5.1.15 we know that there exist two matrices which generate the algebra $M_n(F(\xi_{i,j}^\alpha))$ of all matrices over the field $F(\xi_{i,j}^\alpha)$, hence also two generic matrices generate the algebra $M_n(F(\xi_{i,j}^\alpha))$ as vector space over the function field $F(\xi_{i,j}^\alpha)$. Thus the map $i$ is surjective and so an isomorphism, and $D$ is of dimension $n^2$ over $K$. □

The description of the center of the division ring of fractions of generic matrices over a field $F$ and the computation of its Gelfand–Kirillov dimension will be given in §14.4.1, Theorem 14.4.4, and Corollary 14.4.5.

Of special interest is the case $A = \mathbb{Z}$, and we have the domain $\mathbb{Z}\langle \eta_\alpha \rangle$. It is easy to see that $\mathbb{Q}\langle \eta_\alpha \rangle = \mathbb{Z}\langle \eta_\alpha \rangle \otimes_{\mathbb{Z}} \mathbb{Q}$. In general for a commutative algebra $A$ we have a mapping $\mathbb{Z}\langle \eta_\alpha \rangle \otimes_{\mathbb{Z}} A \to A\langle \eta_\alpha \rangle$, but when for instance $A = \mathbb{Z}/(p)$ is the ground field of characteristic $p$, this map need not be injective **[ADKT00]**, **[Sch85]**.

In Amitsur's original approach he proved that the generic matrices are an Ore domain, so it has a division algebra of quotients. Now the theory is strongly simplified by the study of its center discussed in §10.2, which is a rather remarkable object.

11.3.1.1. *The spectrum of generic matrices.* If $R$ is an $A$-algebra satisfying the identities of $n \times n$ matrices, as for instance the case of a prime PI ring of degree $n$, then $R$ is a quotient of a ring $A\langle\xi_\alpha\rangle$ of generic matrices (usually in infinitely many variables).

If $R$ is any PI algebra, the spectrum of $R$ coincides with the spectrum of $R/N$ where $N$ is its nil radical and, by the Amitsur's theorem, Theorem 11.2.3, $R/N$ satisfies the identities of $n \times n$ matrices for some $n$. Therefore the spectrum of any PI algebra $R$ can be identified to a closed subset of the spectrum of $A\langle\xi_\alpha\rangle$. We see that the spectrum of generic matrices plays a special role, as in commutative algebra that of polynomial rings.

As a first remark let us stress the number $n$ by denoting $A_n\langle\xi_\alpha\rangle$ the generic $n \times n$ matrices. Since $n-1 \times n-1$ matrices satisfy all PI of $n \times n$ matrices, we clearly have that $A_{n-1}\langle\xi_\alpha\rangle = A_n\langle\xi_\alpha\rangle/I_{n-1}$, where $I_{n-1}$ is the image in $A\langle\xi_\alpha\rangle$ of the $T$-ideal of identities of $n-1 \times n-1$ matrices in the free algebra.

REMARK 11.3.2. Thus $\operatorname{Spec}_n(A_n\langle\xi_\alpha\rangle) = \operatorname{Spec}(A_n\langle\xi_\alpha\rangle) \setminus V(I_{n-1})$ and, for all $i \le n$, we have $\operatorname{Spec}_i(A_n\langle\xi_\alpha\rangle) \simeq \operatorname{Spec}_i(A_i\langle\xi_\alpha\rangle)$.

When the number $m$ of generic matrices is finite, we have a theory generalizing the commutative one, which we shall develop in §14.4 where we give a very remarkable structure theory of the spectrum of generic matrices.

**11.3.2. Prime $T$-ideals.** The theorem of Posner has an important application to $T$-ideals. We state it for algebras over infinite fields $F$ first.

THEOREM 11.3.3. *Let $I \subset F\langle X\rangle$ be a nonzero semiprime $T$-ideal. Then $I$ is prime and equals the $T$-ideal of identities of $n \times n$ matrices for some $n$.*

PROOF. $I$ is the $T$-ideal of polynomial identities of $R := F\langle X\rangle/I$, which by hypothesis is a semiprime PI algebra. Let $d$ be the degree of an identity. By Theorem 11.2.3 $R$ embeds into a product of spaces of $h_i \times h_i$ matrices over fields $G_i$ for some list of $h_i \le d/2$, and under this embedding all identities are preserved. Since $F$ is assumed to be an infinite field and each $G_i$ is an extension of $F$, the identities of $M_{h_i}(G_i)$ with coefficients in $F$ coincide with the identiities of $M_{h_i}(F)$, which on the other hand are contained in the identities of $M_k(F)$ for all $k < h_i$. Thus $I$ is the ideal of identities of $M_n(F)$ with coefficients in $F$ where $n$ is the maximum among these $h_i$. □

COROLLARY 11.3.4. *If $F$ is an infinite field and $I \subset F\langle X\rangle$ a nonzero $T$-ideal, then $\sqrt{I}$ equals the $T$-ideal of identities of $n \times n$ matrices for some $n$.*

PROOF. This follows from the previous theorem, since $\sqrt{I}$ is a $T$-ideal, by Corollary 11.1.8. □

Notice that a prime $T$-ideal is also completely prime (Remark 1.1.35) by Theorem 11.3.1.

COROLLARY 11.3.5. *If an algebra $R$ over a commutative ring $A$ satisfies a proper identity (Definition 2.2.33), then it also satisfies a power of a standard identity and a multilinear identity $f$ with coefficients $\pm 1$.*

PROOF. Consider the $T$-ideal $I_f$ in the free algebra $A\langle X\rangle$ over $A$ generated by the identity $f$, and let $N$ be its nil radical. Then $A\langle X\rangle/N$ satisfies the proper identity $f$, that is it is a PI algebra which, by Amitsur's theorem, Theorem 11.2.3, embeds into a finite sum of matrices over commutative rings, in particular it satisfies some standard identity $S_{2n}(x)$ which therefore lies in $N$, hence $S_{2n}(x)^k \in I_f$ for some $k$, that is $S_{2n}(x)^k$ is an identity deduced from $f$. Polarizing this identity, one has a multilinear identity with coefficients $\pm 1$. □

Next assume we work in the free algebra $F\langle X\rangle$ over some infinite field $F$.

COROLLARY 11.3.6. *The radical of a $T$-ideal $Q \subset F\langle X\rangle$ is the ideal of polynomial identities of $k \times k$ matrices, where $k$ is the minimum integer for which there is an exponent $n$ with $S_{2k}(x)^n \in Q$.*

PROOF. Consider the $T$-ideal $P = \sqrt{Q}$, which is the ideal of PI of $h \times h$ matrices for some $h$. If $S_{2k}(x)^n \in Q$, since the free algebra modulo $P$ is a domain, we must have that $S_{2k}(x) \in P$, hence $h \leq k$. Conversely consider $S_{2k}(x)$ as element of the free algebra it is in $P$, but $P/Q$ is a nil ring, so that for some $n$ we have $S_{2h}(x)^n \in Q$. □

Let $A$ be a finite-dimensional algebra over an infinite field $F$. If $A$ is semisimple, then $A = \bigoplus_i S_i$ with $S_i$ simple algebras. Each $S_i$ in turn has some center $G_i$ and, if $\bar{G}_i$ is its algebraic closure, we have $S_i \otimes_{G_i} \bar{G}_i \simeq M_{h_i}(\bar{G}_i)$. We have then

PROPOSITION 11.3.7. *The ideal of polynomial identities of $A$ over $F$ coincides with the ideal of polynomial identities of $M_h(F)$ with $h = \max h_i$.*

PROOF. We have $\mathrm{Id}(A) = \bigcap_i \mathrm{Id}(S_i)$ and

$$\mathrm{Id}(S_i) = \mathrm{Id}(S_i \otimes_F \bar{G}_i) = \mathrm{Id}(M_{h_i}(\bar{G}_i)) = \mathrm{Id}(M_{h_i}(F)).$$

Then if $h < k$, we have $\mathrm{Id}(M_h(F)) \supset \mathrm{Id}(M_k(F))$, and the claim follows. □

In general if $A$ is a finite-dimensional algebra over an infinite field $F$, let $J$ be its radical so $J^m = 0$ for some $m$, and $A/J$ is semisimple.

COROLLARY 11.3.8. *We have $\mathrm{Id}(A/J)^m \subset \mathrm{Id}(A)$ and $\mathrm{Id}(A/J) = \sqrt{\mathrm{Id}(A)}$ is the radical of $\mathrm{Id}(A)$.*

PROOF. If $f \in \mathrm{Id}(A/J)$, then $f$ evaluated in $A$ takes values in $J$ so the first statement follows. Then, since $\mathrm{Id}(A/J)$ is a (completely) prime ideal the second statement also follows. □

## 11.4. Affine algebras

**11.4.1. Lack of Noetherian property.** As we have often announced, finitely generated PI algebras are usually not Noetherian. In fact the following example due to Amitsur (cf. [**Ami70**]) shows that generic matrices $F\langle \xi_\alpha \rangle$ over some field $F$ do not satisfy ascending chain condition of two-sided ideals.

We start from the standard polynomial $St_n(Y_1X, Y_2X, \dots, Y_nX)$ computed in matrix variables $Y_iX$. Suppose we compute this in $n \times n$ matrices and with the $Y_i = \mathrm{diag}(y_{i,1}, \dots, y_{i,n})$ diagonal matrices.

Since this is a multilinear and alternating function in the $n$ variables $Y_i$ which are $n$-dimensional vectors, the same argument as in Lemma 10.2.9 shows that $\mathrm{St}_n(Y_1X, Y_2X, \dots, Y_nX) = \det(Y_1, \dots, Y_n)P(X)$ for some matrix valued equivariant function $P(X)$. We claim that all entries of $P(X)$ are not identically 0, as functions of $X$. For this it is enough to compute it on suitable matrices, for instance,

$$X = \sum_{i=1}^{n-1} e_{i,i+1} + e_{n,1},\ Y_i = e_{i,i} \implies Y_iX = \begin{cases} e_{i,i+1} & \text{if } i < n, \\ e_{n,1} & \text{if } i = n. \end{cases}$$

Thus the only nonzero products of these matrices are $Y_iXY_{i+1}X \cdots Y_{i-1}X = e_{i,i}$, so we get $P(X) = \sum_{i=1}^n \pm e_{i,i}$. In fact we see that the diagonal entries of $P(X)$ are all different from 0 and a similar computation shows that all entries of $P(X)$ are different from 0.

We now work with the algebra $F\langle X, Y\rangle$ generated by two generic $n \times n$ matrices, $n \geq 2$, and consider the polynomials

$$P_k(X, Y) := \mathrm{St}_n(Y^{n-1+k}X, Y^{n-2}X, Y^{n-3}X, \dots, YX, X).$$

THEOREM 11.4.1 (Amitsur). *For all $k$ we have that $P_{k+1}(X, Y)$ does not belong to the two-sided ideal generated by $P_0(X, Y), P_1(X, Y), \dots, P_k(X, Y)$.*

PROOF. One can conjugate the generic matrix $Y$ to be a diagonal matrix, with entries $y_1, \dots, y_n$ (in some extension of the field of the matrix entries). Then by the previous discussion we have that

$$\mathrm{St}_n(Y^{i_1}X, Y^{i_2}X, \dots, Y^{i_{n-1}}X, Y^{i_n}X) = A_{i_1,\dots,i_n}(y_1, \dots, y_n)P(X),$$

where $A_{i_1,\dots,i_n}(y_1, \dots, y_n)$ is the alternating function $\sum_{\sigma \in S_n} \epsilon_\sigma y_{\sigma(1)}^{i_1} \cdots y_{\sigma(n)}^{i_n}$ and $P(X)$ is a nonzero function of $X$.

Since $i_1 > i_2 > \cdots > i_n$, we can write $(i_1, i_2, \dots, i_n) = \lambda + \rho$ with the notations of Lemma 3.2.12. We can apply the theory of Schur functions, Definition 6.3.2, and see that $A_{i_1,\dots,i_n}(y_1, \dots, y_n) = S_\lambda(Y)V(Y)$. In particular we have

$$P_k(X, Y) = S_{(k,0,0,\dots,0)}(Y)V(Y)P(X).$$

Denote by $S_k(Y) := S_{(k,0,0,\dots,0)}(Y)$. By degree considerations $P_k(X, Y)$ belongs to the two-sided ideal generated by $P_j(X, Y)$, $j < k$, only if it is a linear combination of elements of the type $Y^hP_j(X, Y)Y^i = S_j(Y)V(Y)Y^hP(X)Y^i$, with $h + i + j = k$. This is in fact impossible; look at the $(n, n)$ entry of the matrix $Y^hP(X)Y^i$, which equals $y_n^{h+i}P(X)_{(n,n)}$.

If such a linear combination existed, comparing the $(n, n)$ elements, we have thus a linear combination $\sum_{j<k} c_j y_n^{k-j} S_j(Y)V(Y)P(X)_{(n,n)} = S_k(Y)V(Y)P(X)_{(n,n)}$ which would imply that $\sum_{j<k} c_j y_n^{k-j} S_j(Y) = S_k(Y)$. This is impossible since the left-hand side is not a symmetric function of the variables $y_i$, in fact setting $y_n = 0$, the left-hand side vanishes while the right-hand side does not, by the theory of Schur functions Lemma 3.2.16. □

#### 11.4.2. Hilbert Nullstellensatz. We start with

PROPOSITION 11.4.2. *Let $R = F\langle a_1, \dots, a_k\rangle$ be a finitely generated PI algebra, with 1, over a field $F$. If $R$ is a simple algebra, then $\dim_F R < \infty$.*

PROOF. Let $\mathcal{Z}$ be the center of $R$, and let $e_1, \ldots, e_{h^2}$ be a basis of $R$ over $\mathcal{Z}$ which exists by Corollary 11.2.2. There are finitely many elements $g_1, \ldots, g_N \in \mathcal{Z}$ which contain the coordinates of the elements $a_i$ in the basis $e_j$ as well as the coordinates of the multiplication table of the basis elements.

Let $H = F[g_1, \ldots, g_N] \subset \mathcal{Z}$ be the subring of $\mathcal{Z}$ generated by these elements over $F$. Then clearly $S := \bigoplus_{i=1}^{h^2} He_i$ is a subalgebra of $R$ and $a_i \in S$, $\forall i$.

Since $R$ is generated by the elements $a_i$ it follows that $S = R \implies H = \mathcal{Z}$. We can apply the usual result of commutative algebra which states that if a field $\mathcal{Z}$ is a finitely generated algebra over a subfield $F$, then $\dim_F \mathcal{Z} < \infty$. From this the claim follows. □

EXERCISE 11.4.3. In the previous proposition assume that $R$ is finitely generated over $\mathbb{Z}$. Prove that $R$ is finite.

THEOREM 11.4.4 (Nullstellensatz). *Let $R = F\langle a_1, \ldots, a_k\rangle$ be a finitely generated PI algebra, with* 1, *over a field $F$. Then:*

(1) *Each maximal ideal of $R$ has finite codimension.*
(2) *Each irreducible $R$-module is finite dimensional.*
(3) *The Jacobson radical of $R$ is a nil ideal.*

PROOF. (1) This is the content of Proposition 11.4.2.

(2) If $M$ is an irreducible $R$-module, then it is an $R/\operatorname{Ann}(M)$-module where $\operatorname{Ann}(M)$ the annihilator of $M$ is a primitive ideal. By Theorem 11.2.1 $R/\operatorname{Ann}(M)$ is a simple algebra, hence from (1) it is finite dimensional and clearly an irreducible module over a finite-dimensional algebra is finite dimensional.

(3) We may assume that the nil radical of $R$ equals 0 and have to prove that $\{0\}$ is an intersection of maximal ideals of $R$. We know that this is true for $R[x]$ by Theorem 1.1.32 and the PI condition. So take a maximal ideal $M$ of $R[x]$, we know then that $\dim_F R[x]/M < \infty$ and then $\dim_F R/R \cap M < \infty$.

We claim that also $R \cap M$ is a maximal ideal, and this will suffice to finish the proof. Since $\dim_F R/R \cap M < \infty$ it is enough to prove that $R/R \cap M$ is a prime ring. Now since $R[x]/M$ is obtained from $R/R \cap M$ by adding an element in the center, this follows immediately (cf. Exercise 1.1.38). □

REMARK 11.4.5. A consequence of the previous theorem is that when dealing with the spectrum of a finitely generated PI algebra $R$, it is sufficient to consider its maximum spectrum $\operatorname{Spec}_{\max}(R) \subset \operatorname{Spec}(R)$ formed by the maximal ideals of $R$ with the induced topology.

Since a prime ideal $P$ is the intersection of the maximal ideals $\mathfrak{m}$ with $\mathfrak{m} \supset P$, this implies that in $\operatorname{Spec}(R)$, a point $P$ is closed if and only if $P$ is a maximal ideal. For any closed subset $X$ of $\operatorname{Spec}(R)$, the set of closed points in $X$ is dense in $X$. Therefore we have a 1–1 correspondence between the closed subsets of $\operatorname{Spec}(R)$ and the closed subsets of $\operatorname{Spec}_{\max}(R)$.

A much more difficult result is to prove that the Jacobson radical of $R$ is a nilpotent ideal, as will be shown in Chapter 16.

## 11.5. Representable algebras

In this section we present results of Anan'in [**Ana92**], Wehrfritz [**Weh76**], Beidar [**Beĭ86**], and Rowen and Small [**RS15**]. We refer to this last paper for a general survey of related problems and counterexamples. All rings are assumed to have a 1.

**11.5.1. Representable algebras.** We start from a definition introduced by Malćev [**Mal43**] in the year 1942.

DEFINITION 11.5.1. We say that a ring $R$ is *representable* if there is an embedding $R \to M_n(K)$, into the algebra of $n \times n$ matrices over some field $K$.

A subring of a representable ring is representable but there are examples of quotients of representable rings which are not representable, see Example 11.5.23.

Notice that a representable ring is necessarily a PI ring. On the other hand from that example we see that there exist PI rings, even finitely generated over fields, which are not representable, therefore the issue of when a PI ring is representable is of considerable interest.

The problem is of interest even for a commutative ring $R$, and we shall see that if such a ring is Noetherian and contains a field, then it is representable. The most general result in this line is of Rowen and Small [**RS15**]:

THEOREM 11.5.2. *A Noetherian PI algebra $R = A\langle x_1, \dots, x_m\rangle$ finitely generated over a commutative Noetherian ring $A$ containing a field $F$ is representable.*

As we shall see this theorem is proved by reducing to the case in which $A$ is a field, Anan'in's theorem, Theorem 11.5.17. This reduction is essentially parallel to the method of proving the representability theorem for a commutative Noetherian algebra. Let us first discuss this reduction.

We say that an ideal $I$ of a ring $R$ is *indecomposable* if it is not the intersection $I = A \cap B$ of two ideals properly containing $I$.

If $R$ satisfies the ascending chain condition on two-sided ideals, then every ideal $I$ is a finite intersection of indecomposable ideals. In fact otherwise there is $I$ maximal with the property of not satisfying this. But then $I$ cannot be indecomposable, so $I = A \cap B$ and both $A$ and $B$ are a finite intersection of indecomposable ideals by the maximality of $I$, a contradiction.

Therefore we may reduce in Theorem 11.5.2 to the case in which the ideal 0 is indecomposable.

LEMMA 11.5.3. *Let $R$ be an algebra over a commutative Noetherian ring $A$ satisfying the ascending chain condition on two-sided ideals and such that the ideal 0 is indecomposable. Then every $z \in A$ is either nilpotent or it is not a zero divisor in $R$.*

PROOF. Consider for each $k$ the annihilator $A_{z^k}(R) := \{r \in R \mid z^k r = 0\}$. Then $A_{z^k}(R)$ is an ideal and $A_{z^k}(R) \subset A_{z^{k+1}}(R)$, so by hypothesis there is some $k$ for which $A_{z^k}(R) = A_{z^{k+i}}(R)$, $\forall i$. We claim that $A_{z^k}(R) \cap z^k R = \{0\}$. In fact if $z^k r \in A_{z^k}(R)$, we have $z^{2k} r = 0$, but since $A_{z^k}(R) = A_{z^{2k}}(R)$, we have also $z^k r = 0$.

Now by the assumpiton on the ideal $\{0\}$, we must have either $A_{z^k}(R) = \{0\}$ or $z^k R = \{0\}$. In the first case $z$ is not a zero divisor, while in the second it is nilpotent. □

Let us then set $N$ to be the nilpotent radical of $A$. By Lemma 11.5.3 $N$ is a prime ideal and furthermore the set $S := A \setminus N$ is formed of nonzero divisors in $R$. This implies that the localization $\bar{R} := R_S$ contains $R$, that $\bar{A} := A_S$ is a local ring with maximal ideal $\bar{N} := N_S$ nilpotent and some residue field $K := \bar{A}/\bar{N}$.

At this point we invoke a theorem of I. S. Cohen **[Coh46]** (valid for complete local rings containing a field $F$), which ensures that $K$ can be lifted to a subfield $K \subset \bar{A}$, and in this case $\bar{A}$ is of some finite dimension $n$ over $K$.

At this point if $R = A$ is commutative, we are done since $\bar{A}$ embeds in $M_n(K)$. Otherwise if $a_1, \dots, a_k$ are generators of $\bar{A}$ over $K$, we have that

$$R = A\langle x_1, \dots, x_m \rangle = K\langle a_1, \dots a_k, x_1, \dots, x_m \rangle$$

is PI and finitely generated over $K$, the required reduction.

We deduce a result of Wehrfritz **[Weh76]** and Beidar **[Beĭ86]**:

COROLLARY 11.5.4. *Let $R$ be an algebra which is a finite module over a commutative Noetherian $F$-algebra ($F$ a field). Then $R$ is representable.*

**11.5.2. Anan'in's theorem.** We start with a proposition, whose proof will be completed in the next paragraph, from Corollary 11.5.17.

PROPOSITION 11.5.5. *For an algebra $R = F\langle a_1, \dots, a_k \rangle$, finitely generated over a field $F$, the following conditions are equivalent:*

(1) *$R$ is representable.*
(2) *$R$ can be embedded in an $F$-algebra $S$ which is a finitely generated module over its center $\mathcal{Z}$.*

PROOF. Clearly (1) implies (2), so it is enough to show that (2) implies (1).

Let $R = F\langle a_1, \dots, a_k \rangle \subset S := \sum_{i=1}^{\ell} \mathcal{Z} m_i$. We have then $a_i = \sum_{j=1}^{\ell} z_{i,j} m_j$ and also $m_i m_j = \sum_{h=1}^{\ell} z_h^{i,j} m_h$ for some elements $z_{i,j},\ z_h^{i,j} \in \mathcal{Z}$. Let thus

$$C = F[z_{i,j},\ z_h^{i,j}] \subset \mathcal{Z},\ B = \sum_{i=1}^{\ell} C m_i.$$

Clearly $B$ is an algebra over $C$ satisfying the hypotheses of Corollary 11.5.17, hence it is representable. Since $R \subset B$, we have that also $R$ is representable. □

The first positive result is due to Malćev **[Mal43]** who proved that a commutative algebra finitely generated over a field $F$ is representable. This will now be a special case of the result of Anan'in's theorem, Theorem 11.5.17.

In fact from this theorem Malćev deduces the following criterion which we leave as exercise.

EXERCISE 11.5.6. A necessary and sufficient condition for an algebra $R$ finitely generated over a field $F$ to be representable is the following:

There is a positive integer $n$ and a family of ideals $I_\alpha$ of $R$ so that

$$\forall \alpha,\ \dim_F R/I_\alpha \le n, \quad \bigcap_\alpha I_\alpha = \{0\}.$$

Before we proceed, let us make a simple remark.

Take an algebra $R \subset M_n(A)$ for some commutative ring $A$, every element $a \in R$ satisfies its characteristic polynomial. This has the following formal implication: for each $a \in R$, choose $n$ commuting variables $t_1(a), \dots, t_n(a)$, consider finally in the polynomial ring $R[t_i(a)],\ i = 1, \dots, n,\ a \in R$, the ideal $I$ generated by all the elements $H_a = a^{n+1} + t_1(a)a^n + \cdots + t_n(a)a$. Then $R$ embeds in $R[t_i(a)]/I$.

One has also a partial converse of this, a corollary of Proposition 11.5.5 whose proof we leave as exercise:

COROLLARY 11.5.7. *If $R$ is a finitely generated PI algebra and $R$ embeds in the algebra $R[t_i(a)]/I$, then $R$ embeds in some matrix algebra $M_N(A)$.*

PROOF. Apply Theorem 8.2.2 of §8.2.1. Inside $R[t_i(a)]/I$, we have the subalgebra $S$ generated over $R$ by the finite number of elements $t_i(a)$, where $a$ is a monomial in the generators of $R$ of degree $< d$, and $d$ is the degree of an identity of $R$. Then $S$ is representable by Theorem 11.5.17 and, since by hypothesis $R \subset S$, also $R$ is representable. □

**11.5.3. Representable modules.** In order to study representable algebras, it is useful to extend the notion to modules and to prove first some results in this case.

DEFINITION 11.5.8. Let $R$ be an algebra, and let $M$ be a left $R$-module. We say that $M$ is *representable* if we can embed $M \to \hat{M}$ where $\hat{M}$ is an $R$-module but also a finite-dimensional vector space over some field $K$ commuting with $R$.

Equivalently $\hat{M}$ is an $K \otimes_A R$-module, finite dimensional over $K$.

Let us now assume that $R$ is an algebra over a field $F$ and $M$ is an $R$-module.

LEMMA 11.5.9. *$M$ is representable if and only if there is a field $K \supset F$ and a subspace $V$ of $K \otimes_F M$ which is also an $R$ submodule and satisfies $V \cap M = 0$ and $\dim_K K \otimes_F M/V < \infty$.*

*If $M$ is a finitely generated module over $R$, and $R$ is a finitely generated algebra over $F$, we can take $K$ finitely generated over $F$.*

PROOF. Assume there is such a $K$ and $V$. By assumption we have an embedding of $M$ in the $K \otimes_F R$-module $K \otimes_F M/V$, which is finite dimensional, so $M$ is representable.

Conversely assume we have en embedding of $M$ in a $K \otimes_F R$-module $\hat{M}$ which is finite dimensional over $K$. We have a homomorphism (of $K \otimes_F R$-modules) of $K \otimes_F M$ to $\hat{M}$ whose kernel $V$ satisfies the conditions of the lemma.

As for the second part, if $M \subset \hat{M}$, with $\hat{M}$ a $K \otimes_F R$-module and a finite-dimensional $K$ space, consider a basis $v_1, \dots, v_m$ of $\hat{M}$ and a finite set of generators $b_1, \dots, b_\ell$ of $M$ as an $R = F[a_1, \dots, a_k]$-module.

Write $b_i = \sum_{j=1}^{m} \lambda_{i,j} v_j$, then one can take as $K$ the field generated by the elements $\lambda_{i,j}$ and by the entries of the matrices representing the elements $a_i$ in the given basis. □

DEFINITION 11.5.10. Let $R$ be an algebra over a field $F$. A module $M$ over $R$ is *residually finite dimensional*, or simply residually finite, if the intersection of all submodules of finite codimension (over $F$) of $M$ equals 0.

It is *subdirectly irreducible* if the intersection of all proper submodules of $M$ is different from $\{0\}$.

REMARK 11.5.11.

(1) If $M$ is residually finite and $N \subset M$ is a submodule, then $N$ is residually finite.
(2) If we have a family $M_\alpha$ of residually finite modules, then also $\prod_\alpha M_\alpha$ is residually finite.
(3) If we have a family of submodules $M_\alpha \subset M$ such that $M/M_\alpha$ is residually finite, then also $M/\bigcap_\alpha M_\alpha$ is residually finite.

(4) If $N \subset M$ is a submodule of finite codimension and $N$ is residually finite, also $M$ is residually finite.
(5) If a module $M$ is subdirectly irreducible, the intersection of all proper submodules of $M$ is an irreducible submodule $S$ contained in all proper submodules.
(6) If a module $M$ is subdirectly irreducible and $S \subsetneq N \subset M$ is a submodule, then $N$ is subdirectly irreducible.

LEMMA 11.5.12. *Let $M$ be a residually finite $R$-module, and let $N \subset M$ be a finite dimensional (over the field $F$) submodule. Then there is a submodule $P \subset M$ of finite codimension with $P \cap N = \{0\}$.*

PROOF. We consider the submodules $M_i$ of finite codimension of $M$ which intersect into 0. Thus $\bigcap_i (M_i \cap N) = 0$. But since $N$ is finite dimensional, there is a finite intersection of $M_i$ which intersects $N$ in $\{0\}$. We then take as $P$ the finite intersection of these $M_i$ which is clearly also of finite codimension. □

Before we prove Proposition 11.5.14, let us do a special case.

LEMMA 11.5.13. *Let $R = F\langle a_1, \ldots, a_k\rangle$ be a finitely generated prime PI algebra with center $\mathcal{Z}$, and let $B$ be a finitely generated left $R$-module which is torsion free over $\mathcal{Z}$. Then $B$ is representable and residually finite.*

PROOF. Let $K$ be the field of fractions of the center $\mathcal{Z}$ of $R$, by Theorem 11.2.6, the algebra $K \otimes_{\mathcal{Z}} R$ contains $R$, is simple and finite dimensional over its center $K$. Since $B$ is torsion free over $\mathcal{Z}$, it follows that $B$ embeds into $K \otimes_{\mathcal{Z}} B$ which is a finitely generated module over the simple algebra $K \otimes_{\mathcal{Z}} R$, hence it is finite dimensional over $K$. In particular $B$ is representable.

Now choose: a basis $u_i$ of $K \otimes_{\mathcal{Z}} R$ and one $v_j$ of $K \otimes_{\mathcal{Z}} B$ over $K$, generators $a_j$ for $R = F\langle a_1, \ldots, a_m\rangle$ over $F$ and $b_h \in B$ so that $B = \sum_h Rb_h$.

We can then write $a_j = \sum_i \mu_{j,i} u_i$, $b_j = \sum_s \nu_{s,j} v_s$, and the multiplication tables $u_i u_j = \sum_h \lambda_{i,j,h} u_h$, $u_i v_j = \sum_h \phi_{i,j,h} v_h$, with $\lambda_{i,j,h}, \mu_{j,h}, \nu_{s,j}, \phi_{i,j,h} \in K$. Clearly by inverting a suitable nonzero element $c \in \mathcal{Z}$, we have that

$$\lambda_{i,j,h}, \mu_{j,h}, \nu_{s,j}, \phi_{i,j,h} \in \mathcal{Z}[c^{-1}], \quad u_i \in \mathcal{Z}[c^{-1}] \otimes_{\mathcal{Z}} R, \quad v_j \in \mathcal{Z}[c^{-1}] \otimes_{\mathcal{Z}} B.$$

It follows that

$$\bar{R} := \mathcal{Z}[c^{-1}] \otimes_{\mathcal{Z}} R = \bigoplus_i \mathcal{Z}[c^{-1}] u_i, \quad \bar{B} := \mathcal{Z}[c^{-1}] \otimes_{\mathcal{Z}} B = \bigoplus_{i=1}^{k} \mathcal{Z}[c^{-1}] v_i,$$

where $\bar{R} \supset R$ is an algebra and a free module over $\mathcal{Z}[c^{-1}]$ with basis the elements $u_i$ and $\bar{B} \supset B$ is an $\bar{R}$-module and free over $\mathcal{Z}[c^{-1}]$ with basis the elements $v_j$.

Moreover, we claim that the algebra $\mathcal{Z}[c^{-1}]$ is generated over $F$ by the elements $\lambda_{i,j,h}, \mu_{j,h}, \nu_{s,j}, \phi_{i,j,h}, c^{-1}$. In fact if $\bar{\mathcal{Z}} \subset \mathcal{Z}$ is the ring generated by these elements, we have that $R$ embeds in $\bigoplus_i \bar{\mathcal{Z}} u_i$, and this last ring has center $\bar{\mathcal{Z}}$ which implies that $\mathcal{Z}[c^{-1}] \subset \bar{\mathcal{Z}}$, hence the equality.

By the usual commutative Hilbert Nullstellensatz the finitely generated domain $\mathcal{Z}[c^{-1}]$ is residually finite since $\mathcal{Z}[c^{-1}]/\mathfrak{m}$ is finite dimensional for every maximal ideal and the intersection of all maximal ideals is $\{0\}$.

Now $\mathfrak{m}\bar{B} = \bigoplus_{i=1}^k \mathfrak{m} v_i$ is a $\bar{R}$ submodule, $B/(\mathfrak{m}\bar{B}) = \bigoplus_{i=1}^k (\mathcal{Z}[c^{-1}]/\mathfrak{m}) v_i$ is finite dimensional, and $\bigcap_{\mathfrak{m}} \mathfrak{m}\bar{B} = 0$. Thus $\bar{B} = \bigoplus_{i=1}^k \mathcal{Z}[c^{-1}] v_i$ is residually finite as $\bar{R}$-module. Therefore $B \subset \bigoplus_{i=1}^k \mathcal{Z}[c^{-1}] v_i$ is residually finite as $R$-module. □

At this point one should notice that there are prime Noetherian algebras which are not PI, even finitely generated domains, which have no finite-dimensional modules, for instance the Weyl algebra $\mathbb{Q}\langle x, y\rangle/(xy - yx - 1) = \mathbb{Q}[x, \frac{\partial}{\partial x}]$.

So one sees how the PI hypothesis must enter in the proof, through the structure theorems of Kaplansky, Posner, and the Nullstellensatz.

PROPOSITION 11.5.14. *Let $R$ be a finitely generated Noetherian PI algebra over a field $F$. Then every Noetherian left $R$-module $M$ is residually finite.*

PROOF. The proof is in two steps, the first is a general Noetherian induction which allows us to reduce to the case where $R$ is prime and $M$ is a faithful module. Then one has to verify the conditions of Lemma 11.5.13, that is $M$ torsion free over the center $\mathcal{Z}$ of $R$.

STEP 1. Assume by contradiction that there is a Noetherian left $R$-module $M$ which is not residually finite, and take a maximal submodule $N$ of $M$ such that $M/N$ is still not residually finite. Thus we may replace $M$ with $M/N$ and assume that $M$ is not residually finite, while all proper quotients $M/N$ are residually finite.

By Remark 11.5.11(3) it follows that $M$ must be subdirectly irreducible, thus the intersection of all proper submodules of $M$ is a nonzero irreducible submodule $S$, moreover $M/S$ is residually finite, hence $S$ is the intersection of the submodules of finite codimension. By the Hilbert Nullstellensatz, since $R$ is finitely generated and $S$ is irreducible, we have that $S$ is finite dimensional over $F$ (Theorem 11.4.4(2)).

We now define $I(M) \subset R$ to be the set of all elements $r \in R$ such that $RrM$ is finite dimensional over $F$. Clearly $I(M)$ is an ideal of $R$. In particular $I(M)$ contains the annihilator of $M$.

Again by Noetherian induction we may assume that $M$, with the previous properties, has been chosen in such a way that $I(M)$ is maximal. Let

$$M_0 := \mathcal{A}_{I(M)}(M) = \{m \in M \,|\, am = 0, \ \forall a \in I(M)\}.$$

Since $I(M)$ is a two-sided ideal, $M_0$ is an $R$ submodule with $I(M)M_0 = 0$.

Now by Noetherian hypothesis $I(M) = \sum_{i=1}^{k} Ra_i$.

Consider, for each $a_i$ the map $m \to a_i m$, $M \to a_i M \subset Ra_i M$ with kernel $\mathcal{A}_{a_i}(M) := \{m \in M \,|\, a_i m = 0\}$. Since $Ra_i M$ is finite dimensional, it follows that $\mathcal{A}_{a_i}(M)$ is a subspace of $M$ of finite codimension.

Then $M_0 = \bigcap_{i=1}^{k} \mathcal{A}_{a_i}(M)$ has finite codimension and, by Remark 11.5.11(6), $M_0$ is also an infinite-dimensional subdirectly irreducible module.

Clearly $I(M_0) \supset I(M)$, thus, by the maximality assumption on $I(M)$, we have $I(M_0) = I(M)$. We can thus replace $M$ with $M_0$ and then assume that $I(M)M = 0$, that is $I(M)$ equals the annihilator of $M$.

We claim that $I(M)$ is a prime ideal of the ring $R$. Otherwise there are two ideals $X, Y \subset R$ strictly larger that $I(M)$ with $XY \subset I(M)$, and then $XYM = 0$.

If $YM$ is finite dimensional, we have that $Y \subset I(M_0) = I(M)$, instead if $YM$ is infinite dimensional, since $S$ is finite dimensional and it is contained in all proper submodules, we have $S \subsetneq YM$. By Remark 11.5.11(6) we have again that $YM$ is subdirectly irreducible.

Clearly $I(M) \subset I(YM)$, so we have $I(M) = I(YM)$, again by the maximality of $I(M)$. Since now $X \subset I(YM)$, we deduce $X \subset I(M)$, as desired.

Finally we can replace $R$ with $R/I(M)$ and thus assume that $R$ is prime as algebra and $M$ is a faithful $R$-module.

STEP 2. By Lemma 11.5.13, in order to have a contradiction, it remains to prove that $M$ is torsion free over the domain $\mathcal{Z}$ center of $R$.

For each $c \in \mathcal{Z} \setminus \{0\}$ and $k \in \mathbb{N}$, the set $P_k := \{m \in M \mid c^k m = 0\}$ is an $R$ submodule. We have $P_k \subset P_{k+1}$, hence by the Noetherian condition there is a minimum $n$ such that $P_n = P_{n+1}$, that is $\{b \in M \mid c^n b = 0\} = \{b \in M \mid c^{n+1} b = 0\}$.

Assume, by contradiction, that there is a $c \in \mathcal{Z} \setminus \{0\}$ such that the $R$ submodule $P := \{b \in M \mid cb = 0\}$ is nonzero. Since $P$ is a submodule, we have that $P$ contains $S$, in other words $cS = 0$.

Since $M$ is a faithful module, for some $b \in M$ we have $c^n b \neq 0$, so the submodule $c^n M$ is nonzero, hence it contains $S$, that is every element of $S$ is of the form $c^n b$. Now $c^{n+1} b = c(c^n b) = 0$. But, by the choice of $n$, we have that $c^{n+1} b = 0$ implies $c^n b = 0$, hence $S = 0$, a contradiction. □

LEMMA 11.5.15. *Let $R$ be a Noetherian algebra over a field $F$, and let $K \supset F$ be a finitely generated field. Then $K \otimes_F R$ is also Noetherian.*

PROOF. Let $K = F(c_1, \dots, c_h)$. Then $K \otimes_F R$ is obtained from $F[c_1, \dots, c_h] \otimes_F R$ by inverting the central multiplicative set of nonzero elements of $F[c_1, \dots, c_h]$.

By the basis theorem $F[c_1, \dots, c_h] \otimes_F R$ is Noetherian as well as any central localization. □

THEOREM 11.5.16. *Let $R$ be a Noetherian finitely generated PI algebra. Then every Noetherian left $R$-module $M$ is representable.*

PROOF. Suppose by contradiction that this is not true. Let $B$ be a finitely generated nonrepresentable module, we may assume, by Noetherian induction, that every proper quotient of $B$ is representable and among these modules we may choose $B$ so that $L(B) := \{r \in R \mid rB = 0\}$ is maximal.

Now we claim that all nonzero submodules of $B$ are also nonrepresentable. Let $G$ be such a submodule, if it is representable we can find a finitely generated field extension $K \supset F$ so that $G_K := K \otimes_F G$ has a finite codimension submodule $W$ over $R_K = K \otimes_F R$ with $W \cap G = 0$.

Now by Lemma 11.5.15 the algebra $R_K$ is Noetherian, and clearly finitely generated over $K$ and PI. The $R_K$-module $B_K/W$ is finitely generated over $R_K$, so by Proposition 11.5.14 it is residually finite (as $K$ vector space).

By Lemma 11.5.12 there is a submodule $P/W \subset B_K/W$ of finite codimension with $P/W \cap G_K/W = 0$, that is $P \cap G_K = W$.

Now $B/(B \cap P)$ is embedded in $B_K/P$, which is finite dimensional, hence it is representable.

On the other hand $(P \cap B) \cap G \subset P \cap G = P \cap G_K \cap G = W \cap G = 0$. Thus $B$ embeds in $B/(B \cap P) \oplus B/G$. Both summands are representable, and we have a contradiction.

Now we follow a similar reasoning as in Proposition 11.5.14. First prove that $L(B)$ is a prime ideal, take two ideals $XY \subset L(B)$, hence $XYB = 0$. If $YB = 0$, then $Y \subset L(B)$, if $YB \neq 0$, then this is a nonzero submodule which we have seen is not representable.

Again by maximality $L(YB) = L(B)$, hence $X \subset L(B)$.

We now can assume that $B$ is a faithful $R$-module and $R$ is a prime algebra. By Lemma 11.5.13 it is enough to show that $B$ is torsion free for the center $\mathcal{Z}$ of $R$. Otherwise take $0 \neq c \in \mathcal{Z}$ and suppose that $G := \{b \in B \mid cb = 0\}$ is nonzero. By what we have seen, $G$ is not representable but now $c \in L(G)$, hence $L(G) \supsetneq L(B)$ contradicts the maximality of the choice of $B$. □

THEOREM 11.5.17. *Let $R$ be a Noetherian PI algebra finitely generated over a field $F$. Then $R$ is representable.*

PROOF. Consider $R$ as module over itself, then there is an $R_K = K \otimes_F R$-module, finite dimensional over the field $K$, in which $R$ embeds; hence $R$ embeds in the $K$-linear transformations of this vector space. □

COROLLARY 11.5.18. *Let $R$ be an algebra over an infinite field $F$ and be representable. Then $R$ is PI equivalent to an algebra finite dimensional over $F$.*

PROOF. Consider $R \subset M_n(K)$. We may assume that $K$ is algebraically closed. Consider $B = KR \subset M_n(K)$. We have that $B$ is finite dimensional over $K$, and clearly $R$ and $B$ are PI equivalent as $F$-algebras.

If $J$ is the nilpotent radical of $B$, by Wedderburn's principal theorem, Theorem 1.3.13, we have a decomposition $B = \bigoplus_i M_{k_i}(K) \oplus J$. Let $u_1, \dots, u_s$ be a linear basis of $J$ over $K$. Then we easily see that the subalgebra $C$ of $B$, generated over $F$ by $\bigoplus_i M_{k_i}(F)$ and all the elements $u_i$, is finite dimensional, over $F$. We have $B = CK$ and finally $C$ is PI equivalent, as $F$-algebra, to $B$ hence to $R$. □

REMARK 11.5.19. In Anan'in's paper he assumes that the modules over $R$ are indeed bimodules, with respect to some other ring $R$ on which there are no assumptions (so they could be $\mathrm{End}_R(M)$). The proof easily extends to show that when we represent such modules, we can keep the bimodule structure.

**11.5.4. Universal embedding revisited.** Given a representable algebra $R$ and an integer $n$ for which $R$ can be embedded into the algebra on $n \times n$ matrices over some commutative ring $A$, we know that, by Proposition 3.4.8, the universal map is an embedding $j : R \to M_n(F_n(R))$.

We then have the universal trace algebra $\hat{R} = [M_n(F_n(R))]^{\mathrm{GL}(n,F)}$ obtained from Theorem 3.4.24. In characteristic 0 it is obtained by adding to $j(R)$ all the traces of all its elements. If $R$ is finitely generated over $F$, then $\hat{R}$ is a finite module over the finitely generated algebra of invariants $F_n(R)^{\mathrm{GL}(n,F)}$. Of course in characteristic $p > 0$ it is not generated only by traces. If we take an ideal $I$ of $R$, we then see easily that $R/I$ is representable in $n \times n$ matrices, if and only if the ideal $\hat{I}$ generated by $I$ in $[M_n(F_n(R))]^{\mathrm{GL}(n,F)}$ has the property that $\hat{I} \cap R = I$.

This in general need not be true. On the other hand it is possible that still $R/I$ is representable possibly with a larger matrix ring.

**11.5.5. Negative results.** We have discussed up to now some sufficient criteria for representability in matrices. On the other hand we should also discuss some necessary conditions and show by examples that it does not seem possible to give a completely ring-theoretical characterization of rings $R$ which can be embedded into a matrix algebra $M_n(K)$ over a field $K$.

A necessary condition of course is that $R$ should satisfy all identities of $n \times n$ matrices. One natural class to investigate is finitely generated algebras $R$ over a field $F$ which satisfy all polynomial identities of $n \times n$ matrices. By Amitsur's embedding theorem, Theorem 11.2.3, it easily follows that a finitely generated semiprime PI algebra is representable, as we shall soon see the condition on the radical is quite essential.

One finds quickly another condition, a condition on certain special ideals:

DEFINITION 11.5.20. Given a set $X \subset R$, its *left annihilator* (resp., *right annihilator*) is the left (resp., right) ideal,

$$\ell_X(R) := \{a \in R \mid aX = 0\}, \qquad r_X(R) := \{a \in R \mid Xa = 0\}. \tag{11.6}$$

A left (resp., right) ideal is called an *annihilator ideal* if it is of the form $\ell_X(R)$ for some $X$ (similar for right ideals).

EXAMPLE 11.5.21. In the algebra $M_n(K)$, where $K$ is a field, every left ideal is a left annihilator.

In fact given a left ideal $I$ of $M_n(K)$, since $M_n(K)$ is a semisimple module over itself, one has a direct sum decomposition $M_n(K) = I \oplus J$ with $J$ some other left ideal. Then $1 = e + f,\ e \in I,\ f \in J$ and $ef = e - e^2 \in J \cap I = 0$. It follows that $e^2 = e$ and $I = M_n(K)e$ is the left annihilator of $f$, since if $af = 0$, we have $a = ae \in I$. In particular a left ideal $I$ is the annihilator of a subspace $fK^n$ of $K^n$.

Of course the left annihilator is a left ideal and the right annihilator a right ideal. Moreover a left ideal $I$ is a left annihilator if and only if $I = \ell_{r_I(R)}(R)$, finally clearly if $X \subset Y$, we have $\ell_Y(R) \subset \ell_X(R)$.

The following remark is immediate. Let $R \subset S$ be an inclusion of rings and $X \subset R$, then $\ell_X(R) = R \cap \ell_X(S)$. We deduce the following simple fact:

PROPOSITION 11.5.22. *If $R \subset S := M_n(K)$, with $K$ a field, any strictly increasing (or decreasing) sequence $0 \subsetneq I_1 \subsetneq I_2 \subsetneq \cdots \subsetneq I_k \subsetneq R$ of left (or right) annihilator ideals consists of at most $k = n - 1$ terms.*

PROOF. By the previous remark the sequence of left ideals $\ell_{r_{I_i}(R)}(S)$ of $S$ is strictly increasing but in the algebra $M_n(K)$ by Example 11.5.21 it is clear that strictly increasing chains of left ideals correspond to strictly decreasing sequences of subspaces of $K^n$, in particular a sequence of the previous type can have at most $n - 1$ elements. □

The first examples we want to discuss are due to Amitsur **[Ami70]**, Small **[Sma71]**, and Irving and Small **[IS86]** in which they refine the same idea. That is to use Proposition 11.5.22 in order to show that various possible embedding conditions are not sufficient. On the other hand, if one is in characteristic 0 and adds the trace as part of the structure, there is an extremely precise embedding theorem, Theorem 14.2.1.

We start with Small's example which is quite easy to describe.

EXAMPLE 11.5.23. A finitely generated algebra over a field $F$ satisfies all the identities of $4 \times 4$ matrices but has an infinite descending chain of right annihilators (so it is not representable).

Take a field $F$ and construct the rings $F[t],\ F[t, t^{-1}]$ of polynomials and Laurent polynomials respectively.

In the ring of $2 \times 2$ matrices over $F[t, t^{-1}]$, consider the subring

$$R := \begin{vmatrix} F[t,t^{-1}] & F[t,t^{-1}] \\ 0 & F[t] \end{vmatrix} = \left\{ \begin{vmatrix} a & b \\ 0 & c \end{vmatrix},\quad a, b \in F[t, t^{-1}],\quad c \in F[t] \right\}. \tag{11.7}$$

LEMMA 11.5.24. *$R$ is generated over $F$ by the elements*

$$\begin{vmatrix} t & 0 \\ 0 & 0 \end{vmatrix}, \begin{vmatrix} t^{-1} & 0 \\ 0 & 0 \end{vmatrix}, \begin{vmatrix} 0 & 0 \\ 0 & t \end{vmatrix}, \begin{vmatrix} 0 & 1 \\ 0 & 0 \end{vmatrix}.$$

*$R$ is left Noetherian, $R$ is not right Noetherian.*

PROOF. The first part is immediate. Then $R$ as a left module over the commutative Noetherian ring $F[t,t^{-1}] \oplus F[t]$ is generated by two elements $1, \begin{vmatrix} 0 & 1 \\ 0 & 0 \end{vmatrix}$, hence it is Noetherian. On the other hand any $F[t]$ submodule $M \subset F[t,t^{-1}]$ gives rise to a right ideal $\begin{vmatrix} 0 & M \\ 0 & 0 \end{vmatrix}$, and clearly $F[t,t^{-1}]$ as an $F[t]$-module is not Noetherian.

In particular consider the right ideal

$$I := \begin{vmatrix} 0 & F[t] \\ 0 & 0 \end{vmatrix}$$

and the elements $y_i := \begin{vmatrix} 0 & t^{-i} \\ 0 & 0 \end{vmatrix}$, $i = 1, 2, \ldots.$ Set $D_i := \{a \in R \mid y_i a \in I\}$, the set $D_i$ is the right ideal

$$D_i = \begin{vmatrix} F[t,t^{-1}] & F[t,t^{-1}] \\ 0 & t^i F[t] \end{vmatrix}.$$

We now construct, starting from $R$ in the ring of $2 \times 2$ matrices over $R$ the subring

$$B := \begin{vmatrix} F & R \\ 0 & R \end{vmatrix} = \left\{ \begin{vmatrix} a & b \\ 0 & c \end{vmatrix}, \quad a \in F,\ b, c \in R \right\}. \tag{11.8}$$

$B$ can be embedded into $4 \times 4$ matrices over $F[t,t^{-1}]$ and is also clearly finitely generated as algebra over $F$.

Now quite clearly in $B$ if $J$ is a right ideal of $R$, the set $\hat{J} := \begin{vmatrix} 0 & J \\ 0 & 0 \end{vmatrix}$ is a two-sided ideal. We thus apply this to the previous ideal $I$ and consider the quotient algebra $B/\hat{I}$. This is a finitely generated algebra over $F$ which satisfies all polynomial identities of $4 \times 4$ matrices. $B$ can be thought of also as triangular matrices $\begin{vmatrix} F & R/I \\ 0 & R \end{vmatrix}$ where the algebra structure is induced by the *$F, R$-bimodule structure* of $R/I$.

Let now $z_i$ be the class in $B/\hat{I}$ of the element $\begin{vmatrix} 0 & y_i \\ 0 & 0 \end{vmatrix}$. The right annihilator of $z_i$ in $B/\hat{I}$ is given by the classes of the elements

$$\begin{vmatrix} a & b \\ 0 & c \end{vmatrix} \in B \mid \begin{vmatrix} 0 & y_i c \\ 0 & 0 \end{vmatrix} = \begin{vmatrix} 0 & y_i \\ 0 & 0 \end{vmatrix} \begin{vmatrix} a & b \\ 0 & c \end{vmatrix} \in \hat{I} \implies r_{z_i}(B/\hat{I}) = \begin{vmatrix} F & R/I \\ 0 & D_i \end{vmatrix}.$$

The right ideals $D_i$, and so also the right annihilator ideals which we have constructed, form an infinite strictly descending chain. Thus, by Proposition 11.5.22, we have that $B/\hat{I}$ in *not representable*. □

This example is not a graded algebra but in Irving **[Irv83]** lots of interesting examples which are graded are produced.

Notice that the Jacobson radical of $B/\hat{I}$ is $J = \begin{vmatrix} 0 & R/I \\ 0 & 0 \end{vmatrix}$ and $J^2 = 0$.

The fact that the Jacobson radical of a finitely generated PI algebra is nilpotent follows from some deep results of Razmyslov, Braun, and Kemer, which will be discussed in Chapter 16.

It remains a question of whether, if we add the condition that chains of annihilators are bounded, this implies representability. The answer is again no, as shown by examples of Irving and Small **[IS86]**.

At present the only known necessary and sufficient condition for the representability of an algebra is that given by Malćev in Exercise 11.5.6.

As we shall see, for algebras with trace, there is instead a very general theorem, Theorem 14.2.1, at least in characteristic 0.

CHAPTER 12

# Invariants and trace identities

*In this chapter we start to use invariant theory for matrices in a systematic way. We prove the first and second fundamental theorem for matrix invariants in characteristic 0 (for the general case we refer to* **[DCP17]***). The second fundamental theorem then can be interpreted in terms of trace identities for matrices. We then start to discuss general Cayley–Hamilton algebras. We close the chapter with the proof by Formanek of the nonvanishing of a remarkable polynomial due to Regev, which gives a noncommutative discriminant for matrices.*

## 12.1. Invariants of matrices

**12.1.1. FFT for matrices.** We are interested now in the following problem. Let $F$ be a field of characteristic 0, and describe the ring of invariants of the action of the general linear group $\mathrm{GL}(n, F)$ acting by simultaneous conjugation on $m$ copies of the space $M_n(F)$ of square matrices. That is the action,

$$g(X_1, \dots, X_m)g^{-1} := (gX_1g^{-1}, \dots, gX_mg^{-1}). \tag{12.1}$$

In intrinsic language, we have an $n$-dimensional vector space $V$ over $F$ and $\mathrm{GL}(V)$ acts on $\mathrm{End}(V)^{\oplus m}$, and we want to describe the ring of polynomial functions on $\mathrm{End}(V)^{\oplus m}$ which are invariant under the action of $\mathrm{GL}(V)$.

The main theorem is that the ring of invariants of $m,n \times n$ matrices coincides with the ring $\mathcal{T}_{n,m}$ of Definition 3.3.23. It is generated by the coefficients of the characteristic polynomials of polynomials in generic matrices. We prove this in characteristic 0 and see **[DCP17]** for the general case.

It is convenient to work on countably many matrix variables. We denote by $\mathcal{T}_n = \mathcal{T}_{n,\infty}$ or $\mathcal{T}_n(X)$ if we want to specify the list $X$ of variables.

DEFINITION 12.1.1. The space of multilinear invariants of $m$, $n \times n$ matrices will be denoted by $\mathcal{T}_n(m)$.

By the Aronhold method in §3.1.1.5, that is using polarization and restitution, we describe $\mathcal{T}_n$ by reducing to study multilinear invariants $\mathcal{T}_n(m)$ for all $m$.

We will denote by $(X_1, X_2, \dots, X_m)$ an $m$-tuple of $n \times n$ matrices.

We need to interpret the results of Chapter 6, §6.1.1. Theorem 6.1.1 and Lemma 6.3.19 can be reformulated as follows.

Start with the multilinear invariants of $m$ matrices, i.e., the invariant elements of the dual of $\mathrm{End}(V)^{\otimes m}$.

LEMMA 12.1.2. *The space $\mathcal{T}_n(m)$ of multilinear invariants of $m$ matrices is identified with $\mathrm{End}_{\mathrm{GL}(V)}(V^{\otimes m})$, and it is linearly spanned by the functions*

$$T_\sigma(X_1, X_2, \dots, X_m) := \mathrm{tr}(\sigma^{-1} \circ X_1 \otimes X_2 \otimes \cdots \otimes X_m),\ \sigma \in S_m. \tag{12.2}$$

*If* $\sigma = (i_1 i_2 \cdots i_h) \cdots (j_1 j_2 \cdots j_\ell)(s_1 s_2 \cdots s_t)$ *is the cycle decomposition of* $\sigma$, *then we have that*

$$\begin{aligned} T_\sigma(X_1, X_2, \dots, X_m) \\ = \operatorname{tr}(X_{i_1} X_{i_2} \cdots X_{i_h}) \cdots \operatorname{tr}(X_{j_1} X_{j_2} \cdots X_{j_\ell}) \operatorname{tr}(X_{s_1} X_{s_2} \cdots X_{s_t}). \end{aligned} \tag{12.3}$$

PROOF. We first remark that the dual of $\operatorname{End}(V)^{\otimes m}$ can be identified, in a $\operatorname{GL}(V)$ equivariant way (Definition 1.3.19), to $\operatorname{End}(V)^{\otimes m}$ by the pairing formula

$$\begin{aligned} \langle A_1 \otimes A_2 \cdots \otimes A_m | B_1 \otimes B_2 \cdots \otimes B_m \rangle &:= \operatorname{tr}(A_1 \otimes A_2 \cdots \otimes A_m \circ B_1 \otimes B_2 \cdots \otimes B_m) \\ &= \prod \operatorname{tr}(A_i B_i). \end{aligned}$$

Under this isomorphism the multilinear invariants of matrices are identified with the $\operatorname{GL}(V)$ invariants of $\operatorname{End}(V)^{\otimes m}$ which in turn are spanned by the elements of the symmetric group and hence by the elements of formula (12.2).

As for formula (12.3), since the identity is multilinear, it is enough to prove it on the decomposable tensors of $\operatorname{End}(V) = V \otimes V^*$ which are the endomorphisms of rank 1 $u \otimes \phi : v \mapsto \langle \phi \mid v \rangle u$.

So given $X_i := u_i \otimes \phi_i$ and an element $\sigma \in S_m$ in the symmetric group, we need to understand in matrix formulas the invariants $\prod_{i=1}^m \langle \phi \mid u_{\sigma(i)} \rangle$. We need to use the rules

$$u \otimes \phi \circ v \otimes \psi = u \otimes \langle \phi \mid v \rangle \psi, \quad \operatorname{tr}(u \otimes \phi) = \langle \phi \mid u \rangle,$$

from which the formula easily follows. □

THEOREM 12.1.3 (FFT for matrices, I). *The ring* $\mathcal{T}_n = \mathcal{T}_n(X)$ *of invariants of matrices under simultaneous conjugation is generated by the elements*

$$\operatorname{tr}(X_{i_1} X_{i_2} \dots X_{i_{k-1}} X_{i_k}). \tag{12.4}$$

This formula means that we take all possible noncommutative monomials in the $X_i$ and form their traces.

PROOF. The ring of invariants of matrices contains the ring generated by the traces of monomials and both rings are stable under polarization and restitution, hence, by the Aronhold method in §3.1.1.5, it is enough to prove that they coincide on multilinear elements, and this is the content of the previous lemma. □

REMARK 12.1.4. By the invariant theory of Hilbert the ring of invariants in finitely many matrix variables is finitely generated, in our case we have also deduced this fact from Shirshov's theorem (cf. Proposition 2.4.10).

It is an interesting and open problem, for $n > 3$, to find a minimum set of generators for the ring of invariants $\mathcal{T}_n$. For $n = 2$ we have the explicit description of Chapter 9.

In general $\mathcal{T}_n$ is a graded algebra (for its Poincaré series, see **[VdB91]**), so to understand a minimal set of generators it is equivalent to describe the space $\mathcal{T}_{n,+}/\mathcal{T}_{n,+}^2$, where $\mathcal{T}_{n,+}$ is the ideal of invariants with no constant term.

This is of course a representation of the linear group $\operatorname{GL}_\infty$ or the algebra of polarization and restitution operators. By Capelli theory one may thus restrict to analyze in this space $\mathcal{T}_{n,+}/\mathcal{T}_{n,+}^2$ the spaces of multilinear elements, or the spaces spanned by highest weight vectors. This can be related to a problem on the symmetric group. In fact for each $i$ the space $\mathcal{T}_n(i)$ of multilinear elements of degree

$i$ (in the first $i$ variables) of $\mathcal{T}_{n,+}$ coincides with the image of the algebra of the symmetric group $S_i$, by the map $\sigma \mapsto T_\sigma$.

By Theorem 6.1.3 this has as kernel the ideal $I_{n+1}$ generated by the antisymmetrizer $\sum_{\sigma \in S_{n+1}} \epsilon_\sigma \sigma$.

By formula (12.3) it follows that the space of multilinear elements of degree $i$ of $\mathcal{T}_{n,+}^2$ is the image of the span of permutations which are not full cycles. Let us call this subspace $Q_i$, and by $P_i$ the span of the full cycles so that $F[S_i] = P_i \oplus Q_i$. Therefore the space of multilinear elements of degree $i$ can be identified with

$$V_i := F[S_i]/Q_i + I_{n+1} = P_i/P_i \cap (Q_i + I_{n+1}). \tag{12.5}$$

For $n = 3$ this has been computed, by a computer assisted method, by Abeasis and Pittaluga **[AP89]**. Denoting by $S_\lambda$ the irreducible representation of $\mathrm{GL}(m, F)$ associated to a partition $\lambda = (\lambda_1 \geq \lambda_2 \geq \cdots \geq \lambda_k)$ (Definition 6.3.2), the authors obtain the following decomposition for $3 \times 3$ matrices:

$$\begin{aligned} V_3 = & S_{(1)} \oplus S_{(1^2)} \oplus S_{(1^3)} \oplus S_{(3)} \oplus S_{(2^2)} \\ & \oplus S_{(3,1)} \oplus S_{(3,1^2)} \oplus S_{(3,2)} \oplus S_{(5)} \oplus S_{(2^3)} \oplus S_{(4,1^2)}. \end{aligned}$$

In fact one can estimate a priori the degree of the monomials of formula (12.4) necessary to generate all the invariants. At present the best estimate is $n^2$, following by a result of Razmyslov which we shall prove in Theorem 12.2.13.

**12.1.2. Equivariant maps.** Invariants of matrices should always be considered together with a more general concept, that of *equivariant maps*, Definition 1.3.19. That is, we consider for any $n, k$ the space of polynomial maps $f : M_n(F)^k \to M_n(F)$ which are equivariant with respect to the conjugation action of the linear group, that is that satisfy

$$f(gX_1g^{-1}, \ldots, gX_kg^{-1}) = gf(X_1, \ldots, X_k)g^{-1}. \tag{12.6}$$

Several remarks are in order. First of all invariants are among these maps by identifying $F$ with the scalar matrices. Next one can sum and multiply these maps which hence form an algebra. The coordinate elements $(X_1, \ldots, X_k) \mapsto X_i$ belong to this algebra. If $f(X_1, \ldots, X_k)$ is an equivariant map, then $\operatorname{tr} f(X_1, \ldots, X_k)$ is an invariant, and finally if we are considering matrices with some prescribed involution $X \mapsto X^*$, which can be either orthogonal or symplectic, we have that if $f(X_1, \ldots, X_k)$ is an equivariant map, with respect to the orthogonal or symplectic group, then $f(X_1, \ldots, X_k)^*$ is also an equivariant map.

Now there is a simple way to connect equivariant maps with invariants.

That is if $f(X_1, \ldots, X_k)$ is an equivariant map and $X_{k+1}$ is a further variable, then the function $\operatorname{tr}(f(X_1, \ldots, X_k)X_{k+1})$ is an invariant which is linear in $X_{k+1}$. Conversely starting from an invariant $\mathtt{f}(X_1, \ldots, X_k, X_{k+1})$ which is linear in $X_{k+1}$, one can reconstruct an equivariant map $f(X_1, \ldots, X_k)$ such that

$$\mathtt{f}(X_1, \ldots, X_k, X_{k+1}) = \operatorname{tr}(f(X_1, \ldots, X_k)X_{k+1}).$$

For the basic invariants $T_\sigma$ of formula (12.3), we take $m = k + 1$, and we may assume that the last cycle ends with $s_t = k + 1$, so the last factor is of the form $\operatorname{tr}((X_{s_1}X_{s_2} \ldots X_{s_{t-1}})X_{k+1})$. Hence we have that

$$T_\sigma(X_1, X_2, \ldots, X_{k+1}) = \operatorname{tr}(\psi_\sigma(X_1, X_2, \ldots, X_k)X_{k+1}), \tag{12.7}$$

where $\psi_\sigma(X_1, X_2, \ldots, X_k)$ is the equivariant map given by the formula

$$\psi_\sigma(X_1, X_2, \ldots, X_k) = \operatorname{tr}(X_{i_1}X_{i_2}\cdots X_{i_h})\cdots\operatorname{tr}(X_{j_1}X_{j_2}\cdots X_{j_\ell})X_{s_1}X_{s_2}\cdots X_{s_{t-1}}. \tag{12.8}$$

Notice that this formalism is compatible with the symbolic formulas given in (2.3.16). As an example if $\sigma = (1,5,3)(4,2) = (2,4)(3,1,5)$, we have $\psi_\sigma(X_1, \ldots, X_4) = \operatorname{tr}(X_2X_4)X_3X_1$.

We deduce

THEOREM 12.1.5 (FFT for matrices, II). *The multilinear maps of $m$ matrix variables to matrices, equivariant under simultaneous conjugation, are of the form*

$$\psi_\sigma(X_1, X_2, \ldots, X_m),\ \sigma \in S_{m+1}.$$

*The ring of equivariant maps of matrices under simultaneous conjugation is generated over the ring of invariants by the monomials*

$$X_{i_1}X_{i_2}\cdots X_{i_{m-1}}X_{i_m}.$$

PROOF. The first part is the result of the previous discussion, and the second, since we are assuming we are working over a field of characteristic 0, follows from the method of polarization and restitution. □

DEFINITION 12.1.6. The space of multilinear equivariant maps from $m$, $n \times n$ matrices to matrices will be denoted by $T_n(m)$.

By the method of generic elements we deduce (cf. Proposition 2.3.9)

COROLLARY 12.1.7. *The ring of equivariant maps for matrices equals the free algebra with trace modulo the trace identities of matrices. It equals $F\langle\xi\rangle T_n$.*

REMARK 12.1.8. When we consider $n \times n$ matrices $M_n(F)$ as an algebra with 1, then the identity $\operatorname{tr}(1) - n = 0$ is to be considered as one of the trace identities.

Our next task is to understand thus the $T$-ideal kernel of the map of the free algebra to the ring of equivariant maps, that is the $T$-ideal of *trace identities*.

12.1.2.1. *Proof of Theorem* 3.4.27. We can now prove Theorem 3.4.27 stated on page 107. Let $R$ be any $\mathbb{Q}$-algebra presented as a quotient of a suitable free algebra $\mathbb{Q}\langle X\rangle$. We have, with the notations of §3.4.3, the universal ring of Definition 3.4.6 $\mathbb{Q}_n(R)$ and a commutative diagram, as in (12.9), with the vertical arrows surjective.

$$\begin{array}{ccc} \mathbb{Q}\langle X\rangle & \xrightarrow{j_{\mathbb{Q}\langle X\rangle}} & M_n(\mathbb{Q}_n(\mathbb{Q}\langle X\rangle)) \\ \downarrow f & & \downarrow M_n(\mathbb{Q}_n(f)) \\ R & \xrightarrow{j_R} & M_n(\mathbb{Q}_n(R)) \end{array} \tag{12.9}$$

Recall that, for a free algebra $\mathbb{Q}\langle X\rangle$ the universal map into $n \times n$ matrices consists in sending the variables to the generic matrices. So that the universal ring $\mathbb{Q}_n(\mathbb{Q}\langle X\rangle)$ coincides with the polynomial ring in the entries of the generic matrices and the space $M_n(\mathbb{Q}_n(\mathbb{Q}\langle X\rangle))$ with the space of polynomial maps from $X$-tuples of matrices to matrices.

The action of $\mathrm{GL}(n, \mathbb{Q})$, defined in Theorem 3.4.24 is the same as that of this section on polynomial maps (verify this), so that the algebra of invariants

$M_n(\mathbb{Q}_n(\mathbb{Q}\langle X\rangle))^{\mathrm{GL}(n,\mathbb{Q})}$ equals the algebra of equivariant maps described by Theorem 12.1.5. Now since the linear group $\mathrm{GL}(n,\mathbb{Q})$ is linearly reductive by Definition 14.1.6, and the map $M_n(\mathbb{Q}_n(f))$ is equivariant, by Remark 14.1.8 it induces a surjective map at the level of invariants, from which the claim follows easily.

**12.1.3. Trace identities.** *Trace identities and their theory were introduced and developed—independently—by Procesi* **[Pro76a]** *and Razmyslov* **[Raz74a]**. *It was later found that trace identities implicitly appeared in earlier works, in particular on continuum mechanics, especially in dimension* 3, *in the work of Spencer* **[Spe71a]** *and Rivlin* **[Spe09]**. *From another point of view one should mention in particular Kostant's paper* **[Kos58]**.

We now use the second fundamental theorem, Theorem 6.1.3, to understand the relations between the invariants that we have constructed. For this we will again use the language of universal algebra. We have to view the algebras constructed as quotients of the corresponding free algebras, and we have to deduce some information on the kernel of this quotient map.

Recall that in an algebra $R$ with some extra operations an ideal $I$ must be stable under these operations so that $R/I$ can inherit the structure. In an algebra with trace or involution, we assume that $I$ is stable under the trace or the involution. The theory of involutions is postponed to Chapter 13.

With the notations of §2.3.2, let us call by $\mathcal{F}_T = \mathcal{T}_X\langle X\rangle$ the free algebra with trace in variables $X$, and $\mathcal{T}_X$ the trace ring.

DEFINITION 12.1.9. We shall denote by $R_X(n)$ the free algebra modulo the trace identities of $n \times n$ matrices.

By $\mathcal{T}_n(X) = \mathcal{T}_n$ we shall denote the algebra of traces of this algebra.

REMARK 12.1.10. Observe that the spaces $T_n(m)$, $\mathcal{T}_n(m)$ (Definitions 12.1.6 and 12.1.1) are obtained from the corresponding symbolic space $T_X(m)$, $\mathcal{T}_X(m)$ defined in 2.3.14 and by formulas (2.16) by *substituting* or *evaluating* the variables $x_i$ in the generic matrices $X_i$.

This evaluation is also an interpretation of formulas (2.16), (12.2), and (12.7).

We will do the same for orthogonal and symplectic involution and define $R^o_X(n)$, $R^s_X(n)$. We can also use finitely many variables $x_1, \dots, x_m$, and then we use the symbol $F_m$ and $R_m(n) = R_X(n)$, and so on. We have the canonical maps (compatible with trace and, when it applies, with the involution)

$$\pi : \mathcal{F}_T \to R_X(n), \; \pi^o : \mathcal{F}_T^i \to R^o_X(n), \; \pi^s : \mathcal{F}_T^i \to R^s_X(n).$$

We have thus:

*The ideal of trace identities of matrices.*

*The ideal of trace identities of matrices with orthogonal involution.*

*The ideal of trace identities of matrices with symplectic involution.*

In order to describe these ideals, we follow the strategy of exhibiting special identities, construct the $T$-ideal generated by these identities, and then prove that the proposed ideal coincides with the $T$-ideal of all identities by applying Corollary 2.2.44, that is we show that the two ideals coincide on multilinear elements. For this we use the classical *second fundamental theorem, SST*, in the various cases for the various classical groups.

We are going to prove in fact that they are generated as $T$-ideals by some special identities.

We will use Proposition 2.3.16 and formula (2.18) which gives linear isomorphisms $T$, $\Psi$ between group algebras and the spaces $\mathcal{T}_X(m)$, $T_X(m)$ of multilinear trace polynomials in $m$ variables.

The multilinear elements of degree $m$ (in the first variables) are contained in $R_m(n)$ and, by Theorem 12.1.5, are of the form

$$\sum_{\sigma \in S_{m+1}} a_\sigma \psi_\sigma(X_1, X_2, \ldots, X_m).$$

From Lemma 12.1.2 and the second fundamental theorem, Theorem 6.1.3, we have

THEOREM 12.1.11. *The map $\pi \circ T : \mathbb{C}[S_m] \to \mathcal{T}_X(m) \to \mathcal{T}_n(m)$,*

$$\pi \circ T : \sum_{\sigma \in S_m} a_\sigma \sigma \to \sum_{\sigma \in S_m} a_\sigma T_\sigma(X_1, X_2, \ldots, X_m) \tag{12.10}$$

*has as image the multilinear trace invariants for $m$, $n \times n$ matrices, as kernel the ideal $I_{n+1,m}$ generated by the antisymmetrizer in $n+1$ elements.*

*The map*

(12.11)

$$\pi \circ \Psi : \mathbb{C}[S_{m+1}] \to T_X(m) \to T_n(m), \quad \sum_{\sigma \in S_{m+1}} a_\sigma \sigma \to \sum_{\sigma \in S_{m+1}} a_\sigma \psi_\sigma(X_1, X_2, \ldots, X_m)$$

*has as image the multilinear trace polynomial maps for $m$, $n \times n$ matrices, as kernel the ideal generated by the antisymmetrizer in $n+1$ elements.*

*Thus this ideal $I_{n+1,m+1}$ of $\mathbb{C}[S_{m+1}]$ is identified with the space of multilinear trace identities of $n \times n$ matrices, in $m$ variables.*

PROOF. The first part follows from the commutative diagram

$$\begin{array}{ccc} \mathbb{C}[S_m] & \xrightarrow{\ j\ } & \mathrm{End}_{GL(V)}(V^{\otimes m}) \\ 1\big\downarrow & & \simeq\big\downarrow \textit{Lemma } 12.1.2 \\ \mathbb{C}[S_m] & \xrightarrow[(12.2)]{\pi \circ T} & \mathcal{T}_n(m) \end{array} \tag{12.12}$$

and the description of the kernel of $j$ given by Theorem 6.1.3. The second part is just an interpretation of the first by formula (2.17). □

Thus, first of all, the multilinear trace identities appear only for $m \geq n$ and for $m = n$ there is a unique identity (up to scalars).

Thus the first step consists in identifying the identities $T_{n+1}(x_1, \ldots, x_n, x_{n+1})$, $A_n(x_1, \ldots, x_n)$ corresponding to the antisymmetrizer

$$T_{n+1}(x_1, \ldots, x_{n+1}) := \sum_{\sigma \in S_{n+1}} \epsilon_\sigma T_\sigma(x_1, x_2, \ldots, x_{n+1}), \tag{12.13}$$

$$A_n(x_1, \ldots, x_n) := \sum_{\sigma \in S_{n+1}} \epsilon_\sigma \psi_\sigma(x_1, x_2, \ldots, x_n), \tag{12.14}$$

with $\epsilon_\sigma$ the sign of the permutation and

$$T_{n+1}(x_1, x_2, \ldots, x_{n+1}) = t(A_n(x_1, \ldots, x_n)x_{n+1}).$$

For this recall that there is a canonical identity, homogeneous of degree $n$ in 1 variable, called the *Cayley–Hamilton identity*.

Consider the characteristic polynomial of $X$, $\chi_X(t) := \det(t - X)$ (3.20), and we have the *Cayley–Hamilton theorem* $\chi_X(X) = 0$.

If we write $\chi_X(t) = \sum_{i=0}^n (-1)^i \sigma_i(X) t^{n-i}$, the coefficient $\sigma_i(X)$ is the $i$th elementary symmetric function in the eigenvalues $\alpha_j$ of the matrix $X$.

We want to interpret this as a trace identity. Remark thus that $\operatorname{tr}(X^i) = \sum_{j=1}^n \alpha_j^i$ is the $i$th Newton function in the eigenvalues $\alpha_j$ of $X$. Hence, by Corollary 3.3.3 we can interpret each coefficient $(-1)^i \sigma_i(X)$ of the characteristic polynomial as a well-defined polynomial in the elements $\operatorname{tr}(X^i)$.

Thus we can consider, using the universal formula (3.21) for each $\sigma_i(x)$,

$$\mathrm{CH}_n(x) := x^n - \sigma_1(x)x^{n-1} + \sigma_2(x)x^{n-2} + \cdots + (-1)^n \sigma_n(x) \tag{12.15}$$

as a formal element of the free algebra with trace (in one variable).

If we fully polarize this element, we get a multilinear trace identity, denoted $\mathrm{CH}(x_1, \dots, x_n)$, for $n \times n$ matrices, whose terms not containing traces arise from the polarization of $x^n$ and are thus of the form $\sum_{\tau \in S_n} x_{\tau(1)} x_{\tau(2)} \cdots x_{\tau(n)}$.

By the uniqueness of the multilinear identity in degree $n$, we must have that the polarized Cayley–Hamilton identity is a scalar multiple of the identity corresponding to the antisymmetrizer, and to compute the scalar we may look in the two identities at the terms not containing a trace.

Clearly $x_1 x_2 \cdots x_n = \psi_{(1\,2 \cdots n\, n+1)}$ and $\epsilon_{(1\,2 \cdots n\, n+1)} = (-1)^n$, and thus we have finally (see also Lew [**Lew66**]):

PROPOSITION 12.1.12.

$$\mathrm{CH}(x_1, \dots, x_n) = (-1)^n A_n(x_1, \dots, x_n) = (-1)^n \sum_{\sigma \in S_{n+1}} \epsilon_\sigma \psi_\sigma(x_1, x_2, \dots, x_n).$$

*With the notations of Proposition* 2.3.16 *we have*

$$\mathrm{CH}(x_1, \dots, x_n) = (-1)^n \Psi\Big(\sum_{\sigma \in S_{n+1}} \epsilon_\sigma \sigma\Big). \tag{12.16}$$

EXAMPLE $n = 2$ (*polarize* $\mathrm{CH}_2(x)$).

$$\mathrm{CH}_2(x) = x^2 - t(x)x + \det(x) = x^2 - t(x)x + \frac{1}{2}(t(x)^2 - t(x^2)). \tag{12.17}$$

$$\begin{aligned} A_2(x_1, x_2) &= x_1x_2 + x_2x_1 - t(x_1)x_2 - t(x_2)x_1 - t(x_1x_2) + t(x_1)t(x_2) \\ &= \mathrm{CH}_2(x_1 + x_2) - \mathrm{CH}_2(x_1) - \mathrm{CH}_2(x_2). \end{aligned}$$

It is particularly interesting to see what is the polarized form of the determinant. The terms corresponding to the determinant correspond exactly to the sum over all the permutations which fix $n+1$. Similarly the terms of degree $n-k$ in the noncommutative variables correspond to $(-1)^k \sigma_k(x) x^{n-k}$. Therefore we deduce

COROLLARY 12.1.13. *For each* $k \le n$ *the polarized form of* $\sigma_k(X)$ *is the expression*

$$T_k(x_1, \dots, x_k) = \sum_{\sigma \in S_k} \epsilon_\sigma T_\sigma(x_1, \dots, x_k). \tag{12.18}$$

REMARK 12.1.14. Formula (12.18) is independent of $n$. It vanishes when computed in $n \times n$ matrices with $n < k$.

Let us look at some implications for relations among traces.

EXAMPLE 12.1.15 ($n = 2$). $T_3$ equals

$$\begin{aligned} &t(x_1x_2x_3) + t(x_2x_1x_3) - t(x_1)t(x_2x_3) \\ &\qquad - t(x_2)t(x_1x_3) - t(x_1x_2)t(x_3) + t(x_1)t(x_2)t(x_3). \end{aligned}$$

Observe that when we restitute $T_{n+1}$, we get

$$t(\mathrm{CH}_n(x)x) = t(x^{n+1}) - t(x)t(x^n) + \cdots + (-1)^n \det(x)t(x) = \frac{T_{n+1}(x, x, \ldots, x)}{(n+1)!}.$$

The vanishing of this expression for $n \times n$ matrices is precisely the identity which expresses the $n+1$ power sum $t(x^{n+1})$ of $n$ indeterminates as a polynomial in the lower power sums, e.g., $n = 2$,

$$t(x\,\mathrm{CH}_2(x)) = t(x^3) - t(x)t(x^2) + \frac{1}{2}(t(x)^2 - t(x^2))t(x).$$

On $2 \times 2$ matrices $\mathrm{tr}(x\,\mathrm{CH}_2(x)) = 0$ implies

$$t(x^3) = \frac{3}{2}t(x)t(x^2) - \frac{1}{2}t(x)^3, \tag{12.19}$$

which expresses the fact that calling $a_1, a_2$ the two eigenvalues of a $2 \times 2$ matrix

$$a_1^3 + a_2^3 = \frac{3}{2}(a_1 + a_2)(a_1^2 + a_2^2) - \frac{1}{2}(a_1 + a_2)^3$$

an identity of symmetric functions.

COROLLARY 12.1.16. *In an algebra, over* $\mathbb{Q}$, *with trace satisfying* $\mathrm{CH}_n$ *for every element* $x$, *we have* $\sigma_{n+1}(x) = 0$.

PROOF. In the algebra of symmetric functions over $\mathbb{Q}$ imposing $e_{n+1} = 0$ is equivalent to imposing formula (3.15) (see examples (3.17)), which expresses the $n+1$ power sum $\psi_{n+1}$ of $n$ indeterminates as a polynomial in the lower power sums.

Similarly for matrix invariants imposing $\sigma_{n+1} = 0$ is equivalent to imposing the formula of $\mathrm{tr}(X^{n+1})$ in terms of $\mathrm{tr}(X^i)$, $i \leq n$ of $n \times n$ matrices $X$. □

Before we pass to the general case, we should remark that any substitution map in $\mathcal{T}_X\langle X\rangle$ sends the trace ring $\mathcal{T}_X$ into itself. Thus it also makes sense to speak of a $T$-ideal in $\mathcal{T}_X$. In particular we have the $T$-ideal of trace relations, kernel of the evaluation map of Tr into $\mathcal{T}_n(X)$. We wish to prove

THEOREM 12.1.17 (SFT).

(i) *The T-ideal of trace relations, kernel of the evaluation map of* $\mathcal{T}(X)$ *into* $\mathcal{T}_n(X)$ *is generated (as a T-ideal) by the trace relation* $T_{n+1}$.

(ii) *The T-ideal of (trace) identities of* $n \times n$ *matrices, in the free algebra without* 1, *is generated (as a T-ideal) by the Cayley–Hamilton identity.*

*In the free algebra with a* 1 *we have to add the identity* $t(1) = n$.

(iii) *The algebra* $R_X(n)$ *of Definition* 12.1.9 *is thus the free algebra with trace modulo the Cayley–Hamilton identity and* $t(1) = n$.

Before we prove this theorem, let us show how from this one can deduce a theorem of Ziplies **[Zip87]**, **[Zip89]** and Vaccarino **[Vac08]**, **[Vac09]**. We turn to the notations of §4.2.2, $\mathbb{Q}\langle X\rangle$ is the free algebra in the variables $X$.

We construct the algebra $S^n(\mathbb{Q}\langle X\rangle)$ and take its maximal abelian quotient, that is the quotient modulo the ideal generated by commutators. We call this algebra $A^n(\mathbb{Q}\langle X\rangle)$.

We have seen in Theorem 4.2.12 that $S^n(\mathbb{Z}\langle X\rangle)$ (hence also $S^n(\mathbb{Q}\langle X\rangle)$) is generated by the elements $\sigma_q(M)$, where $M$ runs over primitive monomials while

$S^n(\mathbb{Q}\langle X\rangle)$ is generated by the elements $\sigma_1(M)$ where $M$ runs over all monomials. By Proposition 4.2.16 in $A^n(\mathbb{Z}\langle X\rangle)$ we have $\sigma_q(AB) = \sigma_q(BA)$ for all $q$ and all pairs of monomials $A, B$. Moreover in $A^n(\mathbb{Q}\langle X\rangle)$ we have $\sigma_{n+1}(A) = 0$ where $\sigma_{n+1}(A)$ is computed as usual through the elements $\sigma_1(A^i)$. These relations are indeed by Theorem 12.1.17 the defining relations of the trace algebra $\mathcal{T}_n(X)$ over $X$. Consider now the homomorphism $\eta : \mathbb{Q}\langle X\rangle \to \mathbb{Q}\langle\xi\rangle$ to the algebra of generic matrices, $\eta : x_i \to \xi_i$, followed by the determinant

$$\mathrm{Det} : \mathbb{Q}\langle X\rangle \xrightarrow[\eta]{} \mathbb{Q}\langle\xi\rangle \xrightarrow[\det]{} \mathcal{T}_n(X) \tag{12.20}$$

The map Det is a multiplicative polynomial map of degree $n$, hence, by Proposition 4.1.11, it factors to a map

$$\begin{array}{ccc} \mathbb{Q}\langle X\rangle & \xrightarrow{\ \eta\ } & \mathbb{Q}\langle\xi\rangle \\ \downarrow & & \downarrow \det \\ A^n(\mathbb{Q}\langle X\rangle) & \xrightarrow{\ \mathcal{D}\ } & \mathcal{T}_n(X) \end{array}$$

and we have

COROLLARY 12.1.18 (Ziplies and Vaccarino). *The map $\mathcal{D}$ is an isomorphism.*

PROOF. By the FFT and Theorem 4.2.12 the map $\mathcal{D}$ is surjective. On the other hand by Theorems 12.1.17 and Proposition 4.2.16 the defining relations of the algebra $\mathcal{T}_n(X)$ are contained in the defining relations of the algebra $A^n(\mathbb{Q}\langle X\rangle)$, thus the relations must be equal, and this proves the claim. □

PROOF OF THEOREM 12.1.17. From all the remarks made, it is sufficient to prove that a multilinear trace relation (resp., identity) (in the first $m$ variables) is in the $T$-ideal generated by $T_{n+1}$ (resp., $A_n$).

Let us first look at trace relations. By the description of trace identities, it is enough to look at the relations of the type $T(\tau(\sum_{\sigma\in S_{n+1}} \epsilon_\sigma\sigma)\gamma)$, $\tau, \gamma \in S_{m+1}$.

We write such a relation as $T(\tau(\sum_{\sigma\in S_{n+1}} \epsilon_\sigma\sigma))\gamma\tau\tau^{-1})$. We have seen that conjugation corresponds to permutation of variables, an operation allowed in the $T$-ideal, thus we can assume that $\tau = 1$.

We start with the following remark ($m \geq n$).

*Splitting the cycles.* Every permutation $\gamma$ in $S_{m+1}$ can be written as a product $\gamma = \alpha \circ \beta$ where $\alpha \in S_{n+1}$ and, in each cycle of $\beta$, there is at most one element in $1, 2, \ldots, n+1$.

This is a property of permutations based on the observation already made in Remark 12.3.11 that for $a, b, c, \ldots$ numbers and $A, B, C, \ldots$ strings of numbers, so that the followings are permutations, we have

$$(a\ A\ b\ B\ c\ C\ \cdots\ e\ E) = (a\ b\ c \cdots) \circ (A\ a)(B\ b)(C\ c) \cdots (E\ e). \tag{12.21}$$

$$\text{e.g., } A = 7, 8, 4;\ B = 6, 3:\ a = 1,\ b = 2,$$

$$(1, 7, 8, 4, 2, 6, 3) = (1, 2) \circ (7, 8, 4, 1)(6, 3, 2).$$

From the remark it follows that, since $\alpha \in S_{n+1}$,

$$(\sum_{\sigma\in S_{n+1}} \epsilon_\sigma\sigma) \circ \gamma = (\sum_{\sigma\in S_{n+1}} \epsilon_\sigma\sigma) \circ \alpha \circ \beta = \pm(\sum_{\sigma\in S_{n+1}} \epsilon_\sigma\sigma) \circ \beta,$$

we can assume that $\gamma$ has the property that in each cycle of $\gamma$ there is at most one element in $1, 2, \ldots, n+1$.

Assume that $\gamma$ satisfies the previous property, and let $\sigma \in S_{n+1}$. The cycle decomposition of the permutation $\sigma \circ \gamma$ is obtained by formally substituting in the cycles of $\sigma$, to every element $a \in [1, \ldots, n+1]$ the cycle of $\gamma$ containing $a$ as a word (written formally with $a$ at the end), and then adding all the cycles of $\gamma$ not containing elements $\leq n+1$.

We interpret this operation in terms of the corresponding trace elements $T_{\sigma\gamma}(x_1, \ldots, x_{m+1})$ and $T_\sigma(x_1, \ldots, x_{n+1})$. The resulting element $T_{\sigma\gamma}(x_1, \ldots, x_{m+1})$ is the product of a trace element corresponding to the cycles of $\gamma$ not containing elements $\leq n+1$ and the element obtained by substituting in $T_\sigma(x_1, \ldots, x_{n+1})$, for the variables $x_i$, the monomials $M_i x_i = x_{j_1} x_{j_2} \cdots x_{j_k} x_i$ read off from the cycle $(j_1, j_2, \ldots, j_k, i)$ containing $i$ in the cycle decomposition of $\gamma$, as in formula (12.21). As a result we have proved that a trace relation is in the $T$-ideal generated by $T_{n+1}$.

Let us pass now to trace identities. First we remark that by the definition of ideal in a trace ring, the relation $T_{n+1}$ is a consequence of the Cayley–Hamilton identity.

A trace polynomial $f(x_1, x_2, \ldots, x_m)$ is a trace identity for matrices if and only if $t(f(x_1, x_2, \ldots, x_m)x_{m+1})$ is a trace relation. Thus from the previous argument we have that $t(f(x_1, x_2, \ldots, x_m)x_{m+1})$ is a linear combination of elements of type

$$X\, T_{n+1}(M_1, \ldots, M_{n+1}) = X\, t(A_n(M_1, \ldots, M_n)M_{n+1}),$$

where $X$ is some trace expression multilinear in the variables not appearing in $t(A_n(M_1, \ldots, M_n)M_{n+1})$.

Now we have to consider two cases. The variable $x_{m+1}$ appears in $X$ or in one of the $M_i$. In the first case we have $X = t(Bx_{m+1})$, with $B$ some trace polynomial, hence

$$X\, T_{n+1}(M_1, \ldots, M_{n+1}) = t(T_{n+1}(M_1, \ldots, M_{n+1})Bx_{m+1}),$$

and $T_{n+1}(M_1, \ldots, M_{n+1})B$ is a consequence of the Cayley–Hamilton identity.

In the second, due to the antisymmetry of $T_{n+1}$, we can assume that $x_{m+1}$ appears in $M_{n+1} = Bx_{m+1}C$. Hence

$$\begin{aligned} X\, t(A_n(M_1, \ldots, M_n)M_{n+1}) &= X\, t(A_n(M_1, \ldots, M_n)Bx_{m+1}C) \\ &= t(CXA_n(M_1, \ldots, M_n)Bx_{m+1}). \end{aligned}$$

$CXA_n(M_1, \ldots, M_n)B$ is also clearly a consequence of the Cayley–Hamilton identity. □

REMARK 12.1.19. The proof works with the free algebra without 1, or (as we leave to the reader) also for the free algebra with a 1 when we normalize $t(1) = n$.

If we leave $t(1)$ indeterminate and then impose the Cayley–Hamilton identity, we shall see in Proposition 12.2.9 that we actually obtain the direct sum of the relatively free algebras $R(i)$, $i \leq n$.

**12.1.4. Cocharacters of invariants and trace maps.** Both spaces $\mathcal{T}_n(m)$ of multilinear invariants of $m$, $n \times n$ matrices and $T_n(m)$ of multilinear maps of $m$ $n \times n$ matrices to matrices equivariant under simultaneous conjugation are representations of the group $S_m$, and so we may speak of their character.

Moreover we have identified $T_n(m)$ with $\mathcal{T}_n(m+1)$, and the identification is compatible with the action of $S_m \subset S_{m+1}$, hence the character of $T_n(m)$ is obtained by branching the character of $\mathcal{T}_n(m+1)$ from $S_{m+1}$ to $S_m$.

Finally the character of $\mathcal{T}_n(m)$ is, by the first and second fundamental theorems, that of the quotient of the group algebra $F[S_m]$, considered as representation of $S_m$ under conjugation, modulo the ideal $I_{n+1}$. In this sense we may also refer to these characters as the *cocharacters* of invariants or equivariant maps of matrices. Thus the cocharacter of invariants of matrices is given, as representation of $S_m$, by the formula, taking $W$ a vector space of dimension $n$:

$$\text{End}_{\text{GL}(W)}(W^{\otimes m}) = \bigoplus_{\lambda \vdash m,\ \text{ht}(\lambda) \leq n} M_\lambda \otimes M_\lambda. \tag{12.22}$$

Of course the structure of the Kroneker product $M_\lambda \otimes M_\lambda$, as representation of $S_m$, is quite complex. Nevertheless one has the following

PROPOSITION 12.1.20. *The representation $M_\mu$ of $S_m$ which appears in either $\mathcal{T}_n(m)$ or $T_n(m)$ have height $\leq n^2$.*

PROOF. By the branching rules it is enough to do this for $\mathcal{T}_n(m)$. In this case this follows immediately from Corollary 6.3.24 applied to $V^* = \text{End}(W) = M_n(F)$. □

## 12.2. Representations of algebras with trace

**12.2.1. The universal map.** We can now extend Definition 3.4.1. If $R$ is an algebra with trace, by an $n$-dimensional representation over a commutative ring $B$ one means a homomorphism $\phi : R \to M_n(B)$ which is compatible with traces, where matrices have the standard trace. The discussion of §3.4.1 can be repeated verbatim. Assume $B$ is an algebra over a field $F$ of characteristic 0. We have again a representable functor, a universal (trace preserving) map $R \xrightarrow{i_R} M_n(B_n(R))$, an action of $\text{GL}(n, F)$ on $B_n(R)$, and a map

$$R \xrightarrow{i_R} M_n(B_n(R))^{\text{GL}(n,F)}.$$

We shall call $B_n(R)$ the *coordinate ring of the n-dimensional representations of R*, and the map $i_R$ is the *generic n-dimensional representation of R*. We will use the notation $B_n(R)$ to emphasise that now we work with algebras with trace.

We shall see (in Corollary 14.2.2 of Theorem 14.2.1) that this map $i_R$ is *always surjective*, and its kernel is formed by the *obvious elements*, the evaluations $\chi_a^n(a)$ of the Cayley–Hamilton identity (12.23).

Observe that if $R$ has a 1, then if $i_R$ is nonzero, then the subring $\mathbb{Q}[\text{tr}(1)]$ has a homomorphism to $\mathbb{Q}$ mapping $\text{tr}(1)$ to $n$.

**12.2.2. Cayley–Hamilton algebras.** The basic algebraic restriction which we know for the algebra of $n \times n$ matrices over a commutative ring $B$ is the Cayley–Hamilton theorem:

Every matrix $M$ satisfies its characteristic polynomial $\chi_M(\lambda) := \det(\lambda - M)$.

The main remark that allows us to pass to the formal theory is Theorem 3.2.4, which in characteristic 0 gives universal polynomials $F_i(t_1, \ldots, t_i)$ with rational coefficients, given by formula (3.15) and such that Corollary 3.3.3 and formula (3.20):

$$\chi_M(\lambda) = \lambda^n + \sum_{i=1}^{n} (-1)^i F_i(\mathrm{tr}(M), \dots, \mathrm{tr}(M^i))\lambda^{n-i}.$$

At this point we can formally define, in an algebra with trace $R$ over $\mathbb{Q}$, for every element $a$ a formal $n$-characteristic polynomial,

$$\chi_a^n(\lambda) := \lambda^n + \sum_{i=1}^{n} (-1)^i F_i(t(a), \dots, t(a^i))\lambda^{n-i}. \tag{12.23}$$

With this definition the usual Cayley–Hamilton theorem for $n \times n$ matrices tells us that, given any element $a$, the element $\chi_a^n(a)$ vanishes in every $n$-dimensional representation or equivalently it is in the kernel of the universal map. Thus we are led to make the following

DEFINITION 12.2.1. An algebra $R$ over $\mathbb{Q}$, equipped with a trace, is said to be an $n$-Cayley–Hamilton algebra, or to satisfy the $n$th Cayley–Hamilton identity if

$$\chi_a^n(a) = 0, \ \forall a \in R.$$

Equivalently $R$ satisfies the multilinear trace identity $\mathrm{CH}(x_1, \dots, x_n)$ of formula (12.16).

REMARK 12.2.2. There is a subtle point to be clarified. In this definition we need not assume that $R$ is an algebra with a 1; this is actually a relevant point for the theory since we may want to apply this also to nil algebras.

For instance if $R$ is an algebra in which every element $x$ satisfies $x^n = 0$, we think of it as a Cayley–Hamilton algebra in which the trace is identically 0.

Of course by the axioms all elements $t(a)$ are in the center and the *constant term* of the formal characteristic polynomial is in the trace algebra.

REMARK 12.2.3. Given a commutative $\mathbb{Q}$-algebra $A$, a subalgebra $R \subset M_n(A)$ closed under trace is clearly an $n$-Cayley–Hamilton algebra.

A rather remarkable fact, proved in Theorem 14.2.1, is that every $n$-Cayley–Hamilton algebra is of this form. In fact its universal map into $n \times n$ matrices is an embedding.

**Norms.**

The theory of algebras with trace is essentially a theory in characteristic 0. In general it is possible to start with a notion of a *norm* on a $A$-algebra.

DEFINITION 12.2.4. Given $n \in \mathbb{N}$, an *$n$-norm* in an $A$-algebra $R$ is a homogenenous multiplicative polynomial map, of degree $n$ to the center of $R$, satisfying the identity $N(N(a)b) = N(a)^n N(b)$.

Algebra with a norm of a given degree $n$ forms a category, and a free algebra can be defined by first considering the universal multiplicative map from the free algebra $\mathcal{F} := A\langle X\rangle$ to the abelianization of the symmetric tensors $\mathcal{A} := \left((A\langle X\rangle)^{\otimes n}\right)^{S_n}$ and then extending the coefficients taking the free algebra $\mathcal{F}_{\mathcal{A}} := \mathcal{A}\langle X\rangle$ with norm extended as polynomial map.

From a norm one can define a trace by formally constructing a *characteristic polynomial* $N(\lambda - a)$. Then one has the more general notion of Cayley–Hamilton algebra as

DEFINITION 12.2.5. An algebra with an $n$-norm $R$ is a Cayley–Hamilton algebra if each element $a \in R$ satisfies its characteristic polynomial.

We shall not treat this general case since at the moment the theory is not well developed.

Let us show that the definition of Cayley–Hamilton algebra using trace, in characteristic 0 is equivalent to the Definition 12.2.5 given using norms.

So let us consider an algebra with trace satisfying the hypothesis of Definition 12.2.1. We then define the norm of $R$ using the trace and the formula for the $n$th elementary symmetric function $\sigma_n$. We need to prove that $\sigma_n$ is a norm and that the characteristic polynomial defined by formula (12.23) equals that constructed from this norm. Since $R$ is a quotient of the free algebra with trace modulo the Cayley–Hamilton identity, it is enough to prove this statement for this latter algebra.

Now by Theorem 12.1.17, this algebra is the algebra of generic matrices with trace added, and in this algebra the trace corresponds to the usual trace of matrices in its universal embedding as well as the characteristic polynomial. Finally the norm is just the determinant by definition, hence the claim in one direction.

Conversely, assume we have an algebra with norm which is a Cayley–Hamilton algebra according to Definition 12.2.5. By Corollary 12.1.18 the formulas on the norm map are deduced from the formulas of the free algebra which give rise to matrix invariants, hence the expression of the trace obtained by polarizing the norm is related to the norm as the usual trace to the determinant and hence the two characteristic polynomials coincide.

REMARK 12.2.6. Consider the ring of polynomials $\mathbb{Q}[x]$ over $\mathbb{Q}$ equipped with $t : 1 \to x$. This can be thought of as the free algebra with trace over an empty set of variables.

Now impose the $n$th Cayley–Hamilton identity to the element 1. We obtain a polynomial $p_n(x)$. The quotient $\mathbb{Q}[x]/(p_n(x))$ can be thought of as the free $n$-Cayley–Hamilton algebra over an empty set of variables.

PROPOSITION 12.2.7. *We have that* $p_n(x) = (-1)^n (n!)^{-1} \prod_{i=1}^n (x - i)$. *Therefore* $p_n(i) = 0, \forall 1 \le i \le n$.

PROOF. Let us write the $n$ formal characteristic polynomial of formula (12.23) for 1 with $t(1) = x$ an indeterminate. We obtain a polynomial in $t$ and $x$. Setting $t = 1$, we obtain some polynomial $p_n(x)$ of degree $n$ in $x$. Our assumption is that in fact $t(1)$ is a root of this polynomial.

As example for $n = 2$, the polynomial $p_2(x)$ is

$$1 - x - 1/2x + 1/2x^2 = 1 - 3/2x + 1/2x^2 = \frac{1}{2}(x^2 - 3x + 2) = \frac{1}{2}(x-1)(x-2).$$

Now in the space of $n \times n$ matrices we can compute, for each $i = 1, \dots, n$, for $t$ an idempotent of rank $i > 0$ and hence of trace $i$. This idempotent satisfies the Cayley–Hamilton theorem and also $p_n(i)e = 0$, hence $p_n(i) = 0$.

The coefficient $(-1)^n (n!)^{-1}$ arises from restitution, since formula (12.16) is the polarized form of the Cayley–Hamilton identity. □

Observe that

$$\mathbb{Q}[x]/(p_n(x)) = \bigoplus_{i=1}^{n} \mathbb{Q}e_i,\ t(1) = \sum_{i=1}^{n} ie_i,\ e_i^2 = e_i. \tag{12.24}$$

COROLLARY 12.2.8. *If $S$ is an $n$-Cayley–Hamilton algebra with a 1, then*

$$S = \bigoplus_{i=1}^{n} S_i, \quad 1 = \sum_i e_i,\ e_i \in S_i,\ t(e_i) = ie_i.$$

It is then natural to normalize the trace and restrict to $t(1) = n$. The other cases correspond to embeddings of matrices into larger matrices which do not preserve 1.

It is clear that $n$-Cayley–Hamilton algebras form a category. This category has obviously free algebras. Denote by $\bar{R}_X(n)$ the relatively free algebra in the category of trace algebras with a 1 satisfying the $n$th Cayley–Hamilton identity (without imposing $\operatorname{tr}(1) = n$) and by $R_X(n)$ the one with $\operatorname{tr}(1) = n$.

We do this starting from $t(1) = x$, an indeterminate. So we see that the resulting algebra $\bar{R}_X(n)$ contains the subalgebra generated by $x$ modulo the corresponding identity which, by formula (12.24) over $\mathbb{Q}$, is just the direct sum on $n$ copies of $\mathbb{Q}$ in each copy the value of the trace is $i = 1, \dots, n$. So in a natural way each algebra in this category decomposes as direct sum of at most $n$ copies in each we have $t(1) = i$.

We then have

PROPOSITION 12.2.9.

(1) *The algebra $\bar{R}_X(n)$ satisfies the $m$th Cayley–Hamilton identity for all $m \geq n$.*
(2) *The algebra $\bar{R}_X(n) = \bigoplus_{i=1}^n R_X(i)$ is the direct sum of the corresponding relatively free algebras $R_X(i)$, where $t(1)$ is fixed to be equal to $i = 1, \dots, n$.*

PROOF. (1) Observe that from formula (12.15) one has

$$\mathrm{CH}_{n+1}(x) = \mathrm{CH}_n(x)x + (-1)^{n+1}\sigma_{n+1}(x).$$

In an algebra satisfying $\mathrm{CH}_n(x)$, we have from Corollary 12.1.16 that $\sigma_{n+1}(x) = 0$, hence the claim.

(2) By formula (12.24) in $\bar{R}_X(n)$ the identity 1 is a sum of the central idempotents $e_i$ with $\operatorname{tr}(e_i) = i$, hence $\bar{R}_X(n) = \bigoplus_{i=1}^n \bar{R}_X(n)e_i$. Imposing the condition $\operatorname{tr}(1) = i$ is thus equivalent to projecting on the summand $\bar{R}_X(n)e_i$.

Thus, using (1), we need to prove that, in the free algebra with trace the $T$-ideal generated by the $n$-Cayley–Hamilton identity and by $\operatorname{tr}(1) = i$ contains also the $i$-Cayley–Hamilton identity.

Example $n = 2$, $i = 1$,

$$(X+1)^2 - (\operatorname{tr}(X)+1)(X+1) - 1/2\operatorname{tr}((X+1)^2) + 1/2\operatorname{tr}(X+1)^2 = 0$$
$$\implies 2X - \operatorname{tr}(X) - X - \operatorname{tr}(X) + \operatorname{tr}(X) = 0 \implies X = \operatorname{tr}(X)$$

Consider the map from $S_{n+1}$ to $S_n$ which takes the cycle decomposition of $\sigma$ and then drops $n$ from the cycle in which it appears. Then it replaces $n+1$ with $n$. Call $\bar{\sigma}$ the resulting permutation.

Using the terminology of formula (12.14)

$$A_n(x_1, \dots, x_n) := \sum_{\sigma \in S_{n+1}} \epsilon_\sigma \psi_\sigma(x_1, x_2, \dots, x_n),$$

consider $A_n$ as trace identity in the formal algebra with $\operatorname{tr}(1) = i < n$. Compute then the deduced identity

$$A_n(x_1, \dots, x_{n-1}, 1) := \sum_{\sigma \in S_{n+1}} \epsilon_\sigma \psi_\sigma(x_1, x_2, \dots, x_{n-1}, 1) = \bar{A}_n(x_1, \dots, x_{n-1})$$

$$\psi_\sigma(x_1, x_2, \dots, x_{n-1}, 1) = \bar{\psi}_\sigma(x_1, x_2, \dots, x_{n-1}),$$

where in fact

$$\bar{\psi}_\sigma(x_1, x_2, \dots, x_{n-1}) = \begin{cases} \psi_{\bar{\sigma}}(x_1, x_2, \dots, x_{n-1}) & \text{if } \sigma(n) \neq n, \\ i\, \psi_{\bar{\sigma}}(x_1, x_2, \dots, x_{n-1}) & \text{if } \sigma(n) = n, \end{cases}$$

since in the second case $\psi_\sigma(x_1, x_2, \dots, x_n)$ contains a factor $\operatorname{tr}(x_n)$ which is replaced by $\operatorname{tr}(1) = i$.

Given $\tau \in S_n$, we look at all the $\sigma \in S_{n+1}$ with $\bar{\sigma} = \tau$. These are obtained as follows. First substitute in the cycle where $n$ appears, $n$ with $n+1$. We obtain a permutation of $\tilde{\tau} \in S_{n+1}$ fixing $n$. Then the set of $\sigma \in S_{n+1}$ with $\bar{\sigma} = \tau$ consists of $\tilde{\tau}$, fixing $n$, and then $h_1 + h_2 + \cdots + h_k = n$ permutations if $\tau$ has $k$ cycles of lengths $h_i$ obtained by inserting $n$ in all possible ways in one of the cycles of $\tilde{\tau}$.

The contribution $\psi_\tau(x_1, x_2, \dots, x_{n-1})$ comes from all these permutations.

The contribution to $\psi_\tau(x_1, x_2, \dots, x_{n-1})$ coming by inserting $n$ in the cycles is $-n\epsilon_\tau$, since by inserting $n$ into a cycle, we create a permutation of opposite sign. The contribution instead obtained by adding the cycle $(n)$ is $i\epsilon_\tau$. We deduce that

$$\bar{A}_n(x_1, \dots, x_{n-1}) = (n - i)A_{n-1}(x_1, \dots, x_{n-1}).$$

If $i = n$, we have that $A_n(x_1, \dots, x_{n-1}, 1) = 0$, but if $i \neq n$, we have that the element $A_{n-1}(x_1, \dots, x_{n-1})$ is in the $T$-ideal generated by $A_n(x_1, \dots, x_n)$. By induction we finally have that $A_i(x_1, \dots, x_i)$ is also in this $T$-ideal, hence the algebra is $i$ Cayley–Hamilton. □

Under this further constraint now this algebra on the other hand is the one described by Theorem 12.1.17.

COROLLARY 12.2.10. *The algebras $R_m(n)$ of equivariant maps from $m$-tuples of matrices to matrices are free algebras in the category of Cayley–Hamilton algebras with $t(1) = n$.*

**Base change.** If $R$ is an $A$-algebra satisfying the $n$-Cayley–Hamilton identity and $B$ is a commutative $A$ algebra, then $R \otimes_A B$ acquires naturally a $B$-linear trace for which it is also an $n$-Cayley–Hamilton algebra.

12.2.2.1. *A complement on the Cayley–Hamilton algebra.* Assume that $A$ is an algebra with trace $t : A \to A$, and let $J$ be a nilpotent ideal of $A$ with $j^s = 0$, $\forall j \in J$ and such that $t(j) = 0$, $\forall j \in J$.

Then the map $t$ factors to a map that, by abuse of notation, we still call $t : A/J \to A/J$ and $A/J$ with $t$ is also an algebra with trace. Assume further that for some integer $n$ the algebra $A/J, t$ is an $n$-Cayley–Hamilton algebra. Then

LEMMA 12.2.11. *Under the previous hypotheses, $A, \tau$ is an $s \cdot n$-Cayley–Hamilton algebra for the new trace $\tau(a) := s \cdot t(a)$ and $\chi_{sn,a}(t) = \chi_{n,a}(t)^s$.*

PROOF. The main point is to prove that $\chi_{sn,a}(t) = \chi_{n,a}(t)^s$, since by assumption $A/I, t$ is an $n$-Cayley–Hamilton algebra, so $\chi_{n,a}(a) \in J$, $\forall a \in A$, and $t(1) = n$. Then by the assumption on $J$ we have that $\chi_{n,a}(a)^s = 0$ and $\tau(1) = sn$.

So the question is why the polynomial $\chi_{n,a}(t)^s$ equals $\chi_{sn,a}(t)$ for the new trace. This is seen as follows. Since $\chi_{n,a}(a) \in J$, we have $t(a^i\chi_{n,a}(a)) = 0$, hence $t(a^i)$, for $i > n$ is expressed as polynomial in the elements $t(a^j)$, $j \leq n$ by the same formulas as the Newton function $\psi_j(x_1, \dots, x_n)$, $j > n$ is expressed via the Newton function $\psi_j(x_1, \dots, x_n)$, $j \leq n$. Formally we can add $n$ *roots* by formally adding elements $x_i(a)$ which satisfy $\chi_{n,a}(t) = \prod_{i=1}^n (t - x_i(a))$. Then $\chi_{n,a}(t)^s = \prod_{i=1}^n (t - x_i(a))^s$ so the coefficients of $\chi_{n,a}(t)^s$ are expressed as elementary symmetric functions in the elements $x_i(a)$ in which each element is repeated $s$ times. Now the Newton functions $\tau(a^i)$ correspond to $\sum_{i=1}^n s x_i(a)^j = st(a^i)$, and the claim follows. □

In particular let $A$ be finite dimensional over some field $K$, and let $J$ be its radical, $n := \dim_K A/J$, and finally $j^s = 0$, $\forall j \in J$. For an element $a \in A$, let $L_a$ be the left multiplication map $x \mapsto ax$, and let $\bar{L}_a$ be the map induced on $A/J$. Finally define $t(a) := \operatorname{tr}(\bar{L}_a)$. Then the hypotheses of Lemma 12.2.11 are satisfied, and hence $A$ is an $sn$-Cayley–Hamilton algebra with respect to the trace $\tau(a) := s \operatorname{tr}(\bar{L}_a)$.

We shall return to this theory in Chapter 14, §14.1, after developing some features of geometric invariant theory.

**12.2.3. Polynomial identities.** We want to discuss the application of trace identities to polynomial identities of matrices. A polynomial identity is a special type of trace identity in which no traces appear. In fact these identities have been studied before the trace identities although they turn out to be harder to describe. Under the map $\Psi : \mathbb{C}[S_{m+1}] \to F_m$ of formula (2.18) ($F_m$ the free algebra with trace) the elements that correspond to ordinary polynomials without traces are the full cycles $(i_1, i_2, \dots, i_m, m+1)$ and

$$\Psi((i_1, i_2, \dots, i_m, m+1)) = x_{i_1} x_{i_2} \cdots x_{i_m}.$$

Let us define $P_{m+1}$ to be the subspace of $\mathbb{C}[S_{m+1}]$ spanned by the full cycles. Of course we have a canonical identification of $P_{m+1}$ with $\mathbb{C}[S_m]$. Notice that in this identification the left action of $S_m$ on $\mathbb{C}[S_m]$ coincides with the conjugation action of $S_m$ on $P_{m+1}$. Let $I_{n+1}(m+1)$ be the two-sided ideal of $\mathbb{C}[S_{m+1}]$ generated by an antisymmetrizer $\sum_{\pi \in S_{n+1}} \epsilon_\pi \pi$ on $n+1$ elements.

By Theorem 12.1.11 and formula (12.11) this ideal of $\mathbb{C}[S_{m+1}]$ is identified with the space of multilinear trace identities of $n \times n$ matrices, in $m$ variables.

The ideal $I_{n+1}(m+1)$ is the sum of all blocks corresponding to partitions $\lambda$ with height $\geq n+1$.

THEOREM 12.2.12. *The space $M_n(m)$ of multilinear polynomial identities of $n \times n$ matrices of degree $m$ can be identified, under the map $\Psi$, with the space*

$$P_{m+1} \cap I_{n+1}(m+1).$$

PROOF. This is tautological: a polynomial identity is a special case of trace identity. That is, it is a trace identity which lies in $\Psi(P_{m+1})$. □

A more precise description of this space is not known. Of its dimension is known only an asymptotic behaviour.

One simple remark is useful. Consider the automorphism $\zeta$ of the group algebra $\mathbb{C}[S_{m+1}]$ on the group elements by $\zeta(\sigma) := \epsilon_\sigma \sigma$. The map $\zeta$ is multiplication by $(-1)^m$ on $P_{m+1}$ and transforms $I_{n+1}(m+1)$ in the ideal $J_{n+1}(m+1)$ of $\mathbb{C}[S_{m+1}]$ generated by a symmetrizer on $n+1$ elements. It follows that

$$M_n(m) = \Psi(P_{m+1} \cap I_{n+1}(m+1) \cap J_{n+1}(m+1)) .$$

The ideal $I_{n+1}(m+1)$ is the sum of all blocks corresponding to partitions $\lambda$ with height $\geq n+1$ while $J_{n+1}(m+1)$ is the sum of all blocks corresponding to partitions $\lambda$ with first row $\geq n+1$. Thus $I_{n+1}(m+1) \cap J_{n+1}(m+1)$ is the sum of blocks relative to partitions which contain the hook $(n+1, 1^n)$. It follows that $M_n(m) = 0$ if $m < 2n$, which we proved directly in Theorem 10.1.3. As for degree $2n$, $I_{n+1}(2n+1) \cap J_{n+1}(2n+1)$ is the block relative to the hook $(n+1, 1^n)$.

We have the Amitsur–Levitzki identity, and one can prove directly that the space $P_{2n+1} \cap I_{n+1}(2n+1) \cap J_{n+1}(2n+1)$ is one dimensional.

Let us show how the Amitsur–Levitzki identity can be deduced from the Cayley–Hamilton identity. Consider the trace identity

$$\sum_{\sigma \in S_{2n}} \epsilon_\sigma \operatorname{CH}(x_{\sigma(1)}x_{\sigma(2)}, x_{\sigma(3)}x_{\sigma(4)}, \dots, x_{\sigma(2n-1)}x_{\sigma(2n)}).$$

The terms of the Cayley–Hamilton identity which do not contain trace expressions give rise to a multiple (by $n!$) of the Amitsur–Levitzki identity $\operatorname{St}_{2n}(x_1, \dots, x_{2n})$, but the terms which contain trace expressions can be grouped so as to contain each a factor of some type $\operatorname{tr}(S_{2m}(x_1, \dots, x_{2m}))$, $m \leq n$. Now we have already seen, in the first proof of the Amitsur–Levitzki identity, that $\operatorname{tr}(S_{2m}(x_1, \dots, x_{2m})) = 0$ identically as a formal expression.

**12.2.4. Some estimates.** Both the first and second fundamental theorems for matrices are not as precise as the corresponding theorems for vectors and forms. Several things are lacking. The first is a description of a minimal set of generators for the trace invariants. At the moment the best result is given by

THEOREM 12.2.13 (Razmyslov). *The invariants of $n \times n$ matrices are generated by the elements* $\operatorname{tr}(M)$ *where $M$ is a monomial of degree $\leq n^2$.*

*The algebras $R_m(n)$ of equivariant maps from $m$-tuples of matrices to matrices is spanned as a module over the ring on invariants, by monomials of degree $< n^2$.*

PROOF. The precise question is, for which values of $m$ can we express the trace monomial $\operatorname{tr}(X_1X_2 \dots X_{m+1})$ as a product of shorter trace monomials? This in turn is equivalent to asking for which values of $m$, in the free Cayley–Hamilton algebra with trace $R(n)$, the monomial $X_1X_2 \cdots X_m$ lies in the ideal $T(n)^+R(n)$, where as usual $T(n)^+$ denotes the trace invariants without constant term.

Thus the second statement implies the first.

If $\mathcal{F}_T$ denotes the free algebra with trace and $\mathcal{T}$ the trace expressions, by the general correspondence, (12.11), between the symmetric group and the trace expressions, we have that the elements of $S_{m+1}$ which map to $\mathcal{T}^+\mathcal{F}_T$ are not full cycles.

Let $Q_{m+1}$ denote the subspace that they span in $\mathbb{C}[S_{m+1}]$. Let $P_{m+1}$ be the subspace of $\mathbb{C}[S_{m+1}]$ spanned by the full cycles so that $Q_{m+1} \oplus P_{m+1} = \mathbb{C}[S_{m+1}]$. By symmetry, the requirement that modulo the identities of $n \times n$ matrices we have $X_1X_2 \cdots X_m \in T(n)^+R(n)$ is equivalent to requiring that

$$P_{m+1} \subset Q_{m+1} + I_{n+1}(m+1).$$

In other words that $\mathbb{C}[S_{m+1}] = Q_{m+1} + I_{n+1}(m+1)$.

So let us prove that this equality is true for $m = n^2$. Consider the automorphism $\zeta(\sigma) = \epsilon_\sigma \sigma$. The map $\zeta$ preserves $Q_{m+1}$ and maps $I_{n+1}(m+1)$ to $J_{n+1}(m+1)$ the ideal generated by a symmetrizer on $n+1$ elements. Therefore $\zeta(Q_{m+1} + I_{n+1}(m+1)) = Q_{m+1} + J_{n+1}(m+1)$. On the other hand $\zeta = (-1)^m$ on $P_{m+1}$ so that $Q_{m+1} + I_{n+1}(m+1) = Q_{m+1} + I_{n+1}(m+1) + J_{n+1}(m+1)$.

Finally if $m \geq n^2$, every diagram with $m+1$ boxes belongs either to the ideal $I_{n+1}(m+1)$ or to $J_{n+1}(m+1)$ so that $S_{n^2+1} = I_{n+1}(m+1) + J_{n+1}(m+1)$, and the theorem is proved. □

We obtain again a proof with a more precise estimate of Theorem 8.2.5.

COROLLARY 12.2.14. *If $R$ is a finitely generated Cayley–Hamilton algebra over a field $F$, then $R$ is a finite module over its trace algebra $T$ which is a finitely generated algebra.*

**12.2.5. Free nil algebras.** It is interesting that these estimates are related to a theorem in noncommutative algebra known as the Dubnov–Ivanov–Nagata–Higman theorem, see **[DF04]** and **[For90]** for some history. This theorem in various levels of precision is the next statement on nil algebras, Theorem 12.2.16.

To explain the noncommutative algebra, we start with a definition:

DEFINITION 12.2.15. (1) An algebra $R$ (without 1) is said to be nil of exponent $n$ if for every $r \in R$ we have $r^n = 0$.

(2) An algebra $R$ (without 1) is said to be nilpotent of exponent $m$ if $R^m = 0$.

To say that $R$ is nilpotent of exponent $m$ means that every product $r_1 r_2 \cdots r_m = 0$, $r_i \in R$. Of course an algebra $R$ nilpotent of exponent $m$ is also nil of exponent $m$. The converse is not true.

THEOREM 12.2.16.

(i) *The algebra $R_X(n)^+/T_n(X)^+R_X(n)$ is the free nil algebra of exponent $n$ in the variables $X$.*

*In other words, it is the free algebra without* 1 *modulo the $T$-ideal generated by the polynomial identity $z^n = 0$.*

(ii) *The space of multilinear elements of degree $i$ of $R_X(n)^+/T_n(X)^+R_X(n)$ can be identified to the space $F[S_{i+1}]/(Q_{i+1} + I_{n+1}(i+1))$.*

(iii) *In characteristic* 0 *an algebra $R$, nil of exponent $n$, is nilpotent of exponent $\leq n^2$.*

PROOF. (i) In the free algebra with trace without 1 we impose the identity $t(z) = 0$, that is the trace of every element is 0. We obtain the free algebra without trace. Thus the free algebra without 1 modulo the polynomial identity $z^n = 0$ can be seen as the free algebra with trace without 1 modulo the two identities $t(z) = z^n = 0$. Now the Cayley–Hamilton identity plus the identity $t(z) = 0$ becomes $z^n = 0$. Hence the free algebra without 1 modulo the polynomial identity $z^n = 0$ can be seen also as the free algebra with trace without 1 modulo the two trace identities $t(z) = 0$, $\mathrm{CH}_n(z) = 0$.

But clearly this last algebra is $R_X(n)^+/T_n(X)^+R_X(n)$.

(ii) This follows as in the analysis of (12.5) by the map $M \mapsto \mathrm{tr}(MX_{i+1})$ from the space of multilinear equivariant maps in $i$ variables to invariants in $i+1$ variables. Under this map the span of the multilinear elements of $T(n)^+R(n)^+$ corresponds to the span of the multilinear elements of $(T(n)^+)^2$.

(iii) To say that $X_1X_2\cdots X_m$ lies in the ideal $T(n)^+R(n)$ is the same as to say that $X_1X_2\cdots X_m = 0$ in the quotient $R(n)^+/T(n)^+R(n)$. Since this is the free algebra modulo $z^n = 0$, to say that $X_1X_2\cdots X_m = 0$ is equivalent to saying that $X_1X_2\cdots X_m = 0$ is an identity in $R(n)^+/T(n)^+R(n)$ or that $R(n)^+/T(n)^+R(n)$ is nilpotent of exponent $\leq m$.

If the free algebra modulo $z^n = 0$ is nilpotent of exponent $m$, then so is every nil algebra of exponent $n$, since it is a quotient of the free algebra. Finally, from Theorem 12.2.16 we know that $m = n^2$ satisfies the condition that $X_1X_2\cdots X_m$ lies in the ideal $T(n)^+R(n)$. □

REMARK. We have in fact proved, for fixed $n$ (over fields of characteristic 0) that for an integer $m$ the following conditions are equivalent:

(i) $\mathbb{C}[S_{m+1}] = Q_{m+1} + I_{n+1}(m+1)$.
(ii) The invariants of $n \times n$ matrices are generated by the elements $\mathrm{tr}(M)$ where $M$ is a monomial of degree $\leq m$.
(iii) An algebra $R$ nil of exponent $n$ is nilpotent of exponent $\leq m$.
(iv) Such an $m = m(n)$ exists and its minimum value is $\leq n^2$.

It is known by work of Kuzmin [**Kuz75**], that its minimum value is $\geq \binom{n+1}{2}$; see also [**DF04**, Chapter 6]. One may conjecture that it is exactly $\binom{n+1}{2}$. This is true by explicit computations for $n \leq 4$.

## 12.3. The alternating central polynomials

**12.3.1. The discriminant.** We fix a field $F$ and write $M_n = M_n(F)$, $I_n$ for the identity matrix and $G = \mathrm{GL}(n, F)$.

DEFINITION 12.3.1. Given $n^2$, $n \times n$ matrices $\xi_1, \ldots, \xi_{n^2}$, we call their *discriminant* the element $\xi_1 \wedge \cdots \wedge \xi_{n^2} \in \bigwedge^{n^2} M_n$.

If we fix a basis element $u$ of $\bigwedge^{n^2} M_n = Fu$, we have

$$\xi_1 \wedge \cdots \wedge \xi_{n^2} = \Delta(\xi_1, \ldots, \xi_{n^2})u, \tag{12.25}$$

where $\Delta(\xi)$ is a multilinear polynomial alternating in the $n^2$ variables still called the discriminant.

One way of doing this is to take the determinant of the $n^2 \times n^2$ matrix having as rows the matrices $\xi_i$ displayed as an $n^2$ vector.

In general if we consider the determinant $v_1 \wedge v_2 \wedge \cdots \wedge v_m$ of $m$ vectors in a vector space $W$ of dimension $m$, we have under a linear transformation $A : W \to W$ that $Av_1 \wedge Av_2 \wedge \cdots \wedge Av_m = \det(A) v_1 \wedge v_2 \wedge \cdots \wedge v_m$.

In particular $v_1 \wedge v_2 \wedge \cdots \wedge v_m$ is an invariant under the special linear group. In our case $W = M_n$ the group $G$ acting by conjugation is easily seen to be made of linear transformations of $M_n$ of determinant 1, so we have

LEMMA 12.3.2. *$\Delta(\xi)$ is an invariant under conjugation of matrices.*

Now this is part of a more general theory than is the analysis of the algebra, under exterior multiplication, $(\bigwedge M_n^*)^G$ of invariant alternating multilinear functions on $n \times n$ matrices for which we refer to [**BPŠ15**].

To this algebra belong the basic alternating multilinear invariant functions $T_i := \mathrm{tr}(\mathrm{St}_{2i-1}(X_1, \ldots, X_{2i-1}))$, $i = 1, \ldots, n$.

This algebra is the cohomology algebra of the Lie algebra of matrices, a fact which we will not use, the basic result is **[Kos58]**

THEOREM 12.3.3. *$(\bigwedge M_n^*)^G$ is a Hopf algebra and an exterior algebra in its primitive generators, which can be taken to be $T_i$, $i = 1, \ldots, n$.*

It is interesting to cast this theory, which classically belongs to Lie theory since it holds also for simple Lie algebras, in our language in which multilinear invariants of matrices are encoded by elements of the symmetric groups.

If $V$ is a permutation representation of the symmetric group $S_m$, the sign representation of $S_m$ appears in the space spanned by a given orbit if and only if the stabilizer of an element of the orbit is contained in the alternating group.

In particular for the conjugation action of $S_m$ on the group algebra the orbits are the conjugacy classes, indexed by partitions, and it is easily verified that the centralizer of a given permutation is in the alternating group if and only if the partition has all its parts odd and without repetitions. These are all even permutations and coincide with those even permutations whose conjugacy class in the symmetric group splits into two conjugacy classes under the alternating group.

Given one such conjugacy class, one may construct a corresponding antisymmetric element by choosing a representative $\sigma$ and antisymmetrizing it under the conjugation action that is constructing $\sum_{g \in S_n/C_\sigma} \epsilon_g g\sigma g^{-1}$ where $C_\sigma$ is the centralizer of $\sigma$ and we sum over coset representatives. This determines up to sign an antisymetric element associated to this conjugacy class.

Now we pass from the symmetric groups to the corresponding multilinear invariants.

In $S_m$ to the full cycle $\sigma = (1, 2, \ldots, m)$ corresponds $T_\sigma = \mathrm{tr}(X_1 X_2 \cdots X_m)$ and when we antisymmetrize this we have $\mathrm{tr}(\mathrm{St}_m(X_1 X_2 \cdots X_m))$ equal to 0 if $m$ is even. We have our generators of cohomology $T_i = \mathrm{tr}(\mathrm{St}_{2i-1}(X_1, X_2, \ldots, X_{2i-1}))$ if $m = 2i - 1$ is odd.

We define thus $t_i := \sum_{\sigma \in S_{2i-1}} \epsilon_\sigma(\sigma(1), \ldots, \sigma(2i-1)) \in F(S_{2i-1})$ to be the antisymmetrization of $(1, \ldots, 2i-1)$. Of course if we evaluate this formal invariant into $n \times n$ matrices, we know by the Amitsur–Levitzki theorem that it vanishes as soon as $i > n$. One could formalize the language of exterior multiplication in a symbolic way. Let us denote by $A_m$ the space of antiysmmetric elements, under the conjugation action, in $F(S_m)$. We define the exterior product $\bigwedge$ as usual, first we embed $F(S_m) \otimes F(S_p) \to F(S_{m+p})$ in the obvious way by the inclusion $S_m \times S_p \subset S_{m+p}$.

$$A_m \times A_p \to A_{m+p}, \quad a \wedge b := \frac{1}{m!p!} \sum_{g \in S_{m+p}} \epsilon_g g a \otimes b g^{-1}, \tag{12.26}$$

which can also be written in a characteristic free form as sum over coset representatives

$$a \wedge b := \sum_{g \in S_{m+p}/S_m \times S_p} \epsilon_g g a \otimes b g^{-1}.$$

THEOREM 12.3.4. *$\mathbb{A}_\infty := \bigoplus_{i=0}^\infty A_i$ under the previous exterior multiplication is the exterior algebra in the infinitely many generators $t_i$ of degree $2i - 1$.*

The proof is left to the reader.

We can now evaluate all the elements of $\mathbb{A}_\infty$ into matrix invariants of order $n$ by mapping $t_i \mapsto T_i = \mathrm{tr}(\mathrm{St}_{2i-1}(X_1, X_2, \ldots, X_{2i-1}))$, and we know, by the

Amitsur–Levitzki theorem, that all the elements $t_i, i > n$ vanish. The quotient of $\mathbb{A}_\infty$ modulo the ideal in these elements is the exterior algebra in the first $n$ generators $t_1, \dots, t_n$, and we have that

THEOREM 12.3.5. *The algebra, under exterior multiplication, of multilinear alternating invariants of $n \times n$ matrices is the exterior algebra in the generators*

$$T_i = \mathrm{tr}(\mathrm{St}_{2i-1}(X_1, X_2, \dots, X_{2i-1})),\ i = 1, \dots, n.$$

PROOF. We have already seen that this algebra is the quotient of the exterior algebra $\bigwedge(t_1, \dots, t_n)$. It is easy to verify that every proper ideal of $\bigwedge(t_1, \dots, t_n)$ must contain $t_1 \wedge t_2 \wedge \cdots \wedge t_n$, so it is enough to verify that $t_1 \wedge t_2 \wedge \cdots \wedge t_n$ maps to a nonzero invariant. If it maps to 0, we would have no multilinear alternating invariants in degree $n^2$. This is a contradiction since we know that the discriminant is such an invariant, thus $t_1 \wedge t_2 \wedge \cdots \wedge t_n$ maps to a nonzero multiple of the discriminant and in fact to the discriminant by our normalization. □

One way to trivialize $\bigwedge^{n^2} M_n$ is to choose a basis $a_1, \dots, a_{n^2}$ of $M_n$ and set $u := a_1 \wedge \cdots \wedge a_{n^2}$. Then $\Delta(\xi)$ is computed as the determinant of the matrix $A(\xi)$ in which we place the $\xi$ in their coordinates $\xi^i_{h,k}$ as rows or columns using the chosen ordered basis. When we change bases, we change the trivialization by the determinant of the base change. We have just proved that $T_1 \wedge T_2 \wedge \cdots \wedge T_n$ is a nonzero multiple of this determinant. The constant of the change of basis when we take as basis the matrix units can be computed, it is up to a sign **[For87]**

$$C_n := \frac{1!3!5! \cdots (2n-1)!}{1!2! \cdots (n-1)!}. \tag{12.27}$$

From now on we normalize $\Delta := T_1 \wedge T_2 \wedge \cdots \wedge T_n$.

**12.3.2. Alternating central polynomials.** Take a noncommutative polynomial $f(X_1, \dots, X_k)$ in $k$ sets of noncommutative variables $X_1, \dots, X_k$ with each $X_i = X_{i,1}, \dots, X_{i,n^2}$ a set of $n^2$ variables. Assume that $f$ is alternating in each of these sets if variables. We have

PROPOSITION 12.3.6. *$f(X_1, \dots, X_k)$ takes a scalar value $\prod_{i=1}^k \Delta(\xi_i)\alpha$ by evaluation of the $X_i$ in $n \times n$ matrices $\xi_i = \xi_{i,1}, \dots, \xi_{i,n^2}$.*

PROOF. By Lemma 10.2.9 $\prod_{i=1}^k \Delta(\xi_i)$ divides all the entries of $f(\xi_1, \dots, \xi_k)$ as a function on matrices, so $f(\xi_1, \dots, \xi_k) = \prod_{i=1}^k \Delta(\xi_i) U$ where $U$ is independent of the $\xi_i$ so it is just a scalar matrix. But it is also invariant under conjugation, so necessarily we have $U = \alpha I_n$. □

We need to investigate in which case it is $\alpha \neq 0$ so that we are in the presence of a central polynomial.

Let us remark that the first approach would be to consider

$$\sum_{\sigma \in S_{n^2},\ \tau \in S_{n^2}} \epsilon_\sigma \epsilon_\tau x_{\sigma(1)} y_{\tau(1)} x_{\sigma(2)} y_{\tau(2)} \cdots x_{\sigma(n^2)} y_{\tau(n^2)}.$$

In fact this is a PI. As proved by Giambruno and Sehgal **[GS89]**, setting

$$h_{2k}(x,y) := \sum_{\sigma\in S_k,\ \tau\in S_k} \epsilon_\sigma\epsilon_\tau x_{\sigma(1)}y_{\tau(1)}x_{\sigma(2)}y_{\tau(2)}\cdots x_{\sigma(k)}y_{\tau(k)},$$

we have

PROPOSITION 12.3.7. *$h_{2k}(x,y)$ is a PI of $n\times n$ matrices as soon as $k\geq 2n$.*

PROOF. Clearly $h_{2k}(x,y)$ is in the $T$-ideal of $h_{2h}(x,y)$ for $k\geq h$, so it is enough to show that $h_{2(n-1)}(x,y)$ is not a PI, while $h_{2n}(x,y)$ is a PI for $n\times n$ matrices. The first fact is similar to the proof of Theorem 10.1.3, and we omit it. For the second let $A_1,\dots,A_{2n},B_1,\dots,B_{2n}$ be $n\times n$ matrices. Set

$$X_i := \begin{vmatrix}0 & A_i\\ 0 & 0\end{vmatrix},\quad Y_i := \begin{vmatrix}0 & 0\\ B_i & 0\end{vmatrix},\quad i=1,\dots,2n. \tag{12.28}$$

We have that the standard identity $\mathrm{St}_{4n}(X_1,Y_1,\dots,X_{2n},Y_{2n})$ equals to 0.

Now, since $X_iX_j=Y_iY_j=0$ for all $i,j$, the only nonzero terms in this standard identity are

$$\sum_{\sigma\in S_{2n},\ \tau\in S_{2n}} \epsilon_\sigma\epsilon_\tau X_{\sigma(1)}Y_{\tau(1)}X_{\sigma(2)}Y_{\tau(2)}\cdots X_{\sigma(2n)}Y_{\tau(2n)}$$
$$+\sum_{\sigma\in S_{2n},\ \tau\in S_{2n}} \epsilon_\sigma\epsilon_\tau Y_{\sigma(1)}X_{\tau(1)}Y_{\sigma(2)}X_{\tau(2)}\cdots Y_{\sigma(2n)}X_{\tau(2n)}.$$

One easily verifies that this is the block matrix

$$0=\begin{vmatrix}h_{4n}(A,B) & 0\\ 0 & h_{4n}(B,A)\end{vmatrix}.$$

□

Thus the following question arises. Take a monomial linear in the $2n^2$ variables $x_1,\dots,x_{n^2},\ y_1,\dots,y_{n^2}$ and antisymmetrize it with respect to the two sets of variables $x,y$. Can we in this way obtain a nonidentity for $n\times n$ matrices, hence, by Proposition 12.3.7 a nonzero multiple of $\Delta(x)\Delta(y)$?

Clearly the fact that $h_{4n}$ is a PI implies that several of these monomials will produce PIs.

In particular we may assume that our monomial is a product

$$M := m_1(x)n_1(y)m_2(x)n_2(y)\cdots m_k(x)n_k(y) \tag{12.29}$$

with the $m_i(x),\ n_j(y)$ monomials just in $x$ (resp., in $y$) and all nontrivial (that is of degree $\neq 0$ except possibly $m_1$ and $n_k$).

PROBLEM 12.3.8. For which monomials $M$ of (12.29) above does $\mathrm{Alt}_x\,\mathrm{Alt}_y\,M\neq 0$?

There are only partial results to this question. The goal of this section is to show that, if $k=n$ and $\deg m_i=\deg n_i=2i-1$, then the resulting polynomial is not a PI. Then by a variation of the argument used to prove this by Giambruno and Valenti **[GV96]**, show that a necessary condition for $\mathrm{Alt}_x\,\mathrm{Alt}_y\,M\neq 0$ is that the two sequences of degrees of the monomials $m_i$ and $n_j$ be both derived, up to rearrangements, by splitting the sequence of degrees $2i-1$, $i=1,\dots,n$.

In general, take a vector space $V$ of dimension $m$, and consider the multilinear functions, in $mk$ vector variables $v_i\in V$, which are alternating in $k$ groups of $m$ variables. We see immediately that such functions are invariants under the special linear group $\mathrm{SL}(V)=\mathrm{SL}(m,F)$ of elements of determinant 1 in $\mathrm{GL}(V)$.

The $\mathrm{SL}(m, F)$ invariants of $h$ vector variables equals the ring generated by the determinants in any $m$ out of the vector variables; see **[Pro07]**.

It is customary to write

$$[i_1, \ldots, i_m] := \det(v_{i_1}, \ldots, v_{i_m}),$$

the determinant of these vectors. These elements can be used as projective coordinates for the Grassmann variety of $m$ spaces in $h$-dimensional space and are called the Plücker coordinates.

A product of several of these Plücker coordinates is displayed as a rectangular tableau with rows the symbols of the Plücker coordinates.

So they appear only in degree multiples of $m$ and in degree $rm$ they form the irreducible representation of the group $\mathrm{GL}(h, F)$ (which acts by linear combinations of the vector variables).

As shown by Hodge **[HP47]** **[Pro07]** this representation has as its basis the standard (semistandard) tableaux with $r$ rows and $m$ columns.

In this representation we have multilinear elements if and only if $h = km$ and $r = k$. Then the multilinear invariants form the corresponding irreducible representation of the group $S_{km}$, usually written as $M_{(n^2)^k}$ with $k$ columns of length $n^2$.[1]

So we have reached the following point. We have in the space spanned by all noncommutative monomials in $kn^2$ variables, the elements which are obtained by antisymmetrizing $k$ sets of $n^2$ variables, they span some subspace $M_{k,n^2}$ of the free algebra. As polynomials in matrix variables each of these elements produces a scalar multiple of the product of the discriminants which in the language we explained is a tableau in the irreducible representation of the symmetric group associated to the rectangle $k \times n^2$. This mapping is $S_{kn^2}$ equivariant, hence, if it is nonzero, it is surjective, its kernel is formed by the elements of $M^a_{k,h}$ which are polynomial identities. So the main question is to show when this map is nonzero. For $k = 1$ this map is clearly 0, since the only antisymmetric element is the standard identity in $n^2$ variables.

The fact that this map is nonzero for $k \geq 2$ follows from the existence of a remarkable element $F(X, Y)$ discovered by Regev (see (12.30)) and whose nonvanishing on matrices has been proven by a rather ingenious argument by Formanek. This answers the question in a positive way for all $k$ even, but then using the conductor property of this polynomial in Proposition 10.4.9 we shall show the same statement for all $k \geq 2$. So we shall now dedicate ourselves to the task of understanding this construction.

The definition is this. Decompose $n^2 = 1 + 3 + 5 + \cdots + (2n - 1)$ and accordingly decompose the variables $X, Y$ in the list, and construct the monomials $m_i(X), i = 1, \ldots, n$ and similarly $m_i(Y)$ as product in the given order of the given $2i - 1$ variables $X_i$ of the $i$th list, as for instance

$$m_1(X) = X_1, \quad m_2(X) = X_2X_3X_4, \quad m_3(X) = X_5X_6X_7X_8X_9, \ldots.$$

$$m_i(X) = X_{(i-1)^2+1} \cdots X_{i^2}.$$

[1]Unfortunately, in the theory of the symmetric group, antisymmetrization is performed on columns, while in the notation of Hodge it is displayed on rows.

We finally define

$$F(X,Y) := \mathrm{Alt}_X \mathrm{Alt}_Y(m_1(X)m_1(Y)m_2(X)m_2(Y)\cdots m_n(X)m_n(Y)), \tag{12.30}$$

where $\mathrm{Alt}_X$ is the operator of alternation in the variables $X$.

By Proposition 12.3.6 we have that $F(X,Y)$ is a scalar matrix, so we identify it with $n^{-1}\operatorname{tr}(F(X,Y))$ that is

$$\begin{aligned} n\,F(X,Y) &= \operatorname{tr}(\mathrm{Alt}_X \mathrm{Alt}_Y(m_1(X)m_1(Y)m_2(X)m_2(Y)\cdots m_n(X)m_n(Y))) \\ &= \mathrm{Alt}_X \mathrm{Alt}_Y \operatorname{tr}(m_1(X)m_1(Y)m_2(X)m_2(Y)\cdots m_n(X)m_n(Y)). \end{aligned} \tag{12.31}$$

It is then convenient to represent this invariant in the language of the symmetric group as previously explained.

**12.3.3. Formanek's theory.** This theory starts from the following idea. We fix $n \in \mathbb{N}$ and denote, for simplicity, by $C(m)$ the space of multilinear invariants of $n \times n$ matrices, over $\mathbb{Q}$, in $m$ variables, which in Definition 12.1.1 was denoted by $\mathcal{T}_n(m)$.

We know, by Lemma 12.1.2, that this space is a quotient of the group algebra $\mathbb{Q}[S_m]$ via the map which we have introduced in formula (12.3), with the notation $T_\sigma(X_1,\dots,X_m)$,

$$\begin{aligned} T_\sigma(X_1,X_2,\dots,X_m) &= \operatorname{tr}(\sigma^{-1}\circ X_1\otimes X_2\otimes\cdots\otimes X_m) \\ &= \operatorname{tr}(X_{i_1}X_{i_2}\dots X_{i_h})\dots\operatorname{tr}(X_{j_1}X_{j_2}\dots X_{j_\ell})\operatorname{tr}(X_{s_1}X_{s_2}\dots X_{s_t}), \end{aligned} \tag{12.32}$$

where the factors correspond to the cycles of $\sigma$.

When $m \leq n$, we know that the map from $\mathbb{Q}[S_m]$ to $C(m)$ is a linear isomorphism by Theorem 12.1.17.

We define a map, $\Gamma\colon C(n) \to C(n+n^2)$ setting, for $f(X_1,\dots,X_n) \in C(n)$,

$$\Gamma(f)(X_1,\dots,X_n,Y_1,\dots,Y_{n^2}) := \mathrm{Alt}_Y f(X_1m_1(Y),X_2m_2(Y),\dots,X_nm_n(Y)). \tag{12.33}$$

Then we know that this polynomial map is a multiple of the discriminant $\Delta(Y) = T_1(Y)\wedge\cdots\wedge T_n(Y)$, $T_i(Y) = \operatorname{tr}(\mathrm{St}_{2i-1}(Y))$, so there is another invariant polynomial $\Phi(f)(X_1,\dots,X_n) \in C(n)$ defined by

$$\Phi(f)(X_1,\dots,X_n)T_1(Y)\wedge\cdots\wedge T_n(Y) := \Gamma(f)(X_1,\dots,X_n,Y_1,\dots,Y_{n^2}). \tag{12.34}$$

DEFINITION 12.3.9. The map $\Phi : C(n) \to C(n)$ is called the *Formanek transform*.

Since the multilinear invariants of degree $n$ for $n \times n$ matrices have as basis the invariants $T_\sigma, \sigma \in S_n$, in order to understand the Formanek transform it is enough to do it on this basis.

Proposition 12.3.10 tells us that it is enough to compute $\Phi$ on the invariant $T_1$, and then we shall see that this is computed by solving the system of linear equations of formula (12.44). So that for instance from that system for $n = 2$, we get

$$\Phi(T_1) = \frac{1}{3}T_1 - \frac{1}{6}T_{(12)}, \quad \Phi(T_{(12)}) = \frac{1}{3}T_{(12)} - \frac{1}{6}T_1. \tag{12.35}$$

We identify the space $C(n)$ with the group algebra $\mathbb{Q}[S_n]$, and we may define left and right action by $S_n$ through $\rho T_\sigma = T_{\rho\sigma}$, $T_\sigma\rho = T_{\sigma\rho}$.

So we can think of $\Phi$ as a linear map of this group algebra. We will prove, in Theorem 12.3.15, that this is multiplication by $(\sum_{\rho \in S_n} n^{c(\rho)}\rho)^{-1}$ where $c(\rho)$ denotes the number of cycles in $\rho$. The final formula (12.49) will result.

The connection between the map $\Phi$ and the computation of $F(X,Y)$ is the following. By formula (12.33) if $\sigma_0 = (1,2,\dots,n)$ is the full cycle,

(12.36)
$$\Gamma(T_{\sigma_0})(X_1,\dots,X_n,Y_1,\dots,Y_{n^2}) := \mathrm{Alt}_Y \operatorname{tr}(X_1m_1(Y)X_2m_2(Y)\cdots X_nm_n(Y)).$$

By formulas (12.31) and (12.34) we have then

(12.37)
$$\begin{aligned} n\,F(X,Y) &= \mathrm{Alt}_X\,\Gamma(T_{\sigma_0})(m_1(X),\dots,m_n(X),Y_1,\dots,Y_{n^2}) \\ &= \mathrm{Alt}_X\,\Phi(T_{\sigma_0})(m_1(X),\dots,m_n(X))\Delta(Y). \end{aligned}$$

Therefore we need to prove that $\mathrm{Alt}_X\,\Phi(T_{\sigma_0})(m_1(X),\dots,m_n(X)) \neq 0$. In order to do this we need to understand the map $\Phi$, and this is fully explained in Theorem 12.3.15.

The first remarkable and quite hidden property of this transform is

PROPOSITION 12.3.10. *The Formanek transform is equivariant under the left $S_n$ action.*

PROOF. Take the permutation $\delta$ of the last $n^2$ elements $\{n+1,\dots,n+n^2\}$ decomposed into the $n$ cycles $c_i = (i^2-2(i-1)+n,\dots,i^2+n)$ of length $2i-1$.

One easily verifies that $T_{c_i}(Y) = \operatorname{tr}(m_i(Y))$ hence:

(12.38)
$$\Delta(Y) = T_1(Y)\wedge T_2(Y)\wedge\cdots\wedge T_n(Y) = \mathrm{Alt}_Y\,T_\delta(Y).$$

We enlarge each of the cycles $c_i$ to a cycle of length $2i$ which symbolically we write as

$$(i,c_i) = C_i := (i, i^2-2(i-1)+n,\dots,i^2+n) = c_i\circ(i,n+i^2).$$

Set $\tau$ to be the product of the cycles $C_i$.

REMARK 12.3.11. The permutation $\tau$ has the property, easily verified, that when we multiply $\sigma\tau$, $\sigma\in S_n$, the cycles of $\tau$ are *glued* by $\sigma$. A cycle $(i_1,\dots,i_k)$ of $\sigma$ produces the cycle $(i_1,c_{i_1},i_2,c_{i_2},\dots,i_k,c_{i_k})$.[2]

For instance, for $n=2$, we have $\tau = (1,3)(2,4,5,6)$, take $\sigma=(1,2)$, and we have

$$(1,2)(1,3)(2,4,5,6) = (1,3,2,4,5,6).$$

We then remark that, considering the embedding of $S_n \subset S_{n+n^2}$ and denoting by $Y_1 := X_{n+1},\dots,Y_{n^2} := X_{n+n^2}$ the variables relative to the last indices, we have, from formula (12.32) and the gluing rule, the *substitutional rule*,

(12.39)
$$T_\sigma(X_1m_1(Y),X_2m_2(Y),\dots,X_nm_n(Y)) = T_{\sigma\tau}(X_1,\dots,X_n,Y_1,\dots,Y_{n^2}).$$

Thus formula (12.33) becomes

(12.40)
$$\Gamma(T_\sigma)(X,Y) = \mathrm{Alt}_Y\,T_{\sigma\tau}(X,Y).$$

The action of an element $g\in S_{n^2}$ on the invariants $T_\mu$, $\mu\in S_{n+n^2}$, by permuting the variables $Y_1,\dots,Y_{n^2}$, gives, in this group formalism, $T_{g\mu g^{-1}}$, that is conjugation of $\mu$ by $g\in S_{n^2}\subset S_{n+n^2}$.

---

[2]This operation of gluing is the inverse of that of splitting, which we shall use in Theorem 12.1.17.

We apply formula (12.33), defining $\Gamma$, transformed into formula (12.40) by the rule (12.39). So that we have the formula for $\sigma \in S_n$,

$$\Gamma(T_\sigma)(X_1,\dots,X_n,Y_1,\dots,Y_{n^2}) = \sum_{g\in S_{n^2}} \epsilon_g T_{g\sigma\tau g^{-1}} = \sum_{g\in S_{n^2}} \epsilon_g T_{\sigma g\tau g^{-1}}, \tag{12.41}$$

since $S_n$ and $S_{n^2}$ in $S_{n+n^2}$ commute. We deduce, for $\rho \in S_n$ that

$$\Gamma(T_{\rho\sigma}) = \sum_{g\in S_{n^2}} \epsilon_g T_{\rho\sigma g\tau g^{-1}} = \rho\Gamma(T_\sigma), \tag{12.42}$$

as needed. □

Thus in order to understand the Formanek transform $\Phi(T_{\sigma_0})$ for the full cycle, one only needs to understand the transform of $T_1$. This is a quite complex problem. Before we do this, let us make a simple but important remark.

LEMMA 12.3.12.

*Set* $U_\sigma := \mathrm{Alt}_X\, T_\sigma(m_1(X), m_2(X),\dots,m_n(X))$. *Then,*

$$U_\sigma = \begin{cases} T_1(X)\wedge T_2(X)\wedge\cdots\wedge T_n(X) = \Delta(X), & \text{if } \sigma = 1,\\ 0 & \text{if } \sigma\neq 1.\end{cases} \tag{12.43}$$

PROOF. $T_\sigma(m_1(X), m_2(X),\dots,m_n(X)) = T_\gamma(X_1,\dots,X_{n^2})$ where the cycle decomposition of $\gamma$ is easily deduced. A cycle $(i_1,\dots,i_k)$ of $\sigma$ produces a cycle of length $2(\sum i_j) - k$ obtained by substituting each $i_j$ with the cycle of length $2i_j - 1$ given by $((i_j-1)^2+1,\dots,i_j^2)$.

Now $\mathrm{Alt}_X\, T_\gamma(X_1,\dots,X_{n^2})$ is the exterior product of the alternations of the $T_c$ where $c$ are the cycles of $\gamma$. So first of all it is 0 if one of the cycles is even; otherwise it is up to sign the wedge product of the elements $T_{h_i}(X)$, where the $h_i$ are the lengths of the cycles appearing in $\gamma$.

Since, by Theorem 12.3.5, $(\bigwedge M_n^*)^G$ is the exterior algebra in the generators $T_i(X)$, we have that $T_1(X)\wedge T_2(X)\wedge\cdots\wedge T_n(X)$ is the only nonzero product of these elements in degree $n^2$.

This means that $\mathrm{Alt}_X\, T_\gamma(X_1,\dots,X_{n^2}) = 0$ unless the cycles of $\gamma$ are exactly of sizes $1,3,5,7,\dots,2n-1$.

Now it is clear that the only permutation of $S_n$ for which the substitution $T_\sigma(m_1(X), m_2(X),\dots,m_n(X))$ has this cycle decomposition is the identity. □

COROLLARY 12.3.13.

$$n\,F(X,Y) = a_{\sigma_0^{-1}}\Delta(X)\Delta(Y),$$

*where $a_{\sigma_0^{-1}}$ is the coefficient of the full cycle $\sigma_0^{-1}$ in $\Phi(T_1)$.*

PROOF. From formula (12.37) we have that, with $\sigma_0 = (1,2,\dots,n)$,

$$n\,F(X,Y) = \mathrm{Alt}_X\,\Phi(T_{\sigma_0})(m_1(X),\dots,m_n(X))\Delta(Y).$$

Then, by formula (12.43) we deduce that

$$\mathrm{Alt}_X\,\Phi(T_{\sigma_0})(m_1(X),\dots,m_n(X)) = a\Delta(X),$$

where $a$ is the coefficient of the identity in $\Phi(T_{\sigma_0})$. By the equivariance property $\Phi(T_{\sigma_0}) = \sigma_0\Phi(T_1)$, so $a$ is the coefficient of ${\sigma_0}^{-1}$ in $\Phi(T_1)$. □

So we need finally to perform the computation of $a_{\sigma_0^{-1}}$ and prove that it is nonzero.

12.3.3.1. *The value of* $\Phi(T_1)$. Denote by $I$ the identity $n \times n$ matrix. For every permutation $\sigma \in S_n$, we have

LEMMA 12.3.14.

(1) $T_\sigma(I, I, \ldots, I) = n^{c(\sigma)}$ *where* $c(\sigma)$ *is the number of cycles in* $\sigma$.

(2) $\Phi(T_\sigma)(I, I, \ldots, I) = \begin{cases} 1 & \text{if } \sigma = 1, \\ 0 & \text{if } \sigma \neq 1. \end{cases}$

PROOF. (1) $T_\sigma$ is product of $c(\sigma)$ factors. Each evaluated at $I$ gives $\operatorname{tr}(I) = n$.

(2) By definition (formula (12.34)), $\Phi(T_\sigma)(X)\Delta(Y) = \Gamma(T_\sigma)(X, Y)$. It will be enough to compute $\Gamma(T_\sigma)(I, \ldots, I, Y)$.

Use the defining formula (12.34), as transformed by formula (12.41). We remark first the following quite general fact. Take any permutation $\gamma$ on $m$-elements, and consider $T_\gamma(X_1, \ldots, X_m)$ as function on $n \times n$ matrices. If we evaluate $X_m = I$ and define $\bar{\gamma}$ to be the permutation on $m-1$ elements obtained by dropping $m$ in the cycle in which it appears, we have that $T_\gamma(X_1, \ldots, X_{m-1}, I) = T_{\bar{\gamma}}(X_1, \ldots, X_{m-1})$ if $\gamma(m) \neq m$, and if $\gamma$ fixes $m$ it is $T_\gamma(X_1, \ldots, X_{m-1}, I) = nT_{\bar{\gamma}}(X_1, \ldots, X_{m-1})$.

In fact we have that setting in $\operatorname{Tr}(X_{i_1} \cdots X_{i_k} X_m)$, $X_m = I$, the resulting function equals $\operatorname{Tr}(X_{i_1} \cdots X_{i_k})$ if $k \geq 1$, while if $k = 0$, we have $\operatorname{tr}(X_m) \mapsto \operatorname{tr}(I) = n$.

Thus the evaluation $\Gamma(T_\sigma)(X, Y)$ for $X_i = I$, $i = 1, \ldots, n$, is by dropping the first $n$ indices in the element $\sum_{g \in S_{n^2}} \epsilon_g T_{\sigma g \tau g^{-1}}$ and following the previous rule.

Recall that $g \in S_{n^2}$ permutes the last $n^2$ indices so that $\sigma g \tau g^{-1} = g \sigma \tau g^{-1}$, and the operation of dropping the first $n$ indices commutes with conjugation by $g$, that is $\overline{\sigma g \tau g^{-1}} = g \overline{\sigma \tau} g^{-1}$. Clearly, since $g \in S_{n^2}$ the factors of type $\operatorname{Tr}(X_i)$ in the expression of $T_{\sigma g \tau g^{-1}}$ or of $T_{\sigma\tau}$ coincide so that we have finally

$$T_{\sigma g \tau g^{-1}}(I, \ldots, I, Y_1, \ldots, Y_{n^2}) = n^k T_{g \overline{\sigma \tau} g^{-1}}(Y_1, \ldots, Y_{n^2})$$

with $k$ the number of elements in $1, \ldots, n$ fixed by $\sigma\tau$.

It follows that

$$\Gamma(T_\sigma)(I, \ldots, I, Y) = n^k \operatorname{Alt}_Y T_{\overline{\sigma\tau}}(Y_1, \ldots, Y_{n^2}).$$

We know that this antisymmetrization will produce 0 unless $\overline{\sigma\tau}$ is conjugate to the permutation $\delta$, with cycles of all the odd lengths $\leq 2n-1$, which is obtained from $\tau$ by dropping the first $n$ indices.

By Remark 12.3.11, such a permutation is obtained by dropping the first $n$ indices in $\sigma\tau$ only if $\sigma\tau$ has exactly $n$ cycles, that is $\sigma = 1$.

In this case, since $\tau$ does not fix any of the elements $1, 2, \ldots, n$, we have $k = 0$ the power $n^k = n^0 = 1$. The proof is complete. □

Now consider $\Phi(T_1) = \sum_\sigma a_\sigma T_\sigma$, by Proposition 12.3.10 we have, for each $\rho$, $\Phi(T_\rho) = \sum_\sigma a_\sigma T_{\rho\sigma}$. Evaluating both sides for $X_i = I$, we deduce by Lemma 12.3.14 the following system of $n!$ linear equations in the $n!$ unknowns $a_\sigma$,

$$\Phi(T_1) = \sum_\sigma a_\sigma T_\sigma; \quad \sum_\sigma a_\sigma n^{c(\sigma)} = 1, \quad \sum_\sigma a_\sigma n^{c(\rho\sigma)} = 0, \ \forall \rho \neq 1. \tag{12.44}$$

Our final task is to compute $a_{\sigma_0^{-1}}$ from this system of linear equations and show that it is not 0.

12.3.3.2. *A character computation.* Since for any permutation $\rho$, the two permutations $\rho$, $\rho^{-1}$ have the same number of cycles, our first computation from formula (12.44) is

(12.45)

$$(\sum_\sigma a_\sigma\sigma)(\sum_\rho n^{c(\rho)}\rho) = (\sum_\sigma a_\sigma\sigma)(\sum_\rho n^{c(\rho)}\rho^{-1})$$
$$= \sum_{\sigma,\rho} a_\sigma n^{c(\rho)}\sigma\rho^{-1} \stackrel{\rho\mapsto\rho\sigma}{=} \sum_{\sigma,\rho} a_\sigma n^{c(\rho\sigma)}\rho^{-1} = \sum_\rho(\sum_\sigma a_\sigma n^{c(\rho\sigma)})\rho^{-1} \stackrel{by\ (12.44)}{=} 1.$$

Now $\sum_\rho n^{c(\rho)}\rho$ is an element of the center of $\mathbb{Q}[S_n]$, which is the algebra of functions on conjugacy classes; we have seen that it is the inverse of $\sum_\sigma a_\sigma\sigma$. We deduce the following

THEOREM 12.3.15.

(1) $\sum_\sigma a_\sigma\sigma = (\sum_\rho n^{c(\rho)}\rho)^{-1}$ *is a central unit.*
(2) *The Formanek transform* $\Phi$ *is invertible and it is also a right* $S_n$*-module map.*
(3) $\Phi(T_1)(X_1,\dots,X_n)$ *is a symmetric function.*

PROOF. We have just verified (1). As for (2) we use Proposition 12.3.10, and we have $\Phi(\rho) = \sum_\sigma a_\sigma\rho\sigma = \rho(\sum_\sigma a_\sigma\sigma) = (\sum_\sigma a_\sigma\sigma)\rho$.

As for (3) the permutation action on the variables corresponds to conjugation in $S_n$, so this follows from the fact that $(\sum_\sigma a_\sigma\sigma)$ is a class function. □

Denote by $\chi_\lambda$ the irreducible character of the symmetric group $S_n$ associated to the partition $\lambda\vdash n$ and $e_\lambda \in \mathbb{Q}[S_n]$ the corresponding central unit. These elements form a basis of orthogonal idempotents of the center of $\mathbb{Q}[S_n]$.

We can thus write $\sum_\rho n^{c(\rho)}\rho = \sum_{\lambda\vdash n} r_\lambda e_\lambda$, and we have

$$(12.46)\quad \text{(I)}\ \sum_\rho n^{c(\rho)}\rho := \sum_{\lambda\vdash n} r_\lambda e_\lambda, \qquad \text{(II)}\ \sum_\sigma a_\sigma\sigma = (\sum_\rho n^{c(\rho)}\rho)^{-1} = \sum_{\lambda\vdash n} r_\lambda^{-1}e_\lambda.$$

We now exploit the following fact

LEMMA 12.3.16. *The function* $P : \rho \mapsto n^{c(\rho)}$ *is the character of the permutation action of* $S_n$ *on* $V^{\otimes n}$, $\dim_{\mathbb{Q}}(V) = n$.

PROOF. This follows immediately from formula (12.2) and Lemma 12.3.14(1). □

By the Schur–Weyl duality we have $P = \sum_\lambda d_\lambda\chi_\lambda$ where $d_\lambda$ is the dimension of the corresponding irreducible representation of $\mathrm{GL}(V)$.

We can now write a formula for the coefficients $r_\lambda$. By formulas (1.15) and (1.16), we have

$$(12.47)\qquad \text{(I)}\ \ e_\lambda = \frac{\chi_\lambda(1)}{n!}\sum_\sigma \chi_\lambda(\sigma)\sigma, \quad \text{(II)}\ \ \chi_\lambda(e_\mu) = \begin{cases}\chi_\lambda(1) & \text{if } \lambda=\mu,\\ 0 & \text{if } \lambda\neq\mu.\end{cases}$$

Of course $\chi_\lambda(1) \stackrel{(6.21)}{:=} f_\lambda$ is the dimension of the irreducible representation of $S_n$ associated to $\lambda$. One has thus, from formulas (12.46)(I) and (12.47)(II),

$$r_\lambda = \frac{\chi_\lambda(\sum_\rho n^{c(\rho)}\rho)}{\chi_\lambda(1)} = \frac{\sum_\rho P(\rho)\chi_\lambda(\rho)}{\chi_\lambda(1)} = \frac{n!(\chi_\lambda, P)}{\chi_\lambda(1)} = \frac{n!\,d_\lambda}{\chi_\lambda(1)},$$

where $(\chi_\lambda, P)$ is the usual scalar product of characters for which the irreducible characters $\chi_\lambda$ form an orthonormal basis.

So, since $P = \sum_\lambda d_\lambda \chi_\lambda$, we have $(\chi_\lambda, P) = d_\lambda$. The number

$$d_\lambda = \prod_{1 \le i < j \le n} \frac{\lambda_i - \lambda_j + j - i}{j - i} \tag{12.48}$$

is given by Weyl's dimension formula (6.22) while $\chi_\lambda(1) = f_\lambda$ by the hook formula (6.25).

Substituting in formula (12.46) this and also formula (12.47)(I), we finally have the desired formula,

$$\sum_\sigma a_\sigma \sigma = \sum_{\lambda \vdash n} \frac{\chi_\lambda(1)^2}{n!^2 d_\lambda} \sum_\sigma \chi_\lambda(\sigma)\sigma \implies n!^2 a_\sigma = \sum_{\lambda \vdash n} \frac{\chi_\lambda(1)^2 \chi_\lambda(\sigma)}{d_\lambda}. \tag{12.49}$$

We finally need to compute, for $\sigma = \sigma_0$ a full cycle,

$$\sum_{\lambda \vdash n} \frac{\chi_\lambda(1)^2 \chi_\lambda(\sigma_0)}{d_\lambda},$$

and show that it is different from 0.

Recall the basic formula expressing characters through Schur and Newton functions, for a permutation $\sigma = \sigma_\mu$ associated to the partition of cycles $\mu$

$$\text{formula (6.17)}, \qquad \Psi_\mu(X) = \sum_{\lambda \vdash n} \chi_\lambda(\sigma_\mu) S_\lambda(X).$$

In general the numbers $\chi_\lambda(\sigma_\mu)$ can be recursively computed by the Murnagham–Nakayama rule, Theorem 6.4.16, but in our case, $\sigma_0$ a full cycle, so $\mu = n$ and $\Psi_\mu(X) = \psi_n(X)$.

We only need one step of that rule, which says then that $\chi_\lambda(\sigma_0) = 0$ except when $\lambda = k, 1^{n-k}$ is a *hook partition*, with the first row of some length $k$, $1 \le k \le n$ and then the remaining $n - k$ rows of length 1.

In this case we have $\chi_\lambda(\sigma_0) = (-1)^{n-k}$. We thus need to make explicit the integers $d_\lambda, \chi_\lambda(1)$ for such a hook partition.

Instead of Weyl's dimension formula (12.48) there is a more efficient formula, the *hook-content formula* to compute $d_\lambda$, the dimension of the Schur functor $S_\lambda(V)$. Let $s_\lambda(x_1, \dots, x_t)$ be the Schur function of $\lambda$ in $t$ variables. Then $s_\lambda(1, \dots, 1)$ is the number of semistandard tableaux of shape $\lambda$ with indices in $1, \dots, t$ and the dimension of $S_\lambda(W)$ where $\dim_{\mathbb{Q}} W = t$.

Let $\lambda \vdash n$. For $u = (i, j)$ a box of $\lambda$, the content of $u$ is defined as $c_u = j - i$. Also we denote by $h_u$ the hook number of $u$.

The hook-content formula (cf. Stanley **[Sta99**, Corollary 7.21.4]) is

$$s_\lambda(1^t) = \prod_{u \in \lambda} \frac{t + c_u}{h_u} \implies d_\lambda = \prod_{u \in \lambda} \frac{n + c_u}{h_u}.$$

For $\lambda = (k, 1^{n-k})$ and $t = n$, we get that the boxes are

$$u = (1, j),\ j = 1, \dots, k,\ c_u = j - 1,\ h_u = \begin{cases} n & \text{if } j = 1, \\ k - j + 1 & \text{if } j \neq 1, \end{cases}$$

on the first row and, plus $(i+1,1)$, $i = 1,\dots,n-k$ on the first column with $c_u = -i$, $h_u = n-k-i+1$. We deduce

$$\chi_{(k,1^{n-k})}(1) = \frac{n!}{n\prod_{j=2}^{k}(k-j+1)\prod_{i=1}^{n-k}(n-k-i+1)} = \binom{n-1}{n-k},$$

$$d_{(k,1^{n-k})} = \prod_{u\in\lambda}\frac{n+c_u}{h_u} = \frac{(n+k-1)\cdots(k+1)k}{n(k-1)!(n-k)!} = \binom{n+k-1}{n}\binom{n-1}{n-k}.$$

Thus we finally have, substituting in formula (12.49), $\chi_{(k,1^{n-k})}(\sigma_0) = (-1)^{n-k}$ and the previous formulas, and the elementary facts that

$$\begin{aligned}
\sum_{i=0}^{k}(-1)^i\binom{m}{i} &= 1+\sum_{i=1}^{k}(-1)^i\left[\binom{m-1}{i-1}+\binom{m-1}{i}\right]\\
&= (-1)^k\binom{m-1}{k}\binom{n+k-1}{n}^{-1}\binom{n-1}{n-k}\\
&= \frac{n!(n-1)!}{(2n-1)!}\frac{(2n-1)!}{(n+k-1)!(n-k)!}\\
&= \binom{2n-1}{n}^{-1}\binom{2n-1}{n-k},\\
n!^2 a_{\sigma_0} &= \sum_{k=1}^{n}(-1)^{n-k}\binom{n+k-1}{n}^{-1}\binom{n-1}{n-k}\\
&= \binom{2n-1}{n}^{-1}\sum_{i=0}^{n-1}(-1)^i\binom{2n-1}{i},
\end{aligned}$$

$$n!^2 a_{\sigma_0} = (-1)^{n+1}\binom{2n-1}{n}^{-1}\binom{2n-2}{n-1} = (-1)^{n+1}\frac{n}{2n-1} \neq 0. \tag{12.50}$$ □

Compare this with formula (12.35) for $n = 2$.

12.3.3.3. *Corollaries.* The central polynomials $F(X,Y)$ (see formula (12.31)) satisfy the property of Proposition 10.4.9. So they are in the conductor of the inclusion of the algebra of generic matrices into the trace algebra $F\langle\xi\rangle\mathcal{T}_n$. Here $F\langle\xi\rangle$ is the ring of generic matrices and $\mathcal{T}_n$ is the trace algebra, Definition 12.1.1 and Corollary 12.1.7, also denoted by $T_A$ in page 285. Hence they are also in the conductor of the inclusion of their centers. This shows that there is also a common ideal of central polynomials and invariants, and in fact

THEOREM 12.3.17. *The ideal $J = \mathcal{T}_n\Delta(X)\Delta(Y)$ of the trace algebra $\mathcal{T}_n$ of invariants, generated by the polynomial $\Delta(X)\Delta(Y)$, is contained in the center of generic matrices, so it is contained in the conductor of the inclusion of the center of generic matrices into the ring $\mathcal{T}_n$ of invariants.*

In other words all the elements of $J = \mathcal{T}_n\Delta(X)\Delta(Y)$ can be expressed as noncommutative central polynomials, cf. formula (10.9). This gives a remarkable property of the discriminant of a basis as defined in Definition 1.3.16. Over a field $F$ of characteristic 0,

THEOREM 12.3.18. *The discriminant $\det(\operatorname{tr}(u_iu_j))$ of $n^2$ elements $u_1,\dots,u_{n^2}$ in the algebra of generic matrices, which are linearly independent over the center, is a central polynomial in the conductor.*

PROOF. In fact we claim that, up to a nonzero scalar multiple $C$, we have that the determinant $\det(\operatorname{tr}(u_iu_j)) = C\Delta(u_1,\ldots,u_{n^2})^2$.

This can be either proved directly by noticing that the $n^2 \times n^2$ matrix with entries $\operatorname{tr}(u_iu_j)$ is $|u_1,\ldots,u_{n^2}||u_1,\ldots,u_{n^2}|^t$ or by polarizing the discriminant taking two sets of of $n^2$ matrices $u_1,\ldots,u_{n^2}$ and $v_1,\ldots,v_{n^2}$ and noticing that the *joint discriminant* $\det(\operatorname{tr}(u_iv_j))$ is antisymmetric in both the $u$'s and $v$'s so it is a multiple of $\Delta(u_1,\ldots,u_{n^2})\Delta(v_1,\ldots,v_{n^2})$ and clearly by a nonzero constant term. □

Since $\Delta(Z) \in \mathcal{T}_n$, we have $\Delta(X)\Delta(Y)\Delta(Z) \in J$. As a consequence we have that also $\Delta(X)\Delta(Y)\Delta(Z)$ for three sets of variables is in the conductor so it is a central polynomial $F_{3,n}(X,Y,Z)$. A formula for it can be explicitly deduced from formula (10.9).

COROLLARY 12.3.19. *For every $k \geq 2$ and $k$ sets of $n^2$ variables $X_i$, the product $\prod_{i=1}^k \Delta(X_i)$ is given by a central polynomial $F_{k,n}(X_1,\ldots,X_k)$.*

From the discussion at the beginning of this section we have seen that these elements are just the invariants of matrices, thought of as just $n^2$-dimensional vectors, under the multiplication action by the special linear group. These invariants, for any choice of $m$ matrix variables, have a very precise description as the irreducible representation of the linear group $GL(m,F)$ associated to the rectangular diagram with $k$ columns of length $n^2$, and can be put into normal form by standard monomials (cf. **[Pro07]**).

It may be interesting to understand how much of this theory is characteristic free.

A corollary of the previous theory will be used in Chapter 22. Consider $A = F\langle\xi_i\rangle$, the $m$ generic $k \times k$ matrices inside the trace ring $B = \mathcal{T}_k\langle\xi_i\rangle F\langle\xi_i\rangle$ obtained by adding the central ring generated by the traces $\mathcal{T}_k\langle\xi_i\rangle$. According to §12.1.4, one has for each of these algebras a cocharacter, and by Proposition 12.1.20 the multiplicity of a representation $M_\lambda$, in both cases, vanishes as soon as the height of $\lambda$ is $> k^2$. As for height of $\lambda$ equal to $k^2$ we have

THEOREM 12.3.20 (Formanek). *If $\lambda$ is a partition with $k^2$ rows and $\lambda_{k^2} \geq 2$, then the multiplicity $m_\lambda$ (of the cocharacter) for generic matrices $A$ or for the trace algebra $B$ coincide.*

PROOF. We can use for $m_\lambda$ the definition by Proposition 7.1.2(3). It is then enough to show that if $a \in B$ is a $U$-invariant of weight $\lambda$ (that is a highest weight vector by Definition 6.3.6), then $a \in A$. For this we first remark that since $\lambda$ has at least two columns of length $k^2$, we have, by formula (6.19) and Remark 6.3.23 or Theorem 6.3.7(4), that $a = \Delta^2 b$ with $\Delta$, the discriminant, defined by formula (12.25) and $b \in B$. By Theorem 12.3.17 we have that $\Delta^2$ is in $A$, and moreover it is in the conductor of $A$ in $B$, hence $a \in A$. □

CHAPTER 13

# Involutions and matrices

*This chapter is truly a complement; we do not intend to treat this topic in detail.*

## 13.1. Matrices with involutions

**13.1.1. FFT for matrices with involutions.** Recall that an *involution* in an associative $A$-algebra $R$, is an antihomomorphism of order 2, that is a linear map $a \mapsto a^*$ satisfying the properties

$$(a+b)^* = a^* + b^*, \ (ab)^* = b^*a^*, \ (a^*)^* = a. \tag{13.1}$$

Usually one assumes that $\lambda^* = \lambda, \forall \lambda \in A$, although it is also useful to study involutions like the adjoint on complex matrices which induces complex conjugation on the complex numbers.

There are essentially two types of involutions on matrices: orthogonal and symplectic. They correspond to taking a finite-dimensional vector space $V$ with a nondegenerate bilinear form $\{a,b\}$ which is $\epsilon = \pm 1$ symmetric. In the sense that $\{a,b\} = \epsilon\{b,a\}$.

In fact let $V = F^n$, a vector space over a field $F$. A bilinear form is given by a matrix $A$ as $\{u,v\} = u^tAv$ when we write the vectors as columns. It is nondegenerate if $A$ is nonsingular; symmetric if $\{u,v\} = \{v,u\}$ so that $A = A^t$ is a symmetric matrix and antisymmetric if $\{u,v\} = -\{v,u\}$ so that $A = -A^t$ is an antisymmetric or skew symmetric matrix. Such a form induces an involution on matrices $X \mapsto X^*$ given by $\{Xu,v\} = \{u, X^*v\}$, that is

$$(Xu)^tAv = u^tAX^*v \quad \iff X^tA = AX^* \iff X^* = A^{-1}X^tA.$$

Conversely given an involution $X \mapsto X^*$ on matrices, which is the identity on $F$, one has that $X \mapsto (X^*)^t$ is an isomorphism, hence $(X^*)^t = AXA^{-1}$ for some invertible matrix $A$ (Remark 3.4.19), and so $X^* = (A^{-1})^tX^tA^t$. Apply this formula twice to get

$$X = (X^*)^* = (A^{-1})^tAXA^{-1}A^t = (A^{-1}A^t)^{-1}X(A^{-1}A^t).$$

Hence $A^{-1}A^t = \alpha \in F^* \iff A^t = \alpha A \iff A = \alpha A^t \implies A = \alpha^2 A$ which implies $\alpha^2 = 1 \iff \alpha = \pm 1$, and the involution is associated to the form with matrix $A$, symmetric or antisymmetric.

In the symmetric case we denote the form by $(u,v)$ (called a *scalar product*); in the skew symmetric (symplectic case) we denote it by $[u,v]$. In both cases we can identify, using the form, $V = V^*$, $\mathrm{End}(V) = V \otimes V^* = V \otimes V$. On decomposable tensors $u \otimes v$ we use the formula

$$(u \otimes v)(w) = \{v,w\}u \implies (u \otimes v) \circ (w \otimes z) = u \otimes \{v,w\}z.$$

The form induces an involution $X \mapsto X^*$ in $\mathrm{End}(V)$ with the property that

$$\{Xu, v\} = \{u, X^*v\}.$$

We have on decomposable vectors

$$\begin{aligned}\{v,a\}\{u,b\} &= \{u\{v,a\},b\} = \{(u\otimes v)a,b\} = \{a,(u\otimes v)^*b\}\\ &= \epsilon\{a,v\}\{u,b\} = \epsilon\{a,(v\otimes u)b\}\\ &\Rightarrow (u\otimes v)^* = \epsilon\cdot v\otimes u.\end{aligned}$$

If $e_1,\dots,e_n$ is a basis of $V$ and $e^1,\dots,e^n$ is dual with respect to the form in the sense that $\{e^i,e_j\} = \delta^i_j$, then we have $1_V = \sum_{j=1}^n e_j\otimes e^j$.

This implies that $\mathrm{tr}(u\otimes v) = \{v,u\}$. Summarizing

(13.2) $(u\otimes v)\circ(w\otimes z) = u\otimes\{v,w\}z, \quad (u\otimes v)^* = \epsilon\, v\otimes u, \quad \mathrm{tr}(u\otimes v) = \{v,u\}.$

**13.1.2. Trace identities with involutions.** One can deduce a first and second fundamental theorem for invariants of matrices and trace identities also in the case of involutions. For this one needs to recall the first and second fundamental theorems for the orthogonal and the symplectic groups.

One considers thus for a finite-dimensional vector space $V$, $\dim(V) = n$, over a field $F$ (for simplicity of characteristic 0) equipped with a nondegenerate $\epsilon$ symmetric form, the group $G$ of linear transformations preserving the form

$$G := \{g \in \mathrm{GL}(V) \mid \{ga, gb\} = \{a,b\},\ \forall a,b\in V\}.$$

In fact using the involution, we see that $G = \{g\in \mathrm{GL}(V)\mid g^*g = gg^* = 1_V\}$.

When $\epsilon = 1$ the group is the orthogonal group denoted $O(V) = O_n$ while if $\epsilon = -1$ one has that $n$ must be even and the group is the symplectic group denoted $\mathrm{Sp}(V)$.

The first and second fundamental theorems describe the ring of invariants of several copies $V^{\oplus m}$ of the space $V$ under the action of $G$. We usually identify the elements of $V^{\oplus m}$ as $m$-tuples $(v_1,\dots,v_m)$ of vectors $v_i\in V$. For the proof of these theorems, we refer to **[Pro07]**.

THEOREM 13.1.1. *The ring of invariants of $V^{\oplus m}$ is generated by the bilinear products $\{v_i, v_j\}$.*

The second fundamental theorem has a different formulation for the two cases.

In the orthogonal case the elements $\{v_i,v_j\}$ are the entries of an $m\times m$ symmetric matrix.

One easily verifies that such a matrix has rank $\le n = \dim V$ and that in fact any symmetric matrix of rank $\le n$ can be realized in this form. Thus the second fundamental theorem should describe generators of the ideal of functions on the space of symmetric $m\times m$ matrices vanishing on the subvariety of matrices of rank $\le n$. This of course is a proper subvariety if and only if $n < m$.

The coordinate ring of the $m\times m$ symmetric matrices is the polynomial ring $F[x_{i,j}]$ in *symmetric variables* $x_{i,j} = x_{j,i}$, which can be thought of as the entries of a generic symmetric matrix $X$.

THEOREM 13.1.2. *The ideal of the polynomial ring $F[x_{i,j}]$ vanishing on the subvariety of matrices of rank $\le n$ is generated by the determinants of the $n+1\times n+1$ minors of the generic symmetric matrix $X$.*

For the symplectic group for $n = 2k = \dim V$, one has instead as image the subvariety of skew symmetric matrices of rank $\leq n = 2k$. In this case one has a generic skew symmetric matrix $X$ with entries $x_{i,j} = -x_{j,i}, i \neq j$ and should recall that the determinant $\det(A)$ of a skew symmetric $2k \times 2k$ matrix $A$ is the square $\det(A) = \mathrm{pf}(A)^2$ of a polynomial $\mathrm{pf}(A)$ called the *Pfaffian* of $A$:

THEOREM 13.1.3. *The ideal of the polynomial ring $F[x_{i,j}]$ vanishing on the subvariety of skew symmetric matrices of rank $\leq n$ is generated by the Pfaffians of the $n+2 \times n+2$ principal minors of the generic skew symmetric matrix $X$.*

Arguing as in Theorems 12.1.3 and 12.1.5, one then has

THEOREM 13.1.4. *The ring of invariants of $m$ matrices $X_1, \dots, X_m$, under the group $G$ (orthogonal or symplectic) is generated by the elements* $\mathrm{tr}(M)$ *where $M$ runs over all monomials $M$ in the variables $X_1, \dots, X_m$ and their adjoints $X_1^*, \dots, X_m^*$.*

*The ring of equivariant maps of $m$ matrices $X_1, \dots, X_m$ to matrices, under the group $G$ (orthogonal or symplectic) is generated over the ring of invariants, by the matrix variables $X_1, \dots, X_m$ and their adjoints $X_1^*, \dots, X_m^*$.*

PROOF. By Aronhold's method §3.1.1.5, one reduces to study multilinear invariants, and then by identifying $\mathrm{End}(V) = V \otimes V$ the invariants of $m$ matrices $X_i$ can be deduced by setting $X_i = u_i \otimes v_i$ and considering the multilinear invariants on the $2m$ vectors $u_i, v_j$. These invariants are given by products of elements $\{a, b\}$ where $a, b$ run in the entire list.

We understand such a product as a product of traces $\mathrm{tr}(M)$ of monomials in the $X_i = u_i \otimes v_i$ by use of formula (13.2):

To each trace monomial is associated a product $\prod_{i=1}^m \{x_i, y_i\}$, where the list of variables $x_1, \dots, x_m, y_1, \dots, y_m$ is just a reordering of $u_1, \dots, u_m, v_1, \dots, v_m$. Conversely, given such a product, it is easily seen that it comes from a unique trace monomial (in this correspondence one takes care of the formal identities between expressions). As an example (note that the sign depends on the case, orthogonal is $+$ and symplectic is—if the number of $X^*$ is odd):

$$\{u_1, u_4\}\{v_4, v_6\}\{u_6, v_1\}\{u_2, u_5\}\{u_3, v_5\}\{v_2, v_3\} = \pm\, \mathrm{tr}(X_1^* X_4 X_6^*)\, \mathrm{tr}(X_2^* X_5 X_3)$$

$$\{u_1, u_4\}\{v_4, v_6\}\{u_6, v_2\}\{u_2, u_5\}\{u_3, v_5\}\{v_1, v_3\} = \pm\, \mathrm{tr}(X_1^* X_4 X_6^* X_2^* X_5 X_3). \quad \square$$

To each invariant $A$, a monomial into traces of $m$ variables $X_i$, $i = 1, \dots, m+1$, we can, as usual, associate the monomial $B$ in the variables $X_1, \dots, X_m$ such that $A = \mathrm{tr}(B X_{m+1})$.

For instance $\mathrm{tr}(X_1^* X_4 X_6^* X_2^* X_5 X_3) = \mathrm{tr}(X_4^* X_1 X_3^* X_5^* X_2 X_6)$, the corresponding monomial is thus $X_4^* X_1 X_3^* X_5^* X_2$. Extending from monomials to linear combinations, a multilinear trace identity in $m$ variables corresponds to a multilinear trace relation in $m+1$ variables. The second fundamental theorem consists thus in interpreting, in terms of such monomials, the second fundamental theorem for vectors. It is somewhat more complicated than the one for matrices. To do so, it is a question of translating the basic determinants or Pfaffians in terms of trace relations.

As usual we will describe the multilinear relations and deduce the general relations through polarization and restitution.

For us it may be convenient to consider the scalar products between the $u_i, v_j$ as evaluations of variables $x_{ij} = (i|j)$ which satisfy the symmetry condition $(i|j) = \epsilon(j|i)$, with $\epsilon = 1$ in the orthogonal case and $-1$ in the symplectic case. We consider the variables $(i|j)$ as entries of an $\epsilon$ symmetric $2m \times 2m$ matrix.

Consider next the space $M_m$ spanned by the *multilinear monomials*

$$x_{i_1 i_2} x_{i_3 i_4} \cdots x_{i_{2m-1} i_{2m}} = (i_1|i_2)(i_3|i_4) \cdots (i_{2m-1}|i_{2m}),$$

where the index sequence $i_1, i_2, i_3, i_4, \ldots, i_{2m-1}, i_{2m}$ is a permutation of $1, 2, \ldots, 2m$.

Up to sign and symmetry there are $\frac{(2m)!}{2^m m!}$ different monomials.

We have a *coding map* from $M_m$ to the vector space of multilinear pure trace expressions $f(X_1, \ldots, X_{m+1})$ of degree $m+1$ in the variables $X_i$, $X_i^*$ or the noncommutative trace expressions $g(x_1, \ldots, X_m)$ of degree $m$, with $f(X_1, \ldots, X_{m+1}) = \operatorname{tr}(g(X_1, \ldots, X_m) X_{m+1})$.

It is given formally by considering the numbers $i$ as vector symbols, basis of a free module $F$ over the polynomial ring over the variables $(i|j)$. Then set $X_i := (2i-1) \otimes 2i \in F \otimes F \subset \operatorname{End}(F)$, and $(i|j)$ as the scalar product $\langle i|j \rangle$, with the multiplication, trace, and adjoint as in (13.2), deduced by the formulas

$$i \otimes j \circ h \otimes k = i \otimes (j|h)k, \quad \operatorname{tr}(i \otimes j) = \epsilon\,(i|j), \quad (i \otimes j)^* = \epsilon\, j \otimes i.$$

For instance for $m = 2$, we have 15 monomials. We list a few of them for $\epsilon = -1$:

$$\begin{aligned}
(1|2)(3|4)(5|6) &= -\operatorname{tr}(X_1)\operatorname{tr}(X_2)\operatorname{tr}(X_3),\\
(1|2)(3|5)(4|6) &= -\operatorname{tr}(X_1)\operatorname{tr}(X_2^* X_3),\\
(1|2)(3|6)(4|5) &= \operatorname{tr}(X_1)\operatorname{tr}(X_2 X_3),\\
(1|3)(2|5)(4|6) &= -\operatorname{tr}(X_1^* X_2 X_3^*) = -\operatorname{tr}(X_3 X_2^* X_1)
\end{aligned}$$

in the associated noncommutative form,

$$\begin{aligned}
(1|2)(3|4)(5|6) &= -\operatorname{tr}(X_1)\operatorname{tr}(X_2), & (12)(35)(46) &= -\operatorname{tr}(X_1) X_2^*,\\
(1|2)(3|6)(4|5) &= \operatorname{tr}(X_1) X_2, & (1|3)(2|5)(4|6) &= -X_2^* X_1.
\end{aligned}$$

REMARK 13.1.5. Each multilinear monomial in the symbols is decomposed uniquely as a product of *cycles*, which is left to the reader.

Under this coding map, the subspace of relations of invariants corresponds to the trace identities. Thus one can reinterpret the second fundamental theorem as giving basic identities from which all the others can be deduced.

## 13.2. Symplectic and orthogonal case

13.2.0.1. *The symplectic case.* The reinterpretation is particularly simple for the symplectic group. In this case the analogue of the Cayley–Hamilton identity is a *characteristic Pfaffian* equation.

Consider thus $M_{2n}$ the space of $2n \times 2n$ matrices, $J := \begin{vmatrix} 0 & 1_n \\ -1_n & 0 \end{vmatrix}$ the standard matrix of the symplectic involution, and $A^* := -JA^tJ$ the symplectic involution.

Let $M_{2n}^+$ denote the space of symmetric matrices under this involution. Under the map $A \to B := AJ$, we identify $M_{2n}^+$ with the space of skew symmetric matrices, and we can define the *characteristic Pfaffian* as

$$\operatorname{Pf}_A(\lambda) := \operatorname{Pf}((\lambda I - A)J).$$

We leave it to the reader to verify that $\operatorname{Pf}_A(A) = 0$.

*Hint.* A generic symmetric matrix is semisimple, and one can decompose the $2n$-dimensional symplectic space into an orthogonal sum of stable two-dimensional subspaces. On each such subspace $A$ is a scalar.

If we polarize the identity $\mathrm{Pf}_A(A) = 0$, we obtain a multilinear identity, denoted by $P(X_1, \dots, X_n)$, satisfied by matrices which are symmetric under the symplectic involution. In another choice of variables we may write this identity as an identity $P(X_1 + X_1^*, \dots, X_n + X_n^*)$ satisfied by all matrices. This is an identity of degree $n$. On the other hand there is a unique canonical pure trace identity of degree $n+1$ which, under the coding map, comes from the Pfaffian of a $2n+2 \times 2n+2$ matrix of skew indeterminates, thus up to a scalar the polarized characteristic Pfaffian must be a multiple of the noncommutative identity coded by that Pfaffian. For instance, in the trivial case $n = 1$, we get the identity $\mathrm{tr}(X_1) = X_1 + X_1^*$. As for the Cayley–Hamilton identity one may prove:

THEOREM 13.2.1. *All identities for matrices with symplectic involution are a consequence of* $P(X_1 + X_1^*, \dots, X_n + X_n^*)$.

PROOF. We will give the proof for the corresponding trace relations. Consider the matrix of (skew) scalar products:

$$S := \begin{vmatrix} 0 & [u_1, u_2] & \dots [u_1, u_m] & [u_1, v_1] & [u_1, v_2] & \dots [u_1, v_m] \\ [u_2, u_1] & 0 & \dots [u_2, u_m] & [u_2, v_1] & [u_1, v_2] & \dots [u_2, v_m] \\ \dots & \dots & \dots & \dots & \dots & \\ \dots & \dots & \dots & \dots & \dots & \\ [v_1, u_1] & [v_1, u_2] & \dots [v_1, u_m] & 0 & [v_1, v_2] & \dots [v_1, v_m] \\ [v_2, u_1] & [v_2, u_2] & \dots [v_2, u_m] & [v_2, v_1] & 0 & \dots [v_2, v_m] \\ \dots & \dots & \dots & \dots & \dots & \\ \dots & \dots & \dots & \dots & \dots & \\ [v_m, u_1] & [v_m, u_2] & \dots [v_m, u_m] & [v_m, v_1] & [v_m, v_2] & \dots 0 \end{vmatrix}.$$

A typical identity in $m$ variables corresponds to a polynomial

$$[w_1, w_2, \dots, w_{2n+2}][x_1, y_1] \cdots [x_s, y_s]$$

of the following type.

Let $m + 1 = n + 1 + s$, the elements $w_1, w_2, \dots, w_{2n+2}, x_1, y_1, \dots, x_s, y_s$ are given by a permutation of $u_1, \dots, u_m, v_1, \dots, v_m$, and $[w_1, w_2, \dots, w_{2n+2}]$ denotes the Pfaffian of the principal minor of $S$ of row and column indices corresponding to the symbols $w_1, w_2, \dots, w_{2n+2}$. To understand the corresponding trace relation we argue as follows.

(1) Permuting the variables $X_i$ corresponds to permuting simultaneously the $u_i$ and $v_i$ with the same permutation.

(2) Exchanging $X_i$ with $X_i^*$ corresponds to exchanging $u_i$ with $v_i$. Performing these operations, we can reduce our expression to one of type

$$[u_1, \dots, u_k, v_1, \dots, v_k, v_{k+1}, \dots, v_{2n+2-k}][x_1, y_1] \cdots [x_s, y_s].$$

If $k = n + 1$, the Pfaffian $[u_1, \dots, u_{n+1}, v_1, \dots, v_{n+1}]$ corresponds to the identity of the characteristic Pfaffian. The given expression is a product of this relation by trace monomials. Otherwise we look at the symbol $u_{k+1}$ which does not appear in the Pfaffian but appears in one of the terms $[x_i, y_i]$. We have that $u_{k+1}$ is paired with either a term $u_i$ or a term $v_j$. If $u_{k+1}$ is paired with $v_j$, let us use the formula

$$X_j X_{k+1} = u_j \otimes [v_j, u_{k+1}] v_{k+1}.$$

If $u_{k+1}$ is paired with $u_j$, we use the formula

$$X_j^* X_{k+1} = -v_j \otimes [u_j, u_{k+1}] v_{k+1}.$$

In the first case we introduce a new variable $\bar{X}_j := X_j X_{k+1}$. In the second $\bar{X}_j := X_j^* X_{k+1}$, which we write in the form $\bar{X}_j := \bar{u}_j \otimes \bar{v}_j$, with

$$\bar{u}_j := u_j,\ \bar{v}_j := [v_j, u_{k+1}] v_{k+1}, \quad \text{or} \quad \bar{u}_j := -v_j,\ \bar{v}_j := [u_j, u_{k+1}] v_{k+1}.$$

Let us analyze for instance the case $[u_j, u_{k+1}]$, the other being similar. We substitute $[v_j, u_{k+1}]$ inside the Pfaffian:

$$\begin{aligned} &[u_1, \dots, u_k, v_1, \dots, v_k, v_{k+1}, \dots, v_{2n+2-k}][u_j, u_{k+1}] \\ &\quad = [u_1, \dots, u_k, v_1, \dots, v_k, [u_j, u_{k+1}] v_{k+1}, \dots, v_{2n+2-k}] \\ &\quad = [u_1, \dots, u_k, v_1, \dots, v_k, \bar{v}_j, \dots, v_{2n+2-k}]. \end{aligned}$$

We obtain a new formal expression in $m-1$ variables, the variable $X_{k+1}$ has been suppressed while the variable $X_j$ has been substituted with a new variable $\bar{X}_j$. In order to recover the old $m$-variable expression from the new one, one has to substitute $\bar{X}_j$ with $X_j X_{k+1}$ or $X_j^* X_{k+1}$. By induction the theorem is proved. □

**13.2.1. The orthogonal case.** Consider $m$ matrix variables in an $(n-1)$-dimensional vector space with a scalar product, in the form $X_i := u_i \otimes v_i$. The study of their basic relations is more complicated than that of the symplectic case, due to the more complicated form of the second fundamental theorem for orthogonal invariants of $2m$ vectors $u_1, \dots, u_m; v_1, \dots, v_m$.

Consider the matrix of (symmetric) scalar products:

$$S := \begin{vmatrix} (u_1,u_1) & (u_1,u_2) & \dots (u_1,u_m) & (u_1,v_1) & (u_1,v_2) & \dots (u_1,v_m) \\ (u_2,u_1) & (u_2,u_2) & \dots (u_2,u_m) & (u_2,v_1) & (u_1,v_2) & \dots (u_2,v_m) \\ \dots & \dots & \dots & \dots & \dots & \\ \dots & \dots & \dots & \dots & \dots & \\ (u_m,u_1) & (u_m,u_2) & \dots (u_m,u_m) & (u_m,v_1) & (u_m,v_2) & \dots (u_m,v_m) \\ (v_1,u_1) & (v_1,u_2) & \dots (v_1,u_m) & (v_1,v_1) & (v_1,v_2) & \dots (v_1,v_m) \\ (v_2,u_1) & (v_2,u_2) & \dots (v_2,u_m) & (v_2,v_1) & (v_2,v_2) & \dots (v_2,v_m) \\ \dots & \dots & \dots & \dots & \dots & \\ \dots & \dots & \dots & \dots & \dots & \\ (v_m,u_1) & (v_m,u_2) & \dots (v_m,u_m) & (v_m,v_1) & (v_m,v_2) & \dots (v_m,v_m) \end{vmatrix}.$$

A relation may be given by the determinant of any $n \times n$ minor of $S$, times a monomial. Again we have that:

(1) Permuting the variables $X_i$ corresponds to permuing simultaneously the $u_i$ and $v_i$ with the same permutation.

(2) Exchanging $X_i$ with $X_i^t$ corresponds to exchanging $u_i$ with $v_i$.

In this case there are several basic identities to consider, which correspond to the determinants of scalar products, which we may denote by

$$(u_{i_1} u_{i_2} \cdots u_{i_k} v_{j_1} v_{j_2} \cdots v_{j_{n-k}} \mid u_{i_{k+1}} u_{i_{k+2}} \cdots u_{i_n} v_{j_{n+2-k}} v_{j_{n+3-k}} \cdots v_{j_n}).$$

The condition that they represent a multilinear relation in the matrix variables $X_1, \dots, X_n$ means that the indices $i_1, \dots, i_n$ and also the indices $j_1, \dots, j_n$ are permutations of $1, 2, \dots, n$.

Using the previous symmetry laws, we may first exchange all $u_i, v_i$ for all the indices $i$ for which $v_i$ appears on the left but $u_i$ does not appear. After reordering the indices, we end up with a determinant, where $n-k \leq k$,

$$D_{k,n} := (u_1u_2\cdots u_kv_1v_2\cdots v_{n-k} \mid u_{k+1}u_{k+2}\cdots u_nv_{n+1-k}v_{n+2-k}\cdots v_n) \tag{13.3}$$

of the corresponding matrix called $S_{k,n}$.

From the coding map, the above determinant corresponds to some trace relation which we denote by $F_k(X_1,\dots,X_n)$. Unlike the symplectic case, these basic identities do not have enough symmetry to be considered as polarizations of 1-variable identities.

*Symmetries.* Decompose $\{1,2,\dots,n\}$ in three subsets,

$$A := (1,2,\dots,n-k);\ B := (n-k+1,\dots,k);\ C := (k+1,\dots,n). \tag{13.4}$$

Exchanging $u_i, v_i$ , if $i > n-k$, corresponds in the determinant of formula (13.3) to exchanging two rows or, if $i \leq k$, to exchanging two columns. Hence, for such an $i$,

$$(a)\qquad F_k(X_1,\dots,X_i,\dots,X_n) = -F_k(X_1,\dots,X_i^t,\dots,X_n).$$

If $\sigma$ is a permutation of $1,2,\dots,n$ which preserves the subsets $A,B,C$, we have

$$(b)\quad F_k(X_1,\dots,X_i,\dots,X_n) = F_k(X_{\sigma(1)},\dots,X_{\sigma(i)},\dots,X_{\sigma(n)})$$

since it amounts to permuting the same indices of rows and columns in the determinant of formula (13.3) (in case $n=2k$ the second subset is empty). This means that $F_k$ can be viewed as polarization of an identity in three variables $F_k(X,Y,Z)$. Moreover from the first relation the first and third variable can be taken to be skew symmetric variables $\frac{X-X^t}{2}, \frac{Z-Z^t}{2}$.

Finally exchanging all $u_i, v_i$ and transposing the determinant, we get

$$\begin{aligned}&(v_{k+1}v_{k+2}\dots v_nu_{n+2-k}u_{n+3-k}\cdots u_n \mid v_1v_2\dots v_ku_1u_2\dots u_{n-k})\\ &\quad = (u_nu_{n+3-k}\dots u_{n+2-k}v_n\dots v_{k+2}v_{k+1} \mid u_{n-k}\dots u_2u_1v_k\dots v_2v_1) \implies\end{aligned}$$

$$(c)\quad F_k(X_1^t,\dots,X_i^t,\dots,X_n^t) = F_k(X_n,\dots,X_{n-i},\dots,X_1).$$

We need to make a remark when we pass from the trace relation to the noncommutative trace relation. We pick a variable $X_i$ and write

$$F_k(X_1,\dots,X_i,\dots,X_n) = \operatorname{tr}(G_k^i(X_1,\dots,\check{X}_i,\dots,X_n)X_i).$$

It is clear that $G_k^i$ depends on $i$ only up to the symmetry embedded in the previous statements (a), (b), (c).

From the symmetry (b) it appears that we obtain three inequivalent identities $G_k^1, G_k^2, G_k^3$ from the three subsets. From (c) we obtain a different way of deducing the identities associated to the first and the third subset.

$$F_k(X_1,\dots,X_n) = \operatorname{tr}(G_k^1(X_2,\dots,X_n)X_1),$$

gives

$$\begin{aligned}F_k(X_1^t,\dots,X_n^t) &= \operatorname{tr}(G_k^1(X_2^t,\dots,X_n^t)X_1^t)\\ &= \operatorname{tr}(G_k^1(X_2^t,\dots,X_n^t)^tX_1) = \operatorname{tr}(G_k^3(X_n,\dots,X_2)X_1).\end{aligned}$$

Hence

$$G_k^1(X_2^t,\dots,X_n^t)^t = G_k^3(X_n,\dots,X_2).$$

In this sense only $G_k^1, G_k^2$ should be taken as generating identities. One can develop them into formulas for traces as follows.

**13.2.2. Combinatoric of permutations.** Calling $x_1, \ldots, x_n$ (resp., $y_1, \ldots, y_n$) the left and right elements of formula (13.3), one has the expansion of the determinant. Let $B_n$ be the set of bijections between $x_1, \ldots, x_n$ and $y_1, \ldots, y_n$ (which one could also identify as the symmetric group $S_n$ as $\sum_{\sigma \in B_n} \epsilon_\sigma \prod_i (x_i, \sigma(x_i))$, and one has to interpret each term $\prod_i (x_i, \sigma(x_i))$ as a trace monomial. Of course the interpretation depends upon $k$. One can proceed as follows. Let

$$\phi : u_i \mapsto i, v_i \mapsto i, (a_i, b_j) \mapsto (i|j), \ \forall a, b \in \{u, v\}.$$

Thus under the map $\phi$ the symbol $(i|j)$ comes from four different symbols of scalar products $(u_i, u_j)$, $(u_i, v_j)$, $(v_i, u_j)$, $(v_i, v_j)$. The symbols $(i|j)$, are commuting variables, together with the symmetry condition $(i|j) = (j|i)$. On the other hand we have

DEFINITION 13.2.2. A multilinear *scalar* symbol is a product $\prod_{i=1}^n (a_i, b_i)$ in which the elements $a_i, b_i$ form a permutation of $u_1, \ldots, u_n; v_1, \ldots, v_n$. A *special symbol* is a product $\prod_{h=1}^n (i_h | j_h)$, $i_h, j_h \in \{1, \ldots, n\}$, with the special property that each index $i \in \{1, \ldots, n\}$ appears exactly twice in the list $i_h, j_h$, $h = 1, \ldots, n$.

Each of these scalar symbols encodes a multilinear trace monomial $\prod_j \operatorname{tr}(M_j)$ in the variables $X_i$; moreover by the symmetry rules, $\prod_{i=1}^n (a_i, b_i) = \prod_{i=1}^n (b_i, a_i)$ encodes the multilinear trace monomial $\prod_j \operatorname{tr}(M_j^t)$.

DEFINITION 13.2.3. Given a scalar symbol $c$ encoding the multilinear trace monomial $\operatorname{Tr}(c) = \prod_j \operatorname{tr}(M_j)$, its associated partition is the sequence $\lambda_c$ of the lengths of the monomials $M_i$.

To each scalar symbol $\prod_{i=1}^n (a_i, b_i)$, via the map, $\phi$ is associated a special symbol $\prod_{i=1}^n (\phi(a_i), \phi(b_i))$. Then a special symbol arises through $\phi$ from exactly $2^n$ multilinear symbols.

For every subset $T \subset \{1, \ldots, n\}$ with $t$ elements, we may consider a product of type $\prod_{h=1}^t (i_h | j_h)$, $i_h, j_h \in T$, with the special property that each index $i \in T$ appears exactly twice in the list $i_h, j_h$, $h = 1, \ldots, h$. This will be called a *special symbol, supported in* $T$ similarly for scalar symbols. We say by abuse of notation that $a \in T$ if $a = u_i$ or $a = v_i$ with $i \in T$. Let us use the convention

$$u_i^t := v_i, \quad v_i^t := u_i.$$

DEFINITION 13.2.4. A scalar symbol, supported in $T$, $c = \prod_{h=1}^k (a_h, b_h)$, is called a *scalar cycle* if up to reordering its factors and the symmetry $(a_h, b_h) = (b_h, a_h)$ can be brought in the form

$$(a_{i_1}, a_{i_2})(a_{i_2}^t, a_{i_3}) \cdots (a_{i_{k-1}}^t, a_{i_1}^t). \tag{13.5}$$

Equivalently it corresponds to a single trace monomial $\operatorname{Tr}(c) = \operatorname{tr}(M)$.

A special symbol, supported in $T$, $\prod_{h=1}^t (i_h | j_h)$, $i_h, j_h \in T$, is called a *cycle* if up to reordering its factors and the symmetry $(i|j) = (j|i)$ can be brought in the form

$$(i_1, i_2, i_3, \ldots, i_h) := (i_1|i_2)(i_2|i_3) \cdots (i_h|i_1). \tag{13.6}$$

We remark that the scalar cycle (13.5) encodes the matrix invariant $\mathrm{tr}(M)$ with $M = Z_1 \cdots Z_k$, with $Z_h = a^t_{i_h} \otimes a_{i_h}$, a monomial in $X_i, X^t_i$ in which each index $i \in T$ appears exactly once. One easily sees that

PROPOSITION 13.2.5. *Each multilinear symbol decomposes uniquely into a product of cycles over disjoint sets. Under the map $\phi$ a scalar cycle maps to a special cycle.*

Such a symbol will also be represented with the diagram, where each cycle is pictured by a polygon, of which these are two examples for $n = 4$:

(13.7)

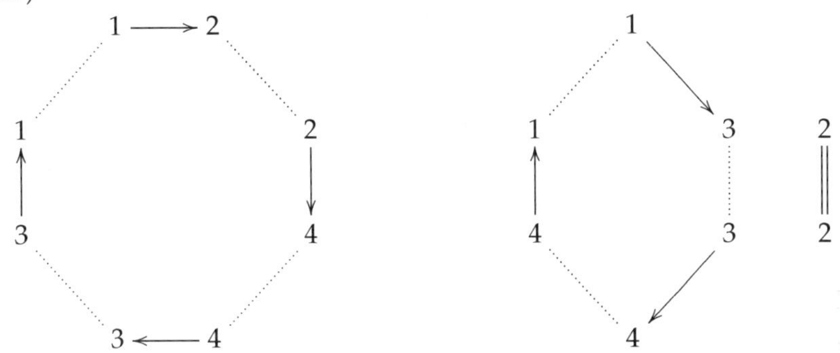


Here the symbol $i \longrightarrow j$ stands for $(i|j)$. A similar picture holds for products $\prod_i (x_i, \sigma(x_i))$ corresponding to the previous special symbols:

(13.8)

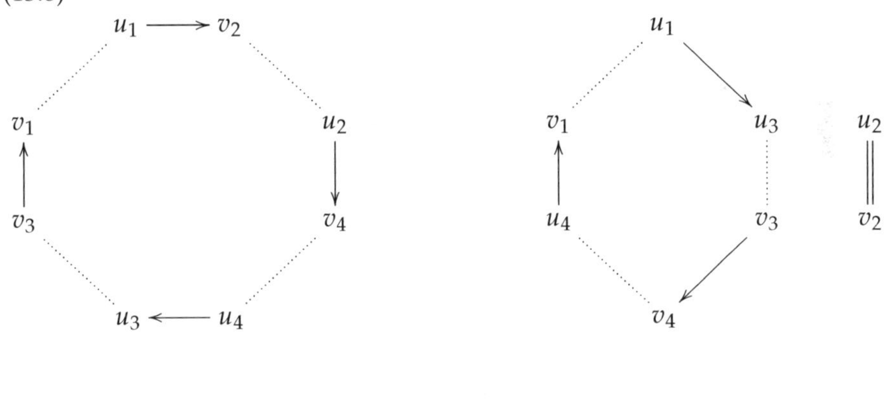


Here the symbols $a_i \longrightarrow b_j$ stand for $(a_i, b_j)$ with $a, b \in \{u, v\}$.

REMARK 13.2.6. From a given special diagram one associates a *canonical multilinear symbol* by replacing in each symbol $i \longrightarrow j$ the symbol $v_i \longrightarrow u_j$.

In terms of matrices the canonical multilinear symbol associated to a cycle $(i_1|i_2)(i_2|i_3) \cdots (i_h|i_1)$ is $\mathrm{tr}(X_{i_1} X_{i_2} \cdots X_{i_h})$.

For instance the canonical symbols corresponding to formula (13.7) are

(13.9)

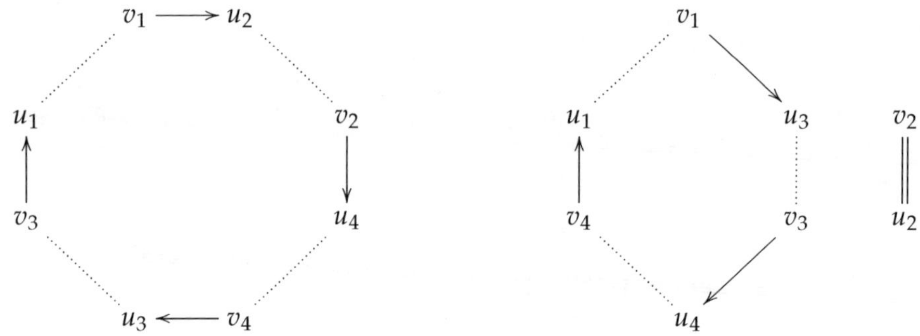


The trace monomials corresponding to (13.9) are

$$\operatorname{tr}(X_1X_2X_4X_3),\ \operatorname{tr}(X_1X_3X_4)\operatorname{tr}(X_2),$$

while the ones corresponding to formula (13.8) are

$$\operatorname{tr}(X_1X_2X_4X_3^t),\ \operatorname{tr}(X_1X_3^tX_4)\operatorname{tr}(X_2).$$

The $2^n$ multilinear symbols associated to a given special symbol are obtained from the canonical symbol by performing in all possible ways exchanges $u_i$ and $v_i$. In term of the corresponding expression in matrix variables, the exchange of $u_i$ and $v_i$ corresponds to substituting $X_i$ with $X_i^t$.

Now set $Y_i := X_i - X_i^t$.

PROPOSITION 13.2.7. *The sum of all multilinear symbols associated to a special cycle* $(i_1|i_2)(i_2|i_3)\cdots(i_h|i_1)$ *each with sign* $(-1)^a$ *where a is the number of exchanges, encodes the trace monomial* $\operatorname{tr}(Y_{i_1}Y_{i_2}\cdots Y_{i_h})$.

PROOF. We prove by induction a stronger statement. Consider the canonical symbol $C$ associated to a special cycle $(i_1|i_2)(i_2|i_3)\cdots(i_h|i_1)$. The sum of all multilinear symbols obtained from $C$ by exchanging $u_i$ with $v_i$ for the first $\ell$ indices $i$, each with sign $(-1)^a$ where $a$ is the number of exchanges, encodes the trace monomial $\operatorname{tr}(Y_{i_1}Y_{i_2}\cdots Y_{i_\ell}X_{i_{\ell+1}}\cdots X_{i_h})$.

In fact if we exchange now $u_{i_{\ell+1}}$ with $v_{i_{\ell+1}}$, we obtain

$$\begin{aligned}\operatorname{tr}(Y_{i_1}Y_{i_2}\cdots Y_{i_\ell}X_{i_{\ell+1}}\cdots X_{i_h}) &- \operatorname{tr}(Y_{i_1}Y_{i_2}\cdots Y_{i_\ell}X^t_{i_{\ell+1}}\cdots X_{i_h})\\ &= \operatorname{tr}(Y_{i_1}Y_{i_2}\cdots Y_{i_\ell}Y_{i_{\ell+1}}X_{i_{\ell+2}}\cdots X_{i_h}).\end{aligned}$$ □

We first need to understand the scalar monomials appearing in $D_{k,n}$, of formula (13.3).

If $k = n$, we easily see that this is the usual trace relation for $n-1 \times n-1$ matrices (12.13). Otherwise for $n/2 \le k < n$ we have identities involving the matrices and their transposes.

For example, for $2 \times 2$ matrices we have $n = 3$ and $k = 2, 3$ for $k = 2$,

$$\begin{pmatrix} & u_1 & u_2 & v_1 \\ & & & \\ u_3 & (u_3, u_1) & (u_3, u_2) & (u_3, v_1) \\ v_2 & (v_2, u_1) & (v_2, u_2) & (v_2, v_1) \\ v_3 & (v_3, u_1) & (v_3, u_2) & (v_3, v_1) \end{pmatrix} \tag{13.10}$$

with determinant

$$(u_3,u_1)(v_2,u_2)(v_3,v_1)-(u_3,u_2)(v_2,u_1)(v_3,v_1)+(u_3,v_1)(v_2,u_1)(v_3,u_2)$$
$$-(u_3,u_1)(v_2,v_1)(v_3,u_2)+(u_3,u_2)(v_2,v_1)(v_3,u_1)-(u_3,v_1)(v_2,u_2)(v_3,u_1)$$

corresponding to

$$\operatorname{tr}(X_3^tX_1)\operatorname{tr}(X_2)-\operatorname{tr}(X_3^tX_2X_1)+\operatorname{tr}(X_3X_2X_1)-\operatorname{tr}(X_3^tX_1X_2^t)+\operatorname{tr}(X_3^tX_2X_1^t)$$
$$-\operatorname{tr}(X_3X_1)\operatorname{tr}(X_2)=-\operatorname{tr}(Y_3X_1)\operatorname{tr}(X_2)+\operatorname{tr}(Y_3X_2X_1)+\operatorname{tr}(Y_3X_1X_2^t)$$
$$=-\operatorname{tr}(Y_3X_1)\operatorname{tr}(X_2)+\operatorname{tr}(Y_3X_2Y_1)=\operatorname{tr}([-\frac{1}{2}\operatorname{tr}(Y_3Y_1)+Y_1Y_3]X_2)$$

is the somewhat trivial identity for skew $2\times 2$ matrices $\frac{1}{2}\operatorname{tr}(Y_3Y_1)=Y_1Y_3$. But for skew $3\times 3$ matrices one has (Corollary 13.2.12) the more complicated identity

$$\operatorname{tr}(Y_2Y_3)Y_1+\operatorname{tr}(Y_1Y_3)Y_2-2(Y_2Y_3Y_1+Y_1Y_3Y_2)=0. \tag{13.11}$$

13.2.2.1. *The case $n=2k$.* In general for $n=2k$ and $h\in\{1,\dots,k\}$ denote by $\bar h:=h+k\in\{k+1,\dots,2k\}$. The development of $D_{k,2k}$ is formed by all the scalar symbols $\prod_{i=1}^{2k}(a_i,b_i)$ with $b_i\in\{1,\dots,k\}$ and $a_i\in\{k+1,\dots,2k\}$. This means that the corresponding trace monomials are products of elements $\operatorname{tr}(Z_1W_1\cdots Z_hW_h)$ where the $Z_j\in\{1,\dots,k\}$ and $W_i\in\{k+1,\dots,2k\}$. That is they belong to special cycles of formula (13.6) of type

$$(i_1,j_1,i_2,j_2,\dots,i_h,j_h),\ i_s\in\{1,\dots,k\}\quad j_s\in\{k+1,\dots,2k\}.$$

DEFINITION 13.2.8. Let $\tau\in S_{2k}$ be the product $\prod_{i=1}^k(i,i+k)$, and let

$$E_{2k}:=\{\sigma\in S_{2k}\mid\sigma(\{1,\dots,k\})=\{k+1,\dots,2k\}.$$

Clearly we have that for $\sigma\in E_{2k}$, $\{1,\dots,k\}=\sigma(\{k+1,\dots,2k\})$ and denoting by $S_k\times S_k$ the subgroup of $S_{2k}$ preserving the two subsets $\{1,\dots,k\}$ and $\{k+1,\dots,2k\}$ we have

$$E_{2k}=\tau\circ S_k\times S_k=S_k\times S_k\circ\tau.$$

Moreover each $\sigma\in E_{2k}$ decomposes into a product of even cycles. Clearly $E_{2k}$ is stable under conjugation by $S_k\times S_k$.

PROPOSITION 13.2.9. *The $S_k\times S_k$ conjugacy classes of $E_{2k}$ are indexed by the partitions of $2k$ with even parts. That is two permutations in $E_{2k}$ are conjugate under $S_k\times S_k$ if and only if their partitions (length of cycles) are the same.*

PROOF. This is essentially clear: let us see that if their partitions are the same they are are conjugate under $S_k\times S_k$. Let us do it for two cycles. In the usual proof $(i_1,j_1,i_2,j_2,\dots,i_h,j_h)$ is conjugate to $(m_1,p_1m_2,p_2,\dots,m_h,p_h)$ using the permutation $\tau: i_s\mapsto m_s;\ j_t\mapsto p_t$, which is clearly in $S_k\times S_k$. □

In particular to a partition $\lambda\vdash 2k$ with even parts, we may choose as representative of the conjugacy class by $S_k\times S_k$ the *ordered* product of cycles of type

$$(h,\bar h,h+1,\overline{h+1},\dots,h+t,\overline{h+t})=(h,h+k,h+1,h+1+k,\dots,h+t,h+t+k).$$

The canonical scalar symbol associated to such a normalized special symbol is the product $C_\lambda$ of

$$\operatorname{tr}(X_hX_{h+k}\cdots X_tX_{t+k})=(v_h,u_{h+k})(v_{h+k},u_{h+1})(v_{h+1},u_{h+k+1})\cdots(v_{t+k},u_h).$$

Denoting by $x_{i,j}$ the entries of the matrix $S_{k,n}$, we need to understand the $\sigma_\lambda$ which gives $C_\lambda$ in the development of $D_{k,2k} = \det(S_{k,n}) = \sum_{\sigma\in S_n} \epsilon_\sigma \prod_{i=1}^n x_{i,\sigma(i)}$. One has

$$(v_h, u_{h+k}) = (u_{h+k}, v_h) = x_{h,h+k}, \ (v_{h+k}, u_{h+1}) = x_{h+k,h+1},$$

so the permutation $\sigma_\lambda$ is

$$\sigma_\lambda(h+a) = k+h+a, \forall a < t, \sigma(k+h+a) = 1+h+a.$$

REMARK 13.2.10. Thus $\sigma_\lambda$ equals its special symbol, its cycles are all even so it has signature $(-1)^{a_\lambda}$ with $a_\lambda$ the number of parts of $\lambda$.

In order to understand $F_k(X_1, \dots, X_{2k})$, we first decompose it as the sum

$$F_k(X_1, \dots, X_{2k}) = \sum_\lambda F_{k,\lambda}(X_1, \dots, X_{2k}),$$

where $F_{k,\lambda}(X_1, \dots, X_{2k})$ is the sum relative to all the scalar symbols with partition $\lambda$.

Recall that in Proposition 2.3.15 we associate to a permutation

$$\sigma = (i_1 i_2 \cdots i_h) \cdots (j_1 j_2 \cdots j_k)(s_1 s_2 \cdots s_t)$$

written in its cycle decomposition, the multilinear trace expression of formula (2.16)

(13.12)

$$T_\sigma(x_1, x_2, \dots, x_m) := \operatorname{tr}(x_{i_1} x_{i_2} \cdots x_{i_h}) \cdots \operatorname{tr}(x_{j_1} x_{j_2} \cdots x_{j_k}) \operatorname{tr}(x_{s_1} x_{s_2} \cdots x_{s_t}).$$

Thus for $\sigma = (i_1, j_1, i_2, j_2, \dots, i_h, j_h) \cdots (p_1, q_1, p_2, q_2, \dots, p_\ell, q_\ell)$ set

$$\phi_\sigma(Y_1, \dots, Y_{2k}) = T_\sigma(Y_1, \dots, Y_{2k}),$$

(13.13)

$$\phi_\sigma(Y_1, \dots, Y_{2k}) := \operatorname{tr}(Y_{i_1} Y_{j_1} Y_{i_2} Y_{j_2} \cdots Y_{i_h} Y_{j_h}) \cdots \operatorname{tr}(Y_{p_1} Y_{q_1} Y_{p_2} Y_{q_2} \cdots Y_{p_\ell} Y_{q_\ell}).$$

For a permutation $\sigma \in E_{2k}$ or a scalar symbol $c$ denote by $\pi(\sigma)$ (resp., $\pi(c)$) its associated partition.

THEOREM 13.2.11. *We have*

$$F_{k,\lambda}(X_1, \dots, X_{2k}) = (-\frac{1}{2})^{a_\lambda} \sum_{\pi(\sigma)=\lambda} \phi_\sigma(Y_1, \dots, Y_{2k}),$$

*hence,*

$$F_k(X_1, \dots, X_{2k}) = \sum_{\sigma\in E_{2k}} (-\frac{1}{2})^{a_{\pi(\sigma)}} \phi_\sigma(Y_1, \dots, Y_{2k}).$$

PROOF. The point is the following. Consider a symbol $c$ with $\pi(c) = s$ encoding a trace monomial $\operatorname{tr}(M_1)t(M_2)\cdots\operatorname{tr}(M_s)$. The same symbol encodes also the different trace monomials obtained from $\operatorname{tr}(M_1)t(M_2)\cdots\operatorname{tr}(M_s)$ by transposing a subset of the $M_j$'s since $\operatorname{tr}(M) = \operatorname{tr}(M^t)$. Thus each of the $2^{2k}$ symbols appearing in $\phi_\sigma(Y_1, \dots, Y_{2k}) = \phi_\sigma(X_1 - X_1^t, \dots, X_n - X_{2k}^t)$ appear also in $2^{\pi(c)}$ different terms of the expansion which thus equals:

$$\sum_{\pi(\sigma)=\lambda} \phi_\sigma(Y_1, \dots, Y_{2k}) = 2^{a_\lambda} \sum_{c \mid \pi(c)=\lambda} \operatorname{Tr}(c).$$

The sign of all these terms is the same, and by Remark 13.2.10 it is $(-1)^{a_\lambda}$. □

We next compute the noncommutative relation associated to $F_k(X_1,\dots,X_{2k})$. For this notice that in formula (13.13) we can assume that $q_\ell = 2k$ and write

$$\phi_\sigma(Y_1,\dots,Y_{2k}) = \mathrm{tr}(\psi_\sigma(Y_1,\dots,Y_{2k-1})Y_{2k},$$

where now $\psi_\sigma(Y_1,\dots,Y_{2k-1})$ is a sum of trace monomials times monomials in the $Y_i$, $i=1,\dots,2k-1$, of odd length.

Now if $M(Y)$ is a monomial of odd length in skew symmetric matrices, we claim that $M(Y)+M^t(Y)$ is skew symmetric. Notice that we are thinking of $M^t(Y)$ as the formal transpose of $M$ evaluated in the $Y$ but now $(M(Y))^t = -M^t(Y)$ since the length is odd.

As a consequence since, in the expansion of $\sum_{\sigma\in E_{2k}}(-\frac{1}{2})^{a_{\pi(\sigma)}}\phi_\sigma(Y_1,\dots,Y_{2k})$, together with each monomial appears also its transpose, we have that the polynomial $\sum_{\sigma\in E_{2k}}(-\frac{1}{2})^{a_{\pi(\sigma)}}\psi_\sigma(Y_1,\dots,Y_{2k-1})$ takes as itsvalues skew symmetric matrices. Since the trace form is nondegenerate on skew symmetric matrices it follows that

COROLLARY 13.2.12. *The polynomial* $\sum_{\sigma\in E_{2k}}(-\frac{1}{2})^{a_{\pi(\sigma)}}\psi_\sigma(Y_1,\dots,Y_{2k-1})$ *is an identity for* $2k-1\times 2k-1$ *skew matrices.*

This gives in particular for $k=2$ the formula (13.11).

13.2.2.2. *General k.* The general case is when in the three subsets $A,B,C$ of formula (13.4) the set $B$ is nonempty. In this case in the development of the determinant $D_{k,n}$ the scalar symbols have still a restriction, that is an index in $A$ is never paired with $A$ and similarly for $C$. So we have

DEFINITION 13.2.13. We set $E_{k,n}$ be the set of permutations $\sigma\in S_n$ such that if $i\in A$, then $\sigma(i)\notin A$, and if $i\in C$, then $\sigma(i)\notin C$.

For the moment we only have an example: $n=4, k=3$. The trace relation is

$$-\frac{1}{2}\mathrm{tr}(Y_4Y_1)\,\mathrm{tr}(X_2)\,\mathrm{tr}(X_3)+\frac{1}{2}\mathrm{tr}(Y_4Y_1)\,\mathrm{tr}(X_3X_2)+\mathrm{tr}(X_3Y_1Y_4)\,\mathrm{tr}(X_2)$$
$$+\mathrm{tr}(X_2Y_1Y_4)\,\mathrm{tr}(X_3)-\mathrm{tr}(Y_4X_3X_2Y_1)-\mathrm{tr}(Y_4X_2X_3Y_1)-\mathrm{tr}(Y_4X_2Y_1X_3^t),$$

written in symmetric form for $X_2, X_3$ as

$$=-\frac{1}{2}\mathrm{tr}(Y_4Y_1)\,\mathrm{tr}(X_2)\,\mathrm{tr}(X_3)+\frac{1}{2}\mathrm{tr}(Y_4Y_1)\,\mathrm{tr}(X_3X_2)+\mathrm{tr}(X_3Y_1Y_4)\,\mathrm{tr}(X_2)$$
$$+\mathrm{tr}(X_2Y_1Y_4)\,\mathrm{tr}(X_3)-\mathrm{tr}(Y_4X_3X_2Y_1)-\mathrm{tr}(Y_4X_2X_3Y_1)$$
$$-\frac{1}{2}\mathrm{tr}(Y_4X_2Y_1X_3^t)-\frac{1}{2}\mathrm{tr}(Y_4X_3Y_1X_2^t).$$

The two noncommutative relations associated are

$$-\frac{1}{2}\mathrm{tr}(Y_3Y_1)\,\mathrm{tr}(X_2)+\frac{1}{2}\mathrm{tr}(Y_3Y_1)X_2+\mathrm{tr}(X_2Y_1Y_3)+\mathrm{tr}(X_2)Y_1Y_3$$
$$-Y_1Y_3X_2-X_2Y_1Y_3-Y_1X_2^tY_3$$

and

$$=-\frac{1}{2}\mathrm{tr}(X_2)\,\mathrm{tr}(X_3)Y_1+\frac{1}{2}\mathrm{tr}(X_3X_2)Y_1+\mathrm{tr}(X_2)Y_1X_3^t+\mathrm{tr}(X_3)Y_1X_2^t$$
$$-Y_1X_2^tX_3^t-Y_1X_3^tX_2^t-\frac{1}{2}X_3Y_1X_2^t-\frac{1}{2}X_2Y_1X_3^t.$$

One can also interpret in a similar way Theorem 13.2.1, as Theorem 12.2.16, on symplectic involution and obtain Theorem thm13-2-14. Denote by $R^s(2n)$ the trace algebra of generic symplectic $2n \times 2n$ matrices (equivariant maps of symplectic matrices) and $T^s(2n)$ the traces (invariants of symplectic matrices) (Theorem 13.1.4). From Theorem 13.2.1 we deduce

THEOREM 13.2.14. *In characteristic* 0,

(i) *The algebra* $R^s(2n)^+/T^s(2n)^+R^s(2n)$ *is the free nil algebra with involution in which every symmetric element is nilpotent of exponent* $n$. *In other words it is the free algebra without one with involution modulo the polynomial identity* $(z+z^*)^n = 0$.

(ii) *If an algebra* $R$ *with involution satisfies* $(z+z^*)^n = 0$, *then it satisfies* $z^{2n} = 0$.

PROOF. The first part is like the case of nil algebras, Theorem 12.2.16. For the second part, it is enough to remark that $2n \times 2n$ symplectic matrices satisfy the Cayley–Hamilton theorem in degree $2n$, which then can be deduced from the Pfaffian identity. □

CHAPTER 14

# A geometric approach

*This chapter has several goals. The first is to study, of a finitely generated PI-algebra $R = F\langle a_1, \dots, a_m\rangle$, its maximal spectrum as some sort of* noncommutative affine variety. *If $\mathfrak{m}$ is a maximal ideal, $F$ is algebraically closed, and $R$ satisfies a polynomial identity of degree $d$, then $R/\mathfrak{m}$ is isomorphic (in a noncanonical way) to $M_h(F)$, $2h \leq d$. This fact induces us to introduce methods of geometric invariant theory.*

*The methods become particularly effective if $R$ is an algebra with trace satisfying the nth Cayley–Hamilton identity. In this case the link with geometry becomes rather strict, as shown in Theorem* 14.2.1 *and in the discussion on page 369 and the following sections.*

## 14.1. Geometric invariant theory

**14.1.1. Hilbert's theory.** In 19th century mathematics one of the main themes of geometry was to understand geometric structures as having groups of symmetries, such as projective, affine, Euclidean, or hyperbolic geometry.

One of the aspects of this study is the study of algebraic invariants. In a simplified treatment, one has a group $G$ of linear transformations on a vector space $V$, then $G$ acts on the space $S(V^*)$ of polynomial functions on $V$ by the formula $(gf)(v) := f(g^{-1}v)$. The space of polynomials invariant under $G$,

$$S(V^*)^G := \{f \in S(V^*) \mid gf = f, \ \forall g \in G\}, \tag{14.1}$$

is a subalgebra, the object of this study. By the main ideas of algebraic geometry we see that the subvariety defined by a set of invariants is stable under the group $G$. The same happens if we take a system of equations stable under $G$, and conversely the ideal associated to a subvariety stable under the group $G$ is also stable under $G$. In particular one may try to describe invariants defining special $G$ stable varieties.

The algebraic part of the theory consists in describing the algebra $S(V^*)^G$ of invariant polynomials. There are several obvious questions in this description, and we shall discuss some of them and refer to Kraft **[Kra84]** for a general treatment.

The geometric part of the study is the decomposition of $V$ into the *G-orbits*, where the orbit of an element $v$ is as usual the set $Gv := \{gv \mid g \in G\}$.

For instance if $G = \mathrm{GL}(n, \mathbb{C})$ acts by conjugation on $M_n(\mathbb{C})$, the orbits are the conjugacy classes, classified by the *Jordan canonical form*.

Clearly an invariant function is constant on any given orbit, and one would like to think of $S(V^*)^G$ as *an algebra parametrizing the G-orbits*. That is, we may ask if the set of $G$-orbits can be given the structure of affine variety with coordinate algebra $S(V^*)^G$, so that the inclusion $S(V^*)^G \subset S(V^*)$ corresponds to the map which associates to a point of $V$ its orbit. This is true if $G$ is a finite group, but as soon as $G$ is infinite several subtle questions arise.

This question has two aspects, the first, known as *Hilbert's* 14*th Problem*, is to prove that the algebra $S(V^*)^G$ is finitely generated.

In the 19th century theory of *binary forms*, where the group is $\mathrm{SL}(2,\mathbb{C})$, this was the great achievement of Gordan, using a *symbolic method*.

If $S(V^*)^G$ is finitely generated, then it is the coordinate ring of an affine variety, which by definition is the *quotient variety* $V//G$. The next problem is to study the map $V \to V//G$ induced by the inclusion $S(V^*)^G \subset S(V^*)$.

It has been shown by Nagata that in general the answer to Hilbert's 14th Problem is negative but it is positive for a class of groups, the *reductive groups*.

Among these are all classical groups, which have a particularly interesting role, especially in our treatment.

In contemporary language one sets the problem using *linear algebraic groups*. These are the subgroups of the group $\mathrm{GL}(n,F)$ of invertible matrices which are closed under the Zariski topology, that is they are defined by polynomial equations. We shall recall some basic facts on these groups by referring to the standard texts on the theory, such as **[Bor69]**, **[Hum80]**, **[Spr09]**, or **[Pro07]**, for a very short and basic introduction.

In fact it is clear that the elements of $S(V^*)$ which are invariant under $G$ are also invariant under the Zariski closure of $G$, this is by definition a linear algebraic group. In our treatment only classical groups will arise, that is the general linear group $\mathrm{GL}(n,F)$ of all invertible matrices over $F$, the special linear group $\mathrm{SL}(n,F)$ of matrices of determinant 1, the projective, orthogonal and symplectic groups. All these groups are reductive.[1]

Denote by $A := F[x_{i,j}]$ the algebra of polynomial functions on $n \times n$ matrices, that is polynomials in the variable entries $x_{i,j}$, $i,j = 1,\dots,n$. The group $\mathrm{GL}(n,F)$ can also be viewed as the subset of $M_n(F) \times F$ of pairs $a, c$ with $\det(a)c = 1$ (and embedded in $\mathrm{SL}(n+1,F)$). As every affine algebraic variety $\mathrm{GL}(n,F)$ has a *ring of regular functions*, in our case it is the ring $A[d^{-1}] = F[x_{i,j}][d^{-1}]$ of the rational functions $fd^{-k}$ with $f \in A$ while $d$, which may appear at the denominator, is the determinant.

DEFINITION 14.1.1. For a subgroup $G \subset \mathrm{GL}(n,F)$, the ring $\mathcal{O}(G)$ of *regular functions* is the restriction to $G$ of the regular functions $A[d^{-1}] = F[x_{i,j}][d^{-1}]$ of $\mathrm{GL}(n,F)$. It is thus by construction isomorphic to $A[d^{-1}]/I$ where $I$ is the ideal of functions vanishing on $G$.

DEFINITION 14.1.2. A finite-dimensional representation $\rho : G \to \mathrm{GL}(N,F)$ of a linear algebraic group $G$ is said to be *rational* if the matrix entries $\rho(g)_{i,j}$ of the elements $\rho(g)$ are regular functions of $g$.

An arbitrary representation is said to be rational if it is a sum of rational representations.

EXAMPLE 14.1.3. An example of a *nonrational* representation is for $\mathrm{GL}(n,\mathbb{C})$ the map $X \mapsto \bar{X}$ (the conjugate).

The simplest examples of algebraic groups are the *tori* $T_n$, that is the groups of invertible diagonal matrices. The coordinate ring of an $n$-dimensional torus is the ring of Laurent polynomials $F[x_1^{\pm 1},\dots,x_n^{\pm 1}]$.

[1] A reductive group is one with connected solvable radical a torus. In characteristic 0 this is equivalent to being linearly reductive (cf. Definition 14.1.6).

A rational representation of a torus decomposes as a direct sum of one-dimensional representations, which are *characters*, that is maps:

(14.2) $$(x_1, \dots, x_n) \mapsto x_1^{h_1} x_2^{h_2} \cdots x_n^{h_n}, \quad (h_1, h_2, \dots, h_n) \in \mathbb{Z}^n.$$

If $G \subset \mathrm{GL}(n, F) = \mathrm{GL}(V)$ is an algebraic subgroup then by definition $V$ is rational.

The main fact is that the coordinate ring $\mathcal{O}(G)$ is a rational representation of $G$ both for the left and for the right actions

$$^g f(h) := f(g^{-1}h), \quad f^g(h) := f(hg).$$

THEOREM 14.1.4. *Every finite-dimensional rational representation of G embeds in a finite direct sum of copies of* $\mathcal{O}(G)$.

EXERCISE 14.1.5. If $V, W$ are rational representations of $G$, so are $V \oplus W$, $V \otimes W$, $V^*$, as well as $R = S(V^*) = \mathcal{P}(V)$ the ring of polynomial functions on $V$.

If $W \subset V$ is a subrepresentation of a rational representation $V$, then $W$ and $V/W$ are rational.

In characteristic 0 all classical groups, that is the general linear group $\mathrm{GL}(n, F)$, the special linear group $\mathrm{SL}(n, F)$, the orthogonal and special orthogonal groups and the symplectic groups have a very special property—they are *linearly reductive*. The main definition is thus

DEFINITION 14.1.6. An algebraic group $G$ is *linearly reductive* if all its rational representations are semisimple.

REMARK 14.1.7. From Theorem 14.1.4 it follows that an algebraic group $G$ is linearly reductive if and only if its coordinate ring $\mathcal{O}(G)$ is a direct sum of irreducible representations under the left (or equivalently the right) action.

In particular tori are linearly reductive in all characteristics.

To the contrary the groups $B_n, U_n$ of upper triangular (resp., strictly upper triangular) matrices are not reductive. As one can see, their defining representation is not semisimple and they are solvable but not tori.

In positive characteristic the only connected linearly reductive groups are tori.

In characteristic 0, linearly reductive groups can be fully classified and are strictly connected with simple Lie algebras and compact Lie groups. In fact one characterization of a linearly reductive group over $\mathbb{C}$ is that it should contain a Zariski dense compact subgroup, as for instance the unitary group inside the general linear group $\mathrm{GL}(n, \mathbb{C})$.[2] A simple criterion, when $F = \mathbb{C}$, to check that an algebraic subgroup $G \subset \mathrm{GL}(n, \mathbb{C})$ is linearly reductive is to verify that it is self-adjoint, that is $X \in G \implies \bar{X}^t \in G$ (cf. **[Pro07]**).

Suppose then that we have a rational representation $V$ of a linearly reductive group $G$. By the theory of semisimple modules, the space $V$ decomposes as a direct sum of its isotypic components. In particular we may consider the decomposition $V = V^G \oplus V_G$, where $V^G$ denote the space of invariants, that is the isotypic component of the trivial representation, and $V_G$ is the sum of all the other isotypic components. The projection $\pi_G : V \to V^G$ with kernel $V_G$ is often called the *Reynolds operator*. It has the basic property that

[2]This is one form of *Weyl's unitary trick*.

REMARK 14.1.8. Given any $G$-equivariant map $f : V \to W$ between two rational representations, the corresponding diagram commutes.

$$\begin{array}{ccc} V & \xrightarrow{f} & W \\ \pi_G \downarrow & & \pi_G \downarrow \\ V^G & \xrightarrow{f} & W^G \end{array} \tag{14.3}$$

Moreover if $f$ is surjective, since $f$ preserves the isotypic components, it is surjective on each component, in particular it is surjective on invariants.

In particular suppose that $V = R$ also has an algebra structure for which $G$ acts as a group of automorphisms. Then the invariants $R^G$ form a subalgebra and, given $a \in R^G$, the two maps of left and right multiplication by $a$ are $G$ equivariant, thus for the Reynolds operator $\pi_G$ we have the Reynolds identities

$$\pi_G(ab) = a\,\pi_G(b), \quad \pi_G(ba) = \pi_G(b)a, \quad \forall b \in R,\ a \in R^G. \tag{14.4}$$

A particularly important case is when $R = \mathcal{P}(V) = S(V^*)$ is the ring of polynomial functions on a finite-dimensional rational representation $V$. In this case we have Hilbert's theory.

The main technical result is a very simple *going up lemma*:

LEMMA 14.1.9. *Let $R$ be a commutative algebra which is a rational representation of a linearly reductive group $G$, and let $I$ be an ideal of the algebra $R^G$ of invariants. Then*

$$IR \cap R^G = I.$$

PROOF. Let $a = \sum_i u_i b_i \in IR \cap R^G$, where $u_i \in I, b_i \in R$. We apply the Reynold's identities and get

$$a = \pi_G(a) = \sum_i u_i \pi_G(b_i) \in I. \tag{14.5}$$

□

When $R$ is noncommutative, then the statement is no more true for two-sided ideals, nevertheless we shall see that in a special and important case this can be extended, see Theorem 14.2.1.

Proposition 14.1.10 and Theorem 14.1.12 form essentially Hilbert's proof of finite generation of invariants of forms. This and the other arguments of Hilbert, like the basis theorem needed for this proof, were major breakthroughs in algebra and in a sense the birth of modern commutative algebra (Hilbert did not use the Reynold's operator but working only with the linear group acting on forms, he used Cayley's $\Omega$ process, the determinant of the matrix of differential operators $\frac{\partial}{\partial x_{i,j}}$).

PROPOSITION 14.1.10. *Let $S = \bigoplus_{i=0}^{\infty} S_i$ be a graded commutative algebra over a field $F$, and let $S_0 = F$, $S_+ = \bigoplus_{i=1}^{\infty} S_i$. If $u_i$ is a basis of $S_+$ modulo $S_+^2$ over the field $F$, then the $u_i$ form a minimal set of generators for the algebra $S$.*

*The elements $u_i$ can be taken from any set of generators of $S_+$ as ideal of $S$.*

PROOF. Denote by $S'$ the subalgebra of $S$ generated by the $u_i$. We want to show that $S = S'$.

This is, by induction on the degree of elements, $a \in S$. Assume we have shown that all elements of degree less than $k$ are in the subalgebra generated by the $u_i$, and let $a \in S_k$.

We first can find a linear combination $\sum_i f_i u_i,\ f_i \in F$ with $b := a - \sum_i f_i u_i \in S_+^2$. Then $b = \sum_j c_j d_j$ where the elements $c_j, d_j$ have degree less than $k$ and hence can be expressed as polynomials in the $u_i$, hence $b \in S'$ and also $a = b + \sum_i f_i u_i \in S'$ thus by induction $S = S'$ in all degrees.

This also shows that the $u_i$ are generators of $S_+$ as ideal.

conversely if $s_+ = \sum_i u_i s$, we have $s_+ = \sum_i f u_i + \sum_i u_i s_+ = \sum_i f u_i + s_+^2$. □

REMARK 14.1.11. The ring of invariants $R^G$ of the action of a group $G$ on a graded algebra and which preserves the degree is a graded subalgebra.

In particular this applies to $R = \mathcal{P}(V)$, the ring of polynomial functions on a finite-dimensional rational representation $V$ of a group $G$.

THEOREM 14.1.12. *Let $R = \mathcal{P}(V)$ be the ring of polynomial functions on a finite-dimensional rational representation $V$ of a linearly reductive group $G$. Then the ring of invariants $R^G$ is finitely generated.*

PROOF. We have seen that $R^G$ is a graded subalgebra of $R$, a polynomial ring. Let $I \subset R^G$ be the space of invariants with no constant term. We have that $IR$, by the Hilbert basis theorem, is a finitely generated ideal and the generators $u_1, \dots, u_k$ can be taken in $I$.

Reasoning as in Lemma 14.1.9, given $u \in I$, we write $u = \sum_i u_i b_i,\ b_i \in R$ and see that $u = \sum_i u_i \pi_G(b_i)$ (where $\pi_G$ is the Reynold's operator), hence $I$ is a finitely generated ideal in $R^G$. We now apply Proposition 14.1.10. □

In Hilbert's theory this means that the ring of invariants $S := \mathcal{P}(V)^G$ can be viewed as the coordinate ring of an affine algebraic variety, which will be denoted by $V//G$, and the inclusion $S \subset \mathcal{P}(V)$ defines a *quotient map* $\pi : V \to V//G$.

More generally, let $W \subset V$ be a subvariety stable under the action of $G$. Then if $I_W \subset \mathcal{P}(V)$ is the ideal of polynomial functions on $V$ vanishing on $W$, we have that $\mathcal{P}(V)/I_W$ is the coordinate ring of the affine variety $W$, and $G$ acts on $I_W$ and the quotient. By Remark 14.1.8 the projection $\pi$ induces a surjective map on invariants. $\pi : \mathcal{P}(V)^G \to (\mathcal{P}(V)/I_W)^G$.

Thus we have also that the ring $(\mathcal{P}(V)/I_W)^G$ is finitely generated. It is thus again the coordinate ring of an affine variety $W//G$ which is indeed a subvariety of $V//G$.

The next question that Hilbert addresses is what can one say about the variety $V//G$.

Of course we have that if $a, b \in V$ are in the same $G$-orbit, then all the invariants take the same values on $a, b$, and so $\pi(a) = \pi(b)$. On the other hand this is true also if $b$ is in the closure of the $G$-orbit of $a$, so under $\pi$ the closure of a $G$-orbit maps to a single point. Here the main result is

THEOREM 14.1.13. *The map $\pi : V \to V//G$ is surjective. For every $p \in V//G$ the fiber $\pi^{-1}(p)$ contains one and only one closed orbit of $G$.*

PROOF. For the first statement let $m_p$ be the maximal ideal of $p$. The fiber $\pi^{-1}(p)$ of $p$ is the subvariety defined by the ideal $m_p\mathcal{P}(V)$ and to say that it is nonempty, that is $\pi$ is surjective, is equivalent as saying that $m_p\mathcal{P}(V) \neq \mathcal{P}(V)$ is a proper ideal. This is ensured by Lemma 14.1.9.

Let us pass to the second statement. For this we use some simple properties of algebraic varieties, that is consider the subvariety $\pi^{-1}(p)$ which is clearly stable under $G$. If in this variety we take a $G$ orbit $O$ of minimal dimension, then $O$ is closed; otherwise the complement of $O$ in its closure would contain orbits of lower dimension. Hence we have that any fibre contains a closed orbit. Now take two distinct, hence disjoint, closed orbits $O_1, O_2$ and let $I_1, I_2$ be their ideals of definition. Since these two subvarieties are disjoint, we must have that $I_1 + I_2 = \mathcal{P}(V)$, in other words we can write $1 = a_1 + a_2$ with $a_i \in I_i$. We apply the Reynold's operator and then see that we can take both $a_i \in \mathcal{P}(V)^G$. This implies that the two projections $p_i = \pi(O_i)$ (which are two points) must be distinct since $a_1$ vanishes on $p_1$ and $a_2$ vanishes on $p_2$ so $a_1 = 1$ on $p_2$ and $a_2 = 1$ on $p_1$. □

We have developed for simplicity the theory as Hilbert did in characteristic 0. Due to the work of Nagata, Mumford, and others, the theory can be developed also in positive characteristic, where the notion of linear reductivity has to be replaced by that of geometric reductivity. This condition is what we have used in the proof of Theorem 14.1.13, that is we now take as a condition that given two distinct closed orbits there is an invariant which vanishes on one and not on the other.

The fact that a reductive group is also geometrically reductive was conjectured by Mumford **[Mum65]**, and it has been known as *Mumford's conjecture* and was proved by Haboush **[Hab75]**. There is a proof using some noncommutative algebra, in the case of the linear group by Formanek and Procesi **[FP76]**.

The interested reader may consult **[Nag60]**, **[Nag63]**, **[Spr09]**, among others.

By their theory it still follows that the ring of invariants is finitely generated and that the quotient variety parametrizes the closed orbits, but the Reynold's identities are no more available and the theory is much more difficult to develop.

In our treatment our prime object is the study of finite-dimensional representations of a free algebra, and this will be done in the next section.

**14.1.2. The variety of semisimple representations.** We now go back to one of our main themes. Consider the space of $k$-tuples of matrices $M_n(F)^k$ where for simplicity we shall now assume that $F$ is algebraically closed (but not necessarily of characteristic 0).

We think of the space $M_n(F)^k$ as the *space of $n$-dimensional representations of the free algebra $F\langle x_1, \ldots, x_k\rangle$ in $k$ generators*, where a representation $\rho : F\langle x_1, \ldots, x_k\rangle \to M_n(F)$ corresponds to the $k$-tuple $(a_i := \rho(x_i))$.

Now the linear group $\mathrm{GL}(n, F)$ acts by simultaneous conjugation. In fact this action is trivial by the scalar matrices and hence should be thought of as an action of the *projective linear group*

$$G := \mathrm{GL}(n, F)/F^* = \mathrm{PGL}(n, F).$$

REMARK 14.1.14.

(1) Clearly two $k$-tuples are in the same orbit if and only if the two representations are isomorphic. In other words the study of isomorphism classes of representations is the same as the study of $G$-orbits.
(2) Notice that in the language of orbits, if $\rho : F\langle x_1, \ldots, x_k\rangle \to M_n(F)$ is a representation, i.e., $\rho \in M_n(F)^k$, its stabilizer is the set of elements of $G$ for which $g\rho g^{-1} = \rho$. That is the elements of $G$ which are in the commutant or centralizer of the representation.

Thus a preliminary fact we should understand is which representations correspond to closed orbits. This has been shown by M. Artin **[Art69]**.

First let us recall that given any module $V$, over an $F$-algebra $R$, so that $V$ is of finite dimension $n$ over the field $F$, we can construct a Jordan–Hölder series $V = V_1 \supset V_2 \supset \cdots \supset V_m \supset V_{m+1} = 0$ of submodules $V_i$ so that $V_i/V_{i+1}$ is always irreducible. The semisimple representation $\bigoplus_i V_i/V_{i+1}$ does not depend, up to isomorphism, on the chain chosen; this is the Jordan–Hölder Theorem. It will be called the *semisimple module* associated to the given module. When we fix a basis of $\bigoplus_i V_i/V_{i+1}$, we have associated to this module a semisimple representation whose orbit is independent of the basis.

THEOREM 14.1.15. *A semisimple representation associated to any given representation is in the closure of its orbit. A representation is in a closed orbit if and only if it is semisimple.*

PROOF. Take a representation of the free algebra given by a $k$-tuple $(a_1, \dots, a_k)$ of $n \times n$ matrices. We want to show that the semisimple representation associated via a Jordan–Hölder chain is in the closure of the orbit. For this we take a basis adapted to the Jordan–Hölder chain so that all the matrices $a_i$ have a block triangular form

$$\begin{pmatrix} A_{1,1} & A_{1,2} & \cdots & A_{1,m} \\ 0 & A_{2,2} & \cdots & A_{1,m} \\ \dots & \dots & \dots & \dots \\ 0 & 0 & \cdots & A_{m,m} \end{pmatrix},$$

where the matrices $A_{i,i}$ of size $\dim(V_i/V_{i+1})$ represent the action on the space $V_i/V_{i+1}$ which is irreducible. Take now a block diagonal matrix $T$ with entry $t^{m-i+1}$ on the $i$th block.

We have that $T$ on the block $A_{i,j}$ acts by multiplication by $t^{j-i}$:

$$T \begin{pmatrix} A_{1,1} & A_{1,2} & \cdots & A_{1,m} \\ 0 & A_{2,2} & \cdots & A_{1,m} \\ \dots & \dots & \dots & \dots \\ 0 & 0 & \cdots & A_{m,m} \end{pmatrix} T^{-1} = \begin{pmatrix} A_{1,1} & tA_{1,2} & \cdots & t^{m-1}A_{1,m} \\ 0 & A_{2,2} & \cdots & t^{m-2}A_{2,m} \\ \dots & \dots & \dots & \dots \\ 0 & 0 & \cdots & A_{m,m} \end{pmatrix}.$$

As long as $t \neq 0$ this is the conjugation action, and we stay in the orbit. But by the formula it is clear that in the closure of the orbit we have the value of this map at $t = 0$ which is the block diagonal matrix

$$\begin{pmatrix} A_{1,1} & 0 & \cdots & 0 \\ 0 & A_{2,2} & \cdots & 0 \\ \dots & \dots & \dots & \dots \\ 0 & 0 & \cdots & A_{m,m} \end{pmatrix}$$

representing the semisimple representation $\bigoplus_i V_i/V_{i+1}$. This proves the first part.

As for the second part it is enough to show that two distinct semisimple representations cannot be in the closure of the same orbit (this is some kind of Jordan–Hölder statement).

In fact this can be viewed as a statement on reductive groups, where closed orbits can be characterized by the fact that the stabilizer is also reductive. We give a more ring-theoretical proof as follows. Take a semisimple representation $\rho_1$. If by contradiction its orbit is not closed, in its closure we can find other orbits and by the previous argument finally another orbit of a semisimple representation $\rho_2$. Clearly $\rho_1, \rho_2$ cannot be distinguished by invariants. So the theorem will follow from

LEMMA 14.1.16. *If $\rho_1, \rho_2$ are nonisomorphic semisimple n-dimensional representations of the free algebra $F\langle X\rangle$, then they can be distinguished by the values of the characteristic polynomials of the elements $\rho_1(a), \rho_2(a)$ for some elements $a \in F\langle X\rangle$.*

PROOF. We first remark that any element $a = a(x_1, \dots, x_k) \in F\langle X\rangle$ is a noncommutative polynomial and $\rho_1(a), \rho_2(a)$ are evaluations of this polynomial in two different $k$-tuples of matrices.

When we evaluate $a$ into $k$-tuples of matrices $(\xi_1, \dots, \xi_k)$, with the evaluation map $e : x_i \mapsto \xi_i$, and then compute the resulting characteristic polynomial, we get a polynomial $\det(t - e(a)) = \det(t - a(\xi_1, \dots, \xi_k))$ whose coefficients are $n$-invariant functions on $M_n(F)^k$ (depending on $a$).

A semisimple representation $\rho$ consists in choosing a $k$-tuple $(a_1, \dots, a_k)$ of matrices generating a semisimple subalgebra $S := \rho(F\langle X\rangle) \subset M_n(F)$.

Observe that the characteristic polynomial of $\rho(a) \in S$ equals $t^n$, that is all its coefficients vanish if and only if $\rho(a)$ is nilpotent.

We claim that the kernel of $\rho$ is formed by the elements $a \in F\langle X\rangle$ with the property that, for all $b \in F\langle X\rangle$, the characteristic polynomial of $\rho(ab) \in S$ equals $t^n$, that is all its coefficients vanish. In fact if $a$ is in the kernel, $\rho(ab) = 0$, but if $a$ has the previous property, then $\rho(a)S$ is a nil ideal of $S$ hence, since $S$ is semisimple, it is zero.

So first of all the values of the characteristic polynomial of the elements $\rho(a)$ determine the kernel of $\rho$ and hence determine at least $S$ as algebra. We need to show that they determine also the isomorphism class of the $n$-dimensional representation.

By Wedderburn's theorem $S = \bigoplus_{i=1}^h \mathrm{End}(V_i)$ with $V_i$ of some dimension $k_i$. A semisimple faithful representation of $S$, of dimension $n$, is up to isomorphism given by taking a direct sum $\bigoplus_{i=1}^h V_i^{\oplus p_i}$ with $0 < p_i$, $\sum_i p_i k_i = n$.

Therefore we only need to show that the invariants determine the multiplicities $p_i$. For this notice that if $e_i$ is the identity element of $\mathrm{End}(V_i)$, on the previous representation it is an idempotent of rank $p_i k_i$, hence its characteristic polynomial is $t^{n-p_ik_i}(t-1)^{p_ik_i}$ and hence determines $p_ik_i$. Having fixed $S$, the multiplicities $p_i$ are determined by the values of the characteristic polynomials of the elements $e_i$. □

□

REMARK 14.1.17. As soon as $k \geq 2$, a *generic* $k$-tuple of $n \times n$ matrices generates the algebra $M_n(F)$, and it is easy to see that such $k$-tuples form a nonempty open set (cf. Theorem 14.3.8). This is the open set of irreducible representations. By Remark 14.1.14(2) and by Schur's lemma, the orbit of an irreducible representation is isomorphic to $\mathrm{GL}(n, F)/F^* = \mathrm{PGL}(n, F)$.

It is a standard result in commutative algebra (or algebraic geometry), that given a dominant map $\pi : V \to W$ between two irreducible varieties (for instance a surjective map), the dimension of $V$ equals $\dim W + \dim(\pi^{-1}(p))$ where $p$ is a generic point of $W$.

In the case of group actions, if $G$ is a reductive group acting on an irreducible variety $V$, we have defined the *quotient* $V//G$ as the variety with coordinate ring the ring of $G$-invariants.

If the generic orbit of $G$ on $V$ is closed, it equals by Hilbert's theory the generic fiber, and so the dimension of the ring of invariants equals the dimension of $V$ minus the dimension of the generic orbit. We deduce

THEOREM 14.1.18. *The ring of invariants of $k \geq 2$-tuples of $n \times n$ matrices, denoted $\mathcal{T}_n(X)$, is the coordinate ring of an irreducible affine variety*

$$V_n(F) := M_n(F)^k // \operatorname{GL}(n, F), \tag{14.6}$$

*of dimension $(k-1)n^2 + 1$ whose points parametrize isomorphism classes of semisimple representations of dimension $n$ of the free algebra in $k$-variables.*

PROOF. This variety is the quotient of the affine space $M_n(F)^m$, which has dimension $mn^2$, by conjugation action. In this action the scalars of $\operatorname{GL}(n, F)$ act trivially so it is really an action of the projective linear group which has dimension $n^2 - 1$.

If $m \geq 2$, a generic element of $M_n(F)^m$ corresponds to an irreducible representation of the free algebra. We have seen that its orbit is closed, and by Schur's lemma the stabilizer under $\operatorname{GL}(n, F)$ is $F^*$. So the stabilizer under $\operatorname{PGL}(n)$ is trivial, and a generic orbit has dimension $n^2 - 1$.

We can thus apply the previous discussion, and we get that the dimension of the quotient is $mn^2 - (n^2 - 1) = (m-1)n^2 + 1$. □

In positive characteristic we have to invoke the theory of Nagata and Mumford and the fact that the linear group is geometrically reductive. On the other hand this statement follows also from the discussion which shall appear in §14.3.3.

Of course, by the invariant theory of a single matrix, if $k = 1$, we have that $M_n(F)/\operatorname{GL}(n, F)$ is an $n$-dimensional affine space, the generic orbit is that of a matrix with distinct eigenvalues with stabilizer an $n$-dimensional torus.

Now consider any finitely generated algebra $R = F\langle a_1, \ldots, a_k \rangle$ which we can thus view as quotient of the free algebra $R = F\langle x_1, \ldots, x_k \rangle / I$. The representations of dimension $n$ of $R$ are thus those representations of the free algebra, that is elements of $M_n(F)^k$, which factor through the kernel $I$. The condition to factor through $I$ defines a subvariety of $M_n(F)^k$. This is the variety defined by the ideal $J \subset F[\xi]$, generated by all entries of the image of $I$, in the matrix ring $M_n(F[\xi])$, given by the map of the free algebra to the algebra of generic matrices.

This ideal $J$ defines the universal ring $F_n(R) = F[\xi]/J$ of Definition 3.4.6.

Notice though that in general $J$ is not the full ideal associated to this variety, that is it is not equal to its radical. Nevertheless we can see that, in characteristic 0, the ring of invariants $F_n(R)^{\operatorname{GL}(n,F)}$ is a quotient of the ring of invariants $\mathcal{T}_n(X)$.

The ring of invariants $F_n(R)^{\operatorname{GL}(n,F)}$ is finitely generated and, although it may have a nonzero nilpotent radical, its associated variety is the set of equivalence classes of semisimple $n$-dimensional representations of $R$.

**Smooth and singular points.** As with any algebraic variety one might study some of its geometric properties. Using Luna's étale slice theorem **[Lun73]**, it is possible to prove

THEOREM 14.1.19. *Except in the case $k = 1$ or $k = n = 2$, where the ring of invariants is a polynomial algebra in five variables (see Example 14.3.2), the variety of formula (14.6) $V_n(F) := M_n(F)^k // \operatorname{GL}(n, F)$ is smooth exactly in the points corresponding to irreducible representations.*

Since these methods go beyond the tools used in this book, we refer for this and related questions to **[LBP87]**, **[LBVdB88]**, **[SS85]**, **[VdB89]**, **[VdB91]**.

## 14.2. The universal embedding into matrices

**14.2.1. The representability of CH algebras.** Recall that for the geometric Hilbert theory, Reynold's operator (14.4) plays a major role in the going up lemma, Lemma 14.1.9. We need a variation of that lemma which applies also to noncommutative algebras and ideals. We have in mind algebras $R$ with trace such that $G$ acts as an automorphism, respecting the trace, that is $\mathrm{tr} : R \to R$ is $G$-equivariant. In this case one has besides the usual identities (14.4),

$$\pi_G(ab) = a\pi_G(b),\ \pi_G(ba) = \pi_G(b)a,\ \pi_G(\mathrm{tr}(b)) = \mathrm{tr}(\pi_G(b)),\quad \forall b \in R,\ \forall a \in R^G. \tag{14.7}$$

Using ideas similar to that of Lemma 14.1.9 we draw a major consequence of Theorem 12.1.17 for Cayley–Hamilton algebras of Definition 12.2.1:

THEOREM 14.2.1. *Let $R$ be an algebra, with 1, with trace, and over a field $F$ of characteristic 0 satisfying the Cayley–Hamilton identity of degree $n$ and $\mathrm{tr}(1) = n$. Then the universal mapping $i_R$ (given by formula (3.52) with $F$ in place of $A$) into $n \times n$ matrices is an isomorphism of $R$ with the invariant algebra*

$$i_R : R \xrightarrow{\cong} M_n(F_n(R))^{\mathrm{GL}(n,F)}. \tag{14.8}$$

PROOF. Denote $G = \mathrm{GL}(n, F)$. Let $R$ be an algebra with trace satisfying the $n$-CH identity. We present $R = R_X(n)/I$, where $R_X(n)$ is a corresponding free algebra with trace modulo the Cayley–Hamilton identity of Theorem 12.1.17(iii) (or Corollary 12.2.10), which we may assume to be in infinitely many variables $X$.

By Corollary 12.1.7 the algebra $R_X(n)$ is embedded in $B := M_n(F[\xi])$ and in fact $R_X(n) = B^G$ is the ring of invariants under conjugation. The action of $G$ on $B$ satisfies all the properties leading to formula (14.7).

We have that $BIB = M_n(J)$ for some $G$ stable ideal $J$ (Lemma 2.6.2), and thus by linear reductivity a surjective map $\pi : R = R_X(n)/I \mapsto M_n(F[\xi]/J)^G$.

$B$ is multigraded and $\pi_G$ preserves the multidegree. We need only to show that $BIB \cap R(n) = I$ since this implies that $\pi$ is also injective hence an isomorphism.

Let $a = \sum a_i u_i b_i \in BIB \cap R_X(n), u_i \in I, a_i, b_i \in B$. Choose a variable $x \in X$ which does not appear in any of these elements. We then have

$$\mathrm{tr}(ax) = \mathrm{tr}(\sum a_i u_i b_i x) = \mathrm{tr}(\sum b_i x a_i u_i).$$

Applying $\pi_G$, we obtain

$$\mathrm{tr}(ax) = \mathrm{tr}(\pi_G(\sum b_i x a_i u_i)) = \mathrm{tr}(\pi_G(\sum b_i x a_i) u_i).$$

Now for each $i$ the element $\pi_G(b_i x a_i)$ is an invariant linear in $x$, so by the first fundamental theorem, Theorem 12.1.3, we have

$$\pi_G(b_i x a_i) = \sum_j s_{i,j} x t_{i,j} + \sum_k \mathrm{tr}(m_{i,k} x) n_{i,k}$$

with $s_{i,j}, t_{i,j}, m_{i,k}, n_{i,k} \in R(n)$ independent of $x$. Thus

$$\begin{aligned}\operatorname{tr}(ax) &= \operatorname{tr}(\sum_i(\sum_j s_{i,j}xt_{i,j} + \sum_k \operatorname{tr}(m_{i,k}x)n_{i,k})u_i) \\ &= \operatorname{tr}(\sum_i\sum_j t_{i,j}u_is_{i,j}x + \sum_k \operatorname{tr}(n_{i,k}u_i)m_{i,k}x),\end{aligned}$$

hence

$$a = \sum_i\sum_j t_{i,j}u_is_{i,j} + \sum_{i,k} \operatorname{tr}(n_{i,k}u_i)m_{i,k}.$$

Since $u_i \in I$ and $I$ is closed under trace, we have finally $a \in I$, as desired. □

COROLLARY 14.2.2. *The universal map $i_R : R \xrightarrow{i_R} M_n(B_n(R))^{\mathrm{GL}(n,F)}$ of an algebra with trace is surjective, and its kernel is the ideal $I_n(R)$ (closed under trace) of $R$ generated by all the evaluations $\chi_n(a)$ of the nth Cayley–Hamilton polynomial.*

PROOF. Clearly the map $i_R$ factors through the quotient $R/I_n(R)$, and it gives the universal map of this algebra which is a CH algebra by construction. □

**Two categories.** This theorem in particular establishes a 1–1 correspondence between $n$-Cayley–Hamilton algebras and some commutative algebras equipped with an action of the group $\mathrm{PGL}(n, F)$, that is $R \mapsto B_n(R)$, with $B_n(R)$ the universal ring. In fact given any commutative ring $A$ equipped with an action of the group $\mathrm{PGL}(n, F)$, we may associate to it an $n$-Cayley–Hamilton algebra by considering the diagonal action of $\mathrm{PGL}(n, F)$ on $M_n(F) \otimes A = M_n(A)$ and taking the invariant algebra $R := M_n(A)^{\mathrm{PGL}(n,F)}$. It is easy to see that this is not an inverse to the previous map but only an adjoint. In fact one easily sees, from the universal properties that there is a $\mathrm{PGL}(n, F)$ equivariant map from $B_n(R)$ to $A$, but this is not necessarily an isomorphism.

For instance take as $A = A_k$ the coordinate ring of the closure of the conjugacy class of some nilpotent $n \times n$ matrix $X$ with $X^k = 0$. There are in general several such nilpotent classes in the subvariety of nilpotent matrices $X^k = 0$ corresponding to different Jordan blocks, each of nilpotency $\leq k$ and at least one of nilpotency $= k$.

EXERCISE 14.2.3. Prove that the CH algebra associated to $A_k$ is $F[x]/(x^k)$. Its associated universal commutative ring is not reduced as soon as $k > 1$, and it has as its associated variety all nilpotent matrices with $X^k = 0$.

## 14.3. Semisimple representations of CH algebras

*In the remaining part of this chapter, for simplicity and as otherwise stated, we assume that $F$ is an algebraically closed field of characteristic* 0.

**14.3.1. A close look at semisimple representations.** As in Definition 12.1.9 we shall denote by $R_m(n) = R_X(n)$ the free trace algebra, in $m$ variables $X$, modulo the trace identities of $n \times n$ matrices and $\mathcal{T}_n(X)$ the algebra of traces over an algebraically closed field $F$. We drop $m$ when we take infinitely many variables.

It is interesting to analyze more closely the variety of semisimple representations $V_n(m) = M_n(F)^m//G$ of formula (14.6). Given a point $p \in V_n(m)$ this is given by a maximal ideal $m_p$ of $\mathcal{T}_n(X)$, and it corresponds to an equivalence class of semisimple representations of $F\langle x_1, x_2, \ldots, x_m\rangle$, the closed orbit in the fiber

$\pi^{-1}(p)$. An explicit representation $\phi$ of $F\langle x_1, x_2, \dots, x_m\rangle$ in the fiber of $p$ is given by a maximal ideal $M_\phi$ in the coordinate ring $F[\xi] = F[\xi_{i,j}^{(\alpha)}]$ of $m$-tuples of $n \times n$ matrices lying over $m_p$, and the representation is given by the evaluation,

$$\phi : R_X(n) \xrightarrow{i} M_n[F[\xi]] \longrightarrow M_n[F[\xi]/M_\phi] = M_n[F].$$

To say that $\phi$ is a semisimple representation is equivalent to saying that the image of $\phi$ is a semisimple algebra $\bigoplus_j M_{h_j}[F] \subset M_n[F]$.

Take now any Cayley–Hamilton algebra $R$ finitely generated over $F$, and let $T$ be its trace algebra. $R = R_m(n)/I$ is the quotient of the free Cayley–Hamilton algebra $R_m(n)$ modulo a trace ideal $I$ and correspondingly $T$ is the quotient of $\mathcal{T}_n(X)$ modulo $I \cap \mathcal{T}_n(X)$.

By Theorem 14.2.1 and functoriality, we have the following commutative diagram.

$$\begin{array}{ccccc} R_X(n) & \xrightarrow[\cong]{i} & M_n[F[\xi]]^{\mathrm{GL}(n,F)} & \longrightarrow & M_n[F[\xi]] \\ \downarrow & & \downarrow & & \downarrow \\ R & \xrightarrow[\cong]{i_R} & M_n[B_n(R)]^{\mathrm{GL}(n,F)} & \longrightarrow & M_n[B_n(R)] \end{array} \tag{14.9}$$

The ring $B_n(R)$ is a quotient $B_n(R) = F[\xi]/J$ and need not be reduced; nevertheless it defines a $\mathrm{GL}(n, F)$ stable subvariety of $M_n(F)^m$ made of the (trace) $n$-dimensional representations of $R$, furthermore $T = B_n(R)^{\mathrm{GL}(n,F)}$.

Again explicitly, a homomorphism $\psi : F[\xi] \to B_n(R) \to F$ gives maximal ideals $M_\psi \subset B_n(R)$, $M'_\psi \subset F[\xi]$ and the following commutative diagram.

(14.10)

$$\begin{array}{ccccccc} R_X(n) & \xrightarrow[\cong]{i} & M_n[F[\xi]]^{\mathrm{GL}(n,F)} & \longrightarrow & M_P[F[\xi]] & \longrightarrow & M_n[F[\xi]/M'_\psi] = M_n[F] \\ \downarrow & & \downarrow & & \downarrow & & \downarrow{\scriptstyle 1} \\ R & \xrightarrow[\cong]{i_R} & M_n[B_n(R)]^{\mathrm{GL}(n,F)} & \longrightarrow & M_n[B_n(R)] & \longrightarrow & M_n[B_n(R)/M_\psi] = M_n[F] \end{array}$$

If $\psi : F[\xi] \to B_n(R) \to F$ gives a point in a closed orbit, then the corresponding representation of $R$ is semisimple. It follows,

THEOREM 14.3.1. *The algebraic variety associated to the ring $T$ parametrizes isomorphism classes of* (*trace compatible*) *semisimple representations of $R$.*

We want to analyze closely the quotient map $\pi : M_n(F)^m \to V_n(m)$ associated to the inclusion of rings $\mathcal{T}_n(X) \subset F[\xi]$. Given a point $p \in V_n(m)$, corresponding to a maximal ideal $m_p \subset \mathcal{T}_n(X)$, the preimage $\pi^{-1}(p)$ is a $\mathrm{GL}(n, F)$ stable subvariety defined by the ideal $m_p F[\xi]$ and containing a unique closed orbit, corresponding to some semisimple representation of the free algebra.

Consider the algebra

$$C_n(p) := R_m(n)/m_p R_m(n). \tag{14.11}$$

EXAMPLE 14.3.2. Take $p := \{0\}$ the *worst point* of the quotient, that is the point associated to the maximal ideal $\mathcal{T}_n^+(X)$. The zero-fiber $\pi^{-1}(\{0\})$ is the set of $m$-tuples of matrices generating a nilpotent subalgebra, the closed orbit is the trivial representation $x_i \mapsto 0$.

By Theorem 12.2.16 we see that $C_n(\{0\})$ is obtained from the free nil algebra of exponent $n$ in $m$ variables by adding 1. From this one easily sees that the ideal generated by $\mathcal{T}_n^+(X)$ in $F[\xi]$ is not semiprime.

For each point $p$ and for $C_n(p)$, the diagram (14.9) becomes

$$\begin{array}{ccccc} R_X(n) & \xrightarrow[\cong]{i} & M_n[F[\xi]]^{\mathrm{GL}(n,F)} & \longrightarrow & M_n[F[\xi]] \\ \downarrow & & \downarrow & & \downarrow \\ C_n(p) & \xrightarrow[\cong]{i_R} & M_n[F[\xi]/m_pF[\xi]]^{\mathrm{GL}(n,F)} & \longrightarrow & M_n[F[\xi]/m_pF[\xi]]. \end{array} \tag{14.12}$$

The algebra $C_n(p)$ has as $n$-dimensional representations the preimage $\pi^{-1}(p)$ and a unique semisimple representation. It is by construction a finitely generated (over the trace algebra) Cayley–Hamilton algebra with trace and the trace takes values in $F$, so we need to start with analyzing this picture and in particular describe the nature of the semisimple representation in terms of the structure of $C_n(p)$.

LEMMA 14.3.3. *Let $a$ be a matrix with entries in a commutative ring $A$. If $a$ is nilpotent,* $\mathrm{tr}(a)$ *is also nilpotent.*

PROOF. Since in a commutative ring $A$ the set of nilpotent elements is the intersection of the prime ideals, we will reduce to the case in which $A$ is a domain. Hence we can embed it into a field. For matrices over a field the trace of a nilpotent matrix is 0. □

PROPOSITION 14.3.4. *Let $R$ be an nth-Cayley–Hamilton algebra, and let $r \in R$ be a nilpotent element. Then* $\mathrm{tr}(r)$ *is nilpotent.*

PROOF. By Theorem 14.2.1 we can embed $R$ into matrices over a commutative ring so that the trace becomes the ordinary trace. Hence the statement follows from the previous lemma. □

PROPOSITION 14.3.5. *Let $R$ be an nth-Cayley–Hamilton algebra over an algebraically closed field $F$ with trace values in $F$ and finitely generated over $F$. Denote by $t$ the trace, then:*[3]

(1) *$R$ is finite-dimensional over $F$.*
(2) *The Jacobson radical $J$ of $R$ is the kernel of the trace form $t(ab)$. So $t$ factors through $R/J$ which is also an nth-Cayley–Hamilton algebra with nondegenerate trace form.*
(3) *$R/J = \bigoplus_{i=1}^s M_{k_i}(F)$, and there exists positive integers $h_i$ with $n = \sum_{i=1}^s h_i k_i$ such that given $a_i \in M_{k_i}(F)$ and* $\mathrm{tr}(a_i)$ *the ordinary trace, we have*

$$t(a_1, a_2, \dots, a_s) = \sum_{i=1}^s h_i \,\mathrm{tr}(a_i).$$

PROOF. (1) The finite dimensionality follows from Theorem 12.2.13 and Corollary 12.2.14.

(2) From Proposition 14.3.4 it follows that the trace $t$ vanishes on the radical $J$ of $R$ which is a nilpotent ideal. Conversely, let $N$ be the kernel of the trace form. It is clearly an ideal since $t(ax) = 0,\ \forall x$ implies $t(uavx) = t(avxu) = 0,\ \forall x$.

[3] One could drop the hypothesis that $R$ is finitely generated and still get (2), (3) but not (1).

Given $a \in N$, the (formal) trace $t(a^i)$ is 0 for all $i$. Therefore the Cayley–Hamilton polynomial of $a$ is $\chi_a^n(t) = t^n$, hence $a^n = 0$ and every element $a \in N$ is nilpotent, so $N \subset J$. The fact that $R/J$ is still a Cayley–Hamilton algebra is clear.

(3) We can assume now that $R$ is semisimple, $R = \bigoplus_{i=1}^s M_{k_i}(F)$ and the trace form is nondegenerate. By the previous discussion, if we present $R$ as a quotient of the free Cayley–Hamilton algebra $R_m(n)$, we have the following commutative diagram.

$$\begin{array}{ccc} R_m(n) & \xrightarrow{tr} & \mathcal{T}_n(X) \\ \downarrow & & \downarrow p \\ R & \xrightarrow{t} & F \end{array}$$

Thus we have a point $p$ in $V_n(m)$ which can be lifted to a semisimple representation of $R_m(n)$ factoring through $R$.

Now $R = \bigoplus_{i=1}^s M_{k_i}(F)$ and an $n$-dimensional representation of this algebra is of the form $\bigoplus h_i F^{k_i}$, $h_i \in \mathbb{N}$, with trace exactly $t(a_1, a_2, \ldots, a_s) = \sum_{i=1}^s h_i \operatorname{tr}(a_i)$. Since the trace form is nondegenerate, we must have $h_i > 0$ for all $i$. □

The previous proposition applies thus to the algebras $C_n(p)$ of formula (14.11). We need to understand a special case, when $\phi$ is irreducible.

PROPOSITION 14.3.6. *Let $R$ be an $n$-Cayley–Hamilton algebra with trace values in $F$ and Jacobson radical $J$. If $R/J = M_n(F)$, then $J = 0$.*

PROOF. Let us first remark that such an $R$ satisfies the hypotheses of Artin's theorem, hence it is an Azumaya algebra of dimension $n^2$ over its center $A$. Let $j : R \to M_n(B)$ be the universal embedding. By Corollary 10.4.3 we have that $j(A) \subset B$ so, since $j$ by definition is compatible with trace, we have for $a \in A$ that $\operatorname{tr}(a) = na$, since by hypothesis the trace has values in $F$, we have $A = F$. We thus conclude by Theorem 5.4.32. □

PROPOSITION 14.3.7. *If $\phi$ is irreducible, then*

$$C_n(p) := R_m(n)/m_p R_m(n) = M_n(F).$$

PROOF. By Proposition 14.3.5 we have that $C_n(p)/J = M_n(F)$ where $J$ is the radical. Then $J = 0$ by Proposition 14.3.6, from the previous analysis, and Proposition 1.3.17 it follows that □

THEOREM 14.3.8. *The set of points $p$ in which $C_n(p) = M_n(F)$ is the open set defined by the nonvanishing of at least one of the* discriminants $\det(\operatorname{tr}(u_i u_j))$ *where the elements $u_i$ vary on all $n^2$-tuples of elements of $R_m(n)$.*

*This set is empty if and only if $m = 1$. For $m = 2$ one takes a diagonal matrix with distinct entries and the matrix of a cyclic permutation which generates $M_n(F)$.*

In fact by Theorem 12.3.18 $\det(\operatorname{tr}(u_i u_j))$ is the value of a central polynomial.

THEOREM 14.3.9. *If $C_n(p) = M_n(F)$, the localized algebra $R_m(n)_{m_p}$ is an Azumaya algebra over the local ring $\mathcal{T}_n(X)_{m_p}$.*

In geometric language, the points $m_p$ of the spectrum of $\mathcal{T}_n(X)$ where $R_m(n)_{m_p}$ is an Azumaya algebra are exactly the points over which the quotient map

$$M_n(F)^m \xrightarrow{\pi} M_n(F)^m // \mathrm{GL}(n, F)$$

is a *principal bundle* over the projective linear group. In fact this can be viewed as a special case of M. Artin's characterization of Azumaya algebras by the polynomial identities in Theorem 10.3.2.

One easily obtains the following

COROLLARY 14.3.10. *Let $R$ be an $n$-Cayley–Hamilton algebra finitely generated over an algebraically closed field $F$. If $R$ is a simple algebra, then $R = M_h(F)$ for some $h$ dividing $n = hk$ and $t(a) = k \cdot \mathrm{tr}(a)$.*

PROOF. From the previous theory there is at least one semisimple representation of $R$, but since $R$ is simple it follows that this is injective, hence $R$ is finite dimensional and the values of the trace are in $F$. The statement follows from Proposition 14.3.5. □

Let us finish this paragraph by pointing out the relationship, for Cayley–Hamilton algebras, between trace representations and irreducible representations. We assume as before that $R$ is a Cayley–Hamilton algebra finitely generated over an algebraically closed field $F$ with center $Z$. Let $V$ be an irreducible representation of $R$, and let $L$ be the ideal of $R$ annihilator of $V$. By Kaplansky's theorem it follows that $R/L$ is a simple algebra, and from the Hilbert Nullstellensatz (Theorem 11.4.4) $R/L = M_k(F)$ with $V$ isomorphic to $F^k$.

It follows that associated to $V$ we have a *central character* $\chi_V : Z \to F$ so that $zv = \chi_V(z)(v)$, $\forall z \in Z, \forall v \in V$. This central character, restricted to the trace ring $T(R)$ induces also a maximal ideal $m$ of $T(R)$ and a point $P$ of the spectrum of semisimple trace representations. $V$ is also an irreducible module for the trace algebra $R/mR$ which, by Proposition 14.3.5 is finite dimensional. Of course the action of $R/mR$ on $V$ is trivial on its Jacobson radical, hence $V$ is one of the irreducible modules appearing in the semisimple representation associated to the point $P$.

Thus the set $\mathrm{Spec}_{\mathrm{irr}}(R)$ of equivalence classes of irreducible representations of $R$ is identified with the maximal spectrum of $R$. Denote by $V_{ss}(R) = \mathrm{Spec}(T(R))$ the equivalence classes of semisimple trace representations. We have

THEOREM 14.3.11. *There is a map*

$$\pi : \mathrm{Spec}(R) = \mathrm{Spec}_{\mathrm{irr}}(R) \to V_{ss}(R) = \mathrm{Spec}(T(R)).$$

(i) *$\pi$ is surjective.*
(ii) *The fiber $\pi^{-1}(P)$ is the finite set of irreducible representations appearing in the semisimple representation associated to $P$.*

One should notice the case of generic $n \times n$ matrices $F\langle \xi_\alpha \rangle$ with the associated Cayley–Hamilton algebra $R_m(n) = F\langle \xi_\alpha \rangle \mathcal{T}_n(X)$. In this case, besides the spectrum of $F\langle \xi_\alpha \rangle \mathcal{T}_n(X)$ and $\mathcal{T}_n(X)$, we can also analyze the map from the irreducible representations of $F\langle \xi_\alpha \rangle \mathcal{T}_n(X)$ to those of $F\langle \xi_\alpha \rangle$. It is clear, since $F\langle \xi_\alpha \rangle \mathcal{T}_n(X)$ is obtained from $F\langle \xi_\alpha \rangle$ by adding elements in the center, that an irreducible map $F\langle \xi_\alpha \rangle \mathcal{T}_n(X) \to M_i(F)$ restricts to an irreducible map of $F\langle \xi_\alpha \rangle$.

Conversely we claim that each $i$-dimensional irreducible representation $V$ of $F\langle \xi_\alpha \rangle$ extends in many different ways to $F\langle \xi_\alpha \rangle \mathcal{T}_n(X)$. In fact take any semisimple representation of the free algebra of dimension $n - i$. Summing to this the representation $V$, we have an $n$-dimensional semisimple representation which is thus a representation of $F\langle \xi_\alpha \rangle \mathcal{T}_n(X)$ containing $V$. The maximal ideal of $\mathcal{T}_n(X)$ kernel of this representation is associated to the semisimple representation chosen. As soon as $i < n$, this gives a very large fiber of the map of spectra. For the $n$-dimensional irreducible representations, the spectra of the three algebras are isomorphic.

14.3.1.1. *The example of* $m = 1$. This is a simple but instructive example. The algebra $R_n(1)$ is the commutative algebra generated by a single element $x$ over the polynomial ring in $n$ variables $\sigma_i(x)$ modulo the unique identity, the characteristic polynomial of $x$ that is the polynomial in formula (12.15). In this case the quotient variety is the affine space $F^n$ with coordinates the values of the $\sigma_i(x)$. Given values $\sigma_i(x) = a_i,\ p = (a_1, \dots, a_n)$ the algebra $C_n(p)$ is the polynomial ring $F[x]$ modulo the polynomial $P_p(x) := x^n - a_1x^{n-1} + a_2x^{n-2} + \cdots + (-1)^n a_n$.

The fiber $\pi^{-1}(p)$ consists of all the matrices with characteristic polynomial $P_p(x)$, among which there is the conjugacy class of semisimple matrices having as eigenvalues the roots of this polynomial.

In this case it is well known that the map $\pi$ has reduced and irreducible fibers all of dimension $n^2 - n$, that is the ideal generated by $m_p$ in $F[\xi_{i,j}]$ is a prime ideal, in fact a complete intersection. For instance for the 0 point one has the variety of nilpotent matrices whose defining ideal is $\sigma_i(x) = 0,\ \forall i$.

**14.3.2. An instructive example.** We want to discuss in full detail the very special case (due to Formanek, Halpin, and Li **[FHL81]**) of two generic $2 \times 2$ matrices. Call these two matrices $X, Y$. We have as in Exercise 9.1.1, the Cayley–Hamilton relations

$$X^2 - \operatorname{tr}(X)X + \det(X) = 0,\ Y^2 - \operatorname{tr}(Y)Y + \det(Y) = 0. \tag{14.13}$$

Substituting to $X$ the element $X + Y$ and using formulas (14.13) and (3.35), we obtain

$$XY + YX - \operatorname{tr}(X)Y - \operatorname{tr}(Y)X + \operatorname{tr}(X)\operatorname{tr}(Y) - \operatorname{tr}(XY) = 0. \tag{14.14}$$

From this one easily proves that the ring $T$ of invariants of 2, $2 \times 2$ matrices is the polynomial ring in five variables

$$a = \operatorname{tr}(X), \quad b = \operatorname{tr}(Y), \quad c = \det(X), \quad d = \det(Y), \quad e = \operatorname{tr}(XY).$$

The trace ring $T\langle X, Y\rangle$ is a free module over $T$ with basis

$$1,\ X,\ Y,\ XY. \tag{14.15}$$

Taking the trace of formulas (14.13) and (14.14), we have

$$\operatorname{tr}(X^2) = \operatorname{tr}(X)^2 - 2\det(X), \qquad \operatorname{tr}(X^2Y) = \operatorname{tr}(X)\operatorname{tr}(XY) - \det(X)\operatorname{tr}(Y),$$
$$\operatorname{tr}((XY)^2) = \operatorname{tr}(XY)^2 - 2\det(X)\det(Y).$$

The matrix of the trace form of the basis (14.15) is thus

$$\begin{vmatrix} \operatorname{tr}(1) & \operatorname{tr}(X) & \operatorname{tr}(Y) & \operatorname{tr}(XY) \\ \operatorname{tr}(X) & \operatorname{tr}(X^2) & \operatorname{tr}(XY) & \operatorname{tr}(X^2Y) \\ \operatorname{tr}(Y) & \operatorname{tr}(YX) & \operatorname{tr}(Y^2) & \operatorname{tr}(XY^2) \\ \operatorname{tr}(XY) & \operatorname{tr}(X^2Y) & \operatorname{tr}(XY^2) & \operatorname{tr}((XY)^2) \end{vmatrix} = \begin{vmatrix} 2 & a & b & e \\ a & a^2 - 2c & e & ae - bc \\ b & e & b^2 - 2d & be - ad \\ e & ae - bc & be - ad & e^2 - 2cd \end{vmatrix} \tag{14.16}$$

with determinant the discriminant of the basis

$$\Delta = -(b^2c + a^2d - 4cd - abe + e^2)^2.$$

On the other hand $[X,Y]^2 = -\det([X,Y])$ is a central polynomial, which in fact equals

$$[X,Y]^2 = \frac{1}{2}\operatorname{tr}([X,Y]^2) = \operatorname{tr}((XY)^2) - \operatorname{tr}(X^2Y^2),$$

$$\operatorname{tr}((XY)^2) = \operatorname{tr}(XY)^2 - 2\det(X)\det(Y) = e^2 - 2cd,$$

$$\begin{aligned}\operatorname{tr}(X^2Y^2) &= \operatorname{tr}((\operatorname{tr}(X)X - \det(X)(\operatorname{tr}(Y)Y - \det(Y))\\ &= \operatorname{tr}(X)\operatorname{tr}(Y)\operatorname{tr}(XY) - \operatorname{tr}(X^2)\det(Y) - \operatorname{tr}(Y^2)\det(X) + 2\det(X)\det(Y)\\ &= abe - a^2d - b^2c + 2cd,\end{aligned}$$

and thus

$$\Delta = -\det([X,Y])^2 = -[X,Y]^4. \tag{14.17}$$

We deduce that the open set of pairs of $2 \times 2$ matrices which generate the entire matrix ring is the set where $\det([X,Y]) \neq 0$.

The hypersurface $H$ defined by $\det([X,Y]) = 0$ is irreducible and is formed of course by the set of pairs of $2 \times 2$ matrices which do not generate the entire matrix ring. At the level of the quotient variety, which is the five-dimensional space of parameters $a, b, c, d, e$, this hypersurface parametrizes semisimple but not simple representations. Such a representation in a basis consists of two diagonal matrices: $X$ of eigenvalues $x_1, x_2$ and $Y$ of eigenvalues $y_1, y_2$. These four parameters are not invariants. In fact the invariants at such a point are

$$a = x_1 + x_2, \quad b = y_1 + y_2, \quad c = x_1x_2, \quad d = y_1y_2, \quad e = x_1y_1 + x_2y_2,$$

which of course are related by the equation $b^2c + a^2d - 4cd - abe + e^2 = 0$.

The spectrum of $R$ instead consists of two parts: $\operatorname{Spec}_2(R)$ is the five-dimensional open set of the quotient variety where $\det([X,Y]) \neq 0$; while $\operatorname{Spec}_1(R)$ is the quotient of $R$ modulo the identities of $1 \times 1$ matrices, that is the commutator ideal. This quotient is just the commutative polynomial ring in two variables and the spectrum is a two-dimensional affine space.

How this space is glued to the five-dimensional spectrum is not very clear. One way of understanding this would be to try to understand curves or even germs of curves in five-space converging to a point of $\operatorname{Spec}_1(R)$.

On the other hand the spectrum of $RT$ can be understood as follows.

There is the open part of irreducible two-dimensional representations which is the same as $\operatorname{Spec}_2(R)$. As for $\operatorname{Spec}_1(RT)$, we have that an irreducible representation of $T$ of dimension 1 appears in a given semisimple representation which corresponds to a point in the hypersurface $H$ defined by $\det([X,Y]) = 0$.

This hypersurface has an open set formed by those semisimple representations, the direct sum of two nonisomorphic representations, and finally the closed set of the ones that correspond to a single irreducible representation with multiplicity 2, which is easily seen to form a plane $F^2$. Thus the spectrum of $RT$ over $H \setminus F^2$ is a double covering while it is homeomorphic over $F^2$.

The reader should analyze in this case the algebras $C_2(p)$ of formula (14.11), in particular $C_2(\{0\})$, which are all four-dimensional.

**14.3.3. A characteristic free approach.** By the theorem of Donkin **[Don92]** the ring $\mathcal{T}_n\langle\xi\rangle$ generated by the coefficients of the characteristic polynomial of the elements of $F_m\langle\xi_1,\dots,\xi_k\rangle$ is the ring of invariants of the projective group acting on the ring of polynomial functions on $M_i(F)^k$. By Amitsur **[Ami79]** this is in fact generated by the coefficients of the characteristic polynomial of the primitive monomials in the $\xi_i$. In characteristic 0 this equals the trace ring which we have denoted by $\mathcal{T}_n(X)$.

Since this group is geometrically reductive also in positive characteristic, we have that the quotient map parametrizes orbits of semisimple representations. A point in the quotient variety corresponds to a maximal ideal $\underline{m}$ of $T$. We shall prove this directly without resorting to geometric invariant theory in the next discussion.

So we have several algebras and their spectra. Let us concentrate on the maximal spectrum, which is enough for finitely generated algebras.

In our setting the relevant rings are the extensions

$$\mathcal{T}_n\langle\xi\rangle \subset F[\xi_{i,j}^t] \subset M_m(F[\xi_{i,j}^t]), \quad R = F_m\langle\xi_1,\dots,\xi_k\rangle \subset RT,$$

on the other hand, as will be clear soon, $R\mathcal{T}_n\langle\xi\rangle \subset M_m(F[\xi_{i,j}^t])$ is not an extension. Let us denote for simplicity of notations

$$A := F[\xi_{i,j}^t].$$

By Corollary 8.2.5 we know that $\mathcal{T}_n\langle\xi\rangle$ is a finitely generated algebra over $F$ and $R\mathcal{T}_n\langle\xi\rangle$ is a finitely generated module over $\mathcal{T}_n\langle\xi\rangle$.

The inclusion $A \subset M_m(A)$ determines an isomorphism of spectra and the maximal spectrum is identified to the space $M_m(F)^k$.

The spectrum of $\mathcal{T}_n\langle\xi\rangle$ parametrizes the equivalence classes of semisimple representations of the free algebra, we have seen this in §14.1.2 in characteristic 0, but we can prove this in general even without invoking geometric invariant theory.

The inclusion $\mathcal{T}_n\langle\xi\rangle \subset A$ induces a mapping of varieties (i.e., maximal ideals). The variety associated to $A$ is by definition $M_m(F)^k$, while we claim (and have proved in characteristic 0 using geometric invariant theory) that the variety associated to $\mathcal{T}_n\langle\xi\rangle$ is what we have called $M_m(F)^k//G$, which parametrizes semisimple representations of $R$.

This means, as we have seen in §14.1.2, that the (quotient) map of varieties $M_m(F)^k \to M_m(F)^k//G$ induced by the inclusion $\mathcal{T}_n\langle\xi\rangle \subset A$ is surjective, and the fiber of each point of $M_m(F)^k//G$ contains a unique closed orbit which thus is the orbit of a semisimple representation Theorem 14.1.15.

The construction in characteristic 0 has been done using the theory of Cayley–Hamilton algebras. Here we present what could be a piece of that theory, which at the moment is only partial, in a characteristic-free form using the more general Definition 12.2.1.

A semisimple representation $\rho : R \to M_m(F)$ extends naturally to a homomorphism $\rho' : R\mathcal{T}_n\langle\xi\rangle \to M_m(F)$. The restriction of $\rho'$ to $\mathcal{T}_n\langle\xi\rangle$ maps to $F$ and, by definition, given any element $r \in R$ or even $r \in R\mathcal{T}_n\langle\xi\rangle$, we have that the characteristic polynomial of $r$, which is a polynomial with coefficients in $\mathcal{T}_n\langle\xi\rangle$, maps to the characteristic polynomial of $\rho'(r)$.

In particular $\rho'(\det(r)) = \det(\rho'(r))$. The image of $\rho$ is by assumption a semisimple algebra hence of the form $\bigoplus_{i=1}^u M_{h_i}(F)$ embedded in $M_m(F)$ as block diagonal matrices. By the theory of semisimple algebras the space $F^m$ is a direct sum of the various irreducible representations $F^{h_i}$ of the blocks $M_{h_i}(F)$ each appearing with some multiplicity $k_i$, so that $m = \sum_i h_i k_i$ and $t = \sum_i k_i$ blocks.

Moreover, if $r \in R$ maps to the $t$ blocks $(\bar{r}_1, \dots, \bar{r}_t)$ we have that $\det(r)$ maps to $\prod_{i=1}^u \det(\bar{r}_i)^{k_i}$. Finally clearly the image of $R$ coincides with the image of $R\mathcal{T}_n\langle\xi\rangle$, since $\mathcal{T}_n\langle\xi\rangle$ maps to $F$.

Conversely take a maximal ideal $\underline{m}$ of the algebra $\mathcal{T}_n\langle\xi\rangle$ and consider the algebra $\bar{R} := R\mathcal{T}_n\langle\xi\rangle/\underline{m}R\mathcal{T}_n\langle\xi\rangle$. We have $\mathcal{T}_n\langle\xi\rangle/\underline{m} = F$ and then, since $R\mathcal{T}_n\langle\xi\rangle$ is a finite module over $\mathcal{T}_n\langle\xi\rangle$, $\bar{R}$ is a finite-dimensional algebra over $F$. Hence its nil radical $N$ is nilpotent and, since $F$ is algebraically closed, we have $\bar{R}/N = \bigoplus_{i=1}^u M_{h_i}(F)$ for some integers $h_i$.

If we compose the map $\det : R\mathcal{T}_n\langle\xi\rangle \to \mathcal{T}_n\langle\xi\rangle \to F$ since it is a polynomial map and $R\mathcal{T}_n\langle\xi\rangle/\underline{m}R\mathcal{T}_n\langle\xi\rangle = R\mathcal{T}_n\langle\xi\rangle \otimes_{\mathcal{T}_n\langle\xi\rangle} F$, we have that det defines a multiplicative polynomial map, homogeneous of degree $m$ which we still call $\det : \bar{R} \to F$. Moreover we can define in the same way the characteristic polynomial of elements in $\bar{R}$, which is also the image of the characteristic polynomial of the elements in $R$ and each element in $\bar{R}$ satisfies its characteristic polynomial.

Now we use two elementary facts from the theory of algebraic groups. Let $\bar{R}^*$ be the multiplicative group of the algebra $\bar{R}$. Consider the exact sequence $0 \to N \to \bar{R} \to \bigoplus_{i=1}^u M_{h_i}(F) \to 0$. The multiplicative group of $\bigoplus_{i=1}^u M_{h_i}(F)$ is the product $\prod_{i=1}^u \mathrm{GL}(h_i, F)$, and we have an exact sequence of homomorphisms

$$1 \to K \to \bar{R}^* \to \prod_{i=1}^u \mathrm{GL}(h_i, F) \to 1.$$

The group $K$ is a unipotent group since it is formed by elements $1 + n,\ n \in N$. The map $\det : \bar{R}^* \to F^*$ is a character and, here we use a basic fact, it is trivial when restricted to any unipotent group, thus this character factors through $K$ and gives a character of the product $\prod_{i=1}^u \mathrm{GL}(h_i, F)$.

The second elementary fact is that a character of the product $\prod_{i=1}^u \mathrm{GL}(h_i, F)$ is necessarily of the form $(a_1, \dots, a_u) \mapsto \prod_{i=1}^u \det(a_i)^{n_i},\ n_i \in \mathbb{Z}$.

The third fact that we need is that det is a polynomial map, hence defined on all of $\bar{R}$ and not only on $\bar{R}^*$, hence this implies that the exponents $n_i \in \mathbb{N}$. Finally since det is homogeneous of degree $m$, we have $m = \sum_i h_i n_i$.

From this formula it follows that the norm of an element $r \in \bar{R}$ depends only on its class module $N$, hence the same norm $\prod_{i=1}^u \det(a_i)^{n_i}$ defines a norm in $\bar{R}/N = \bigoplus_{i=1}^u M_{h_i}(F)$ such that every element of this last algebra satisfies the characteristic polynomial deduced from this norm. If $e_i$ is the unit element of the summand $M_{h_i}(F)$, we have thus that the characteristic polynomial of $(a_1, \dots, a_u)$ is the norm of $t1 - (a_1, \dots, a_u) = (te_1 - a_1, \dots, te_u - a_u)$ is $\prod_{i=1}^u \det(te_i - a_i)^{n_i}$.

The final fact we need to prove is that all $n_i > 0$. For this we use the fact that each element of $\bar{R}$ satisfies its characteristic polynomial.

We have seen also each element of $\bar{R}/N = \bigoplus_{i=1}^u M_{h_i}(F)$ satisfies its characteristic polynomial, which for an element $(a_1, \dots, a_t)$ is $\prod_{i=1}^u \det(te_i - a_i)^{n_i}$. If we had, say, $n_1 = 0$, we would have

$$\det(te_1 - a_1)^{n_1} = 1 \implies \prod_{i=1}^u \det(te_i - a_i)^{n_i} = \prod_{i=2}^u \det(te_i - a_i)^{n_i},$$

so the characteristic polynomial of the element $e_1 = (1,0,0,\dots,0)$ is $t^{\sum_i n_i}$, which is not satisfied by $e_1$.

We have thus now shown that every maximal ideal corresponds to a unique class of semisimple representation, so the claim is complete.

The inclusions of $R, \mathcal{T}_n\langle\xi\rangle$ in $R\mathcal{T}_n\langle\xi\rangle$ are less geometric.

By the Nullstellensatz (Theorem 11.4.4) since $R\mathcal{T}_n\langle\xi\rangle$ is a finitely generated algebra over $F$ given a maximal ideal $M$ of $R\mathcal{T}_n\langle\xi\rangle$, we have that $R\mathcal{T}_n\langle\xi\rangle/M$ is a finite-dimensional simple algebra and, since we are assuming $F$ is algebraically closed, we have that $R\mathcal{T}_n\langle\xi\rangle/M \cong M_i(F)$ for some $i$.

Since $\mathcal{T}_n\langle\xi\rangle$ is in the center of $R\mathcal{T}_n\langle\xi\rangle$ its image in $R\mathcal{T}_n\langle\xi\rangle/M \cong M_i(F)$ must be $F$, hence $M$ intersects $\mathcal{T}_n\langle\xi\rangle$ in a maximal ideal, $\underline{m}$ which we have seen corresponds to a semisimple representation $\rho$. Thus $M$ corresponds to one of the maximal ideals of the algebra $\bar{R} = R\mathcal{T}_n\langle\xi\rangle/\underline{m}R\mathcal{T}_n\langle\xi\rangle$, hence one of the maximal ideals of $\bar{R}/N$ where $N$ is the radical of $\bar{R}$.

By the previous discussion $\bar{R}/N = \bigoplus_{i=1}^t M_{h_i}(F)$ and the blocks $M_{h_i}(F)$ correspond to the irreducible representations.

We have thus that $R\mathcal{T}_n\langle\xi\rangle/M \cong M_i(F)$ has as irreducible representation (on the space $F^i$), one of the irreducible components of $\rho$.

The same maximal ideal gives an irreducible representation of $R$. In other words the fiber for the map $\operatorname{Spec}(R) \to \operatorname{Spec}(\mathcal{T}_n\langle\xi\rangle)$ of a point $\rho$ corresponding to a semisimple representation is the finite set of irreducible representations appearing in $\rho$.

While the fiber of the map $\operatorname{Spec}(\mathcal{T}_n\langle\xi\rangle R) \to \operatorname{Spec}(R)$ associates to an irreducible representation $\sigma$ all pairs $\sigma, \rho$, where $\rho$ corresponds to a semisimple representation in which appears $\sigma$.

## 14.4. Geometry of generic matrices

We can now complete the discussion of the spectrum of generic matrices which we started in §11.3.1.1, page 299.

We can view the next theorem as a precise analysis, in this case, of the general Theorem 11.2.17 for all PI rings.

In order to simplify our treatment, we assume we will work over an algebraically closed field $F$. The notations are $F_n\langle\xi_1,\dots,\xi_m\rangle = F\langle\xi_\alpha\rangle$ the ring generated by $m$ generic $n \times n$ matrices, and $\mathcal{Z} = F \oplus \mathcal{Z}_+$ the center of $F_n\langle\xi_1,\dots,\xi_m\rangle$. By $\mathcal{T}_n\langle\xi_1,\dots,\xi_m\rangle$ we denote the *trace ring* generated by the coefficients of the characteristic polynomials of the primitive monomials in the generic matrices $\xi_i$ (cf. Theorem 3.3.28).

We have the commutative diagram of inclusions,

$$\begin{array}{ccc} F \oplus \mathcal{Z}_+ & \longrightarrow & \mathcal{T}_n\langle\xi_1,\dots,\xi_m\rangle \\ {\scriptstyle i}\downarrow & & \downarrow \\ F_n\langle\xi_\alpha\rangle & \xrightarrow{\ j\ } & F_n\langle\xi_\alpha\rangle\mathcal{T}_n\langle\xi_\alpha\rangle \end{array}, \tag{14.18}$$

which induces maps on the spectra. Since we are in the case where $F$ is algebraically closed and $m$ is finite, by Remark 11.4.5 we may consider only the maximal spectrum which for $\mathcal{T}_n\langle\xi_\alpha\rangle$ are the points of the affine algebraic variety $M_n(F)^m//\operatorname{GL}(n,F)$.

THEOREM 14.4.1.

(1) *A maximal ideal of* $F\langle\xi_\alpha\rangle$ *contains* $\mathcal{Z}_+$ *if and only if it is of PI degree* $< n$.
(2) *The radical of* $\mathcal{Z}_+F\langle\xi_\alpha\rangle$ *is the ideal of PIs of* $n-1\times n-1$ *matrices.*
(3) *The subvariety of* $M_n(F)^m//\,\mathrm{GL}(n,F)$ *defined by* $\mathcal{Z}_+$ *is that of nonirreducible semisimple representations.*
(4) *If* $a\in\mathcal{Z}_+$, *we have that*

$$F_n\langle\xi_\alpha\rangle[a^{-1}] = F_n\langle\xi_\alpha\rangle\mathcal{T}_n\langle\xi_\alpha\rangle[a^{-1}]$$

*is a rank* $n^2$ *Azumaya algebra.*
(5) *The maps*

$$\mathrm{Spec}_n(F_n\langle\xi_\alpha\rangle\mathcal{T}_n\langle\xi_\alpha\rangle) \xrightarrow[\simeq]{j^*} \mathrm{Spec}_n(F_n\langle\xi_\alpha\rangle) \xrightarrow[\simeq]{i^*} \mathrm{Spec}(F\oplus\mathcal{Z}_+)\setminus\{0\}$$

*are all homeomorphisms.*
(6) *The* central character *gives a map* $\mathrm{Spec}_n(F_n\langle\xi_\alpha\rangle\mathcal{T}_n\langle\xi_\alpha\rangle)\to\mathrm{Spec}(\mathcal{T}_n\langle\xi_\alpha\rangle)$ *which is an isomorphism with the open set parametrizing isomorphism classes of irrreducible n-dimensional representations of* $F\langle\xi_\alpha\rangle$.

PROOF. (1) If $\mathfrak{m}$ is a maximal ideal of $R$, we have, by Proposition 11.4.2, since $F$ is algebraically closed, that $R/\mathfrak{m}\simeq M_i(F)$, $i\le n$. Now by Proposition 10.2.2 each element of $\mathcal{Z}_+$ is a PI on $n-1\times n-1$ matrices and it does not vanish on $n\times n$ matrices, hence the claim.

(2) and (3) follow from (1).

(4) By Corollary 10.3.5 the two algebras are both Azumaya of rank $n^2$. In order to show that they coincide it is enough to see that $\mathcal{T}_n\langle\xi_\alpha\rangle$ is contained in the center $\mathcal{Z}[a^{-1}]$ of $F_n\langle\xi_\alpha\rangle[a^{-1}]$.

By definition $\mathcal{T}_n\langle\xi_\alpha\rangle$ is generated by the coefficients of the characteristic polynomials of elements in $F_n\langle\xi_\alpha\rangle\subset M_n[F[\xi_\alpha]]$.

We have that $F_n\langle\xi_\alpha\rangle[a^{-1}]\subset M_n[F[\xi_\alpha][a^{-1}]]$ is an inclusion of Azumaya algebras of rank $n^2$, hence, by Corollary 5.4.29,

$$M_n[F[\xi_\alpha][a^{-1}]] = F_n\langle\xi_\alpha\rangle[a^{-1}]\otimes_{\mathcal{Z}[a^{-1}]}F[\xi_\alpha][a^{-1}].$$

Then the claim follows from Proposition 5.4.41.

(5) The three spaces are covered by the open sets where $a\neq 0$ for $a\in\mathcal{Z}^+$ so it is enough to prove the statement *locally* on one of these open sets. Now for the first map this is clear since the two algebras are now the same. As for the second map we may invoke Corollary 5.4.29 which establishes the stronger 1–1 correspondence between two-sided ideals of an Azumaya algebra and its center.

(6) This follows as in (5) by considering that the open set parametrizing isomorphism classes of irrreducible $n$-dimensional representations of $F\langle x_\alpha\rangle$ is covered by the affine open sets where $a\neq 0$ for $a\in\mathcal{Z}^+$.

On each one of these open sets we have that $\mathcal{Z}[a^{-1}] = \mathcal{T}_n\langle\xi_\alpha\rangle[a^{-1}]$ (by the previous step), hence the claim. □

Notice that we have proved in particular

COROLLARY 14.4.2. *For each* $i\le n$, *the spectrum* $\mathrm{Spec}_i(F_n\langle\xi_1,\dots,\xi_m\rangle)$ *is homeomorphic to the smooth open set of* $M_i(F)^m//\,\mathrm{GL}(i,F)$ *parametrizing isomorphism classes of irreducible i-dimensional representations of the free algebra* $F\langle x_1,\dots,x_m\rangle$.

In fact one can prove **[LBP87]** that, except for the case $i = m = 2$ this open set coincides with the smooth locus of the quotient variety $M_i(F)^m // \mathrm{GL}(i, F)$.

REMARK 14.4.3. If $R = F\langle a_1, \dots, a_m\rangle$ is a finitely generated algebra, a quotient of the algebra of generic matrices $F\langle \xi_1, \dots, \xi_m\rangle$, then for each $i$ the set $\mathrm{Spec}_i(R)$ is a closed subset of $\mathrm{Spec}_i(F\langle \xi_1, \dots, \xi_m\rangle)$, that is it is a quasi-affine variety in the usual sense. On the other hand the topology of the full spectrum is not that of an algebraic variety even for generic matrices.

A partial description of this topology follows from Theorem 14.3.11. Another result on this topology is Schelter's theorem on the catenary property for such algebras, Theorem 15.2.12.

**14.4.1. The dimension of generic matrices.** We now consider the algebra of $m$ generic $n \times n$ matrices $F\langle \xi_1, \dots, \xi_m\rangle$. We have seen that it has a quotient division algebra $D$ of degree $n$, i.e., of dimension $n^2$ over its center $\mathcal{Z}$. We claim

THEOREM 14.4.4. *The field $\mathcal{Z}$ is the field of fractions of the ring $T$ of invariants of $\mathrm{GL}(n, F)$ acting on $M_n(F)^m$. The algebra $F\langle \xi_1, \dots, \xi_m\rangle T$ is a finitely generated module over $T$.*

PROOF. We have a commutative diagram of inclusions

$$\begin{array}{ccc} F\langle \xi_1, \dots, \xi_m\rangle & \longrightarrow & D \\ \downarrow & & \downarrow \\ M_n[F[\xi^a_{i,j}]] & \xrightarrow{\ i\ } & M_n[F(\xi^a_{i,j})] \end{array} \tag{14.19}$$

with $F(\xi^a_{i,j})$ the field of rational functions quotient of the polynomial ring $F[\xi^a_{i,j}]$. Since $D$ has dimension $n^2$ over its center, by Corollary 5.4.29 we have

$$M_n[F(\xi^a_{i,j})] = D \otimes_{\mathcal{Z}} F(\xi^a_{i,j}),$$

and by Theorem 5.4.40 all the coefficients of the characteristic polynomial of an element $a \in D$ lie in $\mathcal{Z}$. By Theorem 12.1.3 in characteristic 0 or by Donkin's theorem in general, the coefficients of the characteristic polynomial of elements $a \in F\langle \xi_1, \dots, \xi_m\rangle$ generate the ring $T$ of invariants and contain the center $C$ of $F\langle \xi_1, \dots, \xi_m\rangle$, so the two rings $C, T$ have the same field of fractions $\mathcal{Z}$.

The fact that the algebra $F\langle \xi_1, \dots, \xi_m\rangle T$ is a finitely generated module over $T$ follows from Theorem 8.2.2. □

COROLLARY 14.4.5. *The Gelfand–Kirillov dimension of generic matrices equals the dimension of the variety $M_n(F)^m // \mathrm{GL}(n, F)$ (or of the corresponding ring $T$ of matrix invariants) and is $(m-1)n^2 + 1$ as soon as $m > 1$.*

PROOF. From Corollary 11.2.9 this dimension equals the transcendence degree of the field $\mathcal{Z}$ of fractions of $T$.

If $F$ is algebraically closed, this equals the dimension of the variety of equivalence classes of semisimple representations $M_n(F)^m // \mathrm{GL}(n, F)$ which by Theorem 14.1.18 is $(m-1)n^2 + 1$ as soon as $m > 1$. □

14.4.1.1. *A birational approach.* The dimension of an irreducible variety equals the transcendence degree of the field of fractions of the coordinate ring of a variety; in other words it is a *birational invariant.* Thus we want to investigate in a more direct way this field in the case of matrices.

Let us denote by $Z_{m,n}$ the field center of the ring of quotients of the ring of $m$ generic $n \times n$ matrices, which we have seen is also the field of fractions of the coordinare ring of $M_n(F)^m // \mathrm{GL}(n, F)$. A challenging question is *whether* $Z_{m,n}$ *is a field of rational functions.* It is easy to reduce this question to $m = 2$.

LEMMA 14.4.6. $Z_{m,n}$ *is a field of rational functions over* $Z_{2,n}$.

PROOF. Take the two generic matrices $X, Y$ so that, by Lemma 5.1.15, the elements $X^iY^j$, $0 \leq i < n$, $0 \leq j < n$ form a basis of $D$ over its center $Z$. For any other generic matrix $Z$, we then have $Z = \sum_{i,j} c_{i,j} X^i Y^j$, $c_{i,j} \in Z$.

One can compute these elements by solving the system of $n^2$ equations, in the $n^2$ unkowns $c_{i,j}$

$$\mathrm{tr}(ZX^hY^k) = \sum_{i,j} c_{i,j}\, \mathrm{tr}(X^iY^jX^hY^k). \tag{14.20}$$

Then we see that $Z$ is the field of fractions of the following algebra. We take the algebra of invariants of two generic matrices $X, Y$ add the inverse of the determinant of the system (14.20), that is the discriminant of the trace form for the basis $X^iY^j$. So in order to have the remaining $m - 2$ generic matrices it is enough to add to the invariants of the first two matrices the $(m-2)n^2$ invariants $\mathrm{tr}(Z_iX^hY^k)$ for each $Z_i$. All invariants of the generic matrices are then in the resulting polynomial ring. Since the transcendence degree of $Z_{m,n}$ over $Z_{2,n}$ is $(m-2)n^2$, the claim follows. The field $Z_{m,n}$ is a field of rational functions over $Z_{2,n}$ in the $(m-2)n^2$ invariants $\mathrm{tr}(Z_iX^hY^k)$. □

The study of the field $Z_{2,n}$ is based on the fact, which follows from geometric invariant theory, that it can be identified with the field of the algebra of invariants of the semidirect product $S_n \ltimes T$ where $T$ is the torus of diagonal matrices acting on the space $\bar{\mathcal{O}}$ of pairs $\delta, Y$ where $\delta$ is a diagonal matrix with nonzero entries and $Y$ is a matrix with nonzero entries $\xi_{h,k}$.

The coordinate ring of $\mathcal{O}$ is the ring $A := F[\xi_{h,k}^{\pm 1}, d_1^{\pm 1}, \ldots, d_n^{\pm 1}]$. The torus $T$ acts only on $B := F[\xi_{h,k}^{\pm 1}]$.

In more algebraic terms $B$ is the group algebra of the free abelian group $M$ with basis elements $\xi_{h,k}$. We then see that

LEMMA 14.4.7. *The ring* $B = F[M]^T$ *of* $T$ *invariants equals the group algebra* $F[N]$. *Where the group* $N$ *is the kernel of the homomorphism from* $\pi : M \to \mathbb{Z}^n$ *given by* $\pi : \xi_{h,k} \mapsto e_h - e_k$.

PROOF. The elements of $M$ are by definition a basis of $B$ over $F$ and each such element $\prod(\xi_{h,k})^{n_{h,k}}$ is in fact an eigenvector for the $T$ action of eigenvalue $\prod(t_h t_k^{-1})^{n_{h,k}} = t^{\sum n_{h,k}(e_h - e_k)}$ where we set $t_h := t^{e_h}$. The ring of invariants under $T$ has as basis the monomials of weight 0, hence the claim. Notice that $\pi$ maps surjectively to the subgroup $\mathbb{Z}^{n,0}$ of $\mathbb{Z}^n$ of the elements in which the sum of the coordinates is 0. In particular $N$ is a free abelian group of rank $n^2 - n + 1$. □

Thus the final question is to understand the invariant theory of $S_n$ on $N \oplus \mathbb{Z}^n$, a free abelian group of rank $n^2+1$.

QUESTION. Is $Z_{2,n}$ a field of rational functions over the base field?

One can of course try several related, possibly weaker questions. This is connected with a general question of rationality for quotient varieties and, as we have seen, to a problem on the symmetric group.

The answer is known to be positive for $n \leq 4$ by work of Formanek **[For80]**. On the other hand there are counterexamples of finite group actions on lattices for which the corresponding field of invariant is not rational (Saltman **[Sal84]** and **[Sal85]**).

The problem is connected with the following problem (Lenstra):

Is there an exact sequence of $S_n$-modules $0 \to N \to P_1 \to P_2 \to 0$ where $P_1, P_2$ are permutation modules? Notice that, in Lemma 14.4.7, $M$ is a permutation module but the image of $\pi$ is not such a module.

**14.4.2. Applications, the crossed product problem.** We have seen that the ring of generic $n \times n$ matrices $R_{m,n} = F\langle \xi_1, \dots, \xi_m \rangle$ is a free algebra in the variety of algebras satisfying the PI of $n \times n$ matrices, its division ring of quotients $D_{m,n}$ can be considered in some form also as *free among all division algebras $D$ which have dimension $n^2$ over the center $Z$*, which is assumed to contain $F$.

We state this property as follows. Let us give any finite number of elements $a_1, \dots, a_k \in D_{m,n}$ (that is rational functions) and a division ring $D$ of dimension $n^2$ over the center $Z$ containing $F$, then

PROPOSITION 14.4.8. *There are homomorphisms from the subalgebra, of $D_{m,n}$, $S := F\langle \xi_1, \dots, \xi_m, a_1, \dots, a_k \rangle$ generated by the $a_i$ and the generic matrices, to $D$. In fact a generic specialization of the $\xi_i$ to $D$ extends to $S$.*

PROOF. Since we know that $D_{m,n}$ is obtained from $R_{m,n}$ by inverting its central elements, there is a central polynomial $f$ such that $S \subset R_{m,n}[f^{-1}]$, so it is enough to prove the statement for $R_{m,n}[f^{-1}]$. We know that $f$ does not vanish on $D$ and takes values in the center, so any evaluation of the generic matrices in $D$ for which $f$ is not 0, an open set, extends to $R_{m,n}[f^{-1}]$. □

This implies that if we can find some elements in the generic division ring which satisfy some special properties, we can do the same in any division ring.

In particular Amitsur used this fact to give a counterexample to the famous *crossed product problem*. This is a topic with a long history which comes from number theory.

Let us recall some facts. Let $D$ be a division algebra of dimension $n^2$ over its center $Z$. We have already recalled that any maximal subfield $K$ of $D$ has dimension $n$ over $Z$ and $D$ has dimension $n$ over $K$.

One can prove also that there is always a separable maximal subfield. The crossed product problem asked if there is also a maximal subfield which is a Galois extension of $Z$.

The interest in this question comes from the Skolem–Noether theorem.

Given any element $g$ of the Galois group $G$ of $K$ over $Z$, there is an element $a_g \in D$ which normalizes $K$ and induces $g$ as an inner automorphism,

$$g(k) = a_g k a_g^{-1}, \ \forall k \in K.$$

Of course $a_g$ is not unique, since $K$ is a maximal subfield, it equals its own commutant. So $a_g$ can be changed by multiplication by any element $h \in K^*$, and $ha_g$ acts in the same way. Thus one has that

$$a_h a_g = k(h,g)a_{hg},\ k(h,g) \in K^*,\ \forall h,g \in G.$$

By the associative law one has the constraint

$$\begin{aligned}(a_h a_g)a_\ell &= k(h,g)a_{hg}a_\ell = k(h,g)k(hg,\ell)a_{(hg)\ell}\\ &= a_h(a_g a_\ell) = a_h k(g,\ell)a_{g\ell} = k(g,\ell)^h a_h a_{g\ell} = k(g,\ell)^h k(h,gl)a_{h(g\ell)}\\ &\implies k(h,g)k(hg,\ell) = k(g,\ell)^h k(h,gl).\end{aligned}$$

This is expressed in the language of *cohomology* by saying that $k(h,g)$ is a 2-cocycle of $G$ with values in $K^*$. When we change the representatives $a_g$ to $\bar{a}_g := k(g)a_g$, we have

$$\bar{a}_h\bar{a}_g = k(h)a_h k(g)a_g = k(h)k(g)^h a_{hg} = k(h)k(g)^h k(hg)^{-1}\bar{a}_{hg}.$$

In the language of *cohomology* we have changed the cocycle by a coboundary and indeed we have identified a cohomology class in $H^2(G,K^*)$ describing the cross product.

Assume $Z$ is a *global field*, that is a finite extension of $\mathbb{Q}$ or of $F_q(x)$ in characteristic $p$, where $F_q$ is the field with $q = p^h$ elements.

A deep theorem of class field theory gives that any finite-dimensional division algebra with center $Z$ has a maximal subfield with cyclic Galois group.

It has then been long asked if every finite-dimensional division algebra with some center $Z$ has a maximal subfield which is Galois over $Z$. Then Amitsur observed

PROPOSITION 14.4.9. *If the division algebra $D_{m,n}, m \geq 2$ (over some base field $F$) has a maximal subfield which is Galois over $Z_{m,n}$, with some Galois group $G$, then every division algebra $D$ of dimension $n^2$ over its center $Z$ and $F \subset Z$ has a maximal subfield which is Galois over $Z$, with the same Galois group $G$.*

PROOF. This follows easily from Proposition 14.4.8.

In fact if $D_{m,n}$ has this property, the maximal subfield $K$ is generated by some element $x$ and has basis $1, x, \dots, x^{n-1}$ over $Z_{m,n}$.

Then there are elements $a_g$, $g \in G$ which satisfy all the previous algebraic identities, in particular

$$a_g x a_g^{-1} = \sum_{i=0}^{n-1} \lambda_{i,g}x^i,\ \lambda_{i,g} \in Z_{m,n}, \quad \text{and} \quad a_g a_h = (\sum_{i=0}^{n-1} \lambda_{i,g,h}x^i)a_{gh}.$$

Finally the elements $a_g x^i$, $g \in G$, $i = 0, \dots, n-1$ form a basis of $D_{m,n}$ over $Z_{m,n}$. This is expressed by imposing that the determinant of the trace form on these elements is nonzero.

All these conditions can be specialized in $D$ by some evaluation of the generic matrices which is nonzero on the common denominators, as central polynomials, of the finitely many elements involved in these identities.

Then one easily sees that the image of $x$ generates the desired Galois extension where the Galois group is induced by conjugation by the images of the elements $a_g$, and thus it is isomorphic to $G$ since it has the same multiplication table. □

From this proposition Amitsur easily deduced, in characteristic 0, that if the square $p^2$ of an odd prime $p$ divides $n$ or 8 divides $n$, then $D_{m,n}$ is not a cross product, thus giving a negative answer, for these $n$, to the cross product problem **[Ami72]**.

His construction consists in exhibiting two special fields. For each there is a division algebra of dimension $n^2$ with such a field as its center. But in one case all Galois extensions of degree $p^2$ are cyclic, while in the other none is cyclic. Thus the two division algebras cannot have the same Galois group for a maximal subfield. The construction was extended in positive characteristic by M. M. Schacher and L. W. Small **[SS73]** and by several other authors.

The case $n = p$ is open for $p \geq 5$. This statement, when $F$ contains the $p$ roots of 1, is equivalent to showing that there exists a noncommutative polynomial $f$ which is not an identity and is not central on $p \times p$ matrices, and such that $f^p$ is central. For instance for $n = 2$ the polynomial $[x, y]$ has this property.

## 14.5. Using Cayley–Hamilton algebras

*We want to develop some basic theory which shows that Cayley–Hamilton algebras give an effective theory of finite-dimensional representations generalizing the theory of commutative algebras. For an extensive treatment the reader is referred to Procesi* **[Pro–nd]**.

**14.5.1. Basic constructions with Cayley–Hamilton algebras.** Let $R$ be an $n$th Cayley–Hamilton finitely generated algebra over an algebraically closed field $F$, with trace $t$ and $t(R) = A$. By Theorem 8.2.5 and Corollary 12.2.14 $A$ is also a finitely generated algebra over $F$. By Theorem 14.3.1 the (closed) points of $V(A)$ parametrize semisimple representations of dimension $n$ of $R$.

Fix a positive integer $r$ and change the trace by taking the new trace $\tau = rt_{R/A}$.

PROPOSITION 14.5.1. *The algebra $R$ with trace $\tau := rt$ is an $(rn)$th-Cayley–Hamilton algebra. A point $p \in V(A)$ determines an $n$-dimensional semisimple representation $M_p^t$ compatible with the trace $t$ and also an $rn$-dimensional semisimple representation $M_p^\tau$ compatible with the trace $\tau$. We have*

$$M_p^\tau = rM_p^t = (M_p^t)^{\oplus r}.$$

PROOF. Let us show first that, if $R$ is a $n$th-Cayley–Hamilton algebra with trace $t$, the algebra $R$ with trace $\tau := rt$ is an $(rn)$-th-Cayley–Hamilton algebra. We know that $R$ embeds in a trace compatible way into $n \times n$ matrices over a commutative ring $A$. Then the natural diagonal embedding of $M_n(A)$ into $M_{rn}(A)$ gives the claim.

Now we have that $V(A)$ parametrizes also semisimple representations (compatible with the new trace) of dimension $rn$. This is done as follows, a point $p \in V(A)$ gives a maximal ideal $\mathfrak{m}$ of $A$ and $R/\mathfrak{m}R$ is finite dimensional with radical $J$. The semisimple representation is the unique one of $(R/\mathfrak{m}R)/J$ with the given trace. We have an obvious map from the variety of $n$-dimensional representations compatible with the reduced trace to the variety of $rn$-dimensional representations compatible with trace $\tau$; it is simply the map that associates to a representation $M$ its direct sum $M^{\oplus r}$.

From this the statement is clear. ∎

There are different ways of defining the direct sum of two trace algebras.

First if we have $R_1, R_2$ with trace algebras $T_1, T_2$ and each $n$-Cayley–Hamilton, the algebra $R_1 \oplus R_2$ with trace $t(r_1, r_2) = (t(r_1) \oplus t(r_2)) \in T_1 \oplus T_2$ is $n$-Cayley–Hamilton with trace algebra $T_1 \oplus T_2$. Conversely,

PROPOSITION 14.5.2. *Let $R$ be an $n$-Cayley–Hamilton algebra with trace algebra a ring $C$ direct sum of subrings $C = \bigoplus_{i=1}^{s} C_i$.*

*Setting $e_i$ to be the unit of $C_i$, we have a decomposition $R = \bigoplus_{i=1}^{s} R_i$, with $R_i := Re_i$. Then the trace $t$ maps $R_i$ to $C_i$ and each $R_i$ is an $n$-Cayley–Hamilton algebra.*

PROOF. This depends on the fact that trace is $C$-linear so that $t(re_i) = t(r)e_i$ and so $t(e_i) = e_i t(1) = ne_i$. □

Let now $R_1$ and $R_2$ be two trace algebras over $A$ which are Cayley–Hamilton for two integers $n_1, n_2$, and $t(R_1) = t(R_2) = A$. Then a point $p \in V(A)$ determines a semisimple representation $M_p^1$ of dimension $n_1$ of $R_1$ and a semisimple representation $M_p^2$ of dimension $n_2$ of $R_2$.

PROPOSITION 14.5.3. *The algebra $R := R_1 \oplus R_2$ with trace*

$$t(r_1, r_2) := t(r_1) + t(r_2)$$

*is an nth-Cayley–Hamilton for $n := n_1 + n_2$, and $t(R) = A$, a point $p \in V(A)$, determines also a semisimple representation $M_p$ of dimension $n = n_1 + n_2$ of $R$ and*

$$M_p = M_p^1 \oplus M_p^2.$$

The proof is similar to that of the previous proposition, and it is omitted.

Finally let now $R_1$ and $R_2$ be two trace algebras over $F$ which are Cayley–Hamilton for two integers $n_1, n_2$, and $t(R_1) = A_1$, $t(R_2) = A_2$. Then a point $p \in V(A_1)$ determines a semisimple representation $M_p^1$ of dimension $n_1$ of $R_1$, and a point $q \in V(A_2)$ determines a semisimple representation $M_q^2$ of dimension $n_2$ of $R_2$.

PROPOSITION 14.5.4. *The algebra $R := R_1 \otimes R_2$ with trace*

$$t(r_1 \otimes r_2) := t(r_1) \otimes t(r_2)$$

*is an nth-Cayley–Hamilton algebra for $n := n_1 n_2$, and $t(R) = A := A_1 \otimes A_2$.*

*A point $(p, q) \in V(A) = V(A_1) \times V(A_2)$ determines also a semisimple representation $M_p$ of dimension $n = n_1 n_2$ of $R$ and*

$$M_p = M_p^1 \otimes M_q^2.$$

Again the proof is similar.

**14.5.2. The reduced trace.** Let $R$ be a prime algebra over a commutative ring $A$ and assume that $A \subset R$ and $R$ is an $A$-module of finite type. In particular $R$ is a prime PI algebra and $A$ is contained in the center $\mathcal{Z}$ of $R$, a domain with some field of fractions $G$.

We are in the settings of Posner's theorem (Theorem 11.2.6). The algebra $S := R \otimes_{\mathcal{Z}} G$ is the ring of fractions of $R$. $S$ is a finite-dimensional simple algebra isomorphic to $M_k(D)$ where $D$ is a division ring finite dimensional over $G$.

If $F$ is the field of fractions of $A$, then $R \subset R \otimes_A F \subset S$. In fact, since $R \otimes_A F$ is a prime algebra finite dimensional over $F$, it is simple. Therefore we have $R \otimes_A F = R \otimes_{\mathcal{Z}} G = S$, and $D$ is also finite dimensional over $F$.

The center $G$ of $S$ is also the center of $D$ and $\dim_G D = h^2$; moreover, if $\bar{G}$ is an algebraic closure of $G$, we have $M_k(D) \otimes_G \bar{G} = M_{hk}(\bar{G})$.

If the finite extension $F \subset G$ is separable, as happens in characteristic 0 and if $p := [G : F] = \dim_F G$, we also have

$$G \otimes_F \bar{G} = \bar{G}^p \implies S \otimes_F \bar{G} = M_k(D) \otimes_F \bar{G} = M_{hk}(\bar{G})^{\oplus p}.$$

We define the number $hkp$ to be the *degree* of $S$ over $F$, and we let

$$hkp := [S : F].$$

If $a \in S$, we have that $a \otimes 1 \in M_k(D) \otimes_F \bar{G}$ is a $p$-tuple of matrices $(A_1, \dots, A_p)$. One defines the *reduced trace* of $a$ to be the sum

$$t(a) = t_{S/F}(a) := \sum_{i=1}^{p} \operatorname{tr}(A_i).$$

A standard argument of Galois theory, as in Theorem 5.4.40, shows that $t(a) \in F$, in characteristic 0. This can be even more easily seen as follows.

Consider the $F$-linear operator $a^L : S \to S$, $a^L(b) := ab$. Let us compute its trace. This can be done in $M_k(D) \otimes_F \bar{G} = M_{hk}(\bar{G})^{\oplus p}$ where $a^L = (A_1^L, \dots, A_p^L)$ and so

$$\operatorname{tr}(a^L) = \sum_{i=1}^{p} \operatorname{tr}(A_i^L) = hk \sum_{i=1}^{p} \operatorname{tr}(A_i) \qquad \implies t_{S/F}(a) = \frac{1}{hk} \operatorname{tr}(a^L).$$

THEOREM 14.5.5. *If $S = R \otimes_A F$ as before and $A$ is integrally closed, we have that the reduced trace $t_{S/F}$ maps $R$ into $A$, so we will denote by $t_{R/A}$ the induced trace. If $R, S$ are $\mathbb{Q}$-algebras with their reduced trace, then they are $n$-Cayley–Hamilton algebras of degree $n = hkp = [S : F]$.*

PROOF. We have a natural representation of $M_{hk}(\bar{G})^{\oplus p}$ as block diagonal $hkp$ matrices for which the reduced trace is the trace. If $R$ is a finite $A$-module, any element of $R$ is integral over $A$, thus its eigenvalues are also integral so that the trace is integral over $A$. If $A$ is integrally closed, then the trace takes values in $A$.

The Cayley–Hamilton identity is that of its matrix representation. □

The importance of the reduced trace comes from the next result. $R, A, F, S$ are as before, $n := [S : F]$ the degree, and we are assuming characteristic 0.

THEOREM 14.5.6. *If $\tau : R \to A$ is any trace for which $R$ is an $m$-Cayley–Hamilton algebra, then there is a positive integer $r$ for which*

$$m = rn, \qquad \tau = r\, t_{R/A}.$$

PROOF. Clearly $\tau$ extends to a trace on $S$ with values in $F$ for which $S$ is an $m$-Cayley–Hamilton algebra.

Let $G$ be a finite Galois extension of $F$ for which $S \otimes_F G = M_{hk}(G)^{\oplus p}$, and $\tau$ extends to a $G$-valued trace on $S \otimes_F G = M_{hk}(G)^{\oplus p}$ which is invariant under the Galois group and for which $S \otimes_F G$ is an $m$-Cayley–Hamilton algebra. Passing to the algebraic closure, we can now apply Proposition 14.3.5(3) where we know that, by invariance under the Galois group, all the integers $h_i$ must be equal to some positive integer $r$. The formula follows from the definitions. □

## 14.6. The unramified locus and restriction maps

**14.6.1. The composition theorem.** Let us go back to the previous setting with $R$ a prime algebra over a $\mathbb{Q}$-algebra $A$, and an $A$-module of finite type.

Let $Z$ be the center of $R$, and let $F$ be the field of fractions of $A$. One has that $S = R \otimes_A F$ is a simple algebra with center $G = Z \otimes_A F$.

If we further assume that $A$, $Z$ are integrally closed and $A$ is Noetherian, we have that $Z$ is an $A$-module of finite type and for the reduced traces we have the formulas,

$$t_{R/A} = t_{Z/A} \circ t_{R/Z}, \quad [S : F] = [S : G][G : F].$$

Assume now that $A$ is a finitely generated algebra over an algebraically closed field $F$. Since we are assuming that $R$ is a finite $A$-module, it follows that also $Z$ is a finite $A$-module, hence also $Z$ is a finitely generated algebra over $F$.

We thus have two varieties $V(A)$, $V(Z)$, the associated affine variety of $A, Z$, respectively. Finally the inclusion $A \subset Z$ defines a morphism of algebraic varieties $\pi : V(Z) \to V(A)$ of degree $p = [G : F]$.

Now the variety $V(A)$ parametrizes semisimple representations (compatible with the reduced trace $t_{R/A}$) of $R$ of dimension $n = [S : F]$. Moreover, we can also use $V(A)$ to parameterize semisimple representations (compatible with the reduced trace $t_{Z/A}$) of $Z$ of dimension $p = [G : F]$.

The affine variety $V(Z)$ parametrizes semisimple representations (compatible with the reduced trace $t_{R/Z}$) of $R$ of dimension $m = [S : G]$.

We want to put all these things together.

Given a point $Q \in V(A)$ denote by $N_Q$ the corresponding $mp$-dimensional semisimple representation of $R$. Given a point $P \in V(Z)$, denote by $M_P$ the corresponding $m$-dimensional semisimple representation of $R$.

First of all an irreducible representation of $Z$ is one-dimensional and corresponds to a point $P \in V(Z)$, a semisimple representation corresponds to a positive cycle $\sum h_i P_i$ of degree $p = \sum_i h_i$. Proposition 14.3.5 implies

PROPOSITION 14.6.1. *Given a point $Q \in V(A)$, we have for the associated semisimple representation $\sum_{i=1}^{s} h_i P_i$ of $Z$, that the points $P_i$ are exactly the points in the fiber $\pi^{-1}(Q)$. So we may identify formally $\sum_{i=1}^{s} h_i P_i$ with the cycle $[\pi^{-1}(P)]$.*

In general a fiber need not have exactly $p$ points but it can have $s \leq p$ points.

In terms of algebras, $Q$ corresponds to a maximal ideal $m$ of $A$ and the points $P_i$ to the maximal ideals of $Z/mZ$. This is a finite-dimensional commutative algebra, and so $\bar{Z} := Z/mZ = \bigoplus_{i=1}^{s} Z_i$ where $Z_i$ is a local ring supported in the point $P_i$, the trace $t_{Z/A}$ induces a trace $t_{\bar{Z}/k}$ on $\bar{Z}$ making this still a $p$-Cayley–Hamilton algebra.

Let $\bar{n}_i$ be the maximal ideal of $Z_i$, and let $n_i$ be the corresponding maximal ideal of $Z$. We have $Z_i/\bar{n}_i = k$ and again Proposition 14.3.5 implies that the induced trace $t_{\bar{Z}/k}$ on $\bar{Z}$ decomposes as the sum of *local factors* $h_i t_i$ where $t_i : Z_i \to Z_i/\bar{n}_i = k$ is the projection and $\sum_i h_i = p$. In particular, if $e_i$ is the idempotent unit of $Z_i$, we have $t_{\bar{Z}/k}(e_i) = h_i$.

Passing to the algebra $R$, we have a direct sum decomposition.

$$\bar{R} := R/mR = R \otimes_A A/m = R \otimes_Z (Z \otimes_A A/m) = \bigoplus_{i=1}^{s} R \otimes_Z Z_i = \bigoplus_{i=1}^{s} \bar{R}_i.$$

As for traces, we know that $t_{R/A} = t_{Z/A} \circ t_{R/Z}$, and modulo $m$ we get traces $t_{\bar{R}/k} = t_{\bar{Z}/k} \circ t_{\bar{R}/\bar{Z}}$. If $e_i \in Z_i$ is the idempotent identity of $Z_i$, we have $\bar{R}_i = \bar{R}e_i$, thus $t_{\bar{R}/\bar{Z}}$ restricts, by Proposition 14.5.2, to $\bar{R}_i$ to a $Z_i$ trace and $t_{\bar{R}/\bar{Z}} = \bigoplus_{i=1}^{s} t_{\bar{R}_i/Z_i}$. Similarly $t_{\bar{Z}/k} = \bigoplus_{i=1}^{s} t_{Z_i/k}$ and

$$t_{\bar{R}/k} = \bigoplus_{i=1}^{s} t_{Z_i/k} \circ t_{\bar{R}_i/Z_i}.$$

The algebra $R_i$ with the trace $t_{\bar{R}_i/Z_i}$ is an $m = [S : G]$-Cayley–Hamilton algebra. If $\bar{n}_i$ is the maximal ideal of $Z_i$, the unique point of $\mathrm{Spec}(Z_i)$ given by $Z_i \to Z_i/\bar{n}_i = k$ corresponds to some semisimple representation $M_i$. As a representation of $R$, $M = M_{P_i}$, where $P_i$ corresponds to the maximal ideal $n_i$ since we have that $R/n_iR = \bar{R}_i/\bar{n}_i\bar{R}_i$. If we denote by $\bar{t}_{\bar{R}_i/Z_i}$ the image of $t_{\bar{R}_i/Z_i}$ modulo $\bar{n}_i$, we have

$$h_i\bar{t}_{\bar{R}_i/Z_i} = t_{\bar{Z}_i/k} \circ t_{\bar{R}_i/Z_i}.$$

It follows that the semisimple representation of the algebra $\bar{R}_i$ relative to the trace $t_{\bar{Z}_i/k} \circ t_{\bar{R}_i/Z_i}$ is $h_iM_i$. Thus the semisimple representation of $\bar{R}$ relative to the trace $t_{\bar{R}/k} = \bigoplus_{i=1}^{s} t_{Z_i/k} \circ t_{\bar{R}_i/Z_i}$ is $\bigoplus_i h_iM_{P_i}$, (cf. Propositions 14.5.1 and 14.5.3).

We have proved

THEOREM 14.6.2. *Given a point $Q \in V(A)$ and its cycle $\sum_i h_iP_i$ in $V(Z)$, we have*

$$N_Q = \bigoplus_i h_iM_{P_i}.$$

**14.6.2. The unramified locus.** Let $R$ be an algebra with trace finitely generated over an algebraically closed field $F$ satisfying the $n$th-Cayley–Hamilton identity with the trace algebra $t(R)$.

DEFINITION 14.6.3. If $P \in V(t(R))$ with maximal ideal $m_P$, we say that $R$ is *unramified* at $P$ if the quotient algebra $R(P) := R \otimes_{t(R)} k = R/m_PR$ is a semisimple algebra.

The condition to be unramified at $P$ is implicit in Proposition 14.3.5. The trace map for $R$ induces the trace $t : R(P) \to k$ for $R(P)$ and the bilinear trace form $t(ab)$. It follows immediately from Proposition 14.3.5 that

PROPOSITION 14.6.4. *The radical $J$ of $R(P)$ is the kernel of the trace form $t(ab)$.*

Let us see what the meaning is of this statement in the case where $R \supset A$ is a prime algebra over an algebra $A$, finitely generated over an algebraically closed field $F$. Let $F$ be the field of fractions of $A$, $S = R \otimes_A F$, $G$ the center of $S$. Assume that $Z := R \cap G$, the center of $R$, is integrally closed. If $m := [S : G]$, the map $\frac{t_{R/Z}}{m}$ is a projection on $Z$ so that $R = Z \oplus R^0$ and $Z$ is a direct summand.

With the previous notations, $V(A)$ (the associated affine variety) parametrizes semisimple representations (compatible with the reduced trace $t_{R/A}$) of $R$ of dimension $n = [S : F]$. $Z$ is a finitely generated algebra over $F$, and its associated affine variety $V(Z)$ parameterizes semisimple representations (compatible with the reduced trace $t_{R/Z}$) of $R$ of dimension $m = [S : G]$. Denote by $\pi : V(Z) \to V(A)$.

If $Q \in V(A)$ corresponds to a maximal ideal $m_Q$ and the quotient algebra $R(Q) := R \otimes_A A/m_Q$ is semisimple, we have that

$$R \otimes_A A/m_Q = Z \otimes_A A/m_Q \oplus R^0 \otimes_A A/m_Q.$$

Since $Z(Q) := Z \otimes_A A/m_Q$ is in the center of $R(Q) := R \otimes_A A/m_Q$, we have that $Z(Q)$ is semisimple, in other words the scheme theoretic fiber $\pi^{-1}(Q)$ is reduced, $Z \otimes_A A/m_Q = \bigoplus_i Z/n_i$, and $Z/n_i = k$ is a point $P_i$ in the fiber of $Q$. We also have $R(P_i) = R \oplus Z/n_i$ and $R \otimes_A A/m_Q = \bigoplus_i R(P_i)$, hence $R(P_i)$ is semisimple. The converse is also clear.

The commutative algebra $Z(Q)$ is semisimple if and only if it is reduced, i.e., the fiber of $Q$ under the map $\pi$ is reduced, which in our case implies that $\pi$ is étale in the points of this fiber. Now we know that $R$ is a finite module over $Z$ and its generic dimension is $m^2$. The dimension of $R(P)$ over $F = Z(P)$, $P \in V(Z)$, is a semicontinuous function, and we always have $\dim_{Z(P)} R(P) \geq m^2$. If $R(P)$ is not simple of dimension $m^2$, from Proposition 14.3.5(3) it follows that if $J$ is the radical of $R(P)$, we have $\dim_k R(P) < m^2$. Hence we have a dichotomy: either $R(P) = M_m(F)$, that is to say that $P$ corresponds to an irreducible representation; or $R(P)$ is not semisimple.

PROPOSITION 14.6.5. *Let $W^0$ be the open set of $V(Z)$ made of points $P$ where $R(P) = M_m(F)$ (the irreducible representations), let $V^0$ be the maximal open set of $V(A)$ with $\pi^{-1}(V^0) \subset W^0$, and let $V^1$ be the open set of $V(A)$ where $Z(Q)$ is reduced. Then the set of points of $V$ where $R(Q)$ is semisimple is $V^0 \cap V^1$.*

The set of points $V^0 \cap V^1$ of $V$ where $R(Q)$ is semisimple is called the *unramified locus* of the $A$-algebra $R$.

From our analysis it follows that, if $m = [S : G]$, $p = [G : F]$ over a point $Q$ in the unramified locus

$$N_Q = \bigoplus_{i=1}^{p} M_{P_i}$$

decomposes as the direct sum of the $p$ irreducible representations $M_{P_i}$ supported at the $p$ distinct points of the fiber $\pi^{-1}(Q)$.

**14.6.3. Restriction maps.** We come now to the final application of the previous theory. The setting we have in mind appears naturally for quantum groups at roots of 1 and their subgroups; see **[DCPRR05]**.

First we need a lemma. Given a prime algebra $R$ finite over $A$ with center $Z$, let $F$ be the quotient field of $A$, and let $F \subset L$ be an extension field.

LEMMA 14.6.6. *The following are equivalent:*

(i) *$R \otimes_A L$ is a simple algebra.*
(ii) *The algebra $Z \otimes_A L$ is a field.*

PROOF. Let $F$ be the quotient field of $A$. We have that $S := R \otimes_A F$ is a simple algebra with center the field $W := Z \otimes_A F$, and that

$$R \otimes_A L = (R \otimes_A F) \otimes_F L = S \otimes_W (W \otimes_F L).$$

Since $S$ is a simple algebra with center $W$, we have by Proposition 5.4.28 that $S \otimes_W (W \otimes_F L)$ is Azumaya over $U := W \otimes_F L$, hence it is simple by Corollary 5.4.31 if and only if $U$ is a field. Finally $W \otimes_F L = (Z \otimes_A F) \otimes_F L = Z \otimes_A L$. □

Assume that we have two prime algebras $R_1 \subset R_2$ over two commutative rings $A_1 \subset A_2 \subset R_2$. Assume as in the previous paragraph that each $R_i$ is finitely generated as an $A_i$ module and that the two rings $A_i$ are integrally closed. We thus have the two reduced traces $t_{R_i/A_i}$, and we want to discuss the compatibility of these traces. In general one can see by simple examples that there is no compatibility. Let us thus make the basic assumption of *compatibility* (with trace). We let $F_i$ be the quotient field of $A_i$ and consider $S_i := R_i \otimes_{A_i} F_i$.

LEMMA 14.6.7. *The following two conditions are equivalent:*

(i) $R_1 \otimes_{A_1} F_2$ *is a simple algebra.*
(ii) *The algebra* $Z_1 \otimes_{A_1} F_2$ *is a field.*
*In this case the map* $i : R_1 \otimes_{A_1} F_2 = S_1 \otimes_{F_1} F_2 \to S_2$ *is injective.*
*These conditions are satisfied if:*
(iii) *The algebra* $Z_1 \otimes_{A_1} A_2$ *is a domain.*

PROOF. The equivalence of the first two conditions is the content of Lemma 14.6.6. It is clear that (iii) implies (ii) since if $Z_1 \otimes_{A_1} A_2$ is a domain, $Z_1 \otimes_{A_1} F_2$ is its quotient field. □

DEFINITION 14.6.8. Consider two algebras $R_i$ over $A_i$, $i = 1, 2$, $R_1 \subset R_2$ and $A_1 \subset A_2 \subset R_2$. We say that they are compatible if the previous two equivalent conditions are satisfied.

In the examples we will study, we will usually verify (iii).

REMARK 14.6.9. If $R_2$ is a domain, then $R_1 \subset R_2$ is compatible with the inclusions $A_1 \subset A_2 \subset R_2$ if and only if the map $i : R_1 \otimes_{A_1} F_2 = S_1 \otimes_{F_1} F_2 \to S_2$ is injective.

PROOF. If $R_2$ is a domain, so is $S_2$, and so if $i$ is injective, we must have that $Z_1 \otimes_{A_1} F_2$ is a field. □

EXAMPLE 14.6.10.

(1) If the extension $F_1 \subset F_2$ is unirational, then $Z_1 \otimes_{A_1} F_2$ is a field.
(2) If $A_1 = Z_1$ is the center of $R_1$, then $Z_1 \otimes_{A_1} F_2 = F_2$ is a field.

Let us show (1). We have $F_2 \subset F_1(t_1, \dots, t_m)$, so

$$Z_1 \otimes_{A_1} F_2 \subset Z_1 \otimes_{A_1} F_1(t_1, \dots, t_m) = (Z \otimes_{A_1} F_1)(t_1, \dots, t_m)$$

is a field.

THEOREM 14.6.11. *Given two compatible algebras* $R_1 \subset R_2$, *we have that for a positive integer* $r$,

$$r[R_1 : A_1] = [R_2 : A_2], \qquad r\, t_{R_1/A_1} = t_{R_2/A_2} \quad \text{on } R_1.$$

PROOF. By the hypotheses made, one can reduce the computation to the two algebras $S_1 \otimes_{F_1} F_2 \subset S_2$ over $F_2$. In this case we know that the reduced trace $t_{S_2/F_2}$ restricted to $S_1 \otimes_{F_1} F_2$ makes it a Cayley–Hamilton trace algebra. Since by assumption $S_1 \otimes_{F_1} F_2$ is simple, one can then apply Theorem 14.5.6. □

Let us now assume we are in the geometric case in which $A_1, A_2$ are further assumed to be finitely generated over an algebraically closed field $k$. If $V(A_1), V(A_2)$ are the two associated affine varieties parametrizing semisimple representations, we have an induced map $\pi : V(A_2) \to V(A_1)$. If $Q \in V(A_2)$ and $M$ is a representation of $R_2$ over $Q$, we then see that $M$ is also a representation of $R_1$ over $\pi(Q)$ but for $r$ times the reduced trace. If $M$ is semisimple as an $R_2$-module, it may well be that it is not semisimple as an $R_1$-module.

THEOREM 14.6.12. *Given two compatible algebras $R_1 \subset R_2$ as before, $Q \in V(A_2)$. $M_Q$ is the corresponding semisimple representation of $R_2$ of dimension $[R_2 : A_2]$, $M_{\pi(Q)}$ is the corresponding semisimple representation of $R_1$ of dimension $[R_1 : A_1]$. We have that the restriction of $M_Q$ to $R_1$ is a trace representation for $r\, t_{R_1/A_1}$, its associated semisimple representation is $rM_{\pi(Q)} = M_{\pi(Q)}^{\oplus r}$.*

*If $\pi(Q)$ lies in the unramified locus of $R_1$ (as an $A_1$ algebra), we have that the restriction of $M_Q$ to $R_1$ is the semisimple representation $rM_{\pi(Q)}$.*

PROOF. Everything follows from the previous discussions except for the last point. Let $m \subset A_2$ be the maximal ideal associated to $Q$ and $m' := m \cap A_1$. By definition of the unramified locus, the algebra $R_1/m'R_1$ is a semisimple algebra for which every representation is semisimple. $\square$

CHAPTER 15

# Spectrum and dimension

*In this chapter we usually assume that R is a PI algebra with a 1. The theme of this chapter is to analyze how various standard facts of commutative algebra may generalize to PI rings, with always in mind the geometric picture of the spectrum as related to representations of finite dimension.*

*We only present a small portion of a large number of results, mostly due to M. Artin and W. Schelter.*

## 15.1. Krull dimension

**15.1.1. Integral extensions.** We follow strictly Artin and Schelter ([**AS81b**], [**AS87**], and [**Sch77a**]), and we assume that all rings have a 1. For the notion of extension we refer to Definition 11.2.12.

Recall that we have introduced, given a ring $R$, the algebra of generalized polynomials which in the special case on one variable is the free product $R * \mathbb{Z}\langle x\rangle$.

DEFINITION 15.1.1. We say that a generalized polynomial $f(x)$ is *monic* if $f(x) = x^m + g(x)$, where $g(x)$ is a generalized polynomial of degree $< m$.

Notice that the terms of $g(x)$ may be in general of the form $r_1 x^{h_1} r_2 x^{h_2} \cdots x^{h_i} r_i$.

DEFINITION 15.1.2. Given two rings $R \subset S$, we say that an element $t \in S$ is *integral* over $R$ if it satisfies a monic generalized polynomial.

We say that an extension $R \subset S$ is *integral*, if every element $s \in S$ is integral over $R$.

REMARK 15.1.3. If the subring $R[t]$, generated by $R$ and $t$, is a finite $R$ bimodule, then $t$ is integral over $R$. In fact then $R[t]$ is spanned by a finite number of generalized polynomials which have some degree $< n$, and then $t^n$ is a linear combination of these with coefficients in $R$ on right or left still a generalized polynomial of the same degree.

In general if $R \subset S$ is an extension, there is no maximal integral extension of $R$ contained in $S$ according to the previous definition, even if $R$ is in the center of $S$ and $S$ is PI. For instance in $R = F$ a field and $S = M_2(F[x])$, then the two elements

$$a = \begin{pmatrix} 0 & 0 \\ x & 0 \end{pmatrix}, \quad b = \begin{pmatrix} 0 & x \\ 0 & 0 \end{pmatrix}$$

both generate an integral extension, but $a + b$ and $ba$ are not integral over $F$.

We now want to introduce the notion of *finite extension*, first remark:

LEMMA 15.1.4. *If $R \subset S$ is an extension, then $S$ is a finite left $R$-module if and only if it is a finite right module and in both cases a linear set of generators can be taken from elements of $S^R$. In this case we say that $S$ is a finite extension of $R$.*

PROOF. Suppose that $S = \sum_i Ru_i$. Since $S = RS^R$ (cf. Definition 11.2.12), we can write $u_i = \sum_j r_{i,j} v_j$ for some $v_j \in S^R$, hence clearly $S = \sum_j Rv_j = \sum_j v_j R$. □

LEMMA 15.1.5. *If $R_1 \subset R_2$ and $R_2 \subset R_3$ are finite extensions, then $R_1 \subset R_3$ is also a finite extension.*

PROOF. We have $R_2 = \sum_i R_1 u_i$ where $u_i$ commutes with $R_1$ and similarly $R_3 = \sum_j R_2 v_j$, then $R_3 = \sum_{i,j} R_1 u_i v_j$ and clearly $u_i v_j$ commutes with $R_1$. □

By Shirshov's theorem if $S$ is a PI algebra, integral over some central ring $A$, then every finitely generated subalgebra is a finite module over $A$.

LEMMA 15.1.6. *If $S = RB$ is a PI extension, $B$ is in the centralizer of $R$ in $S$, and $B \supset A$ is integral over the center $A$ of $R$, then $S$ is integral over $R$.*

PROOF. Let $t \in S$. We write then $t = \sum_i r_i u_i$ with $u_i \in B$. The subring generated by $t$ over $R$ is contained in $RB'$ where $B'$ is the subring of $B$ generated by the elements $u_i$, which by Remark 15.1.3 is a finite $A$-module. Hence $RB'$ is a finite left or right $R$-module and the claim follows from Lemma 15.1.4 and Remark 15.1.3. □

### 15.1.2. Krull dimension.

DEFINITION 15.1.7. The *Krull dimension* of a ring $R$ is the supremum of the lengths of all possible strictly increasing chains of prime ideals

$$P_1 \subsetneq P_2 \subsetneq \cdots \subsetneq P_{k-1} \subsetneq P_k.$$

In general Krull dimension is infinite (sometimes one may even give it a sense as ordinal numbers, but we do not want to treat this here). In classical algebraic geometry, when $R$ is the coordinate ring of an irreducible affine variety $V$, one sees that any increasing maximal chain of prime ideals, corresponding to a maximal chain of decreasing irreducible varieties, has as length the dimension of $V$, which can be defined in several different ways (cf. §8.3).

One possible definition of dimension is the transcendence degree of the function field of the variety over the base field.

15.1.2.1. *Algebras over Noetherian rings.* Let us consider an algebra $R$ over a commutative ring $A$.

DEFINITION 15.1.8. Given an integer $m \in \mathbb{N}$, we say that an ideal $I$ of $R$ is *$m$-representable*, if $R/I$ can be embedded in a matrix algebra $M_m(B)$ where $B$ is a commutative $A$-algebra.

In this section we shall concentrate our study on algebras $R = A\langle a_1, \dots, a_k\rangle$ finitely generated over a commutative Noetherian ring $A$.

THEOREM 15.1.9. *If $R = A\langle a_1, \dots, a_k\rangle$ is finitely generated over a commutative Noetherian ring $A$, then it satisfies the ascending chain condition on $m$-representable ideals.*

PROOF. Recall the construction of §3.4.1 and Proposition 3.4.12.

For any $A$-algebra $S$ we have the universal map $j_{m,S} : S \to M_m(A_m(S))$ into $m \times m$ matrices over a commutative $A$-algebra which we have denoted $A_m(S)$ in §3.4.1; this construction is functorial.

The algebra $S$ can be embedded in a matrix algebra $M_m(B)$, where $B$ is a commutative $A$-algebra, if and only if $j_{m,S}$ is injective. Moreover if $S$ is finitely generated over $A$, so is $A_m(S)$.

Finally, if $\pi : S_1 \to S_2$ is a map, we have an induced map $\pi_m : A_m(S_1) \to A_m(S_2)$, and if $\pi$ is surjective so is $\pi_m$.

It follows that each $m$-representable ideal $I$ of $R$ is associated to an ideal $I_m$ of an $A_m(R)$ kernel of the map $\pi_m : A_m(R) \to A_m(R/I)$ associated to the quotient map $\pi : R \to R/I$. The correspondence $I \mapsto I_m$ is injective and order preserving. Since $A_m(R)$ is finitely generated over $A$, it is Noetherian, hence the claim follows. □

15.1.2.2. *PI algebras over Noetherian rings.* From now on we concentrate on the class of algebras $R = A\langle a_1, \dots, a_k \rangle$ finitely generated over a commutative Noetherian ring $A$ and PI. In Chapter 16 we shall prove that the nil radical of these algebras is nilpotent Theorem 16.1.1. If $A$ is a field, the algebra satisfies the identities of $n \times n$ matrices for some $n$ (see Corollary 16.3.4). For the moment we prove some preliminary facts, assuming for instance the property that an algebra satisfies the identities of $n \times n$ matrices. When we start to study some properties of its spectrum, we will use systematically the results of Theorem 11.2.17.

PROPOSITION 15.1.10. *If $R = A\langle a_1, \dots, a_k \rangle$ is a PI algebra, finitely generated over a commutative Noetherian ring $A$, then $R$ satisfies the ascending chain condition on semiprime ideals.*

PROOF. Assume $R$ satisfies an identity of degree $2d$ or $2d + 1$, and let $I$ be a semiprime ideal. Then by Amitsur's theorem (Theorem 11.2.3), $R/I$ can be embedded in a direct sum of rings of $h_i \times h_i$ matrices over commutative rings for $h_i \leq d$.

All these matrix algebras can be embedded in a single matrix algebra $M_m(B)$ where $B$ is a commutative $A$-algebra and $m$ is a common multiple of the $h_i$, for instance $m = d!$. So all semiprime ideals are $m$-representable, and the claim follows from Theorem 15.1.9. □

REMARK 15.1.11. Proposition 15.1.10 can be stated in a *geometric fashion*, by saying that the topology of the spectrum of $R$ is *Noetherian*, in the sense that *every descending chain of closed sets must be stationary.*

An important step is the study of a prime PI algebra $R = A\langle a_1, \dots, a_k \rangle$ finitely generated over a commutative Noetherian ring $A$. In this case an essential tool is the embedding of $R$ into the *trace ring* $RT$ obtained from $R$ by adding all the coefficients of the characteristic polynomials of the elements of $R$. From Theorem 8.2.5 we have

PROPOSITION 15.1.12. *The trace ring $RT$ is a finite module over $T$ which is a finitely generated algebra over $A$. Hence $RT$ is a Noetherian ring.*

THEOREM 15.1.13. *Let $R = A\langle a_1, \dots, a_k \rangle$ be a prime PI algebra finitely generated over a commutative Noetherian ring $A$. For every ideal $I$ of $R$ we have that its radical $\sqrt{I}$ is an intersection of finitely many primes and it is nilpotent modulo $I$.*

PROOF. If the statement is not true for some $I$, then we may consider the set of all prime ideals $P$ of $R$ such that the statement is not true for $R/P$ and, by Proposition 15.1.10, there is a maximal such $P$. Reducing to this case, we may assume that the statement is true for all rings $R/P$ where $P \neq 0$ is a prime ideal.

We now will prove that the statement is also true for the prime ideal $\{0\}$ reaching a contradiction.

From Proposition 15.1.12 the trace ring $RT$ is a Noetherian ring, so for this ring the statement is true by Corollary 1.4.8.

The ring $RT$ is a prime algebra of some degree $d$. We have seen, in Proposition 11.2.10, that there is a nonzero ideal $I$ of $R$ which is also an ideal of $RT$, i.e., a nonzero *conductor*. For instance for a nonzero evaluation $c \in R$ of a central polynomial for $d \times d$ matrices there is a power $c^k$ in the conductor, Theorem 10.4.11.

Then let $J$ be an ideal of $R$ for which we want to analyze the statement. Then $IJ$ is an ideal of $R$, which is also an ideal of $RT$. So, as we have already remarked, $RT$ is Noetherian. In $RT$ our statement is valid, and thus we have in $RT$ that $\sqrt{IJ} = \hat{P}_1 \cap \hat{P}_2 \cap \cdots \cap \hat{P}_k$ with the $\hat{P}_i$ prime ideals of $RT$ and $(\hat{P}_1 \hat{P}_2 \cdots \hat{P}_k)^N \subset IJ$ for some $N$. Then $P_i = R \cap \hat{P}_i$ is a prime ideal of $R$, $\sqrt{IJ} = P_1 \cap P_2 \cap \cdots \cap P_k$ is a decomposition in $R$ and $(P_1 P_2 \cdots P_k)^N \subset IJ$.

The prime ideals $P_i$ are nonzero so, by our Noetherian induction (cf. Remark 1.4.3), in the rings $R/P_i$ the theorem is valid.

So, setting $J_i \subset R/P_i$ the image of $J$ in $R/P_i$, we have that $\sqrt{J_i} = \bigcap_h \bar{Q}_{i,h}$ where $\bar{Q}_{i,h}$ is the image in $R/P_i$ of a prime ideal $Q_{i,h} \supset P_i$.

We then have in $R$ that $\sqrt{J + P_i} = \bigcap_h Q_{i,h}$ and there are integers $n_i$ such that $(\bigcap_h \bar{Q}_{i,h})^{n_i} \subset J_i$ or $(\bigcap_h Q_{i,h})^{n_i} \subset J + P_i$.

We next claim that

$$\sqrt{J} = \bigcap_i \sqrt{J + P_i} = \bigcap_{i,h} Q_{i,h}.$$

In fact $\sqrt{J + P_i} = \bigcap_h Q_{i,h}$ is by definition and clearly $\sqrt{J} \subset \bigcap_i \sqrt{J + P_i}$. In order to prove the equality $\sqrt{J} = \bigcap_i \sqrt{J + P_i}$, it is thus enough to prove that, for some exponent $M$, we have $(\bigcap_{i,h} Q_{i,h})^M \subset J$.

We have seen that $(P_1 P_2 \cdots P_k)^N \subset IJ$, so, letting $M := hk \max(n_i) N$,

$$(\bigcap_{i,h} Q_{i,h})^M \subset (\prod_{i=1}^{k} (\prod_{i,h} Q_{i,h})^{n_i})^N \subset (\prod_{i=1}^{k} (J + P_i))^N \subset J + (P_1 P_2 \cdots P_k)^N = J. \quad \square$$

COROLLARY 15.1.14. *Let $R = A\langle a_1, \ldots, a_k \rangle$ be a PI algebra finitely generated over a commutative Noetherian ring $A$. Then its nil radical $N$ is the intersection of finitely many minimal primes. If $R$ is the quotient of the ring of $k$ generic $n \times n$ matrices for some $n$, then $N$ is nilpotent.*

PROOF. By Amitsur's theorem (Theorem 11.2.3), $R/N$ satisfies the identities of $n \times n$ matrices for some $n$. Since $R$ and $R/N$ have the same spectrum, it is enough to prove the theorem assuming that $R$ satisfies these identities. First assume that $A$ is a domain. We have that $R = S/I$, where $S = A\langle \xi_1, \ldots, \xi_k \rangle$, is the algebra of $n \times n$ generic matrices, which is a domain and hence prime, to which we can apply Theorem 15.1.13.

Now let us prove the general case. Denote by $N$ the nil radical of $R$. If $\underline{n}$ is the nil radical of $A$, since $A$ is Noetherian, we have that $\underline{n}$ is the intersection of finitely many minimal primes $\underline{p}_i$ of $A$ and it is nilpotent (see Corollary 1.4.8). Thus $\underline{n}R$ is nilpotent and we can reduce modulo this ideal and assume that $\underline{n} = 0 = \bigcap_{i=1}^k \underline{p}_i$.

Each algebra $R/\underline{p}_i R$ is the quotient of the ring of $k$ generic $n \times n$ matrices over the Noetherian domain $A/\underline{p}_i$, so we can apply to it Theorem 15.1.13. Let $\bar{N}_i = N_i/\underline{p}_i R$ be the nil radical of $R/\underline{p}_i R$. It is the intersection of finitely many minimal primes $\bar{Q}_{i,h}$, $h = 1, \dots, t_i$, images of primes $Q_{i,h}$, and it is nilpotent.

Moreover we have that the image $\bar{N}$ of $N$ in $R/\underline{p}_i R$ is contained in $\bar{N}_i$, hence $N \subset N_i$ for all $i$ or $N \subset \bigcap_i N_i$. We claim that $N = \bigcap_{i,h} Q_{i,h}$ and that it is nilpotent.

In fact by construction $N \subset \bigcap_i N_i = \bigcap_{i,h} Q_{i,h}$, so it is enough, in order to prove all claims, to prove that $\bigcap_{i,h} Q_{i,h}$ is nilpotent.

We have as in the proof of Theorem 15.1.13, that $N_i^{n_i} \subset \underline{p}_i R$ implies

$$(\bigcap_{i,h} Q_{i,h})^{\sum_i t_i n_i} \subset (\prod_{i=1}^k (\prod_h Q_{i,h})^{n_i}) \subset \prod_{i=1}^k \underline{p}_i R = 0. \qquad \square$$

## 15.2. A theorem of Schelter

**15.2.1. Chain conditions.** The aim of this section is to show a theorem of Schelter extending to PI rings a well-known property of the coordinate rings of irreducible varieties, the *catenary* property.

DEFINITION 15.2.1. A ring $R$ is *catenary* if all maximal chains of prime ideals of $R$ have the same finite length.

THEOREM 15.2.2. *A prime PI ring $R$ finitely generated over a field $F$ is catenary. Moreover this length, which is the Krull dimension of $R$, equals the transcendence degree of the center of $R$ over $F$.*

First we need some results on integral extensions.

There are various definitions of integrality for noncommutative rings, but the simplest is when $R$ contains a central subring $A$. Then $R$ is integral over $A$ if it satisfies the usual condition, every element $r \in R$ satisfies a monic polynomial with coefficients in $A$.

THEOREM 15.2.3 (Going down). *Let $R$ be a prime PI ring, integral over an integrally closed central subring $A$ (a domain). Suppose that we have two prime ideals $\mathfrak{p}_0 \subset \mathfrak{p}_1 \subset A$ and a prime ideal $P_1$ of $R$ such that $P_1 \cap A = \mathfrak{p}_1$. Then there is a prime ideal $P_0 \subset P_1$ of $R$ such that $P_0 \cap A = \mathfrak{p}_0$.*

LEMMA 15.2.4. *With the hypotheses of Theorem 15.2.3, if $\mathfrak{n}$ is an ideal of $A$, every element of $\mathfrak{n}R$ satisfies a monic polynomial over $A$ with all coefficients besides the leading one in $\mathfrak{n}$.*

PROOF. Take an element $x = \sum_i n_i r_i$ with $n_i \in \mathfrak{n}$ and consider the subring $R'$ generated over $A$ by the elements $n_i r_i$. By Theorem 8.2.1 we have that $R'$ is a finite $A$-module and $x \in R'$.

Every monomial in the elements $n_i r_i$ is of the form $nr$, $n \in \mathfrak{n}$, $r \in R$.

Since we have some equation $r^k + a_1 r^{k-1} + \cdots + a_k = 0$, $a_i \in A$, we also have $(nr)^k + na_1(nr)^{k-1} + \cdots + n^k a_k = 0$. This implies that $nr$ is nilpotent modulo $\mathfrak{n}R'$. By the local nilpotence theorem, Theorem 8.2.2(2), we deduce that every element of $R'$ is nilpotent modulo $\mathfrak{n}R'$, thus for some $t > 0$ we have $x^t \in \mathfrak{n}R'$. Therefore, in some finite set of generators $s_1, \ldots, s_k$ of $R'$ over $A$, we have the matrix of multiplication by $x^t$ of the form $x^t s_i = \sum_j n_{i,j} s_j$, $n_{i,j} \in \mathfrak{n}$. Then $x^t$ satisfies the characteristic polynomial of the matrix $(n_{i,j})$ which has the desired property. □

PROOF OF THEOREM 15.2.3. Consider the set $\mathcal{C}$ of elements of $R$ which are regular, that is nonzero divisors, modulo the prime ideal $P_1$. Let

$$S := (A \setminus \mathfrak{p}_0)\mathcal{C} = \{ac \mid a \in A \setminus \mathfrak{p}_0,\ c \in \mathcal{C}\}.$$

We claim that $R\mathfrak{p}_0 \cap S = \emptyset$. Suppose by contradiction that we have elements $a \in A \setminus \mathfrak{p}_0$, $c \in \mathcal{C}$ such that $s := ac \in R\mathfrak{p}_0$. By Lemma 15.2.4, the element $s$ satisfies a monic polynomial $p(x) = x^m + p_1 x^{m-1} + \cdots + p_m$ with the coefficients $p_i$ in $\mathfrak{p}_0$.

Let $F$ be the field of fractions of $A$. Since $R$ is torsion free over $A$, we have $R \subset F \otimes_A R$ and $s$ satisfies a minimal polynomial $g(x) = x^n + a_1 x^{n-1} + \cdots + a_n$ over the field $F$. Since $A$ is integrally closed and $s$ is integral over $A$, we have that also $g(x)$ has coefficients in $A$. Then $g(x)$ divides $p(x)$ in $A[x]$. Since $p(x) = x^m$ modulo $\mathfrak{p}_0$, we have that also all the the coefficients $a_i$ are in $\mathfrak{p}_0$.

The polynomial $h(x) := x^n + a^{-1}a_1 x^{n-1} + \cdots + a^{-n}a_n$ is the minimal polynomial of $c = a^{-1}s$ over the field $F$. Since we know that $c$ is integral over $A$, we must have that $h(x)$ has coefficients in $A$, i.e., $a^{-i}a_i \in A$. Since $a_i \in \mathfrak{p}_0$, $a \notin \mathfrak{p}_0$, it follows that $a^{-i}a_i \in \mathfrak{p}_0$. Thus $c^n \in \mathfrak{p}_0 R \subseteq P_1$, a contradiction.

Then take an ideal $P_0$ maximal with the property that $P_0 \cap S = \emptyset$. Since $S$ is a multiplicative set, we have that $P_0$ is prime. Now we claim that $P_0 \subset P_1$; otherwise $P_0 + P_1$ intersects $S$. But if $a + b \in S$, $b \in P_1$, we have $a \in S$, so $P_0$ intersects $S$, a contradiction.

Now $P_0 \cap A \supset \mathfrak{p}_0$ and $P_0 \cap (A \setminus \mathfrak{p}_0) = \emptyset$, hence $P_0 \cap A = \mathfrak{p}_0$. □

For a commutative algebra $T$, which is a domain and finitely generated over a field $F$, we use the standard theory coming from affine geometry, that the Krull dimension of $T$ equals the transcendence degree of its field of fractions. The catenary property can be expressed by the fact that if $\mathfrak{p} \subset T$ is a nonzero minimal prime, then $\dim T/\mathfrak{p} = \dim T - 1$.

LEMMA 15.2.5. *Let $R = A\langle a_1, \ldots, a_m\rangle$ be a prime PI algebra over $A$, a Noetherian domain. Let $RT$ be the trace ring of $R$. If $P \neq 0$ is a nonzero minimal prime of $R$, there is a prime $Q$ of $RT$ with $Q \cap R = P$.*

*We have $R/P \subset RT/Q$ both prime rings and of the same PI degree.*

PROOF. By Proposition 15.1.12 the ring $RT$ is Noetherian. Take a nonzero evaluation $a$ of a central polynomial, in the center of $R$ (which is a domain over which $R$ is torsion free). By Theorem 10.4.11 we have some $m$ so that $a^m R$ is an ideal in $RT$. Then $a^m P$ is a nonzero ideal in $RT$.

Take an ideal $Q$ in $RT$ which is maximal with respect to the property that $a^m P \subset Q \cap R \subset P$. We claim that $Q$ is prime and that $Q \cap R = P$.

In fact if we prove the first statement, the second follows by the minimality of $P$ since $Q \cap R$ is a nonzero prime ideal contained in $P$.

If by contradiction $Q$ is not prime, we would have two ideals $I, J$ properly containing $Q$ with $IJ \subset Q$. We have $(I \cap R)(J \cap R) \subset Q \cap R \subset P$ so, since $P$ is prime, either $I \cap R \subset P$ or $J \cap R \subset P$. In either case we contradict the maximality of $Q$.

We have $R/P \subset RT/Q$ both prime rings and of the same PI degree by Proposition 11.2.19. □

REMARK 15.2.6. The following example by M. Artin shows that the previous statement is not true for a height 2 prime. Let $S = F[u, v, v^{-1}]$ and

$$R = \begin{pmatrix} S & uS \\ S & F[v] + uS \end{pmatrix}, \quad P := \begin{pmatrix} S & uS \\ S & F[v]v + uS \end{pmatrix}, \quad S/P \simeq F.$$

Check that $R$ is prime and finitely generated over $F$. The ideal $P$ is prime but does not lift since

$$v^{-1} = \operatorname{tr}\begin{pmatrix} v^{-1} & 0 \\ 0 & 0 \end{pmatrix} \in T, \quad \begin{pmatrix} 0 & 0 \\ 0 & v \end{pmatrix} \in P$$

$$v^{-1}\begin{pmatrix} 0 & 0 \\ 0 & v \end{pmatrix} = \begin{pmatrix} 0 & 0 \\ 0 & 1 \end{pmatrix} \notin P.$$

**15.2.2. Integral and central extensions.** The next theorem generalizes a classical result of Krull for commutative rings.

THEOREM 15.2.7. *Let $R \subset S$ be an integral extension of PI rings.*

(1) (Going-up) *If $P_1 \subset P_2$ are prime ideals in $R$ and $Q_1$ is a prime ideal of $S$ with $Q_1 \cap R = P_1$, there is a prime ideal $Q_2$ of $S$ with $Q_1 \subset Q_2$ and $Q_2 \cap R = P_2$.*
(2) (Incomparability) *If $Q_1, Q_2$ are two different primes in $S$ with $Q_1 \cap R = Q_2 \cap R$, then these two primes are not comparable, i.e., $Q_1 \not\subset Q_2$, $Q_2 \not\subset Q_1$.*
(3) (Lying-over) *For any prime $P$ of $R$ there is a prime $Q$ of $S$ with $Q \cap R = P$.*

PROOF. (1) Passing modulo $Q_1$, we may assume that $Q_1 = 0$, so $S$ and $R$ are prime PI rings. Call $P = P_2$, then take an ideal $Q$ in $S$ maximal with respect to the property that $Q \cap R \subset P$. By Lemma 1.1.40 $Q$ exists, and it is a prime ideal. We want to show that $Q \cap R = P$.

Passing modulo $Q$, we arrive at the following reduction: $R \subset S$ is an integral extension of prime PI rings, $P$ is a prime ideal of $R$, and every nonzero ideal $I$ of $S$ is such that $I \cap R \not\subset P$. We need to show that $P = 0$.

Assume by contradiction that $P \neq 0$. Take a central element $z \in P$ as in Proposition 11.2.5; $z$ is also central in $S$ since $S$ is an extension of $R$ so generated by $R$ and its centralizer. We may consider the ideal $I = zS$. Take then any nonzero element $t \in S$ such that $zt \in I \cap R$. By hypothesis, $t$ satisfies some generalized polynomial $t^n + p(t) = 0$ with *coefficients* in $R$. Then multiplying this by $z^n$, we get that since $p(t)$ has degree $< n$ in $t$ and $z$ is in the center, $z^n p(t) \in zR \subset P$.

Thus each element of $I \cap R$ is in the radical of $P$, but since $R$ is a PI ring and $P$ is prime, the radical of $P$ is $P$, and we have that $I \cap R \subset P$, a contradiction.

(2) Assume that $Q_1 \supsetneq Q_2$. Passing modulo $Q_2$, we may assume that $S$ is prime $Q$ a prime ideal with $Q \cap R = 0$, and we need to prove that $Q = 0$. Otherwise, $Q$ contains some central element $z$ which then satisfies an equation $z^n + r_1 z^{n-1} + \cdots + r_n = 0$, $r_i \in R$, and we may assume this has minimal degree. Thus $r_n \in Q \cap R = 0$, so the equation is $z(z^{n-1} + \cdots) = 0$. But $z$ is not a zero divisor, so we deduce an equation of lower degree, a contradiction.

(3) This is a consequence of (1). Take $Q$ maximal in $S$ with $P_0 := Q \cap R \subset P$. The usual argument shows that $Q$ is prime, so $P_0$ is prime, and we may apply (1). □

One of the main corollaries of this theorem is the fact that

COROLLARY 15.2.8. *Krull dimension is preserved under integral extensions of PI rings.*

PROOF. In fact any chain of distinct prime ideals $Q_1 \subsetneq Q_2 \subsetneq \cdots \subsetneq Q_r$ in $S$ gives rise to a chain of distinct prime ideals $P_1 \subsetneq P_2 \subsetneq \cdots \subsetneq P_r$ in $R$ with $P_i = Q_i \cap R$. Conversely by the going-up theorem (Theorem 15.2.7(1)), any chain of distinct prime ideals $P_1 \subsetneq P_2 \subsetneq \cdots \subsetneq P_r$ in $R$ can be lifted to a chain of primes in $S$. □

**15.2.3. Prime PI rings are catenary.** We come now to the main result from Schelter [**Sch77a**]. Here for an algebra $R$ over a field $F$ by $\dim R$ we mean the GK dimension of $R$ over $F$, cf. §8.3.

EXERCISE 15.2.9. For an algebra $R$ which is a finite module over its center $A$, we have that $\dim R = \dim A$.

DEFINITION 15.2.10. For a prime ideal $P$ of a ring $R$, we define its height $\mathrm{ht}(P)$, as the supremum of the lengths of all the descending chains of prime ideals $P = P_0 \supsetneq P_1 \supsetneq P_2 \supseteq \cdots \supsetneq P_k$.

In particular the Krull dimension (Definition 15.1.7) of $R$ is the supremum of all the heights of all prime ideals.

In Corollary 11.2.9 for a prime PI algebra $R$ over a field $F$ we have seen that its GK dimension $\dim R$ equals the transcendence degree of the center of its ring of fractions over $F$. For the ring of generic matrices, this has been computed in Corollary 14.4.5.

LEMMA 15.2.11. *Let $R$ be a prime PI ring finitely generated over a field $F$ and a finite module over its center. If $0 \neq Q \subset R$ is a prime ideal, then there is a prime ideal $Q_0 \subset Q$ such that*

$$\text{PI-deg}(R/Q_0) = \text{PI-deg}(R) \quad \textit{and} \quad \dim R/Q_0 = \dim R/Q + 1 \tag{15.1}$$

*(and thus $Q$ is minimal over $Q_0$).*

PROOF. We first reduce to the case where $F$ is algebraically closed passing to $R \otimes_F \bar{F}$ where $\bar{F}$ is the algebraic closure. By Lemma 15.1.6, $R \otimes_F \bar{F}$ is integral over $R$, so it has the same Krull dimension of $R$. On the other hand in general $R \otimes_F \bar{F}$ is not prime, and in positive characteristic it may even have a nilpotent ideal.

Nevertheless, by the going-up theorem (Theorem 15.2.7(1)), we can lift $Q \supset \{0\}$ to a pair of prime ideals $Q' \supset J$ of $R \otimes_F \bar{F}$ with $Q' \cap R = Q$, $J \cap R = \{0\}$.

Set $R' := R \otimes_F \bar{F}/J$, and suppose $Q_0'$ in $R'$ satisfies the claim of Lemma 15.2.11 for $Q'/J$. Set $Q_0 := Q_0' \cap R$. We have the central extension $R/Q_0 \subset R'/Q_0'$, so these have the same PI degree which by hypothesis on $Q_0'$ is the PI degree of $R'$ which equals that of $R$. Furthermore since $R/Q_0 \subset R'/Q_0'$ is an integral extension, they have the same GK dimension, so if (15.1) holds for $R'$ it also holds for $R$.

So now our assumptions are that $F$ is algebraically closed, $R$ is prime and a finite module over its affine center $A$; moreover, $Q \cap A$ is a prime ideal.

If $R/Q$ has the same PI degree as $R$, then we are in the part of the spectrum of $R$ which is homeomorphic to an open set of the spectrum of the commutative ring $A$, and then the claim follows from the same statement for $A$.

So we may assume that the PI degree of $R/Q$ is strictly smaller than that of $R$, in particular all central polynomials for $R$ vanish in $R/Q$, that is, take all values in $Q$.

We apply a strong theorem of commutative algebra, Noether normalization as in Bourbaki [**Bou64**, Theorem 1, §3.1].

By this theorem we may find a polynomial ring $F[X_1, \ldots, X_m] \subset A$, such that $A$ is a finite module over this polynomial ring and

$$Q \cap F[X_1, \ldots, X_m] = (X_1, \ldots, X_r)$$

is the ideal generated by $X_1, \ldots, X_r$ for some $r$.

So $m = \dim R$ and $R/Q$ is a finite module over

$$F[X_1, \ldots, X_m]/(Q \cap F[X_1, \ldots, X_m]) = F[X_{r+1}, \ldots, X_m],$$

which is a polynomial ring in $m - r$ variables. In particular $m - r = \dim R/Q$. If $r = 1$, then we are done: $m - 1 = \dim R/Q$ so we could take $Q_0 = 0$.

Let $Z$ be the ideal of $A$ generated by all evaluations of a central polynomial for $R$. We have seen that $Z \subset Q$.

Since $A \supset F[X_1, \ldots, X_m]$ is an integral extension, every nonzero ideal of $A$ intersects $F[X_1, \ldots, X_m]$ in a nonzero ideal, so we may take $j \in Z \cap F[X_1, \ldots, X_m]$, $j \neq 0$.

Since $F$ is infinite and $r \geq 2$, the polynomial $j$ does not vanish on some point outside the linear subspace $U := \{X_1 = X_2 = \cdots = X_r = 0\}$, hence also does not vanish on some subspace of codimension $r - 1$ of $F^m$ containing $U$.

By a linear change of variables this linear subspace could be given by the equations $X_1 = X_2 = \cdots = X_{r-1} = 0$. Now $(X_1, \ldots, X_{r-1}) \subset (X_1, \ldots, X_r)$ is a prime ideal of $F[X_1, \ldots, X_m]$, so, by the going-down theorem, there is in $\operatorname{Spec}(R)$ a prime ideal $Q_0 \subset Q$, with $Q_0$ lying over $(X_1, \ldots, X_{r-1})$. Now, by construction, $j \notin (X_1, \ldots, X_{r-1})$, so $j \notin Q_0$; thus, since $j$ is the value of a central polynomial, $\text{PI-deg}(R[T]/Q_0) = \text{PI-deg}(R)$. We claim that $Q_0$ is the required ideal:

$$\begin{aligned}\dim R/Q_0 &= \dim A/A \cap Q_0 = \dim F[X_1, \ldots, X_m]/(X_1, \ldots, X_{r-1}) \\ &= m - r + 1 = \dim R/Q + 1.\end{aligned}$$

□

The final theorem can be thus formulated as:

THEOREM 15.2.12. *If $R = F\langle a_1, \ldots, a_t \rangle$ is a prime ring satisfying a polynomial identity, $F$ is a field, and $P$ is a prime ideal of $R$, then* $\dim R = \dim(R/P) + \operatorname{ht}(P)$.

PROOF. By induction it is enough to prove this statement when $P$ has height 1. If $\text{PI-deg}(R/P) = \text{PI-deg}(R)$, then we are in the part of the spectrum which is open in an affine variety of dimension $\dim R$ so this follows from affine geometry. We may thus assume that $\text{PI-deg}(R/P) < \text{PI-deg}(R)$.

By Lemma 15.2.5 we can lift $P$ to a prime ideal $Q$ in $RT$, so $Q \cap R = P$.

By Lemma 1.1.40 $Q$ is maximal among ideals $I$ with $I \cap R = P$.

LEMMA 15.2.13. $\dim RT/Q = \dim R/P$.

PROOF. Consider the embedding of rings of fractions $\mathcal{Q}(R/P) \subset \mathcal{Q}(RT/Q)$, and let $E$ be the center of $\mathcal{Q}(R/P)$. Then take the ring $ERT/Q \subset \mathcal{Q}(RT/Q)$ obtained by localizing $RT/Q$ at the nonzero central elements of $R/P$ which are clearly central also in $RT/Q$.

We claim that $ERT/Q$ is a simple ring, and thus it coincides with $\mathcal{Q}(RT/Q)$.

Take a nonzero two-sided ideal $X$ of $ERT/Q$. We have $X \cap RT/Q$ is a nonzero ideal and thus, by maximality of $Q$ that $X \cap R/P \neq 0$. So by Corollary 11.2.7 $X \cap R/P$ meets the center of $R/P$ and contains a nonzero element of $E$. Since $E$ is a field, $X$ is a whole ring.

Moreover $ERT/Q$ is finitely generated over $E$ so by the Nullstellensatz (Theorem 11.4.4) it is also finite dimensional over $E$. This means that the trascendence degree of the center of $\mathcal{Q}(RT/Q)$ so the equality on dimensions follows. □

Now by Lemma 15.2.11 there is a prime ideal $Q_0 \subset Q$ of $RT$ such that

$$\text{PI-deg}(RT/Q_0) = \text{PI-deg}(RT), \quad \dim RT/Q_0 = \dim RT/Q + 1. \tag{15.2}$$

Since, by assumption, $\text{PI-deg}(R/P) < \text{PI-deg}(R) = \text{PI-deg}(RT)$, we cannot have $Q_0 \cap R = P$. So, since $\text{ht}(P) = 1$, we must have $Q_0 \cap R = 0$.

We claim that $Q_0 = 0$, in fact every element of $RT$ is a fraction $a^{-1}b$ with $a, b \in R$ and $a$ in the center of $R$. So, if we had a nonzero element $a^{-1}b \in Q_0$, we would have $b \in Q_0 \cap R = 0$, a contradiction.

So finally formula (15.2) gives

$$\dim R = \dim RT = \dim RT/Q + 1 \overset{15.2.13}{=} \dim R/P + 1,$$

as desired. □

# Part 4

# The relatively free algebras

CHAPTER 16

# The nilpotent radical

*This chapter has several aims. First we prove Theorem* 16.1.1, *with a long history behind, stating the nilpotency of the nil radical of a finitely generated PI algebra over a commutative Noetherian ring. The rest of the chapter is on the structure of T-ideals. In fact the semigroup of T-ideals is free by Theorem* 16.4.3 *of Bergman and Lewin and the theorem of Lewin and its application by Giambruno Zaicev to the T-ideal of block triangular matrices in Theorem* 16.3.2.

## 16.1. The Razmyslov–Braun–Kemer theorem

**16.1.1. The goal.** A basic result in the theory of rings with polynomial identities is the following, first proved by Razmyslov when $A$ is a field and $S$ is assumed to satisfy a Capelli identity (cf. formula (7.5)).

THEOREM 16.1.1 (Braun). *If $S$ is a finitely generated algebra over a commutative Noetherian ring $A$ satisfying a proper polynomial identity, then its nil radical is nilpotent.*

The coefficient ring is just commutative Noetherian, if the base ring is a field and the nil and the Jacobson radical coincide. We can, and will, assume that the algebras have a 1. We have already seen a substantial result in this direction in Corollary 15.1.14 where this theorem has been proved under the hypothesis that $S$ the quotient of the algebra of $k$ generic $n \times n$ matrices for some $n$.

This result has been presented at various levels of generality:

(1) Assuming that $S$ satisfies the identities of $n \times n$ matrices for some $n$.
(2) Assuming that $S$ satisfies a Capelli identity.
(3) The general case.

In the end, at least for algebras over fields, we prove, using results of Lewin and of Giambruno and Zaicev that the condition satisfying identities of $n \times n$ matrices is not restrictive; see Corollary 16.3.4. This requires that we first prove Theorem 16.1.1.

Also the issue of the characteristic is important, and characteristic 0 is the first case to be analyzed. The seminal paper, the one which contains most of the ideas, is due to Razmyslov **[Raz74]**. In this paper he assumes that the algebra satisfies a Capelli identity. Kemer **[Kem80]** has shown that a finitely generated algebra satisfying a polynomial identity always satisfies a Capelli identity.

Later Braun **[Bra84]** presents the general result in all characteristics, and later Zubrilin **[Zub95b]** presents possible generalizations to nonassociative algebras.

The main problem is that for PI algebras there is no Hilbert theorem on the finite basis of ideals. In fact a deep theorem of Kemer gives *finite bases* of $T$-ideals, but we first need to prove the nilpotent property.

We may notice a general result:

PROPOSITION 16.1.2 (Latyshev [**Lat77**]). *Let $R = A\langle a_1, \dots, a_k\rangle$ be a finitely generated PI algebra over a commutative ring $A$, and let $I = \sum_{j=1}^{h} Rt_jR$ be a finitely generated nil ideal. Then $I$ is nilpotent.*

PROOF. Let $d$ be the degree of a PI which we may assume, by Corollary 11.3.5, with coefficients $\pm 1$.

We apply Shirshov's theorem to the words in the elements

$$\underline{b} := a_1, \dots, a_k, t_1, \dots, t_h.$$

Each such word $w_0(\underline{b})$ is a linear combination with integer coefficients of the monomials $w(\underline{b})$, where $w$ runs over the words which are $d$-indecomposable and have the same degree and content as $w_0(\underline{b})$ (Corollary 8.1.10).

Then, by Shirshov's height theorem (Theorem 8.1.7), given $d$ and $\ell = k + h$, there exists an $s \in \mathbb{N}$, dependent on $d, \ell$ such that every $d$-indecomposable word $w$ in $\ell$ letters is of the form $w = u_1^{k_1} u_2^{k_2} \cdots u_j^{k_j}$, $j \le s$, and the $u_i$ are words in the elements $\underline{b}$ with $|u_i| \le d - 1$.

Consider the set $\mathcal{T}$ of the words $u$, $|u| \le d - 1$ which contain at least one element $t_j$, and thus lie in $I$. Let $z := |\mathcal{T}|$ the cardinality.

Such words are, by hypothesis, nilpotent, and we have thus some uniform bound $N$, so that $u^{N+1} = 0$, $\forall u \in \mathcal{T}$. Thus a nonzero word $u_1^{k_1} u_2^{k_2} \cdots u_j^{k_j}$ has $k_i \le N$ whenever $u_i \in \mathcal{T}$. The ideal $I$ is spanned by words $u_1^{k_1} u_2^{k_2} \cdots u_j^{k_j}$ in which at least one of the $u_i$ appearing is in $\mathcal{T}$, i.e., contains a $t$, and $I^M$ is spanned by words whose content contains at least $M$ entries of $t$. These $M$ entries $t_i$ are distributed in the $z$ words of length $\le d - 1$ in $\mathcal{T}$. Each of these words can appear with exponent at most $N$, hence if $M > Nz(d-1)$, we have $I^M = 0$. □

We shall use Proposition 16.1.2 systematically as follows. We have a PI algebra $R = A\langle a_1, \dots, a_k\rangle$ finitely generated over a commutative Noetherian ring $A$, and we wish to prove that its nil radical $N$ is nilpotent. If $I = \sum_{j=1}^{h} Rt_jR$ is a finitely generated nil ideal, it is then enough to prove that the nil radical of $R/I$ is nilpotent. This will allow us to assume that many desired relations, which hold modulo $N$, can be assumed to hold in $R$.

16.1.1.1. *Following Razmyslov.* Some of the basic ideas in the proof of Theorem 16.1.1 come from the seminal paper of Razmyslov [**Raz74b**]. In particular some peculiar role of the Capelli element (7.5) or of the central polynomial $h(x, y)$ of definition (10.19) is derived from the Capelli element.

In general let $f(x_1, \dots, x_{n^2}, \underline{y})$ be a polynomial alternating in the variables $x_i$ and not an identity of $n \times n$ matrices. For $f$ one has the analogue of formula (10.12). We deduce that the following identities hold on $n \times n$ matrices over a commutative ring:

$$\begin{aligned} f(zx_1, \dots, zx_{n^2}, \underline{y}) &= f(x_1z, \dots, x_{n^2}z, \underline{y}), \\ f(zwx_1, \dots, zwx_{n^2}, \underline{y}) &= f(wzx_1, \dots, wzx_{n^2}, \underline{y}) \\ &= f(zx_1w, \dots, zx_{n^2}w, \underline{y}) = f(wx_1z, \dots, wx_{n^2}z, \underline{y}). \end{aligned} \tag{16.1}$$

Moreover, when we polarize formula (10.12), we have by formula (10.13) and for $h = 1, \dots, n^2$, polynomials $f_h(x, y, z) = \sigma_{h;n}(z) f(x, y)$ on matrices. Notice that each $f_h$ is still alternating in the variables $x_i$.

The element $(-1)^h \sigma_{h;n}(z)$ is the coefficient of the term of degree $n^2 - h$ of the characteristic polynomial of left multiplication by $z$ on matrices. We deduce, from the Cayley–Hamilton identity, another identity on $n \times n$ matrices,

$$\sum_{h=0}^{n^2} z^{n^2-h}(-1)^h f_h(x, y, z) = [\sum_{h=0}^{n^2} z^{n^2-h}(-1)^h \sigma_{h;n}(z)] f(x, y) = 0. \tag{16.2}$$

Let $Y$ be a finite set of variables, and let $U = \mathbb{Z}\langle Y \rangle / J$ be the free algebra modulo the $T$-ideal $J$ generated by the identities of formulas (16.1) and (16.2) for the polynomials $C_{n^2}$ and $C_{n^2,h}$ with $h = 1, \dots, n^2$. We deduce the main

THEOREM 16.1.3. *If $Y$ is finite, then there is a $d \in \mathbb{N}$ so that $U$ satisfies all identities of the form $f \cdot C_{n^2}$ and $f \cdot h(x, y)$ (i.e., $h(x, y)$ of formula* (10.19)*) where $f$ is any PI of $d \times d$ matrices in the variables $Y$ (disjoint from the variables appearing in $C_{n^2}$ and $h$).*

PROOF. Consider the free algebra $A\langle Z \rangle$ in countably or finitely many variables over $A$. Single out the variables $x_1, \dots, x_{n^2}$ and the $y_0, \dots, y_{n^2}$ appearing in the Capelli polynomial $y_0 C_{n^2}(x_1, \dots, x_{n^2}, y_1, \dots, y_{n^2-1}) y_{n^2}$. Call $Y$ the remaining variables.

We let $L$ be the $A$-submodule spanned by those elements which are multilinear and alternating in the variables $x_i$. By definition $L$ has as basis all possible evaluations $M_0 C_{n^2}(x_1, \dots, x_{n^2}, M_1, \dots, M_{n^2-1}) M_{n^2}$ of the Capelli polynomial $y_0 C_{n^2}(x_1, \dots, x_{n^2}, y_1, \dots, y_{n^2-1}) y_{n^2}$ for the variables $y_i$, $i = 0, \dots, n^2$, with monomials $M_i$ in the variables $Z \setminus \{x_1, \dots, x_{n^2}\}$. We have that the substitution map $x_i \mapsto a x_i$ induces a linear map $\sigma_a$ on $L$:

$$\begin{aligned} &\sigma_a(M_0 C_{n^2}(x_1, \dots, x_{n^2}, M_1, \dots, M_{n^2-1}) M_{n^2}) \\ &\quad := M_0 C_{n^2}(a x_1, \dots, a x_{n^2}, M_1, \dots, M_{n^2-1}) M_{n^2} \\ &\quad = M_0 a C_{n^2}(x_1, \dots, x_{n^2}, a M_1, \dots, a M_{n^2-1}) M_{n^2}. \end{aligned}$$

The map $a \mapsto \sigma_a$ is a homogeneous polynomial map of degree $n^2$ from $A\langle Y \rangle$ to $A$-linear maps on $L$. This map polarizes to give linear maps $\tau_h(a)$, $h = 1, \dots, n^2$.

Now consider the quotient $A\langle Z \rangle / I$ of this free algebra modulo the polynomial identities given by formula (16.1) and the left-hand side of formula (16.2) applied to $f = C_{n^2}$ and to all the $f_h = C_{n^2,h}$. We see that the maps $\sigma_a$ and $\tau_h(a)$ induce on the image $\bar{L}$ of the group $L$ linear maps depending only on the class of $a$ in the image $\mathcal{F}_Y := A\langle Y \rangle / (I \cap A\langle Y \rangle)$ of the free algebra $A\langle Y \rangle$ in $A\langle Z \rangle / I$. If $a \in A\langle Y \rangle$, the fact that equations (16.1) are identities implies that these maps generate a commutative algebra of maps on $\bar{L}$ which commute by multiplication $\ell(a)$ on the left by elements $a \in \mathcal{F}_Y$. Moreover, formula (16.2), which holds by hypothesis, together with the previous formulas tells us that each $a \in \mathcal{F}_Y$, as operator $\ell(a)$ on $\bar{L}$, satisfies a Cayley–Hamilton identity $\sum_{h=0}^{n^2} \ell(a)^{n^2-h} \tau_h(a) = 0$.

Let $Y$ be finite, say $|Y| = k$. Consider the commutative $A$-algebra $T_Y$ generated by the operators $\tau_h(a)$, $h = 1, \dots, n^2$, $a \in A\langle Y \rangle$ (which by the theory developed in §4.1 is finitely generated). By the formula $\sum_{h=0}^{n^2} \ell(a)^{n^2-h} \tau_h(a) = 0$ and Shirshov's theorem, the algebra generated by $T_Y$ and the elements $\ell(a)$, $a \in A\langle Y \rangle$ is a finite module over $T_Y$. In particular, Corollary 10.1.2, it satisfies the identities of $d \times d$

matrices, with coefficients in $A$, for some $d$. In particular if $f(y_{n^2+1}, \dots, y_{n^2+k})$ is a polynomial identity of $d \times d$ matrices in the variables $Y$ (disjoint from the ones appearing in the polynomial $C_{n^2}$), we have $f(y_{n^2+1}, \dots, y_{n^2+k})f = 0$ for every $f \in \bar{L}$. In particular for $k = C_{n^2}$

$$f(y_{n^2+1}, \dots, y_{n^2+k})y_0 C_{n^2}(x_1, \dots, x_{n^2}, y_1, \dots, y_{n^2-1})y_{n^2} = 0.$$

Since $I$ is a $T$-ideal, this is thus a polynomial identity of $A\langle Z\rangle/I$, from which one deduces also, by Proposition 10.2.20, the identity $f(y_{n^2+1}, \dots, y_{n^2+k})h(x, y)$. $A\langle Z\rangle/I$ contains the algebra $A\langle Y\rangle/(I \cap A\langle Y\rangle)$. Then, by Proposition 2.2.14, $J = (I \cap A\langle Y\rangle)$, and this last algebra is in fact $U$; hence the claim. □

REMARK 16.1.4. If we work over $\mathbb{Q}$, we may take $d = n$ and $Y$ countable.

In fact if we are in characteristic 0, formula (16.2) can be replaced by an identity of degree $n$ in $z$, that is

$$\sum_{h=0}^{n} z^{n-h}(-1)^h \bar{k}_h(x, y, z) = [\sum_{h=0}^{n} z^{n-h}(-1)^h \sigma_h(z)]f(x, y) = 0. \tag{16.3}$$

This is a Cayley–Hamilton algebra for $n \times n$ matrices (see §12.2.2), and in fact one can prove that it is isomorphic to the ring of $n \times n$ generic matrices with trace. This follows from the conductor property of $k$ (Proposition 10.4.9) and the fact that the generic matrices with trace form an integral domain.

One could finish the proof of Razmyslov from this, but we will go directly to Braun's theorem.

**16.1.2. Following Braun.** We follow Rowen's presentation now, cf. **[Row88]**. We let $R = A\langle a_1, \dots, a_k\rangle$ be a PI algebra finitely generated over a commutative Noetherian ring $A$ with nil radical $N$. We start with a variation of Proposition 11.2.5:

LEMMA 16.1.5. *Let $I$ be a nonzero ideal of $R/N$. Then $I$ contains a nonzero element of the center of $R/N$.*

PROOF. By Corollary 15.1.14 we have that $N = P_1 \cap \dots \cap P_k$ is the intersection of its minimal primes, so $R/N \subset \bigoplus_i R/P_i$. Thus, for some $P_i$, we have $I \not\subset P_i$.

Among the primes $P_i$, for which $I \not\subset P_i$, we may choose one $P$ with maximal PI degree $h$.

From Proposition 11.2.5, given a central polynomial $f$ for $h \times h$ matrices, there is an evaluation $\bar{f}$ of $f$ in $I$ which is nonzero in the image $\bar{I} \subset R/P$. Now if the image of $I$ in some $R/P_j$ is nonzero by assumption, the PI degree of this $R/P_j$ is $\leq h$ so that the evaluation of $f$ in $I$ modulo such $P_j$ is either 0 or central. Hence $\bar{f} \in I \subset \bigoplus_i R/P_i$ is central in $\bigoplus_i R/P_i$, hence also in $R/N$. □

Suppose the theorem is not true, and we try to find some kind of *minimal counterexample*. Assume there is a ring $R = A\langle a_1, \dots, a_k\rangle$ with nil radical $N$ not nilpotent:

(1) Consider the set of ideals $I$ of $R$ such that $\sqrt{I}$ is not nilpotent over $I$. By ACC (Proposition 15.1.10) on semiprime ideals, there is one for which $\sqrt{I}$ is maximal. If we replace $R$ with $R/I$, we still have a counterexample.

We may thus assume that if $J$ is an ideal of $R$ such that $\sqrt{J} \neq N$, then $\sqrt{J}$ is nilpotent over $J$.

(2) Next if $I$ is not a nil ideal of $R$, that is $I \not\subset N$, since $N + I \subset \sqrt{I}$, by the previous reduction we have that for some $k$, $(N+I)^k \subset I$, hence $N^k \subset I$.
(3) Let $R$ be a counterexample (i.e., N is not nilpotent), and let $I \subset R$ be a nilpotent ideal. Then $R/I$ is also a counterexample.
(4) Given an element $z \in R$ which is central modulo $N$, we may reduce to the case in which $z$ is in the center. In fact the ideal generated by the commutators $[z, a_i]$ is a finitely generated nil ideal, hence nilpotent by Proposition 16.1.2 of Latyshev. So we can reduce modulo this ideal. Modulo this ideal the image of $z$ is central.
(5) Take now $z \notin N$ in the center of $R$. Then if the nil radical of $R/\operatorname{Ann}(z)$ is nilpotent so is $N$.

In fact by item (2), we have for some $k$, $N^k \subset zR$. Hence, if $N^t \subset \operatorname{Ann}(z)$ we have $N^{k+t} \subset zR \cdot \operatorname{Ann}(z) = 0$.
(6) We reduce to the case $R/N$ prime of some PI degree $n$.

In fact if $R/N$ is not a prime ring, there are two ideals $A_1, A_2 \supsetneq N$ with $A_1 A_2 \subset N$. By Lemma 16.1.5 there are two elements $z_1, z_2 \notin N$ with $z_1 \in A_1, z_2 \in A_2$, hence $z_1 z_2 \in N$. By (4) we may assume that $z_1, z_2$ are also central in $R$. So, for some $m$, we have $0 = (z_1 z_2)^m = z_1^m z_2^m$.

By item (2), we have some exponents for which $N^{k_1} \subset Rz_1$, $N^{k_2} \subset Rz_2$, hence finally

$$N^{m(k_1+k_2)} \subset (Rz_1)^m (Rz_2)^m = 0.$$

The last step will be based on Artin's theorem and Azumaya algebras.

16.1.2.1. *Some generalities relating to Azumaya algebras.* The first step which suggests the possible use of Azumaya algebras is given by

PROPOSITION 16.1.6. *Let $R = A\langle a_1, \dots, a_k \rangle$ be an algebra over a commutative Noetherian ring $A$, and let $I$ be an ideal of $R$ such that $R/I$ is a finitely generated module over its center $Z$. Then:*

(1) *$Z$ is an algebra finitely generated over $A$.*
(2) *$I$ is a finitely generated ideal of $R$.*

PROOF. (1) follows from Proposition 8.2.4 applied to $R/I$.

(2) Take elements $u_1, \dots, u_m \in R$ so that their images $\bar{u}_i \in R/I$ form linear generators of $R/I$ over its center $Z$. We have $\bar{a}_i = \sum \bar{\nu}_i^{(j)} \bar{u}_j$ and the multiplication table $\bar{u}_i \bar{u}_j = \sum_k \bar{\mu}_{i,j}^{(\ell)} \bar{u}_k$, where $\bar{\nu}_i^{(j)}, \bar{\mu}_{i,j}^{(\ell)}$ are in $Z$, and they are images of elements $\nu_i^{(j)}, \mu_{i,j}^{(\ell)}$ in $R$. Next we have that the images of the commutators $[\mu_{i,j}^{(\ell)}, a_t]$, $[\nu_i^j, a_t]$ in $R/I$ are 0, hence we have that

$$a_i - \sum \nu_i^{(j)} u_j,\ u_i u_j - \sum_k \mu_{i,j}^{(\ell)} u_k,\ [\mu_{i,j}^{(\ell)}, a_t],\ [\nu_i^j, a_t] \in I. \tag{16.4}$$

If $J$ is the ideal generated by the finite list of formula (16.4), we clearly have that $R/J$ is a finitely generated module over the finitely generated commutative algebra $S := A[\tilde{\nu}_i^{(j)}, \tilde{\mu}_{i,j}^{(\ell)}]$, with $\tilde{\nu}_i^{(j)}, \tilde{\mu}_{i,j}^{(\ell)}$ the classes of $\nu_i^{(j)}, \mu_{i,j}^{(\ell)}$ modulo $J$.

Hence $R/J$ is Noetherian so that $I/J$ is finitely generated. Since $J$ is finitely generated, so is $I$. □

Let $R = A\langle a_1, \dots, a_k \rangle$ be a PI algebra over a commutative Noetherian ring $A$, let $N$ be its nil radical, and assume that $R/N$ is a prime algebra of PI degree $n$.

We may choose an element $\lambda \in R$, evaluation of a given multilinear central polynomial for $n \times n$ matrices, such that in $R/N$ we have that $\lambda \neq 0$. Then, by Corollary 10.3.5, $R/N[\lambda^{-1}]$ is prime and Azumaya of rank $n^2$ over its center. Under these hypotheses we have

PROPOSITION 16.1.7. *There is a finitely generated ideal $I \subset N$ such that in $R/I$, we have that $\lambda$ is central and $R/I[\lambda^{-1}] = R/N[\lambda^{-1}]$ is Azumaya.*

PROOF. First, since $\lambda$ is central modulo $N$, we have that $[\lambda, a_i] \in N, \forall i$.

So take $J \subset N$ to be the ideal generated by the elements $[\lambda, a_i]$. In $R/J$ the element $\lambda$ is central. We may thus replace $R$ with $R/J$ and assume directly that $\lambda$ is central. Consider then $R[\lambda^{-1}]$ and its quotient $R/N[\lambda^{-1}]$. By Proposition 16.1.7 the kernel of this quotient map is a finitely generated ideal, and by multiplying its generators by some powers of $\lambda$, we may assume that it is generated by a finite number of elements in $R$. Take the ideal $I$ generated by these elements. By construction we have that $R/I[\lambda^{-1}] = R/N[\lambda^{-1}]$. Since $R/N$ is prime we have that $R/N \subset R/N[\lambda^{-1}]$, hence $I \subset N$. □

We arrive at the final reduction. If there is a counterexample $R = A\langle a_1, \dots, a_k \rangle$ to Theorem 16.1.1, then we may assume that $R/N$ is prime and, replacing $R$ with $R/I$ of Proposition 16.1.7, we may assume that there is a central element $\lambda \in R$ with $R[\lambda^{-1}] = R/N[\lambda^{-1}]$ Azumaya.

16.1.2.2. *Constructing a large finite module over the center.* We now will start to explore how to use the previous reduction. Let us start by recalling some standard facts that we have already presented in §5.4. Let $R$ be an algebra (with a 1) over its center $Z$ and consider $R \otimes_Z R^{\mathrm{op}}$ where $R^{\mathrm{op}}$ denotes the opposite of $R$. We have (cf. Remark 5.4.1) a homomorphism

$$\eta_R : R \otimes_Z R^{\mathrm{op}} \to \mathrm{End}_Z(R), \ \eta_R(a \otimes b)(c) := acb.$$

An element $r \otimes 1$ maps to the left multiplication $L_r : x \mapsto rx$ while an element $1 \otimes r$ maps to the right multiplication $R_r : x \mapsto xr$. In particular for an element $u = x \otimes 1 - 1 \otimes x$, we have

$$\eta_R(u) = ad(x), \ ad(x) = L_x - R_x; \ ad(x) : r \mapsto [x, r].$$

Inside $\mathrm{End}_Z(R)$ we have $\hom_Z(R, Z)$. Given an element $u \in R \otimes_Z R^{\mathrm{op}}$ if for all $x \in R$, we have $(x \otimes 1 - 1 \otimes x)u = 0$, then $u$ maps, under $\eta_R$, to an element of $\hom_Z(R, Z)$; see Lemma 5.4.3. In fact one easily sees that

LEMMA 16.1.8. *Let $R$ be generated (as an algebra) over $Z$ by elements $a_1, \dots, a_k$.*

(1) *A necessary and sufficient condition for an element $v \in R \otimes_Z R^{\mathrm{op}}$ to have $\eta_R(v) \in \hom_Z(R, Z)$ is that $\eta_R((a_i \otimes 1 - 1 \otimes a_i)v) = 0, \ \forall i$.*
(2) *If $\eta_R(v) \in \hom_Z(R, Z)$ given $u \in R \otimes_Z R^{\mathrm{op}}$, setting $b := \eta_R(u)(1)$, we have*

$$\eta_R(uv) = L_b \eta_R(v) = R_b \eta_R(v),$$

$$\Longrightarrow \eta_R(uv)(r) = b\, \eta_R(v)(r) = \eta_R(v)(r)b, \ \forall r \in R. \tag{16.5}$$

PROOF. (1) In fact as in Lemma 5.4.3, given $a \in R$,

$$0 = \eta_R((a \otimes 1 - 1 \otimes a)v) \iff a \cdot \eta_R(v)(r) = \eta_R(v)(r)a, \ \forall r \in R.$$

Clearly $\eta_R(v) \in \hom_Z(R, Z)$ if and only if for all $r \in R$, we have that $\eta_R(v)(r)$ is in the center. This is true if and only if $\eta_R(v)(r)$ commutes with the elements $a_i$.

(2) As for the second part, $\eta_R(uv) = \eta_R(u) \circ \eta_R(v)$. We apply this identity to an element $r \in R$, and we have $\eta_R(uv)(r) = \eta_R(u)(\eta_R(v)(r))$. Since $\eta_R(v)(r) \in Z$, we have, by $Z$ linearity of $\eta_R(u)$, that

$$\eta_R(u)(\eta_R(v)(r)) = \eta_R(u)(1 \cdot \eta_R(v)(r)) = \eta_R(u)(1)\eta_R(v)(r). \qquad \square$$

We now need

LEMMA 16.1.9. *Let $R = A\langle a_1, \dots, a_k\rangle$ be an algebra with center $Z$, and let $\lambda \in Z$ be such that $R[\lambda^{-1}]$ is Azumaya. Then there exist elements $u_i, v_i \in R^e = R \otimes_Z R^{\mathrm{op}}$, $i = 1, \dots, h$ and $f \in \mathbb{N}$ with $(a_i \otimes 1 - 1 \otimes a_i)v_j = 0$, $\forall i, j$, and*

$$(16.6) \qquad \lambda^f = \sum_{i=1}^{h} u_i v_i, \quad \text{in} \quad R^e = R \otimes_Z R^{\mathrm{op}}.$$

PROOF. By Proposition 1.5.3 the center $C$ of $R[\lambda^{-1}]$ is the image of $Z[\lambda^{-1}]$.

Recall that we have, for an Azumaya algebra $S$ over its center $C$, an isomorphism $\eta_S : S^e := S \otimes_C S^{\mathrm{op}} \cong \mathrm{End}_C(S)$ given by $\eta(a \otimes b)(c) = acb$. Under this isomorphism the subspace $\hom_C(S, C)$ of $\mathrm{End}_C(S)$ is identified to the subset

$$(S^e)^S := \{u \in S \otimes_C S^{\mathrm{op}} \mid (x \otimes 1 - 1 \otimes x)u = 0, \ \forall x \in S\}.$$

For an Azumaya algebra, one has that $S^e = S^e(S^e)^S$ (Theorem 5.4.25(b)), so one can find elements $\bar{u}_i \in S^e$, $\bar{v}_i \in (S^e)^S$ with $1 = \sum_i \bar{u}_i \bar{v}_i$.

We apply this to $S := R[\lambda^{-1}]$ and call $j : R \otimes_Z R^{\mathrm{op}} \to S \otimes_C S^{\mathrm{op}}$ the natural map. Notice that by the previous observation, $C = Z[\lambda^{-1}]$, the map $j$ is a localization, that is $(R \otimes_Z R^{\mathrm{op}})[\lambda^{-1}] = S^e = S \otimes_C S^{\mathrm{op}}$, Proposition 1.5.4.

Multiplying by a power $\lambda^h$, we have elements $c_i, b_i \in R^e = R \otimes_Z R^{\mathrm{op}}$ such that $j(c_i) = \lambda^h \bar{u}_i$, $j(b_i) = \lambda^h \bar{v}_i$, $\forall i$, and we deduce that $j(\lambda)^{2h} = j(\sum_i c_i b_i)$. We have by construction that $(a_i \otimes 1 - 1 \otimes a_i) j(b_i) = 0$. Hence, by multiplying by a power $\lambda^m$, we may have $(a_i \otimes 1 - 1 \otimes a_i)\lambda^m b_i = 0$ and $\lambda^{2(h+m)} = \sum_i \lambda^m c_i \lambda^m b_i$ in $R^e$.

In the end setting $u_i = \lambda^m c_i, v_i = \lambda^m b_i$ and $f = 2(h+m)$, we have the identity $\lambda^f = \sum_j u_i v_j$, where $(a_i \otimes 1 - 1 \otimes a_i)v_j = 0$, $\forall i, j$ in $R \otimes_Z R^{\mathrm{op}}$. $\square$

PROPOSITION 16.1.10. *Let $R = A\langle a_1, \dots, a_k\rangle$ be an algebra, and let $\lambda \in R$ be a central element, such that $R[\lambda^{-1}]$ is Azumaya. Let $Z$ be the center of $R$. Then there is an exponent $q \in \mathbb{N}$ and elements $b_1, \dots, b_h \in R$ such that $\sum_i Z b_i$ is a subring and $\lambda^q R \subset \sum_i Z b_i$.*

PROOF. Take $f, u_i, v_i$ as in Lemma 16.1.9 and formula (16.6). Then set

$$F_i(r) := \eta_R(v_i)(r), \quad c_i := \eta_R(u_i)(1).$$

The maps $r \mapsto F_i(r) := \eta_R(v_i)(r)$ are $Z$-linear maps to the center $Z$ and $\forall r \in R$,

$$(16.7) \qquad \lambda^f r = \eta_R(\lambda^f)(r) = \sum_i \eta_R(u_i v_i)(r)$$

$$= \sum_i \eta_R(u_i) \circ \eta_R(v_i)(r) \overset{(16.5)}{=} \sum_i c_i \eta_R(v_i)(r) = \sum_i c_i F_i(r) \in \sum_i c_i Z.$$

Thus $\lambda^f R \subset \sum_i c_i Z$ and the elements $c_i$ are linear generators of $R[\lambda^{-1}]$ over $Z[\lambda^{-1}]$.

So $c_ic_j = \sum_\ell c_{i,j}^{(\ell)} c_\ell$, for some $c_{i,j}^{(\ell)} \in Z[\lambda^{-1}]$. Multiplying by some $\lambda^r$, we have that $\lambda^r c_{i,j}^{(\ell)}$ are images of elements in $m_{i,j}^{(\ell)} \in Z$. Hence, setting $g_i := \lambda^r c_i \in R$, we have, in $R[\lambda^{-1}]$, the identities $g_ig_j = \sum_\ell m_{i,j}^{(\ell)} g_\ell$, $m_{i,j}^{(\ell)} \in Z$.

Finally this means that, by multiplying by a further power $\lambda^a$ of $\lambda$, the identities $\lambda^a g_ig_j = \lambda^a \sum_\ell m_{i,j}^{(\ell)} g_\ell$ hold in $R$.

Therefore, setting $b_j := \lambda^a g_j = \lambda^{a+r} c_j$, we have in $R$, since $\lambda^f R \subset \sum_i c_i Z$,

$$b_ib_j = \sum_\ell p_{i,j}^{(\ell)} b_\ell, \; p_{i,j}^{(\ell)} = \lambda^a m_{i,j}^{(\ell)} \in Z, \quad \lambda^q R \subset \sum_i Zb_i, \; q = f + a + r. \tag{16.8}$$

This shows that $\sum_{i=1}^h Zb_i$ is a subring of the ring $R$ satisfying the required condition $\lambda^q R \subset \sum_{i=1}^h Zb_i$. □

16.1.2.3. *Adding polynomial identities.* The next step is to prove

LEMMA 16.1.11. *Given any finite list of homogeneous polynomial identities $f_i$ on $n \times n$ matrices, we can further reduce $R$ so to assume that it satisfies all these identities.*

PROOF. The verbal ideal generated by all evaluations in $R$ of these identities is contained by assumption in $N$. It is not necessarily finitely generated, since a priori it is generated by infinitely many evaluations. So we cannot directly apply the proposition of Latyshev (Proposition 16.1.2), and we need a further reduction.

Let $Z$ be the center of $R$. By Proposition 16.1.10 we may further assume that there is an element $\lambda \notin N$, $\lambda \in Z$, an exponent $q \in \mathbb{N}$, and elements $b_1, \dots, b_h \in R$ such that $\sum_{i=1}^h Zb_i$ is a subring and $\lambda^q R \subset \sum_{i=1}^h Zb_i$.

Let us show how from this statement one can deduce Lemma 16.1.11.

It is enough to do it for a single identity $g(x_1, \dots, x_s)$ of $n \times n$ matrices homogeneous of some degree $p$. Evaluate in $g$ the elements $x_i$ for elements $\sum_{j=1}^r t_{i,j} y_j$, where the $t_{i,j}$ are commutative variables. We obtain a polynomial $\sum_\alpha t^\alpha g_\alpha(y)$ and the coefficients $g_\alpha$ of this, as polynomial in the $t_{i,j}$ are still homogeneous polynomial identities on $n \times n$ matrices in the variables $y_j$. So when we evaluate, in the identities $g_\alpha$, the $y_i$ in the elements $b_i$, we obtain finitely many elements in the nil radical $N$. Hence, by Proposition 16.1.2, we may factor the ideal generated by these elements. As a result we may assume that when we now evaluate $g$ in $\sum_i Zb_i$, we have 0. Hence $g$ vanishes identically on $\lambda^q R$.

Since $g$ has degree $p$, this means that such an identity computed in $R$, multiplied by $\lambda^{q \cdot p}$ is 0.

Next by reduction (5) we can pass modulo the annihilator of this power of $\lambda$ and get that $g$ can be assumed to be an identity of $R$. □

16.1.2.4. *Completing the proof of Theorem* 16.1.1. We now apply Lemma 16.1.11 and assume that $R$ satisfies the identities of formulas (16.1) and (16.2) for $f = C_{n^2}$. We then apply Theorem 16.1.3 to $R$.

In fact in the construction of $\lambda$ we may take it as a value of the central polynomial, for $n \times n$ matrices, $h(x_1, \dots, x_{n^2}, y)$ of formula (10.19). This choice of $\lambda$ allows us to apply Theorem 16.1.3 to $R$ and deduce that for some $d$ the verbal ideal $\mathcal{V}_d(R)$ of $d \times d$ matrices in $k$ variables multiplied by $\lambda$ equals 0.

At this point we can finish the proof of Theorem 16.1.1. The algebra $R$ modulo the annihilator of $\lambda$ is a quotient of the algebra of $k$, $d \times d$ generic matrices with coefficients in $A$. This implies, by Corollary 15.1.14, that for some $p$ we have $N^p \subset \mathcal{V}_d(R) \subset \mathrm{Ann}(\lambda)$, but then, by reduction (5), we are finished.

## 16.2. The theorem of Lewin

*In this section we assume all algebras to be over some field F and to have a* 1.

**16.2.1. Free bimodules.** Given two $T$-ideals $\Gamma_1, \Gamma_2$, their product is still a $T$-ideal. If we present the two $T$-ideals as the ideals $\mathrm{Id}[A], \mathrm{Id}[B]$ of PI of two algebras $A, B$, it is quite natural to ask if one can construct in an explicit way an algebra $C$ such that

$$\mathrm{Id}[A]\,\mathrm{Id}[B] = \mathrm{Id}[C].$$

This is the content of the theorem of Lewin which has quite important applications to representation theory, as we shall see presently.

As usual let us denote by $F\langle X\rangle$ the free algebra in a countable number of variables $X = \{x_1, x_2, \dots, x_i, \dots\}$.

We use the language of *bimodules*. Given two $F$-algebras $A, B$ (with a 1) we make the following

DEFINITION 16.2.1. An $A, B$ bimodule $M$ is an $F$ vector space with a left $A$ action and a right $B$ action with $(am)b = a(mb)$, $\forall a \in A,\ b \in B,\ m \in M$.

Alternatively, we my also think of $M$ as an $A \otimes B^{\mathrm{op}}$ left module.

A basic bimodule is $A \otimes B$, with $B$ acting on the right, and a *free bimodule* is a direct sum of copies of $A \otimes B$. When $A = B$, we shall also speak of an $A$-bimodule.

LEMMA 16.2.2. *Let $U \subset A$, $V \subset B$ be two ideals, and let $M$ be an $A, B$ bimodule. Then $UM + MV$ is a subbimodule and $M/(UM + MV)$ is a bimodule over $A/U$, $B/V$. If $M$ is a free $A, B$ bimodule, then $M/(UM + MV)$ is a free $A/U$, $B/V$ bimodule.*

PROOF. The first part is straightforward. As for the second remark that for two algebras $A, B$ with ideals $U \subset A, V \subset B$, we have

$$(A \otimes B)/(U(A \otimes B) + (A \otimes B)V) = (A \otimes B)/(U \otimes B + A \otimes V) = A/U \otimes B/V.$$

Hence if $M$ is a free $A, B$ bimodule, then $M = \bigoplus_i Am_iB$ where each summand $Am_iB$ is isomorphic to $A \otimes B$. We have that $UM + MV = \bigoplus_i (Um_iB + Am_iV)$ so $M/(UM + MV) = \bigoplus_i Am_iB/(Um_iB + Am_iV)$ is again a free bimodule, with basis the classes $\overline{m}_i$ of the elements $m_i$ over the two algebras $A/U$, $B/V$. □

Next construct the free $F\langle X\rangle$ bimodule $\mathcal{R} := \bigoplus_{i=1}^{\infty} F\langle X\rangle r_i F\langle X\rangle$ having as basis a countable number of elements $R_0 := \{r_1, r_2, \dots, r_i, \dots\}$.

This bimodule comes equipped with a *derivation* $\delta : F\langle X\rangle \to \mathcal{R}$ satisfying

$$\delta(x_i) := r_i, \qquad \delta(fg) = \delta(f)g + f\delta(g). \tag{16.9}$$

In fact by recursion one has

$$\delta(u_1 \cdots u_n) = \sum_{i=1}^{n} u_1 \cdots u_{i-1}\delta(u_i)u_{i+1} \cdots u_n,$$

which applied to a monomial gives the explicit formula on each monomial. As an example,

$$\delta(x_2^2x_1) = r_2x_2x_1 + x_2r_2x_1 + x_2^2r_1.$$

We can also define these notions as a *differential calculus*. Consider the free algebras $F\langle X\rangle$ and $F\langle X \cup R_0\rangle$, and the map $i : f(x) \mapsto f(x+r)$. Then

$$i(f) = f(x_1 + r_1, x_2 + r_2, \dots, x_i + r_i, \dots) = \sum_j \delta_j f(x, r),$$

where $\delta_j f(x, r)$ is formed by the elements of degree $j$ in $r$. The elements of degree 1 in $r$ can be thought of as *differentials* and are clearly a free bimodule over $F\langle X\rangle$ with basis the elements $r_i$ (cf. **[Fox53]**).

Then our derivation $\delta$ equals the *differential* $\delta_1$.

16.2.1.1. *The idealizer construction.* Whenever we have a bimodule $M$ over two algebras $A, B$, one can form the *idealizer* construction, that is consider $M$ as an ideal, with $M^2 = 0$, of an algebra containing $A, B$ and inducing the bimodule structure. This is done by considering the algebra of $2 \times 2$ matrices

$$\begin{pmatrix} A & M \\ 0 & B \end{pmatrix} = \left\{ \begin{pmatrix} a & m \\ 0 & b \end{pmatrix}, \quad a \in A,\ m \in M,\ b \in B \right\}. \tag{16.10}$$

The multiplication of these matrices is via the bimodule structure, and it is equivalent to such a structure:

$$\begin{pmatrix} a_1 & m_1 \\ 0 & b_1 \end{pmatrix} \begin{pmatrix} a_2 & m_2 \\ 0 & b_2 \end{pmatrix} = \begin{pmatrix} a_1 a_2 & a_1 m_2 + m_1 b_2 \\ 0 & b_1 b_2 \end{pmatrix}. \tag{16.11}$$

Let us then apply this to the $F\langle X\rangle$ bimodule $\mathcal{R}$. We deduce a linear map

(16.12)

$$\psi : F\langle X\rangle \to \begin{pmatrix} F\langle X\rangle & \mathcal{R} \\ 0 & F\langle X\rangle \end{pmatrix}, \quad \psi(f) := \begin{pmatrix} f & \delta(f) \\ 0 & f \end{pmatrix}, \quad \psi(x_i) := \begin{pmatrix} x_i & r_i \\ 0 & x_i \end{pmatrix}.$$

The fact that $\delta$ is a derivation is in fact equivalent to the fact that $\psi$ is an algebra homomorphism.

Now take two ideals $U, V$ in $F\langle X\rangle$, then $U\mathcal{R} + \mathcal{R}V$ is a subbimodule of $\mathcal{R}$ and, by Lemma 16.2.2, $\mathcal{R}/(U\mathcal{R} + \mathcal{R}V)$ is again a free bimodule over the two quotient algebras $A := F\langle X\rangle/U$, $B := F\langle X\rangle/V$ with basis the classes of the elements $r_i$, which, by abuse of notation, we still denote by $r_i$.

Then $\begin{pmatrix} U & U\mathcal{R} + \mathcal{R}V \\ 0 & V \end{pmatrix}$, is an ideal of $\begin{pmatrix} F\langle X\rangle & \mathcal{R} \\ 0 & F\langle X\rangle \end{pmatrix}$, and the quotient

$$\begin{pmatrix} F\langle X\rangle/U & \mathcal{R}/(U\mathcal{R} + \mathcal{R}V) \\ 0 & F\langle X\rangle/V \end{pmatrix}$$

is the idealizer construction for this quotient bimodule. Let $\pi$ denote the quotient map.

Thus finally we can consider the composition $\phi := \pi \circ \psi$,

$$\phi : F\langle X\rangle \to \begin{pmatrix} F\langle X\rangle/U & \mathcal{R}/(U\mathcal{R} + \mathcal{R}V) \\ 0 & F\langle X\rangle/V \end{pmatrix}, \quad \phi(f) = \begin{pmatrix} \bar{f} & \overline{\delta(f)} \\ 0 & \bar{f} \end{pmatrix}, \tag{16.13}$$

where the bar denotes the class of an element. These constructions go back to Magnus **[Mag39]**; see also Fox **[Fox53]**. Our first goal is to prove the theorem of Lewin **[Lew74]**, Theorem 16.2.7, that if $F$ is a field, the kernel of $\phi$ is $UV$.

EXAMPLE 16.2.3. Let $U = V$ be the commutator ideal so that $F\langle X\rangle/U = F[X]$ is the *commutative ring of polynomials* in the variables $x_i$. Then $\mathcal{R}/(U\mathcal{R} + \mathcal{R}V)$ is the free module over $F[X] \otimes F[X]$ in the infinite set of generators $\delta(x_i)$.

In this example, given two variables $x, y$, the commutator $[x, y]$ maps to

$$\begin{pmatrix} 0 & \bar{\delta}([x,y]) \\ 0 & 0 \end{pmatrix}, \quad \delta([x,y]) = (x \otimes 1 - 1 \otimes x)\delta(y) - (y \otimes 1 - 1 \otimes y)\delta(x).$$

A product of two commutators maps to 0.

16.2.1.2. *Lexicographic order and free ideal rings.* We need some combinatorial tools. We have already ordered the variables, and we denote by $W$ the set of all monomials, i.e., words, a basis of $F\langle X\rangle$ over $F$.

We totally order the monomials by the *degree lexicographic order.*[1] If $a, b$ are two monomials, we set $a < b$ if either $\deg(a) < \deg(b)$, or, if the two degrees are equal, we set $a < b$ if this happens in the usual lexicographic order.

The main remark on this order is that if $a, b, c$ are monomials, then $a < b$ if and only if $ca < cb$ or $ac < bc$. This is clear for the degrees and if $a, b$ have the same degree, it is an easy property of the lexicographic order.

Thus, given any polynomial $f$ in $F\langle X\rangle$, this has a unique *leading term*, formed by a *leading monomial* that is the highest among the monomials appearing in $f$ with a nonzero coefficient, and a *leading coefficient*.

We denote the leading monomial by $\hat{f}$ and

REMARK 16.2.4. If $f, g$ are two nonzero elements of $F\langle X\rangle$, the leading term of $fg$ is the product of the two leading terms of $f, g$. In particular for the leading monomial we have $\widehat{(fg)} = \hat{f}\hat{g}$.

One use of this combinatorics is to prove that when $F$ is a field, then $F\langle X\rangle$ is a *free ideal ring* (or f.i.r in the language of P.M. Cohn **[Coh06]**). That is, every left (resp., right) ideal of $F\langle X\rangle$ is a free left (resp., right) $F\langle X\rangle$-module. In fact we have an almost canonical way of choosing a basis.

Let thus $I$ be any nontrivial left ideal of $F\langle X\rangle$. Define a subset of monomials $C_I \subset W$ by the following requirement: $c \in C_I$ if and only if we have that

(i) $c$ is a leading monomial $c = \hat{v}$ for some $v \in I$, and
(ii) $c$ is *minimal* in the sense that $c$ cannot be factored as $c = b\hat{w}$ with $w \in I$ and $b \neq 1$.

Thus for every $u \in I$, we see by a simple induction that we can decompose $\hat{u} = b\hat{w}$ with $\hat{w} \in C_I$ (and possibly $b = 1$).

For each $c \in C_I$ choose an element $a_c \in I$ with leading monomial $c$ (i.e., $c = \hat{a}_c$) and leading coefficient 1.

THEOREM 16.2.5. *When $F$ is a field, the elements $a_c \in I,\ c \in C_I$ form a basis of $I$ as left module over $F\langle X\rangle$.*

PROOF. The first remark is that if we choose for each $c \in C_I$ an element $a_c \in I$ such that $c = \hat{a}_c$, we have that these elements $a_c$ generate $I$ as left ideal.

In fact given any $u \in I$, we have that either $\hat{u} \in C_I$ or $\hat{u} = b\hat{w}$ with $\hat{w} \in C_I$.

In either case subtracting a left multiple of $a_u$ or of $a_{\hat{w}}$ to $u$, we have an element in $I$ with strictly lower leading monomial, and we can complete the argument by induction.

We next prove that these elements are linearly independent. This is done again by considering leading terms. When $b \in W$ the leading monomial of $ba_c$ is in fact

[1] This is different than the partial order used in Shirshov's theorem, Theorem 8.1.1. For example, here $x_2 < x_1x_2$, but $x_2 \succ x_1x_2$ in Shirshov's theorem.

$bc$, and if $u \in F\langle X\rangle$ by Remark 16.2.4 the leading monomial of $ua_c$ is $\hat{u}\hat{a}_c = \hat{u}c$. Assume that $\sum_{c\in C_I} u_c a_c = 0$, $u_c \in F\langle X\rangle$. For each $c$ the leading term of $u_c a_c$ is $\hat{u}_c c$. Consider in this sum the terms which have as leading term the highest of all the ones appearing. Suppose we have two terms $c_1, c_2$ giving this highest term $b_1c_1 = b_2c_2$. If $\deg(c_1) = \deg(c_2)$, we must have $c_1 = c_2$ a contradiction; otherwise suppose that $\deg(c_1) < \deg(c_2)$, and we get that $c_2 = bc_1$, $b \neq 1$ which is also a contradiction of the choice of $C_I$. That is, this highest term just comes from a single term and therefore cannot give 0. □

EXERCISE 16.2.6. If $I = F\langle X\rangle_+$ we have $C_I = X$.
What is $C_I$ for $I = F\langle X\rangle_+^k$?

16.2.1.3. *The theorem of Lewin.* We assume that $F$ is a field and use the notations of the previous paragraphs.

THEOREM 16.2.7 (Lewin, **[Lew74]**). *The kernel of* $\phi$ *(formula* (16.13)*) is* $UV$.

PROOF. We may assume that $U, V$ are both nontrivial ideals. Then

$$\ker(\phi) = \{x \in U \cap V \mid \delta(x) \in U\mathcal{R} + \mathcal{R}V\}.$$

Clearly $UV \subset \ker(\phi)$, since $UV \subset U \cap V$ and

$$\delta(uv) = \delta(u)v + u\delta(v) \in \mathcal{R}V + U\mathcal{R}.$$

So the point is to show that $\ker(\phi) \subset UV$.

Take $C := C_V$ and the elements $a_c$, $c \in C$, associated to the ideal $V$, as in the previous discussion. An element $v \in V$ is uniquely of the form $v = \sum_{c\in C} b_c a_c$ with $b_c \in F\langle X\rangle$ by the fact that $a_c$ is a basis of $V$. We need to show that if $v \in \ker(\phi)$, then all $b_c \in U$. We first find a basis of $U\mathcal{R} + \mathcal{R}V$, formula (16.15).

Since $V$ is a nontrivial ideal of $F\langle X\rangle$, every $c \in C$ is a monomial of positive degree, and so we can write it in the form $x_i d_c$ with $x_i \in X$ a variable and $d_c \in W$. By the minimality property defining $C$, we have that $d_c \notin WC$.

Let us denote for convenience $x_i = x_c$ and $r_i = \delta(x_i) = r_c$.

Denote by $R_0 := \{r_i \mid i = 1, 2, \dots\}$ the basis of $\mathcal{R}$ as free bimodule. Notice that the set $R_0W = \{r_i a\}$, $r_i \in R_0$, $a \in W$, is a basis of $R_0F\langle X\rangle$ as $F$-vector space and of $\mathcal{R}$ as free left module over $F\langle X\rangle$. We next decompose this set $R_0W$ into three parts:

$$\begin{aligned} W_1 &:= \{r_c d_c \mid c \in C\}, \\ W_2 &:= \{rbc \mid r \in R_0,\ b \in W,\ c \in C\}, \\ W_3 &:= \{rb \mid r \in R_0,\ b \in W,\ rb \notin W_1 \cup W_2\}. \end{aligned} \tag{16.14}$$

One has to remark first that $W_1 \cap W_2 = \emptyset$. In fact suppose that we had $r_c d_c = rbe$, $e \in C$. This implies $r_c = r$ and $d_c = be$, and hence $c = x_c d_c = x_c be$, but $e \in C$ contradicts the minimality assumption on $c$.

We next consider the degree lexicographic order also on the basis $WR_0W$ of $\mathcal{R}$ over $F$ by considering the variables $X$ as preceding the elements $r_i$. Thus, for a monomial $M = x_{i_1} \cdots x_{i_k}$, the leading term of $\delta(M)$ is $r_{i_1} x_{i_2} \cdots x_{i_k}$. It is easily verified that for any polynomial $a$, the leading term of $\delta(a)$ equals the leading term of $\delta(\hat{a})$, with $\hat{a}$ the leading term of $a$.

Now modify the basis $R_0W$ of $\mathcal{R}$ as free left module over $F\langle X\rangle$ as follows. Since $\hat{a}_c = x_c d_c$, we have $\delta(a_c) = r_c d_c + x_c\delta(d_c) + LT$, where $LT$ denotes a sum of terms strictly lower than $r_c d_c$ in the lexicographic order. Therefore, if we replace

$r_c d_c \in W_1$ by $\delta(a_c)$, we have performed a triangular change of basis in the given order, as left modules. Call $K_1$ this new part of a basis, replacing $W_1$. As for the elements in $W_2$ we can do a similar thing. Take an element $rbc$, then $bc$ is the leading monomial of $ba_c$, so by replacing $rbc$ by $rba_c$, we also perform a triangular change of basis and call $K_2$ this new part of a basis, replacing $W_2$. We keep $W_3$ fixed and call $K := (K_1, K_2, K_3 = W_3)$ this new basis of $\mathcal{R}$ as left $F\langle X\rangle$-module.

Now by Theorem 16.2.5 we have that $K_2 := \{rba_c,\ b \in W,\ c \in C\}$ is a basis of $R_0 V$ as vector space over $F$, hence it is also a basis of $\mathcal{R}V = F\langle X\rangle R_0 V$ as left $F\langle X\rangle$-module, $\mathcal{R}V = F\langle X\rangle K_2$. Since $K$ is a basis of $\mathcal{R}$ as left $F\langle X\rangle$-module, we have $U\mathcal{R} = UK_1 \oplus UK_2 \oplus UK_3$, thus finally

$$U\mathcal{R} + \mathcal{R}V = UK_1 \oplus F\langle X\rangle K_2 \oplus UK_3. \tag{16.15}$$

We are now ready to finish the proof. Let $v \in \ker(\phi) \subset U \cap V$. Then $\delta(v) \in U\mathcal{R} + \mathcal{R}V$ by the definition of $\phi$.

Write $v = \sum_{c\in C} b_c a_c$ by the fact that $a_c$ is a basis of $V$ as left $F\langle X\rangle$ module. Applying the property of derivations, we have

$$\delta(v) = A_v + B_v, \quad A_v = \sum_{c\in C} \delta(b_c)a_c,\ B_v = \sum_{c\in C} b_c\delta(a_c).$$

Next clearly $A_v \in \mathcal{R}V$ and $B_v \in F\langle X\rangle K_1$ so, since $\delta(v) \in U\mathcal{R} + \mathcal{R}V$ and $B_v = \delta(v) - A_v$, we deduce that

$$B_v \in (U\mathcal{R} + \mathcal{R}V) \cap F\langle X\rangle K_1.$$

But formula (16.15) implies then that $(U\mathcal{R} + \mathcal{R}V) \cap F\langle X\rangle K_1 = UK_1$ so $B_v \in UK_1$. Since the elements $\delta(a_c) \in K_1$ are a basis of the left module that they generate and $\sum_{c\in C} b_c\delta(a_c) \in UK_1$, we have that $b_c \in U$ for all $c$ and thus $v = \sum_{c\in C} b_c a_c \in UV$, as desired. □

16.2.1.4. *Consequences for T-ideals.* We apply now the previous theory when $U, V$ are two $T$-ideals and $F$ is a field. Then we have that $UV$ is also a $T$-ideal, and

COROLLARY 16.2.8 (of Theorem 16.2.7). *The relatively free algebra $F\langle X\rangle/UV$ is isomorphic to the subalgebra of* $\begin{pmatrix} F\langle X\rangle/U & \mathcal{R}/(U\mathcal{R}+\mathcal{R}V) \\ 0 & F\langle X\rangle/V \end{pmatrix}$ *generated by the elements* $\phi(x_i) = \begin{pmatrix} a_i & w_i \\ 0 & b_i \end{pmatrix} = \begin{pmatrix} \bar{x}_i & \bar{r}_i \\ 0 & \bar{x}_i \end{pmatrix}$ *(cf. formula (16.12)), where by definition $a_i, b_i$ are the classes of $x_i$ modulo $U, V$, respectively, and $w_i$ is the class of $r_i$ modulo $U\mathcal{R} + \mathcal{R}V$.*

PROOF. The kernel of $\phi$ is $UV$, thus the relatively free algebra $F\langle X\rangle/UV$ is isomorphic to the image of $\phi$. □

## 16.3. *T*-ideals of identities of block-triangular matrices

**16.3.1. Identities of block-triangular matrices.** We can now provide one of the main applications of Theorem 16.2.7 due to Giambruno and Zaicev **[GZ03a]**, that is the description of the $T$-ideals of identities of block-triangular matrices.[2]

Such algebras are strictly related to the Jordan Hölder theory.

Given a vector space $V$ of dimension $n$ over an infinite field $F$, consider a *partial flag* of subspaces $V_0 = 0 \subsetneq V_1 \subsetneq V_2 \subsetneq \cdots \subsetneq V_m = V$ of dimensions $d_1, d_1 + d_2, \ldots, d_1 + d_2 + \cdots + d_m = n$.

[2]These are called *parabolic algebras* in Lie theory.

DEFINITION 16.3.1. The matrix algebra $UT(d_1, d_2, \dots, d_m)$ is the algebra of linear transformations of $V$ which preserve all the subspaces $V_i$.

Thus the $V_i$ are submodules for $UT(d_1, d_2, \dots, d_m)$ and give a Jordan Hölder series, in fact in this case these are the only submodules. If one fixes a basis of $V$ whose intersection with each $V_i$ is a basis of $V_i$, one exhibits the algebra $UT(d_1, d_2, \dots, d_m)$ as *block upper-triangular* $n \times n$ *matrices*. A block $A_{i,j}$, $i, j \le m$ is a $d_i \times d_j$ matrix and further $A_{i,j} = 0$ if $i > j$. The Jacobson radical $J$ of this algebra is clearly the set of elements with $A_{i,i} = 0$, in other words the linear transformations $T : V \to V$ with the property that $T(V_i) \subset V_{i-1}$ for all $i$. Hence we deduce that $J^m = 0$.

Denote by $\Gamma_n$ the ideal of PI of $n \times n$ matrices and by $\Gamma_{d_1, d_2, \dots, d_m}$ the ideal of (stable) PI of $UT(d_1, d_2, \dots, d_m)$ in a given set of variables $X$.[3] Then

THEOREM 16.3.2 (Giambruno and Zaicev **[GZ03a]**). $\Gamma_{d_1, d_2, \dots, d_m} = \Gamma_{d_1} \Gamma_{d_2} \cdots \Gamma_{d_m}$.

PROOF. For simplicity of notation denote by

$$\underline{d} := \{d_1, \dots, d_i, \dots\}, \quad UT(d_1, d_2, \dots, d_m) = UT(\underline{d}).$$

We will prove the theorem by induction on $m$. Hence it is enough to show by induction that

$$\Gamma_{d_1, d_2, \dots, d_m} = \Gamma_{d_1, d_2, \dots, d_{m-1}} \Gamma_{d_m}.$$

First let us show that $\Gamma_{\underline{d}} := \Gamma_{d_1, d_2, \dots, d_m} \supset \Gamma_{d_1} \Gamma_{d_2} \cdots \Gamma_{d_m}$, that is that $\Gamma_{d_1} \Gamma_{d_2} \cdots \Gamma_{d_m}$ vanishes on the algebra $UT(\underline{d})$. For this remark if we compute an element of $\Gamma_{d_i}$ on this algebra, we obtain a matrix $T_i$ which has 0 on the $i, i$ diagonal block. Since in block multiplication $\begin{pmatrix} 0 & a \\ 0 & b \end{pmatrix} \begin{pmatrix} x & y \\ 0 & 0 \end{pmatrix} = 0$, by induction $T_1 T_2 \cdots T_m = 0$.

We now apply the method of generic elements. Consider the polynomial ring $F[\xi_{i,j}^{(t)}]$ in the variables which represent coordinates of matrices $\xi^{(t)}$, $t = 1, \dots, N$ in $UT(\underline{d})$, that is the coordinates of the vector space $UT(\underline{d})^{\mathbb{N}}$.

In other words the numbers $\underline{d} := \{d_1, \dots, d_i, \dots\}$ determine a (simple to describe) set of matrix units $e_{i,j}$, $(i, j) \in E_{\underline{d}}$ so

$$F[\xi]_{\underline{d}} := F[\xi_{i,j}^{(t)}], \ t \in \{1, 2, \dots, N\}, \ (i, j) \in E_{\underline{d}}.$$

Inside the algebra $UT(\underline{d}) \otimes_F F[\xi_{i,j}^{(t)}]$, construct the subalgebra generated by its generic elements $\xi_t(\underline{d}) = \xi_t := \sum_{(i,j) \in E_{\underline{d}}} \xi_{i,j}^{(t)} e_{i,j}$. This, by definition and since $F$ is infinite, is a copy of the free algebra modulo the ideal $\Gamma_{\underline{d}}$, and we need to show that this equals the free algebra modulo $\Gamma_{d_1} \Gamma_{d_2} \cdots \Gamma_{d_m}$.

We can of course represent each element $\xi_t(\underline{d})$ as a $2 \times 2$ block matrix $\xi_t = \begin{pmatrix} \eta_t & w_t \\ 0 & \gamma_t \end{pmatrix}$, where the $\eta_t$ are generic elements for the algebra $UT(d_1, d_2, \dots, d_{m-1})$, and the $\gamma_t$ are generic elements for the algebra $UT(d_m) = M_{d_m}(F)$.

We let

$$A := F\langle \eta_t \rangle = F\langle X \rangle / \Gamma_{d_1, d_2, \dots, d_{m-1}}, \ B := F\langle \gamma_t \rangle = F\langle X \rangle / \Gamma_{d_m} \tag{16.16}$$

be the corresponding algebras of generic matrices.

We want to apply Lewin's theorem so we need to show

[3] This applies also if $X$ is finite.

LEMMA 16.3.3. *Let $A, B$ be the subalgebras generated by the elements $\eta_t$ and $\gamma_t$, respectively. Inside $UT(\underline{d}) \otimes_F F[\xi_{i,j}^{(t)}]$, the elements $w_i$ form a basis of a free bimodule $M$ with respect to multiplication by the two subalgebras $A, B$.*

PROOF. Obviously one can divide, for each $t$, the variables $\xi_{i,j}^{(t)}$ into three blocks corresponding to the coordinates of the three nonzero blocks of the preceding representation of $\xi_t$ as $2 \times 2$ block matrix. We have

$$F[\xi]_{d_1,d_2,\dots,d_m} = F[\xi]_{d_1,d_2,\dots,d_{m-1}} \otimes F[\xi]_{d_m} \otimes F[\xi]_{(d_1+d_2+\cdots+d_{m-1};d_m)},$$

where $F[\xi]_{(d_1+d_2+\cdots+d_{m-1};d_m)}$ is the polynomial ring in the variables relative to the elements $w_t$.

So suppose that we have some linear relation $\sum_t a_t w_t b_t = 0$, where $a_t, b_t$ belong to the corresponding algebras $A, B$. Thus the entries of $a_t$ are in $F[\xi]_{d_1,d_2,\dots,d_{m-1}}$, the entries of $b_t$ are in $F[\xi]_{d_m}$, the entries of $w_t$ are the variables generating $F[\xi]_{(d_1+d_2+\cdots+d_{m-1};d_m)}$.

By definition the coordinates of the $w_t$ are all algebraically independent and also independent from the ones appearing in $a_t, b_t$. So choose an index $j \in [1, N]$ and set equal to 0 all the coordinates of all $w_i$ for $i \neq j$. We see that in the relation $\sum_t a_t w_t b_t$ all terms vanish, except for $a_j w_j b_j$ which is unchanged. Since $j$ is arbitrary, we must have $a_j w_j b_j = 0$ for all $j$.

So let us analyze a possible relation $a_t w_t b_t = 0$. Set $r = d_1 + d_2 + \cdots + d_{m-1}$, $s = d_m$ for simplicity of notation. With this notation $w_t$ is an $r \times s$ matrix of variables, which again for simplicity we denote by $u_{i,j}$, $i = 1, \dots, r$, $j = 1, \dots, s$. Also $a_t$ is some $r \times r$ matrix with entries $a_{i,j}$, and $b_t$ is some $s \times s$ matrix with entries $b_{i,j}$.

The condition $a_t w_t b_t = 0$ means that

$$\sum_{j,k} a_{i,j} u_{j,k} b_{k,\ell} = 0, \ \forall i = 1, \dots, r, \ \ell = 1, \dots, s.$$

This identity is in a polynomial ring in which the $u_{j,k}$ are some of the variables, while the elements $a_{i,j}, b_{k,\ell}$ are polynomials in other disjoint variables. By the hypotheses on the independence of the variables, we thus must have $a_{i,j} b_{k,\ell} = 0$ for all indices involved. Now this clearly implies either $a_t = 0$ or $b_t = 0$. □

Using Lemma 16.3.3 we can now finish the proof. We have that the mapping

$$F\langle X \rangle \to UT(\underline{d}) \otimes_F F[\xi_{i,j}^{(t)}], \qquad x_t \mapsto \xi_t = \begin{pmatrix} \eta_t & w_t \\ 0 & \gamma_t \end{pmatrix},$$

which has as kernel the $T$-ideal of identities of $UT(\underline{d})$, coincides with the mapping $\phi : F\langle X \rangle \to \begin{pmatrix} A & M \\ 0 & B \end{pmatrix}$. This last map (cf. formula (16.16)) has as its kernel the product of the two $T$-ideals $\Gamma_{d_1,d_2,\dots,d_{m-1}}, \Gamma_{d_m}$. □

COROLLARY 16.3.4. *A PI algebra $S$, finitely generated over a field $F$, satisfies the identities of $n \times n$ matrices for some $n$.*

PROOF. We present a PI algebra $S$ as the quotient of a relatively free algebra, that is an algebra $R := F\langle X \rangle / \Gamma$ where $\Gamma$ is the $T$-ideal of identities of $S$.

Now the radical $\sqrt{\Gamma}$ is a semiprime $T$-ideal and, hence, the ideal $\Gamma_k$ of identities of $k \times k$ matrices in the variables $X$ for some $k$.

If $S$ is finitely generated, we can take $R := F\langle x_1, \dots, x_m\rangle/\Gamma$ quotient of a finitely generated free algebra. By the Razmyslov–Braun–Kemer theorem for some power $h$, we have $\sqrt{\Gamma}^h = \Gamma_k^h \subset \Gamma$.

So it is enough to prove that $F\langle x_1, \dots, x_m\rangle/\Gamma_k^h$ satisfies the identities of $n \times n$ matrices for some $n$. Now, by Theorem 16.3.2, $F\langle x_1, \dots, x_m\rangle/\Gamma_k^h$ is the relatively free algebra, in $m$ variables, of the block triangular $kh \times kh$ matrices, with $h$ blocks of size $k$ on the diagonal. In particular this algebra satisfies all identities of $n \times n$ matrices for $n = kh$. □

16.3.1.1. *Block triangular matrices.* We want to apply the previous theory to the algebra $T_{r,m}$ of block triangular matrices, with $m$ blocks of size $r$. This algebra is isomorphic to $M_r(F) \otimes T_m$ where $T_m$ are the triangular $m \times m$ matrices. We can thus apply Theorem 2.6.8 to $T_{r,m}$, with $A = T_m$.

Recall that $M_r(F) * F\langle X\rangle = M_r(F\langle X_{r^2}\rangle)$, and we distinguish the variables $X$ of the generalized polynomial identities of $T_{r,m}$, elements of $M_r(F) * F\langle X\rangle$ and the variables $X_{r^2}$ of the ordinary polynomial identities of $T_m$, elements of $F\langle X_{r^2}\rangle$.

The ideal of polynomial identities of $T_m$ is the $m$th power of the commutator ideal $C$, by Theorem 16.3.2. The ideal of generalized polynomial identities of $T_{r,m}$ with coefficients in $M_r(F)$ equals $M_r(C^m)$ by Theorem 2.6.8.

COROLLARY 16.3.5. *The ideal of generalized polynomial identities of $T_{r,m}$, thought of as $M_r(F)$-algebra, is generated by the product of $m$ standard identities (with variables in between),*

(16.17) $$S_{2r}(x_1, \dots, x_{2r})y_1 S_{2r}(x_{2r+1}, \dots, x_{4r})y_2 \cdots y_{m-1} S_{2r}(x_{2(m-1)r+1}, \dots, x_{2mr}).$$

PROOF. We know that the product of formula (16.17) is an identity, and we know that the $T$-ideal of generalized identities equals $M_r(C^m)$.

In formula (16.17), make the following substitutions (as in formula (2.36)). For each $i = 1, \dots, m$, substitute the $2r$ variables $x_{(2i-1)r+1}, \dots, x_{2ir}$ with the elements

(16.18) $$a_i e_{1,1},\ b_i e_{1,2},\ e_{2,2},\ \dots, e_{r,r},\ e_{r,1},\ i = 1, \dots, m,$$

where $a_i, b_i$ are distinct variables in $X_{r^2}$, which commute with the elements $e_{i,j}$. Then substitute the variables $y_i$ by still other variables $c_i$ in $X_{r^2}$.

By the proof of Proposition 2.6.10, we see that this evaluation is

(16.19) $$[a_1, b_1]c_1[a_2, b_2]c_2 \cdots c_{m-1}[a_m, b_m]e_{1,1}.$$

The coefficient of $e_{1,1}$ is the *product of commutators* and thus the identity which generates the $T$-ideal $C^m$ of identities of triangular matrices. Thus the element in (16.19) generates the generalized identelities of $T_{r,m}$. □

## 16.4. The theorem of Bergman and Lewin

**16.4.1. The free semigroup of $T$-ideals.** In Theorem 16.3.2 we have described the $T$-ideal of polynomial identities of block-triangular matrices as a product of the basic prime $T$-ideals of identities of matrices. It is quite remarkable that under product, both ideals and $T$-ideals of the free algebra have a *unique factorization* in a strong noncommutative sense. In particular the factorization of Theorem 16.3.2 is into indecomposable factors.

Recall that the free monoid in some set $A$, called an alphabet, is the set of words in this alphabet. For ideals we want to consider the monoid structure of multiplication of ideals. So we define

DEFINITION 16.4.1. An element $a$ of a monoid $S$ is *indecomposable* if it is not of the form $a = bc, b, c \neq 1$.

An ideal $I$ of an algebra $R$ is *indecomposable* if it is not of the form $I = JK$ with $J, K \supsetneq I$.

Dubreil-Jacotin **[DJ47]** (see also P. M. Cohn **[Coh70]**) shows

THEOREM 16.4.2. *A semigroup $S$ with a 1 is free on the set of its indecomposable elements if and only if:*

(i) *$S$ has right and left cancellation. This means that*

$$ab = ac \implies b = c, \quad ba = ca \implies b = c, \ \forall a, b, c \in S.$$

(ii) *Whenever two elements of $S$ have a common right multiple, one of them is a right multiple of the other, that is*

$$ac = bd \implies \exists u, \ b = au \quad or \quad a = bu.$$

(iii) *Every element of $S$ is a finite product of indecomposable elements.*

PROOF. Since (iii) is satisfied, what we need to prove is that the indecomposable elements behave as an *alphabet*. If we have two products of indecomposable elements which are equal as elements of the semigroup $a_1 a_2 \cdots a_k = b_1 b_2 \cdots b_h$, then we need to prove that $h = k$ and $a_i = b_i, \ \forall i$.

Now $a_1, b_1$ have a common right multiple, so one is a multiple of the other, but since they are indecomposable, we must have $a_1 = b_1$. Next, by cancellation, we have $a_2 \cdots a_k = b_2 \cdots b_h$, and by induction we finish the proof. □

THEOREM 16.4.3 (Bergman and Lewin **[BL75]**).

(1) *The monoid of ideals of the free algebra is the free monoid with generators the indecomposable ideals.*
(2) *In the free algebra in infinitely many variables the factors in the decomposition as product of indecomposable ideals of a $T$-ideal are also $T$-ideals.*

PROOF. Let us see for the moment why (2) follows from (1). Take a $T$-ideal $I$ and write it, according to (1), as a product $I = I_1 I_2 \cdots I_k$ with the $I_j$ indecomposable ideals. For every special endomorphism $\phi$ of the free algebra (Definition 2.2.9), since $\phi$ is surjective for all ideals $I$, we have $\phi(I)$ is an ideal and

$$I = \phi(I) = \phi(I_1)\phi(I_2) \cdots \phi(I_k).$$

So, by unique factorization, $I_j = \phi(I_j), \ \forall j$. We conclude by Proposition 2.2.10 that all the factors $I_j$ are $T$-ideals. □

From now on we let $R := F\langle X \rangle$ be a free algebra in variables $X$ over a field $F$. We then denote by $R_+$ the ideal of elements of positive degree of $R$, and we think of $F = R/R_+$ as an $R$ bimodule. If $X = x$ is a single variable, then $R = F[x]$ is a unique factorization domain, so we may assume that $X$ has at least two variables. In this case we have

LEMMA 16.4.4. *The only two-sided ideal which is principal as left ideal is $R$.*

PROOF. Let $f \in F\langle X\rangle$, and assume that $F\langle X\rangle f$ is a two-sided ideal and $f$ is of degree $> 0$. Then for each variable $x_i$ we must have that $fx_i = a_i f$ for some $a_i \in F\langle X\rangle$.

It follows easily that all the monomials of highest degree in $f$ are of the form $Nx_i$ for some $N$. This is impossible since there are two different $x_i$. □

We shall then prove that the semigroup of ideals of the free algebra satisfies the three properties of Theorem 16.4.2. After some preliminaries, property (i) is proved in Lemma 16.4.6, property (ii) in Lemma 16.4.9, and finally property (iii) is proved in Lemma 16.4.12. After this the theorem follows.

We need some simple formalism on free modules. For a general ring $S$ the rank of a free module is not well defined. For a free algebra $R$ we have that $F = R/R_+$. If a module $M$ is isomorphic to $R^n$, we have that $F \otimes_R M = M/R_+M = F^n$; so the rank is well defined and will be denoted $\dim M$. Moreover if we have an exact sequence of free modules $M_i$, it splits

$$(16.20)\quad 0 \longrightarrow M_1 \longrightarrow M_2 \longrightarrow M_3 \longrightarrow 0 \implies M_2 = M_1 \oplus M_3.$$

Hence $F \otimes_R M_2 = F \otimes_R M_1 \oplus F \otimes_R M_3$, and we have

LEMMA 16.4.5. *If in an exact sequence of free $R$-modules as in formula* (16.20) *two of the modules $M_i$ have finite rank, then all three have finite rank and*

$$\dim M_2 = \dim M_1 + \dim M_3.$$

We now pass to ideals of $R$ following closely Bergman and Lewin **[BL75]**. We need to verify the three properties of Theorem 16.4.2 for the semigroup of ideals of the free algebra.

LEMMA 16.4.6 ((i) Cancellation). *Let $R$ be a right f.i.r.* (*see page 415*), *let $A$ be a nonzero right ideal of $R$, and let $B, B'$ be left ideals of $R$. Then $AB = AB' \implies B = B'$. Likewise, in a left f.i.r., $AB = A'B \implies A = A'$ for $B$ a nonzero left ideal and $A, A'$ right ideals.*

PROOF. $A$ is free as a right $R$-module, say $A = \bigoplus_i a_i R$. Then $AB = \bigoplus_i a_i B$ (as an additive group), and likewise $AB' = \bigoplus_i a_i B'$. Equating coefficients of any $a_i$, we get $B = B'$. □

LEMMA 16.4.7. *Let $I, J$ be two two-sided ideals of a free algebra $R$. Then*

$$\dim IJ = \dim I \dim J.$$

PROOF. Choose a basis $a_i$ and $b_j$ for $I, J$ as right $R$-modules. So $I = \bigoplus_i a_i R$, $J = \bigoplus_j b_j R$, then $IJ = \bigoplus_i a_i J = \bigoplus_{i,j} a_i b_j R$. □

LEMMA 16.4.8. *Let $I, J$ be two ideals of $R$ such that*

$$I + J = R,\ IJ = JI.$$

*Then either $I = R$ or $J = R$.*

PROOF. Note first that

$$I \cap J = (I \cap J)R = (I \cap J)(I + J) \subset IJ + JI = IJ \subset I \cap J,$$

hence $I \cap J = IJ$. Consider the exact sequence as right ideals

$$0 \longrightarrow IJ = I \cap J \longrightarrow I \oplus J \longrightarrow R = I + J \longrightarrow 0.$$

If $I$ and $J$ have finite rank as free modules, setting $p := \dim I$, $q := \dim J$, we deduce by Lemma 16.4.7 and the previous formula that we have

$$pq + 1 = p + q \iff (p-1)(q-1) = 0 \implies p = 1 \text{ or } q = 1.$$

So one of the two ideals is principal, but in $R$ the only principal ideal is $R$ by Lemma 16.4.4.

Assume that both $I, J$ are different from $R$ and that $I$ has infinite rank. Choose a basis $a_i$ and $b_j$ for $I, J$ as right $R$-modules. So $I = \bigoplus_i a_i R$, $J = \bigoplus_j b_j R$. Since $R = I + J$, there is $x \in I$ with $(1-x) \in J$, and we write $x = \sum_i a_i c_i$. Since $I$ has infinite rank, there is an index $i_0$ such that $a_{i_0}$ does not appear in this expansion of $x$. Now $(1-x)a_{i_0} = a_{i_0} - \sum_{i \neq i_0} a_i c_i a_{i_0} \in JI = IJ = \bigoplus_i a_i J$, so since this is a direct sum, the coefficient 1 of $a_{i_0}$ is in $J$, and thus $J = R$. □

We come now to the factorization problem of ideals.

LEMMA 16.4.9 ((ii) Common factor). *Suppose $AB = CD$, where $A$ and $C$ are nonzero right ideals, and $B$ and $D$ nonzero left ideals of $R$. Then there exists a two-sided ideal $U$ of $R$ such that*

$$\text{either} \quad \begin{cases} C = AU \\ B = UD \end{cases} \qquad \text{or} \quad \begin{cases} A = CU \\ D = UB. \end{cases}$$

PROOF.

CASE 1 ($R = A + C$ and $R = B + D$). Multiply the first of these equations on the right by $B$; then we get

$$B = RB = AB + CB = CD + CB = C(D + B) = CR = C. \tag{16.21}$$

Likewise, multiplying by $D$, we get $D = A$. Hence $A$ and $B$ are two-sided ideals satisfying $A + B = R$ and $AB = BA$, so $A$ or $B = R$, by Lemma 16.4.8. The desired conclusion follows, with $U = B$ or $A$.

CASE 2 ($R = A + C$, no assumption on $B + D$). Arguing as in formula (16.21) we get $B = C(B + D)$ and $D = A(B + D)$.

Let $B + D = \bigoplus_{i \in I} Re_i$. Pick any $i_0 \in I$, and denote by $\pi : \bigoplus Re_i \to R$ the *coefficient of* $e_{i_0}$, a left $R$-linear map. Then we see that $\pi(B), \pi(D)$ are two left ideals and $\pi(B) + \pi(D) = R$.

Apply $\pi$, which is linear by left multiplication by $R$, to the formulas

$$B = C(B + D),\ D = A(B + D),\ AB = CD,$$

getting $\pi(B) = C\pi(B + D) = CR = C$. Similarly we have $\pi(D) = A$ and $A\pi(B) = C\pi(D) \implies CA = AC$. By Lemma 16.4.7 either $A$ or $C$ equal $R$.

We conclude as before.

CASE 3 (The general case). We put $A + C = \bigoplus_j f_j R$ and consider as before a coefficient which is a right $R$-linear map $\pi : A + C \to R$ so that $\pi(A)$, $\pi(C)$ are two right ideals with $\pi(A) + \pi(C) = R$. Applying $\pi$ to $AB = CD$, we get $\pi(A)B = \pi(C)D$. So we are in one of the two cases of Case 2.

For instance the case $\pi(A) = R$ implies $B = \pi(C)D = R\pi(C)D$. Set then $U = R\pi(C)$ a two-sided ideal, and we have $B = UD$. As for $AU = C$, this follows from Lemma 16.4.6 since $CD = AB = AUD$. □

We now come to the last part, verifying property (iii). For this, given a left ideal $I$, consider the set $C_I$ of monomials introduced in Theorem 16.2.5:

DEFINITION 16.4.10. We define the degree $d(I)$ of a left ideal of the free algebra to be the minimum degree of the elements in $C_I$.

Of course $d(I) = 0$ if and only if $I = R$.

LEMMA 16.4.11. *If $I, J$ are two ideals we have $C_{IJ} = C_I C_J$.*
*$d(IJ) = d(I) + d(J)$.*

PROOF. For each $c \in C_I$ choose an element $a_c \in I$ with leading term $c$ so that by Theorem 16.2.5 these elements form a basis of $I$ as left ideal. Do the same for $J$, and to a $d \in C_J$ associate a $b_d \in J$. The elements $a_c b_d$ then form a basis of $IJ$, and by Remark 16.2.4 we have that $cd$ is the leading monomial of $a_c b_d$.

So first let us show that $cd \in C_{IJ}$. For this we need to show that it is minimal, that is we need to exclude the existence of an element $x \in IJ$ with some leading term $\bar{x}$ dividing properly $cd$. Now when we write a linear combination $x = \sum_{c,d} u_{c,d} a_c b_d$, we see that its leading term is of the form $\bar{u}_{c,d} cd$ for some $c, d$ in the sum. So a minimal element in $C_{IJ}$ must necessarily be in $C_I C_J$. Conversely if one of the $cd$ were not minimal, $a_c b_d$ should appear as linear combination of some other elements of the basis which is impossible.

The second part follows from the first. □

PROPOSITION 16.4.12 ((iii) Decomposition). *Given a nontrivial ideal $I$ of $R$ with $d(I) = k$, then $I$ is a product of at most $k$ indecomposable factors.*

PROOF. This follows immediately from Lemma 16.4.9. □

This finishes the proof of the three properties characterizing a free semigroup in the case of the semigroup of ideals of a free algebra and, hence, the proof of Theorem 16.4.3.

There is at the moment no characterization of indecomposable $T$-ideals; nevertheless there are some important examples. Clearly a prime $T$-ideal is indecomposable, and these are the ideals of PI of matrices. Later, in §20.3, we shall study a larger class, the *verbally prime ideals,* which are also indecomposable and characterized by Theorem 20.3.5.

**16.4.2. The Poincaré series and cocharacters.** We want to apply the previous theory to prove a theorem of Formanek **[For85]** and one of Berele and Regev **[BR98]**.

Let us go back to the definition of the multigraded Hilbert–Poincaré series in formula (8.12), and let us first compute this for the free algebra $F\langle x_1, \ldots, x_k\rangle$ in $k$ variables.

THEOREM 16.4.13. *The multigraded Hilbert–Poincaré series of the free algebra $F\langle X\rangle = F\langle x_1, \ldots, x_k\rangle$ is*

$$h_{F\langle X\rangle}(\underline{t}) = (1 - \sum_{i=1}^{k} t_i)^{-1} = 1 + \sum_{j=1}^{\infty} (\sum_{i=1}^{k} t_i)^j. \tag{16.22}$$

PROOF. One may work as follows. Consider the formal power series in the completion of $F\langle x_1,\dots,x_k\rangle$ as in §3.3.2.3, formula (3.38),

$$(1-\sum_{i=1}^{k} x_i)^{-1} = 1+\sum_{j=1}^{\infty}(\sum_{i=1}^{k} x_i)^j = \sum_{w\in W(X)} w.$$

This is the *generating sequence of all monomials*. That is, for all $j$, we have $(\sum_{i=1}^k x_i)^j$ is the sum of all the $k^j$ different words in $x_1,\dots,x_k$ of degree $j$. Then by specializing $x_i \mapsto t_i$, one obtains the identity. $\square$

If $I \subset F\langle X\rangle$ is a right (or left) ideal, we know that as $F\langle X\rangle$-module it is free, and as module we have $I \cong I/I\cdot F\langle X\rangle_+ \otimes_F F\langle X\rangle$. If moreover $I$ is multigraded, then $I/I\cdot F\langle X\rangle_+$ also has a multigraded basis, and we have the identity

(16.23) $$h_I(\underline{t}) = h_{I/I\cdot F\langle X\rangle_+}(\underline{t})h_{F\langle X\rangle}(\underline{t}), \quad h_{F\langle X\rangle/I}(\underline{t}) = [1-h_{I/I\cdot F\langle X\rangle_+}(\underline{t})]h_{F\langle X\rangle}(\underline{t}).$$

Finally assume that $I$ is stable under $\mathrm{GL}(k,F)$ (where in the end we shall take the limit case $k=\infty$). Then the multigraded Hilbert–Poincaré series is a symmetric function, and therefore it can be developed as a power series $\sum_\lambda m_\lambda S_\lambda$ in the Schur symmetric functions.

In particular for $F\langle X\rangle$ we have (cf. Theorem 6.3.20 for $\sigma = 1$)

(16.24) $$h_{F\langle X\rangle}(\underline{t}) = (1-S_1)^{-1} = 1+\sum_{j=1}^{\infty} S_1^j = 1+\sum_{j=1}^{\infty}\sum_{\lambda\vdash j} f_\lambda S_\lambda.$$

We may apply this to $T$-ideals $I$ (for algebras with 1). In this case, if $A = F\langle X\rangle/I$, then $I$ is the ideal of identities of $A$, and we write

$$h_A(\underline{t}) = 1+\sum_\lambda m_\lambda(A)S_\lambda,$$

where $m_\lambda(A)$ is the cocharacter.

It is then useful to define a *reduced cocharacter* by setting

(16.25) $$1-h_{I/I\cdot F\langle X\rangle_+}(\underline{t}) = \sum_\lambda \overline{m}_\lambda(A)S_\lambda.$$

We deduce that

(16.26)
$$h_A(\underline{t}) = \left(\sum_\lambda \overline{m}_\lambda(A)S_\lambda\right)(1-S_1)^{-1} \implies \sum_{\lambda\vdash n} m_\lambda(A)S_\lambda = \sum_{j=0}^{n}\sum_{\lambda\vdash j}\overline{m}_\lambda(A)S_\lambda S_1^{n-j}.$$

Setting $R = F\langle X\rangle = F\langle x_1,\dots,x_k\rangle$, we now have

THEOREM 16.4.14. *Let $I_1, I_2$ be two multigraded ideals of $R = F\langle X\rangle$ (for instance two T-ideals). Let $h_{I_1}(\underline{t}), h_{I_2}(\underline{t})$ be their multigraded Hilbert–Poincaré series. Then the multigraded Hilbert–Poincaré series of $I_1I_2$ is*

(16.27) $$h_{I_1I_2}(\underline{t}) = h_{I_1}(\underline{t})h_{I_2}(\underline{t})(1-\sum_{i=1}^{k} t_i).$$

PROOF. Both $I_1$ and $I_2$ are free right $F\langle x_1,\dots,x_k\rangle$-modules with multigraded bases. For a free module $Y = \bigoplus_{b\in B} bF\langle x_1,\dots,x_k\rangle$ with basis $B$, one then has

$$h_Y(\underline{t}) = \sum_{b\in B} h_{b\,F\langle X\rangle}(\underline{t}) = \sum_{b\in B} t^{\deg b}h_{F\langle X\rangle}(\underline{t}) = (\sum_{b\in B} \underline{t}^{\deg b})(1-\sum_{i=1}^{k} t_i)^{-1}.$$

So if $B_1, B_2$ are a basis relative to $I, J$, respectively, we know that $B_1B_2$ is a basis relative to $IJ$. Hence

$$h_{I_1I_2}(\underline{t}) = (\sum_{b\in B_1B_2} \underline{t}^{\deg b})(1-\sum_{i=1}^k t_i)^{-1} = (\sum_{b\in B_1} \underline{t}^{\deg b})(\sum_{b\in B_2} \underline{t}^{\deg b})(1-\sum_{i=1}^k t_i)^{-1}$$
$$= h_{I_1}(\underline{t})h_{I_2}(\underline{t})(1-\sum_{i=1}^k t_i). \qquad \square$$

So let $A_i = R/I_i$ and $A = R/I_1I_2$. We have $h_I(\underline{t}) = h_R(\underline{t}) - h_{R/I}(\underline{t})$.

COROLLARY 16.4.15.

$$h_A(\underline{t}) = h_{A_1}(\underline{t}) + h_{A_2}(\underline{t}) + h_{A_1}(\underline{t})h_{A_2}(\underline{t})(\sum_{i=1}^k t_i - 1). \tag{16.28}$$

PROOF. $h_R(\underline{t})(1-\sum_{i=1}^k t_i) = 1$ implies

$$h_R(\underline{t}) - h_A(\underline{t}) = (h_R(\underline{t}) - h_{A_1}(\underline{t}))(h_R(\underline{t}) - h_{A_2}(\underline{t}))(1-\sum_{i=1}^k t_i)$$
$$= h_R(\underline{t}) - h_{A_1}(\underline{t}) - h_{A_2}(\underline{t}) + h_{A_1}(\underline{t})h_{A_2}(\underline{t})(1-\sum_{i=1}^k t_i). \qquad \square$$

One can apply this to characters when $I_1, I_2$ are stable under the linear group $\mathrm{GL}(k,F)$ (and one may even have $k=\infty$). In this case the Hilbert–Poincaré series are symmetric functions. Thus we develop $h_{A_i}$ as linear combination of Schur functions $S_\lambda$ with nonnegative coefficients. These coefficients give the multiplicity of the corresponding Schur module $S_\lambda(F^k)$ in the decomposition of the algebra $A_i$ under $\mathrm{GL}(k,F)$.

Write

$$h_{A_i} = \sum_\lambda m_{\lambda,i}S_\lambda,\ h_A = \sum_\lambda m_\lambda(A)S_\lambda,$$

$$\sum_\lambda m_\lambda(A)S_\lambda = \sum_\lambda m_{\lambda,1}S_\lambda + \sum_\mu m_{\mu,2}S_\mu + \sum_\lambda m_{\lambda,1}S_\lambda \sum_\mu m_{\mu,2}S_\mu(S_1-1). \tag{16.29}$$

From formula (6.34) and Definition 6.4.18 of the outer tensor product, it follows by denoting $\chi_n(A)$ the cocharacter in dimension $n$ and equating terms of degree $n$ in formula (16.29) that

COROLLARY 16.4.16. *For $n \geq 1$ we have*

(16.30)

$$\chi_n(A) = \chi_n(A_1) + \chi_n(A_2) - \sum_{i=0}^n \chi_i(A_1)\hat{\otimes}\chi_{n-i}(A_2) + \chi_1\hat{\otimes}\sum_{i=0}^{n-1}\chi_i(A_1)\hat{\otimes}\chi_{n-i-1}(A_2)$$
$$= -\sum_{i=1}^{n-1}\chi_i(A_1)\hat{\otimes}\chi_{n-i}(A_2) + \chi_1\hat{\otimes}\sum_{i=0}^{n-1}\chi_i(A_1)\hat{\otimes}\chi_{n-i-1}(A_2).$$

We then also have a consequence for codimension since $c_n(A) = \deg(\chi_n(A))$, and if $\chi_1, \chi_2$ are characters of the symmetric groups $S_i, S_j$, we have

$$\deg(\chi_1\hat{\otimes}\chi_2) = \binom{i+j}{i}\deg(\chi_1)\deg(\chi_2),$$

and from this and formula (16.30), it follows for $n \geq 1$,

$$c_n(A) = -\sum_{i=1}^{n-1} \binom{n}{i} c_i(A_1)c_{n-i}(A_2) + n\sum_{i=0}^{n-1} \binom{n-1}{i} c_i(A_1)c_{n-i-1}(A_2). \tag{16.31}$$

This can be written as an identity of generating functions

$$c_A(t) := \sum_{i=0}^{\infty} \frac{t^i c_i(A)}{i!}, \quad c_A(t) = (t-1)c_{A_1}(t)c_{A_2}(t) + c_{A_1}(t) + c_{A_2}(t). \tag{16.32}$$

EXAMPLE 16.4.17. The generating sequence $b_k(t)$ of the codimension of the full algebra $A_k$ of upper triangular $k \times k$ matrices is

$$b_k(t) = \sum_{h=0}^{\infty} c_n(A_k)\frac{t^n}{n!} = \sum_{h=1}^{k} \binom{k}{h}(t-1)^{h-1}e^{ht}. \tag{16.33}$$

PROOF. By Theorem 16.3.2 of Giambruno and Zaicev, we have that if $I$ denotes the $T$-ideal generated by the commutator $[x, y]$, then $I^k$ is the $T$-ideal of PI of the full algebra $A_k$ of upper triangular $k \times k$ matrices.

Now $A = F\langle x_1, \dots, x_k\rangle/I$ is the polynomial ring so that $c_n(A) = 1$, $\forall n$, and $c_A(t) = e^t$. Letting $b_k := c_{A_k}(t)$, we thus have the recursive formula

$$b_k = (t-1)b_{k-1}e^t + b_{k-1} + e^t = [(t-1)e^t + 1]b_{k-1} + e^t, \; k \geq 1, \; b_0 = 0,$$

from which one easily deduces

$$b_k = e^t \sum_{h=0}^{k-1} ((t-1)e^t + 1)^h = e^t \frac{((t-1)e^t + 1)^k - 1}{((t-1)e^t + 1 - 1)} = \sum_{h=1}^{k} \binom{k}{h}(t-1)^{h-1}e^{ht}.$$

We remark that the first identities of $A_k$ are in degree $2k$ so the series $b_k(t)$ starts with $\sum_{j=0}^{2k-1} t^j$. For the codimension in degree $2k$ the computer gives, for $k = 1, \dots, 12$,

$$1, 18, 630, 37800, 3515400, 471517200, 86497210800, 20841060240000,$$
$$6389869069584000, 2430526127309280000, 1123451899297247520000,$$
$$620296925072660035200000, \dots. \qquad \square$$

CHAPTER 17

# Finite-dimensional and affine PI algebras

*In this chapter, $F$ denotes a fixed field of characteristic* 0. *Unless otherwise specified by an algebra $A$, we mean an algebra over $F$. Hence, unless otherwise specified by a finite-dimensional algebra $A$, we mean finite-dimensional over $F$ not necessarily with a* 1. *Accordingly by $T$-ideal we mean a $T_+$-ideal. This complication is needed for the application to Specht's problem. The goal is to prove Theorem* 17.1.1.

*This is the first and main step towards Kemer's program to solve Specht's problem, Theorem* 20.2.2. *For this we will need a generalization to superalgebras which we discuss in the next chapter, that is,*

THEOREM (Kemer (Theorem 19.7.4)). *A finitely generated PI superalgebra is PI equivalent to a superalgebra finite dimensional over $F$.*

## 17.1. Strategy

The proof of Theorem 19.7.4 is quite involved. In this section we describe the program that we want to follow. We start by formulating the result in the special case of algebras (rather than superalgebras) and in term of finitely generated algebras. This in fact is the heart of the theory, where all the main difficulties appear:

THEOREM 17.1.1. *A finitely generated PI algebra $A$ is PI equivalent to an algebra $B$ finite dimensional over $F$.*

In fact this theorem can be reformulated as follows.

Let $\mathcal{V}$ be the variety generated by a finitely generated PI algebra $A$.

THEOREM 17.1.2. *The relatively free algebras in the variety $\mathcal{V}$ are representable.*

Let us show the equivalence of these two statements.

PROOF (EQUIVALENCE OF THEOREMS 17.1.1 AND 17.1.2). Assume Theorem 17.1.1 so that $\mathcal{V}$ is generated by a finite-dimensional PI algebra $A$. Then, by the method of generic elements (§2.4), relatively free algebras in the variety $\mathcal{V}$ are embedded in $A \otimes_F K$, where $K$ is some function field, and hence are representable.

Conversely, if $A$ is generated by $m$ elements, the variety $\mathcal{V}$, generated by $A$, is also generated by the relatively free algebra $\mathcal{F}_m$ in $\mathcal{V}$ with $m$ generators. In fact, by definition, $\mathcal{F}_m$ generates a subvariety of $\mathcal{V}$ which contains $A$, since $A$ is a quotient of $\mathcal{F}_m$, hence it equals to $\mathcal{V}$. If $\mathcal{F}_m$ is representable, by Corollary 11.5.18 it is PI equivalent to a finite-dimensional algebra over $F$, hence the claim. □

After showing this, we will prove the analogous result for superalgebras. Its proof goes along the same lines. The main difficulty lies in the fact that such an algebra is not Noetherian in general, so the results of §11.5 do not apply. In fact we have seen in example 11.5.23 a finitely generated PI algebra which is **not** representable. Instead we wish to prove that *the relatively free algebras* in the variety are representable. In more detail we will proceed in several steps.

**First step.**

LEMMA 17.1.3. *A T-ideal $\Gamma$ is the ideal of PIs of a finitely generated algebra if and only if it contains the ideal of the identities of some finite-dimensional algebra $A$.*

PROOF. This is a consequence of Corollary 16.3.4. □

**Second step.**

We introduce certain numerical invariants for finitely generated algebras, in particular the *Kemer index* of a $T$-ideal $\Gamma$ containing some Capelli identity, cf. Definition 17.2.2. This index consists of a pair of nonnegative integers, denoted by $\mathrm{ind}(\Gamma) := (\beta(\Gamma), \gamma(\Gamma))$. Hence we can define the Kemer index of a finitely generated algebra $A$, which we also denote by $\mathrm{ind}(A) := (\beta(A), \gamma(A))$, by setting $\mathrm{ind}(A) := \mathrm{ind}(\Gamma)$, where $\Gamma := \mathrm{Id}(A)$.

After ordering lexicographically the indices, we will show that if $\Gamma_1 \subseteq \Gamma_2$, then $\mathrm{ind}(\Gamma_1) \geq \mathrm{ind}(\Gamma_2)$.

Strict inclusion does not imply $\mathrm{ind}(\Gamma_1) > \mathrm{ind}(\Gamma_2)$, nonetheless in this case we will define certain polynomials in $\Gamma_2 \setminus \Gamma_1$, which we will call *Kemer polynomials*, cf. Definition 17.2.5. These will be useful in our analysis for distinguishing $T$-ideals, thanks to the combinatorial properties that they satisfy.

The proof of the theorem is then carried out by induction on the Kemer index.

**Third step.** We will show that a finite-dimensional algebra $A$ satisfies the same identities of a direct sum $A_1 \oplus \cdots \oplus A_m$ of finite-dimensional algebras, called fundamental or basic. For a fundamental algebra $A$ with Jacobson radical $J$, the Kemer index has a simple interpretation (cf. Definitions 17.2.26 and 17.2.31), that is we have $\beta(A) = \dim_F A/J$ and $J^{\gamma(A)} \neq 0, J^{\gamma(A)+1} = 0$.

Hence, given a finite-dimensional algebra $A$ with $\mathrm{Id}(A) \subseteq \Gamma$, chosen fundamental algebras $A_i$ such that $A$ is PI equivalent to $A_1 \oplus \cdots \oplus A_m$, we have $\mathrm{Id}(A) = \bigcap \mathrm{Id}(A_i) \subseteq \Gamma$. Finally $\mathrm{ind}(\Gamma) \leq \mathrm{ind}(A) = \max_i \mathrm{ind}(A_i)$.

We then introduce trace algebras and see that in a precise sense, Kemer polynomials *absorb traces*. This allows us to construct further finite-dimensional algebras whose identities are contained in $\Gamma$, and finally we arrive at producing a finite-dimensional algebra $A$ with

$$\mathrm{Id}(A) \subset \Gamma, \quad \mathrm{ind}(A) = \mathrm{ind}(\Gamma),$$

and the Kemer polynomials of sufficiently high degree of $A$ are the same as those of $\Gamma$ (Proposition 17.3.15).

This will allow us to prove that Kemer polynomials satisfy the *Phoenix property*, Theorem 17.3.18.

**Fourth step.** If we call $\Gamma'$ the $T$-ideal generated by $\Gamma$ and by the Kemer polynomials described in the previous step, we will have $\mathrm{ind}(\Gamma') < \mathrm{ind}(\Gamma)$. Hence, by induction, there exists a finite-dimensional algebra $A'$ such that $\mathrm{Id}(A') = \Gamma'$. Using the Phoenix property in Proposition 17.4.3, we construct an auxiliary representable algebra $B$ satisfying the identities of $\Gamma$ and such that all $\mu$-Kemer polynomials for $\Gamma$ with $\mu$ large are not identities of $A$.

One then finally has that $\Gamma$ is the ideal of identities of $A' \oplus B'$ where $B'$ is PI equivalent to $B$ and finite dimensional over $F$.

This completes the proof of Theorem 17.1.1.

**Fifth step.** At this point we will give the "superanalogue" of all the previous steps.

## 17.2. Kemer's theory

**17.2.1. Kemer index and polynomials.** In this paragraph we want to introduce certain integers associated to $T$-ideals or to PI algebras and to study some of their properties.

We start by associating two integers $\beta(\Gamma), \gamma(\Gamma)$ to a $T$-ideal $\Gamma$.

We always restrict to $T$-ideals which contain all multilinear polynomials alternating in at least $m$ variables for some $m$. When we work with algebras with a 1 and the corresponding $T$-ideals in the free algebra with a 1, this is equivalent to requiring that the $T$-ideal contains some Capelli polynomial $C_m$.

When this hypothesis is satisfied, the $T$-ideal is determined by its intersection with the free algebra in $(m-1)$-variables. Thus these $T$-ideals appear as $T$-ideals of identities of finitely generated algebras (Theorem 7.6.1).

DEFINITION 17.2.1. A polynomial $f(X_1, \dots, X_\mu, Y)$ is a $\mu$-fold $t$-alternating polynomial if it is multilinear and alternating in $\mu$ disjoint sets of variables $X_i$, with $|X_i| = t$ for all $i$, and it may depend on other variables $Y$.

DEFINITION 17.2.2. For every $T$-ideal $\Gamma$ let $\beta(\Gamma)$ be the greatest integer $t$ such that *for every* $\mu$ there exists a $\mu$-fold $t$-alternating polynomial

$$f(X_1, \dots, X_\mu, Y) \notin \Gamma.$$

We call $\gamma(\Gamma)$ the maximum $s$ for which there exists, for all $\mu \in \mathbb{N}$, a polynomial

$$f(X_1, \dots, X_\mu, Z_1, \dots, Z_s, Y) \notin \Gamma,$$

alternating in $\mu$ layers $X_i$ with $\beta(\Gamma)$ elements and in $s$ layers $Z_j$ with $\beta(\Gamma) + 1$ elements.

The pair $(\beta(\Gamma), \gamma(\Gamma))$ is the *Kemer index* of $\Gamma$ denoted $\operatorname{Ind}\Gamma$.

For a finitely generated algebra $A$ we define the Kemer index of $A$, denoted by $\operatorname{Ind}(A)$, to be the Kemer index of the ideal of polynomial identities of $A$.

See Example 17.2.8. We order the Kemer indices lexicographically and then observe that

REMARK 17.2.3. If $\Gamma_1 \subset \Gamma_2$ are two $T$-ideals, we have

$$\operatorname{Ind}\Gamma_1 \geq \operatorname{Ind}\Gamma_2.$$

If $\Gamma = \bigcap_{i=1}^k \Gamma_i$ is an intersection of $T$-ideals, then

$$\operatorname{Ind}\Gamma = \max \operatorname{Ind}\Gamma_i.$$

If $A = \bigoplus_{i=1}^k A_i$ is a direct sum of algebras, then

$$\operatorname{Ind} A = \max \operatorname{Ind} A_i$$

REMARK 17.2.4. Denote by $s := \gamma(\Gamma)$. By definition there is a minimum $\mu_0 = \mu_0(\Gamma)$ such that every polynomial $f(X_1, \dots, X_\mu, Z_1, \dots, Z_{s+1}, Y)$, alternating in $\mu \geq \mu_0$ layers $X_i$ with $\beta(\Gamma)$ elements and in $s+1$ layers $Z_j$ with $\beta(\Gamma)+1$ elements, lies in $\Gamma$.

DEFINITION 17.2.5. A polynomial $f(X_1,\dots,X_\mu,Z_1,\dots,Z_s,Y)\notin\Gamma$ alternating in $\mu$ layers $X_i$, with $\mu>\mu_0$, with $t=\beta(\Gamma)$ elements and in $s=\gamma(\Gamma)$ layers $Z_j$ with $t+1=\beta(\Gamma)+1$ elements will be called a *$\mu$-Kemer polynomial*.

REMARK 17.2.6. Let $\Gamma_1\subset\Gamma_2$ be two $T$-ideals with the same Kemer index. For $\mu$ large any $\mu$-Kemer polynomial for $\Gamma_2$ is also a $\mu$-Kemer polynomial for $\Gamma_1$.

REMARK 17.2.7. Take a $\mu$-Kemer polynomial $f(X_1,\dots,X_\mu,Z_1,\dots,Z_s,Y)$ which is also linear in a variable $w$ not in the layers $X_i,Z_j$. If we fix one of the layers $X_i$, add to it the variable $w$, and alternate in $f$ these $t+1$ variables, we produce a polynomial $\bar f$ alternating in $\mu-1\geq\mu_0$ layers with $t$ elements and in $s+1$ layers with $t+1$ elements. By the definition of $\mu_0$ the polynomial $\bar f$ is always an element of $\Gamma$.

Similarly take a $\mu$-Kemer polynomial $f(X_1,\dots,X_\mu,Z_1,\dots,Z_s,Y)$ with $\mu>\mu_0+1$. Take two of the layers $X_i,X_j$ and take a variable $w$ out of the layer $X_j$. When we alternate the $t+1$ variables $(X_i,w)$, we necessarily obtain always an element of $\Gamma$.

In fact in this polynomial we have now added to the $s$ layers $Z_j$, alternating in $t+1$ variables, a new layer, getting $s+1$ layers alternating in $t+1$ variables. We still keep the $\mu-2\geq\mu_0$ layers $X_h$, $h\neq i,j$ alternating in $t$ variables.

EXAMPLE 17.2.8. If $A=M_n(F)$, then every $(n^2+1)$-alternating polynomial is in $\mathrm{Id}(A)$. Conversely, using Capelli polynomials, defined by formula (7.5), by Remark 10.2.13[1] we have

$$C_{\mu,n^2}(X_1,\dots,X_\mu,Y):=\prod_{i=1}^{\mu}C_{n^2}(x_{i,1},\dots,x_{i,n^2},y_{i,1},\dots,y_{i,n^2},y_{i,n^2+1}),\tag{17.1}$$

$$X_i:=\{x_{i,1},\dots,x_{i,n^2}\}$$

evaluated in suitable $e_{ij}$'s can take any value $e_{hk}$ for every $\mu$ so, in particular, it is not an identity. Hence $\beta(M_n(F))=n^2$, $\gamma(M_n(F))=0$.

These polynomials are Kemer for $M_n(F)$. On the other hand there are even *minimal degree* Kemer polynomials, in the sense that they depend only on the variables of the layers in which they are alternating. These are the central polynomials of Regev and Formanek, Corollary 12.3.19, $F_{k,n_i}(X):=F_{k,n_i}(X_1^{(i)},\dots,X_k^{(i)})$, $k\geq 2$ which take as value the product $\prod_{j=1}^k\Delta(X_j^{(i)})$.

As we already said, the proof will be carried out by induction, so let us see the basis of induction.

PROPOSITION 17.2.9. *A T-ideal $\Gamma\subset F\langle X\rangle$ has Kemer index $(0,n)$ if and only if $F\langle X\rangle/\Gamma$ is nilpotent of degree $n+1$.*

PROOF. This is indeed immediate, since for a layer with just one variable the condition to be alternating is void. So Kemer index $(0,n)$ means that there are multilinear polynomials of degree $n$ not in $\Gamma$ but all multilinear polynomials of degree $n+1$ are in $\Gamma$.

Notice then that if $X$ is finite, the algebra $F\langle X\rangle/\Gamma$ is finite dimensional. □

---

[1]Here the product is taken in an ordered way.

**17.2.2. An interpretation through cocharacters.** We discuss a connection between Kemer index and cocharacters.

Using the notations of §7.1.1, denote $V_n$ the space of multilinear polynomials in $n$-variables identified with the group algebra $F[S_n]$.

Let $\lambda \vdash n$ be a diagram with columns $k_1, \dots, k_i$, let $T$ be a tableau of shape $\lambda$, let $C_T \equiv \prod_j S_{k_j}$ be its column subgroup, and let $A_T = \frac{1}{\prod_j k_j!} a_T$, $a_T = \sum_{g \in C_T} \epsilon_g g$ be the corresponding antisymmetrizer. We have $A_T^2 = A_T$.

Recall that, for two partitions of $n$ with at most $d$ rows $\mu = (k_1, \dots, k_d)$ and $\nu = (h_1, \dots, h_d)$, we may consider the dominance order $\mu \geq \nu$. That is

$$k_1 + k_2 + \cdots + k_a \geq h_1 + h_2 + \cdots + h_a$$
$$\iff \quad k_d + k_{d-1} + \cdots + k_{d-a} \leq h_d + h_{d-1} + \cdots + h_{d-a}, \ \forall a.$$

First one has

REMARK 17.2.10. Let $T$ be a tableau of shape $\lambda \vdash n$. The shapes $\mu$ for which we have $A_T M_\mu \neq 0$ are all the partitions $\mu \leq \lambda$ in the dominance order.

PROOF. First one observes that the condition $A_T M_\mu \neq 0$ is equivalent to $A_T F[S_n] e_\mu \neq 0$ ($e_\mu$ the Young symmetrizer). This, by applying the anti-isomorphism $\tau : \sigma \mapsto \sigma^{-1}$, is also equivalent to $\tau(e_\mu) F[S_n] A_T \neq 0$. Since $\tau(e_\mu)$ is also a Young symmetrizer for $M_\mu$, this is equivalent to the fact that $M_\mu$ appears in the decomposition of $F[S_n] A_T$. By Young's rule (Theorem 6.2.10) and the dual of Remark 6.2.11, this is equivalent to $\mu \leq \lambda$. □

It follows that in a finite-dimensional representation $P = \bigoplus_{\mu \vdash n} m_\mu M_\mu$, we have a nonzero element $f \in P$ with $A_T f = f$ if and only if there is a $\mu \leq \lambda$ with $m_\mu \neq 0$.

We may apply this to Kemer's polynomials. For each $n$, take an $S_n$ stable complement $P_n$ to $F[S_n] \cap \mathrm{Id}(A)$, hence $P_n \oplus F[S_n] \cap \mathrm{Id}(A) = F[S_n]$.

Decompose $P_n = \bigoplus_{\mu \vdash n} m_\mu M_\mu$ with $m_\mu$ the cocharacters.

Let us use the following notation. Given a partition $\lambda \vdash m$ and $n > m$, denote by $\lambda(n) \vdash n$ the partition which coincides with $\lambda$ on all rows except the first one, where $\lambda(n)_1 = \lambda_1 + (n - m)$.

Given two integers $t, s \in \mathbb{N}$, let $\nu_{t,s,k} = \nu$ be a partition with conjugate partition

$$\check{\nu} = ((t+1)^s, t^k) \tag{17.2}$$

and thus rows, $\nu_i = 0$, $\forall i > t+1$, $\nu_{t+1} = s$ and $\nu_i = k + s$, $\forall\, 1 \leq i \leq t$.

DEFINITION 17.2.11. We will say that, a nonzero multilinear polynomial

$$f = f(X_1, \dots, X_k, Z_1, \dots, Z_s, Y) \in P_n$$

is of type $\nu_{t,s,k}(n)$ if the disjoint sets of variables $X_1, \dots, X_k, Z_1, \dots, Z_s, Y$ satisfy

$$|X_1| = \cdots = |X_k| = t, |Z_1| = \cdots = |Z_s| = t + 1,$$

and $f$ is alternating on each set $\{X_i\}$, and $Z_j$ with $1 \leq i \leq k$; $1 \leq j \leq s$.

PROPOSITION 17.2.12. *If $f \in P_m$ is a nonzero polynomial of type $\nu_{t,s,k}(n)$, then $f$ is a sum of polynomials $f_i$ of type $\nu_{t,s,k}(n)$, each generating in $P_n$ an irreducible representation of $S_n$ isomorphic to some $M_{\lambda_i}$, with $\lambda_i \leq \nu_{t,s,k}(n)$.*

*Conversely if $\lambda \leq \nu_{t,s,k}(n)$, $M_\lambda$ contains nonzero polynomials of type $\nu_{t,s,k}(n)$.*

PROOF. Consider a diagram $\nu = \nu_{t,s,k}(n)$ previously defined in this subsection. We obtain a tableau $T$ by placing in each column of $\nu$, of length $t+1$, the indices of the blocks $Z_i$ and then on the columns of length $t$ those of the blocks $X_i$. The condition of antisymmetry for the given blocks of $f$ is that $A_T f = f \iff a_T f = cf$, with $a_T$ the antisymmetrizer in the columns of $T$ and $c = (t+1)!^s t!^k$.

Now we apply Remark 17.2.10 to the decomposition $P_n = \bigoplus_{\mu \vdash n} m_\mu M_\mu$ into irreducibles $M_\mu$, and this implies that $f \in \bigoplus_{\mu \vdash n,\ \mu \leq \nu} m_\mu M_\mu$.

Conversely if we have a copy of $M_\mu$, $\mu \leq \nu$ in $P_n$, we have an element $f \in M_\mu$, $f \neq 0, A_T f = f$. This is a polynomial of type $\nu_{t,s,k}(n)$. □

COROLLARY 17.2.13. *The following statements are equivalent for an algebra $A$.*

(1) *There is a multilinear polynomial of type $\nu_{t,s,k}(n)$ nonvanishing in $A$.*
(2) *There is a partition $\lambda \vdash n$ with $\lambda \leq \nu_{t,s,k}(n)$ in the dominance order and for the cocharacter $m_\lambda(A) \neq 0$.*

Let $A$ be an algebra satisfying a Capelli identity. Then we know that its cocharacters are contained in a strip of some height $d$.

THEOREM 17.2.14. *The algebra $A$ satisfying a Capelli identity has Kemer index $(t,s)$ $(t > 0)$ if and only if the two following conditions are satisfied:*

(1) *There is an integer $\mu_0$ so that if $\lambda \vdash n$ is a partition with $\lambda_t \geq \mu_0 + s + 1$ and $m_\lambda(A) \neq 0$, then $\lambda_{t+1} + \cdots + \lambda_d \leq s$.*
(2) *For every $k$, there is a partition $\lambda \vdash n$ with $\lambda_t \geq k + s + 1$ and $\lambda_{t+1} + \cdots + \lambda_d = s$ so that $m_\lambda(A) \neq 0$.*

PROOF. From Corollary 17.2.13, if (2) is satisfied, then for every $k$ there are nonvanishing polynomials $f$ alternating in $k$ layers with $t$ elements and $s$ with $t+1$ elements.

So the Kemer index is $(t,s)$, and the previous $f$ are Kemer polynomials, provided we show that for $k$ large there is no nonvanishing polynomial alternating in $k$ layers with $t$ elements and in $s+1$ layers with $t+1$ elements.

If one such element exists, then $m_\lambda(A) \neq 0$ for partitions $\lambda \vdash n$ which are $\leq \nu_{t,s+1,k}(n)$ for all $k$. In particular for $i = t+1$, we have $\sum_{j=t+1}^d \lambda_j \geq \nu_{t+1} = s+1$.

If we assume (1), we have that $\lambda_t < \mu_0 + s + 1$.

But for $i = t$, we have $(d-t+1)\lambda_t \geq \sum_{j=t}^d \lambda_j \geq \nu_t + \nu_{t+1} = k + 2s + 2$. This implies that $\lambda_t \geq \frac{k+2s+2}{d-t+1}$, so $k + 2s + 2 < (d-t+1)(\mu_0 + s + 1)$ puts a bound on $k$.

(1) Assume now that $A$ has Kemer index $(t,s)$. Take $\mu_0$ as in Remark 17.2.4. If $\lambda_t \geq \mu_0 + s + 1$, then also $\lambda_i \geq \mu_0 + s + 1$, $\forall i \leq t$. If furthermore $\lambda_{t+1} + \cdots + \lambda_d > s$, we see that $\lambda \leq \nu_{t,s+1,\mu_0}(n)$ and then by Lemma 17.1.3 $m_\lambda \neq 0$. This implies the existence of a polynomial contradicting the choice of $\mu_0$ as in Remark 17.2.4.

(2) Under the previous conditions assume $\lambda_t \geq \mu_0 + s + 1$ and $\lambda \geq \nu_{t,s,\mu_0}(n)$. Then $\lambda_{t+1} + \cdots + \lambda_d \geq s$ so, by (1), $\lambda_{t+1} + \cdots + \lambda_d = s$. □

REMARK 17.2.15. In the construction of Kemer polynomials, Theorem 17.2.41, the number of extra variables $Y$ can be kept bounded by some fixed number $r$.

COROLLARY 17.2.16. *The partition $\lambda$ with $m_\lambda \neq 0$ of Theorem 17.2.14 and Remark 17.2.10 contains the rectangle $t \times (s+k)$. Observe that outside this rectangle we have $\leq s + r$ boxes, a fixed number.*

**17.2.3. Identities of finite-dimensional algebras.** Infinitely many different, that is nonisomorphic, algebras may have the same polynomial identities, and in general there is no clear representative to choose. We will see that every finite-dimensional algebra is PI equivalent to a finite direct sum of some special algebras called *fundamental* (or *basic*); see Definition 17.2.26. They have the property that their Kemer index has a direct interpretation from the algebra structure.

17.2.3.1. *Split algebras.* Let $A$ be a finite-dimensional algebra. We can decompose $A = \bar{A} \oplus J$, where $\bar{A}$ is semisimple (isomorphic to $A/J$) and $J$ is the radical. The algebra $\bar{A}$ is the direct sum of simple algebras $\bar{A}_i$.

DEFINITION 17.2.17. We say that $A$ is *split* if each $\bar{A}_i = M_{n_i}(F)$ is a matrix algebra over $F$.

THEOREM 17.2.18. *Every algebra $A$ finite dimensional over an infinite field $F$ is PI equivalent, over $F$, to a split algebra.*

PROOF. Given any finite-dimensional algebra $A$, there is a finite-dimensional field extension $F \subset G$ so that $\overline{A \otimes_F G} = \bigoplus_i M_{n_i}(G)$ is a direct sum of matrix algebras over $G$ and $A \otimes_F G$ is PI equivalent to $A$ over $F$. Then split $A \otimes_F G = \bigoplus_i M_{n_i}(G) \oplus J$. The $F$ subalgebra $B := \bigoplus_i M_{n_i}(F) \oplus J$ is finite dimensional over $F$ and clearly PI equivalent to $A \otimes_F G$ and so also to $A$. □

17.2.3.2. *Elementary evaluations.* Since our objective is to study algebras up to PI equivalence, from now on when we consider a finite-dimensional algebra $A$ over $F$, we always assume that it is split.

First of all we fix some notation.

Let $J :=$ rad $A$ be the Jacobson radical of $A$, and let $\bar{A} := A/J$.

By Wedderburn's principal theorem (Theorem 1.3.13), if $A$ is a finite-dimensional algebra, we can decompose $A = \bar{A} \oplus J$ where $\bar{A}$ is semisimple (isomorphic to $A/J$) and $J$ is the radical.

Since $A$ is split, the semidirect sum decomposition $A = \bar{A} \oplus J$ given by Wedderburn's principal theorem has the form:

$$A = \bar{A} \oplus J = R_1 \oplus \cdots \oplus R_q \oplus J, \tag{17.3}$$

where, for every $i$, $R_i$ is isomorphic to $M_{n_i}(F)$ for some $n_i$. We denote by $e_i$ the identity of $R_i$. Clearly if $i \neq j$, we have $e_i \bar{A} e_j = 0$, hence $e_i A e_j = e_i J e_j$.

If $A$ is an algebra with a 1, we have $1 = \sum_i e_i$ and we can consider the *Pierce decomposition*,

$$A = \bigoplus_{i,j=1}^{q} e_i A e_j = \bigoplus_{i=1}^{q} e_i A e_i \bigoplus \bigoplus_{i \neq j=1}^{q} e_i J e_j. \tag{17.4}$$

If $A$ does not have a 1 and we add 1 to $A$, we still may consider the idempotent $e_0 := 1 - \sum_{i=1}^{q} e_i$ in $A \oplus \cdot 1$. We have still

$$A = \bigoplus_{i,j=0}^{q} e_i A e_j = \bigoplus_{i=0}^{q} e_i A e_i \bigoplus \bigoplus_{i \neq j=0}^{q} e_i J e_j. \tag{17.5}$$

REMARK 17.2.19. Observe that, if we add a 1 to an algebra $A$ which already has a 1, call it $1_A$, we have $\sum_{i=1}^{q} e_i = 1_A$ and we get $e_0 A = A e_0 = 0$.

We can visualize this with a graph, which we shall call the *Pierce graph* of $A$.

DEFINITION 17.2.20. The *Pierce graph* of $A$ has vertices $\overset{i}{\circ}$ corresponding to the indices $i \in \{0, 1, \ldots, q\}$ (or $i \in \{1, \ldots, q\}$ if $A$ has a 1). Two distinct vertices $i, j$ are connected by an oriented edge $\overset{i}{\circ}\!\longrightarrow\!\overset{j}{\circ}$ if and only if $e_i A e_j \neq 0$.

EXAMPLE 17.2.21. Let $A$ be the algebra of upper triangular $q \times q$ matrices over $F$. Then $\bar{A} = \bigoplus_{i=1}^{q} F e_{i,i}$ is the algebra of diagonal matrices, $e_i = e_{i,i}$, and $J$ are strictly upper triangular with 0 on the diagonal. The Pierce graph is

$$\overset{1}{\circ}\longrightarrow\overset{2}{\circ}\longrightarrow\overset{3}{\circ}\longrightarrow\overset{4}{\circ}\cdots\cdots\overset{q-2}{\circ}\longrightarrow\overset{q-1}{\circ}\longrightarrow\overset{q}{\circ}.$$

REMARK 17.2.22. Notice that if we have a nonzero product $a_1 a_2 \cdots a_m \neq 0$ and $a_i \in e_{k_i} A e_{h_i}$, we must have $h_i = k_{i+1}$, $\forall i = 1, 2, \ldots, m-1$, and these two indices are joined by an oriented edge. Such a product corresponds to an *oriented path* in this graph.

To $a_i$ we associate the oriented edge $\overset{k_i}{\circ}\!\longrightarrow\!\overset{h_i}{\circ}$, where if $k_i = h_i = e$, we represent this part of the path by a loop at the vertex $e$.

Given a multilinear polynomial $f(x_1, \ldots, x_t, y_1, \ldots, y_m)$ every evaluation of $f$ in $A$ is a sum of evaluations in which each variable is computed in one of the summands $e_i A e_j$ of the Pierce decomposition. Finally if $R_i = M_{n_i}(F)$, we can further decompose each $e_i = \sum_{j=1}^{n_i} e_{j,j}^{(i)}$ and this refines the Pierce decomposition

$$e_i A e_k = \sum_{j=1}^{n_i} e_{j,j}^{(i)} A \sum_{h=1}^{n_k} e_{h,h}^{(k)} = \sum_{j=1}^{n_i} e_{j,j}^{(i)} \bar{A} \sum_{h=1}^{n_k} e_{h,h}^{(k)} \oplus \sum_{j=1}^{n_i} e_{j,j}^{(i)} J \sum_{h=1}^{n_k} e_{h,h}^{(k)}.$$

Moreover recall that if $j \neq k$, we have $e_j \bar{A} e_k = 0 \implies e_j A e_k \subseteq J$, hence $e_j A e_k = e_j J e_k$.

DEFINITION 17.2.23. An evaluation $\eta : F\langle \mathbb{X} \rangle \to A$, in which each element $\eta(x_i)$ is in one of the summands $e_{j,j}^{(i)} \bar{A} e_{h,h}^{(i)}$ or in $e_{j,j}^{(i)} J e_{h,h}^{(k)}$ of the Pierce decomposition, will be called *elementary*.

For a variable $x$, if $\eta(x) \in \bar{A}$ (resp., if $\eta(x) \in J$), we will say that $x$ is a *semisimple variable* (resp., a *radical variable*) for the evaluation $\eta$.

A polynomial $f(x_1, \ldots, x_t, y_1, \ldots, y_m)$, linear in the $x_i$'s is an identity of $A$, if it vanishes under evaluation of every $x_i$ in $\bar{A}$ or in $J$. We will call the first one a **semisimple evaluation** and the second one a **radical evaluation**. We may also restrict to elementary evaluations.

For the semisimple evaluations we can assume that every element is a matrix unit of some simple component.

Using either (17.4) or (17.5), we can assume that every radical evaluation is in $e_i J e_k$ for suitable $i$ and $k$, and we will say that such **evaluation** is **of type** $(i, k)$.

An elementary radical evaluation of type $(i, k)$ is an evaluation in $e_{j,j}^{(i)} J e_{h,h}^{(k)}$ for some $j, h$.

DEFINITION 17.2.24. Set $t_A := \dim_F A/J = \dim_F \bar{A}$, and let $s_A + 1$ be the nilpotency index of $J$, that is $J^{s_A} \neq 0$, $J^{s_A+1} = 0$.

We call the pair $t_A, s_A$ the *$(t, s)$-index of $A$*.

PROPOSITION 17.2.25.

$$\operatorname{Ind} A = (\beta(A), \gamma(A)) \leq (t_A, s_A) \quad \textit{in lexicographic order.}$$

PROOF. Let $f(x_1, \ldots, x_t, y_1, \ldots, y_m)$ be a polynomial alternating in the $t$ variables $x_i$'s. If $f$ does not vanish through the evaluations $x_i \mapsto \bar{x}_i$ and these evaluations are all semisimple, that is $\bar{x}_i \in \bar{A}$, then we must have $t \leq \dim_F \bar{A}$.

In other words, if $t > \dim_F \bar{A}$, then any nonzero evaluation of $f$ must involve a radical evaluation for at least one of $x_1, \ldots, x_t$.

Now we may have at most $s_A$ radical evaluations, otherwise all the products are 0. In particular we cannot have more than $s_A$ layers alternating in $> t_A$ variables, so $\beta(A) \leq t_A$. If $\beta(A) = t_A$, by the same reason $\gamma(A) \leq s_A$. □

DEFINITION 17.2.26. We say that a finite-dimensional algebra $A$ is *basic* if its Kemer index coincides with the $(t,s)$-index.

Let $(\beta(A), \gamma(A)) = \operatorname{Ind} A$ be the Kemer index of $A$, that is of the $T$-ideal of identities of $A$. We have seen in Proposition 17.2.25 that $\operatorname{Ind} A \leq (t_A, s_A)$ in lexicographic order. The main result of this section, Proposition 17.2.34, is that there is a a finite-dimensional algebra $A' = \bigoplus_i A'_i$, which is PI equivalent to $A$ and for which $(t_{A'_i}, s_{A'_i}) = \operatorname{Ind} A'_i$, $\forall i$.

17.2.3.3. *A free finite-dimensional algebra.* Let $A = \bar{A} \oplus J$ be a finite-dimensional algebra decomposed into its semisimple part and nilpotent radical. Let $h := \dim_F J$, and let $t_A, s_A$ be its $(t,s)$-index, i.e., $t_A = \dim_F \bar{A}$, $J^{s_A} \neq 0$, $J^{s_A+1} = 0$, of course $s_A \leq h$.

Consider the free product $A' := \bar{A} * F_+\langle x_1, \ldots, x_h\rangle$ (cf. Definition 2.1.17) in the category of algebras without 1. The algebra $A' = \bar{A} \oplus \mathbb{J}$, and $\mathbb{J}$ is the ideal generated by the variables $x_i$. Furthermore $A'$ is a graded algebra with $\bar{A}$ its degree 0 part, the $x_i$ in degree 1, and $\mathbb{J}$ the graded ideal of elements of degree $\geq 1$.

DEFINITION 17.2.27. Define $\mathcal{A}_{s_A}$ to be $A'/(\mathbb{J}^{s_A+1} + I)$ with $I$ the verbal ideal (Definition 2.2.46) generated by the evaluation in $A'$ of all the PIs of the algebra $A$.

Observe that the PIs of $A$ vanish on $\bar{A}$ so $I \subset \mathbb{J}$. Furthermore $I$ is generated by the evaluations of multilinear identities so that $I$ (and hence also $\mathbb{J}^{s_A+1} + I$) are graded ideals. Hence $\mathcal{A}_{s_A}$ is a graded algebra.

By definition of free product, given elements $\underline{a} := \{a_1, \ldots, a_h\}, a_i \in A$, there is a unique map $\pi_{\underline{a}} : A' \to A$ which is the identity on $\bar{A}$ and maps $x_i$ to $a_i$.

LEMMA 17.2.28. *If $\underline{a} := \{a_1, \ldots, a_h\}, a_i \in J$, the map $\pi_{\underline{a}} : A' \to A$ factors through the ideal $\mathbb{J}^{s_A+1} + I$ to a homomorphism $\pi_{\underline{a}} : \mathcal{A}_{s_A} \to A$.*

PROOF. By definition $I$ maps to 0 in $A$. Since $a_i \in J$ we have that $\mathbb{J}$ maps to $J$ so $\mathbb{J}^{s_A+1}$ maps to $J^{s_A+1} = 0$. □

LEMMA 17.2.29.

(i) *$A$ is a quotient of $\mathcal{A}_{s_A}$.*
(ii) *The algebra $\mathcal{A}_{s_A}$ is finite dimensional and satisfies the same PIs as $A$.*
(iii) *The $(t,s)$-index of $\mathcal{A}_{s_A}$ equals that of $A$.*

PROOF. (i) For the first statement choose the $\underline{a} := \{a_1, \ldots, a_h\}$ generators of $J$. Then $\pi_{\underline{a}} : \mathcal{A}_{s_A} \to A$ is surjective.

(ii) $\mathcal{A}_{s_A}$ is finite dimensional since it is a quotient of $A'/\mathbb{J}^{s_A+1}$ which is finite dimensional by construction. By construction $\mathcal{A}_{s_A}$ satisfies all the PIs of $A$. Since $A$ is a quotient of $\mathcal{A}_{s_A}$ the converse is also true.

(iii) By construction $I \subset \mathbb{J}$ implies (still denoting by $\mathbb{J}$ the same ideal in $\mathcal{A}_{s_A}$) that $\mathcal{A}_{s_A}/\mathbb{J} = \bar{A}$. Finally since we have a surjective map $\mathcal{A}_{s_A} \to A$, we have $\mathbb{J}^{s_A} \neq 0$, so the $(t,s)$-index of $\mathcal{A}_{s_A}$ equals that of $A$. □

By construction the algebra $\mathcal{A}_{s_A}$ satisfies the following *universal property*:

PROPOSITION 17.2.30. *Let $f(a_1, \dots, a_m)$ be an elementary evaluation of a multilinear polynomial $f$ in $A$, with $a_1, \dots, a_k$ semisimple variables and the remaining radical variables. This evaluation factors through the evaluation*

$$f(a_1, \dots, a_k, b_1, \dots, b_{m-k}) \in \mathcal{A}_{s_A}, \ b_i \quad \textit{of degree } 1,$$

*where under the homomorphism $\pi_{\underline{a}} : \mathcal{A}_{s_A} \to A$, which is surjective from the elements of degree 1 to $J$, the elements $b_i$ map to the elements $a_{k+i}$.*

We now set $A_0 := \mathcal{A}_{s_A}/\mathbb{J}^{s_A}$, and we have a PI algebra of $(t,s)$-index $(t_A, s_A - 1)$ which satisfies all the PIs of $A$ and possibly other PIs. It is a *PI-simplification of* $A$.

**17.2.4. Reduced and fundamental algebras.** With the previous notations (see formula (17.3)) consider the quotient map $\pi : A \to A/J = \bigoplus_{i=1}^q R_i$, and for $1 \leq j \leq q$ let

$$R^{(j)} := \bigoplus_{i \neq j,\, i=1}^{q} R_i \quad \text{and} \quad A_j := \pi^{-1} R^{(j)}. \tag{17.6}$$

We have

$$\mathrm{Id}(A) \subset \bigcap_{i=1}^{q} \mathrm{Id}(A_i) \cap \mathrm{Id}(A_0).$$

If we have that $\mathrm{Id}(A) = \bigcap_{i=1}^q \mathrm{Id}(A_i) \cap \mathrm{Id}(A_0)$, then $A$ is PI equivalent to $\bigoplus_{i=1}^q A_i \oplus A_0$. Notice that all the $(t,s)$-indices of these algebras are strictly less than the $(t,s)$-index of $A$, in fact $t_{A_i} < t_A$ while $t_{A_0} = t_A$, $s_{A_0} = s_A - 1$.

DEFINITION 17.2.31.

(1) We say that $A$ is *reduced* or, equivalently, *full* if either $A$ is simple or

$$\mathrm{Id}(A) \subsetneq \bigcap_{i=1}^{q} \mathrm{Id}(A_i). \tag{17.7}$$

A polynomial $f \in \bigcap_{i=1}^q \mathrm{Id}(A_i) \setminus \mathrm{Id}(A)$ will be called *full*.

(2) We say that $A$ is *fundamental* if

$$\mathrm{Id}(A) \subsetneq \bigcap_{i=1}^{q} \mathrm{Id}(A_i) \cap \mathrm{Id}(A_0). \tag{17.8}$$

A polynomial $f \in \bigcap_{i=1}^q \mathrm{Id}(A_i) \cap \mathrm{Id}(A_0) \setminus \mathrm{Id}(A)$ will be called *fundamental*.

Of course a fundamental algebra is also reduced and a fundamental polynomial, in particular, is full.

REMARK 17.2.32. Let $f$ be a fundamental polynomial for $A$. If $a$ is in the $T$-ideal generated by $f$ and not an identity of $A$, then $f$ is also a fundamental polynomial for $A$.

EXAMPLE 17.2.33. In Remark 17.2.40 we will show that an algebra of block triangular matrices is fundamental.

We have already discussed, in Definition 16.3.1, the matrix algebra $UT(d_1, d_2, \dots, d_m)$, defined as the algebra fixing a *partial flag* of subspaces $V_i$. Its radical is the ideal of elements mapping $V_i$ to $V_{i-1}$, an ideal nilpotent of degree $m$. A semisimple part can be chosen by decomposing $V_i = W_i \oplus V_{i-1}$ and considering the semisimple algebra of block diagonal matrices $\bigoplus_{i=1}^m \operatorname{End}(W_i)$.

Its $(t,s)$-index is thus $\sum_i d_i^2, m-1$, and by part (3) of Remark 17.2.40 this also equals the Kemer index.

In this case there are quite explicit Kemer polynomials given by alternating the polynomial of formula (17.9).

**Warning.** If $A$ is nilpotent, then it is automatically fundamental! Any nonvanishing polynomial is also fundamental. So from now on we will treat nonnilpotent algebras.

PROPOSITION 17.2.34. *Every finite-dimensional algebra $A$ is PI equivalent to a finite direct sum of fundamental algebras.*

PROOF. This is immediate. If $A$ is not fundamental, then it is PI equivalent to $\bigoplus_{i=1}^q A_i \oplus A_0$ a direct sum of algebras with strictly lower $(t,s)$-indices. Then we finish by induction. □

Let us now see the properties of a multilinear polynomial $f(x_1, \dots, x_m)$ to be full or fundamental as a condition on elementary (Definition 17.2.23) evaluations. Let us first make

DEFINITION 17.2.35. We say that a multilinear polynomial, nonvanishing on $A$, has *Property K* if every nonzero elementary evaluation in $A$ has exactly $s_A$ radical evaluations.

Let us call, by abuse of notation, *semisimple* (resp., *radical*) a variable which, in an elementary evaluation, has been substituted to a semisimple (resp., radical) element. Take a multilinear polynomial $f(x_1, \dots, x_m)$ which is not a PI of $A$.

LEMMA 17.2.36.

(1) *The polynomial $f$ is full if and only if, for every nonzero elementary evaluation $f(a_1, \dots, a_m) \neq 0$, we have that for all $i = 1, \dots, q$, there is at least one semisimple variable $x_j$ evaluated in $a_j \in R_i$.*
(2) *If $f$ is full, then it is fundamental if and only if it satisfies property K.*

PROOF. (1) Assume $f$ is full according to Definition 17.2.31. If there is no $a_i$ in some $R_j$, the polynomial $f$ is in fact evaluated in the algebra $A_j$ and hence, by definition of full polynomial, it would be zero, a contradiction. The converse is also clear.

(2) Assume $f$ is fundamental according to Definition 17.2.31. First of all we know that we can have at most $s_A$ radical evaluations since $J^{s_A+1} = 0$, so the issue is to show that we cannot have less. By the universal property of $\mathcal{A}_{s_A}$ (Proposition 17.2.30), this evaluation factors through an evaluation of $f$ in $\mathcal{A}_{s_A}$ mapping the radical variables to corresponding elements $b_i$ (of degree 1). If we have fewer than $s_A - 1$ radical variables, this evaluation gives rise to a nonzero element of degree $\leq s_A - 1 \in \mathcal{A}_{s_A}$. On the other hand, when we take the projection from $\mathcal{A}_{s_A}$ to

$A_0 = \mathcal{A}_{s_A-1}$, this element must map to zero since $f$ is by assumption a PI of $A_0$. This is not possible since by definition $\mathcal{A}_{s_A}$ and $A_0 = \mathcal{A}_{s_A-1}$ coincide up to degree $s_A - 1$. The converse is also clear. □

From this let us see how to prove that for a reduced algebra $A$, the first Kemer index equals $t_A$ and for a fundamental algebra $A$ the Kemer index equals the $(t, s)$-index. We shall do this by constructing directly Kemer polynomials out of fundamental polynomials.

17.2.4.1. *Reduced algebras.*

LEMMA 17.2.37. *Let $A = \bar{A} + J$ be a finite-dimensional algebra with $t = \dim A/J$ and $J^r = 0$. If $f$ is a multilinear polynomial which is alternating in $r$ layers of variables, each with at least $t + 1$ elements, then $f$ is a PI of $A$.*

PROOF. It is enough to show that $f$ vanishes for all elementary evaluations. Suppose this is not the case. Then, in such an evaluation, each layer cannot be evaluated only in the semisimple part, since this has dimension $t$ and every multilinear alternating function in $t + 1$ variables vanishes. So for each layer we have an evaluation in $J$, but then since we have $r$ layers, the evaluation lies in $J^r = 0$, a contradiction. □

PROPOSITION 17.2.38. *Let $A = \bar{A} + J$ be a finite-dimensional algebra. Let $\bar{A} = R_1 \oplus \cdots \oplus R_k$, $k \geq 1$, a direct sum of simple algebras. $A$ is reduced if and only if either $k = 1$ or we have $R_1 J R_2 J \cdots J R_k \neq 0$, possibly after reordering the $R_i$.*

PROOF. If $A$ is reduced, by definition there exists a full polynomial $f$. If we take a nonzero elementary evaluation of $f$, we necessarily have that for each $i$ there is a variable evaluated in $R_i$; see Lemma 17.2.36(1). Then one of the monomials appearing in $f$ is nonzero for this evaluation.

In this monomial some of the variables are substituted by elements in $R_i$, in some order which we may (by abuse of notation) consider the given one. Since if $i \neq j$, we have $R_i R_j = 0$, between an element $a_i \in R_i$ and the next one $a_{i+1} \in R_{i+1}$ there must necessarily be an evaluation in $J$. This implies the claim.

The converse depends on Lemma 17.2.39 in which we shall see that there is a polynomial $f$ which is alternating in as many layers with $t$ elements as we want and is not an identity on $A$. Then by Lemma 17.2.37 this is an identity on all the algebras $A_i$ of formula (17.6), and this proves the claim. □

Let $A = \bar{A} + J$ be a finite-dimensional split algebra over a field $F$,

$$\dim \bar{A} = \dim(R_1 \oplus \cdots \oplus R_k) = t.$$

Assume that $R_1 J R_2 J \cdots J R_k \neq 0$. If $e_i$ denotes the unit element of $R_i$ since $R_i = e_i R_i e_i$ and $R_i J \subset J$, we deduce that $e_1 J e_2 J \cdots J e_k \neq 0$. Then there exist elements $u_i \in J$, $i = 1, \ldots, k-1$, which in fact we may assume to be in $e_i J e_{i+1}$, with $e_1 u_1 e_2 u_2 \cdots u_{k-1} e_k \neq 0$.

Take now any integer $d$ such that $d + k > 2$. For each $i$ consider a Kemer polynomial for $M_{n_i}(F)$. For instance the central polynomial, which exists by Corollary 12.3.19, $F_{n_i}(X^{(i)})$, $X^{(i)} := (X_1^{(i)}, \ldots, X_{d+k-1}^{(i)})$, $(|X_j^{(i)}| = n_i^2)$. This polynomial takes as value the scalar matrix product of $\Delta(X_j^{(i)})$.

Alternatively take a product of Capelli polynomials as in formula (17.1). Set $X = (X^{(1)}, \dots X^{(k)})$. Let

(17.9)
$$f(X, z_1, \dots, z_{k-1}) := F_{n_1}(X^{(1)}) z_1 F_{n_2}(X^{(2)}) z_2 \cdots F_{n_{k-1}}(X^{(k-1)}) z_{k-1} F_{n_k}(X^{(k)}).$$

Evaluate this polynomial in $A$ by evaluating $F_{n_i}(X^{(i)})$ in $R_i$ so that its value is $e_i$ and $z_j$ in $u_j$. We see that its value is $e_1 u_1 e_2 u_2 \cdots u_{k-1} e_k \neq 0$. We call $\varphi$ this evaluation (if we choose a different Kemer polynomial, the argument slightly changes).

Set

$$X_j = X_j^{(1)} \cup \cdots \cup X_j^{(k)}, \quad 1 \leq j \leq d+k-1, \quad Z_j := X_j \cup \{z_j\}, \quad 1 \leq j \leq k-1.$$

Notice that $X_i$, $i = k, \dots, d+k-1$ has $t = \sum_{i=1}^{k} n_i^2$ elements and $Z_j$ has $t+1$ elements.

Define $\tilde{f}(X, z_1, z_2, \dots, z_{k-1}) = \tilde{f}(Z_1, Z_2, \dots, Z_{k-1}, X)$ as the polynomial obtained from $f$ by alternating the $k-1$ layers $Z_j$ in $t+1$ elements and the $d$ layers $X_i$, in $t$ elements.

We remark that since each $f$ is already alternating in the layers $X_j^{(i)}$, $j = 1, \dots, d$, when we perform the alternation of $X_j$ or $Z_j$, we may sum only on the coset representatives modulo the product of the symmetric groups fixing the layers $X_j^{(i)}$.

LEMMA 17.2.39. *$\tilde{f}(Z_1, Z_2, \dots, Z_{k-1}, X)$ is*

(1) *$d$-fold $t$-alternating in the layers $X_i$, $i = k, \dots, d+k-1$;*
(2) *$k-1$-fold $t+1$-alternating in the layers $Z_j$;*
(3) *and is not an identity of $A$.*

PROOF. Let us start with a simple remark. Suppose we take, in each $R_i$, some set $S_i$ of elements and consider a monomial $M$ which contains in its factors all the elements of all the $S_i$ and, further, all the elements $u_i \in e_i J e_{i+1}$.

Suppose that $M \neq 0$.

When we read the monomial from left to right and encounter two successive elements $u_i, u_j$, we see that in between we only have elements of the sets $S_h$, but we cannot have two elements from two distinct sets since $S_h S_j = 0$, $\forall h \neq j$. So we must have only elements from some $S_h$. On the other hand $u_i S_h = 0$ unless $h = i+1$, and $S_h u_j = 0$ unless $h = j$. Thus, since $M \neq 0$, we deduce that the elements $u_i$ appear in increasing order from left to right and all the elements of $S_{i+1}$ appear between $u_i, u_{i+1}$.

To prove that $\tilde{f}$ is not an identity of $A$, we show that the same evaluation $\varphi(\tilde{f}) \neq 0$. The evaluation $\varphi(f)$ is a linear combination of monomials $M$ of the previous type. So it is enough to verify that all the terms of the alternation of $f$, corresponding to coset representatives different from the identity, vanish. This follows from the fact that they always produce monomials in which either some exchange in the order of the $u_i$ has been performed or some mixing has occurred between the $S_j$. Then we get $\varphi(\tilde{f}) = \varphi(f) \neq 0$. □

REMARK 17.2.40.

(1) Central polynomials are not essential although they give polynomials of minimal degree. One could also use the products of Capelli polynomials as in formula (17.1), $C_{\nu,n_i^2}(X_1^{(i)},\dots,X_\nu^{(i)},Y_i)$.
(2) What we have proved is that, for a reduced algebra $A$, the first Kemer index is $t = \dim A/J$ and the second Kemer index is at least $k-1$, where $k$ is the number of simple algebras summands of $A/J$. In the case in which $J^k = 0$, this is then the Kemer index by Proposition 17.2.25.
(3) This happens in particular if $A$ is the algebra of block-triangular matrices for some choice of $k$ blocks. Hence $A$ is basic so that by Theorem 17.2.41 $A$ is fundamental.
(4) In Theorem 19.1.7 we shall also see that $\bigwedge V$, where $\dim V = 2h$, is fundamental (but when $\dim V = 2h+1$ it is not fundamental).

17.2.4.2. *Constructing Kemer polynomials.* Let $A = \bar{A} \oplus J$, $\bar{A} = \bigoplus_{i=1}^q R_i$, where $R_k = M_{n_k}(F)$, be a fundamental algebra.

In the literature the next result is usually presented divided in two parts, called the first and second Kemer lemma, but we prefer to prove it all at the same time. Recall the definitions of basic (Definition 17.2.26) and fundamental algebras (Definition 17.2.31).

THEOREM 17.2.41.

(i) *An algebra $A$ is fundamental if and only if it is basic (Definition* 17.2.26*), that is the Kemer index equals the $(t,s)$-index.*
(ii) *In this case, given any fundamental polynomial $f$, we have $\mu$-Kemer polynomials in the T-ideal $\langle f\rangle$ generated by $f$ for every $\mu$.*

PROOF. In one direction the statement is immediate. Assume the Kemer index equals the $(t,s)$-index. If $A$ is not fundamental, it is PI equivalent to a direct sum of algebras with strictly lower $(t,s)$-index. Now we have seen that the Kemer index of a direct sum is the maximum of the Kemer indices and, by Proposition 17.2.25, the Kemer index is always less than or equal to the $(t,s)$-index. This gives a contradiction.

Let us assume $A$ is fundamental, and let us now prove the converse. If $s_A = 0$ that is, if $A = \bigoplus_i M_{h_i}(F)$ is semisimple, then $A$ satisfies the same PI as its largest matrix block. Thus $A$ is fundamental if and only if it is a single matrix algebra $M_n(F)$ and every nonzero polynomial is fundamental. By Remark 17.2.8, the Kemer index is $n^2, 0$, and we have seen how to construct Kemer polynomials in formula (17.1). Finally, since the algebra of generic matrices is a domain, the product $fg$ of any nonzero polynomial $f$ with a Kemer polynomial $g$ in variables disjoint from those of $f$ is clearly Kemer.

If $t_A = 0$, the algebra is nilpotent, then $A^{s_A} \neq 0$, $A^{s_A+1} = 0$ by definition; this has been discussed in Proposition 17.2.9. Then $x_1x_2\cdots x_{s_A}$ is a Kemer polynomial. As for part (ii) any fundamental polynomial is automatically Kemer. So we assume $t_A > 0$, $s_A > 0$. Let $f \in F\langle \mathbb{X}\rangle$ be a fundamental polynomial for $A$, and choose some nonzero elementary evaluation $\eta : F\langle \mathbb{X}\rangle \to A$,

$$\eta(f) = f(r_1,\dots,r_{s_A},b_1,\dots,b_m) \neq 0. \tag{17.10}$$

We may write $f$ as $f(z_1, \dots, z_{s_A}, y_1, \dots, y_m)$ in order to stress that for the $s_A$ variables $z_i$'s, we have, in the given evaluation, radical evaluations $\eta(z_i) = r_i$ (Property K, cf. Lemma 17.2.36) and for the $y_i$ semisimple evaluations $\eta(y_i) = b_i$ (observe that this only means that $b_i \in R_j$ for some $j$ and **not** that $b_i$ is semisimple as element). We have $r_k \in e_{l_k} J e_{l'_k}$ for $1 \leq k \leq s_A$, where the indices $l_k, l'_k$ are indices of vertices of the Pierce graph (including possibly 0 if the algebra is without 1), while the $b_i$ are elements of one of the algebras $R_j$. Since $f$ is fundamental and the evaluation nonzero, all $R_j$ appear.

The idea is to evaluate the $z_k$'s in more complex expressions in $F\langle \mathbb{X} \rangle$ involving again central polynomials in order to create new layers.

From formula (17.10) it follows that there is some monomial, in the elements $r_1, \dots, r_{s_A}, b_1, \dots, b_m$, appearing in the evaluation of $f$, which is nonzero and, by Remark 17.2.22, to this monomial is associated an oriented path $\Pi$ in the Pierce graph defined in Definition 17.2.20, such that the edge corresponding to $r_k \in e_{l_k} J e_{l'_k}$ has source $e_{l_k}$ and target $e_{l'_k}$. We mark this edge with the variable $z_k$. The source $e_{l_k}$ and target $e_{l'_k}$ can be either different vertices, in which case we may call $z_k$ a *bridge variable* or equal, in which case we may call $z_k$ a *loop variable*. Notice that the edges corresponding to the semisimple variables $b_i$ are necessarily loops, originating and ending in $j$, if $b_i \in R_j$. Moreover, since $f$ is full, this path visits all the vertices $1, \dots, q$ of the graph (the graph may also have a vertex 0 if the algebra is without 1 and the path may or may not visit this vertex).

We need to distinguish the cases $q > 1$ and $q = 1$.

(i) $q > 1$. This implies that for each $i = \{1, \dots, q\}$ we may choose an edge, of the path $\Pi$, which joins the vertex $i$ with some other vertex $i' \neq i$. This edge then corresponds, that is it is marked, to a radical variable $z_{t_i}$. It is possible that in this way we choose the same variable twice, associated to the two distinct end points of its edge. Let us call these variables *selected*; notice that they are all bridge variables and at most $q$.

Set $s_A = s$ for simplicity, and let $\nu := \mu + s$. We choose, for each $i = 1, \dots, q$, disjoint lists $X^{(i)} := (X_1^{(i)}, \dots, X_\nu^{(i)})$ of $\nu$ layers $X_k^{(i)}$, $k = 1, \dots, \nu$, each with $n_i^2$ variables. With these we build (see Corollary 12.3.19) the $q$ polynomials $F_{\nu, n_i}(X^{(i)}) = F_{\nu, n_i}(X_1^{(i)}, \dots, X_\nu^{(i)})$, each central for the matrix algebra $R_i$. Denote by $X$ the entire list of the $X_k^{(i)}$, a list of $\nu \cdot t_A$ variables.

We then make a substitution, in the polynomial $f$, only on the selected variables $z_t$. That is we construct a map $\psi : F\langle \mathbb{X} \rangle \to F\langle \mathbb{X} \rangle$ which is the identity on all the variables except the selected $z_t$ and build the polynomial $\psi(f)$. We may distinguish the three cases, namely if $z_t$ is associated to only one index $i$ as source, or as target, or finally it is associated to some $i$ as source, and to some $j \neq i$ as target. For each case we set, respectively, $\psi(z_t)$ to be

(17.11)

$$\psi(z_t) := \{u_i F_{\nu, n_i}(X^{(i)}) v_i z_t,\ z_t u_i F_{\nu, n_i}(X^{(i)}) v_i,\ u_i F_{\nu, n_i}(X^{(i)}) v_i z_t u_j F_{\nu, n_j}(X^{(j)}) v_j\}$$

with the variables $X^{(k)}$ and the two extra variables $u_k, v_k$ for each $k = 1, \dots, q$, all different and not appearing in $f$.

We then obtain a new multilinear polynomial $\psi(f)$, in the $T$-ideal generated by the polynomial $f$, in which we have *loaded* the selected variables with the $q$ polynomials $u_i F_{\nu, n_i}(X^{(i)}) v_i$, one for each semisimple algebra $R_i$.

By Corollary 12.3.19 we may also extend the evaluation $\eta$ to all the new variables $X^{(i)}, u_i, v_i$ so that $\eta$ on the variables $X^{(i)}$ takes values in $R_i$ and for which $\eta(F_{\nu,n_i}(X^{(i)})) = e_i$ the unit element of $R_i$. We also evaluate $\eta(u_i) = \eta(v_i) = e_i$. Under this evaluation clearly $\eta(\psi(z_t)) = \eta(z_t)$ so that $\eta(\psi(f)) = \eta(f) \neq 0$.

By the definition of the polynomials $F_{\nu,n_i}(X_1^{(i)}, \ldots, X_\nu^{(i)})$, we have that $\psi(f)$ is antisymmetric for the group $H := \prod_{j=1}^{\nu} \prod_{i=1}^{q} S_{X_j^{(i)}}$.

We now consider the $\nu$ layers $X_j := (X_j^{(1)}, \ldots, X_j^{(q)})$, $j = 1, \ldots, \nu$, each with $t_A = \sum_{i=1}^{q} n_i^2$ elements. We want to take the last $s$ layers $X_{\mu+i}$, adjoin to $X_{\mu+i}$ the variable $z_i$, and set $Z_i := \{X_{\mu+i}, z_i\}$, $i = 1, \ldots, s$. We have $s$ layers with $t_A + 1$ elements.

We then alternate in the previous polynomial $\psi(f)$ independently each $X_j$, $j = 1, \ldots, \mu$, and each $Z_i$, so that we have now a polynomial $g$ (in the $T$-ideal of $f$) with $\mu$ alternating layers $X_i$, $i = 1, \ldots, \mu$, with $t_A$ elements and $s$ alternating layers $Z_i := \{X_{\mu+i}, z_i\}$, $i = 1, \ldots, s$, with $t_A + 1$ elements. If we can prove that $g$ is not a PI of $A$, then $g$ is the desired Kemer polynomial.

If $G := \prod_{i=1}^{\mu} S_{X_i} \times \prod_{j=1}^{s} S_{Z_j} = S_t^{\mu} \times S_{t+1}^{s}$, we have that $g$ is a sum over cosets $G/H$,

$$g = \sum_{\sigma \in G/H} \epsilon_\sigma \sigma \psi(f). \tag{17.12}$$

We claim that when we perform on $g$ the same evaluation,

$$\eta(g) = \sum_{\sigma \in G/H} \epsilon_\sigma \eta(\sigma \psi(f)) \neq 0. \tag{17.13}$$

Since by construction $\eta(\psi(f)) = \eta(f) \neq 0$, it is enough to show that all terms $\eta(\sigma(\psi(f)))$, $\sigma \neq 1$, vanish. So the value of $\eta(g) = \eta(\psi(f)) = \eta(f)$ is nonzero.

REMARK 17.2.42. We have $\eta(\sigma\psi(f)) = \eta'(f)$, where $\eta'$ is the evaluation of $f$ with $y_i \mapsto \eta'(y_i) = \eta(y_i)$ and $z_t \mapsto \eta'(z_t) := \eta(\sigma(\psi(z_t)))$.

If $z_t$ is not a selected variable, $\psi(z_t) = z_t$ so that $z_t \mapsto \eta'(z_t) = \eta(\sigma(z_t))$.

REMARK 17.2.43. Notice that in this evaluation $\eta'$ of $f$, all the variables $y_i$ remain unchanged, hence semisimple, while one or more of the previous radical variables $z_t$ may also become semisimple. If this happens for at least one of them, then by property K this evaluation vanishes.

Let us choose a $\sigma \neq 1$. We may distinguish three cases.

(1) Assume $\sigma(z_t) = z_t$ for all radical variables, so $\sigma$ permutes only the variables $X$. Since $\sigma \neq 1$, for some $1 \leq k \leq \nu$, in one of the polynomials $u_i F_{\nu,n_i}(X^{(i)}) v_i$ one variable in some $X_k^{(i)}$ is substituted by some variable in some $X_k^{(j)}$ with $j \neq i$. In the evaluation $\eta(\sigma(u_i F_{\nu,n_i}(X^{(i)}) v_i))$ one variable in $X^{(i)}$ is evaluated in an element in $R_j$ with $j \neq i$. This evaluation vanishes.

This is because all the variables $X$ are evaluated in semisimple variables, so if in $X^{(i)}$ one of the variables is evaluated in some $R_j$ all of them must be evaluated in $R_j$ in order for the polynomial $F_{\nu,n_i}(X^{(i)})$ to be nonzero. Then, since $u_i, v_i$ are evaluated in $e_i$, if $j \neq i$, we have an evaluation in $e_i R_j e_i = 0$.

(2) If $\sigma(z_t) \neq z_t$ for a not selected variable, we have $\sigma(z_t) \in X$, so $\eta'(z_t) = \eta(\sigma(z_t))$ is a semisimple evaluation; therefore, the evaluation $\eta'(f)$ vanishes by Remark 17.2.43.

(3) Say $z_1, \dots, z_k$ are the selected variables exchanged with elements $x_i \in X$. If for one of these variables $z_\ell$ we have $\eta'(z_\ell) = \eta(\sigma(\psi(z_\ell)) = 0$, we are done, the evaluation $\eta'(f)$ of the term vanishes. The same happens if one of the elements $\eta'(z_\ell) = \eta(\sigma(\psi(z_\ell))$ is a semisimple evaluation by Remark 17.2.43. Otherwise take one of the $z_\ell$, $1 \leq \ell \leq k$, and for instance assume that $\psi(z_\ell) = u_j F_{\nu,n_j}(X^{(j)}) v_j z_\ell$ so that $\sigma(\psi(z_\ell)) = u_j \sigma(F_{\nu,n_j}(X^{(j)})) v_j \sigma(z_\ell)$ (the other cases are similar). We are assuming $\sigma(z_\ell) \neq z_\ell$, so $\sigma(z_\ell) \in X$ has, under $\eta$, a semisimple evaluation. The corresponding evaluation $\eta'(z_\ell)$ of $z_\ell$,

$$\begin{aligned}\eta'(z_\ell) &= \eta(\sigma(\psi(z_\ell))) = \eta(u_j \sigma(F_{\nu,n_j}(X^{(j)})) v_j) \eta(\sigma(z_\ell)) \\ &= e_j \eta(\sigma(F_{\nu,n_j}(X^{(j)}))) e_j \eta(\sigma(z_\ell)),\end{aligned}$$

is semisimple or 0, unless the loading factor $\sigma(F_{\nu,n_j}(X^{(j)}))$ has under $\eta$ a nonzero radical evaluation in $e_j J e_j$ and $\eta(\sigma(z_\ell)) \in R_j$.

In order for this to happen, one has to have substituted some of the variables $z_1, \dots, z_k$ in $F_{\nu,n_j}(X^{(j)})$ so that $\eta(\sigma(F_{\nu,n_j}(X^{(j)}))) \in e_j A e_j$. Since these variables, which are exchanged, are all bridge variables, if only one such variable $z_t$ with $\eta(z_t) \in e_a J e_b$, $a \neq b$ appears in $\sigma(X^{(j)})$, then, since all the other elements of $\sigma(X^{(j)})$ are in $X$, we have that $\eta(\sigma(F_{\nu,n_j}(X^{(j)})))$ is either 0 or in $e_a J e_b$, $a \neq b$. Thus $\eta(\sigma(F_{\nu,n_j}(X^{(j)}))) \in e_j A e_j$ implies that at least two of the variables $z_j$, $j = 1, \dots, k$, are substituted into the loading factor $F_{\nu,n_j}(X^{(j)})$ of each $z_i$, $i = 1, \dots, k$.

This is not possible for all the $k$ exchanged variables since we need at least two exchanged variables for each of them. Since the loading factors of distinct selected variables involve disjoint variables $X$, we need at least $2k$ variables to perform such a substitution, and we have at our disposal only $k$ variables.

Thus the evaluation $\eta'$ of $f$ has $< s$ radical evaluations so, by property K (Definition 17.2.35), this evaluation $\eta'(f) = \eta(\sigma\psi(f))$ vanishes, as desired.

(ii) $q = 1$, $\bar{A} = M_n(F)$. In this case we start from just one list $X := (X_1, \dots, X_\nu)$ of $\nu$ layers $X_k$, $k = 1, \dots, \nu$, each with $n^2$ variables. We have two subcases. First assume that the algebra $A$ has a 1 (the identity of $\bar{A} = M_n(F)$).

Consider the polynomial $F_{\nu,n}(X) f$. We have some evaluation of $F_{\nu,n}(X)$ in $\bar{A}$ equal to 1 which thus, multiplied by the given evaluation of $f$, does not vanish. By construction this is alternating in the $\nu$ layers $X_i$.

Now, as soon as we alternate one of the $s$ layers $Z_i = (X_{\mu+i}, z_i)$, we are *exporting* a radical variable outside the polynomial $f$ and *importing* in $f$ a semisimple variable instead, and hence we get that the value of $f$ is zero by property K.

Finally assume $q = 1$ and the algebra $A$ does not have a 1. We have the identity $e_1$ of the semisimple part of $A$, and we have added the idempotent $e_0$ so that $e_0 + e_1 = 1$.

If all the variables appearing in $f$ are evaluated in $e_1 A e_1$, we can replace $A$ with $e_1 A e_1$ and we are in the previous case of an algebra with a 1. Otherwise, by hypothesis the polynomial is full, at least one variable is evaluated in $e_1 A e_1$. It is not possible that the remaining variables are all evaluated in $e_0 A e_0$ since $e_1 e_0 = e_0 e_1 = 0$; thus at least 1 variable $z_0$ has an evaluation in $e_0 A e_1$ (or $e_1 A e_0$).

This is necessarily a radical variable, and we can then attach the polynomial $uF_{\nu,n}(X)v$ to the right of this variable $z_0$ (resp., to the left). We have a polynomial $\psi(f)$, and we may extend the evaluation of $f$ to the evaluation $e_1$ of $F_{\nu,n}(X)$, as in the previous case, and to $u, v = e_1$ so to be still nonzero.

We now perform the alternation of the $s$ layers $Z_i$, $i = 1, \ldots, s$, that is of $X_{\mu+i}$ with the corresponding radical variables $z_i$. We want to prove that the same evaluation is still nonzero. As before it is enough to prove that all terms of the alternation different from the term associated to the identity vanish under this evaluation.

When in the alternation some variables $z_i \neq z_0$ are substituted in $X_{\mu+i}$, we have exported semisimple elements in place of $z_i$ while $z_0$ absorbs the nilpotent evaluation, we apply again property K. Finally if $z_0$ is the only variable substituted in $X_{\mu+i}$ the polynomial $uF_{\nu,n}(X)v$ under this substitution vanishes. □

REMARK 17.2.44. From now on we will use only the word "fundamental" and no more "basic".

Theorem 17.2.41 tells us that if $A$ is fundamental and $f$ is a fundamental polynomial of $A$, then the $T$-ideal generated by $f$ contains a Kemer polynomial for $A$. On the other hand we have

PROPOSITION 17.2.45. *A Kemer polynomial, with $\mu$ sufficiently large, for a fundamental algebra $A$ is fundamental.*

PROOF. Each of the algebras $A_i$ and $A_0$, of formulas (17.6) and (17.8), has Kemer index strictly lower than that, $(t, s)$ of $A$. For each thus there is an integer $\nu_0(A_i)$ for which every polynomial alternating in at least $\nu_0(A_i)$ layers with $t$ elements and $s$ layers with $t+1$ elements is a PI. Therefore a Kemer polynomial for $A$, for $\mu > \max(\mu_0(A_i))$, is a PI for all these algebras, and, by definition, it is not an identity for $A$. □

*A complement.* We then introduce

DEFINITION 17.2.46. A $T$-ideal $I$ is called *primary* if it is the ideal of identities of a fundamental algebra.

A $T$-ideal $I$ is *irreducible* if it is not the intersection $I = J_1 \cap J_2$ of two $T$-ideals $J_1, J_2$ properly containing $I$.

Using the basic theorem, to be proved later Corollary 20.2.3, that $T$-ideals satisfy the ascending chain condition, we can prove

PROPOSITION 17.2.47. *Every irreducible $T$-ideal containing a Capelli list is primary and every $T$-ideal containing a Capelli identity is a finite intersection of primary $T$-ideals.*

PROOF. By Noetherian induction every $T$-ideal is a finite intersection of irreducible $T$-ideals; otherwise there is a maximum one which does not have this property, and we quickly have a contradiction.

If a $T$-ideal $I$ contains a Capelli list it is the $T$-ideal of PIs of a finite-dimensional algebra. This algebra, by Proposition 17.2.34, is PI equivalent to a direct sum of fundamental algebras. Hence $I$ is the intersection of the ideals of polynomial identities of these algebras, and if it is irreducible it must coincide with the ideal of polynomial identities of one summand, a fundamental algebra. □

As in the theory of primary decomposition we may define an *irredundant* decomposition $I = J_1 \cap \cdots \cap J_k$ of a $T$-ideal $I$ in primary ideals. In a similar way every finite-dimensional algebra $B$ is PI equivalent to a direct sum $A = \bigoplus_i A_i$ of fundamental algebras which is *irredundant* in the sense that $A$ is not PI equivalent to any proper algebra $\bigoplus_{i \neq j} A_i$. We call this an *irredundant sum of fundamental algebras*.

It is NOT true that the $T$-ideal of PI of a fundamental algebra is irreducible. The following example has been suggested to us by Belov.

Consider the two fundamental algebras $A_{1,2}$, $A_{2,1}$ of block upper-triangular $3 \times 3$ matrices stabilizing a partial flag formed by a subspace of dimension 1 or 2, respectively. They both have a semisimple part isomorphic to $M_2(F) \oplus F$. By the theorem of Giambruno and Zaicev (Theorem16.3.2) the two $T$-ideals of PI are $I_2 I_1$ and $I_1 I_2$, respectively, where $I_k$ denotes the ideal of identities of $k \times k$ matrices. By the theorem of Bergman and Lewin (Theorem 16.4.3) these two ideals are different.

Now in their direct sum $A_{1,2} \oplus A_{2,1}$, consider the algebra $L$ which on the diagonal has equal entries in the two two-by-two and in the two one-by-one blocks. It is easy to see that $L$ is PI equivalent to $A_{1,2} \oplus A_{2,1}$ and that it is fundamental. Its $T$-ideal is not irreducible but it is $I_2 I_1 \cap I_1 I_2$.

## 17.3. The trace algebra

*In this section we shall use heavily the general results of* §2.4.2, *sometimes without explicit reference.*

**17.3.1. The role of traces.** Let $A$ be a fundamental algebra with index $(t, s)$.

Given any set $X$ of variables, consider the $X$-tuples $\bar{A}^X$ of elements of $\bar{A}$, that is the maps $X \to \bar{A}$. Given any element $f(x_1, \dots, x_i, \dots)$ of the free algebra $F\langle X \rangle$, an element $(a_i)_{i \in X} \in A^X$ gives rise to an evaluation map $x_i \mapsto a_i \in \bar{A}$. Denote by $\bar{f}(x_1, \dots, x_i, \dots)$ the corresponding function on $\bar{A}^X$, with values in $\bar{A}$. By taking the trace or the determinant of left multiplication, we have a function $\operatorname{tr}(\bar{f}(x_1, \dots, x_i, \dots)) \in F$, resp., $\det(\bar{f}(x_1, \dots, x_i, \dots))$, on $\bar{A}^X$. Sometimes we write just $f$ instead of $\bar{f}$.

This discussion extends to a direct sum $A = \bigoplus_{i=1}^q A_i$ of fundamental algebras (with radical $J_i$), all with the same Kemer index $(t, s)$. When we take an evaluation $f(x_1, x_2, \dots, x_t, \bar{Y})$ in $A$, this of course equals the direct sum of the evaluations $f(x_1^{(i)}, x_2^{(i)}, \dots, x_t^{(i)}, Y^{(i)})$, where $x_j^{(i)}$ is the component in $A_i$ of the evaluation of $x_j$ in $A$ (same for $Y^{(i)}$). For any $z \in A$, set $\bar{z}_i$ the image of $z$ in $\bar{A}_i$ and set

$$D(z) := (\det(\bar{z}_1), \dots, \det(\bar{z}_q)) \in F^q. \tag{17.14}$$

With $\det(\bar{z}_i)$ the determinant of left multiplication of the class $\bar{z}_i$ of $z$ on the summand $\bar{A}_i$ of $\bar{A}$. In fact since all the summands $\bar{A}_i$ have the same dimension $t$, we have that $\bar{A}$ is a free module of rank $t$ over $F^q$ so multiplications by elements of $\bar{A}$ can be treated as $t \times t$ matrices, now over $F^q$, and $D(z)$ is the determinant of such a matrix. Then of course by introducing a commuting variable $\lambda$, we have

$$D(\lambda - z) = \lambda^t - \operatorname{tr}(\bar{z})\lambda^{-1} + \cdots + (-1)^t D(z). \tag{17.15}$$

This is the characteristic polynomial of left multiplication by $\bar{z}$ on $\bar{A}$, as free $F^q$ module.

DEFINITION 17.3.1. Denote by $\mathcal{T}_A(X) = \mathcal{T}_{\bar{A}}(X)$ the algebra generated by the functions on $X$-tuples of elements of $\bar{A}$ given by the trace $\operatorname{tr}(\bar{f}(x_1, \dots, x_i, \dots))$. Or equivalently generated by $D(f(x_1, \dots, x_i, \dots)) = \det(\bar{f}(x_1, \dots, x_i, \dots))$.

REMARK 17.3.2. Observe that since $D(w_1)D(w_2) = D(w_1w_2)$, the algebra $\mathcal{T}_A(X)$ is linearly spanned by the elements $D(f), f \in \mathcal{F}_A(X)$.

REMARK 17.3.3. Recall that the theorem of Zieplies and Vaccarino (Corollary 12.1.18) tells us that a $\mathcal{T}_t(X)$-module structure on a space $M$ is equivalent to giving a multiplicative polynomial map $\rho$ of degree $t$ from $F\langle X\rangle$ to a commutative algebra of linear operators on $M$. The map $z \mapsto (\det(\bar{z}_1), \dots, \det(\bar{z}_q))$, $z \in \mathcal{F}(X)$, is a multiplicative map homogeneous of degree $t$. This is a different explanation of the fact that we still have that $\mathcal{T}_A(X)$ is a quotient of $\mathcal{T}_t(X)$.

Choosing a basis of $\bar{A}$ over $F^q$, we have that left multiplication by elements of $\bar{A}$ are given by $t \times t$ matrices over $F^q$. Therefore $\mathcal{T}_A(X)$ is a quotient of the algebra of invariants $\mathcal{T}_t(X)$ of $t \times t$ matrices. This quotient depends upon the structure of $\bar{A}$ (see §18.1.3.2 for some discussion of this dependence). When $A$ is fundamental, then $q = 1$.

We then can construct the algebra $\mathcal{T}_A(X)\mathcal{F}_A(X)$. This is an algebra of polynomial functions in the variables $X$, that is from $A^X$ to $A$.

This is an algebra with trace having $\mathcal{T}_A(X)$ as trace algebra. One should remark that any substitution $X \to \mathcal{F}_A(X)$ (or even $X \to \mathcal{T}_A(X)\mathcal{F}_A(X)$) extends to a homomorphism of algebras with trace, a substitution. In other words this can be considered as the *relatively free* algebra with trace of the trace algebra $A$.

The main reason to introduce the algebra $\mathcal{T}_A(X)\mathcal{F}_A(X)$ is that when $X$ is finite it is the Noetherian theorem, Theorem 17.3.5, which $\mathcal{F}_A(X)$ is not. The second reason has to do with Kemer polynomials (§17.3.1.2).

We discuss this issue using the method of generic elements.

17.3.1.1. *Generic elements.* In commutative algebra a polynomial is at the same time a symbolic expression and a function. Here we replace the language as functions on $A$ with the equivalent symbolic language. The advantage is that we can also use rational functions without having to care on which set they can be evaluated as functions.

In §2.4 and in particular in §2.4.2 we have discussed the method of generic elements for algebras, finite modules over some commutative algebra $C$ finite dimensional over a field $F$ and equipped with a trace with values in $C$.

We want to apply this method to an algebra $A = \bigoplus_{i=1}^q A_i$ direct sum of fundamental algebras, with the same Kemer index $(t, s)$. Let $J = \bigoplus J_i$ be the radical of $A$, and let $A/J = \bigoplus_{i=1}^q \bar{A}_i = \bigoplus_{i=1}^k A_i/J_i$. In this case $A$ and $\bar{A}$ are considered also as a module over $C = F^q$. Notice that $\bar{A}$ is free over $C$ of rank $t$.

Let $\dim_F A = n$, and fix a basis $a_1, \dots, a_n$ of $A$ over $F$ union of bases of the summands $A_i$. We may also choose, for each summand $A_i$ decomposed as $\bar{A}_i \oplus J_i$, the basis to be formed of a basis of $J_i$ and one of $\bar{A}_i$. Choose an indexing set $I$ usually $\mathbb{N}$ or a finite subset of this. Let $\xi_i^{(\alpha)}$, $i = 1, \dots, n$; $\alpha \in I$, denote the variables from which one builds the generic elements $\xi_\alpha = \sum_{i=1}^n \xi_i^{(\alpha)} a_i$. The generic element $\xi_\alpha$ decomposes as a sum $\xi_\alpha = \sum_{i=1}^q \xi_{\alpha,i}$ with $\xi_{\alpha,i}$ generic for $A_i$. The algebra of generic elements is

$$\mathcal{F}_A(X) := F\langle \xi_\alpha \rangle \subset \bigoplus_{i=1}^q F\langle \xi_{\alpha,i} \rangle = \bigoplus_{i=1}^q \mathcal{F}_{A_i}(X),$$

where by $F\langle \xi_\alpha \rangle$ we denote the subalgebra of $A[\xi_i^{(\alpha)}]$ over $F$ generated by the elements $\xi_\alpha$ (with or without 1 depending on the assumptions on $A$).

We may also introduce the field of rational functions in the variables $\xi_i^{(\alpha)}$, denoted by $L := F(\xi_i^{(\alpha)})$. We have by definition $F\langle \xi_{(\alpha)} \rangle \subset A[\xi_i^{(\alpha)}]$, $i = 1, \dots, n; \alpha \in I$, the polynomial ring, and of course

$$A[\xi_i^{(\alpha)}] = \bigoplus_{j=1}^{q} A_j[\xi_{\alpha,j}] \subset \bigoplus_{j=1}^{q} A_j \otimes_F L, \ F\langle \xi_{\alpha,j} \rangle \subset A_j[\xi_i^{(\alpha)}] \subset A_j \otimes_F L.$$

Recall that the map $F\langle x_1, \dots, x_m, \dots \rangle \to F\langle \xi_1, \dots, \xi_m, \dots \rangle$, $x_i \mapsto \xi_i$, has as kernel the $T$-ideal of polynomial identities of $A$ in the variables $X := \{x_\alpha \mid \alpha \in I\}$. The algebra $F\langle \xi_1, \dots, \xi_m, \dots \rangle$ is isomorphic to the relatively free algebra $\mathcal{F}_A(X)$ of $A$.

For each $i = 1, \dots, q$, we also have that $\mathcal{F}_{A_i}(X) = F\langle \xi_{1,i}, \dots, \xi_{m,i}, \dots \rangle$ is the relatively free algebra for $A_i$.

Notice that the radical of $A \otimes_F L$ is $J \otimes_F L$, and modulo the radical this is $\bar{A} \otimes_F L$ which is a direct sum of matrix algebras $\bigoplus_{j=1}^{r} M_{n_j}(L)$.

We consider $A$ and $A \otimes L$ as a direct sum of algebras with trace. The trace is the one defined by formula (17.15). For $a \in A$, we have $t(a) := (t_1(a), \dots, t_q(a)) \in F^q$; for $a \in A \otimes L$, we have $t(a) \in L^{\oplus q}$.

Notice that if the algebra has a 1 we have $L^q \subset A \otimes_F L$, and the map $a \mapsto t(a)$ satisfies all the properties of an abstract trace, according to Definition 2.3.1 (Exercise 2.3.5). If we do not assume a 1, we use Remark 2.3.6; in particular $t(t(a)b) = t(a)t(b)$ still holds.

PROPOSITION 17.3.4. *We have according to Remark* 2.3.18 (*and formula* (17.14))

$$\mathcal{T}_A(X) := F[t(a)]|_{a \in \mathcal{F}_A} \subset L^{\oplus q}. \tag{17.16}$$

*The algebra $\mathcal{T}_A(X)$, generated over $F$ by all the elements $t(a)$, $a \in \mathcal{F}_A(X)$, is the pure trace algebra of $A$. The algebra $\mathcal{T}_A(X)\mathcal{F}_A(X) \subset A \otimes L$ is the relatively free algebra of $A$ as trace algebra, that is the free algebra with trace modulo the trace identities of $A$.*

From now on we assume $X$ is fixed and drop the symbol $X$. We have a variation of the results of §2.4.2 and in particular of Theorem 2.4.8 and Remark 2.4.9, Proposition 2.4.10, and Theorem 8.2.5 that

THEOREM 17.3.5. *If $X$ is finite, $\mathcal{T}_A = \mathcal{T}_A(X)$ is a finitely generated $F$-algebra and $\mathcal{T}_A\mathcal{F}_A$ is a finitely generated module over $\mathcal{T}_A$.*

PROOF. We first claim that every element $a \in \mathcal{F}_A$ satisfies a monic polynomial with coefficients in $\mathcal{T}_A$. In fact the coefficients are polynomials in the elements $t(a^j)$ for $j \leq t$.

Since $\bar{A}$ is a free $F^q$ module of dimension $t$ (the first Kemer index), the projection $\bar{a}$ of $a$ in $\bar{A} \otimes_F L$ satisfies, formula (17.15), its characteristic polynomial $D(\lambda - a)$. This is the evaluation in $a$ of a universal trace expression in $x$.

Now every element of the kernel of $p : A \to \bar{A}$ is nilpotent of some degree $\leq s$, and finally we deduce that (we need to add a factor $a$ since we are not assuming the existence of a 1)

$$D(\lambda - a)^s a = 0, \quad \forall a \in \mathcal{T}_A\mathcal{F}_A.$$

We can then conclude by applying Theorem 8.2.1 and Remark 8.2.3. We have a Shirshov basis for $\mathcal{F}_A$. Then, since we know that $t(t(a)b) = t(a)t(b)$, we deduce that $\mathcal{T}_A$ is generated by the traces $t(M)$ where $M$ is a monomial in the Shirshov basis with exponents less than the degree of $ts+1$. Finally $\mathcal{T}_A\mathcal{F}_A$ is spanned over $\mathcal{T}_A$ by this finite number of monomials. □

REMARK 17.3.6. If X is countable, we may also consider the algebras

$$\mathcal{T}_A(X), \mathcal{T}_A(X)\mathcal{F}_A(X)$$

as *finitely generated* as in classical invariant theory. That is using the Capelli identity, one know that there are finitely many generators provided we also use linear changes on variables, that is special automorphisms.

REMARK 17.3.7. By definition $\mathcal{F}_A$ is isomorphic to the relatively free algebra in the variety generated by $A$. On the other hand $\mathcal{T}_A\mathcal{F}_A$ depends upon $A$ through the trace identities of $A$ thought of as a trace algebra, and $\mathcal{T}_A\mathcal{F}_A$ is a relatively free algebra (cf. Theorem 2.4.8 and Remark 2.4.9). It is interesting to understand how much this is dependent on $A$, and this issue will be treated in Theorem 18.1.16.

17.3.1.2. *Kemer polynomials absorb traces.* We now connect the previous construction with Kemer polynomials. We start with a fundamental algebra $A$ with Kemer index $(t,s)$. By property K (Definition 17.2.35), every elementary nonzero evaluation of a Kemer polynomial $g$ has all the radical evaluations in the big layers (alternating in $t+1$ variables). In other words, no matter what are the evaluations in the big layers, if one evaluates in the radical a variable in a small layer or outside the layers, one gets 0. That is

PROPOSITION 17.3.8. *The evaluation of a Kemer polynomial $g$ outside the big layers factors modulo the radical.*

So let us fix one of the small layers of the variables say $x_1,\dots,x_t$, and denote for simplicity by $Y$ the remaining variables. Having fixed some evaluation $\bar{Y}$ of the variables $Y$ outside the chosen small layer of variables, we deduce that the map $A^t\to A$, given by $(x_1,\dots,x_t)\mapsto g(x_1,x_2,\dots,x_t,\bar{Y})$, factors through a multilinear alternating map from $(A/J)^t$ to $A$ (in $t$ variables),

$$(x_1,\dots,x_t)\mapsto(\bar{x}_1,\dots,\bar{x}_t)\mapsto g(\bar{x}_1,\dots,\bar{x}_t\bar{Y}),$$

where by $\bar{x}$ we denote the class in $A/J$ of an element $x\in A$.

Since $\dim A/J=t$ by Lemma 10.2.9 we deduce, fixing a basis of $\bar{A}=A/J$ (or just an element of $0\neq v\in\bigwedge^t(\bar{A})$) that this map has the form

(17.17)
$$g(x_1,x_2,\dots,x_t,\bar{Y})=\det(\bar{x}_1,\dots,\bar{x}_t)u(\bar{Y}),\ \det(\bar{x}_1,\dots,\bar{x}_t)v=\bar{x}_1\wedge\cdots\wedge\bar{x}_t.$$

REMARK 17.3.9.

(1) The map $\bar{Y}\mapsto u(\bar{Y})$ is no more given by a noncommutative polynomial in the variables $\bar{Y}$, but it is only a polynomial map.
(2) Since $g$ is Kemer and there are $s$ radical evaluations among the elements $\bar{Y}$, we also have $u(\bar{Y})\in J^s$. Further $Ju(\bar{Y})=0$ hence we also have $\forall x\in A$,

$$xg(x_1,x_2,\dots,x_t,\bar{Y})=\bar{x}\det(\bar{x}_1,\dots,\bar{x}_t)u(\bar{Y}).$$

As a consequence of (17.17) we deduce, by Proposition 10.2.10, the important identity as functions on $A$,

$$g(zx_1,zx_2,\dots,zx_t,Y)=\det(\bar{z})g(x_1,x_2,\dots,x_t,Y), \tag{17.18}$$

where $\det(\bar{z})$ means the determinant of left multiplication of $\bar{z}$ on $\bar{A} := A/J$. Notice that this identity is independent on the chosen layer.

When we polarize this identity, we have, on the right-hand side, the characteristic polynomial (of left multiplication of $\bar{z}$ on $A/J$),

$$g(\lambda - z)x_1, (\lambda - z)x_2, \ldots, (\lambda - z)x_t, Y) = \det((\lambda - \bar{z}))g(x_1, x_2, \ldots, x_t, Y). \tag{17.19}$$

Equating the coefficients of the powers of $\lambda$, we have in particular

$$\operatorname{tr}(\bar{z})g(x_1, \ldots, x_t, Y) = \sum_{k=1}^{t} g(x_1, \ldots, x_{k-1}, zx_k, x_{k+1}, \ldots, x_t, Y), \tag{17.20}$$

where $\operatorname{tr}(\bar{z})$ is again the trace of left multiplication in $\bar{A} \simeq A/J$.

Assume now $A = \bigoplus_{i=1}^{q} A_i$ is a direct sum of fundamental algebras (with radical $J_i$) all with the same Kemer index $(t,s)$. With these notations, if $\mu$ is sufficiently large, a polynomial $g$ is a $\mu$-Kemer polynomial for $A$ if and only if it is a $\mu$-Kemer polynomial for one of the algebras $A_i$. But $g$ can also be a PI for some of the $A_j$.

As in the previous paragraph we think of $A$ as a free module over the direct sum of $q$ copies of $F$.

We obtain a formula analogous to (17.18) (cf. formula (17.14)):

$$\begin{aligned} g(zx_1, zx_2, \ldots, zx_t, Y) &= \sum_{i=1}^{q} \det(\bar{z}_i) g(x_1^{(i)}, x_2^{(i)}, \ldots, x_t^{(i)}, Y^{(i)}) \\ &= D(z)g(x_1, x_2, \ldots, x_t, Y),\ D(z) := (\det(\bar{z}_1), \ldots, \det(\bar{z}_q)) \in F^q. \end{aligned} \tag{17.21}$$

We now observe that the left-hand side of formula (17.18) and hence also the right-hand side of this formula is still alternating in all the layers in which $g$ is alternating. Since for generic $z \in A$, we have $\operatorname{tr}(\bar{z}) \neq 0$, this is not an identity of $A$, so it is again a Kemer polynomial.

THEOREM 17.3.10. *Let $g(x_1, x_2, \ldots, x_t, Y)$ be a $\mu$ Kemer polynomial for $A$. The T-ideal $K_A(f) \subset \mathcal{F}_A(X) \subset \mathcal{T}_A(X)\mathcal{F}_A(X)$ generated by $g$, in the relatively free algebra of $A$, is an ideal (in fact a T-ideal) of $\mathcal{T}_A(X)\mathcal{F}_A(X)$.*

PROOF. The right-hand side of formula (17.21) equals $D(z)g(x_1, x_2, \ldots, x_t, Y)$. Since the algebra $\mathcal{T}_A(X)\mathcal{F}_A(X)$ is closed under substitutions, which are trace preserving homomorphisms, it follows that formula (17.21) can also be treated as a trace identity. So take an element $u \in K_A(f)$, obtained by substitution of the variables $x_1, x_2, \ldots, x_t, Y$ with elements in $\mathcal{F}(X)$ from $g(x_1, x_2, \ldots, x_t, Y)$.

The function $D(w)g$ is obtained from the polynomial $g(zx_1, zx_2, \ldots, zx_t, Y)$ by the same substitution of the variables $x_1, x_2, \ldots, x_t, Y$ and $z \mapsto w$. So $D(w)g$ lies in $K_A(f)$. The elements $D(w)$, $w \in \mathcal{F}_A(X)$, span linearly $\mathcal{T}_A(X)$, Remark 17.3.2, so the claim is proved. □

REMARK 17.3.11 (WARNING). The fact that $K_A(f)$ is a module over some algebra $\mathcal{T}_A(X)$ has been proved using a particular algebra $A$. But the left-hand side of formula (17.21) makes no reference to $A$. Any finite-dimensional algebra $B$ PI equivalent to $A$ will produce a module structure. A priori the two commutative algebras $\mathcal{T}_A(X), \mathcal{T}_B(X)$ need not be isomorphic and $K_A(f)$ need not be faithful under them. What is true is that they both induce the same commutative algebra of operators on $K_A(f)$.

In fact, both are quotients of the algebra $\mathcal{T}_t(X)$ of invariants of $t \times t$ matrices. We may just think of $K_A(f)$ as a module over this last algebra, without reference to a particular choice of $A$. The annihilator in $\mathcal{T}_t(X)$ of the module $K_A(f)$ is, on the other hand, intrinsic and thus a subtle PI invariant of $A$; see §18.1.3.2.

With $A$ as before we thus have

COROLLARY 17.3.12. *The T-ideal $I_S$ of $\mathcal{F}_A$ generated by all evaluations of any set $S$ of $\mu$-Kemer polynomials for $\mu$ large is a $\mathcal{T}_A$ submodule and thus a common ideal in $\mathcal{F}_A$ and $\mathcal{T}_A\mathcal{F}_A$.*

COROLLARY 17.3.13. *The T-ideal generated by $\mu$-Kemer polynomials, $\mu > \mu_0$ is a bimodule over the relatively free trace algebra $\mathcal{F}_{\bar{A}}(X)\mathcal{T}_{\bar{A}}(X)$ of $\bar{A}$ by right and left multiplication.*

$$\mathcal{F}_{\bar{A}}(X)\mathcal{T}_{\bar{A}}(X) \subset \bigoplus_i \mathcal{F}_{\bar{A}_i}(X)\mathcal{T}_{\bar{A}_i}(X).$$

PROOF. When we multiply a Kemer polynomial by any element $z$ of the free algebra as function, this depends only on the value of $z$ modulo the radical. This together with the previous theorem give the result. □

The main consequence of these last statements is for $A$ an arbitrary finite-dimensional algebra, and $\mathcal{F}_A$ its relatively free algebra.

THEOREM 17.3.14. *The algebra $\mathcal{F}_A/I_S$, where $I_S$ is an ideal generated by all evaluations of any set $S$ of $\mu$-Kemer polynomials for $\mu$ large, is representable hence PI equivalent to a finite-dimensional algebra.*

PROOF. By the usual argument using a Capelli polynomial, Corollary 7.3.7, we may reduce to the case in which $\mathcal{F}_A$ is in finitely many variables.

We may assume up to PI equivalence, that $A = \bigoplus_i A_i$ is a direct sum of fundamental algebras. Decompose this direct sum as $B \oplus C$ where $B$ is the direct sum of fundamental algebras with the same Kemer index as $A$ and $C$ of smaller Kemer index. We then embed $i : \mathcal{F}_A \subset \mathcal{F}_B \oplus \mathcal{F}_C$. Next the Kemer $\mu$-polynomials for high $\mu$ are PI of $C$ so we have that, under the inclusion $i$, the ideal $I_S \subset \mathcal{F}_B$. Hence $\mathcal{F}_A/I_S \subset \mathcal{F}_B/I_S \oplus \mathcal{F}_C$. Since clearly $\mathcal{F}_C$ is representable, it is enough to show that $\mathcal{F}_B/I_S$ is representable. By Corollary 17.3.12 we have that $I_S$ is also an ideal in $\mathcal{T}_B\mathcal{F}_B$ so that $\mathcal{F}_B/I_S \subset \mathcal{T}_B\mathcal{F}_B/I_S$. This last algebra is a finite module over a finitely generated algebra, hence, by Proposition 11.5.5, it is representable. So also the subalgebra $\mathcal{F}_B/I_S$ is representable.

By Corollary 11.5.18 it follows that $\mathcal{F}_A/I_S$ is PI equivalent to some finite-dimensional $F$-algebra. □

Finally for $\Gamma$ the $T$-ideal of a finitely generated PI algebra we have

PROPOSITION 17.3.15.

(i) *There exists a finite-dimensional algebra $A$ such that $\Gamma \supseteq \mathrm{Id}(A)$ and $\mathrm{ind}(A) = \mathrm{ind}(\Gamma)$.*

(ii) *Let $I_\mu$ denote the T-ideal in $\mathcal{F}_A$ generated by the $\mu$ Kemer polynomials of $A$.*

*We can choose $A$ and $\mu$ big enough that $(\Gamma/\mathrm{Id}(A)) \cap I_\mu = 0$. Thus that we may say that $\Gamma$ and $\mathrm{Id}(A)$ have the same Kemer $\mu$-polynomials.*

PROOF. We know, Lemma 17.1.3, that there are finite-dimensional algebras $A$ with $\Gamma \supseteq \mathrm{Id}(A)$. Take $A$ of previous type with minimal Kemer index $\kappa_0 \geq \kappa$. Set $\bar{\Gamma} := \Gamma/\mathrm{Id}(A) \subset \mathcal{F}_A$.

Let $I_\mu$ be the $T$-ideal in $\mathcal{F}_A$ generated by Kemer $\mu$-polynomials of $A$ for $\mu$ large. If $\kappa_0 > \kappa$, we have that the ideal $I_\mu \subset \bar{\Gamma}$. But, by Theorem 17.3.14, $\mathcal{F}_A/I_\mu$ is the relatively free algebra of a finite-dimensional algebra $A'$ with Kemer index $< \kappa_0$, so we may replace $A$ by $A'$, a contradiction.

Once we have that $A$ satisfies (i) we can easily modify it so to satisfy (ii) for the Kemer polynomials.

Apply Theorem 17.3.14 to $\mathcal{F}_A/\bar{\Gamma} \cap I_\mu$ which again is the relatively free algebra of a finite-dimensional algebra $A'$ with $\Gamma \supseteq \mathrm{Id}(A')$ and $\mathrm{ind}(A') = \mathrm{ind}(\Gamma)$.

By Remark 17.2.6 the $T$-ideal generated by the Kemer $\mu$-polynomials of $A'$ is $\bar{I}_\mu = I_\mu/\bar{\Gamma} \cap I_\mu$. Finally by construction $\Gamma/\mathrm{Id}(A') = \bar{\Gamma}/\bar{\Gamma} \cap I_\mu$ and

$$(\Gamma/\mathrm{Id}(A')) \cap \bar{I}_\mu = 0. \qquad \square$$

17.3.1.3. *A Cayley–Hamilton theorem.* Let $A$ be a finite-dimensional algebra of Kemer index $(t, s)$. Take a $\mu$-Kemer polynomial $g(Y, w) \in \mathcal{F}_A(X)$, $\mu > \mu_0$, with $Y$ the variables containing the layers and depending also linearly on some extra variable $w$. Then we have the following Cayley–Hamilton theorem which will be used repeatedly later:

THEOREM 17.3.16. *For every element $a \in \mathcal{F}_A(X)$ we have*

$$g(Y, a^{t+j}) + \sum_{i=1}^{t} (-1)^i \sigma_i(a) g(Y, a^{t+j-i}) = 0, \ \forall j \geq 0. \tag{17.22}$$

PROOF. We may assume that $A = \bigoplus_i A_i$ is a direct sum of fundamental algebras. On all the summands which have strictly lower Kemer index, already the polynomial $g$ vanishes. Restricting to the other summands, we may assume that $A$ is fundamental and so $\bar{A} = A/J$ has dimension $t$. By Proposition 17.3.8 any evaluation of $g$ in $A$ factors through evaluation of $w$ in $\bar{A}$, that is as a function we have

$$g(Y, a^{t+j}) + \sum_{i=1}^{t} (-1)^i \sigma_i(a) g(Y, a^{t+j-i}) = g(Y, \bar{a}^{t+j} + \sum_{i=1}^{t} (-1)^i \sigma_i(\bar{a}) \bar{a}^{t+j-i}) = 0,$$

since $\bar{a}^{t+j} + \sum_{i=1}^{t} (-1)^i \sigma_i(\bar{a}) \bar{a}^{t+j-i} = 0$ by the definition of $\sigma_i(a)$ and the usual Cayley–Hamilton theorem applied to multiplication by $a$ on $\bar{A}$. $\square$

**17.3.2. The Phoenix property.** Let $\Gamma = \mathrm{Id}(A) \subset F_+\langle X \rangle$ be the $T$-ideal of identities of a finitely generated algebra $A$. For a set $\mathcal{S} \subset F_+\langle X \rangle$ denote by $\langle \mathcal{S} \rangle$ the $T$-ideal generated by $\mathcal{S}$.

DEFINITION 17.3.17. We say that a set $S \subset F_+\langle X \rangle \setminus \Gamma$ satisfies the Phoenix property, relative to $\Gamma$, if given any $a \in \langle \mathcal{S} \rangle$, with $a \notin \Gamma$, then $S \cap \langle a \rangle \neq \emptyset$.

A further step for the proof of Kemer's representability is the following. Let $\Gamma$ be as before.

THEOREM 17.3.18. *The set $S$ of $\mu$-Kemer polynomials for $\Gamma$, for all $\mu$ sufficiently large, satisfies the Phoenix property relative to $\Gamma$.*

PROOF. By definition $S \subset F_+\langle X \rangle \setminus \Gamma$. By Proposition 17.3.15 there is a finite-dimensional algebra $A = \bigoplus_i A_i$, a direct sum of fundamental algebras $A_i$, such that $\Gamma \supset \mathrm{Id}(A)$ and $\Gamma$ and $\mathrm{Id}(A)$ have the same Kemer index and $\mu$-Kemer polynomials for $\mu$ large.

The Phoenix property relative to $\Gamma$ follows from the same property relative to $\mathrm{Id}(A)$. In fact if $g$ is a $\mu$-Kemer polynomial for $\Gamma$, by hypothesis it is also one for $A$. If $a \in \langle f \rangle \notin \Gamma$, then $a \notin \mathrm{Id}(A)$, so if $\mathrm{Id}(A)$ satisfies the Phoenix property there is a Kemer polynomial for $\mathrm{Id}(A)$ in $\langle a \rangle$. But by hypothesis this is also a Kemer polynomial for $\Gamma$.

Let us prove thus the Phoenix property for $\mathrm{Id}(A)$.

If $g$ is a $\mu$-Kemer polynomial for $A$, it is also fundamental for the algebras $A_i$ with the same Kemer index as $A$ where it is nonzero (Proposition 17.2.45), and it is a PI of all the summands $A_j$ with lower Kemer index. From Remark 17.2.32 it follows that if a polynomial $a \in \langle f \rangle$ is not a PI of $A$, then it is also fundamental for one algebra $A_{i_0}$ of the same list of $A_i$ which has the same Kemer index as $A$.

When we multilinearize $a$, we have also a fundamental polynomial for $A_{i_0}$, and the Phoenix property follows from Theorem 17.2.41. This shows that in the $T$-ideal generated by a fundamental polynomial, we can find $\mu$-Kemer polynomials for all large $\mu$. □

### 17.4. The representability theorem, Theorem 17.1.1

**17.4.1. A crucial lemma.** Let $\Gamma$ be a $T$-ideal in countably many variables $\mathbb{X}$ with Kemer index $(m, s)$, and let $\mu_0$ be as in Remark 17.2.4.

Take a $\mu + 1$-Kemer polynomial $g$, for $\Gamma$, for $\mu \geq \mu_0$ and single out one of the small layers $(x_1, \dots, x_m;)$ so $f = g(x; Z)$.

Let $A$ be a finite-dimensional algebra, which exists by Proposition 17.3.15 with $\Gamma \supseteq \mathrm{Id}(A), \mathrm{ind}(A) = \mathrm{ind}(\Gamma)$, and set $\bar{\Gamma} := \Gamma / \mathrm{Id}(A) \subset \mathcal{F}_A$.

Hence $\Gamma$ and $\mathrm{Id}(A)$ have the same Kemer $\mu$-polynomials in the sense of Proposition 17.3.15(ii). That is the $T$-ideal generated by the $\mu$-Kemer polynomials of $\Gamma$ is identified modulo $\Gamma$ to the $T$-ideal generated by the $\mu$-Kemer polynomials of $A$ in the relatively free algebra of $A$. So we treat it as functions on $A$.

We then restrict the variables to a finite list

$$X = U \cup W; \; W := Z \cup \{x_1, \dots, x_m\}, \tag{17.23}$$

$|U| > 2r$ and $\Gamma$ contains the Capelli list $\mathcal{C}_r$, for some $r$, and $|Z| = \mu_0 m + s(m+1)$.

We decompose $Z$ into $\mu_0$ layers $Z_1, \dots, Z_{\mu_0}$ with $m$ elements and $s$ layers $\bar{Z}_1, \dots, \bar{Z}_s$ with $m+1$ elements.

The condition on $|U|$ ensures that every polynomial $f$ in any number of variables which is not in $\Gamma$ has a nonzero evaluation in $F\langle U \rangle / (F\langle U \rangle \cap \Gamma)$.

By abuse of notation we still denote by $\Gamma$ the $T$-ideal $F\langle X \rangle \cap \Gamma$ in $F\langle X \rangle$.

DEFINITION 17.4.1. We take for $\mathbb{M} \subset F\langle X \rangle / \Gamma$ the space formed by the polynomials which are alternating and multilinear in the $\mu_0$ small layers and the $s$ big layers in which we have decomposed the variables in $Z$ and the small layer formed by $\{x_1, \dots, x_m\}$, as in formula (17.23). They may depend on the other variables $U$ in any form.

The space $\mathbb{M}$ is contained in the space of Kemer $\mu_0$-polynomials for $\Gamma$, and so if we choose $\mu_0$ sufficiently large, we may identify it with a corresponding space of functions on $A$ (Proposition 17.3.15). Corollary 17.3.10 or formula (17.21) shows that $\mathbb{M}$ is also a module under $\mathcal{T}_{\bar{A}}(U)$ (Definition 17.3.1). In fact if in formula (17.21), applied to an element of $\mathbb{M}$, we take for $z$ an element of $F\langle U \rangle$, we still have an element of $\mathbb{M}$.

In particular a $\mu$-Kemer polynomial $g(\mathbb{X})$ with $\mu > \mu_0 + 1$ has a nonzero evaluation in $F\langle U\rangle/(F\langle U\rangle \cap \Gamma)$. Clearly we may factor this through an evaluation in which one small layer, say $\{\mathtt{x}_1, \dots, \mathtt{x}_m\}$, is substituted with $\{x_1, \dots, x_m\}$, then another $\mu_0$ small layers and the $s$ big layers, in a 1–1 way, in the corresponding layers of the variables $Z$ and the remaining variables in $F\langle U\rangle$. This evaluation is nonzero in $\mathbb{M}$, we call this a *special evaluation* denoted $\bar{g}(X) \in F\langle X\rangle/\Gamma$.

Observe that such an evaluation is also a Kemer polynomial, although in the variables $X$.

We apply the Cayley–Hamilton theorem, Theorem 17.3.16, to Kemer polynomials in $\mathbb{M}$.

A CRUCIAL LEMMA. We now arrive at the crucial lemma. The hypotheses on $\Gamma$ and the notations $\mathbb{M}$ are those of the previous section.

Denote for simplicity $R := F\langle X\rangle/\Gamma$, $S := F_+\langle U\rangle/\Gamma \cap F_+\langle U\rangle \subset R$ (cf. Lemma 2.1.15). Now consider the trace algebra $\mathcal{T}_{\bar{A}}(U)$ (quotient of generic $m \times m$ matrices) in the variables $U$, and let $\mathfrak{R} = \mathcal{T}_{\bar{A}}(U) \otimes_F R$. We identify $R$ with $1 \otimes R$.

The ideal $K_\mu \subset R$ generated by $\mu$-Kemer polynomials is already a $\mathcal{T}_{\bar{A}}(U)$ module, thus we have a map,

$$\pi : \mathcal{T}_{\bar{A}}(U) \otimes K_\mu \to K_\mu = 1 \otimes K_\mu;\ \pi(\tau \otimes f) := \tau \cdot f = 1 \otimes \tau \cdot f. \tag{17.24}$$

The elements $\tau \otimes f - \tau \cdot f$ are in the kernel of $\pi$. $\pi$ is the identity on $K_\mu = 1 \otimes K_\mu$.

For each $a \in S$, let $H_a = a^{m+1} - \sigma_1(a) \otimes a^m + \cdots \pm \sigma_m(a) \otimes a$, where $t^m + \sum_{i=1}^m (-1)^i \sigma_i(a) t^{m-i} \in \mathcal{T}_{\bar{A}}(U)[t]$ is the characteristic polynomial of the left multiplication by $\bar{a}$ on $\bar{A}$ (as $F^q$ algebra). Consider finally in the algebra $\mathfrak{R}$ the ideal $I$ generated by the the elements $H_a$ for all $a \in S$.

LEMMA 17.4.2. *A nonzero element $f$ of $\mathbb{M}$ remains nonzero in $\mathfrak{R}/I$.*

PROOF. Assume by contradiction that $f \in \mathbb{M}$ lies in $I$. Then there are elements $f_{i,a}, u_{i,a} \in R$ and $\tau_{i,a} \in \mathcal{T}_{\bar{A}}(U)$ such that

$$\begin{aligned} f = \sum_{i,a} (\tau_{i,a} \otimes f_{i,a}) H_a u_{i,a} &= \sum_{i,a} \tau_{i,a} \otimes f_{i,a} a^{m+1} u_{i,a} \\ &+ \sum_{i,a} \sum_{j=1}^{m} (-1)^j \tau_{i,a} \sigma_j(a) \otimes f_{i,a} a^{m+1-j} u_{i,a}. \end{aligned} \tag{17.25}$$

The left-hand side of this equality is homogeneous of degree 1 in all the variables $W$.

We may first develop each $f_{i,a}$ and each $u_{i,a}$ into its homogeneous components with respect to $W$ so, since $H_a$ is independent of $W$, we may assume that each term $f_{i,a} H_a u_{i,a}$ is homogeneous in $W$, and finally since $\mathfrak{R}$ is multigraded in these variables, we may assume that each term $f_{i,a} H_a u_{i,a}$ is homogeneous of degree 1 in all the variables $W$. Using an auxiliary variable $w$, this term is the evaluation, for $w = H_a$, of the term $f_{i,a} w u_{i,a}$, homogeneous of degree 1 in all the variables $W \cup \{w\}$.

As next step we may apply the alternation operators Alt on both sides of formula (17.25) to all the layers, small and big of $Z \cup \{x_1, \dots, x_m\}$.

The left-hand side remains fixed. As for the right-hand side, since the elements $\tau_{i,a}$, $H_a$ do not depend on the variables in which we alternate, each $f_{i,a} H_a u_{i,a}$, is replaced by the evaluation, for $w = H_a$, of the alternation $\mathtt{f}_{i,a}(\dots, w) := \mathrm{Alt}(f_{i,a} w u_{i,a})$ of $f_{i,a} w u_{i,a}$.

Therefore it follows that $\mathtt{f}_{i,a}(\dots,w) = \mathrm{Alt}(f_{i,a}wu_{i,a})$ is a $\mu$-Kemer polynomial, or 0, and this is an identity inside $\mathcal{T}_{\bar{A}}(U)\otimes K_\mu$. We then apply the map $\pi$ of formula (17.24) and see that the right-hand side vanishes by Theorem 17.3.16. □

**17.4.2. An auxiliary algebra.** Before finishing the proof of the main representability theorem, we need a final

PROPOSITION 17.4.3. *Let $\Gamma = \mathrm{Id}(R)$ be the T-ideal of PIs of a PI algebra R finitely generated over F with Kemer index $(m,s)$. Then, there exists an algebra A which is finite dimensional over F with the following properties:*

(1) $\Gamma \subset \mathrm{Id}(A)$.
(2) *All $\mu$-Kemer polynomials for $\Gamma$, with $\mu$ large, are not identities of A. In particular $\Gamma$ and A have the same Kemer index.*

PROOF. First we restrict to $X = U \cup W$, finitely many variables, as defined in formula (17.23), and so we may assume $R = F\langle X\rangle/(\Gamma \cap F\langle X\rangle)$.

We can thus apply Lemma 17.4.2. Take the Shirshov basis for the algebra $R = F\langle X\rangle/\Gamma$ (Definition 8.1.12) associated to the generators $X$. We divide this basis into two parts: the elements $a_1,\dots,a_k$ give a Shirshov basis for the subalgebra generated by the elements $U$, and the other $b_i$ containing at least one of the remaining variables $W$. By definition these elements are monomials, hence multihomogeneous in the variables $W$.

By definition every $\mu$-Kemer polynomial with $\mu > \mu_0 + 2$ has a nonzero evaluation in $\mathbb{M} \subset F\langle X\rangle/\Gamma$ defined in Definition 17.4.1. We construct a new algebra $\tilde{R}$ as follows. Let $R[t_{j,i}]$, $j = 1,\dots,m$, $i = 1,\dots,k$, be the algebra of polynomials over $R$ in $km$ variables. For each $i = 1,\dots,k$, set $\tilde{H}_{a_i} := {a_i}^{m+1} + t_{1,i}{a_i}^m + \cdots + t_{m,i}a_i$, and define

$$\tilde{R} := R[t_{j,i}]/J,\ J = \langle \tilde{H}_{a_i}\rangle.$$

So $J$ is the ideal generated by the elements $\tilde{H}_{a_i}$. Modulo $J$ the elements $a_i$ of the Shirshov basis become integral over $F[t_{j,i}]$.

Since the elements $t_{j,i}$ are indeterminates over $F$, we have a homomorphism $\rho : R[t_{j,i}] \to \mathfrak{R} = \mathcal{T}_{\bar{A}}(U) \otimes_F R$ mapping $t_{j,i} \mapsto (-1)^j\sigma_j(a_i)$. Under this homomorphism $\tilde{H}_{a_i}$ is mapped to $H_{a_i} = a_i^{m+1} - \sigma_1(a_i)\otimes a_i^m + \cdots \pm \sigma_m(a_i)\otimes a_i$. By the definition of the ideal $I \subset \mathfrak{R}$, we have $\lambda(J) \subset I$, so $\lambda$ factors to a map of $\tilde{R}$ to $\mathfrak{R}/I$.

By Lemma 17.4.2, we deduce that the space $\mathbb{M}$ projects isomorphically to its image in $\tilde{R}$ so every $\mu$-Kemer polynomial has a nonzero evaluation in the image $\mathbb{M} \subset \tilde{R}$.

Now consider the ideal $L$ of $\tilde{R}$ sum of all the elements which are homogeneous of degree at least 2 in at least one of the variables $W$, and let $B := \tilde{R}/L$. The elements $b_j$ of the Shirshov basis satisfy $b_j^2 \in L$, hence $b_j^2 = 0$ in $B$. It follows that $B$ is a finite module over $F[t_{j,i}]$ since now all the elements of the Shirshov basis have become integral, moreover no special evaluation of a Kemer polynomial lies in $L$, by homogeneity. Therefore the algebra $B$ is representable so PI equivalent to a finite-dimensional algebra $A$ which satisfies all the requirements of Proposition 17.4.3. □

**17.4.3. Proof of Theorem 17.1.1.** We prove Theorem 17.1.1 by induction on the Kemer index. We take a $T$-ideal $\Gamma \subset F_+\langle X\rangle$ of identities of a finitely generated algebra $X$, a countable set of variables. We know then that it contains a Capelli list

(Definition 7.3.3), and it is generated as $T$-ideal by its intersection with a finitely generated free algebra $F_+\langle\Lambda\rangle$ ($\Lambda \subset X$ finite), in particular it is the $T$-ideal of identities of the finitely generated algebra $F_+\langle\Lambda\rangle/(\Gamma \cap F_+\langle\Lambda\rangle)$ (Theorem 7.3.9).

Consider its Kemer index. If this index is $(0,s)$, then $F_+\langle\Lambda\rangle/\Gamma \cap F_+\langle\Lambda\rangle$ is finite dimensional, and there is nothing to prove.

Assume the index of $\Gamma$ is $(m,s)$ with $m > 0$. By Proposition 17.4.3 we have a finite-dimensional algebra $A$ and some $\mu > 0$ which satisfies $\Gamma \subset \mathrm{Id}(A)$. $A$ and $\Gamma$ have the same Kemer index and all Kemer polynomials, for $\Gamma$, with at least $\mu$ layers with $m$ elements are not identities of $A$.

With this $\mu$ consider the $T$-ideal $\Gamma' := \Gamma + K$ where $K$ is the $T$-ideal of $F_+\langle X\rangle$ generated by all evaluations of Kemer polynomials $\mathcal{S}$ for $\Gamma$ with at least $\mu$ layers with $m$ elements.

Since $\Gamma' \supset \Gamma$, it still contains a Capelli list so (Theorem 7.3.9) $\Gamma'$ is the ideal of identities of a finitely generated algebra, and we clearly have that the Kemer index of $\Gamma'$ is strictly less than that of $\Gamma$ since all evaluations of Kemer polynomials $\mathcal{S}$ for $\Gamma$ with at least $\mu$ layers with $m$ elements are in $\Gamma'$. By induction $\Gamma'$ is the $T$-ideal of identities of a finite-dimensional algebra $B$.

We claim that $\Gamma$ is the ideal of identities of $A \oplus B$. In other words we need to prove that

$$\Gamma = \mathrm{Id}(A) \cap \mathrm{Id}(B) = \mathrm{Id}(A) \cap \Gamma'.$$

We have by construction that $\Gamma \subset \mathrm{Id}(A) \cap \Gamma'$. So let $f \in \mathrm{Id}(A) \cap \Gamma'$, and suppose that $f \notin \Gamma$. Since $f \in \Gamma'$ we have that there is a $g$ with $f - g \in \Gamma$, and $g \in K$ is in the $T$-ideal generated by the $\mu$-Kemer polynomials $\mathcal{S}$ of $\Gamma$. By the Phoenix property of §17.3.2 we have that there is a Kemer polynomial $g'$ for $\Gamma$ in the $T$-ideal generated by $g$. Since $f \in \mathrm{Id}(A) \cap \Gamma'$, we also have $g' \in \mathrm{Id}(A) \cap \Gamma'$. But then $g'$ is an identity of $A$, a contradiction, since $A$ is constructed in such a way that the $\mu$ Kemer polynomials of $\Gamma$ are not PI of $A$. This proves Theorem 17.1.1. □

## 17.5. The abstract Cayley–Hamilton theorem

**17.5.1. A theorem of Zubrilin.** *This section is a complement. We present one of the main ideas in the papers of Razmyslov on the fact that the nil radical of a finitely generated PI algebra is nilpotent in the form given by his student Zubrilin* **[Zub95b]**.

17.5.1.1. *Antisymmetry.* A $\mu$-Kemer polynomial $g$, for a $T$-ideal $\Gamma$ of Kemer index $(t,s)$, has the following property. Take $t+1$ variables appearing in $g$ linearly, and outside the big layers, and alternate $g$ in these variables. Then provided $\mu$ is sufficiently large (for instance $\mu > \mu_0 + t + 1$), the resulting polynomial lies in $\Gamma$. In fact this polynomial has now $s+1$ big layers and at least $\mu - t - 1$ small layers.

In this section we show that this is the essential property which has been used in the proof of Theorem 17.3.16 and its corollary.

In this section we work with algebras over general commutative rings $A$. In this setting the notion of symmetry is delicate. When we have an action of the symmetric group $S_n$ on some $A$-module $M$, we say that an element $a \in M$ is *antisymmetric* if $\sigma(a) = \epsilon_\sigma a$ for all $\sigma \in S_n$, and we denote $M^a$ the space of antisymmetric elements.

A way of constructing antisymmetric elements is, given $b \in M$, to form the sum $\mathrm{Alt}(b) := \sum_{\sigma \in S_n} \epsilon_\sigma \sigma(b)$. If $n!$ is invertible in the coefficient ring, then this can be made into a projection $\frac{\mathrm{Alt}}{n!}$ of $M \to M^a$. Otherwise it is easy to give examples in which not every element of $M^a$ is of the form $\text{Alt}(b)$ for some $b$. The most trivial is $M = M^a$, the one-dimensional representation of $S_n$ given by the sign; here the alternation operator is 0 if the characteristic $p$ is $\leq n$.

REMARK 17.5.1. When we consider, in a free algebra, a polynomial which is multilinear and antisymmetric in some of its variables, it can always be written as a linear combination of antisymmetrizations of monomials in those variables.

17.5.1.2. *The abstract Cayley–Hamilton theorem.* We start by proving an abstract version of Theorem 17.3.16, extracting some formal properties of Kemer polynomials.

Let $\Gamma$ be a $T$-ideal, and let $g(x_1, \dots, x_m, x_{m+1}, y) \notin \Gamma$ be a polynomial multilinear in the variables $x_1, \dots, x_m, x_{m+1}$ and skew symmetric (alternated) in $x_1, \dots, x_m$ (and may depend on other variables $y$).

A special case may be when $g(x_1, \dots, x_m, x_{m+1}, y) = x_{m+1}h(x_1, \dots, x_m, y)$.

**Assume that:** if we first substitute in $g$, $x_{m+1} \mapsto zx_{m+1}$ with $z$ some extra variable, and alternate the polynomial $g(x_1, \dots, x_m, zx_{m+1}, y)$ in all the variables $x_1, \dots, x_m, x_{m+1}$, we have that the resulting polynomial is in $\Gamma$.

Of course the same holds by replacing $z$ with any polynomial, and in particular with $z^t$ for some $t > 0$. If $\Gamma$ is a $T$-ideal in the category of algebras without 1, we assume that this is true also for $t = 0$.

Since $g$ is already skew symmetric in $x_1, \dots, x_m$, we define the alternation as the sum $\tilde{g} := \sum_{\sigma \in S_{m+1}/S_m} \epsilon_\sigma g(x_{\sigma(1)}, \dots, x_{\sigma(m)}, zx_{\sigma(m+1)}, y)$ on coset representatives of $S_{m+1}/S_m$. As coset representatives of $S_{m+1}/S_m$ we can take the identity 1, which has sign $+1$ and the $m$ transpositions $(i, m+1)$, $i = 1, \dots, m$, which have sign $-1$. Then we have that our assumption can be written as

(17.26)

$$g(x_1, \dots, x_m, zx_{m+1}, y) - \sum_{j=1}^{m} g(x_1, \dots, x_{j-1}, x_{m+1}, x_{j+1}, \dots, x_m, zx_j, y) \in \Gamma.$$

For any polynomial $h(x_1, \dots, x_m, W)$ multilinear in $x_1, \dots, x_m$ ($W$ some other variables), we can define the polarized form, relative to the variables $x_1, \dots, x_m$. We use a parameter $\lambda$ commuting with the variables

$$h((\lambda + z)x_1, \dots, (\lambda + z)x_m, W) = \sum_{q=0}^{m} \lambda^{m-q} \sigma_{q,z} h(x_1, \dots, x_m, W),$$

where

$$\sigma_{q,z} h := \sum_{1 \leq i_1 < \cdots < i_q \leq m} h(x_1, \dots, zx_{i_1}, \dots, zx_{i_q}, \dots, x_m, W), \; 0 \leq q \leq m. \tag{17.27}$$

Notice that $\sigma_{0,z}h = h$ and $\sigma_{q,z}$ is a linear operator on functions which are multilinear in the variables $x_1, \dots, x_m$. Since the formula is symmetric in the $x$, it commutes with permutations of the variables $x_1, \dots, x_m$.

In a similar way define the polarized form only for the variables $x_2, \dots, x_m$,

$$\tau_{q,z} h := \sum_{2 \leq i_1 < \cdots < i_q \leq m} h(x_1, \dots, zx_{i_1}, \dots, zx_{i_q}, \dots, x_m, W), \; 1 \leq q \leq m - 1. \tag{17.28}$$

So, for all $h$ and $1 \leq q \leq m-1$, we have

$$\sigma_{q,z}h(x_1,\dots,x_m,W) = \tau_{q,z}h(x_1,\dots,x_m,W) + \tau_{q-1,z}h(zx_1,x_2,\dots,x_m,W)$$

$$\text{and} \qquad \sigma_{0,z}h(x_1,\dots,x_m,W) = \tau_{0,z}h(x_1,\dots,x_m,W), \tag{17.29}$$

$$\sigma_{m,z}h(x_1,\dots,x_m,W) = \tau_{m-1,z}h(zx_1,x_2,\dots,x_m,W).$$

THEOREM 17.5.2 (Abstract Cayley–Hamilton theorem). *For $g$ satisfying formula (17.26) take a variable $z$. Then*

$$\sum_{q=0}^{m}(-1)^q\sigma_{q,z}g(x_1,\dots,x_m,z^{m-q}x_{m+1},y) \in \Gamma. \tag{17.30}$$

PROOF. Denote for simplicity by

$$G_j(\dots,x_j) := g(x_1,\dots,x_{j-1},x_{m+1},x_{j+1},\dots,x_m,x_j,y).$$

Clearly, if $h = h(x_1,\dots,x_m,W) \in \Gamma$, we have that $\sigma_{q,z}h \in \Gamma$ thus for all $q$ and for all $t$, we have by formula (17.26):

$$\sigma_{q,z}g(x_1,\dots,x_m,z^tx_{m+1},y) = \sum_{j=1}^{m}\sigma_{q,z}G_j(\dots,z^tx_j), \qquad \text{modulo } \Gamma. \tag{17.31}$$

We substitute this identity in formula (17.30). Then proving equation (17.30) is equivalent to proving

$$\sum_{q=0}^{m}(-1)^q\sum_{j=1}^{m}\sigma_{q,z}G_j(\dots,z^{m-q}x_j) \in \Gamma. \tag{17.32}$$

This follows from the stronger condition, that we now show, for all $j$,

$$\sum_{q=0}^{m}(-1)^q\sigma_{q,z}G_j(\dots,z^{m-q}x_j) = 0. \tag{17.33}$$

By symmetry it is enough to do this for $j=1$. Denote by $G(\dots,x_1) = G_1(\dots,x_1)$. By formula (17.29) we have, for $1 \leq q \leq m-1$,

$$\sigma_{q,z}G(\dots,z^{m-q}x_1) = \tau_{q,z}G(\dots,z^{m-q}x_1) + \tau_{q-1,z}G(\dots,z^{m-q+1}x_1),$$

$$\sigma_{0,z}G(\dots,z^mx_1) = \tau_{0,z}G(\dots,z^mx_1), \quad \sigma_{m,z}G(\dots,x_1) = \tau_{m-1,z}G(\dots,zx_1).$$

We get that the left-hand side of formula (17.33) is

$$\tau_{0,z}G(\dots,z^mx_1) + \sum_{q=1}^{m-1}(-1)^q\Big(\tau_{q,z}G(\dots,z^{m-q}x_1) + \tau_{q-1,z}G(\dots,z^{m-q+1}x_1)\Big)$$
$$+(-1)^m\tau_{m-1,z}G(\dots,zx_1) = 0.$$ □

EXERCISE 17.5.3. Deduce from the abstract Cayley–Hamilton theorem the ordinary one.

*Hint.* Use the fact that generic matrices are a domain and use the Capelli polynomial.

17.5.1.3. *Combinatorial statement on the symmetric group.* Consider the symmetric group $S_{2m}$ on $X \cup Y$, with $X = \{1, \ldots, m\}$ and $Y = \{m+1, \ldots, 2m\}$. For any subset $Z \subset X$ define the element $T(Z) \in \mathbb{Z}[S_{2m}]$ as

$$T(Z) := \sum_{\sigma \in S_{2m} | \sigma(Z) \subseteq Y} \epsilon_\sigma \sigma. \tag{17.34}$$

Of course $T(\emptyset) = \sum_{\sigma \in S_{2m}} \epsilon_\sigma \sigma$. Notice that $\{\sigma \in S_{2m} \mid \sigma(X) = X\} \simeq S_m \times S_m$.

PROPOSITION 17.5.4. *We have the identity*

$$\sum_{Z \subseteq X} (-1)^{|Z|} T(Z) = \sum_{\sigma(X)=X} \epsilon_\sigma \sigma = \sum_{(\sigma,\tau) \in S_m \times S_m} \epsilon_\sigma \epsilon_\tau \sigma\tau. \tag{17.35}$$

PROOF. For $\sigma \in S_{2m}$ let $Z(\sigma) = \{i \in X \mid \sigma(i) \in Y\}$. We have

$$\sum_{Z \subseteq X} (-1)^{|Z|} T(Z) = \sum_{Z \subseteq X} (-1)^{|Z|} \sum_{\sigma \,|\, \sigma(Z) \subseteq Y} \epsilon_\sigma \sigma = \sum_{\sigma \in S_{2m}} \Big[ \sum_{Z \subset Z(\sigma)} (-1)^{|Z|} \Big] \epsilon_\sigma \sigma.$$

For any finite set $A$ we have $\sum_{Z \subseteq A} (-1)^{|Z|} = 1$ when $A = \emptyset$ and $= 0$ otherwise. Clearly $Z(\sigma) = \emptyset$ if and only if $\sigma(X) = X$, and the claim follows. □

It is better to write formula (17.35) singling out the term $(-1)^{|X|} T(X)$ and reformulate it as in formula (17.36) which will be used in Lemma 17.5.6.

The element $T(X)$ is the sum, with signs, of the $(m!)^2$ permutations in $S_{2m}$ with $\sigma(X) = Y$, $\sigma(Y) = X$.

These permutations are in fact the full coset $(S_m \times S_m)\gamma = \gamma(S_m \times S_m)$ of the subgroup $S_m \times S_m$ times the permutation $\gamma = \prod_{i=1}^m (i, m+i)$. The permutation $\gamma$ is the product of $m$ transpositions, hence it has sign $(-1)^m$ and $\gamma^2 = 1$. Thus $(-1)^{|X|} T(X) = \sum_{\sigma(X)=X} \epsilon_\sigma \sigma \gamma$.

We deduce

$$\sum_{\sigma(X)=X} \epsilon_\sigma \sigma - \Big( \sum_{\sigma(X)=X} \epsilon_\sigma \sigma \Big) \gamma = \sum_{Z \subsetneq X} (-1)^{|Z|} T(Z). \tag{17.36}$$

It is useful to give a more explicit description of each $T(Z)$.

For any $Z \subset X$, set $U_Z := \{\sigma \in S_{2m} \mid \sigma(Z) \subset Y\}$ so that $T(Z) = \sum_{\sigma \in U_Z} \epsilon_\sigma \sigma$. Consider the group $G_Z := \{\sigma \in S_{2m} \mid \sigma(Z) \subset Z\}$, this equals the product of the group $S_Z$ of permutations of $Z$ times the group $S_{Y \cup X \setminus Z}$ of permutations of $Y \cup X \setminus Z$. Multiplication to the right by elements of $G_Z$ clearly preserves the set $U_Z$.

We may choose a set $R_Z$ of representatives of the orbits of this action, for instance for each subset $Z' \subset Y$ with $|Z| = |Z'|$ the permutation, *a shuffle*, which maps $Z$ to $Z'$ in increasing order and the same for the complement.

We thus finally have

PROPOSITION 17.5.5.

$$T(Z) = \sum_{\tau \in R_Z} \epsilon_\tau \tau \Big( \sum_{\sigma \in S_Z} \epsilon_\sigma \sigma \Big) \Big( \sum_{\sigma \in S_{Y \cup X \setminus Z}} \epsilon_\sigma \sigma \Big). \tag{17.37}$$

*Application to PI theory.* Let $g(x; y) := g(x_1, \ldots, x_m; y_1, \ldots, y_m)$ be a polynomial multilinear and alternating separately in the variables $x_1, \ldots, x_m$ and $y_1, \ldots, y_m$ and possibly depending on other variables $U$. Set $\mathrm{Alt}_{m,m} := \sum_{(\sigma,\tau) \in S_m \times S_m} \epsilon_\sigma \epsilon_\tau \sigma \tau$, where the first $S_m$ permutes the variables $x_1, \ldots, x_m$ and the second the variables

$y_1, \dots, y_m$. By definition there is a polynomial $g(x_1, \dots, x_m; y_1, \dots, y_m)$, multilinear in the variables $x$ and in the variables $y$, such that we have

$$g(x_1, \dots, x_m; y_1, \dots, y_m) = \mathrm{Alt}_{m,m}\, g(x_1, \dots, x_m; y_1, \dots, y_m). \tag{17.38}$$

LEMMA 17.5.6. *We have*

$$g(x_1, \dots, x_m; y_1, \dots, y_m) - g(y_1, \dots, y_m; x_1, \dots, x_m) \tag{17.39}$$

*vanishes modulo the T-ideal $\Gamma$ obtained from $g(x; y)$ by alternating any $m+1$ variables among the $x_1, \dots, x_m; y_1, \dots, y_m$.*

PROOF. We have the action of the symmetric group $S_{2m}$ permuting the variables $x_1, \dots, x_m; y_1, \dots, y_m$ and thus acting on the polynomial $g(x; y)$. The two sides of formula (17.36) thus give two equal operators in the group algebra $\mathbb{Z}[S_{2m}]$, which when applied to $g(x; y)$ give two equal polynomials.

$$\Big( \sum_{\sigma(X)=X} \epsilon_\sigma \sigma - \Big( \sum_{\sigma(X)=X} \epsilon_\sigma \sigma \Big) \gamma \Big) g(x; y) = \Big( \sum_{Z \subsetneq X} (-1)^{|Z|} T(Z) \Big) g(x; y). \tag{17.40}$$

By the definition of $\gamma$ we have $\gamma g(x; y) = g(y; x)$ and the left-hand side of formula (17.40) is the polynomial of formula (17.39).

We claim that the right-hand side is in $\Gamma$. In fact, by formula (17.37), each $T(Z)$, $Z \subsetneq X$, is the product of the operator $\sum_{\tau \in R_Z} \epsilon_\tau \tau (\sum_{\sigma \in S_Z} \epsilon_\sigma \sigma)$ by $(\sum_{\sigma \in S_{Y \cup X \setminus Z}} \epsilon_\sigma \sigma)$ which alternates at least $m+1$ variables $Y \cup (X \setminus Z)$ from $x_1, \dots, x_m; y_1, \dots, y_m$ in $g(x; y)$, so by hypothesis it maps $g(x; y)$ into $\Gamma$. $\square$

Let $g(x; y) := g(x_1, \dots, x_m; y_1, \dots, y_m)$ be a polynomial multilinear in the variables $x_1, \dots, x_m$ and $y_1, \dots, y_m$ and possibly depending on other variables $U$, and a $T$-ideal $\Gamma$.

DEFINITION 17.5.7. We say that the polynomial $g(x; y)$ satisfies Property A relative to $\Gamma$ if the following holds.

(1) There is a polynomial $g(x; y)$ with

$$g(x_1, \dots, x_m; y_1, \dots, y_m) = \mathrm{Alt}_{m,m}\, g(x_1, \dots, x_m; y_1, \dots, y_m).$$

(2) Take two extra variables $u, v$, then the polynomials obtained from

$$g(vx; uy), g(vx; y), g(x; uy), g(x; y),$$

by alternating $m+1$ among the variables $x_1, \dots, x_m; y_1, \dots, y_m$ (the last three follow from the first in the category of algebras with 1), lie in $\Gamma$.

PROPOSITION 17.5.8. *Assume that the polynomial $g(x; y)$ satisfies Property A relative to a T-ideal $\Gamma$. Modulo the T-ideal $\Gamma$ we have*

$$g(ux_1, \dots, ux_m; vy_1, \dots, vy_m) = g(vx_1, \dots, vx_m; uy_1, \dots, uy_m) \tag{17.41}$$

$$= g(uvx_1, \dots, uvx_m; y_1, \dots, y_m) = g(vux_1, \dots, vux_m; y_1, \dots, y_m). \tag{17.42}$$

PROOF. We first show formula (17.41). All computations are done modulo $\Gamma$. We have $g(x; y) = \mathrm{Alt}_{m,m}\, g(x; y)$ and $g(vx, uy) = \mathrm{Alt}_{m,m}\, g(vx, uy)$. By Lemma 17.5.6 applied to the two previous cases, $g(x; y) = g(y; x)$ and $g(vx; uy) = g(vy; ux)$. Substituting in the first $x \mapsto ux$, $y \mapsto vy$ implies

$$g(ux; vy) = g(vy; ux) = g(vx; uy), \tag{17.43}$$

and we have formula (17.41). Similarly

$$(17.44)\qquad \text{(i)}\quad g(ux;y)=g(x;uy),\quad \text{(ii)}\quad g(vx;y)=g(x;vy).$$

We now substitute in formula (17.44)(i) $x\mapsto vx$ and in (ii) $x\mapsto ux$ and get

$$(17.45)\qquad g(uvx;y)\overset{(17.44)\text{ (i)}}{=}g(vx;uy)\overset{(17.43)}{=}g(ux;vy)\overset{(17.44)\text{ (ii)}}{=}g(vux;y).$$

The last equation proves formula (17.42). □

### 17.5.2. Connection with the theorem of Ziplies and Vaccarino.

17.5.2.1. *Some operators.* We start by working over a commutative ring $A$ of coefficients. Later we will assume that $A=F$ is a field of characteristic 0. Fix a $T$-ideal $\Gamma\subset A\langle X\rangle$ with $X$ some set of variables. We decompose the variables as $X=U\cup\{x_1,\dots,x_m,y_1,\dots,y_m\}$.

Consider the subset $M\subset A\langle X\rangle$ of all polynomials $g(x,y,U)\in A\langle X\rangle$ that satisfy Property A, of Definition 17.5.7, with respect to $\Gamma$.

PROPOSITION 17.5.9.

(1) *$M$ is a subspace.*
(2) *$M$ is stable under all endomorphisms of $A\langle X\rangle$ that send variables in $U$ into $A\langle U\rangle$.*
(3) *$M$ is also stable under all endomorphisms of $A\langle X\rangle$ which fix the variables in $U$ and for some $a,b\in A\langle U\rangle$ map $x_i\mapsto ax_i$, $\forall i$ and $y_i\mapsto by_i$, $\forall i$.*

PROOF. (1) This is essentially clear if $f_1,f_2$ satisfy Property A, of Definition 17.5.7, with respect to $\Gamma$. We first have $f_1=\mathrm{Alt}_{m,m}g_1$, $f_2=\mathrm{Alt}_{m,m}g_2$ so we deduce $\alpha f_1+\beta f_2=\mathrm{Alt}_{m,m}g$, $g=\alpha f_1+\beta f_2$.

One needs to prove that $g$ satisfies the second condition but this is really obvious.

(2) If $\Phi$ is such an endomorphism, it clearly commutes with antisymmetrization of the variables $x,y$ so that $\Phi(f)=\mathrm{Alt}_{m,m}\Phi(g)$. The fact that $\Phi(g)$ satisfies Definition 17.5.7(2) depends on the fact that $\Gamma$ is a $T$-ideal.

(3) This follows in a similar way as 2). Let us check just one of the conditions. Take $\bar g(x;y):=g(ax_1,\dots,ax_m,by_1,\dots by_m,U)$. We want to show that by alternating $m+1$ among the variables $x_1,\dots,x_m;y_1,\dots,y_m$ the polynomial obtained from $\bar g(ux,vy)$ is in $\Gamma$. But in fact $\bar g(ux,vy)=g(aux,bvy)$, so again this follows from the fact that $\Gamma$ is a $T$-ideal. □

We consider now the image $\bar M$ of $M$ modulo $\Gamma$. We may apply Proposition 17.5.8 to all elements $g(x,y,U)\in M$. We have thus on $\bar M$ the operators $\tau(u)$, $u\in A\langle U\rangle$ defined by

$$(17.46)\qquad \tau(u)g(x,y,U)=g(ux,y,U)\overset{\mathrm{mod}\ \Gamma}{=}g(x,uy,U).$$

By inspection of formula (17.46) we see that

PROPOSITION 17.5.10. *The map $u\mapsto\tau(u)$, as function of $u$, is a multiplicative polynomial map of degree $m$ from the free algebra $A\langle U\rangle$ to the opposite of the algebra $\mathrm{End}_A(\bar M)$ of operators on $\bar M$. By formula (17.42) its image is a commutative algebra of operators on $\bar M$.*

As with any such multiplicative map we can apply the theory developed in §4.2 and Proposition 4.1.11. Thus $\tau$ factors as

$$\begin{array}{ccc} A\langle U\rangle & \xrightarrow{u\mapsto u^{\otimes m}} & (A\langle U\rangle^{\otimes m})^{S_m}/I \\ & \searrow^{\tau} & \downarrow{\scriptstyle \bar\tau} \\ & & \mathrm{End}_A(\bar M) \end{array} \qquad \begin{array}{c} (A\langle U\rangle^{\otimes m})^{S_m} \overset{\text{Def. 4.1.10}}{=} \mathcal{S}^m(A\langle U\rangle) \\ \\ (A\langle U\rangle^{\otimes m})^{S_m}/I \overset{\text{Def. 4.2.14}}{=} \mathbb{A}^m_A(A\langle U\rangle) \end{array}$$

with $I$ the ideal generated by commutators.

PROPOSITION 17.5.11. *The map $u \mapsto \tau(u)$ induces a homomorphism $\bar\tau$ of the algebra $\mathbb{A}^m_A(A\langle U\rangle)$ to the algebra of linear operators on $\bar M$ with $\bar\tau(u^{\otimes m}) = \tau(u)$.*

By Theorem 4.2.12 the algebra $\mathcal{S}^m(A\langle U\rangle)$ is generated by the elements $\sigma_q(P)$, $q \leq m$, as $P$ runs over all primitive monomials in $U$. We have (see page 122) that $\mathbb{A}^m_A(A\langle U\rangle)$ is canonically isomorphic to the algebra of invariants of $m\times m$ matrices in variables $U$. Moreover an element $g(U) \in A\langle U\rangle$ maps into the corresponding polynomial $g(\bar U)$ in generic matrices $\bar U$, while $\sigma_q(g(U))$, $q \leq m$, maps into the $q$th coefficient of the characteristic polynomial of the corresponding polyomial $g(\bar U)$ in generic matrices.

Given $z \in A\langle U\rangle$ the element $\sigma_h(z)$ maps to $\tau_h(z)$ defined through the polarization $\tau(1+tz) = 1 + \sum_{h=1}^m t^h \tau_h(z)$,

$$\sum_{1\leq i_i<\cdots<i_h\leq m} g(x_1,\dots,zx_{i_1},\dots,zx_{i_h},\dots x_m,y) = \tau_h(z)g(x_1,\dots,x_m,y). \tag{17.47}$$

DEFINITION 17.5.12. We define $\mathcal{A}$ to be the algebra of linear operators on $\bar M$ image of the homomorphism of $\mathbb{A}^m_A(A\langle U\rangle)$ induced by the map $u\mapsto\tau(u)$, to the linear operators $\mathrm{End}_A(\bar M)$ on $\bar M$.

We deduce by Theorem 4.2.12 that

THEOREM 17.5.13. *$\mathcal{A}$ is a commutative algebra isomorphic to the quotient of the algebra of invariants of $m\times m$ matrices in variables $U$. It is generated by the elements $\tau_q(N)$ where $N$ runs over all primitive monomials and $1 \leq q \leq m$ which correspond to the invariants $\sigma_q(N)$.*

PROOF. For $u, v \in A\langle U\rangle$ we have that $\tau(u)\tau(v) = \tau(v)\tau(u)$ by formula (17.42). Since the elements $\tau(u)$ commute, the map factors modulo the abelianization $\mathbb{A}^m(A\langle U\rangle)$ of $\mathcal{S}^m(A\langle U\rangle)$. We then apply Theorems 4.2.12 and 4.2.20. □

17.5.2.2. *Kemer polynomials.* We finish by explaining what we started with, that is the hypotheses of Lemma 17.5.6 and Proposition 17.5.8, for some $T$-ideal $\Gamma$, are naturally satisfied by Kemer's polynomials for $\Gamma$. So the previous discussion is an alternative to the use of an auxiliary finite-dimensional algebra as in §17.3.1.

Let $\Gamma$ be a $T$-ideal with Kemer index $(m,s)$, and let $\mu_0$ be as in Remark 17.2.4. Take a $\mu+2$-Kemer polynomial $g$, for $\Gamma$, for $\mu \geq \mu_0$ and single out two of the small layers $(x_1,\dots,x_m; y_1,\dots,y_m)$ so $f = g(x;y,Z) = g(x;y)$.

PROPOSITION 17.5.14. *The polynomial $g(x;y)$ satisfies Property A, of Definition 17.5.7, relative to $\Gamma$.*

PROOF. Use the two small layers $(x_1, \dots, x_m; y_1, \dots, y_m)$ in $g$ in order to write as in formula (17.38)

$$g(x_1, \dots, x_m; y_1, \dots, y_m, Z) = \mathrm{Alt}_{m,m}\, g(x_1, \dots, x_m; y_1, \dots, y_m, Z).$$

Clearly $g(x_1, \dots, x_m; y_1, \dots, y_m, Z)$ is a $\mu$-Kemer polynomial. Its layers involve variables $Z$ disjoint from the variables $(x; y) = (x_1, \dots, x_m; y_1, \dots, y_m)$, the same is true for $g(vy; ux), g(vy; x), g(y; ux)$.

Now if we apply alternation to $m+1$ out of these variables $(x; y)$, in one of these polynomials we obtain a polynomial with $\mu$-alternating layers with $m$ elements and $s+1$ alternating layers with $m+1$ elements. By the definition of $\mu_0$ this polynomial lies in $\Gamma$. □

CHAPTER 18

# The relatively free algebras

*In this chapter we apply the theory developed in the previous chapter so to give various properties of the relatively free algebra associated to a finite-dimensional algebra $A$, with a particular eye at the case in which $A$ is fundamental.*

*We will need some basic facts of commutative algebra and invariant theory which are collected in §18.2. Some of these facts will later play a role in the study of the growth of the codimension in Chapter 22.*

## 18.1. Rationality and a canonical filtration

**18.1.1. Relation with Cayley–Hamilton algebra.** Let $A$ be an algebra presented as quotient $F\langle X\rangle/I$ of a free algebra. The condition that $A$ may be embedded in $n \times n$ matrices over some commutative ring can be formulated in different ways. For one thing it means that the universal morphism into $n \times n$ matrices over a commutative ring is injective. In characteristic 0, the condition can be formulated as follows. Consider the free algebra with trace $T\langle X\rangle$ and there the ideal $\mathrm{CH}_n$ generated by the $n$ Cayley–Hamilton polynomial, we have

PROPOSITION 18.1.1. *Let $A = F\langle X\rangle/I$ be presented as quotient of a free algebra, and let $F$ be a field of characteristic 0. Then $A$ can be embedded in $n \times n$ matrices over some commutative ring if and only if, denoting by $I_n$ the ideal in $T\langle X\rangle$ generated by $I$ and $\mathrm{CH}_n$, we have $I_n \cap F\langle X\rangle = I$.*

PROOF. If $A$ can be embedded into $M_n(B)$ with $B$ a commutative ring, the mapping $F\langle X\rangle \to A = F\langle X\rangle/I \to M_n(B)$ extends to a map of $T\langle X\rangle$ to $M_n(B)$ which factors through the ideal generated by $\mathrm{CH}_n$. The converse is the statement of Theorem 14.2.1. □

**18.1.2. The structure of relatively free algebras.** The approach in this section follows closely **[Pro16]**. Let thus $R := F\langle X\rangle/I$ be a relatively free algebra in a finite number $m$ of variables $X$.

We have proved that $I$ is the $T$-ideal of identities (in $m$ variables) of a finite-dimensional algebra $A = \bigoplus_i A_i$, a direct sum of fundamental algebras $A_i$. So, choosing such an $A$, we can write $R = \mathcal{F}_A(X)$ where we shall identify $\mathcal{F}_A(X)$ to the algebra of generic elements of $A$.

We can decompose the direct sum $A = \bigoplus_i A_i$ in two parts: $A = B \oplus C$, where $B$ is the direct sum of the $A_i$ with the same Kemer index as $A$; while $C$ is the sum of those of strictly lower Kemer index.

We have $\mathrm{Id}(A) = \mathrm{Id}(B) \cap \mathrm{Id}(C)$ and so, dropping $X$ for simplicity, we have an embedding $\mathcal{F}_A \subset \mathcal{F}_B \oplus \mathcal{F}_C$ of the corresponding relatively free algebras.

We can apply to $B$ Theorem 17.3.5 and embed $\mathcal{F}_B \subset \mathcal{F}_B\mathcal{T}_B$ in a finitely generated module over a finitely generated commutative algebra $\mathcal{T}_B$, (both graded).

Let $K_0 \subset \mathcal{F}_A$ be the $T$-ideal generated by the $\mu$-Kemer polynomials of $A$ for sufficiently large $\mu$. Since these polynomials are PI of $C$ it follows that, under the embedding $\mathcal{F}_A \subset \mathcal{F}_B \oplus \mathcal{F}_C$, the ideal $K_0$ maps isomorphically to the corresponding ideal in $\mathcal{F}_B$ which, by Theorem 17.3.14, is a finite module over $\mathcal{T}_B$.

The algebra $R_1 := R/K_0$ is also a relatively free algebra with strictly lower Kemer index so we can repeat the previous construction. Let $K_1 \subset R$ be the $T$-ideal with $K_0 \subset K_1$, and let $K_1/K_0$ be the $T$-ideal in $R_1$ generated by the corresponding Kemer polynomials of the analogous direct sums of finite-dimensional fundamental algebras.

If we iterate the construction, since at each step the Kemer index strictly decreases, we must stop after a finite number of steps.

THEOREM 18.1.2.

(1) *We have a filtration* $0 \subset K_0 \subset K_1 \subset \cdots \subset K_u = R$ *of T-ideals, such that* $K_{i+1}/K_i$ *is the T-ideal in* $R_i := R/K_i$ *generated by the corresponding* $\mu$*-Kemer polynomials for* $\mu$ *suitably large.*
(2) *Each algebra* $R_i$ *can be embedded* $R_i \subset R_i\mathcal{T}_{R_i}$ *in a finitely generated module over a finitely generated commutative algebra* $\mathcal{T}_{R_i}$.
(3) $K_{i+1}/K_i$ *is an ideal of* $R_i\mathcal{T}_{R_i}$*, in particular a finitely generated module over the finitely generated commutative algebra* $\mathcal{T}_{R_i}$.

REMARK 18.1.3. As in Remark 17.3.6 if $X$ is countable, we may also consider the algebras $\mathcal{T}_{R_i}$, $R_i\mathcal{T}_{R_i}$ as *finitely generated* as in classical invariant theory. That is using the Capelli identity one knows that there are finitely many generators provided we also use linear changes on variables, that is special automorphisms. We will say that they are *T-finitely generated*.

COROLLARY 18.1.4 (Belov [**Bel97**]). *If* $R := F\langle X\rangle/I$ *is a relatively free algebra in a finite number of variables X, let* $R(i)$ *be its part of degree i, its Hilbert series*

$$H_R(t) := \sum_{k=0}^{\infty} \dim(R(i))\, t^i$$

*is a rational function of the form*

$$\frac{p(t)}{\prod_{j=1}^{N}(1-t^{h_j})}, \quad h_j \in \mathbb{N}, \quad p(t) \in \mathbb{Z}[t]. \tag{18.1}$$

PROOF. Clearly $H_R(t) = \sum_{i=0}^{u-1} H_{K_{i+1}/K_i}(t)$. We know that $K_{i+1}/K_i$ is a finitely generated module over a finitely generated graded algebra $\mathcal{T}_{R_i}$. If $\mathcal{T}_{R_i}$ is generated by some elements $a_1, \dots, a_m$ of degrees $h_i$ by commutative algebra, one has that (cf. Theorem 8.3.2):

$$H_{K_{i+1}/K_i}(t) = \frac{p_i(t)}{\prod_{j=1}^{m}(1-t^{h_j})}, \quad p_i(t) \in \mathbb{Z}[t].$$

Summing this finite number of rational functions, we have the result:

$$H_R(t) = \sum_i H_{K_{i+1}/K_i}(t) = \sum_i \frac{p_i(t)}{\prod_{j=1}^{m}(1-t^{h_j})}. \tag{18.2}$$ □

In Theorem 18.1.19 and Corollary 18.1.21 we will apply a deeper geometric analysis in order to compute the dimension of the relatively free algebras from the Hilbert series.

One can generalize these results by considering a relatively free algebra in $k$ variables $X$ as multigraded by the degrees of the variables $X$. Then we write its generating series of the multigrading

$$H_R(x) = \sum_{h_1,\cdots,h_k} \dim(R_{h_1\cdots h_k}) t_1^{h_1} \cdots t_k^{h_k}. \tag{18.3}$$

This is of course the graded character of the induced action of the torus of diagonal matrices (which is a standard choice for a maximal torus of the general linear group $\mathrm{GL}(k,F)$) acting linearly on the space of variables.

The series of formula (18.3) should be interpreted in terms of the representation theory of $\mathrm{GL}(k,F)$. In each degree $d$ the homogeneous part $R_d$ of the algebra $R$ is some quotient of the representation $V^{\otimes d}$, $\dim V = k$, so

$$R_d = \bigoplus_{\lambda \vdash d} m_\lambda(R) S_\lambda(V)$$

and, for $\lambda \vdash d$, $m_\lambda(R) \leq f_\lambda$, with $f_\lambda$ the dimension of the corresponding irreducible representation of the symmetric group $S_d$ (Theorem 6.3.4).

The character of $S_\lambda(V)$ is the corresponding Schur function $S_\lambda(t_1,\ldots,t_k)$ (a symmetric function), so that finally we have

$$H_R(x) = \sum_d \sum_{\lambda \vdash d} m_\lambda(R) S_\lambda(t_1,\ldots,t_k). \tag{18.4}$$

We have seen, in Proposition 7.1.2 and Theorem 7.3.10, that, as soon as $k$ is sufficiently large so that the $T$-ideal of $R$ in $k$ variables contains a Capelli list, the numbers $m_\lambda(R)$ are independent of $k$ and are the multiplicities of the cocharacters of $R$. Thus it will also be interesting to write directly a generating function for these multiplicities,

$$\bar{H}_R(x) = \sum_d \sum_{\lambda \vdash d} m_\lambda t^\lambda. \tag{18.5}$$

In the next paragraphs, using the theory of highest weight vectors, we shall show that also these two functions (of formulas (18.4) and (18.5)) are rational and of special type. We will connect this to the theory of partition functions.

Finally in Theorem 18.3.3 we will apply this theory to give a precise quantitative result on the growth of the colength of $R$ (Definition 7.1.1(3)).

### 18.1.3. A close look at the filtration.

REMARK 18.1.5. The filtration given by Theorem 18.1.2 is almost canonical. In fact at each step we have used the algebras $A = B \oplus C$ only to estimate a value for $\mu$ so to choose only those $\mu$-Kemer polynomials which vanish on $C$. On the other hand a more careful analysis of formulas (17.21) and (17.20) together with the formalism of Corollary 12.1.18 shows that all the constructions may be made intrinsic and hence canonical.

This is now our objective.

18.1.3.1. *Kemer polynomials.* With the notations of the previous section let us choose a $T$-ideal $\Gamma \subset F\langle X\rangle$ of finite Kemer index $(t,s)$. For $\mu \geq \mu_0$ let $K_\mu$ denote the $T$-ideal of $F\langle X\rangle/\Gamma$ generated by $\mu$ Kemer polynomials. We have, by definition, *substitutional maps* for every $\Phi : X \to F\langle X\rangle$ and an induced homomorphism $\bar{\Phi}$ : $F\langle X\rangle \to F\langle X\rangle$ which induces homomorphisms

$$\bar{\Phi} : F\langle X\rangle/\Gamma \to F\langle X\rangle/\Gamma, \quad \bar{\Phi} : K_\mu \to K_\mu.$$

On the other hand we also have the formal trace algebra $\mathcal{T}_X$ of Definition 2.3.12 and its quotient $\mathcal{T}_t(X)$ of invariants of $t \times t$ matrices in which also the substitutional maps $\Phi : X \to F\langle X\rangle$ induce homomorphisms.

Recall that the theorem of Zieplies and Vaccarino (Corollary 12.1.18) tells us that a $\mathcal{T}_t(X)$-module structure on a space $M$ is equivalent to giving a multiplicative polynomial map $\rho$ of degree $t$ from $F\langle X\rangle$ to a commutative algebra of linear operators on $M$. Under this map, given $u \in F\langle X\rangle$, the action of $\rho(u)$ equals the action of $\det(\bar{u})$, where $\bar{u}$ is the evaluation of $u$ in generic matrices. We claim

THEOREM 18.1.6. *There is a minimal $\mu_1 \geq \mu_0$ such that for $\mu \geq \mu_1$, we have that $K_\mu$ is naturally a $\mathcal{T}_t(X)$-module with an action commuting with multiplication by $F\langle X\rangle/\Gamma$ and is compatible with substitutions. That is for all $a \in \mathcal{T}_t(X)$, $b \in K_\mu$, $\Phi : X \to F\langle X\rangle$, we have $\bar{\Phi}(a \cdot b) = \bar{\Phi}(a) \cdot \bar{\Phi}(b)$.*

*Let $\rho$ denote the map of $F\langle X\rangle$ to linear operators on $K_\mu$ determining this module structure. If $b = f(x_1, \dots, x_t, Z)$ is a $\mu$ Kemer polynomial with $x_1, \dots, x_t$, one of its small layers, and $z \in F\langle X\rangle$ we have*

$$\rho(z)b = f(zx_1, \dots, zx_t, Z). \tag{18.6}$$

PROOF. By definition $K_\mu$ is the ideal deduced from Kemer polynomials by all possible substitutions of variables, so if there is such a module structure, it is clearly completely determined by formula (18.6).

The existence of such $\mu$ depends upon the discussion at the beginning of §18.1.2. By that discussion $F\langle X\rangle/\Gamma$ is the $T$-ideal of PIs of $A = B \oplus C$, with $B$ a direct sum of fundamental algebras all with first Kemer index $t$, and $C$ has strictly lower Kemer index. For some large $\mu$, $K_\mu$ embeds in $\mathcal{F}_B$.

The fact that such formula extends to a module action follows from the fact $\mathcal{F}_B$ which embeds into its algebra with trace $\mathcal{T}_B(X)\mathcal{F}_B(X)$ which is by construction (cf. Proposition 17.3.4) stable under substitutions and where $\mathcal{T}_B(X)$, by formula (17.21) and Corollary 17.3.10, is a quotient of $\mathcal{T}_t(X)$ (see also Theorem 17.3.10). □

REMARK 18.1.7. We thus have that $\mu_1$ is canonically associated to the $T$-ideal $I$, and we can take $K_{\mu_1}$ as the first step of the filtration.

For instance if $I$ is the $T$-ideal of commutative algebras, the relatively free algebra is the ring $F_+[X]$ of commutative polynomials (without 1) in the variables $X$, and the ring $\mathcal{T}_1(X)$ is the ring $F[X]$ of commutative polynomials with 1 in the variables $X$. We have $\mu_1 = \mu_0 = 1$, and $K_1 = F_+[X]$ is of course a module over $F[X]$ and exhausts the filtration.

More generally let $I_n$ be the $T$-ideal of identities of $n \times n$ matrices. Again one has $\mu_1 = \mu_0 = 1$ and $K_1$ is the $T$-ideal (modulo $I_n$) generated by the Capelli polynomial $C_{n^2}$. In this case $\mathcal{T}_{M_n(F)}(X)$ equals $\mathcal{T}_n(X)$ the invariants of $n \times n$ matrices, which appears as quotient of $\mathcal{T}_{n^2}(X)$ by the usual embedding by left multiplication of $M_n(F)$ into $M_{n^2}(F)$. Instead of formula (18.6) one should invoke formula (10.12).

There is finally one further normalization which is useful. If $\mu \geq \mu_1$, we have seen that $K_\mu$ is canonically a $\mathcal{T}_t(X)$-module. Let $I_\mu \subset \mathcal{T}_t(X)$ be the annihilator ideal of $K_\mu$. If $\nu > \mu \geq \mu_1$, we have $K_\nu \subset K_\mu$, hence $I_\nu \supset I_\mu$. Since $\mathcal{T}_t(X)$ is, for $X$ finite, a finitely generated algebra, it is Noetherian so at some value $\mu_2 \geq \mu_1$, we have $I_\nu = I_{\mu_2}$, $\forall \nu \geq \mu_2$.

A priori this $\mu_2$ depends on $X$, but notice that if $X = \{x_1, \dots, x_m\}$ the spaces $\mathcal{T}_t(x_1, \dots, x_m)$, $K_\mu(x_1, \dots, x_m)$, $I_\mu(x_1, \dots, x_m)$ are all stable under the action of the group $\mathrm{GL}(m, F)$ acting linearly on $\mathrm{span}(x_1, \dots, x_m)$. By the Capelli identity there is a $k$ such that all these spaces, as representations of $\mathrm{GL}(m, F)$, are sums of Schur modules $S_\lambda$ with height $\leq k$, hence they are determined by the first $k$ variables.

So $\mu_1, \mu_2$ are determined by choosing $X = (x_1, \dots, x_k)$.

18.1.3.2. *The varieties* $W_{n_1,\dots,n_q}$. Given a $T$-ideal $\Gamma$ of finite Kemer index $(t, s)$, we have thus constructed in an intrinsic form a commutative algebra $\mathcal{T}_\Gamma$ acting faithfully on $K_\mu$, $\mu \geq \mu_1$.

By Theorem 18.1.6 if $\Gamma$ is the ideal of identities of a finite-dimensional algebra $A$, then $\mathcal{T}_\Gamma$ equals the algebra $\mathcal{T}_A$ defined by Definition 17.3.1 if $A$ is fundamental and in formula (17.16) for a general $A$. $\mathcal{T}_\Gamma = \mathcal{T}_A$ is a natural quotient of the algebra $\mathcal{T}_t(X)$ of invariants of $m = |X|$-tuples of $t \times t$ matrices.

Let us first analyze the case in which $A$ is fundamental. We write as usual

$$A = \bar{A} \oplus J, \ \bar{A} = \bigoplus_{i=1}^{q} A_i, \quad A_i = M_{n_i}(F), \quad t = \sum_{i=1}^{q} n_i^2.$$

Here $J$ is the radical and $\bar{A}$ is the semisimple part.

Recall that if $|X| = m$, we have that $\mathcal{T}_t(X)$ is the coordinate ring of the quotient variety $V_t(m) := M_t(F)^m // \mathrm{PGL}(t)$ (formula (14.6)) whose points, by Corollary 14.1.18, parametrize equivalence classes of semisimple representations of dimension $t$ of the free algebra in $m$ variables, that is the closed orbits of the space of $m$-tuples of $t \times t$ matrices under conjugation.

Since $\mathcal{T}_A(X)$ is a quotient of $\mathcal{T}_t(X)$, it is the coordinate ring of a subvariety of the quotient variety $V_t(m)$. We want to investigate the nature of this subvariety.

Let us make first an elementary remark. If $X$ is a $t \times t$ matrix, the trace of the left multiplication $L_X : a \mapsto Xa$ on the space $M_t(F)$ equals $t \cdot \mathrm{tr}(X)$.

If $A = M_t(F)$, we have that the algebra $\mathcal{T}_A(X) = \mathcal{T}_t(X)$, since we are assuming characteristic 0 and the algebra $\mathcal{T}_t(X)$ is also generated by the traces of the monomials in the left multiplications on $M_t(F)$. In this case, however, we are considering the variety associated to $\mathcal{T}_A(X)$ as a subvariety of that of semisimple $t^2$-dimensional representations. This is easily seen to arise from the fact that as representation by left multiplication this $t^2$-dimensional representation is $M_t(F) = (F^t)^{\oplus t}$, that is the $t$-dimensional representation with multiplicity $t$, so the embedding of $V_t(m) \to V_{t^2}(m)$ which associates $V^{\oplus t}$ to a $t$-dimensional semisimple representation $V$ (this is similar to Proposition 14.5.1).

Given a semisimple algebra $\bar{A} := \bigoplus_{i=1}^{q} M_{n_i}(F)$, set $\sum_i n_i^2 = t$. We may consider $\mathrm{End}_F(\bar{A}) = M_t(F)$ as the algebra of $t \times t$ matrices. Left multiplication of $\bar{A}$ on itself induces thus an inclusion $j : \bar{A} \to \mathrm{End}_F(\bar{A}) = M_t(F)$, its image is a subalgebra of block diagonal matrices in which each block $M_{n_i}(F)$ appears with multiplicity $n_i$.

If $a = (a_1, \ldots, a_q) \in \bigoplus_{i=1}^q A_i = \bar{A}$, we have that $L_a = \bigoplus_{i=1}^q L_{a_i}$ so

$$\operatorname{tr}(L_a) = \sum_{i=1}^q \operatorname{tr}(L_{a_i}) = \sum_{i=1}^q n_i \operatorname{tr}(a_i).$$

Thus, when we restrict the invariants $\mathcal{T}_t(X)$ to this subalgebra, we see that when $z$ is some polynomial in $X$ the restriction of the function $\operatorname{tr}(L_z)$ to this subalgebra equals $\sum_{i=1}^q n_i \operatorname{tr}_i(z)$, where $\operatorname{tr}_i(z)$ is in the algebra $\mathcal{T}_{n_i}(X)$. This means that $\mathcal{T}_t(X)$ maps to the $G = \prod_{i=1}^q \operatorname{PGL}(n_i, F)$ invariants, that is the coordinate ring of $\prod_{i=1}^q V_{n_i}(m)$ which is $\mathcal{T}_{n_1}(X) \otimes \cdots \otimes \mathcal{T}_{n_q}(X)$. Let us define by $\mathcal{T}_{\bar{A}}(X)$ the image of this homomorphism.

We have shown that as $\prod_{i=1}^q V_{n_i}(m)$ is the quotient of $m$ copies of the space $\bigoplus_{i=1}^q M_{n_i}(F)$ under the group $G = \prod_{i=1}^q \operatorname{PGL}(n_i, F)$ acting by conjugation, we have the commutative diagram, where the two maps $\pi_G, \pi_{\operatorname{PGL}(t,F)}$ are quotients under the two groups

$$(18.7)\qquad \begin{array}{ccc} (\bigoplus_{i=1}^q M_{n_i}(F))^m & \xrightarrow{\ j\ } & M_t(F)^m \\ \pi_G \big\downarrow & & \pi_{\operatorname{PGL}(t,F)} \big\downarrow \\ \prod_{i=1}^q V_{n_i}(m) & \xrightarrow{\ \bar{j}\ } & V_t(m). \end{array}$$

We then define $W_{n_1,\ldots,n_q}$ to be the image of the map $\bar{j}$. From the diagram and the definitions it follows that

PROPOSITION 18.1.8. *$W_{n_1,\ldots,n_q}$ is the set of equivalence classes of semisimple representations which are direct sum of $q$ semisimple representations $N_i$ of dimension $n_i$, for each $i = 1, \ldots, q$ each with multiplicity $n_i$. Of course, generically, each $N_i$ is irreducible.*

We first claim that this image is closed, that is $W_{n_1,\ldots,n_q}$ is a subvariety, for this by standard arguments of commutative algebra it is sufficient to prove that $\bar{j}$ is a *finite map* to its image, the content of Lemma 18.1.9.

Denote for simplicity by $A_i = \mathcal{T}_{n_i}(X)$ the ring of invariants of $m$-tuples of $n_i \times n_i$ matrices, the coordinate ring of the variety $V_{n_i}(m)$. Thus $A_1 \otimes A_2 \otimes \cdots \otimes A_q$ is the coordinate ring of the variety $\prod_{i=1}^q V_{n_i}(m)$. We claim that

LEMMA 18.1.9. *$A_1 \otimes A_2 \otimes \cdots \otimes A_q$ is integral over $\mathcal{T}_{\bar{A}}(X)$. In fact we have generators which are integral of degree bounded by some number independent of $m$.*

PROOF. Consider a polynomial

$$(18.8)\qquad f(t) = x^t - a_1 x^{t-1} + a_2 x^{t-2} + (-1)^t a_t$$

with roots $x_1, \ldots, x_t$, so that the elements $a_i$ are the elementary symmetric functions in the $x_i$. We may even take the $x_i$ as indeterminates. Given an integer $k \in \mathbb{N}$ and a set $S \subset \{1, \ldots, t\}$ with $h$ elements, set

$$X_S^k := \sum_{i \in S} x_i^k.$$

Consider next the polynomial of degree $N := \binom{t}{h}$

$$(18.9)\qquad x^N + \sum_{i=1}^N (-1)^i b_i x^{N-i} := \prod_{S \subset \{1,\ldots,t\},\ |S|=h} (x - X_S^k).$$

The coefficients $b_i$ of this polynomial are clearly symmetric functions in the variables $x_j$, so they are expressible as polynomials in the elements $a_i$. In fact $b_i$ is a polynomial of degree $ki$ in the variables $x_j$ so it is a polynomial of this weight when we give to $a_i$ the weight $i$.

We apply this construction to our setting. Each $A_i$ is generated by the traces $\mathrm{tr}(M_i)$ of the monomials $M$ acting on the $i$th summand.

For a given monomial $M$ consider as $f(t)$ of formula (18.8) the characteristtic polynomial of left multiplication by $M$. It is a polynomial with coefficients in $\mathcal{T}_{\bar{A}}(X)$ by the universal formulas expressing the elementary symmetric functions $a_i$ in the eigenvalues of $M$ through the elements $\mathrm{tr}(M^k)$, $k = 1, \dots, t$. Thus the element $\mathrm{tr}(M_i)$ is a sum of $n_i$ eigenvalues out of the $t$ eigenvalues of the monomial $M$. We deduce, from formula (18.9), a monic polynomial of degree $\binom{t}{n_i}$ satisfied by $\mathrm{tr}(M_i)$ with coefficients polynomials in the elements $\mathrm{tr}(M^k)$, $k = 1, \dots, t$, thus in $\mathcal{T}_{\bar{A}}(X)$. This proves that $A_1 \otimes A_2 \otimes \cdots \otimes A_q$ is integral over $\mathcal{T}_{\bar{A}}(X)$ and the bound on the generators. □

COROLLARY 18.1.10. *The two varieties* $\prod_{i=1}^q V_{n_i}(m)$, *and* $W_{n_1,\dots,n_q}$ *have the same dimension* $\sum_{i=1}^q (mn_i^2 - (n_i^2 - 1)) = (m-1)t + q$.

PROOF. This follows from Lemma 18.1.9 and Theorem 14.1.18. □

REMARK 18.1.11. The map $j : \prod_{i=1}^q V_{n_i}(m) \to W_{n_1,\dots,n_q}$ is surjective but it is almost never an isomorphism in general. In fact it is not even birational. To compute its degree, let us analyze its generic fiber.

The generic fiber is obtained when in $\prod_{i=1}^q V_{n_i}(m)$ and in $W_{n_1,\dots,n_q}$ we have that all the summands $N_i$ are irreducible and not just semisimple. In this case if we have several indices $i$ with the same $n_i$, then in the expression $\bigoplus_i n_i N_i$ we may permute the indices of the irreducible representations of the same dimension so they come from different ways of arranging them in the factors $V_{n_i}(m)$. Thus if we have certain multiplicities $h_1, \dots, h_k$ of the different indices $n_i$, we see that the generic fiber is formed by $\prod_i h_i!$ points.

In fact even if there are no multiplicities, so the map $j$ is birational, the same argument may show that in nongeneric fibers we may perform some of these permutations, and so the map is not usually bijective.

EXAMPLE 18.1.12. $t = 2 = 1 + 1$, and the variety $V_1(m)$ is just affine space with its coordinate ring the polynomial ring $F[x_1, \dots, x_m]$.

The variety $V_1(m) \times V_1(m)$ is a $2m$-dimensional affine space with its coordinate ring the polynomial ring $F[x_1, \dots, x_m, y_1, \dots, y_m]$. The algebra $\mathcal{T}_{\bar{A}}(X)$ is the following. Take a monomial in the elements $\xi_i := (x_i, y_i)$ which should be thought of as diagonal $2 \times 2$ matrices. Its trace is the sum of the same monomial computed once in the $x_i$ and once in $y_i$. It is a symmetric function in the exchange $\tau$ between $x_i, y_i$. The map is generically 2–1, and the image of coordinate rings is the ring of $\tau$ invariants.

REMARK 18.1.13. As in Remark 18.1.7, in the special case $q = 1, \bar{A} = M_n(F)$, $t = n^2$, we see that $\bar{j}$ is an isomorphism of $V_n(m)$ with $W_n(m)$. The embedding of $V_n(m)$ into $V_{n^2}(m)$ consists in associating to a semisimple representation $N$ of dimension $n$ of the free algebra, the semisimple representation $N^n$ (cf. Proposition 14.5.1).

18.1.3.3. *The support of Kemer polynomials.* We now apply the previous discussion to Kemer polynomials. First let us take a fundamental algebra $A$ with semisimple part $\bar{A} = \bigoplus_{i=1}^{q} M_{n_i^2}(F)$, $t = \sum_i n_i^2$. We have constructed the map from $\mathcal{T}_t(X)$ to $\mathcal{T}_A$.

LEMMA 18.1.14. *The algebra $\mathcal{T}_A$ is the coordinate ring of the irreducible subvariety $W_{n_1,\dots,n_q}$ of $M_t(F)^m // \mathrm{PGL}(t,F)$.*

PROOF. By definition $\mathcal{T}_A$ is the algebra generated by the traces of $m$-tuples of elements of $\bar{A} = \bigoplus_{i=1}^{q} M_{n_i^2}(F)$, $t = \sum_i n_i^2$, acting on itself by left multiplication so the lemma follows from the previous discussion as summarized by formula (18.7). □

COROLLARY 18.1.15. *Two fundamental algebras, which have the same Kemer polynomials (in particular if they are PI equivalent), have an isomorphic semisimple part.*

PROOF. This follows from the fact that the varieties $W_{n_1,\dots,n_q}$ are all distinct since they parametrize different sets of semisimple representations; Proposition 18.1.8. □

Let $A = \bigoplus_{i=1}^{u} A_i$ be a direct sum of fundamental algebras all with the same Kemer index. To each $A_i$ we have associated the irreducible subvariety of the quotient $M_t(F)^m // \mathrm{PGL}(t,F)$:

$$W_i := W_{n_1,\dots,n_{q_i}}, \quad \bar{A}_i = \bigoplus_{j=1}^{q_i} M_{n_j}(F).$$

From formula (17.16) it follows that the commutative algebra $\mathcal{T}_A(X)$, quotient of $\mathcal{T}_t(X)$ embeds into the product $\prod_{i=1}^{u} \mathcal{T}_{A_i}(X)$, while each $\mathcal{T}_{A_i}(X)$ is also a quotient of $\mathcal{T}_A(X)$. Therefore we have the rather interesting fact:

THEOREM 18.1.16. *The image of the algebra $\mathcal{T}_t(X)$ acting on the space of Kemer polynomials $K_{\mu_1}$ is the coordinate ring of a, possibly reducible, subvariety of the quotient $M_t(F)^m // \mathrm{PGL}(t,F)$ union of the subvarieties $W_i$, $i = 1,\dots,u$, where $W_i$ is the variety associated to $\mathcal{T}_{A_i}(X)$. The module $K_{\mu_1}$ of Kemer polynomials is supported on this subvariety.*

Notice that $\mathcal{T}_t(X)$ is an object intrinsically defined and then also $\bigcup_i W_i$ is intrinsic being the support of $K_{\mu_1}$. The subvarieties $W_i = W_{n_1,\dots,n_{q_i}}$ reflect the structure of the semisimple parts of the fundamental algebras $A_i$ which may appear as summands of maximal Kemer index of an algebra $A$ having as identities the given $T$-ideal.

There are some subtle points in this construction. It is possible that some variety $W_i$ is contained in another $W_j$. This gives an *embedded component* which may not be visible just by the structure of the module $K_{\mu_1}$ but depends on the embedding $K_{\mu_1} \subset \bigoplus_i K_i$, which in turn depends on the particular decomposition of the ideal $\mathrm{Id}_A = \bigcap_{i=1}^{u} \mathrm{Id}_{A_i}$. This is worth further investigation.

A specific element of $K_{\mu_1}$ vanishes on one of the varieties $W_i$ if and only if it is a polynomial identity for the corresponding summands $A_i$.

This answers at least partially the question posed in Remark 18.1.5, on how intrinsic are the constructions associated to a particular choice of the algebra $A$ having the chosen $T$-ideal as ideal of polynomial identities.

We shall thus make use of the following refinement of the Kemer index:

DEFINITION 18.1.17. Given a $T$-ideal $\Gamma$ of identities of a fundamental algebra, we take as *index* of $\Gamma$ the pair $\bar{A}, s$, where $s$ is the second Kemer index (while the first is $\dim \bar{A}$).

18.1.3.4. *The dimension of relatively free algebras.* From the theory developed one can extract some further properties of these rational functions. We apply the results and definitions of §8.3 to the noncommutative relatively free algebras.

We take as definition of dimension for a relatively free algebra, the one given by Definition 8.3.5 for its Hilbert series, which measures growth, and by Proposition 8.3.10 it equals the Gelfand–Kirillov dimension.

Analyze first a fundamental algebra $A$, with semisimple part $\bar{A}=\bigoplus_{i=1}^{q} M_{n_i}(F)$.

Take the relatively free algebra in $m$ variables $\mathcal{F}_A(m) := F\langle \xi_1, \ldots, \xi_m \rangle$ for $A$. We have an inclusion of $\mathcal{F}_A(m) \subset \mathcal{F}_A(m)\mathcal{T}_A(m)$ and also an inclusion of the ideal $K_A \subset \mathcal{F}_A(m)$ generated by Kemer polynomials. Since both $K_A$ and $\mathcal{F}_A(m)\mathcal{T}_A(m)$ are finitely generated torsion free modules over $\mathcal{T}_A(m)$, it follows that the Gel'fand–Kirillov dimension of the algebras $K_A$, $\mathcal{F}_A(m)$, $\mathcal{F}_A(m)\mathcal{T}_A(m)$, $\mathcal{T}_A(m)$ are all equal, and can be computed by any of the four Hilbert series. We have computed the dimension $\dim \mathcal{T}_A(m) = (m-1)t + q$; here $t = \sum_{i=1}^{q} n_i^2$ is also the first Kemer index.

PROPOSITION 18.1.18. *The GK dimension of the relatively free algebra in $m$ variables for a fundamental algebra $A$ is $(m-1)t+q$, where $t$ is the first Kemer index. The GK dimension of the relatively free algebra in $m$ variables for a direct sum $A = \bigoplus_i A_i$ of fundamental algebras is the maximum of the GK dimension of the relatively free algebras in $m$ variables for the fundamental algebras $A_i$.*

PROOF. The first part we have just proved. As for the second remark that the relatively free algebra in $m$ variables for a direct sum $\bigoplus_i A_i$ is contained in the direct sum of the relatively free algebras in $m$ variables for the summands $A_i$, so its dimension is smaller than or equal to the maximum of the dimensions of the algebras relative to the summands. On the other hand each relatively free algebra in $m$ variables for the summands $A_i$ is also a quotient of the relatively free algebras in $m$ variables for the direct sum, so the claim follows. □

Consider again a fundamental algebra with semisimple part $\bar{A} = \bigoplus_{i=1}^{q} M_{n_i}(F)$ and $R$ its relatively free algebra in $m$ variables. We can also approach this in a more intrinsic form using the filtration of $R$. By Theorem 18.1.16 and the remark following it the dimension is the maximum of the dimensions of the modules $K_{i+1}/K_i$. Each one is supported in a union of varieties $W_{n_1,\ldots,n_k}$, but on each of these subvarieties the module is torsion free, so it has the same dimension as its support. Thus we deduce

THEOREM 18.1.19. *The dimension of $\mathcal{F}_A(m)$ is the maximum dimension of the varieties $W_{n_1,\ldots,n_k}$.*

Now the dimension of $W_{n_1,\ldots,n_k}$ is $mt - \sum_{i=1}^{q}(n_i^2 - 1) = (m-1)t + q$, so we see that as $m$ grows the maximum is obtained from the factors $R_i$ for which the first Kemer index equals the Kemer index $t$ of $R$ and among these the one with maximum $q$.

DEFINITION 18.1.20. The integer $q$ is an invariant of the $T$-ideal, called $q$ *invariant*.

COROLLARY 18.1.21. *The dimension of $\mathcal{F}_A(m)$ for $m$ large is $(m-1)t+q$ where $t$ is the first Kemer index and $q$ is the $q$ invariant.*

Observe that when $A = M_n(F)$, $\mathcal{F}_A(m)$ is the ring of generic $n \times n$ matrices, then $t = n^2$ and $q = 1$, the formula is valid for all $m \geq 2$. Since block-triangular matrices are a fundamental algebra, the formula is valid for them and for all $m \geq 2$, with $q$ equal to the number of blocks and $t$ to the dimension of the semisimple part.

Proposition 18.1.18 was proved for upper block-triangular matrices by A. Giambruno and M. Zaicev in **[GZ03a]**.

**18.1.4. Cocharacters.** In Theorem 7.3.13 we have seen that the multiplicity of a cocharacter in the case of $2 \times 2$ matrices has the following property. We have a finite partition of the set of allowed $\lambda$ so that on each set the multiplicity is a polynomial function in the parts (or differences) of $\lambda$.

In Theorem 18.3.1 we want to show that this is a general phenomenon for finite-dimensional algebras, as shown by Berele **[Ber06]** by extending the theorem of Belov **[Bel97]**. This requires some standard preliminaries.

## 18.2. Complements of commutative algebra and invariant theory

**18.2.1. Partition functions.** The notion of a partition function, which we previously discussed for a sequence of positive integers (Definition 8.3.3), can be given in general for a sequence of integral vectors $A := \{a_1, \dots, a_m\}$, $a_i \in \mathbb{Z}^p$, for which there is a linear function $f$ with $f(a_i) > 0$, $\forall i$.[1]

Then one defines the partition function on $\mathbb{Z}^p$,

$$b \in \mathbb{Z}^p, \quad P_A(b) = \#\{t_1, \dots, t_m \in \mathbb{N} \mid \sum_{i=1}^{m} t_i a_i = b\}.$$

Of course $P_A(b) = 0$ unless $b \in C(A) := \{\sum_i x_i a_i \mid x_i \in \mathbb{R}^+\}$, the *positive cone* generated by the elements $a_i$. The assumption $f(a_i) > 0$, $\forall i$, means that $C(A)$ is *pointed*, that is it does not contain any line (only half lines).

REMARK 18.2.1. When $p = 1$ and the elements $a_i \in \mathbb{N}$, the partition function goes back to Euler. Even in this special case this function is quite complex. On each coset, modulo the least common multiple of the $a_i$, it coincides (on $\mathbb{N}$) with a polynomial with rational coefficients of degree $m-1$ (Proposition 8.3.6). The leading terms of these polynomials coincide and equal a formula for a volume of a variable polytope.

A similar but subtler theory holds in higher dimension **[DCP11]**.

It is customary to express the partition function by a generating function; this is a series in $p$ variables $x_1, \dots, x_p$. When $b = (b_1, \dots, b_p) \in \mathbb{Z}^p$, we set $x^b := \prod_{i=1}^{p} x_i^{b_i}$, and then by setting

$$\mathcal{P}_A = \sum_{b \in \mathbb{Z}^p} P_A(b) x^b,$$

[1] This restriction is essential, otherwise the value of the partition function is $\infty$.

one sees that

$$\mathcal{P}_A = \frac{1}{\prod_{i=1}^m (1 - x^{a_i})}.$$

This formal series is also truly convergent on some region of space. We can interpret this in terms of graded algebras. Let $R_A = F[y_1, \dots, y_m]$ be the polynomial algebra in $m$ variables $y_i$ to which we give a $\mathbb{Z}^p$ grading by assigning to $y_i$ the degree $a_i$. For every graded vector space $V = \bigoplus_{a \in \mathbb{Z}^p} V_a$, for which $\dim(V_a) < \infty$, $\forall a$, we can define its *graded Hilbert series*

$$H_V = \sum_{a \in \mathbb{Z}^p} \dim(V_a) x^a.$$

One has to be careful when manipulating such series since in general a product of two such series makes no sense. If we have two graded vector spaces $V, W$ with the previous restriction on graded dimensions, in general $V \otimes W$ does not satisfy this restriction. The product makes sense if we restrict to series supported in a given pointed cone; in this case we have that $H_{V \otimes W} = H_V H_W$.

In general let us consider a finitely generated graded $R_A$-module $M$. Then we have the generalization of Theorem 8.3.2,

LEMMA 18.2.2. *The partition function of $A$ coincides with the Hilbert series of $R_A$. The Hilbert series of $M$ has the form*

$$H_M = \frac{p(x)}{\prod_{i=1}^m (1 - x^{a_i})}, \quad p(x) \in \mathbb{Z}[x_1^{\pm 1}, \dots, x_p^{\pm 1}], \tag{18.10}$$

*that is $p(x)$ is a finite linear combination with integer coefficients of Laurent monomials. If $A \subset \mathbb{N}^p$ and $M$ is graded in $\mathbb{N}^p$, then the elements $p(x)$ are polynomials.*

PROOF. The standard proof is by induction on $m$. We prove the second part since the first is a special case. If $m = 0$, then $M$ is finite dimensional, it has a basis $v_1, \dots, v_u$ of vectors with some degrees $c_i$, and its Hilbert series is $\sum_{j=1}^u x^{c_i}$. Otherwise consider the exact sequence

$$0 \longrightarrow M_0 \longrightarrow M \xrightarrow{y_m} M \longrightarrow M_1 \longrightarrow 0,$$

where the middle map is multiplication by $y_m$. We have that $M_0, M_1$ are both modules over $F[y_1, \dots, y_{m-1}]$ for which we apply induction. One easily sees that

$$0 = H_{M_0} - H_M + x^{a_m} H_M - H_{M_1} \implies H_M = \frac{1}{1 - x^{a_m}} (H_{M_0} - H_{M_1}),$$

which by induction implies the claim. □

Notice that this also explains easily why the partition function is given by the Hilbert series $H_{R_A}$.

18.2.1.1. *Partition function.* Let $A$ be as before; the theory is explained in detail in **[DCP11]**. We want to decompose the cone $C(A)$ into *big cells* and define its singular and regular points.

The set of *singular* points $C_s(A)$ is the union of all cones $C(B)$ for $B$ any subset of $A$ which does not span $\mathbb{R}^p$.

The big cells are defined as the connected components of the open set of $C(A)$ of *regular points* which is obtained when removing from $C(A)$ the singular points $C_s(A)$. One verifies that each big cell is the interior of a cone, the intersection of simplicial cones generated by bases extracted from $A$.

One finally needs the notion of *quasi-polynomial* on $\mathbb{Z}^p$. This is a function $f$ on $\mathbb{Z}^p$ for which there exists a subgroup $\Lambda \subset \mathbb{Z}^p$ of finite index, so that $f$, restricted to each coset of $\Lambda$ in $\mathbb{Z}^p$, coincides with some polynomial.

The theory of partition functions plays a major role a finite-dimensional space of quasi-polynomials introduced by Dahmen and Micchelli. Let us recall their theory **[DM88]** and see also Stanley **[Sta82]**.

- Given a list of integral vectors $A := \{a_1, \ldots, a_m\}$, $a_i \in \mathbb{Z}^p$ spanning $\mathbb{R}^p$, we call a subset $Y$ of $A$ a *cocircuit* if it is minimal with the property that $A \setminus Y$ does not span $\mathbb{R}^m$. The set of all cocircuits of $A$ will be denoted by $\mathcal{E}(A)$.
- To $A$ is associated a remarkable convex polytope, the *zonotope*
$$Z(A) := \{\sum_i t_i a_i \,|\, 0 \le t_i \le 1\}.$$
- The faces of this zonotope come in opposite pairs corresponding to the cocircuits of $A$. Moreover $Z(A)$ can be paved by parallelepipeds associated to all the bases of $\mathbb{R}^p$ which can be extracted from $A$. Each of these parallelepipeds has as its volume a positive integer, the absolute value of the determinant of the corresponding basis elements.
- For a list of integral vectors $Y$ we then define the difference operator
$$\nabla_Y := \prod_{y \in Y} \nabla_y, \quad \nabla_y f(x) = f(x) - f(x - y).$$
- The space $\mathrm{DM}(A)$ of Dahmen and Micchelli is the space of integral valued functions on $\mathbb{Z}^p$ which are solutions of the system of difference equations $\nabla_Y f = 0$, $Y \in \mathcal{E}(A)$.
- The space $\mathrm{DM}(A)$ is a space of quasi-polynomials of degree $m - 1$, it is a free abelian group of dimension $\delta(A)$ the weighted number of bases extracted from $A$, or the volume of $Z(A)$.
- In fact if we take a generic shift $a - Z(A)$, with $a \in \mathbb{R}^p$, we have that $a - Z(A) \cap \mathbb{Z}^p$ has exactly $\delta(A)$ elements, and the restriction of $\mathrm{DM}(A)$ to $a - Z(A) \cap \mathbb{Z}^p$ is an isomorphism with the space of integral valued functions on $a - Z(A) \cap \mathbb{Z}^p$.

Assuming that $A$ spans $\mathbb{R}^p$, the main theorem on partition functions (see **[DCP11]**) is

THEOREM 18.2.3. *$P_A$ is supported on the intersection of the lattice $\mathbb{Z}^p$ with the cone $C(A)$. Given a big cell $\mathfrak{c}$ of $C(A)$ and a point $a \in \mathfrak{c}$ very close to 0, one has that $a - Z(A) \cap \mathbb{Z}^p$ intersects $C(A)$ only in 0. $P_A$ coincides on each big cell $\mathfrak{c}$ with the quasi-polynomial in the space* $\mathrm{DM}(A)$ *which is 1 at 0 and equals 0 in the other points of $a - Z(A) \cap \mathbb{Z}^p$.*

Finally the partition function may be interpreted as counting the number of integral points in the variable convex compact polytope
$$V(b) := \{(t_1, \ldots, t_m) \,|\, t_i \in \mathbb{R}^+, \quad \sum_i t_i a_i = b\}.$$

Thus the partition function is asymptotic to the volume of this variable polytope. This volume function, denoted by $T_A(b)$, is a *spline*, a piecewise polynomial function on $C(A)$, which is polynomial in each big cell of degree $m - 1$. Again there is a remarkable theory behind these functions, cf. **[DCP11]**.

18.2.1.2. *Nice rational functions.* We want to pass from formula (18.10) expressing the graded dimension of some relatively free algebra to a formula expressing directly the multiplicities $m_\lambda$. We do this following Berele **[Ber06]**.

Let us say that:

DEFINITION 18.2.4. A rational function $f(x_1,\dots,x_p)$ in $p$ variables $x_1,\dots,x_p$ is *nice* if the denominator is a product of factors of type $1-x^{\underline{h}}$, where, for a $p$-tuple $\underline{h}=(h_1,\dots,h_p)$, we denote $x^{\underline{h}}=x_1^{h_1}\cdots x_p^{h_p}$. If furthermore $f(x_1,\dots,x_p)$ is symmetric and we expand

$$\prod_{\underline{h}}\frac{1}{1-x^{\underline{h}}}=\prod_{\underline{h}}\sum_{k=0}^{\infty}x^{k\underline{h}},$$

we express $f(x_1,\dots,x_p)$ in a symmetric power series which can thus be developed in the basis of Schur functions and have

$$f(x_1,\dots,x_p)=\sum_{\lambda}m_\lambda S_\lambda(x_1,\dots,x_p).$$

One can observe that from formula (3.18) and the Definition 3.18 of Schur functions there follows

LEMMA 18.2.5. *$m_{h_1,\dots,h_p}$ is the coefficient of $x_1^{h_1+p-1}x_2^{h_2+p-2}\cdots x_p$ in the product $f(x_1,\dots,x_p)V(x_1,\dots,x_p)$, where $V(x_1,\dots,x_p)$ is the Vandermonde determinant.*

It is then interesting to have a generating series directly for the $m_\lambda$; we will discuss this in the next paragraph.

18.2.1.3. *U invariants.* In order to apply the previous theory to cocharacters, we need to recall some basic facts on highest weight vectors and $U$ invariants, where $U$ is the subgroup of the linear group of upper-triangular matrices with a 1 on the diagonal.

Recall that we have shown in Remark 7.1.2 that the multiplicity $m_\lambda$ of a cocharacter equals the multiplicity of the highest weight vectors of weight $\lambda$ in the relatively free algebra in $k$ variables, if we are considering an algebra of dimension $k$.

Recall that a highest weight vector is an element of a given representation of $G=\mathrm{GL}(k,F)$, which is invariant under $U$, and it is a weight vector under the maximal torus of diagonal matrices $(t_1,\dots,t_k)$.

The highest weight is then a *dominant weight* sum of the fundamental weights $\omega_i:=\prod_{j=1}^{i}t_j$, which is the highest weight of $\bigwedge^i F^k$. If $m_i$ is the number of columns of the partition of length $i$, then the corresponding weight is $\sum_i m_i\omega_i$, that is the character $\prod_{i=1}^{k}(\prod_{j=1}^{i}t_j)^{m_i}$, we identify partitions with dominant weights and thus write $\lambda=\sum_i m_i\omega_i$, the highest weight of the Schur module $S_\lambda(F^k)$.

By Theorem 18.1.2 the Hilbert series of $R$ is the sum of the contributions of the finitely many factors $K_{i+1}/K_i$ which are all modules over finitely generated algebras. Moreover these are all stable under the action of the linear group $G=\mathrm{GL}(k,F)$ so that we have some decomposition into irreducible representations

$$K_{i+1}/K_i=\bigoplus_{\lambda}S_\lambda(F^k)^{\oplus m_{i,\lambda}}. \tag{18.11}$$

These multiplicities $m_{i,\lambda}$ can be computed by taking the vector space of $U$ invariants, $(K_{i+1}/K_i)^U = \bigoplus_\lambda (S_\lambda(F^k)^U)^{\oplus m_{i,\lambda}}$.

Since $\dim S_\lambda(F^k)^U = 1$ is generated by a single vector of weight $\lambda$, we have that the graded Hilbert series of $(K_{i+1}/K_i)^U$, which is $\mathbb{N}^k$ graded by the coordinates $m_i$ of the dominant weights, is the generating function of the multiplicities $m_{i,\lambda}$.

So our aim is to prove that $(K_{i+1}/K_i)^U$ is a finitely generated graded module over a graded polynomial algebra in finitely many variables corresponding to the fundamental weights. Then we can apply Lemma 18.2.2 and Theorem 18.2.3 which properly interpreted give the desired result.

This actually is a standard fact on reductive groups following Grosshans **[Gro73]** and **[BBHM63]**.

Consider the coordinate ring $F(\mathrm{GL}(k, F))$ of the linear group, concretely it is $F[x_{i,j}, d^{-1}]$, the polynomial ring over $F$ in $n^2$ variables $x_{i,j}$ with the determinant $d$ of the generic matrix $(x_{i,j})$ inverted.

On this algebra the group $\mathrm{GL}(k, F)$ acts both on the right and on the left and the polynomial ring $F[x_{i,j}]$. We have the Cauchy decomposition (formula (6.18)) $F[x_{i,j}] = \bigoplus_\lambda S_\lambda(F^k) \otimes S_\lambda((F^k)^*)$.

This is a special case of a general fact for a linearly reductive algebraic group $G$. The algebra $F(G)$ of functions on $G$ decomposes as $F(G) = \bigoplus V_i \otimes V_i^*$, where $V_i$ runs over all rational irreducible representations of $G$.

It is a standard fact for any reductive group that if we fix a Borel subgroup $B$ with unipotent radical $U$, the algebra $F(G)^U$ (where $U$ acts on the right) is a finitely generated algebra over which $G$ acts on the left.

This algebra is the coordinate ring of an affine variety acted by $G$, which contains as open orbit $G/U$. Moreover this variety is normal and with rational singularities **[Pop86]** (not needed in this book).

LEMMA 18.2.6. *For every rational representation $V$ of a linear algebraic group $G$, we have the isomorphism*

$$\pi : [F(G) \otimes V]^G \simeq V, \quad \textit{with} \quad \pi(f(g) \otimes v) := f(1)v, \tag{18.12}$$

*where $G$ acts diagonally on $F(G) \otimes V$ with the left action on $F(G)$.*

PROOF. The space of polynomial maps from $G$ to $V$ is $F(G) \otimes V$. The element $\sum_i f_i(g) \otimes v_i$ corresponds to the map $\phi : g \mapsto \sum_i f_i(g)v_i$ and finally $\phi(1) = \sum_i f_i(1)v_i$. The given action is

$$x \sum_i f_i \otimes v_i = \sum_i {}^x f_i \otimes x v_i, \quad g \mapsto \sum_i f_i(x^{-1}g) x v_i = x\phi(x^{-1}g).$$

Saying that $\phi$ is invariant under this action is equivalent to saying that $\phi$ is $G$ equivariant, that is $\phi(xg) = x\phi(g)$. Thus we have $\phi(g) = \phi(g1) = g\phi(1)$; in other words $\phi$ is determined by the vector $\phi(1) \in V$. Conversely given any vector $v \in V$, the map $g \mapsto gv$ is $G$ equivariant, so we have a canonical identification $\pi[F(G) \otimes V]^G \equiv V$, $f(g) \otimes v \mapsto f(1)v$. □

Moreover the map $\pi$ is $G$ equivariant if, on $[F(G) \otimes V]^G$, we use the $G$ action on the right of $F(G)$, since for an equivariant map $\phi$ we have $\phi^h(g) := \phi(gh)$ (induced from right action), which maps to $\phi^h(1) = \phi(h) = h\phi(1)$.

COROLLARY 18.2.7. *For any subgroup $H \subset G$ we have*

$$[F(G)^H \otimes V]^G \overset{\pi}{\simeq} V^H, \tag{18.13}$$

*where $G$ acts diagonally on $F(G)^H \otimes V$ and with the left action on $F(G)^H$. In fact this isomorphism is given by restricting to $(F(G)^H \otimes V)^G$ the explicit map $\pi : V \otimes F(G) \to V$, $\pi : v \otimes f(g) \mapsto vf(1)$. If $V$ is an algebra, with $G$ acting as automorphisms, this identification is an isomorphism of algebras.*

PROOF. This depends upon the fact that the right action of $H$ and the left action of $G$ commute. As for the second statement it is enough to observe that $\pi$ is a homomorphism of algebras. □

The invariant theory for $\mathcal{U}$ is not ruled by Hilbert's theory, since $\mathcal{U}$ is not reductive. Using the previous corollary we thus replace $\mathcal{U}$ with $G$ and obtain in this way the desired statement on finite generation. In fact a simple argument, as in Hilbert theory, shows the following.

Let $M$ be a module finitely generated over a finitely generated commutative algebra $A$. Assume that a linearly reductive group $G$ acts on $M$, and on $A$ by automorphisms in a compatible way, that is $g(am) = g(a)g(m)$, $a \in A$, $m \in M$. Then

LEMMA 18.2.8. *The space of invariants $M^G$ is finitely generated as a module over the algebra $A^G$.*

PROOF. Consider the $A$ submodule $AM^G$ of $M$. It is finitely generated by some elements $m_1, \dots, m_k \in M^G$. Thus if $u \in M^G$, we have $u = \sum_i a_i m_i$, $a_i \in A$. Now the map $A^k \to M$ given by $\sum_i a_i m_i$ is $G$ equivariant, so it commutes with the projection to the invariants (in $A$ called the Reynolds operator $R$), so $u = \sum_i R(a_i) m_i$, $R(a_i) \in A^G$. □

At this point we only have to apply the theory of graded modules over graded algebras, here the grading is by the semigroup of dominant weights (which one can identify to $\mathbb{N}^k$), and we use the fact that the algebra $F(G)^U$ is finitely generated by elements which have as weight the fundamental weights. Thus if $V$ is a finitely generated module over the finitely generated algebra $A$ with $G$ action, we have that $V \otimes F(G)^U$ is a finitely generated module over the finitely generated algebra $A \otimes F(G)^U$ with diagonal $G$ action. We deduce

THEOREM 18.2.9. *$V^U$ is a finitely generated module over the finitely generated algebra $A^U$.*

In the cases we are interested in the group $G = \mathrm{GL}(k, F)$, so we are assuming that $A, V$ are polynomial representations of $G$ (6.3.1), which we can expand as direct sum of Schur modules.

COROLLARY 18.2.10. *If $V$ and $A$ are as before and $V = \bigoplus_\lambda m_\lambda S_\lambda(F^k)$, the generating function $\sum_\lambda m_\lambda t^\lambda$ is a rational function, in the variables $t_i := t^{\omega_i}$, with denominator a product of $1 - t^{\mu_i} = 1 - t^{\sum_i m_i \omega_i} = 1 - \prod_i t_i^{m_i}$. The $\mu_i = \sum_i m_i \omega_i$, $m_i \in \mathbb{N}$, are the dominant weights of some finite set of irreducible representations generating $A$ as algebra.*

PROOF. If $V = \bigoplus_\lambda m_\lambda S_\lambda(F^k)$, we have $V^U = \bigoplus_\lambda m_\lambda S_\lambda(F^k)^U$, and $S_\lambda(F^k)^U$ is one dimensional generated by a vector of weight $\lambda$ so $\sum_\lambda m_\lambda t^\lambda$ is the Hilbert series of the graded module $V^U$ (graded by dominant weights). Then compute the generating function of $V^U$ using its identification with $(A \otimes F(\mathrm{GL}(k, F))^U)^G$, and then apply Lemmas 18.2.2 and 18.2.8. We only need remark that the torus $T$ acts on $(A \otimes F(\mathrm{GL}(k, F))^U)^G$ by acting on $F(\mathrm{GL}(k, F))$ on the right and under the identification the weight is preserved. □

In the algebras that we are studying and the various graded objects associated with them, we have the action of $\mathrm{GL}(k, F)$ which is in fact a polynomial action, that is no inverse of the determinant appears.

The action of the torus of diagonal matrices $t_1, \dots, t_k$ determines the weight decomposition and so the multigrading and the grading.

The fundamental weight $\omega_i = t_1 t_2 \cdots t_i$ has degree $i$, so the ordinary Hilbert series of $V^U$, $\sum_n \dim V_n^U$ is given by substituting $t_i := t^{\omega_i} \mapsto t^i$ so that $t^{\sum_i m_i \omega_i} \mapsto t^{\sum_i m_i i}$.

Notice that in degree $n$ the dimension of $V_n^U$ equals the length or number of irreducible components in which $V_n$ decomposes.

COROLLARY 18.2.11. *The length of $V_n$ is a rational function with numerator a polynomial in $\mathbb{Z}[t]$ and denominator a product of factors of type $1 - t^{h_i}$. The same statement holds for the colength of a PI algebra satisfying some Capelli identity.*

**18.2.2. Invariants of several copies of $V$.** Let $V$ be a vector space of some dimension $k$, and let $G$ be a semisimple group acting on $V$. The invariants of $m$ copies $V^m$ under the action of $G$ are also a representation of $\mathrm{GL}(m, F)$, in fact from Cauchy's formula (6.18) we have

(18.14)
$$S[[(V^*)^m] = \bigoplus_\lambda S_\lambda(V^*) \otimes S_\lambda(F^m) \implies S[[(V^*)^m]^G = \bigoplus_\lambda S_\lambda(V^*)^G \otimes S_\lambda(F^m).$$

If we are interested in understanding the multiplicity with which a given representation $S_\lambda(F^m)$ appears, we know that it equal to $\dim S_\lambda(V^*)^G$, provided that $m$ is larger than the height of $\lambda$ but it may appear only if the height of $\lambda$ is $\leq \dim V$. Thus this multiplicity stabilizes for $m \geq \dim V$. On the other hand if $U$ is the unipotent group of $\mathrm{SL}(k, F)$ of upper-triangular matrices with 1 on the diagonal, we have

$$S[[(V^*)^k]^{G \times U} = \bigoplus_\lambda S_\lambda(V^*)^G \otimes S_\lambda(F^k)^U, \ \dim S_\lambda(F^k)^U = 1.$$

Thus the generating function of these multiplicities is the generating function of $S[[(V^*)^m]^{G \times U}$.

In particular for growth we need to compute the dimension of the algebra $S[[(V^*)^m]^{G \times U}$. This we compute as follows. From the previous section we have that

$$S[[(V^*)^k]^{G \times U} = (S[[(V^*)^k] \otimes F(\mathrm{SL}(k))^U)^{G \times \mathrm{SL}(k)}. \tag{18.15}$$

THEOREM 18.2.12. *If $G$ acts faithfully on $V$ and $G \times \mathrm{SL}(k)$ acts freely on a nonempty open set of the variety $V^k \times \mathrm{SL}(k)/U$, the dimension of $S[[(V^*)^k]^{G \times U}$ is*

$$\frac{k^2 + k}{2} - \dim(G). \tag{18.16}$$

PROOF. From formula (18.15) this dimension is the dimension of the quotient variety of $G \times \mathrm{SL}(k)$ acting on the variety $V^k \times W$, where $W$ is the variety of coordinate ring $F(\mathrm{SL}(k))^U$ which contains $\mathrm{SL}(k)/U$ as dense open set, so its dimension is $(k^2-1) - \frac{k^2-k}{2} = \frac{k^2+k-2}{2}$.

The group $\mathrm{SL}(k, F)$ is semisimple and simply connected. So by a theorem of Popov (cf. [**Pop74**, Corollary of Proposition 1]) the coordinate ring of $\mathrm{SL}(k, F)$ is factorial, then, since $U$ is a connected group, also the ring $F(\mathrm{SL}(k))^U$ is factorial, hence the variety $V^k \times W$ is factorial.

For a semisimple group $H$ acting on an irreducible affine variety $X$ which is also factorial and with the generic orbit equal to $H$ or just with finite stabilizer, one knows, by another theorem of Popov, cf. [**Pop70**], that the generic orbit is closed, so it equals the generic fiber and, by Hilbert's theory, the quotient variety has dimension $\dim W - \dim H$. Thus we have in our case $\dim W = k^2 + \frac{k^2+k-2}{2}$ and since $H = G \times \mathrm{SL}(k)$ $\dim H = \dim G + k^2 - 1$. The formula follows. □

From this formula we see that only for somewhat small $G$ may we have this strong condition. The following criterion is then useful:

PROPOSITION 18.2.13. *If $G$ acts faithfully on a space $V$ of dimension $k$ and the generic stabilizer of the action of $G$ on $V$ is a torus, $G \times \mathrm{SL}(k)$ acts freely on a nonempty open set of $V^k \times \mathrm{SL}(k)/U$ and $G \times U$ acts freely on a nonempty open set of $V^k$.*

PROOF. In fact let us look at the stabilizer of $((v_1, \dots, v_k), U)$. It is the subgroup of $G \times U$ stabilizing $(v_1, \dots, v_k)$. Then, for a generic choice of $(v_1, \dots, v_k)$, this is the stabilizer of a generic orbit of $G \times U$ on $V^k$ so it is enough to show that this is trivial.

The action of $(g, u)$ on a vector $(v_1, \dots, v_k)$ is the action of $u$ on $(gv_1, \dots, gv_k)$. Moreover $u = 1 + \Lambda$ is some triangular matrix with $\lambda_{j,i} \neq 0, \implies j < i$, so finally

$$(g, u)(v_1, \dots, v_k) = (w_1, \dots, w_k), \quad w_i = gv_i + \sum_{j<i} \lambda_{j,i} g v_j. \tag{18.17}$$

Hence if $(g, u)$ stabilizes the vector $(v_1, \dots, v_k)$, we must have $v_i = w_i \ \forall i$.

In particular $v_1 = gv_1, v_2 = gv_2 + \lambda v_1$. So since $v_1$ is generic, $g$ is a semisimple element being in a torus. Decompose $V = V^g \oplus V_g$, the invariants and a stable complement, and write each $v_i = a_i + b_i$, $a_i \in V^g$, $b_i \in V_g$.

Consider thus $a_2 + b_2 = v_2 = gv_2 + \lambda v_1 = a_2 + gb_2 + \lambda a_1$. This implies that $b_2 - gb_2 = \lambda a_1 \in V_g \cap V^g = \{0\}$ implies $b_2 - gb_2 = 0$, but this implies $b_2 \in V^g$, hence $b_2 = 0$. Since $b_2$ is generic in $V_g$, this implies $V_g = 0$ or $g = 1$. Thus from formula (18.17), we deduce $\sum_{j<i} \lambda_{j,i} v_j = 0, \ \forall i$.

Then since $(v_1, \dots, v_k)$ are generic, they are linearly independent, and this implies $\lambda_{j,i} = 0$ and also $u = 1$. □

## 18.3. Applications to PI algebras

**18.3.1. Cocharacters.** We are now ready to apply the previous ideas to cocharacters, and we deduce

THEOREM 18.3.1. *Given a PI algebra $A$, denote by $m_\lambda(A)$ the multiplicity of $\chi_\lambda$ in the cocharacter of $A$. The generating series $\Xi_A := \sum_{\lambda, \mathrm{ht}(\lambda) \le k} m_\lambda(A) t^\lambda$ is a rational* *function as in formula* (18.10),

$$\Xi_A = \frac{\sum_{j=1}^{q} n_j t^{b_j}}{\prod_{i=1}^{m}(1-t^{a_i})}, \quad n_j \in \mathbb{Z}, b_j,\ a_i \in \mathbb{Z}^k, \tag{18.18}$$

PROOF. This follows from the fact that $m_\lambda(A)$ equals the multiplicity of $S_\lambda(V)$ in the relatively free algebra generated by $A$ over a vector space $V$ with $\dim V = k$. One then applies the standard filtration and the basic Corollary 18.2.10. □

18.3.1.1. *Colength.* We want to investigate now the colength of $R = \bigoplus_{n=0}^{\infty} R_n$, where $R$ is the relatively free algebra of some finite-dimensional algebra and $R_n$ the part of degree $n$. By definition (Definition 7.1.13) the colength is the function $\mathrm{col}_R(n)$ of $n$ which measures the number of irreducible representations of $\mathrm{GL}(k, F)$ decomposing the part $R_n$ of degree $n$ (if $R$ is relatively free in $k$ variables).

Arguing as for the dimension the colength of $R$ is the sum of the colengths of the factors $K_{i+1}/K_i$, i.e., $\mathrm{col}_R(n) = \sum_i \mathrm{col}_{K_{i+1}/K_i}(n)$.

Each one of these factors is a finite torsion-free module over some finitely generated algebra $T_{R_i}$ which is the coordinate ring of one of the varieties $W$. By Lemma 18.1.9 in turn we may replace it with the ring of invariants $T_A$ of several copies of a semisimple algebra $A = \bigoplus_{i=1}^{q} M_{n_i}(F)$ under its automorphism group $G = \prod_i \mathrm{PSL}(n_i)$. So we may apply Corollary 18.2.10, and we have that the generating series of the colength is a rational function with denominator a product of factors $1 - t^{h_i}$.

This implies that the colength $\mathrm{col}_R(n)$ stabilizes to a quasi-polynomial whose degree equals the dimension of the algebra $T_{R_i}^U$. Since $T_A$ is a finite module over $T_{R_i}$, by Theorem 18.2.9, $T_A^U$ is also a finite module over $T_{R_i}^U$, and hence the dimension of $T_{R_i}^U$ equals that of $T_A^U$.

The algebra $T_A^U$ is that coming from formula (18.15) for $V = A = \bigoplus_{i=1}^{q} M_{n_i}(F)$ under the group $G = \prod_i \mathrm{PSL}(n_i)$ which acts faithfully on $\bar{A}$, and it is semisimple. Moreover the generic element of $A$ is a list of matrices each with distinct eigenvalues so that the stabilizer is a product of maximal tori and hence reductive. We thus have verified all the properties of Proposition 18.2.13. We can thus apply Theorem 18.2.12, and then the formula for the dimension of $T_R^U$ is given by formula (18.16), where $k = \dim A = t = \sum_i n_i^2$ while $\dim G = \sum_{i=1}^{q}(n_i^2 - 1) = t - q$.

We finally get for the quotient the dimension

PROPOSITION 18.3.2. *If* $A = \bigoplus_{i=1}^{q} M_{n_i}(F)$, $G = \prod_{i=1}^{q} \mathrm{PSL}(n_i)$, *the dimension of the algebra of formula* (18.15), *equals* $\frac{t^2-t}{2} + q$.

PROOF. Here we apply formula (18.15) with $k = t$ and $\dim G = t - q$. □

THEOREM 18.3.3. *The degree of the colength* $\mathrm{col}_R(n)$ *of* $R$ *(relatively free in* $m$ *variables for a fundamental algebra) is the maximum of the degrees of the colengths of the factors* $K_{i+1}/K_i$. *When* $m$ *is sufficiently large, this is independent of* $m$ *and* $\mathrm{col}_R(n)$ *is a quasi-polynomial of degree* $\frac{t^2-t}{2} + q - 1$, *where* $t$ *is the first Kemer index of* $R$ *while* $q$ *is the* $q$ *invariant* (*Definition* 18.1.20).

The colength was computed by Berele **[Ber10]** for minimal varieties (prime product algebras in his terminology), hence in particular for upper block-triangular matrices.

18.3.1.2. *Proper versus ordinary cocharacters.* We now discuss a complement, based on the ideas developed in §7.4. If $A$ satisfies the Capelli identity $C_k$, one has that all nonzero $m_\lambda$ have height $\leq k$. Thus the previous identity rather than being computed in the formal ring of symmetric functions can be computed in the ring of symmetric functions in $k$ variables $x_1, \ldots, x_k$. Then $\sum_\lambda m_\lambda(A) S_\lambda(x_1, \ldots, x_k)$ is the generating function for the multigraded relative free algebra in $k$ variables, while

$$\sum_i S_i(x_1, \ldots, x_k) = \frac{1}{\prod_{i=1}^k (1 - x_i)}.$$

From formula (7.16) we deduce

COROLLARY 18.3.4. *$\sum_\lambda \overline{m}_\lambda(A) S_\lambda(x_1, \ldots, x_k)$ is also a nice symmetric rational function.*

## 18.4. Model algebras

**18.4.1. The canonical model of fundamental algebras.** We want to discuss now the problem of choosing special fundamental algebras in the PI equivalence classes. In Corollary 18.1.15 we have seen that two PI-equivalent fundamental algebras have isomorphic semisimple parts. Thus it is natural to study fundamental algebras $A$ with a given fixed semisimple part $\bar{A} = \bigoplus_{i=1}^q M_{n_i}(F)$ so that $\beta(A) = t = \sum_{i=1}^q n_i^2$.

REMARK 18.4.1. From Lemma 17.2.36 it follows that the second Kemer index $\gamma(A)$, which for a fundamental algebra coincides with the maximum $s$ for which $J^s \neq 0$, must be $\geq q - 1$. The case $\gamma(A) = q - 1$ is attained by all possible algebras of upper-triangular matrices with semisimple part $\bar{A}$.

So our present goal is to analyze fundamental algebras with index (cf. Definition 18.1.17) $(\bar{A}, s)$, and $s \geq q - 1$ (and thus Kemer index $t = \dim(\bar{A}), s$). We use now Definition 17.2.46 and Proposition 17.2.47.

Consider a $T$-ideal $I$ of identities of a fundamental algebra $A = \bar{A} \oplus J$. We have seen in Corollary 18.1.15 that the semisimple part $\bar{A}$ is determined by $I$. We want to construct a canonical fundamental algebra having semisimple part $\bar{A}$ and $I$ as $T$-ideal of identities.

First we construct a universal object. Take the free product $\bar{A} \star F\langle X\rangle$, for $X = \{x_1, \ldots, x_m\}$. We assume $m \geq s$. We now take this free product modulo the ideal of elements of degree $\geq s + 1$ in the variables $X$, and we call $\mathcal{F}_{\bar{A},s}(X)$ the resulting algebra.

$\mathcal{F}_{\bar{A},s}(X)$ is a finite-dimensional algebra with semisimple part $\bar{A}$ and Jacobson radical $J$ of nilpotency $s + 1$ generated by the $x_i$. It satisfies a universal property among such algebras.

REMARK 18.4.2 (Universal property). Given a finite-dimensional algebra $A$ with semisimple part $\bar{A}$ and Jacobson radical $J$ of nilpotency $s + 1$, any map $X \to J$ extends to a unique homomorphism of $\mathcal{F}_{\bar{A},s}(X)$ to $A$ which is the identity on $\bar{A}$.

18.4.1.1. *Description of $\mathcal{F}_{\bar{A},s}(X)$.* Let $V$ be the vector space with basis the elements $x_i$. In degree $h$ the algebra $\mathcal{F}_{\bar{A},s}(X) = \mathcal{F}_{\bar{A},s}(V)$ can be described as follows. Consider

DEFINITION 18.4.3. Let $\mathcal{M}$ be the monoid in two generators $a, b$ with relation $a^2 = a$. Elements of this monoid correspond to words in $a, b$ in which $aa$ never appears as subword. When we multiply two such words, if we have the factor $aa$ appearing we reduce it by the rule to $a$.

Now to such a word $w$ we associate a tensor product $T_w$ of $\bar{A}$ whenever we have an $a$ and $V$ when we have a $b$. For instance:

$$w = abbab \mapsto T_w = \bar{A} \otimes V \otimes V \otimes \bar{A} \otimes V.$$

The multiplication of two such words $w_1 w_2$ according to the previous rule, induces a multiplication $T_{w_1} T_{w_2} \subset T_{w_1 w_2}$. We take the corresponding tensor product, and if we get a factor $\bar{A} \otimes \bar{A}$, we replace it by $\bar{A}$ by multiplication.

Then we see that $\mathcal{F}_{\bar{A},s}(X)$ is the direct sum $\oplus T_w$, where $w$ runs over all the words of previous type with at most $s$ appearances of $b$. It is a graded algebra over this monoid, truncated at degree $s$ in $b$.

As representation of $\mathrm{GL}(m, F) \times G = \mathrm{GL}(V) \times G$ $(G = \prod_{i=1}^{q} \mathrm{PSL}(n_i))$, $\mathcal{F}_{\bar{A},s}(V)$ in degree $h$ is a direct sum of $c_{i,j}$ spaces each isomorphic to $\bar{A}^{\otimes i} \otimes V^{\otimes h}$ where $c_{i,j}$ is the number of words in $\mathcal{M}$ with $i$ times $a$ and $j$ times $b$. The summands correspond to the type of elements in the free product which are monomials in $j$ elements of $V$ and $i$ elements of $\bar{A}$.

COROLLARY 18.4.4. *Having fixed s, the* $\mathrm{GL}(V)$ *invariant subspaces of* $\mathcal{F}_{\bar{A},s}(V)$ *all intersect* $\mathcal{F}_{\bar{A},s}(V_s)$ *where* $V_s$ *is the subspace of dimension s spanned by the first s variables.*

PROOF. This is a property of each $V^{\otimes h}$, $h \leq s$. A $\mathrm{GL}(V)$ invariant subspace is generated by its highest weight vectors which in $V^{\otimes h}$ depend on the first $h$ elements of a chosen basis of $V$ (the variables). □

Corollary 18.4.4 gives another justification for restricting our analysis to the algebras $\mathcal{F}_{\bar{A},s}$ (with $X$ a set of $s$ variables).

PROPOSITION 18.4.5. *When* $s \geq q - 1$, *we have that* $\mathcal{F}_{\bar{A},s}$ *is a fundamental algebra with index* $(\bar{A}, s)$.

PROOF. This can be done by following the proof of Kemer's lemma and proving the existence of Kemer's polynomials. We just sketch the proof.

If $\bar{A} = \bigoplus_{i=1}^{q} M_{n_i}(F)$, take, for every $k$ and for each $i = 1, \dots, q$, a product $P_i$ of $k + s$ Capelli polynomials in $n_i^2$ variables $X_{i,j}, i = 1, \dots, q;\ j = 1, \dots, k + s$. When evaluated in suitable matrix units, each $P_i$ is nonzero.

Take the product $M_0 P_1 M_1 P_2 \cdots M_{q-1} P_q M_q$ of all these polynomials where the $M_i$ are monomials in $s$ variables $z_1, \dots, z_s$ so that each variable appears once and $M_i$ for $i = 1, \dots, q - 1$ has degree at least 1. Here we use $s \geq q - 1$.

Then form $k$ layers $X_j := \bigcup_i X_{i,j},\ j = 1, \dots, k$, with $t = \sum_i n_i^2$ elements and $s$ layers with $t + 1$ elements $Z_j := z_j, \bigcup_i X_{i,j+k},\ j = 1, \dots, s$. Then alternate all these layers independently obtaining a new polynomial $f$. Looking carefully at the description of $\mathcal{F}_{\bar{A},s}$, one has a nonzero evaluation of $f$ by evaluating the $z_j$ in the elements $x_j \in \mathcal{F}_{\bar{A},s}$ and the other variables in suitable matrix units. The polynomial $f$ is thus Kemer and gives the Kemer index $(t, s)$. □

DEFINITION 18.4.6. The algebra $\mathcal{F}_{\bar{A},s}$ is the *universal fundamental algebra* for the pair $\bar{A}, s$.

We remark that $\mathcal{F}_{\bar{A},s}$ is defined for all $s$. But only if $s \geq q-1$ do we have that $\mathcal{F}_{\bar{A},s}(X)$ is fundamental by Proposition 18.4.5.

18.4.1.2. *The algebra $\mathcal{A}_s(X)$ associated to a fundamental algebra $A$.* Given a finite-dimensional algebra $A$ with given $(t,s)$-index, we define $\mathcal{A}_s(X)$ to be $\mathcal{F}_{\bar{A},s}(X)$ modulo the ideal generated by all polynomial identities of $A$. All these PI vanish on $\bar{A}$, hence they take values in the ideal generated by the $x_i$. Therefore the Jacobson radical $J_s$ of $\mathcal{A}_s(X)$ is the ideal generated by the $x_i$, and its semisimple part is $\bar{A}$.

By construction and Remark 18.4.2, given any list of elements $a_1, \dots, a_m \in J$, there is a morphism $\pi : \mathcal{A}_s(X) \to A$ which is the identity on $\bar{A}$ and maps $x_i \mapsto a_i$.

For $m$ sufficiently large this morphism may be chosen so that the radical $J_s$ maps surjectively to the radical $J$, thus also the map of $\mathcal{A}_s(X)$ to $A$ is surjective. Hence $\mathcal{A}_s(X)$ satisfies the same PI as $A$.

DEFINITION 18.4.7. We now define $\mathcal{F}_{\bar{A},s}$ and $\mathcal{A}_s$ to be the algebras $\mathcal{F}_{\bar{A},s}(X)$ and $\mathcal{A}_s(X)$, where $X = \{x_1, \dots, x_s\}$ is formed by $s$ variables.

LEMMA 18.4.8. *$\mathcal{A}_s$ satisfies the same identities as $A$.*

PROOF. By construction all identities of $A$ are satisfied by $\mathcal{A}_s$, so we need to prove the converse. Take a multilinear polynomial $f$ which is not a PI of $A$. We need to show that it does not vanish on $\mathcal{A}_s$.

There is a substitution of $f$ in $A$, which we may assume to be elementary, (cf. Definition 17.2.23), which is nonzero.

In this substitution at most $s$ variables are in the radical and the others are in some matrix units of $\bar{A}$.

We now take the same substitution for matrix units in $\bar{A}$ and the remaining variables, which we may call $y_1, \dots, y_k$, $k \leq s$. We substitute in $x_1, \dots, x_k \in X \subset \mathcal{A}_s$. By the universal property the evaluation of $f$ in $A$ factors through this evaluation in $\mathcal{A}_s$, which therefore is different from 0. □

We now apply this general discussion to fundamental algebras.

LEMMA 18.4.9. *Let $B, A = B/I$ be two finite-dimensional algebras, let $J$ be the radical of $B$, and let $I \subset J$, so $A, B$ have the same semisimple part $\bar{A}$. Assume that $A, B$ have the same nilpotency index and $A$ is fundamental. Then $B$ is fundamental and with the same Kemer index as $A$.*

PROOF. The assumption implies that $A, B$ have the same $t, s$ index. Then the statement follows from Theorem 17.2.41 since by hypothesis the Kemer index of $A$ equals the $t, s$ index. On the other hand clearly the Kemer index of $B$, which is less than or equal to its $t, s$ index, cannot be less than the Kemer index of $A$. □

PROPOSITION 18.4.10. *If $A$ is fundamental, $\mathcal{F}_{\bar{A},s}$ and $\mathcal{A}_s$ are fundamental algebras with index $(\bar{A}, s)$.*

PROOF. For $\mathcal{F}_{\bar{A},s}$ this is the content of Proposition 18.4.5 and then the statement follows from Theorem 17.2.41, since by construction its Kemer index equals the $(t,s)$-index. □

DEFINITION 18.4.11. We call $\mathcal{A}_s$ the *canonical model* of $A$.

Notice a property of this model, if $f(y_1, \ldots, y_k)$ is a multilinear noncommutative polynomial. In order to verify if it is a PI of $\mathcal{A}_s$, one has to choose in all possible ways a subset of the variables $y$ with $h \leq s$ elements, evaluate these variables in the canonical elements $x_1, \ldots, x_h$, and then evaluate all the other variables into matrix units of $\bar{A}$. If all these finitely many evaluations are 0, then $f$ is a PI. This follows easily from the universal property of Remark 18.4.2.

We remark also that by construction the nilpotent subalgebra of $\mathcal{F}_{\bar{A},s}(X)$, generated by the $x_i$, is a relatively free nilpotent algebra.

QUESTION. When is the $T$-ideal of identities of $\mathcal{F}_{\bar{A},s}$ irreducible, and how is it described?

REMARK 18.4.12. The algebra $R$ of block upper-triangular matrices relative to a decomposition of a vector space $V = \bigoplus_{i=1}^{q} V_i$, $\dim V_i = n_i$ can be presented as a quotient of $\mathcal{F}_{\bar{A},q-1}$ as follows. $R$ is described as the direct sum $\bigoplus_{i \leq j} \hom(V_i, V_j)$. $\bar{A} = \bigoplus_i \hom(V_i, V_i)$ and $\hom(V_i, V_j)$ is an irreducible module under $\hom(V_i, V_i)$ $\oplus \hom(V_j, V_j)$. Take the elements $E_{i,i+1}$ to be a nonzero matrix in the corresponding block $\hom(V_i, V_{i+1})$. It is easily seen that the homomorphism mapping $\mathcal{F}_{\bar{A},q-1}$ to $R$, which is the identity on $\bar{A}$ and maps $x_i \to E_{i,i+1}$, is surjective.

CHAPTER 19

# Identities and superalgebras

*Flipping a diagram along the diagonal corresponds to tensoring an irreducible representation of the symmetric group with the sign representation. In a sense one exchanges symmetry and antisymmetry.*

*The $(k, \ell)$ hook theorem, Theorem 7.5.1, suggests that there is a corresponding duality, flipping the diagrams for polynomial identities. This in fact can be made into a full theory, but in order to do this one has to extend the theory to superalgebras, that is algebras graded modulo 2.*

*The infinite Grassmann algebra plays the role of the sign representation. The operation of flipping the diagram corresponds to some extent to taking a tensor product with the infinite Grassmann algebra.*

*The theory has remarkable consequences for PI theory, which we will discuss as Kemer's theory. The main goals are*

THEREOM 19.7.4. *Every PI superalgebra $R$ finitely generated over a field $F$, is $PI_2$ equivalent to a superalgebra finite dimensional over $F$.*

THEREOM 19.8.1. *Every nontrivial variety of algebras is generated by the Grassmann envelope of a finite-dimensional superalgebra.*

## 19.1. The Grassmann algebra

Let $F$ be a field of characteristic different from 2.

DEFINITION 19.1.1. The Grassmann algebra over $F$ is the algebra generated by 1 and a countable set of elements $e_1, e_2, \dots,$ called *Grassmann variables* subject to the condition $e_ie_j = -e_je_i$, for all $i, j \geq 1$. In symbols

$$G = \langle 1, e_1, e_2, \dots \mid e_ie_j = -e_je_i, \text{ for all } i, j \geq 1 \rangle.$$

In fact $G$ is the exterior algebra $\bigwedge V$, where $V$ is the infinite-dimensional vector space with basis the elements $e_i$. In the literature on exterior algebras the product is denoted by $\wedge$ as $e_i \wedge e_j$, but in our treatment we use the usual notation $e_ie_j$. The standard theory of exterior algebras shows that $G$ has a basis consisting of the monomials of the type

$$e_{i_1}e_{i_2} \cdots e_{i_k},\ 1 \leq i_1 < \cdots < i_k.$$

The algebra $G$ has a natural structure of superalgebra (cf. Definition 19.2.1) $G = G_0 \oplus G_1$, where $G_0$ is the linear span of the monomials in the $e_i$'s of even length, and $G_1$ is the linear span of the monomials of odd length.

Notice that $G_0$ coincides with the center of $G$ while for two elements $a, b \in G_1$, we have $ab = -ba$. Hence $[G, G] = [G_1, G_1] \subseteq G_0$. It follows that $G$ satisfies the PI algebra

$$[x, y, z] := [[x, y], z] = 0. \tag{19.1}$$

REMARK 19.1.2. The Grassmann algebra in infinitely many variables over a field of characteristic 0 does not satisfy any standard identity.

PROOF. We have $\mathrm{St}_k(e_1, \dots, e_k) = k! e_1 e_2 \cdots e_k$. □

Next we are going to show that $\mathrm{Id}(G) = \langle [x, y, z] \rangle_T$, is the $T$-ideal generated by $[x, y, z]$. We start with the following

LEMMA 19.1.3. $[x_1, x_2][x_3, x_4] + [x_1, x_3][x_2, x_4] \in \langle [x, y, z] \rangle_T$.

PROOF. We have, $[xy, z] = xyz - zxy = xyz - xzy + xzy - zxy = x[y, z] + [x, z]y$, hence

$$\begin{aligned}[x_1x_4, x_2, x_3] &= [x_1[x_4, x_2], x_3] + [[x_1, x_2]x_4, x_3] \\ &= x_1[x_4, x_2, x_3] + [x_1, x_3][x_4, x_2] + [x_1, x_2][x_4, x_3] + [x_1, x_2, x_3]x_4\end{aligned}$$

implies

$$[x_1, x_3][x_2, x_4] + [x_1, x_2][x_3, x_4] = -[x_1x_4, x_2, x_3] + x_1[x_4, x_2, x_3] + [x_1, x_2, x_3]x_4.$$

The right-hand side is clearly in $\langle [x, y, z] \rangle_T$, so the claim follows. □

By identifying variables $x_1 = x_3$ and $x_4 = x_3$, we get

COROLLARY 19.1.4. $[x_1, x_2][x_1, x_3] \in \langle [x, y, z] \rangle_T$.

For the next theorem we assume the characteristic of the base field $F$ to be 0.

THEOREM 19.1.5 (Krakowski and Regev; Olsson and Regev). *Let $c_n(G)$ denote the nth codimension of the infinite-dimensional Grassmann algebra $G$, and let $\chi_n(G)$ be its nth cocharacter. Then*

(1) *The T-ideal of the identities of G is generated—as a T-ideal—by the single polynomial $[[x_1, x_2], x_3]$.*
(2) (*cf.* **[KR73]**)

$$c_n(G) = 2^{n-1}, \quad \textit{and}$$

(3) (*cf.* **[OR76]**)

$$\chi_n(G) = \sum_{j=1}^{n} \chi_{(j, 1^{n-j})},$$

*with $(j, 1^{n-j})$ a hook partition.*

PROOF. Given $n$ variables $x_1, \dots, x_n$ and a set $S$ of $k$ out of the variables $x_i$ such that $n - k = 2h$ is even, we construct the following polynomial $P_S$ (or $P_S^n$ if we want to recall $n$). We have in increasing order

$$S = \{i_1 < i_2 < \cdots < i_k\}, \ [1, \dots, n] \setminus S = \{j_1 < j_2 < \cdots < j_{2h}\}.$$

We then set

$$P_S = x_{i_1} x_{i_2} \cdots x_{i_k} [x_{j_1}, x_{j_2}] \cdots [x_{j_{2h-1}}, x_{j_{2h}}], \tag{19.2}$$

e.g.,

$$P_{3,5}^6 = x_3x_5[x_1, x_2][x_4, x_6], \qquad P_{1,3,5}^7 = x_1x_3x_5[x_2, x_4][x_6, x_7].$$

This produces exactly $2^{n-1}$ choices of elements $P_S$.

We have already checked that $[x, y, z] \in \mathrm{Id}(G)$. Then every left normed commutator of length $\geq 3$ is in $\mathrm{Id}(G)$.

By Proposition 7.4.4, and Corollary 19.1.4, a repeated application of Lemma 19.1.3 shows that every multilinear polynomial $f(x_1, x_2, \ldots, x_n)$ can be written, modulo the $T$-ideal, as a linear combination of the $2^{n-1}$ elements $P_S$.

Now we claim that these elements are linearly independent as functions on the Grassmann algebra. This will prove parts (1) and (2).

Assume $\sum_S \lambda_S P_S = 0$ on $G$. We want to prove, by descending induction on $k := |S|$, that $\lambda_S = 0$, $\forall S$.

Let us start with $S = \{1, \ldots, n\}$. We substitute the $x_i$ with disjoint even elements as, for instance, $x_i \mapsto e_{2i-1}e_{2i}$. Then all polynomials $P_S$ with $|S| < n$ vanish since they contain at least one commutator in which a variable is evaluated in the center, so we deduce that $\lambda_{\{1,\ldots,n\}} = 0$.

Assume by induction that $\lambda_S = 0$ for all $k = |S| > h$ and $n - h$ even.

Take any subset $A \subset \{1, \ldots, n\}$ with $h$ elements, and substitute these $h$ variables with even elements $e_i e_j$ and the remaining with odd elements $e_k$ so that no index is repeated. Then all the terms $P_R$ with $|R| \leq h$ different from $P_S$ vanish since one of the variables corresponding to $A$ is inside a commutator. We deduce that also $\lambda_A = 0$.

This proves (1) and (2). As for (3) we have to understand the action of the symmetric group on the span $W$ of the elements $P_S^n$ in $V_n / V_n \cap \mathrm{Id}(G)$.

Since $S_n$ is generated by the simple transpositions $s_i := (i, i+1), i = 1, \ldots, n-1$, let us first analyze the action of these elements on $P_S$.

For $n = 2m$ or $n = 2m + 1$, we filter the space $W$ by setting

$$W_k := \{\operatorname{span}\langle P_S \rangle \mid |S| \leq n - 2k\},\ k = 0, \ldots, m.$$

We have $W_k \supset W_{k+1}$ and claim that all the spaces $W_k$ are stable under the symmetric group.

Let $|S| = h = n - 2k$. If $i, i+1$ are both not in $S$, we have two possibilities: if the commutator $[x_i, x_{i+1}]$ is a factor of $P_S$, then $s_i P_S = -P_S$; otherwise, in the expression of $P_S = M[x_a, x_i][x_{i+1}, x_b]N$ will appear the two factors $[x_a, x_i][x_{i+1}, x_b]$ with $a < i$, $i + 1 < b$. Using the identity of Lemma 19.1.3, we see that

$$s_i P_S = M[x_a, x_{i+1}][x_i, x_b]N = -P_S, \text{ modulo } \mathrm{Id}(G).$$

If $i, i+1 \in S$, we see that $s_i(x_i x_{i+1}) = x_{i+1}x_i = x_i x_{i+1} - [x_i, x_{i+1}]$, and $s_i P_S = P_S + Q$, where $Q$ is in $W_{k+1}$.

Finally let us assume that one of the two indices is in $S$ and the other is not, say for instance $i \in S$. Then by the very definition it is easy to see that $s_i P_S = P_{s_i S}$ where $s_i S$ is obtained from $S$ by replacing $i$ with $i + 1$ (or vice versa).

This shows immediately that all the spaces $W_k$ are stable under the symmetric group. Moreover, for a given $h = n - 2k$, we see that the representation of the symmetric group $W_k / W_{k+1}$ has a basis of the classes of the elements $P_S$ with $|S| = h$ and the action of the symmetric group is the one just described,

$$s_i P_S = \begin{cases} P_S & \text{if } \{i, i+1\} \subset S \\ -P_S & \text{if } \{i, i+1\} \cap S = \emptyset \\ P_{s_i S} & \text{if } |\{i, i+1\} \cap S| = 1. \end{cases} \tag{19.3}$$

One easily sees that this is the representation induced from $S_h \times S_{n-h}$ to $S_n$ from the one-dimensional representation which is trivial on $S_h$ and the sign on $S_{n-h}$.

Now one knows that given a representation of a symmetric group $S_k$, one associates to it the Frobenius character $\sum_\lambda m_\lambda S_\lambda$ where the $S_\lambda$ are the Schur functions. When you do induction of the tensor product of two representations of $S_k, S_h$ to $S_{k+h}$, by the Frobenius formula, the Frobenius character of the induced representation is the product of the two corresponding Frobenius characters (cf. (6.34)).

In our case the product of the two Schur functions $S_{(n-2k)}$ and $S_{(1^{2k})}$ is given by Pieri's formula (Proposition 3.2.18). This gives exactly the two hooks $(h, 1^{2k})$ and $(h+1, 1^{2k-1})$, where $h = n - 2k$, except in the degenerate cases when $h = 0$ (if $n = 2k$ is even) which gives the sign, or $h = n$ giving the trivial representation.

Even in positive characteristic one may show that there is a filtration of $W_k$ with associated grading of the usual forms of the hook representations.

In characteristic 0 all terms are distinct and add to Theorem 19.1.5(3). □

Notice that the proof works in all characteristics $\neq 2$, although only in characteristic 0 can we deduce from the given filtration an actual decomposition into irreducible components.

In case one considers the Grassmann algebra without 1 in characteristic 0, the above theorem still holds, but in positive characteristic the above result is no longer valid. It can be shown that over an infinite field of characteristic $p > 2$, the $T$-ideal of identities of such an algebra is generated by $[x, y, z]$ and $x^p$ (**[GK01]**).

### 19.1.1. Finite-dimensional Grassmann algebras.

REMARK 19.1.6. Recall the following relations (Lemma 7.3.11). Let $\alpha \in F$ be a scalar. Then, if $k$ is odd,

$$\mathrm{St}_k(x_1, \dots, x_{i-1}, \alpha, x_{i+1}, \dots, x_k) = (-1)^{i+1} \alpha \, \mathrm{St}_{k-1}(x_1, \dots, x_{i-1}, x_{i+1}, \dots, x_k),$$

whereas if $k$ is even

$$\mathrm{St}_k(x_1, \dots, x_i, \alpha, x_{i+1}, \dots, x_k) = 0. \tag{19.4}$$

Let $G_k$ be the finite-dimensional Grassmann algebra with 1 on a $k$-dimensional vector space with basis $e_1, \dots, e_k$. We want to prove

THEOREM 19.1.7. *$G_{2h}$ and $G_{2h+1}$ are PI equivalent. $G_{2h}$ is a fundamental algebra.*

PROOF. Recall that the $n$th cocharacter of the infinite-dimensional Grassmann algebra $G$ is

$$\chi_n(G) = \sum_{j=1}^{n} \chi_{(j, 1^{n-j})}.$$

Since $\mathrm{Id}(G_k) \supseteq \mathrm{Id}(G)$, then $\chi_n(G_k) \leq \chi_n(G)$ (the multiplicities in $\chi_n(G_k)$ are bounded by the multiplicities in $\chi_n(G)$). We claim that

$$\chi_n(G_k) = \sum_{j=n-k+1}^{n} \chi_{(j, 1^{n-j})}, \text{ if } k \text{ is odd}$$

and

$$\chi_n(G_k) = \sum_{j=n-k}^{n} \chi_{(j, 1^{n-j})}, \text{ if } k \text{ is even.}$$

In fact, for every $1 \leq r \leq k$, consider the tableau whose first column is $1, 2, \dots, r$, and first row $1, r+1, \dots, n$, and the multilinear polynomial $f_r$ corresponding, in the identification with the algebra of the symmetric group, to the essential idempotent, Remark 6.2.4, product of the symmetrizer on the first row and the antisymmetrizer on the first column.

$$f_r = \sum_{\sigma \in S_{n-r+1}} \mathrm{St}_r(x_{\sigma(1)}, x_2, \ldots, x_r) x_{\sigma(r+1)} \cdots x_{\sigma(n)},$$

where $S_{n-r+1}$ acts on $1, r+1, \ldots, n$. In the space $V_n$ of multilinear polynomials, $f_r$ generates under the symmetric group $S_n$ a copy of the corresponding hook representation.

If $r$ is even, we consider the evaluation $\varphi$ such that $\varphi(x_i) = e_i$, $1 \le i \le r$, and $\varphi(x_s) = 1, r+1 \le s \le n$. Then, by Remark 19.1.2 and formula (19.4),

$$\varphi(f_r) = (n-r)!\varphi(\mathrm{St}_r(e_1, \ldots, e_r)) = (n-r)!r!e_1 \cdots e_r \neq 0.$$

Whereas if $r$ is odd, we consider the evaluation $\varphi$ such that $\varphi(x_1) = 1$, $\varphi(x_i) = e_i$, $2 \le i \le r$, and $\varphi(x_s) = 1, r+1 \le s \le n$. Then

$$\varphi(f_r) = (n-r+1)!\varphi(\mathrm{St}_{r-1}(e_2, \ldots, e_r)) = (n-r+1)!(r-1)!e_2 \cdots e_r \neq 0.$$

Notice that if $k$ is even, such an evaluation gives also $\varphi(f_{k+1}) \neq 0$.

Thus $\chi_{(n-r+1,1^{r-1})}$ appears in the decomposition of $\chi_n(G_k)$, for all $1 \le r \le k$, if $k$ is odd and for all $1 \le r \le k+1$, if $k$ is even.

Now, by Remark 19.1.6, $\mathrm{St}_r$ is an identity of $G_k$, for any $r > k$ in the odd case, and for any $r > k+1$ in the even case. So for these values, $f_r$ vanishes, due to the idempotent property. This means that the corresponding hook representations are also identities. This proves the claim.

Consider the Wedderburn–Malćev decomposition of $G_k$,

$$G_k = F + J,$$

where $J$ is the algebra $G_k$ without 1. We have $J^k \neq 0$ and $J^{k+1} = 0$.

We can now apply Theorem 17.2.14. By looking at the decomposition of the cocharacter of $G_k$, we deduce that its Kemer index is $(1, k)$ if $k$ is even and $(1, k-1)$ if $k$ is odd. Since the $(t, s)$-index of $G_k$ is $(1, k)$, we deduce that if $k$ is even $G_k$ is a fundamental algebra. □

Next we construct Kemer polynomials for $G_{2h}$ (and by PI equivalence also for $G_{2h+1}$). For every $n > k$ define

$$g_{n,k} := \left(\prod_{j=1}^{k} \mathrm{Alt}_{j,k+j}\right) [\mathrm{St}_k(x_1, x_2, \ldots, x_k) x_{k+1} x_{k+2} \cdots x_{2k} \cdots x_n],$$

where $\mathrm{Alt}_{a,b} := 1 - (a, b)$ is the alternation on the corresponding variables $x_a, x_b$.

Let $\varphi$ be the evaluation such that $\varphi(x_i) = e_i$, $1 \le i \le k$, and $\varphi(x_s) = 1$, $k+1 \le s \le n$. Taking into account the above two remarks when $k = 2h$ is even, we get

$$\varphi(g_{n,k}) = \mathrm{St}_k(e_1, \ldots, e_k) 1 \cdots 1 = k! e_1 \cdots e_k \neq 0,$$

and $g_{n,k}$ is a Kemer polynomial for $G_{2h}$.

If $k$ is odd, we let $\varphi$ be the evaluation such that $\varphi(x_i) = e_i$, $1 \le i \le k-1$, $\varphi(x_s) = 1$, $k \le s \le n$. Then $\varphi(g_{n,k-1}) = (k-1)!e_1 \cdots e_{k-1} \neq 0$, and $g_{n,k-1}$ is a Kemer polynomial in this case, as predicted by PI equivalence.

EXERCISE 19.1.8. Prove that the $T$-ideal of the Grassmann algebra $G_k := \bigwedge V$, where $\dim_F V = k$, is generated by the identity $[x, y, z]$ and $\mathrm{St}_{2h}(x_1, \ldots, x_{2h})$, where $h$, is the minimum integer such that $2h > k$.

Equivalently,

$$\mathrm{Id}(G_{2k}) = \mathrm{Id}(G_{2k+1}) = \langle [x_1, x_2, x_3], [x_1, x_2] \cdots [x_{2k+1}, x_{2k+2}] \rangle_T.$$

19.1.1.1. *Complement.* Observe that in characteristic 0 the free algebra $F\langle X\rangle/T$, where $T$ is the $T$-ideal generated by $[x, y, z]$, decomposes as

$$F\langle X\rangle/T = \bigoplus_{h,k} S_{(h,1^k)}(W),$$

where $W$ is the vector space with basis the variables $X$. Thus we have that only hook representations appear and each appears with multiplicity 1.

This implies that any subspace of $\bigoplus_{h,k} S_{((h,1^k))}(W)$, which is invariant under the group $\mathrm{GL}(W)$ of linear substitutions of variables, is a direct sum of a subset of the summands $S_{h,1^k}(W)$. In particular a $T$-ideal $I$ of $\bigoplus_{h,k} S_{(h,1^k)}(W)$ corresponds to some special subset.

Now we have two possible notions of $T$-ideals, one in the category of algebras with a 1. In this case $I$ is generated by proper polynomials, and one sees that it is necessarily generated by some standard identity. Thus $I$ is the $T$-ideal of identities of $G_n$ for some $n$.

EXERCISE 19.1.9. We work in the category of algebras without 1 and then describe the $T$-ideal of $\bigoplus_{h,k} S_{((h,1^k))}(W)$ generated by one of the summands $S_{(h,1^k)}(W)$. In particular we show that it contains a standard identity. Describe combinatorially all $T$-ideals of $\bigoplus_{h,k} S_{(h,1^k)}(W)$.

19.1.1.2. *Grassmann algebras and standard identities.* Next we want to study the relation between the Grassmann algebra and the standard polynomials.

THEOREM 19.1.10. *A variety $\mathcal{V}$ satisfies a standard identity if and only if $G \notin \mathcal{V}$.*

PROOF. By Remark 19.1.2 the algebra $G$ does not satisfy a standard identity of any degree. It follows that if $\mathrm{St}_m(x_1, \ldots, x_m)$ is a multilinear identity of $\mathcal{V}$ for some $m \geq 1$, then $G \notin \mathcal{V}$. Conversely, assume that $f(x_1, \ldots, x_n)$ is a multilinear identity of $G$. We must have

$$\sum_{\sigma \in S_n} \epsilon_\sigma \sigma f(x_1, \ldots, x_n) = \sum_{\sigma \in S_n} \epsilon_\sigma f(x_{\sigma(1)}, \ldots, x_{\sigma(n)}) = 0,$$

since any antisymmetric polynomial is a multiple of the standard identity and no standard identity is satisfied by $G$.

Now if $G \notin \mathcal{V}$, there is a multilinear polynomial $g(x_1, \ldots, x_n) \in \mathrm{Id}(\mathcal{V})$ which is not an identity of $G$. By the proof of Theorem 19.1.5 we may write, with $P_S$ defined in (19.2),

$$g(x_1, \ldots, x_n) = \sum_S \alpha_S P_S + f(x_1, \ldots, x_n),$$

where $f(x_1, \ldots, x_n) \in \mathrm{Id}(G)$, and not all $\alpha_S$ vanish. If $n = 2q$ and

$$g(x_1, \ldots, x_{2q}) = \alpha[x_1, x_2] \cdots [x_{2q-1}, x_{2q}] + f(x_1, \ldots, x_{2q}), \tag{19.5}$$

consider the antisymmetrizer

$$\mathrm{Alt}_{2q} := \sum_{\sigma \in S_{2q}} \epsilon_\sigma \sigma \in F[S_{2q}].$$

We have already remarked that this applied to $f$ gives 0.

On the other hand, since $\mathrm{Alt}_{2q}$ antisymmetrizes the $2q$ variables $x_1, \ldots, x_{2q}$, we get that

$$\mathrm{Alt}_{2q}[x_1, x_2] \cdots [x_{2q-1}, x_{2q}] = 2^q \, \mathrm{St}_{2q}(x_1, \ldots, x_{2q}).$$

Hence we deduce that $\mathrm{Alt}_{2q}\, g = 2^q \, \mathrm{St}_{2q}$ is an identity of $\mathcal{V}$. This completes the proof of the theorem in this case.

Otherwise there is a maximum $k$ and some subset $S$ with $k$ elements, so that the coefficient $\alpha_S \neq 0$. We take these $k$ variables in $S$, call them $y_1, \ldots, y_k$, and substitute each one of them by a commutator in new variables. We see that in each $P_{S'}$, $S' \neq S$, a variable appearing inside a commutator is substituted by a commutator, so all terms of $g$ except $P_S$ are mapped into identities of $G$. Instead $P_S$ is mapped into a nonzero multiple of a product of commutators, where we apply the previous analysis. □

## 19.2. Superalgebras

19.2.0.1. *Basic formalism.* A $\mathbb{Z}_2$-*graded* space $M$ is just a vector space decomposed as direct sum of two subspaces: $M = M_0 \oplus M_1$, indexed by $0, 1$. A subspace $N$ of a $\mathbb{Z}_2$-graded space $M = M_0 \oplus M_1$ is called *graded* if $N = N_0 \oplus N_1$, where $N_0 = N \cap M_0$ and $N_1 = N \cap M_1$.

DEFINITION 19.2.1. An algebra $A$ is a *superalgebra* if it is a $\mathbb{Z}_2$-graded space $A = A_0 \oplus A_1$ such that

$$A_0^2 + A_1^2 \subseteq A_0, \; A_0 A_1 + A_1 A_0 \subseteq A_1.$$

An element $a \in A$ is *homogeneous* if $a \in A_0$ (homogeneous of degree 0) or $a \in A_1$ (homogeneous of degree 1).

Superalgebras form a category, with *graded morphisms*. A graded morphism $\phi : A = A_0 \oplus A_1 \to B = B_0 \oplus B_1$ is such that $\phi(A_0) \subset B_0$, $\phi(B_1) \subset B_1$.

A subalgebra or an ideal is graded if it is a graded subspace, the image of a graded morphism is a graded subalgebra, and the kernel a graded ideal. The usual homomorphism theorems hold.

In superalgebras there are also the notions of being *graded commutative* and of a *super tensor product*.

DEFINITION 19.2.2. A superalgebra $A = A_0 \oplus A_1$ is *graded commutative* if given two homogeneous elements $a, b \in A$, we have $ab = (-1)^{\deg(a)\deg(b)} ba$.

The tensor product $A \otimes B$ of $A = A_0 \oplus A_1$, $B = B_0 \oplus B_1$ is the algebra $A \otimes B$ graded by $(A \otimes B)_0 := A_0 \otimes B_0 \oplus A_1 \otimes B_1$ and $(A \otimes B)_1 := A_0 \otimes B_1 \oplus A_1 \otimes B_0$.

The graded, or super, tensor product $A \otimes_{\mathrm{su}} B$ is instead obtained by twisting the multiplication in $A \otimes B$ using the rule that the multiplication between homogeneous elements is given by

$$(a \otimes b)(c \otimes d) = (-1)^{\deg(b)\deg(c)}(ac \otimes bd). \tag{19.6}$$

REMARK 19.2.3.

(1) The Grassmann algebra is graded commutative.

(2) If $A = A_0$ or $B = B_0$, then the tensor product and the super tensor product coincide.
(3) The super tensor product is associative,
$$(A \otimes_{\text{su}} B) \otimes_{\text{su}} C = A \otimes_{\text{su}} (B \otimes_{\text{su}} C).$$
(4) The mapping $a \otimes b \mapsto (-1)^{\deg(a)\deg(b)} b \otimes a$ is an isomorphism between $A \otimes_{\text{su}} B$ and $B \otimes_{\text{su}} A$.

$A \otimes B$ and $A \otimes_{\text{su}} B$ are usually not isomorphic, as we shall see in several examples. Nevertheless we have

LEMMA 19.2.4. *If $F$ contains an element $u$ with $u^2 = -1$, then $(A \otimes B)_0$ is isomorphic to $(A \otimes_{\text{su}} B)_0$.*

PROOF. We have $(A \otimes B)_0 = A_0 \otimes B_0 \oplus A_1 \otimes B_1$. Consider the map $j$ which sends $a \otimes b$ to $a \otimes_{\text{su}} b$ if $a \otimes b \in A_0 \otimes B_0$ and to $ua \otimes_{\text{su}} b$ if $a \otimes b \in A_1 \otimes B_1$.

For simplicity call $M_0 := A_0 \otimes B_0,\ M_1 := A_1 \otimes B_1$. We claim that this is the desired isomorphism. In fact this map is clearly bijective, and for two elements $x = a \otimes b, y = c \otimes d$ we have that

$$j(xy) = \begin{cases} ac \otimes_{\text{su}} bd, & \text{if } x, y \in M_0,\ x, y \in M_1, \\ u\, ac \otimes_{\text{su}} bd, & \text{if } x \in M_0,\ y \in M_1 \text{ or } y \in M_0,\ x \in M_1, \end{cases}$$

$$j(x)j(y) = \begin{cases} ac \otimes_{\text{su}} bd, & \text{if } x, y \in M_0, \\ -(a \otimes_{\text{su}} b)\,(c \otimes_{\text{su}} d) = ac \otimes_{\text{su}} bd, & \text{if } x, y \in M_1, \\ u\, ac \otimes_{\text{su}} bd, & \text{if } x \in M_0,\ y \in M_1 \text{ or } y \in M_0,\ x \in M_1. \end{cases}$$ □

Every algebra $A$ which is graded by $\mathbb{N}$ or $\mathbb{Z}$ produces a superalgebra by setting $A_0$ to be the sum of the even parts and $A_1$ the sum of the odd parts. So for instance a polynomial ring can be thought of as superalgebra, which is commutative but not graded commutative, unless we give to each variable an even degree.

EXAMPLE 19.2.5. Given a vector space $V$, the graded algebra $\bigwedge V = \bigoplus_{i=0}^{\infty} \bigwedge^i V$ is noncommutative but graded commutative as a superalgebra. We have for two vector spaces $V, W$,
$$\bigwedge(V \oplus W) = \bigwedge(V) \otimes_{\text{su}} \bigwedge(W).$$
In particular for the Grassmann algebra $G = \bigwedge V$, where $V$ is infinite dimensional we have an isomorphism $G \otimes_{\text{su}} G \equiv G$. It is easy to see that $G \otimes G$ is not isomorphic to $G$ and in fact it does not even satisfy the same polynomial identities (cf. Lemma 20.3.15).

19.2.0.2. *Group gradings.* Given a finite group $\Gamma$, we have two different ways of connecting $\Gamma$ with a ring structure $R$. One is just the notion of algebra $R$ with a $\Gamma$ action by automorphisms. The other is the notion of $\Gamma$-grading of $R$, that is a direct sum decomposition
$$R = \bigoplus_{g \in \Gamma} R_g, \quad R_g R_h \subset R_{gh}.$$
If $\Gamma$ is abelian, there is a *dual group* $\hat{\Gamma} := \{\chi : \Gamma \to \mathbb{Q}/\mathbb{Z}\}$, where $\chi$ is a homomorphism and $\mathbb{Q}/\mathbb{Z}$ is usually identified to the group of roots of 1 in $\mathbb{C}^*$.

It is well known that $\hat{\Gamma}$ is isomorphic, in a noncanonical way, to $\Gamma$ and $\hat{\hat{\Gamma}} = \Gamma$. Then, if $\Gamma$ is a finite abelian group and $R$ is an algebra over a field of characteristic

0, containing the $|\Gamma|$ roots of 1, a $\Gamma$ grading on $R$ is equivalent to a $\hat{\Gamma}$ action on $R$, where by definition

$$\chi.a = \chi(g)a, \forall a \in R_g, \ \chi \in \hat{\Gamma}.$$

Let $R$ be a split semisimple finite-dimensional algebra (Definition 17.2.17) over a field $F$ with a $\Gamma$ action. Then $R = \bigoplus_i M_{n_i}(F)$ and $\Gamma$ permutes the summands $M_{n_i}(F)$. Given one such orbit of summands, the stabilizer of one of the summands, say $M_n(F)$, is a subgroup $H$ acting by automorphisms on $M_n(F)$.

Such an action is a *projective representation* of $H$, which by the theory of Schur multipliers is induced by a linear representation of an associated finite group $\tilde{H}$. In characteristic 0, such a representation is defined clearly over some finite extension of $\mathbb{Q}$, so any such algebra $R$ over $F$, an algebraically closed field of characteristic 0, is obtained by field extension from a similar algebra defined over the algebraic closure $\bar{\mathbb{Q}}$ of $\mathbb{Q}$.

Both $\Gamma$ graded algebras and algebras with a $\Gamma$ action form categories with free algebras. In the case of an abelian group $\Gamma$ the previously established connection (between algebras with a $\hat{\Gamma}$ actions and algebras with a $\Gamma$ grading) is an isomorphism of categories.

For $\Gamma$ actions, the free algebra in a set $X$ of generators is just the free algebra in the set $X \times \Gamma$, where $\Gamma$ acts on the second factors. Concretely, the variables will be called $y_{i,g}$, $i = 1, \dots, m$, $g \in \Gamma$. In the abelian group case there is a simple way to pass from these to the graded variables.

In characteristic different from 2 let us be more explicit for this duality between $\mathbb{Z}_2 := \mathbb{Z}/(2)$-gradings and automorphisms of order 2 on an algebra $A$. First we may consider the *graded variables*

$$x_i := \frac{y_{i,0} + y_{i,1}}{2}, \ z_j := \frac{y_{i,0} - y_{i,1}}{2}.$$

If $A = A_0 \oplus A_1$ is a superalgebra and $A_1 \neq 0$, then the map $\sigma : A \to A$ such that $a_0 + a_1 \mapsto a_0 - a_1, a_i \in A_i$, is an automorphism of order 2. Otherwise $A = A_0$ and $\sigma$ is the identity. Conversely, if $\sigma \in \mathrm{Aut}(A)$ has order 2, then $A = A_0 \oplus A_1$, where $A_0$ and $A_1$ are the eigenspaces relative to the eigenvalues 1 and $-1$, respectively; i.e., $A_0 = \{a + \sigma(a) \mid a \in A\}$ and $A_1 = \{a - \sigma(a) \mid a \in A\}$.

COROLLARY 19.2.6. *Over a field $F$ of characteristic different from 2, the category of superalgebras is isomorphic to the category of algebras equipped with a $\mathbb{Z}/(2)$ action.*

It follows in particular that in the category of superalgebras one can define the product of any family of superalgebras. Notice that if $A, B$ are two superalgebras described by two automorphisms of order 2, $\tau_1, \tau_2$, the automorphism $\tau := \tau_1 \otimes \tau_2$ of $A \otimes B$ has order 2 and gives the tensor product grading to $A \otimes B$. It is easily verified that it is also an automorphism of $A \otimes_{\mathrm{su}} B$.

Let $J(A)$ be the Jacobson radical of the superalgebra $A$. Since $J(A)$ is stable under automorphisms, we get the following

THEOREM 19.2.7. *The Jacobson radical of a superalgebra $A$ over a field $F$ of characteristic different from 2 is a graded ideal.*

DEFINITION 19.2.8. As for algebras we say that a superalgebra $A$ is *simple* if $A^2 \neq \{0\}$ and $A$ does not contain proper graded ideals.

19.2.0.3. *Endomorphisms of graded space.*

DEFINITION 19.2.9. Given two $\mathbb{Z}_2$-graded vector spaces $V = V_0 \oplus V_1$ and $W = W_0 \oplus W_1$, a linear map $f : V \to W$ is of degree 0 if $f(V_0) \subset W_0$, $f(V_1) \subset W_1$ and of degree 1, if $f(V_0) \subset W_1$, $f(V_1) \subset W_0$.

Take a basis of $V$ in which the first part is a basis of $V_0$ and the second part is a basis of $V_1$; similarly for $W$. Then write the linear maps in block matrix form. The elements of degree 0 (resp., 1) are

$$\hom(V,W)_0 := \begin{vmatrix} A_{0,0} & 0 \\ 0 & A_{1,1} \end{vmatrix}, \quad \hom(V,W)_1 := \begin{vmatrix} 0 & A_{0,1} \\ A_{1,0} & 0 \end{vmatrix}. \tag{19.7}$$

In particular if $V = W$, we have that $\mathrm{End}(V)$ acquires the structure of a superalgebra. In this case the automorphism of order 2 is conjugation by the element $\begin{vmatrix} 1 & 0 \\ 0 & -1 \end{vmatrix}$, that is the identity on $V_0$ and $-1$ on $V_1$.

Notice that if $F(1)$ is the one-dimensional vector space in degree 1, we have that $F(1) \otimes V$, which as space is just $V$, now has $F(1) \otimes V_0$ in degree 1 and $F(1) \otimes V_1$ in degree 0. On the other hand $\mathrm{End}(F(1)) = F$ in degree 0, and we see that $\mathrm{End}(V)$ is isomorphic to $\mathrm{End}(F(1) \otimes V)$.

DEFINITION 19.2.10. We denote by $M_{k,l}$ the superalgebra $\mathrm{End}(V_0 \oplus V_1)$, where $\dim V_0 = k$, $\dim V_1 = l$.

In other words if $n = k + l$, we have $M_n(F)$ with grading $(M_{k,l}(F)_0, M_{k,l}(F)_1)$, given by formula (19.7), the degree of the endomorphisms.

Now, given four graded vector spaces $V, V', W, W'$, we have the notion of graded tensor product of linear maps $f : V \to V', g : W \to W'$ by

$$f \otimes_{\mathrm{su}} g(v \otimes w) := (-1)^{\deg(g)\deg(v)} f(v) \otimes g(w).$$

In the block description of $f, g$, we have

$$f_{i,j} \otimes_{\mathrm{su}} g_{h,k} = (-1)^{(h+k)k} f_{i,j} \otimes g_{h,k}.$$

PROPOSITION 19.2.11. *Given two finite-dimensional $\mathbb{Z}_2$-graded vector spaces $V$, $W$, we have an isomorphism*

$$j : \mathrm{End}(V) \otimes_{\mathrm{su}} \mathrm{End}(W) \to \mathrm{End}(V \otimes W),\ j(A \otimes B) := A \otimes_{\mathrm{su}} B.$$

Notice that in this case we have also $\mathrm{End}(V) \otimes \mathrm{End}(W) \simeq \mathrm{End}(V \otimes W)$, so in fact $\mathrm{End}(V) \otimes \mathrm{End}(W)$ is isomorphic to $\mathrm{End}(V) \otimes_{\mathrm{su}} \mathrm{End}(W)$.

19.2.0.4. *The regular representation.* In the ordinary theory of algebras we have for an algebra $A$ the regular representation. It is given by left multiplication on $A$, $a \mapsto L_a$, $L_a(x) = ax$, and the centralizer of the left action is given by right action of the opposite. Next if $A \subset B$ is an inclusion of superalgebras, the (super)-centralizer of $A$ is the set of elements of $B$ which super commute with $A$.

That is, $\{b \in B \mid ab = (-1)^{\deg(a)\deg(b)} ba,\ \forall a \in A\}$.

Define the *opposite* $A^{\mathrm{op}}$ of a superalgebra as the algebra structure on $A$ given by $a \cdot b := (-1)^{\deg(a)\deg(b)} ba$ and the *envelope* of a superalgebra (as in the theory of separable algebras) by

$$A^{\mathrm{e}} := A \otimes_{\mathrm{su}} A^{\mathrm{op}}.$$

The right action then of the opposite of $A$ on $A$ should be defined with a sign, by $R_a(x) := (-1)^{\deg(a)\deg(x)}xa$. In this way we have a homomorphism of $A^{\mathrm{op}}$ into $\mathrm{End}(A)$, and $A^{\mathrm{op}}$ is the super centralizer of the left action and conversely.

If $A$ is super commutative, the left and right actions coincide.

Define next a map $j : A^{\mathrm{e}} \to \mathrm{End}(A)$ on graded elements by

$$j(a \otimes b)c := (-1)^{\deg(b)\deg(c)}acb.$$

One easily verifies that this is a homomorphism of superalgebras, and then in the previous example we have that this is an isomorphism.

EXERCISE 19.2.12. Verify that $(A \otimes_{\mathrm{su}} B)^{\mathrm{e}}$ is isomorphic to $A^{\mathrm{e}} \otimes_{\mathrm{su}} B^{\mathrm{e}}$.

EXAMPLE 19.2.13. Consider the superalgebra $A := F + cF$ with $c^2 = 1$ and $A_0 = F$, $A_1 = cF$. In terms of the automorphism description, this is $F \oplus F$ with exchange automorphism and $c = (1, -1)$. Consider its regular representation given by (left) multiplication. This embeds $A$ into $\mathrm{End}(A)$ as the centralizer of the multiplication by $c$.

It is easily verified that $A \otimes A = A \oplus A$.

Consider instead $A \otimes_{\mathrm{su}} A$, which has as its basis $1, x = c \otimes 1, y = 1 \otimes c$, $xy = c \otimes c$, with $1, xy$ in degree 0 and $x, y$ in degree 1.

The multiplication rules given by the super tensor product structure are

$$x^2 = y^2 = 1,\ xy = -yx,\ (xy)^2 = -1.$$

As algebra this is isomorphic to $2 \times 2$ matrices by setting

$$x = \begin{vmatrix} 1 & 0 \\ 0 & -1 \end{vmatrix},\quad y = \begin{vmatrix} 0 & 1 \\ 1 & 0 \end{vmatrix},\quad xy = \begin{vmatrix} 0 & 1 \\ -1 & 0 \end{vmatrix}.$$

The automorphism of order 2 inducing the superalgebra structure can be taken as conjugation by $xy$ so $A \otimes_{\mathrm{su}} A = M_{1,1}$.

**19.2.1. Simple superalgebras.** Our next goal is the classification of finite-dimensional simple superalgebras. For this we need a standard lemma which is a special case of the Noether–Skolem theorem.

LEMMA 19.2.14. *If $F$ is a field, every $F$-automorphisms $\sigma$ of $M_n(F)$ is inner, i.e., of the form $X \mapsto YXY^{-1}$, where $Y$ is an invertible matrix.*

PROOF. Consider $F^n$ as an $M_n(F)$-module in the usual way and also in a *twisted* way by defining $(X, v) \mapsto \sigma(X)v$, $X \in M_n(F)$, $v \in F^n$.

Since up to isomorphism $M_n(F)$ has only one irreducible module, there is an isomorphism $Y : F^n \to F^n$ between the two modules, that is $YX = \sigma(X)Y$, for all $X \in M_n(F)$. □

THEOREM 19.2.15. *A finite-dimensional simple superalgebra over an algebraically closed field $F$ of characteristic different from 2 is isomorphic to one of the following superalgebras:*

(1) *$M_n(F + cF)$ where $c^2 = 1$, with grading $(M_n(F), cM_n(F))$, $n \geq 1$;*
(2) *the superalgebras $M_{k,l}(F)$;*
(3) *in particular for $l = 0$, we have $M_n(F)$, with grading $(M_n(F), 0)$, $n \geq 1$.*

PROOF. If $\sigma$ is the automorphism inducing the $\mathbb{Z}_2$-structure on $A$, the simplicity of $A$ as a superalgebra says that $A$ has no proper $\sigma$-stable ideals.

We first claim that $A$ is a semisimple algebra. In fact the radical $J(A)$ is stable under all automorphisms of $A$ so in particular it is a $\sigma$-stable ideal. By hypothesis then $J(A) = 0$ or $J(A) = A$. But, since $J(A)$ is a nilpotent ideal this implies that $J(A)^2 \neq J(A)$ is a proper ideal or zero. Hence it must be zero. Thus, if $J(A) = A$, we have $A^2 = 0$, which is excluded in the definition of simple superalgebra.

We claim that either $A$ is simple as an algebra or $A = B \oplus B^\sigma$, for some simple subalgebra $B$ of $A$. Suppose $A$ is not simple as an algebra, and let $B$ be a minimal ideal of $A$. Clearly $B^\sigma$ is still a minimal ideal of $A$ and $B^\sigma \neq B$ for the $\sigma$-simplicity of $A$. Since $B \oplus B^\sigma$ is a $\sigma$-stable ideal of $A$, we get that $A = B \oplus B^\sigma$, as claimed.

By the Wedderburn theorem, Theorem 1.1.14, since $F$ is algebraically closed, a finite-dimensional simple algebra is a full matrix algebra $M_n(F)$, we thus get that either $A \cong M_n(F)$ or $A \cong M_n(F) \oplus M_n(F)$, and $\sigma$ exchanges these two summands.

Suppose first that the second possibility occurs. Then we can write

$$M_n(F) \oplus M_n(F)^\sigma = M_1 \oplus M_{-1},$$

where

$$M_1 = \{(a, a) \mid a \in M_n(F)\} \subseteq M_n(F) \oplus M_n(F)^\sigma$$

is the eigenspace of $\sigma$ relative to the eigenvalue 1, while

$$M_{-1} = \{(b, -b) \mid b \in M_n(F)\} \subseteq M_n(F) \oplus M_n(F)^\sigma$$

is the eigenspace of $\sigma$ relative to the eigenvalue $-1$. Notice that $M_1 \cong M_n(F)$, and $M_{-1} = cM_1$, with

$$c = \begin{pmatrix} I & 0 \\ 0 & -I \end{pmatrix},$$

where $I$ is the $n \times n$ identity matrix. We have proved that $A$ is a superalgebra of type (1), and it is also clear that $A = M_n(F) \otimes (F + cF)$.

Now suppose that $A$ is isomorphic to $M_n(F)$, as algebra. Fixing such an isomorphism $j : A \to M_n(F)$, we have a corresponding isomorphism, $j \circ \sigma \circ j^{-1}$ of $M_n(F)$ which by abuse of notation we still denote by $\sigma$. Thus $A$ is isomorphic as superalgebra to $M_n(F)$ with the induced automorphism $\sigma$.

If $\sigma$ is the identity, we are in case (3), $\ell = 0$. In any case by Lemma 19.2.14, $\sigma(X) = YXY^{-1}$ is the inner automorphism induced by a matrix $Y$. Notice that $Y$ is not uniquely determined since any nonzero scalar multiple of $Y$ induces the same automorphism.

Also $\sigma^2$ is the identity, and it is induced by the inner automorphism associated to $Y^2$, hence $Y^2$ is central in $M_n(F)$ and, so, $Y^2 = \alpha I$, for some nonzero scalar $\alpha$. Since $F$ is algebraically closed, we can write $\alpha = \beta^2$, for some $\beta \in F$. Hence, by eventually multiplying $Y$ by $\beta^{-1}$, we may assume that $Y^2 = I$.

It follows that the eigenvalues of $Y$ are 1 and $-1$, and such a matrix is conjugate to one of the form

$$\begin{pmatrix} I_k & 0 \\ 0 & -I_l \end{pmatrix}, \tag{19.8}$$

where $I_k$ and $I_l$ denote the identity matrices of orders $k$ and $l$, respectively.

We can still multiply $Y$ with $\pm 1$, so, eventually multiplying by $-1$, the matrix $Y$ is of the form (19.8) with $1 \le l \le k$, $k + l = n$. Now we have

$$\sigma(X) = YXY^{-1} = \begin{pmatrix} I_k & 0 \\ 0 & -I_l \end{pmatrix} \begin{pmatrix} X_{11} & X_{12} \\ X_{21} & X_{22} \end{pmatrix} \begin{pmatrix} I_k & 0 \\ 0 & -I_l \end{pmatrix}$$
$$= \begin{pmatrix} X_{11} & -X_{12} \\ -X_{21} & X_{22} \end{pmatrix},$$

where the $X_{ij}$'s are block matrices of the corresponding order. Translating the above relations into the corresponding language of gradings, we have that $A_0$ is the eigenspace of the eigenvalue 1 while $A_1$ is the eigenspace of the eigenvalue $-1$, respectively, the spaces of matrices

$$\begin{pmatrix} X_{11} & 0 \\ 0 & X_{22} \end{pmatrix}, \quad \begin{pmatrix} 0 & X_{12} \\ X_{21} & 0 \end{pmatrix}.$$

We obtain that $M_n(F)$ has a superalgebra structure as described in the case (3). This completes the proof of the theorem. □

REMARK 19.2.16. The previous three types are in fact defined for any field. We call them *split simple superalgebras.*

The reader can prove that $F$ is of characteristic 0 and $A$ is a simple superalgebra finite dimensional over $F$. Then there is a finite field extension $F \subset G$ such that $G \otimes_F A$ is a direct sum of split simple superalgebras.

For $F$ not necessarily algebraically closed one still has

THEOREM 19.2.17. *A finite-dimensional semisimple superalgebra over a field $F$ of characteristic different from 2 is a finite direct sum of simple superalgebras.*

PROOF. By the Wedderburn theorem, we can write $A = B_1 \oplus \cdots \oplus B_m$ where the $B_i$'s are all the minimal two-sided ideals of $A$. Let $\sigma$ be the automorphism determined by the superstructure of $A$. Then for every $i$, $B_i^\sigma$ is still a minimal ideal of $A$. Hence either $B_i^\sigma = B_i$ or $B_i^\sigma = B_j$, for some $j \neq i$. In the first case $B_i$ is a simple superalgebra, and in the second case $B_i \oplus B_j$ is a simple superalgebra, and we get the desired decomposition. □

The study of simple superalgebras over $F$ can be performed in a similar way to the algebraically closed case, but requires a closer analysis of automorphisms of order 2 of simple algebras.

There is a theorem analogous to the Wedderburn–Malćev theorem for superalgebras, a consequence of Theorem 1.3.20 for $G = \mathbb{Z}_2$.

THEOREM 19.2.18. *Let $A$ be a finite-dimensional superalgebra over a field of characteristic 0. Then $A = \bar{A} \oplus J$, where $\bar{A} = \bar{A}_0 \oplus \bar{A}_1$ is a maximal semisimple graded subalgebra of $A$ and $J = J_0 \oplus J_1$ is the Jacobson radical of $A$.*

19.2.1.1. *Tensor and super tensor products of simple algebras.* We need to understand the behavior of the previously described simple superalgebras under tensor product and super tensor product.

Take two of these superalgebras $A_1, A_2$ described through the automorphisms $\tau_1, \tau_2$. Then their tensor product is $A_1 \otimes A_2$ with the $\mathbb{Z}_2$ grading induced by the automorphism $\tau_1 \otimes \tau_2$.

We also have the notion of super tensor product, $A_1 \otimes_{su} A_2$. We take the graded vector space $A_1 \otimes A_2$ with the product given by formula (19.6). There are several cases to analyze but we can use the commutativity of tensor and super tensor products:

(1) $A_1 = M_k(F)$, then the analysis of $M_k(F) \otimes A = M_k(F) \otimes_{su} A$ for $A$ a simple superalgebra divides into three cases:
   (i) $A = M_h(F)$, then $M_k(F) \otimes M_h(F) = M_{hk}(F)$.
   (ii) $A = M_h(F + cF)$.

$$M_k(F) \otimes M_h(F + cF) = M_k(F) \otimes M_h(F) \otimes (F + cF) = M_{kh}(F + cF).$$

   (iii) $A = \mathrm{End}(W) = M_{a,b}$, for a graded vector space $W$. Think in more intrinsic terms as $M_k(F) = \mathrm{End}(V_0)$, the endomorphisms of a trivially graded vector space $\dim V_0 = k$, then for a graded vector space $W = W_0 \oplus W_1$ with $\dim W_0 = a$, $\dim W_1 = b$. We have

$$\mathrm{End}(V_0) \otimes \mathrm{End}(W) = \mathrm{End}(V_0) \otimes_{su} \mathrm{End}(W) = \mathrm{End}(V_0 \otimes W) = M_{ka,kb}.$$

(2) When we have $A_1 = M_k(F) \oplus M_k(F) \cong M_k(F + cF) = M_k(F) \otimes (F + cF)$ with the exchange automorphism, we have that

$$A_1 \otimes A_2 = M_k(F) \otimes (F + cF) \otimes A_2, \ A_1 \otimes_{su} A_2 \cong M_k(F) \otimes [(F + cF) \otimes_{su} A_2],$$

so we only need to understand $(F + cF) \otimes A_2$, $(F + cF) \otimes_{su} A_2$.
   (i) If $A_2 = M_h(F) \otimes (F + cF)$ (as algebra use Exercise 19.2.13), then

$$A_1 \otimes A_2 = M_{hk}(F) \otimes ((F + cF) \otimes (F + cF)) \cong M_{hk}(F) \otimes ((F + cF) \oplus (F + cF)),$$

$$A_1 \otimes_{su} A_2 = M_{hk}(F) \otimes M_{1,1} = M_{hk,hk}.$$

   (ii) If $A_2 = M_{l,k}$, then since $F + cF = F \oplus F$ we have the isomorphisms as algebras $M_{l,k} \otimes (F \oplus F) = M_{l,k} \oplus M_{l,k} = M_{l+k} \oplus M_{l+k}$. To understand the superalgebra structure, notice that the automorphism exchanges the two summands so that by the proof of Theorem 19.2.15,

$$M_{l,k} \otimes (F + cF) \cong M_{l+k}(F + cF).$$

   As for $M_{l,k} \otimes_{su} (F + cF)$, a direct computation shows that it is isomorphic to $M_{l,k} \otimes (F + cF) \cong M_{l+k}(F + cF)$.

(3) It only remains to show the case in which both $A_1 = M_{k,l}$ and $A_2 = M_{h,m}$. In this case we have already shown, in Proposition 19.2.11, that $M_{k,l} \otimes M_{h,m}$ and $M_{k,l} \otimes_{su} M_{h,m}$ are both isomorphic to $M_{hk+lm,km+lh}$.

It follows that

PROPOSITION 19.2.19.

$$M_k(F + cF) \otimes_{su} M_{h,m} = M_{k(h+m)}(F + cF)$$
$$M_k(F + cF) \otimes_{su} M_h(F + cF) \cong M_{kh,kh}$$
$$M_{k,l} \otimes_{su} M_{h,m} \cong M_{hk+lm,km+lh}.$$

In particular we have

$$M_{k,l} \otimes_{su} M_{1,1} \cong M_{k+l,k+l}$$

$$(F + cF)^{\otimes_{su} 2k} \cong M_{2^{k-1},2^{k-1}}, \quad (F + cF)^{\otimes_{su} 2k+1} \cong M_{2^k}(F + cF).$$

When $A_1 = M_k(F)$ and $\tau_1 = \mathrm{Id}$, then $A_1 = M_{k,0}$ so

$$M_{k,0} \otimes_{su} M_{h,m} \cong M_{kh,km}.$$

## 19.3. Graded identities

**19.3.1. $T_2$-ideals.** From now on the field $F$ is assumed to be of characteristic 0 and often also algebraically closed. Let $F\langle X\rangle$ be the free associative algebra with identity over $F$, freely generated by an infinite countable set $X$. We write $X$ as

$$X = Y \cup Z, \ Y \cap Z = \emptyset, \ Y, Z \text{ countable.}$$

We call $Y$ the *even variables* and $Z$ the *odd variables*. From now on we denote by $x$ and $x_i$ the elements of $X$, by $y$ and $y_i$ the elements of $Y$, and by $z$ and $z_i$ the elements of $Z$. The elements of the algebra $F\langle X\rangle$ are therefore polynomials, and they will be denoted by $f(x_1,\dots,x_n)$. Alternatively, we use the notation $f(y_1,\dots,y_k;z_1,\dots,z_m)$ to denote a polynomial where the variables $y_1,\dots,y_k$ are in $Y$ and the variables $z_1,\dots,z_m$ are in $Z$. Denote by $\mathcal{F}_0$ the subspace of $F\langle X\rangle$ generated by the monomials in the variables of $X$ that have even total degree in the variables of $Z$, and with $\mathcal{F}_1$ the subspace generated by the monomials of odd total degree in the variables of $Z$. Then we clearly have

$$F\langle X\rangle = \mathcal{F}_0 \oplus \mathcal{F}_1$$

and $F\langle X\rangle$ is a superalgebra with grading $(\mathcal{F}_0, \mathcal{F}_1)$, which we denote by $F\langle Y, Z\rangle$. We will call superpolynomials the elements of $F\langle Y, Z\rangle$.

Clearly $F\langle Y, Z\rangle$ is a *free superalgebra*, in the sense that given any superalgebra $A$, the graded morphisms from $F\langle Y, Z\rangle$ to $A$ are in 1–1 correspondence with the pairs of maps $Y \to A_0$, $Z \to A_1$.

DEFINITION 19.3.1. We say that an ideal $I$ of $F\langle X\rangle$ is a $T_2$-ideal if it is invariant under the endomorphisms $\varphi$ of $F\langle X\rangle$ such that $\varphi(\mathcal{F}_0) \subseteq \mathcal{F}_0$ and $\varphi(\mathcal{F}_1) \subseteq \mathcal{F}_1$, i.e.,

$$\begin{aligned}&f(y_1,\dots,y_k;z_1,\dots,z_m) \in I \implies \text{ for every}\\ &a_1,\dots,a_k \in \mathcal{F}_0 \text{ and } b_1,\dots,b_m \in \mathcal{F}_1, \quad f(a_1,\dots,a_k;b_1,\dots,b_m) \in I.\end{aligned}$$

Clearly a $T$-ideal is also a $T_2$-ideal.

DEFINITION 19.3.2. We say that a superalgebra $A = A_0 \oplus A_1$ satisfies a graded identity (or a superidentity) $f(y_1,\dots,y_k;z_1,\dots,z_m)$ if, for every choice of elements $a_1,\dots,a_k \in A_0$ and $b_1,\dots,b_m \in A_1$, the equality $f(a_1,\dots,a_k;b_1,\dots,b_m) = 0$ holds.

Clearly the ideal of identities of an associative algebra $A$ is a $T$-ideal, while the ideal of graded identities of a superalgebra $A$ is a $T_2$-ideal, which we will denote by $\mathrm{Id}_2(A)$.

The notion of variety of algebras that we introduced in the previous chapters applies also to superalgebras, if we consider as homomorphisms the maps that preserve the grading. Hence we define a variety of superalgebras in a way analogous to the varieties of algebras (§2.2).

There is clearly a correspondence between $T_2$-ideals and varieties of superalgebras analogous to the one between $T$-ideals and varieties of algebras. We have a characterization of varieties as in Proposition 2.2.24 and Theorem 2.2.25.

PROPOSITION 19.3.3. *Let $\mathcal{V}$ be a variety of superalgebras.*

(1) *If $R_i$ where $i \in I$ is a family of superalgebras in $\mathcal{V}$, then $\prod_{i\in I} R_i \in \mathcal{V}$.*
(2) *If $R$ is a superalgebra in $\mathcal{V}$ and $S \subset R$ is a subsuperalgebra, then $S \in \mathcal{V}$.*
(3) *If $R$ is a superalgebra in $\mathcal{V}$ and $S = R/I$ is a quotient superalgebra (by a graded ideal $I$), then $S \in \mathcal{V}$.*

As in Proposition 2.2.25 we have that the previous properties characterize varieties of superalgebras.

THEOREM 19.3.4. *A full subcategory $\mathcal{V}$ of the category of superalgebras which satisfies properties* (1), (2), (3) *of Proposition* 19.3.3 *is a variety of superalgebras.*

We leave it to the reader to verify that both statements are proved in a way which is identical to the case of algebras.

As for algebras one needs also to distinguish these notions for superalgebras with or without 1. In the second case the free superalgebra has to be taken without 1 so that its endomorphisms are restricted. We can substitute variables only with polynomials with no constant terms. So we have a weaker condition (and a larger class) of $T_2$-ideals.

REMARK 19.3.5. Arguing as in Proposition 2.2.38, it can be shown that the $T_2$-ideals are homogeneous in all the variables.

EXAMPLE 19.3.6 (Exercise). The Grassmann algebra is the relatively free algebra in odd generators in the variety of super-commutative superalgebras.

If $I_1$ and $I_2$ are graded ideals we can define the $T_2$-equivalence, which we will denote by $\sim_{T_2}$, as in Definition 2.2.4.

Analogously to Proposition 2.2.16 we can prove that if $I_1$ and $I_2$ are graded ideals, then $I_1 \sim_{T_2} I_2$ if and only if the variety of superalgebras generated by $F\langle X\rangle/I_1$ is equal to the variety of superalgebras generated by $F\langle X\rangle/I_2$.

Also for the $T_2$-ideals the multilinear identities play a central role.

LEMMA 19.3.7. *In characteristic* 0, *every $T_2$-ideal is generated as a $T_2$-ideal by its multilinear polynomials.*

PROOF. This lemma follows directly from Corollary 2.2.44, by observing that in the linearization process and its inverse of restitution, one can use substitutions that correspond to endomorphisms preserving the grading. □

We need an analogue of the theorem of Leron and Vapne, Theorem 7.2.3, where for simplicity, we shall say that two superalgebras $A, B$ are $PI_2$ equivalent if they satisfy the same super identities.

THEOREM 19.3.8. *Let $A, A', B, B'$ be superalgebras over a field. Then if $A$ is $PI_2$ equivalent to $A'$ and $B$ is $PI_2$ equivalent to $B'$, we have that $A \otimes_{su} B$ is $PI_2$ equivalent to $A' \otimes_{su} B'$ (all equivalences and tensor product are as superalgebras).*

PROOF. Let $U_1 = F\langle X\rangle/I_1$, $U_2 = F\langle X\rangle/I_2$ be the relatively free algebras in some large number of variables for $A, A'$ and $B, B'$, respectively, so that these algebras are quotients of their relatively free algebras. $I_1, I_2$ are the two $T_2$-ideals of identities of the two pairs of algebras.

It is then enough to show that $A \otimes_{su} B$ (and hence also $A' \otimes_{su} B'$) is $PI_2$ equivalent to $U_1 \otimes_{su} U_2$.

In fact it is enough to do it in two steps and prove for instance that $A \otimes_{su} B$ is $PI_2$ equivalent to $U_1 \otimes_{su} B$, and then, in the same way, that $U_1 \otimes_{su} B$ is $PI_2$ equivalent to $U_1 \otimes_{su} U_2$.

Now let $\mathcal{I}$ be the set of homomorphisms (as superalgebras) of $U_1$ to $A$, and we deduce a mapping $j : U_1 \to A^{\mathcal{I}}$ given by evaluations of all these homomorphisms.

Recall that $A^{\mathcal{I}}$ has the structure of superalgebra induced by the automorphism of order 2. Since $I_1$ is the ideal of superidentities of $A$, we have that $j : U_1 \to A^{\mathcal{I}}$ is injective, so $U_1 \otimes_{\text{su}} B$ injects into $(A^{\mathcal{I}}) \otimes_{\text{su}} B$ which in turn injects into $(A \otimes_{\text{su}} B)^{\mathcal{I}}$ as superalgebra

Therefore $U_1 \otimes_{\text{su}} B$ satisfies all PIs of $A \otimes_{\text{su}} B$. But now $A$ is a quotient of $U_1$, so we have also that the converse $A \otimes_{\text{su}} B$ satisfies all super PIs of $U_1 \otimes_{\text{su}} B$, and the claim follows. □

**19.3.2. Super-codimension and cocharacters.** One may define the super-codimension and cocharacter in a straightforward fashion.

Given two positive integers $a, b$, we have

DEFINITION 19.3.9. We define $V_{a,b}$ to be the space of multilinear polynomials in $a$ even and $b$ odd variables.

Observe that the group $S_a \times S_b$ acts on this space by permuting the two types of variables.

A representation of this is the following. To a monomial in $a$ even and $b$ odd variables we associate the sequence of $+.-$ of the sign of the variables. If $n = a + b$, we have $\binom{n}{a} = \binom{n}{b}$ such sequences which distribute the monomials into disjoint sets, each of which have exactly $a!b!$ monomials so that $\dim V_{a,b} = \binom{n}{a}a!b! = n!$.

A representation of $S_a \times S_b$, thus $V_{a,b}$, is the direct sum of $\binom{n}{a}$ copies of the regular representation $F[S_a] \times F[S_b]$. Thus it decomposes as a direct sum of the representations $M_\lambda \otimes M_\mu$, $\lambda \vdash a$, $\mu \vdash b$ each with multiplicity $\binom{n}{a} f_\lambda f_\mu$.

$$V_{a,b} = \bigoplus_{\lambda \vdash a,\ \mu \vdash b} \binom{n}{a} f_\lambda f_\mu M_\lambda \otimes M_\mu. \tag{19.9}$$

Given a $T_2$ ideal $I$ (the superidentities of some superalgebra $R$) inside $V_{a,b}$, we have the subspace $V_{a,b} \cap I$ of multilinear superidentities of $R$ of degrees $a, b$ which is clearly a representation of $S_a \times S_b$, so we may consider

DEFINITION 19.3.10. We define the dimension of $V_{a,b}/(V_{a,b} \cap I)$ as super codimension (it depends on the two parameters $a, b$) and

$$V_{a,b}/(V_{a,b} \cap I) = \bigoplus_{\lambda \vdash a,\ \mu \vdash b} m_{\lambda,\mu} M_\lambda \otimes M_\mu; \tag{19.10}$$

we may call $\sum_{\lambda,\mu} m_{\lambda,\mu} \chi_\lambda \otimes \chi_\mu$ the *supercocharacter*.

19.3.2.1. *Algebras and superalgebras.* Every superalgebra $A = A_0 \oplus A_1$ can be considered as just algebra by forgetting the grading. Conversely there are several ways of constructing superalgebras from algebras. The trivial construction is to take an algebra $A$ and grade it with $A_0 = A$, $A_1 = 0$. In this superalgebra the super identities are the identities of $A$ in the variables $y$ and the identity $z = 0$. There is a subtler construction in which the $T_2$-ideal of superidentities is also a $T$-ideal so each superidentity is in fact a polynomial identity.

EXAMPLE 19.3.11. Let $R$ be an algebra. We make $S = R \oplus R$ into a superalgebra by setting $S_0 = \{(r, r),\ r \in R\}$, $S_1 = \{(r, -r),\ r \in R\}$.

Let us show that $T(R) = T_2(S)$. In fact every PI of $R$ is also a PI of $S$, hence also a graded PI. Let us take a multihomogeneous graded PI, $f(y_1, \ldots, y_k; z_1, \ldots, z_m)$.

When we evaluate $y_i = (r_i, r_i)$ and $z_j = (s_j, -s_j)$, we have as coordinates $(f(r_1, \dots, r_k; s_1, \dots, s_m), (-1)^t f(r_1, \dots, r_k; s_1, \dots, s_m))$, where $t$ is the degree in the variables $z$, hence the claim.

Notice that this algebra is isomorphic to $R[\mathbb{Z}_2]$, the group algebra of $\mathbb{Z}_2$ over $R$.

REMARK 19.3.12. A superalgebra $A = A_0 \oplus A_1$ satisfies an ordinary multilinear identity $g \equiv 0$ of degree $n$ if and only if $A$ as superalgebra satisfies all the $2^n$ graded identities $h \equiv 0$, where every $h$ is obtained from $g$ by substituting $k$ variables, $0 \leq k \leq n$, with even variables and all the other $n - k$ variables with odd variables.

REMARK 19.3.13. Given a variety $\mathcal{V}$ of algebras, consider the class $\tilde{\mathcal{V}}$ of superalgebras, whose associated algebra is in $\mathcal{V}$. Clearly $\tilde{\mathcal{V}}$ satisfies the properties of Proposition 19.3.3, and so it is also a variety of superalgebras, which by abuse of notation we still denote by $\mathcal{V}$. In this case the associated $T_2$-ideal is a $T$-ideal, the $T$-ideal of $\mathcal{V}$, and hence each of its multilinear elements satisfies the property of Remark 19.3.12.

In other words, any graded identity of a superalgebra $A \in \mathcal{V}$ is also a polynomial identity for $A$ as algebra.

It is easy to see that, by Example 19.3.11, if $R$ generates the variety $\mathcal{V}$ as algebras, then $S = R \oplus R$ generates the same variety but as superalgebras.

19.3.2.2. $S_2$ *ideals.* As in the case of $S$-ideals of Definition 2.2.35, we may consider the ideals of $F\langle X\rangle$ which are stable under linear substitutions of variables which preserve the two subspaces spanned by the two sets of variables $Y, Z$. That is we consider those ideals which are stable under the product of the two infinite linear groups of these two subspaces. These ideals will be called $S_2$ ideals.

## 19.4. The role of the Grassmann algebra

In this paragraph $F$ is a field of characteristic 0, although the definitions hold in general.

**19.4.1. The $*$ transform.** Consider a polynomial $f(y_1, \dots, y_m; z_1, \dots, z_n) \in F\langle X\rangle$, with $F\langle X\rangle = F\langle Y, Z\rangle$ the free superalgebra. Suppose that $f$ is linear in each of the odd variables $z_1, \dots, z_n$ appearing in $f$. Then we can write $f$ in the form

$$f = \sum_u \sum_{\sigma \in S_n} \alpha_{\sigma,u} u_1 z_{\sigma(1)} u_2 z_{\sigma(2)} \cdots u_n z_{\sigma(n)} u_{n+1},$$

where the $u_i$'s are words, possibly empty, in the variables of $Y$ and $\alpha_{\sigma,u} \in F$.

We apply to $f$ the operator of multiplying by the sign of $\sigma$ the term associated to $\sigma$ and get a new polynomial that we shall denote $f^*$. Then $f^*$ is defined as

$$f^* := \sum_u \sum_{\sigma \in S_n} \epsilon_\sigma \alpha_{\sigma,u} u_1 z_{\sigma(1)} u_2 z_{\sigma(2)} \cdots u_n z_{\sigma(n)} u_{n+1}.$$

If we assume that the variables $Z$ are well ordered, this definition continues to hold whenever we take a polynomial which is multilinear in some finite set of variables in $Z$.

In fact

EXERCISE 19.4.1.

(1) $f^*$ can be defined by adding Grassmann variables $e_i$ (commuting with the variables $X$), for each $z_i$ by the formula

$$f(y_1,\dots,y_k;z_1\otimes e_1,\dots,z_m\otimes e_m)=f^*(y_1,\dots,y_k;z_1,\dots,z_m)\otimes e_1\cdots e_m. \tag{19.11}$$

(2) With the notations of §19.3.2 consider the space $V_{a,b}$ of polynomials which are multilinear in $a$ even and $b$ odd variables decomposed as in formula (19.9),

$$V_{a,b}=\bigoplus_{\lambda\vdash a,\ \mu\vdash b}\binom{n}{a}f_\lambda f_\mu M_\lambda\otimes M_\mu.$$

On this space the operator $*$ maps each representation $M_\lambda\otimes M_\mu$ to $M_\lambda\otimes(M_\mu\otimes\mathrm{sgn})=M_\lambda\otimes M_{\check\mu}$ where sgn is the sign representation and $\check\mu$ is the dual partition of $\mu$.

The basic properties of the linear operator $*$ are

LEMMA 19.4.2. *If $f$ and $g$ are multilinear polynomials in disjoint variables of $Z$, then $(fg)^*=f^*g^*$. Moreover $(f^*)^*=f$.*

PROOF. This is immediate from formula (19.11). □

Let $P$ be the graded space of all multilinear polynomials in the algebra $F\langle X\rangle$, and let $I$ be a $T_2$-ideal. Then

DEFINITION 19.4.3. We denote by $I^*$ the $T_2$-ideal generated by the set $(I\cap P)^*$.

LEMMA 19.4.4. *If $I$, $I_1$, and $I_2$ are $T_2$-ideals, we have*

(i) $I^*\cap P=(I\cap P)^*$;
(ii) $(I_1\cap I_2)^*=I_1^*\cap I_2^*$;
(iii) $(I_1+I_2)^*=I_1^*+I_2^*$;
(iv) $I^{**}=I$.

PROOF. First we show that for every $T_2$-ideal $I$, we have $I^*\cap P=(I\cap P)^*$. Since the map $*$ preserves $P$, we clearly have $(I\cap P)^*\subset I^*\cap P$. For the opposite inclusion, if $f\in I^*\cap P$, then $f$ can be written as a linear combination of multilinear polynomials of the form $ahb$, where the polynomial $h$ is $h=g(M_1,\dots,M_k;N_1,\dots,N_m)$ and $g\in(I\cap P)^*$. Hence $g=k^*(y_1,\dots,y_k;z_1,\dots,z_m)$, $k\in I\cap P$. The $M_i$ are monomials of even and the $N_j$ monomials of odd degree in $z$, and $a,b\in F\langle X\rangle$.

Now we claim that if $\phi(y,z):=k(M_1,\dots,M_k;N_1,\dots,N_m)$, we have that

$$h=g(M_1,\dots,M_k;N_1,\dots,N_m)=\pm\phi^*(y,z).$$

In fact, when we substitute to the variables $z_j$ the elements $z_j\otimes e_j$, we have that $M_i\mapsto M_i\otimes u_i$, $N_j\mapsto N_j\otimes v_j$ where $u_i$ is a product of an even number of $e_j$, while $v_j$ is a product of an odd number of $e_j$. These $v_j$ also behave as Grassmann variables and $v_1\cdots v_n=\pm e_1\cdots e_k$ where $k$ is the degree of $h$ in the variables $z$. Then since $\phi(y,z)\in I\cap P$, we have $h\in(I\cap P)^*$.

We have $ahb=(a^*h^*b^*)^*$. Since $h^*\in I\cap P$, it follows that $ahb\in(I\cap P)^*$. Then $f\in(I\cap P)^*$. Therefore $I^*\cap P\subseteq(I\cap P)^*$.

From the equality we just proved, it follows that for each $T_2$-ideal $I_1, I_2$

$$(I_1 \cap I_2)^* \cap P = (I_1 \cap I_2 \cap P)^* = (I_1 \cap P)^* \cap (I_2 \cap P)^* = I_1^* \cap I_2^* \cap P.$$

Since the ground field $F$ has characteristic 0, Lemma 19.3.7 says that for any two $T_2$-ideals $J_1$ and $J_2$, the equality $J_1 \cap P = J_2 \cap P$ implies $J_1 = J_2$. From this and the last equality it follows that $(I_1 \cap I_2)^* = I_1^* \cap I_2^*$.

Using the first equality again, we obtain

$$\begin{aligned}(I_1 + I_2)^* \cap P &= ((I_1 + I_2) \cap P)^* = ((I_1 \cap P) + (I_2 \cap P))^* \\ &= (I_1 \cap P)^* + (I_2 \cap P)^* = (I_1^* + I_2^*) \cap P,\end{aligned}$$

which implies $(I_1 + I_2)^* = I_1^* + I_2^*$.

Finally

$$I^{**} \cap P = (I^* \cap P)^* = (I \cap P)^{**} = I \cap P,$$

hence $I^{**} = I$. □

LEMMA 19.4.5. *If $I_1$ and $I_2$ are $T_2$-ideals, then*

$$(I_1 I_2)^* = I_1^* I_2^*.$$

PROOF. Let $f \in I_1 I_2 \cap P$. From the homogeneity of the $T_2$-ideals it follows that the polynomial $f$ can be written as a linear combination of multilinear polynomials of the form $g_1 g_2$, where $g_1 \in I_1$ and $g_2 \in I_2$. But then we have $(g_1 g_2)^* = g_1^* g_2^*$, hence $f^* \in I_1^* I_2^*$. Therefore $(I_1 I_2)^* \subseteq I_1^* I_2^*$ for every pair of $T_2$-ideals $I_1$ and $I_2$. Substituting $I_1^*$ to $I_1$ and $I_2^*$ to $I_2$ in this inclusion, we obtain $(I_1^* I_2^*)^* \subseteq I_1^{**} I_2^{**} = I_1 I_2$; therefore $I_1^* I_2^* \subseteq (I_1 I_2)^*$. The lemma is proved. □

**19.4.2. The Grassmann envelope.** Next we introduce the so-called *Grassmann envelope* of a superalgebra that will play a central role in the theory.

DEFINITION 19.4.6. Let $A = A_0 \oplus A_1$ be a superalgebra, and let $G = G_0 \oplus G_1$ be the Grassmann algebra with its standard $\mathbb{Z}_2$-grading. The algebra

$$G(A) = (A_0 \otimes G_0) \oplus (A_1 \otimes G_1) = (A \otimes G)_0$$

is called the *Grassmann envelope* of the superalgebra $A$.

Notice that $G(A)$ also has a structure of superalgebra where $G(A)_0 = A_0 \otimes G_0$ and $G(A)_1 = A_1 \otimes G_1$. Also, if we repeat the construction, we obtain

$$G(G(A)) = (A_0 \otimes G_0 \otimes G_0) \oplus (A_1 \otimes G_1 \otimes G_1),$$

which is contained in $A \otimes B$ where $B = G_0 \otimes G_0 \oplus G_1 \otimes G_1$ is a commutative algebra. Hence,

PROPOSITION 19.4.7. *$A$ and $G(G(A))$ have the same ordinary identities, i.e., we have $\mathrm{Id}(A) = \mathrm{Id}(G(G(A)))$ and $\mathrm{Id}_2(A) = \mathrm{Id}_2(G(G(A)))$ also as superalgebras.*

EXAMPLE 19.4.8. If $R$ is an algebra and $A = A_0 \oplus A_1$ a superalgebra, then $A \otimes R = A_0 \otimes R \oplus A_1 \otimes R$ is a superalgebra and

$$G(A \otimes R) = G(A) \otimes R.$$

Consider the superalgebra $S := R \oplus R$ of Example 19.3.11. We have that $S = F[c] \otimes R$ where $F[c] = F \oplus cF$, $c^2 = 1$.

The algebra $G(S) = G(F[c]) \otimes R = (G_0 \oplus c\,G_1) \otimes R$ can be identified to $G \otimes R$, since $G_0 \oplus c\,G_1$ is isomorphic to $G$ via the map $a + cb \mapsto a + b$.

LEMMA 19.4.9. *Let $f(y_1,\dots,y_k,z_1,\dots,z_m)$ be a multilinear polynomial. When we evaluate $f$ in elements of a Grassmann envelope $G(A)$, $y_i \mapsto a_i \otimes g_i$, $z_j \mapsto b_j \otimes h_j$ with $g_i \in G_0$, $h_j \in G_1$, we have*

$$f(a_1 \otimes g_1,\dots,a_k \otimes g_k; b_1 \otimes h_1,\dots,b_m \otimes h_m) = f^*(a_1,\dots,a_k,b_1,\dots,b_m) \otimes g_1 \cdots g_k h_1 \cdots h_m. \tag{19.12}$$

PROOF. $f$ is $G_0$-linear since $G_0$ is in the center, hence the claim follows from formula (19.11). □

Let $A$ be a superalgebra, and let $G(A)$ be its Grassmann envelope regarded as a superalgebra. Then

LEMMA 19.4.10. *A multilinear polynomial $f$ is a graded identity of $A$ if and only if $f^*$ is a graded identity of $G(A)$. In other words*

$$\mathrm{Id}_2(A)^* = \mathrm{Id}_2(G(A)). \tag{19.13}$$

PROOF. From formula (19.12), since the elements $g_i, h_j$ can be taken arbitrarily, it follows that a multilinear polynomial $g^*$ is a graded identity of $A$ if and only if $g$ is a graded identity of $G(A)$. Since $(f^*)^* = f$, the claim follows. □

DEFINITION 19.4.11. Let $\mathcal{V}$ be a variety of superalgebras. Denote by $\mathcal{V}^*$ the class of all superalgebras $A = A_0 \oplus A_1$ such that $G(A) \in \mathcal{V}$.

We have the following theorem.

THEOREM 19.4.12. *For every variety of superalgebras $\mathcal{V}$ the class $\mathcal{V}^*$ is a variety of superalgebras and $\mathcal{V} = (\mathcal{V}^*)^*$. If $I$ is the $T_2$-ideal of identities of $\mathcal{V}$, then $I^*$ is the $T_2$-ideal of identities of $\mathcal{V}^*$.*

PROOF. By Lemma 19.4.10, given a superalgebra $A = A_0 \oplus A_1$, $G(A)$ satisfies the multilinear graded identity $f \equiv 0$ if and only if $A$ satisfies the multilinear graded identity $f^* \equiv 0$.

It follows from Lemmas 19.4.2 and 19.4.4 that the superalgebras $A$, such that $G(A) \in \mathcal{V}$, coincide with the variety defined by the $T_2$-ideal $I^*$.

Conversely given any superalgebra $A$, we know, by Proposition 19.4.7, that $A$ is equivalent to $G(G(A))$, so $G(A) \in \mathcal{V}^*$ if and only if we have $A \in \mathcal{V}$. This proves that $\mathcal{V} = (\mathcal{V}^*)^*$ and the theorem. □

COROLLARY 19.4.13. *If $\mathcal{V}^*$ is generated by a superalgebra $A$, then $\mathcal{V}$ is generated by the superalgebra $G(A)$.*

PROOF. This follows from Lemma 19.4.10 and Theorem 19.4.12. □

This theorem and Corollary 19.4.13 can be applied also in the special case in which $\mathcal{V}$ is a variety of algebras, that is its $T_2$-ideal $I$ is also a $T$-ideal; see Remark 19.3.13.

In this case, if $A$ generates $\mathcal{V}^*$, as in Corollary 19.4.13, we have that $G(A)$ generates $\mathcal{V}$ also as a variety of algebras. In fact if $f$ is an identity of $G(A)$, it is also a multilinear graded identity. So $f^* \in I^*$ is a multilinear graded identity of $A$, hence $f \in I$ is an identity in the variety $\mathcal{V}$, by Remark 19.3.13.

Observe that, if $\mathcal{V}$ is a variety of algebras, it is not necessarily true that $\mathcal{V}^*$ is also a variety of algebras. Take as example $\mathcal{V}$ the variety of commutative algebras.

The ideal $I$ is generated by the commutative law. Now write the commutative law separating the variables as

$$y_1y_2 - y_2y_1,\ y_1z_1 - z_1y_1,\ z_1z_2 - z_2z_1.$$

When we take the $T_2$ ideal $I^*$, we see that it is generated by

$$y_1y_2 - y_2y_1,\ y_1z_1 - z_1y_1,\ z_1z_2 + z_2z_1. \tag{19.14}$$

REMARK 19.4.14. If $\mathcal{V}$ is the variety of commutative algebras, the variety $\mathcal{V}^*$ is the variety of (super)-commutative superalgebras generated by the Grassmann (super)-algebra.

This $T_2$-ideal is not a $T$-ideal otherwise, by Remark 19.3.12; from the identities (19.14) one would also deduce $y_1y_2 + y_2y_1$ and so also $y_1y_2$.

PROPOSITION 19.4.15.

(1) *A superalgebra $A$ satisfies a polynomial identity if and only if $G(A)$ also satisfies polynomial identities.*
(2) *If a $T_2$-ideal $I$ contains a nonzero $T$-ideal $J$, then also the ideal $I^*$ contains a nonzero $T$-ideal.*

PROOF. (1) If $A$ satisfies a polynomial identity, since $G$ is a PI algebra it follows from the theorem of Regev (Theorem 7.2.2) that the $T$-ideal of identities of $A \otimes G$ is nonzero. Since $G(A) \subset A \otimes G$, the claim follows. The converse follows since $A$ is PI equivalent to $G(G(A))$.

(2) The ideal $I$ is the ideal of graded identities of the superalgebra $A = F\langle X\rangle/I$. By Lemma 19.4.10 $I^*$ is the ideal of graded identities of the superalgebra $A_0 \otimes G_0 \oplus A_1 \otimes G_1$. Now the ideal $I^*$ contains the ideal $J'$ of (ordinary) identities of $A \otimes G$. Again by the Regev theorem, since $J \neq 0$ is an ideal of identities of $A$, we have $J' \neq 0$. □

Next we shall prove a technical lemma. Recall that $V_n \cong F[S_n]$ is the space of multilinear polynomials in $x_1, \dots, x_n$ and $H(k,l)$ is the infinite hook of arm $k$ and leg $l$ (formula (6.26)).

LEMMA 19.4.16. *Let $M$ be an irreducible $S_n$-submodule of $V_n$ corresponding to a partition $\lambda \vdash n$. If $\lambda \in H(k,l)$, then there exists a decomposition of the set of variables $X_n = \{x_1, \dots, x_n\}$ into a disjoint union*

$$X_n = X_1^+ \cup \dots \cup X_{k'}^+ \cup X_1^- \cup \dots \cup X_{l'}^-, \tag{19.15}$$

*with $k' \le k, l' \le l$ such that $M$ can be generated by a polynomial $f = f(x_1, \dots, x_n)$ which is symmetric on each set of variables $X_i^+, 1 \le i \le k'$, and alternating on each set of variables $X_j^-, 1 \le j \le l'$.*

PROOF. We know that for each tableau $T_\lambda$ of shape $\lambda$, the module $M$ is isomorphic to $F[S_n]e_{T_\lambda}$ where $e_{T_\lambda}$ is the corresponding Young symmetrizer.

By formula (cf. (6.4)) it follows then that $M$ is generated by a polynomial of the type $f'(x_1, \dots, x_n) = e_{T_\lambda}g(x_1, \dots, x_n)$. Since $e_{T_\lambda} = s_{T_\lambda}a_{T_\lambda}/p(\lambda)$ and $s_{T_\lambda}$ is the symmetrizer on rows, the polynomial $f'$ is symmetric on each set of variables corresponding to the rows of $T_\lambda$.

Since $\lambda \in H(k,l)$, the diagram $\lambda$ has $l' \le l$ columns of height $> k$ and $k' \le k$ rows not contained in the first $l'$ columns. The tableau $T_\lambda$ can be thought of as a filling of $\lambda$ with the variables $X_n$.

Then $X_1^-, \ldots, X_{l'}^-$ will be the subsets of $X_n$ corresponding to the $l'$ columns of $T_\lambda$ of height $> k$. Also we let $X_1^+, \ldots, X_{k'}^+$ be the subsets of $X_n$ corresponding to the truncated rows of $T_\lambda$ outside the first $l'$ columns.

Denote by $H$ the subgroup of $S_n$ of all permutations stabilizing the first $l'$ columns of $T_\lambda$. This subgroup clearly commutes with the subgroup $K$ stabilizing the truncated rows $X_1^+, \ldots, X_{k'}^+$ outside the first $l'$ columns. Thus the polynomial $f := (\sum_{\tau \in H} (\text{sgn } \tau)\tau) f'$ is still symmetric under the group $K$ and antisymmetric under $H$ by construction. We claim that $f$ is nonzero. So, since $M = F[S_n]f'$ is irreducible, we also have $F[S_n]f = M$, and $f$ has the desired properties.

In order to prove that $f \neq 0$, recall that $e_{T_\lambda} = p(\lambda)^{-1} s_{T_\lambda} a_{T_\lambda}$ is an idempotent, so $f' = e_{T_\lambda} f'$ where $a_{T_\lambda}$ antisymmetrizes all columns of $\lambda$.

Hence $a_{T_\lambda} (\sum_{\tau \in H} (\text{sgn } \tau)\tau) = |H| a_{T_\lambda}$ which implies $e_{T_\lambda} f = |H| f' \neq 0$. □

19.4.2.1. *The main theorem.* Before we state and prove the main theorem of this section, Theorem 19.4.18, we need some preliminaries. Let $\mathcal{V}$ be a nontrivial variety of associative algebras over a field $F$ of characteristic 0. Consider the supervariety $\mathcal{V}^*$ and its relatively free superalgebra $\mathcal{F}$ in $k$ free even generators $u_1, \ldots, u_k$, and $l$ free odd generators $v_1, \ldots, v_l$.

LEMMA 19.4.17. *Let $h \leq k$, $\ell \leq l$, and let*

$$f = f(Y_1, \ldots, Y_h, Z_1, \ldots, Z_\ell)$$

*be a graded multilinear polynomial which is symmetric on each set of even variables $Y_i, 1 \leq i \leq h$, and alternating on each set of odd variables $Z_j, 1 \leq j \leq \ell$. If $f^* \in \mathrm{Id}_2(\mathcal{F})$, then $f^* \in \mathrm{Id}_2(\mathcal{V}^*)$ and $f$, regarded as a polynomial in ungraded variables, is an identity of $\mathcal{V}$.*

PROOF. Let $|Y_i| = m_i$, $1 \leq i \leq h$, $|Z_j| = n_j$, $1 \leq j \leq \ell$. Notice that, by the definition of the map $*$, the polynomial $f^*$ is *symmetric* on each of the sets $Y_1, \ldots, Y_h, Z_1, \ldots, Z_\ell$.

Thus $f^*$ is, up to a nonzero scalar factor, the polarization of the polynomial $p(y_1, \ldots, y_h, z_1, \ldots, z_\ell)$ obtained from $f^*$ by identifying all the variables of each of the sets $Y_i$ with a variable $y_i$ and all the variables of each of the sets $Z_j$ with a variable $z_j$ (restitution). The $y_i$'s are even variables, the $z_j$'s are odd variables, and $p$ is homogeneous of degree $m_i$ in $y_i$ and $n_j$ in $z_j$.

Since $f^* \in \mathrm{Id}_2(\mathcal{F})$, we also have that $p(y_1, \ldots, y_h, z_1, \ldots, z_\ell) \in \mathrm{Id}_2(\mathcal{F})$, hence $p(u_1, \ldots, u_h, v_1, \ldots, v_\ell) = 0$. Since $\mathcal{F}$ is a relatively free algebra, this implies that the polynomial $p(y_1, \ldots, y_h, z_1, \ldots, z_\ell)$ is a graded identity of $\mathcal{V}^*$.

Then, since $f^*$ is obtained from $p$ by polarization, it is also a graded identity of $\mathcal{V}^*$. Thus $f$ is a graded identity of $\mathcal{V}$. But since $\mathcal{V}$ is a variety of algebras, this is also an ordinary identity. □

The following theorem is the main result of this section but it will be replaced later by the much stronger Theorem 19.8.1.

THEOREM 19.4.18. *Every nontrivial variety $\mathcal{V}$ of associative algebras over a field of characteristic 0 is generated by the Grassmann envelope of a finitely generated PI superalgebra.*

PROOF. According to Remark 19.3.13, we can think of $\mathcal{V}$ also as variety of superalgebras.

Let $k$ and $l$ be integers such that the cocharacter $\chi_n(\mathcal{V})$ lies in the hook $H(k,l)$ for every $n \geq 1$. Let $\mathcal{F}_{k,l} = F\langle u_1, \dots, u_k; v_1, \dots, v_l \rangle$ be the relatively free superalgebra of the supervariety $\mathcal{V}^*$ in $k$ free even generators $u_1, \dots, u_k$, and let $l$ be the free odd generators $v_1, \dots, v_l$. The theorem is thus a consequence of the following more precise

THEOREM 19.4.19. *Assume that the cocharacters of a variety $\mathcal{V}$ of associative algebras are contained in the hook $H(k,l)$. Then the variety $\mathcal{V}$ is generated by the Grassmann envelope $G(\mathcal{F})$ of the algebra $\mathcal{F}_{k,l}$.*

Clearly $G(\mathcal{F}) = \mathcal{F}_0 \otimes G_0 \oplus \mathcal{F}_1 \otimes G_1 \in \mathcal{V}$ by Definition 19.4.11 of $\mathcal{V}^*$. It remains to show that all the identities of $G(\mathcal{F})$ are identities of $\mathcal{V}$.

Let $f \equiv 0$ be a multilinear identity of $G(\mathcal{F})$ of degree $n$, and let $M = F[S_n]f$ be the $S_n$-module generated by $f$. $M$ is formed by identities of $G(\mathcal{F})$, and we need to show that $M \subseteq \mathrm{Id}(\mathcal{V}) \cap V_n$. By complete reducibility $M$ is a sum of irreducibles, hence we may assume that $M$ is an $S_n$-irreducible module corresponding, say, to a partition $\lambda \vdash n$.

If $\lambda \notin H(k,l)$, since $\chi_n(\mathcal{V})$ lies in the hook $H(k,l)$, then $M \subseteq \mathrm{Id}(\mathcal{V})$, and we are done. Therefore we may assume that $\lambda \in H(k,l)$.

By Lemma 19.4.16, $M$ is generated by a polynomial $f = f(x_1, \dots, x_n)$ which is symmetric on $h \leq k$ disjoint sets $X_i^+, 1 \leq i \leq h$, and is alternating on $\ell \leq l$ disjoint sets $X_j^-, 1 \leq j \leq \ell$ and $\bigcup_{i,j}(X_i^+ \cup X_j^-) = \{x_1, \dots, x_n\}$.

We rewrite $f = f(X_1^+, \dots, X_h^+, X_1^-, \dots, X_\ell^-)$.

Now we regard the variables of the sets $X_i^+$ as even variables, and we call $X_i^+ = Y_i$, and the variables of the sets $X_j^-$ as odd variables $X_j^- = Z_j$. By doing so, we regard $f = f(Y_1, \dots, Y_h, Z_1, \dots, Z_\ell) \in \mathrm{Id}_2(G(\mathcal{F}))$ as a graded identity of the superalgebra $G(\mathcal{F})$. Since $f \in \mathrm{Id}_2(G(\mathcal{F}))$, by Lemma 19.4.10 $f^* \in \mathrm{Id}_2(\mathcal{F})$.

We can now apply Lemma 19.4.17 and conclude that $f \in \mathrm{Id}(\mathcal{V})$. $\square$

**19.4.3. The colength is polynomially bounded.** We can now complete the proof of Theorem 8.3.15. Let $\mathcal{V}$ be a nontrivial variety of algebras over $F$ of characteristic 0. Our next aim is to prove that the multiplicities in the $n$th cocharacter of $\mathcal{V}$ are polynomially bounded. We shall prove more: that the colength sequence of $\mathcal{V}$ is polynomially bounded.

LEMMA 19.4.20. *Let $\mathcal{V}$ be a nontrivial variety of algebras over a field $F$ of characteristic 0, and let $\chi_n(\mathcal{V}) = \sum_{\lambda \vdash n} m_\lambda \chi_\lambda$ be its $n$th cocharacter. Then there exist constants $C$ and $t$ such that $m_\lambda \leq Cn^t$, for all $n \geq 1$.*

PROOF. By Theorem 19.4.19 we have $\mathcal{V} = \mathrm{var}(G(L))$, where we denote by $L = F\langle u_1, \dots, u_k; v_1, \dots, v_l \rangle$ the relatively free superalgebra in $k$ free even generators and $l$ free odd generators of $\mathcal{V}^*$. For every $n \geq 1$, let $L_n$ be the subspace of $L$ generated by the monomials in the free generators of degree at most $n$.

Since $L$ is finitely generated and PI, by Proposition 8.3.13 $\dim L_n \leq Cn^t$ for some constants $C, t$. Hence, in order to finish the proof, it is enough to show that $m_\lambda \leq \dim L_n$ for all $n$ and for all partitions $\lambda \vdash n$.

Suppose by contradiction that $m_\lambda = d > \dim L_n$ for some $n \geq 1$ and some partition $\lambda \vdash n$. Notice that since the cocharacter of $\mathcal{V}$ lies in the infinite hook $H(k,l)$, then $\lambda \in H(k,l)$.

Thus $m_\lambda > d$ says that there exists a direct sum of $d$ irreducible $S_n$-submodules $M_i \subset V_n$ whose character is $\chi_\lambda$ such that $(M_1 \oplus \cdots \oplus M_d) \cap \mathrm{Id}(\mathcal{V}) = 0$.

Now, by Lemma 19.4.16 there exists a decomposition of the set of variables $X_n = \{x_1, \ldots, x_n\}$ into a disjoint union $X_n = X_1^+ \cup \cdots \cup X_{k'}^+ \cup X_1^- \cup \cdots \cup X_{l'}^-$, with $k' \leq k, l' \leq l$ such that, each $M_r$, $r = 1, \ldots, d$, can be generated by a polynomial $f_r = f_r(x_1, \ldots, x_n)$ which is symmetric on each set $X_i^+, 1 \leq i \leq k'$, and alternating on each set $X_j^-, 1 \leq j \leq l'$. By eventually multiplying on the left by permutations $\rho_r \in S_n$, we may assume that all the polynomials $f_1, \ldots, f_d$ have the same properties of symmetrization and alternation on the same sets of variables.

By assumption no nonzero linear combination of $f_1, \ldots, f_d$ is a polynomial identity of $\mathcal{V}$.

Next we consider each polynomial $f_r$ as a graded polynomial by regarding the variables of $X_1^+ \cup \cdots \cup X_k^+$ as even and the variables of $X_1^- \cup \cdots \cup X_l^-$ as odd. Then we can consider the polynomials $f_r^*$, $1 \leq r \leq d$, and their evaluation $\varphi$ in $L$ such that $\varphi(X_j^+) = u_j, 1 \leq j \leq k$, and $\varphi(X_j^-) = v_j, 1 \leq j \leq l$.

Now, since $\deg f_r = n$, then $\varphi(f_r) \in L_n$. On the other hand $\dim L_n < d$ implies that the elements $\varphi(f_1^*), \ldots, \varphi(f_d^*)$ are linearly dependent over $F$. It follows that there exist (not all zero) scalars $\alpha_1, \ldots, \alpha_d$ such that $\varphi(\alpha_1 f_1^* + \cdots + \alpha_d f_d^*) = 0$. Since $L = F\langle u_1, \ldots, u_k; v_1, \ldots, v_l \rangle$ is relatively free in $\mathcal{V}^*$, this says that $\alpha_1 f_1^* + \cdots + \alpha_d f_d^*$ is a graded identity of $L$. But then by Lemma 19.4.17 $\alpha_1 f_1 + \cdots + \alpha_d f_d$ is an identity of $\mathcal{V}$, and this is a contradiction to our assumption on the $f_i$. □

Let $\mathcal{V}$ be a variety of algebras over a field $F$ of characteristic 0, and let $\chi_n(\mathcal{V}) = \sum_{\lambda \vdash n} m_\lambda \chi_\lambda$ be its $n$th cocharacter. Recall that

$$l_n(\mathcal{V}) = \sum_{\lambda \vdash n} m_\lambda, \ n = 1, 2, \ldots$$

is called the colength sequence of $\mathcal{V}$.

THEOREM 19.4.21. *If $\mathcal{V}$ is a nontrivial variety, its colength sequence is polynomially bounded.*

PROOF. Recall that the $n$th cocharacter of $\mathcal{V}$ lies in the hook $H(k, l)$. Hence in light of Lemma 19.4.20 it is enough to bound polynomially the number of partitions of $n$ lying in $H(k, l)$. If $\lambda \in H(k, l)$, the first $k$ rows of $\lambda$ have length bounded by $n$, hence we have at most $n^k$ choices. Also we have at most $n^l$ choices for the length of the first $l$ columns. Hence the total number of partitions $\lambda \in H(k, l)$ is bounded by $n^{k+l}$. □

**19.4.4. Generic elements.** Recall that a superalgebra $A = A^{(0)} \oplus A^{(1)}$ is *supercommutative* if, for any homogenous elements $a, b \in A$, we have $ab = (-1)^{|a||b|} ba$. For instance, the Grassmann algebra with its usual $\mathbb{Z}_2$-grading is an example of supercommutative algebra. The free supercommutative algebra of countable rank $S$ is defined by its universal property: we let $\Xi = \{\xi_{i,j}\}, \mho = \{\eta_{i,j}\}$ be countable sets, then

$$S = F\langle \Xi, \mho \rangle$$

is the algebra with 1 generated by $\Xi, \mho$ over $F$ subject to the condition that the elements of $\Xi$ are central and those of $\mho$ anticommute among them.

The free superalgebra has a natural $\mathbb{Z}_2$-grading $S = S^{(0)} \oplus S^{(1)}$ by requiring that the variables $\xi_{i,j}$ are even and the variables $\eta_{i,j}$ are odd.

Clearly $S \cong F[\Xi] \otimes G$, $S^{(0)} = F[\Xi] \otimes G_0$, $S^{(1)} = F[\Xi] \otimes G_1$, where $G$ is the Grassmann algebra generated by the $\{\eta_{i,j}\}$.

One can define the superenvelope of a superalgebra $A = A^{(0)} \oplus A^{(1)}$ to be

$$S(A) = S^{(0)} \otimes A^{(0)} \oplus S^{(1)} \otimes A^{(1)} = G(A) \otimes F[\Xi].$$

REMARK 19.4.22. For any superalgebra $A$, $S(A)$ and $G(A)$ satisfy the same superidentities.

Let $A = A_0 \oplus A_1$ be a finite-dimensional superalgebra (over some infinite field $F$), and let $G(A)$ be its Grassmann envelope. We claim that we have also in this case, for every $m = 1, \dots, \infty$, an algebra of $m$ generic elements, contained in the superenvelope $S(A)$, which is isomorphic to the relatively free algebra for $G(A)$ in $m$ variables.

Namely, fix a basis $a_1, \dots, a_h$ of $A_0$ and $b_1, \dots, b_k$ of $A_1$, and then fix some number of variables $m$.

Choose $mh$ polynomial (i.e., commutative) variables $\xi_{i,j}$, $i = 1, \dots, m$, $j = 1, \dots, h$, and $mk$ Grassmann variables $\eta_{i,j}$, $i = 1, \dots, m$, $j = 1, \dots, k$, out of the infinite list of Grassmann variables generating $G$. Then set

$$\xi_i := \sum_{j=1}^{h} \xi_{i,j} \otimes a_j + \sum_{\ell=1}^{k} \eta_{i,\ell} \otimes b_\ell \in S(A) \tag{19.16}$$

and the subalgebra $\mathcal{G}$ that they generate. Consider the homomorphism $\pi$ of the free algebra $F\langle x_1, \dots, x_m \rangle$ to the algebra generated by the $\xi_i$ mapping $x_i \mapsto \xi_i$. Then we claim that

PROPOSITION 19.4.23. *The kernel of $\pi$ is the ideal of polynomial identities in $m$ variables for $G(A)$.*

PROOF. Evaluate a noncommutative polynomial $f(x_1, \dots, x_m)$ in $G(A)$ by substituting each $x_i$ with some element $u_i = \sum_{j=1}^{h} \alpha_{i,j} \otimes a_j + \sum_{\ell=1}^{k} \beta_{i,\ell} \otimes b_\ell$, with $\alpha_{i,j} \in G_0$, $\beta_{i,\ell} \in G_1$.

Consider then the homomorphism $\rho : A \otimes F[\xi_{i,j}, \eta_{i,\ell}] \to A \otimes G$ which is the identity on $A$ and maps $\xi_{i,j} \mapsto \alpha_{i,j}$, $\eta_{i,\ell} \mapsto \beta_{i,\ell}$. This maps $\xi_i$ to $u_i$ and $f(u_1, \dots, u_m) = \rho(f(\xi_1, \dots, \xi_m))$. Thus if $f(x_1, \dots, x_m)$ vanishes when computed on the generic elements $\xi_i$, it is a polynomial identity for $G(A)$. The converse depends upon the fact that a nonzero polynomial in the variables $\xi_{i,j}$ does not vanish identically on $G_0$. □

## 19.5. Finitely generated PI superalgebras

*In this section we repeat, for superalgebras, the theory developed in Chapter 17 for algebras. This will eventually lead to the theorems of Kemer having as its goal the solution of the Specht problem.*

**19.5.1. Representable superalgebras.** We start from a simple remark:

PROPOSITION 19.5.1. *If $R$ is a superalgebra which can be embedded in a finite-dimensional algebra over some field $K$, then $R$ can be embedded as superalgebra in a finite-dimensional superalgebra over $K$. We say it is super-representable.*

PROOF. We may assume $i : R \to A - M_n(K)$ is the given embedding, and we consider the automorphism of order 2, $\tau : R \to R$ defining the superalgebra structure. We then embed $R \to A \oplus A$ by $j : r \mapsto (i(r), i(\tau(r)))$ and give to $A \oplus A$ the superalgebra structure $\tau : (a, b) \mapsto (b, a)$. We see that the constructed embedding $j$ satisfies the required properties. □

COROLLARY 19.5.2. *If $R$ is a superalgebra which is a finite module over a finitely generated commutative algebra of degree* 0, *over a field $F$, then $R$ is super-representable.*

PROOF. This follows from Theorem 11.5.17 and Proposition 19.5.1. □

Notice that the analogue of Corollary 11.5.18 holds for superalgebras, due to Theorem 19.2.18. That is:

COROLLARY 19.5.3. *A representable superalgebra over a field $F$ is $PI_2$ equivalent to a superalgebra finite dimensional over $F$.*

PROOF. We may assume $K$ is algebraically closed. If $R \subset M_n(K) \oplus M_n(K)$ as superalgebra, consider $RK$ a superalgebra, finite dimensional over $K$, $\mathrm{PI}_2$ equivalent to $R$. Modulo the radical, it is of the form $S \otimes_F K$ with $S$ a semisimple superalgebra finite dimensional over $F$. Then take homogeneous generators of its radical and $S$ together with these generators generates a superalgebra $R'$ finite dimensional over $F$ with $RK = R'K$, hence the claim. □

As in Theorem 17.2.18 we have the notion of split finite-dimensional superalgebra, as one for which $\bar{A}$, the semisimple part, is isomorphic to a direct sum of split simple algebras. With essentially the same proof we have

THEOREM 19.5.4. *Every superalgebra finite dimensional over a field $F$* (*of characteristic not* 2) *is* $\mathrm{PI}_2$ *equivalent to a finite-dimensional split superalgebra.*

**19.5.2. The Kemer superindex.** We assume $F$ is of characteristic 0. As in Definition 17.2.17 we may replace any finite-dimensional superalgebra with one which is split as a superalgebra, Theorem 19.5.4.

In this section we do the last step of our program, i.e., we give a *super* analogue of the results that we have seen until now.

LEMMA 19.5.5. *For every* PI-*superalgebra $A$, finitely generated over $F$, there exists a superalgebra $B$, finite dimensional over $F$, such that* $\mathrm{Id}_2(B) \subseteq \mathrm{Id}_2(A)$.

PROOF. Let $A$ be a finitely generated superalgebra with ordinary PI. Then $\mathrm{Id}_2(A)$ contains the $T$-ideal $\mathrm{Id}(A)$ of ordinary identities of $A$, $\mathrm{Id}_2(A) \supseteq \mathrm{Id}(A)$. By Lemma 17.1.3 there is a finite-dimensional algebra $B$ such that $\mathrm{Id}(B)$ is contained in $\mathrm{Id}(A)$. Now consider the algebra $B \otimes F[\mathbb{Z}_2] = B \oplus B$ as a superalgebra. By Example 19.3.11 $\mathrm{Id}(B) = \mathrm{Id}_2(B \otimes F[\mathbb{Z}_2])$. Thus $\mathrm{Id}_2(A)$ contains $\mathrm{Id}_2(B \otimes F[\mathbb{Z}_2])$ and $B \otimes F[\mathbb{Z}_2]$ is finite dimensional. □

Now we give all the definitions, analogous to the ones that we have seen until now in this chapter, for superpolynomials, i.e., for the elements of the free superalgebra $F\langle Y, Z\rangle$ where $Y$ are the degree 0 variables and $Z$ are the degree 1 variables.

DEFINITION 19.5.6. The first Kemer superindex of a $T_2$-ideal $\Gamma$ is the maximum (if it exists), in the lexicographic order, of the pairs of numbers $t_0, t_1$ for which the following statement holds. For all $\mu \in \mathbb{N}$ there is a superpolynomial $f \notin \Gamma$ which is *$\mu$-fold $(t_0, t_1)$-alternating*. This means that there exist sets of variables $X_i = Y_i \cup Z_i$

with $Y_i \subseteq Y$, $Z_i \subseteq Z$, $1 \le i \le \mu$, $|Y_i| = t_0$ and $|Z_i| = t_1$, such that $f$ is alternating in each $Y_i$ and each $Z_i$.

We shall apply this notion in particular to finitely generated superalgebras which satisfy some ordinary PI, hence they satisfy some Capelli identity so that this maximum is certainly achieved.

As for the second Kemer superindex of a $T_2$-ideal $\Gamma$, it is better to use a Young diagram, in the form of *Ferrer's diagram*. That is a finite set $S$ of pairs $(a,b) \in \mathbb{N}^2$ such that, if $(a,b) \in S$ and $(a',b')$ is such that $a' \le a$, $b' \le b$, then $(a',b') \in S$. Below is an example.

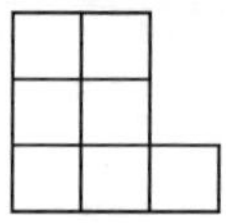

$$= \begin{matrix} \cdot & \cdot & \\ \cdot & \cdot & \\ \cdot & \cdot & \cdot \end{matrix} = \begin{matrix} (0,2) & (1,2) & \\ (0,1) & (1,1) & \\ (0,0) & (1,0) & (2,0) \end{matrix}$$

In particular for such a diagram one has the *corners*, that is the coordinates $a, b$ of a box which if removed still gives a diagram; see page 183. In the example we have two corners, $(2,0), (1,2)$.

Assume that a $T_2$-ideal $\Gamma$ has first Kemer superindex $(t_0, t_1)$. Consider thus the Ferrer's diagram formed by all pairs $(a,b)$ for which, for all $\mu \in \mathbb{N}$, there is a superpolynomial $f \notin \Gamma$ which is $\mu$-fold $(t_0,t_1)$-alternating, and moreover we also have that it is $a$-fold alternating in layers of degree 0 variables with $t_0+1$ elements and $b$-fold alternating in layers of degree 1 variables with $t_1+1$ elements.

DEFINITION 19.5.7. This set of pairs is finite, and it gives a Young diagram $D$ which will be called the *second Kemer superindex*.

We also denote $\beta(\Gamma) = (t_0,t_1)$, $\gamma(\Gamma) = D$ and by $\mathrm{ind}_2(\Gamma) = (\beta(\Gamma), D)$ which we identify to the triple $(t_0, t_1, D)$. We order lexicographically these triples where $D_1 \le D_2$ means that $D_1 \subset D_2$.

If $\Gamma = \mathrm{Id}_2(A)$, we write $\mathrm{ind}_2(\Gamma) = \mathrm{ind}_2(A) = (\beta(A), \gamma(A))$.

Clearly, if $\mathrm{Id}_2(A_1) \subseteq \mathrm{Id}_2(A_2)$, then $\mathrm{ind}_2(A_1) \ge \mathrm{ind}_2(A_2)$.

REMARK 19.5.8. For a finite-dimensional superalgebra $A$ with radical $J$, we have also two integers $t_{A,0}, t_{A,1}$ called the *first algebra superindex of* $A$ and a *Young diagram* $D_A$ *of* $A$ called the *second algebra superindex of* $A$:

$$t_{A,0} = \dim(A/J)_0,\ t_{A,1} = \dim(A/J)_1.$$

To define $D_A$, take all possible pairs $(a,b)$, $a, b = 0, 1, \dots$ for which there is a nonzero product of $a$ elements in $J_0$ and $b$ elements in $J_1$. This is clearly a finite set of pairs and it corresponds to a Young diagram, called the *Young diagram of* $A$ denoted by $D_A$. If $A = \bigoplus_i A_i$, then

$$(t_{A,0}, t_{A,1}) = \sum_i (t_{A_i,0}, t_{A_i,1}), \quad D_A = \bigcup_i D_{A_i}. \tag{19.17}$$

REMARK 19.5.9. The same proof as in Proposition 17.2.25 shows that the first Kemer superindex of $A$ is always less than or equal to the first algebra superindex of $A$. Moreover the second Kemer superindex is a diagram $D \subseteq D_A$.

We shall soon see that of particular importance is the case where $D$ is a rectangle. In this case $D$ is determined by its upper right corner, a pair of integers $a, b$, which will then be referred to as the second Kemer superindex.

In this case the Kemer superindex will be denoted by $(t_0, t_1, a, b)$.

We use similar notation for the algebra superindex.

19.5.2.1. *Kemer superpolynomials.* Let $A$ be an algebra with first Kemer superindex $(t_0, t_1)$ and second index $D$. First, as in Remark 17.2.4 there is some $\mu_0 \in \mathbb{N}$ such that each superpolynomial, that is multilinear and alternating in $\mu_0$ layers with $t_0$ even and $t_1$ odd variables and also alternating in $a$ layers with $t_0 + 1$ even and $b$ layers with $t_1 + 1$ odd variables with $(a, b) \notin D$, is a superidentity of $A$. Then

DEFINITION 19.5.10. A superpolynomial multilinear and alternating in $\mu \geq \mu_0$ layers with $t_0$ even and $t_1$ odd variables and also alternating in $a$ layers with $t_0 + 1$ even and $b$ layers with $t_1 + 1$ odd variables with $(a, b)$ a corner of $D$ will be called a *Kemer superpolynomial of the algebra A*.

By definition of the Kemer superindex such polynomials exist for all $\mu \geq \mu_0$.

**19.5.3. The case of simple superalgebras.** First of all we should analyze these numbers for simple finite-dimensional superalgebras. We have two nontrivial types.

TYPE 1. The first type is $A = M_n(F) \oplus M_n(F)$, where the 0 part is the set $(a, a)$, $a \in M_n(F)$, while the 1 degree part is the set $(a, -a)$, $a \in M_n(F)$. We have $t_{A,0} = t_{A,1} = n^2$. Thus the first algebra superindex of $A$ is $(n^2, n^2)$. The second algebra superindex is $(0, 0)$.

We may take $Y, Z$ as superpolynomials alternating in $h \geq 2$ sets of $n^2$ variables, for instance the product of two Regev polynomials given by Corollary 12.3.19, respectively, in variables in $Y$ and $Z$.

$$F^{(1)}_{h,n}(Y_1, \dots, Y_h, Z_1, \dots, Z_h) := F_{h,n}(Y_1, \dots, Y_h)F_{h,n}(Z_1, \dots, Z_h).$$

REMARK 19.5.11. If the evaluation of $F_{h,n}(X_1, \dots, X_h)$ in matrices $a^i_{j,k}$ is 1, then the evaluation in the pairs $(a^i_{j,k}, a^i_{j,k})$ gives $1, 1$ and that in the pairs $(a^i_{j,k}, -a^i_{j,k})$ gives $1, (-1)^{hn^2}$.

DEFINITION 19.5.12. We define

(19.18)

$$F^{(1)}_{A,h,n}(Y_1, \dots, Y_h, Z_1, \dots, Z_h) := F_{h,n}(Y_1, \dots, Y_h)F_{h,n}(Z_1, \dots, Z_h), \text{ if } hn^2 \text{ even,}$$
$$F^{(1)}_{A,h,n}(Y_1, \dots, Y_h, Z_1, \dots, Z_h, Z) := F_{h,n}(Y_1, \dots, Y_h)F_{h,n}(Z_1, \dots, Z_h)Z, \text{ if } hn^2 \text{ odd,}$$

where $Z$ is some single odd variable.

In both cases we have thus an even polynomial which has an evaluation that by the previous remark gives as result the identity 1 of the simple superalgebra.

From Definition 19.5.7 it follows that the Kemer superindex of $A$ equals the algebra superindex of $A$, i.e., $(n^2, n^2, 0, 0)$.

TYPE 2. The second type is $A = M_{k,\ell}(F)$ with

$$t_{A,0} = k^2 + \ell^2, \quad t_{A,1} = 2k\ell,$$

the matrix units $e_{i,j}$ are distributed into a basis of $A_0$ and a basis of $A_1$.

By Corollary 12.3.19 if we evaluate each layer $X_i$ of the central polynomial $F_n(X_1, \dots, X_h)$ in the $n^2$ elements of the basis $e_{i,j}$, we have a nonzero multiple of the identity. This means that we can arbitrarily assign a sign $+$ to $k^2 + \ell^2$ of the $n^2$ variables and call them $Y_i$, and $-$ to the remaining $2k\ell$ and call them $Z_i$ variables.

In order to remember this, let us denote this block by $(Y, Z)_i$ and the corresponding superpolynomial by

$$F^{(2)}_{A,h}(Y_1, \dots, Y_h, Z_1, \dots, Z_h) := F_n((Y, Z)_1, \dots, (Y, Z)_h). \tag{19.19}$$

Since the polynomial is alternating in each block $(Y, Z)_i$, it is clearly alternating in the separate homogeneous components.

Again it follows that the superindex of $A$ equals the $(t_0, t_1)$ superindex which equals $k^2 + \ell^2, 2k\ell, 0, 0$.

19.5.3.1. *Fundamental superalgebras.* We construct fundamental superalgebras in a similar way to what we did for algebras. Start from a finite-dimensional superalgebra $A$ with radical $J$ and semisimple part $\bar{A} := A/J = \bigoplus_{i=1}^q R_i$ with the $R_i$ simple superalgebras. Consider as in §17.2.4 the quotient map $\pi : A \to A/J = \bigoplus_{i=1}^q R_i$ and for $1 \leq j \leq q$ let

$$R^{(j)} := \bigoplus_{i \neq j,\, i=1}^{q} R_i \quad \text{and} \quad A_j := \pi^{-1} R^{(j)}.$$

We then perform the construction of §17.2.3.3 by choosing instead of the variables $x_i$ both variables $x_i$ of degree 0 and $z_j$ of degree 1.

Take the free product $A' := \bar{A} * F\langle x_1, \dots, x_a, z_1, \dots, z_b \rangle$ with the obvious structure of superalgebra modulo the verbal ideal generated by the superidentities of $A$. The algebra $A'$ is bigraded by the two sets of variables $x, y$. We have a homomorphism of $A'$ to $\bar{A}$ evaluating the $x, z$ in 0, and we let $J'$ be its kernel.

For $a, b$ sufficiently large we have that $A$ is a quotient of $A'$ by a map which is the identity on $\bar{A}$ and which maps the elements $x_i$ to a basis of $J_0$ and $z_j$ to a basis of $J_1$.

Consider the Young diagram $Y := D_A$ defined in Remark 19.5.8 for the algebra $A$. Define $\mathcal{A}_Y$ as the quotient of $A'$ modulo the bigraded ideal generated by the products involving $u$ elements of $J'_0$ and $v$ elements of $J'_1$ for all pairs $(u, v) \notin Y$, that is the elements of bidegree $(u, v)$ for all pairs $(u, v) \notin Y$. We have that $A$ is a quotient of $\mathcal{A}_Y$ which is finite dimensional with semisimple part $\bar{A}$ and with radical the ideal generated by the classes of the elements $x_i, z_j$, which have degree 1 in the natural grading of $\mathcal{A}_Y$.

By construction we have that $\mathcal{A}_Y$ and $A$ are PI equivalent. They also have the same algebra superindex since they have the same semisimple part and also the same algebra Young diagram by construction.

Then take the various subdiagrams $Y_i$ obtained from $Y$ by removing some corner $(u_i, v_i)$. To these diagrams are associated the superalgebras $\mathcal{A}_{Y_i}$ quotient of $\mathcal{A}_Y$ modulo the ideal generated by all elements of bidegree $(u_i, v_i)$ (which is contained in the Jacobson radical).

Now the algebra Young diagram of $\mathcal{A}_{Y_i}$ is in fact $Y_i$.

DEFINITION 19.5.13. We say that a superalgebra $A$ is *fundamental* if

$$\mathrm{Id}_2(A) \subsetneq \bigcap_{i=1}^{q} \mathrm{Id}_2(A_i) \bigcap_j \mathrm{Id}_2(\mathcal{A}_{Y_j}). \tag{19.20}$$

A multilinear superpolynomial

$$f \in \bigcap_{i=1}^{q} \mathrm{Id}_2(A_i) \bigcap_j \mathrm{Id}_2(\mathcal{A}_{Y_j}) \setminus \mathrm{Id}_2(A) \tag{19.21}$$

will be called *fundamental.*

Note that all the algebras in the right-hand side of formula (19.20) have an algebra superindex which is strictly lower than the algebra superindex of A.

Clearly an argument similar to that of Proposition 17.2.34 shows that

PROPOSITION 19.5.14.

(1) *Every finite-dimensional superalgebra $A$ is* PI *super equivalent to a direct sum of fundamental superalgebras.*
(2) *Assume that $A$ is a fundamental superalgebra.*
  (a) *Given a multilinear fundamental superpolynomial $f$, for $A$, a nonzero elementary evaluation of $f$ visits all the $R_i$ and if it has $u$ (resp., $v$) evaluations in $J_0$ (resp., in $J_1$), then $(u, v)$ is a corner of the Young diagram.*
  (b) *Finally the Young diagram $Y$ has a unique corner, that is it is a rectangle.*

PROOF. (1) This is by induction on the algebra superindex.

(2)(a) Assume $f$ is fundamental according to Definition 19.5.13, formula (19.21), in particular it is multilinear. First of all a nonzero elementary evaluation of $f$ must visit all the simple blocks $\bar{A}_i$, otherwise it would be an evaluation in one of the algebras $A_i$, hence 0.

Assume that in this evaluation we have $u$ even (resp., $v$ odd) radical variables. By the universal property of $\mathcal{A}_Y$ (as in Proposition 17.2.30), this evaluation factors through an evaluation of $f$ in $\mathcal{A}_Y$ mapping the $(u, v)$ radical variables to corresponding elements $x_i, z_j$ (of degree 1).

Note that this evaluation produces a nonzero element in $\mathcal{A}_Y$ of bidegree $u, v$. If we have that $(u, v)$ is not a corner, then $(u, v)$ is contained in each subdiagram $Y_i$ and our evaluation remains nonzero when we map into $\mathcal{A}_{Y_i}$, since $\mathcal{A}_Y$ and $\mathcal{A}_{Y_i}$ coincide in bidegree $(u, v)$. But this is false since $f$ is, by assumption, a PI of $\mathcal{A}_{Y_i}$.

(2)(b) Suppose by contradiction that there is another corner $(u', v')$, and let $\mathcal{A}_{Y'}$ be the algebra $\mathcal{A}_Y$ modulo the part in bidegree $(u', v')$. Then by hypothesis $f$ is a PI of $\mathcal{A}_{Y'}$, but this is a contradiction since $\mathcal{A}_{Y'}$ and $\mathcal{A}_Y$ coincide in bidegree $(u, v)$. □

**19.5.4. Kemer superindex of fundamental superalgebras.** There is an analogue of Theorem 17.2.41 for superalgebras.

THEOREM 19.5.15.

(1) *A finite-dimensional superalgebra $A$ is fundamental if and only if:*
  (a) *The first Kemer superindex equals the first algebra superindex.*

$$\beta(A) = (t_{A,0}, t_{A,1}) = (\dim_F \bar{A}_0, \dim_F \bar{A}_1).$$

  (b) *The second Kemer superindex $\gamma(A)$ is a rectangle (see Remark 19.5.9) and coincides with the diagram $D_A$ of $A$ (cf. Remark 19.5.8).*
(2) *In this case, given any fundamental superpolynomial $f$, we have $\mu$-Kemer polynomials in the $T_2$-ideal $\langle f \rangle$ generated by $f$ for every $\mu$.*

PROOF. The proof is really a variation of the proof of Theorem 17.2.41, and we include it for convenience of the reader.

In one direction the statement is immediate, as in the proof of Theorem 17.2.41. Assume the Kemer superindex equals the algebra superindex.

If $A$ is not fundamental, it is PI equivalent to a direct sum of algebras with strictly lower algebra superindex. Now we have seen that the Kemer superindex of a direct sum is the maximum of the Kemer superindices and, by Remark 19.5.9 (as in Proposition 17.2.25), that the Kemer superindex is always less than or equal to the algebra superindex; this gives a contradiction.

Let us now prove the converse. Assume $A$ is fundamental. We know (for any algebra) that the Kemer superindex of $A$ is less than or equal to the algebra superindex. We want to show they are equal. Since $A$ is fundamental, its algebra Young diagram is a rectangle with corner $(\sigma_0, \sigma_1)$. If $\sigma_0 = \sigma_1 = 0$, then $A = \bigoplus_i R_i$ is semisimple, with $R_i$ simple superalgebras. The condition of (19.20) easily implies that we must have $A$ simple. We have seen how to construct Kemer superpolynomials in §19.5.3, and there we proved the required equality.

If $t_{A,0} = 0$, then also $t_{A,1} = 0$ and the algebra is nilpotent. Then, by the definition of $(\sigma_0, \sigma_1)$, there is a monomial $y_1 y_2 \cdots y_{\sigma_0+\sigma_1}$ with $\sigma_0$ variables $y_i = x_i$ of degree 0 and $\sigma_1$ variables $y_i = z_i$ of degree 1 which is nonzero on $A$, so it is a Kemer superpolynomial. As for part (2) any fundamental superpolynomial is automatically Kemer.

So we assume $\bar{A} = \bigoplus_{i=1}^q R_i$ and $\dim(R_i)_0 = t_{i,0}$, $\dim(R_i)_1 = t_{i,1}$, so that $t_{A,0} = \sum_{i=1}^q t_{i,0} > 0$, $t_{A,1} = \sum_{i=1}^q t_{i,1} \geq 0$. Denote by $e_i$ the unit element of $R_i$ (of degree 0). Let $g$ be a fundamental superpolynomial for $A$, and choose some nonzero elementary evaluation,

$$g(r_1, \ldots, r_{\sigma_0}, u_1, \ldots, u_{\sigma_1}, b_1, \ldots, b_m) \neq 0, \tag{19.22}$$

where $(\sigma_0, \sigma_1)$ is the unique corner of $D_A$ and where, in the given evaluation, we have elementary radical substitutions $r_k \in J_0, u_j \in J_1$ (property K, cf. Lemma 17.2.36) and semisimple substitutions $b_h \in \bar{A}$ (of degree 0 or 1).

Call $m_k, n_j, y_h$ the variables being substitutes to $r_k, u_j, b_h$, respectively, which we call *even and odd radical and semisimple variables*.

Then $r_k \in e_{\ell_k} J_0 e_{\ell'_k}$ and $u_j \in e_{\ell_j} J_1 e_{\ell'_j}$ for $1 \leq k \leq \sigma_0$, $1 \leq j \leq \sigma_1$, while the $b_h$ are elements of the algebras $R_i$ and, since $g$ is fundamental and the evaluation nonzero, all $R_i$ appear.

From formula (19.22) it follows that there is some nonzero monomial, in the elements $r_1, \ldots, r_{\sigma_0}, u_1, \ldots, u_{\sigma_1}, b_1, \ldots, b_m$, appearing in the evaluation of $g$.

By Remark 17.2.22, to this monomial is associated an oriented path $\Pi$ in the Pierce graph defined in a similar way as Definition 17.2.20 using the decomposition into orthogonal idempotents of $\bar{A}_0$. In $\Pi$ the edge corresponding to $r_k \in e_{\ell_k} J_0 e_{\ell'_k}$ has source $e_{\ell_k}$ and target $e_{\ell'_k}$, similarly for $u_j \in e_{\ell_j} J_1 e_{\ell'_j}$. We mark this edge with the *radical variable* $m_k$ *or* $n_j$ associated to the element $r_k$ (resp., $u_j$).

Moreover, since $g$ is full, this path visits all the vertices $1, \ldots, q$ of the graph (the graph may also have a vertex 0 if the algebra is without 1 and the path may or may not visit this vertex).

We need to distinguish the cases $q > 1$ and $q = 1$.

(1) $q > 1$. This implies that for each $i \in \{1, \ldots, q\}$, corresponding to a semisimple block, we may choose an edge of the path $\Pi$ which joins the vertex $i$ with

some other vertex $i' \neq i$. This edge then corresponds, that is it is marked, to a radical variable $w_i$ (even or odd) associated to some element $r_{v_i}$ or $u_{v_i}$. It is possible that in this way we choose the same edge, and hence the same variable twice, associated to the two distinct end points of this edge.

Set $\max(\sigma_0, \sigma_1) = s$ for simplicity, and let $\nu := \mu + s$. We make a substitution only on the variables $w_j$ previously chosen.

We distinguish the three cases; namely if $w_j$ is associated to only one index $j$ as source, or is associated to only one index $j$ as target or finally it is associated to some $j$ as source, and to some $d \neq j$ as target, i.e., $w_j = w_d$.

We then substitute, in the polynomial $g$, each variable $w_j$, respectively, with

$$u_j F^{(i)}_{R_j,\nu} y_j w_j, \quad w_j x_j F^{(i)}_{R_j,\nu} y_j, \quad x_j F^{(i)}_{R_j,\nu} y_j w_j u_d F^{(i)}_{R_d,\nu} v_d, \quad \text{if } w_j = w_d,$$

where for $F^{(i)}_{A,\nu} = F^{(i)}_n(Y, Z)$, $i = 1, 2$, we take the superpolynomial (of formulas (19.18) or (19.19)) associated to the type $i = 1, 2$ of the simple superalgebra $R_j$ in which it must be evaluated.

These polynomials are taken in disjoint sets of variables. Finally the $x_j, y_j$ are even variables to be evaluated in the unit element $e_j$ of $R_j$.

Notice that in this way we have added, in the polynomial $g$, for each index $j = 1, \dots, q$, an even superpolynomial (of formulas (19.18) or (19.19)) nonidentity of the simple superalgebra $R_j$ which is alternating in $\nu$ even layers $Y_f^{(j)}$, each with $t_{j,0}$ elements, and in $\nu$ odd layers $Z_f^{(j)}$, $f = 1, \dots, \nu$, each with $t_{j,1}$ elements.

We then obtain some new polynomial $\tilde{g}$ in the $T_2$-ideal generated by $g$.

We have a nonzero evaluation of $\tilde{g}$ which extends the previous evaluation.

In fact we may take the one in which the inserted superpolynomials $F^{(i)}_{R_j,\nu}$, containing the new variables, take as value the identity $e_j$ of the corresponding simple block $R_j$.

Consider the following layers of variables

$$Y_f := \{Y_f^{(1)}, Y_f^{(2)}, \dots, Y_f^{(q)}\}, \qquad Z_f := \{Z_f^{(1)}, Z_f^{(2)}, \dots, Z_f^{(q)}\}, \quad f = 1, \dots, \nu.$$

Each $Y_f$ (resp., $Z_f$), is a layer with $t_{A,0} = \sum_{j=1}^q t_{j,0}$ variables (resp., $t_{A,1} = \sum_{j=1}^q t_{j,1}$ variables).

We now alternate independently each layer $Y_f$ and $Z_f$ (as in Lemma 17.2.39). That is we take the sum with sign on all coset representatives of $S_{t_{A,0}}$ with respect to the subgroup $\prod_{i=1}^q S_{t_{i,0}}$ and the same for the $Z$ variables. We may do this since the polynomial we start from is already alternating in each layer $Y_f^{(j)}$.

At this point we need to show that the constructed polynomial is not a PI of $A$. The argument is the same as for Theorem 17.2.41, so we omit it.

We still need to construct the big sets. The radical variables are either even or odd. We must add even radical variables to a small set of even variables and similarly for odd radicals. Note that the radicals that were chosen above or, more generally, the radical variables that bridge different super simple components (of course there may be more than the just chosen above) do not cause any problem since any nontrivial permutation of them with semisimple variables yields zero. So the main issue is to alternate the radical variables that are surrounded by the identity of the same super simple component, say $i$. For these we need to worry only where the alternation is with semisimple elements of the super simple algebra $R_i$ (other alternations yield zero). Now, an alternation of a radical variable surrounded by the identity of $R_i$ yields an evaluation of the original polynomial $p$ with less radical values (two or more radical variables are joined) and this vanishes because the polynomial $p$ is fundamental. Indeed, the corresponding point $(\sigma'_0, \sigma'_1)$, where $\sigma'_0$ (resp., $\sigma'_0$) is the number of even (resp., odd) of radical evaluations, is in $D_A$ and is clearly not a corner there (in fact $(\sigma'_0, \sigma'_1) \neq (\sigma_0, \sigma_1)$ which is the unique corner of $D_A$). It follows that the evaluation factors through an evaluation in the algebra $A_{Y_0}$ obtained by removing from $D_A$ the corner $(\sigma_0, \sigma_1)$. Since the polynomial is an identity of $A_{Y_0}$, we get zero.

As a result we have that all nontrivial permutations of the radical elements with semisimple elements yield zero. Thus we have constructed a Kemer polynomial in the case where $A$ has a unit and $q > 1$. In the case where $A$ has a unit and $q = 1$, we multiply the polynomial by 1 and insert there small sets (even and odd). If we exchange a radical variable with an external semisimple variable, we obtain zero since we are reducing the number of radical evaluations in the polynomial.

(2) $q = 1$. The argument is the same as for Theorem 17.2.41, so we omit it. □

We have an analogue of Proposition 17.2.45:

PROPOSITION 19.5.16. *A Kemer superpolynomial, with $\mu$ sufficiently large, for a fundamental superalgebra $A$ is fundamental.*

19.5.4.1. *Even index.* Given a superalgebra $A$ we may define its *even* Kemer index as in the usual case but only for layers of even variables. We then have also the even $(t_0, s_0)$ index by setting $t_0 := \dim \bar{A}_0$ and $J_0^s \neq 0,\ J_0^{s+1} = 0$.

PROPOSITION 19.5.17. *If $A$ is fundamental with Kemer superindex $(t_0, t_1, a, b)$ (see Remark 19.5.9), then its even index is $(t_0, a)$.*

PROOF. The only point is to prove that $s = a$. In fact by the definition of $(a, b)$ we have that $J_0^a \neq 0$ but also $J_0^{a+1} = 0$, otherwise the box $(a+1, 0)$ should be in the corresponding Young diagram. □

## 19.6. The trace algebra

**19.6.1. The role of traces.** Let $A$ be a fundamental superalgebra with Kemer superindex $(t_0, t_1, s, u)$ and so even index $(t_0, s)$. Let $J$ be its radical and $\bar{A} := A/J$ which we assume to be split (Theorem 19.5.4). If $a \in A$, denote by $\bar{a} \in \bar{A}$ its class modulo $J$.

Given any set $X := (Y, Z)$ of even and odd variables, consider the $Y, Z$-tuples $A_0^Y \times A_1^Z$ of elements of $A$, that is the graded maps $X \to A$, which induce $Y, Z$-tuples $\bar{A}_0^Y \times \bar{A}_1^Z$ of elements of $\bar{A}$, that is the graded maps $X \to \bar{A}$.

Given any even element $f(y_1, \dots, y_i, \dots; z_1, \dots)$ of the free algebra $F\langle Y, Z\rangle$, an element $(a_i)_{i\in X} \in A_0^Y \times A_1^Z$ gives rise to an evaluation map $x_i \mapsto \bar{a}_i \in \bar{A}$. Denote by $\bar{f}(x_1, \dots, x_i, \dots)$ the corresponding function on $\bar{A}_0^Y \times \bar{A}_1^Z$, with values in $\bar{A}_0$.

By taking the trace or the determinant of left multiplication on $\bar{A}_0$, we have a function $\mathrm{tr}_0(\bar{f}(x_1, \dots, x_i, \dots)) \in F$ (resp., $\det_0(\bar{f}(x_1, \dots, x_i, \dots))$) on $\bar{A}_0^Y \times \bar{A}_1^Z$. Sometimes we write just $f$ instead of $\bar{f}$.

DEFINITION 19.6.1. Denote by $\mathcal{T}_A^{\mathrm{su}}(X) = \mathcal{T}_{\bar{A}}^{\mathrm{su}}(X)$ the algebra generated by the functions on $X$-tuples of elements of $\bar{A}$ given by the trace $\mathrm{tr}_0(\bar{f}(x_1, \dots, x_i, \dots))$

where $f$ is an even superpolynomial. Or it is generated equivalently by $D_0(f(x_1,\dots,x_i,\dots)) = \det_0(\bar{f}(x_1,\dots,x_i,\dots))$.

This discussion extends to a direct sum $A = \bigoplus_{i=1}^q A^{(i)}$ of fundamental superalgebras (with radical $J_i$) all with the same Kemer superindex $(t,s,a,b)$. When we take an evaluation $f(x_1,x_2,\dots,x_t,\bar{W})$ in $A$ this of course equals the direct sum of the evaluations $f(x_1^{(i)},x_2^{(i)},\dots,x_t^{(i)},W^{(i)})$, where $x_j^{(i)}$ is the component in $A^{(i)}$ of the evaluation of $x_j$ in $A$ (same for $W^{(i)}$). For any $y \in A_0$, set $\bar{y}_i$ the image of $y$ in $\bar{A}_0^{(i)}$ and set

$$D(y) := (\det(\bar{y}_1),\dots,\det(\bar{y}_q)) \in F^q. \tag{19.23}$$

With $\det(\bar{y}_i)$ the determinant of left multiplication of the class $\bar{y}_i$ of $y$ on the summand $\bar{A}_0^{(i)}$ of $\bar{A}_0$. In fact since all the summands $\bar{A}_0^{(i)}$ have the same dimension $t$, we have that $\bar{A}_0$ is a free module of rank $t$ over $F^q$, so multiplications by elements of $\bar{A}_0$ can be treated as $t \times t$ matrices, now over $F^q$ and $D_0(y) \in F^q$ is the determinant of such a matrix. Then by introducing a commuting variable $\lambda$, we have

$$D_0(\lambda - y) = \lambda^t - \operatorname{tr}_0(\bar{y})\lambda^{-1} + \cdots + (-1)^t D_0(y). \tag{19.24}$$

This is the characteristic polynomial, with coefficients in $F^q$, of left multiplication by $\bar{y}$ on $\bar{A}_0$, as free $F^q$ module.

We can construct the algebra $\mathcal{T}_A^{\mathrm{su}}(X)\mathcal{F}_A^{\mathrm{su}}(X)$, an algebra of superpolynomial functions in the variables $X$. That is graded polynomials from $A_0^Y \times A_1^Z$ to $A$.

The 0 degree subalgebra $\mathcal{T}_A^{\mathrm{su}}(X)\mathcal{F}_A^{\mathrm{su}}(X)_0$ is an algebra with trace having $\mathcal{T}_A^{\mathrm{su}}(X)$ as trace algebra. One should remark that any graded substitution $X \to \mathcal{F}_A^{\mathrm{su}}(X)$ (or even $X \to \mathcal{T}_A^{\mathrm{su}}(X)\mathcal{F}_A^{\mathrm{su}}(X)$) extends, in the 0 component, to a homomorphism of algebras with trace, a substitution.

REMARK 19.6.2. One could extend to superalgebras the theory of algebras with trace by defining the trace as a map only on even elements.

The formalism allows us to define the free superalgebra with trace, and then $\mathcal{T}_A^{\mathrm{su}}(X)\mathcal{F}_A^{\mathrm{su}}(X)$ can be considered as the *relatively free* algebra with trace of the trace algebra $A$.

The main reason to introduce the algebra $\mathcal{T}_A^{\mathrm{su}}(X)\mathcal{F}_A^{\mathrm{su}}(X)$ is that when $X$ is finite it is the Noetherian theorem, Theorem 17.3.5, which $\mathcal{F}_A^{\mathrm{su}}(X)$ is not. The second reason has to do with Kemer superpolynomials; see §17.3.1.2.

One can discuss this issue using the method of generic elements exactly as in §17.3.1.1, separating the generic elements of degree 0, $\xi_i$ from those of degree 1, $\eta_i$.

The map $F\langle y_1,\dots,y_m,\dots;z_1,\dots\rangle \to F\langle \xi_1,\dots,\xi_m,\dots;\eta_1,\dots\rangle$, $y_i \mapsto \xi_i$, $z_j \mapsto \eta_j$ has as kernel the $T_2$-ideal of superpolynomial identities of $A$ in the corresponding variables $X := \{Y,Z\}$. The algebra $F\langle \xi_1,\dots,\xi_m,\dots;\eta_1,\dots\rangle$ is isomorphic to the relatively free superalgebra $\mathcal{F}_A^{\mathrm{su}}(X)$ of $A$. We can decompose $\xi_i = \bigoplus_{j=1}^q \xi_{j,i}$, $\eta_i = \bigoplus_{j=1}^q \eta_{j,i}$ with $\xi_{j,i}$, $\eta_{j,i}$ generic for $A_0^{(i)}, A_1^{(i)}$. For each $i = 1,\dots,k$, we also have that $\mathcal{F}_{A^{(i)}}^{\mathrm{su}}(X) = F\langle \xi_{1,i},\dots,\xi_{m,i},\dots;\eta_{1,i},\dots,\rangle$ is the relatively free superalgebra for $A^{(i)}$.

As in §17.3.1.1 let $L$ denote the field of rational functions in the variables $\xi,\eta$ appearing. $A \otimes_F L = A_0 \otimes_F L \oplus A_1 \otimes_F L$ is a finite-dimensional superalgebra over

$L$. Notice that the radical of $A \otimes_F L$ is $J \otimes_F L$ and modulo the radical, this is $\bar{A} \otimes_F L$ which is a direct sum of simple superalgebras $\bigoplus_{j=1}^r S_j$ of the possible three types.

We consider $A$ and $A \otimes L$ as superalgebras with trace, Remark 19.6.2. For $A$ the trace has values in $F^q$; for $A \otimes L$ the trace has values in $L^q$.

PROPOSITION 19.6.3. *We have according to Remark* 2.3.18 (*and formula* (17.14))

$$\mathcal{T}_A^{\mathrm{su}}(X) := F[t_0(a)]|_{a \in \mathcal{F}_A^{\mathrm{su}}(X)_0} \subset L^{\oplus q}. \tag{19.25}$$

*The algebra $\mathcal{T}_A^{\mathrm{su}}(X)$, generated over $F$ by all the elements $t_0(a)$, $a \in \mathcal{F}_A^{\mathrm{su}}(X)_0$, is the pure trace algebra of $A$. The algebra $\mathcal{T}_A^{\mathrm{su}}(X)\mathcal{F}_A^{\mathrm{su}}(X) \subset A \otimes L$ is the relatively free algebra of $A$ as trace superalgebra.*

From now on we assume $X$ is fixed and drop the symbol $X$. We have a variation of the results of §2.4.2 and in particular of Theorem 2.4.8 and Remark 2.4.9, Proposition 2.4.10 and Theorem 8.2.5 that

THEOREM 19.6.4. *If $X$ is finite, $\mathcal{T}_A^{\mathrm{su}} = \mathcal{T}_A^{\mathrm{su}}(X)$ is a finitely generated $F$-algebra and $\mathcal{T}_A^{\mathrm{su}}\mathcal{F}_A^{\mathrm{su}}$ is a finitely generated module over $\mathcal{T}_A^{\mathrm{su}}$.*

PROOF. We first claim that every homogeneous element $a \in \mathcal{F}_A^{\mathrm{su}}$ satisfies a monic polynomial with coefficients in $\mathcal{T}_A^{\mathrm{su}}$. In fact if $a$ is odd, we have $a^2$ is even, and it is enough to prove this for even elements. So take $a \in \mathcal{F}_A^{\mathrm{su}}(X)_0$.

Since $\bar{A}_0$ is a free $F^q$ module of dimension $t$ (the first even Kemer index), the projection $\bar{a}$ of $a$ in $\bar{A}_0 \otimes_F L$ satisfies (by formula (17.15)) its characteristic polynomial $D_0(\lambda - a)$.

Now every element of the kernel of $p : A_0 \to \bar{A}_0$ is nilpotent of some degree $\leq s$, and finally we deduce (we need to add a factor of $a$ since we are not assuming the existence of a 1) that

$$D_0(\lambda - a)^s a = 0, \quad \forall a \in \mathcal{T}_A^{\mathrm{su}}\mathcal{F}_A^{\mathrm{su}}|_0.$$

We can then conclude by applying Theorem 8.2.1 and Remark 8.2.3. We have a Shirshov basis for $\mathcal{F}_A^{\mathrm{su}}$ made of homogeneous elements. Then, since we know that $t_0(t_0(a)b) = t_0(a)t_0(b)$ for $a, b$ even, we deduce that $\mathcal{T}_A^{\mathrm{su}}$ is generated by the traces $t_0(M)$ where $M$ is an even monomial in the Shirshov basis with exponents less than the degree of $ts + 1$. Finally $\mathcal{T}_A^{\mathrm{su}}\mathcal{F}_A^{\mathrm{su}}$ is spanned over $\mathcal{T}_A^{\mathrm{su}}$ by this finite number of monomials. □

19.6.1.1. *Kemer superpolynomials absorb traces.* This is basically identical to the earlier §17.3.1.2. There is one small point to notice.

REMARK 19.6.5. We have seen that, the Kemer superindex of a fundamental superalgebra is a $4^{tuple}$ of numbers $(t_0, t_1, a, b)$. Now the lexicographic order of such $4^{tuples}$ is a total order contrary to the partial order of superindices induced by inclusion of diagrams. Therefore if $A$ is a finite-dimensional superalgebra $\mathrm{PI}_2$ equivalent to a direct sum of fundamental superalgebras $A_i$, it is possible to select the maximum of these superindices and consider the superpolynomials relative to this maximum. For $\mu$ large they vanish on the summands which do not attain this maximum.

In general when we speak of Kemer superpolynomials for $A$ we will implicitly assume that they have this property. In other words, according to Definition 19.5.10, they have a nonzero evaluation in the lowest corner of the diagram.

Let us now assume that $A$ is a direct sum of $q$ fundamental superalgebras, all with the same Kemer superindex. We have the analogue of Proposition 17.3.8:

PROPOSITION 19.6.6. *The evaluation of a Kemer superpolynomial $g$ outside the big layers factors modulo the radical.*

So let $g(y_1, y_2, \dots, y_t, \bar{W})$ be a Kemer superpolynomial, and let $y_1, y_2, \dots, y_t$ be a small even layer. We have that $\bar{A}_0$ is a free $F^q$ module and $\bigwedge^t \bar{A}_0 = F^q v$ (the exterior product is as $F^q$ module):

$$(19.26) \qquad g(y_1, y_2, \dots, y_t, \bar{W}) = \det{}_0(\bar{y}_1, \dots, \bar{y}_t) u(\bar{W}),$$
$$\det{}_0(\bar{y}_1, \dots, \bar{y}_t) v = \bar{y}_1 \wedge \dots \wedge \bar{y}_t.$$

As a consequence of (19.26) we deduce that

$$(19.27) \qquad g(uy_1, uy_2, \dots, uy_t, W) = \det{}_0(\bar{u}) g(y_1, y_2, \dots, y_t, W), \ u \in A_0,$$

where $\det{}_0(\bar{u})$ means the determinant of left multiplication of $\bar{u}$ on $\bar{A}_0 := A_0/J_0$ as free $F^q$ module. Notice that this identity is independent on the chosen layer.

With $A$ as before we thus have

COROLLARY 19.6.7. *The $T_2$-ideal $I_S$ of $\mathcal{F}_A^{\mathrm{su}}$ generated by all evaluations of any set $S$ of $\mu$-Kemer superpolynomials for $\mu$ large is a $\mathcal{T}_A^{\mathrm{su}}$ submodule and thus a common ideal in $\mathcal{F}_A^{\mathrm{su}}$ and $\mathcal{T}_A^{\mathrm{su}} \mathcal{F}_A^{\mathrm{su}}$.*

COROLLARY 19.6.8. *The $T_2$-ideal generated by $\mu$-Kemer superpolynomials $\mu > \mu_0$ is a bimodule over the relatively free trace algebra $\mathcal{F}_{\bar{A}}^{\mathrm{su}}(X) \mathcal{T}_{\bar{A}}^{\mathrm{su}}(X)$ of $\bar{A}$ by right and left multiplication.*

$$\mathcal{F}_{\bar{A}}^{\mathrm{su}}(X) \mathcal{T}_{\bar{A}}^{\mathrm{su}}(X) \subset \bigoplus_i \mathcal{F}_{\bar{A}^{(i)}}^{\mathrm{su}}(X) \mathcal{T}_{\bar{A}^{(i)}}^{\mathrm{su}}(X).$$

PROOF. When we multiply a Kemer polynomial by any element $z$ of the free algebra as function this depends only on the value of $z$ modulo the radical. This together with Theorem 19.6.4 give the result. □

The main consequence of these last statements is for $A$ an arbitrary finite-dimensional superalgebra and $\mathcal{F}_A^{\mathrm{su}}$ its relatively free superalgebra.

THEOREM 19.6.9. *The algebra $\mathcal{F}_A^{\mathrm{su}}/I_S$, where $I_S$ is an ideal generated by all evaluations of any set $S$ of $\mu$-Kemer superpolynomials for $\mu$ large, is representable, hence* PI *equivalent to a finite-dimensional algebra.*

PROOF. Here by Kemer superpolynomials we mean those of Remark 19.6.5. We separate $B \oplus C$ with $B$ a sum of fundamental superalgebras with maximal Kemer superindex, and then the proof is the same as in Theorem 17.3.14.

By Corollary 11.5.18 it follows that $\mathcal{F}_A^{\mathrm{su}}/I_S$ is $\mathrm{PI}_2$ equivalent to some finite-dimensional $F$-superalgebra. □

Finally, for $\Gamma$ the $T_2$-ideal of a finitely generated PI superalgebra we have

PROPOSITION 19.6.10.

(i) *There exists a finite-dimensional superalgebra $A$ such that $\Gamma \supseteq \mathrm{Id}_2(A)$ and $\mathrm{ind}_2(A) = \mathrm{ind}_2(\Gamma)$.*
(ii) *Let $I_\mu$ denote the $T_2$-ideal in $\mathcal{F}_A^{\mathrm{su}}$ generated by the $\mu$ Kemer superpolynomials of $A$.*

*We can choose $A$ and $\mu$ big enough so that $(\Gamma/\mathrm{Id}_2(A)) \cap I_\mu = 0$, thus we may say that $\Gamma$ and $\mathrm{Id}_2(A)$ have the same Kemer $\mu$-superpolynomials.*

PROOF. The proof is really identical to that of Proposition 17.3.15 using the previous results. One starts by decomposing $A = B \oplus C$ with $A$ fundamental superalgebras all with the same Kemer superindex, and $B$ a sum of fundamental superalgebras with strictly inferior Kemer superindex. □

19.6.1.2. *A Cayley–Hamilton theorem.* We may repeat the Theorem in §17.3.1.3 with the same proof. Let $A$ be a finite-dimensional superalgebra of first even Kemer index $t$. Take a $\mu$-Kemer superpolynomial $g(X, w) \in \mathcal{F}_A^{\mathrm{su}}(X)$, $\mu > \mu_0$, with $X$ the variables containing the layers and depending also linearly on some extra even variable $w$.

THEOREM 19.6.11. *For every element $a \in \mathcal{F}_A^{\mathrm{su}}(X)_0$, we have*

$$g(X, a^{t+j}) + \sum_{i=1}^{t} (-1)^i \sigma_i(a) g(X, a^{t+j-i}) = 0, \ \forall j \geq 0. \tag{19.28}$$

**19.6.2. The Phoenix property.** This is the same as in §17.3.2. Let $\Gamma = \mathrm{Id}_2(A) \subset F_+\langle X\rangle$ be the $T_2$-ideal of identities of a finitely generated superalgebra $A$.

THEOREM 19.6.12. *The set $S$ of $\mu$-Kemer superpolynomials for $\Gamma$ for all $\mu$ sufficiently large satisfies the Phoenix property relative to $\Gamma$.*

## 19.7. The representability theorem, Theorem 19.7.4

**19.7.1. A crucial lemma.** This is also very similar to the analogue §17.4.1. Let $\Gamma$ be a $T_2$-ideal in countably many variables $\mathbb{X}$ with even Kemer index $(m, s)$, and let $\mu_0$ be as in Remark 17.2.4.

Take a $\mu+1$-Kemer superpolynomial $f$, for $\Gamma$, for $\mu \geq \mu_0$, and single out one of the even small layers $(y_1, \ldots, y_m)$; so $f = f(y; W)$. Let $A$ be a finite-dimensional superalgebra, which exists by Proposition 19.6.10 with $\Gamma \supseteq \mathrm{Id}_2(A), \mathrm{ind}_2(A) = \mathrm{ind}_2(\Gamma)$, and set $\bar{\Gamma} := \Gamma/\mathrm{Id}_2(A) \subset \mathcal{F}_A^{\mathrm{su}}$.

Hence $\Gamma$ and $\mathrm{Id}_2(A)$ have the same Kemer $\mu$-polynomials in the sense of Proposition 19.6.10(ii). That is the $T_2$-ideal generated by the $\mu$-Kemer superpolynomials of $\Gamma$ is identified modulo $\Gamma$ to the $T_2$-ideal generated by the $\mu$-Kemer superpolynomials of $A$ in the relatively free superalgebra of $A$. So we treat it as functions on $A$.

We then restrict the variables to a finite list

$$X = U \cup W; \ W := Y \cup \{y_1, \ldots, y_m\}, \ Y \text{ even}, \tag{19.29}$$

similarly to formula (17.23).

We decompose $Y$ into $\mu_0$ small layers $Y_1, \ldots, Y_{\mu_0}$ with $m$ elements and $s$ big layers $\bar{Y}_1, \ldots, \bar{Y}_s$ with $m+1$ elements.

The condition on $|U|$ ensures that $\Gamma$ that every superpolynomial $f$ in any number of variables which is not in $\Gamma$ has a nonzero evaluation in $F\langle U\rangle/(F\langle U\rangle \cap \Gamma)$.

By abuse of notation we still denote by $\Gamma$ the $T_2$-ideal $F\langle X\rangle \cap \Gamma$ in $F\langle X\rangle$.

DEFINITION 19.7.1. We take for $\mathbb{M} \subset F\langle X\rangle/\Gamma$ the space formed by the superpolynomials which are alternating and multilinear in the $\mu_0$ small layers and the $s$ big layers in which we have decomposed the variables in $Y$ and the small layer formed by $\{y_1, \ldots, y_m\}$, as in formula (17.23). This is similar in a sufficiently large number of odd layers. They may depend on the other variables $U$ in any form.

Formula (19.27) shows that $\mathbb{M}$ is also a module under $\mathcal{T}^{\text{su}}_{\bar{A}}(U)$ (Definition 19.6.1). In fact if in formula (19.27), applied to an element of $\mathbb{M}$, we take for $u$ an element of $F\langle U\rangle$, we still have an element of $\mathbb{M}$.

In particular a $\mu$-Kemer superpolynomial $f(\mathbb{X})$ with $\mu > \mu_0 + 1$ has a nonzero evaluation in $F\langle U\rangle/(F\langle U\rangle \cap \Gamma)$. Clearly we may factor this through an evaluation in which one small layer, say $\{\mathrm{y}_1, \ldots, \mathrm{y}_m\}$ is substituted with $\{y_1, \ldots, y_m\}$, then another $\mu_0$ small layers and the $s$ big layers, in a 1–1 way, in the corresponding layers of the variables $Y$ and the remaining variables in $F\langle U\rangle$. This evaluation is nonzero in $\mathbb{M}$; we call this a *special evaluation* denoted $\bar{f}(X) \in F\langle X\rangle/\Gamma$.

Observe that such an evaluation is also a Kemer superpolynomial although in the variables $X$. We apply the Cayley–Hamilton theorem, Theorem 17.3.16, to Kemer superpolynomials in $\mathbb{M}$.

**A crucial lemma.** We now arrive at the crucial lemma. The hypotheses on $\Gamma$ and the notations $\mathbb{M}$ are those of the previous section.

Denote for simplicity $R := F\langle X\rangle/\Gamma$, $S := F_+\langle U\rangle/\Gamma \cap F_+\langle U\rangle \subset R$ (cf. Remark 2.1.15).

Now consider the trace algebra $\mathcal{T}^{\text{su}}_{\bar{A}}(U)$ (quotient of generic $m \times m$ matrices) in the variables $U$, and let $\mathfrak{R} = \mathcal{T}^{\text{su}}_{\bar{A}}(U) \otimes_F R$. We identify $R$ with $1 \otimes R$.

The ideal $K_\mu \subset R$ generated by $\mu$-Kemer superpolynomials is already a $\mathcal{T}^{\text{su}}_{\bar{A}}(U)$ module, thus we have a map as in formula (17.24).

For each $a \in S_0$, let $H_a = a^{m+1} - \sigma_1(a) \otimes a^m + \cdots \pm \sigma_m(a) \otimes a$, where $t^m + \sum_{i=1}^m (-1)^i \sigma_i(a) t^{m-i} \in \mathcal{T}^{\text{su}}_{\bar{A}}(U)[t]$ the characteristic polynomial of the left multiplication by $\bar{a}$ on $\bar{A}_0$ (as $F^q$ algebra).

Consider finally in the algebra $\mathfrak{R}$ the graded ideal $I$ generated by the elements $H_a$ for all $a \in S_0$.

LEMMA 19.7.2. *A nonzero element $f$ of $\mathbb{M}$ remains nonzero in $\mathfrak{R}/I$.*

PROOF. As in Lemma 17.4.2. □

### 19.7.2. An auxiliary algebra.

As in §17.4.2 we need

PROPOSITION 19.7.3. *Let $\Gamma = \mathrm{Id}_2(R)$ be the $T_2$-ideal of* $\mathrm{PI}_2$*'s of a* PI *superalgebra $R$ finitely generated over $F$ with first coordinate of the Kemer superindex $m$. Then, there exists a superalgebra $A$ which is finite dimensional over $F$ with the following properties:*

(1) $\Gamma \subset \mathrm{Id}_2(A)$.
(2) *All $\mu$-Kemer superpolynomials for $\Gamma$, with $\mu$ large, are not identities of $A$. In particular $\Gamma$ and $A$ have the same Kemer superindex.*

PROOF. First we restrict to $X = U \cup W$, finitely many variables, as defined in formula (19.29), and so we may assume $R = F\langle X\rangle/(\Gamma \cap F\langle X\rangle)$.

We can thus apply Lemma 19.7.2. Take the Shirshov basis for the algebra $R = F\langle X\rangle/\Gamma$ (Definition 8.1.12) of homogeneous elements. We divide this basis into three parts: the elements $a_1, \ldots, a_k$ are even and give a Shirshov basis for the even part of the subalgebra generated by the elements $U$, the $c_1, \ldots, c_h$ are odd and complete with the $a_i$ a Shirshov basis for the subalgebra generated by the elements $U$. The other $b_i$ containing at least one of the remaining variables $W$. By definition these elements are monomials, hence multihomogeneous in the variables $W$.

By definition every $\mu$-Kemer superpolynomial with $\mu > \mu_0 + 2$ has a nonzero evaluation in $\mathbb{M} \subset F\langle X\rangle/\Gamma$ defined in Definition 17.4.1. We construct a new superalgebra $\tilde{R}$ as follows. Let $R[t_{j,i}]$, $j = 1, \dots, m$, $i = 1, \dots, k+h$, be the algebra of polynomials over $R$ in $(k+h)m$ variables. For each $i = 1, \dots, k$; $j = 1, \dots, h$, set

$$\tilde{H}_{a_i} := a_i{}^{m+1} + t_{1,i}a_i{}^m + \cdots + t_{m,i}a_i, \; \tilde{H}_{c_i^2} := c_i{}^{2(m+1)} + t_{1,k+i}c_i{}^{2m} + \cdots + t_{m,k+i}c_i{}^2,$$

and define

$$\tilde{R} := R[t_{j,i}]/J, \; J = \langle \tilde{H}_{a_i}, \; \tilde{H}_{c_i^2}\rangle.$$

So $J$ is the ideal generated by the elements $\tilde{H}_{a_i}, \tilde{H}_{c_i^2}$. Modulo $J$ the elements $a_i, c_j$ of the Shirshov basis become integral over $F[t_{j,i}]$.

Since the elements $t_{j,i}$ are indeterminates over the field $F$, we have a homomorphism $\rho : R[t_{j,i}] \to \mathfrak{R} = \mathcal{T}_{\bar{A}}^{\mathrm{su}}(U) \otimes_F R$ mapping $t_{j,i} \mapsto (-1)^j\sigma_j(a_i), i \le k$, and $t_{j,k+i} \mapsto (-1)^j\sigma_j(c_i^2), i \le h$. Under this homomorphism $\tilde{H}_{a_i}$ is mapped to $H_{a_i} = a_i^{m+1} - \sigma_1(a_i) \otimes a_i^m + \cdots \pm \sigma_m(a_i) \otimes a_i$, similarly for $\tilde{H}_{c_i^2}$. By the definition of the ideal $I \subset \mathfrak{R}$, we have $\lambda(J) \subset I$ so $\lambda$ factors to a map of $\tilde{R}$ to $\mathfrak{R}/I$.

By Lemma 19.7.2, we deduce that the space $\mathbb{M}$ projects isomorphically to its image in $\tilde{R}$, so every $\mu$-Kemer superpolynomial has a nonzero evaluation in the image $\mathbb{M} \subset \tilde{R}$.

Now consider the graded ideal $G$ of $\tilde{R}$, sum of all the elements which are homogeneous of degree at least 2 in at least one of the variables $W$, and let $B := \tilde{R}/G$.

The elements $b_j$ of the Shirshov basis satisfy $b_j^2 \in G$, hence $b_j^2 = 0$ in $B$. It follows that $B$ is a finite module over $F[t_{j,i}]$ since now all the elements of the Shirshov basis have become integral, moreover no special evaluation of a Kemer superpolynomial lies in $G$, by homogeneity. Therefore the superalgebra $B$ is representable so $\mathrm{PI}_2$-equivalent to a finite-dimensional superalgebra $A$, which satisfies all the requirements of Proposition 19.7.3. □

### 19.7.3. Proof of Theorem 19.7.4.

THEOREM 19.7.4. *Every* PI *superalgebra $R$, finitely generated over a field $F$, is $PI_2$ equivalent to a superalgebra finite dimensional over $F$.*

We prove this theorem as we did for Theorem 17.1.1, by induction on the Kemer superindex. Take a $T_2$-ideal $\Gamma \subset F_+\langle X\rangle$ of identities of a finitely generated PI superalgebra, $X$ a countable set of variables. We know then that $\Gamma$ contains a Capelli list 7.3.3 and it is generated as $T_2$-ideal by its intersection with a finitely generated free superalgebra $F_+\langle\Lambda\rangle$ ($\Lambda \subset X$ finite), in particular it is the $T_2$-ideal of identities of the finitely generated algebra $F_+\langle\Lambda\rangle/(\Gamma \cap F_+\langle\Lambda\rangle)$ (Theorem 7.3.9).

Consider its even Kemer index. If this index is $(0, s)$, then $F_+\langle\Lambda\rangle/\Gamma \cap F_+\langle\Lambda\rangle$ is finite dimensional, and there is nothing to prove.

Assume the even index of $\Gamma$ is $(m, s)$ with $m > 0$. By Proposition 19.7.3 we have a finite-dimensional superalgebra $A$ and some $\mu > 0$ which satisfies $\Gamma \subset \mathrm{Id}_2(A)$. $A$ and $\Gamma$ have the same Kemer superindex and all Kemer superpolynomials, for $\Gamma$, with at least $\mu$ layers (even and odd) with $m$ elements that are not identities of $A$.

With this $\mu$ consider the $T_2$-ideal $\Gamma' := \Gamma + K$ where $K$ is the $T_2$-ideal of $F_+\langle X\rangle$ generated by all evaluations of the set $\mathcal{S}$ of Kemer superpolynomials for $\Gamma$ with at least $\mu$ even layers with $m$ elements.

Since $\Gamma' \supset \Gamma$, it still contains a Capelli list (Theorem 7.3.9) so $\Gamma'$ is the ideal of identities of a finitely generated superalgebra. We clearly have that the Kemer superindex of $\Gamma'$ is strictly less than that of $\Gamma$ since all evaluations of the set $\mathcal{S}$ of Kemer superpolynomials for $\Gamma$ with at least $\mu$ even layers with $m$ elements are in $\Gamma'$. By induction on the Kemer superindex, $\Gamma'$ is the $T_2$-ideal of identities of a finite-dimensional superalgebra $B$.

We claim that $\Gamma$ is the ideal of identities of $A \oplus B$. In other words we need to prove that

$$\Gamma = \mathrm{Id}_2(A) \cap \mathrm{Id}_2(B) = \mathrm{Id}_2(A) \cap \Gamma'.$$

We have by construction that $\Gamma \subset \mathrm{Id}_2(A) \cap \Gamma'$. So let $f \in \mathrm{Id}_2(A) \cap \Gamma'$, and suppose that $f \notin \Gamma$. Since $f \in \Gamma'$ we have that there is a $g$ with $f - g \in \Gamma$ and $g \in K$ is in the $T_2$-ideal generated by the $\mu$-Kemer superpolynomials $\mathcal{S}$ of $\Gamma$. By the Phoenix property of §17.3.2 we have that there is a Kemer superpolynomial $g'$ for $\Gamma$ in the $T_2$-ideal generated by $g$. Since $f \in \mathrm{Id}_2(A) \cap \Gamma'$ we also have $g' \in \mathrm{Id}_2(A) \cap \Gamma'$. But then $g'$ is an identity of $A$, a contradiction, since $A$ is constructed in such a way that the $\mu$-Kemer superpolynomials of $\Gamma$ are not $\mathrm{PI}_2$ of $A$. This proves the theorem. □

## 19.8. Grassmann envelope and finite-dimensional superalgebras

THEOREM 19.8.1. *Every nontrivial variety of algebras is generated by the Grassmann envelope of a finite-dimensional superalgebra.*

PROOF. Let $\mathcal{V}$ be a nontrivial variety of associative algebras. By Theorem 19.4.18,

$$\mathcal{V} = \mathrm{var}(A_0 \otimes G_0 \oplus A_1 \otimes G_1),$$

where $A = A_0 \oplus A_1$ is a finitely generated PI-superalgebra and $G = G_0 \oplus G_1$ is the Grassmann algebra. By Theorem 19.7.4 the variety of superalgebras generated by $A$ is also generated by a finite-dimensional superalgebra $B = B_0 \oplus B_1$, hence $\mathrm{Id}_2(A) = \mathrm{Id}_2(B)$, and the statement follows from Corollary 19.4.13. □

Recall that a free graded commutative superalgebra is a *polynomial ring*

$$F[\xi, \eta] = F[\xi] \otimes \bigwedge(\oplus_j F\eta_j)$$

in some commutative variables $\xi_i$ of degree 0 and some Grassmann variables (Definition 19.1.1) $\eta_j$ of degree 1.

This algebra satisfies the identity $[x, y, z] = 0$.

COROLLARY 19.8.2. *Any relatively free algebra in an arbitrary number of variables of a nontrivial variety $\mathcal{V}$ is embeddable in a matrix algebra of finite order over a free graded commutative superalgebra.*

PROOF. Let $B$ be a relatively free algebra of a nontrivial variety $\mathcal{V}$. By Theorem 19.8.1 $\mathcal{V} = \mathrm{var}(A_0 \otimes G_0 \oplus A_1 \otimes G_1)$ for some finite-dimensional superalgebra $A = A_0 \oplus A_1$ and $G = \bigwedge(\bigoplus_j F\eta_j)$. Then, by Proposition 19.4.23, the elements $\xi_i$, $i = 1, \ldots, m$ of formula (19.16) which are contained in $A \otimes_F F[\xi_{i,j}] \otimes G$ generate the free algebra $F\langle x_1, \ldots, x_m\rangle$ modulo the PIs of $\mathcal{V}$, that is they generate an algebra isomorphic to $B$. Clearly since $A$ is finite dimensional, it embeds into a matrix algebra, hence the claim. □

In particular such a relatively free algebra is embeddable into the algebra of $m \times m$ matrices, for some $m$, over the Grassmann algebra over some field $K \supset F$. We may say that it is *superrepresentable*.

CHAPTER 20

# The Specht problem

*In this chapter, algebras and superalgebras are over a field $F$ of characteristic 0. Sometimes for simplicity we assume $F$ is algebraically closed. From the representability theorem for superalgebras, Theorem 19.7.4, we prove that every nontrivial variety of algebras is generated by the Grassmann envelope of a finite-dimensional superalgebra. Moreover, if the variety does not contain the Grassmann algebra $G$, then it is generated by a finite-dimensional algebra.*

*These results will allow us to prove the following embedding theorem for algebras and a solution of Specht problem, both results due to Kemer:*

(1) *Any relatively free* PI *algebra can be embedded in a matrix algebra over the relatively free algebra satisfying the identity $[x_1, x_2, x_3] = 0$.*
(2) *We will give a positive answer to Specht's problem by proving that the ideal of identities of everyalgebra over a field of characteristic 0 is finitely generated as a T-ideal (proposed by Specht in* **[Spe50]**).

## 20.1. Standard and Capelli

LEMMA 20.1.1. *Let $A = A_0 \oplus A_1$ be a finite-dimensional simple superalgebra with $A_1 \neq 0$. Then $G(A)$ does not satisfy any standard identity.*

PROOF. By the classification theorem (Theorem 19.2.15), we have two cases.

TYPE 1. $A = M_k(F[c])$ so $G(A) = M_k(G)$, for some $k \geq 1$, then since $G \subseteq M_k(G)$, $G(A)$ does not satisfy any standard identity by Remark 19.1.2.

TYPE 2. $A = M_{k,l}(F)$ and $G(A) = M_{k,l}(G)$, for some $k \geq l \geq 1$. We claim that also in this case $M_{k,l}(G)$ does not satisfy any standard identity.

Clearly it is enough to prove that $M_{1,1}(G)$ has this property. To this end, recalling that $G$ is generated by the $e_i$'s, we compute

$$e_{11}\,\mathrm{St}_{2r}(e_1e_{12}, e_2e_{21}, \dots, e_{2r-1}e_{12}, e_{2r}e_{21}) = 2(r!)(e_1 \cdots e_{2r})e_{11} \neq 0.$$

This is similar for standard identities of odd degree. □

With the notations of Theorem 19.2.18, let $A = A_0 \oplus A_1$ be a finite-dimensional superalgebra, and let $A = \bar{A} \oplus J$, $A_0 = \bar{A}_0 \oplus J_0$, $A_1 = \bar{A}_1 \oplus J_1$ be a Wedderburn decomposition. Then we have

LEMMA 20.1.2. *The Grassmann algebra $G$ is contained in the variety generated by the Grassmann envelope $G(A)$ of $A$ if and only if $A_1 \neq J_1$ is not nilpotent.*

PROOF. The semisimple superalgebra $\bar{A} = \bigoplus_{i=1}^{q} B(i)$ decomposes as a direct sum of simple superalgebras $B(i)$. By the classification, if $\bar{A}_1 \neq 0$, then for a summand $B = B(i_0)$, we must have $B_1 \neq 0$ so $B$ is of either Type 1 or 2 (see

Theorem 19.2.15). In both cases $B_1$ generates $B$, so one has that $\bar{A}_1$ is not nilpotent. It is thus clear that $A_1$ is nilpotent if and only if $A_1 = J_1$.

Assume first that $A_1^k = 0$, then also $(A_1 \otimes G_1)^k = 0$. We claim that the standard polynomial $\mathrm{St}_m$ of degree $m \geq \dim A_0 + (\dim A_0 + 1)k$ vanishes on $G(A)$. In fact, by multilinearity, it is enough to prove that $\mathrm{St}_m$ vanishes when computed on elements in either $A_0 \otimes G_0$ or $A_1 \otimes G_1$. Clearly $\mathrm{St}_m$ vanishes if at least $\dim A_0 + 1$ variables are evaluated in $A_0 \otimes G_0$. So let us assume that at most $\dim A_0$ variables are evaluated in $A_0 \otimes G_0$. Hence we may assume that at least $(\dim A_0 + 1)k$ variables are evaluated in $A_1 \otimes G_1$. Since each monomial has $m$ terms, we must have in it at least one sequence of $k$ variables all computed in $A_1 \otimes G_1$, so each monomial of $\mathrm{St}_m$ vanishes. Therefore, by Theorem 19.1.10, $G$ is not contained in the variety generated by $G(A)$.

If $A_1$ is not nilpotent, then the semisimple part $\bar{A}$ has $\bar{A}_1 \neq 0$, hence there is a simple superalgebra $B \subset \bar{A} \subset A$, with $B_1 \neq 0$. By Lemma 20.1.1 we have then that $G(B)$ does not satisfy any standard identity. Hence by Theorem 19.1.10 the Grassmann algebra $G$ is contained in the variety generated by $G(A)$. □

THEOREM 20.1.3. *Let $I \subset F_+\langle X\rangle$ be a $T$-ideal, and let $X$ be a countable list of variables. Let $\sqrt{I}$ be its nil radical. Then the following statements are equivalent:*

(1) *$I$ is the ideal of identities of a finite-dimensional algebra $A$.*
(2) *$I$ contains a Capelli identity (or a Capelli list, Exercise 7.3.4).*
(3) *$I$ contains a standard identity.*
(4) *There is some $m \in \mathbb{N}$ with $(\sqrt{I})^m \subset I$.*

PROOF. Clearly (1) implies (2) which implies (3). Also (1) implies (4) since, if $J$ is the radical of $A$, we have $J^m = 0$ for some $m$, then we have that $\sqrt{I} = \mathrm{Id}(A/J)$ and $(\sqrt{I})^m \subset I$; see Corollary 11.3.8.

Assume (3). By Theorem 19.8.1 every nontrivial variety of algebras is generated by the Grassmann envelope $R = A_0 \otimes G_0 \oplus A_1 \otimes G_1$ of a finite-dimensional superalgebra which, by Remark 19.4.22 is PI equivalent to the superenvelope $S(A) = S^{(0)} \otimes A_0 \oplus S^{(1)} \otimes A_1 = G(A) \otimes F[\Xi]$.

Take then $A = A_0 \oplus A_1$ such that the $T$-ideal of $G(A)$ is $I$. By Lemma 20.1.2 we must have that $A_1 = J_1$ is nilpotent.

Consider the relatively free algebra $\mathcal{F} \simeq F_+\langle X\rangle/I$ of $G(A)$ generated by elements $\xi_i + \eta_i$ which is contained in the superenvelope $S(A)$ by Proposition 19.4.23. The relatively free algebra $\mathcal{F}$ maps to the relatively free algebra of $A_0$ generated by the elements $\xi_i$, and let $\bar{K} = K/I$ be the kernel of this map. We have that $\bar{K}$ is contained in the ideal $S(J)$, the superenvelope of the Jacobson radical $J$ of $A$, which is nilpotent. Thus, for some $h$ we have $K^h \subset I$.

Since $I \subset K \subset \sqrt{I}$ we have $\sqrt{I} = \sqrt{K}$. On the other hand $K$ is the ideal of identities of the finite-dimensional algebra $A_0$ so that, by the first part, $(\sqrt{I})^m \subset K$, for some $m$. Thus $(\sqrt{I})^{h \cdot m} \subset I$ and (4) follows.

(2) implies (1) is the basic Theorem 17.1.1. Hence we complete the proof by proving (4) implies (2). In fact either $\sqrt{I} = F_+\langle X\rangle$ or $\sqrt{I}$ is the ideal of PIs of $n \times n$ matrices, for some $n$. In the first case $I$ is the $T$-ideal of the Grassmann envelope of a nilpotent superalgebra, which is also nilpotent as algebra. Thus $I$ contains all polynomials of degree $\geq m$, for some $m$. In the last case we have that $(\sqrt{I})^m$ is

the ideal of PIs of block-triangular matrices, and by Theorem 16.3.2 it contains a Capelli list. □

COROLLARY 20.1.4. *If a variety $\mathcal{V}$ does not contain the Grassmann algebra G, then it is generated by a finite-dimensional algebra, and every relatively free algebra of the variety $\mathcal{V}$ is representable.*

PROOF. If $\mathcal{V}$ does not contain the Grassmann algebra, by Theorem 19.1.10, $\mathcal{V}$ satisfies a standard identity, hence it is generated by a finite-dimensional algebra. Thus by the equivalence of Theorems 17.1.1 and 17.1.2, each relatively free algebra of the variety $\mathcal{V}$ is representable. □

## 20.2. Solution of the Specht's problem

We start with a lemma.

LEMMA 20.2.1. *Let $A = A_0 \oplus A_1$ be a finite-dimensional superalgebra with Jacobson radical $J = J(A)$. Then*

$$A_0 = A_0^2 + A_1^2 + (A_0 \cap J) \tag{20.1}$$

*and*

$$A_1 = A_0A_1 + A_1A_0 + (A_1 \cap J). \tag{20.2}$$

PROOF. As $A$ is a finite-dimensional superalgebra, by Theorems 19.2.18 and 19.2.7, we have $A = \bar{A} + J$, $A_0 = \bar{A}_0 \oplus J_0$, and $A_1 = \bar{A}_1 \oplus J_1$, where $\bar{A} = \bar{A}_0 \oplus \bar{A}_1$ is a semisimple superalgebra ($J_0 = A_0 \cap J$, $J_1 = A_1 \cap J$). Since $\bar{A}$ is an algebra with unit $1 \in \bar{A}_0$, $\bar{A}^2 = \bar{A}$. The required equalities easily follow. □

Next we prove the main theorem. We are working in the category of algebras in which we do not assume the existence of 1 and thus in $F_+\langle X\rangle$, the free algebra without 1 and corresponding $T$-ideals.

THEOREM 20.2.2 (Kemer). *Every associative* PI *algebra over a field of characteristic 0 has a finite basis of identities.*

PROOF. Suppose by contradiction that there is an algebra $E$ which has an infinite basis of identities $f_1 = 0$, $f_2 = 0, \dots$. That is we assume that the chain of $T$-ideals generated by these polynomials

$$\langle f_1\rangle_T \subsetneq \langle f_1\rangle_T + \langle f_2\rangle_T \subsetneq \cdots \tag{20.3}$$

is strictly increasing. As $F$ has characteristic 0, we may assume that each $f_i$ is multilinear. We may also assume, passing to a subsequence, that

$$\deg f_1 < \deg f_2 < \cdots. \tag{20.4}$$

Denote by $\langle f_i\rangle_T^+$ the $T$-ideal generated by all multilinear polynomials $f \in \langle f_i\rangle_T$ of degree strictly greater than $n_i := \deg f_i$. In particular

$$f_i(x_1, \dots, x_kx, x_{k+1}, \dots, x_{n_i}) \in \langle f_i\rangle_T^+, \tag{20.5}$$

for every $x \in X$ and every $k = 1, \dots, n_i$.

Consider the $T$-ideal $\Gamma = \langle f_1\rangle_T^+ + \langle f_2\rangle_T^+ + \cdots$. We claim that for every $i$ we have $f_i \notin \Gamma$.

Indeed, if $f_i \in \Gamma$ for some $i$, then formula (20.4) implies that $f_i \in \langle f_1\rangle_T + \cdots + \langle f_{i-1}\rangle_T$. It would follow that the chain (20.3) is not strictly increasing.

By Theorem 19.8.1 we have $\Gamma = \mathrm{Id}(B)$, where $B = A_0 \otimes G_0 \oplus A_1 \otimes G_1$ is the Grassmann envelope of a finite-dimensional superalgebra $A = A_0 \oplus A_1$.

From formulas (20.1) and (20.2) of Lemma 20.2.1 we have, with $J$ the radical of $A$,

$$B \subset (A_0^2 + A_1^2) \otimes G_0 + (A_0A_1 + A_1A_0) \otimes G_1 + J \otimes G. \tag{20.6}$$

We have that $G_{+,0} := G_1^2$ is the part of degree $\geq 2$ of $G_0$. We also have

$$B^2 = A_0^2 \otimes G_0 + A_1^2 \otimes G_1^2 + (A_0A_1 + A_1A_0) \otimes G_1$$

$$\implies (A_0^2 + A_1^2) \otimes G_{+,0} + (A_0A_1 + A_1A_0) \otimes G_1 + J \otimes G \subset B^2 + J \otimes G. \tag{20.7}$$

formulas (20.6) and (20.7) imply

$$BG_1^2 = BG_{+,0} \subseteq B^2 + J \otimes G. \tag{20.8}$$

As $\langle f_i \rangle_T^+ \subseteq \Gamma = \mathrm{Id}(B)$, we deduce from (20.5) that $B$ satisfies every identity of the form

$$f_i(x_1, \ldots, x_k x, x_{k+1}, \ldots, x_n) = 0,$$

i.e., we have the equality

$$f_i(B, \ldots, B, B^2, B, \ldots, B) = \{0\}.$$

From this and (20.8) it follows that

$$f_i(BG_{+,0}, \ldots, BG_{+,0}) \subseteq f_i(J \otimes G, \ldots, J \otimes G) \subseteq J^{\deg f_i} \otimes G.$$

But if the degree of $f_i$ is strictly greater than the index of nilpotency of the ideal $J$, then $f_i \in \mathrm{Id}(BG_{+,0})$.

Now notice that the algebra $BG_{+,0} \cong B \otimes G_{+,0}$ has the same ideal of identities of $B$. In fact, for every $n$ we can find $n$ elements $a_i \in G_{+,0}$ with $\prod_{i=1}^n a_i \neq 0$ so that for every homogeneous multilinear polynomial $f(x_1, \ldots, x_n)$ we have $f(b_1a_1, \ldots, b_na_n) = f(b_1, \ldots, b_n)a_1a_2 \cdots a_n$ equals 0, for $b_i \in B$, if and only if $f$ is a PI in $BG_{+,0}$.

Finally we obtained that $f_i \in \mathrm{Id}(B)$, which contradicts the fact that $f_i \notin \Gamma$. The theorem is proved. □

COROLLARY 20.2.3. *T-ideals satisfy the ascending chain condition.*

COROLLARY 20.2.4. *Every finitely generated* PI *superalgebra has a finite basis of graded identities.*

PROOF. Let $A$ be a finitely generated PI superalgebra. By Theorem 19.7.4 $\mathrm{Id}_2(A) = \mathrm{Id}_2(B)$ for some finite-dimensional superalgebra $B = B_0 \oplus B_1$. Hence $B$, and so $A$, satisfies the Capelli polynomials $C_{k+1}(y; x)$ and $C_{l+1}(z; x)$, the first alternating in $k+1$, $k = \dim B_0$ even variables $y$, and the second alternating in $l+1$, $l = \dim B_1$ odd variables $z$.

Suppose that $A$ has an infinite basis of identities $f_1, f_2, \ldots$, and assume, as we may, that the polynomials $f_i$ are multilinear and $\deg f_1 < \deg f_2 < \cdots$. As in Theorem 20.2.2, we consider the $T_2$-ideal

$$\Gamma = \{f_1\}_{T_2}^+ + \{f_2\}_{T_2}^+ + \cdots,$$

where $\langle f_i \rangle_T^+$ is the $T_2$-ideal generated by all multilinear polynomials $f \in \langle f_i \rangle_{T_2}$ of degree greater than $\deg f_i$.

We have that $f_i \notin \Gamma$ for every $i$. Also, $C_{k+2}(y;x)$, $C_{l+2}(z;x) \in \Gamma$ and this implies, arguing as in Theorem 7.3.9, that $\Gamma$ is the $T_2$-ideal of identities of a superalgebra $C$ generated by $k+1$ elements of homogeneous degree 0 and $l+1$ elements of degree 1. Then by Theorem 19.7.4 $\mathrm{Id}_2(C) = \mathrm{Id}_2(D)$ for some finite-dimensional superalgebra $D = D_0 \oplus D_1$. As $\{f_i\}_{T_2}^+ \subseteq \Gamma$, we have the equalities

$$f_i(D_0, \ldots, D_0^2 + D_1^2, \ldots, D_0, D_1, \ldots, D_1) = \{0\},$$

$$f_i(D_0, \ldots, D_0, D_1, \ldots, D_0 D_1 + D_1 D_0, \ldots, D_1) = \{0\}.$$

As in Theorem 20.2.2, from this we deduce

$$f_i(D_0, \ldots, D_0, D_1, \ldots, D_1) \subseteq J^{\deg f_i}.$$

Hence if the degree of $f_i$ is greater than the index of nilpotency of $J$, then $f_i \in \mathrm{Id}_2(D) = \Gamma$, a contradiction. □

## 20.3. Verbally prime $T$-ideals

We have seen in Theorem 11.3.3 that the only $T$-ideals of the free algebra which are prime are the ideals of identities of $n \times n$ matrices. We may on the other hand relax the requirement as follows.

DEFINITION 20.3.1. A $T$-ideal $I$ is *verbally prime* if, whenever we have two $T$-ideals $U, V$ with $UV \subseteq I$, then either $U$ or $V$ is contained in $I$. We shall say that a variety $\mathcal{V}$ is verbally prime, if its $T$-ideal of identities is verbally prime.

REMARK 20.3.2. Notice that an equivalent condition for a $T$-ideal of an algebra $R$ to be verbally prime is *the product of every nonidentity polynomial (multilinear) with disjoint variables is a nonidentity.*

We need

REMARK 20.3.3. If a variety $\mathcal{V}$ is generated by an algebra $R$, then $\mathcal{V}$ is verbally prime if and only if whenever we have two nonzero verbal ideals $U(R), V(R)$ of $R$ (obtained by evaluating the $T$-ideals $U, V$ in $R$), we have $U(R)V(R) \neq 0$.

PROOF. This is essentially tautological since $U(R)V(R) = (UV)(R)$ is $\{0\}$ if and only if $UV$ is contained in the $T$-ideal $\mathrm{Id}(R)$ of identities of $R$. □

Next we are going to show that in characteristic 0 the only proper verbally prime $T$-ideals are the ideals of identities of the Grassmann envelope of a finite-dimensional simple superalgebra. Recalling the classification of the simple superalgebras given in Theorem 19.2.15, we get that their Grassmann envelope is one of the following algebras

(1) $M_k(G_0) \sim_{\mathrm{PI}} M_k(F), k \geq 1$;
(2) $G(M_k(F + cF)) = M_k(G), k \geq 1$;

(3)

(20.9)

$$G(M_{k,l}(F)) = M_{k,l}(G) = \begin{array}{c} \\ k \\ \\ \\ l \\ \\ \end{array}\left(\begin{array}{ccc|ccc} \multicolumn{3}{c|}{k} & \multicolumn{3}{c}{l} \\ G_0 & \cdots & G_0 & G_1 & \cdots & G_1 \\ \vdots & & \vdots & \vdots & & \vdots \\ G_0 & \cdots & G_0 & G_1 & \cdots & G_1 \\ \hline G_1 & \cdots & G_1 & G_0 & \cdots & G_0 \\ \vdots & & \vdots & \vdots & & \vdots \\ G_1 & \cdots & G_1 & G_0 & \cdots & G_0 \end{array}\right), \quad k \geq l \geq 1.$$

We remark that type (1) is obtained from the trivial superalgebra $M_{k,0}(F)$; type (2) is obtained by noticing that

$$G(M_k(F \oplus cF)) = M_k(F) \otimes G_0 \oplus cM_k(F) \otimes G_1 \cong M_k(G_0) \oplus M_k(G_1) = M_k(G).$$

LEMMA 20.3.4. *Let $R$ be a* PI *algebra, and let $I \subset R$ be a nilpotent ideal, $I^k = 0$. Then* $\mathrm{Id}(R/I)^k \subset \mathrm{Id}(R)$.

PROOF. This is essentially trivial since if $f(x_1, \dots, x_m) \in \mathrm{Id}(R/I)$, this polynomial evaluated in $R$ takes values in $I$, so the product of $k$ such polynomials evaluated in $R$ is 0. □

THEOREM 20.3.5. *Let $F$ be a field of characteristic* 0. *The only proper verbally prime $T$-ideals of $F\langle X\rangle$ are*

(1) $\mathrm{Id}(M_k(F)), k \geq 1$;
(2) $\mathrm{Id}(M_k(G)), k \geq 1$;
(3) $\mathrm{Id}(M_{k,l}(G)), k \geq l \geq 1$.

PROOF. Let $P$ be a proper verbally prime $T$-ideal. From Theorem 19.8.1 we know that $P = \mathrm{Id}(G(A))$, for some finite-dimensional superalgebra $A$.

Write $A = B \oplus J$ where $B = B_0 \oplus B_1$ is a semisimple superalgebra and where $J = J_0 \oplus J_1$ is the Jacobson radical of $A$. Then $G(A) = G(B) \oplus G(J)$, a vector space decomposition, $G(B)$ is a subalgebra, and $G(J) = G_0 \otimes J_0 \oplus G_1 \otimes J_1$ is a nilpotent ideal. From Lemma 20.3.4 it follows that for some exponent $h$ we have

$$\mathrm{Id}(G(B))^h \subseteq \mathrm{Id}(G(A)). \tag{20.10}$$

Since $P = \mathrm{Id}(G(A))$ is verbally prime from formula (20.10) it follows that $\mathrm{Id}(G(B)) \subseteq P$. But clearly $P$ vanishes on $G(B)$, so $P \subseteq \mathrm{Id}(G(B))$ and $P = \mathrm{Id}(G(B))$.

Now, let $B = B_1 \oplus \cdots \oplus B_s$, a direct sum of simple superalgebras.

Then $\mathrm{Id}(G(B)) = \mathrm{Id}(G(B_1)) \cap \cdots \cap \mathrm{Id}(G(B_s)) \supseteq \mathrm{Id}(G(B_1)) \cdots \mathrm{Id}(G(B_s))$ and, so, $\mathrm{Id}(G(B)) = \mathrm{Id}(G(B_i))$, for some $i$, being verbally prime.

We have proved that $P$ is the ideal of identities of one of the algebras listed in Theorem 19.2.15. Next we need to show that each of these $T$-ideals is verbally prime.

In case (1) we know that $P$ is the ideal of identities of $M_k(F)$ and it is even prime.

For the other cases we shall use Remark 20.3.3. We need to prove that if $I, J$ are two verbal ideals with $IJ = 0$, one of the two must be 0. We do this in the following.

LEMMA 20.3.6. *If $P = \mathrm{Id}(A)$ where $A = M_n(F)$ or $M_{k,l}(G)$ or $M_n(G)$, then $P$ is verbally prime.*

PROOF. Let $U, V$ be $T$-ideals not contained in $P$, and let $f \in U, g \in V$ be such that $f, g \notin P$. Then there exist $a_1, \dots, a_n, b_1, \dots, b_m \in A$ such that

$$f(a_1, \dots, a_n) = a \neq 0 \quad \text{and} \quad g(b_1, \dots, b_m) = b \neq 0.$$

In case $A = M_{k,l}(G)$ or $M_n(G)$, we may assume that the two sets of elements $\{a_1, \dots, a_n\}$ and $\{b_1, \dots, b_m\}$ depend on disjoint sets of generators $e_i$'s of the Grassmann algebra $G$.

Since in both cases $a, b$ are matrices, at least one entry in position $(i, j)$ for $a$ and $(h, k)$ for $b$ is not zero. Then we take $x_1, x_2$ and $y_1, y_2$ multiples of the matrix units and in disjoint sets of generators $e_i$'s, with the property that $x_1 a x_2 = \alpha_1 e_{1,1}, y_1 b y_2 = \alpha_2 e_{1,1}$ have a nonzero entry only in say the $(1, 1)$ position. We can also assume that these two entries $\alpha_1, \alpha_2$ are in disjoint sets of the generators $e_i$ of $G$. Then $x_1 a x_2 y_1 b y_2 = \alpha_1 \alpha_2 e_{1,1} \neq 0$, since in the Grassmann algebra the product of two nonzero elements in disjoint sets of generators is nonzero. Since $x_1 a x_2 y_1 b y_2$ is the evaluation of an element in $UV$ we have $UV \not\subset P$ as desired. □

20.3.0.1. *Invariant ideals.* It is clear that a verbal ideal of an algebra $A$ is closed under all automorphisms of the algebra. In particular one could consider the Aut-prime algebras as an interesting object.

DEFINITION 20.3.7. An algebra $A$ is Aut-prime if for for any two ideals $I, J$, invariant under all automorphisms of $A$ with $IJ = 0$, we have that one of the two must be 0.

We shall show that the algebras of Theorem 20.3.5 have this property. In fact we shall classify all the invariant ideals of these algebras. This can be useful for establishing further properties of these algebras.

CASE 1. We have matrices which are a simple algebra.

CASE 2. We know (cf. Lemma 2.6.2) that any ideal $I$ of a matrix algebra with a 1, in particular of $M_n(G)$, is of the form $M_n(E)$ where $E$ is an ideal of $G$. If $I$ is a verbal ideal it is furthermore closed under all automorphisms of $M_n(G)$ so that $E$ is closed under all automorphisms of $G$.

Now $G$ is the exterior algebra $\bigwedge V$ where $V$ is the infinite-dimensional vector space with basis the elements $e_i$. So in particular we have the linear maps of $V$, the group $\mathrm{GL}(V)$, which give a group of automorphisms of $\bigwedge V$. In this group there are the scalar maps, and an ideal invariant under scaling $e_i \mapsto \lambda e_i$ is just a homogeneous ideal. But (cf. Remark 6.3.5) for each $i$ the space $\bigwedge^i V$ is an irreducible representation of $\mathrm{GL}(V)$, so $E \cap \bigwedge^i V$ is either 0 or $\bigwedge^i V$. Since $\bigwedge^i V$ generates as ideal the part of $\bigwedge V$ of degree $\geq i$, it follows that the only $\mathrm{GL}(V)$ invariant ideals of $\bigwedge V$ are the ideals $E_i := \bigoplus_{k=i}^{\infty} \bigwedge^k V$, so in $M_n(G)$ the ideals $M_n(E_i)$. We have $E_a E_b = E_{a+b}$, in particular the product of two nonzero invariant ideals is nonzero so $\bigwedge V$, and hence $M_n(G)$, is Aut-prime.

CASE 3. Let $A = M_{k,l}(G)$, and assume that we have two verbal ideals $I, J$ with $IJ = 0$. Again we know that a verbal ideal $I$ is closed under all automorphisms of the algebra $A$. In particular it must be invariant under the automorphism of order 2 defining the superalgebra structure of $M_{k,l}(G)$ and under all automorphisms induced again by $\mathrm{GL}(V)$.

Let us understand these special ideals. In particular we have $I = I_0 \oplus I_1$ where

$$I_0 = I \cap M_{k,l}(G)_0 = (I \cap M_k(G_0)) \oplus (I \cap M_l(G_0)).$$

The ideal is stable under $\mathrm{GL}(V)$, and $G_0 := \bigoplus_{k=0}^{\infty} \bigwedge^{2k} V$ is a direct sum of irreducible representations. We define thus $E_{0,i} := \bigoplus_{k=i}^{\infty} \bigwedge^{2k} V$, and these are the only $\mathrm{GL}(V)$ invariant ideals of $G_0$.

Let us put by convention $E_{0,\infty} := 0$. We deduce that we must have that $I_0$ is the direct sum of two ideals

$$I_0 = M_k(E_{0,i}) \oplus M_l(E_{0,j}),\ i, j \in \mathbb{N} \cup \{\infty\},$$

a priori one or both of them could be 0.

Next $I_1 \subset A_1 = \hom(F^k, F^l) \otimes G_1 \oplus \hom(F^l, F^k) \otimes G_1$.

Now we consider another group of automorphisms. The product of the two linear groups $\mathrm{GL}(k, F) \times \mathrm{GL}(l, F)$ induce inner automorphisms of $M_{k,l}(F)$ which preserve the superalgebra structure. Under this group both spaces $\hom(F^k, F^l)$, and $\hom(F^l, F^k)$ are nonisomorphic irreducible representations.

It follows that under the group $\mathrm{GL}(k, F) \times \mathrm{GL}(l, F) \times \mathrm{GL}(V)$ the representations $\hom(F^k, F^l) \otimes \bigwedge^{2p+1} V, \hom(F^l, F^k) \otimes \bigwedge^{2p+1} V$ are nonisomorphic and irreducible.

Thus $I_1$ must be a sum of some of these representations, but since $I = I_0 \oplus I_1$ is an ideal, if $I_1$ contains for instance $\hom(F^k, F^l) \otimes \bigwedge^{2p+1} V$, then it also contains $\hom(F^k, F^l) \otimes \bigwedge^{2q+1} V,\ \forall q \geq p$.

Thus $I_1$ must be of the form $\hom(F^k, F^l) \otimes E_{1,a} \oplus \hom(F^l, F^k) \otimes E_{1,b}$, for some $a, b$, where by $E_{1,a}$ we denote $E_{1,a} := \bigoplus_{k=a}^{\infty} \bigwedge^{2k+1} V$ (with $E_{1,\infty} = 0$).

So a $\mathrm{GL}(k, F) \times \mathrm{GL}(l, F) \times \mathrm{GL}(V)$-invariant ideal is of the form

$$M_k(E_{0,i}) \oplus M_l(E_{0,j}) \oplus \hom(F^k, F^l) \otimes E_{1,a} \oplus \hom(F^l, F^k) \otimes E_{1,b},$$

for suitable $i, j, a, b$. Now we claim that the constraints are

$$a \leq i \leq a+1,\ b \leq i \leq b+1, a \leq j \leq a+1,\ b \leq j \leq b+1.$$

In fact,

$$M_k(E_{0,i}) \hom(F^k, F^l) \otimes G_1 = \hom(F^k, F^l) \otimes E_{1,i} \implies i \geq a;$$

$$\hom(F^l, F^k) \otimes G_1 \hom(F^k, F^l) \otimes E_{1,a} = M_l(E_{0,a+1}) \implies j \leq a+1;$$

$$M_l(E_{0,j}) \hom(F^l, F^k) \otimes G_1 = \hom(F^l, F^k) \otimes E_{1,j} \implies j \geq b;$$

$$\hom(F^k, F^l) \otimes E_{1,a} \hom(F^l, F^k) \otimes G_1 = M_k(E_{0,a+1}) \implies i \leq a+1.$$

Similarly $j \leq b+1$, so any nonzero invariant ideal contains an ideal of the form

$$K_c = M_k(E_{0,c}) \oplus M_l(E_{0,c}) \oplus \hom(F^k, F^l) \otimes E_{1,c} \oplus \hom(F^l, F^k) \otimes E_{1,c},$$

where $c = \max(a, b, i, j)$. Since $K_c K_d = K_{c+d} \neq 0$ this proves that $M_{k,l}(G)$ is Aut-prime.

EXERCISE 20.3.8 (Problem). Determine which of the possible invariant ideals are also verbal ideals.

*Hint.* Use various standard polynomials.

### 20.3.1. The radical of verbally prime ideals.

THEOREM 20.3.9. *Let $P$ be a verbally prime ideal of $F\langle X\rangle$. Then $\mathrm{Id}(M_n(F)) \subseteq P$, $n \geq 1$, if and only if $P = \mathrm{Id}(M_k(F))$, for some $k \leq n$.*

PROOF. If $P = \mathrm{Id}(M_k(F))$, then clearly $\mathrm{Id}(M_n(F)) \subseteq P$ if and only if $k \leq n$. In fact, if $k \leq n$, $M_k(F)$ is a subalgebra of $M_n(F)$, and so $\mathrm{Id}(M_n(F)) \subseteq P$; whereas, if $k > n$, $\mathrm{St}_{2n}$ is an identity of $M_n(F)$ but not of $M_k(F)$.

The other cases follow from Lemma 20.1.1. □

We need a simple remark on the Grassmann algebra.

LEMMA 20.3.10. *Let $x_1, \ldots, x_m \in G_1$, then the ideal $I$ generated by the $x_i$ is nilpotent of degree $\leq m+1$. If we take $x_1, \ldots, x_m \in (G \otimes G)_1 = G_0 \otimes G_1 \oplus G_1 \otimes G_0$, then the ideal $I$ generated by the $x_i$ is nilpotent of degree $\leq 2m+1$.*

PROOF. First if $a, b \in G_1$, we have $ab = -ba$, in particular $a^2 = 0$. So take $m+1$ elements $z_k := \sum_{j=1}^{m} x_j y_{k,j} \in I$, $y_{k,j} \in G$. When we expand $z_1 z_2 \cdots z_{m+1}$, we have in each monomial appearing some $x_i^2$ (up to reordering the product) so $I^{m+1} = 0$.

In the second case write $x_i = a_i + b_i$, $a_i \in G_0 \otimes G_1$, $b_i \in G_1 \otimes G_0$, and take $2m+1$ elements $z_k := \sum_{j=1}^{2m} x_j y_{k,j} \in I$. When we expand $z_1 z_2 \cdots z_{2m+1}$ we have in each monomial appearing (up to reordering the product) at least $m+1$ factors of type $a_i$ or $m+1$ factors of type $b_i$. This is zero by the same argument as in the previous case, so $I^{2m+1} = 0$. □

LEMMA 20.3.11 (Berele [**Ber19**]).

(1) $\mathrm{St}_h \notin \mathrm{Id}(M_n(G)), \mathrm{St}_h \notin \mathrm{Id}(M_{k,l}(G))$ *for all $h, n, k, l$.*

(2) $\mathrm{St}_{2n}^{n^3+1} \in \mathrm{Id}(M_n(G))$.
*We have $\mathrm{St}_k^l \notin \mathrm{Id}(M_n(G))$, for all $l \geq 1$ and $k \leq 2n-1$.*

(3) $\mathrm{St}_{2k}^{4k^2 l+1} \in \mathrm{Id}(M_{k,l}(G))$.
*We have $\mathrm{St}_{2k-1}^m \notin \mathrm{Id}(M_{k,l}(G))$, for any $m \geq 1$.*

PROOF. As we have seen in the proof of Theorem 20.3.9, if neither $M_{k,l}(G)$ nor $M_n(G) \supseteq G$ satisfy any standard identity, then (1) is clear. Also, since $M_n(G) \supseteq M_n(G_0)$ and $M_n(G_0)$ has the same identities as $M_n(F)$, it follows that $M_n(G)$ does not satisfy $St_{2n-1}^l$, for any $l \geq 1$.

Let $\varphi$ be an evaluation of $\mathrm{St}_{2n}$ in $M_n(G)$. For $1 \leq i \leq 2n$, let $\varphi(x_i) = a_i + b_i$, where $a_i \in M_n(G_0)$ and $b_i \in M_n(G_1)$. Then, by the Amitsur–Levitzki theorem

$$\varphi(\mathrm{St}_{2n}) = \mathrm{St}_{2n}(a_1, \ldots, a_{2n}) + \sum \mathrm{St}_{2n}(c_1, \ldots, c_{2n}) \stackrel{10.1.4}{=} \sum \mathrm{St}_{2n}(c_1, \ldots, c_{2n}),$$

where $c_i = a_i$ or $b_i$ and the sum is over all $2n$-tuple $(c_1, \ldots, c_{2n})$ with at least one $c_i = b_i$.

If precisely one $c_i$ is $b_i$, say $c_1 = b_1$, then by writing $b_1 = \sum u_{ij} e_{ij}$, with $u_{ij} \in G_1$, we get that

$$\mathrm{St}_{2n}(b_1, a_2, \ldots, a_{2n}) = \sum u_{ij} \mathrm{St}_{2n}(e_{ij}, a_2, \ldots, a_{2n}) = 0,$$

by the Amitsur–Levitzki theorem since $u_{ij}$ commutes with all the $a_i \in M_n(G_0)$.

Hence we may consider $\sum \mathrm{St}_{2n}(c_1, \ldots, c_{2n})$ where at least two $c_i$ belong to $M_n(G_1)$. It follows that the matrix entries of $\sum \mathrm{St}_{2n}(c_1, \ldots, c_{2n})$ belong to $I^2$, where $I$ is the two-sided ideal of $G$ generated by the $2n^3$ entries of the elements $b_i$.

From Lemma 20.3.10 this is nilpotent of index at most $2n^3+1$.

Since $\varphi(St_{2n}) \in M_n(I^2)$, it follows that $\mathrm{St}_{2n}^{n^3+1} \in \mathrm{Id}(M_n(G))$. This proves (2).

Also, as in the previous case, since $M_{k,l}(G) \supseteq M_k(G_0)$, then $M_{k,l}(G)$ does not satisfy $\mathrm{St}_{2k-1}^m$, for any $m \geq 1$.

The proof that $\mathrm{St}_{2k}^{4k^2l+1} \in \mathrm{Id}(M_{k,l}(G))$ also works as in the previous case by choosing an evaluation $\varphi$ of $\mathrm{St}_{2k}$ in $M_{k,l}(G)$ such that $\varphi(x_i) = a_i + b_i$, where $a_i \in (M_{k,l}(G))_0$ and $b_i \in (M_{k,l}(G))_1$, $1 \leq i \leq 2k$.

Since $k \geq l \geq 1$, then $\mathrm{St}_{2k}(a_1, \dots, a_{2k}) = 0$ and $\varphi(\mathrm{St}_{2k}) \in M_{k+l}(I)$, where $I$ is the two-sided ideal of $G$ generated by the $2k(2kl)$ entries of the elements $b_i$. The conclusion follows as above. □

These estimates are not sharp, for instance we have

PROPOSITION 20.3.12. *In $M_{1,1}(G)$ we have $[X,Y]^2 \neq 0,\ [X,Y]^3 = 0$.*

PROOF. Let $X, Y \in M_{1,1}(G)$,

$$X = \begin{vmatrix} a_1 & b_1 \\ c_1 & d_1 \end{vmatrix}, \quad Y = \begin{vmatrix} a_2 & b_2 \\ c_2 & d_2 \end{vmatrix},$$

so $a_i, d_i \in G_0$, $b_i, c_i \in G_1$. We exploit systematically the fact that $G_0$ is the center, and if $u, v \in G_1$, $uv = -vu$ in particular $u^2 = 0$. One easily verifies that

$$[X,Y] = A + B, \quad A = \begin{vmatrix} z & 0 \\ 0 & z \end{vmatrix}, B = \begin{vmatrix} 0 & x \\ y & 0 \end{vmatrix},$$

where

$$z = b_1c_2 - b_2c_1, \quad x = b_1(d_2 - a_2) + b_2(a_1 - d_1), \quad y = c_1(a_2 - d_2) + c_2(d_1 - a_1).$$

Now the remarkable fact that $A$ is a *scalar* with values in $G_0$ implies that

$$[X,Y]^2 = (A+B)^2 = A^2 + 2AB + B^2 = \begin{vmatrix} z^2 + xy & 2zx \\ 2zy & z^2 - xy \end{vmatrix} \neq 0,$$

$$[X,Y]^3 = (A+B)^3 = A^3 + 3A^2B + 3AB^2 + B^3 = \begin{vmatrix} z^3 + 3zxy & 3z^2x + xyx \\ 3z^2y + yxy & z^3 - 3zxy \end{vmatrix},$$

and one checks that $z^2 = -2b_1c_2b_2c_1$ and $z^3 = z^2x = z^2y = xyx = yxy = 0$. □

Recall that the radical of a nonzero $T$-ideal is the $T$-ideal $I_k$ of identities of $k \times k$ matrices for some $k$ (Corollary 11.3.4). We thus set

DEFINITION 20.3.13. The *radical index* of a PI algebra $A$, denoted by $\mathrm{ri}(A)$ is the integer $k$ for which $\sqrt{\mathrm{Id}(A)} = I_k$.

A simple way to compute the radical index of $A$ is from the remark that $\mathrm{ri}(A)$ is the minimum $k$ for which $A$ satisfies some power $\mathrm{St}_{2k}^n$ of the standard identity in $2k$ variables.

COROLLARY 20.3.14. *The radical index of the verbally prime algebras is $n$ for the algebra $M_n(G)$ and $k$ for$M_{k,l}(G)$.*

PROOF. We apply Corollary 11.3.6 and Lemma 20.3.11. □

We want next to show that all these constructed verbally prime ideals are distinct. By the previous corollary all the $M_n(G)$ have different identities, it remains to distinguish $M_{k,l}(G)$ from $M_k(G)$ and $M_{k,l}(G)$ from $M_{k,l'}(G)$ for $l \neq l'$, since the index $k$ is determined as the minimum $h$ for which $St_{2h}$ to some power is an identity.

For this we shall distinguish them by taking a further tensor product with $G$. Then use the Leron–Vapne theorem, Theorem 7.2.3, so that if $A \otimes G$ is not PI equivalent to $B \otimes G$, then $A$ is not PI equivalent to $B$.

LEMMA 20.3.15. $G \otimes G \sim_{\text{PI}} M_{1,1}(G)$.

PROOF. We prove the lemma by first showing that $G \otimes G$ generates a verbally prime variety, so $G \otimes G$ is PI equivalent to one of the algebras of the list in Theorem 20.3.5. Then we see that such a variety is also generated by $M_{1,1}(G)$, by excluding all other cases using some explicit identities satisfied by $G \otimes G$.

The proof that $\text{Id}(G \otimes G)$ is a verbally prime $T$-ideal is similar to that of Lemma 20.3.6. Take $U, V$ $T$-ideals not contained in $\text{Id}(G \otimes G)$, and let $f \in U, g \in V$ be such that $f, g \notin \text{Id}(G \otimes G)$. Then there exist $a_1, \dots, a_n, b_1, \dots, b_m \in G \otimes G$ such that $f(a_1, \dots, a_n) = a \neq 0$ and $g(b_1, \dots, b_m) = b \neq 0$. We may take the two sets of elements $\{a_1, \dots, a_n\}$ and $\{b_1, \dots, b_m\}$ in such a way that they depend on disjoint sets of generators $e_i$'s of the Grassmann algebra $G$.

But then one can find $c \in G \otimes G$ such that $acb \neq 0$. Hence $UV \not\subseteq \text{Id}(G \otimes G)$ and $\text{Id}(G \otimes G)$ is a verbally prime $T$-ideal.

Clearly $x^k$ and $[x, y]$ are not identities of $G \otimes G$, for any $k \geq 1$. If we show that $[x, y]^t \in \text{Id}(G \otimes G)$, for some $t > 1$, then, since $[x, y] = \text{St}_2$, from Lemma 20.3.11 it will follow that $\text{Id}(G \otimes G) = \text{Id}(G)$ or $\text{Id}(M_{1,1})$.

In fact write $x = \sum_{i,j=0,1} x_{i,j}, y = \sum_{i,j=0,1} y_{i,j}$, where $i, j$ indicate the degrees in the two factors.

One checks immediately that $[x, y]$ is in the ideal generated by the elements $x_{0,1}, y_{0,1}, x_{1,0}, y_{1,0}$. Now both $x_{0,1}, y_{0,1}$ and $x_{1,0}, y_{1,0}$ generate nilpotent ideals of order 3 so it follows that $[x, y]^5 = 0$.

We finish the proof by remarking that $G \otimes G$ does not satisfy $[[x, y], z] = 0$, so it is not PI equivalent to $G$. □

From Proposition 20.3.12 we finally deduce

COROLLARY 20.3.16. *$G \otimes G$ satisfies $[x, y]^3 = 0$ and not $[x, y]^2 = 0$.*

EXERCISE 20.3.17. The algebras $G \otimes G$ and $M_{1,1}(G)$ satisfy $[[[x_1, x_2], [y_1, y_2]], z]$ but no identity of degree 4.

PROPOSITION 20.3.18. *The algebras $M_n(G) \otimes G$ and $M_{k,l}(G) \otimes G$ are verbally prime and of radical index $n, k + l$, respectively.*

PROOF. The proof that these algebras are verbally prime follows the same lines as in Lemma 20.3.6, by noticing that these algebras are contained in the matrix algebra $M_n(G \otimes G)$, where $n = k + l$ in the second case.

As for the radical index, notice that in both cases the matrices we obtain contain all the $n \times n$ matrices which in each entry has either all possible elements of $G_0 \otimes G_0$ or all possible elements of $G_1 \otimes G_1$. If we then evaluate $\text{St}_k$ with $k < 2n$ on these matrices, we clearly have that there are nonzero evaluations, as in the proof of Lemma 20.3.11. On the other hand when we evaluate $\text{St}_{2n}$ we can argue as in the same lemma, but now the entries are in $G \otimes G$, so we use the second part of Lemma 20.3.10 and deduce that $\mathrm{St}_{2n}^{8n^3+1} = 0$. $\square$

COROLLARY 20.3.19. *The* PI *equivalence type of a verbally prime, but not prime,* PI *algebra $A$ is determined by the pair of indices* $(\mathrm{ri}(A),\ \mathrm{ri}(A \otimes G))$. *This pair of indeces is* $(n, n)$ *for* $M_n(G)$ *and* $(k, k+l)$ *for* $M_{k,l}(G)$.

THEOREM 20.3.20 (Kemer).

(1) $M_n(G) \otimes G \sim_{\mathrm{PI}} M_{n,n}(G)$.
(2) $M_{k,l}(G) \otimes G \sim_{\mathrm{PI}} M_{k+l}(G)$.
(3) $M_{k,l}(G) \otimes M_{k',l'}(G) \sim_{\mathrm{PI}} M_{kk'+ll',kl'+lk'}(G)$.

PROOF. (1) We have

$$M_n(G) \otimes G = M_n(G \otimes G) \overset{7.2.3}{\sim}_{\mathrm{PI}} M_n(M_{1,1}(G)) = M_{n,n}(G).$$

(2) By Corollary 20.3.19 if $A = M_{k,l}(G) \otimes G$, $B = M_{k+l}(G)$ it is enough to check that $(\mathrm{ri}(A),\ \mathrm{ri}(A \otimes G)) = (\mathrm{ri}(B),\ \mathrm{ri}(B \otimes G))$. By Proposition 20.3.18 we have $\mathrm{ri}(A) = \mathrm{ri}(B) = k + l$, as for $\mathrm{ri}(B \otimes G)$ by the same proposition it is also $k + l$.

Now $A \otimes G = M_{k,l}(G) \otimes G \otimes G \overset{20.3.15}{\sim}_{\mathrm{PI}} M_{k,l}(G) \otimes M_{1,1}(G)$. It is enough to show that $M_{k,l}(G) \otimes M_{1,1}(G)$ has radical index $k + l$. We have to look more closely at the isomorphism $M_{k,l} \otimes M_{1,1} = M_{k+l,k+l}$.

If we follow this isomorphism, we see that $M_{k,l}(G) \otimes M_{1,1}(G)$ can be embedded into $2(k+l) \times 2(k+l)$ matrices, so that the entries in the 0 degree part are in $(G \otimes G)_0$ and in the degree 1 part in $(G \otimes G)_1$. Then arguing as in Lemma 20.3.10, we see that the radical degree is indeed $k + l$.

(3) The algebra $M_{k,l}(G) \otimes M_{k',l'}(G)$ differs from $M_{kk'+ll',kl'+lk'}(G)$ by the fact that in formula (20.9) the entries in $G_0$ are replaced by entries in $H_0 := G_0 \otimes G_0 \oplus G_1 \otimes G_1$, while the entries in $G_1$ are replaced by entries in $H_1 := G_0 \otimes G_1 \oplus G_1 \otimes G_0$. So we have the inclusions,

$$M_{kk'+ll',kl'+lk'}(G) \subset M_{k,l}(G) \otimes M_{k',l'}(G) \subset M_{kk'+ll',kl'+lk'}(H). \tag{20.11}$$

We have $H = H_0 \oplus H_1 \supset G$ is a supercommutative superalgebra, so also a quotient of $G$. Thus clearly $M_{kk'+ll',kl'+lk'}(G)$ and $M_{kk'+ll',kl'+lk'}(H)$ are PI equivalent and the claim follows. $\square$

$\square$

CHAPTER 21

# The PI-exponent

*This chapter is devoted to the proof of Theorem 7.1.7 by Giambruno and Zaicev. For every* PI *algebra $A$ over a field of characteristic* 0*, the limit*

$$\lim_{n\to\infty} c_n(A)^{1/n} = \exp(A)$$

*exists and is an integer, denoted* $\exp(A)$*, called the* PI *exponent of $A$.*

## 21.1. The asymptotic formula

*In fact in this and the next chapters we shall study deeper properties of the asymptotic behavior of the function $c_n(A)$.*

DEFINITION 21.1.1. Recall that two positive functions $f(n), g(n)$, $n \in \mathbb{N}$ (or also $n \in \mathbb{R}^+$) are *asymptotic*, and we write $f(n) \simeq g(n)$ if $\lim_{n\to\infty} f(n)/g(n) = 1$.
The relation $f(n) \simeq g(n)$ is clearly an equivalence.

For $2 \times 2$ matrices, by formula (9.66), we have that $c_n(M_2(F))$ is asymptotic to $C_{n+1}$, the $(n+1)$-th Catalan number. One can see from formula (22.55) applied to $\binom{2n}{n}$, the asymptotic $C_n \simeq \frac{1}{\sqrt{\pi}} n^{-3/2} 4^n \implies c_n(M_2(F)) \simeq \frac{1}{\sqrt{\pi}} n^{-3/2} 4^{n+1}$.
We will prove the following sequence of results by various authors:

THEOREM 21.1.2.

(1) (Regev)

$$c_n(M_k(F)) \simeq C_k n^{\frac{1-k^2}{2}} k^{2(n+1)}, \tag{21.1}$$

*where the constant*

$$C_k = \left(\frac{1}{\sqrt{2\pi}}\right)^{k-1} \left(\frac{1}{2}\right)^{\frac{(k^2-1)}{2}} \cdot 1!2! \cdots (k-1)! \cdot k^{\frac{k^2}{2}}. \tag{21.2}$$

(2) (Giambruno and Zaicev) *For any* PI *algebra $A$ there exist constants $C_1 > 0, C_2, t_1, t_2$ such that*

$$C_1 n^{t_1} d^n \leq c_n(A) \leq C_2 n^{t_2} d^n, \tag{21.3}$$

*and $d = \exp(A)$ is a nonnegative integer. Also, if $A$ is finite dimensional the exponent equals the first Kemer index.*

(3) (Aljadeff, Janssens, and Karasik) *Let $A = \bar{A} \oplus J$ be a fundamental algebra with $\bar{A} = \bigoplus_{i=1}^q A_i$, $A_i = \mathtt{M}_{d_i}(F)$, and let $(d, s)$ be its Kemer index, where $d = \dim \bar{A}$ and $s$ is equal to that integer $s$ such that $J^s \neq 0$, $J^{s+1} = 0$. Then there exist two positive constants $C_1, C_2$ such that, $\forall n$,*

$$C_1 n^{-\frac{d-q}{2}+s} d^n \leq c_n(A) \leq C_2 n^{-\frac{d-q}{2}+s} d^n. \tag{21.4}$$

(4) (Formanek) *The trace algebra associated to A has the same codimension growth as A.*
(5) (Berele and Regev) *Finally, if A has a* 1*, in* (21.3) *we have that* $t_1 = t_2$ *is an integer or half integer and* $C_1 = C_2$.

**21.1.1. Introduction to codimension growth.** We have seen in Theorem 7.1.6 that the codimension $c_n(A)$ of a PI algebra is exponentially bounded.

An explicit knowledge of the codimension $c_n(A)$ is known only in few cases: in particular for $A$ the Grassmann algebra, we have the formula $c_n(A) = 2^{n-1}$ (Theorem 19.1.5); for the algebra of $k \times k$ upper triangular matrices we have formula (16.33); and for $2 \times 2$ matrices we have formula (9.66).

From formula (16.31) and Theorem 16.3.2 one can deduce explicit formulas for the codimension of an algebra of block triangular matrices with blocks of size 1,2.

In general, even for the $k \times k$ matrix algebras, except for $k = 2$, an explicit knowledge of the codimension $c_n(A)$ seems to be an impossible task. One theme which has been vastly investigated is thus to describe it asymptotically.

In particular Amitsur has conjectured the following.

**The exponent.** For every PI algebra $A$ we have that $\lim_{n\to\infty} c_n(A)^{\frac{1}{n}}$ exists and it is a positive integer, it is called the *exponent* of $A$.

This conjecture has been proved by Giambruno and Zaicev **[GZ99]**, and this is one of the themes of this chapter.

They proved

THEOREM 21.1.3. *There exist constants* $C_1 > 0, C_2 > 0$*, and* $t_1 \leq t_2 \in \mathbb{R}$ *such that*

$$C_1 n^{t_1} d^n \leq c_n(A) \leq C_2 n^{t_2} d^n, \tag{21.5}$$

*where* $d = \lim_{n\to\infty} c_n(A)^{1/n} = \exp(A)$ *is an integer, the* exponent.

PROOF. The integer $d$ is described in a precise way in Theorem 21.2.2. The proof is obtained by combining Lemma 21.2.3, which gives an estimate from above, and Lemma 21.2.6 which gives an estimate from below. ☐

By the methods developed first by Berele and Regev and then by Berele **[Ber08a]**, one can prove

THEOREM 21.1.4. *If A has a* 1*, then* $t_1 = t_2 \in \frac{1}{2}\mathbb{Z}$*,* $C_1 = C_2$*. That is, there exists an integer or half integer h and a positive constant C so that*

$$c_n(A) \simeq C n^h d^n. \tag{21.6}$$

In fact for algebras satisfying a Capelli identity, we have the more precise result (23.17) describing the number $h$ (as for $d$, see Theorem 21.2.2). At the moment one of the obstacles to obtaining such a precise result in the general case is the fact that we do not have a precise estimate for the codimension growth for the Grassmann envelope of a simple superalgebra.

Based on some steps of the proof of this result, Giambruno and Zaicev **[GZ14]** show that for any $A$ (with or without a 1) the two exponents $h_1, h_2$ of formula (21.5) are equal and an integer or half integer $h$, that is

$$C_1 n^h d^n \leq c_n(A) \leq C_2 n^h d^n, \quad h \in \frac{1}{2}\mathbb{Z}. \tag{21.7}$$

Usually the exponent can be computed, while for the number $h$ one only knows its existence and crude estimates. In fact even for the Grassmann envelope of a simple superalgebra the number $h$ is not known.

For $A$ a finite-dimensional fundamental algebra $h$ is computed by Aljadeff, Janssens, and Karasik, see formula (21.4). For a general finite-dimensional algebra the computation of the pair $d, h$ can be at least theoretically reduced to that of fundamental algebras due to Proposition 17.2.34 and the following proposition whose proof is left to the reader.

PROPOSITION 21.1.5. *If $R = \bigoplus_{i=1}^k R_i$, then*

$$c_n(R_i) \leq c_n(R) \leq \sum_{i=1}^{k} c_n(R_i). \tag{21.8}$$

As a corollary one has for $R = \bigoplus_{i=1}^k R_i$,

COROLLARY 21.1.6. *If the codimension growth of $R_i$ is given by formula* (21.7) *with $d = d_i$, $h = h_i$, then the codimension growth of $R$ is also given by a formula like* (21.7) *with $(d, h)$ maximal in the lexicographic order among the pairs $(d_i, h_i)$.*

As for the constant $C$ it is known only for matrices and, as a consequence, for block-triangular matrices $UT(d_1, d_2, \ldots, d_q)$ (Definition 16.3.1); see Corollary 23.2.3.

In this and next two chapters we shall present all of these results except the very general one of Berele; the statements are summarized in Theorem 21.1.2.

When $A$ is finite dimensional with 1, we prove the statement of Theorem 21.1.4, in Theorem 21.1.2(5).

As a consequence of Theorem 21.1.3 we recover a result of Kemer, that the codimensions either grow exponentially or, if $d = 1$, are polynomially bounded (no intermediate growth) **[Kem78]**.

### 21.1.2. Generalities on codimension growth.

THEOREM 21.1.7. *The sequence of codimensions of an associative* PI *algebra $R$ is eventually nondecreasing, i.e., $c_{n+1}(R) \geq c_n(R)$, for all $n$ large enough.*

PROOF. Let us remark first that if $R$ has a 1, or more generally if there is no $r \in R$, $r \neq 0$ such that $rR = 0$ (or $Rr = 0$), then the statement is trivial since if $f(x_1, \ldots, x_n)$ is not a PI for $R$, then $x_{n+1} f(x_1, \ldots, x_n)$ (or $f(x_1, \ldots, x_n) x_{n+1}$) is also not a PI for $R$.

For the general case assume, as we may, that the algebra is the Grassmann envelope $G(A)$ of a finite-dimensional superalgebra $A = A_0 \oplus A_1$. Write $A = \bar{A} + J$, where $\bar{A}$ is a maximal semisimple $\mathbb{Z}_2$-graded subalgebra, $J$ is the Jacobson radical of $A$, and let $t \geq 1$ be such that $J^t = 0$. We shall prove that $c_n(G(A)) \leq c_{n+1}(G(A))$ as soon as $n \geq t$.

Since $J$ is nilpotent, we may clearly assume that $\bar{A} \neq 0$. Let $e$ be the unit element of $\bar{A}$, and let $T_e : A \to A$ be the right multiplication by $e$. Clearly $T_e$ has 0 and 1 as eigenvalues, and we decompose $A$ into the sum of the corresponding eigenspaces $A = A^0 \oplus A^1$. Since $ae = a$, for any $a \in \bar{A}$, $\bar{A} \subseteq A^1$ and let $J = J^0 \oplus J^1$ be the decomposition of $J$ into the corresponding eigenspaces. Hence $A^0 = J^0$, $A^1 = \bar{A} + J^1$, where $xe = x$ for $x \in A^1$, and $xe = 0$ for $x \in A^0 = J^0$. Since $e$ is of degree 0 (in the $\mathbb{Z}_2$ gradation of the superalgebra) both $A^0$ and $A^1$ are $\mathbb{Z}_2$ graded so that $G(A) = G(A^0) \oplus G(A^1)$ and, since $A^0 \subset J$, we have $G(A^0)^t = 0$.

Given a multilinear polynomial $f(x_1, \ldots, x_n) \in V_n$, we define

$$\tilde{f}(x_1, \ldots x_n, x_{n+1}) := \sum_{j=1}^{n} f(x_1, \ldots, x_j x_{n+1}, \ldots, x_n) \in V_{n+1}. \tag{21.9}$$

If $f \in \mathrm{Id}(G(A))$, also $\tilde{f} \in \mathrm{Id}(G(A))$, so the linear map $f \mapsto \tilde{f}$ induces a linear map

$$V_n / V_n \cap \mathrm{Id}(G(A)) \to V_{n+1} / V_{n+1} \cap \mathrm{Id}(G(A)).$$

It is thus enough to show that this map is injective for $n \geq t$.

So take $f(x_1, \ldots, x_n) \in V_n \setminus V_n \cap \mathrm{Id}(G(A)$ and $n \geq t$. Since $f$ is not an identity of $G(A)$, we can find $a_1, \ldots, a_n$, each either in $A^0$ or in $A^1$, and suitable homogeneous elements $g_1, \ldots, g_n$ of the Grassmann algebra $G$ such that

$$f(a_1 \otimes g_1, \ldots, a_n \otimes g_n) \neq 0. \tag{21.10}$$

Now

$$\tilde{f}(a_1 \otimes g_1, \ldots, a_n \otimes g_n, e \otimes 1) = \sum_{j=1}^{n} f(a_1 \otimes g_1, \ldots, (a_j \otimes g_j)(e \otimes 1), \ldots, a_n \otimes g_n).$$

We have

$$(a_j \otimes g_j)(e \otimes 1) = a_j e \otimes g_j = \begin{cases} a_j \otimes g_j & \text{if } a_j \in A^1 \\ 0 & \text{if } a_j \in A^0. \end{cases}$$

Therefore $\tilde{f}(a_1 \otimes g_1, \ldots, a_n \otimes g_n, e \otimes 1) = p f(a_1 \otimes g_1, \ldots, a_n \otimes g_n)$, where $p$ is the number of indices $j$ with $a_j \in A^1$. Since $n \geq t$ and $f(a_1 \otimes g_1, \ldots, a_n \otimes g_n) \neq 0$, we cannot have that all the $a_j \in A^0$, hence $p > 0$.

Therefore we have $\tilde{f}(a_1 \otimes g_1, \ldots, a_n \otimes g_n, e \otimes 1) \neq 0$, that is $\tilde{f}$ is not an identity. □

Next we construct a PI algebra $R$ whose sequence of codimensions has a prescribed finite number of inequalities $c_n(R) > c_{n+1}(R)$.

REMARK 21.1.8. For any integer $k \geq 1$, there exists an associative PI algebra $R$ such that $c_{n_1}(R) > c_{n_1+1}(R), \ldots, c_{n_k}(R) > c_{n_k+1}(R)$, for some $n_1 < \cdots < n_k$.

PROOF. Let $F_0[y]$ be the ring of polynomials in the variable $y$ with zero constant term, and define $R_q = F_0[y]/(y^{q+1})$. Then $R_q^q \neq 0$ and $R_q^{q+1} = 0$.

Fix an integer $r > 1$ and choose integers $n_1 < \cdots < n_k$ as follows: For $1 \leq j \leq k+1$ write $A_j = M_{k+r-j}(R_{n_j})$. Clearly

$$c_n(A_j) = \begin{cases} c_n(M_{k+r-j}(F)), & \text{if } n \leq n_j \\ 0, & \text{if } n > n_j. \end{cases}$$

Hence $A_j$ has the same multilinear identities of degree $\leq n_j$ as $M_{k+r-j}(F)$. Moreover, since the codimensions $c_n(M_t(F))$ grow exponentially as $t^2$, we can find integers $n_1 < \cdots < n_{k+1}$ such that

$$c_{n_j}(A_j) > c_{n_j+1}(A_{j+1}), \tag{21.11}$$

for all $j = 1, \ldots, k$.

Set $R = A_1 \oplus \cdots \oplus A_{k+1}$. Then from (21.11) it follows that

$$c_{n_j}(R) = c_{n_j}(A_j) > c_{n_j+1}(A_{j+1}) = c_{n_j+1}(R),$$

for all $j = 1, \ldots, k$. □

## 21.2. The exponent of an associative PI algebra

Here we want to find upper and lower bounds of the codimensions of an arbitrary PI algebra. By Theorem 19.8.1 we may assume that our algebra is $G(A)$, the Grassmann envelope of a finite-dimensional superalgebra $A = A_0 \oplus A_1$, and the base field $F$ is algebraically closed.

Write $A = \bar{A} + J$ where $\bar{A}$ is a semisimple superalgebra, $\bar{A} = B_1 \oplus \cdots \oplus B_k$, a direct sum of simple superalgebras and $J$ is the radical. In this setting we have an analogue of Definition 17.2.31 and Proposition 17.2.38.

DEFINITION 21.2.1. We say that a semisimple superalgebra $A_1 \oplus \cdots \oplus A_h \subseteq \bar{A}$, where $A_1, \ldots, A_h \in \{B_1, \ldots, B_k\}$ are distinct, is *admissible in* $A$, if the product $A_1 J A_2 J \cdots J A_h \neq 0$, i.e., $A_1 \oplus \cdots \oplus A_h + J$, is a reduced or full superalgebra.

We shall prove

THEOREM 21.2.2. *The* PI *exponent of* $G(A)$ *equals the maximal dimension* $d$ *of an admissible subalgebra of* $\bar{A}$.

The strategy to prove this theorem consists in establishing both an upper and a lower bound $\ell_A(n) \leq c_n(G(A)) \leq u_A(n)$ with the property that

$$\lim_{n\to\infty} \ell_A(n)^{1/n} = \lim_{n\to\infty} u_A(n)^{1/n} = d.$$

**21.2.1. An upper bound for the codimensions.** In order to bound from above the codimensions of $G(A)$, we start by fixing a basis of the finite-dimensional superalgebra $A = A_0 \oplus A_1 = \bar{A} + J$ formed by $a_1, \ldots, a_h$ a basis of $A_0$ and $b_1, \ldots, b_k$ a basis of $A_1$. Also fix some number of variables $n$.

We shall prove

LEMMA 21.2.3. $c_n(G(A)) \leq Cn^s d^n$, *for some constant* $C$, *where* $d$ *is the maximal dimension of an admissible subsuperalgebra of* $\bar{A}$ *and* $J^s \neq 0$, $J^{s+1} = 0$.

PROOF. Choose $nh$ polynomial variables $\xi_{i,j}$, $i = 1, \ldots, n$, $j = 1, \ldots, h$, and $nk$ Grassmann variables $\eta_{i,j}$, $i = 1, \ldots, n$, $j = 1, \ldots, k$ (Definition 19.1.1). Then set $G$ to be the Grassmann algebra generated by the $\{\eta_{i,j}\}$ and

$$\xi_i := \sum_{j=1}^{h} \xi_{i,j} a_j + \sum_{\ell=1}^{k} \eta_{i,\ell} b_\ell \in A \otimes G[\xi_{i,j}].$$

Denote by $\mathcal{G}$ the subalgebra generated by the elements $\xi_i$.

Consider the homomorphism $\pi$ of the free algebra $F\langle x_1, \ldots, x_n\rangle$ to the algebra generated by the $\xi_i$ mapping $x_i \mapsto \xi_i$. By Proposition 19.4.23 the kernel of $\pi$ is the ideal of polynomial identities in $n$ variables for $G(A)$.

Therefore the codimension $c_n(G(A))$ equals the dimension of the span of the $n!$ monomials $\xi_{\sigma(1)} \cdots \xi_{\sigma(n)}$, $\sigma \in S_n$.

Now each of these monomials, expanded in the basis of $A$, is a linear combination of *multilinear* monomials in the variables $\Xi : \{\xi_{i,j}, \eta_{h,\ell}\}$, that is elements of type

$$u\Gamma, \quad \Gamma = \gamma_{1,i_1}\gamma_{2,i_2} \cdots \gamma_{n,i_n}, \tag{21.12}$$

with $u \in A$ an element of the chosen basis and $\gamma_{k,i_k}$ equal either to $\xi_{k,i_k}$ or to $\eta_{k,i_k}$.

Thus we shall estimate the codimension $c_n(G(A))$ through an estimate of the number of possible monomials $\Gamma$ of this last type which can appear as coefficients of $\xi_{\sigma(1)} \cdots \xi_{\sigma(n)}$ in a chosen basis. For this we first choose our basis wisely.

Decompose $A = \bar{A} \oplus J$ with $J$ the radical, $\bar{A} = \bigoplus B_i$ the semisimple part written as direct sum of simple superalgebras, and choose our basis as a basis of $J$ and a basis for each of the $B_i$. Thus, when we take one of the variables $\xi_{i,j}, \eta_{i,\ell}$, it makes sense to say that it belongs to $J$ (call it a radical variable) or to one of the $B_i$. Notice that, among the variables $\Xi$, we have thus $n \dim J$ radical variables.

Given this, take a nonzero element $\Gamma = \Gamma_r \Gamma_s$ of type (21.12) of degree $n$ where $\Gamma_r$ contains $i$ radical variables and $\Gamma_s$ contains $n - i$ semisimple variables. There are $\binom{n}{i}$ subsets of possible $i$ indices $j$ with $\gamma_{j,i_j}$ a radical variable. For each of these we have necessarily $i \le s$ (since $J^{s+1} = 0$). This gives rise to at most $\binom{n}{i} \dim J^i$ possible monomials $\Gamma_r$ with $i$ radical variables.

Next such a monomial involves $m := n - i$ variables from all of some distinct simple components $B_{i_1}, \ldots, B_{i_t}$, and this is possible only if, for some order of these simple components, $B_1 J B_2 J \cdots J B_t \neq 0$, that is $B = B_1 \oplus \cdots \oplus B_t \subseteq \bar{A}$ is admissible.

Each admissible subalgebra $B$ of dimension $d_B$ has a basis with $d_0$ elements of degree 0 and $d_1$ of degree 1 summing $d_0 + d_1 = d_B$, and it contributes thus for each $i$ a list of monomials in $d_0$ polynomial and $d_1$ Grassmann variables as above.

We now make a simple remark. Suppose we are given $n$ sets of polynomial variables $\xi_{i,j}$, $i = 1, \ldots, n$, $j = 1, \ldots, d_0$, and $n$ sets of Grassmann variables $\eta_{i,j}$, $i = 1, \ldots, n$, $j = 1, \ldots, d_1$, and consider multilinear monomials $\Gamma = \gamma_{1,i_1}\gamma_{2,i_2} \cdots \gamma_{n,i_n}$, $\gamma_{i,j} \in \{\xi_{i,j}, \eta_{i,j}\}$. Thus this set of monomials, up to sign, coincides with a list of $d^m$ monomials.

Thus by the previous remark, such an admissible subalgebra $B$ may contribute at most $\sum_{i=0}^{s} \binom{n}{i} d_B^{n-i} \le C n^s d_B^n$ monomials all together if we have $r$ admissible subalgebras, and if we set $d := \max d_B$, we estimate the codimension by $c_n(G(A)) \le C' n^s d^n$. □

**21.2.2. An estimate from below.** We start with the following.

21.2.2.1. *A combinatorial lemma.*

DEFINITION 21.2.4. Given integers $k, l, t \ge 0$, define the following partition

$$h(k, l, t) = ((l + t)^k, l^t).$$

Hence $h(k, l, t) \vdash (k + l)t + kl$ is the following hook-shaped diagram.

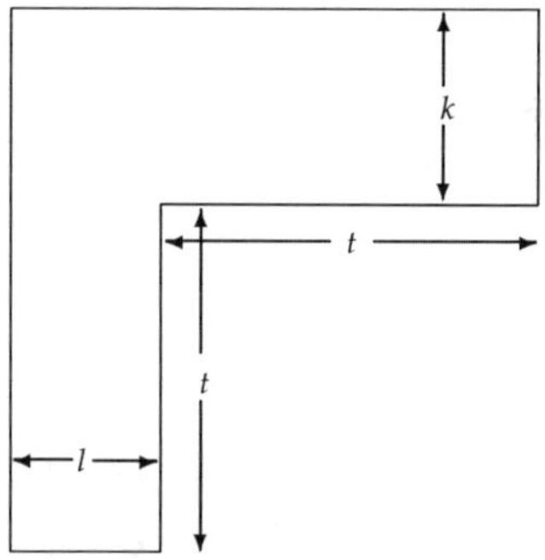

Recall that for a partition $\lambda \vdash n, f_\lambda = \chi_\lambda(1)$ is the degree of the irreducible $S_n$-character associated to $\lambda$.

LEMMA 21.2.5. *Let $k, l \geq 0$ be fixed integers. If $h(k,l,t) \vdash n = t(k+l)+kl$, then*

$$f_{h(k,l,t)} \underset{n\to\infty}{\simeq} at^{\frac{1-(k^2+l^2)}{2}}(k+l)^n \simeq a'n^{\frac{1-(k^2+l^2)}{2}}(k+l)^n, \tag{21.13}$$

*for some (explicit) constants $a, a'$.*

PROOF. We shall apply the hook formula. Let $\Pi = \prod h_{ij}$ be the product of the hook numbers of $h(k,l,t)$. We decompose the diagram into three rectangles and write $\Pi = \Pi_1\Pi_2\Pi_3$, where

$$\Pi_1 = \prod_{i-1}^{k}\prod_{j=l+1}^{l+t} h_{ij}, \quad \Pi_2 = \prod_{i=k+1}^{k+t}\prod_{j=1}^{l} h_{ij}, \quad \Pi_3 = \prod_{\substack{1\le i\le k\\ 1\le j\le l}} h_{ij}.$$

Then

$$\Pi_1 = t!\frac{(t+1)!}{1!}\frac{(t+2)!}{2!}\cdots\frac{(t+k-1)!}{(k-1)!},$$
$$\Pi_2 = t!\frac{(t+1)!}{1!}\frac{(t+2)!}{2!}\cdots\frac{(t+l-1)!}{(l-1)!}$$

and

$$\Pi_3 = (2t+1)(2t+2)\cdots(2t+k)\cdot(2t+2)(2t+3)\cdots(2t+k+1)\cdot$$
$$\cdots(2t+l)(2t+l+1)\cdots(2t+l+k-1) \simeq (2t)^{kl}.$$

A convenient form of that formula is the following.

Let $m \to \infty$, and let $b$ be fixed. Then (as will also appear in formula (22.25)) using Stirling's formula (22.37) $n! \simeq \sqrt{2\pi n}\,\left(\frac{n}{e}\right)^n$, we have

$$(m+b)! = m^b\prod_{j=1}^{b}(1+\frac{j}{m})m! \simeq \sqrt{2\pi}\cdot m^b(\frac{m}{e})^m\sqrt{m}. \tag{21.14}$$

$$\Pi_1 \simeq \prod_{i=0}^{k-1}\left[\frac{\sqrt{2\pi}}{i!}\cdot t^i(\frac{t}{e})^t\sqrt{t}\right] = c_1e^{-kt}t^{kt+\frac{k(k-1)}{2}}\sqrt{t}^k = c_1e^{-kt}t^{kt+\frac{k^2}{2}}.$$

Similarly for $\Pi_2$.

$$\prod h_{ij} = \Pi_1\Pi_2\Pi_3 \simeq c_2t^{\frac{k^2+l^2}{2}}t^{kl}\frac{t^{t(k+l)}}{e^{t(k+l)}}, \quad c_2 = 2^{kl}\prod_{i=0}^{k-1}\frac{\sqrt{2\pi}}{i!}\prod_{i=0}^{l-1}\frac{\sqrt{2\pi}}{i!}.$$

Then, since $n = t(k+l)+kl$, we obtain as $t \to \infty$,

$$n! \simeq \sqrt{2\pi}\cdot t^{kl}\frac{(t(k+l))^{t(k+l)}}{e^{t(k+l)}}\sqrt{t} = \sqrt{2\pi}\cdot t^{kl+\frac{1}{2}}\frac{t^{t(k+l)}}{e^{t(k+l)}}(k+l)^{t(k+l)}.$$

Hence

$$f_{h(k,l,t)} = \frac{n!}{\prod h_{ij}} \simeq a\cdot t^{\frac{1-(k^2+l^2)}{2}}\cdot(k+l)^{t(k+l)}.$$

Finally, replacing $t \to \infty$ with $n = t(k+l)+kl \to \infty$, we obtain formula (21.13),

$$f_{h(k,l,t)} \underset{n\to\infty}{\simeq} at^{\frac{1-(k^2+l^2)}{2}}(k+l)^n \simeq a'n^{\frac{1-(k^2+l^2)}{2}}(k+l)^n$$

for some constants $a, a'$, as required. □

21.2.2.2. *The estimate.* We now can prove the main lemma.

LEMMA 21.2.6. *$c_n(G(A)) \geq Cn^a d^n$, for some constants $C, a$, where $d$ is the maximal dimension of an admissible subsuperalgebra of $\bar{A}$.*

We may assume that $A$ is fundamental and use a variation of Corollary 17.2.16 for superalgebras.

From the discussion in §19.5.4, if $A$ is a fundamental superalgebra with algebra index $(s_A, t_A)$ $d = s_A + t_A = \dim \bar{A}$, then for all $n$ large there are Kemer superpolynomials of total degree $n$ of the form $f(Y_1, \ldots, Y_\mu; Z_1, \ldots, Z_\mu, Y, Z)$, with the $Y_i$ even layers alternating in $s_A$ variables and the $Z_j$ odd layers alternating in $t_A$ variables (there is also the possible degenerate case where $t_A = 0$) with the number of extra variables $Y, Z$ bounded by some fixed numbers $r_1, r_2$.

Then the proof of Corollary 17.2.16 applies separately to the even and odd variables. We get for the supercocharacter defined in formula (19.10)

COROLLARY 21.2.7. *There are two partitions $\lambda \vdash a, \mu \vdash b$ with $m_{\lambda,\mu} \neq 0$, so that $\lambda$ contains a rectangle $s_A \times k$ and $\mu$ contains a rectangle $t_A \times k$. Outside these two rectangles we have a bounded number $r'$ of boxes. In particular $a = s_A k + \alpha, b = t_A k + \beta$, $\alpha + \beta = r'$ and $n = (s_A + t_A)k + r'$.*

Then if we apply the $*$ transform (see §19.4.1) to such a representation $M_\lambda \otimes M_\mu$ of $S_a \times S_b$ appearing in the supercocharacter, by Exercise 19.4.1, we obtain

$$(M_\lambda \otimes M_\mu)^* = M_\lambda \otimes M_{\check{\mu}},$$

and now the partition $\check{\mu}$ dual (or conjugate) to $\mu$ contains the transpose rectangle $k \times t_A$.

By Lemma 19.4.10 this means that in the supercocharacter of the Grassmann envelope $G(A)$ of the superalgebra $A$ in degree $n$, the character of $M_\lambda \otimes M_{\check{\mu}}$ appears with nonzero multiplicity, that is the elements of $M_\lambda \otimes M_{\check{\mu}}$ are not superidentities, in particular they are not polynomial identities.

For the usual cocharacter of $G(A)$ this implies that there is an irreducible representation $M_\gamma$ of $S_n$ with $m_\gamma \neq 0$ (the usual cocharacter), which, restricted to $S_a \times S_b$ contains $M_\lambda \otimes M_{\check{\mu}}$. This is equivalent to saying that $M_\gamma$ appears in the induced representation $\mathrm{Ind}_{S_a \times S_b}^{S_n} M_\lambda \otimes M_{\check{\mu}}$.

Finally this induced representation, restricted to $S_{(s_A+t_A)k}$, contains the induced representation, formula (1.17),

$$\mathbb{M} := \mathrm{Ind}_{S_{ks_A} \times S_{kt_A}}^{S_{(s_A+t_A)k}} M_{(k^{s_A})} \otimes M_{(t_A{}^k)}. \tag{21.15}$$

Recall that, with the symbol $(a^b)$, we denote the partition with $b$ rows of length $a$, identified, as diagram, to a rectangle. Thus the irreducible representation $M_\gamma$ of $S_n$ (which gives a contribution to the cocharacter), restricted to $S_{(s_A+t_A)k}$ contains some irreducible representation $M_\delta$ appearing in the induced representation $\mathbb{M}$.

Thus $c_n(G(A)) \geq \dim M_\gamma \geq \dim M_\delta$, implies that the codimension of the Grassmann envelope of the superalgebra $A$ in degree $n$ is larger than $\dim M_\delta$. With $M_\delta$ some irreducible representation appearing in formula (21.15).

In order to estimate $\dim M_\delta$, we apply the Frobenius character. Under this theory to an irreducible representation $M_\delta$ corresponds the Schur function $S_\delta$ and to $\mathbb{M}$ the product of the two Schur functions $S_{(k^{s_A})} S_{(t_A{}^k)}$ (formula (6.34)).

We have thus that the Schur function $S_\delta$, corresponding to $M_\delta$, appears in the product of the two Schur functions $S_{(k^{s_A})}S_{({t_A}^k)}$.

This has been discussed in Theorem 6.4.20, the Littlewood–Richardson rule.

Of this rule we only need the easy part (i), which implies that, as diagrams, $\delta \supset ({t_A}^k \cup k^{s_A})$ the union of the two diagrams.

Notice that the partition obtained as the union of the two rectangles $({t_A}^k)$ and $(k^{s_A})$ equals the hook-shaped partition $(k^{s_A}, {t_A}^{k-s_A})$.

We can finally deduce

COROLLARY 21.2.8. *The codimension of the Grassmann envelope of the superalgebra $A$ in degree $n = (s_A + t_A)k + r'$ is larger than the dimension of $M_{(k^{s_A}, {t_A}^{k-s_A})}$.*

*Proof of Lemma* 21.2.6 We claim that Corollary 21.2.8 gives, as in Lemma 21.2.5, an estimate from below of the codimension as $> Cn^a(s_A + t_A)^n$. In fact clearly by inspection on the simple superalgebras, we always have $t_A \leq s_A$, therefore the hook partition $(k^{s_A}, {t_A}^{k-s_A})$ contains the partition $h(s_A, t_A, k - s_A) \vdash n - f$ of Definition 21.2.4, where $f \leq r' + s_A - t_A$.

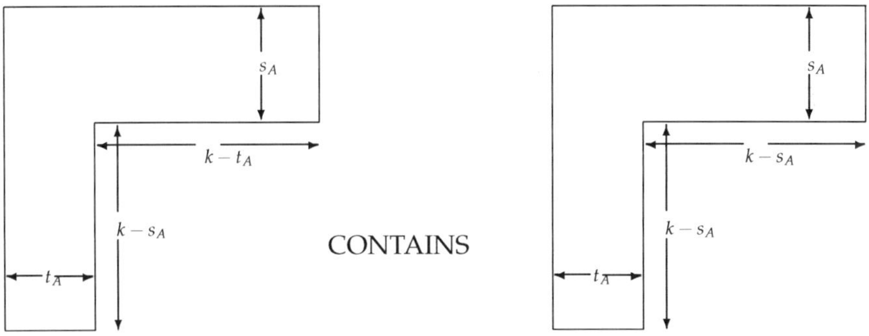


The claim then follows from Lemma 21.2.5.

*Proof of Theorem* 21.1.3. This is obtained by combining Lemmas 21.2.3 and 21.2.6.

## 21.3. Growth of central polynomials

As for polynomial identities, also for central polynomials (see §10.2), one can introduce codimensions and measure their growth. The study was started by Regev; see **[Reg16]**.

We modify Definition 10.2.1 as follows. A polynomial $f \in F\langle X\rangle$ is a central polynomial for an algebra $A$ if it takes values in the center of $A$. In case $f$ takes a nonzero value, we say that $f$ is a proper central polynomial of $A$.

The set $\mathrm{Id}^z(A)$ of central polynomials of $A$ is a $T$-space, i.e., a vector space invariant under the endomorphisms of $F\langle X\rangle$.

We have the notion of central codimension $c_n^z(A) = \dim \frac{V_n}{V_n \cap \mathrm{Id}^z(A)}$ and of proper central codimension $\delta_n(A) = \dim \frac{V_n \cap \mathrm{Id}^z(A)}{V_n \cap \mathrm{Id}(A)}$.

The relation between the usual codimensions and the above two is clear:

$$c_n(A) = \delta_n(A) + c_n^z(A). \tag{21.16}$$

An analogous more general relation also holds between the corresponding cocharacters. From formula (21.16) it follows that if $A$ is a PI algebra, also the two sequences $\delta_n(A)$ and $c_n^z(A)$ are exponentially bounded.

It turns out that the exponential growth of the two sequences can be computed for any PI algebra **[Reg16]**, **[GZ19]**. Assume, as we may, that our algebra is the Grassmann envelope $G(A)$ of a finite-dimensional superalgebra $A$ over an algebraically closed field $F$. Write $A = \bar{A} + J$, where $\bar{A} = A_1 \oplus \cdots \oplus A_m$, with $A_1, \ldots, A_m$ simple superalgebras and $J$ the Jacobson radical.

DEFINITION 21.3.1. Let $B = A_{i_1} \oplus \cdots \oplus A_{i_k} \subseteq \bar{A}$ be a semisimple subsuperalgebra. Then $B$ is a *centrally admissible* subalgebra for the envelope $G(A)$ if there exists a multilinear central polynomial $f = f(x_1, \ldots, x_s)$ of $G(A)$ with $s \geq k$, such that $f(a_1, \ldots, a_k, b_1, \ldots, b_{s-k}) \neq 0$, for some $a_1 \in G(A_{i_1}), \ldots, a_k \in G(A_{i_k})$, $b_1, \ldots, b_{s-k} \in G(A)$.

THEOREM 21.3.2. *If $G(A)$ has centrally admissible subalgebras, then for all $n \geq 1$*

$$C_1 n^{t_1} d^n \leq \delta_n(G(A)) \leq C_2 n^{t_2} d^n,$$

*for some constants $C_1 > 0, C_2, t_1, t_2$, where $d$ is the maximal dimension of a centrally admissible subalgebra of $G(A)$.*

*If $G(A)$ has proper central polynomials but there are no centrally admissible subalgebras for $G(A)$, then $\delta_n(G(A)) = 0$, for all $n$ large enough.*

From Theorem 21.3.2 it immediately follows that if $R$ is any PI algebra over a field of characteristic 0, then the proper central exponent

$$\exp^\delta(R) = \lim_{n\to\infty} \delta_n(R)^{1/n}$$

exists and is a nonnegative integer. Also $\exp^\delta(R) \leq \exp(R)$.

Regarding the sequence $c_n^z(R)$, the result is

THEOREM 21.3.3. *If $R$ is any* PI *algebra over a field of characteristic* 0, *its central exponent*

$$\exp^z(R) = \lim_{n\to\infty} c_n^z(R)^{1/n}$$

*exists. Moreover* $\exp^z(R) = \exp(R)$, *provided* $\exp(R) \geq 2$. *When* $\exp(R) = 1$, *then* $\exp^z(R) = 0$ *or* 1.

## 21.4. Beyond associative algebras

The notion of polynomial identity is quite general and applies to any algebraic structure where there is a symbolic calculus. In particular to nonnecessarily associative algebras, by this we mean just a vector space $A$ over a field $F$ with a bilinear map $m : A \otimes_F A \to A$, called the multiplication.

In this case the notion of polynomial identity and codimension is through the use of nonassociative polynomials. Of course one may look at special varieties as Lie or Jordan algebras. The theory is much more complicated and quite beyond the purpose of this book. We may mention that even for Lie and Jordan algebras there are examples in which the codimension is not exponentially bounded **[Vol84]**, **[Dre87]**, **[GZ11]**, although it is easy to see that this is so for finite-dimensional algebras **[BD02]**, **[GZ10b]**.

In this case one has

THEOREM 21.4.1. *If $A$ is a nonnecessarily associative finite-dimensional algebra $A$, then* $c_n(A) \leq (\dim A)^{n+1}$.

The exponent of finite-dimensional Lie or Jordan algebras exists **[Zaĭ02]**, **[GSZ11]** and is clearly bounded from above by $\dim A$.

For general nonnecessarily associative algebras, the PI exponent exists **[GZ12]**, provided they are simple.

THEOREM 21.4.2. *If $A$ is a nonnecessarily associative finite-dimensional simple algebra $A$, then* $\exp(A)$ *exists.*

For simple associative or Lie or Jordan algebras, the equality $\exp(A) = \dim A$ holds, provided that the base field is algebraically closed. More generally this equality holds provided that $A$ has a suitable bilinear form **[GSZ11]**.

COROLLARY 21.4.3. *Let $A$ be a finite-dimensional simple algebra over an algebraically closed field of characteristic* 0. *If $A$ is a Lie algebra or a unitary noncommutative Jordan algebra, then* $\exp(A) = \dim A$.

PROOF. If A is a semisimple Lie algebra, the Killing form is nondegenerate. If A is a semisimple unitary noncommutative Jordan algebra, by **[Sch66**, pp. 141–142**]** $\alpha(x, y) = \operatorname{tr}(R_{xy+yx} + L_{xy+yx})$ is a nondegenerate bilinear form. □

The equality $\exp(A) = \dim A$ of Corollary 21.4.3 is no more valid even for simple Lie superalgebras over an algebraically closed field. Since simple Lie superalgebras are simple in a nongraded sense, by the above discussion their PI exponent exists. Nevertheless it can be shown **[GZ12]** that for the infinite family of Lie superalgebras of type $b(t)$, $t \geq 3$, the PI exponent is strictly less than the dimension:

Recall that $b(t)$ is the set of matrices

$$\begin{pmatrix} A & B \\ C & -A^T \end{pmatrix},$$

where $A, B, C \in M_t(F)$, $B^T = B, C^T = -C$, and $\operatorname{tr} A = 0$. Then $b(t) = L = L_0 \oplus L_1$, where

$$L_0 = \{\begin{pmatrix} A & 0 \\ 0 & -A^T \end{pmatrix} \mid A \in M_t(F), \operatorname{tr}(A) = 0\}$$

and

$$L_1 = \{\begin{pmatrix} 0 & B \\ C & 0 \end{pmatrix} \mid B^T = B, C^T = -C \in M_t(F)\}$$

is a simple Lie superalgebra for $t \geq 3$.

REMARK 21.4.4. The number of possible brackets of a nonassociative monomial of length $n$ is the $n$th Catalan number $\frac{1}{n}\binom{2n-2}{n-1} \leq 4^n$. It follows that if the cocharacter of an algebra is contained in a hook, then the corresponding codimensions are exponentially bounded (this shows that the codimensions of finite-dimensional algebras are exponentially bounded, see Theorem 21.4.1).

Even when the codimensions of a nonassociative PI algebra are exponentially bounded, the exponential behavior can be very wild.

By exploiting the combinatorial properties of periodic or Sturmian words, one can construct several counterexamples. The basic properties of these words can be found for instance in **[Lot02]**.

Recall that if $w = w_1w_2\cdots$ is an infinite word in the alphabet $\{0,1\}$, the complexity of $w$ is the function $Comp : \mathbb{N} \to \mathbb{N}$ such that $Comp_w(n)$ is the number of distinct subwords of $w$ of length $n$. If $Comp_w(n) = T$, for all $n \geq T$, then $w$ is a periodic word of period $T$. In general $Comp_w$ is an increasing function and $Comp_w(n) \geq n+1$, for all $n \geq 1$. If $Comp_w(n) = n+1$, for all $n \geq 1$, then $w$ is called a Sturmian word. for such words the limit

$$\pi(w) = \lim_{n\to\infty} \frac{w_{k+1} + \cdots + w_{k+n}}{n}$$

exists and is called the slope of $w$. A well-known result on the combinatorics of words asserts that given any real number $\alpha$, $0 < \alpha < 1$, if $\alpha$ is rational, there exists a periodic word $w$ whose slope is $\alpha$, whereas if $\alpha$ is irrational, there exists a Sturmian word whose slope is $\alpha$.

We recall a construction given in **[GMZ08]** and **[GMZ06]**.

Let $K = (k_1, k_2, \ldots)$ be a sequence of positive integers, and let $A(K)$ be the algebra over $F$ with basis given by the set

$$\{a, b\} \cup Z_1 \cup Z_2 \cup \cdots,$$

where

$$Z_i = \{z_j^{(i)} \mid 1 \leq j \leq k_i\}, \quad i = 1, 2, \ldots,$$

and the multiplication table is given by

$$z_1^{(i)}a = z_2^{(i)}, \ldots, z_{k_i-1}^{(i)}a = z_{k_i}^{(i)}, \ z_{k_i}^{(i)}a = 0, \quad i = 1, 2\ldots,$$

if $k_i \geq 2$, while $z_1^{(i)}a = 0$ if $k_i = 1$. Also

$$z_{k_i}^{(i)}b = z_1^{(i+1)}, \quad i = 1, 2, \ldots,$$

and all other products are equal to zero.

It is easily checked that

$$x_1(x_2x_3) \equiv 0 \tag{21.17}$$

is an identity of $A(K)$.

Now let $w = w_1w_2\cdots$ be an infinite word in the alphabet $\{0,1\}$. Given an integer $m \geq 2$, let $K_{m,w} = (k_1, k_2, \ldots)$ be the sequence defined by

$$k_i = \begin{cases} m, & \text{if } w_i = 0 \\ m+1, & \text{if } w_i = 1, \end{cases}$$

and write $A(m, w) = A(K_{m,w})$.

A basic result proved for these algebras **[GMZ08]** says that if $w$ is an infinite periodic or Sturmian word with slope $\alpha$, then $\exp(A(m, w))$ exists and equals

$$\frac{1}{\beta^\beta(1-\beta)^{1-\beta}},$$

where $\beta = \frac{1}{m+\alpha}$.

Out of this **[GMZ08]** one can prove

THEOREM 21.4.5. *For any real number $\alpha > 1$, there exists an algebra $A_\alpha$ such that*

$$\exp(A_\alpha) = \alpha.$$

Even if the codimensions are exponentially bounded, the existence of the PI exponent is not granted **[Zai14]**. By the above methods, for any real number $\alpha > 1$, one can construct an algebra $A_\alpha$ such that

$$\liminf_{n\to\infty} \sqrt[n]{c_n(A_\alpha)} = 1, \quad \limsup_{n\to\infty} \sqrt[n]{c_n(A_\alpha)} = \alpha.$$

What about intermediate growth of the codimensions? Lie algebras do not have intermediate growth **[Mis96]**, but this is no more the case for more general nonassociative algebras. In fact, by using the construction of the algebras $A(K)$ given above **[GMZ06]**, one can prove the following

THEOREM 21.4.6. *For any real number $\beta \in \mathbb{R}$ with $0 < \beta < 1$, there is an algebra $A$ such that*

$$\lim_{n\to\infty} \log_n \log_n(c_n(A)) = \beta.$$

*Hence $c_n(A) \simeq n^{n^\beta}$, asymptotically.*

## 21.5. Beyond the PI exponent

The Grassmann algebra $G$ and the algebra $UT(1,1)$ of $2 \times 2$ upper triangular matrices over $F$ have PI exponent equal to 2. Now, by using the characterization of the exponent, it is easy to prove that any $T$-ideal properly containing $\mathrm{Id}(G)$ or $\mathrm{Id}(UT(1,1))$ has polynomial growth of the codimensions. Actually,

PROPOSITION 21.5.1. *These are the only two $T$-ideals with such property.*

PROOF. Let $\mathcal{V}$ be a variety such that $G, UT(1,1) \notin \mathcal{V}$. We shall prove that $\exp(\mathcal{V}) \leq 1$, and this will do.

Assume, as we may, that the base field $F$ is algebraically closed. Since $G \notin \mathcal{V}$, by Corollary 20.1.4 we have $\mathcal{V} = \mathrm{var}(A)$ with $A$ a finite dimensional algebra. Also, since $UT(1,1) \notin \mathcal{V}$, then either $A = J$ or $A = F_1 \oplus \cdots \oplus F_q + J$, where $F_i \cong F$ and $F_i J F_k = 0$, for all $i \neq k$. This says that $\exp(\mathcal{V}) = \exp(A) \leq 1$. □

In the language of varieties of algebras this proposition says that $\mathrm{var}(G)$ and $\mathrm{var}(UT(1,1))$ are the only two varieties of almost polynomial growth.

One generalizes this phenomenon as follows.

DEFINITION 21.5.2. A variety of algebras $\mathcal{V}$ is called minimal of exponent $d \geq 2$ if any proper subvariety has exponent strictly smaller that $d$.

Hence $G$ and $UT(1,1)$ generate the only two minimal varieties of exponent two. With more effort one could actually prove that any verbally prime variety is minimal.

In order to state the general result, we need to introduce a definition.

Let $A$ be a finite-dimensional simple superalgebra. Let $e_1, \ldots, e_n \in A_0$ be orthogonal idempotents such that $e_1 + \cdots + e_n = 1$ and for every $i = 1, \ldots, n$, $Ae_i$ $(e_iA)$ is a minimal left (resp., right) graded ideal of $A$. Such idempotents are called the *minimal graded idempotents* of the simple superalgebra $A$.

Now let $A = \bar{A} + J$ be a finite-dimensional superalgebra over an algebraically closed field, and let $\bar{A} = A_1 \oplus \cdots \oplus A_q$ with the $A_i$'s graded simple.

DEFINITION 21.5.3. $A = \bar{A} + J$ is a minimal superalgebra if there exist homogeneous elements $w_{12}, \ldots, w_{q-1,q} \in J$ and minimal graded idempotents $e_1 \in A_1, \ldots, e_q \in A_q$ such that

$$e_i w_{i,i+1} = w_{i,i+1} e_{i+1} = w_{i,i+1}, \quad i = 1, \ldots, q-1,$$
$$w_{12} w_{23} \cdots w_{q-1,q} \neq 0,$$

and $w_{12}, \ldots, w_{q-1,q}$ generate $J$ as a two-sided ideal of $A$.

The classification of minimal varieties (**[GZ03a]**, **[GZ03b]**) is given in

THEOREM 21.5.4. *The following properties are equivalent.*

(1) *$\mathcal{V}$ is a minimal variety of exponent $d \geq 2$;*
(2) *$\mathrm{Id}(\mathcal{V})$ is a product of verbally prime T-ideals;*
(3) *$\mathcal{V} = \mathrm{var}(G(A))$ for some minimal superalgebra $A$ such that $\dim \bar{A} = d$.*

As a corollary we get that the number of minimal varieties of given exponent $d \geq 2$ is finite.

We remark that if $A$ is a minimal superalgebra with trivial $\mathbb{Z}_2$-grading, then $A \cong UT(d_1, \ldots, d_q)$, where $\dim A_i = d_i$, $1 \leq i \leq q$. Hence $\mathrm{var}(UT(d_1, \ldots, d_q))$ are the only minimal varieties generated by a finite-dimensional algebra.

Another application of the exponent is the following. Given a polynomial $f \in F\langle X \rangle$, let $\mathrm{var}(f)$ be the variety of algebras whose $T$-ideal is generated by $f$.

DEFINITION 21.5.5. The exponent $\exp(f)$ of the polynomial $f$ is the exponent of the variety $\mathrm{var}(f)$.

The following theorem is proved in **[BR01]** and says that the standard polynomial is of maximal exponent among the homogeneous polynomials of the same degree.

THEOREM 21.5.6.

(1) $\exp(\mathrm{St}_{2n}) = \exp(\mathrm{St}_{2n+1}) = n^2$;
(2) *If $f$ is a homogeneous polynomial of degree $n \geq 4$, then $\exp(f) \leq [\frac{n}{2}]^2 = \exp(\mathrm{St}_n)$.*

Another interesting issue is how close are the $T$-ideal generated by the standard identity of degree $2k$ and the $T$-ideal of identity of $k \times k$ matrices. We have (see **[GZ03c]**)

THEOREM 21.5.7. *If $F$ is algebraically closed, then $\mathrm{var}(\mathrm{St}_{2k}) = \mathrm{var}(M_k(F) \oplus B)$, where $B$ is a finite-dimensional algebra with $\exp(B) < k^2$. Hence*

$$c_n(\mathrm{St}_{2k}) \simeq c_n(M_k(F)).$$

There are analogous results for the Capelli polynomial.

CHAPTER 22

# Codimension growth for matrices

*This chapter deals with the asymptotic formulas for the codimension of a matrix algebra. It differs from the rest of the book in that the methods require some nontrivial analytic estimates, mostly inspired from methods of probability theory.*

## 22.1. Codimension growth for matrices

**22.1.1. Some background.** In 1981 Regev gave a precise description of the asymptotic behavior of $c_n(M_k(F))$. His theorem is

THEOREM 22.1.1.

$$c_n(M_k(F)) \simeq C_k n^{\frac{1-k^2}{2}} k^{2(n+1)}, \tag{22.1}$$

*where* $C_k$ *is the constant*

$$C_k = \Big(\frac{1}{\sqrt{2\pi}}\Big)^{k-1} \Big(\frac{1}{2}\Big)^{\frac{(k^2-1)}{2}} \cdot 1!2! \cdots (k-1)! \cdot k^{\frac{k^2}{2}}. \tag{22.2}$$

In particular from the easier result, already proved in Theorem 21.2.2, it also follows that

COROLLARY 22.1.2. *The exponent of* $M_k(F)$ *equals* $k^2$.

Theorem 22.1.1 is achieved in several steps.

One can also introduce the *trace codimensions*, $\mathrm{Td}_n(M_k(F)), \mathrm{td}_n(M_k(F))$, which are, respectively, the dimension of the space of multilinear expressions of $k \times k$ matrices and their traces of degree $n$, and the dimension of the space $[(\mathrm{End}(V)^*]^{\otimes n}$ of multilinear invariants of $k \times k$ matrices, $\dim_F V = k$.

By the basic Theorem 12.1.11 one has

(1)

$$\mathrm{Td}_n(M_k(F)) = \mathrm{td}_{n+1}(M_k(F)), \tag{22.3}$$

since the space of multilinear expressions of $k \times k$ matrices and their traces of degree $n$ is identified with the space $[(\mathrm{End}(V)^*]^{\otimes n+1}$ of multilinear invariants of $k \times k$ matrices of degree $n+1$, $\dim V = k$.

(2) In turn this is the endomorphism algebra

$$\mathrm{End}_{GL(V)} V^{\otimes n+1}, \ \dim V = k.$$

By the Schur–Weyl duality this algebra is spanned by the symmetric group $S_{n+1}$. It is then isomorphic to $\bigoplus_{\lambda \vdash n+1 \mid \mathrm{ht}(\lambda) \le k} M_\lambda \otimes M_\lambda^*$ where $M_\lambda$ is the irreducible representation with Young diagram $\lambda$, Theorem 6.1.3(2).

(3) Thus if $f_\lambda := \dim M_\lambda$, we have

$$\mathrm{Td}_n(M_k(F)) = \sum_{\lambda \vdash n+1 \mid \mathrm{ht}(\lambda) \le k} f_\lambda^2. \tag{22.4}$$

Of course we have $c_n(M_k(F)) \leq \mathrm{Td}_n(M_k(F))$. By Formanek, ([**For84**, Theorem 16]), we have

THEOREM 22.1.3.

$$\mathrm{Td}_n(M_k(F)) \simeq c_n(M_k(F)). \tag{22.5}$$

PROOF. Since $M_k(F)$ satisfies the Capelli polynomial $C_{k^2+1}$ we know, from Corollary 7.3.6 that the multiplicities $m_\lambda$ vanish as soon as $\lambda$ has at least $k^2+1$ rows. The multiplicity of a representation $M_\lambda$, in $\mathrm{Td}_n(M_k(F))$ by Proposition 12.1.20, also vanishes as soon as the height of $\lambda$ is $> k^2$. Therefore, from Theorem 12.3.20, the two multiplicities of cocharacters may differ only for partitions $\lambda$ with $\lambda_{k^2} \leq 1$. Thus to prove that Theorem 12.3.20 implies Theorem 22.1.3, that is formula (22.5), it is enough to show that the contribution to the asymptotic behavior of partitions with the $k^2$ row bounded by some fixed number is negligible since it is $< Cn^t(k^2-1)^n$. This follows easily from Lemma 21.2.5. ☐

Therefore formula (22.1) is equivalent to

$$\sum_{\lambda \vdash n+1 \,|\, \mathrm{ht}(\lambda) \leq k} f_\lambda^2 = \mathrm{Td}_n(M_k(F)) = \mathrm{td}_{n+1}(M_k(F)) \simeq C_k n^{\frac{1-k^2}{2}} k^{2(n+1)}. \tag{22.6}$$

If we are only interested in the exponent, the proof is quite easy. We first have that, since $\mathrm{End}_{\mathrm{GL}(V)} V^{\otimes n} \subset \mathrm{End}\, V^{\otimes n}$,

$$\mathrm{Td}_n(M_k(F)) = \dim \mathrm{End}_{\mathrm{GL}(V)} V^{\otimes n} \leq k^{2n}.$$

Next let $S_\lambda(V), M_\lambda$ be the irreducible $\mathrm{GL}(V), S_n$-module appearing in the decomposition (6.8) of $V^{\otimes n}$ and $d_\lambda(V), f_\lambda$ their dimension. We have $\sum_{\lambda \vdash n \,|\, \mathrm{ht}(\lambda) \leq k} d_\lambda(V) f_\lambda = k^n$. Moreover by Cauchy's formula (6.18) we have

$$\bigoplus_{\lambda \vdash n} S_\lambda(V) \otimes S_\lambda(V) = S^n[V \otimes V]$$

which implies

$$\sum_{\lambda \vdash n} d_\lambda(V)^2 = \dim S^n[V \otimes V] = \binom{n+k^2-1}{k^2-1}.$$

Then by Cauchy's formula and the Cauchy–Schwarz inequality we get

$$k^{2n} = \Big(\sum_{\lambda \vdash n \,|\, \mathrm{ht}(\lambda) \leq k} d_\lambda(V) f_\lambda\Big)^2 \leq \mathrm{Td}_n(M_k(F)) \binom{n+k^2-1}{k^2-1}$$

$$\implies \mathrm{Td}_n(M_k(F)) \geq k^{2n} \binom{n+k^2-1}{k^2-1}^{-1}.$$

From this Corollary 22.1.2 follows.

22.1.1.1. *Beyond matrices.* The case of an algebra $R$ satisfying a Capelli identity will be discussed in §23.1.1. For the general case, the result is not as precise as the previous case, and was already discussed in §21.2.

**22.1.2. The central limit theorem.** *We are inspired by a standard theorem of probability. In fact we shall not use this directly, so this section serves only as motivation and can be skipped. For basic facts we refer to* **[Dur10]**.

Recall that an $\mathbb{R}^k$-valued *random variable* is a measurable map $X : \Pi \to \mathbb{R}^k$, where $\Pi$ is a set with a probability measure $dp$. Given a (measurable) set $A \subset \mathbb{R}^k$, we set

$$P(X \in A) := \text{measure}\, X^{-1}(A).$$

In general one assumes that $X \in L^2$ (i.e., $\int |X|^2 dp < \infty$) and then

DEFINITION 22.1.4. The *expected value* $E(X) =: \int_\Pi X(p)dp \in \mathbb{R}^k$ is called *the mean*, and $E(|X - E(X)|^2) \in \mathbb{R}^+$ is the *variance*.

For two random variables $X, Y$ (on the same space $\Pi$ with values in $R^k$) and $(X, Y)$ their scalar product, one defines the *covariance*

$$\operatorname{cov}(X, Y) := E[(X - E(X), Y - E(Y))] = E((X, Y)) - (E(X), E(Y)).$$

Given a finite-dimensional real vector space $V$ and $X : \Pi \to V$ a random variable (when $V = \mathbb{R}^k$, in coordinates $X = (x_1, \dots, x_k)$), we have the positive semidefinite quadratic form $Q_X^*(u) := E((u, X - E(X))^2)$ (intrinsically defined for $u$ in the dual space $V^*$) and in dual coordinates $u = (u_1, \dots, u_k)$,

$$\begin{aligned} Q_X^*(u) &:= E((u, X - E(X))^2) \\ &= E(\sum_i u_i^2 (x_i - E(x_i))^2 + 2 \sum_{i<j} u_i u_j (x_i - E(x_i))(x_j - E(x_j))), \end{aligned}$$

with positive semidefinite *covariant matrix*,

$$\Sigma_X := (\operatorname{cov}(x_i, x_j)). \tag{22.7}$$

If this quadratic form is positive, then it induces a positive quadratic form $Q(v)$ on $V$, a Euclidean structure, with matrix $\Sigma_X^{-1}$ in the coordinates and a translation invariant measure $dv = dx_1 \cdots dx_k$ in coordinates. Hence it induces the *Gaussian, or normal, distribution*, i.e., the probability measure on $V$,

$$\mathcal{N}_k(0, \Sigma_X) = \frac{1}{\sqrt{(2\pi)^k |Q|^{-1}}} e^{-\frac{Q(v)}{2}} dv, \quad |Q| = \det(\Sigma_X^{-1}). \tag{22.8}$$

The identity map $1_V : V \to V$, thought of as random variable for this probability space, has the matrix $\Sigma_X$ as covariant matrix.

Two $\mathbb{R}^k$-valued random variables $X, Y$ are said to be *independent* if

$$\forall A, B \subset \mathbb{R}^k \text{ (measurable sets)} \mid P((X, Y) \in A \times B) = P(X \in A) \cdot P(Y \in B).$$

In particular for two independent real-valued random variables one clearly has

$$E(XY) = E(X)E(Y) \implies \operatorname{cov}(X, Y) = 0.$$

Given a random variable $X$, one can produce $n$ *replicas* $\mathbf{X_1}, \dots, \mathbf{X_n}$ of $X$ (defined on $\Pi^n$) which are independent. Then one constructs the *average* random variable

$$\bar{\mathbf{X}}_\mathbf{n} := 1/n \sum_{i=1}^n \mathbf{X_i}.$$

If $\mu$ denotes the common expectation value $E(\mathbf{X_i})$, we have still $E(\bar{\mathbf{X}}_\mathbf{n}) = \mu$.

Let us next remark that if $X$ is a real-valued random variable with $E(X) = 0$ and variance $\sigma^2$ the variance of $\bar{\mathbf{X}}_\mathbf{n}$ is $\sigma^2/n$, it is natural in general to consider the random variables $\sqrt{n}(\bar{\mathbf{X}}_{\mathbf{n}} - \mu)$ which have always the same variance $\sigma^2$ and see if, as $n$ goes to $\infty$, they converge. In fact one of the main theormes of probability theory is that they converge to the Gaussian distribution of variance $\sigma^2$.

There is a similar analysis for vector variables. The multivariate *central limit theorem* states that

$$\sqrt{n}\,(\bar{\mathbf{X}}_n - \mu) \overset{D}{\to} \mathcal{N}_k(0, \Sigma_X), \tag{22.9}$$

where the covariance matrix $\Sigma_X$ is defined in formula (22.7) and the convergence is in Law, that is if $f(v)$ is a continuous function on $V$ such that

$$\frac{1}{\sqrt{(2\pi)^k |Q|^{-1}}} \int_V |f(v)| e^{-\frac{Q(v)}{2}} dv < \infty,$$

one has

$$\lim_{n\to\infty} E(f(\sqrt{n}(\bar{\mathbf{X}}_{\mathbf{n}} - \mu))) = \frac{1}{\sqrt{(2\pi)^k |Q|^{-1}}} \int_V f(v) e^{-\frac{Q(v)}{2}} dv < \infty. \tag{22.10}$$

22.1.2.1. *Using the central limit theorem.* In our case the treatment is truly combinatorial, and we start with $\mathbb{R}^k$ with canonical basis $e_1, \dots, e_k$.

DEFINITION 22.1.5. Denote by $X$ a random variable which takes only the values $e_1, \dots, e_k$ with equal probability $1/k$.

In this example we have that $E(X) = \frac{1}{k}\sum_{i=1}^k e_i := \mu$. The random variable $X - \mu$ has mean 0 and takes values in the vector lattice $(\mathbb{Z}^k - \mu) \cap U$ with the space $U$ defined by

$$U = \{(x_1, \dots, x_k) \in \mathbb{R}^k \mid \sum_i x_i = 0\}. \tag{22.11}$$

On $U$ one has the standard Euclidean structure induced by the norm $\sum_{i=1}^k x_i^2$ on $\mathbb{R}^k$ and thus a Euclidean measure $dy = dy_1 \cdots dy_{k-1}$ in any system of coordinates $(y_1, \dots, y_{k-1})$ with respect to some orthonormal basis. As for the covariance matrix we have

$$E((u, X - E(X))^2) = \frac{1}{k}\sum_{i=1}^k (u_i - \frac{1}{k}\sum_j u_j)^2 = \frac{1}{k}\left(\sum_i u_i^2 - \frac{1}{k}(\sum_j u_j)^2\right).$$

Using the Euclidean structure, we identify $U$ with its dual. Restricted to $U$, we have $\sum_j u_j = 0$, so

$$E((u, X - E(X))^2) = \frac{1}{k}|u|^2, \quad u \in U. \tag{22.12}$$

LEMMA 22.1.6. *The $(k-1)$-dimensional lattice $\mathbb{Z}^k \cap U$ divides $U$ into cells of volume $\sqrt{k}$ with respect to the Euclidean measure.*

PROOF. A basis of this lattice is given by the elements $u_i := e_i - e_k$, $i = 1, \dots, k-1$. Thus the volume of a cell is the absolute value of the determinant of the matrix $Z$ with rows the coordinates of these vectors in an orthonormal basis of $U$. This is the square root of the determinant of the matrix $X = ZZ^t$ of scalar products $(u_i, u_j) = 1, i \neq j, (u_i, u_i) = 2$. We have $X = 1 + Y$ is the identity matrix plus the matrix $Y$ all of 1, which has eigenvalues $k-1, 0, \dots, 0$. Hence $X$ has eigenvalues $k, 1, \dots, 1$ and determinant $k$. □

The quadratic form $Q(x) = k|x|^2$ on the $(k-1)$-dimensional space $U$ has constant diagonal matrix $k$ in an orthonormal basis. Then formula (22.8) becomes

LEMMA 22.1.7. *We have*

$$\int_U e^{-\frac{k}{2}\sum_{i=1}^k x_i^2} dy = \left(\int_{-\infty}^{\infty} e^{-k\frac{t^2}{2}} dt\right)^{k-1} = \sqrt{\frac{2\pi}{k}}^{k-1}. \tag{22.13}$$

PROOF. In an orthonormal basis of coordinates $(y_1, \dots, y_{k-1})$ for $U$ we have $\sum_{i=1}^k x_i^2 = \sum_{i=1}^{k-1} y_i^2$, so

$$\int_U e^{-\frac{k}{2}\sum_{i=1}^k x_i^2} dy = \int_{\mathbb{R}^{k-1}} e^{-\frac{k}{2}\sum_{i=1}^{k-1} y_i^2} dy_1 \cdots dy_{k-1} = \sqrt{\frac{2\pi}{k}}^{k-1}. \qquad \square$$

For every $n$ consider the set

$$L_k(n) := \{(\ell_1, \ell_2, \dots, \ell_k),\ \ell_i \in \mathbb{N},\ \sum_i \ell_i = n\}.$$

If we take independent replicas $X_i$ of $X$, we have the random variable $\frac{1}{n}\sum_{i=1}^n X_i$ taking values only on the set $\frac{1}{n}L_k(n)$, moreover

$$P(\sum_{i=1}^n X_i = (a_1, \dots, a_k)) = \binom{n}{a_1, \dots, a_k} k^{-n}. \tag{22.14}$$

The random variable appearing in the central limit theorem is thus given by $\sqrt{n}(\frac{1}{n}\sum_{i=1}^n X_i - \mu)$ and takes values in $\frac{1}{\sqrt{n}}L_k(n) - \sqrt{n}\mu$. The central limit theorem in our case can be stated as follows.

Given $(h_1, \dots, h_k) \in \mathbb{R}^k \mid \sum_i h_i = n$, following formula (22.9), we write

$$h_j = \frac{n}{k} + c_j\sqrt{n} \iff c_j = \frac{1}{\sqrt{n}}(h_j - \frac{n}{k}) \tag{22.15}$$

and set

$$\mathsf{c} := (c_1, \dots, c_k), \quad (h_1, \dots, h_k) = \rho(c_1, \dots, c_k) = \rho(\mathsf{c}) := \sqrt{n}\cdot \mathsf{c} + n \cdot \mu. \tag{22.16}$$

Then the set $\Gamma_n$ of elements $\mathsf{c} := (c_1, \dots, c_k) \in U$ with the corresponding

$$\rho(\mathsf{c}) = (h_1, \dots, h_k) \in \mathbb{Z}^k, \quad (\sum_i h_i = n),$$

is the lattice $\sqrt{n}^{-1}[\mathbb{Z}^n - n\,\mu] \cap U$ which, by Lemma 22.1.6 divides $U$ into cells of volume $n^{-\frac{k-1}{2}}\sqrt{k}$ with respect to the Euclidean measure. Set

$$\bar{L}_k(n) := \{\mathsf{c} \in U \mid \rho(\mathsf{c}) = (\ell_1, \ell_2, \dots, \ell_k),\ \ell_i \in \mathbb{N},\ \sum_i \ell_i = n\}. \tag{22.17}$$

Denote by $\bar{L}_k(n) \subset \Gamma_n \subset U$ the set of elements $\mathsf{c} \in U$ with the corresponding $\rho(\mathsf{c}) = (h_1, \dots, h_k) \in L_k(n)$.

THEOREM 22.1.8 (Central limit theorem). *For every continuous function $f(\mathsf{c})$, where the variable $\mathsf{c} = (c_1, \dots, c_k)$ is in $U$, with*

$$\int_U |f(u)| e^{-\frac{k}{2}|u|^2} dy < \infty,$$

*using formula* (22.12) *and* (22.8) *or* (22.13), *one has*

(22.18) $$\lim_{n\to\infty}\sum_{\mathrm{c}\in\bar{L}_k(n)} f(\mathrm{c})\binom{n}{\rho(\mathrm{c})}/k^n = \left(\sqrt{\frac{k}{2\pi}}\right)^{k-1}\int_U f(u)e^{-\frac{k}{2}|u|^2}dy.$$

A remark is in order. Set $|\mathrm{c}|^2 := \sum_{i=1}^k c_i^2$. By the theory of the Riemann integral, using a sum over the lattice $\Gamma_n$ one also has

$$\lim_{n\to\infty}\sqrt{k}\sum_{\mathrm{c}\in\Gamma_n} f(\mathrm{c})e^{-\frac{k}{2}\sum_{i=1}^k c_i^2}n^{-\frac{k-1}{2}} = \int_U f(u)e^{-\frac{k}{2}|u|^2}dy.$$

But by the fast decay at infinity of the Gaussian function, also

(22.19) $$\lim_{n\to\infty}\sqrt{k}\sum_{\mathrm{c}\in\bar{L}_k(n)} f(\mathrm{c})e^{-\frac{k}{2}\sum_{i=1}^k c_i^2}n^{-\frac{k-1}{2}} = \int_U f(u)e^{-\frac{k}{2}|u|^2}dy.$$

This means that, comparing formulas (22.18) and (22.19), we should look for some close correlation, as $n\to\infty$, between $\sqrt{k}e^{-\frac{k|\mathrm{c}|^2}{2}}n^{-\frac{k-1}{2}}$ and $\left(\sqrt{\frac{k}{2\pi}}\right)^{1-k}\binom{n}{\rho(\mathrm{c})}/k^n$. That is some statement close to (but stronger than)

(22.20) $$\sqrt{2\pi}^{1-k}k^{\frac{k}{2}}k^n e^{-\frac{k|\mathrm{c}|^2}{2}}n^{-\frac{k-1}{2}} \simeq \binom{n}{\rho(\mathrm{c})},$$

$$\iff \sqrt{2\pi}^{1-k}k^{\frac{k}{2}}k^n e^{-\frac{k}{2n}|h|^2+\frac{n}{2}}n^{-\frac{k-1}{2}} \simeq \binom{n}{h_1,\dots,h_k}.$$

A precise statement will be shown in formula (22.55) to prove Theorem 22.2.8.

## 22.2. The codimension estimate for matrices

*This section is devoted to the proof of formula* (22.6).

**22.2.1. Partitions and the symmetric group.** The partitions $\lambda\vdash n$ of $n$ index the irreducible representations of $S_n$. Recall that we denote $M_\lambda$ the irreducible representation corresponding to $\lambda$ and $f_\lambda := \dim M_\lambda$.

DEFINITION 22.2.1. Let $P_k(n) := \{\lambda\vdash n \mid ht(\lambda)\le k\}$. That is

(22.21) $$P_k(n) := \{(r_1,\dots,r_k)\in\mathbb{N}^k \mid r_1\ge\cdots\ge r_k,\ \sum_{i=1}^k r_i = n\},$$
$$\Pi_k(n) := \{(r_1,\dots,r_k)\in\mathbb{N}^k \mid r_1>\cdots> r_k,\ \sum_{i=1}^k r_i = n\}.$$

Let (the definition comes from Lie theory **[Pro07]**)

$$\rho_k := (k-1,k-2,\dots,2,1,0)\in\Pi_k\left(\frac{k(k-1)}{2}\right).$$

LEMMA 22.2.2. *The mapping* $\lambda=(r_1,\dots,r_k)\mapsto\lambda+\rho_k$ *is a 1–1 correspondence between* $P_k(n)$ *and the set* $\Pi_k(m)$, *with* $m := n+k(k-1)/2$.

PROOF. This simple exercise is left to the reader. □

Recall the hook formula or Young–Frobenius formula (6.4.2) for $f_\lambda$.

Let $\lambda = (r_1, \ldots, r_k)$, $\lambda + \rho_k = (\ell_1, \ldots, \ell_k)$, $\ell_i = r_i + k - i$.

$$f_\lambda = n! \frac{\Delta(\ell_1, \ldots, \ell_k)}{\ell_1! \cdots \ell_k!} = \frac{n!}{m!}\binom{m}{\ell_1, \ldots, \ell_k}\Delta(\ell_1, \ldots, \ell_k), \tag{22.22}$$

where

$$\Delta(\ell_1, \ldots, \ell_k) := \prod_{1 \le i < j \le k} (\ell_i - \ell_j) = \prod_{1 \le i < j \le k} (r_i - r_j - i + j). \tag{22.23}$$

For $\mu = (\ell_1, \ldots, \ell_k) \in P_k(m)$, set

$$g_\mu := \binom{m}{\ell_1, \ldots, \ell_k}\Delta(\ell_1, \ldots, \ell_k). \tag{22.24}$$

REMARK 22.2.3. Notice that $g_\mu = 0$ for $\mu \in P_k(m) \setminus \Pi_k(m)$. For every pair of $u, v \in \mathbb{N}$, we have

$$\frac{(u+v)!}{u!} = \prod_{j=1}^{v}(u+j) = u^v \prod_{j=1}^{v}(1 + \frac{j}{u}), \tag{22.25}$$

therefore

$$\boxed{f_\lambda = n^{-k(k-1)/2} \prod_{1 \le j \le k(k-1)/2} (1 + \frac{j}{n})^{-1} g_{\lambda+\rho_k}}. \tag{22.26}$$

**The first problem.** For $\beta > 0$, in particular $\beta = 2$, we are interested in the asymptotic behavior of

$$c_\beta(k,n) := \sum_{\lambda \in P_k(n)} f_\lambda^\beta = n^{-\frac{\beta k(k-1)}{2}} \prod_{1 \le j \le k(k-1)/2} (1 + \frac{j}{n})^{-\beta} \sum_{\lambda \in P_k(n)} g_{\lambda+\rho_k}^\beta. \tag{22.27}$$

Since $\prod_{1 \le j \le k(k-1)/2}(1 + \frac{j}{n})^{-\beta}$ tends to 1 as $n \to \infty$, we may equivalently study the asymptotics of

$$C_\beta(k,m) := \sum_{\lambda \in P_k(m)} g_\lambda^\beta \simeq n^{\frac{\beta k(k-1)}{2}} c_\beta(k,n), \quad m = n + \frac{k(k-1)}{2}. \tag{22.28}$$

Observe that, if $\lambda = \rho(\mathrm{c})$ (see Definition 22.16), we have $g_\lambda = |\Delta(\rho(\mathrm{c}))|\binom{m}{\rho(\mathrm{c})}$.

Now $|\Delta(\rho(\mathrm{c}))|^\beta \binom{m}{\rho(\mathrm{c})}^\beta$ is symmetric with respect to the group $S_k$ permuting the variables $c_i$ and it is 0 when two of the $c_i$ are equal. The open set of the vector space $U$ where the $c_i$ are all distinct, decomposes into $k!$ connected components, one of which is $c_1 > \cdots > c_k$, permuted in a simply transitive way by the symmetric group $S_k$.

Thus formula (22.28) becomes, with $\bar{L}_k(m)$ defined in (22.17),

$$C_\beta(k,m) = \sum_{\lambda \in P_k(m)} g_\lambda^\beta = \frac{1}{k!} \sum_{\mathrm{c} \in \bar{L}_k(m)} |\Delta(\rho(\mathrm{c}))|^\beta \binom{m}{\rho(\mathrm{c})}^\beta, \ m = n + \frac{k(k-1)}{2}. \tag{22.29}$$

Next clearly if $\rho(\mathrm{c}) = \lambda \in P_k(m)$, we have

$$|\Delta(\rho(\mathrm{c}))| = m^{\frac{k(k-1)}{4}} |\Delta(\mathrm{c})|,$$

so, *if in the asymptotics we can substitute*, as in formula (22.20), we deduce

$$C_\beta(k,m) \simeq \frac{1}{k!}(\sqrt{2\pi}^{1-k}k^{\frac{k}{2}})^\beta \sum_{c\in \bar{L}_k(m)} m^{\beta\frac{k(k-1)}{4}}|\Delta(c)|^\beta e^{-\frac{\beta k}{2}|c|^2} m^{-\beta\frac{k-1}{2}} k^{\beta m}$$

$$(22.30) \qquad = F_\beta(k,m) \sum_{c\in \bar{L}_k(m)} |\Delta(c)|^\beta e^{-\frac{\beta k}{2}|c|^2}\sqrt{k}m^{-\frac{k-1}{2}},$$

$$F_\beta(k,m) := \frac{1}{\sqrt{k}k!}(\sqrt{2\pi}^{1-k}k^{\frac{k}{2}})^\beta m^{\beta\frac{(k-2)(k-1)}{4}+\frac{k-1}{2}}k^{\beta m}.$$

We then have that

$$\lim_{m\to\infty} \sum_{c\in \bar{L}_k(m)} |\Delta(c)|^\beta e^{-\frac{\beta k}{2}|c|^2}\sqrt{k}\, m^{-\frac{k-1}{2}} = I(k,\beta)$$

$$(22.31) \qquad = \int_U |\Delta(c)|^\beta e^{-\frac{\beta k}{2}|c|^2} dy < \infty$$

is a Riemann sum converging to the integral $I(k,\beta)$ on $U = \mathbb{R}^{k-1}$ which one can compute from the Mehta and Selberg integral in formula (22.65). In particular

$$I(k,2) = (2\pi)^{\frac{k-1}{2}}(2k)^{-(k^2-1)/2}\prod_{j=1}^{k} j!.$$

So finally

THEOREM 22.2.4.

$$(22.32) \quad C_\beta(k,m) \simeq A_{k,\beta}m^{\beta\frac{(k-2)(k-1)}{4}+\frac{k-1}{2}}k^{\beta m}, \quad A_{k,\beta} = \frac{1}{k!\sqrt{k}}(\sqrt{2\pi}^{1-k}k^{\frac{k}{2}})^\beta I(k,\beta).$$

Furthermore $m^{\beta\frac{(k-2)(k-1)}{4}+\frac{k-1}{2}} \simeq n^{\beta\frac{(k-2)(k-1)}{4}+\frac{k-1}{2}}$, hence the two formulas (22.32) and (22.28) imply,

$$(22.33) \qquad c_\beta(k,m) \simeq B_{k,\beta}n^{-\frac{\beta(k+2)(k-1)}{4}+\frac{k-1}{2}}k^{\beta n}, \quad B_{k,\beta} = A_{k,\beta}k^{\frac{\beta k(k-1)}{2}}.$$

For the case of the codimension of $k\times k$ matrices we have $\beta = 2$. Since we have by formula (22.4) $c_2(k,n+1) = \mathrm{Td}_n(M_k(F)) \simeq c_n(M_k(F))$, we have, as corollary, Theorem 22.1.1:

COROLLARY 22.2.5.

$$(22.34) \qquad c_n(M_k(F)) \simeq C_k n^{-\frac{(k+1)(k-1)}{2}}k^{2(n+1)}.$$

$$(22.35) \qquad C_k = B_{k,2} = (1/\sqrt{2\pi})^{(k-1)}\cdot 2^{-(k^2-1)/2}k^{k^2/2}\prod_{j=1}^{k-1} j!$$

The problem is thus to justify formula (22.30). In fact as we shall see in Theorem 22.2.12, one can substitute as in formula (22.20) only around the **center** of the distribution, which we will fix to be $k|c_i| < K\sqrt{\log n}$ with $K > 0$ a suitable constant. Then we must prove that the error made in the **tail** is small with $m$. This is a probabilistic argument which we do in the next section.

**22.2.2. Some analytic estimates.** We use the standard notation $O(z^j)$, $|z| < c < 1, j > 0$, to denote any function $f(z)$ of $z$ for which in the range $|z| < c$ one has $|f(z)| < K|z|^j$ for some $K > 0$. This applies also if $z = n^{-1}$, $n \in \mathbb{N}, n \gg 0$. If the function depends on parameters, we assume $K$ independent of the parameters.

REMARK 22.2.6. This will be used mainly for $\log(1+x) = \sum_{i=1}^{\infty}(-1)^{i+1}\frac{x^i}{i}$, $|x| < 1$ and $e^x$.

REMARK 22.2.7. We use systematically the fact that if $f(z) = \sum_{j=0}^{\infty} a_j z^j$ is a converging power series for $|z| < c'$, then for every $c < \min(1, c')$ one has $\sum_{j=h}^{\infty} a_j z^j = z^h \sum_{j=0}^{\infty} a_{j+h} z^j = O(z^h)$, $\forall h$, $|z| < c$.

Our first goal is to prove a precise formulation of formula (22.20):

THEOREM 22.2.8. *If $k \max(|c_j|) < K\sqrt{\log n}$ with $K > 0$, we have*

$$\left| e^{-\frac{k|c|^2}{2}} - \sqrt{2\pi}^{k-1} k^{-\frac{k}{2}} k^{-n} n^{\frac{k-1}{2}} \binom{n}{\rho(c)} \right| \leq e^{-\frac{k|c|^2}{2}} O(\log(n)^{3/2} n^{-\frac{1}{2}}). \tag{22.36}$$

This will be done in two steps. The first is Lemma 22.2.9, which produces formula (22.47) for the second term of formula (22.36), and it is a straightforward application of Stirling's formula.

The second is Theorem 22.2.12, which completes the proof.

22.2.2.1. *Stirling's formula.* A basic tool here is Stirling's formula, which gives the asymptotic of $n!$. We shall use it in the form:

$$\text{(i)} \quad n! = \sqrt{2\pi n}\left(\frac{n}{e}\right)^n e^{\lambda_n}, \quad \text{(ii)} \quad \frac{1}{12n+1} < \lambda_n < \frac{1}{12n}. \tag{22.37}$$

There are several proofs of this formula. If one does not need to specify the value $\sqrt{2\pi}$ of the constant, the proof is quite elementary, and it already appears in de Moivre's 1730 *Miscellanea analytica*, where he also has the first results approximating binomial coefficients through the normal curve.

A simple proof consists in computing $\log(n!) = \sum_{i=1}^{n-1} \log(i+1)$ and considering $\log(p+1)$ as the area of a rectangle with base $[p, p+1]$ and height $\log(p+1)$. This area is divided into three parts, $\log(p+1) = A_p + B_p - \epsilon_p$, the part $A_p$ under the curve $\log(x)$, and that above the curve, which is the difference between the area $B_p$ of a right triangle of legs $1, (\log(p+1) - \log(p))$ minus the area of the *small* part $\epsilon_p$ between this curve and the *secant* hypothenuse of the previous triangle. The computation is by elementary integrals using the fact that a primitive of $\log(x)$ is $x\log(x) - x$.

The total contribution of the areas $\sum_p A_p$, $p = 1, \ldots, n-1$ is

$$\int_1^n \log(x)dx = [x\log(x) - x]_1^n = n\log n - n + 1. \tag{22.38}$$

The total contribution of the areas of the $B_p$, $p = 1, \ldots, n-1$ is

$$\sum_{p=1}^{n-1} \frac{1}{2}(\log(p+1) - \log(p)) = \frac{\log n}{2}. \tag{22.39}$$

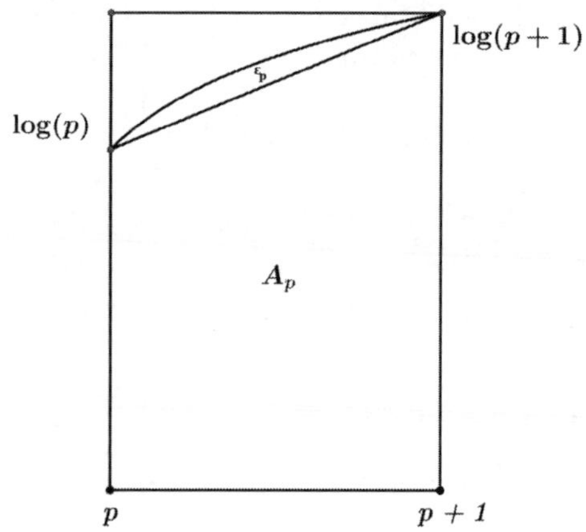


FIGURE 1. The three areas

It remains the contribution of the parts $\epsilon_p$ which is
(22.40)
$$\sum_{p=1}^{n-1}[\int_p^{p+1}\log(x)dx - \frac{1}{2}(\log(p+1)+\log(p))] = \sum_{p=1}^{n-1}[(p+\frac{1}{2})\log(\frac{p+1}{p})-1].$$

Thus the total of formulas (22.38), (22.39), and (22.40) is

$$n\log n - n + \frac{1}{2}\log n + 1 - \sum_{p=1}^{n-1}[(p+\frac{1}{2})\log(\frac{p+1}{p})-1].$$

Since by formula (22.42) we have $(p+\frac{1}{2})\log(\frac{p+1}{p}) - 1 > 0$, the asymptotic formula is equivalent to proving that

$$1 - \sum_{p=1}^{\infty}[(p+\frac{1}{2})\log(\frac{p+1}{p})-1] = \log(\sqrt{2\pi}). \tag{22.41}$$

The computation of formula (22.41) is usually done by different approaches to formula (22.37), and we leave it to the textbooks, or see **[Rom00]**.

Note that by the next formula (22.43), the series converges to a number $y$ with $0 < y < 1 - 1/12$. This gives formula (22.37) with the unknown constant $e^{1-y}$ in place of the explicit $\sqrt{2\pi}$.

For the equality one has to estimate the error for $n$,

$$\lambda_n := \sum_{p=n}^{\infty}[(p+\frac{1}{2})\log(\frac{p+1}{p})-1].$$

We follow Robbins **[Rob55]**. Recall the series for $x < 1$,

$$\log(\frac{1+x}{1-x}) = \log(1+x) - \log(1-x) = 2(x + \frac{x^3}{3} + \frac{x^5}{5} + \frac{x^7}{7} + \cdots).$$

Therefore setting $x = (2p+1)^{-1}$, we have $\frac{p+1}{p} = \frac{1+x}{1-x}$ and

$$(p+\frac{1}{2})\log(\frac{p+1}{p}) - 1 = \frac{1}{x}(x + \frac{x^3}{3} + \frac{x^5}{5} + \frac{x^7}{7} + \cdots) - 1 = \frac{x^2}{3} + \frac{x^4}{5} + \frac{x^6}{7} + \cdots$$

$$\frac{x^2}{3} + \frac{x^4}{5} + \frac{x^6}{7} + \cdots < \frac{x^2}{3}(\sum_{i=0}^{\infty} x^{2i}) = \frac{x^2}{3(1-x^2)} = \frac{1}{12}\left(\frac{1}{p} - \frac{1}{p+1}\right). \tag{22.42}$$

$$\implies \lambda_n < \frac{1}{12}\sum_{p=n}^{\infty}\left(\frac{1}{p} - \frac{1}{p+1}\right) = \frac{1}{12n}. \tag{22.43}$$

$$\frac{x^2}{3} + \frac{x^4}{5} + \frac{x^6}{7} + \cdots > \frac{x^2}{3}(\sum_{i=0}^{\infty}(\frac{x^2}{3})^i) = \frac{x^2}{3(1-\frac{x^2}{3})} > \frac{1}{12}\left(\frac{1}{p+\frac{1}{12}} - \frac{1}{p+1+\frac{1}{12}}\right),$$

$$\implies \lambda_n > \frac{1}{12}\sum_{p=n}^{\infty}\left(\frac{1}{p+\frac{1}{12}} - \frac{1}{p+1+\frac{1}{12}}\right) = \frac{1}{12n+1}.$$

This finishes the proof of formula (22.37)

22.2.2.2. *Using Stirling's formula.* One deduces that if $h_i > 0$,

$$\binom{n}{h_1,\ldots,h_k} = \sqrt{2\pi}^{1-k}\sqrt{\frac{n}{\prod_i h_i}}\left(\frac{n}{e}\right)^n \cdot \prod_i \left(\frac{e}{h_i}\right)^{h_i} e^{\lambda_n - \sum \lambda_{h_i}}$$

$$= \sqrt{2\pi}^{1-k} n^{n+\frac{1}{2}} \cdot \prod_i \left(\frac{1}{h_i}\right)^{h_i+\frac{1}{2}} e^{\lambda_n - \sum \lambda_{h_i}}$$

$$= \sqrt{2\pi}^{1-k} n^{\frac{(-k+1)}{2}} \cdot \prod_i \left(\frac{n}{h_i}\right)^{h_i+\frac{1}{2}} e^{\lambda_n - \sum \lambda_{h_i}}. \tag{22.44}$$

By induction on $k$ we have that the contribution of the partitions with $k-1$ rows to formula (22.6) is negligible, and we may assume thus $h_i \geq 1$ for all $i$.

Write, as in formula (22.15), $h_j = \frac{n}{k} + c_j\sqrt{n},\ 1 \leq h_j \leq n$ so that

$$c_1 + \cdots + c_k = 0 \quad \text{with} \quad -\frac{\sqrt{n}}{k} + \sqrt{n}^{-1} \leq c_j < \sqrt{n},\ j = 1,\ldots,k. \tag{22.45}$$

We deduce

$$\frac{h_j}{n} = \frac{1}{k}(1 + \frac{kc_j}{\sqrt{n}}) \implies \prod_j \left(\frac{n}{h_j}\right)^{h_j+\frac{1}{2}} = k^{n+\frac{k}{2}}\prod_j(1+\frac{kc_j}{\sqrt{n}})^{-\frac{n}{k}-c_j\sqrt{n}-\frac{1}{2}}. \tag{22.46}$$

Hence from formulas (22.44) and (22.46) we deduce

LEMMA 22.2.9. *Setting* $G(k,n) := \sqrt{2\pi}^{k-1}k^{-n-\frac{k}{2}}n^{\frac{k-1}{2}}$, *we have*

$$G(k,n)\binom{n}{h_1,\ldots,h_k} = \prod_j(1+\frac{kc_j}{\sqrt{n}})^{-\frac{n}{k}-c_j\sqrt{n}-\frac{1}{2}}e^{\lambda_n - \sum\lambda_{h_i}}, \tag{22.47}$$

*where for any* $x \in \mathbb{N}$ *by* $\lambda_x$ *we mean the specific number with*

$$\frac{1}{12x+1} < \lambda_x < \frac{1}{12x}$$

*appearing in Stirling's formula.*

22.2.2.3. *Computing* $\lim_{n\to\infty}\prod_j(1+\frac{kc_j}{\sqrt{n}})^{-\frac{n}{k}-c_j\sqrt{n}-\frac{1}{2}} = e^{-k\frac{\sum_{i=1}^k c_i^2}{2}}$. A basic fact of analysis is

$$\lim_{n\to\infty}(1+\frac{c}{n})^n = e^c \quad\text{or}\quad \lim_{n\to\infty}(1+\frac{c}{n})^{an+b} = e^{ac},\ \forall a,b\in\mathbb{R}. \tag{22.48}$$

We then have

PROPOSITION 22.2.10. *Let* $c_1,\dots,c_k$, $\sum_i c_i = 0$, *and set* $|\mathsf{c}|^2 := \sum_{i=1}^k c_i^2$. *Then*

$$\lim_{n\to\infty}\prod_{i=1}^k(1+\frac{c_i}{n})^{n^2} = e^{-\frac{|\mathsf{c}|^2}{2}}. \tag{22.49}$$

PROOF. For instance for $k=2$ this is

$$\lim_{n\to\infty}(1+\frac{c}{n})^{n^2}(1-\frac{c}{n})^{n^2} = \lim_{n\to\infty}(1-\frac{c^2}{n^2})^{n^2} = e^{-c^2} = e^{-\frac{c^2+(-c)^2}{2}}.$$

In general, taking logarithms, we see that since $\sum_{i=1}^k c_i = 0$, we have

$$n^2\sum_{i=1}^k\log(1+\frac{c_i}{n}) = n^2\left(\sum_{i=1}^k\frac{c_i}{n}-\frac{1}{2}\sum_{i=1}^k\frac{c_i^2}{n^2}+O(1/n^3)\right) = -\frac{1}{2}\sum_{i=1}^k c_i^2+O(1/n),$$

hence the claim. □

COROLLARY 22.2.11. *If* $\sum_i c_i = 0$, *we have*

$$\lim_{n\to\infty}\prod_j(1+\frac{kc_j}{\sqrt{n}})^{-\frac{n}{k}-c_j\sqrt{n}-\frac{1}{2}} = e^{-k\frac{\sum_{i=1}^k c_i^2}{2}}. \tag{22.50}$$

PROOF. By formula (22.49) we have

$$\lim_{n\to\infty}\prod_j(1+\frac{kc_j}{\sqrt{n}})^{-\frac{n}{k}} = \lim_{n\to\infty}\prod_j(1+\frac{kc_j}{n})^{-\frac{n^2}{k}} = e^{k^{-1}\frac{\sum_{i=1}^k(kc_i)^2}{2}} = e^{k\frac{\sum_{i=1}^k c_i^2}{2}},$$

while by formula (22.48),

$$\lim_{n\to\infty}(1+\frac{kc_j}{\sqrt{n}})^{-c_j\sqrt{n}-\frac{1}{2}} = e^{-kc_j^2},$$

hence, since $e^{k\frac{\sum_{i=1}^k c_i^2}{2}}\prod_j e^{-kc_j^2} = e^{-k\frac{\sum_{i=1}^k c_i^2}{2}}$, the claim. □

22.2.2.4. *Estimate of the error.* We now give a quantitative form of Corollary 22.2.11.

THEOREM 22.2.12. *If* $\max(k|c_j|) < K\sqrt{\log(n)}$, *we have*

$$\prod_j(1+\frac{kc_j}{\sqrt{n}})^{-\frac{n}{k}-c_j\sqrt{n}-\frac{1}{2}}e^{\lambda_n-\sum_i\lambda_{h_i}} = e^{-k\frac{|\mathsf{c}|^2}{2}+O(\log(n)^{3/2}n^{-\frac{1}{2}})}. \tag{22.51}$$

PROOF. Let $z := \sqrt{\frac{\log(n)}{n}}$, and notice that since $k\max(|c_j|) < K\sqrt{\log(n)}$, we have

$$\frac{1}{h_j} = \frac{k}{n}(1+kc_j\sqrt{n}^{-1})^{-1} = \frac{k}{n}(1+O(\sqrt{\log(n)/n}))^{-1} = O(1/n).$$

Thus from formula (22.37)(ii) we have $\lambda_n - \sum_i \lambda_{h_i} = O(1/n)$. Now, by Remark 22.2.6,
(22.52)

$$\log(1+\frac{k\,c_j}{\sqrt{n}}) = O(z) = \frac{k\,c_j}{\sqrt{n}} + O(z^2), \quad \sum_{j=1}^{k} \log(1+\frac{k\,c_j}{\sqrt{n}}) = -\frac{k^2|c|^2}{2n} + O(z^3).$$

Set

$$\log\left(\prod_j (1+\frac{k\,c_j}{\sqrt{n}})^{-\frac{n}{k}-c_j\sqrt{n}-\frac{1}{2}}\right) := F(n,k,c) = A+B+C.$$

$$A = -\frac{n}{k}\sum_{j=1}^{k}\log(1+\frac{k\,c_j}{\sqrt{n}}),\ B = -\sum_{j=1}^{k} c_j\sqrt{n}\log(1+\frac{k\,c_j}{\sqrt{n}}), \tag{22.53}$$

$$C = -\frac{1}{2}\sum_{j=1}^{k}\log(1+\frac{k\,c_j}{\sqrt{n}}).$$

From formula (22.52) we estimate

$$A = -\frac{n}{k}(-\frac{k^2|c|^2}{2n} + O(z^3)) = \frac{k|c|^2}{2} + O(n\cdot z^3)) = \frac{k|c|^2}{2} + O(\sqrt{\frac{\log(n)^3}{n}}),$$

$$B = -\sum_{j=1}^{k} c_j\sqrt{n}(\frac{k\,c_j}{\sqrt{n}} + O(z^2)) = -k|c|^2 + O(\frac{\log(n)}{\sqrt{n}}), \quad C = O(\frac{\log(n)}{n}).$$

$$\implies F(n,k,c) = -\frac{k|c|^2}{2} + O(\sqrt{\frac{\log(n)^3}{n}}), \tag{22.54}$$

from which formula (22.51) follows. □

From this Theorem 22.2.8, that is formula (22.36), follows, and we finally have in the same range that

$$\binom{n}{h_1,\ldots,h_k} = \sqrt{2\pi}^{1-k} k^{\frac{k}{2}} k^n n^{\frac{1-k}{2}} e^{-k\frac{|c|^2}{2} + O(\log(n)^{\frac{3}{2}} n^{-\frac{1}{2}})}. \tag{22.55}$$

REMARK 22.2.13. The strong constraint $k\max(|c_j|) < K\sqrt{\log(n)}$ will be sufficient for the rest of the proof. We can also impose a weaker constraint $\max(|c_j|) < n^{\frac{1}{2}+\alpha}$, $-\frac{1}{2} < \alpha < -\frac{1}{3}$. Arguing in the same way with $z = n^{\alpha}$, we see that we have

$$\prod_j (1+\frac{k\,c_j}{\sqrt{n}})^{-\frac{n}{k}-c_j\sqrt{n}-\frac{1}{2}} e^{\lambda_n - \sum_i \lambda_{h_i}} = e^{-k\frac{|c|^2}{2} + O(n^{1+3\alpha})}. \tag{22.56}$$

From formulas (22.47) and (22.51) we finally have in the same range that

$$\binom{n}{h_1,\ldots,h_k} = \sqrt{2\pi}^{1-k} k^{\frac{k}{2}} k^n n^{\frac{1-k}{2}} e^{-k\frac{|c|^2}{2} + O(n^{1+3\alpha})},\ -\frac{1}{2} < 1+3\alpha < 0. \tag{22.57}$$

22.2.2.5. *An upper bound.* We need next an upper bound for the product

$$\prod_{j=1}^{k}\left(1+\frac{c_jk}{\sqrt{n}}\right)^{-n/k-c_j\sqrt{n}-\frac{1}{2}},$$

where

$$c_1+\cdots+c_k=0 \quad \text{with} \quad \sqrt{n}^{-1}(1-\frac{n}{k})\leq c_j<\sqrt{n}, \quad j=1,\ldots,k.$$

We start with an elementary lemma:

LEMMA 22.2.14.

(1) *Let* $-1<x\leq 1$. *Then* $(1+x)\log(1+x)\geq x+x^2/3$.
(2) *Let* $1<x$. *Then* $(1+x)\cdot\log(1+x)>x+\frac{1}{3x}$.

PROOF. (1) Expanding in series $(1+x)\cdot\log(1+x)-(x+x^2/3)$, we have

$$\sum_{i=1}^{\infty}(-1)^{i+1}\frac{x^i}{i}+\sum_{i=2}^{\infty}(-1)^{i}\frac{x^i}{i-1}-\left(x+\frac{x^2}{3}\right)$$
$$=\frac{x^2}{6}(1-x)+\sum_{i=2}^{\infty}\left(\frac{x^{2i}}{2i(2i-1)}-\frac{x^{2i+1}}{2i(2i+1)}\right).$$

Since each term

$$\frac{x^{2i}}{2i(2i-1)}-\frac{x^{2i+1}}{2i(2i+1)}=\frac{x^{2i}}{2i}\left(\frac{1}{2i-1}-\frac{x}{2i+1}\right)$$

is positive, the claim follows.

(2) We leave this as exercise. Note that $2\log 2\geq 1.4>1+1/3$ and, by computing the derivative, we have that $(1+x)\log(1+x)-x-\frac{1}{3x}$ is an increasing function. □

LEMMA 22.2.15. *Let* $c_1+\cdots+c_k=0$ *with* $\sqrt{n}^{-1}(1-\frac{n}{k})\leq c_j<\sqrt{n}$, *for* $j=1,\ldots,k$. *Then there is a constant* $\gamma>0$ *with*

$$\prod_{j=1}^{k}\left(1+\frac{c_jk}{\sqrt{n}}\right)^{n/k+c_j\sqrt{n}+\frac{1}{2}}\geq\gamma n^{-\frac{k}{2}}e^{\frac{k}{3k^2}(c_1^2+\cdots+c_k^2)},\ \gamma>0. \tag{22.58}$$

PROOF. Denote $d_j=c_jk/\sqrt{n}$. Then from the hypotheses on the elements $c_j$, we have $\frac{k}{n}-1\leq d_j<k$ and $d_1+\cdots+d_k=0$. We split the left-hand side of (22.58) as

$$\prod_{j=1}^{k}\left(1+\frac{c_jk}{\sqrt{n}}\right)^{n/k+c_j\sqrt{n}+\frac{1}{2}}=\prod_{j=1}^{k}(1+d_j)^{\frac{n}{k}(1+d_j)}\cdot\prod_{j=1}^{k}(1+d_j)^{\frac{1}{2}}$$

and estimate the two factors separately. From Lemma 22.2.15 and the two constraints on the $d_j$, it follows that $(1+d_j)\log(1+d_j) > d_j + \frac{d_j^2}{3k^3}$, hence,

$$\prod_{j=1}^{k}(1+d_j)^{\frac{n}{k}(1+d_j)} = \exp\sum_{j=1}^{k}\frac{n}{k}(1+d_j)\log(1+d_j)$$

$$\geq \exp\frac{n}{k}\sum_{j=1}^{k}\left(d_j+\frac{d_j^2}{3k^3}\right) \overset{\sum_j d_j=0}{=} \exp\frac{n}{k}\sum_{j=1}^{k}\frac{d_j^2}{3k^3}$$

$$= \exp\frac{n}{3k^4}\sum_{j=1}^{k}\frac{c_j^2k^2}{n} = \exp\frac{1}{3k^2}(c_1^2+\cdots+c_k^2). \tag{22.59}$$

We have $\prod_{j=1}^{k}(1+d_j)^{\frac{1}{2}} \geq (\frac{k}{n})^{\frac{k}{2}}$ since $d_j+1\geq\frac{k}{n}$. From this we have the desired inequality (22.58). □

We show next that this upper bound is enough to *control the tail* (see page 562).

22.2.2.6. *The main theorem.* We have set $G(k,n) := \sqrt{2\pi}^{k-1}k^{-\frac{k}{2}}k^{-n}n^{\frac{k-1}{2}}$ in Lemma 22.2.9. The asymptotic behavior of $C_\beta(k,m)$ is thus reduced to that of

$$\tilde{C}_\beta(k,m) := k!m^{-\beta\frac{k(k-1)}{4}}C_\beta(k,m)G(k,m)^\beta\sqrt{k}m^{\frac{1-k}{2}} \quad m = n+\frac{k(k-1)}{2},$$

once we prove

THEOREM 22.2.16.

$$\lim_{m\to\infty}\tilde{C}_\beta(k,m) = \int_U|\Delta(\mathrm{c})|^\beta e^{-\frac{\beta k|\mathrm{c}|^2}{2}}dy < \infty. \tag{22.60}$$

PROOF. Recall formula (22.29) with $m = n+\frac{k(k-1)}{2}$,

$$C_\beta(k,m) = \frac{1}{k!}\sum_{\mathrm{c}\in\bar{L}_k(m)}|\Delta(\rho(\mathrm{c}))|^\beta\binom{m}{\rho(\mathrm{c})}^\beta = \frac{1}{k!}m^{\beta\frac{k(k-1)}{4}}\sum_{\mathrm{c}\in\bar{L}_k(m)}|\Delta(\mathrm{c})|^\beta\binom{m}{\rho(\mathrm{c})}^\beta,$$

so

$$\tilde{C}_\beta(k,m) = \sum_{\mathrm{c}\in\bar{L}_k(m)}|\Delta(\mathrm{c})|^\beta\left(G(k,m)\binom{m}{\rho(\mathrm{c})}\right)^\beta\sqrt{k}m^{\frac{1-k}{2}}$$

$$\overset{(22.47)}{=} \sum_{\mathrm{c}\in\bar{L}_k(m)}|\Delta(\mathrm{c})|^\beta\left(\prod_j(1+\frac{kc_j}{\sqrt{m}})^{-\frac{m}{k}-c_j\sqrt{m}-\frac{1}{2}}e^{\lambda_m-\sum\lambda_{h_i}}\right)^\beta\sqrt{k}m^{\frac{1-k}{2}}. \tag{22.61}$$

Here $\bar{L}_k(m) := \{(c_1,\dots,c_k)\in U \mid h_j = \frac{m}{k}+c_j\sqrt{m}\in\mathbb{N}\}$, defined in formula (22.17), is a lattice in $U$ which divides $U$ into cells of volume $\sqrt{k}m^{\frac{1-k}{2}}$ (Lemma 22.1.6).

Let us split this sum into two parts,

$$\tilde{C}_\beta(k,m) = \tilde{C}_\beta^{(1)}(k,m)+\tilde{C}_\beta^{(2)}(k,m),$$

where the first includes all terms for which $C := k\max(|c_i|) < K\sqrt{\log m}$, the second is the remainder.

(1) By formula (22.51) the first sum can be replaced by

$$\tilde{C}_\beta^{(1)}(k,m) := \sum_{\mathrm{c}\in \bar{L}_k(m)\,|\,C<K\sqrt{\log m}} |\Delta(\mathrm{c})|^\beta \left(\prod_j (1+\frac{kc_j}{\sqrt{m}})^{-\frac{m}{k}-c_j\sqrt{m}-\frac{1}{2}} e^{\lambda_m-\sum\lambda_{h_i}}\right)^\beta \sqrt{k}\, m^{\frac{1-k}{2}}$$

$$(22.62)\qquad = \sum_{\mathrm{c}\in \bar{L}_k(m)\,|\,C<K\sqrt{\log m}} |\Delta(\mathrm{c})|^\beta e^{-\frac{\beta k|\mathrm{c}|^2}{2}+O(\log(m)^{\frac{3}{2}} m^{-\frac{1}{2}})} \sqrt{k}\, m^{\frac{1-k}{2}}.$$

Now set $R_\beta(k,m) := \sum_{\mathrm{c}\in \bar{L}_k(m)\,|\,C<K\sqrt{\log m}} |\Delta(\mathrm{c})|^\beta e^{-\frac{\beta k|\mathrm{c}|^2}{2}} \sqrt{k}\, m^{\frac{1-k}{2}}$. Hence,

$$(22.63)\qquad \tilde{C}_\beta^{(1)}(k,m) = R_\beta(k,m) e^{O(\log(m)^{\frac{3}{2}} m^{-\frac{1}{2}})}.$$

This sum $R_\beta(k,m)$ is a Riemann sum relative to a bounded part of the lattice of covolume $\sqrt{k}\, m^{\frac{1-k}{2}}$ which as $m\to\infty$ covers the entire space. Since the integrand has extremely fast decay at infinity, this Riemann sum converges to the right-hand side of formula (22.60), a finite positive number. Since from formula (22.63),

$$\begin{aligned} |\tilde{C}_\beta^{(1)}(k,m) - R_\beta(k,m)| &= |R_\beta(k,m)(1-e^{O(\log(m)^{\frac{3}{2}} m^{-\frac{1}{2}})})| \\ &\leq R_\beta(k,m) O(\log(m)^{\frac{3}{2}} m^{-\frac{1}{2}}), \end{aligned}$$

we have that also $\tilde{C}_\beta^{(1)}(k,m)$ converges to the same limit.

(2) As for the second sum we can majorize $|\tilde{C}_\beta^{(2)}(k,m)|$ by

$$\sum_{i=1}^k \sum_{\mathrm{c}\in \bar{L}_k(m)\,|\,k|c_i|\geq K\sqrt{\log m}} |\Delta(\mathrm{c})|^\beta \left(\prod_j (1+\frac{kc_j}{\sqrt{m}})^{-\frac{m}{k}-c_j\sqrt{m}-\frac{1}{2}} e^{\lambda_m-\sum\lambda_{h_i}}\right)^\beta \sqrt{k}\, m^{\frac{1-k}{2}}.$$

Let us use the symbol *cost* for a constant $> 0$ which at each step is updated.

For each of these $k$ terms we can combine the inequality (22.58) with

$$|\Delta(\mathrm{c})|^\beta < cost\cdot m^{\beta\frac{k(k-1)}{4}}, \quad \left(e^{\lambda_m-\sum\lambda_{h_i}}\right)^\beta < e^{(k+1)\beta},$$

getting the estimate, by symmetry on the indices $i$,

$$|\tilde{C}_\beta^{(2)}(k,m)| \leq cost\cdot m^{\beta(\frac{k}{2}+\frac{k(k-1)}{4})} \sum_{\mathrm{c}\in \bar{L}_k(m)\,|\,k|c_1|\geq K\sqrt{\log m}} e^{-\frac{\beta}{3k^2}|\mathrm{c}|^2} \sqrt{k}\, m^{\frac{1-k}{2}}.$$

As for this last Riemann sum, due to the very fast decay of the integrand at infinity, one can verify that it is approximately the corresponding integral

$$\sum_{\mathrm{c}\in \bar{L}_k(m)\,|\,k|c_1|\geq K\sqrt{\log m}} e^{-\frac{\beta}{3k^2}|\mathrm{c}|^2} \sqrt{k}\, m^{\frac{1-k}{2}} \sim \int_{\mathrm{c}\in U\,|\,k|c_1|\geq K\sqrt{\log m}} e^{-\frac{\beta}{3k^2}|\mathrm{c}|^2} dy$$

(with $dy$ the standard Euclidean measure of $U$ of definition (22.11)).

The coordinate $c_1$ is a linear function on $U$. Let $u_1, u_2, \ldots, u_{k-1}$ be an orthonormal basis of $U$ such that $u_2, \ldots, u_{k-1}$ is a basis of the kernel of $c_1$, and let $a = c_1(u_1)$ be a nonzero number. In the coordinates $\sum_{i=1}^{k-1} x_i u_i$ the region

$B_m := \{c \in U \mid k|c_1| \geq K\sqrt{\log m}\}$ is that where $|x_1| \geq |ak|^{-1}K\sqrt{\log m}$ so set $H := |ak|^{-1}K$.

$$\int_{B_m} e^{-\frac{\beta}{3k^2}|c|^2} dy = \int_{\mathbb{R}^{k-2}} e^{-\frac{\beta}{3k^2}\sum_{j\neq 1} x_j^2} \prod_{j\neq 1} dx_j \int_{H\sqrt{\log m}}^{\infty} e^{-\frac{\beta}{3k^2}x_1^2} dx_1$$

$$= cost \cdot \int_{H\sqrt{\log m}}^{\infty} e^{-\frac{\beta}{3k^2}x_1^2} dx_1 < cost \cdot e^{-\frac{\beta}{3k^2}H^2\log(m)} = cost \cdot m^{-\frac{\beta H^2}{3k^2}},$$

because as soon as $y > 1$

$$\int_y^{\infty} e^{-ax^2}dx < \int_y^{\infty} xe^{-ax^2}dx \overset{z:=\sqrt{a}x}{=} a^{-1}\int_{\sqrt{a}y}^{\infty} ze^{-z^2}dz = \frac{1}{2a}\int_{ay^2}^{\infty} e^{-u}du = \frac{1}{2a}e^{-ay^2}.$$

Apply to $a = \frac{\beta}{3k^2}$, $y = H\sqrt{\log m}$ so

$$ay^2 = \frac{\beta}{3k^2}H^2\log m \implies e^{-\frac{\beta H^2}{3k^2}\log m} = m^{-\frac{\beta H^2}{3k^2}}.$$

So if we fix $H$ so that $\frac{k(k+1)}{4} - \frac{H^2}{3k^2} < 0$, we have that $\lim_{m\to\infty} \tilde{C}_\beta^{(2)}(k,m) = 0$, as desired. □

**22.2.3. The Mehta and Selberg integrals.** It is possible to compute explicitly the integral $I(k,\beta)$ of formulas (22.31) and (22.60),

$$I(k,\beta) = \int_U \left(|\Delta(u)| \cdot e^{-\frac{k}{2}|u|^2}\right)^{\beta} du.$$

We can use *Mehta's integral* $M(k,\gamma)$ which is

$$\frac{1}{(2\pi)^{k/2}}\int_{-\infty}^{\infty}\cdots\int_{-\infty}^{\infty}\prod_{i=1}^{k} e^{-t_i^2/2}\prod_{1\leq i<j\leq k}|t_i - t_j|^{2\gamma}\,dt_1\cdots dt_k = \prod_{j=1}^{k}\frac{\Gamma(1+j\gamma)}{\Gamma(1+\gamma)}.$$

It is the partition function for a gas of point charges moving on a line that are attracted to the origin (Mehta 2004). Its value can be deduced from that of the Selberg integral **[Sel44]**. This was conjectured by Mehta and Dyson (1963), who were unaware of Selberg's earlier work.

Setting $t_i := (k\beta)^{\frac{1}{2}}x_i$, and $|\Delta(x)| := \prod_{1\leq i<j\leq k}|x_i - x_j|$, we get

$$M(k,\beta/2) := \frac{(k\beta)^{k/2+\beta k(k-1)/4}}{(2\pi)^{k/2}}\int_{-\infty}^{\infty}\cdots\int_{-\infty}^{\infty} e^{-k\beta\frac{\sum_{i=1}^{k}x_i^2}{2}}|\Delta(x)|^{\beta}\,dx_1\cdots dx_k.$$

We reduce $I(k,\beta)$ to Metha's integral by the following standard argument.

We have the orthogonal decomposition $\mathbb{R}^k = \mathbb{R}j \oplus U$, $j = (\frac{1}{\sqrt{k}},\ldots,\frac{1}{\sqrt{k}})$. The function $\Delta(x_1,\ldots,x_k) = \Delta(xj+u) = \Delta(u)$ is independent of $x$ and

$$e^{-\frac{k}{2}(x_1^2+\cdots+x_k^2)} = e^{-\frac{k}{2}|xj+u|^2} = e^{-\frac{k}{2}x^2}e^{-\frac{k}{2}|u|^2}, \qquad \sqrt{\frac{\pi}{a}} = \int_{-\infty}^{+\infty} e^{-ax^2}\,dx.$$

Thus, setting $v := xj + u$ and $dv$ the Euclidean measure in $\mathbb{R}^k$ we have $dv = dxdu$, so

$$(2\pi)^{k/2}\sqrt{k\beta}^{-k-\beta k(k-1)/2} M(k,\beta/2) = \int_{\mathbb{R}^n}\left[|\Delta(x)|\cdot e^{-\frac{k}{2}|x|^2}\right]^\beta dx_1\cdots dx_k$$

$$= \int_{-\infty}^{\infty} e^{-\beta\frac{k}{2}x^2}dx \int_U \left[|\Delta(u)|\cdot e^{-\frac{k}{2}|u|^2)}\right]^\beta du$$

$$= \sqrt{\frac{2\pi}{\beta k}}\int_U \left[|\Delta(u)|\cdot e^{-\frac{k}{2}|u|^2)}\right]^\beta du = \sqrt{\frac{2\pi}{k\beta}}\, I(k,\beta). \tag{22.64}$$

Therefore,

THEOREM 22.2.17.

$$I(k,\beta) = (2\pi)^{\frac{k-1}{2}}(k\beta)^{-\frac{k}{2}-\frac{\beta k(k-1)}{4}+\frac{1}{2}}\prod_{j=1}^{k}\frac{\Gamma(1+j\beta/2)}{\Gamma(1+\beta/2)}. \tag{22.65}$$

*Plugging this into formula* (22.32) *of Theorem* 22.2.4*, we get the explicit asymptotics of* $C_\beta(k,m)$.

In particular

$$I(k,2) = (2\pi)^{\frac{k-1}{2}}(2k)^{-(k^2-1)/2}\prod_{j=1}^{k} j!$$

from which, by formula (22.32) we get the final formula (22.35),

$$B_{k,2} = (1/\sqrt{2\pi})^{(k-1)}\cdot 2^{-(k^2-1)/2}k^{k^2/2}\prod_{j=1}^{k-1} j!\,. \tag{22.66}$$

Example $k = 2, \beta = 2$. From our considerations $c_n(M_2(F))$ is asymptotic to $S_2(2, n+1)$. From formula (9.66) $c_n(M_2(F))$ is asymptotic to $C_{n+1}$ a Catalan number. Asymptotically, the Catalan numbers grow as

$$C_{n+1} \simeq \frac{4^{n+1}}{(n+1)^{3/2}\sqrt{\pi}},$$

so from our formulas

$$S_2(2,n+1) \simeq C_{2,2}\cdot\left(\frac{1}{n+1}\right)^{3/2}\cdot 4^{n+1},$$

and from formula (22.35) we have $C_{2,2} = \frac{1}{\sqrt{\pi}}$, as desired.

CHAPTER 23

# Codimension growth for algebras satisfying a Capelli identity

*This chapter deals with the asymptotic formulas for the codimension of a PI algebra. In this chapter all rings are assumed to be algebras over a field F of characteristic 0 satisfying a Capelli identity. For a PI algebra satisfying a Capelli identity, we may approach the codimension growth through the cocharacters.*

## 23.1. PI algebras satisfying a Capelli identity

**23.1.1. Preliminaries.** Let $R$ be an algebra satisfying a Capelli identity. Then by Theorem 17.1.1 and Proposition 17.2.34 $R$ is PI equivalent to a direct sum of basic or fundamental algebras $A_i$. In view of formula (21.4) and of Proposition 21.1.5 we may thus reduce the study to that of a fundamental algebra. Our goal is to prove the statement of formula (21.4) in Theorem 21.1.2.

Consider a fundamental (or basic) algebra $A = \bar{A} \oplus J$ with, $\bar{A} = \bigoplus_{i=1}^{q} A_i$, and $A_i = M_{d_i}(F)$. Let $J$ be the radical of $A$, with $J^s \neq 0,\ J^{s+1} = 0$. By Corollary 18.1.15 the three invariants $d = \sum_{i=1}^{q} d_i^2 = \dim \bar{A}, s, q$ depend only upon the polynomial identities ($(d, s)$ being the Kemer index).

One can define for any possible function $p(\lambda)$, $\lambda \vdash n$, taking positive integer values on partitions a sort of *virtual codimension* and a *virtual exponent* (if it exists)

$$c_n(p) := \sum_{\lambda \vdash n} p(\lambda) f_\lambda, \quad e(p) := \lim_{n\to\infty} (c_n(p))^{\frac{1}{n}}. \tag{23.1}$$

In this general setting the question becomes to relate the codimension growth with special properties of the function $p(\lambda)$. This is useful since it allows a lot of possible inductions.

In general this appears as follows.

23.1.1.1. *Polynomial representations.* Take a polynomial representation $W$ of degree $n$ of the linear group $\mathrm{GL}(k, F)$ with $k \gg n$. Then $W$ decomposes as

$$W = \bigoplus_{\lambda \vdash n} p_\lambda S_\lambda(F^k).$$

Here $S_\lambda(F^k)$ denotes the usual Schur functor or irreducible representation of the linear group $\mathrm{GL}(k, F)$ of highest weight $\lambda$, and $\mathrm{p}_\lambda \in \mathbb{N}$ denotes the multiplicity of the irreducible representation $S_\lambda(F^k)$ in $W$.

On this representation acts the group $D := \{t_1, \ldots .t_k\}$ of diagonal matrices, and $W$ has a basis of weight vectors $v$ relative to characters $\prod_{i=1}^{k} t_i^{h_i}$, with $h_i \in \mathbb{N}$, $\sum_i h_i = n$.

For every $s \leq k$ consider the subspace $W_s$ spanned by the weight vectors $v$ relative to characters $\prod_{i=1}^{s} t_i^{h_i}$. One has $[S_\lambda(F^k)]_s = S_\lambda(F^s)$, and thus

$$W_s = \bigoplus_{\lambda \vdash n} p_\lambda S_\lambda(F^s)$$

as representation of the subgroup $\mathrm{GL}(s,F) \subset \mathrm{GL}(k,F)$ fixing the last $k-s$ basis vectors $e_i$.

One also has the *space of multilinear elements* that is the subspace $W^{mult}$ spanned by the weight vectors $v$ relative to characters $\prod_{i=1}^{n} t_i$. Of course this space is contained in $W_n$. This space is a representation of the symmetric group $S_n \subset \mathrm{GL}(n,F)$ and one has also that (Corollary 6.3.11)

$$|\lambda| = n, \quad S_\lambda(F^n)^{\mathrm{mult}} = M_\lambda, \tag{23.2}$$

with $M_\lambda$ the irreducible representation of $S_n$ associated to $\lambda$.

REMARK 23.1.1. The choice of the first $n$ indices is not relevant. If we take any set $S$ of $n$ indices in $1,\dots,k$, the subspace of $W$ spanned by the weight vectors of weight $\prod_{i\in S} t_i$ is isomorphic, as representation of $S_n$, to $W^{\mathrm{mult}}$.

Thus

LEMMA 23.1.2.

$$\dim W^{\mathrm{mult}} = \sum_{\lambda \vdash n} p_\lambda f_\lambda,\ f_\lambda = \dim M_\lambda.$$

Finally let $U(k) \subset \mathrm{GL}(k,F)$ be the group of upper triangular matrices with 1 on the diagonal. We have by Theorem 6.3.7:

(i) The space $S_\lambda(F^k)^{U(k)} = Fu_\lambda$ is one dimensional generated by a *highest weight vector, denoted by* $u_\lambda$.
(ii) The group of diagonal matrices $D := \{t_1,\dots,t_k\}$ acts on $S_\lambda(F^k)^{U(k)} = Fu_\lambda$ by the character $t_1^{\ell_1}\cdots t_k^{\ell_k}$, where the $\ell_i$ are the rows of $\lambda$.
(iii) In particular if $\lambda$ has $p \leq k$ rows, we see that $u_\lambda \in S_\lambda(F^p)^{U(p)}$.

Therefore

$$W^{U(k)} = \bigoplus_{\lambda \vdash n} p_\lambda S_\lambda(F^k)^{U(k)}, \quad S_\lambda(F^k)^{U(k)} = Fu_\lambda. \tag{23.3}$$

LEMMA 23.1.3. *The multiplicity* $p_\lambda$ *equals the dimension of the subspace* $W^{\lambda,U(k)}$ *of* $W^{U(k)}$ *of vectors of weight* $\lambda$. *If* $\mathrm{ht}(\lambda) \leq p$, *we have* $W^{\lambda,U(k)} = W_p^{\lambda,U(p)}$.

The notion of codimension is a special case of a notion associated to a graded $\mathbb{W} = \bigoplus_{n=0}^{\infty} \mathbb{W}(n)$ representation of the limit group $\mathrm{GL}_\infty(F) = \bigcup_{k=1}^{\infty} \mathrm{GL}(k,F)$.

Notice that for such a group it makes sense to consider $S_\lambda(F^\infty) = \bigcup_{k=1}^{\infty} S_\lambda(F^k)$ irreducible under $\mathrm{GL}_\infty(F)$.

Assume that for all $n$ one has that

$$\mathbb{W}(n) = \bigoplus_{\lambda \vdash n} p_\lambda S_\lambda(F^\infty),\ p_\lambda < \infty, \tag{23.4}$$

then one can define a *$\mu$-dimension* as the dimension

$$\mu_n(\mathbb{W}) := \dim \mathbb{W}(n)^{\mathrm{mult}}. \tag{23.5}$$

When $R$ is a PI algebra, its codimension $c_n(R)$ equals the dimension of the space of multilinear elements of degree $n$ in the relatively free algebra in $n$ variables $\mathbb{W} = F\langle \xi_1, \dots, \xi_n \rangle$ in the variety generated by $R$. Hence

$$c_n(R) = \mu_n(\mathbb{W}).$$

The algebra $F\langle \xi_1, \dots, \xi_n \rangle$ is the quotient of the free algebra $F\langle x_1, \dots, x_k \rangle$ modulo the $T$-ideal of polynomial identities of $R$. The space $\mathbb{W}(n)^{\text{mult}}$, of its multilinear elements of degree $n$, is the quotient of the corresponding space of multilinear elements of degree $n$ in the free algebra $F\langle x_1, \dots, x_k \rangle$, identified to the group algebra of $S_n$, modulo the space of multilinear polynomial identities of $R$. Hence the name codimension of $R$.

In general it is better not to use the word *codimension* since there is no bigger space of which the multilinear elements are a quotient.

EXAMPLE 23.1.4. As in Corollary 6.3.24, if a group $G$ acts on a finite-dimensional vector space $V$, then it acts on $(V^*)^{\oplus \infty}$. We have that $\mathbb{W} := S[(V^*)^{\oplus \infty}]^G$ (the ring of invariants of countably many copies of the representation $V$) is a graded representation of $\mathrm{GL}_\infty(F)$. In this case $\mathbb{W}(n)^{\text{mult}}$ is the space of multilinear invariants of $n$ copies of $V$.

An important case for us is the case $V = M_k(F)$ and $G = \mathrm{GL}(k, F)$ acting on $V$ by conjugation. Then $\mathbb{W}(n)^{\text{mult}}$ is identified with the quotient of the group algebra $F[S_n]$ modulo the ideal $I_{k+1}$ generated by the antisymmetrizer in $k+1$ elements; see §12.1.4.

Under the hypotheses on $\mathbb{W}$ of formula (23.4) one can also define the *character sequence* (generalizing the notion of cocharacter). Each $W(n)$ has as Frobenius character the symmetric function $\sum_\lambda p_\lambda S_\lambda$ with $S_\lambda$ the Schur function in the limit algebra of symmetric functions.

Let us simplify the notations of Definition 6.4.9, and denote, for a given $k$, by $\Pi_k := H(k, 0)$ (resp., $\Pi_k(n) := H(k, 0, n)$) the set of all partitions (resp., of partitions of $n$), with at most $k$ rows $\lambda_1, \lambda_2, \dots, \lambda_k$. In our cases we will be able to use Capelli-type arguments to show that in our graded space decomposition, given by formula (23.4), we further have

$$\mathbb{W}(n) = \bigoplus_{\lambda \in \Pi_k(n)} p_\lambda S_\lambda(F^\infty), \; p_\lambda < \infty. \tag{23.6}$$

DEFINITION 23.1.5. If $\mathbb{W}$ satisfies property (23.6), we shall say that it has *height* $\leq k$.

We deduce the

PROPOSITION 23.1.6. *Let $\mathbb{W}$ be a graded $\mathrm{GL}_\infty(F)$ representation of height $\leq k$. Then*

$$\mu_n(\mathbb{W}) = \sum_{\lambda \in \Pi_k(n)} p_\lambda f_\lambda. \tag{23.7}$$

*The multiplicity $p_\lambda$ equals the dimension of the subspace of the space $\mathbb{W}_k^{U(k)}$ formed by vectors of weight $\lambda$.*

Notice the formula for $\mu$-dimension under tensor product:

PROPOSITION 23.1.7.

$$\mu_n(\mathbb{W}_1 \otimes \mathbb{W}_2) = \sum_{k=0}^{n} \binom{n}{k} \mu_k(\mathbb{W}_1)\mu_{n-k}(\mathbb{W}_2). \tag{23.8}$$

PROOF. We have

$$\mathbb{W}_1 \otimes \mathbb{W}_2(n) = \bigoplus_{k=0}^{n} \mathbb{W}_1(k) \otimes \mathbb{W}_2(n-k).$$

The space of multilinear elements of $\mathbb{W}_1(k) \otimes \mathbb{W}_2(n-k)$ has as basis the elements $u \otimes v$ where $u, v$ are weight vectors so that the (total) weight of $u \otimes v$ is $\prod_{i=1}^{n} t_i$. Thus there is a subset $S \subset (1, \dots, n)$ of cardinality $k$ so that $u$ is a weight vector of weight $\prod_{i \in S} t_i$ and $v$ is a weight vector of *complementary* weight $\prod_{i \in \{1,\dots,n\} \setminus S} t_i$.

By Remark 23.1.1, for each such subset $S$, the span of the $u \otimes v$ of previous type is isomorphic to $\mathbb{W}_1(k)^{\text{mult}} \otimes \mathbb{W}_2(n-k)^{\text{mult}}$, hence the claim. □

We will also need a special consequence of the previous proposition.

PROPOSITION 23.1.8. *Assume that $\mu_n(\mathbb{W}_1) = 0$ for $n \geq i$ and that $\mu_n(\mathbb{W}_2)$ is an increasing function. Then there is a positive constant $C$ such that*

$$\forall n, \quad \mu_n(\mathbb{W}_1 \otimes \mathbb{W}_2) \leq C\binom{n}{i}\mu_n(\mathbb{W}_2). \tag{23.9}$$

23.1.1.2. *Estimates.* We shall apply Proposition 23.1.8 to $\mu$-dimension using the following

LEMMA 23.1.9 (Beckner and Regev, **[BR98]**). *Let $(\alpha_1, \dots, \alpha_q) \in \mathbb{R}_+^q$ be positive numbers, and let $(e_1, \dots, e_q) \in \mathbb{R}^q$. Set $|\alpha| = \sum_j \alpha_j$, $p_j := \alpha_j/|\alpha|$. For $h_1 + \cdots + h_q = n$, let us denote by $\underline{h} := h_1, \dots, h_q$ and $|\underline{h}| = n$*

$$\sum_{|\underline{h}|=n,} \binom{n}{\underline{h}} \prod_{i=1}^{q} \alpha_i^{h_i} h_i^{e_i} = |\alpha|^n \sum_{|\underline{h}|=n} \binom{n}{\underline{h}} \prod_{i=1}^{q} p_i^{h_i} h_i^{e_i} \simeq |\alpha|^n n^{\sum_i e_i} \prod_{j=1}^{q} p_j^{e_j}, \tag{23.10}$$

$$e.g., \quad q = 2,\ (\alpha_1 + \alpha_2)^n n^{e_1+e_2} \frac{\alpha_1^{e_1}\alpha_2^{e_2}}{(\alpha_1+\alpha_2)^{e_1+e_2}}. \tag{23.11}$$

SKETCH. This follows the same methods as in §22.2.2. The difference is that by setting $p_i := \alpha_i/|\alpha|$, one has that $\sum_i p_i = 1$, and $p_i$ is the probability of the random variable $X$ to take the value $e_i$.

Then the probability $\binom{n}{\underline{h}} \prod_{i=1}^{q} p_i^{h_i}$ of the vector $\underline{h} = (h_1, \dots, h_q)$ sum of $n$ independent replicas of $X$ with $\sum_i h_i = n$ tends to be concentrated around the vector $n(p_1, \dots, p_q)$ so that

$$|\alpha|^n \sum_{|\underline{h}|=n} \binom{n}{\underline{h}} \prod_{i=1}^{q} p_i^{h_i} h_i^{e_i} \simeq |\alpha|^n \left( \sum_{|\underline{h}|=n} \binom{n}{\underline{h}} \prod_{i=1}^{q} p_i^{h_i} \right) \prod_{i=1}^{q} (np_i)^{e_i} = |\alpha|^n n^{\sum e_i} \prod_{i=1}^{q} p_i^{e_i}.$$

A complete justification of the last step can be given by using Stirling's formula directly. Let us sketch this for $q = 2$ using Stirling's formula,

$$\binom{n}{i} p_1^i p_2^{n-i} = \sqrt{2\pi}^{-1} n^{-1/2} \cdot e^{\lambda_n - \lambda_i - \lambda_{n-i}} \left( \left(\frac{n}{i}\right)^{i+1/2} \left(\frac{n}{n-i}\right)^{n-i+1/2} p_1^i p_2^{n-i} \right).$$

We set

$$\frac{i}{n} = p_1 + \frac{d_i}{\sqrt{n}}, \quad \frac{n-i}{n} = p_2 - \frac{d_i}{\sqrt{n}} \quad (d_{i+1} - d_i = \sqrt{n}),$$

$$\begin{aligned} A &:= \log\left(\left(\frac{n}{i}\right)^{i+1/2}\left(\frac{n}{n-i}\right)^{n-i+1/2} p_1^i p_2^{n-i}\right) \\ &= -\log(p_1 + \frac{d_i}{\sqrt{n}})(n(p_1 + \frac{d_i}{\sqrt{n}}) + 1/2) - \log(p_2 - \frac{d_i}{\sqrt{n}})(n(p_2 - \frac{d_i}{\sqrt{n}}) + 1/2) \\ &\quad + n\log(p_1)(p_1 + \frac{d_i}{\sqrt{n}}) + n\log(p_2)(p_2 - \frac{d_i}{\sqrt{n}}), \end{aligned}$$

let $\frac{|d_i|}{p_i\sqrt{n}} < h(n)$, $\lim_{n\to\infty} nh(n)^2 = \infty$, $\lim_{n\to\infty} nh(n)^3 = 0$ , for instance $h(n) = n^{-\alpha}$, $1/3 < \alpha < 1/2$, then

$$\begin{aligned} \log(p_1 + \frac{d_i}{\sqrt{n}}) &= \log(p_1) + \log(1 + \frac{d_i}{p_1\sqrt{n}}) \\ &= \log(p_1) + \frac{d_i}{p_1\sqrt{n}} - \frac{1}{2}(\frac{d_i}{p_1\sqrt{n}})^2 + O(h(n)^3) \\ &\implies A = -(\frac{d_i}{p_1\sqrt{n}} - \frac{1}{2}(\frac{d_i}{p_1\sqrt{n}})^2 + O(h(n)^3)) \\ &\qquad \times (n(p_1 + \frac{d_i}{\sqrt{n}}) + 1/2) \\ &\qquad + (\frac{d_i}{p_2\sqrt{n}} - \frac{1}{2}(\frac{d_i}{p_2\sqrt{n}})^2 + O(h(n)^3)) \\ &\qquad \times (n(p_2 - \frac{d_i}{\sqrt{n}}) + 1/2) - \frac{1}{2}\log(p_1p_2) \\ &= -d_i\sqrt{n} - \frac{d_i^2}{2p_1} + d_i\sqrt{n} - \frac{d_i^2}{2p_2} - \frac{1}{2}\log(p_1p_2) + O(nh(n)^3) \\ &= -\frac{d_i^2}{2p_1p_2} - \frac{1}{2}\log(p_1p_2) + O(nh(n)^3). \end{aligned}$$

From these formulas one estimates

$$\binom{n}{i} p_1^i p_2^{n-i} i^{e_1}(n-i)^{e_2} = \frac{1}{\sqrt{2\pi n}} e^{-\frac{d_i^2}{2p_1p_2} + O(nh(n)^3)} n^{e_1+e_2} p_1^{e_1} p_2^{e_2}(1 + O(h(n)^2)),$$

and the Riemann sum in the range tends to the Gaussian integral equal to 1.

$$d_i = -p_1\sqrt{n} + \frac{i}{\sqrt{n}},$$

$$\lim_{n\to\infty} \sum_{|d_i|<h(n)} \frac{1}{\sqrt{2\pi p_1p_2}} e^{-\frac{d_i^2}{2p_1p_2}} n^{-1/2} = \frac{1}{\sqrt{2\pi p_1p_2}} \int_{-\infty}^{\infty} e^{-\frac{x^2}{2p_1p_2}} dx = 1. \qquad \square$$

We will be interested in sequences $a_n$ which are asymptotic to some $Cn^e\alpha^n$.

Given a sequence $a = \{a_n\}$ we have introduced an equivalence relation setting $a \simeq b$ if and only if $\lim a_n/b_n = 1$.

Given $a = \{a_n\}$ we associate to $a$ a formal power series $f_a = La$ ($L$ is a linear operator)

$$f_a(x) := \sum_{n\geq 0} \frac{a_n}{n!} x^n .$$

For instance $f_{\{c^n\}}(x) = e^{cx}$. The previous equivalence associates an equivalence relation between the associated power series $f_a \simeq f_b$.

Next define the sequence $a \star b$ by

$$(a \star b)_n := \sum_{i,j\geq 0,\, i+j=n} \binom{n}{i} a_i b_j.$$

This is an associative product so that

$$f_a \cdot f_b = f_{a\star b} .$$

We are particularly interested in the sequences $a(\alpha, e) := \{n^e \alpha^n\},\ \alpha > 0$, so that the sequence in formula (23.10) is

$$a(\alpha_1, e_1) \star a(\alpha_2, e_2) \star \cdots \star a(\alpha_q, e_q).$$

For $q = 2$ Lemma 23.1.9 reads as

$$a(\alpha_1, e_1) \star a(\alpha_2, e_2) \simeq \frac{\alpha_1^{e_1} \alpha_2^{e_2}}{(\alpha_1 + \alpha_2)^{e_1+e_2}} a(\alpha_1 + \alpha_2, e_1 + e_2). \tag{23.12}$$

The following analysis has been explained to us by Luca Biasco.

Let us introduce a partial order between sequences

$$a \ll b \quad \iff \quad |a_n| \leq b_n\, ,\ \forall n \geq 0\, ,$$

which induces a partial order between the associated power series $f_a \ll f_b$. Note that

$$f_1 \ll g_1\, ,\ f_2 \ll g_2 \implies f_1 f_2 \ll g_1 g_2\, , \qquad f_1 + f_2 \ll g_1 + g_2\, .$$

Given a sequence $a$ we also define its *majorant sequence* $\hat{a}$ with $\hat{a}_n := |a_n|$. Observe that

$$a \ll b \iff \hat{a} \ll b\, .$$

Observe also that $a \ll \hat{a}$ and $a \ll b$ implies $b = \hat{b}$.

THEOREM 23.1.10. *If*

$$a_n, b_n > 0,\ \ a \simeq a',\ b \simeq b',\ \ \lim_n a_n/(a \star b)_n = \lim_n b_n/(a \star b)_n = 0\, ,$$

*then*

$$a \star b \ \simeq\ a' \star b'\, .$$

PROOF. Observe that $a \simeq a'$ implies the existence of $\bar{a}_n$ such that

$$a'_n = a_n(1 + \bar{a}_n)\, , \qquad \bar{a}_n \to 0\, .$$

Similarly

$$b'_n = b_n(1 + \bar{b}_n)\, , \qquad \bar{b}_n \to 0\, .$$

So let us define

$$g_{a,a'} := L(|\bar{a}_n a_n|)\, ,\ \text{ or } g_{a,a'}(x) := \sum_{n\geq 0} \frac{|\bar{a}_n a_n|}{n!} x^n\, .$$

Note that

$$f_{a'} - f_a \ll g_{a,a'}\, , \qquad f_{b'} - f_b \ll g_{b,b'}\, .$$

Moreover, since $\bar{a}_n \to 0$, $\bar{b}_n \to 0$, and $a_n, b_n > 0$, we have the following simple

LEMMA 23.1.11. *$\forall \epsilon > 0$ there exist two polynomials $P(x), Q(x)$ with positive coefficients such that*

$$f_{a'} - f_a \ll g_{a,a'} \ll \epsilon f_a + P, \qquad f_{b'} - f_b \ll g_{b,b'} \ll \epsilon f_b + Q.$$

Thus

$$\begin{aligned} f_{a'\star b'} - f_{a\star b} &= f_{a'} f_{b'} - f_a f_b \\ &\ll \widehat{(f_{a'} - f_a)} f_b + \widehat{(f_{b'} - f_b)} f_a + \widehat{(f_{a'} - f_a)}\widehat{(f_{b'} - f_b)} \\ &\ll (\epsilon f_a + P) f_b + (\epsilon f_b + Q) f_a + (\epsilon f_a + P)(\epsilon f_b + Q) \\ &= (2\epsilon + \epsilon^2) f_{a\star b} + (1+\epsilon)(P f_b + Q f_a) + PQ. \end{aligned}$$

From this one obtains, using $\lim_n a_n/(a\star b)_n = \lim_n b_n/(a\star b)_n = 0$, that

$$\limsup_{n\to\infty} \left| \frac{(a'\star b')_n}{(a\star b)_n} - 1 \right| \le 2\epsilon + \epsilon^2 .$$

Since $\epsilon$ is arbitrary it follows that

$$\lim_{n\to\infty} \frac{(a'\star b')_n}{(a\star b)_n} = 1 . \qquad \square$$

THEOREM 23.1.12. *Assume that two graded vector spaces $\mathbb{W}_1, \mathbb{W}_2$ have the asymptotics for $\mu$-dimension*

$$\mu_n(\mathbb{W}_i) \simeq a_i n^{e_i} \alpha_i^n, \quad i = 1, 2.$$

*Then for $\mathbb{W}_1 \otimes \mathbb{W}_2$ we have*

$$\mu_n(\mathbb{W}_1 \otimes \mathbb{W}_2) \simeq \frac{a_1 a_2\, \alpha_1^{e_1} \alpha_2^{e_2}}{(\alpha_1+\alpha_2)^{e_1+e_2}} n^{e_1+e_2} (\alpha_1+\alpha_2)^n. \tag{23.13}$$

PROOF. By formula (23.8) we have

$$\mu(\mathbb{W}_1 \otimes \mathbb{W}_2) = \mu(\mathbb{W}_1) \star \mu(\mathbb{W}_2) \tag{23.14}$$

while

$$\{\sum_{k=0}^{n} \binom{n}{k} a_1 k^{e_1} \alpha_1^k a_2 (n-k)^{e_2} \alpha_2^{n-k}\} = \{a_1 n^{e_1} \alpha_1^n\} \star \{a_2 n^{e_2} \alpha_2^n\}.$$

These sequences satisfy the hypotheses of Theorem 23.1.10, and hence the claim. $\square$

The only case of importance for us is the asymptotics of the codimension of $n \times n$ matrices given by formula (22.1).

23.1.1.3. *Codimension and $\mu$-dimension.* The first step is to apply this method to two commutative rings associated to a semisimple algebra $A = \bigoplus_{i=1}^q M_{k_i}(F)$ of dimension $d = \sum_{i=1}^q k_i^2$. On this algebra the group $G := \prod_{i=1}^q \mathrm{GL}(k_i, F)$ acts as group of automorphisms, and for any number of copies we have, denoting by $\mathcal{P}(V) = S[V^*]$ the ring of polynomials on a space $V$

$$\mathcal{P}(A^k)^G = \bigotimes_{i=1}^{q} \mathcal{P}(M_{k_i}(F)^k)^{\mathrm{GL}(k_i, F)}. \tag{23.15}$$

We can thus study the $\mu$-dimension growth of the dimension $\mu_n(\mathcal{P}(A^k)^G)$ of the multilinear invariants of $\mathcal{P}(A^k)^G$, as in formula (23.5), which we denote for simplicity by $\mu_n(A)$.

For $q = 1$ and $A = M_k(F)$ this is given by formula (22.1). From Theorem 23.1.12, applied to $\mathbb{W}_i = \mathcal{P}(M_{k_i}(F)^k)^{GL(k_i,F)}$, and formula (23.13) it follows that

PROPOSITION 23.1.13. *$C_k$ as in formula (21.2):*

$$c_\mu(A) \simeq Cn^{-\frac{d-q}{2}}d^n, \ C = d^{\frac{d-q}{2}}\prod_{i=1}^{q} C_{k_i}k_i^{1-k_i^2}. \tag{23.16}$$

PROOF. We apply formula (23.13) to the factors with asymptotics $C_j n^{-\frac{k_j^2-1}{2}}k_j^{2n}$ where $\alpha_j = k_j^2$ and $e_j = -\frac{k_j^2-1}{2}$ so that $\sum_j \alpha_j = d, \ \sum_j e_j = -\frac{d-q}{2}$. □

REMARK 23.1.14. For a function $c(n) \simeq Cn^\alpha d^n, \ C > 0$, we have for each integer $a$ that $c(n+a) \simeq Cd^a\, c(n)$.

Given two functions $c_1(n) \simeq C_1 n^{\alpha_1}d_1^n, \ C_1 > 0, \ c_2(n) \simeq C_2 n^{\alpha_2}d_2^n, C_2 > 0$, for the function $c_1(n) + c_2(n)$, we have:

(1) $c_1(n) + c_2(n) \simeq c_1(n)$ if $d_1 > d_2$ or if $d_1 = d_2$ and $\alpha_1 > \alpha_2$.
(2) $c_1(n) + c_2(n) \simeq (C_1 + C_2)n^\alpha d^n$ if $d_1 = d_2 = d$ and $\alpha_1 = \alpha_2 = \alpha$.

### 23.1.2. Codimension.

According to Aljadeff, Janssens, and Karasik [**AJK17**].

23.1.2.1. *The main theorem.* Let $R$ be an algebra satisfying a Capelli identity. Then $R$ is PI equivalent to a direct sum of basic or fundamental algebras $A_i$.

For a fundamental algebra $A = \bar{A} \oplus J$ with $\bar{A} = \bigoplus_{i=1}^q A_i, \ A_i = M_{d_i}(F)$ and $J$ the radical with $J^s \neq 0, \ J^{s+1} = 0$, the three invariants $d = \sum_{i=1}^q d_i^2 = \dim \bar{A}, s, q$ depend only upon the polynomial identities ($(d, s)$ being the Kemer index).

Assume thus that $R$ is fundamental. We want to prove

THEOREM 23.1.15. *There exist two positive constants $C_1, C_2$ such that, $\forall n$,*

$$C_1 n^{-\frac{d-q}{2}+s}d^n \leq c_n(R) \leq C_2 n^{-\frac{d-q}{2}+s}d^n. \tag{23.17}$$

REMARK 23.1.16. If $R$ is finite dimensional but not fundamental we know that $R$ is PI equivalent to a direct sum $\bigoplus_i R_i$ with $R_i$ fundamental. We have in general

$$c_n(R_i) \leq c_n(R) \leq \sum_i c_n(R_i).$$

Thus the previous Theorem 23.1.15 can be formulated also for such an $R$ where as $d$ one has to take the maximum of the $d_i$ of the various $R_i$, and then among these take for $-\frac{d-q}{2} + s$ the maximum.

23.1.2.2. *Generalities.* The codimension $c_n(R)$ of an algebra $R$ is the dimension of the space (of dimension $n!$) of multilinear polynomials in $n$ variables modulo the polynomial identities. If $R$ is finite dimensional (over the base field $F$) one can apply the method of generic elements, choose a basis $u_1, \ldots, u_d$ of $R$ and $n$ sets of polynomial variables $\xi_{i,j}, \ i = 1, \ldots, n; \ j = 1, \ldots d$. Define the generic elements

$$X_i := \sum_{j=1}^{d} \xi_{i,j}u_j, \ i = 1, \ldots, n, \ X_i \in R \otimes_F F[\xi_{i,j}] = R[\xi_{i,j}].$$

Then it is immediate that

LEMMA 23.1.17. *$c_n(R)$ equals the dimension, over $F$, of the span of all the products $X_{\sigma(1)} \cdots X_{\sigma(n)}, \sigma \in S_n$.*

Then suppose we *specialize* the generic elements $X_n \mapsto \bar{X}_n$ by specializing the variables $\xi_{i,j} \mapsto \bar{\xi}_{i,j}$. We then have the obvious

LEMMA 23.1.18. *$c_n(R)$ is greater than or equal to the dimension, over $F$, of the span of the products $\bar{X}_{\sigma(1)} \cdots \bar{X}_{\sigma(n)} \in R \otimes_F F[\bar{\xi}_{i,j}], \sigma \in S_n$.*

We will apply this lemma in particular to $n$ generic elements $X_i$ plus some $h$ elements $a_j \in R$ which we may think of as specializations of some further generic elements $X_{n+j}$. Then the linear span of all these monomials in $n+h$ elements has dimension $\le c_{n+h}(R)$.

23.1.2.3. *An estimate from above.* Let $A = \bar{A} \oplus J$ be any finite-dimensional algebra. Let $\bar{A} = \bigoplus_{h=1}^q A_h$, let $A_h = M_{d_h}(F)$, and let $J$ be the radical with $J^s \neq 0$, $J^{s+1} = 0$. Denote by $e_{i,j}(h)$, $i, j = 1, \dots, d_h$, the matrix units of $A_h = M_{d_h}(F)$.

We have a Pierce decomposition associated to the decomposition of $1 \in \bar{A}$, $1 = \sum_{i=1}^q \sum_{j=1}^{d_i} e_{j,j}(i)$ if $A$ has a 1; otherwise adding 1 to $A$, we have to add an orthogonal idempotent $e_0$ so that $1 = e_0 + \sum_{i=1}^q \sum_{j=1}^{d_i} e_{j,j}(i)$.

Choose an *elementary basis* $W := \{w_t,\ t = 1, \dots, p\}$ of $J$, that is we may assume that each $w_t \in e_{j,j}(i) A e_{h,h}(k)$ with $0 \le i, k \le q$, $1 \le j \le d_i$, $1 \le h \le d_k$ (with the possible presence of $e_0 = e_{1,1}(0)$ if $A$ does not have 1).

We then write the generic elements in this basis

$$X_j = \sum_{\ell=1}^q \sum_{h,k=1}^{d_\ell} \xi_{h,k,\ell}^{(j)} e_{h,k}(\ell) + \sum_{t=1}^p \eta_{j,t} w_t \in A \otimes F[\xi, \eta], \ j = 1, \dots, n,$$

with $\xi = \{\xi_{h,k,\ell}^{(j)}\}$ parameters of copies of the coordinates of $\bar{A}$ in the elementary basis and $\eta = \{\eta_{j,t}\}$ of copies of the coordinates of $J$ in the basis $w_t$.

We then have, by Lemma 23.1.17, that the $n$th codimension $c_n(A)$ of $A$ is the dimension of the span $\mathfrak{M}_n \subset A \otimes F[\xi, \eta]$ of the $n!$ monomials $X_{\sigma(1)} \cdots X_{\sigma(n)} \in A \otimes F[\xi, \eta]$ with $\sigma \in S_n$.

We want to prove an estimate on this dimension. For this decompose

$$(23.18) \qquad X_j = \sum_{\ell=1}^q X_j(\ell) + W_j, \ X_j(\ell) := \sum_{h,k=1}^{d_\ell} \xi_{h,k,\ell}^{(j)} e_{h,k}(\ell), \ W_j := \sum_t \eta_{j,t} w_t.$$

We then consider inside $A \otimes F[\xi, \eta]$ the subring $Y$ generated by all the elements $X_j(\ell), \eta_{j,t} w_t$, $j = 1, \dots, n$; $\ell = 1, \dots, q$; $t = 1, \dots, p$.

The algebra $Y$ is multigraded in these variables $(X_j(\ell), \eta_{j,t} w_t)$, so it makes sense to speak of the space $Y_n$ of *multilinear* elements of degree $n$ in $Y$. They are spanned by the products of $n$ out of the list $X_j(\ell), \eta_{j,t} w_t$, in which each index $j \in \{1, \dots, n\}$ appears (exactly once). Of course the space $\mathfrak{M}_n \subset Y_n$ so that we can estimate

$$c_n(A) = \dim_F \mathfrak{M}_n \le \dim_F Y_n.$$

Now, since $J^{s+1} = 0$ and $w_t \in J$, we have that a nonzero monomial in $X_j(\ell), \eta_{j,t} w_t$ can contain at most $s$ out of the elements $\eta_{j,t} w_t$. We thus may consider the space $Y_n$

as the sum of finitely many spaces $Y_n^{\vec{\pi}}$ each determined by the ordered sequence $\vec{\pi}:= w_{t_1}, \dots, w_{t_u}$, $u \le s$, $1 \le t_i \le p$ appearing in the monomial.

Notice that a monomial of this type may appear only if the sequence $\vec{\pi}$ determines in fact a path on the Pierce graph. That is there must exist indices $a_0, a_1, \dots, a_u \in \{1, \dots, q\}$ such that, if $e_i = \sum_{j=1}^{d_i} e_{j,j}(i)$ denotes the unit element of $A_i$, we have

$$w_{t_i} = e_{h_i,h_i}(a_{i-1}) w_{t_i} e_{k_i,k_i}(a_i) \quad \text{and hence} \quad w_{t_i} = e_{a_{i-1}} w_{t_i} e_{a_i}. \tag{23.19}$$

We thus estimate separately the dimension of each subspace $Y_n^{\vec{\pi}}$ of monomials of the type associated to each path $\vec{\pi}$. Let us ignore the case in which $A$ does not have a 1 which follows by the same lines.

In a monomial in $Y_n^{\vec{\pi}}$ appear the elements $\eta_{f(1),t_1} w_{t_1}, \dots, \eta_{f(u),t_u} w_{t_u}$ for some bijective map $f : \{1, \dots, u\} \to S$ to a subset $S$ of $\{1, \dots, n\}$. Then a nonzero monomial has the form (use formula (23.19))

$$\begin{aligned} &\prod_{i=1}^{u} \eta_{f(i),t_i} M_0 w_{t_1} M_1 w_{t_2} M_2 \cdots w_{t_{u-1}} M_{u-1} w_{t_u} M_u \\ &\overset{(23.19)}{=} \prod_{i=1}^{u} \eta_{f(i),t_i} M_0 e_{a_0} w_{t_1} e_{a_1} M_1 e_{a_1} \cdots M_{u-1} e_{a_{u-1}} w_{t_u} e_{a_u} M_u, \end{aligned}$$

hence each $M_i$ is a product of elements out of the list $X_j(a_i)$, $j = 1, \dots, n$.

Moreover the indices $j$ appearing in all these monomials form the complement of $S$ in $\{1, \dots, n\}$ (and each appears only once). Thus

LEMMA 23.1.19. *The dimension of $Y_n^{\vec{\pi}}$ is $\le u!\binom{n}{u} d \le C_0 \cdot n^s d$. Where $d$ is the dimension of the span of the monomials $M_0 w_{t_1} M_1 \cdots M_{u-1} w_{t_u} M_u$.*

PROOF. The number $\binom{n}{u}$ counts the subsets $S$ while $u!$ counts the possible maps $f$, since $u \le s$ the estimate follows. □

We need finally to estimate the dimension of the vector space $\mathbb{W}_{t_1,\dots,t_u}$ span of the monomials $M_0 w_{t_1} M_1 \cdots M_{u-1} w_{t_u} M_u$, where in total the $M_i$ depend on $n - u$ variables $X_j(\ell_j)$, $j \in \{1, \dots, n\} \setminus S$, $\ell_j \in \{1, \dots, q\}$.

This dimension is less than or equal to the sum of the dimension of the span of all the monomials $M_0 w_{t_1} M_1 \cdots M_{u-1} w_{t_u} M_u$ where the $n - u$ variables $X_j(\ell)$ are divided among the indices $\ell$, $1 \le \ell \le q$, with each index $\ell$ appearing some $n_\ell$ times with $\sum_{\ell=1}^{q} n_\ell = n - u$.

Without loss of generality we may assume that $\{1, \dots, n\} \setminus S = \{1, \dots, n-u\}$.

So let us estimate first the dimension of the subspace $\mathbb{W}_{\mathcal{D};t_1,\dots,t_u}$ of $\mathbb{W}_{t_1,\dots,t_u}$, where the $X_j(\ell)$ appear only for $j \in D_\ell$, for a given choice of a decomposition $\mathcal{D} = \{D_1, \dots, D_q\}$ of $\{1, \dots, n-u\}$ into $q$ sets each with $n_\ell$ elements, $1 \le \ell \le q$ (and $\sum_{\ell=1}^{q} n_\ell = n - u$).

LEMMA 23.1.20. *We have for $\mathcal{D} = \{D_1, \dots, D_q\}$, $|D_\ell| = n_\ell$.*

$$\dim \mathbb{W}_{\mathcal{D};t_1,\dots,t_u} \le C \prod_{\ell=1}^{q} n_\ell^{-\frac{d_\ell^2 - 1}{2}} d_\ell^{2n_\ell}. \tag{23.20}$$

*From which one has*

$$\dim \mathbb{W}_{t_1,\dots,t_u} = \bigoplus_{\mathcal{D}} \dim \mathbb{W}_{\mathcal{D};t_1,\dots,t_u} \le C \sum \binom{n-u}{n_1,\dots,n_q} \prod_{j=1}^{q} n_\ell^{-\frac{d_\ell^2-1}{2}} d_\ell^{2n_\ell},$$

*which, applying Lemma 23.1.9 to $\alpha_\ell = d_\ell^2$ and $e_\ell = -\frac{d_\ell^2-1}{2}$, is bounded by*

$$\le C'(n-u)^{-\frac{d-q}{2}} d^{n-u} \le C'' n^{-\frac{d-q}{2}} d^n. \tag{23.21}$$

PROOF. Of course the point is to prove formula (23.20). We further decompose

$$M_0 w_{t_1} M_1 \cdots M_{u-1} w_{t_u} M_u = e_{a_0} M_0 w_{t_1} M_1 \cdots M_{u-1} w_{t_u} M_u e_{a_u}$$
$$= \sum_{h=1}^{d_{a_0}} \sum_{k=1}^{d_{a_u}} e_{h,h}(a_0) M_0 w_{t_1} M_1 \cdots M_{u-1} w_{t_u} M_u e_{k,k}(a_u).$$

Notice that a term

$$e_{h,h}(a_0) M_0 w_{t_1} M_1 \cdots M_{u-1} w_{t_u} M_u e_{k,k}(a_u)$$
$$= F_{h,k}(X_j(\ell))_{\ell=1,\dots,q;\ j\in D_\ell} e_{h,h}(a_0) w_{t_1} \cdots w_{t_u} e_{k,k}(a_u). \tag{23.22}$$

Here $F_{h,k}(X_j(\ell))$ is a multilinear function of the $n-u$ matrix variables $X_j(\ell)$, $j \in D_\ell$ which depends upon the path $\vec{\pi}$, the monomials $M_0, M_1, \dots, M_u$, and the two indices $h, k$. Namely it is a product of matrix coefficients. It is thus enough to show that the dimension of the span $\langle F_{h,k}(X_j(\ell))\rangle$ of these functions $F_{h,k}(X_j(\ell))$ has dimension bounded by the right-hand side of formula (23.20).

We now assume to have fixed $h = h_0, k = k_{u+1}$ and drop the pedex writing $F = F_{h_0,k_{u+1}}$.

Notice that $X_j(a)X_h(b) = 0$ if $a \neq b$. Thus, from formula (23.19), each monomial $M_i$ will give a nonzero contribution only if it depends only on the variables $X_j(a_i)$ ($j \in D_{a_i}$). So we denote $M_i$ by $M_i(X_j(a_i))$, and

$$\begin{aligned} w_{t_i} M_i(X_j(a_i)) w_{t_{i+1}} &= w_{t_i} e_{k_i,k_i}(a_i) M_i(X_j(a_i)) e_{h_{i+1},h_{i+1}}(a_i) w_{t_{i+1}} \\ &= (M_i(X_j(a_i)))_{k_i,h_{i+1}} w_{t_i} e_{k_i,h_{i+1}}(a_i) w_{t_{i+1}} \\ &\implies F(X_j(\ell)) = \prod_{i=0}^{u} (M_i(X_j(a_i)))_{k_i,h_{i+1}}. \end{aligned} \tag{23.23}$$

The notation $M_i(X_j(a_i))$ reminds us that the monomial $M_i$ is computed in the elements $x_j \mapsto X_j(a_i)$, hence it is a matrix in $\mathtt{M}_{d_{a_i}}(F[\xi])$ of which we are taking the $k_i, h_{i+1}$ entry.

We then write, collecting factors relative to the indices $a_i = \ell$, $\forall 1 \le \ell \le q$,

$$F(X_j(\ell)) = \prod_{\ell=1}^{q} F_\ell(X_j(\ell)), \quad \text{with} \quad F_\ell(X_j(\ell)) = \prod_{i \mid a_i=\ell} (M_i(X_j(\ell)))_{k_i,h_{i+1}}. \tag{23.24}$$

For a fixed $\ell$ let $s_1, \dots, s_{f_\ell}$ be the distinct indices $i$ with $a_i = \ell$. Denote for simplicity $\bar{k}_r := k_{s_r}$, $\bar{h}_r := h_{s_r}$, $1 \le r \le f_\ell$, and also $e_{i,j} = e_{i,j}(\ell)$.

We have then

$$F_\ell(X_j(\ell)) = \prod_{r=1}^{f_\ell} \bar{M}(r)_{\bar{k}_r,\bar{h}_{r+1}}, \quad \bar{M}(r) := M_{s_r}(X_j(\ell)),$$

(23.25) $$F_\ell(X_j(\ell))e_{1,1} = e_{1,\bar{k}_1}\bar{M}(1)e_{\bar{h}_2,\bar{k}_2}\bar{M}(2)e_{\bar{h}_3,\bar{k}_3}\cdots e_{\bar{h}_{f_\ell},\bar{k}_{f_\ell}}\bar{M}(f_\ell)e_{\bar{h}_{f_\ell+1},1}.$$

The value of $F_\ell(X_j(\ell))e_{1,1}$ is thus the evaluation of a monomial in $n_\ell$ generic $d_\ell \times d_\ell$ matrices $X_j(\ell)$ which we have added a fixed set of $f_\ell + 1$ matrix units.

Thus if we set $m_\ell = n_\ell + f_\ell + 1$, for each $1 \leq \ell \leq q$, the estimate of the dimension of the span of the coefficients $F_\ell(X_j(\ell))$ is therefore, by Lemma 23.1.18, less than or equal to the codimension

$$\dim\langle F_{h,k}(X_j(\ell))\rangle \leq c_{m_\ell}(M_{d_\ell}(F)) \leq C_\ell m_\ell^{-\frac{d_\ell^2-1}{2}} d_\ell^{2m_\ell} \leq C'_\ell n_\ell^{-\frac{d_\ell^2-1}{2}} d_\ell^{2n_\ell},$$

which is the desired estimate. □

We deduce finally the desired upper bound:

PROPOSITION 23.1.21. *There is a positive constant $\bar{C} > 0$ such that $c_n(R) \leq \bar{C}n^{-\frac{d-q}{2}+s}d^n$.*

PROOF. The codimension is a finite sum of terms each of which, by Lemma 23.1.19, is a product of a factor $\leq \binom{n}{u} \leq C_0 \cdot n^s$, $u \leq s$, and a factor which we have estimated by $C'n^{-\frac{d-q}{2}}d^n$, hence the claim. □

23.1.2.4. *A property of generic matrices.* Before discussing the estimate from below we recall a simple (possibly well known) fact on generic matrices.

Consider a nonzero element $F := f(X_1, \ldots, X_t) \in M_k[F[\xi]]$ in the algebra of $t$, $k \times k$ generic matrices $X_i := (\xi_{s,u}^{(i)})$, where $F[\xi] = F[\xi_{s,u}^{(i)}], i = 1, \ldots, t,\ s, u = 1, \ldots, k$, is the polynomial ring in the $tk^2$ variables $\xi_{s,u}^{(i)}$.

LEMMA 23.1.22. *All the diagonal entries of $F := f(X_1, \ldots, X_t)$ are nonzero polynomials.*

PROOF. Assume by contradiction that some such entry vanishes, for instance $F_{1,1} = 0$. Since the map $f$ is equivariant, we have for every invertible matrix $g \in \mathrm{GL}(k, F)$ that

$$gFg^{-1} = f(gX_1g^{-1}, \ldots, gX_tg^{-1}).$$

That is the entries of $gFg^{-1}$ are obtained from the entries of $F$ by applying the automorphism $\bar{g} : F[\xi] \to F[\xi]$ induced by $g$ on the polynomial ring $F[\xi]$. That is

$$\bar{g}(\xi_{s,u}^{(i)}) = \bar{g}((X_i)_{s,u}) := (gX_ig^{-1})_{s,u} \implies (gFg^{-1})_{s,u} = \bar{g}(F_{s,u}).$$

In particular for all $g \in \mathrm{GL}(k, F)$, we must have the vanishing of the entry $(gFg^{-1})_{1,1} = 0$. We claim that all off-diagonal entries must be nonzero.

In fact, if $F_{i,j} = 0$, $i \neq j$, by applying conjugation by a permutation matrix $\sigma$, then one has that $F_{\sigma(i),\sigma(j)} = 0$, so all off-diagonal entries are 0. But since also $F_{1,1} = 0$, the element $F$ has determinant 0. This is a contradiction since, from Theorem 11.3.1 it follows that all nonzero elements of the algebra of generic matrices are invertible.

Therefore we have $F_{2,1} \neq 0$. Then conjugating $F$ with the matrix $1 + e_{2,1}$, since we have assumed that $F_{1,1} = 0$, one has $[(1 + e_{2,1})F(1 - e_{2,1})]_{1,1} = \bar{g}(F_{1,2}) \neq 0$, where $\bar{g}$ is the automorphism induced by $g = (1 + e_{2,1})$ on $F[\xi]$, a contradiction. □

COROLLARY 23.1.23. *Given any $i = 1, \dots, k$, the linear map $F = f(X_1, \dots, X_t) \mapsto F_{i,i}$ from the algebra of generic matrices to polynomials is injective.*

PROOF. By Lemma 23.1.22, its kernel is 0. □

23.1.2.5. *An estimate from below.* For this estimate we must assume that $R = A = \bigoplus_{i=1}^q A_i \oplus J$ (same notations as before $\bar{A} = \bigoplus_{i=1}^q A_i$, $d = \dim \bar{A}$) is a fundamental algebra.

Next, since $A$ is fundamental, there is a noncommutative polynomial $f$ multilinear and alternating in $s$ layers $x_i = \{x_{i,1}, \dots, x_{i,d+1}\}$ each with $d+1$ variables, and linear in variables $y_1, \dots, y_q$ (and possibly dependent always in a multilinear way on some further variables $E$) with the property that it has a nonzero evaluation. We may assume the evaluation to be elementary and (up to rearranging the indices) $y_i$ is evaluated in $A_i$.

REMARK 23.1.24. For such a nonzero evaluation one has necessarily that since $f$ is alternating in each layer of the $d+1$ variables $x_i$ and $d = \dim \bar{A}$ at least one of the variables in $x_i$ is evaluated in $J$. But since $J^{s+1} = 0$, only one variable is evaluated in some element $h_i \in J$, and we may assume it is the first $z_h := x_{h,1}$, $h = 1, \dots, s$. By the same reason all other variables are evaluated in $\bar{A}$.

Now say that $y_i$ is evaluated in $e_{h_i,k_i}(i)$, replace $y_i$ by $u_i v_i w_i y_i$ obtaining a new polynomial $\tilde{f}$. The evaluation of $f$ can be extended to an evaluation of $\tilde{f}$ by replacing the new variables by

$$u_i, v_i, w_i \mapsto e_{h_i,h_i}(i), \implies u_i v_i w_i y_i \mapsto e_{h_i,k_i}(i).$$

Now consider the polynomial $\tilde{f}$ in which all the variables $u_i, w_i, y_i, x_{h,k}, k > 1, h = 1, \dots, s$, have been evaluated in the same way as for the previous evaluations. The variables $v_i, z_h$ are left not evaluated.

Call $g(v_1, \dots, v_q, z_1, \dots, z_s)$ this generalized polynomial. It is a multilinear expression in these variables and some fixed number $r = r(g)$ of fixed elements of $A$ (in fact one easily sees that $r = |E| + d^s + 3q$). By Remark 23.1.24, the polynomial $g$ has the property that if in a given evaluation of all the variables, we have evaluated some $z_h$ in $\bar{A}$ or some $v_i$ in $J$, then it vanishes.

Moreover, write an element $a \in A$ (or in $A \otimes F[\xi]$) as

$$a = \sum_{i=1}^q a(i) + j(a),\ a(i) \in A_i[\xi] = M_{d_i}(F[\xi]),\ a(i) = (a(i)_{\alpha,\beta}),\ j(a) \in J[\xi].$$

If we evaluate $v_h \mapsto a_h = \sum_{i=1}^q a_h(i) + j(a_h)$ we have

$$\begin{aligned} u_i v_i w_i y_i &\mapsto e_{h_i,h_i}(i)\Big(\sum_{\ell=1}^q a_i(\ell) + j(a_i)\Big) e_{h_i,h_i}(i) e_{h_i,k_i}(i) \\ &= a_i(i)_{h_i,h_i} e_{h_i,k_i}(i) + e_{h_i,h_i}(i) j(a_i) e_{h_i,k_i}(i),\ e_{h_i,h_i}(i) j(a_i) e_{h_i,k_i}(i) \in J[\xi]. \end{aligned}$$

From the properties of $g$ (Remark 23.1.24), we have that $g(v_1, \dots, v_q, z_1, \dots, z_s)$ evaluated in elements $v_h \mapsto a_h = \sum_{i=1}^q a_h(i) + j(a_h)$ and arbitrary elements $z_i$ equals

(23.26)
$$\prod_{i=1}^q a_i(i)_{h_i,h_i} g(e_{h_1,h_1}(1), \dots, e_{h_q,h_q}(q), z_1, \dots, z_s) := \prod_{i=1}^q a_i(i)_{h_i,h_i} G(z_1, \dots, z_s),$$

where $G(z_1,\dots,z_s) := g(e_{h_1,h_1}(1),\dots,e_{h_q,h_q}(q),z_1,\dots,z_s)$, a polynomial map in the elements $z_i$.

Finally, if we substitute for $z_i \mapsto a_i + j_i$, $a_i \in \bar{A}[\xi]$, $j_i \in J[\xi]$, we have

$$G(a_1+j_1,\dots,a_s+j_s) = G(j_1,\dots,j_s), \tag{23.27}$$

and for some choice of $j_1,\dots,j_s$, we have $G(j_1,\dots,j_s) \neq 0$.

Choose $q$ integers $\underline{n} := n_1,\dots,n_q$ adding to $n$, and let $m = n+s$, $n_{q+1} = s$. Let us denote by $b_i := n_1+\cdots+n_i$, $i = 1,\dots,q$, and by convention $b_0 := 0$, $b_{q+1} := m$.

The set $\{1,\dots,m\}$ is partitioned in $q+1$ subsets $\mathcal{P}_{0,\underline{n}} := \{B_1,\dots,B_q,B_{q+1}\}$ with

$$B_i = \{b_{i-1}+1, b_{i-1}+2,\dots,b_i\}.$$

The symmetric group $S_m$ acts transitively on the set $\mathfrak{P}_{\underline{n}}$ of all the partitions $\{C_1,\dots,C_q,C_{q+1}\}$ of $\{1,\dots,m\}$ for which each $C_j$ has $n_j$ elements. The stabilizer of $\mathcal{P}_{0,\underline{n}}$ is the product of the $q+1$ symmetric groups $\prod_{i=1}^{q+1} S_{B_i} = \prod_{i=1}^{q+1} S_{n_i}$, a subgroup of $S_m$ with $\prod_{i=1}^{q+1} n_i!$ elements. Hence $\mathfrak{P}_{\underline{n}}$ has $\frac{m!}{\prod_{i=1}^{q+1} n_i!} = \binom{m}{n_1,\dots,n_q,n_{q+1}}$ elements.

Consider also $m$ variables $x_h$, $h = 1,\dots,m$, and the monomials $M_1,\dots,M_q$ with $M_i = x_{b_{i-1}+1}x_{b_{i-1}+2}\cdots x_{b_i}$ product of the corresponding $n_i$ consecutive variables. Make $g$ into a generalized polynomial of the $m$ variables $x_h$, $h = 1,\dots,m$, by substituting to $v_i$ the monomial $M_i$ and to $z_j$ the variable $x_{n+j}$ call

$$g_{\mathcal{P}_{0,\underline{n}}}(x_1,\dots,x_m) := g(M_1,\dots,M_q,x_{n+1},\dots,x_m).$$

Consider the $m$ generic elements of $A$, decomposed as in formula (23.18)

$$X_j = \sum_{i=1}^{q} X_j(i) + W_j \in A[\xi,\eta],\ j = 1,\dots,m,$$

with $X_j(i)$ generic elements of $A_i = M_{d_i}(F)$, i.e., generic $d_i \times d_i$ matrices and $W_j$ generic elements of $J$.

For $\sigma \in S_m$ consider then $g_{\mathcal{P}_{0,\underline{n}}}(X_{\sigma(1)},\dots,X_{\sigma(m)})$. By formulas (23.26) and (23.27) we have

$$g_{\mathcal{P}_{0,\underline{n}}}(X_{\sigma(1)},\dots,X_{\sigma(m)}) = \prod_{i=1}^{q} M_i^{\sigma}(X)_{h_i,h_i} G(W_{\sigma(n+1)},\dots,W_{\sigma(m)}), \tag{23.28}$$

where

$$M_i^{\sigma}(X) = X_{\sigma(b_{i-1}+1)}(i)X_{\sigma(b_{i-1}+2)}(i)\cdots X_{\sigma(b_i)}(i) \in M_{d_i}(F[\xi]), \tag{23.29}$$

and of course $M_i^{\sigma}(X)_{h_i,h_i}$ is a diagonal entry of the matrix $M_i^{\sigma}(X)$.

Consider the group $H = \prod_{i=1}^{q} S_{n_i} = \prod_{i=1}^{q} S_{B_i}$ defined by $\sigma(\mathcal{P}_{0,\underline{n}}) = \mathcal{P}_{0,\underline{n}}$ and $\sigma(n+i) = n+i$, $i = 1,\dots,s$.

COROLLARY 23.1.25. *The dimension of the vector space spanned by the elements* $g_{\mathcal{P}_{0,\underline{n}}}(X_{\sigma(1)},\dots,X_{\sigma(m)})$, *where* $\sigma$ *runs over* $H$, *equals* $\prod_{i=1}^{q} c_{n_i}(M_{d_i})$.

PROOF. Since by construction $G(X_{n+1},\dots,X_m) = G(W_{n+1},\dots,W_m) \neq 0$, this space, by formula (23.28), is isomorphic to the space spanned by the polynomials $\prod_{i=1}^{q} M_i^{\sigma}(X)_{h_i,h_i}$, $\sigma \in H$.

For each $i = 1, \dots, q$ the polynomial functions $M_i^\sigma(X)_{h_i,h_i}$ depend and are multilinear, only upon the variables $X_j(i),\ j \in B_i$.

Moreover write $\sigma = (\tau_1, \dots, \tau_q),\ \tau_j \in S_{B_j} = S_{n_j}$, then we have $M_i^\sigma(X)_{h_i,h_i} = M_i^{\tau_i}(X)_{h_i,h_i}$, by formula (23.29).

Thus, if $i \neq j$, the polynomial functions $M_i^\sigma(X)_{h_i,h_i},\ M_j^\sigma(X)_{h_j,h_j},\ \sigma \in H$, are in disjoint sets of variables.

Therefore the span of the polynomials $\prod_{i=1}^q M_i^\sigma(X)_{h_i,h_i},\ \sigma \in H$, is the tensor product of the spaces spanned by the polynomials $M_i^{\tau_i}(X)_{h_i,h_i},\ \tau_i \in S_{n_i}$.

This last space, by Lemma 23.1.22, is isomorphic to the span of multilinear products of $n_i$ generic $d_i \times d_i$ matrices, hence, by Lemma 23.1.17, it has dimension $c_{n_i}(M_{d_i})$. □

REMARK 23.1.26. Given a partition $\mathcal{P} = \{C_1, \dots, C_q, C_{q+1}\}$ and a permutation $\tau \in S_m$ with $\tau(\mathcal{P}_{0,\underline{n}}) = \mathcal{P}$, we have that the element $g_{\mathcal{P}_{0,\underline{n}}}(X_{\tau(1)}, \dots, X_{\tau(m)})$ depends, and is multilinear, only upon the elements $X_j(i),\ j \in C_i,\ i = 1, \dots, q$, and $W_j,\ j \in C_{q+1}$. Moreover $\tau H \tau^{-1} = \prod_{i=1}^q S_{C_i}$.

It follows that, denoting the span of the elements $g_{\mathcal{P}_{0,\underline{n}}}(X_{\sigma(1)}, \dots, X_{\sigma(m)}),\ \sigma \in \tau H,\ \tau(\mathcal{P}_{0,\underline{n}}) = \mathcal{P}$, with $\tau$ fixed, by $\mathcal{M}_\mathcal{P}$ we have

LEMMA 23.1.27. *The spaces $\mathcal{M}_\mathcal{P} \subset A[\xi, \eta]$, as $\mathcal{P} = \{C_1, \dots, C_q, C_{q+1}\}$ runs over all partitions of $\{1, \dots, m\}$ (with $q+1$ parts and $|C_{q+1}| = s$), form a direct sum. Moreover $\mathcal{M}_\mathcal{P} = \tau \mathcal{M}_{\mathcal{P}_0}$.*

PROOF. By Remark 23.1.26 these spaces are formed by homogeneous elements of $A[\xi, \eta]$ of different multidegrees in the variables $X_j(i), W_j$. □

Observe next that the direct sum of all these spaces is contained in the space obtained by specialization of the span of multilinear products of $n + r$ generic elements of $A$, for $r = r(g)$ independent of $n$ counts the number of the elements in $A$ appearing in the generalized polynomial $g$. From Lemma 23.1.18 we deduce

$$\sum_{\underline{n}=(n_1,\dots,n_q)\,|\,\sum_i n_i = n}\ \sum_{\mathcal{P} \in \mathfrak{P}_{\underline{n}}} \dim(\mathcal{M}_\mathcal{P}) \le c_{n+t}(A) \tag{23.30}$$

with $t$ some fixed number. By symmetry we have $\dim(\mathcal{M}_\mathcal{P}) = \dim(\mathcal{M}_{\mathcal{P}_{0,\underline{n}}})$ for all $\mathcal{P} \in \mathfrak{P}_{\underline{n}}$, so, since $n_{q+1} = s$, we have

$$\sum_{\mathcal{P} \in \mathfrak{P}_{\underline{n}}} \dim(\mathcal{M}_\mathcal{P}) = \binom{m}{n_1, \dots, n_{q+1}} \dim(\mathcal{M}_{\mathcal{P}_{0,\underline{n}}}) = \binom{m}{s}\binom{n}{n_1, \dots, n_q} \dim(\mathcal{M}_{\mathcal{P}_{0,\underline{n}}})$$
$$\ge C n^s \binom{n}{n_1, \dots, n_q} \dim(\mathcal{M}_{\mathcal{P}_{0,\underline{n}}}).$$

From Corollaries 23.1.23 and 23.1.25 or formula (23.28) it then follows that

$$\dim(\mathcal{M}_{\mathcal{P}_{0,\underline{n}}}) = \prod_{i=1}^q c_{n_i}(M_{d_i}) \ge C \prod_{i=1}^q n_i^{-\frac{d_i^2-1}{2}} d_i^{2n_i}.$$

Hence, using also formula (23.30),

$$c_{n+t}(A) \ge C n^s \sum_{n_1,\dots,n_q\,|\,\sum_i n_i = n} \binom{n}{n_1, \dots, n_q} \prod_{i=1}^q n_i^{-\frac{d_i^2-1}{2}} d_i^{2n_i}. \tag{23.31}$$

By Lemma 23.1.9 of Beckner and Regev **[BR98]** finally

$$\sum_{n_1,\dots,n_q \mid \sum_i n_i = n} \binom{n}{n_1,\dots,n_q} \prod_{i=1}^{q} n_i^{-\frac{d_i^2-1}{2}} d_i^{2n_i} \simeq n^{-\frac{d-q}{2}} d^n, \; d = \sum d_i^2.$$

Hence formula (23.31) yields the conclusion

$$c_m(A) \geq C(m-t)^{-\frac{d-q}{2}+s} d^{m-t} \geq C' m^{-\frac{d-q}{2}+s} d^m. \tag{23.32}$$

**23.1.3. Algebras with 1, Theorem 21.1.2(5).** We want to sketch now (a variation of) the approach by Berele and Regev to Theorem 21.1.2(5). This approach is based on a detailed study of the cocharacters. If $A$ satisfies a Capelli identity $C_k$, we know that $m_\lambda(A) \neq 0$ only for $\lambda, \mathrm{ht}(\lambda) \leq k$. We then use formula (7.9) for the codimension and Theorem 18.3.1.

By this theorem and the theory of partition functions (see §18.2.1.1), it follows that formula (7.9) should be split into a finite sum of contributions as follows.

Let $P_k := \bigcup_n P_k(n)$ be the set of partitions with $k$ parts. First, by formula (18.10) as in Corollary 18.2.10, there exist vectors $X := \{a_1, \dots, a_m\}$ and $\{b_1, \dots, b_q\}$ in the set $P_k$, and integers $n_j$ so that the generating function $\Xi_A := \sum_{\lambda, \mathrm{ht}(\lambda) \leq k} m_\lambda(A) t^\lambda$ for cocharacters is a sum,

$$\Xi_A = \sum_{j=1}^{q} \Xi_{A,j}, \quad \Xi_{A,j} = n_j t^{b_j} \mathcal{P}_X, \tag{23.33}$$

where, if $C_X$ is the cone generated by $X$ in $\mathbb{N}^k$ we have the partition function

$$\mathcal{P}_X = \frac{1}{\prod_{i=1}^m (1 - t^{a_i})} = \sum_{a \in C_X} \mathcal{P}_X(a) t^a.$$

Finally

$$t^{b_j} \mathcal{P}_X = \sum_{a \in C_X} \mathcal{P}_X(a) t^{a+b_i} = \sum_{c \in C_X + b_i} \mathcal{P}_X(c - b_i) t^c.$$

Next the cone $C_X$ is divided into a finite number of polyhedral cones $C_j$ where $\mathcal{P}_X(a)$ is a quasi-polynomial, hence there is finally some positive integer $d$ such that on the intersection of $C_j$ with each coset $\underline{i} + d\mathbb{Z}^k$, $\underline{i} = (i_1, \dots, i_k)$, in $\mathbb{Z}^k$ the function $\mathcal{P}_X(a)$ is a polynomial.

23.1.3.1. *Some geometry of the subdivision.* By a standard theorem on polyhedral cones (see for instance **[DCP11**, §1.2.3**]**) one may further assume that each $C_j$ is *simplicial*, that is there is some basis of partitions $(v_1, \dots, v_k)$ so that

$$C_j := C(v_1, \dots, v_k) = \{\sum_i t_i v_i \mid t_i \geq 0\}. \tag{23.34}$$

In fact we can be more precise, using Theorem 17.2.14, if $A$ satisfies a Capelli identity $C_k$ and has as its first Kemer index $t$ and second Kemer index $s$, we know that for the partitions $\lambda \vdash m$, $\lambda = (\lambda_1, \cdots, \lambda_k)$, for which $m_\lambda \neq 0$, as soon as $m$ is large, we also have $\lambda_{t+1} + \cdots + \lambda_k \leq s$, $\lambda_i = 0$, $\forall i > d$.

Therefore we can a priori say that the codimension is also a finite sum of terms, arising from formula (7.9) where the sequence $\nu := (\lambda_{t+1}, \dots, \lambda_k)$ is fixed. So it depends only on the partition $(\lambda_1, \dots, \lambda_t)$. Denote

$$P_t(\nu) := \{(\lambda_1 \geq \cdots \geq \lambda_k) \mid (\lambda_{t+1}, \dots, \lambda_k) = \nu\}. \tag{23.35}$$

Notice that, if $P_t := \{(\lambda_1 \geq \cdots \geq \lambda_t)\}$ denotes the cone in $\mathbb{N}^t$ of all partitions, we identify $P_t$ with a subset of $P_k$ by adding $k-t$ entries equal to 0. We have

$$P_t(\nu) = P_t + u,\ u = (\lambda_{t+1}, \ldots, \lambda_{t+1}, \lambda_{t+1}, \ldots, \lambda_k).$$

In other words $P_t(\nu)$ is a $t$-dimensional cone in $\mathbb{N}^k$ with vertex $u$.

We can further describe $P_t(\nu) \cap C_j + b_j$ assuming that $C_j$ is given by formula (23.34). Given a partition $a \in P_k$, let us write $a = (a_0, a_1)$ where $a_0$ consists of the first $t$ coordinates and $a_1$ of the remaining coordinates. Then

$$P_t(\nu) \cap C_j + b_j = \{\sum_i t_i v_i + b_j \mid \sum_i t_i v_{i,1} + b_{j,1} = \nu,\ t_i \geq 0\}.$$

Subdivide the indices $i = 1, \ldots, k$ into those, call this set $S$, for which $v_{i,1} \neq 0$ and the complement $S'$.

The set of $t_i,\ i \in S,\ t_i \geq 0$, for which $\sum_{i \in S} t_i v_{i,1} + b_{j,1} = \nu$ is compact so the set $\sum_{i \in S} t_i v_i + b_j$ is also compact and thus contains only finitely many integral points $a_1, \ldots, a_k$. Thus the intersection of $P_t(\nu) \cap C_j + b_j$ with the integral lattice is contained in the finite union of the subsets $C(v_i,\ i \in S') + a_j$, where by construction if $i \in S'$, we have $v_i, \in P_t$.

Summarizing, we can decompose the set $P_t(\nu) \cap C_j + b_j \cap \underline{i} + d\mathbb{Z}^k$ into finitely many sets of type $C(v_1, \ldots, v_t) + a_j \cap \underline{i} + d\mathbb{Z}^k$ with $(v_1, \ldots, v_t)$ a basis in $P_t$.

Finally if $v_1, \ldots, v_h$ is any independent set of $P_t$, let us define

$$C^\circ(v_1, \ldots, v_h) := \{\sum_i t_i v_i \mid t_i > 0\}$$

to be the relative interior of the cone $C(v_1, \ldots, v_h)$. We finally have a disjoint union of sets

$$\alpha := C^\circ(v_1, \ldots, v_h) + \underline{a}$$

covering the set of partitions $\lambda \vdash n$ for $n$ large for which $m_\lambda(A) \neq 0$.

We have that the codimension is a finite sum of terms $c_n(A, \alpha)$ where

$$c_n(A, \alpha) = \sum_{\lambda \vdash n, \lambda \in \alpha} m_\lambda(A) f_\lambda, \quad \alpha = C^\circ(v_1, \ldots, v_h) + \underline{a} \cap \underline{i} + d\mathbb{Z}^k. \tag{23.36}$$

Of course, in this finite sum, only the terms which have exponential growth $t^n$ may contribute to the asymptotic.

23.1.3.2. *The function $f_\lambda$.* With the notations of §22.2 we start by analyzing the function $f_\lambda$ when $\lambda := (\lambda_1, \ldots, \lambda_k) \vdash n \in P_t(\nu)$ where $\nu = (\lambda_{t+1}, \ldots, \lambda_k)$ is fixed.

Let us set $a := \lambda_{t+1} + \cdots + \lambda_k \leq s$, $r := \lambda_1 + \cdots + \lambda_t$ so that $n = r + a$.

We compute $f_\lambda$ by the Young–Frobenius formula (6.4.2), so let $\ell_i := \lambda_i + k - i$, $\mu = (\ell_1, \ldots, \ell_k) := \lambda + \rho_k$, and $m := \ell_1 + \cdots + \ell_k = n + k(k-1)/2$.

The first remark is that by formula (22.26), in formula (23.36) we may replace $f_\lambda$ with $n^{-k(k-1)/2} g_{\lambda+\rho_k} = n^{-k(k-1)/2} g_\mu$ and thus instead of using formula (23.36), we can study the asymptotic of

$$\bar{c}_n(A, \alpha) = \sum_{\lambda \vdash n, \lambda \in \alpha} m_\lambda(A) g_{\lambda+\rho_k}. \tag{23.37}$$

Denote by $\mu_0 := (\ell_1, \ldots, \ell_t)$ and $p := \ell_1 + \cdots + \ell_t$, so

$$p = m - \sum_{j=t+1}^{k} \ell_j = m - a - (k-t)(k-t-1)/2 = m - b = n + c,$$

where we set for simplicity

$$b := a + (k-t)(k-t-1)/2, \quad c = k(k-1)/2 - b.$$

We have $\mu_0 \in \Pi_t(p)$ and we see that, from the formula (22.24) for $g_\mu$,

$$\binom{m}{\ell_1, \dots, \ell_k} = \binom{p}{\ell_1, \dots, \ell_t} \frac{\prod_{j=1}^b (p+j)}{\prod_{j=t+1}^k \ell_j!} = \binom{p}{\ell_1, \dots, \ell_t} p^b \frac{\prod_{j=1}^b (1+\frac{j}{p})}{\prod_{j=t+1}^k \ell_j!}, \tag{23.38}$$

(23.39)

$$\Delta(\ell_1, \dots, \ell_k) = \Delta(\ell_1, \dots, \ell_t) q(\ell_1, \dots, \ell_t),\ q(\ell_1, \dots, \ell_t) := \frac{\prod_{i=1}^t \prod_{j=t+1}^d (\ell_i - \ell_j)}{\prod_{j=t+1}^d \ell_j!}$$

is a polynomial in $\ell_1, \dots, \ell_t$, and finally

$$g_\mu = q(\ell_1, \dots, \ell_t) g_{\mu_0} p^b K(p), \quad K(p) := \frac{\prod_{j=1}^b (1+\frac{j}{p})}{\prod_{j=t+1}^k \ell_j!}.$$

Now $p^b$ is asymptotic to $n^b$ and $K(p)$ tends to the constant $(\prod_{j=t+1}^d \ell_j!)^{-1}$ as $n \to \infty$.

Let us fix one of the sets $\alpha := C^\circ(v_1, \dots, v_h) + \underline{a} \cap \underline{i} + d\mathbb{Z}^k \subset P_t(\nu)$. When $\lambda \in \alpha$, we have that $\ell_1, \dots, \ell_t$ lies in a corresponding region $\alpha'$

$$\begin{aligned} \alpha' &:= \{(\ell_1, \dots, \ell_t) \mid \ell_1, \dots, \ell_k \in \alpha + \rho_k\} \cap \underline{i} + \rho_k + d\mathbb{Z}^k \\ &= C^\circ(u_1, \dots, u_h) \cap \underline{j} + d\mathbb{Z}^t,\ u_i = v_i + \rho_{k,0},\ \underline{j} = \underline{i}_0 + \rho_{k,0}. \end{aligned}$$

In this region $m_\lambda(A)$ is some nonnegative polynomial $A(\ell_1, \dots, \ell_t)$. Therefore the heart of the computation is to study the asymptotics of

$$c_\alpha(n+c) = c_\alpha(p) := \sum_{(\ell_1, \dots, \ell_t) \in \alpha',\ \sum_{i=1}^t \ell_i = p} Q(\ell_1, \dots, \ell_t) \binom{p}{\ell_1, \dots, \ell_t}, \tag{23.40}$$

where $Q(\ell_1, \dots, \ell_t) := A(\ell_1, \dots, \ell_t)\Delta(\ell_1, \dots, \ell_k) q(\ell_1, \dots, \ell_t)$ is also a nonnegative polynomial. This is actually closer to the central limit theorem since there is no exponent $\beta$.

REMARK 23.1.28. Observe that the sum in formula (23.40) is empty, and hence $c_\alpha(p) = 0$ unless $p$ is congruent, modulo $d$, to the sum of the entries of $\underline{j}$.

23.1.3.3. *The asymptotics of* $c_\alpha(p)$. The reader should be convinced that the arguments of §22.2.2 apply in this case. We make the usual substitution

$$(\ell_1, \dots, \ell_t) = (\frac{p}{t} + c_1\sqrt{p}, \dots, \frac{p}{t} + c_t\sqrt{p}),\ (c_1, \dots, c_t) := \rho_p(\ell_1, \dots, \ell_t)$$

with $\rho_p(\ell) := \sqrt{p}^{-1}(\ell - \frac{p}{t})$.

For given $p$ consider the subset $C^\circ(u_1, \dots, u_h)_p$ of $C^\circ(u_1, \dots, u_h)$ of partitions of $p$.

First let us describe $\rho_p(C^\circ(u_1, \dots, u_h)_p)$. For this let us normalize the elements $u_i = (u_{i,1} \geq \cdots \geq u_{i,t} \geq 0)$ so that $\sum_j u_{i,j} = 1$, thus $C^\circ(u_1, \dots, u_h)_p$ is the convex hull of the vectors $pu_1, \dots, pu_h$.

$$\rho_p(u_i) = \sqrt{p}(u_{i,1} - \frac{1}{t}, \dots, u_{i,t} - \frac{1}{t}) := \sqrt{p}(u_i - u_0),\ u_0 := (\frac{1}{t}, \dots, \frac{1}{t}).$$

Setting $\Delta$ to be the convex hull of the vectors $u_i$, we have

$$\rho_p(C^\circ(u_1,\dots,u_h)_p) = \sqrt{p}(\Delta - u_0).$$

The formula (23.40) is now a sum for $(c_1,\dots,c_t) \in \rho_p(\alpha')$ which is the set $\beta$ of points in $\sqrt{p}(\Delta - u_0)$ which lie in the lattice $\sqrt{p}^{-1}(\underline{j} + d\mathbb{Z}^t)$.

Under this change of variables the polynomial $\bar{Q}(\ell_1,\dots,\ell_t)$ gives some polynomial in $\sqrt{p}$ with coefficient polynomials in the $c_i$ with leading term, in $\sqrt{p}$ some polynomial $p^h A_h(c_1,\dots,c_t)$, $h \in \mathbb{Z}/2$, and $A_h(c_1,\dots,c_t))$ is still positive on the region where the sum is taken.

At this point we can substitute, in formula (23.40) with formula (22.30) for the multinomial and obtain, setting $\beta := \sqrt{p}(\Delta - u_0) \cap \sqrt{p}^{-1}(\underline{j} + d\mathbb{Z}^t)$,

$$c_\alpha(p) := p^h t^p K \sum_{(c_1,\dots,c_t)\in\beta} A_h(c_1,\dots,c_t) e^{-\frac{t|c|^2}{2}} \sqrt{t} d p^{-(t-1)/2} \tag{23.41}$$

with $K$ constant. One easily sees that we may assume that $h = t$ so that $\Delta$ is a $(t-1)$-dimensional simplex and the sum in formula (23.41) is, in fact by Remark 23.1.28, either 0 or a Riemann sum relative to a lattice, inside $\sqrt{p}(\Delta - u_0)$, of covolume proportional to $p^{-(t-1)/2}$ if $p \equiv k$, mod $d$ for a given $k$ of the function $A_h(c_1,\dots,c_t)e^{-\frac{t|c|^2}{2}}$.

We have to distinguish the case in which 0 is or is not in $\Delta - u_0$. By the definitions it is easily seen that $u_0 \in \Delta$ if and only of $u_0$ is one of the vertices $u_i$, and we may assume $u_0 = u_1$.

If $0 \notin \Delta - u_0$, one sees that the domain $\sqrt{p}(\Delta - u_0)$ as $p$ goes to infinity tends to infinity and thus, due to the exponential decay of the integrand, this Riemann sum tends to 0. In fact one may estimate its decrease by $Cp^j e^{-p}$ with exponential decrease and thus give terms which can be ignored. The estimate is easily reduced to the estimate of

$$\frac{1}{2}\int_{\sqrt{p}}^{\infty} x^{2j+1} e^{-x^2} dx = \int_p^\infty y^j e^{-y} dy \le Cp^j e^{-p}$$

since the domain $\sqrt{p}(\Delta - u_0)$ is contained in a region where for some $i$ we have $|c_I| \ge c\sqrt{p}$ for some fixed $c$.

On the other hand if $u_0 = u_1$, then the domains $\sqrt{p}(\Delta - u_0)$ are increasing and cover the cone generated by the vectors $u_i - u_0$, $i = 2,\dots,t-1$. In this case the Riemann sum tends to the integral of a positive rapidly decreasing function in this cone which is a (finite) positive number. Thus this contribution to the codimension is $Cp^h t^p$ with $C$ some positive constant.

Given $k = 0,\dots,d-1$ for $p \equiv k$, mod $d$, in a given congruence class, summing all the finite terms of previous type for which the exponent $h$ is maximal, we have that for each $k$ the codimension $c_p(A)$ with $p \equiv k$, mod $d$ is asymptotic to $C_k p^{h_k} t^p$ with $C_k$ some positive constant.

Actually since the codimension is an increasing function (Theorem 21.1.7), one has that the exponents $h_k$ must all be equal, but in general the constants $C_k$ could be distinct. We have proved

THEOREM 23.1.29. *Given an algebra $A$ satisfying a Capelli identity with first Kemer index $t$, there exists a positive integer $d$, an integer or half integer $h$, and $d$ positive* *constants* $C_k, k = 0, \dots, d-1$ *so that*

$$c_p(A) \simeq C_k p^h t^p, \; p \equiv k, \bmod d. \tag{23.42}$$

This theorem, together with the precise estimate of $h$ given in Theorem 23.1.15, strengthens the information of formula (23.17). When $A$ has a 1, one has

THEOREM 23.1.30. *Given an algebra $A$ with 1 satisfying a Capelli identity with first Kemer index $t$, there exists an integer or half integer $h$ and a positive constant $C$ so that*

$$c_p(A) \simeq C p^h t^p. \tag{23.43}$$

In order to prove this theorem one has to recall a theorem of Drensky, that is the cocharacter sequence is *Young derived*, see formula (7.15).

Using Corollary 18.3.4, we see that the function $\overline{m}_\lambda$ has the same properties of being a quasipolynomial in some regions as $m_\lambda$ so the same arguments as before imply

THEOREM 23.1.31. *There exists some positive integer $\bar{d}$, an integer or half integers $\bar{h}_k$, an integer $\bar{d}$, positive constants $\bar{C}_k, k = 0, \dots, \bar{d}-1$ so that*

$$\bar{c}_p(A) \simeq \bar{C}_k p^{\bar{h}_k}(t-1)^p, \; p \equiv k, \bmod \bar{d}. \tag{23.44}$$

PROOF. We only need to prove that the *exponent* is $t-1$. This comes from the fact that $m_\lambda(A) \neq 0$ for large $\lambda$ implies that $\lambda_{t+1} \leq s$. Now these $m_\lambda(A)$ are obtained from the $\bar{m}_\lambda(A)$ by Pieri's rule so that necessarily $\bar{m}_\lambda(A) \neq 0$ for large $\lambda$ implies that $\lambda_t \leq s$.

This gives the growth $\leq t-1$ which is in fact sufficient for the next arguments. □

Now for the codimension, using formula (7.15), we have

$$c_n(A) = \sum_{j=0}^{n} \binom{n}{j} \bar{c}_j(A) = \sum_{k=0}^{d-1} \sum_{\substack{0 \leq j \leq n \\ j \equiv k \bmod d}} \binom{n}{j} \bar{c}_j(A) \tag{23.45}$$

$$\overset{(23.44)}{\simeq} \sum_{k=0}^{d-1} \sum_{\substack{0 \leq j \leq n \\ j \equiv k \bmod d}} \binom{n}{j} \bar{C}_k j^{\bar{h}_k}(t-1)^j.$$

LEMMA 23.1.32. *For each $0 \leq k \leq d-1$ there is a positive constant $a_k$ so that the sum*

$$A_{k,n} := \sum_{\substack{0 \leq j \leq n \\ j \equiv k \bmod d}} \binom{n}{j} \bar{C}_k j^{\bar{h}_k}(t-1)^j$$

*is asymptotic to $a_k n^{\bar{h}_k} t^n$.*

PROOF. Let $\omega$ be a primitive $d$ root of 1 so that

$$\sum_{\beta=0}^{d-1} \omega^{(q-k)\beta} = \begin{cases} 0 \text{ if } q \not\equiv k \bmod d \\ d \text{ if } q \equiv k \bmod d. \end{cases}$$

Write

$$A_{k,n} := \frac{1}{d}\sum_{0\le j\le n}[\sum_{\beta=0}^{d-1}\omega^{(j-k)\beta}]\binom{n}{j}\bar{C}_k j^{\bar{h}_k}(t-1)^j$$

$$(23.46)\qquad = \frac{1}{d}\sum_{\beta=0}^{d-1}\sum_{0\le j\le n}\omega^{-k\beta}\binom{n}{j}\bar{C}_k j^{\bar{h}_k}[\omega^{\beta}(t-1)]^j.$$

We first observe that

$$|\sum_{0\le j\le n}\omega^{-k\beta}\binom{n}{j}j^{\bar{h}_k}[\omega^{\beta}(t-1)]^j| \le \sum_{0\le j\le n}\binom{n}{j}j^{\bar{h}_k}[|\omega^{\beta}|(t-1)]^j,$$

where the equality is only for $\beta = 0$.

We recall the notation $a(\alpha, e) := \{n^e\alpha^n\}$, $\alpha > 0$ (page 578), and we apply formula (23.12) to $a(1,0) \star a(|\omega^{\beta}|(t-1), \bar{h}_k)$ obtaining that

$$\{\sum_{0\le j\le n}\binom{n}{j}j^{\bar{h}_k}[|\omega^{\beta}|(t-1)]^j\} = a(1,0) * a(|\omega^{\beta}|(t-1), \bar{h}_k)$$

$$\simeq Ka(|\omega^{\beta}|(t-1)+1, \bar{h}_k).$$

In the sum of formula (23.46), when $1 \le \beta \le d-1$, we have $|\omega^{\beta}| < 1$ so we deduce $|\omega^{\beta}|(t-1)+1 < t$ and the dominant term is for $\beta = 0$ giving the desired asymptotic. □

From Lemma 23.1.32, Theorem 21.1.2(5) follows immediately. So for a finite-dimensional algebra $R$ with 1, one has an asymptotic $c_n(R) \simeq Cn^h\ell^n$ where, by Remark 23.1.16, the two numbers $h, \ell$ can be computed once we describe a list of fundamental algebras $R_i$ such that $R$ is PI equivalent to $\bigoplus_i R_i$.

REMARK 23.1.33. The restriction in Theorem 21.1.2(5) that $A$ satisfies a Capelli identity was removed by A. Berele; see [**Ber08a**, Theorem 4.22].

## 23.2. Special finite-dimensional algebras

**23.2.1. Block triangular matrices.** By Theorem 16.3.2 we know that the $T$-ideal of polynomial identities of $UT(d_1, d_2, \dots, d_q)$ is the product of $T$-ideals of matrices $\mathrm{Id}(M_{d_1}(F))\,\mathrm{Id}(M_{d_2}(F))\cdots\mathrm{Id}(M_{d_q}(F))$.

We know, by Remark 17.2.40, that this algebra is fundamental, with Kemer index $(d = \sum_{i=1}^q d_i^2, q-1)$. We want to strengthen Theorem 21.1.2 by proving that, in formula (23.17), $C_1 = C_2$ and compute this constant in Corollary 23.2.3.

First, in order to stress the special role of these algebras, we present Proposition 23.2.1, which is of independent interest.

Let $A = \bar{A} + J$ be a finite-dimensional algebra over an algebraically closed field $F$ of characteristic 0.

PROPOSITION 23.2.1. *If $A_1 \oplus \cdots \oplus A_q \subseteq \bar{A}$ is an admissible subalgebra of $A$ (Definition 21.2.1), and $A_i \cong M_{d_i}(F)$, $1 \le i \le q$ ($F$ is algebraically closed), then $A$ contains a subalgebra isomorphic to $UT(d_1, \dots, d_q)$.*

PROOF. Recall that $A_1 \oplus \cdots \oplus A_q$ is admissible if $A_1JA_2J\cdots JA_q \neq 0$, i.e., $A_1 \oplus \cdots \oplus A_q + J$ is reduced or full (Definition 17.2.31, Proposition 17.2.38).

Let $u_1, \dots, u_{q-1} \in J$ be such that $A_1u_1A_2u_2\cdots u_{q-1}A_q \neq 0$.

Set $\delta := d_1 + \cdots + d_q$ and identify $A_1 \oplus \cdots \oplus A_q \cong M_{d_1}(F) \oplus \cdots \oplus M_{d_q}(F)$ with the subalgebra of $M_\delta(F)$ of block-diagonal matrices

$$\begin{pmatrix} M_{d_1}(F) & & 0 \\ & \ddots & \\ 0 & & M_{d_q}(F) \end{pmatrix}.$$

This is done by setting $\delta_0 = 0$ and

$$\delta_k = d_1 + \cdots + d_k,\ 1 \leq k \leq q,$$

so that $A_k \equiv \operatorname{span}\{e_{\alpha\beta} \mid \delta_{k-1} + 1 \leq \alpha, \beta \leq \delta_k\},\ 1 \leq k \leq q$.

In this setting the condition $A_1 u_1 A_2 u_2 \cdots u_{q-1} A_q \neq 0$ says that there exist $e_{\alpha_1\beta_1}, \ldots, e_{\alpha_q\beta_q} \in M_q(F)$ such that

$$e_{\alpha_1\beta_1} u_1 e_{\alpha_2\beta_2} u_2 \cdots u_{q-1} e_{\alpha_q\beta_q} \neq 0, \tag{23.47}$$

and $e_{\alpha_k\beta_k} \in A_k,\ 1 \leq k \leq q$, with $\delta_{k-1} + 1 \leq \alpha_k, \beta_k \leq \delta_k$.

On the other hand the subalgebra $UT(d_1, \ldots, d_q) = \bigoplus_{1 \leq i \leq j \leq q} A_{i,j} \subset M_q(F)$ has as basis the matrix units $e_{i,j}$ with the restriction that if $\delta_{k-1} + 1 \leq i \leq \delta_k$, then $\delta_{k-1} + 1 \leq j$ which distribute into bases of the various blocks $A_{i,j}$. Moreover it is easy to see that each nonzero ideal of $UT(d_1, \ldots, d_q)$ contains the entire block $A_{1,q}$.

In order to prove the lemma, we shall construct, starting from relation (23.47), a set of elements $x_{ij} \in A$ in correspondence with the matrix units of $UT(d_1, \ldots, d_q)$ and such that they satisfy the same multiplication rules as these matrix units.

Now, if $i, j$ are such that $\delta_{k-1} + 1 \leq i, j \leq \delta_k$, for some $1 \leq k \leq q$, define $x_{ij} = e_{ij} \in A_k$. In case $1 \leq i < j \leq q$ are such that

$$\delta_{k-1} + 1 \leq i \leq \delta_k, \quad \delta_{k+s-1} + 1 \leq j \leq \delta_{k+s}$$

with $1 \leq k \leq q, s \geq 1$ and $k + s \leq q$, then define

$$x_{ij} = e_{i\alpha_k} e_{\alpha_k\beta_k} u_k e_{\alpha_{k+1}\beta_{k+1}} u_{k+1} \cdots u_{k+s-1} e_{\alpha_{k+s}\beta_{k+s}} e_{\beta_{k+s} j}.$$

By direct computation we get that

$$x_{ij} x_{jk} = x_{ik}, \qquad x_{ij} x_{tk} = 0 \qquad \text{if} \quad j \neq t, \tag{23.48}$$

therefore these elements span a subalgebra $C$ containing $\bigoplus_{i=1}^q A_i$ and define a surjective homomorphism of $UT(d_1, \ldots, d_q)$ to $C$ by mapping $e_{i,j} \mapsto x_{i,j}$.

In order to show that this homomorphism is in fact an isomorphism, we need to prove that it is injective. If it were not injective, then by Remark 23.1.33 its kernel contains the block $A_{1,m}$, but this is not the case since, by formula (23.47), the element $x_{\alpha_1,\beta_q} \neq 0$ and $e_{\alpha_1,\beta_q} \in A_{1,q}$. □

23.2.1.1. *Estimates.* We shall apply the results of §16.4.2 to codimension. Let $c_n(A)$ denote the codimensions of the PI algebra $A$.

THEOREM 23.2.2. *For $j = 1, \ldots, q$, let $I_j$ be T-ideals, and let $I_j = \operatorname{Id}(A_j) \subset F\langle X\rangle$, such that $c_n(A_j) \simeq a_j n^{e_j} \alpha_j^n$. Let $I = I_1 \cdots I_q$, and let $A$ satisfy $I = \operatorname{Id}(A)$. Then*

$$c_n(A) \simeq a\, n^e \alpha^{n-1}, \tag{23.49}$$

*where* $\alpha = \alpha_1 + \cdots + \alpha_q$, $e = e_1 + \cdots + e_q + q - 1$, *and*

$$a = a_1 \cdots a_q \frac{\alpha_1^{e_1} \cdots \alpha_q^{e_q}}{(\alpha_1 + \cdots + \alpha_q)^e}.$$

PROOF. The general case easily follows by induction from the case $q = 2$, which we now prove.

Estimate the degree of the third summand in (16.30):

$$\deg[\chi_{(1)} \hat{\otimes} \sum_{j=0}^{n-1} \chi_j(A_1) \hat{\otimes} \chi_{n-j-1}(A_2)]$$

$$n \sum_{j=0}^{n-1} \binom{n-1}{j} c_j(A_1) c_{n-j-1}(A_2) \cong n \sum_{j=0}^{n-1} \binom{n-1}{j} a_1 j^{e_1} \alpha_1^j a_2 (n-j-1)^{e_2} \alpha_2^{n-j-1}$$

$$\overset{(23.11)}{\cong} a_1 a_2 \frac{\alpha_1^{e_1} \alpha_2^{e_2}}{(\alpha_1 + \alpha_2)^{e_1+e_2}} (n-1)^{e_1+e_2+1} (\alpha_1 + \alpha_2)^{n-1}$$

$$\cong a n^e \alpha^{n-1}.$$

Similarly, the last summand in (16.30) yields $a \cdot n^{e_1+e_2} \alpha^{n-1}$ which can be discarded since $e_1 + e_2 < e$. Finally, the asymptotics of $c_n(A_j)$, $j = 1, 2$, are given, and obviously both can be discarded. $\square$

COROLLARY 23.2.3. *For the algebras* $R = UT(d_1, d_2, \ldots, d_q)$ *the two constants* $C_1, C_2$ *of Theorem* 21.1.2 *are equal and with value* ($C_k$ *as in formula* (21.2)):

$$c_n(R) \simeq C n^{-\frac{d-q}{2} + q - 1} d^{n-1}, \quad C = d^{\frac{d-q}{2}} \prod_{i=1}^{q} C_{d_i} d_i^{1-d_i^2}, \; d = \sum_{i=1}^{q} d_i^2.$$

PROOF. This is proved by Theorem 16.3.2 and Theorem 23.2.2. $\square$

For instance for $UT(1,1)$ we have the asymptotic $n2^{n-1}$, while from formula (16.33) the exact codimension is $(n-2)2^{n-1} + 2$.

EXERCISE 23.2.4. Let $R$ be the two-dimensional noncommutative algebra without 1,

$$R := \left\{ \begin{vmatrix} a & b \\ 0 & 0 \end{vmatrix} \right\}.$$

Prove that $c_n(R) = n$.

*Hint*. Use the generic elements $\xi_i := \left\{ \begin{vmatrix} a_i & b_i \\ 0 & 0 \end{vmatrix} \right\}$.

APPENDIX A

# The Golod–Shafarevich counterexamples

The $F$-algebra $A$ is *nil* if for every element $a \in A$ there exists a power $p = p(a)$ such that $a^{p(a)} = 0$. In particular such an algebra is algebraic, see Section 8.2.1. Theorem 8.2.1 shows that a PI algebra which is finitely generated and algebraic is finite dimensional. Thus a finitely generated nil PI algebra is finite dimensional. This raises the question, What happens if we remove the assumption that the algebra $A$ is PI?

QUESTION (cf. Kurosch **[Kur41]**). If the $F$-algebra $A$ is finitely generated and nil, is it necessarily finite dimensional?

The celebrated Golod–Shafarevich counterexample shows that there exist algebras $A$ which are finitely generated and nil but are infinite dimensional. Following **[RR09]**, we now give full details of that counterexample. This is done in two main steps. In the first step a certain finitely generated nil algebra $A$ is constructed. The second step consists of the proof that *that* algebra $A$ is infinite dimensional. We remark that usually in the literature, that second step applies techniques of Hilbert series; see **[Gol64]**, **[Gol68]**, **[GŠ64]**, **[Her94]**, **[Row88]**, and **[Vin65]**. The proof presented here of that second step is completely elementary, where Hilbert series considerations are replaced by an induction argument.

**A.0.1. Preliminaries.** Throughout this section $F$ is a field (of any characteristic), $d \geq 2$ is a natural number, $V$ is a vector space of dimension $\dim V = d$, and

$$T = F\langle x_1, \ldots, x_d \rangle \equiv \bigoplus_{n=0}^{\infty} V^{\otimes n}$$

is the free associative noncommutative algebra of the polynomials in $d$ variables. Let

$$T_{\geq 1} \subset T, \qquad T_{\geq 1} = \bigoplus_{n=1}^{\infty} V^{\otimes n}$$

denote the polynomials with no constant term. If $I \subseteq T_{\geq 1}$ is a two-sided ideal, then the quotient algebra $T_{\geq 1}/I$ is finitely generated.

The Golod–Shafarevich counterexample can be stated as follows.

THEOREM A.0.1 (**[GŠ64]**). *There exists a homogeneous two-sided ideal $I \subset T_{\geq 1}$, such that the finitely generated algebra $A = T_{\geq 1}/I$ is both nil and infinite dimensional.*

The proof is given in what follows.

The two-sided ideal $I$ given here is generated by a sequence $H = \{f_1, f_2, \ldots\}$ of homogeneous polynomials, so $I = I_H$. We first define when such a sequence $H$ is a *G.S. sequence* (Definition A.0.3) (G.S. stands for Golod–Shafarevich). Then in Section A.0.5 we construct a G.S. sequence, thus proving that such sequences exist.

Since $T_{\geq 1}/I_H$ is finitely generated, Theorem A.0.1 is a corollary of the following theorem.

THEOREM A.0.2. *Let $H \subset T_{\geq 1}$ be a G.S. sequence, then the quotient algebra $T_{\geq 1}/I_H$ is both nil and infinite dimensional.*

We begin with the following definition.

DEFINITION A.0.3. Let $H = \{f_1, f_2, \dots\} \subset T_{\geq 1}$ be a sequence of *homogeneous* polynomials. We assume that all $\deg f_j \geq 2$. Let $r_n$ be the number of elements of $H$ of degree $n$. Let $I = I_H$ be the two-sided ideal generated in $T_{\geq 1}$ by $H$. Call $H$ a G.S. sequence if the ideal $I_H$ and the corresponding numbers $r_n$ satisfy

(1) For every polynomial $g \in T_{\geq 1}$ there exists $m = m(g)$ such that $g^m \in I$.
(2) For some $\epsilon > 0$ satisfying $d - 2\epsilon > 1$, $\epsilon^2(d-2\epsilon)^{n-2} \geq r_n$ for all $n \geq 2$.

Note that by condition (1) of Definition A.0.3, the finitely generated algebra $A = T_{\geq 1}/I_H$ is nil. The first step in proving Theorem A.0.2 is Theorem A.0.4 **[Gol64]** of §A.0.2, which applies to any homogeneous ideal $I \subseteq T_{\geq 2}$ and which establishes the basic inequality (A.2) below. This inequality, together with condition (2) of Definition A.0.3, imply Proposition A.0.5, which states that $A = T_{\geq 1}/I_H$ is infinite dimensional. The construction in §A.0.5 of G.S. sequences (see Theorem A.0.11) thus completes the proof of Theorem A.0.1.

**A.0.2. The basic inequality.** Recall that $\dim V = d$. Let $T_n \equiv V^{\otimes n} \subset T$ denote the homogeneous polynomials of total degree $n$, so $\dim T_n = d^n$ and

$$T = \bigoplus_{n=0}^{\infty} T_n \equiv \bigoplus_{n=0}^{\infty} V^{\otimes n}.$$

We denote

$$T_{\geq k} = \bigoplus_{n=k}^{\infty} T_n \equiv \bigoplus_{n=k}^{\infty} V^{\otimes n}.$$

Thus $T_{\geq 1} \subseteq T$ are the polynomials with no constant term.

As in Definition A.0.3, $H = \{f_1, f_2, \dots\}$ is a sequence of *homogeneous* polynomials where all $\deg f_j \geq 2$, so $H \subseteq T_{\geq 2}$. Also $r_n$ is the number of elements of $H$ of degree $n$. Thus

$$H = \bigcup_{n=2}^{\infty} H_n, \quad \text{where} \quad H_n = \{f_j \mid \deg f_j = n\} \quad \text{and} \quad |H_n| = r_n.$$

Let $R = \operatorname{span}_F H$, so

$$R = \bigoplus_{n=0}^{\infty} R_n = \bigoplus_{n=2}^{\infty} R_n, \quad \text{where} \quad R_n = \operatorname{span} H_n.$$

Note that $\dim R_n \leq r_n$. Since all $\deg f_j \geq 2$, we have $r_0 = r_1 = 0$ and $R \subseteq T_{\geq 2} \subseteq T_{\geq 1}$. Since $T_{\geq 1} = TT_1$, we have $R \subseteq TT_1$. Let $I = \langle f_1, f_2, \dots \rangle$ be the two-sided ideal generated in $T$ by the sequence $H$, so that

$$I = TRT \subseteq T_{\geq 2} \subseteq T_{\geq 1}.$$

Let $A = T_{\geq 1}/I$ be the corresponding algebra. Since $I$ is generated by homogeneous polynomials of degree $\geq 2$,

$$I = \bigoplus_{n=0}^{\infty} I_n = \bigoplus_{n=2}^{\infty} I_n,$$

where $I_n = I \cap T_n$. Let $B_n \subseteq T_n$ be a complement vector space of $I_n$,

$$T_n = I_n \oplus B_n, \tag{A.1}$$

and denote $b_n = \dim B_n$. Since $I_0 = I_1 = 0$, hence $B_0 = T_0 = F$ and $B_1 = T_1$, so $b_0 = \dim_F F = 1$ and $b_1 = \dim_F T_1 = d$. Denote $B = \bigoplus_{n\geq 0} B_n$, so

$$T = I \oplus B.$$

With these notations, we now prove the basic inequality (A.2).

THEOREM A.0.4 (**[Gol64]**). *With the previous hypotheses and notations we have, for all $n \geq 2$,*

$$b_n \geq db_{n-1} - \sum_{j=0}^{n-2} r_{n-j} b_j. \tag{A.2}$$

PROOF (Following **[Vin65]**). Recall that $R = \operatorname{span}_F\{f_1, f_2, \ldots\}$ and that $T = I \oplus B$. We first show that

$$I = IT_1 + BR. \tag{A.3}$$

Note that $T = T_{\geq 1} \oplus F = TT_1 \oplus F$, hence

$$I = TRT = TR(TT_1 \oplus F) = (TRT)T_1 + TR = IT_1 + TR. \tag{A.4}$$

Now $T = I \oplus B$, $R \subseteq TT_1$ and $IT = I$, hence

$$TR = (I \oplus B)R = IR + BR \subseteq ITT_1 + BR = IT_1 + BR. \tag{A.5}$$

Since $IT_1 + IT_1 = IT_1$, by (A.4) and (A.5)

$$I = IT_1 + TR \subseteq IT_1 + (IT_1 + BR) \subseteq IT_1 + BR. \tag{A.6}$$

Since $I \supseteq IT_1, BR$, together with (A.6), it implies that $I = IT_1 + BR$, proving (A.3).

Taking the $n$th homogeneous component of (A.3) and recalling that $R_0 = R_1 = 0$ yields

$$I_n = I_{n-1}T_1 + \sum_{k=0}^{n} B_{n-k}R_k = I_{n-1}T_1 + \sum_{k=2}^{n} B_{n-k}R_k. \tag{A.7}$$

Note that $\dim(I_{n-1}T_1) \leq (\dim I_{n-1})(\dim T_1) = (\dim I_{n-1})d$ and $\dim(B_{n-k}R_k) \leq b_{n-k}r_k$. Taking dimensions on both sides of (A.7), we obtain

$$\dim I_n \leq (\dim I_{n-1})d + \sum_{k=2}^{n} b_{n-k}r_k. \tag{A.8}$$

Substituting $j = n - k$ in the above sum gives

$$\dim I_n \leq (\dim I_{n-1})d + \sum_{j=0}^{n-2} b_j r_{n-j}. \tag{A.9}$$

By (A.1), $d^n = \dim T_n = \dim I_n + \dim B_n = \dim I_n + b_n$, so $\dim I_n = d^n - b_n$ and similarly $\dim I_{n-1} = d^{n-1} - b_{n-1}$. Substituting this into (A.9) (and cancelling $d^n$) yields the desired inequality (A.2). □

**A.0.3. Infinite dimensionality of the algebra $A = T_{\geq 1}/I$.** The following proposition is at the heart of the Golod-Shafarevich construction.

PROPOSITION A.0.5 (**[Gol64]**). *Let $b_n$ and $r_\ell$ be the sequences in Theorem* A.0.4, *hence in particular satisfying* (A.2), *and recall that $d \geq 2$. Let $\epsilon > 0$ such that $d - 2\epsilon > 1$. If $r_\ell \leq \epsilon^2(d-2\epsilon)^{\ell-2}$ for all $\ell \geq 2$, then all $b_n \geq 1$ (in fact, it follows that the $b_n$'s grow exponentially), and therefore the algebra $A = T_{\geq 1}/I$ is infinite dimensional.*

As remarked already, usually the proof of Proposition A.0.5 applies techniques of Hilbert series (see **[Gol64]**, **[Gol68]**, **[GŠ64]**, **[Her94]**, **[Row88]** and **[Vin65]**). The following is a rather elementary proof of that proposition, just using induction.

PROOF. Let $u = d - 2\epsilon$. Then by assumption, $\epsilon^2 u^{\ell-2} \geq r_\ell$ for each $\ell$. Substitute $\ell = n + 2 - j$, so $\epsilon^2 u^{n-j} \geq r_{n+2-j}$ for $0 \leq j \leq n$. Multiplying by $b_j$ and summing on $j$, we obtain

$$\epsilon^2 \sum_{j=0}^{n} u^{n-j} b_j \geq \sum_{j=0}^{n} r_{n+2-j} b_j. \tag{A.10}$$

CLAIM A.0.6. For all $n \geq 0$, $b_{n+1} \geq \epsilon \sum_{j=0}^{n} u^{n-j} b_j$.

PROOF. By induction on $n$. For $n = 0$, $b_1 = d \geq \frac{d-1}{2} \geq \epsilon = \epsilon b_0$, and so the claim holds.

Now let $n \geq 1$, assume that $b_{n+1} \geq \epsilon \sum_{j=0}^{n} u^{n-j} b_j$, and show that

$$b_{n+2} \geq \epsilon \sum_{j=0}^{n+1} u^{n+1-j} b_j. \tag{A.11}$$

Since $d = 2\epsilon + u$, we have

$$\begin{aligned} db_{n+1} &= (\epsilon + u) b_{n+1} + \epsilon b_{n+1} \geq (\epsilon + u)\epsilon \sum_{j=0}^{n} u^{n-j} b_j + \epsilon b_{n+1} \\ &= \epsilon^2 \sum_{j=0}^{n} u^{n-j} b_j + \epsilon \sum_{j=0}^{n} u^{n+1-j} b_j + \epsilon b_{n+1} \\ &= \epsilon^2 \sum_{j=0}^{n} u^{n-j} b_j + \epsilon \sum_{j=0}^{n+1} u^{n+1-j} b_j. \end{aligned}$$

Therefore by (A.10),

$$db_{n+1} \geq \sum_{j=0}^{n} r_{n+2-j} b_j + \epsilon \sum_{j=0}^{n+1} u^{n+1-j} b_j.$$

By (A.2) with $n + 2$ replacing $n$, this gives

$$b_{n+2} \geq db_{n+1} - \sum_{j=0}^{n} r_{n+2-j} b_j \geq \epsilon \sum_{j=0}^{n+1} u^{n+1-j} b_j.$$

This implies (A.11), and hence completes the proof (by induction) of Claim A.0.6.

□

Now by Claim A.0.6 and by (A.10),

$$\epsilon b_{n+1} \geq \epsilon^2 \sum_{j=0}^{n} u^{n-j} b_j \geq \sum_{j=0}^{n} r_{n+2-j} b_j,$$

and this implies that

$$db_{n+1} = \epsilon b_{n+1} + (d-\epsilon) b_{n+1} \geq \sum_{j=0}^{n} r_{n+2-j} b_j + (d-\epsilon) b_{n+1}. \tag{A.12}$$

Thus

$$b_{n+2} \geq_{\text{by (A.2)}} \geq db_{n+1} - \sum_{j=0}^{n} r_{n+2-j} b_j \geq_{\text{by (A.12)}} \geq (d-\epsilon) b_{n+1}.$$

Since $b_0 = 1$, it follows that for all $n$, $b_n \geq (d-\epsilon)^n$. Note that, writing $A = \bigoplus_{n=1}^{\infty} A_n$ where $A_n = \{t + I | t \in T_n\}$, we have $b_n = \dim A_n$. Thus we have shown that $A$ is infinite dimensional. □

**A.0.4. Toward the construction of G.S. sequences.** Fix some $\epsilon > 0$ satisfying $d - 2\epsilon > 1$. Following Definition A.0.3, we now construct a sequence of homogeneous polynomials of degrees $\geq 2$, $f_1, f_2, \dots \in T_{\geq 2} \subset T_{\geq 1} \subset T = F\langle x_1, \dots, x_d \rangle$, having the following properties:

(i) For every $g \in T_{\geq 1}$ there exists a power $n$ such that $g^n \in I$, where $I = \langle f_1, f_2, \dots \rangle$ is the two-sided ideal generated by the $f_j$'s; and
(ii) Let $r_\ell$ be the number of $f_j$'s of degree $\ell$, then $r_{\ell+2} \leq \varepsilon^2 (d - 2\varepsilon)^\ell$.

Let $A = T_{\geq 1}/I$, then $A$ is generated by the $d$ elements $x_1 + I, \dots, x_d + I$, hence is finitely generated. By (i) here it is nil, and by (ii) together with Proposition A.0.5 it is infinite dimensional.

A.0.4.1. *Preparatory remarks.* Let $q$ and $n$ be positive integers.

Define $I(q,n) := \{(i_1, \dots, i_n) \mid 1 \leq i_1, \dots, i_n \leq q\}$, and define $J(q,n) \subseteq I(q,n)$ via $J(q,n) := \{(i_1, \dots, i_n) \mid 1 \leq i_1 \leq i_2 \leq \cdots \leq i_n \leq q\}$.

REMARK A.0.7. Note that $|I(q,n)| = q^n$. Also, $|J(q,n)| = \binom{n+q-1}{q-1}$; for example, the correspondence between $1 \leq j_1 \leq \cdots \leq j_n \leq q$ and

$$\longleftrightarrow \; 1 \leq j_1 < j_2 + 1 < j_3 + 2 < \cdots < j_n + n - 1 \leq n + q - 1$$

gives a bijection between $J(q,n)$ and the set of $n$-subsets of an $(n+q-1)$-set.

Note that

$$|J(q,n)| = \binom{n+q-1}{q-1} \leq (n+q-1)^{q-1}, \tag{A.13}$$

and the right-hand side is a polynomial in $n$ (of degree $q-1$).

Given $\pi \in S_n$ and $\underline{i} = (i_1, \dots, i_n) \in I(q,n)$, set $\pi(\underline{i}) = (i_{\pi^{-1}(1)}, \dots, i_{\pi^{-1}(n)})$. If $\sigma, \pi \in S_n$, then $\sigma(\pi(\underline{i})) = (\sigma\pi)(\underline{i})$, hence $I(q,n)$ is the disjoint union of the corresponding orbits. Given $\underline{i} \in I(q,n)$, let $O_{\underline{i}} = \text{orbit}(\underline{i}) = \{\pi(\underline{i}) \mid \pi \in S_n\}$ denote the orbit of $\underline{i}$ under the $S_n$ action. Then

$$I(q,n) = \bigcup_{\underline{j} \in J(q,n)} O_{\underline{j}}\,,$$

a disjoint union. If $\underline{j} = (j_1, \ldots, j_n) \in J(q,n)$, then $\underline{j} = (1, \ldots, 1, 2, \ldots, 2, \ldots)$, and we denote $\underline{j} = (1^{\mu_1}, \ldots, q^{\mu_q})$, where $k$ appears $\mu_k$ times in $\underline{j}$. Then $S_{\mu_1} \times \cdots \times S_{\mu_q}$ fixes $\underline{j}$, and

$$|O_{\underline{j}}| = \frac{n!}{\mu_1! \cdots \mu_q!}.$$

For example, $(1,1,1,2,2,2,2) = (1^3, 2^4) \in J(2,7)$, and $|O_{(1^3,2^4)}| = 7!/(3!4!) = 35$.

A.0.4.2. *The order-symmetric polynomials.*

DEFINITION A.0.8. Let $\underline{j} = (j_1, \ldots, j_n) \in J(q,n)$. Then define the *order-symmetric polynomial*

$$p_{\underline{j}}(y_1, \ldots, y_q) = \sum_{\underline{i} \in O_{\underline{j}}} y_{i_1} \cdots y_{i_n},$$

a homogeneous polynomial of degree $n$ in $y_1, \ldots, y_q$.

Note that we can also write

$$p_{\underline{j}}(y_1, \ldots, y_q) = \frac{1}{\mu_1! \cdots \mu_q!} \cdot \sum_{\pi \in S_n} y_{j_{\pi^{-1}(1)}} \cdots y_{j_{\pi^{-1}(n)}}.$$

We have, given commuting variables $\lambda_1, \ldots, \lambda_q$:

$$(\sum_{i=1}^{q} \lambda_i y_i)^n = \sum_{\underline{j} \in J(q,n)} \prod_{i=1}^{q} \lambda_i^{\mu_i} p_{\underline{j}}(y_1, \ldots, y_q). \tag{A.14}$$

Recall that $d \geq 2$. Given $0 < c \in \mathbb{N}$, let $q = d + d^2 + \cdots + d^c$. Then $q$ is the number of monomials of degree between 1 and $c$ in the noncommuting variables $x_1, \ldots, x_d$. Let $\{M_1, \ldots, M_q\}$ be the set of these monomials. Given $0 < n \in \mathbb{N}$ and $\underline{j} \in J(q,n)$, denote

$$h_{\underline{j}}(x) = h_{\underline{j}}(x_1, \ldots, x_d) = p_{\underline{j}}(M_1, \ldots, M_q). \tag{A.15}$$

With these notations we prove

LEMMA A.0.9. *Let $0 < c \in \mathbb{N}$, $q = d + d^2 + \cdots + d^c$, with $\{M_1, \ldots, M_q\}$ the monomials of degree between 1 and $c$. Let $0 < n \in \mathbb{N}$, and let $I \subseteq F\langle x_1 \cdots, x_d \rangle$ be a two-sided ideal containing $h_{\underline{j}}(x)$ for all $\underline{j} \in J(q,n)$, where $h_{\underline{j}}(x)$ are given by* (A.15). *Let $g = g(x_1, \ldots, x_d) \in T_{\geq 1}$ be a polynomial of degree $\leq c$. Then $g^n \in I$.*

PROOF. Since $M_1, \ldots, M_q$ are all the nonconstant monomials (in $x_1, \ldots, x_d$) of degree $\leq c$, and since $\deg g \leq c$, we can write

$$g = \sum_{i=1}^{q} \alpha_i M_i \quad \text{with} \quad \alpha_i \in F.$$

Then from formula (A.14) we have, substituting $\lambda_i$ with $\alpha_i$ and $y_j$ with $M_j$,

$$g^n = (\sum_{i=1}^{q} \alpha_i M_i)^n = \sum_{\underline{j} \in J(q,n)} \prod_{i=1}^{q} \alpha_i^{\mu_i} p_{\underline{j}}(M_1, \ldots, M_q) = \sum_{\underline{j} \in J(q,n)} \prod_{i=1}^{q} \alpha_i^{\mu_i} h_{\underline{j}}(x_1, \ldots, x_d). \tag{A.16}$$

Thus $g^n$ is a linear combination of the polynomials

$$\{h_{\underline{j}}(x_1, \ldots, x_d) \mid \underline{j} \in J(q,n)\}.$$

By assumption all these $h_{\underline{j}}(x)$ are in $I$, hence $g^n \in I$. □

REMARK A.0.10. We saw in Remark A.0.7 that $|J(q,n)| \le (n+q-1)^{q-1}$. The right-hand side is a polynomial in $n$ (of degree $q-1$), hence it grows slower than any exponential function in $n$, in particular slower than $\varepsilon^2 \cdot \alpha^n$, provided $\varepsilon > 0$ and $\alpha > 1$. This implies the following.

Let $d \ge 2, \varepsilon > 0$ such that $d - 2\varepsilon > 1$. Then there exists $n$ large enough such that

$$|J(q,n)| \le \varepsilon^2 \cdot (d-2\varepsilon)^{n-2}. \tag{A.17}$$

**A.0.5. The construction of a G.S. sequence.** We now prove

THEOREM A.0.11. *G.S. sequences exist. Namely, let $d \ge 2$, $\epsilon > o$ such that $d - 2\epsilon > 1$. Then there exists a sequence $f_1, f_2 \ldots \in T$ of homogeneous polynomials of degree $\ge 2$ satisfying conditions* (1) *and* (2) *of Definition* A.0.3.

PROOF. The construction is inductive, starting with the empty sequence. The induction assumption is that by the $k$th step we have chosen two integers $c_k \le c'_k$ and a sequence of homogenous polynomials $f_1, \ldots, f_{m_k}$ of degrees between 2 and $c'_k$. We assume that $f_1, \ldots, f_{m_k}$ satisfy the following

Condition $(c_k)$.

(1) For any $\ell \le c'_k$, the number $r_\ell$ of the elements $f_i$ of degree $\ell$ satisfies $r_{\ell+2} \le \varepsilon^2 \cdot (d-2\varepsilon)^\ell$.
(2) For any $g \in T_{\ge 1}$ of degree $\le c_k$, there exists some power $n$ such that $g^n \in I_k$, where $I_k = \langle f_1, \ldots, f_{m_k} \rangle$.

**The inductive step.**

In the next step (step $k+1$) we choose two integers $c_{k+1} \le c'_{k+1}$ satisfying $c'_k < c_{k+1}$. We then construct another block of polynomials $f_{m_k+1}, \ldots, f_{m_{k+1}}$ having degrees $c'_k < \deg f_j < c'_{k+1}$. In this way we obtain the sequence $f_1, \ldots, f_{m_k}, f_{m_k+1}, \ldots, f_{m_{k+1}}$.

The construction starts with choosing any $c_{k+1}$ such that $c'_k < c_{k+1}$.

Let $q = q_{k+1} = d + d^2 + \cdots + d^{c_{k+1}}$, then choose $n$ large enough such that

$$1 < n, \quad c'_k < n, \quad \text{and} \quad |J(q,n)| < \epsilon^2 (d-2\epsilon)^{n-2}. \tag{A.18}$$

By Remark A.0.10 and by (A.17) such $n$ exists. Next choose $c'_{k+1} = n \cdot c_{k+1}$, so $c_{k+1} < c'_{k+1}$.

Denote $m_{k+1} = |J(q,n)| + m_k$, then let $f_{m_k+1}, f_{m_k+2}, \ldots, f_{m_{k+1}}$ be the $|J(q,n)|$ polynomials

$$h_{\underline{j}}(x_1, \ldots, x_d) = p_{\underline{j}}(M_1, \ldots, M_q), \quad \underline{j} \in J(q,n), \tag{A.19}$$

given by (A.15). Note that for $\underline{j} \in J(q,n)$, $\deg p_{\underline{j}}(y_1, \ldots, y_q) = n$ and all $1 \le \deg M_i \le c_{k+1}$, hence

$$c'_k < n \le \deg p_{\underline{j}}(M_1, \ldots, M_q) \le n \cdot c_{k+1} = c'_{k+1}. \tag{A.20}$$

From now the new sequence from the above two blocks of $f'_j s$:

$$f_1, \ldots, f_{m_k}, f_{m_k+1}, \ldots, f_{m_{k+1}}.$$

We show that this last sequence satisfies Condition $(c_{k+1})$. We also show that $r_\ell$, calculated in step $k$ for $\ell \le c'_k$, remains unchanged when calculated in step $k+1$. Hence the numbers $r_\ell$ are well defined for the resulting infinite sequence $f_1, f_2, \ldots$. This will complete the proof of the theorem.

Note that $\langle f_{m_k+1}, f_{m_k+2}, \ldots, f_{m_{k+1}} \rangle \subseteq \langle f_1, \ldots, f_{m_k}, f_{m_k+1}, \ldots, f_{m_{k+1}} \rangle = I_{k+1}$.

By Lemma A.0.9 with $c = c_{k+1}$, for any polynomial $g \in T_{\geq 1}$ of degree $\leq c_{k+1}$,

$$g^n \in \langle f_{m_k+1}, f_{m_k+2}, \ldots, f_{m_{k+1}} \rangle,$$

hence $g^n \in I_{k+1}$, which is part (2) of Condition $(c_{k+1})$.

By assumption Condition $(c_k)$, the degrees of the polynomials in the first block $f_1, \ldots, f_{m_k}$ are $\leq c'_k$.

By (A.18) $c'_k < n$, and by (A.20), the degrees of $f_{m_k+1}, \ldots, f_{m_{k+1}}$ are between $n$ and $c'_{k+1} = c_{k+1}n$. Recall that for each degree $\ell$, $r_\ell$ is the number of elements of degree $\ell$ in the sequence $f_1, \ldots, f_{m_k}, f_{m_k+1}, \ldots, f_{m_{k+1}}$. Thus for $\ell \leq c'_k$ there is no contribution to $r_\ell$ from the second block $f_{m_k+1}, \ldots, f_{m_{k+1}}$. Therefore these $r_\ell$'s remain unchanged when the process of constructing the $f_j$'s continues.

If $\ell \leq c'_k$, it is given (by the induction assumption) that $r_\ell \leq \varepsilon^2 \cdot (d - 2\varepsilon)^{\ell-2}$.

Similarly, for $\ell > c_k$ there is no contribution to $r_\ell$ from the first block, so $r_\ell$ is calculated on the second block $f_{m_k+1}, \ldots, f_{m_{k+1}}$ only. And since by (A.19) the total number of the $f_j$'s in that second block is $|J(q,n)|$, hence that $r_\ell < |J(q,n)|$. If $\ell > c'_k$, we may assume that $\ell \geq n$ (since there are no $f_j$'s with degree between $c'_k + 1$ and $n$). Then by (A.18)

$$r_\ell \leq |J(q,n)| < \varepsilon^2 \cdot (d - 2\varepsilon)^{n-2} \leq \varepsilon^2 \cdot (d - 2\varepsilon)^{\ell-2}.$$

This completes the inductive step. □

**We summarize.**

We are given $d \geq 2$, $\varepsilon > 0$ such that $d - 2\varepsilon > 1$. We have constructed the infinite sequence $\{f_1, f_2, \ldots\}$ of homogeneous polynomials in $x_1, \ldots, x_d$ of degree $\geq 2$. Let $I = \langle f_1, f_2, \ldots \rangle \subseteq F\langle x_1, \ldots, x_d \rangle$ denote the ideal generated by the $f_j$'s. Let $r_\ell$ denote the number of $f_j$'s of degree $\ell$.

These polynomials satisfy that $r_\ell \leq \varepsilon^2 \cdot (d - 2\varepsilon)^{\ell-2}$ for all $\ell \geq 2$. Then, by Proposition A.0.5, the algebra $T_{\geq 1}/I$ is infinite dimensional. Obviously, it is finitely generated. But by construction, for every $g \in T_{\geq 1}$ there exists $n$ such that $g^n \in I$, hence $T_{\geq 1}/I$ is nil.

# Bibliography


[AP89] S. Abeasis and M. Pittaluga, *On a minimal set of generators for the invariants of* $3 \times 3$ *matrices*, Comm. Algebra **17** (1989), no. 2, 487–499, DOI 10.1080/00927878908823740. MR978487

[Adi79] S. I. Adian, *The Burnside problem and identities in groups*, Ergebnisse der Mathematik und ihrer Grenzgebiete [Results in Mathematics and Related Areas], vol. 95, Springer-Verlag, Berlin-New York, 1979. Translated from the Russian by John Lennox and James Wiegold. MR537580

[Alb39] A. A. Albert, *Structure of Algebras*, American Mathematical Society Colloquium Publications, vol. 24, American Mathematical Society, New York, 1939. MR0000595

[AEGN02] E. Aljadeff, P. Etingof, S. Gelaki, and D. Nikshych, *On twisting of finite-dimensional Hopf algebras*, J. Algebra **256** (2002), no. 2, 484–501, DOI 10.1016/S0021-8693(02)00092-3. MR1939116

[AGLM11] E. Aljadeff, A. Giambruno, and D. La Mattina, *Graded polynomial identities and exponential growth*, J. Reine Angew. Math. **650** (2011), 83–100, DOI 10.1515/CRELLE.2011.004. MR2770557

[AHN10] E. Aljadeff, D. Haile, and M. Natapov, *Graded identities of matrix algebras and the universal graded algebra*, Trans. Amer. Math. Soc. **362** (2010), no. 6, 3125–3147, DOI 10.1090/S0002-9947-10-04811-7. MR2592949

[AJK17] E. Aljadeff, G. Janssens, and Y. Karasik, *The polynomial part of the codimension growth of affine PI algebras*, Adv. Math. **309** (2017), 487–511, DOI 10.1016/j.aim.2017.01.022. MR3607284

[AKBK16] E. Aljadeff, A. Kanel-Belov, and Y. Karasik, *Kemer's theorem for affine PI algebras over a field of characteristic zero*, J. Pure Appl. Algebra **220** (2016), no. 8, 2771–2808, DOI 10.1016/j.jpaa.2015.12.008. MR3471186

[AKB10] E. Aljadeff and A. Kanel-Belov, *Representability and Specht problem for G-graded algebras*, Adv. Math. **225** (2010), no. 5, 2391–2428, DOI 10.1016/j.aim.2010.04.025. MR2680170

[AK08] E. Aljadeff and C. Kassel, *Polynomial identities and noncommutative versal torsors*, Adv. Math. **218** (2008), no. 5, 1453–1495, DOI 10.1016/j.aim.2008.03.014. MR2419929

[Ami51] S. A. Amitsur, *Nil PI-rings*, Proc. Amer. Math. Soc. **2** (1951), 538–540, DOI 10.2307/2032001. MR42383

[Ami56] S. A. Amitsur, *Radicals of polynomial rings*, Canadian J. Math. **8** (1956), 355–361, DOI 10.4153/CJM-1956-040-9. MR78345

[Ami70] S. A. Amitsur, *A noncommutative Hilbert basis theorem and subrings of matrices*, Trans. Amer. Math. Soc. **149** (1970), 133–142, DOI 10.2307/1995665. MR258869

[Ami71] S. A. Amitsur, *A note on PI-rings*, Israel J. Math. **10** (1971), 210–211, DOI 10.1007/BF02771571. MR321967

[Ami72] S. A. Amitsur, *On central division algebras*, Israel J. Math. **12** (1972), 408–420, DOI 10.1007/BF02764632. MR318216

[Ami79] S. A. Amitsur, *On the characteristic polynomial of a sum of matrices*, Linear and Multilinear Algebra **8** (1979/80), no. 3, 177–182, DOI 10.1080/03081088008817315. MR560557

[AL50] A. S. Amitsur and J. Levitzki, *Minimal identities for algebras*, Proc. Amer. Math. Soc. **1** (1950), 449–463, DOI 10.2307/2032312. MR36751

[AP66] S. A. Amitsur and C. Procesi, *Jacobson-rings and Hilbert algebras with polynomial identities*, Ann. Mat. Pura Appl. (4) **71** (1966), 61–72, DOI 10.1007/BF02413733. MR206044

[Ana92] A. Z. Anan'in, *Representability of Noetherian finitely generated algebras*, Arch. Math. (Basel) **59** (1992), no. 1, 1–5, DOI 10.1007/BF01199007. MR1166010

[AAR99] G. E. Andrews, R. Askey, and R. Roy, *Special functions*, Encyclopedia of Mathematics and its Applications, vol. 71, Cambridge University Press, Cambridge, 1999. MR1688958

[Aok83] T. Aoki, *Calcul exponentiel des opérateurs microdifférentiels d'ordre infini. I* (French), Ann. Inst. Fourier (Grenoble) **33** (1983), no. 4, 227–250. MR727529

[Art64] E. Artin, *The gamma function*, Translated by Michael Butler. Athena Series: Selected Topics in Mathematics, Holt, Rinehart and Winston, New York-Toronto-London, 1964. MR0165148

[Art69] M. Artin, *On Azumaya algebras and finite dimensional representations of rings*, J. Algebra **11** (1969), 532–563, DOI 10.1016/0021-8693(69)90091-X. MR242890

[AS79] M. Artin and W. Schelter, *A version of Zariski's main theorem for polynomial identity rings*, Amer. J. Math. **101** (1979), no. 2, 301–330, DOI 10.2307/2373980. MR527994

[AS81a] M. Artin and W. Schelter, *On two-sided modules which are left projective*, J. Algebra **71** (1981), no. 2, 401–421, DOI 10.1016/0021-8693(81)90183-6. MR630605

[AS81b] M. Artin and W. Schelter, *Integral ring homomorphisms*, Adv. in Math. **39** (1981), no. 3, 289–329, DOI 10.1016/0001-8708(81)90005-0. MR614165

[AS87] M. Artin and W. F. Schelter, *Graded algebras of global dimension* 3, Adv. in Math. **66** (1987), no. 2, 171–216, DOI 10.1016/0001-8708(87)90034-X. MR917738

[ADKT00] T. Asparouhov, V. Drensky, P. Koev, and D. Tsiganchev, *Generic* $2 \times 2$ *matrices in positive characteristic*, J. Algebra **225** (2000), no. 1, 451–486, DOI 10.1006/jabr.1999.8143. MR1743670

[AG60a] M. Auslander and O. Goldman, *Maximal orders*, Trans. Amer. Math. Soc. **97** (1960), 1–24, DOI 10.2307/1993361. MR117252

[AG60b] M. Auslander and O. Goldman, *The Brauer group of a commutative ring*, Trans. Amer. Math. Soc. **97** (1960), 367–409, DOI 10.2307/1993378. MR121392

[Azu51] G. Azumaya, *On maximally central algebras*, Nagoya Math. J. **2** (1951), 119–150. MR40287

[Bah91] Yu. A. Bahturin, *Identities*, Enciclopedia of Math. Sci. **18**, Algebra II, Springer-Verlag, Berlin-New York (1991), 107–234.

[BD02] Y. Bahturin and V. Drensky, *Graded polynomial identities of matrices*, Linear Algebra Appl. **357** (2002), 15–34, DOI 10.1016/S0024-3795(02)00356-7. MR1935223

[BZS08] Yu. A. Bakhturin, M. V. Zaĭtsev, and S. K. Segal, *Finite-dimensional simple graded algebras* (Russian, with Russian summary), Mat. Sb. **199** (2008), no. 7, 21–40, DOI 10.1070/SM2008v199n07ABEH003949; English transl., Sb. Math. **199** (2008), no. 7-8, 965–983 (2008). MR2488221

[BZ98] Yu. A. Bakhturin and M. V. Zaĭtsev, *Identities of special Jordan algebras with finite grading* (Russian, with Russian summary), Vestnik Moskov. Univ. Ser. I Mat. Mekh. **2** (1998), 26–29, 73; English transl., Moscow Univ. Math. Bull. **53** (1998), no. 2, 28–31. MR1706152

[BR87] W. Beckner and A. Regev, *Asymptotics and algebraicity of some generating functions*, Adv. in Math. **65** (1987), no. 1, 1–15, DOI 10.1016/0001-8708(87)90015-6. MR893467

[BR98] W. Beckner and A. Regev, *Asymptotic estimates using probability*, Adv. Math. **138** (1998), no. 1, 1–14, DOI 10.1006/aima.1994.1503. MR1645060

[Beĭ86] K. I. Beĭdar, *On A. I. Mal'tsev's theorems on matrix representations of algebras* (Russian), Uspekhi Mat. Nauk **41** (1986), no. 5(251), 161–162. MR878332

[Bel97] A. Ya. Belov, *Rationality of Hilbert series with respect to free algebras* (Russian), Uspekhi Mat. Nauk **52** (1997), no. 2(314), 153–154, DOI 10.1070/RM1997v052n02ABEH001786; English transl., Russian Math. Surveys **52** (1997), no. 2, 394–395. MR1480146

[Bel00] A. Ya. Belov, *Counterexamples to the Specht problem* (Russian, with Russian summary), Mat. Sb. **191** (2000), no. 3, 13–24, DOI 10.1070/SM2000v191n03ABEH000460; English transl., Sb. Math. **191** (2000), no. 3-4, 329–340. MR1773251

[Bel10] A. Ya. Belov, *Local finite basis property and local representability of varieties of associative rings* (Russian, with Russian summary), Izv. Ross. Akad. Nauk Ser. Mat. **74** (2010), no. 1, 3–134, DOI 10.1070/IM2010v074n01ABEH002481; English transl., Izv. Math. **74** (2010), no. 1, 1–126. MR2655238

[BBL97] A. Ya. Belov, V. V. Borisenko, and V. N. Latyshev, *Monomial algebras*, J. Math. Sci. (New York) **87** (1997), no. 3, 3463–3575, DOI 10.1007/BF02355446. Algebra, 4. MR1604202

[BKGRV14] A. Belov-Kanel, A. Giambruno, L. H. Rowen, and U. Vishne, *Zariski closed algebras in varieties of universal algebra*, Algebr. Represent. Theory **17** (2014), no. 6, 1771–1783, DOI 10.1007/s10468-014-9469-8. MR3284328

[BKRV15] A. Belov-Kanel, L. Rowen, and U. Vishne, *Specht's problem for associative affine algebras over commutative Noetherian rings*, Trans. Amer. Math. Soc. **367** (2015), no. 8, 5553–5596, DOI 10.1090/tran/5983. MR3347183

[KBKR16] A. Kanel-Belov, Y. Karasik, and L. H. Rowen, *Computational aspects of polynomial identities. Vol. 1*, 2nd ed., Monographs and Research Notes in Mathematics, CRC Press, Boca Raton, FL, 2016. Kemer's theorems. MR3445601

[KBR05] A. Kanel-Belov and L. H. Rowen, *Computational aspects of polynomial identities*, Research Notes in Mathematics, vol. 9, A K Peters, Ltd., Wellesley, MA, 2005. MR2124127

[Ber82] A. Berele, *Homogeneous polynomial identities*, Israel J. Math. **42** (1982), no. 3, 258–272, DOI 10.1007/BF02802727. MR687131

[Ber85] A. Berele, *Magnum P.I*, Israel J. Math. **51** (1985), no. 1-2, 13–19, DOI 10.1007/BF02772954. MR804472

[Ber93] A. Berele, *Classification theorems for verbally semiprime algebras*, Comm. Algebra **21** (1993), no. 5, 1505–1512, DOI 10.1080/00927879308824633. MR1213969

[Ber94] A. Berele, *Supertraces and matrices over Grassmann algebras*, Adv. Math. **108** (1994), no. 1, 77–90, DOI 10.1006/aima.1994.1066. MR1293582

[Ber97] A. Berele, *Approximate multiplicities in the trace cocharacter sequence of two three-by-three matrices*, Comm. Algebra **25** (1997), no. 6, 1975–1983, DOI 10.1080/00927879708825967. MR1446144

[Ber05] A. Berele, *Colength sequences for matrices*, J. Algebra **283** (2005), no. 2, 700–710, DOI 10.1016/j.jalgebra.2004.09.020. MR2111218

[Ber06] A. Berele, *Applications of Belov's theorem to the cocharacter sequence of p.i. algebras*, J. Algebra **298** (2006), no. 1, 208–214, DOI 10.1016/j.jalgebra.2005.09.011. MR2215124

[Ber08a] A. Berele, *Properties of hook Schur functions with applications to p.i. algebras*, Adv. in Appl. Math. **41** (2008), no. 1, 52–75, DOI 10.1016/j.aam.2007.03.002. MR2419763

[Ber08b] A. Berele, *Maximal multiplicities in cocharacter sequences*, J. Algebra **320** (2008), no. 1, 318–340, DOI 10.1016/j.jalgebra.2008.03.001. MR2417991

[Ber08c] A. Berele, *Properties of hook Schur functions with applications to p.i. algebras*, Adv. in Appl. Math. **41** (2008), no. 1, 52–75, DOI 10.1016/j.aam.2007.03.002. MR2419763

[Ber10] A. Berele, *Bounds on colength and maximal multiplicity sequences*, J. Algebra **324** (2010), no. 12, 3262–3275, DOI 10.1016/j.jalgebra.2010.10.007. MR2735383

[Ber13a] A. Berele, *Using hook Schur functions to compute matrix cocharacters*, Comm. Algebra **41** (2013), no. 3, 1123–1133, DOI 10.1080/00927872.2011.630711. MR3037185

[Ber13b] A. Berele, *Invariant theory for matrices over the Grassmann algebra*, Adv. Math. **237** (2013), 33–61, DOI 10.1016/j.aim.2012.12.021. MR3028573

[Ber14] A. Berele, *Cocharacter sequences are holonomic*, J. Algebra **412** (2014), 150–154, DOI 10.1016/j.jalgebra.2014.02.030. MR3215949

[Ber19] A. Berele, *Powers of standard identities satisfied by verbally prime algebras*, Comm. Algebra **47** (2019), no. 12, 5338–5347, DOI 10.1080/00927872.2019.1618865. MR4019344

[BB99] A. Berele and J. Bergen, *P.I. algebras with Hopf algebra actions*, J. Algebra **214** (1999), no. 2, 636–651, DOI 10.1006/jabr.1998.7707. MR1680597

[BR83] A. Berele and A. Regev, *Applications of hook Young diagrams to P.I. algebras*, J. Algebra **82** (1983), no. 2, 559–567, DOI 10.1016/0021-8693(83)90167-9. MR704771

[BR87] A. Berele and A. Regev, *Hook Young diagrams with applications to combinatorics and to representations of Lie superalgebras*, Adv. in Math. **64** (1987), no. 2, 118–175, DOI 10.1016/0001-8708(87)90007-7. MR884183

[BR98] A. Berele and A. Regev, *Codimensions of products and of intersections of verbally prime T-ideals*, Israel J. Math. **103** (1998), 17–28, DOI 10.1007/BF02762265. MR1613536

[BR01] A. Berele and A. Regev, *Exponential growth for codimensions of some p.i. algebras*, J. Algebra **241** (2001), no. 1, 118–145, DOI 10.1006/jabr.2000.8672. MR1838847

[BR08] A. Berele and A. Regev, *Asymptotic behaviour of codimensions of p. i. algebras satisfying Capelli identities*, Trans. Amer. Math. Soc. **360** (2008), no. 10, 5155–5172, DOI 10.1090/S0002-9947-08-04500-5. MR2415069

[BR85] A. Berele and J. B. Remmel, *Hook flag characters and their combinatorics*, J. Pure Appl. Algebra **35** (1985), no. 3, 225–245, DOI 10.1016/0022-4049(85)90042-8. MR777256

[BS99] A. Berele and J. R. Stembridge, *Denominators for the Poincaré series of invariants of small matrices*, Israel J. Math. **114** (1999), 157–175, DOI 10.1007/BF02785575. MR1738677

[Ber73] G. M. Bergman, *Infinite multiplication of ideals in $\aleph_0$-hereditary rings*, J. Algebra **24** (1973), 56–70, DOI 10.1016/0021-8693(73)90152-X. MR309982

[BL75] G. M. Bergman and J. Lewin, *The semigroup of ideals of a fir is (usually) free*, J. London Math. Soc. (2) **11** (1975), no. 1, 21–31, DOI 10.1112/jlms/s2-11.1.21. MR379554

[BBHM63] A. Białynicki-Birula, G. Hochschild, and G. D. Mostow, *Extensions of representations of algebraic linear groups*, Amer. J. Math. **85** (1963), 131–144, DOI 10.2307/2373191. MR155938

[Bo70] H. Boerner, *Representations of Groups*, North Holland (1970).

[Bor69] A. Borel, *Linear algebraic groups*, Notes taken by Hyman Bass, W. A. Benjamin, Inc., New York-Amsterdam, 1969. MR0251042

[BK76] W. Borho and H. Kraft, *Über die Gelfand-Kirillov-Dimension*, Math. Ann. **220** (1976), no. 1, 1–24, DOI 10.1007/BF01354525. MR412240

[Bou64] N. Bourbaki, *Éléments de mathématique. Fasc. XXX. Algèbre commutative. Chapitre 5: Entiers. Chapitre 6: Valuations* (French), Actualités Scientifiques et Industrielles, No. 1308, Hermann, Paris, 1964. MR0194450

[Bou87] J.-F. Boutot, *Singularités rationnelles et quotients par les groupes réductifs* (French), Invent. Math. **88** (1987), no. 1, 65–68, DOI 10.1007/BF01405091. MR877006

[Bra84] A. Braun, *The nilpotency of the radical in a finitely generated PI ring*, J. Algebra **89** (1984), no. 2, 375–396, DOI 10.1016/0021-8693(84)90224-2. MR751151

[BPŠ15] M. Brešar, C. Procesi, and Š. Špenko, *Quasi-identities on matrices and the Cayley-Hamilton polynomial*, Adv. Math. **280** (2015), 439–471, DOI 10.1016/j.aim.2015.03.021. MR3350226

[Bri12] M. Brion, *Spherical varieties*, Highlights in Lie algebraic methods, Progr. Math., vol. 295, Birkhäuser/Springer, New York, 2012, pp. 3–24, DOI 10.1007/978-0-8176-8274-3_1. MR2866845

[Bro93] R. Brown, *On a conjecture of Dirichlet*, Amer. Math. Soc., Providence, RI, 1993.

[Cap02] A. Capelli, *Lezioni sulla teoria delle forme algebriche*, Napoli 1902.

[CCR03] J. O. Carbonara, L. Carini, and J. B. Remmel, *Trace cocharacters and the Kronecker products of Schur functions*, J. Algebra **260** (2003), no. 2, 631–656, DOI 10.1016/S0021-8693(03)00013-9. MR1967315

[Coh46] I. S. Cohen, *On the structure and ideal theory of complete local rings*, Trans. Amer. Math. Soc. **59** (1946), 54–106, DOI 10.2307/1990313. MR16094

[Coh65] P. M. Cohn, *Universal algebra*, Harper & Row, Publishers, New York-London, 1965. MR0175948

[Coh70] P. M. Cohn, *On a class of rings with inverse weak algorithm*, Math. Z. **117** (1970), 1–6, DOI 10.1007/BF01109821. MR279123

[Coh06] P. M. Cohn, *Free ideal rings and localization in general rings*, New Mathematical Monographs, vol. 3, Cambridge University Press, Cambridge, 2006. MR2246388

[Cze71] A. J. Czerniakiewicz, *Automorphisms of a free associative algebra of rank* 2. *I*, Trans. Amer. Math. Soc. **160** (1971), 393–401, DOI 10.2307/1995814. MR280549

[CR62] C. W. Curtis and I. Reiner, *Representation theory of finite groups and associative algebras*, Pure and Applied Mathematics, Vol. XI, Interscience Publishers, a division of John Wiley & Sons, New York-London, 1962. MR0144979

[DM88] W. Dahmen and C. A. Micchelli, *The number of solutions to linear Diophantine equations and multivariate splines*, Trans. Amer. Math. Soc. **308** (1988), no. 2, 509–532, DOI 10.2307/2001089. MR951619

[DCEP80] C. De Concini, D. Eisenbud, and C. Procesi, *Young diagrams and determinantal varieties*, Invent. Math. **56** (1980), no. 2, 129–165, DOI 10.1007/BF01392548. MR558865

[DCEP82] C. De Concini, D. Eisenbud, and C. Procesi, *Hodge algebras*, Astérisque, vol. 91, Société Mathématique de France, Paris, 1982. With a French summary. MR680936

[DCK90] C. De Concini and V. G. Kac, *Representations of quantum groups at roots of* 1, Operator algebras, unitary representations, enveloping algebras, and invariant theory (Paris, 1989), Progr. Math., vol. 92, Birkhäuser Boston, Boston, MA, 1990, pp. 471–506. MR1103601

[DCKP92] C. De Concini, V. G. Kac, and C. Procesi, *Quantum coadjoint action*, J. Amer. Math. Soc. **5** (1992), no. 1, 151–189, DOI 10.2307/2152754. MR1124981

[DCPP15] C. De Concini, P. Papi, and C. Procesi, *The adjoint representation inside the exterior algebra of a simple Lie algebra*, Adv. Math. **280** (2015), 21–46, DOI 10.1016/j.aim.2015.04.011. MR3350211

[DCP76] C. De Concini and C. Procesi, *A characteristic free approach to invariant theory*, Advances in Math. **21** (1976), no. 3, 330–354, DOI 10.1016/S0001-8708(76)80003-5. MR422314

[dCP97] C. De Concini, C. Procesi, *Quantum Schubert cells*, Algebraic groups and Lie groups, ed. Lehrer, pp. 127-160, (1997).

[DCP11] C. De Concini and C. Procesi, *Topics in hyperplane arrangements, polytopes and box-splines*, Universitext, Springer, New York, 2011. MR2722776

[DCP17] C. De Concini and C. Procesi, *The invariant theory of matrices*, University Lecture Series, vol. 69, American Mathematical Society, Providence, RI, 2017. MR3726879

[DCPRR05] C. De Concini, C. Procesi, N. Reshetikhin, and M. Rosso, *Hopf algebras with trace and representations*, Invent. Math. **161** (2005), no. 1, 1–44, DOI 10.1007/s00222-004-0405-0. MR2178656

[dLV99] A. de Luca and S. Varricchio, *Finiteness and regularity in semigroups and formal languages*, Monographs in Theoretical Computer Science. An EATCS Series, Springer-Verlag, Berlin, 1999. MR1696498

[Deu68] M. Deuring, *Algebren* (German), Zweite, korrigierte auflage. Ergebnisse der Mathematik und ihrer Grenzgebiete, Band 41, Springer-Verlag, Berlin-New York, 1968. MR0228526

[Dic88] W. Dicks, *On a characterization of Azumaya algebras*, Publ. Mat. **32** (1988), no. 2, 165–166, DOI 10.5565/PUBLMAT_32288_03. MR975894

[Dil50] R. P. Dilworth, *A decomposition theorem for partially ordered sets*, Ann. of Math. (2) **51** (1950), 161–166, DOI 10.2307/1969503. MR32578

[Đok07] D. Ž. Đoković, *Poincaré series of some pure and mixed trace algebras of two generic matrices*, J. Algebra **309** (2007), no. 2, 654–671, DOI 10.1016/j.jalgebra.2006.09.018. MR2303199

[DF04] M. Domokos and P. E. Frenkel, *On orthogonal invariants in characteristic 2*, J. Algebra **274** (2004), no. 2, 662–688, DOI 10.1016/S0021-8693(03)00513-1. MR2043371

[Don92] S. Donkin, *Invariants of several matrices*, Invent. Math. **110** (1992), no. 2, 389–401, DOI 10.1007/BF01231338. MR1185589

[Don93] S. Donkin, *Invariant functions on matrices*, Math. Proc. Cambridge Philos. Soc. **113** (1993), no. 1, 23–43, DOI 10.1017/S0305004100075757. MR1188816

[Dre81] V. S. Drenski, *A minimal basis for identities of a second-order matrix algebra over a field of characteristic* 0 (Russian), Algebra i Logika **20** (1981), no. 3, 282–290, 361. MR648317

[Dre74] V. S. Drenski, *Identities in Lie algebras* (Russian), Algebra i Logika **13** (1974), 265–290, 363–364. MR0374220

[Dre81] V. S. Drensky, *Codimensions of T-ideals and Hilbert series of relatively free algebras*, C. R. Acad. Bulgare Sci. **34** (1981), no. 9, 1201–1204. MR649144

[Dre87] V. Drensky, *Polynomial identities for the Jordan algebra of a symmetric bilinear form*, J. Algebra **108** (1987), no. 1, 66–87, DOI 10.1016/0021-8693(87)90122-0. MR887192

[Dre90] V. Drensky, *Polynomial identities for* $2 \times 2$ *matrices*, Acta Appl. Math. **21** (1990), no. 1-2, 137–161, DOI 10.1007/BF00053295. MR1085776

[Dre92] V. Drensky, *Relations for the cocharacter sequences of T-ideals*, Proceedings of the International Conference on Algebra, Part 2 (Novosibirsk, 1989), Contemp. Math., vol. 131, Amer. Math. Soc., Providence, RI, 1992, pp. 285–300. MR1175839

[Dre95] V. Drensky, *New central polynomials for the matrix algebra*, Israel J. Math. **92** (1995), no. 1-3, 235–248, DOI 10.1007/BF02762079. MR1357754

[Dre00] V. Drensky, *Free algebras and PI-algebras: Graduate course in algebra*, Springer-Verlag Singapore, Singapore, 2000. MR1712064

[DF04] V. Drensky and E. Formanek, *Polynomial identity rings*, Advanced Courses in Mathematics. CRM Barcelona, Birkhäuser Verlag, Basel, 2004. MR2064082

[DG03] V. Drensky and G. K. Genov, *Multiplicities of Schur functions in invariants of two* $3 \times 3$ *matrices*, J. Algebra **264** (2003), no. 2, 496–519, DOI 10.1016/S0021-8693(03)00070-X. MR1981418

[DGV06] V. Drensky, G. K. Genov, and A. Valenti, *Multiplicities in the mixed trace cocharacter sequence of two* $3 \times 3$ *matrices*, Internat. J. Algebra Comput. **16** (2006), no. 2, 275–285, DOI 10.1142/S0218196706002974. MR2228513

[DK85] V. Drensky and A. Kasparian, *A new central polynomial for* $3 \times 3$ *matrices*, Comm. Algebra **13** (1985), no. 3, 745–752, DOI 10.1080/00927878508823188. MR773761

[DK87] V. S. Drensky and P. E. Koshlukov, *Weak polynomial identities for a vector space with a symmetric bilinear form* (English, with Bulgarian summary), Mathematics and mathematical education (Bulgarian) (Sunny Beach (Sl"Bryag), 1987), Publ. House Bulgar. Acad. Sci., Sofia, 1987, pp. 213–219. MR949941

[DPC94] V. Drensky and G. M. Piacentini Cattaneo, *A central polynomial of low degree for* $4 \times 4$ *matrices*, J. Algebra **168** (1994), no. 2, 469–478, DOI 10.1006/jabr.1994.1240. MR1292776

[DR93] V. Drensky and T. G. Rashkova, *Weak polynomial identities for the matrix algebras*, Comm. Algebra **21** (1993), no. 10, 3779–3795, DOI 10.1080/00927879308824765. MR1231633

[DV86] V. S. Drenski and L. A. Vladimirova, *Varieties of pairs of algebras with a distributive lattice of subvarieties*, Serdica **12** (1986), no. 2, 166–170. MR867365

[DJ47] M.-L. Dubreil-Jacotin, *Sur l'immersion d'un semi-groupe dans un groupe* (French), C. R. Acad. Sci. Paris **225** (1947), 787–788. MR22224

[Dur10] R. Durrett, *Probability: theory and examples*, 4th ed., Cambridge Series in Statistical and Probabilistic Mathematics, vol. 31, Cambridge University Press, Cambridge, 2010. MR2722836

[Dvi93] Y. Dvir, *On the Kronecker product of $S_n$ characters*, J. Algebra **154** (1993), no. 1, 125–140, DOI 10.1006/jabr.1993.1008. MR1201916

[Eis95] D. Eisenbud, *Commutative algebra: With a view toward algebraic geometry*, Graduate Texts in Mathematics, vol. 150, Springer-Verlag, New York, 1995. MR1322960

[Fer98] D. Ferrand, *Un foncteur norme* (French, with English and French summaries), Bull. Soc. Math. France **126** (1998), no. 1, 1–49. MR1651380

[Fox53] R. H. Fox, *Free differential calculus. I. Derivation in the free group ring*, Ann. of Math. (2) **57** (1953), 547–560, DOI 10.2307/1969736. MR53938

[For72] E. Formanek, *Central polynomials for matrix rings*, J. Algebra **23** (1972), 129–132, DOI 10.1016/0021-8693(72)90050-6. MR302689

[For80] E. Formanek, *The center of the ring of* $4 \times 4$ *generic matrices*, J. Algebra **62** (1980), no. 2, 304–319, DOI 10.1016/0021-8693(80)90184-2. MR563230

[For82] E. Formanek, *The polynomial identities of matrices*, Algebraists' homage: papers in ring theory and related topics (New Haven, Conn., 1981), Contemp. Math., vol. 13, Amer. Math. Soc., Providence, R.I., 1982, pp. 41–79. MR685937

[For84] E. Formanek, *Invariants and the ring of generic matrices*, J. Algebra **89** (1984), no. 1, 178–223, DOI 10.1016/0021-8693(84)90240-0. MR748233

[For85] E. Formanek, *Noncommutative invariant theory*, Group actions on rings (Brunswick, Maine, 1984), Contemp. Math., vol. 43, Amer. Math. Soc., Providence, RI, 1985, pp. 87–119, DOI 10.1090/conm/043/810646. MR810646

[For86] E. Formanek, *Functional equations for character series associated with* $n \times n$ *matrices*, Trans. Amer. Math. Soc. **294** (1986), no. 2, 647–663, DOI 10.2307/2000206. MR825728

[For87] E. Formanek, *A conjecture of Regev about the Capelli polynomial*, J. Algebra **109** (1987), no. 1, 93–114, DOI 10.1016/0021-8693(87)90166-9. MR898339

[For89] E. Formanek, *Polynomial identities and the Cayley-Hamilton theorem*, Math. Intelligencer **11** (1989), no. 1, 37–39, DOI 10.1007/BF03023774. MR979022

[For90] E. Formanek, *The Nagata-Higman theorem*, Acta Appl. Math. **21** (1990), no. 1-2, 185–192, DOI 10.1007/BF00053297. MR1085778

[For91] E. Formanek, *The polynomial identities and invariants of* $n \times n$ *matrices*, CBMS Regional Conference Series in Mathematics, vol. 78, Published for the Conference Board of the Mathematical Sciences, Washington, DC; by the American Mathematical Society, Providence, RI, 1991. MR1088481

[FHL81] E. Formanek, P. Halpin, and W. C. W. Li, *The Poincaré series of the ring of* $2 \times 2$ *generic matrices*, J. Algebra **69** (1981), no. 1, 105–112, DOI 10.1016/0021-8693(81)90130-7. MR613860

[FP76] E. Formanek and C. Procesi, *Mumford's conjecture for the general linear group*, Advances in Math. **19** (1976), no. 3, 292–305, DOI 10.1016/0001-8708(76)90026-8. MR404279

[FW08] P. J. Forrester and S. O. Warnaar, *The importance of the Selberg integral*, Bull. Amer. Math. Soc. (N.S.) **45** (2008), no. 4, 489–534, DOI 10.1090/S0273-0979-08-01221-4. MR2434345

[Frö73] A. Fröhlich, *The Picard group of noncommutative rings, in particular of orders*, Trans. Amer. Math. Soc. **180** (1973), 1–45, DOI 10.2307/1996653. MR318204

[FH91] W. Fulton and J. Harris, *Representation theory*, Graduate Texts in Mathematics, vol. 129, Springer-Verlag, New York, 1991. A first course; Readings in Mathematics. MR1153249

[GR85] A. M. Garsia and J. Remmel, *Shuffles of permutations and the Kronecker product*, Graphs Combin. **1** (1985), no. 3, 217–263, DOI 10.1007/BF02582950. MR951014

[GK66] I. M. Gel′fand and A. A. Kirillov, *On fields connected with the enveloping algebras of Lie algebras* (Russian), Dokl. Akad. Nauk SSSR **167** (1966), 503–505. MR0195912

[GK01] A. Giambruno and P. Koshlukov, *On the identities of the Grassmann algebras in characteristic* $p > 0$, Israel J. Math. **122** (2001), 305–316, DOI 10.1007/BF02809905. MR1826505

[GMZ06] A. Giambruno, S. Mishchenko, and M. Zaicev, *Algebras with intermediate growth of the codimensions*, Adv. in Appl. Math. **37** (2006), no. 3, 360–377, DOI 10.1016/j.aam.2005.02.005. MR2261178

[GMZ08] A. Giambruno, S. Mishchenko, and M. Zaicev, *Codimensions of algebras and growth functions*, Adv. Math. **217** (2008), no. 3, 1027–1052, DOI 10.1016/j.aim.2007.07.008. MR2383893

[GS89] A. Giambruno and S. K. Sehgal, *On a polynomial identity for* $n \times n$ *matrices*, J. Algebra **126** (1989), no. 2, 451–453, DOI 10.1016/0021-8693(89)90312-8. MR1024999

[GSZ11] A. Giambruno, I. Shestakov, and M. Zaicev, *Finite-dimensional non-associative algebras and codimension growth*, Adv. in Appl. Math. **47** (2011), no. 1, 125–139, DOI 10.1016/j.aam.2010.04.007. MR2799615

[GV96] A. Giambruno and A. Valenti, *Central polynomials and matrix invariants*. part A, Israel J. Math. **96** (1996), no. part A, 281–297, DOI 10.1007/BF02785544. MR1432737

[GZ99] A. Giambruno and M. Zaicev, *Exponential codimension growth of PI algebras: an exact estimate*, Adv. Math. **142** (1999), no. 2, 221–243, DOI 10.1006/aima.1998.1790. MR1680198

[GZ03a] A. Giambruno and M. Zaicev, *Minimal varieties of algebras of exponential growth*, Adv. Math. **174** (2003), no. 2, 310–323, DOI 10.1016/S0001-8708(02)00047-6. MR1963697

[GZ03b] A. Giambruno and M. Zaicev, *Codimension growth and minimal superalgebras*, Trans. Amer. Math. Soc. **355** (2003), no. 12, 5091–5117, DOI 10.1090/S0002-9947-03-03360-9. MR1997596

[GZ03c] A. Giambruno and M. Zaicev, *Asymptotics for the standard and the Capelli identities*, Israel J. Math. **135** (2003), 125–145, DOI 10.1007/BF02776053. MR1996399

[GZ05] A. Giambruno and M. Zaicev, *Polynomial identities and asymptotic methods*, Mathematical Surveys and Monographs, vol. 122, American Mathematical Society, Providence, RI, 2005. MR2176105

[GZ10a] A. Giambruno and M. Zaicev, *Codimension growth of special simple Jordan algebras*, Trans. Amer. Math. Soc. **362** (2010), no. 6, 3107–3123, DOI 10.1090/S0002-9947-09-04865-X. MR2592948

[GZ10b] A. Giambruno and M. Zaicev, *Codimension growth of special simple Jordan algebras*, Trans. Amer. Math. Soc. **362** (2010), no. 6, 3107–3123, DOI 10.1090/S0002-9947-09-04865-X. MR2592948

[GZ12] A. Giambruno and M. Zaicev, *On codimension growth of finite-dimensional Lie superalgebras*, J. Lond. Math. Soc. (2) **85** (2012), no. 2, 534–548, DOI 10.1112/jlms/jdr059. MR2901077

[GZ14] A. Giambruno and M. Zaicev, *Growth of polynomial identities: is the sequence of codimensions eventually non-decreasing?*, Bull. Lond. Math. Soc. **46** (2014), no. 4, 771–778, DOI 10.1112/blms/bdu031. MR3239615

[GZ18] A. Giambruno and M. Zaicev, *Central polynomials and growth functions*, Israel J. Math. **226** (2018), no. 1, 15–28, DOI 10.1007/s11856-018-1704-2. MR3819685

[GZ19] A. Giambruno and M. Zaicev, *Central polynomials of associative algebras and their growth*, Proc. Amer. Math. Soc. **147** (2019), no. 3, 909–919, DOI 10.1090/proc/14172. MR3896042

[GZ11] A. Giambruno and E. Zelmanov, *On growth of codimensions of Jordan algebras*, Groups, algebras and applications, Contemp. Math., vol. 537, Amer. Math. Soc., Providence, RI, 2011, pp. 205–210, DOI 10.1090/conm/537/10576. MR2799101

[Gia93] A. Giambruno (ed.), *Recent developments in the theory of algebras with polynomial identities*, Circolo Matematico di Palermo, Palermo, 1993. Papers from the conference held in Palermo, June 15–18, 1992; Rend. Circ. Mat. Palermo (2) Suppl. No. 31 (1993) (1993). MR1244612

[GRZ03] A. Giambruno, A. Regev, and M. Zaicev (eds.), *Polynomial identities and combinatorial methods*, Lecture Notes in Pure and Applied Mathematics, vol. 235, Marcel Dekker, Inc., New York, 2003. MR2017608

[Gol74] A. W. Goldie, *Lectures on quotient rings and rings with polynomial identities*, Mathematisches Institut Giessen, Universität Giessen, Giessen, 1974. Given at the University of Giessen, Summer 1974; Notes by Wolfgang Hamernik; Vorlesungen aus dem Mathematischen Institut Giessen, Heft 1. MR0371943

[Gol64] E. S. Golod, *On nil-algebras and finitely approximable p-groups* (Russian), Izv. Akad. Nauk SSSR Ser. Mat. **28** (1964), 273–276. MR0161878

[Gol68] E. S. Golod, *Some problems of Burnside type* (Russian), Proc. Internat. Congr. Math. (Moscow, 1966), Izdat. "Mir", Moscow, 1968, pp. 284–289. MR0238880

[GŠ64] E. S. Golod and I. R. Šafarevič, *On the class field tower* (Russian), Izv. Akad. Nauk SSSR Ser. Mat. **28** (1964), 261–272. MR0161852

[GW09] R. Goodman and N. R. Wallach, *Symmetry, representations, and invariants*, Graduate Texts in Mathematics, vol. 255, Springer, Dordrecht, 2009. MR2522486

[GW04] K. R. Goodearl and R. B. Warfield Jr., *An introduction to noncommutative Noetherian rings*, 2nd ed., London Mathematical Society Student Texts, vol. 61, Cambridge University Press, Cambridge, 2004. MR2080008

[GB89] D. Gouyou-Beauchamps, *Standard Young tableaux of height* 4 *and* 5, European J. Combin. **10** (1989), no. 1, 69–82, DOI 10.1016/S0195-6698(89)80034-4. MR977181

[GKP94] R. L. Graham, D. E. Knuth, and O. Patashnik, *Concrete mathematics*, 2nd ed., Addison-Wesley Publishing Company, Reading, MA, 1994. A foundation for computer science. MR1397498

[Gro73] F. Grosshans, *Observable groups and Hilbert's fourteenth problem*, Amer. J. Math. **95** (1973), 229–253, DOI 10.2307/2373655. MR325628

[Gro60] A. Grothendieck, *EGA I, Le langage des schemas* 1960 Publications mathématiques, n. 45, Rondpoint Bugeaud - Paris (XVIe)

[Gro64] A. Grothendieck, *Éléments de géométrie algébrique. IV. Étude locale des schémas et des morphismes de schémas. I* (French), Inst. Hautes Études Sci. Publ. Math. **20** (1964), 259. MR173675

[Gro66] A. Grothendieck, *Éléments de géométrie algébrique. IV. Étude locale des schémas et des morphismes de schémas. III*, Inst. Hautes Études Sci. Publ. Math. **28** (1966), 255. MR217086

[Gro68] A. Grothendieck, *Le groupe de Brauer I,II,III*, Dix Exposés sur la Cohomologie des Schemas, North Holland (1968), 46–188.

[Gro95] A. Grothendieck, *Le groupe de Brauer. I. Algèbres d'Azumaya et interprétations diverses [ MR0244269 (39 #5586a)]* (French), Séminaire Bourbaki, Vol. 9, Soc. Math. France, Paris, 1995, pp. Exp. No. 290, 199–219. MR1608798

[Gur64] G. B. Gurevich, *Foundations of the theory of algebraic invariants*, Translated by J. R. M. Radok and A. J. M. Spencer, P. Noordhoff Ltd., Groningen, 1964. MR0183733

[Hab75] W. J. Haboush, *Reductive groups are geometrically reductive*, Ann. of Math. (2) **102** (1975), no. 1, 67–83, DOI 10.2307/1970974. MR382294

[Hal86] M. Hall Jr., *Combinatorial theory*, 2nd ed., Wiley-Interscience Series in Discrete Mathematics, John Wiley & Sons, Inc., New York, 1986. A Wiley-Interscience Publication. MR840216

[Hal83a] P. Halpin, *Some Poincaré series related to identities of* $2 \times 2$ *matrices*, Pacific J. Math. **107** (1983), no. 1, 107–115. MR701811

[Hal83b] P. Halpin, *Central and weak identities for matrices*, Comm. Algebra **11** (1983), no. 19, 2237–2248, DOI 10.1080/00927878308822961. MR714201

[Her94] I. N. Herstein, *Noncommutative rings*, Carus Mathematical Monographs, vol. 15, Mathematical Association of America, Washington, DC, 1994. Reprint of the 1968 original; With an afterword by Lance W. Small. MR1449137

[Hig56] G. Higman, *On a conjecture of Nagata*, Proc. Cambridge Philos. Soc. **52** (1956), 1–4. MR73581

[HR74] M. Hochster and J. L. Roberts, *Rings of invariants of reductive groups acting on regular rings are Cohen-Macaulay*, Advances in Math. **13** (1974), 115–175, DOI 10.1016/0001-8708(74)90067-X. MR347810

[HP47] W. V. D. Hodge and D. Pedoe, *Methods of Algebraic Geometry. Vol. I*, Cambridge, at the University Press; New York, The Macmillan Company, 1947. MR0028055

[Hum80] J. E. Humphreys, *Arithmetic groups*, Lecture Notes in Mathematics, vol. 789, Springer, Berlin, 1980. MR584623

[Hut75] J. P. Hutchinson, *Eulerian graphs and polynomial identities for skew-symmetric matrices*, Canadian J. Math. **27** (1975), no. 3, 590–609, DOI 10.4153/CJM-1975-070-1. MR404052

[Irv83] R. S. Irving, *Affine PI-algebras not embeddable in matrix rings*, J. Algebra **82** (1983), no. 1, 94–101, DOI 10.1016/0021-8693(83)90175-8. MR701038

[IS86] R. S. Irving and L. W. Small, *The embeddability of affine PI-algebras in rings of matrices*, J. Algebra **103** (1986), no. 2, 708–716, DOI 10.1016/0021-8693(86)90162-6. MR864439

[Iva94] S. V. Ivanov, *The free Burnside groups of sufficiently large exponents*, Internat. J. Algebra Comput. **4** (1994), no. 1-2, ii+308, DOI 10.1142/S0218196794000026. MR1283947

[IO96] S. V. Ivanov and A. Yu. Ol′shanskiĭ, *Hyperbolic groups and their quotients of bounded exponents*, Trans. Amer. Math. Soc. **348** (1996), no. 6, 2091–2138, DOI 10.1090/S0002-9947-96-01510-3. MR1327257

[Jac45] N. Jacobson, *Structure theory for algebraic algebras of bounded degree*, Ann. of Math. (2) **46** (1945), 695–707, DOI 10.2307/1969205. MR14083

[Jac56] N. Jacobson, *Structure of rings*, American Mathematical Society, Colloquium Publications, vol. 37, American Mathematical Society, 190 Hope Street, Prov., R. I., 1956. MR0081264

[Jac75] N. Jacobson, PI-*algebras: An introduction*, Lecture Notes in Mathematics, Vol. 441, Springer-Verlag, Berlin-New York, 1975. MR0369421

[Jac89] N. Jacobson, *Basic algebra. II*, 2nd ed., W. H. Freeman and Company, New York, 1989. MR1009787

[Jam78] G. D. James, *The representation theory of the symmetric groups*, Lecture Notes in Mathematics, vol. 682, Springer, Berlin, 1978. MR513828

[JK81] G. James and A. Kerber, *The representation theory of the symmetric group*, Encyclopedia of Mathematics and its Applications, vol. 16, Addison-Wesley Publishing Co., Reading, Mass., 1981. With a foreword by P. M. Cohn; With an introduction by Gilbert de B. Robinson. MR644144

[Jun31] R. Jungen, *Sur les séries de Taylor n'ayant que des singularités algébrico-logarithmiques sur leur cercle de convergence* (French), Comment. Math. Helv. **3** (1931), no. 1, 266–306, DOI 10.1007/BF01601817. MR1509439

[Kap48] I. Kaplansky, *Rings with a polynomial identity*, Bull. Amer. Math. Soc. **54** (1948), 575–580, DOI 10.1090/S0002-9904-1948-09049-8. MR25451

[Kap57] I. Kaplansky, *Problems in the theory of rings. Report of a conference on linear algebras, June, 1956, pp. 1-3*, National Academy of Sciences-National Research Council, Washington, Publ. 502, 1957. MR0096696

[Kap70] I. Kaplansky, *"Problems in the theory of rings" revisited*, Amer. Math. Monthly **77** (1970), 445–454, DOI 10.2307/2317376. MR258865

[Kem78] A. R. Kemer, *The Spechtian nature of T-ideals whose condimensions have power growth* (Russian), Sibirsk. Mat. Ž. **19** (1978), no. 1, 54–69, 237. MR0466190

[Kem80] A. R. Kemer, *Capelli identities and nilpotency of the radical of finitely generated PI-algebra* (Russian), Dokl. Akad. Nauk SSSR **255** (1980), no. 4, 793–797. MR600746

[Kem84] A. R. Kemer, *Varieties and $Z_2$-graded algebras* (Russian), Izv. Akad. Nauk SSSR Ser. Mat. **48** (1984), no. 5, 1042–1059. MR764308

[Kem87] A. R. Kemer, *Finite basability of identities of associative algebras* (Russian), Algebra i Logika **26** (1987), no. 5, 597–641, 650. MR985840

[Kem90] A. R. Kemer, *Identities of finitely generated algebras over an infinite field* (Russian), Izv. Akad. Nauk SSSR Ser. Mat. **54** (1990), no. 4, 726–753, DOI 10.1070/IM1991v037n01ABEH002053; English transl., Math. USSR-Izv. **37** (1991), no. 1, 69–96. MR1073084

[Kem91] A. R. Kemer, *Ideals of identities of associative algebras*, Translations of Mathematical Monographs, vol. 87, American Mathematical Society, Providence, RI, 1991. Translated from the Russian by C. W. Kohls. MR1108620

[Kem95] A. Kemer, *Multilinear identities of the algebras over a field of characteristic p*, Internat. J. Algebra Comput. **5** (1995), no. 2, 189–197, DOI 10.1142/S0218196795000124. MR1328550

[Kem03] A. Kemer, *On some problems in PI-theory in characteristic p connected with dividing by p*, Proceedings of the Third International Algebra Conference (Tainan, 2002), Kluwer Acad. Publ., Dordrecht, 2003, pp. 53–66. MR2026093

[Kem77] G. R. Kempf, *Some quotient varieties have rational singularities*, Michigan Math. J. **24** (1977), no. 3, 347–352. MR491675

[KKM83] A. A. Kirillov, M. L. Kontsevich, and A. I. Molev, *Algebras of intermediate growth* (Russian, with English summary), Akad. Nauk SSSR Inst. Prikl. Mat. Preprint **39** (1983), 19. Translated in Selecta Math. Soviet. **9** (1990), no. 2, 137–153. MR753873

[KK83] A. A. Kirillov and M. L. Kontsevich, *The growth of the Lie algebra generated by two generic vector fields on the line* (Russian, with English summary), Vestnik Moskov. Univ. Ser. I Mat. Mekh. **4** (1983), 15–20. MR713969

[KO72] M. A. Knus and M. Ojanguren, *Sur le polynôme caractéristique et les automorphismes des algèbres d'Azumaya* (French), Ann. Scuola Norm. Sup. Pisa Cl. Sci. (3) **26** (1972), 225–231. MR360659

[KO74a] M.-A. Knus and M. Ojanguren, *Théorie de la descente et algèbres d'Azumaya* (French), Lecture Notes in Mathematics, Vol. 389, Springer-Verlag, Berlin-New York, 1974. MR0417149

[KO74b] M.-A. Knus and M. Ojanguren, *A Mayer-Vietoris sequence for the Brauer group*, J. Pure Appl. Algebra **5** (1974), 345–360, DOI 10.1016/0022-4049(74)90043-7. MR369339

[Kos88] P. Koshlukov, *Polynomial identities for a family of simple Jordan algebras*, Comm. Algebra **16** (1988), no. 7, 1325–1371, DOI 10.1080/00927878808823634. MR941174

[Kos90] A. I. Kostrikin, *Around Burnside*, Ergebnisse der Mathematik und ihrer Grenzgebiete (3) [Results in Mathematics and Related Areas (3)], vol. 20, Springer-Verlag, Berlin, 1990. Translated from the Russian and with a preface by James Wiegold. MR1075416

[Knu70] D. E. Knuth, *Permutations, matrices, and generalized Young tableaux*, Pacific J. Math. **34** (1970), 709–727. MR272654

[Knu98] D. E. Knuth, *The art of computer programming. Vol. 3*, Addison-Wesley, Reading, MA, 1998. Sorting and searching; Second edition [of MR0445948]. MR3077154

[Kos58] B. Kostant, *A theorem of Frobenius, a theorem of Amitsur-Levitski and cohomology theory*, J. Math. Mech. **7** (1958), 237–264, DOI 10.1512/iumj.1958.7.57019. MR0092755

[Kra84] H. Kraft, *Geometrische Methoden in der Invariantentheorie* (German), Aspects of Mathematics, D1, Friedr. Vieweg & Sohn, Braunschweig, 1984. MR768181

[KR73] D. Krakowski and A. Regev, *The polynomial identities of the Grassmann algebra*, Trans. Amer. Math. Soc. **181** (1973), 429–438, DOI 10.2307/1996643. MR325658

[Kur41] A. Kurosch, *Ringtheoretische Probleme, die mit dem Burnsideschen Problem über periodische Gruppen in Zusammenhang stehen* (Russian, with German summary), Bull. Acad. Sci. URSS. Sér. Math. [Izvestia Akad. Nauk SSSR] **5** (1941), 233–240. MR0005723

[Kuz75] E. N. Kuzmin, *On the Nagata-Higman theorem*, in: Mathematical Structures-Computational Mathematics-Mathematical Modeling, Sofia, 1975, pp. 101–107, in Russian.

[Lam91] T. Y. Lam, *A first course in noncommutative rings*, Graduate Texts in Mathematics, vol. 131, Springer-Verlag, New York, 1991. MR1125071

[Lat72] V. N. Latyšev, *On Regev's theorem on identities in a tensor product of* PI*-algebras* (Russian), Uspehi Mat. Nauk **27** (1972), no. 4(166), 213–214. MR0393114

[Lat77] V. N. Latyshev, *Non-matrix varieties of associative algebras*, Dr. of Sci. Thesis, Moscow 1977, 150 pages.

[LBP87] L. Le Bruyn and C. Procesi, *Étale local structure of matrix invariants and concomitants*, Algebraic groups Utrecht 1986, Lecture Notes in Math., vol. 1271, Springer, Berlin, 1987, pp. 143–175, DOI 10.1007/BFb0079236. MR911138

[LBP90] L. Le Bruyn and C. Procesi, *Semisimple representations of quivers*, Trans. Amer. Math. Soc. **317** (1990), no. 2, 585–598, DOI 10.2307/2001477. MR958897

[LBVdB88] L. Le Bruyn and M. Van den Bergh, *Regularity of trace rings of generic matrices*, J. Algebra **117** (1988), no. 1, 19–29, DOI 10.1016/0021-8693(88)90238-4. MR955588

[LB08] L. Le Bruyn, *Noncommutative geometry and Cayley-smooth orders*, Pure and Applied Mathematics (Boca Raton), vol. 290, Chapman & Hall/CRC, Boca Raton, FL, 2008. MR2356702

[LZ12] G. Lehrer and R. Zhang, *The second fundamental theorem of invariant theory for the orthogonal group*, Ann. of Math. (2) **176** (2012), no. 3, 2031–2054, DOI 10.4007/annals.2012.176.3.12. MR2979865

[LV70] U. Leron and A. Vapne, *Polynomial identities of related rings*, Israel J. Math. **8** (1970), 127–137, DOI 10.1007/BF02771307. MR269694

[Lev43] J. Levitzki, *On the radical of a general ring*, Bull. Amer. Math. Soc. **49** (1943), 462–466, DOI 10.1090/S0002-9904-1943-07959-1. MR8072

[Lev46] J. Levitzki, *On a problem of A. Kurosch*, Bull. Amer. Math. Soc. **52** (1946), 1033–1035, DOI 10.1090/S0002-9904-1946-08705-4. MR19600

[Lew66] J. S. Lew, *The generalized Cayley-Hamilton theorem in n dimensions* (English, with French summary), Z. Angew. Math. Phys. **17** (1966), 650–653, DOI 10.1007/BF01597249. MR213381

[Lew67] J. S. Lew, *Reducibility of matrix polynomials and their traces* (English, with French summary), Z. Angew. Math. Phys. **18** (1967), 289–293, DOI 10.1007/BF01596920. MR210725

[Lew74] J. Lewin, *A matrix representation for associative algebras. I, II*, Trans. Amer. Math. Soc. **188** (1974), 293–308; ibid. 188 (1974), 309–317, DOI 10.2307/1996780. MR338081

[Lot97] M. Lothaire, *Combinatorics on words*, Cambridge Mathematical Library, Cambridge University Press, Cambridge, 1997. With a foreword by Roger Lyndon and a preface by Dominique Perrin; Corrected reprint of the 1983 original, with a new preface by Perrin. MR1475463

[Lot02] M. Lothaire, *Algebraic combinatorics on words*, Encyclopedia of Mathematics and Its Applications vol. **90**, Cambridge University Press, Cambridge, 2002.

[Lun73] D. Luna, *Slices étales* (French), Sur les groupes algébriques, Soc. Math. France, Paris, 1973, pp. 81–105. Bull. Soc. Math. France, Paris, Mémoire 33, DOI 10.24033/msmf.110. MR0342523

[Mac95] I. G. Macdonald, *Symmetric functions and Hall polynomials*, 2nd ed., Oxford Mathematical Monographs, The Clarendon Press, Oxford University Press, New York, 1995. With contributions by A. Zelevinsky; Oxford Science Publications. MR1354144

[ML98] S. Mac Lane, *Categories for the working mathematician*, 2nd ed., Graduate Texts in Mathematics, vol. 5, Springer-Verlag, New York, 1998. MR1712872

[Mag39] W. Magnus, *On a theorem of Marshall Hall*, Ann. of Math. (2) **40** (1939), 764–768, DOI 10.2307/1968892. MR262

[Mal42] A. Malćev, *On the representation of an algebra as a direct sum of the radical and a semi-simple subalgebra*, C. R. (Doklady) Acad. Sci. URSS (N.S.) **36** (1942), 42–45. MR0007397

[Mal43] A. Malćev, *On the representations of infinite algebras* (Russian, with English summary), Rec. Math. [Mat. Sbornik] N.S. **13 (55)** (1943), 263–286. MR0011084

[Mat80] H. Matsumura, *Commutative algebra*, 2nd ed., Mathematics Lecture Note Series, vol. 56, Benjamin/Cummings Publishing Co., Inc., Reading, Mass., 1980. MR575344

[Mat89] H. Matsumura, *Commutative ring theory*, 2nd ed., Cambridge Studies in Advanced Mathematics, vol. 8, Cambridge University Press, Cambridge, 1989. Translated from the Japanese by M. Reid. MR1011461

[MR01] J. C. McConnell and J. C. Robson, *Noncommutative Noetherian rings*, Revised edition, Graduate Studies in Mathematics, vol. 30, American Mathematical Society, Providence, RI, 2001. With the cooperation of L. W. Small. MR1811901

[Mis96] S. P. Mishchenko, *Lower bounds on the dimensions of irreducible representations of symmetric groups and of the exponents of the exponential of varieties of Lie algebras* (Russian, with Russian summary), Mat. Sb. **187** (1996), no. 1, 83–94, DOI 10.1070/SM1996v187n01ABEH000101; English transl., Sb. Math. **187** (1996), no. 1, 81–92. MR1380205

[Mor58] K. Morita, *Duality for modules and its applications to the theory of rings with minimum condition*, Sci. Rep. Tokyo Kyoiku Daigaku Sect. A **6** (1958), 83–142. MR96700

[Mor65] K. Morita, *Adjoint pairs of functors and Frobenius extensions*, Sci. Rep. Tokyo Kyoiku Daigaku Sect. A **9** (1965), 40–71 (1965). MR190183

[Mue76] B. J. Mueller, *Rings with polynomial identities*, Universidad Nacional Autónoma de México, Mexico City, 1976. Monografias del Instituto de Matematicas, No. 4. [Monographs of the Institute of Mathematics, No. 4]. MR0472905

[Mum65] D. Mumford, *Geometric invariant theory*, Ergebnisse der Mathematik und ihrer Grenzgebiete, Neue Folge, Band 34, Springer-Verlag, Berlin-New York, 1965. MR0214602

[Nag60] M. Nagata, *On the fourteenth problem of Hilbert*, Proc. Internat. Congress Math. 1958, Cambridge Univ. Press, New York, 1960, pp. 459–462. MR0116056

[Nag63] M. Nagata, *Invariants of a group in an affine ring*, J. Math. Kyoto Univ. **3** (1963/64), 369–377, DOI 10.1215/kjm/1250524787. MR179268

[Ol'91] A. Yu. Ol'shanskiĭ, *Geometry of defining relations in groups*, Mathematics and its Applications (Soviet Series), vol. 70, Kluwer Academic Publishers Group, Dordrecht, 1991. Translated from the 1989 Russian original by Yu. A. Bakhturin. MR1191619

[OR76] J. B. Olsson and A. Regev, *Colength sequence of some T-ideals*, J. Algebra **38** (1976), no. 1, 100–111, DOI 10.1016/0021-8693(76)90247-7. MR409547

[Ore31] O. Ore, *Linear equations in non-commutative fields*, Ann. of Math. (2) **32** (1931), no. 3, 463–477, DOI 10.2307/1968245. MR1503010

[Pap97] C. J. Pappacena, *An upper bound for the length of a finite-dimensional algebra*, J. Algebra **197** (1997), no. 2, 535–545, DOI 10.1006/jabr.1997.7140. MR1483779

[Pie82] R. S. Pierce, *Associative algebras*, Graduate Texts in Mathematics, vol. 88, Springer-Verlag, New York-Berlin, 1982. Studies in the History of Modern Science, 9. MR674652

[Pir96] G. Pirillo, *A proof of Shirshov's theorem*, Adv. Math. **124** (1996), no. 1, 94–99, DOI 10.1006/aima.1996.0079. MR1423200

[Pio08] D. I. Piontkovskiĭ, *On the Kurosh problem in varieties of algebras* (Russian, with English and Russian summaries), Fundam. Prikl. Mat. **14** (2008), no. 5, 171–184, DOI 10.1007/s10958-009-9711-9; English transl., J. Math. Sci. (N.Y.) **163** (2009), no. 6, 743–750. MR2533586

[PY95] V. P. Platonov and V. I. Yanchevskii, *Finite-dimensional division algebras [MR1170354 (93h:16026)]*, Algebra, IX, Encyclopaedia Math. Sci., vol. 77, Springer, Berlin, 1995, pp. 121–239, DOI 10.1007/978-3-662-03235-0_2. Translated from the Russian by P. M. Cohn. MR1392479

[Pop70] V. L. Popov, *Criteria for the stability of the action of a semisimple group on the factorial of a manifold* (Russian), Izv. Akad. Nauk SSSR Ser. Mat. **34** (1970), 523–531. MR0262416

[Pop74] V. L. Popov, *Picard groups of homogeneous spaces of linear algebraic groups and one-dimensional homogeneous vector fiberings* (Russian), Izv. Akad. Nauk SSSR Ser. Mat. **38** (1974), 294–322. MR0357399

[Pop86] V. L. Popov, *Contractions of actions of reductive algebraic groups* (Russian), Mat. Sb. (N.S.) **130(172)** (1986), no. 3, 310–334, 431, DOI 10.1070/SM1987v058n02ABEH003106; English transl., Math. USSR-Sb. **58** (1987), no. 2, 311–335. MR865764

[Pos60] E. C. Posner, *Prime rings satisfying a polynomial identity*, Proc. Amer. Math. Soc. **11** (1960), 180–183, DOI 10.2307/2032951. MR111765

[Pro66a] C. Procesi, *A noncommutative Hilbert Nullstellensatz*, Rend. Mat. e Appl. (5) **25** (1966), no. 1-2, 17–21. MR466129

[Pro66b] C. Procesi, *The Burnside problem*, J. Algebra **4** (1966), 421–425, DOI 10.1016/0021-8693(66)90031-7. MR0212081

[Pro67a] C. Procesi, *Non commutative Jacobson-rings*, Ann. Scuola Norm. Sup. Pisa Cl. Sci. (3) **21** (1967), 281–290. MR224652

[Pro67b] C. Procesi, *Non-commutative affine rings* (English, with Italian summary), Atti Accad. Naz. Lincei Mem. Cl. Sci. Fis. Mat. Natur. Sez. I (8) **8** (1967), 237–255. MR0224657

[PS68] C. Procesi and L. Small, *Endomorphism rings of modules over* PI*-algebras*, Math. Z. **106** (1968), 178–180, DOI 10.1007/BF01110128. MR233846

[Pro68a] C. Procesi, *Sugli anelli non commutativi zero dimensionali con identità polinomiale* (Italian, with English summary), Rend. Circ. Mat. Palermo (2) **17** (1968), 5–12, DOI 10.1007/BF02849545. MR252427

[Pro68b] C. Procesi, *Sulle identità delle algebre semplici* (Italian, with English summary), Rend. Circ. Mat. Palermo (2) **17** (1968), 13–18, DOI 10.1007/BF02849546. MR265402

[Pro72a] C. Procesi, *On the identities of Azumaya algebras*, Ring theory (Proc. Conf., Park City, Utah, 1971), Academic Press, New York, 1972, pp. 287–295. MR0338061

[Pro72b] C. Procesi, *Dipendenza integrale nelle algebre non commutative* (Italian), Symposia Mathematica, Vol. VIII (Convegno sulle Algebre Associative, INDAM, Rome, 1970), Academic Press, London, 1972, pp. 295–308. MR0338062

[Pro72c] C. Procesi, *On a theorem of M. Artin*, J. Algebra **22** (1972), 309–315, DOI 10.1016/0021-8693(72)90148-2. MR302681

[Pro73] C. Procesi, *Rings with polynomial identities*, Marcel Dekker, Inc., New York, 1973. Pure and Applied Mathematics, 17. MR0366968

[Pro74] C. Procesi, *Finite dimensional representations of algebras*, Israel J. Math. **19** (1974), 169–182, DOI 10.1007/BF02756630. MR357507

[Pro76a] C. Procesi, *The invariant theory of $n \times n$ matrices*, Advances in Math. **19** (1976), no. 3, 306–381, DOI 10.1016/0001-8708(76)90027-X. MR419491

[Pro76b] C. Procesi, *The invariants of $n \times n$ matrices*, Bull. Amer. Math. Soc. **82** (1976), no. 6, 891–892, DOI 10.1090/S0002-9904-1976-14196-1. MR419490

[Pro79] C. Procesi, *Trace identities and standard diagrams*, Ring theory (Proc. Antwerp Conf. (NATO Adv. Study Inst.), Univ. Antwerp, Antwerp, 1978), Lecture Notes in Pure and Appl. Math., vol. 51, Dekker, New York, 1979, pp. 191–218. MR563295

[Pro84] C. Procesi, *Computing with $2 \times 2$ matrices*, J. Algebra **87** (1984), no. 2, 342–359, DOI 10.1016/0021-8693(84)90141-8. MR739938

[Pro87] C. Procesi, *A formal inverse to the Cayley-Hamilton theorem*, J. Algebra **107** (1987), no. 1, 63–74, DOI 10.1016/0021-8693(87)90073-1. MR883869

[Pro98] C. Procesi, *Deformations of representations*, Methods in ring theory (Levico Terme, 1997), Lecture Notes in Pure and Appl. Math., vol. 198, Dekker, New York, 1998, pp. 247–276. MR1767983

[Pro07] C. Procesi, *Lie groups: An approach through invariants and representations*, Universitext, Springer, New York, 2007. MR2265844

[Pro16] C. Procesi, *The geometry of polynomial identities* (Russian, with Russian summary), Izv. Ross. Akad. Nauk Ser. Mat. **80** (2016), no. 5, 103–152, DOI 10.4213/im8436; English transl., Izv. Math. **80** (2016), no. 5, 910–953. MR3588807

[Pro–nd] C. Procesi, *T-ideals of Cayley Hamilton algebras*, arXiv:2008.02222.

[Ray70] M. Raynaud, *Anneaux locaux henséliens* (French), Lecture Notes in Mathematics, Vol. 169, Springer-Verlag, Berlin-New York, 1970. MR0277519

[Raz73] Ju. P. Razmyslov, *A certain problem of Kaplansky* (Russian), Izv. Akad. Nauk SSSR Ser. Mat. **37** (1973), 483–501. MR0338063

[Raz74] Ju. P. Razmyslov, *The Jacobson Radical in PI-algebras* (Russian), Algebra i Logika **13** (1974), 337–360, 365. MR0419515

[Raz73] Ju. P. Razmyslov, *The existence of a finite basis for the identities of the matrix algebra of order two over a field of characteristic zero* (Russian), Algebra i Logika **12** (1973), 83–113, 121. MR0340348

[Raz74a] Ju. P. Razmyslov, *Identities with trace in full matrix algebras over a field of characteristic zero* (Russian), Izv. Akad. Nauk SSSR Ser. Mat. **38** (1974), 723–756. MR0506414

[Raz74] Ju. P. Razmyslov, *Trace identities of matrix algebras via a field of characteristic zero*, Math. USSR Izvestia (translation). **8** (1974), 727–760.

[Raz74b] Ju. P. Razmyslov, *Identities with trace in full matrix algebras over a field of characteristic zero* (Russian), Izv. Akad. Nauk SSSR Ser. Mat. **38** (1974), 723–756. MR0506414

[Raz74c] Ju. P. Razmyslov, *Existence of a finite base for certain varieties of algebras* (Russian), Algebra i Logika **13** (1974), no. 6, 685–693, 720. MR0396686

[Raz85] Ju. P. Razmyslov, *Trace identities and central polynomials in matrix superalgebras $M_{n,k}$* (Russian), Mat. Sb. (N.S.) **128(170)** (1985), no. 2, 194–215, 287. MR809485

[Raz92] Ju. P. Razmyslov, *Identities of algebras and their representations*, Proceedings of the International Conference on Algebra, Part 2 (Novosibirsk, 1989), Contemp. Math., vol. 131, Amer. Math. Soc., Providence, RI, 1992, pp. 173–192, DOI 10.1090/conm/131.2/1175831. MR1175831

[Raz94] Ju. P. Razmyslov, *Identities of algebras and their representations*, Translations of Mathematical Monographs, vol. 138, American Mathematical Society, Providence, RI, 1994. Translated from the 1989 Russian original by A. M. Shtern. MR1291603

[RZ94] Ju. P. Razmyslov and K. A. Zubrilin, *Capelli identities and representations of finite type*, Comm. Algebra **22** (1994), no. 14, 5733–5744, DOI 10.1080/00927879408825159. MR1298747

[Reg72] A. Regev, *Existence of identities in $A \otimes B$*, Israel J. Math. **11** (1972), 131–152, DOI 10.1007/BF02762615. MR314893

[Reg78] A. Regev, *The representations of $S_n$ and explicit identities for P. I. algebras*, J. Algebra **51** (1978), no. 1, 25–40, DOI 10.1016/0021-8693(78)90133-3. MR469965

[Reg79] A. Regev, *Algebras satisfying a Capelli identity*, Israel J. Math. **33** (1979), no. 2, 149–154, DOI 10.1007/BF02760555. MR571250

[Reg80a] A. Regev, *The Kronecker product of $S_n$-characters and an $A \otimes B$ theorem for Capelli identities*, J. Algebra **66** (1980), no. 2, 505–510, DOI 10.1016/0021-8693(80)90100-3. MR593607

[Reg80b] A. Regev, *The polynomial identities of matrices in characteristic zero*, Comm. Algebra **8** (1980), no. 15, 1417–1467, DOI 10.1080/00927878008822526. MR584301

[Reg81] A. Regev, *Asymptotic values for degrees associated with strips of Young diagrams*, Adv. in Math. **41** (1981), no. 2, 115–136, DOI 10.1016/0001-8708(81)90012-8. MR625890

[Reg84] A. Regev, *Codimensions and trace codimensions of matrices are asymptotically equal*, Israel J. Math. **47** (1984), no. 2-3, 246–250, DOI 10.1007/BF02760520. MR738172

[Reg94] A. Regev, *Young-derived sequences of $S_n$-characters*, Adv. Math. **106** (1994), no. 2, 169–197, DOI 10.1006/aima.1994.1055. MR1279217

[RA82] A. Regev and S. A. Amitsur, *PI-algebras and their cocharacters*, J. Algebra **78** (1982), no. 1, 248–254, DOI 10.1016/0021-8693(82)90110-7. MR677720

[RR09] A. Regev and A. Regev, *The Golod-Shafarevich counterexample without Hilbert series*, Groups, rings and group rings, Contemp. Math., vol. 499, Amer. Math. Soc., Providence, RI, 2009, pp. 257–264, DOI 10.1090/conm/499/09808. MR2581943

[Reg16] A. Regev, *Growth for the central polynomials*, Comm. Algebra **44** (2016), no. 10, 4411–4421, DOI 10.1080/00927872.2015.1090577. MR3508307

[RV06] Z. Reichstein and N. Vonessen, *Group actions and invariants in algebras of generic matrices*, Adv. in Appl. Math. **37** (2006), no. 4, 481–500, DOI 10.1016/j.aam.2005.08.007. MR2266894

[Reu86] C. Reutenauer, *Mots de Lyndon et un théorème de Shirshov* (French), Ann. Sci. Math. Québec **10** (1986), no. 2, 237–245. MR877397

[Reu93] C. Reutenauer, *Free Lie algebras*, London Mathematical Society Monographs. New Series, vol. 7, The Clarendon Press, Oxford University Press, New York, 1993. Oxford Science Publications. MR1231799

[RS87] C. Reutenauer and M.-P. Schützenberger, *A formula for the determinant of a sum of matrices*, Lett. Math. Phys. **13** (1987), no. 4, 299–302, DOI 10.1007/BF00401158. MR895292

[Riv55] R. S. Rivlin, *Further remarks on the stress-deformation relations for isotropic materials*, J. Rational Mech. Anal. **4** (1955), 681–702, DOI 10.1512/iumj.1955.4.54025. MR71980

[Riv76] R. S. Rivlin, *The application of the theory of invariants to the study of constitutive equations*, Trends in applications of pure mathematics to mechanics (Conf., Univ. Lecce, Lecce, 1975), Pitman, London, 1976, pp. 299–310. Monographs and Studies in Math., Vol. 2. MR0489123

[RS75] R. S. Rivlin and G. F. Smith, *On identities for* $3 \times 3$ *matrices* (English, with Italian summary), Rend. Mat. (6) **8** (1975), no. 2, 345–353. MR387314

[Rob55] H. Robbins, *A remark on Stirling's formula*, Amer. Math. Monthly **62** (1955), 26–29, DOI 10.2307/2308012. MR69328

[Rob63] N. Roby, *Lois polynomes et lois formelles en théorie des modules* (French), Ann. Sci. École Norm. Sup. (3) **80** (1963), 213–348. MR0161887

[Rob80] N. Roby, *Lois polynômes multiplicatives universelles* (French, with English summary), C. R. Acad. Sci. Paris Sér. A-B **290** (1980), no. 19, A869–A871. MR580160

[Rom00] D. Romik, *Stirling's approximation for* $n!$*: the ultimate short proof?*, Amer. Math. Monthly **107** (2000), no. 6, 556–557, DOI 10.2307/2589351. MR1767064

[Ros76] S. Rosset, *A new proof of the Amitsur-Levitski identity*, Israel J. Math. **23** (1976), no. 2, 187–188, DOI 10.1007/BF02756797. MR401804

[Row80] L. H. Rowen, *Polynomial identities in ring theory*, Pure and Applied Mathematics, vol. 84, Academic Press, Inc. [Harcourt Brace Jovanovich, Publishers], New York-London, 1980. MR576061

[Row88] L. H. Rowen, *Ring theory. Vol. II*, Pure and Applied Mathematics, vol. 128, Academic Press, Inc., Boston, MA, 1988. MR945718

[RS15] L. Rowen and L. Small, *Representability of algebras finite over their centers*, J. Algebra **442** (2015), 506–524, DOI 10.1016/j.jalgebra.2014.10.028. MR3395071

[Sag01] B. E. Sagan, *The symmetric group: Representations, combinatorial algorithms, and symmetric functions*, 2nd ed., Graduate Texts in Mathematics, vol. 203, Springer-Verlag, New York, 2001. MR1824028

[Sal84] D. J. Saltman, *Noether's problem over an algebraically closed field*, Invent. Math. **77** (1984), no. 1, 71–84, DOI 10.1007/BF01389135. MR751131

[Sal85] D. J. Saltman, *Groups acting on fields: Noether's problem*, Group actions on rings (Brunswick, Maine, 1984), Contemp. Math., vol. 43, Amer. Math. Soc., Providence, RI, 1985, pp. 267–277, DOI 10.1090/conm/043/810657. MR810657

[SS73] M. M. Schacher and L. W. Small, *Noncrossed products in characteristic* $p$, J. Algebra **24** (1973), 100–103, DOI 10.1016/0021-8693(73)90155-5. MR308185

[Sch66] R. D. Schafer, *An introduction to nonassociative algebras*, Academic Press, New York, 1966.

[Sch75] W. Schelter, *Essential extensions and intersection theorems*, Proc. Amer. Math. Soc. **53** (1975), no. 2, 328–330, DOI 10.2307/2040006. MR389971

[Sch77a] W. Schelter, *Affine PI rings are catenary*, Bull. Amer. Math. Soc. **83** (1977), no. 6, 1309–1310, DOI 10.1090/S0002-9904-1977-14427-3. MR450314

[Sch77b] W. Schelter, *Azumaya algebras and Artin's theorem*, J. Algebra **46** (1977), no. 1, 303–304, DOI 10.1016/0021-8693(77)90409-4. MR444694

[Sch82] W. F. Schelter, *Affine rings satisfying a polynomial identity*, Algebraists' homage: papers in ring theory and related topics (New Haven, Conn., 1981), Contemp. Math., vol. 13, Amer. Math. Soc., Providence, R.I., 1982, pp. 217–221. MR685954

[Sch84] W. F. Schelter, *Smooth affine PI algebras*, Methods in ring theory (Antwerp, 1983), NATO Adv. Sci. Inst. Ser. C Math. Phys. Sci., vol. 129, Reidel, Dordrecht, 1984, pp. 483–488. MR770611

[Sch85] W. F. Schelter, *On a question concerning generic matrices over the integers*, J. Algebra **96** (1985), no. 1, 48–53, DOI 10.1016/0021-8693(85)90038-9. MR808840

[Sch86] W. F. Schelter, *Smooth algebras*, J. Algebra **103** (1986), no. 2, 677–685, DOI 10.1016/0021-8693(86)90160-2. MR864437


[Sch85] W. F. Schelter, *On a question concerning generic matrices over the integers*, J. Algebra **96** (1985), no. 1, 48–53, DOI 10.1016/0021-8693(85)90038-9. MR808840


[Sch61] C. Schensted, *Longest increasing and decreasing subsequences*, Canadian J. Math. **13** (1961), 179–191, DOI 10.4153/CJM-1961-015-3. MR121305

[Schu01] I. Schur, *Uber eine Klasse von Matrizen, die sich einer gegebenen Matrix zuordnen lassen*, Diss. Berlin (1901).

[Sel44] A. Selberg, *Bemerkninger om et multipelt integral*, Nordisc Mat. Tidskr. **26** (1944), 71–78.

[Ser65] J.-P. Serre, *Algèbre locale. Multiplicités* (French), Cours au Collège de France, 1957–1958, rédigé par Pierre Gabriel. Seconde édition, 1965. Lecture Notes in Mathematics, vol. 11, Springer-Verlag, Berlin-New York, 1965. MR0201468

[Shc01a] V. V. Shchigolev, *Finite basis property of T-spaces over fields of characteristic zero* (Russian, with Russian summary), Izv. Ross. Akad. Nauk Ser. Mat. **65** (2001), no. 5, 191–224, DOI 10.1070/IM2001v065n05ABEH000362; English transl., Izv. Math. **65** (2001), no. 5, 1041–1071. MR1874359

[Shc01b] V. V. Shchigolev, *Finite basis property of T-spaces over fields of characteristic zero* (Russian, with Russian summary), Izv. Ross. Akad. Nauk Ser. Mat. **65** (2001), no. 5, 191–224, DOI 10.1070/IM2001v065n05ABEH000362; English transl., Izv. Math. **65** (2001), no. 5, 1041–1071. MR1874359

[She91] I. P. Shestakov, *Superalgebras and counterexamples* (Russian), Sibirsk. Mat. Zh. **32** (1991), no. 6, 187–196, 207, DOI 10.1007/BF00971214; English transl., Siberian Math. J. **32** (1991), no. 6, 1052–1060 (1992). MR1156760

[Šir57a] A. I. Širšov, *On some non-associative null-rings and algebraic algebras* (Russian), Mat. Sb. N.S. **41(83)** (1957), 381–394. MR0089841

[Šir57b] A. I. Širšov, *On rings with identity relations* (Russian), Mat. Sb. N.S. **43(85)** (1957), 277–283. MR0095192

[Šir58] A. I. Širšov, *On the Levitzki problem* (Russian), Dokl. Akad. Nauk SSSR **120** (1958), 41–42. MR0098767

[Sib68] K. S. Sibirskiĭ, *Algebraic invariants of a system of matrices* (Russian), Sibirsk. Mat. Ž. **9** (1968), 152–164. MR0223379

[Sma71] L. W. Small, *An example in PI-rings*, J. Algebra **17** (1971), 434–436, DOI 10.1016/0021-8693(71)90025-1. MR272823

[SS85] L. W. Small and J. T. Stafford, *Homological properties of generic matrix rings*, Israel J. Math. **51** (1985), no. 1-2, 27–32, DOI 10.1007/BF02772956. MR804474

[SSW85] L. W. Small, J. T. Stafford, and R. B. Warfield Jr., *Affine algebras of Gel′fand-Kirillov dimension one are PI*, Math. Proc. Cambridge Philos. Soc. **97** (1985), no. 3, 407–414, DOI 10.1017/S0305004100062976. MR778674

[Smo02] A. Smoktunowicz, *A simple nil ring exists*, Comm. Algebra **30** (2002), no. 1, 27–59, DOI 10.1081/AGB-120006478. MR1880660

[Spe50] W. Specht, *Gesetze in Ringen. I* (German), Math. Z. **52** (1950), 557–589, DOI 10.1007/BF02230710. MR35274

[SR58a] A. J. M. Spencer and R. S. Rivlin, *The theory of matrix polynomials and its application to the mechanics of isotropic continua*, Arch. Rational Mech. Anal. **2** (1958/59), 309–336, DOI 10.1007/BF00277933. MR100600

[SR58b] A. J. M. Spencer and R. S. Rivlin, *Finite integrity bases for five or fewer symmetric* $3 \times 3$ *matrices*, Arch. Rational Mech. Anal. **2** (1958/59), 435–446, DOI 10.1007/BF00277941. MR100601

[Spe61] A. J. M. Spencer, *The invariants of six symmetric* $3 \times 3$ *matrices*, Arch. Rational Mech. Anal. **7** (1961), 64–77, DOI 10.1007/BF00250750. MR120246

[Spe71a] A. J. M. Spencer, *Continuum physics*, Vol. 1, pp. 239–352, Academic Press, New York, 1971.

[Spe71b] A. J. M. Spencer, *Theory of invariants*, A.C. Eringen (Ed.), Continuum Physics, vol. I, Academic Press, San Diego (1971), pp. 240–353.

[Spe09] A. J. M. Spencer, *Ronald Rivlin and invariant theory*, Internat. J. Engrg. Sci. **47** (2009), no. 11-12, 1066–1078, DOI 10.1016/j.ijengsci.2009.01.004. MR2590475

[SR60] A. J. M. Spencer and R. S. Rivlin, *Further results in the theory of matrix polynomials*, Arch. Rational Mech. Anal. **4** (1960), 214–230 (1960), DOI 10.1007/BF00281388. MR109830

[Spr77] T. A. Springer, *Invariant theory*, Lecture Notes in Mathematics, Vol. 585, Springer-Verlag, Berlin-New York, 1977. MR0447428

[Spr09] T. A. Springer, *Linear algebraic groups*, 2nd ed., Modern Birkhäuser Classics, Birkhäuser Boston, Inc., Boston, MA, 2009. MR2458469

[Sta73] R. P. Stanley, *Linear homogeneous Diophantine equations and magic labelings of graphs*, Duke Math. J. **40** (1973), 607–632. MR317970

[Sta82] R. P. Stanley, *Linear Diophantine equations and local cohomology*, Invent. Math. **68** (1982), no. 2, 175–193, DOI 10.1007/BF01394054. MR666158

[Sta83] R. P. Stanley, *Combinatorics and commutative algebra*, Progress in Mathematics, vol. 41, Birkhäuser Boston, Inc., Boston, MA, 1983. MR725505

[Sta97] R. P. Stanley, *Enumerative combinatorics. Vol. 1*, Cambridge Studies in Advanced Mathematics, vol. 49, Cambridge University Press, Cambridge, 1997. With a foreword by Gian-Carlo Rota; Corrected reprint of the 1986 original. MR1442260

[Sta99] R. P. Stanley, *Enumerative combinatorics. Vol. 2*, Cambridge Studies in Advanced Mathematics, vol. 62, Cambridge University Press, Cambridge, 1999. With a foreword by Gian-Carlo Rota and appendix 1 by Sergey Fomin. MR1676282

[ŞVO99] D. Ştefan and F. Van Oystaeyen, *The Wedderburn-Malćev theorem for comodule algebras*, Comm. Algebra **27** (1999), no. 8, 3569–3581, DOI 10.1080/00927879908826648. MR1699590

[Ste65] R. Steinberg, *Regular elements of semisimple algebraic groups*, Inst. Hautes Études Sci. Publ. Math. **25** (1965), 49–80. MR180554

[Stu95] B. Sturmfels, *On vector partition functions*, J. Combin. Theory Ser. A **72** (1995), no. 2, 302–309, DOI 10.1016/0097-3165(95)90067-5. MR1357776

[Svi11] I. Sviridova, *Identities of pi-algebras graded by a finite abelian group*, Comm. Algebra **39** (2011), no. 9, 3462–3490, DOI 10.1080/00927872.2011.593417. MR2845581

[Svi13] I. Sviridova, *Finitely generated algebras with involution and their identities*, J. Algebra **383** (2013), 144–167, DOI 10.1016/j.jalgebra.2013.03.002. MR3037972

[STR93] J. Szigeti, Z. Tuza, and G. Révész, *Eulerian polynomial identities on matrix rings*, J. Algebra **161** (1993), no. 1, 90–101, DOI 10.1006/jabr.1993.1207. MR1245845

[Swa63] R. G. Swan, *An application of graph theory to algebra*, Proc. Amer. Math. Soc. **14** (1963), 367–373, DOI 10.2307/2033801. MR149468

[Taf57] E. J. Taft, *Invariant Wedderburn factors*, Illinois J. Math. **1** (1957), 565–573. MR0098124

[Ter86] Y. Teranishi, *The ring of invariants of matrices*, Nagoya Math. J. **104** (1986), 149–161, DOI 10.1017/S0027763000022728. MR868442

[Ter88] Y. Teranishi, *The Hilbert series of rings of matrix concomitants*, Nagoya Math. J. **111** (1988), 143–156, DOI 10.1017/S0027763000001057. MR961222

[Vac08] F. Vaccarino, *Generalized symmetric functions and invariants of matrices*, Math. Z. **260** (2008), no. 3, 509–526, DOI 10.1007/s00209-007-0285-2. MR2434467

[Vac09] F. Vaccarino, *Homogeneous multiplicative polynomial laws are determinants*, J. Pure Appl. Algebra **213** (2009), no. 7, 1283–1289, DOI 10.1016/j.jpaa.2008.11.036. MR2497575

[VdB89] M. Van den Bergh, *Trace rings of generic matrices are Cohen-Macaulay*, J. Amer. Math. Soc. **2** (1989), no. 4, 775–799, DOI 10.2307/1990894. MR1001850

[VdB91] M. Van den Bergh, *Explicit rational forms for the Poincaré series of the trace rings of generic matrices*, Israel J. Math. **73** (1991), no. 1, 17–31, DOI 10.1007/BF02773421. MR1119924

[Vas99] S. Yu. Vasilovsky, *$Z_n$-graded polynomial identities of the full matrix algebra of order $n$*, Proc. Amer. Math. Soc. **127** (1999), no. 12, 3517–3524, DOI 10.1090/S0002-9939-99-04986-2. MR1616581

[VL70] M. R. Vaughan-Lee, *Varieties of Lie algebras*, Quart. J. Math. Oxford Ser. (2) **21** (1970), 297–308, DOI 10.1093/qmath/21.3.297. MR269710

[Vin65] È. B. Vinberg, *On the theorem concerning the infinite-dimensionality of an associative algebra* (Russian), Izv. Akad. Nauk SSSR Ser. Mat. **29** (1965), 209–214. MR0172892

[Vol84] I. B. Volichenko, *Varieties of Lie algebras with identity $[[X_1, X_2, X_3], [X_4, X_5, X_6]] = 0$ over a field of characteristic zero* (Russian), Sibirsk. Mat. Zh. **25** (1984), no. 3, 40–54. MR746940

[Weh76] B. A. F. Wehrfritz, *Automorphism groups of Noetherian modules over commutative rings*, Arch. Math. (Basel) **27** (1976), no. 3, 276–281, DOI 10.1007/BF01224671. MR409615

[Wey97] H. Weyl, *The classical groups*: *Their invariants and representations*, Princeton Landmarks in Mathematics, Princeton University Press, Princeton, NJ, 1997. Fifteenth printing; Princeton Paperbacks. MR1488158

[Wil94] H. S. Wilf, *generatingfunctionology*, 2nd ed., Academic Press, Inc., Boston, MA, 1994. MR1277813

[Zaĭ02] M. V. Zaĭtsev, *Integrality of exponents of growth of identities of finite-dimensional Lie algebras* (Russian, with Russian summary), Izv. Ross. Akad. Nauk Ser. Mat. **66** (2002), no. 3, 23–48, DOI 10.1070/IM2002v066n03ABEH000386; English transl., Izv. Math. **66** (2002), no. 3, 463–487. MR1921808

[Zai14] M. Zaicev, *On existence of PI-exponents of codimension growth*, Electron. Res. Announc. Math. Sci. **21** (2014), 113–119, DOI 10.3934/era.2014.21.113. MR3260134

[Zel91] E. I. Zel′manov, *Superalgebras and identities*, Algebra and analysis (Kemerovo, 1988), Amer. Math. Soc. Transl. Ser. 2, vol. 148, Amer. Math. Soc., Providence, RI, 1991, pp. 39–46, DOI 10.1090/trans2/148/04. MR1109062

[Zel90] E. I. Zel'manov, *Solution of the restricted Burnside problem for groups of odd exponent* (Russian), Izv. Akad. Nauk SSSR Ser. Mat. **54** (1990), no. 1, 42–59, 221; English transl., Math. USSR-Izv. **36** (1991), no. 1, 41–60. MR1044047

[Zel91] E. I. Zel'manov, *Solution of the restricted Burnside problem for 2-groups* (Russian), Mat. Sb. **182** (1991), no. 4, 568–592; English transl., Math. USSR-Sb. **72** (1992), no. 2, 543–565. MR1119009

[Zel05] E. Zelmanov, *Infinite algebras and pro-p groups*, Infinite groups: geometric, combinatorial and dynamical aspects, Progr. Math., vol. 248, Birkhäuser, Basel, 2005, pp. 403–413, DOI 10.1007/3-7643-7447-0_11. MR2195460

[Zip87] D. Ziplies, *Generators for the divided powers algebra of an algebra and trace identities*, Beiträge Algebra Geom. **24** (1987), 9–27. MR888200

[Zip89] D. Ziplies, *Abelianizing the divided powers algebra of an algebra*, J. Algebra **122** (1989), no. 2, 261–274, DOI 10.1016/0021-8693(89)90215-9. MR999072

[Zub94] A. N. Zubkov, *Endomorphisms of tensor products of exterior powers and Procesi hypothesis*, Comm. Algebra **22** (1994), no. 15, 6385–6399, DOI 10.1080/00927879408825195. MR1303010

[Zub96] A. N. Zubkov, *On a generalization of the Razmyslov-Procesi theorem* (Russian, with Russian summary), Algebra i Logika **35** (1996), no. 4, 433–457, 498, DOI 10.1007/BF02367026; English transl., Algebra and Logic **35** (1996), no. 4, 241–254. MR1444429

[Zub95a] K. A. Zubrilin, *On the class of nilpotency of obstruction for the representability of algebras satisfying Capelli identities* (Russian, with English and Russian summaries), Fundam. Prikl. Mat. **1** (1995), no. 2, 409–430. MR1790973

[Zub95b] K. A. Zubrilin, *Algebras that satisfy the Capelli identities* (Russian, with Russian summary), Mat. Sb. **186** (1995), no. 3, 53–64, DOI 10.1070/SM1995v186n03ABEH000021; English transl., Sb. Math. **186** (1995), no. 3, 359–370. MR1331808

[Zub97] K. A. Zubrilin, *On the largest nilpotent ideal in algebras that satisfy Capelli identities* (Russian, with Russian summary), Mat. Sb. **188** (1997), no. 8, 93–102, DOI 10.1070/SM1997v188n08ABEH000253; English transl., Sb. Math. **188** (1997), no. 8, 1203–1211. MR1481398

# Index

# Index of Symbols

## Selected Published Titles in This Series

66 **Eli Aljadeff, Antonio Giambruno, Claudio Procesi, and Amitai Regev,** Rings with Polynomial Identities and Finite Dimensional Representations of Algebras, 2020
65 **László Lovász,** Graphs and Geometry, 2019
64 **Kathrin Bringmann, Amanda Folsom, Ken Ono, and Larry Rolen,** Harmonic Maass Forms and Mock Modular Forms: Theory and Applications, 2017
63 **Cornelia Druţu and Michael Kapovich,** Geometric Group Theory, 2018
62 **Martin Olsson,** Algebraic Spaces and Stacks, 2016
61 **James Arthur,** The Endoscopic Classification of Representations, 2013
60 **László Lovász,** Large Networks and Graph Limits, 2012
59 **Kai Cieliebak and Yakov Eliashberg,** From Stein to Weinstein and Back, 2012
58 **Freydoon Shahidi,** Eisenstein Series and Automorphic $L$-Functions, 2010
57 **John Friedlander and Henryk Iwaniec,** Opera de Cribro, 2010
56 **Richard Elman, Nikita Karpenko, and Alexander Merkurjev,** The Algebraic and Geometric Theory of Quadratic Forms, 2008
55 **Alain Connes and Matilde Marcolli,** Noncommutative Geometry, Quantum Fields and Motives, 2008
54 **Barry Simon,** Orthogonal Polynomials on the Unit Circle: Part 1: Classical Theory; Part 2: Spectral Theory, 2005
53 **Henryk Iwaniec and Emmanuel Kowalski,** Analytic Number Theory, 2004
52 **Dusa McDuff and Dietmar Salamon,** $J$-holomorphic Curves and Symplectic Topology, Second Edition, 2012
51 **Alexander Beilinson and Vladimir Drinfeld,** Chiral Algebras, 2004
50 **E. B. Dynkin,** Diffusions, Superdiffusions and Partial Differential Equations, 2002
49 **Vladimir V. Chepyzhov and Mark I. Vishik,** Attractors for Equations of Mathematical Physics, 2002
48 **Yoav Benyamini and Joram Lindenstrauss,** Geometric Nonlinear Functional Analysis, 2000
47 **Yuri I. Manin,** Frobenius Manifolds, Quantum Cohomology, and Moduli Spaces, 1999
46 **J. Bourgain,** Global Solutions of Nonlinear Schrödinger Equations, 1999
45 **Nicholas M. Katz and Peter Sarnak,** Random Matrices, Frobenius Eigenvalues, and Monodromy, 1999
44 **Max-Albert Knus, Alexander Merkurjev, Markus Rost, and Jean-Pierre Tignol,** The Book of Involutions, 1998
43 **Luis A. Caffarelli and Xavier Cabré,** Fully Nonlinear Elliptic Equations, 1995
42 **Victor W. Guillemin and Shlomo Sternberg,** Variations on a Theme by Kepler, 1991
40 **R. H. Bing,** The Geometric Topology of 3-Manifolds, 1983
38 **O. Ore,** Theory of Graphs, 1962
37 **N. Jacobson,** Structure of Rings, 1956
36 **Walter Helbig Gottschalk and Gustav Arnold Hedlund,** Topological Dynamics, 1955
32 **R. L. Wilder,** Topology of Manifolds, 1949
31 **E. Hille and R. S. Phillips,** Functional Analysis and Semi-groups, 1996
30 **Tibor Radó,** Length and Area, 1948
29 **A. Weil,** Foundations of Algebraic Geometry, 1946
28 **G. T. Whyburn,** Analytic Topology, 1942
27 **S. Lefschetz,** Algebraic Topology, 1942
26 **N. Levinson,** Gap and Density Theorems, 1940

For a complete list of titles in this series, visit the
AMS Bookstore at **www.ams.org/bookstore/collseries/**.